/QD61.T4VOL.18>C1/

/QD61.T4VOL.18>C1/

TECHNIQUES OF CHEMISTRY

VOLUME XVIII

MICROWAVE MOLECULAR SPECTRA

MICROWAVE MOLECULAR SPECTRA

WALTER GORDY

Duke University
Durham, North Carolina

ROBERT L. COOK

Mississippi State University
Mississippi State, Mississippi

A WILEY-INTERSCIENCE PUBLICATION

JOHN WILEY & SONS

New York • Chichester • Brisbane • Toronto • Singapore

Library of Congress Catalogue Card Number: 84-40367

ISBN 0-471-08681-9

Printed in the United States of America

10 9 8 7 6 5 4 3 2 1

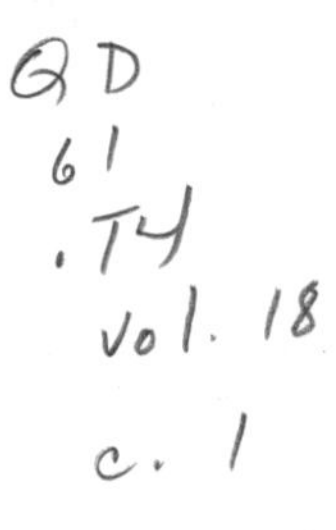

TO THE DUKE MICROWAVERS

The more than one hundred graduate students and research associates who have contributed much to the development of microwave spectroscopy.

INTRODUCTION TO THE SERIES

Techniques of Chemistry is the successor to the Technique of Organic Chemistry Series and its companion—Technique of Inorganic Chemistry. Because many of the methods are employed in all branches of chemical science, the division into techniques for organic and inorganic chemistry has become increasingly artificial. Accordingly, the new series reflects the wider application of techniques, and the component volumes for the most part provide complete treatments of the methods covered. Volumes in which limited areas of application are discussed can be easily recognized by their titles.

Like its predecessors, the series is devoted to a comprehensive presentation of the respective techniques. The authors give the theoretical background for an understanding of the various methods and operations and describe the techniques and tools, their modifications, their merits and limitations, and their handling. It is hoped that the series will contribute to a better understanding and a more rational and effective application of the respective techniques.

Authors and editors hope that readers will find the volumes in this series useful and will communicate to them any criticisms and suggestions for improvements.

ARNOLD WEISSBERGER

Research Laboratories
Eastman Kodak Company
Rochester, New York

PREFACE

Many exciting new developments in microwave spectroscopy have occurred since the earlier edition of this book was published in 1970. The frequency coverage of coherent microwave spectroscopy is still expanding. Although not comparable to the fourfold expansion that occurred in the 17 years preceding the 1970 edition, the millimeter wave range has been extended from 800 GHz to above 1000 GHz since 1970. Subtle new techniques continue to improve the sensitivity and resolving power of microwave spectrometers as well as their applicability to new types of spectra.

Detection of the microwave spectra of nonpolar, spherical-top molecules such as CH_4 and of "forbidden" ΔK transitions in symmetric-top molecules such as PH_3 has been achieved. Microwave "molecular ion" spectroscopy has become a practical reality since 1970. The observation of rotational spectra of weak, hydrogen-bonded complexes and rare gas atom—molecule complexes (van der Waals molecules) has become widespread. New techniques for observation of molecules in highly excited vibrational states have been devised. Significant advances in the theory of complex microwave molecular spectra have been made. In the last 15 years microwave spectroscopy has advanced from observation of molecules in laboratory cells to observation of molecules in interstellar space. An important new field, microwave molecular astronomy, has been created. Microwave sources and techniques have been combined with infrared and optical lasers to form the new and rapidly advancing field of microwave–optical double resonance spectroscopy.

Although the spectra described in the earlier edition are in no sense out of date, the new developments made the revision of *Microwave Molecular Spectra* desirable if not necessary. The basic theory and measurements of microwave spectroscopy have a remarkable durability which results from the high resolution and accuracy of measurement that characterize all coherent radiation spectroscopy of sharp line spectra. Because we could not justifiably delete nor significantly reduce the basic material of the earlier edition, a moderate expansion of the volume was necessary to achieve an adequate coverage of microwave molecular spectra in the 1980s.

As was true for the earlier edition, we are indebted to many people for assistance in the revision of this book. Again, Vida Miller Gordy graciously assisted with every phase of the manuscript preparation. We have benefited by discussions with Frank De Lucia, Eric Herbst, Paul Helminger, and K. V. L. N. Sastry. The Winnewissers—Manfred, Brenda, and Gisbert—have helped to keep us informed about the latest results in the field. M. C. L. Gerry read parts of the manuscript and made helpful suggestions. We are also grateful to other spectroscopists for sending us preprints and reprints describing results obtained in their laboratories. Among them are Lisa Nygaard, E. Tiemann, H. Dreizler, H. D. Rudolph, A. F. Krupnov, and J. L. Destombes.

Finally, we wish to express a tribute to the memory of Dr. William West, editor of the earlier edition. He was a great person, a considerate and competent editor, and a treasured friend.

WALTER GORDY
ROBERT L. COOK

Durham, North Carolina
July 1984

PREFACE TO THE SECOND EDITION

In the 17 years since the first book on microwave spectroscopy was written, the field has developed so extensively that it is not possible to give a comprehensive coverage of all its aspects within a single volume of manageable size. Not only have the applications been increased enormously and the instruments and techniques diversified, but the microwave region itself has also been greatly expanded. The frequency range in which spectral measurements can be made with microwave methods has increased more than fourfold since 1953. Many measurements are now made at submillimeter wavelengths. With the molecular beam maser the already ultrahigh resolution of microwave spectroscopy has been increased by more than an order of magnitude. Rapidly recording, highly sensitive microwave spectrometers have become commercially available. High-speed computers have taken much of the labor out of the analysis and have made the study of complex molecules more feasible. New theoretical developments have increased the possibility of understanding complex spectra and have also increased the usefulness of such spectra. Microwave spectral measurements on short-lived gaseous free radicals and on substances with vaporization temperatures of the order of a thousand degrees are commonly made. Spectral frequencies are measured to accuracies of the order of one part in 10^8. Accurate molecular structures and other properties have been found for molecules numbering into the thousands.

In this volume we have sought to provide a basis for the understanding of microwave spectra in the gas phase. In doing this we have developed the theory from what we considered to be the simplest approach consistent with essential correctness and applicability. Although we make no effort to include all the useful information about molecules that has been derived from the spectra—a hardly achievable goal—we have included a variety of types of information about selected molecules. A reasonably complete listing of the molecular structures derived up to 1969 is given in the appendix. Although the book is a member of the *Technique of Organic Chemistry* series, we have by no means limited the coverage to organic molecules. To do so would have been too wasteful because the theory applicable to organic molecules is generally applicable to inorganic molecules. The book is written for chemical physicists and physical chemists as well as for organic chemists—if indeed such separate classifications are *bona fide* in this age. We have not included a discussion of the determination of nuclear moments and masses, subjects that are perhaps of more interest to physicists than to chemists. Most nuclear moments and isotopic masses that can be measured to advantage with microwave spectra have already been measured. Rather we have treated nuclear hyperfine structure and isotopic shifts of spectral lines with the aim of using them to gain information about molecules. To achieve more thorough coverage of gases we have omitted solid-state studies, which are made primarily with microwave paramagnetic resonance (included in another volume of this series) and liquid-state studies made chiefly through dielectric dispersion and absorption. For the same reason we have omitted discussion of instruments and techniques of measurement.

Nevertheless we have tried to serve the dual purpose of providing a convenient source and reference book for much of the valuable information gained about molecules through microwave spectroscopy and of providing a textbook that explains essential theory for interpretation of the spectra and derivation of information from spectra. Although not written specifically for the purpose, the book can be used as a text for a course or seminar on microwave spectra. A quantitative study of microwave spectroscopy provides numerous, rather elegant examples of the application of quantum mechanics to problems of molecular dynamics. Although the quantum mechanical treatment can in some cases be somewhat complicated in its details, much of the information derivable from microwave spectra of molecules is of such importance that it justifies more than casual attention.

Several people have assisted considerably in the preparation of the volume. Vida Miller Gordy has helped persistently and effectively in the preparation and proofreading of the manuscript. Jean Luffman's typing of the entire manuscript presented us with a beautiful final copy from the many revisions given her in the course of the writing. Janet Jackel drew most of the illustrations. Several research associates and graduate students at Duke University—James Cederberg, Frank De Lucia, Steve Guastafson, Paul Helminger, Sam McKnight, William Oelfke, Edward Pearson, David Straub, Gisbert Winnewisser, Ray Winton, and Fred Wyse—have read sections of the book and made helpful suggestions. To each one we offer our thanks.

WALTER GORDY
ROBERT L. COOK

Durham, North Carolina
April 20, 1969

CONTENTS

CHAPTERS

APPENDIXES

Chapter I

INTRODUCTION

Microwave and radiofrequency spectroscopy are branches of spectroscopy in which the spectral transitions are measured with coherent radiation sources. The radiation sources employed are phase-coherent oscillators which provide energy in a frequency band so narrow that in comparison with most spectral lines they can be considered monochromatic sources. These essentially monochromatic sources are tunable and can be swept over a spectral line to be measured or can be conveniently tuned over wide regions in search of unknown spectral lines. The resolution easily obtainable with these tuned radio electronic oscillators is thousands of times greater than that of conventional infrared spectrometers employing noncoherent sources, with prisms or gratings for dispersion.

Figure 1.1 shows the extent of the microwave region of the electromagnetic spectrum. This region is designated as extending from wavelengths of approximately 30 cm to those of 0.3 mm, or from frequencies of 1 GHz to those of 1000 GHz, or 1 THz. The GHz unit (gigacycles per second), now commonly used by microwave spectroscopists, represents 10^9 Hz, or a thousand megacycles per second.

In the initial period of microwave spectroscopy, measurements were made in the centimeter wave region with oscillators and detectors developed for microwave radar during World War II. The experimental techniques and extensive results for this period are described in the earlier books on microwave spectroscopy [1, 2]. In later years the range of microwave measurements has been extended throughout the millimeter and into the submillimeter wavelengths, to 0.3 mm (1000 GHz or 1 THz).

In 1954, coherent microwave spectral measurements by Burrus and Gordy [3] were made to overlap the far-infrared grating measurements of Genzel and Eckhardt [4]. This extension of microwave spectroscopy into the submillimeter wave region was made possible by a crystal harmonic generator and a

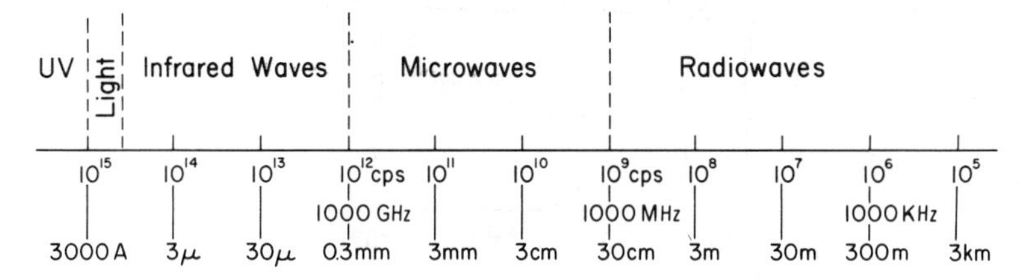

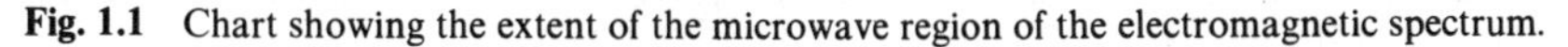
Fig. 1.1 Chart showing the extent of the microwave region of the electromagnetic spectrum.

crystal diode detector designed by King and Gordy [5]. Later refinement of the harmonic generators and improvement in the sensitivity of the submillimeter wave detectors led to further extension of coherent, tunable, microwave spectroscopy [6–9] to frequencies of 800 GHz (0.37 mm) by 1970 [8] and to 1037 GHz (0.38 mm) by 1983 [9]. Since harmonics are exact multiples of the fundamental frequency of the source power, this type of generator carries with it a precise frequency-measuring chain to the highest detectable harmonic. For example, the $J=8 \rightarrow 9$ transition of CO was detected and measured [9] at 1,036,312.35 MHz with a Doppler-limited accuracy of about ± 0.15 MHz. Among the later improvements of the submillimeter wave spectrometers at Duke was the use of ion-bombarded [10] silicon crystals in the harmonic generators and replacement of the crystal diode detector with detectors operating at low temperatures, particularly by an indium–antimonide photoconducting detector [8] and by a silicon bolometer [11].

Very impressive submillimeter wave spectral observations with primary radiation from tunable, submillimeter wave, backward-wave oscillators (BWO's) are now made by A. F. Krupnov [12, 13] and his group working in Gorky, USSR. For detection of the spectra, they employ an acoustic detector that depends on the thermal expansion of the gaseous sample produced by the resonant absorption of the radiation at the frequency of the spectral absorption line. They designate this type of spectrometer by RAD, which signifies "radiospectroscope with acoustic detector." Their experiments, initiated about 1970, have resulted in high-resolution measurements in the submillimeter wave region to frequencies above 1000 GHz, or to wavelengths of 0.3 mm. An illustration of the remarkable performance of this spectrometer [14] is given in Fig. 1.2. A description of the spectrometer with a discussion of its performance is given in reviews by Krupnov and Burenin [12] and by Krupnov [13].

The development of effective microwave instruments and techniques for measurements in the shorter millimeter and the submillimeter wave region

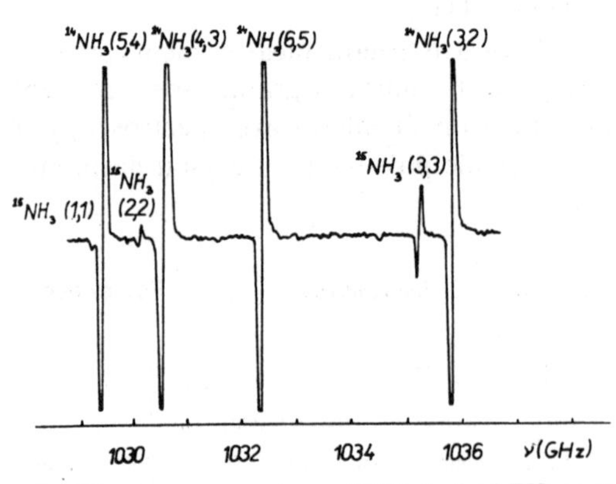

Fig. 1.2 Part of the submillimeter wave spectrum of NH_3 in the 1 THz region (1000 GHz). From Belov et al. [14].

from 3 mm to approximately 0.3 mm (100 to 1000 GHz) has not only increased the coverage of molecular rotational spectra but has also made possible the use of semioptical techniques [15] for the focusing and direction of radiation without any sacrifice in the resolution and accuracy of the microwave source and detector. It has also made possible the use of free-space cells for corrosive or unstable molecules and ions, the use of molecular-beam, high-temperature absorption spectrometers for molecules having high vaporization temperatures, and the use of precision parallel-plate cells for measurement of Stark components. The increase in the abundance and the strength of microwave molecular lines with increase of frequency insures that the last explored region of the electromagnetic spectrum, the submillimeter wave region, will continue to be a region of interest and value to the chemist and the physicist. Experimental methods and descriptions of the results obtained with millimeter and submillimeter wave spectroscopy are available in several reviews [15–19]. Treatments of selected results will be found throughout this volume.

Because of their low intensity, spontaneous emission lines are not observable in the microwave or radiofrequency region. Since the invention of the maser, however, observation of certain microwave spectral lines through stimulated emission has become common. To make such an observation one must in some way upset the Boltzmann distribution so as to obtain an excess population in the upper of two quantum levels involved in the transition. In the first operating maser, Gordon, Zeiger, and Townes [20] accomplished this by removal of the molecules in the lower state through deflection of a molecular beam by an inhomogeneous electric field. Several other methods have since been devised for achievement of excess population in the upper state sufficient for observation of stimulated emission spectroscopy. Perhaps those most widely used are optical pumping [21] and chemical pumping [22].

The exceptional resolution in a beam-maser spectrometer makes it of great advantage in the study of specialized problems. The resolution obtainable in microwave spectroscopy with the molecular beam maser is illustrated [23] by Fig. 1.3, which shows the triplet hyperfine structure due to the deuterium nucleus superimposed upon a hyperfine component of ^{14}N in the $J=0\leftarrow1$ rotational transition of DCN. The total width of this triplet which occurs at a frequency of 72,414 MHz is only 68.7 kHz, and the two closest components are separated by only 23 kHz. These frequencies are measured to an accuracy of better than a kilohertz, or to one part in 10^8. Application of this beam maser has been extended into the upper submillimeter region by Garvey and De Lucia [24] to measure with comparable resolution the hyperfine structure of the $J=0\leftarrow1$ transition of ND_3 at 309 GHz and of the $1_{10}\rightarrow1_{01}$ transition of D_2O at 317 GHz. Developments in techniques and applications of molecular beam masers up to 1975 are described in an excellent review by Lainé [25].

Since the publication of the earlier edition of this book (1970), microwave–optical laser double resonance has become increasingly important in measurement of the rotational structure of excited vibrational and electronic levels of molecules [26–30]. Harold Jones gives a thorough review of the important

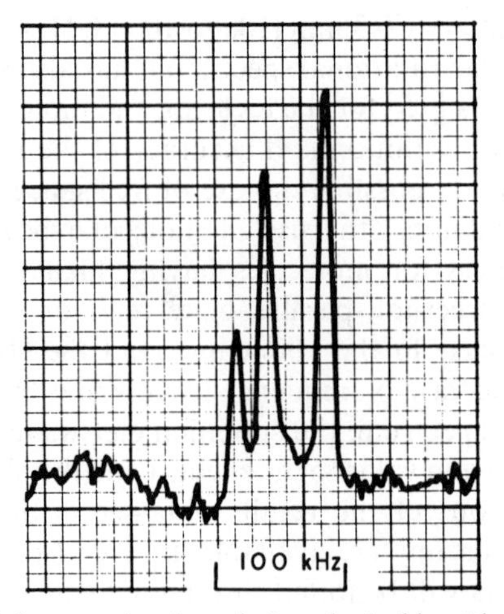

Fig. 1.3 Illustration of the exceptional resolution obtainable with the molecular beam maser. The spectra, observed at 72,414 MHz, represent a part of the hyperfine structure of the $J=0 \rightarrow 1$ transition of DCN. The splitting is due to the nuclear quadrupole coupling of deuterium. From De Lucia and Gordy [23].

infrared–microwave double resonance technique [30]. The development of tunable dye-lasers and the expansion of microwave techniques for the millimeter and submillimeter wave lengths have made microwave–optical double resonance increasingly feasible. An excellent monograph on laser spectroscopy by Demtroder [31] explains experimental techniques and spectral applications. Microwave–microwave double resonance has become a powerful technique for study of special problems such as molecular collision processes. Experimental methods and applications are described in a review by Baker [32]. The use of this technique for identification of spectral transitions in asymmetric rotors is described in Chapter VII, Section 5.

Lines of the strong HCN laser source [33, 34] in the 0.337 mm wave region have been accurately measured in frequency units by Hocker et al. [35, 36], who detected beat notes between the laser frequencies and the harmonics of a calibrated microwave source. A laser source thus calibrated was later used to give measurable beat notes with a laser operating at a higher frequency [37]; the second laser thus calibrated could be used for measurement of beat notes with a laser operating at a still higher frequency, and so on. With this method, optical laser sources have been measured indirectly, but precisely, in absolute frequency units. By this means the microwave harmonic frequency-measuring chain has been projected to the optical region.

Reviews of various types of lasers, with references to numerous monographs on lasers and their various applications, may be found in *Quantum Electronics*, a 1979 volume of *Methods in Experimental Physics* [38].

This book is primarily concerned with interpretation of observed spectra.

Descriptions of instruments and experimental procedures must be obtained from other sources. An excellent treatment (1979) of modern microwave spectrometers and experimental methods is given by Roussy and Chantry [39]. The pulsed, coherent microwave Fourier-transform spectrometer introduced by Ekkers and Flygare [40] holds much promise for subtle observations requiring exceptional sensitivity and resolution [41]. Interferometric methods for non-coherent (black-body) millimeter and submillimeter spectroscopy described by Fleming [42] are convenient and useful for applications where very high resolution is not required.

A technique that has evolved since the earlier edition is the adaptation of the "supersonic free jet" [43] in the study of van der Waals molecules with microwave spectroscopy [41] (Chapter V, Section 9). The method also makes possible observation of very weak hydrogen-bonded complexes that are unstable at ordinary temperatures (Chapter V, section 8) and the study of rotational transitions in normal molecules at very low temperatures, at which they would solidify in a normal absorption cell. The cooling in the supersonic jet is caused by the expansion of a highly pressurized gas as it escapes through a restricting nozzle into an evacuated microwave cavity. The restricting nozzle also collimates the escaping gas into a beam [43].

Applications of microwave spectroscopy to the study of molecular fragments, gaseous free radicals, and molecules in excited electronic states with unpaired electron spins have significantly increased since the earlier edition. Carrington's monograph on the microwave spectroscopy of gaseous free radicals (1974) describes both spectra and experimental methods [44]. The developments in microwave–optical double resonance have greatly facilitated the measurement of paramagnetic transitions of molecules in excited triplet states [45]. Since the initial observations of the microwave rotational spectra of ionized molecules such as CO^+ and HCO^+ by R. C. Woods and his associates [46, 47] in 1975, the spectroscopy of molecular ions has become a rapidly developing component of microwave spectroscopy [48].

The extensive observation of microwave spectral lines of molecules in interstellar space is an exciting development, most of which has occurred since the earlier edition. Lines from molecular free radicals and ions, as well as from stable molecules, are being observed with microwave spectral telescopes, operated primarily in the millimeter wave range. Many of these are emission lines originating through a cosmological laser action. The first molecular species to be detected in interstellar space through measurement of its microwave spectral lines was the OH free radical observed in 1963 [49]. The first of the stable molecules thus to be observed [50] in outer space (1968) was ammonia, NH_3, which, it is interesting to recall, was the first terrestrial molecule to be observed with microwave spectroscopy, the first to control the frequencies of an atomic clock, and the molecule used in the first operating maser. Within a few years after these initial microwave observations of molecules in outer space, microwave spectroscopy had become a powerful new technique for astronomical investigations. Laboratory measurement and identification of the observed interstellar transitions greatly facilitated the astronomical applications. Late reviews of this rapidly developing field are available [51–53].

The interesting microwave spectrum in Fig. 1.4, the $J=0\rightarrow1$ rotational transition of $(NNH)^+$ showing the ^{14}N nuclear quadrupole structure, was recorded from the Orion molecular cloud [54]. This molecular species had not been previously observed in the laboratory. Identification of the species was confirmed by comparison of the observed hyperfine pattern with a theoretically calculated pattern, as shown in the figure. Preliminary observations [55] and a tentative identification [56] had been made earlier.

Discrete spectral transitions that are measured primarily in the radiofrequency region up to 1 GHz (1000 MHz) are nuclear resonances (magnetic and electric quadrupole) and molecular beam resonances (electric and magnetic). Measurable pure rotational spectra of gaseous molecules fall predominantly in the microwave region; electron spin magnetic resonance is most

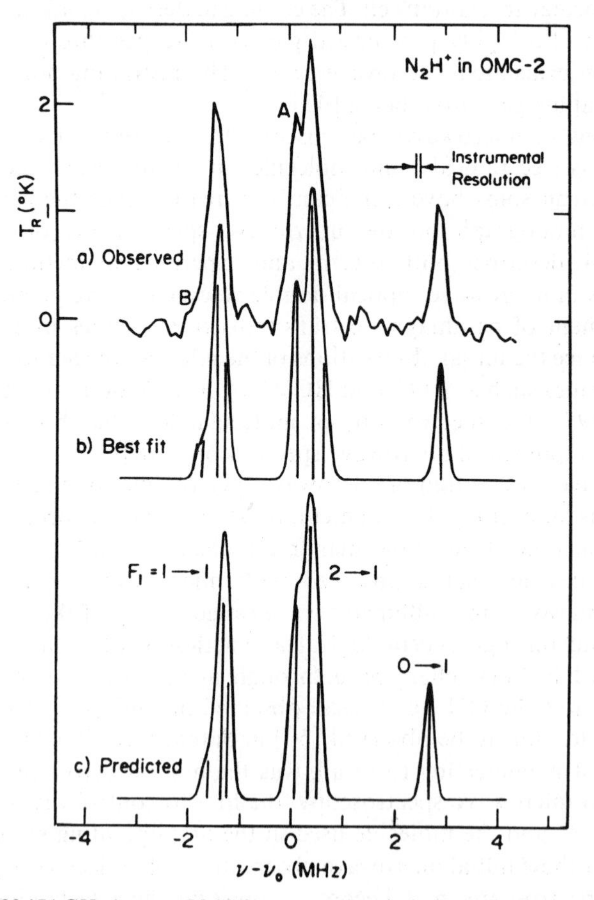

Fig. 1.4 The 93.174 GHz lines in OMC-2, compared with theoretical calculations of the hyperfine structure of the $J=1\rightarrow0$ transition of N_2H^+. From Thaddeus and Turner [54].

advantageously observed in this region. The mathematical methods for treating these types of spectra, the procedures for calculating the energy eigenvalues, line intensities, selectrion rules, and so on, are basically similar. They all include momentum or intrinsic spin operators, dipole moment or direction cosine matrix elements, and state populations that depend on similar statistical mechanisms. The applications described in this volume are primarily rotational spectra of gaseous molecules. Microwave electron spin resonance spectra in condensed matter is treated in a companion volume of this series [57].

The development and increased availability of high-speed computers have paralleled the improvements in techniques and the expanding frequency coverage of microwave spectroscopy. These computers have greatly augmented the contributions in microwave spectroscopy by facilitating the interpretation of complex spectra.

We make no attempt to give a complete or comprehensive treatment of microwave spectroscopy. The field is now so large that such a treatment is impossible in a single volume. We shall limit our coverage mainly to problems of interest to chemists and chemical physicists. The tabulations and references we give are not intended to be complete but to be illustrative of what has been done and what can be done in the field. There are other books that describe instruments and techniques as well as spectra [58–63]. Extensive tabulations of line frequencies measured with microwave spectroscopy are given in a series of volumes published under the auspices of the National Bureau of Standards [64]. Extensive tabulations of molecular structures and constants derived from microwave spectroscopy are available in the Landolt-Börnstein series [65]. *Molecular Spectroscopy*, an annual review published by the Chemical Society (London), covers "Microwave Spectroscopy" from 1970 [66]. A book by Mizushima [67] on the theory of rotating diatomic molecules contains many applications to microwave spectra. A tabulation of the constants of diatomic molecules (to 1978) has been provided by Huber and Herzberg [68]. The valuable book *Modern Aspects of Microwave Spectroscopy*, edited by Chantry, is already referenced several times [13, 30, 32, 39, 42, 52]. Advances in the application of microwave spectroscopy to the area of chemical analysis are reviewed elsewhere [69, 70].

References

1. W. Gordy, W. V. Smith, and R. F. Trambarulo, *Microwave Spectroscopy*, Wiley, New York, 1953. Republication by Dover, New York, 1966.
2. C. H. Townes and A. L. Schawlow, *Microwave Spectroscopy*, McGraw-Hill, New York, 1955.
3. C. A. Burrus and W. Gordy, *Phys. Rev.*, **93**, 897 (1954).
4. L. Genzel and W. Eckhardt, *Z. Physik*, **139**, 592 (1954).
5. W. C. King and W. Gordy, *Phys. Rev.*, **90**, 319 (1953).
6. M. Cowan and W. Gordy, *Phys. Rev.*, **111**, 209 (1958).
7. G. Jones and W. Gordy, *Phys. Rev.*, **135**, A295 (1964).
8. P. Helminger, F. C. De Lucia, and W. Gordy, *Phys. Rev. Lett.*, **25**, 1397 (1970).

9. P. Helminger, J. K. Messer, and F. C. De Lucia, *Appl. Phys. Lett.*, **42**, 309 (1983).
10. R. S. Ohl, P. P. Budenstein, and C. A. Burrus, *Rev. Sci. Instrum.*, **30**, 765 (1959).
11. W. Steinbach and W. Gordy, *Phys. Rev.*, **A8**, 1753 (1973).
12. A. F. Krupnov and A. V. Burenin, "New Methods in Submillimeter Microwave Spectroscopy," in *Molecular Spectroscopy: Modern Research*, Vol. 2, K. N. Rao and C. W. Mathews, Eds., Academic, New York, 1976, pp. 93–126.
13. A. F. Krupnov, "Modern Submillimetre Microwave Scanning Spectrometry," in *Modern Aspects of Microwave Spectroscopy*, G. W. Chantry, Ed., Academic, London, 1979, pp. 217–256.
14. S. P. Belov, L. I. Gershstein, A. F. Krupnov, A. V. Maslovskij, S. Urban, V. Špirko, and D. Papoušek, *J. Mol. Spectrosc.*, **84**, 288 (1980).
15. W. Gordy, "Microwave Spectroscopy in the Region of 4–0.4 Millimetres," in *Pure and Applied Chemistry*. Vol. 11, Butterworths, London, 1965, pp. 403–434.
16. G. Winnewisser, M. Winnewisser, and B. P. Winnewisser, "Millimetre Wave Spectroscopy," in *MTP International Review of Science, Physical Chemistry*, Vol. 3, *Spectroscopy*, D. A. Ramsey, Ed., Butterworths, London, 1972, pp. 241–296.
17. F. C. De Lucia, "Millimeter and Submillimeter Wave Spectroscopy," in *Molecular Spectroscopy: Modern Research*, K. N. Rao and C. W. Mathews, Eds., Vol. 2, Academic, New York, 1976, pp. 69–92.
18. Y. Morino and S. Saito, "Microwave Spectroscopy," in *Molecular Spectroscopy: Modern Research*, K. N. Rao and C. W. Mathews, Eds., Vol. 1, Academic, New York, 1972, pp. 9–28.
19. D. R. Johnson and R. Pearson, Jr., "Microwave Region," in *Methods in Experimental Physics*, Vol. 13, *Spectroscopy*, Part B, D. Williams, Ed., Academic, New York, 1976, pp. 102–133.
20. J. P. Gordon, H. J. Zeiger, and C. H. Townes, *Phys. Rev.*, **99**, 59 (1955).
21. J. Brossel and A. Kastler, *Compt. Rendu Acad. Sci. (Paris)*, **229**, 1213 (1949); F. Bitter, *Phys. Rev.*, **76**, 833 (1949).
22. T. A. Cool, "Chemically Pumped Lasers," in *Methods in Experimental Physics*, Vol. 15, *Quantum Electronics*, Part B, C. L. Tang, Ed., Academic, New York, 1979, pp. 95–142.
23. F. C. De Lucia and W. Gordy, *Phys. Rev.*, **187**, 58 (1969).
24. R. M. Garvey and F. C. De Lucia, *Can. J. Phys.*, **55**, 1115 (1977).
25. D. C. Lainé, "Advances in Molecular Beam Masers," in *Advances in Electronics and Electron Physics*, Vol. 39, L. Marton, Ed., Academic, New York, 1975, pp. 183–251.
26. T. Shimizu and T. Oka, *J. Chem. Phys.*, **53**, 2536 (1970).
27. M. Takami and K. Shimoda, *J. Mol. Spectrosc.*, **59**, 35 (1976).
28. I. Botskor and H. Jones, *J. Mol. Spectrosc.*, **81**, 1 (1980).
29. F. Kohler, H. Jones, and H. D. Rudolph, *J. Mol. Spectrosc.*, **80**, 56 (1980).
30. H. Jones, "Infrared Microwave Double Resonance Techniques," in *Modern Aspects of Microwave Spectroscopy*, G. W. Chantry, Ed., Academic, London, 1979, pp. 123–216.
31. W. Demtroder, *Laser Spectroscopy*, Springer-Verlag, New York, 1980.
32. J. G. Baker, "Microwave–Microwave Double Resonance," in *Modern Aspects of Microwave Spectroscopy*, G. W. Chantry, Ed., Academic, London, 1979, pp. 65–122.
33. H. A. Gebbie, N. W. B. Stone, and F. D. Findlay, *Nature*, **202**, 685 (1964).
34. D. R. Lide and A. G. Maki, *Appl. Phys. Lett.*, **11**, 62 (1967).
35. L. O. Hocker, A. Javan, D. R. Rao, L. Frankel, and T. Sullivan, *Appl. Phys. Lett.*, **10**, 147 (1967).
36. L. O. Hocker and A. Javan, *Phys. Lett.*, **A25**, 489 (1967).
37. L. O. Hocker and A. Javan, *Phys. Lett.*, **A26**, 255 (1968).
38. C. L. Tang, Ed., *Quantum Electronics*, Vol. 15, Part B of *Methods of Experimental Physics*, Academic, New York, 1979.

39. G. Roussy and G. W. Chantry, "Microwave Spectrometers," in *Modern Aspects of Microwave Spectroscopy*, G. W. Chantry, Ed., Academic, London, 1979, pp. 1–63.

40. J. Ekkers and W. H. Flygare, *Rev. Sci. Instrum.*, **47**, 448 (1976).

41. T. J. Balle, E. J. Campbell, M. R. Keenan, and W. H. Flygare, *J. Chem. Phys.*, **72**, 922 (1980).

42. J. W. Fleming, "Interferometric Spectrometry at Millimetre and Submillimetre Wavelengths," in *Modern Aspects of Microwave Spectroscopy*, G. W. Chantry, Ed., Academic, London, 1979, pp. 257–309.

43. D. H. Levy, "The Spectroscopy of Very Cold Gases," *Science*, **214**, 263–269 (1981).

44. A. Carrington, *Microwave Spectroscopy of Free Radicals*, Academic, New York, 1974.

45. M. A. El-Sayed, "Phosphorescence–Microwave Multiple Resonance Spectroscopy," in *MTP International Review of Science, Physical Chemistry*, Vol. 3, *Spectroscopy*, D. A. Ramsey, Ed., Butterworths, London, 1972, pp. 119–153.

46. T. A. Dixon and R. C. Woods, *Phys. Rev. Lett.*, **34**, 61 (1975).

47. R. C. Woods, T. A. Dixon, J. R. Saykally, and P. G. Szanto, *Phys. Rev. Lett.*, **35**, 1269 (1975).

48. R. J. Saykally and R. C. Woods, "High Resolution Spectroscopy of Molecular Ions," *Ann. Rev. Phys. Chem.*, **32**, 403–431 (1981).

49. S. Weinreb, A. H. Barrett, M. L. Meeks, and J. C. Henry, *Nature*, **200**, 829 (1963).

50. A. C. Cheung, D. M. Rank, C. H. Townes, D. D. Thornton, and W. J. Welch, *Phys. Rev. Lett.*, **21**, 1701 (1968).

51. W. B. Somerville, "Interstellar Radio Spectrum Lines," *Rep. Prog. Phys.*, **40**, 483–565 (1977).

52. G. Winnewisser, E. Churchwell, and C. M. Walmsley, "Astrophysics of Interstellar Molecules," in *Modern Aspects of Microwave Spectroscopy*, G. W. Chantry, Ed., Academic, London, 1979, pp. 313–503.

53. A. Carrington and D. A. Ramsey, Eds., *Molecules in Interstellar Space*, The Royal Society, London, 1982.

54. P. Thaddeus and B. F. Turner, *Astrophys. J. (Letters)*, **201**, L25 (1975).

55. B. F. Turner, *Astrophys. J. (Letters)*, **193**, L83 (1974).

56. S. Green, J. A. Montgomery, Jr., and P. Thaddeus, *Astrophys. J. (Letters)*, **193**, L89 (1974).

57. W. Gordy, *Theory and Applications of Electron Spin Resonance*, Wiley-Interscience, New York, 1980.

58. M. W. P. Strandberg, *Microwave Spectroscopy*, Methuen, London, 1954.

59. D. J. E. Ingram, *Spectroscopy at Radio and Microwave Frequencies*, Butterworths, London, 1955. Second Ed., Plenum Press, New York, 1967.

60. T. M. Sugden and C. N. Kenney, *Microwave Spectroscopy of Gases*, Van Nostrand, London, 1965.

61. J. B. Wollrab, *Rotational Spectra and Molecular Structure*, Academic, New York, 1967.

62. J. G. Baker gives an excellent treatment of harmonic generators and semiconductor detectors for millimeter and submillimeter waves in *Spectroscopic Techniques for Far Infra-red, Submillimetre and Millimetre Waves*, D. H. Martin, Ed., North Holland Publishing Co., Amsterdam, 1967, Ch. 5. E. H. Putley and D. H. Martin give a thorough treatment of bolometer detectors as well as point detectors. Op cit., Ch. 4.

63. H. W. Kroto, *Molecular Rotation Spectra*, John Wiley, London, 1975.

64. *Microwave Spectra Tables*, National Bureau of Standards, U.S. Department of Commerce, Washington, D.C. From 1964.

65. Landolt-Börnstein, *Numerical Data and Functional Relationships in Science and Technology, Group II, Atomic and Molecular Physics*, K.-H. Hellwege and A. M. Hellwege, Eds. Vol. 4, *Molecular Constants from Microwave Spectroscopy*, by B. Starck, Springer-Verlag, Berlin, 1967. Vol. 6, (Supplement and Extension to Vol. 4), *Molecular Constants from Microwave, Molecular Beam, and Electron Spin Resonance Spectroscopy*, Contributors: J. Demaison, W. Hüttner, B. Starck/I. Buck, R. Tischer, and M. Winnewisser. Springer-Verlag, Berlin, 1974.

Vol. 7, *Structure Data of Free Polyatomic Molecules*, J. H. Callomon, E. Hirota, K. Kuchitsu, W. J. Lafferty, A. G. Maki, and C. S. Pote, with assistance of I. Buck and B. Starck. Springer-Verlag, Berlin, 1976. Vol. 14 (Supplement to Vols. II/4 and II/6), *Molecular Constants Mostly from Microwave, Molecular Beam, and Electron Resonance Spectroscopy*, J. Demaison, A. Dubrulle, W. Hüttner, and E. Tiemann. Springer-Verlag, Berlin, 1982.

66. *Molecular Spectroscopy: Specialized Periodical Reports*, The Chemical Society, London. "Microwave Spectroscopy," A. C. Ligon and D. J. Millen, Vol. 1 (1973), pp. 1–61; Vol. 2 (1974), pp. 1–99; Vol. 3 (1975), pp. 1–103. J. N. Macdonald and J. Sheridan, Vol. 4 (1976), pp. 1–69; Vol. 5 (1977), pp. 1–59; Vol. 6 (1979), pp. 1–45.
67. M. Mizushima, *The Theory of Rotating Diatomic Molecules*, Wiley, New York, 1975.
68. K. P. Huber and G. Herzberg, *Molecular Spectra and Molecular Structure*, Vol. IV, *Constants of Diatomic Molecules*, Van Nostrand Reinhold, New York, 1979.
69. R. L. Cook and G. E. Jones, "Microwave Spectroscopy," in *Systematic Materials Analysis*, J. H. Richards and R. V. Peterson, Eds., Vol. 2, Academic, New York, 1974.
70. R. Varma and L. W. Hrubesh, "Chemical Analysis by Microwave Rotational Spectroscopy," in *Chemical Analysis*, Vol. 52, P. J. Elving and J. D. Winefordner, Eds., John Wiley, New York, 1979.

Chapter **II**

THEORETICAL ASPECTS OF MOLECULAR ROTATION

1 CLASSICAL ANGULAR MOMENTA AND ROTATIONAL ENERGY

Derivation of the quantum mechanical properties of molecular rotors, including their microwave spectra, begins with the classical expressions for the angular momenta and rotational energy. Likewise, the final derivation of the molecular structure from the observed spectral constants requires a knowledge of classical moments of inertia. Hence we begin with a brief summary of the classical mechanics of rotating bodies.

The classical angular momentum of a rigid system of particles

$$\mathbf{P}=\mathbf{I}\cdot\boldsymbol{\omega} \tag{2.1}$$

where $\boldsymbol{\omega}$ is the angular velocity and $\boldsymbol{I}$ is the moment of inertia tensor which in dyadic notation is written as

$$\begin{aligned}\mathbf{I}=I_{xx}\boldsymbol{ii}+I_{xy}\boldsymbol{ij}+I_{xz}\boldsymbol{ik}\\+I_{yx}\boldsymbol{ji}+I_{yy}\boldsymbol{jj}+I_{yz}\boldsymbol{jk}\\+I_{zx}\boldsymbol{ki}+I_{zy}\boldsymbol{kj}+I_{zz}\boldsymbol{kk}\end{aligned} \tag{2.2}$$

with

$$
\begin{aligned}
I_{xx} &= \sum m(y^2+z^2) \\
I_{yy} &= \sum m(z^2+x^2) \\
I_{zz} &= \sum m(x^2+y^2) \\
I_{xy} &= I_{yx} = -\sum mxy \\
I_{zx} &= I_{xz} = -\sum mxz \\
I_{yz} &= I_{zy} = -\sum myz
\end{aligned} \tag{2.3}
$$

in which m is the mass of a particular particle and x, y, z are its positional coordinates relative to a rectangular coordinate system fixed in the body and with its origin at the center of gravity of the body. The summation is taken over all the particles of the body. The origin of the coordinate system is chosen at the center of mass because this choice allows the total kinetic energy to be written as the sum of the kinetic energy of translational motion of the center of mass plus the kinetic energy of the motion relative to the center of mass. The translational and rotational motions can hence be treated separately. It is always possible to choose the coordinate axes in such a way that the products of inertia vanish, leaving only the diagonal elements, called the principal moments of inertia. The principal moments are the three roots I of the cubic equation

$$
\begin{vmatrix}
I_{xx}-I & I_{xy} & I_{xz} \\
I_{yx} & I_{yy}-I & I_{yz} \\
I_{zx} & I_{zy} & I_{zz}-I
\end{vmatrix} = 0 \tag{2.4}
$$

When the notation x, y, z represents the principal axes system, the components of angular momentum become

$$
P_x = I_x\omega_x, \qquad P_y = I_y\omega_y, \qquad P_z = I_z\omega_z \tag{2.5}
$$

The rotational kinetic energy is

$$
\begin{aligned}
E_r &= \tfrac{1}{2}\boldsymbol{\omega}\cdot\boldsymbol{I}\cdot\boldsymbol{\omega} \\
&= \tfrac{1}{2}(I_{xx}\omega_x^2 + I_{yy}\omega_y^2 + I_{zz}\omega_z^2 + 2I_{xy}\omega_x\omega_y + 2I_{xz}\omega_x\omega_z + 2I_{yz}\omega_y\omega_z)
\end{aligned} \tag{2.6}
$$

which in the principal axes system becomes

$$
\begin{aligned}
E_r &= \tfrac{1}{2}I_x\omega_x^2 + \tfrac{1}{2}I_y\omega_y^2 + \tfrac{1}{2}I_z\omega_z^2 \\
&= \frac{1}{2}\left(\frac{P_x^2}{I_x}\right) + \frac{1}{2}\left(\frac{P_y^2}{I_y}\right) + \frac{1}{2}\left(\frac{P_z^2}{I_z}\right)
\end{aligned} \tag{2.7}
$$

Now suppose that the body is subjected to a torque

$$
\boldsymbol{\tau} = \boldsymbol{i}\tau_X + \boldsymbol{j}\tau_Y + \boldsymbol{k}\tau_Z \tag{2.8}
$$

relative to the space-fixed axes X, Y, Z. The time rate of change of angular momentum relative to the space-fixed axes is equal to the applied torque,

$$\frac{d\mathbf{P}}{dt}=\boldsymbol{i}\left(\frac{dP_X}{dt}\right)+\boldsymbol{j}\left(\frac{dP_Y}{dt}\right)+\boldsymbol{k}\left(\frac{dP_Z}{dt}\right)=\boldsymbol{\tau} \tag{2.9}$$

where X, Y, Z are space-fixed axes. When no torque is applied, it is evident that

$$\mathbf{P}=\boldsymbol{i}P_X+\boldsymbol{j}P_Y+\boldsymbol{k}P_Z=\text{constant} \tag{2.10}$$

also that the components, and hence P_X^2, P_Y^2, and P_Z^2 are each constant. Thus

$$P^2=P_X^2+P_Y^2+P_Z^2=\text{constant} \tag{2.11}$$

The rate of change of the total angular momentum with reference to a system x, y, z, fixed in a rotating body, caused by a torque relative to that system, consists of two parts. One part is due to the time rate of change of the components P_x, P_y, P_z; the other is due to the fact that the body-fixed axes x, y, z are themselves rotating with angular velocities ω_x, ω_y, ω_z. The latter contributes a term $\boldsymbol{\omega}\times\mathbf{P}$ to the rate of change of the angular momentum. Thus

$$\boldsymbol{i}\frac{dP_x}{dt}+\boldsymbol{j}\frac{dP_y}{dt}+\boldsymbol{k}\frac{dP_z}{dt}+\boldsymbol{\omega}\times\mathbf{P}=\boldsymbol{\tau} \tag{2.12}$$

Note that $\boldsymbol{i}, \boldsymbol{j}, \boldsymbol{k}$ are now unit vectors of the body-fixed system. The corresponding component equations are

$$\begin{aligned}
&\frac{dP_x}{dt}+\omega_y P_z-\omega_z P_y=\tau_x\\
&\frac{dP_y}{dt}+\omega_z P_x-\omega_x P_z=\tau_y\\
&\frac{dP_z}{dt}+\omega_x P_y-\omega_y P_x=\tau_z
\end{aligned} \tag{2.13}$$

Now x, y, z are chosen as the principal axes of the body so that $P_x=I_x\omega_x$, $P_y=I_y\omega_y$, $P_z=I_z\omega_z$. Substitution of the values of ω from these equations, upon the assumption that the body is rotating freely, with no torque applied ($\boldsymbol{\tau}=0$) leads to these equations

$$\begin{aligned}
&\frac{dP_x}{dt}+\left(\frac{1}{I_y}-\frac{1}{I_z}\right)P_yP_z=0\\
&\frac{dP_y}{dt}+\left(\frac{1}{I_z}-\frac{1}{I_x}\right)P_zP_x=0\\
&\frac{dP_z}{dt}+\left(\frac{1}{I_x}-\frac{1}{I_y}\right)P_xP_y=0
\end{aligned} \tag{2.14}$$

These are known as Euler's equations of motion. Multiplication of the first by P_x, the second by P_y, and the third by P_z, followed by addition of the three, yields

$$P_x\left(\frac{dP_x}{dt}\right)+P_y\left(\frac{dP_y}{dt}\right)+P_z\left(\frac{dP_z}{dt}\right)=0 \tag{2.15}$$

which upon integration and multiplication by 2 yields

$$P_x^2+P_y^2+P_z^2=\text{constant}=P^2 \tag{2.16}$$

This shows that when no torque is applied, the square of the total angular momentum expressed in the body-fixed axes is constant. Multiplication of the respective equations (2.14) by

$$\frac{P_x}{I_x}, \quad \frac{P_y}{I_y}, \quad \text{and} \quad \frac{P_z}{I_z}$$

followed by addition and integration yields

$$E_r=\tfrac{1}{2}\left(\frac{P_x^2}{I_x}+\frac{P_y^2}{I_y}+\frac{P_z^2}{I_z}\right)=\text{constant} \tag{2.17}$$

which shows that with no torque applied the kinetic energy of rotation remains constant.

In angular momentum space with P_x, P_y, and P_z as the coordinates of a point, it is evident that (2.16) is an equation of a sphere with radius P. In the same momentum coordinates, (2.17) represents an ellipsoid with principal semi-axes of $(2I_xE_r)^{1/2}$, $(2I_yE_r)^{1/2}$, and $(2I_zE_r)^{1/2}$. Since the values of P_x, P_y, and P_z must satisfy both equations, it is evident that end points of **P** can only be along the intersection of the sphere and the ellipsoid. If the sphere and the ellipsoid are to intersect, the values of P must be between those of the minimum and maximum axes of the ellipsoid. This is illustrated by the prolate, symmetric-top case (Fig. 2.1) for which the two smaller semi-axes of the energy ellipsoid are equal ($I_x=I_y$).

The energy ellipsoid is fixed in the body, whereas **P** is fixed in magnitude and in direction in space. Therefore the body must rotate in such a manner that the fixed vector **P** continues to terminate on the surface of the ellipsoid. For a symmetric top, this motion is such that the terminal of **P** traces circles around the symmetrical ellipsoid, as indicated by Fig. 2.1. For the asymmetric

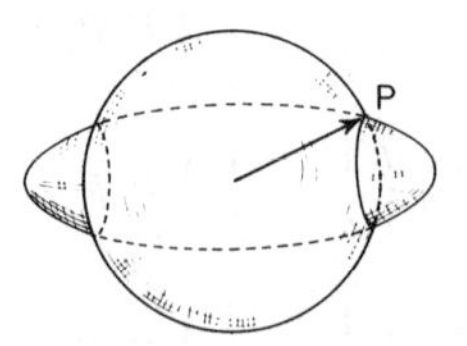

Fig. 2.1 Diagram illustrating the restrictions on the motions of a prolate symmetric top. The allowed values of P are described by the intersection of the momentum sphere (Eq. 2.16) and the energy ellipsoid (Eq. 2.17).

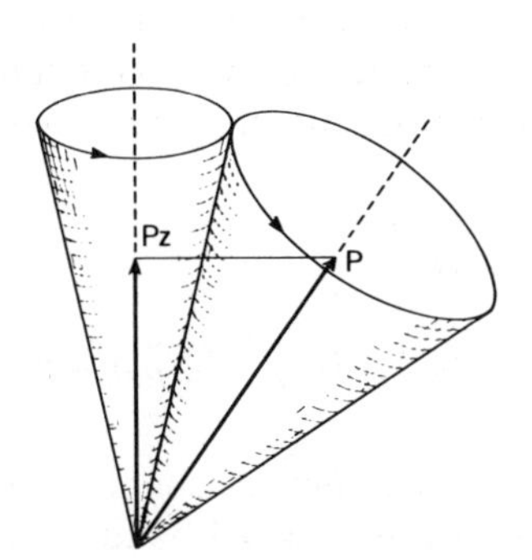

Fig. 2.2 Simulation of the classical motions of a prolate symmetric top.

rotor in which none of the semi-axes are equal, the curve traced out by **P** on the surface of the ellipsoid is more complicated, but the stable motions are still highly restricted.

The rotational motions of a symmetric top are simulated by a cone that rolls without slipping around the surface of a second cone that is fixed in space. The total momentum vector **P** is along the axis of the fixed cone, and the component P_z is along the axis of the rolling cone, as indicated by Fig. 2.2 for the prolate top.

That P_z is a constant of the motion of a symmetric top is easily proved by the substitution of $I_x = I_y$ in (2.14). One obtains

$$\frac{dP_z}{dt} = 0 \qquad \text{hence} \qquad P_z = \text{constant} \tag{2.18}$$

For the asymmetric rotor $I_x \neq I_y \neq I_z$ there is, in contrast, no internal axis about which the rotation is constant, although P^2 is a constant for the asymmetric rotor and is independent of the coordinate system in which it is expressed.

2 ANGULAR MOMENTUM OPERATORS AND MATRIX ELEMENTS

Most of the problems treated in this volume involve angular momentum operators. The characteristic energy levels required for finding the microwave spectral frequencies are eigenvalues of the Hamiltonian operators which usually can be expressed in terms of the angular momentum operators of a particle or system of particles. The matrix elements of these angular momentum operators are therefore useful in finding the characteristic energies of the system, that is, the eigenvalues of the Hamiltonian operator.

The classical angular momentum of a system of particles can be expressed by

$$\mathbf{P} = \sum_n \mathbf{r}_n \times \mathbf{p}_n \tag{2.19}$$

where $\mathbf{p}_n$ is the instantaneous linear momentum $m_n\mathbf{v}_n$ of the nth particle and $\mathbf{r}_n$ is its radius vector from the center of rotation assumed to be fixed in space. Expanded in terms of its components in space-fixed rectangular coordinates X, Y, Z, it is

$$\begin{aligned}\mathbf{P}&=\boldsymbol{i}P_X+\boldsymbol{j}P_Y+\mathbf{k}P_Z\\&=\sum_n\left[\boldsymbol{i}(Yp_Z-Zp_Y)_n+\boldsymbol{j}(Zp_X-Xp_Z)_n+\mathbf{k}(Xp_Y-Yp_X)_n\right]\end{aligned}\tag{2.20}$$

To derive the corresponding quantum mechanical angular momentum operators one substitutes the relations $X\to X$, and so on, and $p_X\to(\hbar/i)(\partial/\partial X)$, and so on. Thus the component angular momentum operators are

$$\begin{aligned}P_X&=\sum_n\frac{\hbar}{i}\left[Y\left(\frac{\partial}{\partial Z}\right)-Z\left(\frac{\partial}{\partial Y}\right)\right]_n\\P_Y&=\sum_n\frac{\hbar}{i}\left[Z\left(\frac{\partial}{\partial X}\right)-X\left(\frac{\partial}{\partial Z}\right)\right]_n\\P_Z&=\sum_n\frac{\hbar}{i}\left[X\left(\frac{\partial}{\partial Y}\right)-Y\left(\frac{\partial}{\partial X}\right)\right]_n\end{aligned}\tag{2.21}$$

where $i=(-1)^{1/2}$ and $\hbar=h/2\pi$. In spherical coordinates these component operators are

$$\begin{aligned}P_X&=\frac{\hbar}{i}\left[-\sin\phi\left(\frac{\partial}{\partial\theta}\right)-\cot\theta\cos\phi\left(\frac{\partial}{\partial\phi}\right)\right]\\P_Y&=\frac{\hbar}{i}\left[\cos\phi\left(\frac{\partial}{\partial\theta}\right)-\cot\theta\sin\phi\left(\frac{\partial}{\partial\phi}\right)\right]\\P_Z&=\frac{\hbar}{i}\left(\frac{\partial}{\partial\phi}\right)\end{aligned}\tag{2.22}$$

and the important operator conjugate to the square of the total angular momentum is

$$P^2=-\hbar^2\left\{\left(\frac{1}{\sin\theta}\right)\left(\frac{\partial}{\partial\theta}\right)\left[\sin\theta\left(\frac{\partial}{\partial\theta}\right)\right]+\frac{1}{\sin^2\theta}\left(\frac{\partial^2}{\partial\phi^2}\right)\right\}\tag{2.23}$$

It is easily shown from the foregoing expression that P^2 commutes with its component operators, for example,

$$P^2P_Z-P_ZP^2=0\tag{2.24}$$

The component operators do not commute among themselves, however. The following commutation rules are easily shown to hold

$$P_XP_Y-P_YP_X=i\hbar P_Z$$

$$P_Y P_Z - P_Z P_Y = i\hbar P_X \tag{2.25}$$
$$P_Z P_X - P_X P_Z = i\hbar P_Y$$

These important commutation relations for the components can be expressed most compactly by the vector equation

$$\mathbf{P} \times \mathbf{P} = i\hbar \mathbf{P} \tag{2.26}$$

Also of convenience for some manipulations are the operators defined by

$$P_+ = P_X + iP_Y \tag{2.27}$$

$$P_- = P_X - iP_Y \tag{2.28}$$

The rules for commutation of these operators are

$$P_+P_- - P_-P_+ = 2\hbar P_Z \tag{2.29}$$

$$P_Z P_+ - P_+ P_Z = \hbar P_+ \tag{2.30}$$

$$P_Z P_- - P_- P_Z = -\hbar P_- \tag{2.31}$$

Since the component operators P_X and P_Y commute with P^2, it is evident that P_+ and P_- commute with P^2.

It is one of the principles of quantum mechanics that operators which commute have common sets of eigenfunctions. Therefore P^2 and P_Z have common eigenfunctions, which we designate as $\psi_{J,M}$. Thus we can write

$$P^2 \psi_{J,M} = k_J \psi_{J,M} \tag{2.32}$$

$$P_Z \psi_{J,M} = k_M \psi_{J,M} \tag{2.33}$$

where k_J and k_M represent temporarily the eigenvalues of P^2 and P_Z corresponding to the eigenstate described by $\psi_{J,M}$. By application of P_+ and P_- (called raising and lowering operators) to (2.33) and various other manipulations [1] the quantized values $k_J = \hbar^2 J(J+1)$ and $k_M = \hbar M$ are obtained, where M and J have integral values and $|M| \leqslant J$. Thus (2.32) and (2.33) can be expressed as

$$P^2 \psi_{J,M} = \hbar^2 J(J+1) \psi_{J,M} \tag{2.34}$$

$$P_Z \psi_{J,M} = \hbar M \psi_{J,M} \tag{2.35}$$

where

$$J = 0, 1, 2, 3, \ldots \quad \text{and} \quad M = J, J-1, J-2, \ldots, -J \tag{2.36}$$

With the commutation rules it can be shown that

$$P_+ \psi_{J,M} = C_+ \psi_{J,M+1} \tag{2.37}$$

$$P_- \psi_{J,M} = C_- \psi_{J,M-1} \tag{2.38}$$

where C_+ and C_- are scalar constants. Thus P_+ and P_- are raising and lowering operators in the sense that operation with P_+ on $\psi_{J,M}$ raises the M subscript by one, whereas operation with P_- lowers it by one. The constant C can be evaluated upon the condition that the wave function be properly normalized. Multiplication of (2.37) and (2.38) by their complex conjugates and integration over all coordinates with the use of (2.34) and (2.35) and the normalizations

$$\int \psi^*_{J,M+1}\psi_{J,M+1}d\tau = 1 \qquad \text{and} \qquad \int \psi^*_{J,M-1}\psi_{J,M-1}\,d\tau = 1 \tag{2.39}$$

yield the values

$$C_+ = i\hbar[J(J+1) - M(M+1)]^{1/2} \tag{2.40}$$

$$C_- = -i\hbar[J(J+1) - M(M-1)]^{1/2} \tag{2.41}$$

Consequently (2.37) and (2.38) become

$$P_+\psi_{J,M} = i\hbar[J(J+1) - M(M+1)]^{1/2}\psi_{J,M+1} \tag{2.42}$$

$$P_-\psi_{J,M} = -i\hbar[J(J+1) - M(M-1)]^{1/2}\psi_{J,M-1} \tag{2.43}$$

The particular choice of the C's as imaginary is discussed at the end of this section. Multiplication of (2.42) by $\psi^*_{J,M+1}$ and (2.43) by $\psi^*_{J,M-1}$ and integration over all the coordinates yield the nonvanishing matrix elements of P_+ and P_- given below, (2.45) and (2.46).

For convenience, we shall use in most applications the abbreviated bracket notation for indication of wave functions and matrix elements. For example, the matrix elements $\int \psi^*_{J,M+1} P_+\psi_{J,M}\,d\tau$ of the operator P_+ in the representation J, M is indicated by

$$\int \psi^*_{J,M+1}P_+\psi_{J,M}\,d\tau \equiv (J,\, M+1|P_+|J,\, M) \tag{2.44}$$

The nonvanishing matrix elements of P_+ and P_- found from (2.42) and (2.43) are

$$(J,\, M+1|P_+|J,\, M) = i\hbar[J(J+1) - M(M+1)]^{1/2} \tag{2.45}$$

$$(J,\, M-1|P_-|J,\, M) = -i\hbar[J+1) - M(M-1)]^{1/2} \tag{2.46}$$

That these are all the nonvanishing matrix elements of P_+ and P_- can be seen by the fact that multiplication by any other member of the orthogonal set of functions, say by $\psi^*_{J,M+2}$, would have reduced the right-hand side of the equation to zero because of the orthogonality requirement, $\int \psi^*_{J,M}\psi_{J',M'}\,d\tau = 0$ when either $J \neq J'$ or $M \neq M'$. All eigenfunctions of Hermitian operators having different eigenvalues must be orthogonal, and the functions $\psi_{J,M}$ are eigenfunctions of the Hermitian operators P^2 and P_Z.

From (2.34) and (2.35) it is evident that the nonvanishing matrix elements of P^2 and P_Z are

$$(J,\, M|P^2|J,\, M) = \hbar^2 J(J+1) \tag{2.47}$$

$$(J,\, M|P_Z|J,\, M) = \hbar M \tag{2.48}$$

It is, of course, always true that the matrix of an operator in the representation of its eigenfunctions is diagonal.

The operators P_X and P_Y are not diagonal in the J, M representation, that is, in the eigenfunctions of P^2 and P_Z, because they do not commute with P_Z. Since P_X or P_Y commutes with P^2, we could have chosen a set of eigenfunctions common to P^2 and P_X, or to P^2 and P_Y, but then the matrix elements of P_Z would not have been diagonal in either of these representations. The choice of the particular pair P^2 and P_Z is, of course, completely arbitrary.

From algebraic manipulation of (2.27), (2.28), (2.45), and (2.46) and with the condition

$$(J, M|P_i|J', M') = (J', M'|P_i|J, M)^* \tag{2.49}$$

that the quantum mechanical operators of all physically real quantities must be Hermitian, the nonvanishing matrix elements of P_X and P_Y in the J, M representation are found to be

$$(J, M|P_Y|J, M\pm 1) = \frac{\hbar}{2}\,[J(J+1) - M(M\pm 1)]^{1/2} \tag{2.50}$$

$$(J, M|P_X|J, M\pm 1) = \mp\frac{i\hbar}{2}\,[J(J+1) - M(M\pm 1)]^{1/2} \tag{2.51}$$

These elements are seen to be diagonal in J, but not in M. The phase choice here is consistent with that usually selected for the body-fixed angular momentum components [see (2.63) and (2.64)], that is, P_Y is real and positive and P_X is imaginary. The choice made for P_X and P_Y in most books on quantum mechanics is just the reverse of this, since $C_\pm$ is chosen to be real rather than imaginary as we have done in (2.40) and (2.41). The choice of phase is, however, completely arbitrary and has no effect on physical observables.

Spin Operators and Matrix Elements

Although intrinsic spin angular momentum has no classical counterpart, the electronic spin vector **S** and the nuclear spin vector **I** are assigned similar angular momentum operators which obey the same commutation rules as those for **P**:

$$\mathbf{S} \times \mathbf{S} = i\hbar\mathbf{S} \tag{2.52}$$

$$\mathbf{I} \times \mathbf{I} = i\hbar\mathbf{I} \tag{2.53}$$

One can then in a similar way find the nonvanishing matrix elements of S^2, I^2, and their various components. These have the same form as those for the operator **P** since they are derived from the same commutation rules. The justification for the assumption of the analogous commutation rules is the test of experience. The consequences of these assumptions are borne out by all measurements of fine or hyperfine structure in atomic or molecular spectra and likewise by electronic spin and nuclear resonance experiments. There is one important difference, however. Whereas the quantum numbers for P_Z have only integral values, those for S_Z or I_Z may take half-integral values also. This difference does not violate in any way the derivations of the angular properties from the commutation rules, which require only that the values of M must differ by integral steps and that the values must be symmetrical about zero. If the smallest numerical value of M is zero, as found for molecular end-over-end rotation, the values of M, hence of J, can be integers only. Note, however, that it is possible to have a ladder of nonintegral values of M separated by integral units and symmetric about zero when—and only when—the nonzero values of M are odd integral multiples of a half. Certain particles or systems of particles are found to have intrinsic spin values which are half-integrals. It follows that $(\frac{1}{2})\hbar$ is the smallest observable component of this momentum. The most notable of such particles is the free electron which has an intrinsic spin of $S=1/2$; hence $M_S=1/2$ and $-1/2$. Organic molecules are commonly observed in triplet states for which $S=1$. Nuclear spin values as high as $I=6$ (for ^{50}V) have been observed, but the most common ones are 9/2 or less.

Matrix elements of the spin operators can be found from (2.47), (2.48), (2.50), and (2.51) by substitution of the I or S for J and of M_I or M_S for M.

The Symmetric-Top Rotor

The symmetric-top rotor has a component of its angular momentum about the internal symmetry axis of inertia which is a constant of the motion (see Section 1). If the body-fixed coordinate system is chosen as x, y, z with z as the axis of symmetry, then the operator P_z will commute with P^2 since both are constants of the motion, that is, are simultaneously defined. Furthermore, P_Z still commutes with P^2 since the latter is independent of the coordinate system employed

$$P^2=P_x^2+P_y^2+P_z^2=P_X^2+P_Y^2+P_Z^2 \tag{2.54}$$

Thus, both P_Z and P_z commute with P^2 and hence with each other and have a common set of eigenfunctions, $\psi_{J,K,M}$. Therefore, in the bracket $\psi_{J,K,M} \equiv |J, K, M)$ notation

$$P^2(J, K, M)=k_J|J, K, M) \tag{2.55}$$

$$P_z|J, K, M)=k_K|J, K, M) \tag{2.56}$$

$$P_Z|J, K, M)=k_M|J, K, M) \tag{2.57}$$

The eigenvalues k_J and k_M must be the same as those previously determined for the rotor in spaced-fixed coordinates without regard to symmetry, or with $k_J=\hbar^2 J(J+1)$ and $k_M=M\hbar$.

In a similar way, the values of k_K can be found from the commutation rules of the angular momentum operators expressed in the internal coordinate system. These rules are similar to those for the space-fixed system except for a change in sign of i. In this system P_z, of course, commutes with P^2, and

$$\begin{aligned} P_xP_y - P_yP_x &= -i\hbar P_z \\ P_yP_z - P_zP_y &= -i\hbar P_x \\ P_zP_x - P_xP_z &= -i\hbar P_y \end{aligned} \tag{2.58}$$

The change in the sign of i has the additional effect of making P_- a raising and P_+ a lowering operator in opposition to the corresponding space-fixed operators.

From these rules the value of k_K is found, as for k_M, to be $K\hbar$, where $|K|$ is an integer equal to, or less than, J. Since P^2 is independent of the coordinate system, k_J is found to be as before, $\hbar^2 J(J+1)$. Thus for the symmetric top the diagonal matrix elements are

$$(J, K, M|P^2|J, K, M) = \hbar^2 J(J+1) \tag{2.59}$$

$$(J, K, M|P_z|J, K, M) = K\hbar \tag{2.60}$$

$$(J, K, M|P_Z|J, K, M) = M\hbar \tag{2.61}$$

where

$$\begin{aligned} J &= 0, 1, 2, \ldots \\ K &= J, J-1, J-2, \ldots, -J \\ M &= J, J-1, J-2, \ldots, -J \end{aligned} \tag{2.62}$$

The matrix elements of P_x and P_y found from the commutation rules in the body-fixed system are independent of M, but we retain the M subscript to indicate the common eigenfunction

$$(J, K, M|P_x|J, K\pm 1, M) = \pm\frac{i\hbar}{2}[J(J+1) - K(K\pm 1)]^{1/2} \tag{2.63}$$

$$(J, K, M|P_y|J, K\pm 1, M) = \frac{\hbar}{2}[J(J+1) - K(K\pm 1)]^{1/2} \tag{2.64}$$

The nonvanishing matrix elements of P_X and P_Y are the same as those already stated in (2.50) and (2.51) since they are independent of the internal coordinates and hence of K.

Squared Operators

In finding the eigenvalues of the Hamiltonian operator we shall often have need of the matrix elements of the squared operators P_x^2, P_y^2, and so on, of the angular momentum operators. These can be found from the matrix elements already given by application of the matrix product rule

$$(J, K, M|P_g^2|J', K', M') = \sum_{J'',K'',M''} (J, K, M|P_g|J'', K'', M'') \times (J'', K'', M''|P_g|J', K', M') \tag{2.65}$$

The matrix elements of P^4 and P_z^2 are:

$$(J, K, M|P^4|J', K', M') = (J, K, M|P^2|J, K, M)^2 = \hbar^4 J^2(J+1)^2 \quad (2.66)$$

and

$$(J, K, M|P_z^2|J, K, M) = K^2\hbar^2 \quad (2.67)$$

However, the matrix elements for P_x^2 and P_y^2 are both diagonal and off-diagonal

$$\begin{aligned}(J, K, M|P_y^2|J', K', M') = {} & [(J, K, M|P_y|J, K+1, M)(J, K+1, M|P_y|J, K, M) \\ & + (J, K, M|P_y|J, K-1, M)(J, K-1, M|P_y|J, K, M)]\delta_{K'K} \\ & + [(J, K, M|P_y|J, K+1, M)(J, K+1, M|P_y|J, K+2, M)]\delta_{K'K+2} \\ & + [(J, K, M|P_y|J, K-1, M)(J, K-1, M|P_y|J, K-2, M)]\delta_{K'K-2} \end{aligned} \quad (2.68)$$

The first two terms on the right are diagonal, and with the aid of the Hermetian property and (2.64) they can be combined to give

$$(J, K, M|P_y^2|J, K, M) = \frac{\hbar^2}{2}[J(J+1) - K^2] \quad (2.69)$$

The last two terms are off-diagonal by two units of K and with (2.64) can be written as

$$\begin{aligned}(J, K, M|P_y^2|J, K\pm 2, M) = {} & \frac{\hbar^2}{4}[J(J+1) - K(K\pm 1)]^{1/2} \\ & \times [J(J+1) - (K\pm 1)(K\pm 2)]^{1/2}\end{aligned} \quad (2.70)$$

where the upper plus signs are to be taken together and the lower minus signs together. In a similar way, the matrix elements of P_x^2 are seen to be

$$(J, K, M|P_x^2|J, K, M) = \frac{\hbar^2}{2}[J(J+1) - K^2] \quad (2.71)$$

$$\begin{aligned}(J, K, M|P_x^2|J, K\pm 2, M) = {} & -\frac{\hbar^2}{4}[J(J+1) - K(K\pm 1)]^{1/2} \\ & \times [J(J+1) - (K\pm 1)(K\pm 2)]^{1/2}\end{aligned} \quad (2.72)$$

3 MATRIX ELEMENTS OF ROTATIONAL HAMILTONIAN OPERATORS

The Hamiltonian operator is obtained from the classical Hamiltonian when the momenta are replaced by their conjugate operators. When no torques are applied, the classical Hamiltonian of the rigid rotor consists of only kinetic energy which can be expressed in terms of the components of angular momentum in the principal axes, as in (2.7). To find the corresponding Hamiltonian operator, one simply substitutes for the P's the conjugate angular momentum operators. In the body-fixed principal axes, x, y, z, this operator is

$$\mathscr{H}_r = \frac{1}{2}\left(\frac{P_x^2}{I_x} + \frac{P_y^2}{I_y} + \frac{P_z^2}{I_z}\right) \tag{2.73}$$

where

$$P_x = \frac{\hbar}{i}\left(y\frac{\partial}{\partial z} - z\frac{\partial}{\partial y}\right), \qquad P_y = \frac{\hbar}{i}\left(z\frac{\partial}{\partial x} - x\frac{\partial}{\partial z}\right), \qquad P_z = \frac{\hbar}{i}\left(x\frac{\partial}{\partial y} - y\frac{\partial}{\partial x}\right) \tag{2.74}$$

The eigenvalues of the Hamiltonian operators represent the quantized energies from which the microwave spectral frequencies are determined. Finding these eigenvalues is therefore one of our most important problems. The Hamiltonian operators dealt with in this volume can usually be expressed in terms of angular momentum operators or intrinsic spin operators which are of the same form as angular momentum operators. For this reason one often makes use of the matrix elements of angular momentum operators and spin operators when finding the energy levels involved in microwave spectral transitions.

If the Hamiltonian operator is found to commute with the angular momentum operators, it will be diagonal in the representation in which those operators are diagonal. Its matrix elements can then be readily found from the known diagonal matrix elements of the angular momentum operators.

As a simple example, let us consider the rigid spherical-top rotor for which the three principal moments of inertia are equal, $I_x = I_y = I_z = I$. The Hamiltonian operator $\mathscr{H}_r$ becomes

$$\mathscr{H}_r = \frac{1}{2I}(P_x^2 + P_y^2 + P_z^2) = \frac{P^2}{2I} \tag{2.75}$$

Since I is a constant, the Hamiltonian obviously commutes with P^2. The nonvanishing matrix elements of $\mathscr{H}_r$, in this case the eigenvalues of $\mathscr{H}_r$, that is, the quantized energy values, are

$$E_J = (J, M|\mathscr{H}_r|J, M) = \frac{1}{2I}(J, M|P^2|J, M) = \frac{\hbar^2 J(J+1)}{2I} \tag{2.76}$$

The Hamiltonian of the symmetric-top rotor commutes with P_z and P_Z as well as with P^2 and is therefore diagonal in the J, K, M representation. This can be easily seen if I_x is set equal to I_y in (2.73); the Hamiltonian for the symmetric top can then be expressed as

$$\mathscr{H}_r = \frac{P^2}{2I_y} + \left(\frac{1}{2I_z} - \frac{1}{2I_y}\right)P_z^2 \tag{2.77}$$

Since P^2 and P_z^2 commute with P_z and are diagonal in the J, K, M representation, $\mathscr{H}_r$ is also diagonal in the same representation, with matrix elements

$$E_{J,K} = (J, K, M|\mathscr{H}_r|J, K, M) = \frac{\hbar^2}{2}\left[\frac{J(J+1)}{I_y} + \left(\frac{1}{I_z} - \frac{1}{I_y}\right)K^2\right] \tag{2.78}$$

which represent the characteristic rotational energies. Note that these diagonal elements of $\mathscr{H}_r$ do not depend on M, in agreement with the classical principle that the rotational energy in the absence of torques is independent of the direction in which the angular momentum vector points in space.

The Hamiltonian operator for the asymmetric-top rotor for which $I_x \neq I_y \neq I_z$ does not commute with the operator P_z or with the other component operators P_x or P_y. It is thus not diagonal in the symmetric-top J, K, M representation. Its matrix elements may be easily found in the symmetric-top eigenfunctions, as is done later; but the resulting matrix will not be diagonal, and the elements will not represent eigenvalues of $\mathscr{H}_r$. However, $\mathscr{H}_r$ does commute with P^2 and P_Z, and hence the matrix will be diagonal in the J and M quantum numbers. In principle, and in practice for J not too high, it is possible to diagonalize the resulting matrix and thus to obtain the eigenvalues of $\mathscr{H}_r$. This is equivalent to the setting up and solving of the secular equation as described in Section 4. For this purpose the matrix elements of $\mathscr{H}_r$ in the J, K, M representation as given below will be needed.

By substitution of the matrix elements of P_x^2, P_y^2, and P_z^2 from (2.67) and (2.69) to (2.72) into the Hamiltonian operator for the asymmetric rotor, (2.73), expressed in the coordinates of its principal axes of inertia, the nonvanishing matrix elements of $\mathscr{H}_r$ are found to be

$$(J, K|\mathscr{H}_r|J, K)=\frac{\hbar^2}{4}\left[J(J+1)\left(\frac{1}{I_x}+\frac{1}{I_y}\right)+K^2\left(\frac{2}{I_z}-\frac{1}{I_x}-\frac{1}{I_y}\right)\right] \quad (2.79)$$

and

$$(J, K|\mathscr{H}_r|J, K\pm 2)=\frac{\hbar^2}{8}(J(J+1)-K(K\pm 1)]^{1/2} \times [J(J+1)-(K\pm 1)(K\pm 2)]^{1/2}\left(\frac{1}{I_y}-\frac{1}{I_x}\right) \quad (2.80)$$

Thus there are diagonal elements of $\mathscr{H}_r$, (2.79), but these do not represent eigenvalues of $\mathscr{H}_r$ because in this representation there are also nonvanishing, off-diagonal elements, (2.80), in K.

4 METHODS OF FINDING EIGENVALUES OF HAMILTONIAN OPERATORS

Certain systems such as symmetric-top rotors have sufficient symmetry that the Schrödinger equation

$$\mathscr{H}\psi=E\psi \quad (2.81)$$

is solvable for the eigenfunctions ψ and the eigenvalues E of the Hamiltonian operators. Alternately, it is possible to find the eigenvalues of $\mathscr{H}$, also the direction cosine matrix elements for such systems from the commutation rules of the component operators as indicated in Sections 3 and 6, even without a specific knowledge of the eigenfunctions.

When the symmetry of the Hamiltonian operator does not allow direct solution of the Schrödinger equation, the customary procedure is to expand the unknown eigenfunction in terms of a known orthogonal set such as those of the symmetric top; for example, let us assume the eigenfunction of (2.81) to be expressed by

$$\psi = \sum_n c_n \psi_n \tag{2.82}$$

where ψ_n represents a member of a normalized orthogonal set of functions and the c_n are weighting constants. Substitution of (2.82) into (2.81) yields

$$\sum_n c_n \mathscr{H} \psi_n = E \sum_n c_n \psi_n \tag{2.83}$$

By multiplication of this equation by the ψ_m^*, or the conjugate of a member of the orthogonal set, and integration over all coordinates, one obtains

$$\sum_n c_n \int \psi_m^* \mathscr{H} \psi_n \, d\tau = E \sum_n c_n \int \psi_m^* \psi_n \, d\tau \tag{2.84}$$

Since the assumed set is orthogonal $\int \psi_m^* \psi_n \, d\tau = 0$ except when $m=n$ and since they are assumed to be normalized, this quantity is unity when $m=n$. For convenience, the matrix of $\mathscr{H}$ may be expressed in the bracket notation $\int \psi_m^* \mathscr{H} \psi_n \, d\tau = (m|\mathscr{H}|n)$. In many texts it is written simply as $\mathscr{H}_{m,n}$. Thus (2.84) can be written as

$$\sum_n c_n [(m|\mathscr{H}|n) - E\delta_{m,n}] = 0 \tag{2.85}$$

where $\delta_{m,n} = 1$ when $m=n$ and $\delta_{m,n} = 0$ when $m \neq n$. The expression (2.85) represents a set of l linear equations containing l unknown coefficients which have a nontrivial solution only if the determinant of the coefficient vanishes, where l is the number of functions in the set. If this determinant is set equal to zero, the secular equation,

$$|(m|\mathscr{H}|n) - E\delta_{m,n}| = 0 \tag{2.86}$$

is obtained. If the matrix elements $(m|\mathscr{H}|n)$ are known or can be found, this equation can, in principle, be solved for the values of E. These values, which are the various roots of the secular equation with E considered as the unknown, represent the eigenvalues of $\mathscr{H}$. By substitution of each of these values of E into (2.87) with the known value of $(m|\mathscr{H}|n)$, ratios of the various coefficients c_n/c_{n-1}, and so on, can be found; and with the auxiliary equation,

$$\sum |c_n|^2 = 1 \tag{2.87}$$

obtained from normalization of the ψ of (2.82), the values of the c_n coefficients can be obtained. Thus the eigenfunction ψ_j corresponding to the particular root j of the secular equation or particular value E_j can be found in terms of the assumed functions ψ_n of some other operator.

It should be noted that a knowledge of the assumed eigenfunctions ψ_n is not actually required for the setting up of the secular equation and hence for the finding of the energy values E_j; only the matrix elements $(m|\mathscr{H}|n)$ are required. These matrix elements (Section 2) can often be found from commutation rules without a specific knowledge of the eigenfunctions. For simplicity we have assumed that only the one quantum number n is required for the labeling of the assumed set of functions preceding; but for the problems considered in this volume additional ones will usually be required.

It is apparent that operators such as $\mathscr{H}$ can be represented by matrices and their eigenvalues can be found by diagonalization of the corresponding matrix. The matrix formulation of quantum mechanics has been developed by Heisenberg. In Appendix A some important features of matrix mechanics pertinent to microwave spectroscopy are reviewed.

A classical example of the application of the secular equation is in the finding of the eigenvalues of the field-free, rigid, asymmetric rotor. As we have seen in Section 3, the matrix elements of the Hamiltonian operator of the asymmetric rotor in the J, K representation, that is, the representation in which the Hamiltonian of the symmetric-top rotor is diagonal, can be readily found. Because these elements are diagonal in J, the secular equation factorizes into subequations corresponding to the different values of J. Since the values of K range in unit steps from J to $-J$, the subdetermined equations have the form

$$\begin{array}{c|cccccc|c}
K/K' & J & J-1 & J-2 & J-3 & \cdots & -J & \\
\hline
J & \mathscr{H}_J^J-E & 0 & \mathscr{H}_J^{J-2} & 0 & \cdots & 0 & \\
J-1 & 0 & \mathscr{H}_{J-1}^{J-1}-E & 0 & \mathscr{H}_{J-1}^{J-3} & \cdots & 0 & \\
J-2 & \mathscr{H}_{J-2}^J & 0 & \mathscr{H}_{J-2}^{J-2}-E & 0 & \cdots & 0 & =0 \\
J-3 & 0 & \mathscr{H}_{J-3}^{J-1} & 0 & \mathscr{H}_{J-3}^{J-3}-E & \cdots & 0 & \\
\cdots & \cdots & \cdots & \cdots & \cdots & \cdots & \cdots & \\
-J & 0 & 0 & 0 & 0 & \cdots & \mathscr{H}_{-J}^{-J}-E &
\end{array} \tag{2.88}$$

in which we have, for convenience, designated $(J, K|\mathscr{H}|J, K')=\mathscr{H}_K^{K'}$. From (2.79) and (2.80) we see that nonvanishing elements of $\mathscr{H}$ occur only for $K'=K$ and for $K'=K\pm 2$. It is evident that the secular determinant has the dimension of $2J+1$ and will therefore have $2J+1$ roots or energy values E corresponding to each value of J. Although the required matrix elements and secular equation can be found easily, solution of this equation becomes increasingly difficult as J increases. It can be solved only with approximation methods and most advantageously with modern computers.

As a more specific illustration, let us solve the secular equation of a rigid asymmetric rotor when $J=1$. From the matrix elements obtained from (2.79) and (2.80) with $A = \hbar^2/2I_x$, $B = \hbar^2/2I_y$, and $C = \hbar^2/(2I_z)$, the secular equation is seen to be

$$\begin{array}{c|ccc|l} K/K' & 1 & 0 & -1 & \\ \hline 1 & \left(\frac{A+B}{2}+C\right)-E & 0 & -\frac{A-B}{2} & \\ 0 & 0 & (A+B)-E & 0 & =0 \\ -1 & -\frac{A-B}{2} & 0 & \left(\frac{A+B}{2}+C\right)-E & \end{array} \quad (2.89)$$

Solution of this cubic equation yields the three values of E as

$$\begin{aligned} E_0 &= A+B \\ E_+ &= B+C \\ E_- &= A+C \end{aligned} \quad (2.90)$$

In the treatment of the higher J values of the asymmetric rotor we change the form of the Hamiltonian to take advantage of the symmetric properties in the reduced energies. These more involved levels are treated in Chapter VII.

5 EIGENFUNCTIONS OF ANGULAR MOMENTUM OPERATORS

The eigenfunction $\psi_{J,M}$ of the angular momentum operators P^2 and P_Z in space-fixed coordinates are most easily found by use of these operators expressed in spherical coordinates. Application of P_Z from (2.22) in (2.35) yields

$$\frac{\hbar}{i}\frac{\partial \psi_{J,M}}{\partial \phi} = M\hbar\psi_{J,M} \quad (2.91)$$

We assume that

$$\psi_{J,M} = \Phi \cdot \Theta \quad (2.92)$$

where Φ is a function of ϕ only and Θ a function of θ only. Then (2.91) can be written

$$\frac{\partial \Phi}{\partial \phi} = iM\Phi \quad (2.93)$$

The solution is:

$$\Phi_M = N_M e^{iM\phi} \quad (2.94)$$

and, with M an integer, Φ is insured of being single valued, that is, $\Phi(\phi) = \Phi(\phi + 2\pi)$. Normalization of the eigenfunction requires

$$\int \Phi * \Phi \, d\tau = N_M^2 \int_0^{2\pi} d\phi = 1 \quad (2.95)$$

Thus $N_M = 1/(2\pi)^{1/2}$ and

$$\psi_{J,M} = (1/2\pi)^{1/2}\, e^{iM\phi}\Theta \tag{2.96}$$

To find the θ-dependent part of the function we apply the operator P^2 from (2.23) in (2.34) and obtain

$$\frac{1}{\sin\theta}\frac{\partial}{\partial\theta}\left(\sin\theta\,\frac{\partial\psi_{J,M}}{\partial\theta}\right)+\frac{1}{\sin^2\theta}\,\frac{\partial^2\psi_{J,M}}{\partial\phi^2}+J(J+1)\psi_{J,M}=0 \tag{2.97}$$

which, upon substitution of $\psi_{J,M}$ from (2.92) and transformation, becomes

$$\frac{1}{\sin\theta}\frac{\partial}{\partial\theta}\left(\sin\theta\,\frac{\partial\Theta}{\partial\theta}\right)-\frac{M^2}{\sin^2\theta}\,\Theta+J(J+1)\Theta=0 \tag{2.98}$$

This is the well-known Legendre equation, solution of which is

$$\Theta_{J,M}=N_{J,M}P_J^{|M|}(\cos\theta) \tag{2.99}$$

where $P_J^{|M|}(\cos\theta)$ represents the associated Legendre polynomials and $N_{J,M}$ is a constant which is determined by the normalizing condition

$$\int \Theta_{JM}^{*}\Theta_{J,M}\,d\theta = N_{J,M}^2 \int [P_J^{|M|}(\cos\theta)]^*[P_J^{|M|}(\cos\theta)]\,d\theta=1 \tag{2.100}$$

to be

$$N_{J,M}=\left[\frac{(2J+1)}{2}\cdot\frac{(J-|M|)!}{(J+|M|)!}\right]^{1/2} \tag{2.101}$$

Combination of these various factors gives the normalized eigenfunction of the angular momentum operators P^2 and P_Z to be

$$\psi_{J,M}=\frac{1}{\sqrt{2\pi}}\left[\frac{2J+1}{2}\cdot\frac{(J-M)!}{(J+M)!}\right]^{1/2} e^{iM\phi}P_J^{|M|}(\cos\theta) \tag{2.102}$$

which are commonly called the surface spherical harmonics.

The operators P^2 and P_Z commute with the Hamiltonian operator of the rigid linear rotor, and the functions of (2.102) are also rotational energy state functions of linear molecules when centrifugal distortion is neglected. The same eigenfunctions $\psi_{J,M}$ are obtained from a solution of the Schrödinger wave equation $\mathscr{H}_r\psi = E\psi$ for the rigid linear molecule.

Subsequent discussion of spin statistics (Chapter III, Section 4) will require a knowledge of the effect on the wave function of an exchange of the nuclei in a linear rotor. This operation is equivalent to the transformation $(\theta, \phi)\rightarrow(\pi-\theta, \phi+\pi)$. Replacement of ϕ by $\phi+\pi$ in (2.102) multiplies the function by $(-1)^M$. The $P_J^{|M|}(\cos\theta)$ is a polynomial of $\cos\theta$ of order $J-|M|$, involving even (odd) powers of $\cos\theta$ if $J-|M|$ is even (odd). Replacement of θ by $\pi-\theta$ hence multiplies the function by $(-1)^{J-|M|}$. The net result is that the total function is multiplied by $(-1)^J$.

To obtain the common eigenfunctions $\psi_{J,K,M}$ for the symmetric-top operators P^2, P_z, and P_Z, one can express the angular momentum operators in

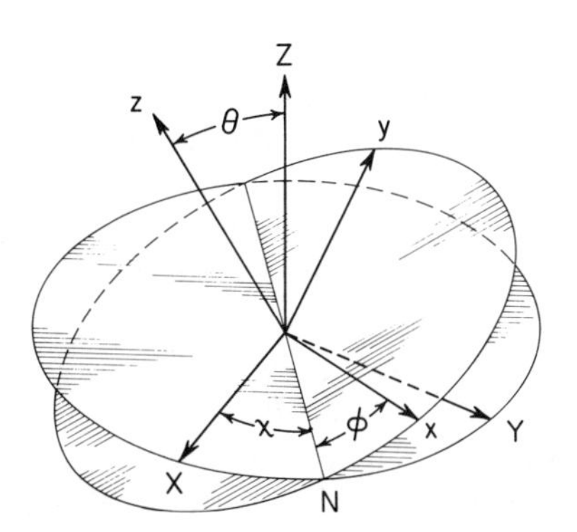

Fig. 2.3 Euler's angles defining the orientation of the body-fixed (x, y, z) axes relative to the space-fixed (X, Y, Z) axes. The line N represents the intersection of the xy plane with the XY plane and is called the line of nodes.

Euler's angles θ, ϕ, χ, and proceed in a manner similar to that followed previously, but the calculations are much more involved. Dennison [2] has obtained the symmetric-top eigenvalues with matrix methods; Reiche and Radmacher [3] and Kronig and Rabi [4], by solution of the Schrödinger equation for the symmetric top. The resulting eigenfunctions are

$$\psi_{J,K,M} = \Theta_{J,K,M}\, e^{iK\phi}\, e^{iM\chi} \tag{2.103}$$

where

$$\Theta_{J,K,M} = N_{J,K,M}\left(\sin\frac{\theta}{2}\right)^{|K-M|}\left(\cos\frac{\theta}{2}\right)^{|K+M|} F\left(\sin^2\frac{\theta}{2}\right) \tag{2.104}$$

The function $F(\sin^2\theta/2)$ is a hypergeometric series, and $N_{J,K,M}$ is a normalizing factor determined by the condition $N_{J,K,M}^2 \int \psi_{J,K,M}\psi_{J,K,M}^* \, d\tau = 1$. Euler's angles are illustrated in Fig. 2.3.

Eigenfunctions of asymmetric rotors can be obtained by expansion of these functions in terms of those for the symmetric top and evaluation of the coefficients in the expansion with the aid of energy values obtained from solution of the secular equation as described in Section 4. However, most of the quantities needed for microwave spectroscopic analysis are obtained from matrix elements derived from commutation rules without a specific knowledge of the eigenfunctions.

6 MATRIX ELEMENTS OF DIPOLE MOMENTS AND DIRECTION COSINES

Particularly for calculating transition probabilities and Stark and Zeeman effects, we shall need the direction cosine matrix elements of body-fixed axes of a rotor relative to space-fixed axes. Microwave spectral transitions are induced by interaction of the electric or magnetic components of the radiation fixed in space with the electric or magnetic dipole components fixed in the rotating body. Likewise, the Stark and Zeeman effects arise from interaction of space-fixed electric or magnetic fields with components of the dipole moment fixed in the rotors. These interactions can be considered as occurring between the spaced-fixed fields with components of the dipole moments resolved along the space-fixed axes, but it is evident that the magnitude of the space-fixed components will depend on the state of rotation. In finding these interactions one needs matrix elements of the components of the dipole moment referred to the space-fixed axes. These elements must be expressed in the eigenfunctions of the particular energy states between which the transition occurs.

In spinning electrons and nuclei, the body-fixed magnetic dipole moments are along the spin axis and are constants in all magnetic resonance experiments because S and I do not change. Thus the components of $\boldsymbol{\mu}$ resolve along space-fixed axes as the components of $\mathbf{S}$ and $\boldsymbol{I}$ and can be readily found from the angular momentum matrix elements of the components of $\mathbf{S}$ and $\boldsymbol{I}$.

The electric dipole moments matrix elements in space-fixed axes of rotating molecules are more complicated than those of electronic or nuclear spin magnetic moments for a number of reasons, the most important of which is that the electric dipole moment of the rotating molecule does not lie along $\mathbf{P}$ as it does along $\mathbf{S}$ or $\boldsymbol{I}$ and hence has matrix elements that are off-diagonal in J, whereas the matrix elements of the spin magnetic moments are diagonal in the corresponding quantum number S or I. The electric dipole moment of the linear molecule is perpendicular to $\mathbf{P}$ and has no matrix elements diagonal in J. It consequently has no resolvable component along a space-fixed axis and hence no first-order Stark effect. Another complication arises from the lower symmetry possible in molecules, especially in the asymmetric rotor which may have dipole components along each of the principal inertial axes fixed in the rotating body and for which the eigenfunctions of high J transitions are exceedingly complicated. Also, in rotating molecules the body-fixed electric dipole components are not independent of the electronic and vibrational states although these latter states do not generally change with the pure rotational transitions observed in the microwave region. Therefore we can treat the electric dipole moment components in the rotating, body-fixed system as constants. The problem is to find the nonvanishing matrix elements resolved on space-fixed axes in the representation that diagonalizes the rotational Hamiltonian operator.

Without loss of generality, we can choose the body-fixed dipole components along the principal axes of inertia. We designate the space-fixed axes by $F = X, Y, Z$, the body-fixed principal axes of inertia by $g = x, y, z$, and the cosine of the angle between F and g as $\Phi_{F,g}$. For example, a molecule with constant dipole moment components in the body-fixed principal axes of μ_x, μ_y, μ_z would have components along the space-fixed Z axes of

$$\mu_Z = \mu_x \Phi_{Zx} + \mu_y \Phi_{Zy} + \mu_z \Phi_{Zz} \tag{2.105}$$

If ψ_r represents the rotational eigenfunctions, the matrix elements of the dipole moment with reference to the space-fixed axes would be

$$\int \psi_r^* \mu_Z \psi_r' \, d\tau = \mu_x \int \psi_r^* \Phi_{Zx} \psi_r' \, d\tau + \mu_y \int \psi_r^* \Phi_{Zy} \psi_r' \, d\tau + \mu_z \int \psi_r^* \Phi_{Zz} \psi_r' \, d\tau \tag{2.106}$$

or, more generally

$$\int \psi_r^* \mu_F \psi_r' \, d\tau = \sum_g \mu_g \int \psi_r^* \Phi_{Fg} \psi_r' \, d\tau \tag{2.107}$$

Thus the matrix elements required are those for the direction cosine Φ_{Fg}. For the linear or symmetric-top molecule, the eigenfunctions are known, and the integrals $\int \psi_r^* \Phi_{Fg} \psi_r' \, d\tau$ can be evaluated in a straightforward manner. For the linear molecule this procedure is relatively simple, but for the symmetric top it is very tedious because of the complex form of the eigenfunctions. These matrix elements can be found in a simpler way from commutation rules between the angular momentum operators and the direction cosines [5], or between the angular momentum operators and the vector operators [1]. The somewhat involved derivations will not be given here, only the needed results.

The nonvanishing direction cosine matrix elements in the symmetric-top representation J, K, M are most useful. They are given in Table 2.1. These elements are separated into factors that depend on the different quantum numbers as follows

$$(J, K, M|\Phi_{Fg}|J', K', M') = (J|\Phi_{Fg}|J')(J, K|\Phi_{Fg}|J', K')(J, M|\Phi_{Fg}|J', M') \tag{2.108}$$

Those for the linear molecule may be obtained from them if K is set equal to zero. Those for the asymmetric rotor are obtained from them by methods [5] similar to those described for the finding of the energy eigenvalues of the asymmetric rotor (see Chapter VII).

As an illustration of the use of the elements of Table 2.1, let us employ them to find the dipole moment matrix elements that correspond to rotational absorption transitions $J \to J+1$ for the symmetric-top molecule in which the permanent dipole moment lies wholly along the axis of symmetry, $\mu = \mu_z$. The accidentally symmetric-top in which $I_x = I_y$ might have μ_x or μ_y which is not zero, but an accidentally symmetric-top molecule is exceedingly rare. Using (2.106) with $\mu_x = 0$, $\mu_y = 0$ and the $J' = J+1$ column of Table 2.1 we find

$$(J, K, M|\mu_Z|J+1, K, M) = \mu \frac{[(J+1)^2 - K^2]^{1/2}[(J+1)^2 - M^2]^{1/2}}{(J+1)[(2J+1)(2J+3)]^{1/2}} \tag{2.109}$$

$$(J, K, M|\mu_X|J+1, K, M \pm 1) = +\frac{i\mu}{2} \frac{[(J+1)^2 - K^2]^{1/2}[(J \pm M+1)(J \pm M+2)]^{1/2}}{(J+1)[(2J+1)(2J+3)]^{1/2}} \tag{2.110}$$

$$(J, K, M|\mu_Y|J+1, K, M \pm 1) = \mp \frac{\mu}{2} \frac{[(J+1)^2 - K^2]^{1/2}[(J \pm M+1)(J \pm M+2)]^{1/2}}{(J+1)[(2J+1)(2J+3)]^{1/2}} \tag{2.111}$$

Table 2.1 Factors of Direction Cosine Matrix Elements[a] of Symmetric-Top Rotors[b]

Matrix Element Factor	*Value of J′*		
	$J+1$	J	$J-1$
$(J\|\Phi_{Fg}\|J')$	$\{4(J+1)[(2J+1)(2J+3)]^{1/2}\}^{-1}$	$[4J(J+1)]^{-1}$	$[4J(4J^2-1)^{1/2}]^{-1}$
$(J, K\|\Phi_{Fz}\|J', K)$	$2[(J+1)^2-K^2]^{1/2}$	$2K$	$-2(J^2-K^2)^{1/2}$
$(J, K\|\Phi_{Fy}\|J', K\pm 1) = \mp i(J, K\|\Phi_{Fx}\|J', K\pm 1)$	$\mp[(J\pm K+1)(J\pm K+2)]^{1/2}$	$[J(J+1)-K(K\pm 1)]^{1/2}$	$\mp[(J\mp K)(J\mp K-1)]^{1/2}$
$(J, M\|\Phi_{Zg}\|J', M)$	$2[(J+1)^2-M^2]^{1/2}$	$2M$	$-2(J^2-M^2)^{1/2}$
$(J, M\|\Phi_{Yg}\|J', M\pm 1) = \pm i(J, M\|\Phi_{Xg}\|J', M\pm 1)$	$\mp[(J\pm M+1)(J\pm M+2)]^{1/2}$	$[J(J+1)-M(M\pm 1)]^{1/2}$	$\mp[(J\mp M)(J\mp M-1)]^{1/2}$

[a]Cross et al. [5].
[b]The matrix elements are obtained from the factors with the relation: $(J, K, M|\Phi_{Fg}|J', K', M') = (J|\Phi_{Fg}|J')(J, K|\Phi_{Fg}|J', K')(J, M|\Phi_{Fg}|J', M')$, $F = X, Y, Z$, and $g = x, y, z$.

Transitions $J \rightarrow J+1$ and $M \rightarrow M+1$, and $M \rightarrow M-1$ are induced by interactions of the radiation with rotating components of the dipole moment in the XY plane. This is evident from the fact that the nonvanishing matrix elements from these two transitions can be expressed as

$$(J, K, M|\mu_Y - i\mu_X|J+1, K, M-1) = \frac{\mu[(J+1)^2 - K^2]^{1/2}[(J-M+1)(J-M+2)]^{1/2}}{(J+1)[(2J+1)(2J+3)]^{1/2}} \tag{2.112}$$

$$(J, K, M|\mu_Y + i\mu_X|J+1, K, M+1) = -\mu\frac{[(J+1)^2 - K^2]^{1/2}[(J+M+1)(J+M+2)]^{1/2}}{(J+1)[(2J+1)(2J+3)]^{1/2}} \tag{2.113}$$

Transitions of the type $M \rightarrow M$ require radiation with the electric field in the Z direction.

The corresponding matrix elements for the linear molecule can be obtained from those for the symmetric top if K is set equal to zero in (2.109)–(2.113) or in Table 2.1.

The transition probabilities for induced absorption or stimulated emission are proportional to the squared dipole moment matrix elements with reference to the space-fixed axis (Chapter III, Section 1) as expressed by the Einstein B coefficient

$$B_{J,K,M \rightarrow J',K',M'} = \frac{8\pi^3}{3h^2} \sum_{F=X,Y,Z} |(J, K, M|\mu_F|J', K', M')|^2 \tag{2.114}$$

This equation provides a basis for the selection rules of dipole absorption or emission and is a factor in the calculation of intensities of spectral lines. Dipole transitions occur only between levels for which the matrix elements expressed by (2.109)–(2.113) do not all vanish. The dipole matrix elements for stimulated emission between the same levels can be found in a similar manner from Table 2.1 or simply by a reversal of the quantum numbers in the various elements, J', K', M' for J, K, M. This follows from the fact that the matrices are Hermitian or that the value of B is the same for induced absorption as for stimulated emission.

It is of interest that dipole matrix elements corresponding to $J \rightarrow J$ occur for the symmetric top, but not for the linear molecule. From Table 2.1 it is seen that when μ is along the z axis as before, the $J \rightarrow J$ elements of the symmetric top are

$$(J, K, M|\mu_Z|J, K, M) = \frac{\mu KM}{J(J+1)} \tag{2.115}$$

$$(J, K, M|\mu_Y - i\mu_X|J, K, M-1) = \mu\frac{K[J(J+1) - M(M-1)]^{1/2}}{J(J+1)} \tag{2.116}$$

$$(J, K, M|\mu_Y + i\mu_X|J, K, M+1) = \mu\frac{K[J(J+1) - M(M+1)]^{1/2}}{J(J+1)} \tag{2.117}$$

From the nonvanishing matrix elements thus obtained from Table 2.1, the selection rules for the genuine symmetric top (μ along z) are found to be

$$\Delta J=0, \pm 1, \qquad \Delta K=0, \qquad \Delta M=0, \pm 1 \tag{2.118}$$

Note, however, that the $J \rightarrow J$, $K \rightarrow K$, $M \rightarrow M$ for the rigid molecule gives rise to a constant, space-fixed component, an eigenvalue of μ along Z that cannot give resonance coupling with an ac electric field. Also, this transition for the strictly rigid top does not connect different energy states and is therefore trivial. When, however, there is an inversion vibration, as for NH_3, the dipole moment does not remain constant but changes with the inversion from $+\mu_Z$ to $-\mu_Z$ and thus can couple through resonance with an ac electric field component along Z which has the same frequency as that of the inversion. The inversion gives rise to a splitting of the rotational levels that have opposite symmetry, plus and minus, with transition probabilities proportional to the square of the foregoing matrix elements for $\Delta J=0$, $\Delta K=0$. It should also be noted that (2.116) and (2.117) indicate that when a field is applied, electric dipole transitions corresponding to $J \rightarrow J$, $K \rightarrow K$, $M \rightarrow M \pm 1$ can be observed between the nondegenerate Stark or Zeeman components of a given rotational state, even when there is no inversion or other vibration.

For an accidentally symmetric top in which μ does not lie exactly along z, the selection rules for K must also include the possibility $\Delta K = \pm 1$, but such cases are rare.

By substitution of $K=0$ in the matrix elements for the symmetric top, the dipole moment matrix elements corresponding to rotational absorption transitions of linear molecules are found to be

$$(J, M|\mu_Z|J+1, M)=\mu\left[\frac{(J+1+M)(J+1-M)}{(2J+1)(2J+3)}\right]^{1/2} \tag{2.119}$$

$$(J, M|\mu_Y - i\mu_X|J+1, M-1)=\mu\left[\frac{(J+1-M)(J+2-M)}{(2J+1)(2J+3)}\right]^{1/2} \tag{2.120}$$

$$(J, M|\mu_Y + i\mu_X|J+1, M+1)=-\mu\left[\frac{(J+1+M)(J+2+M)}{(2J+1)(2J+3)}\right]^{1/2} \tag{2.121}$$

Note that the $J \rightarrow J$ transitions are zero when $K=0$. Those for $J \rightarrow J-1$ are not zero and can be found in a similar way from Table 2.1. From the nonvanishing elements the selection rules for the rigid linear molecular rotor are seen to be

$$\Delta J = \pm 1, \qquad \Delta M=0, \pm 1 \tag{2.122}$$

Transition probabilities for the particular Stark or Zeeman components are proportional to the squares of the foregoing matrix elements. In the field-free molecular rotor, the Stark components are degenerate, and the intensities of the unsplit, pure rotational lines are proportional to the squares of these matrix elements summed over all M values of the final J levels and, if the stimulating radiation is isotropic, over the three coordinates, X, Y, Z. The summation can be achieved simply by addition of the squared component matrices. The M dependency cancels. The result for linear molecules is

$$
\begin{aligned}
|(J|\mu|J+1)|^2 &= \sum_{F=X,Y,Z} \sum_{M'} |(J, M|\mu_F|J+1, M')|^2 \\
&= (J, M|\mu_Z|J+1, M)^2 + \tfrac{1}{2}[(J, M|\mu_Y - i\mu_X|J+1, M-1)^2 \\
&\qquad + (J, M|\mu_Y + i\mu_X|J+1, M+1)^2] \\
&= \mu^2(J+1)/(2J+1) \qquad (2.123)
\end{aligned}
$$

Similarly, the dipole moment matrix elements for the field-free symmetric top must be summed over the M' degeneracies of the final state when the effective transition probabilities for the rotational absorption lines are obtained. The result for the $J, K \rightarrow J+1, K$ transition is

$$
\begin{aligned}
|(J, K|\mu|J+1, K)|^2 &= \sum_{F=X,Y,Z} \sum_{M'} |(J, K, M|\mu_F|J+1, K, M')|^2 \\
&= \mu^2 \frac{(J+1)^2 - K^2}{(J+1)(2J+1)} \qquad (2.124)
\end{aligned}
$$

For the $J \rightarrow J$ or the inversion type of transition the result is

$$
|(J, K^-|\mu|J, K^+)|^2 = \sum_{F=X,Y,Z} \sum_{M'} |(J, K^-, M|\mu_F|J, K^+, M')|^2 = \mu^2 \frac{K^2}{J(J+1)} \qquad (2.125)
$$

The $(+)$ and $(-)$ symbols represent the two inversion states.

When plane-polarized radiation is used, coupling with molecules occurs only along one axis, and the foregoing squared sums must be reduced by $\frac{1}{3}$ for calculation of the line intensities.

Dipole moment matrices and selection rules for the asymmetric rotor are described in Chapter VII.

References

1. E. Feenberg and G. E. Pake, *Notes on the Quantum Theory of Angular Momentum*, Addison-Wesley, Cambridge, Mass., 1953.
2. D. M. Dennison, *Phys. Rev.*, **28**, 318 (1926); *Rev. Mod. Phys.*, **3**, 280 (1931).
3. F. Reiche and H. Rademacher, *Z. Physik*, **39**, 444 (1927).
4. R. de L. Kronig and I. I. Rabi, *Phys. Rev.*, **29**, 262 (1927).
5. P. C. Cross, R. M. Hainer, and G. C. King, *J. Chem. Phys.*, **12**, 210 (1944).

Chapter **III**

MICROWAVE TRANSITIONS —LINE INTENSITIES AND SHAPES

Microwave spectral lines arise from transitions between quantized energy levels of which the separations ΔE correspond to the microwave quanta $h\nu$, where ν is the frequency of the microwave radiation. If the molecule, particle, or system of particles has a dipole moment, electric or magnetic, the radiation field will couple with it to induce transitions when the frequency of the radiation corresponds to the Bohr condition $h\nu = \Delta E$. An electric dipole will couple with the electric vector; a magnetic dipole, with the magnetic component of the radiation. Electric dipole coupling is generally the mechanism for inducing molecular rotational transitions; magnetic dipole coupling, the mechanism for inducing transitions between states of electron spin resonance. Since electric quadrupole transitions and spontaneous emission are too weak in the microwave region to be of interest, we shall be concerned with induced dipole transitions.

1 LINE STRENGTHS

Consider two discrete, nondegenerate levels of a substance between which microwave transitions can occur. For simplicity, we designate each level by only one quantum number, m for the lower level and n for the upper one. To be specific, we assume that the substance is made up of molecules, although the theory described applies equally well to spinning electrons, to nuclei oriented in magnetic fields, or to any quantum mechanical system considered in this volume. Now suppose that the substance is exposed to radiation at the resonant frequency

$$\nu_{mn} = \frac{E_n - E_m}{h} \tag{3.1}$$

A molecule originally in state m will have a probability

$$p_{m\to n} = \rho(\nu_{mn}) B_{m\to n} \tag{3.2}$$

of absorbing a quantum $h\nu_{mn}$ and of undergoing a transition $m \to n$ in unit time, where $\rho(\nu_{mn})$ represents the density of the radiation and $B_{m\to n}$ represents the Einstein coefficient of absorption for the particular transition of the substance. A molecule originally in state n will have a probability

$$p_{m\leftarrow n} = \rho(\nu_{mn}) B_{m\leftarrow n} + A_{m\leftarrow n} \tag{3.3}$$

of emitting a quantum $h\nu_{mn}$ and of undergoing the transition $m \leftarrow n$ in unit time, where $B_{m\leftarrow n}$ is the Einstein coefficient of induced emission and $A_{m\leftarrow n}$ is the Einstein coefficient of spontaneous emission for the transition.

Einstein originally assumed the coefficients of induced emission and absorption to be equal, an assumption later verified by quantum mechanical calculations and by experiments. From time-dependent perturbation theory it can be shown (see, for example, Pauling and Wilson [1]) that for isotropic radiation the probability of absorption in unit time is

$$p_{m\to n} = \left(\frac{8\pi^3}{3h^2}\right) [|(m|\mu_X|n)|^2 + |(m|\mu_Y|n)|^2 + |(m|\mu_Z|n)|^2] \rho(\nu_{mn}) \tag{3.4}$$

where

$$(m|\mu_F|n) = \int \psi_m^* \mu_F \psi_n \, d\tau \tag{3.5}$$

and where $F = X, Y, Z$ represents the matrix elements of the dipole moment component of the molecule or other particle referred to space-fixed axes for the transition $m \to n$. The dipole moment matrix is Hermitian $(m|\mu_F|n) = (n|\mu_F|m)^*$. Thus the Einstein B coefficients are

$$B_{mn} = B_{m\to n} = B_{m\leftarrow n} = \left(\frac{8\pi^3}{3h^2}\right) [|(m|\mu_X|n)|^2 + |(m|\mu_Y|n)|^2 + |(m|\mu_Z|n)|^2] \tag{3.6}$$

The quantities $(m|\mu_F|n)$ are the matrix elements of the dipole moment components resolved on the space-fixed axes in the representation that diagonalizes the energy matrix of the molecule. For most molecules, the dipole moment components can be considered as constant in the rotating body-fixed axes, and the matrix elements may then be expressed only in molecular rotational eigenfunctions. Dipole moment matrix elements for molecular rotational transitions are given in Chapter II, Section 6. Since the B coefficients are equal, we shall hereafter drop the arrow subscripts.

The coefficient of spontaneous emission A is relatively inconsequential at microwave and radiofrequencies. A rigorous derivation is difficult, but the coefficient may be found easily from the B coefficients by use of Planck's radiation law with the assumption that the only mechanism for exchange of energy between levels is through radiation-induced or spontaneous emission. For isolated units in which no collision or like thermal process occurs, this assumption is obviously justifiable, and we assume this idealized condition when finding the coefficient A. It is possible, however, to treat relaxation processes such as collisions as arising from radiation components generated by the collisions that contribute to $\rho(\nu_{mn})$.

If N_m represents the number of molecules per unit volume in the lower state, the number per unit volume undergoing transitions to the upper state in time Δt will be

$$\Delta N_{m\to n} = N_m B_{mn} \rho(\nu_{mn}) \Delta t \tag{3.7}$$

The number per unit volume making the reverse transition in time Δt is

$$\Delta N_{m\leftarrow n} = N_n [B_{mn} \rho(\nu_{mn}) + A_{m\leftarrow n}] \Delta t \tag{3.8}$$

If thermal equilibrium is maintained by these processes, $(\Delta N_{m\to n})/(\Delta t) = (\Delta N_{m\leftarrow n})/(\Delta t)$, and (3.7) with (3.8) yields

$$\frac{N_n}{N_m} = \frac{B_{mn}\rho(\nu_{mn})}{B_{mn}\rho(\nu_{mn}) + A_{m\leftarrow n}} \tag{3.9}$$

At thermal equilibrium the Boltzmann law requires:

$$\frac{N_n}{N_m} = e^{-h\nu_{mn}/kT} \tag{3.10}$$

Combination of the Boltzmann relation with (3.9) yields upon transformation

$$\rho(\nu_{mn}) = \frac{A_{m\leftarrow n}}{B_{mn}} \frac{1}{\exp(h\nu_{mn}/kT) - 1} \tag{3.11}$$

Conformity of this expression with Planck's radiation law

$$\rho(\nu) = \left(\frac{8\pi h\nu^3}{c^3}\right)\left(\frac{1}{\exp(h\nu/kT) - 1}\right) \tag{3.12}$$

requires

$$A_{m\leftarrow n} = \left(\frac{8\pi h\nu_{mn}^3}{c^3}\right) B_{mn} \tag{3.13}$$

With B_{mn} from (3.6), this gives

$$A_{m\leftarrow n} = \frac{64\pi^4 \nu_{mn}^3}{3hc^3} [|(m|\mu_X|n)|^2 + |(m|\mu_Y|n)|^2 + |(m|\mu_Z|n)|^2] \tag{3.14}$$

In the optical region spontaneous emission is of comparable importance with stimulated emission. However, the variation of the A coefficient with the cube of the radiation frequency makes spontaneous emission insignificant in the microwave and radiofrequency regions. In this regions we can neglect it in comparison with other relaxation processes.

Suppose now that the radiation density $\rho(\nu_{mn})$ is the coherent radiation from a controlled microwave or radiofrequency oscillator. If V is the volume of the sample, the total number of molecules undergoing the transition $m \rightarrow n$ per second is $VN_m p_{m\rightarrow n}$. Since each of them absorbs an energy $h\nu_{mn}$, the total energy per second (power) required for this transition is

$$P_{m\rightarrow n} = VN_m p_{m\rightarrow n} h\nu_{mn} = VN_m B_{mn} \rho(\nu_{mn}) h\nu_{mn} \tag{3.15}$$

and the total power returned to the radiation field through stimulated emission is

$$P_{m\leftarrow n} = VN_n B_{mn} \rho(\nu_{mn}) h\nu_{mn} \tag{3.16}$$

The emitted power will be in phase with the coherent radiation that stimulates it and thus will add power of the same frequency and in phase with that from the original source. As a result of these processes, there will be a net change in the coherent power in the radiation field of

$$\Delta P = P_{m\leftarrow n} - P_{m\rightarrow n} = V(N_n - N_m) B_{mn} \rho(\nu_{mn}) h\nu_{mn} \tag{3.17}$$

Usually the population of the lower state is the greater, $N_m > N_n$, and the ΔP of (3.17) is negative, that is, there will be a net absorption of power. If, as in maser spectrometers, the upper state has the greater population, $N_n > N_m$, the ΔP will be positive, that is, there will be a net gain of power or amplification. It is evident, however, that application of resonant coherent power causes more transitions from the state having the greater population and hence tends to equalize the population of the two states. When $N_n = N_m$, there will be no net change in power, $\Delta P = 0$. It is further evident that continuous observation of either absorption or amplification of the applied radiation would require that some other process be operative to preserve a difference in populations, to offset the equalizing tendency of the applied power. Thermal motions that cause interactions between the molecules or particles most commonly serve this function. However, the thermal process always tends to produce a greater population of the lower energy level so that under conditions of thermal equilibrium the states have a difference in population given by the Boltzmann law, (3.10). Clearly, application of coherent resonant power will tend to upset the thermal equilibrium between the two levels so that the relative populations will no longer be given by Boltzmann's law. If, however, the thermal relaxation is rapid as compared with the rate of exchange of energy of the molecules with the applied radiation field, the thermal processes will dominate and thermal equilibrium will be maintained, or nearly so. Since most microwave absorption spectrometers operate under conditions of approximate thermal equilibrium, we shall calculate the peak absorption coefficients for this condition. We can achieve this condition by lowering the applied radiation $\rho(\nu_{mn})$ so as to decrease the probability of radiation-induced transitions or by increasing the rate of thermal relaxation through some process such as an increase of temperature or pressure of the sample.

Under conditions of thermal equilibrium we can employ Boltzmann's law with (3.17) to find the power absorbed

$$P_{abs} = -\Delta P = VN_m(1 - e^{-h\nu_{mn}/kT})B_{mn}\rho(\nu_{mn})h\nu_{mn} \tag{3.18}$$

A quantity more useful to spectroscopists than the absorbed power is the absorption coefficient α defined by

$$\alpha = -\left(\frac{1}{P}\right)\left(\frac{\Delta P}{\Delta x}\right) \tag{3.19}$$

where ΔP represents the power absorbed in an element of cell length Δx where the total power is P.

The element of volume ΔV of the sample can be expressed as $\Delta V = S\Delta x$ where S represents the crosssection of the absorption cell; the energy density in this section as $\rho(\nu) = P/cS$ where P is the power at the input and c is the velocity of propagation; and the number of particles in the lower state as $N_m = NF_m$ where N is the number of particles per unit volume of the sample and F_m is the fraction of these which are in the state m. Substitution of these relations into (3.18) and transformation give

$$\alpha_{mn} = -\frac{1}{P}\frac{\Delta P}{\Delta x} = \frac{NF_m}{c}(1 - e^{-h\nu_{mn}/kT})B_{mn}h\nu_{mn} \tag{3.20}$$

For microwave spectroscopy, where $h\nu < kT$, the expression for the difference in population of the energy states can be expanded in a converging series and α_{mn} expressed as

$$\alpha_{mn} = \frac{NF_m(h\nu_{mn})^2}{ckT}\left(1 - \frac{1}{2}\frac{h\nu_{mn}}{kT} + \cdots\right)B_{mn}$$
$$\approx \frac{NF_m(h\nu_{mn})^2}{ckT}B_{mn} \tag{3.21}$$

In convenient units for numerical computations,

$$\frac{h\nu}{kT} = 1.44\frac{\nu(\text{cm}^{-1})}{T} = 48 \times 10^{-6}\frac{\nu(\text{MHz})}{T} \tag{3.22}$$

where T is measured on the absolute scale (°K).

Except at low temperatures or very high submillimeter wave frequencies, $h\nu_{mn} \ll kT$, and we can negelect higher terms in the expansion and employ the approximate form of (3.21). For example, with measurements made with 1-cm wavelength at room temperature (300°K) neglect of the $h\nu_{mn}/2kT$ term would cause an error in the calculated value of only 0.2%; for measurements at the same frequency, but at liquid helium temperature (4.3°K) the error would be

5%. In the $\frac{1}{2}$-mm region ($\nu=20\ \text{cm}^{-1}$) and at dry ice temperatures, neglect of the higher terms in the expansion would cause an error of the order of 7%. Note that the predicted absorption with this approximation is greater than the actual absorption.

So far we have assumed that frequencies ν_{mn} for transitions between m and n are the same for all particles of the samples. This is, of course, not true; no spectral line is infinitely sharp. The mutual interactions between particles, Doppler broadening, and other factors discussed in Section 2 cause the different molecules in the indicated states to have slightly different energies and transition frequencies. Thus the absorption will be distributed over a range of frequencies, and the absorption coefficient will be a function of the frequencies within the absorption range. This function, called the line-shape function, will vary according to type and degree of perturbation, but most microwave and radiofrequency absorption lines considered in this volume are sharp and have approximately the Lorentzian shape function, (3.44). The formula for absorption coefficients α_ν for isotropic radiation of any frequency ν under conditions of thermal equilibrium can be found from (3.21) upon substitution of B_{mn} and upon multiplication by the properly normalized shape function $S(\nu, \nu_0)$ discussed in Section 2. The resulting formula is

$$\alpha_\nu=\frac{8\pi^3 NF_m\nu^2}{3ckT}\left(1-\frac{1}{2}\frac{h\nu}{kT}\right)[|(m|\mu_X|n)|^2+|(m|\mu_Y|n)|^2+|(m|\mu_Z|n)|^2]S(\nu, \nu_0) \tag{3.23}$$

where ν_0 is the frequency for the peak absorption and ν is any frequency within the range of absorption.

With plane polarized radiation commonly employed in microwave spectroscopy we can without loss of generality choose the coupling vector of the radiation along the Z axis and can drop the X and Y components from the dipole moment matrix elements if we multiply the preceding expression for isotropic radiation by the factor of 3. However, the net result is the same if we keep all components of (3.23) and do not multiply by 3. With the Lorentzian shape function, (3.44), the absorption coefficient is

$$\alpha_\nu=\frac{8\pi^2 NF_m\nu^2}{3ckT}\left(1-\frac{1}{2}\frac{h\nu}{kT}\right)|(m|\mu|n)|^2\frac{\Delta\nu}{(\nu_0-\nu)^2+(\Delta\nu)^2} \tag{3.24}$$

where

F_m = fraction of molecules in the lower state of the transition (Section 3)

N = number of molecules per unit volume

For gaseous samples

$$N=9.68\times 10^{18}p_{\text{mm}}/T \tag{3.25}$$

where p_{mm} is the pressure in mm of Hg and T is the absolute temperature

$\Delta\nu$ = half-width of line measured between half-intensity points

For pressure broadened lines (see Section 2)

$$\Delta\nu \approx 300 p_{\mathrm{mm}} \frac{(\Delta\nu)_1}{T} \tag{3.26}$$

where $(\Delta\nu)_1$ is the line breadth measured at $T = 300°\mathrm{K}$ and 1 mm of Hg pressure.

An important quantity often measured is the absorption coefficient at the resonant frequency ν_0, called the peak absorption coefficient $\alpha_{\max}$. When $\nu = \nu_0$ is substituted in (3.24), the peak absorption coefficient is seen to be

$$\alpha_{\max} = \frac{8\pi^2 N F_m \nu_0^2}{3ckT(\Delta\nu)} \left(1 - \frac{1}{2}\frac{h\nu_0}{kT}\right) |(m|\mu|n)|^2 \tag{3.27}$$

The peak absorptions are useful for quick measurements of relative intensities of lines having similar shape and width. The absolute or integrated intensity defined by

$$I = \int_0^\infty \alpha_\nu \, d\nu \tag{3.28}$$

is independent of the line width.

From the absorption coefficient α_ν one can calculate the power absorbed at the frequency ν for an absorption cell of given length x. Upon integration of (3.19) and evaluation of the constant of integration with the condition that at $x = 0$, $P = P_0$, one obtains

$$P_x = P_0 \exp(-\alpha_\nu x) \tag{3.29}$$

where P_0 is the power at the input and P_x is the power at the output of the cell of length x and the power absorbed is $P_{\mathrm{abs}} = P_0 - P_x$. If the power attenuation constant in the absorption cell walls is α_c, the transmitted power is

$$P_x = P_0 \exp[-(\alpha_\nu + \alpha_c)x] \tag{3.30}$$

For experimental evaluation of the spectral absorption coefficient from the measured ratio P_x/P_0, one can use the integrated expression

$$\alpha_\nu = -\alpha_c - \frac{1}{x} \ln\left(\frac{P_x}{P_0}\right) \tag{3.31}$$

The constant α_c can be measured by the power loss when no absorbing gas is in the cell, but in actual practice it is simpler to measure α_c not by removing the gas from the cell but simply by tuning the oscillator out of the region of absorption of the gas where $\alpha_\nu \approx 0$. Since α_c does not vary significantly with frequency over the region of absorption of the usually sharp spectral line, its effects are not apparent in the typical frequency-sweep spectrometer, which responds only to a variation of power with frequency. The power loss in the cell does, however, affect the sensitivity of the spectrometer, and α_c must be known for an absolute measurement of α_ν.

2 LINE SHAPES AND WIDTHS

In this volume we shall not be concerned with physical theories of line shapes as such, but primarily with the effects of line width and shape on the detectability, resolution, and accuracy of measurement of the line frequency. Most microwave measurements are made on gaseous absorption lines having widths ranging from a few kilohertz to a few megahertz. We shall discuss briefly the factors which determine width of this range.

Natural Line Width

The natural width of a molecular line may be considered to arise from the probability of spontaneous emission which would limit the lifetime in the upper state n if the molecule were completely isolated from interaction with radiation fields or with other molecules or particles. The lifetime in the state and the energy spread of the state is related by the uncertainty principle

$$\Delta t \cdot \Delta E \approx \hbar \tag{3.32}$$

The corresponding frequency spread is

$$\Delta \nu = \frac{\Delta E}{h} \approx \frac{1}{2\pi(\Delta t)} \tag{3.33}$$

For a molecule in a state n that can only make spontaneous transitions to a lower state m, the "natural" lifetime in the state can be taken as

$$\Delta t = \frac{1}{A_{mn}} = \frac{3hc^3}{64\pi^4 \nu_{mn}^3 |(m|\boldsymbol{\mu}|n)|^2} \tag{3.34}$$

where A_{mn} is the probability of spontaneous emission given by (3.14). Substitution of this Δt into (3.33) gives

$$\Delta \nu \approx \frac{32\pi^3 \nu_{mn}^3}{3hc^3} |(m|\mu|n)|^2 \tag{3.35}$$

When the dipole moment is expressed in convenient debye units (1 D$=10^{-18}$ esu), this becomes

$$\Delta \nu \approx 1.86 \times 10^{-38} \nu_{mn}^3 |(m|\mu|n)|^2 \tag{3.36}$$

Typically for polar molecules, the dipole moment matrix elements for rotational transitions are of the order of 1 D. With $|(m|\mu|n)|^2 \approx 1$ and for molecules in ground vibrational and electronic states, the natural half-width $\Delta\nu$ of a rotational line in the 3-mm wave region, $\nu_{mn} = 10^{11}$ Hz, is of the order of 10^{-5} Hz; that of a line in the 3-cm wave region, of the order of 10^{-8} Hz. It is evident that such line widths are entirely negligible in comparison with line width arising from other factors described later.

If a molecule in state m can undergo a transition through spontaneous emission to more than one lower state, the natural line width will depend on the sum of the probabilities of the transition to the lower states. If the lower states are represented by $i = 1, 2, \ldots$,

$$\Delta\nu \approx \frac{1}{2\pi} \sum_i A_{im} \approx \frac{32\pi^3}{3hc^3} \sum_i \nu_{im}^3 |(m|\mu|i)|^2 \tag{3.37}$$

In the observation of microwave transitions between rotational sublevels of excited vibrational or electronic states, the lifetime in the state m is determined predominantly by the probability of spontaneous emission to the ground vibrational or the ground electronic state; and the spread of the level, hence the natural microwave line width, is correspondingly greater. The natural width can be estimated from (3.37) with ν and the dipole moment matrix elements corresponding to the transition to the ground vibrational or ground electronic state; for example, suppose we are observing rotational lines in the first excited vibrational state of HCl. The frequency corresponding to a transition to the ground vibrational state is 8.65×10^{13} Hz. The matrix elements corresponding to this vibrational transition are still of the order of a debye unit. Thus the natural line width, from (3.37), is of the order of 12 or 24 Hz if we consider the line spread of the lower rotational level. This broadening is still smaller than that caused by molecular collisions, Doppler broadening, and so on, in the usual microwave experiment. For molecules or atoms in excited electronic states, however, this "natural broadening" factor can be the dominant one unless the excited electronic state is a metastable state, for which $|(m|\mu|n)|^2$ is zero in the first approximation. For ν_{mn} in the visible region at 4000 Å and $|(m|\mu|n)|^2 \approx 1$, the broadening would be of the order of 24 MHz.

Doppler Broadening

The frequency of radiation absorbed by a molecule in motion depends to some extent on the velocity of the molecule relative to that of the radiation. If ν_0 represents the resonant absorption frequency of the molecule at rest, the frequency that it will absorb when moving with a velocity v relative to that of the radiation will be $\nu = \nu_0(1 + v/c)$, where c is the velocity of the radiation. The change in frequency $\nu_0(v/c)$ is called the Doppler shift of the frequency. For molecules moving in opposite direction to the radiation (v negative), the absorption frequency will be lower than ν_0; for those moving in the direction of the radiation (v positive), the absorption frequency will be higher than ν_0. Molecules moving at right angles to the path of the radiation will have no Doppler shift.

Similar shifts are observed for emission of radiation, with v representing the molecular velocity relative to that of the emitted radiation which is detected. The molecules of a gas at thermal equilibrium will have a Maxwell-Boltzmann distribution of velocities, and the various Doppler shifts of all the molecular frequencies will give rise to a line-shape function

$$S_d(\nu, \nu_0) = S_0 \exp\left[-\frac{mc^2}{2kT}\left(\frac{\nu - \nu_0}{\nu_0}\right)^2\right] \tag{3.38}$$

where k is the Boltzmann constant and T is the absolute temperature. The Doppler line width $2(\Delta\nu)_d$, the width between half-intensity points caused by Doppler broadening alone, is seen from (3.38) to be

$$2(\Delta\nu)_d = \frac{2\nu_0}{c}\left(\frac{2NkT \ln 2}{M}\right)^{1/2} = 7.15 \times 10^{-7}\left(\frac{T}{M}\right)^{1/2} \nu_0 \tag{3.39}$$

where M represents the molecular weight in atomic mass units and N is Avogadro's number. The line breadth $\Delta\nu$ is usually defined as the width between the peak frequency ν_0 and the frequency at half-intensity, so that the line width measured between half-intensity points is $2(\Delta\nu)$.

By solving (3.39) for ν_0 in terms of $(\Delta\nu)_d$ and substituting this value for ν_0 into (3.38), one obtains the more convenient expression

$$S_d(\nu, \nu_0) = S_0 \exp\left\{-(\ln 2)\left[\frac{\nu - \nu_0}{(\Delta\nu)_d}\right]^2\right\} \tag{3.40}$$

for the line-shape function. Substituting this expression for the shape function in (3.23) with N in terms of pressure p from (3.25) when $h\nu \ll kT$ and using $\ln 2 = 0.693$, one obtains

$$\alpha_\nu(p \leqslant p_c) = C\nu^2 p \exp\left\{-0.693\left[\frac{\nu - \nu_0}{(\Delta\nu)_d}\right]^2\right\} \tag{3.41}$$

for the absorption coefficient in the pressure range below p_c above which collision broadening begins to be significant (see Pressure Broadening in Section 2). Here C is a constant independent of the pressure and frequency. At the critical pressure and resonant frequency, the Doppler broadened line has its maximum absorption coefficient

$$(\alpha_{\max})_d = C\nu_0^2 p_c \tag{3.42}$$

as is seen from substitution of $p = p_c$ and $\nu = \nu_0$ into (3.41). Substitution of C from (3.42) into (3.41) yields

$$\left(\frac{\alpha_\nu}{\alpha_{\max}}\right)_d = \frac{p}{p_c}\left(\frac{\nu}{\nu_0}\right)^2 \exp\left\{-0.693\left[\frac{\nu - \nu_0}{(\Delta\nu)_d}\right]^2\right\} \tag{3.43}$$

where $p \leqslant p_c$. The $(\alpha_{\max})_d$ as defined here has the same numerical value as does the $\alpha_{\max}$ for the pressure-broadened line described in Section 2. The function of (3.43) represents the normalized shape function of the Doppler-broadened line. It is plotted in Fig. 3.1 for a few values of p/p_c with an assumed Doppler width $(\Delta\nu)_d = 0.5$ MHz. Note that the width is constant and that the peak intensity decreases directly with the pressure. Since one gains nothing in resolution while losing peak intensity, it is usually undesirable to reduce the pressure below p_c, or that for which the Doppler broadening becomes a dominant factor.

Doppler broadening of gaseous lines is much smaller than pressure broadening under the usual conditions of observation in the centimeter wave region. Because of its increase with frequency, however, Doppler broadening is some-

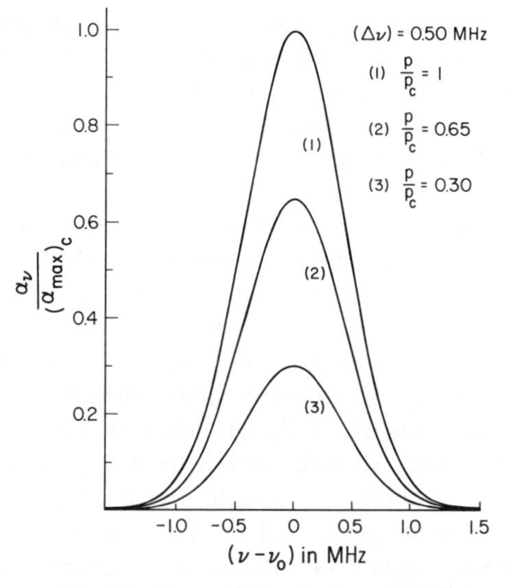

Fig. 3.1 Shape of Doppler-broadened lines.

times the dominant factor in the broadening of submillimeter wave spectral lines. As an example, let us consider NH_3, which has a molecular weight of 17. At $T = 300°K$, the ammonia inversion lines in the centimeter wave region, $\nu_0 \approx 24{,}000$ MHz, have a Doppler width of 72 kHz; the $J = 0 \rightarrow 1$ rotational line [2] of NH_3 with $\nu_0 = 572{,}053$ MHz has a Doppler width $2(\Delta\nu)_d$ of 1.7 MHz. Although this increase of Doppler broadening with increase of ν_0 does not reduce the absolute accuracy of measurement of the resonant frequency, which depends on $\Delta\nu/\nu_0$, it interferes with resolution of the hyperfine structure, the separation of which does not increase with ν_0.

The Doppler broadening can be circumvented or greatly reduced by the use of a molecular beam absorption spectrometer in which the absorbing molecules are sprayed across the path of the microwave radiation. Fortunately, the greater absorption of molecules at higher frequencies makes this method more feasible at the submillimeter wave frequencies where Doppler broadening becomes significant. Since the absolute resolution and accuracy of measurement depend on $\Delta\nu/\nu_0$, it is evident that with a molecular beam spectrometer to reduce $\Delta\nu$, exceptionally high precision can be achieved in the submillimeter wave region with microwave electronic methods.

Pressure Broadening

At the pressures usually employed in the measurement of microwave spectral lines of gases, ~ 1 to 10^{-3} mm of Hg, pressure arising from binary collisions of the molecules is the predominant line-broadening factor. If the collision is considered as abruptly terminating the life of the molecule in a particular rotational energy state, the line width can be calculated from (3.33) with the Δt's representing the times between collisions. The line-shape function would be expected to have the same form as that for the distribution of the collision times between the molecules, which would in turn depend upon the Maxwell-Boltzmann distribution of velocities. If τ is the mean time between collisions that end the lifetime in the state, (3.33) indicates a line breadth of $\Delta\nu = 1/(2\pi\tau)$.

All collisions do not cause transitions. Furthermore, transitions may be induced by long-range resonance interactions in which the molecules do not collide in the usual sense. Also, a molecule can experience a Stark perturbation of its energy levels from the dipole field of a molecule which approaches closely. The pressure broadening thus depends on properties of the molecules as well as on velocity distribution and collision times. Theoretically, the line shape might be used for measurement of dipole moments, but such measurements would be much more difficult to make and the resulting dipole moments much less accurate than those obtained from the Stark effect of rotational lines. However, collisions between polar and nonpolar molecules have been used for approximate measurements of molecular quadrupole moments [3] of nonpolar molecules for which no other measurements are available.

At the low pressures commonly used in microwave measurements where the half-width $\Delta\nu \ll \nu_0$, collision-broadened lines have the Lorentzian shape function [4]:

$$S(\nu, \nu_0) = \frac{1}{\pi}\left[\frac{\Delta\nu}{(\nu_0-\nu)^2+(\Delta\nu)^2}\right] \tag{3.44}$$

where

$\Delta\nu = 1/(2\pi\tau)$

τ = mean time between collisions

ν_0 = the peak resonant frequency

Over the range of very low pressures considered, 1 mm of Hg or less, the mean collision time τ varies inversely as the pressure, and therefore $\Delta\nu$ increases linearly with the pressure. The number of absorbing particles also increases linearly with pressure. Hence, within the range of pressures sufficiently high that molecular collisions are the dominant cause of broadening and sufficiently low that the line width is directly proportional to pressure, we can by substitution of (3.25) and (3.26) into (3.24) express the absorption coefficient as a function of frequency and pressure

$$\alpha_\nu = C\nu^2\left[\frac{(\Delta\nu)_1 p^2}{(\nu-\nu_0)^2+(\Delta\nu)_1^2 p^2}\right] \tag{3.45}$$

where p is the pressure, $(\Delta\nu)_1$ is the line breadth for $p = 1$ mm pressure, $T = 300°\mathrm{K}$, and C is a constant that represents other intensity factors in (3.24) that do not depend on pressure or frequency. Note that the peak absorption coefficient α_{max} obtained when ν equals ν_0 is independent of pressure

$$\alpha_{max} = \frac{C\nu_0^2}{(\Delta\nu)_1} \tag{3.46}$$

This is an important property in the observation of spectra since the detecability of an absorption line generally depends on the peak intensity rather than on the integrated intensity. Thus one gains resolution without losing detectability by decreasing the pressure until other broadening factors that do not depend on pressure (mainly Doppler broadening) become significant. As the pressure is reduced to the point at which the Doppler broadening becomes dominant, the line shape changes to the Maxwell-Boltzmann shape described earlier, and the peak intensity α_{max} does not remain constant but decreases directly with the decrease in pressure.

By substituting $C\nu_0^2 = \alpha_{max}(\Delta\nu)_1$ from (3.46) into (3.45) and rearranging the terms, one obtains

$$\frac{\alpha_\nu}{\alpha_{max}} = \frac{(\nu/\nu_0)^2}{[(\nu-\nu_0)/p(\Delta\nu)_1]^2+1} \tag{3.47}$$

Equation 3.47 represents the normalized-shape function of pressure-broadened lines in the pressure range for which the linewidth is directly proportional to pressure, the range in which most microwave lines of gases are measured. Figure 3.2 shows plots of (3.47) for a typical line-breadth parameter $(\Delta\nu)_1 =$ 10 MHz/mm of Hg and for three different pressures ranging from 1 to 10^{-1} mm of Hg. These curves represent the Lorentzian line shape.

In the higher range of pressures, above 10 mm of Hg, deviations from the Lorentzian-shape function become evident. The more general Van Vleck-

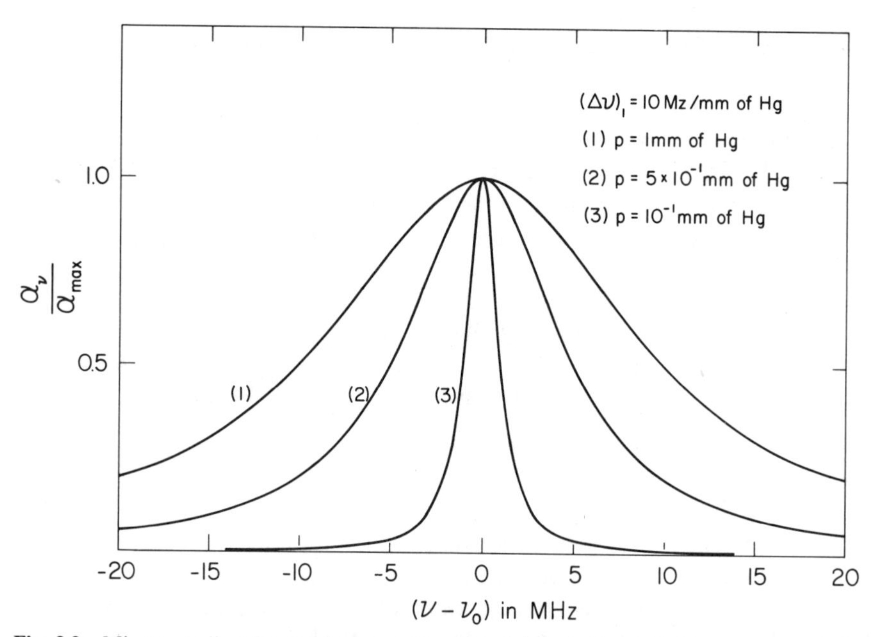

Fig. 3.2 Microwave line shapes of pressure-broadened lines in the typical pressure range where the line width is directly proportional to pressure. The line-shape function is Lorentzian.

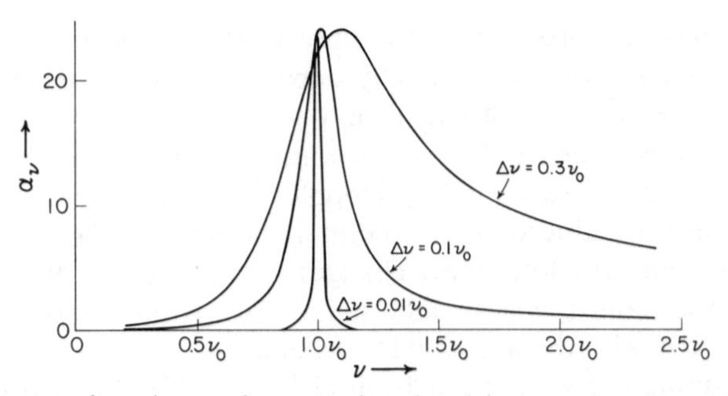

Fig. 3.3 Microwave line shapes of pressure-broadened lines in the relatively high pressure range—the Van Vleck–Weisskopf shape function [7]. From *Microwave Spectroscopy* by Townes and Schawlow. Copyright 1955 by McGraw-Hill, Inc. Used with permission of McGraw-Hill Book Company.

Weisskopf [5] line-shape function

$$S(\nu, \nu_0) = \frac{\nu}{\pi\nu_0}\left[\frac{\Delta\nu}{(\nu_0-\nu)^2+(\Delta\nu)^2}+\frac{\Delta\nu}{(\nu_0+\nu)^2+(\Delta\nu)^2}\right] \tag{3.48}$$

is found to fit satisfactorily the observed microwave line shapes at low gas pressures as well as at relatively high ones, of the order of an atmosphere. Figure 3.3 shows graphically the Van Vleck-Weisskopf line shape with different relative values of $\Delta\nu$ and ν_0. For ν_0 in the microwave region, these curves all correspond to relatively high pressures. For low pressures where $\Delta\nu \ll \nu_0$, the second term of (3.48) becomes very small in comparison with the first, and the expression reduces to the Lorentzian-shape function, (3.44), in the region of significant absorption where $\nu \approx \nu_0$.

In Table 3.1 illustrative line-breadth parameters are given for a few gases. These represent the half-widths of the lines between half-intensity points for a pressure of 1 mm of Hg of the gas. From these examples one can estimate roughly the line width to expect for other, unmeasured lines.

A rigorous treatment of pressure broadening is given by Anderson [6]. His theory relates the line shapes to the various types of intermolecular forces between the colliding molecules and provides a mechanistic understanding of line broadening. Discussion of collision broadening can be found in books on microwave spectroscopy by Townes and Schawlow [7] and by Gordy et al. [8]. A comprehensive review of the subject is given by Birnbaum [9].

Instrumental Distortion

POWER EFFECTS

At low pressures with cells of small volume it is relatively easy to distort the line noticeably by power saturation of the molecules. In the center of the line where the absorption is greatest, the molecules at first absorb power at a rate greater than that at which they can return to the lower state through thermal

Table 3.1 Self-broadening Line-breadth Parameters of Rotational Lines ($T=300°K$)

Molecule	*Transition*		$(\Delta\nu)_1$ *MHz/mm of* Hg	*Reference*
OCS	$J=0\rightarrow1$		6.44	*a*
N_2O	$1\rightarrow2$		5.22	*a*
ClCN	$1\rightarrow2$		25.0	*b*
BrCN	$2\rightarrow3$		22.0	*c*
CH_3F	$0\rightarrow1$		20.0	*d*
	$2\rightarrow3$	$K=0$	17.2	*e*
		$K=1$	17.2	
		$K=2$	17.2	
		$K=3$	17.2	
CHF_3	$0\rightarrow1$		35.1	*f*
	$6\rightarrow7$	$K=3$	9.0	*e*
		$K=4$	9.0	
		$K=5$	9.0	
		$K=6$	9.0	
H_2O	$2_2\rightarrow3_{-2}$		19.1	*g*

[a] B. T. Berendths, Thesis, Catholic Univ. of Nijmegen (Netherlands), 1966. (Quoted by G. Birnbaum [9].)
[b] C. H. Townes, A. N. Holden, and F. R. Merritt, *Phys. Rev.*, **74**, 1113 (1948).
[c] S. L. Srivastava and V. Prakash, *J. Chem. Phys.*, **42**, 3738 (1965).
[d] O. R. Gilliam, H. D. Edwards, and W. Gordy, *Phys. Rev.*, **75**, 1014 (1949).
[e] G. Birnbaum, E. R. Cohn, and J. R. Rusk, *J. Chem. Phys.*, **49**, 5150 (1968).
[f] C. O. Britt and J. E. Boggs, *J. Chem. Phys.* **45**, 3877 (1966).
[g] J. R. Rusk, *J. Chem. Phys.*, **42**, 493 (1965).

relaxation processes. As a result, difference in population of the two states $\Delta N = N_n - N_m$ becomes less than that for thermal equilibrium, and hence the power absorption is less than that which would occur if thermal equilibrium could be maintained. In the event of complete saturation, the amount of power absorbed is determined completely by the rate of relaxation, and a further increase of incident power will cause no further power absorption. Because of its greater probability of absorption, the center of the line becomes saturated more completely than the wings, and a broadening of the line occurs. The power density at which saturation broadening becomes noticeable depends on the Einstein coefficient B or the squared dipole moment matrix elements for the transitions as well as on the pressure of the gas.

For low pressures and incomplete saturation, Karplus and Schwinger [10] have modified the Lorentzian-shape function to take account of power effects. The resulting line-shape function is

$$S(\nu, \nu_0) = \frac{1}{\pi}\left[\frac{\Delta\nu}{(\nu_0-\nu)^2+(\Delta\nu)^2+B_{mn}P/c\pi^2}\right] \tag{3.49}$$

where B_{mn} is the Einstein coefficient and P is the power density. The presence of saturation will decrease α_ν, as is apparent from (3.49) and (3.23).

In the shorter millimeter wave region and particularly in the submillimeter wave region, the absorption coefficients of many molecules are so large that absorption approaching 100% of the power occurs in the center of the line for cell lengths of only a few centimeters. Since the absorption is less in the wings of the line, this nearly complete absorption in the center causes a flattening of the line peak and an increase in the indicated line width. Unlike the power saturation effects, this distortion is not lessened but increased by increase of pressure, but it can be eliminated by a severe decrease in pressure or by a decrease in the effective cell length.

COLLISION WITH CELL WALLS

In an absorption cell of small dimensions, broadening caused by collision with cell walls can become a significant factor when the pressure is so low that the cell dimensions become comparable to the mean free path between the molecules. The half-width caused by cell-wall broadening can be calculated by the formula $\Delta\nu = 1/(2\pi\tau)$, where τ is the mean time between collisions of the molecules with the cell walls. If a, b, and c are dimensions of a rectangular cell, the line width caused by collision with the cell walls [11] is approximately

$$2(\Delta\nu) = 10\left(\frac{1}{a}+\frac{1}{b}+\frac{1}{c}\right)\left(\frac{T}{M}\right)^{1/2} \tag{3.50}$$

where $(2\Delta\nu)$ is in kHz units and a, b, c are in cm units. T is the absolute temperature, and M is the molecular weight. Because the length c is large in the usual waveguide cell, the factor $1/c$ is negligible. Usually the cross-sectional dimensions of the absorption cell are made sufficiently large that the cell-wall broadening is negligible. However, G-band waveguide which supports only the dominant mode for 3-mm wave radiation has the cross-sectional dimensions of approximately 0.2×0.1 cm. In an observation of carbon monoxide gas (molecular weight $M = 28$) at room temperature ($T = 300°$K) in a cell made of G-band waveguide, the cell broadening would be a half-megahertz.

BROADENING IN MOLECULAR BEAMS

In molecular-beam spectrometers in which the molecules undergoing transitions are sprayed across the path of the radiation, there is a broadening that results from the limited time the molecules are in the radiation field. This broadening can also be approximated by the formula $\Delta\nu = 1/(2\pi\tau)$ where τ is the mean time the molecules are in the radiation. If $\bar{v}$ is the mean velocity of the molecules of the beam and d is their path length in the radiation field, $\tau = d/\bar{v}$, and the broadening is

$$\Delta\nu \approx \frac{\bar{v}}{2\pi d} \tag{3.51}$$

If $\bar{v}$ is in cm/sec and d is in cm, $\Delta\nu$ will be in Hz units.

DISTORTION BY MODULATORS, DETECTORS, AND AMPLIFIERS

Spectral lines are often deliberately distorted by Stark or Zeeman modulation so that their sensitivity to detection may be increased and so that they may be distinguished from sharp reflections of electronic resonances caused by impedence mismatch in the microwave components. Similar distortions of the apparent line shape are produced by frequency modulation of the radiation source to aid detection. The most common types of these distorted presentations represent the first and second derivatives of the contour of the actual line. These derivative curves for a line having the typical Lorentzian shape are illustrated in Fig. 3.4. Experimentally, the first derivative curve is obtained by a frequency modulation of the source or spectrum line and the tuning of a phase-lock-in detector to the fundamental frequency of the modulation. The second derivative curve is obtained from a tuning of the phase-lock-in detector to a frequency twice that of the modulation. To prevent further distortion, in both instances the frequency of the modulation as well as its range of variation must be small compared with the line width. The first derivative is the more sensitive form for detection; the second derivative, the more convenient for measurement of the

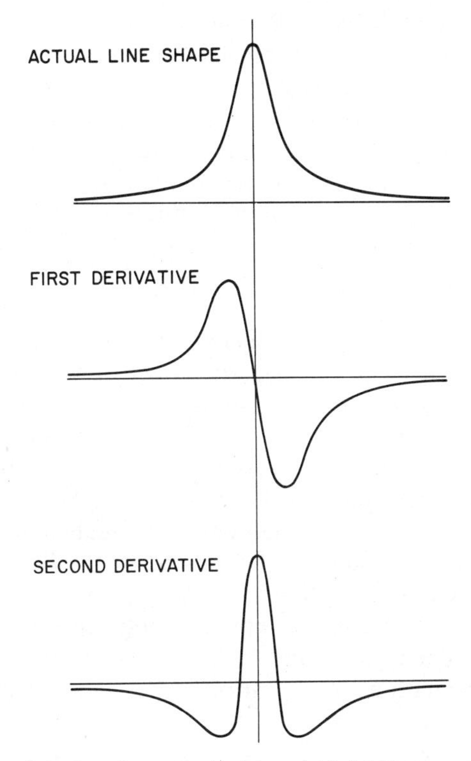

Fig. 3.4 Comparison of the line shape obtained by a high-fidelity recording of a typical microwave line with those obtained by detection systems that display the first or second derivatives of the true line shape function (assumed to be Lorentzian).

line center and the half-width. Note from comparisons in Fig. 3.4 that the second derivative curve has an artificially sharpened peak which occurs at the center of the actual line, also minima which occur approximately at the half-intensity points. The center of the line occurs at a zero-point of the first derivative curve. Broad-banded receivers required for a high-fidelity cathode-ray-scope display of an absorption line like that of Fig. 3.4 are less sensitive than either the first- or the second-derivative detectors, but many microwave spectral lines are so strong that their detection is no problem.

When the rate of frequency modulation is not small compared with the line width, other distortion effects are encountered. This distortion also depends on the type of modulation, that is, whether sine-wave, square-wave, and so on. These effects are discussed by Karplus [12]; also by Townes and Schawlow [13].

In the typical crystal video-type spectrometer [8] the source frequency is swept completely over the spectral line at a constant rate and repeated at a constant rate, usually many times a second. A cathode-ray trace synchronized with this sweep is used to display the line which is detected with a crystal receiver and is amplified with a video-frequency receiver. The degree and type of distortion depend on the rate of frequency modulation relative to the band width of the receiver. A high-fidelity receiver must amplify equally all the Fourier components of the modulated signal. Such a receiver is not the most sensitive one for detection of the lines within thermal noise, nor will it discriminate against the spurious signals caused by impedance mismatch in the microwave components. Generally the choice of sweep-rate is a compromise between fidelity and sensitivity, depending on the strength and width of the line to be measured and the degree of mismatch in the microwave system.

3 POPULATION OF ENERGY STATES

The absorption line strength formula developed previously includes a factor N_m which represents the number of particles per unit volume of sample that are in the lower energy state involved in the transition. For spin resonance—electronic or nuclear—which is not split by interactions with other states, the calculation of this number N_m is relatively simple, but for molecular rotational spectra it can be complicated by the multiplicity of populated states having different energies. Under conditions of thermal equilibrium the relative population of the different states is determined by the Boltzmann law, (3.10), and by the degeneracies of the different states.

Let us assume first a spectrum of nondegenerate energy levels 0, 1, 2, 3, . . . of particles in thermal equilibrium. If N_0 represents the population of the lowest level, the populations of the various other levels relative to that of the lowest level will be

$$N_1 = N_0 e^{-E_1/kT}, \qquad N_2 = N_0 e^{-E_2/kT}, \qquad N_3 = N_0 e^{-E_3/kT} \tag{3.52}$$

where $E_1, E_2, E_3, \ldots$ are the energies of the various levels relative to the energy of the ground state. The total number N of particles per unit volume will be

$$N = N_0 + N_1 + N_2 + N_3 + \cdots = N_0 \sum_i e^{-E_i/kT} \tag{3.53}$$

where the summation is taken over all the states. If the ith level has the degeneracy g_i, the sum total population in all states per unit volume will then be

$$N = N_0 \sum_i g_i e^{-E_i/kT} \tag{3.54}$$

and the number in the state m with degeneracy g_m and energy E_m is

$$N_m = N_0 g_m e^{-E_m/kT} \tag{3.55}$$

The fractional part F_m of the particles in state m is

$$F_m = \frac{N_m}{N} = \frac{g_m e^{-E_m/kT}}{\sum_i g_i e^{-E_i/kT}} = \frac{g_m e^{-E_m/kT}}{Q} \tag{3.56}$$

The denominator in the final expression, designated

$$Q = \sum_i g_i e^{-E_i/kT} \tag{3.57}$$

is called the partition function. It is of great significance in the calculation of spectral-line intensities. Note that N_0 cancels from the ratio and that neither Q nor the numerator contains the actual number of particles in any state. Nevertheless, one can use this factor to obtain immediately the actual number of particles in any given state of known quantity of sample. The Q is a function of all the different forms of energies that the particles or system of particles may have.

Molecular Vibrational and Rotational States

The partition function for free gaseous molecules is a function of the electronic, the vibrational, the rotational, and the nuclear spin states. If interactions between these various states are neglected, it can be expressed as the product

$$Q = Q_e Q_v Q_r Q_n \tag{3.58}$$

Most stable organic molecules at ordinary temperatures are in ground electronic singlet Σ states, and for them we can set $Q_e = 1$. For this reason we are here concerned mainly with the last three factors.

If anharmonicities are neglected, the vibrational partition function, with the energy levels measured with respect to the ground vibrational state, can be expressed as

$$Q_v = \left(\sum_{v_1} e^{-v_1 h\omega_1/kT}\right)\left(\sum_{v_2} e^{-v_2 h\omega_2/kT}\right)\left(\sum_{v_3} e^{-v_3 h\omega_3/kT}\right)\cdots \tag{3.59}$$

where $v_1, v_2, v_3, \ldots$, the vibrational quantum numbers, can each have the values 0, 1, 2, 3, ... and where $\omega_1, \omega_2, \omega_3, \ldots$ are the frequencies of the fundamental modes of vibration. The summation is taken over all values of v_1, v_2, $v_3, \ldots$, and each fundamental mode of vibration is counted separately. This product by evaluation of the summations can be expressed in the more convenient form

$$Q_v = (1 - e^{-h\omega_1/kT})^{-d_1}(1 - e^{-h\omega_2/kT})^{-d_2}(1 - e^{-h\omega_3/kT})^{-d_3} \cdots \tag{3.60}$$

where ω_1, ω_2, $\omega_3, \ldots$ are the fundamental frequencies and each different frequency appears only once and where d_1, d_2, $d_3, \ldots$ are the degeneracies of the fundamental vibrations, that is, the number of modes having the same frequency.

The fraction of molecules in a given vibrational state specified by the set of quantum numbers $v_1, v_2, v_3, \ldots \equiv v$, at a temperature T, is

$$F_v = \frac{d_v e^{-h\Sigma_i v_i \omega_i/kT}}{Q_v} = d_v e^{-h\Sigma_i v_i \omega_i/kT}(1 - e^{-h\omega_1/kT})^{d_1}(1 - e^{-h\omega_2/kT})^{d_2} \cdots \tag{3.61}$$

When, as is customary, the vibrational frequencies are expressed in cm^{-1}, the exponent can be numerically written

$$\frac{h\omega_i}{kT} = 1.439 \frac{\omega_i(\text{cm}^{-1})}{T(°\text{K})} \tag{3.62}$$

The fraction $F_{J,i}$ of molecules in the rotational state $|J, i)$ with rotational energy $E_{J,i}$ and in vibrational state v is given by

$$F_{J,i} = \frac{N_{J,i}}{N} = \frac{F_v g_J g_i e^{-E_{J,i}/kT}}{Q_r} \tag{3.63}$$

where

$$Q_r = \sum_J \sum_i g_J g_i e^{-E_{J,i}/kT}$$

Here J is the usual total angular momentum quantum number, and i signifies any internal quantum numbers such as K of the symmetric top. The degeneracies $g_J = 2J + 1$ (when no external field is applied) associated with the "outer" quantum number M_J are the same for all classes of molecules. The degeneracy factor g_i is that associated with all inner quantum numbers and includes, in addition to the K degeneracy of the symmetric top, all degeneracies caused by nuclear spins, inversion, and internal rotation. Usually, however, the inversion and internal rotation degeneracies are the same for the different levels and cancel from the ratio g_i/Q_r. Except for molecules having certain symmetries described later, the effects of nuclear spins also cancel from the ratio g_i/Q_r and hence need not be included in either. Any nuclear hyperfine splitting of the levels can be neglected in the calculation of the integrated population of the rotational level.

Linear molecules have no K degeneracy; and polar-linear molecules, which are the only ones having an observable rotational spectrum, have no center of symmetry. Therefore g (nuclear) cancels from the ratio g_i/Q_r, and we can set $g_i = 1$ and can drop the i subscripts in (3.63). The rotational energy relative to the lowest level is adequately expressed by the rigid rotor approximation of (5.3). Thus for diatomic or linear polyatomic molecules with no center of symmetry [14]

$$Q_r = \sum_J (2J+1) \exp\left[\frac{-hBJ(J+1)}{kT}\right]$$
$$= \frac{kT}{hB} + \frac{1}{3} + \frac{1}{15}\left(\frac{hB}{kT}\right) + \frac{4}{315}\left(\frac{hB}{kT}\right)^2 + \cdots$$
$$\cong \frac{kT}{hB} \text{ (linear molecules)} \tag{3.64}$$

Therefore, for linear and diatomic molecules in the vibrational state v,

$$F_J = \frac{N_J}{N} \cong F_v \left(\frac{hB}{kT}\right)(2J+1) \exp\left[-\frac{hBJ(J+1)}{kT}\right] \tag{3.65}$$

For a symmetric-top molecule there are degeneracies associated with the internal quantum number K as well as nuclear spin degeneracies g (nuclear) that do not cancel from (3.63) and must therefore be included. Although g (nuclear), which would appear in the definition of Q_r, can depend on the rotational state, in the high-temperature approximation the correct result is obtained if g (nuclear) is replaced by the total statistical weight factor g_n [see (3.74)] divided by the symmetry number σ. By definition of a reduced statistical weight factor g_I which includes the g_n factor, the fraction of molecules in a given rotational state J, K can be expressed as

$$F_{J,K} = \frac{N_{J,K}}{N} = \frac{F_v g_J g_K g_I e^{-E_{J,K}/kT}}{Q_r} \tag{3.66}$$

where F_v = fraction of molecules in the particular vibrational state being considered
$g_J = 2J+1$
$g_K = 1$ for $K=0$ and 2 for $K \neq 0$
g_I = reduced nuclear spin weight factor defined by (3.77)
$E_{J,K} = h[BJ(J+1) + (A-B)K^2]$

and the rotational partition function is defined as

$$Q_r = \frac{1}{\sigma} \sum_{J=0}^{J=\infty} \sum_{K=-J}^{K=J} (2J+1) e^{-E_{J,K}/kT} \tag{3.67}$$

where σ is a measure of the degree of symmetry, discussed further in Section 4. For the usual symmetric top which has C_3 or C_{3v} symmetry, $\sigma = 3$. To a good approximation, the summation can be expressed [14] as

$$Q_r \cong \frac{1}{\sigma}\left[\left(\frac{\pi}{B^2 A}\right)\left(\frac{kT}{h}\right)^3\right]^{1/2} = \left(\frac{5.34 \times 10^6}{\sigma}\right)\left(\frac{T^3}{B^2 A}\right)^{1/2} \text{ (symmetric top)} \tag{3.68}$$

In the last expression the spectral constants B and A are in MHz units.

A similarly approximate formula for the partition function for the asymmetric rotor is

$$Q_r = \left(\frac{5.34 \times 10^6}{\sigma}\right)\left(\frac{T^3}{ABC}\right)^{1/2} \quad \text{(asymmetric top)} \tag{3.69}$$

Furthermore, in (3.66) the $E_{J,\tau}$ term for the particular asymmetric rotor (Chapter VII) must be employed rather than the $E_{J,K}$ term. Since there is no K degeneracy for the asymmetric rotor, $g_K = 1$. Many asymmetric rotors have no symmetry that allows exchange of identical nuclei with a turning of the rigid molecule. For such molecules, $\sigma = 1$ and $g_I = 1$. Effects of symmetry will be described in Section 4.

4 SYMMETRY PROPERTIES

Symmetry operations, such as inversion of all the coordinates at the origin or exchange of identical nuclei which leave the overall wave function of a molecule either unchanged or changed only in sign, have important consequences in the determination of selection rules and statistical weights of energy levels. The classification of the overall wave function on the basis of such symsymmetry operations is indicated as parity. The function that is changed in sign has a different parity from one that is left unchanged by the symmetry operation. Selection rules based on symmetry properties are called parity selection rules or symmetry selection rules.

In determination of parity it can be assumed that eigenfunctions corresponding to various energy operators of the molecule—electronic, vibrational, molecular rotational, and nuclear spin—can be used to represent the complete wave function

$$\psi \text{ overall} = \psi_e \psi_v \psi_r \psi_n \tag{3.70}$$

where ψ_e represents the electronic function, ψ_v the vibrational function, ψ_r the molecular rotational wave function, and ψ_n the nuclear spin function. Later we shall describe various symmetry operations on which the parity classifications are based.

Inversion of the Coordinates

If the coordinates of all the particles of the molecules, expressed in the Cartesian system with the origin at the center of mass, are changed in sign ($x_i \rightarrow -x_i$, $y_i \rightarrow -y_i$, $z_i \rightarrow -z_i$), an equivalent equilibrium configuration of the molecule is achieved. For this operation, called inversion or reflection of the coordinates at the origin, the part of the product $\psi_e \psi_v \psi_r$ must remain unchanged or be changed in sign only. Whether this product function changes sign or not depends on the behavior of each factor with respect to inversion. This part of the wave function does not include the nuclear spin function and is often referred to as the coordinate function although it does include the electron spin.

Generally, reflection of the coordinates at the origin chosen as the center of mass of the molecule does not exchange identical nuclei but may do so in molecules having certain symmetries such as CO_2. However, such linear symmetrical molecules have no dipole moment and hence no observable microwave spectra. The effect of exchange of identical nuclei is described later in this section.

In linear or planar polyatomic molecules in general, the new configuration achieved by inversion of the coordinates can also be achieved by rotation of the molecule. Inversion doubling of the levels does not occur. For nonplanar molecules, however, the new, inverted configuration cannot be achieved by rotation alone about any succession of axes; it can be achieved only if a potential barrier is overcome. When the barrier is very high, stable right and left configurations of molecules called optical isomers may occur. The two inverted configurations for nonplanar molecules (not achievable by rotation) lead to a doubling of all rotational levels, called inversion doubling. For almost all nonplanar molecules, the barrier to inversion is so high that this doubling is not resolvable, and the rotational levels are, in effect, doubly degenerate. The term $g_{\text{inv}}=2$ would occur in both the numerator and the denominator of (3.63); hence it does not affect the term F_{Ji}, the fraction of molecules in a particular rotational state. In a few molecules, most notably ammonia, the barrier to inversion is sufficiently low that the molecule oscillates back and forth between the two equilibrium configurations, giving rise to a measurable separation of the inversion doublets and to an inversion spectrum, as described in Chapter VI, Section 3.

The functions that are unchanged upon reflection of the coordinates at the origin are designated as positive (+) functions; those that are changed in sign, as negative (−) functions. Selection rules that depend on these properties are easily derived. The dipole moment matrix elements between the states, represented by ψ_i and ψ_j, upon which the transition probability depends, can be expressed as

$$\mu_{ij}=\int \psi_i \mu_X \psi_j d\tau + \int \psi_i \mu_Y \psi_j d\tau + \int \psi_i \mu_Z \psi_j d\tau \tag{3.71}$$

Now for a reflection of the coordinates of all the particles at the center of mass, the components μ_X, μ_Y, and μ_Z change sign; and unless the functions ψ_i and ψ_j have opposite parity (opposite sign) with respect to this operation, the integrals will change sign. Since these are definite integrals, they cannot change sign unless they are identically zero. Thus for dipole-induced transitions, only positive and negative functions can combine. The parity selection rules are therefore

$$+\leftrightarrow- \qquad +\not\leftrightarrow+ \qquad -\not\leftrightarrow- \tag{3.72}$$

In practically all nonplanar molecules the inversion doubling is not resolvable, and these selection rules place no restrictions on the intercombinations of the rotational levels because the two components of each level have opposite symmetry.

The rotational levels of linear or planar molecules do not have the inversion degeneracy because the inverted state can be achieved also by rotation. Nevertheless, the alternate rotational levels have opposite parity with respect to the reflection of the coordinates at the center of gravity. This can readily be seen for linear molecules by an examination of the eigenfunctions of (2.102). Reflection at the origin is achieved by the transformation $\theta \rightarrow \pi - \theta$ and $\phi \rightarrow \pi + \phi$. The functions of (3.70) remain unchanged for these transformations when J is even but are changed in sign when J is odd. Therefore the selection rules $+ \leftrightarrow -$, $+ \not\leftrightarrow +$, $- \not\leftrightarrow -$ are the same as those already proved, $J \rightarrow J \pm 1$.

The parity of the coordinate function depends on ψ_e and ψ_v as well as upon ψ_r. Most molecules have completely paired electronic systems with positive ψ_e. Likewise, the ground vibrational states of molecules in general have positive ψ_v. Thus most molecules for which rotational spectra are observed have positive $\psi_e \psi_v$, and the parity of ψ_r determines the parity of the overall coordinate function $\psi_e \psi_v \psi_r$. For these cases of linear molecules, the coordinate functions with $J = 0, 2, 4, \ldots$ are plus, and those with $J = 1, 3, 5, \ldots$ are minus.

Symmetry of Momental Ellipsoid

Classification of the rotational levels on the basis of the symmetry properties of the momental ellipsoid is of importance in determination of the selection rules and line intensities for an asymmetric rotor.

Because of the symmetry of the momental ellipsoid, the probability density ψ_r^2 of a rigid asymmetric rotor does not change for a rotation of π degrees about any of the principal axes of inertia. Conventionally, these axes are designated a, b, c and are chosen so that the principal moments of inertia have the order $I_a < I_b < I_c$. The rotational wave function either is unchanged (is even) or is changed (is odd) in sign by an operation that turns the molecule π degrees about a principal axis. Because a rotation of π about any two principal axes in succession is equivalent to a rotation of π about the third, ψ_r either must be odd with respect to two of the operations and even with respect to the third or must be even with respect to all three. There are thus two classes of functions. Those which are symmetric or even with respect to all three operations are sometimes designated A functions; those which are even with respect to one axis and odd with respect to the other two are designated B functions. Since there are three axes, it is evident that the B functions can be further divided into three subclasses: B_a, B_b, and B_c where the subscripts designate the particular principal axis of inertia a, b, or c with respect to which ψ_r is symmetric. Thus there are four species of functions. In microwave spectroscopy these functions are conventionally designated by two indices which refer to the operation C_2^a about the axis of least moment of inertia a and the operation C_2^c about the axis of greatest moment of inertia c, with the operation about a being designated first. For example, the designation *eo* means that ψ_r is symmetric (even) with respect to a rotation of π about the axis of least moment of inertia a and is antisymmetric (odd) with respect to a similar rotation about the axis of largest moment of inertia c. This function is evidently a B_a function. Similarly, $ee \equiv A$,

$oo \equiv B_b$, and $oe \equiv B_c$. In the $J_{K_{-1}K_1}$ designation of the asymmetric rotor levels, the evenness or oddness of the integral subscripts K_{-1} and K_1 reveals the parity of the levels (see Chapter VII, Section 2). If K_{-1} is an odd integral, the rotational eigenfunction for the state is odd with respect to a rotation of π about a; if K_{-1} is zero or an even integral, the function is even with respect to this operation. The K_1 subscript has a similar relation to this operation about c. Thus a $J_{2,5}$ level has the parity eo; a $J_{2,4}$ level, ee; a $J_{3,1}$ level, oo.

The foregoing symmetry operations C_2^a, C_2^b, C_2^c (with the identity operation E), under which the asymmetric rotor functions are classified, form a symmetry group known as the Four group, see Table 7.4. The symbols A, B_a, B_b, and B_c, introduced to distinguish the symmetry of the functions, correspond to the designations of the symmetry species of the group.

Selection rules based on the symmetry properties of the momental ellipsoid depend on which principal axis has a component of dipole moment. These rules are described in Chapter 7. The symmetry of the rotational levels with respect to rotation of π about a principal axis is of importance in the determination of the parity of the overall wave function with respect to exchange of identical nuclei. It will be discussed in the section to follow.

Effects of Nuclear Spin

For an exchange of identical nuclei the overall wave function of the molecule expressed by (3.70) must either remain unchanged or change sign or phase only. Those which remain unchanged by this operation are designated as symmetric functions and those which change sign as antisymmetric. It is found by experience that for particles having spins of zero or an integer (Bose particles) the overall functions are symmetric, and for particles having spins of half-integers (Fermi particles) the overall functions are antisymmetric with regard to an operation that exchanges the identical particles.

Selection rules based on the symmetric–antisymmetric classication are derived in a manner similar to those of (3.71). If ψ_i and ψ_j represent the overall wave function of two states i and j, the transition dipole moment between the states is

$$\mu_{ij} = \sum_F \int \psi_i \mu_F \psi_j \, d\tau \tag{3.73}$$

where $F = X, Y, Z$.

For an operator that merely exchanges two identical nuclei, the sign of μ_F does not change, and therefore the integral will change sign unless ψ_i and ψ_j have the same symmetry. Since the definite integral cannot change sign unless it is zero, the transition moment μ_{ij} is zero and the transition probability is also zero unless ψ_i and ψ_j have the same parity. Not only the dipole moment but all higher induced moments, quadrupole, and so on, are unchanged in sign for an operation that involves only exchange of identical nuclei. Therefore the parity selection rules based on this classification of the overall functions are

$$\text{symmetric} \leftrightarrow \text{symmetric} \qquad \text{antisymmetric} \leftrightarrow \text{antisymmetric}$$
$$\text{symmetric} \not\leftrightarrow \text{antisymmetric}$$

These rules hold rigorously for all possible mechanisms of induced transitions. However, the selection rules derived on the basis of the nonvanishing of the matrix elements of the dipole moments (Chapter II, Section 6) are sufficient for most applications in microwave spectroscopy.

The presence of identical nuclei in the molecule can have important consequences in the determination of the statistical weights of energy levels and the relative intensities of rotational lines. When there is no symmetry and no nuclear hyperfine splitting, each molecular rotational level will have a degeneracy

$$g_n = (2I_1+1)(2I_2+1)(2I_3+1)\cdots = \prod_i (2I_i+1) \tag{3.74}$$

where I_i represents the spin of the ith nucleus and where the product is taken over all nuclei of the molecule. With no symmetry, however, this product will cancel from the numerator and denominator in (3.63) and thus need not be considered in the calculations of population of the rotational states. Even when there is nuclear coupling which gives rise to a splitting of the rotational levels, it is more convenient to neglect the nuclear splitting in the calculation of the overall, integrated number of molecules in the rotational state and then to take account of the molecular distribution among the hyperfine levels in a separate calculation.

In certain nonlinear molecules there are symmetries that give rise to inequivalent nuclear statistical weights for different rotational levels, and hence these weights do not cancel from the formula for calculation of population of rotational levels. For such molecules the rotational partition function must be divided by a symmetry number σ, defined as the number of indistinguishable positions that can be achieved through simple rotation of the rigid molecule.

The reduction in Q by the factor σ arises from the symmetry properties of the overall wave function for the molecule which must be symmetric for an exchange of identical particles obeying the Bose-Einstein statistics (zero or integral spins) and must be antisymmetric for an exchange of particles obeying Fermi-Dirac statistics (spins of odd integral multiples of $\frac{1}{2}$). In the determination of these symmetries the overall function of the molecule can be expressed as the product of (3.70). Except when the spins of the identical particles are zero, both symmetric and antisymmetric nuclear spin functions ψ_n can be chosen from combinations of those of the identical nuclei. Thus when $I \neq 0$, one can choose ψ_n's that make the product $\psi_r\psi_n$ either symmetric or antisymmetric whatever the symmetry of ψ_r. However, the number of symmetric and antisymmetric nuclear spin functions that can be thus formed are not equivalent, and the weights of the rotational levels that can be matched with them to satisfy the required overall symmetry will not be equal. When $I=0$, the nuclear spin functions are all symmetric, and only symmetric ψ_r can be multiplied with them to give the symmetric, overall function required for Bose particles. In this instance the odd rotational levels will be entirely missing when $\psi_e\psi_v$ is even, and the even rotational levels will be missing when that product is odd.

Although both symmetric and antisymmetric spin functions can be chosen from combinations of the separate spin functions of the identical nuclei, not all of the possible combinations can be paired with a ψ_r to give the proper symmetry of the overall functions. If there are n such identical nuclei, there will be $(2I+1)^n$ possible functions, but on the average only $(2I+1)^n/\sigma$ combinations of them will form the correct overall symmetry with the ψ_r functions. Although the nuclear statistical weight factor for various rotational levels will be different in the summation of the levels for determination of Q_r, we can simply multiply each level by the averaged nuclear degeneracy $(2I+1)^n/\sigma$. This is completely satisfactory except for very light molecules or extremely low temperatures.

Diatomic molecules with identical nuclei have no electric dipole moment that gives rise to an observable microwave spectrum; but an important one, O_2, has a magnetic spin dipole moment that gives rise to an observable microwave spectrum. It provides a simple illustration of the effects of nuclear spin statistics on the population of rotational states. Its ground electronic state ψ_e is antisymmetric for the exchange of the nuclei, and its vibrational states are symmetric. Therefore $\psi_e\psi_v$ is odd. Because O_2 has a spin angular momentum, the rotational quantum number is represented by N rather than J (see Chapter IV, Section 2). The rotational wave functions are symmetric when N is even (0, 2, 4, etc.) and antisymmetric when N is odd (1, 3, 5, etc.). The nuclear spin I of ^{16}O is zero, and hence for $^{16}O^{16}O$, ψ_n is symmetric. Furthermore, the overall wave function must be symmetric for the operation which exchanges the two identical Bose particles (for a rotation of π). Since the product $\psi_e\psi_v$ is antisymmetric, it is obvious that only the ψ_r functions having odd N satisfy this condition. The rotational levels corresponding to even values of N are entirely missing. Since only half of the possible energy levels exist, it is evident that the expression for Q_r must be divided by $\sigma=2$.

When the spin of the identical nuclei of a diatomic molecule is not zero, all the rotational levels will be populated, but with different weights. This condition, which does not give rise to a microwave spectrum, is most simply illustrated by H_2, for which $I=\frac{1}{2}$. Let us designate the individual spin states corresponding to $M_I=+\frac{1}{2}$ as α and those corresponding to $M_I=-\frac{1}{2}$ as β. Obviously the combined spin functions ψ_n for the two-proton system corresponding to $|\alpha\alpha)$ and $|\beta\beta)$ will not change sign and therefore will be symmetric upon exchange of the identical nuclei. The combinations $|\alpha\beta)$ or $|\beta\alpha)$, where the two proton spins differ, cannot be distinguished, and the acceptable functions are independent linear combinations:

$$\psi_s=\frac{1}{\sqrt{2}}\,[|\alpha\beta)+|\beta\alpha)] \tag{3.75}$$

$$\psi_a=\frac{1}{\sqrt{2}}\,[|\alpha\beta)-|\beta\alpha)] \tag{3.76}$$

The first of these is symmetric under proton exchange, and the second is antisymmetric. There are thus three symmetric spin functions and one antisymmetric spin function. The total spin angular momentum quantum number is $T=1$ for the symmetric spin functions and $T=0$ for the antisymmetric spin function. For H_2, $\psi_e\psi_v$ is symmetric, and $\psi_e\psi_v\psi_r$ is symmetric for even J and antisymmetric for odd J. The overall function must be antisymmetric for the operator that exchanges the identical Fermi particles. This requirement can be met by a combination of the odd J rotational levels (odd $\psi_e\psi_v\psi_r$) with the even or symmetric spin states (weight 3) or by a combination of even rotational levels (even $\psi_e\psi_v\psi_r$) with the odd spin states (weight 1). Consequently, all the rotational levels occur, but the ones with $J=1, 3, 5, \ldots$ have three times the weight of those with $J=0, 2, 4, \ldots$. The average weight of the levels $(3+1)/2=2$ is again only one-half the total number of spin states $(2I+1)^2=4$.

For diatomic molecules having identical nuclei with spin I, there are $(I+1)(2I+1)$ symmetric (even) spin functions and $I(2I+1)$ antisymmetric (odd) spin functions. The relative weights of adjacent rotational levels will be $I/(I+1)$, and the average weights of the levels will be $(2I+1)^2/2$. The symmetry of $\psi_e\psi_v$ and the conformity of I to Bose or Fermi statistics (as explained previously) determine whether the J even or the J odd rotational levels will have the greater weight. A summary and values of g (nuclear) are given in Table 3.2.

For determination of nuclear statistical weights in symmetric-top rotors, the rotational levels are conveniently classified under the rotational subgroup of the full point group of the molecule. This rotational subgroup is comprised of the identity operation and the rotational operations that exchange identical nuclei. The rotational subgroup for symmetric tops belonging to the point group C_{3v} is C_3, the symmetry species of which are A and E (see Table 12.1). Similarly, ψ_e and ψ_v may be classified with respect to the given rotational subgroup. Although the calculation of nuclear statistical weights in symmetric polyatomic molecules is appreciably more involved than that for diatomic molecules, nevertheless, it is based on the same principle, requiring the proper symmetry of the overall wave function. Nuclear statistical weights g (nuclear)

Table 3.2 Symmetry of the Eigenfunctions and Nuclear Statistical Weights for Exchange of Two Identical Nuclei

Statistics	*Spin* I	*Total Function* ψ *Overall*	*Coordinate Function* $\psi_e\psi_v\psi_r$	*Spin Function* ψ_n	*Statistical Weight*[a] g(*nuclear*)
Fermi-Dirac	$\frac{1}{2}, \frac{3}{2}, \ldots$	asym	sym	asym	$(2I+1)I$
			asym	sym	$(2I+1)(I+1)$
Bose-Einstein	$0, 1, 2, \ldots$	sym	sym	sym	$(2I+1)(I+1)$
			asym	asym	$(2I+1)I$

[a] $g_I=g$ (nuclear)$/(2I+1)^2$.

for the largest class of symmetrical molecules, those having C_{3v} symmetry (NH_3, CH_3Cl, CH_3CCH, etc.) are listed in Table 3.3. These have been obtained with group theory by Dennison [15] and by Wilson [16]. Other derivations are described in books by Townes and Schawlow [7] and by Sugden and Kenney [17]. Because of the opposite symmetry of the doubly degenerate components of the levels when $K \neq 0$, the same statistical weight formulas hold for either Bose or Fermi particles. However, when $K=0$, the weights for the different inversion levels depend on the nature of the nuclear statistics as indicated in Table 3.3. These formulas hold for symmetric vibrational and electronic states, $\psi_e \psi_v$ even. They also hold for molecules in degenerate vibrational states with ψ_e even if K is replaced by $K-l$. If the inversion doubling is not resolved, the statistical weights would be the sum of those for the sublevels.

Table 3.3 Nuclear Statistical Weights[a] for Symmetric-top Molecules Having C_{3v} Symmetry

Statistics	*J*	*K*	*Inversion Level*	*Statistical Weight g(nuclear)*
		When Inversion Levels Are Separated[b]		
Either	Even or odd	Not 0 and multiple of 3	Either	$\frac{1}{3}(2I+1)(4I^2+4I+3)$
Either	Even or odd	Not 0 and not multiple of 3	Either	$\frac{1}{3}(2I+1)(4I^2+4I)$
Fermi-Dirac Bose-Einstein	Even Odd	0	Lower	$\frac{1}{3}(2I+1)(2I-1)I$
Fermi-Dirac Bose-Einstein	Even Odd	0	Upper	$\frac{1}{3}(2I+1)(2I+3)(I+1)$
Fermi-Dirac Bose-Einstein	Odd Even	0	Lower	$\frac{1}{3}(2I+1)(2I+3)(I+1)$
Fermi-Dirac Bose-Einstein	Odd Even	0	Upper	$\frac{1}{3}(2I+1)(2I-1)I$
		When Inversion Levels Are Not Separated[c]		
Either	Even	0 or multiple of 3		$\frac{1}{3}(2I+1)(4I^2+4I+3)$
Either	Even or odd	Not multiple of 3		$\frac{1}{3}(2I+1)(4I^2+4I)$

[a]These statistical weights are for rotational lines of symmetric-top molecules with three off-axis atoms having nuclear spin I when the molecule is in a nondegenerate vibrational level and in $^1\Sigma$ electronic state. The values apply for a degenerate vibrational level if K is replaced by $K-l$. The reduced statistical weight g_I is obtained from division of the g (nuclear) by the total number of spin functions $(2I+1)^3$.

[b]In (3.66), $g_K=1$ for all K.

[c]In (3.66), $g_K=1$ for $K=0$ and $g_K=2$ for $K \neq 0$.

Theoretically, the averaged nuclear statistical weight factor involved in Q_r is g_n/σ where g_n is the total spin degeneracy and σ is the symmetry number as explained previously. However, it is more convenient in numerical calculations to omit the factor g_n from the expression for Q_r and to employ a reduced nuclear statistical weight g_I in the numerator of (3.66). This reduced factor is defined as

$$g_I = \frac{g(\text{nuclear})}{g_n} \tag{3.77}$$

Values of g_I for symmetric-top molecules having C_{3v} and C_3 symmetry are also given in Chapter VI.

A number of asymmetric rotors studied with microwave spectroscopy have two identical nuclei and C_{2v} symmetry. Examples are H_2O and SO_2. The identical nuclei can be exchanged and the molecule changed into an indistinguishable configuration by rotation of π degrees about the symmetry axis. With respect to this operation, there are $(I+1)(2I+1)$ symmetric and $I(2I+1)$ antisymmetric nuclear spin functions, just as for diatomic molecules. If $\psi_e\psi_v$ is positive, the symmetric nuclear spin functions must be paired with the even rotational levels when $I=0, 1, 2, \ldots$ (Bose-Einstein statistics) and with the odd rotational levels when $I=\frac{1}{2}, \frac{3}{2}, \frac{5}{2}, \ldots$ (Fermi-Dirac statistics). Thus far the conditions are exactly those which were described for the diatomic molecule. The difference in the two cases arises not from the nuclear statistics but from the greater difficulty in determination of the parity (odd or even character) of the wave functions ψ_r of the asymmetric rotor with respect to the operation which exchanges the identical nuclei. This procedure will be briefly described.

To determine the effects of nuclear statistical weights on a particular level, one must know which of the axes, a, b, or c, is the symmetry axis and must then ascertain the parity of ψ_r with respect to this axis. One then chooses the nuclear spin functions ψ_n which with the particular ψ_r will give the correct symmetry for the overall wave function ψ. For illustration, let us choose a molecule with C_{2v} symmetry for which $\psi_e\psi_v$ is even and the symmetry axis is the intermediate axis b. If spins of the two identical nuclei are zero or integrals (Bose-Einstein statistics), the overall wave function must be symmetric with respect to a rotation of π about the symmetry axis b. Now the level $J_{4,5}$ will be antisymmetric (odd) with respect to this operation; since the $\psi_e\psi_v$ is even, only the antisymmetric nuclear spin function will give the required overall symmetry (even). The nuclear statistical weight for the level will then be $I(2I+1)$. When $I=0$ it is evident that the level does not occur. If the symmetry axis had been the a axis, the symmetric spin function would have been required and the weights would have been $(I+1)(2I+1)$. For nuclei obeying the Fermi-Dirac statistics, $I=\frac{1}{2}, \frac{3}{2}, \ldots$, these weights would be reversed, that is, weights $(I+1)(2I+1)$ when b is the symmetry axis and $I(2I+1)$ when a is the symmetry axis. To obtain the corresponding reduced statistical weights, one divides these numbers by $(2I+1)^2$.

When there is more than one pair of equivalent nuclei, the statistical weights can be obtained by an extension of the foregoing discussion. For a molecule such as CH_2F_2, a rotation about the twofold axis of symmetry simultaneously exchanges the two hydrogens and the two fluorines. The resultant statistics, whether Bose-Einstein or Fermi-Dirac, depends on the statistics of the individual pairs and can be obtained by consideration of the effect of exchanging each pair separately. Exchange of the two hydrogens which are fermions must change the sign of the total wave function. The same holds for the exchange of two fluorines. Therefore a simultaneous exchange of both pairs of nuclei must leave the total wave function for CH_2F_2 unchanged, and the resultant statistics is Bose-Einstein statistics. On the other hand, for CD_2F_2, where deuterium is a boson, the resultant statistics is Fermi-Dirac; and the nuclear spin functions must be combined with the coordinate function in such a way as to yield a total wave function which changes sign for a rotation about the twofold axis.

For one pair of identical nuclei having spins I_1 there are a total of $(2I_1+1)^2$ spin functions, of which $(I_1+1)(2I_1+1)$ are symmetric spin functions and $I_1(2I_1+1)$ are antisymmetric. Likewise, the other pair of identical nuclei having spins I_2 have in all $(2I_2+1)^2$ spin functions, of which $(I_2+1)(2I_2+1)$ are symmetric and $I_2(2I_2+1)$ are antisymmetric. The total spin functions for the system of two pairs of identical nuclei will be product functions composed of the spin functions associated with I_1 and those associated with I_2. Products composed of the symmetric spin functions of the two types of nuclei will give total spin functions which are symmetric. Similarly, products of antisymmetric spin functions give symmetric total spin functions. There are thus for the composite system the following number of symmetric spin functions

$$\begin{aligned} g^s(\text{nuclear}) &= (I_1+1)(2I_1+1)(I_2+1)(2I_2+1) + I_1(2I_1+1)I_2(2I_2+1) \\ &= (2I_1+1)(2I_2+1)(2I_1I_2+I_1+I_2+1) \end{aligned} \tag{3.78}$$

If the symmetric spin functions of one type of nuclei are combined with the antisymmetric spin functions of the other, the total spin functions will be antisymmetric. The total number of such combinations will be

$$\begin{aligned} g^a(\text{nuclear}) &= (I_1+1)(2I_1+1)I_2(2I_2+1) + I_1(2I_1+1)(I_2+1)(2I_2+1) \\ &= (2I_1+1)(2I_2+1)(2I_1I_2+I_1+I_2) \end{aligned} \tag{3.79}$$

The total number of spin functions in all will be $(2I_1+1)^2(2I_2+1)^2$. For the case of CH_2F_2 with $\psi_e\psi_v$ even and the symmetry axis the b axis, we find from (3.78) and (3.79) that the *ee* and *oo* levels will have a statistical weight of 10 while the *eo* and *oe* levels have a statistical weight of 6.

An instructive illustration of the effects of nuclear statistics on the relative intensities of the lines of asymmetric rotors is provided by Fig. 3.5, which represents a number of lines corresponding to different J values for the ${}^RQ_{K_{-1}=0}$ branch of D_2S_2 observed by Winnewisser et al. [18]. For the $D^{32}S^{32}SD$ species it is readily apparent that the lines alternate in intensity in the ratio 1:2. Those for J even are the stronger set. This intensity alternation is due to the nuclear statistics of the identical D nuclei which are exchanged by a rota-

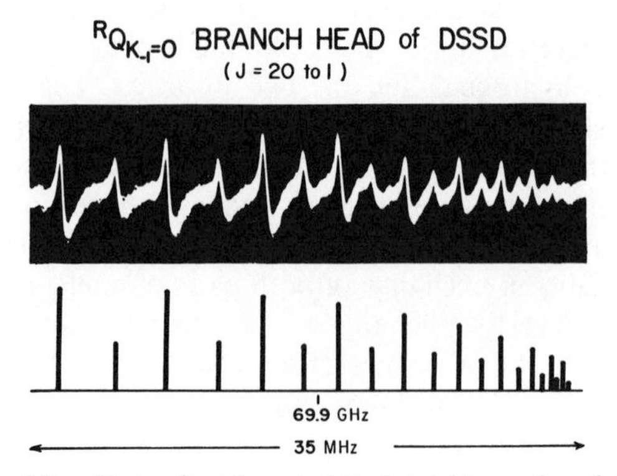

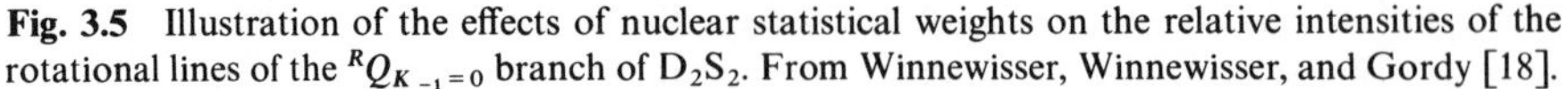

Fig. 3.5 Illustration of the effects of nuclear statistical weights on the relative intensities of the rotational lines of the $^{R}Q_{K_{-1}=0}$ branch of D_2S_2. From Winnewisser, Winnewisser, and Gordy [18].

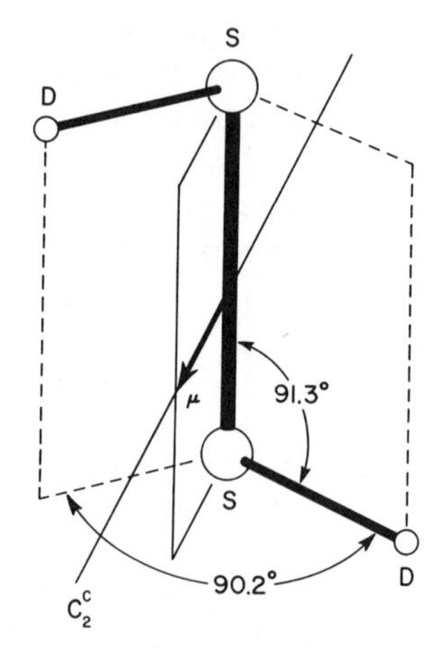

Fig. 3.6 Structural diagram of the D_2S_2 molecule.

tion of π degrees about the c axis, which is perpendicular to the SS bond as is indicated by Fig. 3.6. The operation C_2^c also exchanges two identical ^{32}S nuclei, but because they have zero spins the resultant spin of the identical nuclei exchanged by the operation is $I=1$, and the overall wave function must conform to the Bose-Einstein statistics. The molecules observed are in symmetric ground electronic and ground vibrational states. Thus $\psi_e\psi_v$ is even, and the product $\psi_r\psi_n$ must be symmetric with respect to the operation C_2^c which rotates the molecule about the symmetry axis c. This means that only even spin functions can combine with the even ψ_r and that only odd spin functions can combine with the odd ψ_r.

The D_2S_2 molecule is an almost accidentally prolate symmetric top with $I_b \approx I_c$. In the symmetric-top approximation the observed transitions correspond to $J \rightarrow J$ and $K=0 \rightarrow 1$. Classically, this corresponds to a rotation of the deuteriums and the dipole moment about the SS bond. In the asymmetric-top $J_{K_{-1}K_1}$ notation, the transitions are $J_{K_{-1}=0,K_1=J} \rightarrow J_{K_{-1}=1,K_1=J}$. Because the dipole moment is along c, the parity of K_1 does not change, and the allowed transitions are $eo \rightarrow oo$ and $ee \rightarrow oe$. The first type is between wave functions ψ_r that are antisymmetric with respect to C_2^c (K_1 odd), and the second is between functions that are symmetric with respect to this operation (K_1 even). Also note that for the $J_{0,K_1=J}$ levels or the $J_{K_{-1}=1,K_1=J}$ levels the parity of K_1 is even when J is even and odd when J is odd. Thus to achieve an overall wave function that is symmetric with respect to C_2^c we must combine the symmetric nuclear spin functions, weight $(I+1)(2I+1)=6$, with even J levels and must combine the antisymmetric nuclear spin functions, weight $I(2I+1)=3$, with the odd J levels. Therefore the transitions involving levels of even J are expected to have twice the strength of the transitions involving levels of odd J, in agreement with the observations.

Substitution of ^{34}S for ^{32}S in one position, to form $D^{32}S^{34}SD$, destroys the symmetry and the alternation in the line intensities. Although ^{34}S, like ^{32}S, has $I=0$, the difference in mass causes the center of mass to shift so that the nuclei are no longer exchanged by a rotation of π about a principal axis. Since $I=\frac{1}{2}$ for H, the $H^{32}S^{32}SH$ obeys the Fermi-Dirac statistics, and the alternation of the line intensity is $(I+1)/I=3:1$, with those for J odd being the stronger set. This case was also observed by Winnewisser et al. [18]. Table 3.2 indicates the nuclear statistical weights for exchange of two identical nuclei of any spin value I.

References

1. L. Pauling and E. B. Wilson, *Introduction to Quantum Mechanics*, McGraw-Hill, New York, 1935, Chap. XI.
2. P. Helminger and W. Gordy, *Bull. Am. Phys. Soc.*, **11**, 543 (1967).
3. W. V. Smith and R. R. Howard, *Phys. Rev.*, **79**, 132 (1950); C. Greenhow and W. V. Smith. *J. Chem. Phys.*, **19**, 1298 (1951).
4. H. A. Lorentz, *Proc. Amsterdam Acad.*, **8**, 591 (1906).
5. J. H. Van Vleck and V. F. Weisskopf, *Rev. Mod. Phys.*, **17**, 227 (1945). See also H. Fröhlich, *Nature*, **157**, 478 (1946).
6. P. W. Anderson, *Phys. Rev.*, **76**, 647 (1949). See also H. Margenau, *Phys. Rev.*, **76**, 1423 (1949).
7. C. H. Townes and A. L. Schawlow, *Microwave Spectroscopy*, McGraw-Hill, New York, 1955.
8. W. Gordy, W. V. Smith, and R. F. Trambarulo, *Microwave Spectroscopy*, Wiley, New York, 1953. Republication by Dover, New York, 1966.
9. G. Birnbaum, "Microwave Pressure Broadening" in *Intermolecular Forces*, J. O. Hirschfelder, Ed. (*Adv. Chem. Phys.*, **12**), Wiley, New York, 1967, pp. 487–548.
10. R. Karplus and J. Schwinger, *Phys. Rev.*, **73**, 1020 (1948).
11. R. H. Johnson and M. W. P. Strandberg, *Phys. Rev.*, **86**, 811 (1952).
12. R. Karplus, *Phys. Rev.*, **73**, 1027 (1948).
13. See Chap. 10 of [7].
14. G. Herzberg, *Infrared and Raman Spectra of Polyatomic Molecules*, Van Nostrand, New York, 1945. Chap. V.
15. D. M. Dennison, *Rev. Mod. Phys.*, **3**, 280 (1931).

16. E. B. Wilson, Jr., *J. Chem. Phys.*, **3**, 276 (1935).
17. T. M. Sugden and C. N. Kenney, *Microwave Spectroscopy of Gases*, Van Nostrand, London, 1965.
18. G. Winnewisser, M. Winnewisser, and W. Gordy, *J. Chem. Phys.*, **49**, 3465 (1968).

Chapter **IV**

DIATOMIC MOLECULES

1 MOLECULES WITHOUT ELECTRONIC ANGULAR MOMENTUM

Gross Rotational Spectrum—Rigid Rotor Approximation

Most stable diatomic molecules have a $^1\Sigma$ ground electronic state with no unbalanced electronic angular momentum. When there is no nuclear coupling, the angular momentum arises wholly from the end-over-end rotation. If small effects caused by centrifugal distortion and by interaction between vibration and rotation are negelected and if no external field is applied, the rotational spectra of these molecules are simple. To provide an understanding of the gross rotational spectra of diatomic molecules we shall first describe the rigid rotor approximation.

The Hamiltonian for the rigid rotor is given by (2.73). For diatomic, or linear polyatomic, molecules in $^1\Sigma$ states, the principal moment of inertia about the molecular bond axis is zero, and the moments of inertia for rotation in two orthogonal planes about an axis normal to the bond are equal. Thus

$I_a=0$, and from (2.5), $P_z=P_a=0$; $I_b=I_c=I$, and

$$\mathscr{H}_r=\frac{1}{2I}(P_x^2+P_y^2)=\frac{P^2}{2I} \tag{4.1}$$

Therefore, from (2.59),

$$E_J=\frac{\hbar^2}{2I}J(J+1)=hBJ(J+1) \tag{4.2}$$

With the selection rules $\Delta J=\pm 1$ for dipole transitions (see Chapter II, Section 6), the rotational absorption lines are

$$\nu_r=\frac{E_{J+1}-E_J}{h}=2B(J+1) \tag{4.3}$$

where $J=0, 1, 2, \ldots$ and where

$$B=\frac{h}{8\pi^2 I} \tag{4.4}$$

is the rotational spectral constant. The gross rotational spectrum then consists of a harmonic series of lines having frequencies

$$\nu=2B, 4B, 6B, \ldots \tag{4.5}$$

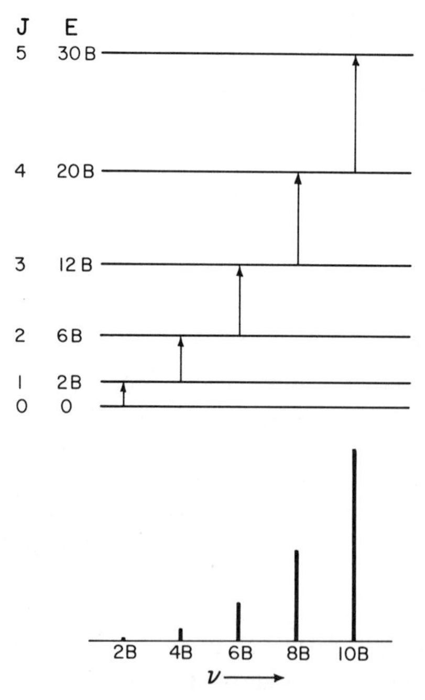

Fig. 4.1 Diagram indicating the sequence of the rotational levels and corresponding absorption lines for a diatomic molecule. This diagram also applies to linear polyatomic molecules discussed in Chapter 5.

The first few energy levels, transitions, and corresponding absorption lines are indicated in Fig. 4.1. The heights of the bars indicate the relative intensities that are assumed to increase as the cube of the frequency (see Section 3).

Note that the lowest rotational frequency corresponds to $2B$. Thus, if B is too large, that is, I too small, the molecule might have its rotational spectrum originating in the infrared region outside the region of observation with microwave techniques. However, all the diatomic molecules, except the lightest halogen hydride HF, for which $\nu(J=0\rightarrow1)=2B=1233$ GHz or $\lambda=0.243$ mm, have their rotational spectra originating in the region presently observable with microwave spectroscopy. The light H_2 molecule has no dipole moment and no microwave spectrum. Free radicals such as CH and OH have B values comparable to that of HF, but they have Λ doublet spectra (see Section 2) which can be observed at microwave frequencies. Lithium hydride has its lowest rotational frequency in the 0.68 mm wave region. The $J=0\rightarrow1$ transition of LiD in the 1.2-mm wave region [1] and that of HCl in the 0.48-mm region [2] have been measured. Essentially all diatomic molecules are now accessible to microwave spectral measurements although the lighter diatomic hydrides have only one or two rotational lines falling in the microwave region.

Nonrigid Molecules—The Vibrating Rotor

Diatomic molecules in $^1\Sigma$ ground states (no electronic angular momentum) are sufficiently simple that they can be treated as distortable vibrating rotors. A wave equation can be set up and solved for their vibrational and rotational energies. The solution yields directly the effects of centrifugal distortion and of interaction between vibration and rotation, which for polyatomic molecules must be calculated with perturbation theory. To obtain such a solution, however, one must know or assume the potential function for the interaction of the two atoms. We shall give the spectral constants obtained from solution of the wave equation with the familiar Morse potential function [3] and with the Dunham potential function [4], which is a series expansion having many adjustable potential parameters. Only the essential results will be given here. Descriptions of the solution of the equations, which are rather lengthy, can be found elsewhere [4–6].

In spherical coordinates the wave equation describing the rotationvibration motion of a nonrigid diatomic molecule, without interaction of electronic angular momentum ($^1\Sigma$ electronic states) and without nuclear couplings, is

$$\frac{1}{r^2}\frac{\partial}{\partial r}\left(r^2\frac{\partial\psi}{\partial r}\right)+\frac{1}{r^2\sin\theta}\frac{\partial}{\partial\theta}\left(\sin\theta\frac{\partial\psi}{\partial\theta}\right)+\frac{1}{r^2\sin^2\theta}\left(\frac{\partial^2\psi}{\partial\phi^2}\right)$$

$$+\frac{8\pi^2\mu}{h^2}[E-U(r)]\psi=0 \tag{4.6}$$

where the origin of the coordinate system is chosen at the center of mass and where r represents the internuclear distance. The reduced mass is

$$\mu=\frac{m_1m_2}{m_1+m_2} \tag{4.7}$$

where m_1 and m_2 are the masses of the individual atoms. With the assumption of a suitable potential function $U(r)$, it is possible to obtain a detailed solution of the equation and to find the eigenfunctions and the energy eigenvalues.

The most commonly used potential function is the Morse potential,

$$U(r)=\mathscr{D}[1-e^{-a(r-r_e)}]^2 \tag{4.8}$$

where $\mathscr{D}$ is the energy of dissociation measured from the bottom of the potential well, r_e is the equilibrium distance between the nuclei, and a is a constant. This function is represented approximately by the curve of Fig. 4.2. Solution of (4.6) with a Morse-type potential has been obtained by Morse [3] for $J=0$ and by Pekeris [6] for higher values of J. The energy eigenvalues obtained from the solution can be expressed in the form

$$\frac{E_{v,J}}{h}=\omega_e(v+\tfrac{1}{2})-\omega_e x_e(v+\tfrac{1}{2})^2+B_vJ(J+1)-D_v J^2(J+1)^2+H_v J^3(J+1)^3+\cdots \tag{4.9}$$

in which v represents the vibrational quantum number $v=0, 1, 2, \ldots$, J represents the rotational quantum number $J=0, 1, 2, \ldots$; ω_e is the fundamental vibrational frequency; and $\omega_e x_e$ is the anharmonicity constant. The vibrational transitions are not observable in the microwave region. We are concerned here with pure rotational transitions occurring within a given vibrational state and observed in absorption for which the selection rules are $\Delta v=0$ and $J\rightarrow J+1$. Application of these rules with the Bohr postulate, $h\nu=\Delta E$, to (4.9) yields the rotational frequencies

$$\nu=2B_v(J+1)-4D_v(J+1)^3+H_v(J+1)^3[(J+2)^3-J^3]+\cdots \tag{4.10}$$

$$B_v=B_e-\alpha_e(v+\tfrac{1}{2})+\gamma_e(v+\tfrac{1}{2})^2+\cdots \tag{4.11}$$

$$D_v=D_e+\beta_e(v+\tfrac{1}{2})+\cdots \tag{4.12}$$

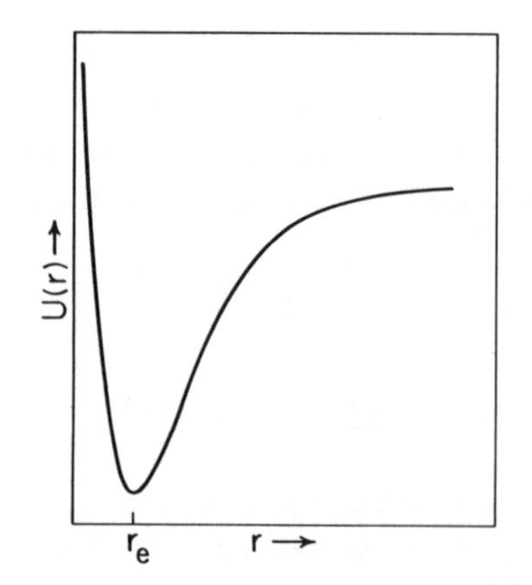

Fig. 4.2 Graph of the typical Morse potential function.

$$H_v \approx H_e = \frac{2D_e}{3\omega_e^2}(12B_e^2 - \omega_e\alpha_e) \tag{4.13}$$

The α_e and γ_e represent rotation–vibration interaction constants. The γ_e, which is small as compared with the α_e, can be neglected in some applications. The D_v represents the first-order centrifugal stretching term, and the term $\beta_e(v+\frac{1}{2})$ corrects for the effects of vibration on the centrifugal stretching constant. Because the higher-order centrifugal stretching correction (the term in H_v) is so small that its effects are barely observable in the microwave region, it is usually neglected.

For measurement of D_v for any given vibrational state, at least two rotational lines of molecules in that state must be measured. When H_v is not negligible, additional lines must be measured. For accurate measurement of D_v, one of the measured lines should be of a relatively high J transition, preferably in the shorter millimeter or submillimeter range where the centrifugal stretching effects are large. For an increase in accuracy and as a check on the consistency of the results, measurements are often made on many different transitions. An illustration is given in Table 4.1 of the use of (4.10) for evaluation of B_0 and D_0 for $^{12}C^{16}O$ from precise measurements of the first seven rotational transitions of the ground vibrational state. Effects of the higher-order terms at these frequencies are too small to permit a meaningful evaluation of H_0. Note that these lowest rotational frequencies of CO fall primarily in the submillimeter wave region. Millimeter or submillimeter wave measurements on all the hydrogen halides have now been made. The resulting B_0 and D_0 values are listed in Table 4.2.

Table 4.1 Rotational Frequencies and Derived Rotational Constants for $^{12}C^{16}O$ in the Ground Vibrational State

Transition	*Observed Frequency (MHz)*	*Calculated Frequency (MHz)*	$\nu_{obs}-\nu_{calc}$ *(MHz)*
$0\to1$	115,271.204[a]	115,271.206	−0.002
$1\to2$	230,537.974[b]	230,538.005	−0.031
$2\to3$	345,795.989[c]	345,795.993	−0.004
$3\to4$	461,040.811[c]	461,040.764	+0.047
$4\to5$	576,267.934[c]	576,267.910	+0.024
$5\to6$	691,472.978[c]	691,473.027	−0.049
$6\to7$	806,651.719[c]	806,651.708	+0.011

Derived Rotational Constants

$B_0 = 57{,}635.970 \pm 0.006$ MHz
$D_0 = 0.18358 \pm 0.00014$

[a]Rosenblum et al. [9].
[b]M. Cowan and W. Gordy, *Bull. Am. Phys. Soc.*, **2**, 212 (1957).
[c]P. Helminger, F. C. De Lucia, and W. Gordy, *Phys. Rev. Lett.*, **25**, 1397 (1970).

Table 4.2 Rotational Constants B_0 and D_0 and Equilibrium Internuclear Distances r_e for the Hydrogen Halides[a]

Halides	B_0 *(MHz)*	D_0 *(MHz)*	r_e
$^2H^{19}F$	325,584.98(30)	17.64	0.916914
$^1H^{35}Cl$	312,989.297(20)	15.863	1.2745991
$^1H^{37}Cl$	312,519.121(20)	15.788	1.2745990
$^2H^{35}Cl$	161,656.238(14)	4.196(3)	1.2745990
$^2H^{37}Cl$	161,183.122(16)	4.162(3)	1.2745998
$^3H^{35}Cl$	111,075.84	1.977	1.2745997
$^3H^{37}Cl$	110,601.62	1.960	1.2745985
$^1H^{79}Br$	250,358.51(15)	10.44(6)	1.4144691
$^1H^{81}Br$	250,280.58(15)	10.44(6)	1.4144705
$^2H^{79}Br$	127,357.639(12)	2.6529(14)	1.4144698
$^2H^{81}Br$	127,279.757(17)	2.6479(20)	1.4144698
$^3H^{79}Br$	86,251.993	1.234	1.4144705
$^3H^{81}Br$	86,174.078	1.232	1.4144691
$^1H^{127}I$	192,657.577(19)	6.203(3)	1.609018
$^2H^{127}I$	97,537.092(9)	1.578(1)	1.609018
$^3H^{127}I$	65,752.305	0.7150	1.609018

[a]These constants were derived by De Lucia, Helminger, and Gordy [*Phys. Rev.*, **A3**, 1849 (1971)] from their submillimeter wave measurements with the aid of earlier millimeter wave measurements of the tritium halides and with some of the D_0 values derived from infrared data, as cited by them. The r_e values have higher-order corrections applied (Chapter XIII, Section 6). The last three figures given for r_e have only relative significance because of the limited accuracy of Planck's constant.

If B_e, α_e, and γ_e are to be obtained entirely from microwave rotational spectra, measurements must be made on rotational lines in three vibrational states. Enough lines must be measured for each vibrational state so that the effects of centrifugal stretching on B_v can be determined. Normally, the ground state and the first two excited vibrational states ($v=0$, 1, and 2) are observed, and the corresponding rotational constants B_0, B_1, and B_2 are obtained. From (4.11) one finds that

$$\begin{aligned} B_e &= \tfrac{1}{8}(15B_0 - 10B_1 + 3B_2) \\ \alpha_e &= 2B_0 - 3B_1 + B_2 \\ \gamma_e &= \tfrac{1}{2}(B_0 - 2B_1 + B_2) \end{aligned} \tag{4.14}$$

Because of low population the second vibrational state for molecules with high ω_e may not be observable. Then γ_e may be neglected, and reasonably accurate values of B_e and α_e may be obtained from B_0 and B_1 only.

Theoretically,

$$B_e = \frac{h}{(8\pi^2 I_e)} = \frac{h}{(8\pi^2 \mu r_e^2)} \tag{4.15}$$

where r_e represents the equilibrium internuclear distance, the separation of the nuclei corresponding to the minimum on the potential energy curve (Fig. 4.2), and where I_e is the moment of inertia that the molecule would have if the nuclei actually were separated by the distance r_e with an effective reduced mass of μ. Thus a determination of B_e from the spectral data provides an evaluation of r_e. However, one should be aware of the fact that B_e obtained as just described depends on an assumed potential function and neglects the effects caused by electronic motions and displacements. Although gross effects of electronic interaction with rotation occur only when there is an unbalanced electronic spin or orbital momentum as described in Section 2, there are slight interactions between the electronic and rotational motions even for molecules in $^1\Sigma$ states. Because of unequal sharing of the bonding electrons and the distortion of the electronic clouds of each atom by the chemical bonding, the effective reduced mass is not exactly that obtained by substitution of the atomic masses of the neutral atoms into (4.7). These slight electronic effects on r_e are treated in Chapter XI, Section 7 and Chapter XIII, Section 6. Corrections for these higher-order effects were included in the calculation of the r_e values for the hydrogen halides listed in Table 4.2. As a result, the r_e values listed for the different isotopic species of the same molecule are the same, within the accuracy of the measurements. The largest effects on the interatomic distances are due to the vibrational motions and to centrifugal stretching by rotation. These effects are taken into account in the preceding theory. The determination of B_e with the assumption of a different potential function, the Dunham series function, is described later.

The equilibrium values B_e, I_e, and r_e have the advantage of being independent of the vibrational motions, and hence r_e is independent of isotopic substitution. However, these values are purely theoretical. Because of the asymmetry in the potential curve they do not correspond to averaged values of the internuclear distance for any eigenstate. Perhaps of more real significance for the chemist are the averaged values I^* and $\langle r \rangle$ for the ground vibrational state and for a particular isotopic species of the molecule. Evaluation of these from B_0 is discussed in Chapter XIII.

The vibrational constants ω_e and $\omega_e x_e$ are related to the rotational constants B_e, α_e, and D_e through the auxiliary relations derived by Pekeris [6]

$$\omega_e^2 = \frac{4B_e^3}{D_e} \tag{4.16}$$

$$\omega_e x_e = B_e \left(\frac{\alpha_e \omega_e}{6B_e^2} + 1 \right)^2 \tag{4.17}$$

with the assumption of the Morse potential function. Also, this assumption leads to the prediction,

$$\mathscr{D} = \frac{\omega_e^2}{4\omega_e x_e} \tag{4.18}$$

for the dissociation energy of the molecule. Equation 4.18 provides only an approximate value of $\mathscr{D}$. For the lighter diatomic molecules for which ω_e

is large, the sparse population of excited vibrational states makes it difficult to obtain enough microwave data for a complete evaluation of B_e, α_e, and D_e. When this is true, (4.16) and (4.17) make possible the use of ω_e and $\omega_e x_e$ from vibrational spectra with microwave data on the ground vibrational state for an accurate determination of these constants. For such light molecules, the vibrational frequencies fall in the near-infrared region where ω_e and $\omega_e x_e$ can be measured very accurately. In contrast, ω_e and $\omega_e x_e$ for the heavier molecules cannot be measured very accurately with optical spectroscopy, but the smallness of ω_e and consequently the greater population of their excited vibrational states make possible the measurements of rotational transitions in a sufficient number of vibrational states for precise evaluation of B_e, α_e, and D_e from microwave data only. Then, (4.16) and (4.17) give the vibrational frequency ω_e and the anharmonicity constant $\omega_e x_e$. The values of these vibrational constants for the heavier molecules are usually more precise when measured in this manner from the rotational spectra than when measured directly in the infrared region. For such molecules, however, it is advantageous to apply the more elaborate Dunham theory, which is described next.

Dunham's Solution

Employing the Wentzel-Kramers-Brillouin method, Dunham [4] has solved the wave equation for the diatomic molecule with the assumption of a series expansion of the potential function in the form

$$U(\xi)=a_0\xi^2(1+a_1\xi+a_2\xi^2+a_3\xi^3+\cdots)+B_eJ(J+1)(1-2\xi+3\xi^2-4\xi^3+\cdots) \tag{4.19}$$

where $\xi=(r-r_e)/r_e$ and where $a_0, a_1, a_2, \ldots$ are potential constants. The resulting expression which he obtained for the energy levels is

$$\begin{aligned}\frac{E_{vJ}}{h}&=\sum_{l,m} Y_{l,m}(v+\tfrac{1}{2})^l J^m(J+1)^m\\&=Y_{10}(v+\tfrac{1}{2})+Y_{20}(v+\tfrac{1}{2})^2+Y_{01}J(J+1)\\&\quad+Y_{11}(v+\tfrac{1}{2})J(J+1)+Y_{21}(v+\tfrac{1}{2})^2J(J+1)\\&\quad+Y_{31}(v+\tfrac{1}{2})^3J(J+1)+Y_{02}J^2(J+1)^2\\&\quad+Y_{12}(v+\tfrac{1}{2})J^2(J+1)^2+Y_{03}J^3(J+1)^3+\cdots\end{aligned} \tag{4.20}$$

where the Y's are observable constants known as Dunham's constants. Usually only five of the Dunham constants are required for a fitting of the observable microwave rotational spectra. To seven terms, the rotational frequency formula is

$$\begin{aligned}\nu=&\,2Y_{01}(J+1)+2Y_{11}(v+\tfrac{1}{2})(J+1)+2Y_{21}(v+\tfrac{1}{2})^2(J+1)\\&+2Y_{31}(v+\tfrac{1}{2})^3(J+1)+4Y_{02}(J+1)^3+4Y_{12}(v+\tfrac{1}{2})(J+1)^3\\&+Y_{03}(J+1)^3[(J+2)^3-J^3]+\cdots\end{aligned} \tag{4.21}$$

One must measure at least seven spectral lines to evaluate the seven constants, but in practice it is desirable to measure more than this and to obtain a least-squares fitting of the formula to the measured frequency. For this task it is desirable to use a high-speed computer. The terms in Y_{31} and Y_{03} are very small and often are negligible.

With the possible exception of Y_{01}, the Dunham constants are equivalent within the accuracy of the usual microwave measurements to the more familiar constants of (4.9)–(4.13). The relationships are

$$Y_{01} \approx B_e, \quad Y_{02} = -D_e, \quad Y_{03} = H_e, \quad Y_{11} = -\alpha_e, \quad Y_{12} = -\beta_e$$
$$Y_{21} = \gamma_e, \quad Y_{10} = \omega_e, \quad \text{and} \quad Y_{20} = -\omega_e x_e \tag{4.22}$$

From a comparison of (4.11) and (4.21) it is evident that

$$B_v = Y_{01} + Y_{11}(v+\tfrac{1}{2}) + Y_{21}(v+\tfrac{1}{2})^2 + Y_{31}(v+\tfrac{1}{2})^3 + \cdots \tag{4.23}$$

The more precise relationship of Y_{01} and B_e is

$$Y_{01} = B_e\left[1 + C_{01}\left(\frac{B_e^2}{\omega_e^2}\right)\right] \tag{4.24}$$

where

$$C_{01} = Y_{10}^2\left(\frac{Y_{21}}{4Y_{01}^3}\right) + 16a_1\left(\frac{Y_{20}}{3Y_{01}}\right) - 8a_1 - 6a_1^2 + 4a_1^3 \tag{4.25}$$

Dunham gives equations relating the various Y's to the potential constants a_0, a_1, a_2, and so on. Generally, these equations are rather involved, and all need not be repeated here. The expressions most often needed are those which give the first few potential constants in terms of the Y's that can be obtained directly from the measured spectral lines. The first four potential constants can be evaluated from

$$a_0 = \frac{\omega_e^2}{4B_e} = \frac{B_e^2}{D_e} \approx -\frac{Y_{01}^2}{Y_{02}}$$

$$a_1 = \frac{Y_{11}Y_{10}}{6Y_{01}^2} - 1 = \frac{Y_{11}}{3(-Y_{02}Y_{01})^{1/2}} - 1$$

$$a_2 = \frac{Y_{12}}{6}\left(\frac{Y_{01}}{-Y_{02}^3}\right)^{1/2} + \frac{9}{8}a_1(2+a_1) + \frac{19}{8}$$

$$a_3 = -\frac{2}{15}\frac{Y_{21}}{Y_{02}} + \frac{a_2}{5}(3+13a_1) - \frac{a_1}{2}[4+3a_1(1+a_1)] - 1 \tag{4.26}$$

The vibrational constants Y_{10} and Y_{20} are not obtained directly from the rotational frequencies but can be derived with use of the relations,

$$Y_{10} = 2\left(\frac{B_e^3}{-Y_{02}}\right)^{1/2} \approx 2\left(\frac{Y_{01}^3}{-Y_{02}}\right)^{1/2}$$

$$Y_{20} = \tfrac{3}{2}B_e(a_2 - \tfrac{5}{4}a_1^2) \tag{4.27}$$

Higher-order Corrections of the Theory

There are three small deviations of Y_{01} from the equilibrium constant B_e which are detectable with highly precise microwave measurements for some diatomic molecules in $^1\Sigma_0$ ground states. These are generally designated as δ_1, δ_2, and δ_3 in the probable order of their decreasing significance. The δ_3, called Dunham's correction, is included in Dunham theory and is calculable with (4.24). Mechanistically, it arises from an anharmonicity generated by rotation–vibration interactions. The deviations δ_1 and δ_2, not included in Dunham's solutions, result, in principle, from the inadequacy of the Born-Oppenheimer approximation [7] employed by Dunham. Mechanistically, δ_1 may be ascribed to electron-cloud distortions and to electronic components in the total angular momentum induced by the end-over-end rotation of the molecule. These effects also give rise to the molecular rotational g factor for $^1\Sigma$ molecules, and the δ_1 correction may be calculated from the experimentally measured g_{mol}, as is shown in Chapter XI, Section 7 and Chapter XIII, Section 6 (See Eq. 11.115). The deviation δ_2, although related to δ_1, differs from it mechanistically. The induced, second-order, electronic component in the total angular momentum (which contributes to δ_1) causes the rotating internuclear frame to wobble, thus inducing a slight stretching force that in diatomic molecules is along the internuclear axis. The theory for this effect, known as wobble-stretching, was first derived by Van Vleck [8] and was further developed and applied by Rosenblum et al. [9] in their early microwave spectral measurements of CO. Applications to other molecules are described in Chapter XIII, Section 6.

In summary, the precise B_e value for a diatomic molecule in a $^1\Sigma$ electronic ground state is related to Y_{01} by

$$B_e = Y_{01} + \delta_1 + \delta_2 + \delta_3 \tag{4.28}$$

where δ_1 is the correction for electronic distortion and electronic angular momentum components (L-uncoupling) described in Chapter XI, Section 7, δ_2 is the wobble-stretch correction described in Chapter XIII, Section 6, and δ_3 is the Dunham correction given by (4.24). In calculation of δ_1 for diatomic molecules with (11.115), Y_{01} may be substituted for B_{eff} and $B_e - (\delta_2 + \delta_3)$ for B. Thus, for this case (11.115) becomes

$$B_e - (\delta_2 + \delta_3) = \frac{Y_{01}}{1 + (m/M_p)g} \cong Y_{01} + \delta_1 \tag{4.29}$$

and hence

$$\delta_1 = -\left(\frac{m}{M_p}\right) g Y_{01} = -\frac{g Y_{01}}{1836} \tag{4.30}$$

There is no direct method for calculation of δ_2, but because of its $1/\mu^2$ dependence it may be obtained from reduced mass ratios when more than one isotopic species of the molecule is measured (See Chapter XIII, Section 6).

Because of the proportionality relationship,

$$Y_{lm} \propto \left(\frac{1}{\mu}\right)^{(l+2m)/2} \tag{4.31}$$

between spectral constants Y_{lm} and the reduced mass $\mu = m_1 m_2/(m_1 + m_2)$ of the molecule, the very useful relationship

$$\frac{Y_{lm}}{Y'_{lm}} = \left(\frac{\mu'}{\mu}\right)^{(l+2m)/2} \tag{4.32}$$

exists between the corresponding constants Y_{lm} and Y'_{lm} and the reduced masses μ and μ' of different isotopic species of the same diatomic molecule. The relationship holds to the accuracy of most microwave measurements except for Y_{01} which does not always correspond to B_e within the accuracy of the measurements. The relationship can be used for calculation of Y_{lm} values for isotopic species of low abundance for which the rotational lines cannot be measured as completely or as accurately as those of a more abundant species. When sufficient lines can be accurately measured for different isotopic species, the relationship can be used to give accurate isotopic mass ratios of atoms. In this way, precise isotopic ratios have been obtained for some of the heavier elements [10–12].

Applications of Dunham's Theory

Successful application of Dunham's theory requires measurements on different rotational transitions in at least two, and preferably more, excited vibrational states. For an accurate evaluation of the stretching constant, molecules should be measured over a range of J values that includes some relatively high ones. It is evident that a wide frequency coverage is needed, including millimeter or even submillimeter waves, and that the required range of data is not easily obtainable for molecules with high values of B and ω. Nevertheless, sufficient data have been obtained for useful application of Dunham's theory to the alkali halides, including the lightest of the group ^{6}LiF. Table 4.3 illustrates the close fitting of the frequencies calculated with (4.21) to observed rotational lines of three vibrational states measured over a wide range of millimeter and submillimeter wave frequencies. The six Dunham constants obtained from this fitting are those listed for ^{6}LiF in Table 4.4. The Y values of Table 4.4 have been used with (4.26) for calculation of the first four potential constants in the series expansion of (4.19). The resulting constants are listed in Table 4.5. With (4.19) these constants provide accurate description of the potential curve in the region where $|r - r_e| \ll r$. The series of (4.19) does not converge rapidly when $|r - r_e|$ is large, and the description becomes less precise. In Table 4.4, note the accuracy of the indirectly measured vibrational constants, ω_e and $\omega_e x_e$.

During the early years of microwave spectroscopy, the lower J rotational transitions of the mixed diatomic halides, ClF, BrF, BrCl, ICl, and IBr, were measured in the centimeter wave region. These measurements provided valuable

Table 4.3 Comparison of Observed Frequencies of ^{6}LiF with Those Calculated from (4.21) with Constants from Table 4.4

Vibration v	*Transition* $J \to J+1$	*Measured Frequency*[a] *(MHz)*	*Calculated Frequency*[a] *(MHz)*	*Difference*
0	0→1	89,740.46[b]	89,740.47	−0.01
0	1→2	179,470.35	179,470.37	−0.02
0	2→3	269,179.18	269,179.12	0.06
0	3→4	358,856.19	358,856.16	0.03
0	4→5	448,491.07	448,490.92	0.15
0	5→6	538,072.65	538,072.83	−0.18
1	0→1	88,319.18[b]	88,319.19	−0.01
1	1→2	176,627.91	176,627.92	−0.01
1	2→3	264,915.79	264,915.74	0.05
1	3→4	353,172.23	353,172.19	0.04
1	4→5	441,386.83	441,386.82	0.01
2	0→1	86,921.20[b]	86,921.20	0.00
2	1→2	173,832.04	173,832.06	−0.02
2	2→3	260,722.24	260,722.23	0.01
2	3→4	347,581.39	347,581.38	0.01
3	1→2	171,082.27	171,082.26	0.01
3	2→3	256,597.84	256,597.82	0.02
3	3→4	342,082.66	342,082.71	−0.05

[a]All measured frequencies (except those indicated by *b*) and all calculated frequencies are taken from Pearson and Gordy [1].

[b]L. Wharton, W. Klemperer, L. P. Gold, J. J. Gallagher, and V. E. Derr, *J. Chem. Phys.*, **38**, 1203 (1963).

information about nuclear coupling, but the transitions measured were insufficient to permit accurate evaluation of B_e, r_e, and other molecular constants. In 1980, Willis and Clark [13] measured the higher millimeter wave rotational transitions of the entire group in excited as well as ground vibrational states. Some of the halides were heated to temperatures of several hundred degrees to populate the higher vibrational states. The spectral constants obtained for the different isotopic species appear in the various tables of the paper. The molecular constants obtained, in addition to the Dunham Y's, include accurate Dunham potential constants, a_0, a_1, a_2, a_3, and r_e values for the entire group. References to the earlier microwave papers on these molecules may also be found in their paper. Tabulation of the earlier measurements of frequencies and the constants derived from them are given by Lovas and Tiemann [14].

Observations at High Temperatures and High Frequencies

The development of high-temperature cells for microwave spectroscopy has made possible the study of many diatomic molecules which have too low

Table 4.4 Observed and Derived Molecular Constants of Lithium Fluoride[a]

Constants	$^6Li^{19}F$	$^7Li^{19}F$
Observed		
$Y_{01}(\approx B_e)$ MHz	45,230.848(13)	40,329.808(14)
$Y_{11}(=-\alpha_e)$ MHz	−722.417(28)	−608.182(29)
$Y_{21}(=\gamma_e)$ MHz	5.918(16)	4.670(17)
$Y_{31}(=\delta_e)$ MHz	−0.212(28)	−0.0104(28)
$Y_{02}(=-D_e)$ kHz	−442.95(16)	−352.36(17)
$Y_{12}(=-\beta_e)$ kHz	4.81(12)	3.73(9)
Derived		
ω_e cm^{-1}	964.24(18)	910.25(22)
$\omega_e\chi_e$ cm^{-1}	9.136(45)	8.104(42)
B_e MHz	45,230.556	40,329.768
r_e Å	1.563857(21)	1.563857(21)
a_0 cm^{-1}	154,063(57)	153,976(74)
a_1 cm^{-1}	−2.70127(38)	−2.70062(49)
a_2 cm^{-1}	5.084(17)	5.101(18)
a_3 cm^{-1}	−7.85(15)	−7.98(15)

[a]Pearson and Gordy [1].

Table 4.5 Molecular Constants of Lithium and Sodium Hydroxides

Constants	^6LiD[a]	^7LiD[a]	NaH[b]	NaD[b]
Y_{01}(MHz)	131,615.07(4)	126,905.36(4)	146,999.10(30)	76,659.59(20)
Y_{11}(MHz)	−2,898.90(4)	−2,744.61(4)	−4,108.99(80)	−1,546.94(40)
Y_{21}(MHz)			32.83(50)	8.76(20)
Y_{31}(MHz)			−0.96(7)	−0.16(4)
Y_{02}(MHz)			−10.307(20)	−2.802(6)
B_e(MHz)	131,673(4)	126,961(4)	147,076(10)	76,680(2)
r_e(Å)	1.59490(2)		1.88654(10)	

[a]From Pearson and Gordy [1].
[b]From Sastry et al. [32].

a vapor pressure for observation at room temperature. Even the light diatomic molecule LiD must be heated to a temperature of a few hundred degrees C before it has a vapor pressure adequate for microwave measurements. The earliest high-temperature microwave spectrometer, developed by Townes and his co-workers [15], operated at centimeter wavelengths and employed a waveguide cell heated to the desired temperature. A number of spectrometers have since been constructed with heated Stark modulation cells which operate very effectively in the centimeter and upper millimeter wave regions. Particularly noteworthy is the work of Lide [16] and his co-workers at the Bureau of Standards, also that by Fitzky [17], Törring [18], and Hoeft [19] in Germany. High-temperature spectrometers of great convenience and effectiveness have been developed for the shorter millimeter and submillimeter wavelengths [20 – 27].

Many of the nonvolatile substances now being studied cannot be observed simply by vaporization of the substance in a hot cell, as was done for the alkali halides [15, 20, 21]. At the elevated temperatures required for evaporation, the molecules usually dissociate, if not completely, so much that the products of dissociation prevent detection of the much lower concentrations of the undissociated molecules. Special techniques have been developed for production of the molecules in the vapor state by reactions that occur within the absorption cell. An example of these techniques is the flowing of halide gas over hot chips of copper metal to produce the copper halides [28]. Measurable SiO vapor was produced by the heating of homogeneous mixtures of silicon dioxide to 1350°C directly within the absorption cell [10]. In the hot cell developed in the laboratory of M. Winnewisser at Giessen [23, 24], the metal atoms are vaporized into the microwave absorption chamber where they interact with flowing gases to produce the desired product such as BaO [29] or BaS [25].

The advantage of the higher frequencies for high-temperature spectroscopy lies partly in the greater ease with which short waves can be focused with horns and lenses. This, in turn, makes practical the use of free-space cells in which the molecules are simply vaporized into, or across, the path of the focused radiation. A further advantage comes from the rapid increase of absorption of the molecule with increase of frequency, which makes possible the use of a much shorter absorbing path. Fig. 4.3, showing the spectra of ^{6}LiF, illustrates the detection and display of absorption lines at high temperature in the submillimeter wave region. These spectra were obtained when the submillimeter waves were simply focused with horns and lenses through the central region of a cylindrical, stainless steel pipe, the central region of which was heated to vaporize the salt. Usually the rotational transitions observed in the millimeter and submillimeter wave regions are of such high J that the nuclear hyperfine structure treated in Chapter IX is not resolvable.

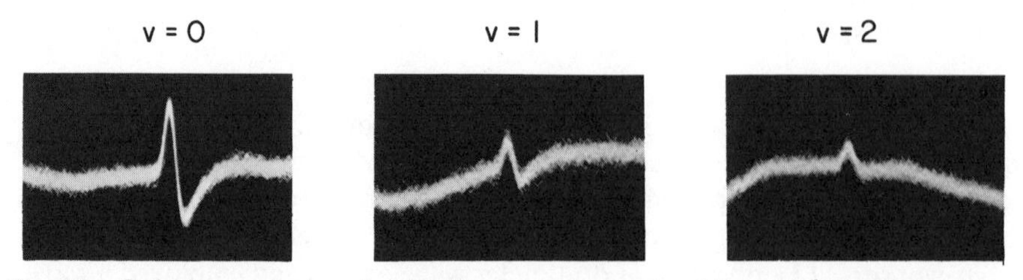

Fig. 4.3 Cathode ray display of submillimeter wave rotational lines of ^{6}LiF observed at a temperature of 950°K. The lines represent the $J=3\rightarrow4$ transition at $\lambda=0.85$ in different vibrational states as indicated. The line widths are approximately 1.6 MHz. From Pearson and Gordy [1].

Molecular Constants from High-temperature Spectroscopy

Many diatomic molecules in $^1\Sigma$ ground states have been studied with microwave rotational spectroscopy at high temperatures. These include the light molecules LiD and NaH and many heavier ones such as AgCl, TiCl, PbO, and GeS. The different isotopic species were measured for most of them. From these measurements were evaluated such molecular properties as moments of inertia, internuclear distances, fundamental vibrational frequencies, anharmonicities, potential constants, centrifugal distortion constants, molecular dipole moments, and nuclear quadrupole couplings. Although not all such quantities were evaluated for all the molecules studied, the moments of inertia and internuclear distances were obtained for each one. The dipole moments and nuclear couplings of some of these molecules will be treated in later chapters. Also, the evaluation of the internuclear distances will be described in Chapter XIII.

Probably the most comprehensive investigations are those made on the alkali halides, the first group of molecules to be observed at high temperatures with microwave spectroscopy [15]. Measurements on these molecules have been extended into the millimeter and submillimeter regions with spectrometers designed to reduce the excessive pressure and Doppler broadening at the elevated temperatures required for vaporization of these halides. By application of Dunham's theory, vibrational as well as rotational constants, internuclear distances, and potential constants were obtained. As an illustration, the constants for LiF are given in Table 4.4. Those for the other alkali halides may be found in the literature [21, 22, 30, 31].

Although the rotational transitions of LiD were measured [1] with molecules produced by vaporization of solid lithium deuteride in a hot cell, early effort to observe the heavier alkali hydrides in this way failed, apparently because of rapid dissociation of the molecules at the high temperature required for their vaporization. Sastry et al. [32] have now succeeded in measuring sharp lines of NaH and NaD by producing the molecules directly in the microwave absorption cell from flowing hydrogen gas and sodium vapor that were caused to react by a glow discharge. The important molecular constants they obtained are listed in Table 4.5, together with those previously measured for LiD. Higher-order corrections (earlier in Section 1) were applied for determination of r_e and B_e values.

Thorough studies of the diatomic halides of silver, copper, aluminum, and bismuth have now been made. Some of the molecular constants derived from these measurements are listed in Table 4.6. Other constants may be obtained from the original sources cited in the table. At the temperatures required for vaporization of these substances, excited vibrational levels are significantly populated. This makes possible the measurement of rotational transitions for several vibrational states and accurate evaluation of the rotation–vibration interactions, the vibrational frequencies, and the anharmonic constants. Note that ω_e for these molecules is obtained to 5 or 6 significant figures and $\omega_e x_e$ to 3 or 4 figures. As an example, the rotational structure of 14 vibrational states was measured for aluminum iodide [33].

Table 4.6 Selected Molecular Constants of Diatomic Halides of Silver, Copper, Aluminum, and Bismuth

Diatomic Halide	$Y_{01} \approx B_e$ (MHz)	ω_e (cm^{-1})	$\omega_e\chi_e$ (cm^{-1})	r_e (Å)	Ref.
$^{107}Ag^{19}F$	7,965.545(9)	514.6(111)	2.95(9)	1.983171(23)	a
$^{107}Ag^{35}Cl$	3,686.9639(6)	353.52(4)	1.169(3)	2.28078(31)	b,c,d
$^{107}Ag^{79}Br$	1,943.6420(15)			2.393100(3)	d,e
$^{107}Ag^{127}I$	1,345.1103(25)			2.544611(3)	d
$^{63}Cu^{19}F$	11,374.214(20)	622.65	3.95	1.744923(20)	a
$^{63}Cu^{35}Cl$	5,343.768(20)	417.599(42)	1.617(6)	2.051177(8)	f
$^{63}Cu^{79}Br$	3,055.707(16)	314.816(12)	0.955(1)	2.173435(6)	g
$^{63}Cu^{127}I$	2,197.102(2)	264.897(18)	0.715(2)	2.338317(2)	h
$^{27}Al^{19}F$	16,562.930(6)	802.85(25)	4.86(6)	1.65436(2)	i,j,k
$^{27}Al^{35}Cl$	7,313.206(4)	481.67(14)	2.07(4)	2.13011(3)	l,j,k
$^{27}Al^{79}Br$	4,772.825(2)	378.19(6)	1.327(9)	2.2980(3)	l
$^{27}Al^{127}I$	3,528.5533(8)	316.25(2)	0.981(2)	2.53709(3)	l
$^{209}Bi^{35}Cl$	2,761.8538(13)	308.18	1.09	2.47155(7)	m
$^{209}Bi^{79}Br$	1,295.5609(12)	209.62	0.52	2.60953(7)	n
$^{209}Bi^{127}I$	816.11943(13)	164.12	0.321	2.80053(8)	o

[a]F. J. Hoeft, F. J. Lovas, E. Tiemann, and T. Törring, *Z. Naturforsch.*, **25a**, 35 (1970).
[b]Pearson and Gordy [22].
[c]L. C. Krisher and W. G. Norris, *J. Chem. Phys.*, **44**, 391 (1966).
[d]J. Hoeft, F. J. Lovas, E. Tiemann, and T. Törring, *Z. Naturforsch.*, **26a**, 240 (1971).
[e]L. C. Krisher and W. G. Norris, *J. Chem. Phys.*, **44**, 974 (1966).
[f]Manson et al. [28].
[g]E. L. Manson, F. C. De Lucia, and W. Gordy, *J. Chem. Phys.*, **63**, 2724 (1975).
[h]E. L. Manson, F. C. De Lucia, and W. Gordy, *J. Chem. Phys.*, **62**, 4796 (1975).
[i]F. C. Wyse, W. Gordy, and E. F. Pearson, *J. Chem. Phys.*, **52**, 3887 (1970).
[j]J. Hoeft, F. J. Lovas, E. Tiemann, and T. Törring, *Z. Naturforsch.*, **25a**, 1029 (1970).
[k]D. R. Lide, *J. Chem. Phys.*, **38**, 2027 (1963); **42**, 1013 (1965).
[l]Wyse and Gordy [33].
[m]Kuijpers et al. [27].
[n]P. Kuijpers and A. Dymanus, *Chem. Phys. Lett.*, **39**, 217 (1976).
[o]Kuijpers et al. [26].

Microwave spectra of the alkaline earth oxides, CaO, SrO, and BaO, likewise the sulfide BaS, have been observed. For each of them, numerous rotational transitions of several vibrational states were measured in the millimeter wave range. An example of the observed spectra, that for BaS, is shown in Fig. 4.4 (on page 90). Accurate values of the molecular constants were obtained, some of which are listed in Table 4.7.

Pure rotational spectra of a large number of diatomic molecules formed by combinations of elements of Group IV with elements of Group VI have been measured for several vibrational states. All these substances, except those formed with carbon, are solids at room temperature and require elevated temperature for production of measurable vapors. Some of the molecular constants obtained are listed in Table 4.8 with references to the original sources, from which more complete results may be learned. Note that all of the very heavy lead group, PbO, PbS, PbSe, and PbTe, have been measured. It is of interest that maser emission lines corresponding to the first three rotational transitions of SiO in the first excited vibrational state, $v=1$, and the $J=1\rightarrow 0$ transition in the $v=2$ state have been observed in a large number of astronomical sources [34]. Laboratory measurements aided in the identification of these interstellar lines. Rotational lines of the related molecule CO have also been detected [35] from interstellar sources.

The interesting diatomic molecule PN, first observed with optical spectroscopy [36] in 1933, was not measured with microwave spectroscopy [37, 38] until 1972, following its observation in the $J=1$ rotational state with molecular beam electric resonance (MBER) [39] in 1971. Millimeter wave rotational transitions were measured [37] for the ground state and for four excited vibrational states at temperatures of approximately 800°C. The MBER measurements yielded accurate hyperfine coupling constants and the precise dipole moment, 2.7471 D. The millimeter wave measurement yielded accurate Dunham Y's, potential constants a_0, a_1, a_2, a_3, and the equilibrium values $B_e=23{,}578.34(8)$ MHz and $r_e=1.49086(2)$ Å.

Table 4.7 Molecular Constants of Some Alkaline Earth Oxides and Sulfides

Constants	$^{40}Ca^{16}O$[a]	$^{88}Sr^{16}O$[b]	$^{138}Ba^{16}O$[b,c]	$^{138}Ba^{32}S$[d,e]
Observed				
Y_{01}(MHz)	13,324.3578(32)	10,132.5841(16)	9,371.9403(12)	3,097.28216(26)
Y_{11}(MHz)	−99.3110(28)	−65.8012(26)	−41.7482(21)	−9.44620(33)
Y_{21}(MHz)			−0.12522(60)	−0.013323(66)
Y_{02}(kHz)	−19.623(33)	−10.848(11)	−8.1555(68)	−0.918568(63)
Y_{12}(kHz)	−0.203(30)	−0.108(19)	−0.024(10)	0.001554(93)
Derived				
ω_e(cm^{-1})	732.48(62)	653.30(33)	670.24(28)	379.42
$\omega_e\chi_e$(cm^{-1})	4.71(14)	3.74(14)	2.26(10)	0.8710
r_e(Å)	1.822203(68)	1.919809(95)	1.939630(7)	2.5073184(15)

[a]R. A. Creswell, W. H. Hocking, and E. F. Pearson, *Chem. Phys. Lett.*, **48**, 369 (1977).
[b]Hocking et al. [29].
[c]J. Hoeft, F. J. Lovas, E. Tiemann, and T. Törring, *Z. Naturforsch.*, **25a**, 1750 (1970).
[d]Helms et al. [25].
[e]E. Tiemann, C. Ryzlewicz, and T. Törring, *Z. Naturforsch.*, **31a**, 128 (1976).

Table 4.8 Selected Molecular Constants of Diatomic Molecules Formed by Si, Ge, Sn, and Pb with O, S, Se, and Te

Diatomic Molecules	$Y_{01}(\cong B_e)(MHz)$	$Y_{11}(MHz)$	$-Y_{02}(kHz)$	$\omega_e(cm^{-1})$	$\omega_e\chi_e(cm^{-1})$	r_e(Å)	*Ref*
$^{28}Si^{16}O$	21,787.453(11)	151.026(11)	29.38(13)	1252(3)	5.96(71)	1.50973(4)	*a,b*
$^{28}Si^{32}S$	9,099.5365(12)	44.1616(11)	5.997(59)			1.929320(57)	*c,d*
$^{28}Si^{80}Se$	5,756.365(11)	23.286(10)	2.524			2.058326(60)	*e*
$^{74}Ge^{16}O$	14,560.872(16)	92.306(27)				1.624647(5)	*f,g*
$^{74}Ge^{32}S$	5,593.1019(22)	22.4569(19)	2.41(11)	569(13)	1.723(53)	2.0120772(12)	*h*
$^{74}Ge^{80}Se$	2,888.218(3)	8.669(2)	0.60(6)			2.134632(63)	*i*
$^{74}Ge^{130}Te$	1,958.7903(15)	5.1702(12)	0.353(33)	308(15)	0.62(4)	2.3401556(26)	*j*
$^{120}Sn^{16}O$	10,664.189(17)	64.243(37)	7.98(2)	882	3.93	1.832198(38)	*k*
$^{120}Sn^{32}S$	4,103.0013(12)	15.1585(17)	1.272(32)	491.6(63)	1.412(25)	2.2090172(7)	*l*
$^{120}Sn^{80}Se$	1,948.584(5)	5.111(5)	0.33(8)			2.325603(70)	*m*
$^{120}Sn^{130}Te$	1,273.4936(10)	2.8609(10)	0.165(10)	237.6(75)	0.408(17)	2.5228035(28)	*n*
$^{208}Pb^{16}O$	9,212.791(12)	57.405(14)				1.921813(60)	*o*
$^{208}Pb^{32}S$	3,487.1435(24)	13.0373(52)	1.012(27)	431.8(58)	1.277(24)	2.2868535(13)	*l*
$^{208}Pb^{80}Se$	1,516.9358(19)	3.8952(15)	0.210(19)	272.3(50)	0.552(14)	2.402223(32)	*p*
$^{208}Pb^{130}Te$	938.5824(18)	2.02160(72)	0.080(16)	214(22)	0.374(55)	2.594969(13)	*q*

[a]T. Törring, *Z. Naturforsch.*, **23a**, 777 (1968).
[b]Manson et al. [10].
[c]E. Tiemann, E. F. Renwanz, J. Hoeft, and T. Törring, *Z. Naturforsch.*, **27a**, 1566 (1972).
[d]J. Hoeft, *Z. Naturforsch.*, **20a**, 1327 (1965).
[e]J. Hoeft, *Z. Naturforsch.*, **20a**, 1122 (1965).
[f]R. Honerjager and R. Tischer, *Z. Naturforsch.*, **28a**, 1374 (1973).
[g]T. Törring, *Z. Naturforsch.*, **21a**, 287 (1966).
[h]J. Hoeft, F. J. Lovas, E. Tiemann, R. Tischer, and T. Törring, *Z. Naturforsch.*, **24a**, 1217 (1969).
[i]J. Hoeft, *Z. Naturforsch.*, **21a**, 1240 (1966).
[j]J. Hoeft and H.-P. Nolting, *Z. Naturforsch.*, **22a**, 1121 (1967).
[k]T. Törring, *Z. Naturforsch.*, **22a**, 1234 (1967).
[l]Hoeft et al. [11].
[m]J. Hoeft, *Z. Naturforsch.*, **21a**, 437 (1966).
[n]J. Hoeft and E. Tiemann, *Z. Naturforsch.*, **23a**, 1034 (1968).
[o]Torring [18].
[p]J. Hoeft and K. Manns, *Z. Naturforsch.*, **21a**, 1884 (1966).
[q]Tiemann et al. [12].

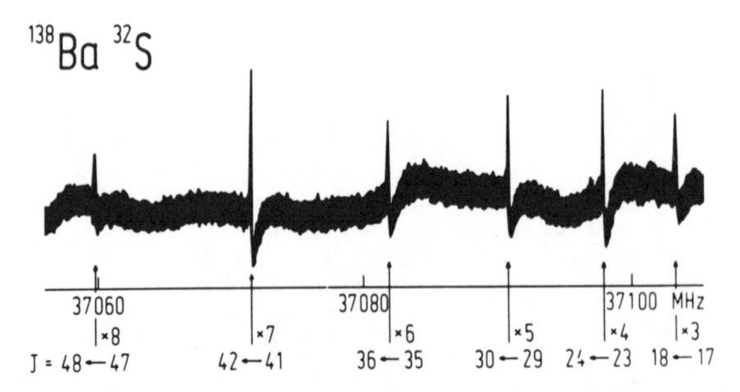

Fig. 4.4 Video oscilloscope display of a sequence of rotational transitions in $^{138}Ba^{32}S$. The third through the eighth harmonic of the klystron fundamental are shown covering simultaneously portions of the spectrum between 110 and 297 GHz. From Helms, Winnewisser and Winnewisser [25].

2 MOLECULES WITH ELECTRONIC ANGULAR MOMENTUM

Because of the interaction of the electronic angular momentum with the molecular rotation, molecules having unbalanced electronic angular spin or orbital momentum cannot be treated with the theory presented in Section 1. The nature of the microwave spectra of such molecules depends on the type of electronic state and upon the degree of coupling among the various angular momentum vectors (electronic spin, orbital angular momentum, and molecular end-over-end rotation). There are only a few stable diatomic molecules with unbalanced angular momentum, but with the microwave techniques now available the spectra of a number of unstable diatomic radicals such as CH and CN may be observable in the microwave region. The unstable species OH and OD have been successfully investigated, as have the moderately unstable molecules NO, SO, and ClO. A number of others have been observed with paramagnetic resonance, as described in Chapter XI.

If a diatomic molecule has orbital electronic angular momentum $L\neq 0$, this will be coupled strongly to the internuclear axis by the electrical forces of the chemical bond. As a result, the **L** vector will not be fixed in space but will precess rapidly about the internuclear axis. Thus **L** itself is not a constant of the motion and is not defined as it is in free atoms. However, the components of the orbital angular momentum are resolved along the bond axis and have definite quantized values. The electrical forces of the bond can be regarded as giving rise to a strong internal Stark effect which lifts the L degeneracy and resolves the components M_L along the bond axis. Because this is a second-order effect, the states corresponding to $+M_L$ and $-M_L$ are degenerate, and only the magnitude of M_L is required in labeling the states of different energy. Conventionally, the magnitudes are designated by Λ and have the values

$$\Lambda = |M_L| = 0, 1, 2, \ldots, L \tag{4.33}$$

Molecules for which $\Lambda=0$ have Σ states; those for which $\Lambda=1$ have Π states; those for which $\Lambda=2$ have Δ states, and so on.

Electrical forces do not couple directly to the resultant electronic spin angular momentum **S** since its associated moment is a purely magnetic dipole, but when $\Lambda \neq 0$, the magnetic field generated by the orbital motion which is directed along the bond axis will interact with the spin magnetic moment and, in the absence of other magnetic fields such as that generated by molecular rotation, will resolve it into quantized components along the bond axis. This internal "Zeeman" splitting of the S levels is a first-order effect, and the levels corresponding to the plus and minus components are not degenerate, as are those of L. The components of S along the axis, in units of $\hbar$, are designated by Σ (not to be confused with the Σ state), the possible values of which are

$$\Sigma = S, S-1, \ldots, -S \tag{4.34}$$

When $\Lambda = 0$, Σ is not defined. The resultant angular momentum along the molecular axis (in units of $\hbar$) is

$$\Omega = |\Lambda + \Sigma| \tag{4.35}$$

where Ω is the quantum number of the total electronic angular momentum for the nonrotating molecule corresponding to the quantum number J for the atom. Even when there is molecular rotation, Ω is still defined in some molecules.

Conventionally, J is used for designation of the quantum number for the total angular momentum of the molecule (or of the free atom) exclusive of nuclear spin. When there is nuclear coupling, the total angular momentum including nuclear spin is designated by **F**. Thus for an atom, $\mathbf{J} = \mathbf{L} + \mathbf{S}$; but for a molecule, **J** represents the sum of the electronic angular momentum and the angular momentum for the end-over-end rotation of the molecule, which is designated by **O**. When the molecule has no electronic angular momentum—when in the $^1\Sigma$ state—$J = O$, and only the quantum number J is required. The symbol **N** is used to designate the angular momentum exclusive of spin, that is, $\mathbf{N} = \mathbf{\Lambda} + \mathbf{O}$.

As we have observed, the term symbols Σ, Π, Δ, Φ, and so on, are employed to denote, respectively, the value of $\Lambda = 0, 1, 2, 3, \ldots$. To this is added as a left superscript the multiplicity $2S+1$ and as a right subscript the value of $\Lambda + \Sigma$. Note that $\Lambda + \Sigma$ may be negative in comparison to Ω. For $\Lambda = 1$, $S = 1$, we have the states $^3\Pi_0$, $^3\Pi_1$, $^3\Pi_2$.

Hund's Coupling Cases

CASE (*a*)

When $\Lambda \neq 0$ and the magnetic field generated by the rotation is not large, **S** as well as **L** is resolved along the molecular axis, and Ω is a good quantum number even when $O \neq 0$. Then **O** and $\mathbf{\Omega}$ form a resultant **J**, as indicated in the vector diagram of Fig. 4.5. This condition of coupling in which both **L** and **S** are resolved along the molecular axis in the presence of rotation, is known as Hund's case (*a*).

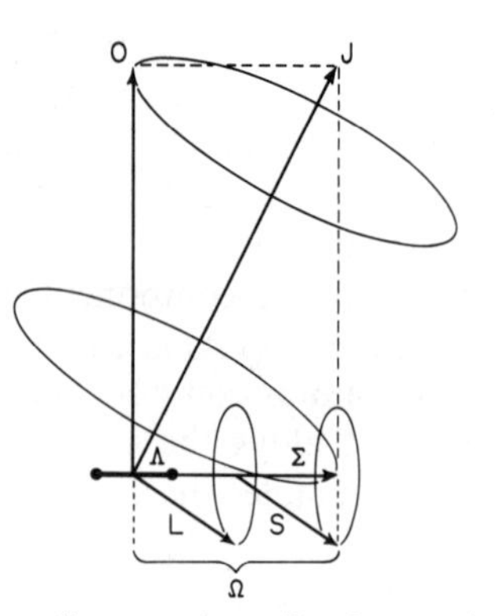

Fig. 4.5 Vector diagram of coupling for Hund's case (a).

It is evident from the vector diagram that for this case the molecular axis will precess about the direction of **J**, as does the symmetry axis z for the polyatomic symmetric-top molecule (Chapter V). Indeed, the eigenvalues for the rotational energy can be found by the same methods as those for the symmetric top: let I_α be the moment of inertia associated with the orbital angular momentum about the bond axis, and let I_b be the moment of inertia for rotation about an axis normal to the bond. The rotational Hamiltonian will then be

$$\mathscr{H}_r = \frac{1}{2}\left(\frac{\mathbf{O}^2}{I_b} + \frac{\mathbf{\Omega}^2}{I_a}\right) \tag{4.36}$$

However, from the vector diagram, Fig. 4.5,

$$\mathbf{O}^2 = \mathbf{J}^2 - \mathbf{\Omega}^2 \tag{4.37}$$

Hence,

$$\mathscr{H}_r = \frac{1}{2}\left[\frac{\mathbf{J}^2}{I_b} + \mathbf{\Omega}^2\left(\frac{1}{I_a} - \frac{1}{I_b}\right)\right] \tag{5.38}$$

Thus $\mathscr{H}_r$ is diagonal in the representation in which $\mathbf{J}^2$ and $\mathbf{\Omega}^2$ are diagonal. Therefore,

$$\begin{aligned} E_{J,\Omega} &= \frac{\hbar^2}{2}\left[\frac{J(J+1)}{I_b} + \Omega^2\left(\frac{1}{I_a} - \frac{1}{I_b}\right)\right] \\ &= h[B_v J(J+1) + \Omega^2(A-B)] \end{aligned} \tag{4.39}$$

where

$$J = \Omega, \Omega+1, \Omega+2, \ldots \tag{4.40}$$

Thus values of J begin with Ω, and levels for $J<\Omega$ do not occur.

The selection rules for dipole-induced transitions are

$$\Delta J = 0, \pm 1, \qquad \Delta\Omega = 0, \pm 1 \tag{4.41}$$

However, those transitions corresponding to a change in the electronic quantum number $\Delta\Omega = \pm 1$ represent such large changes in energy that the frequencies do not fall in the microwave region. The transitions that give rise to observable microwave absorption frequencies are therefore

$$\Delta J = 1, \qquad \Delta\Omega = 0 \tag{4.42}$$

The corresponding frequencies are

$$\nu = 2B_v(J+1) \tag{4.43}$$

This formula is the same as that obtained for the rigid diatomic molecule or the linear polyatomic molecule without electronic angular momentum Superficially, the electronic motions appear to have no effect on the frequencies, but this is not true. The values of the quantum number J are no longer restricted to integral values beginning with zero. The lowest value of J is not zero but Ω; and when Ω is a half-integral, J is also a half-integral.

Equation 4.43 applies only to the idealized coupling case (*a*), which never holds strictly. There are noticeable effects on the microwave spectra which are due to interactions of the end-over-end rotation and to centrifugal stretching. Before describing these effects we shall consider the second important coupling case known as Hund's case (*b*).

CASE (*b*)

When $\Lambda = 0$ but $S \neq 0$, the electronic momentum is due to spin only. Since there is no orbital field to couple the spin moment to the internuclear axis, the spin moment becomes coupled to the axis of rotation **N** through a weak magnetic field generated by the end-over-end rotation of the molecule. For certain light molecules such as OH, the molecular field generated by the rotation becomes so large for high rotational states that the spin **S** is resolved along **J** rather than along the molecular axis, even if $\Lambda \neq 0$. The idealized coupling case in which **S** is coupled to **N** to form a resultant **J**, as illustrated by the vector diagram of Fig. 4.6, is known as Hund's case (*b*). When $\Lambda = 0$, **N** becomes equivalent to **O**, and for molecules in multiplet Σ states such as O_2 the spin is coupled to **N**, which is now perpendicular to the internuclear axis, to form the rotational axis **J**. The significant numbers are

$$N = \Lambda, \Lambda+1, \Lambda+2, \ldots \tag{4.44}$$

and

$$J = N+S, N+S-1, N+S-2, \ldots, |N-S| \tag{4.45}$$

OTHER CASES

The molecules that have been investigated with microwave spectroscopy, as well as most of those likely to be investigated, are in Π or Σ states which can be classified as approximately Hund's case (*a*) or (*b*), or else as intermediate between these two cases. Hund proposes other idealized cases such as those

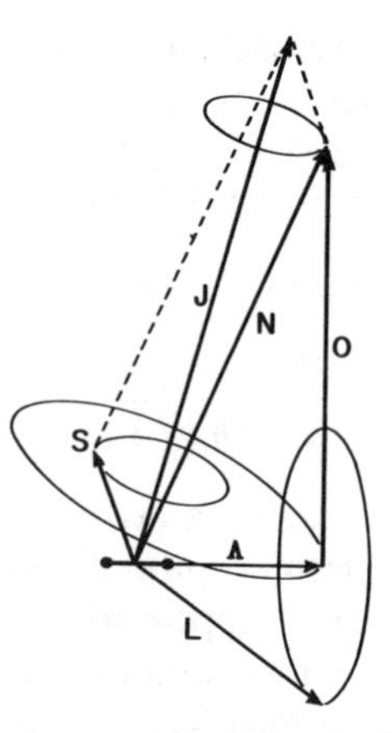

Fig. 4.6 Vector diagram of coupling for Hund's case (b).

in which both **L** and **S** are coupled to form a resultant which is, in turn, coupled to the molecular axis or to the axis of rotation. These infrequent cases are described by Herzberg in his treatise on diatomic molecules [40].

Molecules in the $^2\Sigma$ State

Molecules having electronic spin, but no orbital momentum and no strong nuclear coupling, approximate closely Hund's case (*b*). The spin is coupled to the rotational axis through interaction of the spin magnetic moment with the weak magnetic field generated by the molecular rotation. To a good approximation, this field is directly proportional to the end-over-end rotational momentum **N**, and the spin magnetic moment is $g_s\beta\mathbf{S}$. When there is more than one unpaired electron in the molecule, the individual spins are coupled to form a resultant spin greater than $\frac{1}{2}$ and multiplet states greater than the doublet Σ considered in this section. Although the coupling model still conforms to that of Fig. 4.7, the Hamiltonians of the higher spin states, $^3\Sigma$, $^4\Sigma$, and so on, are complicated by a spin–spin interaction considered for $^3\Sigma$ states in the next section. The $^2\Sigma$ molecules treated in this section have only one unpaired electron and hence no electron spin–spin interaction.

The Hamiltonian operator for the rotating $^2\Sigma^+$ molecule, including the spin–rotation interaction and the first term in the centrifugal distortion perturbation energy but not the nuclear interactions, may be expressed as

$$\mathscr{H} = B_v\mathbf{N}^2 - D_v\mathbf{N}^4 + \gamma_v\mathbf{S}\cdot\mathbf{N} \tag{4.46}$$

Here **N** is the end-over-end rotational operator, **S** is the electron spin operator; B_v is the rotational constant, D_v the centrifugal distortion constant, and γ_v the spin–rotation coupling constant for the vibrational state v. Without nuclear moments, the angular momentum **J**, the vector sum of **N** and **S**, is the total angular momentum. From the vector diagram of Fig. 4.7 it is apparent that

$$\mathbf{J}^2 = \mathbf{N}^2 + \mathbf{S}^2 - 2\mathbf{S}\cdot\mathbf{N}$$

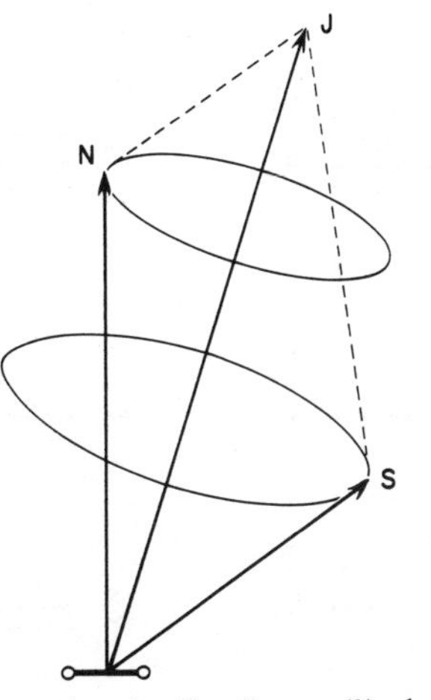

Fig. 4.7 Vector diagram of coupling for Hund's case (*b*) when molecules are in Σ states.

and thus

$$\mathbf{S}\cdot\mathbf{N}=\tfrac{1}{2}(\mathbf{J}^2-\mathbf{N}^2-\mathbf{S}^2) \tag{4.47}$$

Therefore, the $\mathscr{H}$ operator of (4.46) can be expressed as

$$\mathscr{H}=(B_v-D_v\mathbf{N}^2)\mathbf{N}^2+\tfrac{1}{2}\gamma_v(\mathbf{J}^2-\mathbf{N}^2-\mathbf{S}^2) \tag{4.48}$$

For the assumed coupling model the Hamiltonian, (4.48), is diagonal in J, N, S, and the diagonal elements yield the energies

$$E_{J,N,S}=B_vN(N+1)-D_vN^2(N+1)^2+\tfrac{1}{2}\gamma_v[J(J+1)-N(N+1)-S(S+1)] \tag{4.49}$$

in which the quantum numbers are

$$N=0,\ 1,\ 2,\ 3,\ \ldots$$
$$S=\pm\tfrac{1}{2}$$
$$J=N+\tfrac{1}{2}\quad\text{and}\quad N-\tfrac{1}{2}$$

Substitution of these values of S and J in (4.49) gives the two sets of levels

$$\left(\frac{E_N}{h}\right)_{J=N+(1/2)}=B_vN(N+1)-D_vN^2(N+1)^2+\frac{\gamma_vN}{2} \tag{4.50}$$

$$\left(\frac{E_N}{h}\right)_{J=N-(1/2)}=B_vN(N+1)-D_vN^2(N+1)^2-\frac{\gamma_v}{2}(N+1) \tag{4.51}$$

The sets of levels given by these expressions are illustrated in Fig. 4.8*a* which shows that the rotational levels are split into doublets by the spin rotation interaction and that the lowest level corresponds to $J=\frac{1}{2}$.

Selection rules for electric dipole, rotational transitions are

$$\Delta N=\pm1,\qquad \Delta J=0,\ \pm1$$

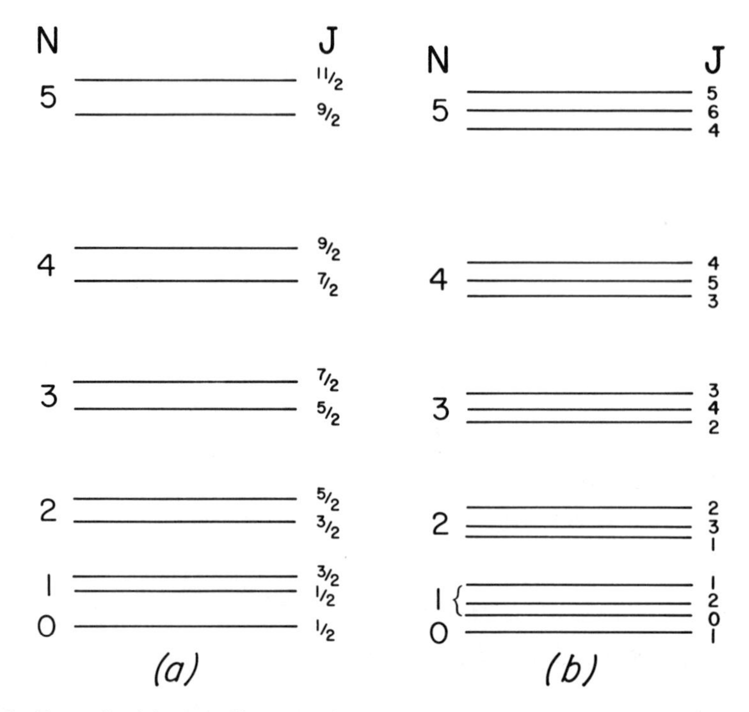

Fig. 4.8 Energy level diagram illustrating fine structure of rotational levels of diatomic molecules: (*a*) in $^2\Sigma$ electronic states and (*b*) in $^3\Sigma$ electronic states.

Thus the lowest rotational transition, $N=0\rightarrow1$, is a doublet; higher transitions are triplets, but the $\Delta J=0$ component is weak and diminishes rapidly with increasing N. Only the doublet corresponding to $\Delta J=1$ is likely to be detected for the higher N transitions. The relative intensities of these fine-structure components can be found from Appendix I by substitution of N for J, J for F, and S for $I=\frac{1}{2}$.

An important application of the above theory is the analysis of the observed microwave transitions of the molecular ion $^{12}C^{16}O^+$ which has a $^2\Sigma^+$ electronic ground state and no nuclear spin moments. The $N=0\rightarrow1$ transition of CO^+, reported by Dixon and Woods [41] in 1975, was the first microwave spectral observation of molecular ions in the laboratory. This observation aided in the detection of a pair of lines from interstellar space tentatively assigned as fine-structure components of the $N=1\rightarrow2$ transition of CO^+ and confirmed later by measurement of these components of the $N=1\rightarrow2$ transition in the laboratory by Sastry et al. [42]. The $N=0\rightarrow1$ doublet of $^{12}C^{16}O^+$, accurately measured by Piltch et al. [43], falls at 117,692.29 MHz ($J=\frac{1}{2}\rightarrow\frac{1}{2}$) and 118,101.89 MHz ($J=\frac{1}{2}\rightarrow\frac{3}{2}$) for ions in the ground vibrational state. The somewhat lower frequencies for the corresponding transitions for the $v=1$ and $v=2$ vibrational states have been measured by Bogey et al. [44] at Lille.

Sastry et al. [42] measured the fine-structure lines for the three transitions $N=1\rightarrow2$, $2\rightarrow3$, $3\rightarrow4$, which extend into the submillimeter frequencies to 471,952.34 MHz. These various measurements were analyzed by the investigators to give the rotational and fine-structure constants of (4.50) and (4.51) listed for $^{12}C^{16}0^+$ in Table 4.9.

The analysis of the $X^2\Sigma^+$ spectrum of $^{12}C^{14}N$ and of $^{28}Si^{14}N$ involves the ^{14}N hyperfine structure caused by nuclear magnetic and nuclear quadrupole coupling. Microwave rotational transitions of CN were first measured for interstellar molecules [45], and the first analysis of the ^{14}N hyperfine structure was made of the interstellar spectrum by Penzias, Wilson, and Jefferts [46]. The effective Hamiltonian including the hyperfine interaction is

$$\mathscr{H} = \mathscr{H}_{N,S-N} + \mathscr{H}_{\mathrm{MHS}} + \mathscr{H}_Q \tag{4.52}$$

where $\mathscr{H}_{N,S-N}$ is the rotational and spin–rotation $\mathscr{H}$ of (4.46), $\mathscr{H}_{\mathrm{MHS}}$ is the operator for the nuclear magnetic interaction, and $\mathscr{H}_Q$ is that for nuclear quadrupole interactions.

The nuclear magnetic operator of (4.52) has the form

$$\mathscr{H}_{\mathrm{MHS}} = b\mathbf{I}\cdot\mathbf{S} + cI_zS_z + C_I\mathbf{N}\cdot\mathbf{I} \tag{4.53}$$

where b, c, and C_I are the nuclear magnetic coupling constants as designated by Frosch and Foley [47], who originally worked out the theory of the magnetic hyperfine structure. The nuclear quadrupole interaction has the general form of that described for axially symmetric $^1\Sigma$ molecules in Chapter IX, but its specific form depends on the nature of the coupling vectors. Usually the quadrupole coupling is very weak compared with the magnetic interactions and can be treated with first-order perturbation theory.

The nature of the solution of the $\mathscr{H}$ of (4.52) depends on the relative magnitudes of the various coupling terms. When the nuclear coupling terms are at least moderately smaller than the spin–rotation term, the coupling scheme $\mathbf{N}+\mathbf{S}=\mathbf{J}$ may still be used, but now $\mathbf{J}$ couples with the nuclear spin $\mathbf{I}$ to form the resultant $\mathbf{F}=\mathbf{J}+\mathbf{I}$, as illustrated in Fig. 9.3. These conditions hold for both $^{12}C^{14}N$ and $^{28}Si^{14}N$.

Dixon and Woods [48], who were first to measure microwave rotational transitions of CN in the laboratory, tabulated matrix elements for the various terms in $\mathscr{H}$ which one may use to obtain the energy matrix required for a precise solution for its energies. Solution of the rather complex secular equation that results can be achieved with numerical computer methods. The $\mathscr{H}_{\mathrm{MHS}}$ for $^{28}Si^{14}N$ is like that for $^{12}C^{14}N$, and the methods for its solution are similar [49]. The rotational and coupling constants for both these molecules are listed in Table 4.9.

The hyperfine structure of the species $^{13}C^{16}O^+$, which has ^{13}C magnetic hyperfine structure but no nuclear quadrupole interaction, has been measured and analyzed by Piltch et al. [43]. In this ion the ^{13}C magnetic coupling is much stronger than the spin–rotation interaction, $b_0=14.58$ MHz compared

Table 4.9 Molecular Constants (in MHz) from Microwave Spectra of Some Molecules in $^2\Sigma^+$ Ground States

Constant	$^{12}C^{16}O^+$	$^{12}C^{14}N$	$^{28}Si^{14}N$
B_0	58,983.040(12)[a]	56,693.097(2)[d,e]	21,827.799(11)[f]
D_0	0.1896(5)[a]	0.1914(2)(Y_{02})[e]	0.0358(4)[f]
γ_0	273.01(5)[a]	217.486(5)[d,e]	505.11(4)[f]
B_1	58,412.72(4)[b]	56,171.110(5)[d,e]	
γ_1	271.27(11)	215.071(12)[d,e]	
	$^{13}C^{16}O^+$		
b_0	1,458(16)[c]	−33.966(10)[d]	19.46(21)[f]
c_0	144.6(21)[c]	60.329(20)[d]	94.47(27)[f]
$eQq_0(^{14}N)$		−1.31(3)[d]	3.05(23)[f]

[a]Sastry et al. [42].
[b]Bogey et al. [44].
[c]Piltch et al. [43].
[d]Dixon and Woods [48].
[e]D. D. Skatrud, F. C. De Lucia, G. A. Blake, and K. V. L. N. Sastry, *J. Mol. Spectrosc.*, **99**, 35 (1983).
[f]Saito et al. [49].

with $\gamma_0 = 260$ MHz. As a result, the $\mathbf{S} \cdot \mathbf{N}$ coupling is broken, and the combinations used in the analysis are

$$\mathbf{I} + \mathbf{S} = \mathbf{F}_2 \qquad \mathbf{N} + \mathbf{F}_2 = \mathbf{F} \tag{4.54}$$

No **J** resultant is formed. The matrix elements of the various operators in the appropriate basis functions and a simplified secular equation for solution of this type of problem is given in the original paper [43]. The ^{13}C coupling constants b_0 and c_0 are listed in Table 4.9.

Molecules in $^3\Sigma$ States

When there is more than one unpaired electron, as in such states as $^3\Sigma$, $^4\Sigma$, and so on, there is an additional magnetic term that is not in $^2\Sigma$ states, one arising from spin–spin interaction of the unpaired electrons. The most important cases are those for $^3\Sigma$ states which will be described later.

The rotational levels of molecules in the $^3\Sigma$ state are each split into a spin-triplet, sometimes called a ρ-type triplet, corresponding to the three possible space orientations of the electronic spin vector **S**. With molecular rotational momentum again designated by **N**, the resultant angular momentum is $\mathbf{J} = \mathbf{N} + \mathbf{S}$, as shown in Fig. 4.7. Since $S = 1$, the possible values of the total angular quantum number J are

$$J = N + 1, \qquad J = N, \qquad J = N - 1 \tag{4.55}$$

where N is the integral quantum number for the molecular rotation. Thus for any given N (except $N = 0$) there are three levels: $N+1$, N, and $N-1$.

By substitution of the three J values of (4.55) into Eq. (4.49) the values for the spin–rotation interaction are found to be $\gamma(N+1)$ for $J=N+1$, 0 for $J=N$, and $-\gamma N$ for $J=N-1$. Here a $(-\gamma)$ has been subtracted from each of the energy terms. This will not affect the calculated frequencies if γ may be regarded as a constant.

The dipole–dipole interaction between the spins of two unpaired electrons was first worked out by Kramers [50]. The interaction is proportional to $(3\cos^2\theta-1)$, where θ, when averaged, is the angle between the molecular axis and **S** axis. Hebb [51] showed that, in addition to the dipole–dipole interaction, a coupling of the ground electronic states with excited states having electronic angular momentum also contributes to the magnetic interaction. This term has the same angular dependence as that of the direct spin–spin interaction, and the two are represented by a common interaction constant λ. The latter interaction varies inversely as the energy difference between the higher states and the ground state; it is large when a Π state is near the ground Σ state. In the first-order approximation, to which Hund's case (*b*) applies, the relevant energies of the $^3\Sigma$ state molecules are

$$\frac{E_{J=N+1}}{h}=B_vN(N+1)+\gamma(N+1)-\frac{2\lambda(N+1)}{2N+3} \tag{4.56}$$

$$\frac{E_{J=N}}{h}=B_vN(N+1) \tag{4.57}$$

$$\frac{E_{J=N-1}}{h}=B_vN(N+1)-\gamma N-\frac{2\lambda N}{2N-1} \tag{4.58}$$

where B_v is the rotational constant, γ is the spin–rotation constant, and λ is the spin–spin interaction constant. Figure 4.8*b* shows the order of the energy levels when the triplet splitting is much smaller than the rotational energies.

The spin–spin interaction tends to align the spins along the internuclear axis as in Hund's case (*a*). Therefore Hund's case (*b*) will not apply rigorously for Σ states with more than one unpaired electron. For this reason, (4.56–4.58), which are based on the assumption of case (*b*), are not applicable for fitting accurate microwave measurements. More precise formulas are obtained when a partial decoupling of **S** from **N** is taken into account and the secular equation is solved for the coupling intermediate between Hund's cases (*b*) and (*a*). A solution of this kind, first obtained by Schlapp [52], has been improved by Miller and Townes [53], Mizushima and Hill [54], Tinkham and Strandberg [55], and others [56–58]. Development of the theory for finding the energies and transition frequencies will be described briefly.

The Hamiltonian operator for a diatomic molecule in a triplet electronic ground state may be expressed as

$$\mathscr{H}=\mathscr{H}_{\text{rot}}+\mathscr{H}_{S-N}+\mathscr{H}_{S-S} \tag{4.59}$$

Here

$$\mathscr{H}_{\text{rot}}=B_v\mathbf{N}^2-D_v\mathbf{N}^4=\mathbf{N}^2(B_v-D_v\mathbf{N}^2) \tag{4.60}$$

is the operator for the end-over-end rotation, including $D_v\mathbf{N}^4$, that for energies of centrifugal distortion; B_v is the rotational constant for the particular vibrational state^v, and D_v is the centrifugal stretching constant. For simplicity, we omit the subscript v on the constants in the following development. The second term

$$\mathscr{H}_{S-N}=\gamma^{(0)}\mathbf{N}\cdot\mathbf{S}+\gamma^{(1)}(\mathbf{N}\cdot\mathbf{S})\mathbf{N}^2=\mathbf{N}\cdot\mathbf{S}(\gamma^{(0)}+\gamma^{(1)}\mathbf{N}^2) \tag{4.61}$$

is the spin–rotation operator in which the last term accounts for the change in the spin–rotation energy caused by centrifugal distortion. The $\gamma^{(0)}$ and $\gamma^{(1)}$ signify the first-order and second-order spin–rotation coupling constants which are sometimes signified by μ_0 and μ_1. The last term

$$\mathscr{H}_{S-S}=\tfrac{2}{3}\lambda(3S_z^2-\mathbf{S}^2) \tag{4.62}$$

is the spin–spin operator arising from interaction of the magnetic dipole moments of the two unpaired electrons. This interaction is assumed to be axially symmetric about the molecular axis z. The constant λ signifies the spin–spin coupling constant. Effects of centrifugal stretching on this term are included later, in the coefficient λ, by (4.66)–(4.68).

To obtain the characteristic energies of $\mathscr{H}$, one must first find the nonvanishing matrix element $(\phi|\mathscr{H}|\phi')$ in a suitable representation and then set up and solve the corresponding secular equation, such as (2.85). Since $\mathbf{J}=\mathbf{N}+\mathbf{S}$ (see Fig. 4.7), the logical choice is $\phi)=|J, M_JN, S)$. The finding of nonvanishing elements of $\mathscr{H}_{\text{rot}}$, (4.60), and $\mathscr{H}_{S-N}$, (4.61), in this representation poses no problem for it is evident that both expressions are diagonal in this expansion. However, finding those of $\mathscr{H}_{S-S}$, (4.62), is difficult because neither S_z nor S_z^2 is diagonal in this representation. To solve the problem, one can transform S_z to space-fixed axes X, Y, Z and employ known matrix elements of S_X, S_Y, and S_Z (Chapter II, Section 2) with those of the transforming direction cosine elements. Probably the simplest procedure for finding these marix elements is to express $\mathscr{H}$ or $\mathscr{H}_{S-S}$ in spherical tensor notation [57] and use standard formulas already developed in the theories of irreducible tensor operators [59] to obtain the desired matrix elements. This method is described in Chapter XV. The resulting matrix elements are [60]

$$(\phi'|\mathscr{H}_{\text{rot}}|\phi)=[BN(N+1)-DN^2(N+1)^2]\delta_{N,N'} \tag{4.63}$$

$$(\phi'\mathscr{H}_{S-N}|\phi)=\frac{1}{2}[\gamma^{(0)}+\gamma^{(1)}N(N+1)][J(J+1)-N(N+1)-S(S+1)]\delta_{N,N'} \tag{4.64}$$

$$\begin{aligned}(\phi'|\mathscr{H}_{S-S}|\phi)=&-\tfrac{2}{3}\sqrt{30}\lambda(-1)^{S+J+N}\begin{Bmatrix}J & N & S\\ 2 & S & N\end{Bmatrix}\sqrt{\frac{N(N+1)(2N+1)}{(2N-1)(2N+3)}}\,\delta_{N,N'}\\ &2\sqrt{5}\lambda(-1)^{S+J+N}\begin{Bmatrix}J & N+2 & S\\ 2 & S & N\end{Bmatrix}\sqrt{\frac{(N+1)(N+2)}{(2N+3)}}\,\delta_{N+2,N'}\\ &2\sqrt{5}\lambda(-1)^{S+J+N}\begin{Bmatrix}J & N-2 & S\\ 2 & S & N\end{Bmatrix}\sqrt{\frac{N(N-1)}{(2N-1)}}\,\delta_{N-2,N'}\end{aligned} \tag{4.65}$$

For simplicity, the effects of centrifugal stretching on the spin–spin interaction have not been included in $\mathscr{H}_{S\cdot S}$. By comparison with (4.60) and (4.61), we include these stretching effects in the coefficient λ as

$$\lambda = \lambda^{(0)} + \lambda^{(1)}\mathbf{N}^2 \tag{4.66}$$

which, for the term diagonal in N [first term on the right of (4.65)], becomes

$$\lambda = \lambda^{(0)} + \lambda^{(1)}N(N+1) \qquad N=N' \tag{4.67}$$

For the last two terms of (4.65), which are off-diagonal by 2 in the rotational quantum number N, this coefficient expansion is [54]

$$\lambda = \lambda^{(0)} + \lambda^{(1)}(J^2+J+1) \quad \begin{cases} N'=J-1 \\ N'=J+1 \end{cases} \tag{4.68}$$

The term in the curly brackets of (4.65) is the six-j symbol [59], which has a definite numerical value for a given set of numbers in brackets. With the restricted values of N,

$$N = J-1, \qquad J, \qquad J+1$$

the nonvanishing matrix elements of the $\mathscr{H}$ of (4.59) are found from (4.63)–(4.65), with (4.67) and (4.68), to be

$$a = \mathscr{H}_{J-1,J-1} = BJ(J-1) - DJ^2(J+1)^2 + \gamma^{(0)}(J-1) + \gamma^{(1)}J(J-1)^2 + [\tfrac{2}{3} - 2J/(2J+1)][\lambda^{(0)} + \lambda^{(1)}J(J-1)] \tag{4.69}$$

$$b = \mathscr{H}_{J,J} = BJ(J+1) - DJ^2(J+1)^2 - [\gamma^{(0)} + \gamma^{(1)}J(J+1)] + \tfrac{2}{3}[\lambda^{(0)} + \lambda^{(1)}J(J+1)] \tag{4.70}$$

$$c = \mathscr{H}_{J+1,J+1} = B(J+1)(J+2) - D(J+1)^2(J+2)^2 - [\gamma^{(0)}(J+2) + \gamma^{(1)}(J+1)(J+2)^2] + [\tfrac{2}{3} - 2(J+1)/(2J+1)][\lambda^{(0)} + \lambda^{(1)}(J+1)(J+2)] \tag{4.71}$$

$$d = \mathscr{H}_{J-1,J+1} = \mathscr{H}_{J+1,J-1} = 2[\sqrt{J(J+1}/(2J+1)][\lambda^{(0)} + \lambda^{(1)}(J^2+J+1)] \tag{4.72}$$

Because $\mathscr{H}$ is diagonal in J, the energy determinant for each value of J, except $J=0$, factors into a $3\times 3'$ determinant to give a secular equation of the form

$$\begin{array}{c|ccc|} N/N' & J-1 & J & J+1 \\ \hline J-1 & a-E & 0 & d \\ J & 0 & b-E & 0 \\ J+1 & d & 0 & c-E \end{array} = 0 \tag{4.73}$$

where a, b, c, and d are given by (4.69)–(4.72). For the special case of $J=0$, $N=N'=1$, the energy equation reduces to

$$\mathscr{H}_{11} - E = 0 \tag{4.74}$$

Thus the $J=0$ level is unsplit and has the energy value

$$E(J=0) = \mathscr{H}_{11} = 2(B-2D) - 2(\gamma^{(0)} + 2\gamma^{(1)}) - \tfrac{4}{3}(\lambda^{(0)} + 2\lambda^{(1)})J=0 \tag{4.75}$$

Solution of the secular equation (4.73) yields the triplet energy values for each value of J, except zero, to be

$$E_0(J)=b \tag{4.76}$$

$$E_{\pm}(J)=\tfrac{1}{2}(a+c)\pm\tfrac{1}{2}[(a-c)^2+4d^2]^{1/2} \qquad J\neq 0 \tag{4.77}$$

In some papers, μ_0, μ_1, B_0, and B_1 are used for $\gamma^{(0)}$, $\gamma^{(1)}$, B, and $-D$, respectively, as employed here. Because of admixture of the rotational states for $N=J\pm1$ with those for $N'=N\pm2=J=1$, N is not a good quantum number except when $N=J$. For this reason the pseudo-quantum number n is often used to designate the sequence of the triplet rotational levels. However, we shall continue to use N to indicate this sequence with the understanding that it is not, strictly speaking, a good quantum number.

The frequencies of the allowed rotational transitions conforming to the selection rules, $N\rightarrow N+2$, $\Delta J=0$, ±1, and calculated from the energy equations (4.75)–(4.77) may be expressed [61] as

$$\nu_R(N)=(2B-2\gamma^{(1)}+\tfrac{4}{3}\lambda^{(1)})(N+1)-4D(N+3)^3 \tag{4.78}$$

The frequencies of the fine structure transitions conforming to the selection rules $\Delta N=0$, $\Delta J=\pm1$ may likewise be expressed as

$$\nu_{\pm}(N)=\lambda^{(0)}+\tfrac{1}{2}\gamma^{(0)}-4\gamma^{(1)}+(\lambda^{(1)}+\tfrac{5}{2}\gamma^{(1)})(k^2+k+2) \\ \pm\{f(k)-(2k+1)[B-2D(k^2+k+1)-\gamma^{(1)}+\tfrac{2}{3}\lambda^{(1)}]\} \tag{4.79}$$

where

$$f(k)^2=F^2+k(k+1)G^2 \tag{4.80}$$

in which

$$F=(2k+1)[(B-\tfrac{1}{2}\gamma^{(0)})-(2D+\tfrac{1}{2}\gamma^{(1)})(k^2+k+1)-\tfrac{3}{2}\gamma^{(1)}]-\frac{\lambda^{(0)}+\tfrac{1}{3}\lambda^{(1)}(7k^2+7k+4)}{2k+1} \tag{4.81}$$

$$G=\frac{2\lambda^{(0)}+2\lambda^{(1)}(k^2+k+1)}{2k+1} \tag{4.82}$$

and

$$k=N\pm1 \tag{4.83}$$

The transitions between the lower energy levels corresponding to these frequencies are indicated in the diagrams of Fig. 4.9 for $^{16}O_2$ and $^{18}O_2$ on the left and for $^{16}O^{18}O$ on the right. Note that $\nu_+(N)$ corresponds to transitions of $\Delta N=0$, $(J=N+1)\rightarrow(J=N)$ and that $\nu_-(N)$ corresponds to $\Delta N=0$, $(J=N-1)\rightarrow(J=N)$.

For reasons of symmetry, explained in Chapter III, Section 4, the alternate rotational levels of $^{16}O_2$ and $^{18}O_2$, or any diatomic molecule having identical nuclei with zero spins, are missing. Because the ground electronic state of

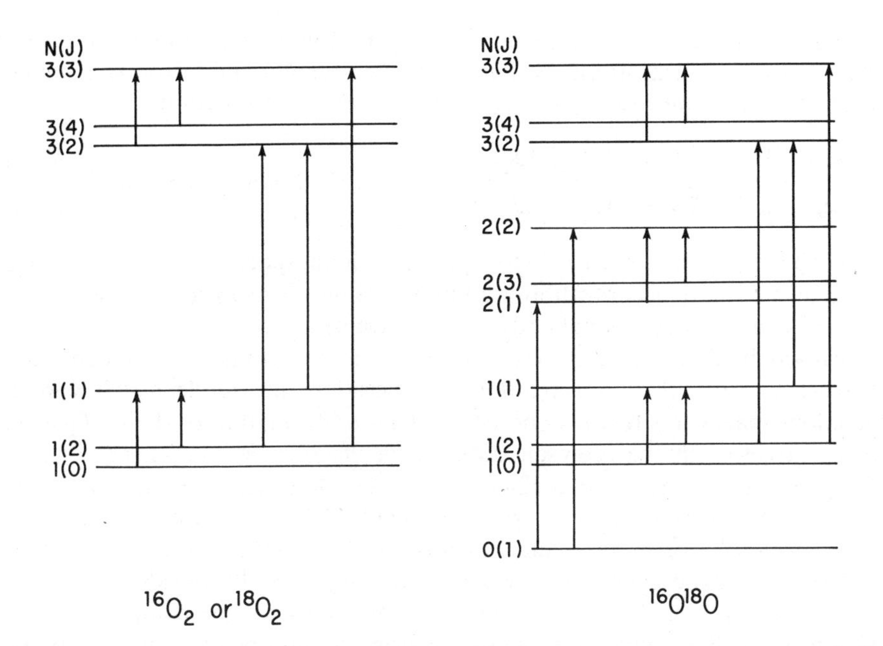

Fig. 4.9 Magnetic dipole transitions of oxygen for the lower $N(J)$ levels. From Steinbach [60].

oxygen $^3\Sigma_g^-$ is antisymmetric and the ground vibrational state and the nuclear spin state are symmetric, only antisymmetric rotational states can make the overall wave function symmetric, as required. The rotational states having odd rotational numbers, $N=1, 3, 5, \ldots$, are antisymmetric, and thus those states with N even do not exist for ^{16}O and ^{18}O. Note in Fig. 4.9 that there are no levels corresponding to $N=0, 2, \ldots$, in the diagram on the left. However, $^{16}O^{18}O$ does not have identical nuclei, and hence its rotational levels are not thus restricted. Also note in Fig. 4.9 that there are no missing levels in the diagram to the right.

The selection rules and transition probabilities depend on the matrix elements of the transition moments, electric or magnetic, expressed by (3.6), Chapter III, Section 1. A molecule in a triplet electronic state always has a magnetic spin moment and may also have an electric dipole moment. Symmetric diatomic molecules such as $^{16}O_2$ have only the magnetic dipole moment, whereas an asymmetric molecule in a triplet state, such as SO, has an electric as well as a magnetic dipole moment and may undergo both magnetic and electric dipole transitions. Selection rules for triplet state molecules depend only on the nonvanishing of the matrix elements of the transition moments. For magnetic dipole transitions, the selection rules are:

$$\text{series 1.} \quad \Delta N=0, \qquad \Delta J=\pm 1 \qquad (4.84)$$

magnetic dipole transitions

$$\text{series 2.} \quad \Delta N=\pm 2, \qquad \Delta J=0, \mp 1 \qquad (4.85)$$

Series 1 are transitions between the triplet components of rotational levels.

Series 2 with $\Delta J=0$ correspond to pure rotational transitions, and with $\Delta J \pm 1$ consist of transitions between both rotational and electron spin states. Electric dipole transitions give rise to two additional series of lines:

$$\text{series 3.} \quad \Delta N = \pm 1, \quad \Delta J = \pm 1 \qquad \text{electric dipole transitions} \tag{4.86}$$

$$\text{series 4.} \quad \Delta N = \pm 1, \quad \Delta J = 0 \tag{4.87}$$

Series 3 corresponds essentially to a pure rotational spectrum since the change in rotational state occurs without a flipping of the electron spin. Series 4 consists of combined rotational and electronic transitions.

Although obtaining the matrix elements of the transition moments for triplet state molecules is a straightforward operation, the complexity of the wave functions makes it a tedious one which requires evaluation of the coefficients or eigenvectors in the representation which diagonalizes the energy matrix. For the triplet state these coefficients may be found from solution of (2.85) with the matrix elements of $\mathscr{H}$ given by (4.69)–(4.72) and the energy values E found from solution of the secular equation (4.73). It is evident that the resulting expressions for the eigenvectors will be quite complex and the evaluation of the matrix elements of the squared transition moments will be complicated. By making use of the theory of irreducible spherical tensor operators and computer evaluations, Steinbach [60] has calculated eigenvectors and magnetic dipole transition probabilities for molecular oxygen—$^{16}O_2$, $^{18}O_2$, and $^{16}O^{18}O$. The resulting quantities, which require too much space for reproduction here, may be found in Steinbach's Ph.D. dissertation.

Because of the slight displacement of the center of mass from the internuclear center in $^{16}O^{18}O$, there should be a very slight electric dipole moment in this species, which may give rise to rotational transitions corresponding to the selection rules expressed by (4.61) and (4.62). To our knowledge, none of these transitions have yet been detected. Steinbach, however, has also calculated the probable frequencies and relative transition probabilities in terms of the unknown moment μ_z for transitions up to $6(7)\to 7(8)$. These calculations should be helpful to anyone trying to detect these transitions.

The microwave spectrum of the important oxygen molecule has been extensively investigated. The fine-structure transition of ^{16}O fall primarily in the 5 to 6 mm wavelength range although one transition, $N=1$, $J=0\to 1$, occurs in the 2.5 mm wave region. This series, particularly for $^{16}O_2$ because of its significant influence on atmospheric microwave transmission, was measured many times in the early years of microwave spectroscopy. The results are described in earlier books on microwave spectra. In efforts toward more precise molecular constants, the fine-structure frequencies of $^{16}O_2$ were remeasured with high precision by Zimmerer and Mizushima [62] and by West and Mizushima [63]; the results were analyzed by Welsh and Mizushima [64]. Frequencies of the fine-structure series for $^{18}O_2$ and $^{16}O^{18}O$ have been measured precisely by Steinbach and Gordy [57, 61]. Table 4.10 shows the close correlation of these measurements with frequencies calculated from (4.78) and (4.79).

Table 4.10 Comparison of Observed and Calculated Microwave Frequencies of $^{18}O_2$ and $^{16}O^{18}O$

Transition	*Observed Frequency (MHz)*	*Predicted Frequency (MHz)*	*Difference*
	$^{18}O_2$ *Frequencies*[a]		
$\nu_+(1)$	57,239.907(20)	57,239.952(12)	−0.045
$\nu_+(3)$	58,899.771(27)	58,899.732(12)	+0.039
$\nu_+(5)$	59,811.414(34)	59,811.404(15)	+0.010
$\nu_+(7)$	60,505.782(62)	60,505.840(47)	−0.058
$\nu_-(3)$	61,529.854(32)	61,529.864(18)	−0.010
$\nu_-(5)$	59,871.473(23)	59,871.464(19)	+0.009
$\nu_-(7)$	58,962.067(24)	58,962.047(15)	+0.020
$\nu_-(9)$	58,270.727(9)	58,270.794(46)	−0.067
1(2)→3(2)	378,831.51(10)	378,831.51(2)	0.000
	$^{16}O^{18}O$ *Frequencies*[b]		
$\nu_+(3)$	58,656.447(65)	58,656.313(27)	+0.134
$\nu_+(6)$	60,094.935(17)	60,095.012(31)	−0.077
$\nu_+(8)$	60,801.121(58)	60,800.64(64)	+0.157
$\nu_-(4)$	60,861.117(114)	60,861.240(39)	−0.123
$\nu_-(5)$	60,105.612(77)	60,105.615(45)	−0.003
$\nu_-(6)$	59,539.420(56)	59,539.445(42)	−0.025
$\nu_-(7)$	59,074.760(67)	59,074.722(30)	+0.038
$\nu_-(8)$	58,670.780(73)	58,670.743(33)	+0.037
$\nu_-(9)$	58,306.003(55)	58,306.058(95)	−0.055
0(1)→2(1)	233,946.179(61)	233,946.178(52)	+0.001

[a]From Steinbach and Gordy [57].
[b]From Steinbach and Gordy [61].

Rotational transitions corresponding to the frequencies of (4.78) are essential for precise evaluation of the molecular parameters. Since these transitions fall in the submillimeter wave region and have rather low transition probabilities, only a few have been measured with microwave methods. The first of these, the $N(J)=1(2)\rightarrow 3(2)$ transition of $^{16}O_2$, was measured by McKnight and Gordy [65] at 424,763.80(20) MHz. This transition has also been measured with microwave techniques for $^{18}O_2$ and the $N(J)=0(1)\rightarrow 2(1)$ transition for $^{16}O^{18}O$. The frequencies are listed in Table 4.10. Several higher rotational transitions of $^{16}O_2$ have been measured with laser magnetic resonance techniques [66, 67].

The molecular parameters are listed in Table 4.11 for the three isotopic species $^{16}O_2$, $^{18}O_2$, and $^{16}O^{18}O$ in the ground electronic state $^3\Sigma_g^-$ and the ground vibrational state, $v=0$. For $^{18}O_2$ and $^{16}O^{18}O$, they were obtained from a least-squares, computer fitting of (4.78) and (4.79) to the fine structure frequencies and the one rotational transition as shown in Table 4.10. For

Table 4.11 Some Derived Molecular Constants of Oxygen in the $^3\Sigma_g^-$, $v=0$ Ground State[a]

Constant	$^{16}O_2$	$^{16}O^{18}O$	$^{18}O_2$
B_0 (MHz)	43,100.460(6)	40,707.408(10)	38,313.730(7)
D_0 (MHz)	0.14501(27)	0.129	0.115
$\lambda_0^{(0)}$ (MHz)	59,501.341(4)	59,499.097(43)	59,496.698(11)
$\lambda_0^{(1)}$ (MHz)	0.05848(2)	0.05312(80)	0.05211(42)
$\gamma_0^{(0)}$ (MHz)	−252.586(1)	−238.4888(7)	−224.439(3)
$\gamma_0^{(1)}$ (MHz)	−0.000247(1)	−0.000619(116)	−0.000351(54)
B_e (MHz)	43,344.7(7.5)	40,931.7(6.9)	38,518.6(6.3)
r_e (Å)	1.207433(106)	1.207429(103)	1.207427(100)

[a]From Steinbach and Gordy [61]. Some of the parameter designations are changed from those of the original source.

$^{16}O_2$, they were similarly calculated from the fine structure frequencies reported by Welsh and Mizushima [64] together with the remeasured $N(J)=1(2)\rightarrow 3(2)$ rotational frequency, 424,763.21(10) MHz.

The equilibrium values of B_e given in Table 4.11 were derived from the B_0 values with the first-order correction

$$B_e'=B_0+\tfrac{1}{2}\alpha \tag{4.88}$$

for vibration–rotation distortions, (13.103), and with the correction

$$B_e=(1-g_r)B_e' \tag{4.89}$$

for electronic-cloud distortions described in Chapter XI, Section 7, (11.115). The α value used for $^{16}O_2$ was 478(15) MHz, derived from a reanalysis by Albritten et al. [68]. of the optical measurements by Babcock and Herzberg [69]. Values of α for $^{18}O_2$ and $^{16}O^{18}O$ were obtained from this $^{16}O_2$ value by use of isotopic ratio relations. The value of $g_r=-1.25(8)\times 10^{-4}$ for $^{16}O_2$, observed by Evenson and Mizushima [67], was used for all three isotopic species.

The fine-structure spectrum of $^{16}O_2$ in the first excited vibrational state has been measured by Amano and Hirota [70]. They combined their results with the frequencies for the ground vibrational state measured by Mizushima and associates [62–64] to obtain the equilibrium values of the molecular constants listed in Table 4.12. They also measured some fine structure frequencies for $^{16}O^{18}O$ in the ground vibrational state.

Another diatomic molecule having a triplet sigma ground state is sulfur monoxide. It also has been studied extensively. Because SO has an electric dipole moment as well as a magnetic spin moment, it can undergo electric dipole transitions, (4.61)–(4.62), as well as magnetic dipole transitions, (4.59)–(4.60). Unlike $^{16}O_2$, it has no missing rotational levels since its nuclei are dissimilar. Thus, its microwave spectrum is rich in variety. Several of its electric dipole

Table 4.12 Equilibrium Constants of $^{16}O_2$ in the $^3\Sigma_g^-$ Ground Electronic State[a]

Constant	*Value*
B_e	43,336.2(15) MHz
λ_e	59,429.08(11) MHz
γ_e	−252.265(11) MHz
r_e	1.20748(5) Å

[a]From Amano and Hirota [70].

transitions occurring in the millimeter wave region were measured by Winnewisser et al. [71] and some in the centimeter wave region by Powell and Lide [72]. At least one of the magnetic dipole transitions has been measured [73]. The electric dipole transitions were used to obtain rather accurate values of B_0 and of the coupling constants λ and γ reported in the earlier edition of this book. Clark and De Lucia [74] and Amano et al. [75] have now observed electric dipole transitions of $^{32}S^{16}O$ in the first excited vibrational state as well as additional ones in the ground vibrational state. From their measurements they were able to derive equilibrium values for the molecular constants from microwave measurements only. For the ground vibrational state, Clark and De Lucia measured many new lines of higher frequency in the millimeter and in the submillimeter wave range, to 609,959.62 MHz. They combined these measurements with those made earlier in the lower frequency region, a total of 56 transitions, in a highly accurate recalculation of the molecular constants of $^{32}S^{16}O$ in the $^3\Sigma_g^-$, $v=0$ ground state. The results are shown in Table

Table 4.13 Molecular Constants of $^{32}S^{16}O$ in the $^3\Sigma_g^-$ Ground State[a]

Constant	$v=0$	$v=1$
B_v (MHz)	21,523.56	21,351.58(40)
D_v (MHz)	0.03399(4)	(0.034)[b]
$\lambda_v^{(0)}$ (MHz)	158,254.387(26)	159,204.7(9.4)
$\lambda_v^{(1)}$ (MHz)	0.305(2)	(0.305)[b]
$\gamma_v^{(0)}$ (MHz)	−168.342(10)	−171.5(12)
r_v (Å)	1.4840383(1)	1.4900031(10)
B_e (MHz)	21,609.552(21)	
λ_e (MHz)	157,779.2(47)	
r_e (Å)	1.481026(1)	

[a]From Clark and De Lucia [74].
[b]Assumed to be the same as for $v=0$ state.

4.13. Although only five transitions were observed for the $v=1$ state, these were adequate for calculation of the constants of the $v=1$ state, with the assumption that the centrifugal stretching effects are the same as those for the $v=0$ state. From the combined measurements, the equilibrium constants as listed in Table 4.13 were obtained. It is of interest to compare the coupling constants λ and γ of SO with those of O_2.

Molecules in Π and Δ States

The heavier molecules in Π states or in Δ states approximate closely Hund's case (*a*). There are, however, deviations from this idealized case which cause noticeable effects on the microwave spectra of moderately heavy molecules such as NO. The light diatomic hydrides in Π states generally have coupling intermediate between that for case (*a*) and case (*b*); often for high rotational states they approximate that for case (*b*).

Deviation from Hund's case (*a*) is caused by interaction of the molecular rotation with the electronic motions. This interaction leads to a partial spin-uncoupling as well as *L*-uncoupling from the molecular axis. The spin-uncoupling leads to a change in the effective B_v value and a shift in the rotational levels, which increases with the rotational energy. The *L*-uncoupling lifts the $\pm\Lambda$ degeneracy ($\Lambda \neq 0$) and leads to a doublet splitting (Λ splitting) of the levels as well as to a slight displacement. In the first order approximation, the displacement caused by *L*-uncoupling can be neglected and the Λ doublet splitting assumed to be symmetrical about the unperturbed level. The Λ doubling likewise increases with the molecular rotation.

Spin uncoupling from the molecular axis or Λ results from an admixing of the spin states by the rotation and in the first-order approximation is inversely proportional to the separation of the spin states. For $^2\Pi$ states it results from a slight admixing of the $^2\Pi_{1/2}$ and the $^2\Pi_{3/2}$ states via the rotation. When $2JB \ll |\Lambda A|$ (small spin-uncoupling), effects of spin-uncoupling can be taken into account most simply by substitution of B_{eff} for B in (4.2). The energy perturbed by spin-uncoupling is then

$$E_J = hB_{\text{eff}}J(J+1) \tag{4.90}$$

where $J=\Omega, \Omega+1, \Omega+2, \ldots$. In the first-order approximation [76, 77] for spin doublet states $^2\Pi$ or $^2\Delta$,

$$B_{\text{eff}} = B_v\left(1 \pm \frac{B_v}{\Lambda A} + \cdots\right) \tag{4.91}$$

and for spin triplet states $^3\Pi$ or $^3\Delta$ with $\Sigma=0$ ($^3\Pi_1$ or $^3\Delta_2$) $B_{\text{eff}}=B_v$ while for $\Sigma=\pm1$

$$B_{\text{eff}} = B_v\left(1 \pm \frac{2B_v}{\Lambda A} + \cdots\right) \tag{4.92}$$

Here A is the spin orbit coupling constant ($A\mathbf{S}\cdot\mathbf{\Lambda}$) which is equivalent to the energy separation of the spin states. For $A>0$ the so-called regular case is obtained where the energies of the component states are in the order of their $(\Lambda+\Sigma)$ values; and for $A<0$ the inverted case is obtained where the spin states are inverted. As an example, NO in its $^2\Pi$ ground state has a fine structure constant A of 122 cm^{-1}, which represents in cm^{-1} units the difference in energy between the lower $^2\Pi_{1/2}$ and the upper $^2\Pi_{3/2}$ spin states. For NO, $B=1.70$ cm^{-1}, and thus the first-order correction term B/A equals 0.014. When A is positive, as for NO, the rotational levels for the $^2\Pi_{1/2}$ state are lowered, and those for the $^2\Pi_{3/2}$ state are raised by the spin uncoupling.

The L-uncoupling from the molecular axis is caused by the magnetic field generated by the end-over-end rotation which induces an admixing of the ground Π states (or ground Δ states) with the nearby Σ states. This interaction gives rise to a splitting of the doubly degenerate orbital states which can be represented in zero-order by the linear combinations

$$\psi_+=\frac{1}{\sqrt{2}}[\psi_e(\Lambda)+\psi_e(-\Lambda)] \tag{4.93}$$

$$\psi_-=\frac{1}{\sqrt{2}}[\psi_e(\Lambda)-\psi_e(-\Lambda)] \tag{4.94}$$

The two electronic states thus have opposite symmetry, indicated by Π^+ and Π^- or Δ^+ and Δ^-. In Mulliken's notation they are indicated as c or d states or as Π^c and Π^d, and so on. Generally only one Σ state lies low enough to cause significant splitting. Each substate Π^+ or Π^- has a set of rotational levels associated with it, and each J level is hence split into two closely spaced components. The Λ doublet splitting has been calculated with first-order perturbation theory by Van Vleck [78] and by Mulliken and Christy [79]. Dousmanis et al. [80] have extended the calculations to higher order for Π states. Table 4.14 gives a summary of Van Vleck's calculations of the first-order Λ doublet splitting for different ground states and for different coupling cases (b) or (a) of the unperturbed state.

Because the single state $^1\Pi$ or $^1\Delta$ have no spin complications, the Λ doublet splitting is relatively simple. That for the $^1\Delta$ state, which has the coupling constant $48B_v^4/\nu_e^3$, is so small for low J values that it can be neglected.

The most important states for the microwave spectroscopist are the $^2\Pi$ states. These approximate Hund's case (a) for the heavier molecules and case (b) for the light diatomic hydrides except for low rotational states. Note from Table 4.14 that for case (a) the doubling constant for the $^2\Pi_{3/2}$ state is smaller by a factor of $2(B_v/A)^2$ than that for the $^2\Pi_{1/2}$ state, but that its Λ splitting increases rapidly with rotation, approximately as J^3, whereas the splitting for the $^2\Pi_{1/2}$ state increases approximately as the first power of J. For the $J=\frac{1}{2}\rightarrow\frac{3}{2}$ transition of NO in the $^2\Pi_{1/2}$ state the Λ doublet splitting is 355 MHz [81], whereas for the $J=\frac{3}{2}\rightarrow\frac{5}{2}$ transition for the $^2\Pi_{3/2}$ state it is only 2.7 MHz [82].

For molecules which approximate closely Hund's case (a), $2BJ\ll|\Lambda A|$,

Table 4.14 Theoretical Λ-Type Doubling

State	*Coupling Case*	E_Λ *(approximate Λ-type splitting of levels)*[a]	*Approximate Theoretical Coupling Constants*[b]
$^1\Pi$		$q_\Lambda J(J+1)$	$q_\Lambda = \dfrac{4B^2}{\nu_e}$
$^2\Pi$	Case (*b*)	$q_\Lambda N(N+1)$	
$^2\Pi$	Case (*a*) $\Omega=\frac{1}{2}$	$q_\Lambda^a(J+\frac{1}{2})$	$q_\Lambda^a = \dfrac{4AB}{\nu_e}$
	$\Omega=\frac{3}{2}$	$q_\Lambda^b(J^2-\frac{1}{4})(J+\frac{3}{2})$	$q_\Lambda^b = 2q_\Lambda^a\left(\dfrac{B}{A}\right)^2$
$^3\Pi$	Case (*b*)	$q_\Lambda N(N+1)$	
$^3\Pi$	Case (*a*) $\Omega=0$	q_Λ^c	$q_\Lambda^c = \dfrac{2A^2}{\nu_e}$
	$\Omega=1$	$q_\Lambda J(J+1)$	
	$\Omega=2$	~ 0	
$^1\Delta$		$q_\Lambda^d J(J^2-1)(J+2)$	$q_\Lambda^d = \dfrac{48B^4}{\nu_e^3}$

[a]Van Vleck [78].
[b]ν_e represents the transition frequency between the ground level and the lowest Σ state; *A*, the spin-orbit coupling constant; *B*, the rotational constant. *A*, *B*, and q_Λ are in frequency units.

the *L*-uncoupling energy splitting for $^2\Pi_{1/2}$ states is given to a good approximation (see Table 4.14) by

$$\frac{E_\Lambda}{h} = q_\Lambda^a(J+\tfrac{1}{2}) \tag{4.95}$$

For the $^2\Pi_{3/2}$ states it is given by

$$\frac{E_\Lambda}{h} = q_\Lambda^b(J-\tfrac{1}{2})(J+\tfrac{1}{2})(J+\tfrac{3}{2}) \tag{4.96}$$

where q_Λ^a is the doubling constant for the $^2\Pi_{1/2}$ state and $q_\Lambda^b \approx 2(B_v/A)^2 q_\Lambda^a$ is that for the $^2\Pi_{3/2}$ state. Approximately, $q_\Lambda^a = 4B_vA/\nu_e$, where ν_e is the transition frequency between the $^2\Pi_{1/2}$ state and the nearest state with which the admixing occurs to cause the Λ doubling. Interaction with other higher Σ states is neglected.

Centrifugal distortion constants for the $^2\Pi$ state, calculated by Almy and Horsfall [83] and by Dixon [84], for both $^2\Pi_{1/2}$ and $^2\Pi_{3/2}$ states, can be represented with sufficient accuracy by the terms $-D_v[(J-\frac{1}{2})(J+\frac{1}{2})^2(J+\frac{3}{2})+1]$ where D_v is the centrifugal stretching constant $D_v \approx 4B_v^3/\omega^2$.

By summation of the various terms, the perturbed rotational levels when $2JB \ll A$ are found to be for the $^2\Pi_{1/2}$ state

$$\frac{E_J^{\pm}}{h} = B_v\left(1 - \frac{B_v}{A}\right)J(J+1) \pm \frac{1}{2}q_\Lambda^a(J+\tfrac{1}{2}) - D_v[(J-\tfrac{1}{2})(J+\tfrac{1}{2})^2(J+\tfrac{3}{2})+1] \quad (4.97)$$

where

$$J = \tfrac{1}{2}, \tfrac{3}{2}, \tfrac{5}{2}, \ldots$$

For the $^2\Pi_{3/2}$ state they are found to be

$$\frac{E_J^{\pm}}{h} = B_v\left(1 + \frac{B_v}{A}\right)J(J+1) \pm \frac{1}{2}q_\Lambda^b(J-\tfrac{1}{2})(J+\tfrac{1}{2})(J+\tfrac{3}{2})$$
$$-D_v[(J-\tfrac{1}{2})(J+\tfrac{1}{2})^2(J+\tfrac{3}{2})+1] \quad (4.98)$$

where

$$J = \tfrac{3}{2}, \tfrac{5}{2}, \tfrac{7}{2}, \ldots$$

The + sign yields the upper Λ-doublet level and the − sign yields the lower level. (These signs should not be confused with the + or − symmetry properties of the levels.) These equations are adequate for heavy molecules which generally have large spin-orbit couplings and small B_v values. They apply reasonably well to the semiheavy molecule NO, but accurate evaluation of the molecular constants of NO from the precise microwave measurements which have been made on this molecule requires more rigorous theory [85].

Light diatomic hydrides such as CH or OH have large B_v values and small spin-orbit coupling constants so that $2JB > |\Lambda A|$ except for low J values. Usually their couplings are intermediate between case (*a*) and case (*b*), and at high J values they closely approach case (*b*). For doublet states ($S=\tfrac{1}{2}$), Hill and Van Vleck [76] have solved the secular equation for the intermediate spin-coupling case. The resulting equations, which do not include effects of L-uncoupling for a regular $^2\Pi$ state are
for $^2\Pi_{1/2}$, $\Lambda = 1$

$$\frac{E_J}{h} = B_v\left\{(J+\tfrac{1}{2})^2 - \Lambda^2 - \frac{1}{2}\left[4(J+\tfrac{1}{2})^2 + \frac{A}{B_v^2}(A-4B_v)\Lambda^2\right]^{1/2}\right\} \quad (4.99)$$

and for $^2\Pi_{3/2}$, $\Lambda = 1$

$$\frac{E_J}{h} = B_v\left\{(J+\tfrac{1}{2})^2 - \Lambda^2 + \frac{1}{2}\left[4(J+\tfrac{1}{2})^2 + \frac{A}{B_v^2}(A-4B_v)\Lambda^2\right]^{1/2}\right\} \quad (4.100)$$

where A is again the spin-orbit coupling constant and B_v is the rotational constant for the particular vibrational state. For an inverted $^2\Pi$ state the state designations should be reversed. The effects of centrifugal distortion are not included. To a good approximation, the same term in D_v as that of (4.97) can be employed. For large A/B_v the square root may be expanded; and, if terms independent of J are omitted, the foregoing equations reduce to those given previously for small spin uncoupling, (4.90) and (4.91). Figure 4.10 gives an energy level diagram showing the transition from case (*a*) to case (*b*).

Van Vleck [78] has also derived an expression for the Λ doubling energy for the intermediate field case of $^2\Pi$ states. The rather involved expression for E_Λ is

$$\frac{E_\Lambda}{h}=\frac{q_\Lambda}{2}(J+\tfrac{1}{2})\left[\left(2+\frac{A'}{B'}\right)\left(1+\frac{2-A/B}{X}\right)+\frac{4(J+\frac{3}{2})(J-\frac{1}{2})}{X}\right] \tag{4.101}$$

where

$$X=\pm\left[\frac{A}{B}\left(\frac{A}{B}-4\right)+4(J+\tfrac{1}{2})^2\right]^{1/2}$$

and $A'/B' \approx A/B$, where again B is the rotational constant and A is the fine structure constant. For $A>0$ the positive sign in the definition of X yields the splitting of the $^2\Pi_{3/2}$ state and the negative sign, the splitting of the $^2\Pi_{1/2}$ state. For $A<0$ just the reverse is true.

For diatomic molecules in $^2\Pi$ states and in fixed vibrational states, $\Delta v=0$, the selection rules for electric dipole transitions which correspond to microwave frequencies are

$$J\rightarrow J\pm 1, \qquad N\rightarrow N\pm 1, \qquad +\rightarrow -$$

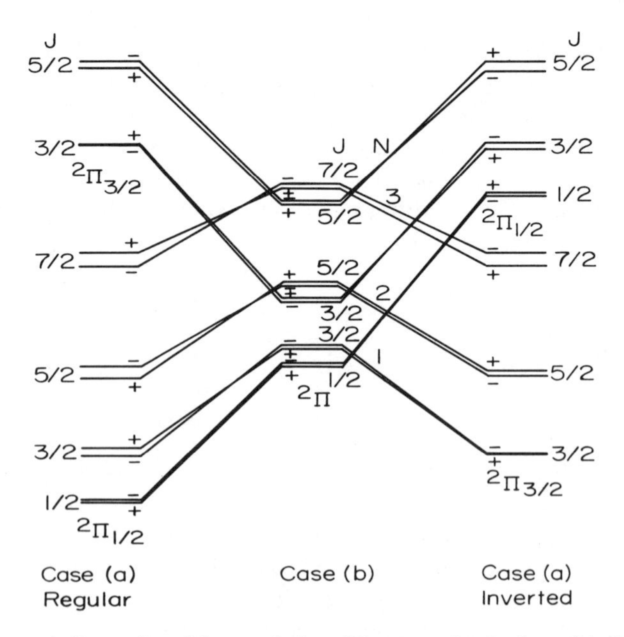

Fig. 4.10 Schematic illustration of the correlation of the energy levels of case (**b**) with those of the regular and inverted case (*a*) for the $^2\Pi$ states. The Λ-type doubling is exaggerated. (After G. W. King, *Spectroscopy and Molecular Structure*, Holt, Rinehart and Winston, New York, 1964.)

In the last expression the + and − signs refer to the symmetry or antisymmetry, respectively, of the overall wave function (see Chapter III, Section 4). The Λ doublets of the rotational levels have opposite symmetry, and the symmetry of the upper and lower component alternates, as illustrated in Fig. 4.10. There are hence two sets of rotational lines for a $^2\Pi$ state corresponding to $J, \Pi_L \rightarrow J+1, \Pi_L$ and $J, \Pi_U \rightarrow J+1, \Pi_U$ where L and U designate, respectively, the lower and upper doublet states of Π. The rotational line frequencies are

$$\nu_r = \frac{E_{J+1}^{\pm} - E_J^{\pm}}{h} \tag{4.102}$$

Direct transitions between the Λ doublet levels of a given rotational state corresponding to the selection rules $J \rightarrow J$ or $N \rightarrow N$ and $\Pi_L \rightarrow \Pi_U$ are also possible. The squared matrix elements for these internal transitions are for Hund's case (*a*)

$$|(J, \Omega, \Pi_L|\mu|J, \Omega, \Pi_U)|^2 = \frac{\mu^2\Omega^2}{J(J+1)} \tag{4.103}$$

where μ is the dipole moment. For Hund's case (*b*) the transition matrix elements are

$$|(J, \Lambda, \Pi_L|\mu|J, \Lambda, \Pi_U)|^2 = \frac{\mu^2\Lambda^2}{J(J+1)} \tag{4.104}$$

Since the dipole moment is along the molecular axis, these matrix elements can be easily derived from the vector models, Figs. 4.5 and 4.6. They may be obtained from (2.125) by identification of Λ or Ω with K. Dousmanis et al [80] derived the transition matrix elements for intermediate coupling cases. Frequencies corresponding to the Λ-type doublet spectra are

$$\nu_\Lambda = \frac{(E_\Lambda^+)_J - (E_\Lambda^-)_J}{h} \tag{4.105}$$

and can be calculated from the various expressions for E_Λ. For example, in the limiting Hund's case (*a*) approximation of the $^2\Pi$ state, the Λ doublet transitions have the frequencies

$$\nu_\Lambda = q_\Lambda^a(J+\tfrac{1}{2}) \qquad \text{for } ^2\Pi_{1/2} \tag{4.106}$$

and

$$\nu_\Lambda = q_\Lambda^b(J-\tfrac{1}{2})(J+\tfrac{1}{2})(J+\tfrac{3}{2}) \qquad \text{for } ^2\Pi_{3/2} \tag{4.107}$$

Figure 4.11 shows Λ doublet separations of different J levels of some diatomic hydrides in Π states, as calculated by Mulliken and Christy [79]. The $\Delta\nu_{dc}$ represents the frequencies (here in cm^{-1} units) of Λ doublet transitions. Note that many of the predicted frequencies fall in the millimeter or observable submillimeter wave regions 1 to 23 cm^{-1}.

For the lighter diatomic hydrides, CH, OH and so on, the Λ doublet spectra are the only spectra observable in the microwave region because the B_v values

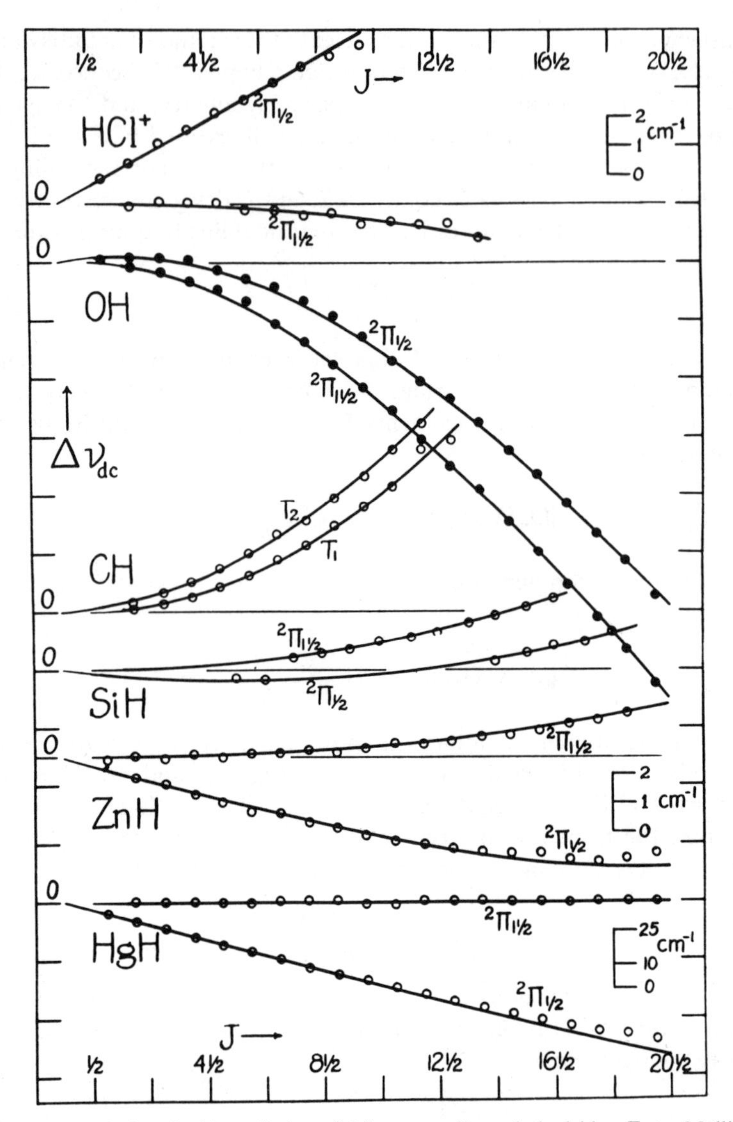

Fig. 4.11 Theoretical Λ doublet splitting of *J* for some diatomic hydrides. From Mulliken and Christy [79].

for these molecules are so large that the rotational transitions fall in the infrared region outside the present range of frequencies observable with microwave techniques. For OH, with $B_0 = 555$ GHz, the lowest rotational transition occurs in the shorter submillimeter wave region, but its Λ doublet spectrum spreads over the millimeter region and into the centimeter region. Chemists may find much use for these Λ doublet spectra in future study of light gaseous free radicals. The difficulty in observing such spectra is due primarily to their instability, a difficulty which is lessened with use of the free-space absorption cells now possible in millimeter wave spectroscopy.

Townes and his collaborators [80] were first to observe microwave lines corresponding to direct transitions between Λ doublets of OH and OD which have $^2\Pi$ ground states. Although these molecules approach closely Hund's case (*b*) for high rotational states, they have intermediate coupling for the lower rotational states for which the lines were observed. In addition to the various electronic effects, these spectra have doublet hyperfine splitting caused by H, or triplet splitting caused by D. For interpretation of the spectra, Dousmanis et al. [80] have modified Van Vleck's [78] intermediate coupling theory and have included the theory for the nuclear coupling. The resulting formulas are too complex for inclusion here. They can be used profitably for detailed analyses of similar spectra. Because the rotational constant for ^{16}OH, $B_v=555{,}040$ MHz as obtained from ultraviolet spectroscopy, is very large, only the Λ doublet spectrum occurs in the microwave region. Originally, the observed lines of ^{16}OH were fitted with (4.101) to an accuracy of about 40 MHz, with the values $q_\Lambda=1159$ MHz, $A'/B'=-6.073$, and $A/B=-7.547$. With the more involved theory, Dousmanis et al. were able to obtain a somewhat better fitting with two Λ doublet constants designated as α_p and β_p having values of -2361 and 576 MHz, respectively. From this analysis they obtained $A/B=-7.444\pm0.017$ for ^{16}OH. For OD they obtained:

$$A/B=13.954\pm0.032, \qquad \alpha_p=-1549 \text{ MHz}, \qquad \beta_p=162 \text{ MHz}$$

Since this earlier work Coxon [86], from an analysis of the combined optical and microwave spectra of ^{16}OH, has derived accurate values for the spin-orbit coupling constant A_0 and the rotational constant B_0 for the ground state $^2\Pi$ and $v=0$. Perhaps even more accurate is the determination of these constants from laser magnetic resonance measurements of rotational transitions [87–89]. These lead to the values [89], $A_0=-4{,}168.63913(78)$ MHz and $B_0=555.66097(11)$ MHz, which yield $A_0/B_0=-7.5201$.

Microwave doublet transitions led to the detection of OH radicals in interstellar space, the first interstellar molecular species to be discovered with microwave spectroscopy [90]. Since that discovery, Λ doublets of OH and OD have been observed from many astronomical sources and have been useful for estimation of D/H ratios for astronomical mapping, for the study of chemical mechanisms in interstellar space and other astronomical phenomena [91, 92].

Lambda doublet transitions have now been detected from the important species CH in interstellar space [93, 94], even though no such transitions, to our knowledge, have yet been observed from CH molecules in the laboratory. There seems little doubt that the Λ doublet spectrum of CH will prove to be valuable in future astronomical observations, as OH has already proved to be. It is important for microwave spectroscopists to persist in their efforts to measure this spectrum for earth-based CH.

Microwave lines of the $J=\frac{1}{2}\rightarrow\frac{3}{2}$ transition of ^{14}N^{16}O in the $^2\Pi_{1/2}$ ground state have been observed by Burrus and Gordy [81] and by Gallagher et al.

[95]. Gallagher and Johnson [85] extended the measurements to ^{15}NO, the $J=\frac{3}{2}\rightarrow\frac{5}{2}$ transition; Favero et al. [82] observed the $J=\frac{3}{2}\rightarrow\frac{5}{2}$ transition of ^{14}NO in the $^2\Pi_{3/2}$ state. Intermediate coupling theory similar to that of Dousmanis et al. [80] was applied to the latter measurements.

Microwave measurements of the two lowest rotational transitions of NS, the $J=\frac{1}{2}\rightarrow\frac{3}{2}$ and $\frac{3}{2}\rightarrow\frac{5}{2}$ transitions, which fall in the millimeter wave region, have been carried out by Amano et al. [96] for both the $^2\Pi_{1/2}$ and $^2\Pi_{3/2}$ electronic ground states. Accurate measurements of the Λ doublet splittings of the $^2\Pi_{1/2}$ ground state were obtained for both $^{14}N^{32}S$ and $^{14}N^{34}S$. For $^{14}N^{32}S$, Amano et al. found $B_0=23{,}156.01(16)$ MHz, and with α_e^B from band spectra they calculated the equilibrium value, $r_e=1.4938(2)$ Å. They also measured the ^{14}N nuclear coupling and the molecular dipole moments. Their frequency measurements were no doubt helpful in the discovery of NS in interstellar space through detection of the doublet emission lines of the $^2\Pi_{1/2}$ state by Kuper et al. [97] and by Gottlieb et al. [98].

Millimeter wave rotational transitions have been measured by Amano et al. [99] for the unstable molecule ClO, which has the interesting feature of an inverted $^2\Pi$ ground state, with the $^2\Pi_{3/2}$ below the $^2\Pi_{1/2}$ state. The $J=\frac{3}{2}\rightarrow\frac{5}{2}$ rotational transition of the even more unstable species BrO has been detected by Powell and Johnson [100] for the $^2\Pi_{3/2}$, which is also the ground electronic state. Rotational transitions of the transient CF radical in both $^2\Pi_{1/2}$ and $^2\Pi_{3/2}$ states have been measured by Saito et al. [101].

The first microwave spectral observation of a molecule in an excited electronic state is that made by Saito [102] of the $J=2\rightarrow3$ rotational transition of SO in the metastable $^1\Delta$ state, which is approximately 6,350 cm^{-1} above the $^3\Sigma$ ground state. Later, Clark and De Lucia [74] confirmed this assignment and extended the measurements to three other rotational transitions in the 100 to 300 GHz range. The $^1\Delta$ state conforms closely to Hund's case (*a*), Fig, 4.5. For the $^1\Delta$ state, $\Sigma=0$, $\Lambda=2$, and thus $\Omega=2$. If the centrifugal stretching and Λ doubling are neglected, the allowed rotational energies are given by (4.39). This equation, with the selection rules of (4.42), leads to the approximate formula of (4.43) for the rotational frequencies. This frequency formula is like that for a rigid diatomic molecule in a $^1\Sigma$ state; however, the lowest J values are restricted by (4.40) to $J=\Omega=2$. Thus the lowest rotational transition is the $J=2\rightarrow3$, which for $^{12}S^{16}O$ was found to occur at 127,770 MHz.

The Λ-type doublet splitting of the rotational levels for the $^1\Delta$ state (see bottom entry in Table 4.14) is given by $q_\Lambda J(J^2-1)(J+2)$, and the approximate value of q_Λ is $48B^4/\nu_e^3$. The latter expression gives the negligible value of 10^{-6} MHz for the $^1\Delta$ state of SO. In agreement, no evidence for Λ doublet splitting was observed. Thus for the transitions measured, the Λ doublet splitting can be neglected and the observed rotational frequencies be expressed by

$$\nu=2B_v(J+1)-4D_v(J+1)^3 \tag{4.108}$$

where the last term is the first-order correction for centrifugal distortion and $J=2, 3, 4, 5, \ldots$. For $^{32}S^{16}O$ in the $^1\Delta$, $v=0$, state, Clark and DeLucia [74] obtained: $B_0 = 21{,}295.405(24)$ MHz, $D_0 = 0.0350(4)$ MHz, and $r_0 = 1.491967(8)$ Å. It is interesting to compare these values with those for the $^3\Sigma$ ground state in Table 4.13.

3 LINE INTENSITIES

Most gaseous absorption lines are measured in the pressure range where the intensity at the center of the line is independent of pressure (Chapter III, Section 2). Under these conditions the absorption coefficient at the center of the line α_{max}, where the absorption is greatest, gives the best measure of the detectability of the line with the usual spectrometers and provides a convenient comparison of the relative intensities of different lines measured within the pressure range 10^{-3} to 100 mm of Hg, where the peak absorptions are constant.

The formula for the peak intensities of the field-free diatomic molecule in $^1\Sigma$ states can be obtained from (3.27) if $|(m|\mu|n)|^2$ is replaced by the squared matrix elements $|(J|\mu|J+1)|^2$ from (2.123) and if F_m is replaced by F_J from (3.65). For calculation of intensities, the rigid rotor approximation is entirely adequate. With use of $\nu_0 = 2B(J+1)$ and with some rearrangements, the formula for α_{max} specialized to diatomic and linear polyatomic molecules becomes

$$\alpha_{max} = \frac{4\pi^2 N F_v h \mu^2 \nu_0^3}{3c(kT)^2 \Delta\nu} \left(1 - \frac{1}{2}\frac{h\nu_0}{kT}\right) e^{-(Jh\nu_0)/(2kT)} \tag{4.109}$$

With $N = 9.68 \times 10^{18} i_c p_{mm}/T$, from (3.25), with $\Delta\nu = 300 p_{mm}(\Delta\nu)_1/T$, from (3.26), and with substitution of numerical values for the constants, (4.109) reduced to a form convenient for calculations becomes

$$\alpha_{max}(\text{cm}^{-1}) = \frac{4.94 \times 10^{-3} i_c F_v \mu^2 \nu_0^3}{(\Delta\nu)_1 T^2} \left(1 - \frac{0.024\nu_0}{T}\right) e^{-0.024J\nu_0/T} \tag{4.110}$$

where ν_0 is the line frequency in GHz (MHz $\times 10^{-3}$) where $(\Delta\nu)_1$ is the line breadth in MHz per mm of Hg pressure at 300°K, μ is the dipole moment in debye units, T is the absolute temperature in Kelvin units, i_c is the isotopic concentration of the particular species observed, and F_v is the fraction of molecules in the particular vibrational state observed.

Equation (4.109) or (4.110) is adequate for the entire microwave region up to the highest submillimeter wave frequencies so far measured ($\lambda = 0.33$ mm or $\nu = 1000$ GHz). For the millimeter and centimeter wave region the term 0.024 ν/T in parentheses can be dropped. Except for molecules with very small B_0 values, for which J becomes very large for the frequencies observed, the exponential term can be expanded to give the more convenient expression

$$\alpha_{max}(\text{cm}^{-1}) = \frac{4.94 \times 10^{-3} i_c F_v \mu^2 \nu_0^3}{(\Delta\nu)_1 T^2} \left(1 - \frac{0.024J\nu_0}{T}\right) \tag{4.111}$$

which is usually applicable to frequencies of about 100 GHz ($\lambda = 3$ mm) or for higher frequencies when $J\nu \ll T$. For the centimeter wave region the simpler expression

$$\alpha_{\max}(\text{cm}^{-1}) = \frac{4.94 \times 10^{-3} i_c F_v \mu^2 \nu_0^3}{(\Delta\nu)_1 T^2} \tag{4.112}$$

given in the earlier microwave texts is adequate.

Since there is only one fundamental vibrational frequency and hence no degenerate modes, (3.61) for F_v is particularly simple for diatomic molecules. With $d=1$ and with Q_v from (3.60),

$$F_v = \frac{e^{-hv\omega/(kT)}}{Q_v} = e^{-hv\omega/(kT)}(1 - e^{-h\omega/(kT)})$$
$$= e^{-1.44v\omega/T}(1 - e^{-1.44\omega/T}) \tag{4.113}$$

where v is the vibrational quantum number and ω is the fundamental or lowest vibrational frequency. In the last form, ω is in cm^{-1} and T, the absolute temperature, in Kelvin units. In the ground vibrational state $v=0$ and

$$F_{v=0} = 1 - e^{-1.44\omega/T} \tag{4.114}$$

In Figure 4.12 we have plotted the calculated peak absorption coefficients of CO as a function of frequency for the ground vibrational state. For CO, $B_0 = 57{,}896$ MHz, GHz$=1.93$ cm^{-1}, and $\nu_{J=0\to1} = 115.8$ GHz. The dipole moment $\mu = 0.112$ D has been measured with millimeter-wave spectroscopy by Burrus [103]. Because of its low dipole moment we have assumed that $(\Delta\nu)_1$ has a value of 3 MHz, only slightly greater than that for O_2, and have negelected possible variation of $(\Delta\nu)_1$ with frequency. From optical spectroscopy, $\omega = 2170$ cm^{-1}, and therefore from (4.114) $F_{v=0} = 0.995$. For the most abundant

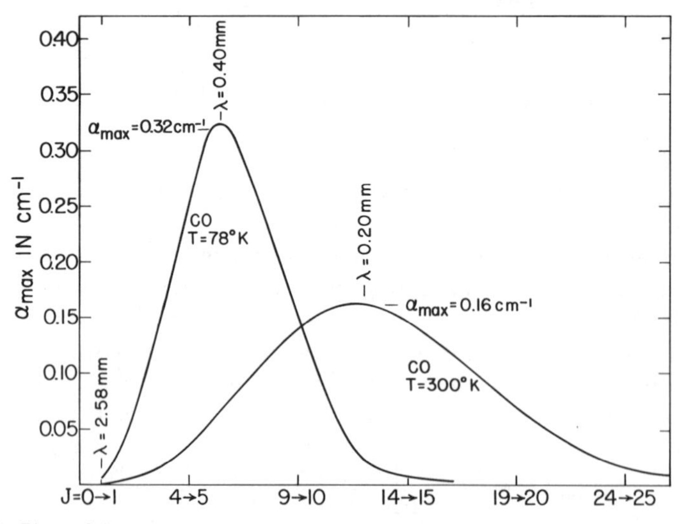

Fig. 4.12 Plots of the peak absorption coefficients of pressure-broadened, rotational absorption lines of CO in the ground vibrational state at two different temperatures.

species $i_c \approx 0.99$. With these constants, (4.110) was used for calculation of the absorption coefficients for $T=300°K$ and for $T=78°K$. At the latter temperature CO continues to have significant vapor pressure; it has been measured with the absorption cell under liquid nitrogen. Note that at 78°K the highest observed microwave frequency, for CO the $J=6 \rightarrow 7$ transition at 806,651.7 MHz($\lambda=0.37$), falls in the region of greatest absorption, whereas for 300°K the region of greatest absorption falls in the infrared region not yet reachable with microwave techniques.

Figure 4.13 provides a comparison of the variation of rotational line peak intensity for a heavy CsI diatomic molecule and a relatively light CsF diatomic molecule at elevated temperatures. For heavier diatomic molecules with B_0 of 1000 MHz or less, the maximum absorption is reached in the region now covered with microwave measurements even at temperatures appreciably above 300°K. For example, CsI, with $B_0=707$ MHz, has its lines of maximum strength occurring in the 1.0-mm wave region, even when $T=900°K$, to which it must be heated if it is to have measurable vapor. The dipole moment of CsI is large [8], 12.1 D; in calculation of the CsI curve of Fig. 4.13 the value of $(\Delta\nu)_1$ was assumed to be 20 MHz. Any error in the assumed value of $(\Delta\nu)_1$ does not alter the shape of the curve if $(\Delta\nu)_1$ does not vary significantly with frequency. Figure 4.13 represents the ground vibrational state for which $F_v=0.24$.

For $T=78°K$ and $J=5 \rightarrow 6$, $\alpha_{max}=0.32$ cm^{-1} for CO. With a cell length x of 10 cm, the fractional power absorption at this frequency is

$$\frac{P_{abs}}{P_0}=1-e^{-0.32x}=0.96 \tag{4.115}$$

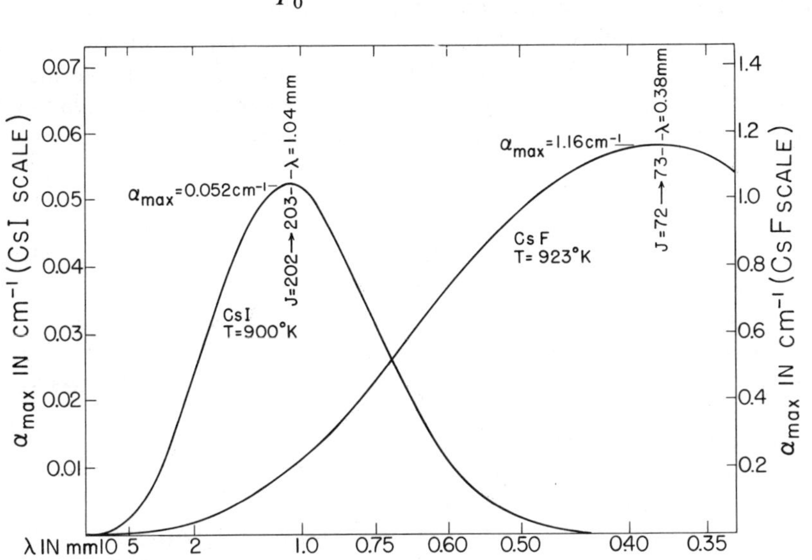

Fig. 4.13 Plots of the peak absorption coefficients of pressure-broadened rotational absorption lines of CsF and CsI at their approximate vaporization temperatures.

Thus despite its small dipole moment, CO would absorb 96% of the energy at the peak of the $J=5 \rightarrow 6$ transition with a cell length of only 10 cm when $T=78°K$. Cesium iodide at $T=900°K$ in the ground vibrational state has an α_{max} of 0.052 in the 10-mm wave region; with an effective cell length of 10 cm it would absorb 40% of the power at the peak of the rotational lines. Although the CsI lines have nuclear quadrupole hyperfine structure, the splitting is not resolvable for the high J values of lines observed in the millimetre wave region.

From Figs. 4.12 and 4.13 it is evident that the lower millimeter to upper submillimeter region, with wavelengths from about a half millimeter to five millimeters, is the optimum one for detection of rotational spectra of the heavy and moderately heavy diatomic molecules—the metallic halides, PbS, GeS, SnSe, etc.—whereas the submillimeter region is the optimum one for the lighter molecules such as CO, CN, LiF, or the halogen and metallic hydrides. One can readily estimate the region of optimum absorption for any temperature and B value from the simple formula

$$J_{opt} \approx 5.5 \left(\frac{T}{B}\right)^{1/2} \tag{4.116}$$

described in Chapter V, Section 4. Here J_{opt} represents approximately the lower rotational quantum number of the strongest rotational absorption line when the rotational constant B is in GHz and the absolute temperature T is in Kelvin units.

When the rotational transition has fine or hyperfine structure or Stark or Zeeman splitting, the foregoing formulas can be used for calculation of the intensities of the overall multiplet, and special formulas (see Chapters IX and X) can be used for calculation of the relative intensities of the components of the multiplet. However, for transitions involving changes in the electronic quantum numbers, such as the Λ doublet spectra, different dipole moment matrix elements must be employed. The matrix elements chosen depend on the nature of the electronic state and the special coupling case.

References

1. E. Pearson and W. Gordy, *Phys. Rev.*, **177**, 59 (1969).
2. G. Jones and W. Gordy, *Phys. Rev.*, **136**, A1229 (1964).
3. P. M. Morse, *Phys. Rev.*, **34**, 57 (1929).
4. J. L. Dunham, *Phys. Rev.*, **41**, 721 (1932).
5. L. Pauling and E. B. Wilson, *Introduction to Quantum Mechanics*, McGraw-Hill, New York, 1935.
6. C. L. Pekeris, *Phys. Rev.*, **45**, 98 (1934).
7. W. Born and R. Oppenheimer, *Ann. Physik*, **84**, 457 (1927).
8. J. H. Vleck, *J. Chem. Phys.*, **4**, 327 (1936).
9. B. Rosenblum, A. H. Nethercot, and C. H. Townes, *Phys. Rev.*, **109**, 400 (1958).
10. E. L. Manson, W. W. Clark, F. C. De Lucia, and W. Gordy, *Phys. Rev.*, **A15**, 223 (1977).
11. J. Hoeft, F. J. Lovas, E. Tiemann, R. Tischer, and T. Törring, *Z. Naturforsch.*, **24a**, 1222 (1969).

12. E. Tiemann, J. Hoeft, and B. Shenk, *Z. Naturforsch.*, **24a**, 787 (1969).
13. R. E. Willis and W. W. Clark, *J. Chem. Phys.*, **72**, 4946 (1980).
14. F. E. Lovas and E. Tiemann, *J. Phys. Chem. Ref. Data*, **3**, 609 (1974).
15. A. Honig, M. Mandel, M. L. Stitch, and C. H. Townes, *Phys. Rev.*, **96**, 629 (1954); M. L. Stitch, A. Honig, and C. H. Townes, *Rev. Sci. Instrum.* **25**, 759 (1954).
16. D. R. Lide, P. Cahill, and L. P. Gold, *J. Chem. Phys.*, **40**, 156 (1964).
17. H. G. Fitzky, *Z. Physik*, **151**, 351 (1958).
18. T. Törring, *Z. Naturforsch.*, **19a**, 1426 (1964).
19. J. Hoeft, *Z. Physik*, **163**, 262 (1961); *Z. Naturforsch.*, **19a**, 1134 (1964); **20a**, 826 (1965).
20. A. K. Garrison and W. Gordy, *Phys. Rev.*, **108**, 899 (1957).
21. J. R. Rusk and W. Gordy, *Phys. Rev.*, **127**, 817 (1962).
22. E. Pearson and W. Gordy, *Phys. Rev.*, **152**, 42 (1966).
23. M. Winnewisser, *Z. Angew. Phys.*, **30**, 359 (1971).
24. M. Winnewisser and B. P. Winnewisser, *Z. Naturforsch.*, **29a**, 633 (1974).
25. D. A. Helms, M. Winnewisser, and G. Winnewisser, *J. Phys. Chem.*, **84**, 1758 (1980).
26. P. Kuijpers, T. Törring, and A. Dymanus, *Chem. Phys.*, **12**, 309 (1976).
27. P. Kuijpers, T. Törring and A. Dymanus, *Chem. Phys.*, **18**, 401 (1976).
28. E. L. Manson, F. C. De Lucia, and W. Gordy, *J. Chem. Phys.*, **62**, 1040 (1975).
29. W. H. Hocking, E. F. Pearson, R. A. Creswell, and G. Winnewisser, *J. Chem. Phys.*, **68**, 1128 (1978).
30. S. E. Veazey and W. Gordy, *Phys. Rev.*, **138**, A1303 (1965).
31. P. L. Clouser and W. Gordy, *Phys. Rev.*, **134**, A863 (1964).
32. K. V. L. N. Sastry, E. Herbst, and F. C. De Lucia, *J. Chem. Phys.*, **75**, 4753 (1981).
33. F. C. Wyse and W. Gordy, *J. Chem. Phys.*, **56**, 2130 (1972).
34. L. E. Snyder and D. Buhl, *Astrophys. J.*, **197**, 329 (1975).
35. R. W. Wilson, K. B. Jefferts, and A. A. Penzias, *Astrophys. J.*, (*Lett.*) **161**, L43 (1970).
36. J. Curry, L. Herzberg, and G. Herzberg, *Z. Physik*, **86**, 348 (1933).
37. F. C. Wyse, E. L. Manson, and W. Gordy, *J. Chem. Phys.*, **57**, 1106 (1972).
38. J. Hoeft, E. Tiemann, and T. Torring, *Z. Naturforsch.*, **27a**, 703 (1972).
39. J. Raymonda and W. Klemperer, *J. Chem. Phys.*, **55**, 232 (1971).
40. G. Herzberg, *Molecular Spectra and Molecular Structure. I. Spectra of Diatomic Molecules*, 2nd ed., Van Nostrand, New York, 1950.
41. T. A. Dixon and R. C. Woods, *Phys. Rev. Lett.*, **34**, 61 (1975).
42. K. V. L. N. Sastry, P. Helminger, E. Herbst, and F. C. De Lucia, *Astrophys. J.*, **250**, L91 (1981).
43. N. D. Piltch, P. G. Szanto, T. G. Anderson, C. S. Gudeman, T. A. Dixon, and R. C. Woods, *J. Chem. Phys.*, **76**, 3385 (1982).
44. M. Bogey, C. Demuynck, and J. L. Destombes, *Mol. Phys.*, **46**, 679 (1982).
45. K. B. Jefferts, A. A. Penzias, and R. W. Wilson, *Astrophys. J.*, **161**, L87 (1970).
46. A. A. Penzias, R. W. Wilson, and K. B. Jefferts, *Phys. Rev.*, **32**, 707 (1974).
47. R. A. Frosch and H. M. Foley, *Phys. Rev.*, **88**, 1337 (1952).
48. T. A. Dixon and R. C. Woods, *J. Chem. Phys.*, **67**, 3956 (1977).
49. S. Saito, Y. Endo, and E. Hirota, *J. Chem. Phys.*, **78**, 6447 (1983).
50. H. A. Kramers, *Z. Physik*, **53**, 422 (1929).
51. M. H. Hebb, *Phys. Rev.*, **49**, 610 (1936).
52. B. Schlapp, *Phys. Rev.*, **51**, 342 (1937).
53. S. L. Miller and C. H. Townes, *Phys. Rev.*, **90**, 537 (1953).

54. M. Mizushima and R. M. Hill, *Phys. Rev.*, **93**, 745 (1954).
55. M. Tinkham and M. W. P. Strandberg, *Phys. Rev.*, **97**, 937 (1955); 951 (1955).
56. R. Tischer, *Z. Naturforsch.*, **22a**, 1711 (1967).
57. W. Steinbach and W. Gordy, *Phys. Rev.*, **A8**, 1753 (1973).
58. M. Mizushima, *The Theory of Rotating Diatomic Molecules*, Wiley, New York, 1975.
59. R. L. Cook and F. C. De Lucia, *Am. J. Phys.*, **39**, 1433 (1971).
60. W. Steinbach, Ph.D. dissertation, Duke University, 1974.
61. W. Steinbach and W. Gordy, *Phys. Rev.*, **A11**, 729 (1975).
62. R. W. Zimmerer and M. Mizushima, *Phys. Rev.*, **121**, 152 (1961).
63. B. G. West and M. Mizushima, *Phys. Rev.*, **143**, 31 (1966).
64. W. M. Welsh and M. Mizhushima, *Phys. Rev.*, **A5**, 2692 (1972).
65. J. S. McKnight and W. Gordy, *Phys. Rev. Lett.*, **21**, 1787 (1968).
66. K. M. Evenson, H. P. Broida, J. S. Wells, R. J. Mahler, and M. Mizushima, *Phys. Rev. Lett.*, **21**, 1038 (1968).
67. K. M. Evenson and M. Mizushima, *Phys. Rev.*, **A6**, 2197 (1972).
68. D. L. Albritton, W. J. Harrop, A. L. Schmeltekopf, and R. M. Zare, *J. Mol. Spectrosc.*, **46**, 103 (1973).
69. H. D. Babcock and G. Herzberg, *Astrophys. J.*, **108**, 167 (1948).
70. T. Amano and E. Hirota, *J. Mol. Spectrosc.*, **53**, 346 (1974).
71. M. Winnewisser, K. V. L. N. Sastry, R. L. Cook, and W. Gordy, *J. Chem. Phys.*, **41**, 1687 (1964).
72. F. X. Powell and D. R. Lide, *J. Chem. Phys.*, **41**, 1413 (1964).
73. Measurement by M. Winnewisser, reported by W. Gordy, "Microwave Spectroscopy in the Region of 4–0.4 Millimetres," in *Molecular Spectroscopy VIII*, Butterworths, London, 1965.
74. W. W. Clark and F. C. De Lucia, *J. Mol. Spectrosc.*, **60**, 332 (1976).
75. T. Amano, E. Hirota, and Y. Morino, *J. Phys. Soc. Jap.*, **22**, 399 (1967).
76. E. L. Hill and J. H. Van Vleck, *Phys. Rev.*, **32**, 250 (1923).
77. R. S. Mulliken, *Rev. Mod. Phys.*, **2**, 60 (1930); **3**, 89 (1931).
78. J. H. Van Vleck, *Phys. Rev.*, **33**, 467 (1929).
79. R. S. Mulliken and A. Christy, *Phys. Rev.*, **38**, 87 (1931).
80. G. C. Dousmanis, T. M. Sanders, and C. H. Townes, *Phys. Rev.*, **100**, 1735 (1955).
81. C. A. Burrus and W. Gordy, *Phys. Rev.*, **92**, 1437 (1953).
82. P. G. Favero, A. M. Mirri, and W. Gordy, *Phys. Rev.*, **114**, 1534 (1959).
83. G. M. Almy and R. B. Horsfall, *Phys. Rev.*, **51**, 491 (1937).
84. R. N. Dixon, *Phil. Trans. R. Soc.* **252A**, 165 (1960).
85. J. J. Gallagher and C. M. Johnson, *Phys. Rev.*, **103**, 1727 (1956).
86. J. A. Coxon, *Can. J. Phys.*, **58**, 933 (1980).
87. K. M. Evenson, J. S. Wells, and H. E. Radford, *Phys. Rev. Lett.*, **25**, 199 (1970).
88. M. Mizushima, *Phys. Rev.*, **A5**, 143 (1972).
89. M. Brown, C. M. L. Kerr, F. D. Wayne, K. M. Evenson, and H. E. Radford, *J. Mol. Spectrosc.*, **86**, 544 (1981).
90. S. Weinreb, A. H. Barrett, M. L. Meeks, and J. C. Henry, *Nature*, **200**, 829 (1963).
91. L. E. Snyder, "Molecules in Space," in *MTP International Review of Science*, *Physical Chemistry*, Vol. 3, *Spectroscopy*, D. A. Ramsey, Ed., Butterworths, London, 1972, pp. 193–240.
92. G. Winnewisser, E. Churchwell, and C. M. Walksley, "Astrophysics of Interstellar Molecules," in *Modern Aspects of Microwave Spectroscopy*, G. W. Chantry, Ed., Academic, London, 1979, pp. 313–503.

93. B. E. Turner and B. Zukerman, *Astrophys. J.*, **187**, L59 (1973).
94. O. E. H. Rydbeck, J. Ellder, and W. M. Irvine, *Astron. Astrophys.*, **33**, 315 (1974).
95. J. J. Gallagher, F. D. Bedard, and C. M. Johnson, *Phys. Rev.*, **93**, 729 (1954).
96. T. Amano, S. Saito, E. Hirota, and Y. Morino, *J. Mol. Spectrosc.*, **32**, 97 (1969).
97. T. B. H. Kuiper, B. Zukerman, R. K. Kakar, and E. N. Rodriguez-Kuiper, *Astrophys. J.*, **200**, L151 (1975).
98. C. A. Gottlieb, J. A. Ball, E. W. Gottlieb, C. J. Lada, and H. Penfield, *Astrophys. J.*, **200**, L147 (1975).
99. T. Amano, S. Saito, E. Hirota, Y. Morino, D. R. Johnson, and F. X. Powell, *J. Mol. Spectrosc.*, **30**, 275 (1969).
100. F. X. Powell and D. R. Johnson, *J. Chem. Phys.*, **50**, 4596 (1969).
101. S. Saito, Y. Endo, M. Takami, and E. Hirota, *J. Chem. Phys.*, **78**, 116 (1983).
102. S. Saito, J. Chem. Phys., **53**, 2544 (1970).
103. C. A. Burrus, *J. Chem. Phys.*, **28**, 427 (1958).

Chapter **V**

LINEAR POLYATOMIC MOLECULES

Because the potential functions of nonrigid polyatomic molecules are more complicated than those of diatomic molecules, it is generally not feasible to solve their wave equations by the same methods as were used for the diatomic molecules. Instead, the customary procedure is to find a first solution for the polyatomic molecule with the assumption that it is a completely rigid, field-free rotor and then to treat centrifugal distortion, Stark and Zeeman interactions with perturbation theory. For these calculations, specific eigenfunctions are not required; only the matrix elements of angular momentum operators and the direction cosine matrix elements, as given in Chapter II, are required. These elements can be derived from matrix methods without specific knowledge of the eigenfunctions. Thus one need not solve the wave equation to find the required spectral formulas and selection rules. Even though the wave equation for a rigid, linear rotor is readily solvable, the desired eigenvalues of the rotational energy can be obtained more easily from the matrix elements of the angular momentum operators derived from the commutation rules of these operators. In this chapter we shall treat only unperturbed, field-free rotors, leaving the Stark and Zeeman effects and the nuclear hyperfine structure to later chapters. Methods for calculation of molecular structures from the spectra are given in Chapter XIII.

1 THEORETICAL FORMULATIONS: ENERGIES AND FREQUENCIES

Molecules in the Ground Vibrational State

For a rigid, linear, polyatomic molecule with no resultant electronic angular momentum, the angular momentum about the figure axis a is zero, just as for the diatomic molecule, and $I_b = I_c = I$. Therefore the rotational Hamiltonian can be expressed as

$$\mathscr{H}_0 = \frac{P^2}{2I} \tag{5.1}$$

where I is the moment of inertia and P^2 is the operator conjugate to the square of the total angular momentum. Since the nonvanishing matrix elements of the operator P^2 are $(J|P^2|J) = \hbar^2 J(J+1)$ (see Chapter II, Section 2) the Hamiltonian operator $\mathscr{H}_0$ has only diagonal elements, and the energy eigenvalues are therefore

$$E_J^0 = (J|\mathscr{H}_0|J) = \left(\frac{1}{2I}\right)(J|P^2|J) = \frac{h^2 J(J+1)}{8\pi^2 I} \tag{5.2}$$

With the substitution of the spectral constant $B \equiv h/8\pi^2 I$, this becomes

$$E_J^0 = hBJ(J+1) \tag{5.3}$$

The bracket notation signifies $(J|\mathscr{H}_0|J) = \int \psi_J^{0*} \mathscr{H}_0 \psi_J^0 \, d\tau$ where ψ_J^0 represents the eigenfunctions of the unperturbed operator $\mathscr{H}_0$. Since $\mathscr{H}_0$, P^2, and P_Z are commuting operators, they have a common set of eigenfunctions ψ_{JM}^0. The eigenfunctions for the rigid diatomic or linear polyatomic molecule are the same as those of (2.102) and can be obtained from solution of Schrödinger's wave equation. Furthermore, the component of the angular momentum of the linear rotor that can be resolved along a fixed axis by a Stark or Zeeman field is

$$P_Z = \frac{h}{2\pi} M_J \tag{5.4}$$

The quantum number of the angular momentum, as was found earlier, can take only the integral values

$$J=0, 1, 2, 3, \ldots \tag{5.5}$$

and M_J, which measures in units of $\hbar$ the components of the angular momentum along a space-fixed axis, can take the values

$$M_J=J, J-1, J-2, \ldots, -J \tag{5.6}$$

The selection rules for dipole absorption of radiation (see Chapter II, Section 6), based on the condition that the matrix elements of the molecular dipole moment $\int \psi_{JM}\mu\psi_{J'M'}\,d\tau$ must not vanish, are found to be

$$J'=J\pm 1, \qquad M_{J'}=M_J \text{ or } M_J\pm 1 \tag{5.7}$$

just as they are for diatomic molecules. The selection rules for M_J are of interest only when an external field is applied (see Chapter X). The transitions $J\rightarrow J+1$ correspond to absorption of radiation of the field-free rotor. Hence the absorption line frequencies are

$$\nu=2B(J+1) \tag{5.8}$$

where the spectral constant B is expressed in frequency units.

From (5.8) it is seen that the rotational frequencies of the rigid, linear, polyatomic molecule, like those of the diatomic molecule, form a series of $2B$, $4B$, $6B$, and so on. The lowest rotational frequency $2B$ for most polyatomic molecules, falls in the centimeter wave region, although for HCN it occurs in the 3-mm wave region.

The corresponding Hamiltonian for the nonrigid rotor can be written as

$$\mathscr{H}=\mathscr{H}_0+\mathscr{H}_d \tag{5.9}$$

where $\mathscr{H}_d$ represents the centrifugal distortion energy. For the linear molecule there is only one axis of distortion, and $(P_x^2+P_y^2)^2=P^4$, $P_z^2=0$. With these conditions, (8.35) for $\mathscr{H}_d$ takes the simple form

$$\mathscr{H}_d=-\left(\frac{h}{\hbar^4}\right)DP^4 \tag{5.10}$$

in which D is a constant. The first-order perturbation energy is the average of $\mathscr{H}_d$ over the eigenfunction ψ_J^0 of the unperturbed Hamiltonian $\mathscr{H}_0$, or

$$E^{(1)}=(J|\mathscr{H}_d|J)=-\left(\frac{h}{\hbar^4}\right)D(J|P^4|J)=-hDJ^2(J+1)^2 \tag{5.11}$$

where D, the centrifugal stretching constant, depends on the potential functions of the various bonds and can, in principle, be expressed as a function of the vibrational frequencies or force constants. For a triatomic linear molecule, Nielsen [1] has shown that

$$D=4B^3\left(\frac{\zeta_{21}^2}{\omega_3^2}+\frac{\zeta_{23}^2}{\omega_1^2}\right) \tag{5.12}$$

where ω_1 and ω_3 are the bond-stretching frequencies and ζ_{21} and ζ_{23} are Coriolis coupling constants.

The next higher-order term in the expression has the form, $H_0 J^3(J+1)^3$, where $H_0 \approx H_e \ll D_0$. Thus the rotational energy for the linear molecules in the ground vibrational state may be expressed as

$$\frac{E_J}{h} = B_0 J(J+1) - D_0 J^2(J+1)^2 + H_0 J^3(J+1)^3 \tag{5.13}$$

With the selection rules for dipole absorption of radiation, $J \rightarrow J+1$, the rotational frequencies are

$$\nu = 2B_0(J+1) - 4D_0(J+1)^3 + H_0(J+1)^3[(J+2)^3 - J^3] \tag{5.14}$$

This equation is found to be adequate for most microwave measurements of linear molecules in the ground vibrational state. It has the same form as (4.10) for diatomic molecules.

Molecules in Excited Vibrational States

Theory for the rotational spectra of linear polyatomic molecules in excited vibrational states is more complicated than that for diatomic molecules, not only because of the greater number of vibrating modes but also because of the special type of interaction of rotation with the degenerate bending modes of vibration.

A linear molecule with n atoms has $(3n-5)$ modes of vibration. Not all these modes have different frequencies ω_i, however. The bending modes are always doubly degenerate; that is, there are two bending modes of vibration, both of which have the same frequency; and in molecules with $n > 3$ there is more than one degenerate mode. A nonsymmetric, triatomic linear molecule such as OCS or HCN has two nondegenerate parallel modes v_1 and v_3 and a doubly degenerate bending mode v_2, as indicated in Fig. 5.1. The parallel modes correspond essentially to the stretching vibrations of the two bonds, XY and YZ.

As in diatomic molecules, the averaged or effective bond lengths tend to be slightly longer in the excited parallel modes because of the asymmetry in the Morse-type potential curve (anharmonicity). This leads to a slightly increasing moment of inertia and a consequent lowering of B_v and the rotational frequency with increasing vibrational quantum number for the parallel modes. In contrast, linear molecules in bending vibrational modes are slightly shorter, on the average, than are the molecules of the ground vibrational states. Their effective B_v and their rotational frequencies are expected therefore to be slightly higher than the corresponding ones for molecules in the ground vibrational states.

Except for deviations caused by l-type resonances or Fermi resonances (discussed later in this section), the effects of the vibration on B_v are to first order given by

$$B_v = B_e - \sum_i \alpha_i \left(v_i + \frac{d_i}{2} \right) \tag{5.15}$$

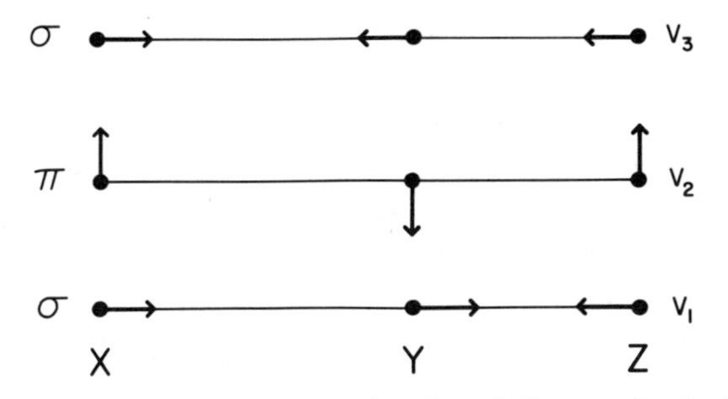

Fig. 5.1 Indication of the vibrational modes of a triatomic linear molecule. The bending mode v_2 is doubly degenerate because the molecule is free to bend in two orthogonal planes.

where v_i is the vibrational quantum number for the ith excited mode, d_i is the degeneracy of that mode, and α_i is a small interaction constant that gives the first-order correction of B_v for the ith mode. For reasons given previously, α_i is generally positive for parallel modes and negative for perpendicular modes. Vibrational modes having the same fundamental frequency, degenerative modes, are counted only once in the summation. For nondegenerate modes, $d_i=1$; for doubly degenerate modes, $d_i=2$, and so on. Note that B_v for the ground vibrational state differs from B_e by $\Sigma\alpha_i d_i/2$. Thus the evaluation of B_e to first order requires measurement of rotational lines of the molecule in at least one excited vibrational level for each of the fundamental modes. Interactions among the different modes of excitation give rise to higher-order cross terms included in (5.25).

Likewise, the centrifugal stretching constants differ slightly for the excited vibrational states. To first order, this difference is expressed by

$$D_v = D_e + \sum_i \beta_i \left(v_i + \frac{d_i}{2} \right) \tag{5.16}$$

Since D_v is already a very small constant in comparison with B_v, the changes in the higher-order stretching constant H_v with vibrational state are usually negligible.

In addition to the shifts in the lines described above, there are other effects of interaction of rotation with the degenerate bending vibrational modes. If z is chosen along the figure axis, the degenerate vibrations of the nonrotating molecules can be considered as orthogonal bending motions of the molecule in the xz and yz planes. In the rotating molecule the degeneracy is lifted by the interaction between vibration and rotation. The vibrational states can be represented as combinations of the two idealized states of the nonrotating molecule. Two such orthogonal vibrations with a 90° phase difference are equivalent to rotations, clockwise and counterclockwise, about the figure axis z. Thus a degenerate bending vibrational state with vibrational quantum number v_i can be regarded as having components of angular momentum $p_z = l\hbar$ with rotational quantum numbers

$$l_i = v_i,\ v_i - 2,\ v_i - 4,\ \ldots,\ -v_i \tag{5.17}$$

For a first excited degenerate bending mode, $v_i=1$ and $l_i=\pm 1$; for a second, $v_i=2$, and $l_i=0, \pm 2$; for a third, $v_i=3$, $l_i=\pm 1, \pm 3$, and so on. The various vibrational states are indicated by a sequence of their vibrational quantum numbers. For example, the vibrational states of a triatomic molecule would be indicated by $v_1 v_2^{|l|} v_3$, where the subscripts indicate the fundamental modes specified in Fig. 5.1 and the corresponding v numbers indicate their states of excitation. Examples are

$$0\ 1^1\ 0, \qquad 0\ 2^0\ 0, \qquad 0\ 2^2\ 0, \qquad 1\ 1^1\ 0.$$

When more than one bending vibrational state is excited, the quantum number for the total vibrational angular momentum, $l=\sum_i l_i$, characterizes the component of angular momentum about the molecular axis. Since the angular momentum thus generated is about the figure axis, l is analogous to the K quantum number of the symmetric top or the Λ component in diatomic molecules with electronic orbital momentum. Following a proposal by Mulliken [2], the vibrational states of linear polyatomic molecules are often designated as are those of diatomic molecules, with the value of $|l|$ replacing that for Λ, that is, vibrational states $l=0$ are designated as Σ states; with $|l|=1$, as π states; with $|l|=2$, as Δ states, and so on.

In an excited degenerate bending mode, the rotating linear molecule has an effective moment of inertia I_a as well as a component of angular momentum about the figure axis. As for the rigid symmetric top (to be discussed in Chapter VI), its rotational Hamiltonian can be expressed as

$$\mathscr{H}_r=\frac{P^2}{2I_b}+\frac{1}{2}\left(\frac{1}{I_a}-\frac{1}{I_b}\right)p_z^2 \tag{5.18}$$

Since p_z^2 has eigenvalues $\hbar^2 l^2$, the allowed energy values are

$$\frac{E_r}{h}=B_v J(J+1)+(A_v-B_v)l^2 \tag{5.19}$$

The term $A_v l^2$ is usually treated as vibrational energy, and the rotational energy is expressed by

$$\frac{E_r}{h}=B_v[J(J+1)-l^2] \tag{5.20}$$

For pure rotational transitions, the selection rules are $J\to J+1$ and $l\to l$; the rotational frequencies in this approximation are

$$\nu=2B_v(J+1) \tag{5.21}$$

where J represents the total angular momentum quantum number including the vibrational angular momentum. Thus J has the values

$$J=|l|, |l|+1, |l|+2, \ldots \tag{5.22}$$

Except for the small change in B_v given by (5.15), (5.21) is like that for the rigid molecule in the ground vibrational state. There is, however, an important effect which is due to the fact that J must now include l. In an excited bending mode when $|l|=1$, the lowest value of J is 1. As a result, the line corresponding to $J=0\rightarrow1$ of the ground vibrational state does not occur. For a higher excited vibrational state when $l=2$, the first two rotational states are missing and the $J=0\rightarrow1$ and $1\rightarrow2$ levels are missing, and so on. In addition to these effects and the l-doubling effects to be described later, there are slight effects of the vibrational angular momentum in the centrifugal stretching term which are altered to $-D_v[J(J+1)-l^2]^2$ and $H_v[J(J+1)-l^2]^3$.

Because of the Coriolis coupling force proportional to $\mathbf{v}\times\boldsymbol{\omega}$ between the vibrational motion $\mathbf{v}$ and the angular rotational motion $\boldsymbol{\omega}$ in orthogonal planes, the $\pm l$ degeneracy of (5.19) is lifted, and a doublet splitting of the rotational lines (called l doubling) is produced. As a result of this interaction, the effective moment of inertia generated about the z axis is not quite symmetric, and the molecule behaves like a slightly asymmetric rotor in which there is a small splitting of the K levels which are degenerate in the corresponding symmetric top. For the Π state, $|l|=1$, and the l splitting ΔE_l is approximately proportional to the square of the angular momentum and is expressed by [3, 4]

$$(\Delta E)_{|l|=1}=h\left(\frac{q_i}{2}\right)(v_i+1)J(J+1) \tag{5.23}$$

where v_i is the vibrational quantum number of the ith degenerate bending mode and q_i is the coupling constant of the vibration and rotation for this mode when $|l|=1$.

For linear molecules in degenerate bending modes when the splitting for $|l|=1$ is included, the rotational energies may be expressed as

$$\frac{E_r^{\pm}}{h}=B_v[J(J+1)-l^2]\pm\left(\frac{q_i}{4}\right)(v_i+1)J(J+1)-D_v[J(J+1)-l^2]^2 + H_v[J(J+1)-l^2]^3 \tag{5.24}$$

If the lower and upper component levels of an l doublet are signified by subscripts L and U, respectively, the possible transitions are J, $l_U\rightarrow J+1$, l_U and J, $l_L\rightarrow J+1$, l_L. The squared dipole matrix elements, upon which the probabilities of these transitions depend, may be obtained from (2.124) by replacement of K with l_L or l_U. With the omission of the usually negligible term in H_v, the rotational frequencies are

$$\nu_{\pm}=2B_v(J+1)\pm\tfrac{1}{2}q_i(v_i+1)(J+1)-4D_v(J+1)[(J+1)^2-l^2] \tag{5.25}$$

where

$$B_v=B_e-\sum_i \alpha_i\left(v_i+\tfrac{1}{2}d_i\right)+\sum_{ij}\gamma_{ij}\left(v_i+\tfrac{1}{2}d_i\right)\left(v_j+\tfrac{1}{2}d_j\right) + \sum_{ijk}\varepsilon_{ijk}\left(v_i+\tfrac{1}{2}d_i\right)\left(v_j+\tfrac{1}{2}d_j\right)\left(v_k+\tfrac{1}{2}d_k\right)+\gamma_{ll}l^2 \tag{5.26}$$

and

$$D_v = D_e + \sum_i \beta_i(v_i + \tfrac{1}{2}d_i) \tag{5.27}$$

In (5.26) the factors α_i, γ_{ij}, ε_{ijk}, and γ_{ll} are interaction constants which are a measure of the effects of the vibration–rotation interactions on B_v, and $l = \sum_i l_i$ is the quantum number for the total vibrational angular momentum about the molecular axis.

In measurements on the lower vibrational states, those most often investigated, the terms in α_i and γ_{ij} in the expansion of B_v, (5.26), are usually adequate for a fitting of the observed spectra. However, De Lucia and Helminger [3], who measured rotational transitions of 12 excited vibrational states of HCN in active laser plasmas, found the term ε_{ijk} of (5.26) to be required for a fitting of the data of some relatively low excited states. For example, they found that ε_{222} makes a significant contribution to the B_v value for the lower bending modes of HCN. The various parameters of (5.26) that they evaluated from their measurements are reproduced in Table 5.1.

The theory of the interactions of rotation with bending vibrations in linear molecules has been developed by Nielsen and Shaffer [4, 5]. For the linear XYZ molecules with $|l| = 1$ for the degenerate bending mode v_2 they derive the following expression for the coupling constant in (5.23).

$$q_2 = 2\frac{B_e^2}{\omega_2}\left[1 + 4\sum_i \zeta_{2i}^2\left(\frac{\omega_2^2}{\omega_i^2 - \omega_2^2}\right)\right] \approx 2.6\frac{B_e^2}{\omega_2} \tag{5.28}$$

Table 5.1 Equilibrium Rotational Constants of $H^{12}C^{14}N$ as Derived from (5.26) with Experimental Values of B_v for 12 Excited Vibrational States[a]

Constant	*Value (MHz)*	*Standard Deviation* σ
B_e	44,511.62	0.03
α_1	300.028	0.03
α_2	−108.369	0.05
α_3	312.983	0.03
γ_{11}	−0.893	0.009
γ_{12}	−3.583	0.033
γ_{13}	6.448	0.027
γ_{22}	0.762	0.027
γ_{23}	6.332	0.033
γ_{33}	−4.521	0.018
ε_{222}	0.094	0.003
γ_{ll}	−6.224	0.006

[a]From De Lucia and Helminger [3].

where ω_2 signifies the vibrational frequency of the degenerate bending mode and ω_i, the frequencies of other vibrational modes, and where ζ_{2i}, the Coriolis coupling constants, are parameters that depend on the force constants, atomic mass, and so on. Expressions for ζ_{2i}, which are relatively involved, may be found in the original papers. When $|l|>1$, the l splitting is much less than when $|l|=1$ and is usually not observable when $|l|>2$. For the alkali hydroxides, Kuijpers et al. [6] found the l doubling to be measurable for $|l|=1$ and 2, but not detectable for $|l|=3$.

Bogey and Bauer [7], who measured rotational transitions of very high vibrational states of OCS, excited through active transfer from N_2, used the theoretical formalism of Amat et al. [8] for their analysis. They give particular formulas for the frequencies of the $v_i=3$, 4, and 5 degenerate bending modes with the allowed values of $|l|$ for each mode. These formulas should be useful in the analysis of the microwave spectra of other linear triatomic molecules in highly excited bending modes, but they require too much space for reproduction here.

An observational scan [9] displaying the $J=2\rightarrow3$ rotational line of FCP in seven different vibrational states, shown in Fig. 5.2 illustrates the nature of the experimental observations from which vibrational effects on rotational spectra are evaluated.

l-Type Doublet Spectra

As Shulman and Townes [10] proved, direct transitions between l-type doublet levels $l_L\leftrightarrow l_U$ with no change in J can be observed. The squared dipole moment matrix elements upon which the transition probabilities depend are

$$|(J, l_L|\mu|J, l_U)|^2=\frac{\mu^2 l^2}{J(J+1)} \tag{5.29}$$

These elements are obtainable from (2.125) by replacement of K by l. For light molecules such as HCN, these transitions correspond to frequencies in the microwave region and provide the most precise measurements available of the coupling constant q. When q is measured in frequency units, the l-doublet transitions, according to Eq. (5.23), correspond to frequencies of

$$(\nu_l)_i=\left(\frac{q_i}{2}\right)(v_i+1)J(J+1) \tag{5.30}$$

For a given molecule q_i is approximately constant; therefore for a given bending mode v_i, the l-doublet frequencies should be proportional to $J(J+1)$. That this is nearly true was shown by the early measurements of Shulman and Townes [10] on HCN. However, measurable deviations [11, 12] from this formula occur as J becomes large. Nielsen and Amat [13] have shown that this deviation can be attributed to a change of q_i by centrifugal distortion. To first order, the dependence of q_i on J is expressed by

$$q_i=q_i^{(0)}-q_i^{(1)}J(J+1) \tag{5.31}$$

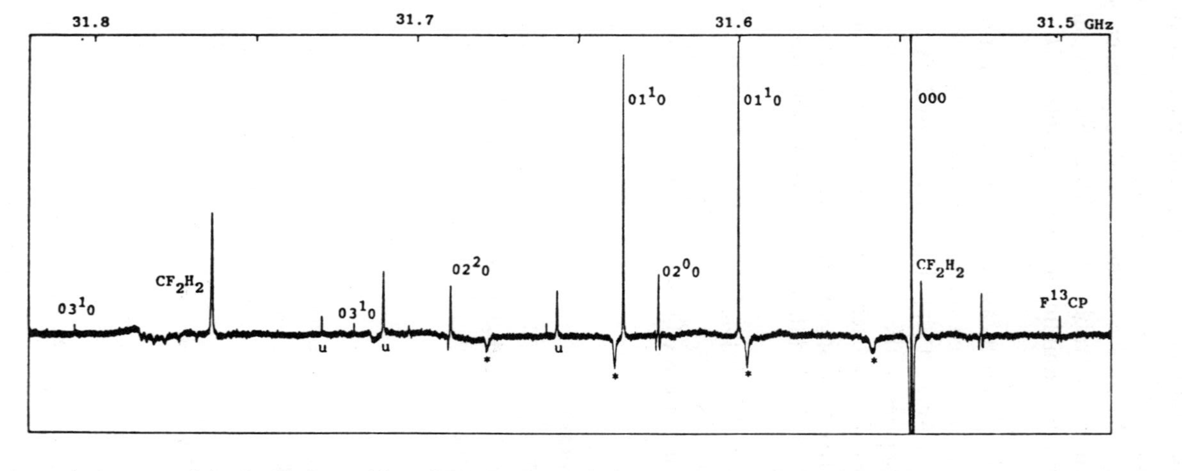

Fig. 5.2 Moderate resolution scan of the $J=3\leftarrow2$ transition of $FC\equiv P$ observed with a $2800\ V\ cm^{-1}$ Stark modulation. Unassigned lines are labeled *u* and the *l*-doublet Stark lobes by an asterisk. Note that CF_2H_2 impurity lines are also detected as indicated. From Kroto, Nixon, and Simmons [9].

so that the l-type doublet frequencies are expressed by

$$(\nu_l)_i=\left(\frac{q_i^{(0)}}{2}\right)(v_i+1)J(J+1)-\left(\frac{q_i^{(1)}}{2}\right)(v_i+1)J^2(J+1)^2 \tag{5.32}$$

Törring [12] found that the several l-type doublet transition frequencies which they observed for the degenerate bending mode v_2 in HCN and DCN when $v_2=1$ and $|l|=1$, could be fitted to the formulas with ν_l in kHz

$$\nu_l(H^{12}C^{12}N)=224{,}478\,J(J+1)-0.002666\,J^2(J+1)^2 \tag{5.33}$$

and

$$\nu_l(D^{12}C^{12}N)=186{,}192\,J(J+1)-0.00220\,J^2(J+1)^2 \tag{5.34}$$

Winnewisser and Bodenseh [14] have found that still higher order terms were required for satisfactory agreement with their precise measurements on l-type doublet transitions of the degenerate bending vibrations $v_4=1$ and $v_5=1$ of HCNO over a wide range of J values. They employed the expression

$$(\nu_l)_i=q_i^{(0)}J(J+1)-q_i^{(1)}J^2(J+1)^2+q_i^{(2)}J^3(J+1)^3-q_i^{(3)}J^4(J+1)^4+\cdots \tag{5.35}$$

which was indicated by the theoretical work of Ramadier and Amat [15]. The fitting of the observed frequencies with those calculated from (5.35) was very close. The values obtained for the various q's are listed in Table 5.2. Figure 5.3 shows plots of the observed l-doublet transition frequencies as a function of J. The v_4 and v_5 for HCNO correspond to the HCN and the CNO bending modes, respectively.

Later investigations [16–20] of the interesting molecule HCNC (fulminic acid) provide a basis for understanding the anomalies of the doublet splitting described above. Yamada et al. [18] have shown that the abnormal splitting arises partly from accidental resonances among some excited vibrational levels. Near coincidences were found to occur between levels of two bending modes, v_4 and v_5, and the lowest-lying stretching mode, v_3. The accidental resonances were found to be less normal and the l-type splitting more nearly normal [19]

Table 5.2 l-Type Doublet Constants of $H^{12}C^{14}N^{16}O$ Obtained by Application of (5.35) to the l-Type Doublet Spectra for the $v_5=1$ and the $v_4=1$ Vibrational States[a]. Values are given in MHz units

$v_5=1$	$v_4=1$
$q_5^{(0)}=34.6391\pm0.0001$	$q_4^{(0)}=23.6722\pm0.0009$
$q_5^{(1)}=(0.1623\pm0.0002)\ 10^{-3}$	$q_4^{(1)}=(0.6139\pm0.0032)\ 10^{-3}$
$q_5^{(2)}=(1.00\pm0.11)\ 10^{-9}$	$q_4^{(2)}=(0.1417\pm0.0034)\ 10^{-6}$
	$q_4^{(3)}=(0.199\pm0.012)\ 10^{-10}$

[a]From Winnewisser and Bodenseh [14].

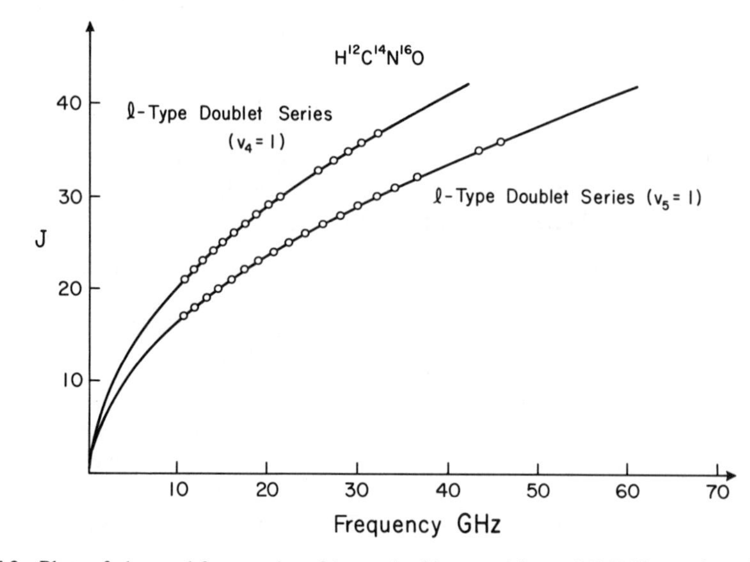

Fig. 5.3 Plots of observed frequencies of *l*-type doublet transitions of HCNO as a function of *J*. From Winnewisser and Bodenseh [14].

in DCNO than in HCNO. The abnormalities are also enhanced by the quasi-linearity of the molecule (Section 7). Bunker et al. [20] conclude that the equilibrium structure of HCNO is linear but that in the zero-point vibrational state and in the v_1 and v_2 vibrational states the HCN bending potential has its minimum off the internuclear axis. This would seem to be the cause of its quasi-linear behavior.

Table 5.3 shows a comparison of the $q_l^{(0)}$ values for the first excited bending modes ($v_i = 1$) for a few linear molecules.

Fermi Resonance

In addition to the effects of vibrations on rotational spectra, described earlier, there are smaller and less frequently encountered perturbations caused by Fermi resonance between the two vibrational levels of the same symmetry that have nearly the same energy. As Fermi predicted [21], such an interaction gives rise to a repulsion which further separates the vibrational levels. Such displacements in the vibrational levels will be accompanied by changes in the α_i and B_v values.

Nuclear Effects

Practically all stable linear molecules have singlet sigma electronic ground states and hence no molecular magnetic field except a weak one generated by the molecular rotation. For this reason most linear molecules observed with microwave spectroscopy do not have a resolvable nuclear magnetic hyperfine structure. However, hyperfine structure due to nuclear quadrupole

Table 5.3 *l*-Type Doublet Constants for First Excited Bending Modes ($v_i=1$) of Some Linear Molecules. Values are Given in MHz Units

Molecule[a]	q_2	*Ref.*	*Molecule*[a]	q_2	*Ref.*
HCN	224.47	b	OCS	6.344	e
DCN	186.19	b	^{79}BrCN	3.915	c
NNO	23.73	c	OC^{80}Se	3.172	c
FCN	19.85	d	ICN	2.688	e
^{35}ClCN	7.459	c	SCTe	0.659	f
	q_4	q_5			
HCCF	19.12	12.57			d
DCCF	15.33	13.26			d
	q_5	q_6	q_7		
HCCCN	2.56	3.57	6.54		d
DCCCN	2.68	3.10	5.97		d

[a]Atomic symbols are for the most abundant species where not specified.
[b]T. Törring [12].
[c]C. A. Burrus and W. Gordy, *Phys. Rev.*, **101**, 599 (1956).
[d]J. K. Tyler and J. Sheridan, *Trans. Faraday Soc.*, **59**, 2661 (1963).
[e]C. H. Townes, A. N. Holden, and F. R. Merritt, *Phys. Rev.*, **74**, 1113 (1948).
[f]W. A. Hardy and G. Silvey, *Phys. Rev.*, **95**, 385 (1954).

coupling is often resolved in the microwave rotational spectra. Nuclear quadrupole moments are electric moments, and their coupling to the molecular frame depends on the molecular electric field gradient. Thus large nuclear quadrupole splittings can occur for molecules in "nonmagnetic" singlet sigma states, but only nuclei with spins greater than $\frac{1}{2}$ can have quadrupole moments. For this reason the rotational spectra of many linear organic molecules do not have resolvable hyperfine structure.

When there is nuclear hyperfine splitting of the rotational lines, it must be analyzed, and the hypothetical unsplit rotational frequencies corresponding to those which would occur if there were no nuclear splitting must be found before the preceding formulas are applied for calculation of the spectral constant B and the corresponding moment of inertia. Methods for analysis of the hyperfine structure are given in Chapter IX. Because of its rapid decrease with increase of J, the hyperfine splitting is frequently negligible for transitions observed at millimeter wave frequencies and almost always negligible for those observed in the submillimeter wave region.

The effect of nuclear isotopic substitution is to change the moment of inertia of the molecule and hence to shift all the rotational lines. Each isotopic species of a given molecule has a complete rotational spectrum that with microwave spectrometers can be easily resolved from that of other isotopic species except when the isotopic difference occurs only in an atom that happens to occur at, or very near, the center of gravity of the molecule. The different isotopic species are very useful in the determination of the interatomic distances of polyatomic molecules from microwave spectroscopy. See Chapter XIII. The rotational transitions of an isotopic species have their particular rotational spectra and nuclear hyperfine structure, which must be analyzed separately.

2 LINE INTENSITIES

The formula for calculation of the rotational absorption coefficients of a linear polyatomic molecule is the same as that for a diatomic molecule (Chapter IV, Section 3). The molecule OCS is often used for tuning or monitoring of spectrometers. Furthermore, its dipole moment, B value, and other constants, upon which the absorption coefficient depends, are in the middle range of those of linear polyatomic molecules. As an example, we shall calculate the peak absorption coefficients of its microwave rotational lines for the ground vibrational state. For this purpose we apply (4.110), for which pressure-broadened lines are assumed to be in the range where the peak intensities are independent of pressure. In the computation it is more convenient to express α_{max} as a function of J. Hence we substitute $\nu_0=2B(J+1)$ into (4.110) and obtain

$$\alpha_{max}=X(J+1)^3\left[1-\frac{0.048B(J+1)}{T}\right]e^{-0.048BJ(J+1)/T} \tag{5.36}$$

where

$$X=\alpha_{max}(0\to1)=\frac{3.95\times10^{-2}i_cF_v\mu^2B^3}{(\Delta\nu)_1T^2} \tag{5.37}$$

In this equation the spectral constant B is expressed in GHz (kilomegahertz), but the line breadth parameter $(\Delta\nu)_1$ is in MHz per mm of Hg at 300°K. For OCS, $B=6.08$ GHz and $(\Delta\nu)_1=6.44$ MHz. The isotopic concentration of the species is indicated by i_c. We shall choose T as 195°K, the temperature of dry ice. For $^{16}O^{12}C^{32}S$ in its natural abundance, $i_c=0.94$. The dipole moment $\mu=0.709$ D. There are three fundamental vibrational modes, but only the degenerate bending mode $\omega_2=527$ cm^{-1} and the CS stretching mode $\omega_1=$ 859 cm^{-1} have significant populations at $T=195$°K. The CO stretching mode $\omega_3=2079$ cm^{-1}. Therefore the population of the ground vibrational state given by (3.56) for $v=0$ is

$$F_{v=0}=(1-e^{-1.44\omega_1/T})^{d_1}(1-e^{-1.44\omega_2/T})^{d_2}(1-e^{-1.44\omega_3/T})^{d_3}=0.96 \tag{5.38}$$

With these values

$$\alpha_{max}(0\to1)=1.65\times10^{-5}\text{ cm}^{-1}$$

and

$$\alpha_{max}(\text{OCS})=1.65\times10^{-5}(J+1)^3[1-15\times10^{-4}(J+1)]\times e^{-15\times10^{-4}J(J+1)} \tag{5.39}$$

This function is plotted in Fig. 5.4. The graph shows that OCS at dry-ice temperature has its strongest rotational absorption lines in the region of $\lambda = 0.77$ mm, for which $J = 31 \rightarrow 32$, approximately. The frequency for this strongest transition is $\nu_{opt} = 390$ GHz, and the absorption coefficient at the line peak $\alpha_{max}(31 \rightarrow 32) = 1.15 \times 10^{-1}$ cm^{-1}. In comparison with such a large absorption, the losses to the cell walls are negligible, and the power absorption fractional in a cell 20 cm in length is

$$\frac{P_{abs}}{P_0} = 1 - e^{-\alpha x} = 1 - e^{-2.3} = 0.90 \tag{5.40}$$

Thus, in the optimum region 90% of the power is absorbed in a cell only 20 cm in length. Although an absorption coefficient of 1.65×10^{-5} cm^{-1} (that for the $J = 0 \rightarrow 1$ line) can be detected easily with conventional microwave spectrometers, the $J = 0 \rightarrow 1$ line of the isotopic species ^{18}OCS in its natural abundance of 0.2% would have a peak absorption coefficient of only 3.3×10^{-8} cm^{-1} and could be detected with only the more sensitive spectrometers. In the region of optimum absorption, $\lambda = 0.77$ mm, the peak absorption coefficient for this rare isotopic species is 2.3×10^{-4} cm^{-1}, and the lines can be readily detected without isotopic concentration.

Optimum Regions for Observations

Now that the microwave region has been extended well into the submillimeter wave region, it covers the frequency range where most molecules have their strongest rotational lines. With such coverage available, it is of advantage to know in what regions the different molecules have their optimum detectability. This is particularly desirable when one is searching for lines of molecules

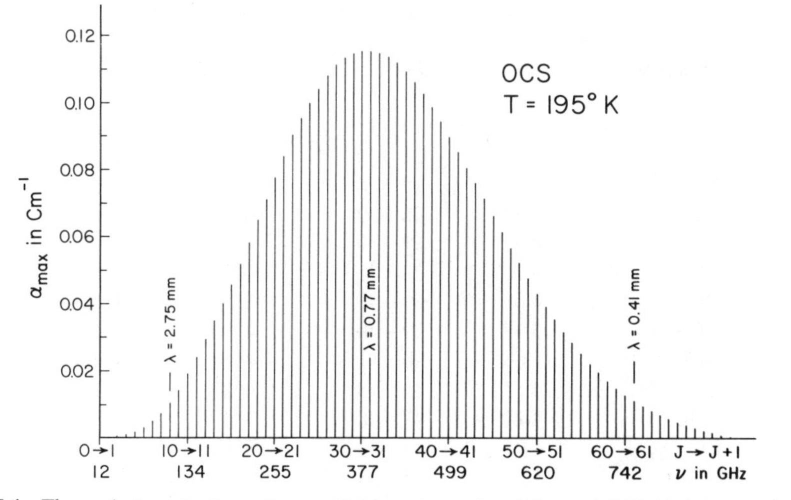

Fig. 5.4 Theoretical peak absorption coefficients of rotational lines of OCS in the ground vibrational state.

with very small dipole moments, for lines of rare isotopic species, or for rotational lines of molecules in sparsely populated vibrational states. To find the J value of the strongest $J \rightarrow J+1$ transition we differentiate α_{max} with respect to J in (5.36), set the derivative equal to zero, and solve for J. With the omission of some very small terms, the result is

$$J_{opt} \text{ (for strongest line)} \approx 5.5 \left(\frac{T\,(\mathrm{K}^\circ)}{B\,(\mathrm{GHz})} \right)^{1/2} \tag{5.41}$$

Having obtained J_{opt}, one can readily find the optimum frequency region from the relation

$$\nu \text{ (for strongest line)} = 2B(J_{opt}+1) \tag{5.42}$$

It is apparent from examination of Table 5.4 and the similar one for symmetric-top molecules in Chapter VI that the optimum region for observation of most molecular rotational spectra is already accessible to the high-resolution, high-precision methods of microwave spectroscopy. Only the very lightest linear polyatomic molecules, HCN and HCP, have their greatest absorption coefficient in the infrared region, beyond the range of microwave spectrometers, and by observing these molecules at reduced temperatures one can shift their region of optimum absorption to the edge of the microwave region. Although one can obtain better resolution of nuclear hyperfine structure on low J transitions, which usually fall in the lower frequency microwave region, the optimum region for observation of microwave rotational spectra is that from about 5 to 0.5 mm wavelength.

Table 5.4 *J* Transitions and Frequencies for Regions of Greatest Absorption Line Strengths of Linear Molecules at $T=300°$K

Molecule	*B (GHz)*	$J \rightarrow J+1$	ν_{opt} *(GHz)*	λ_{opt}*(mm)*
HCN	44.31	14→15	1330	0.23
HCP	19.97	21→22	877	0.34
NNO	12.56	27→28	703	0.43
HCNO	11.47	28→29	665	0.45
FCN	10.55	29→30	634	0.47
HCCF	9.70	30→31	600	0.50
OCS	6.08	39→40	496	0.60
^{35}ClCN	5.97	39→40	478	0.63
HCC^{35}Cl	5.68	40→41	465	0.65
HCCCN	4.55	44→45	400	0.75
^{79}BrCN	4.12	47→48	394	0.76
OC^{78}Se	4.04	47→48	388	0.77
ICN	3.22	53→54	348	0.86
SCTe	1.56	76→77	240	1.25

3 MEASUREMENTS WITH SPECIAL HIGH-RESOLUTION TECHNIQUES

The Doppler-broadened line width, which increases directly with the transition frequency [see (3.39)], becomes an important factor in limiting the accuracy of transitions measured with conventional spectrometers at high frequencies. Three effective devices for circumventing effects of Doppler broadening are the molecular beam maser, the collimated molecular beam spectrometer, and the Lamb-dip method.

A section of the DCN spectrum in the shorter millimeter wave region, shown in Fig. 1.3, demonstrates the remarkable resolution of the beam maser. In an ordinary spectrometer, the well-resolved hyperfine structure of the deuterium, shown in this figure within a span of 100 kHz, would have been completely masked by Doppler broadening. The values of B_0 and D_0 for $H^{12}C^{14}N$ that are obtained from these beam measurements [22] are listed later, in Table 5.7. The frequencies of the $J=0\leftarrow1$ and $1\leftarrow2$ transitions from which they are derived were measured to a precision of one part in 10^8. The first HCN maser operation was achieved on the $J=0-1$ transition by Marcuse [23], who did not use it for accurate measurement of the molecular constants.

A collimated molecular beam absorption spectrometer was developed [24, 25] in the early years of microwave spectroscopy for reduction of line broadening in the alkali halides at their high vaporization temperatures. Since that time, Dymanus and his associates at Nijmegen [26–28] have developed a Stark-modulation, molecular-beam absorption spectrometer that is operable for gases at ordinary or lowered temperatures. It has resolution comparable to that of the beam maser and offers the advantage of being simpler to operate.

Another effective means of combating the limitations to measurement caused by Doppler broadening is the Lamb-dip method [29–31]. It does not require collimated molecular beams, but it does require sufficient source power to saturate the molecules in the center of the Doppler-broadened line and an absorption cell in which the source power is so reflected as to pass through the absorbing molecules in opposite directions. A relatively simple qualitative explanation for the dip can be given. Because the molecules absorbing at the center of the Doppler band have zero velocity relative to that of the transmitted radiation, they absorb, at the same resonant frequency ν_0, the power traveling in either the forward or the reverse direction. In contrast, molecules having a velocity v relative to the forward direction will have a velocity $-v$ relative to the reverse direction and will thus absorb the forward and the reverse radiation at different frequencies, $\nu_0+\Delta\nu$ and $\nu_0-\Delta\nu$, on opposite sides of the center frequency ν_0. Thus the molecules absorbing at the *center* frequency ν_0 are subjected to twice as much resonant power as are those absorbing at slightly higher or slightly lower frequencies. If the power can be critically adjusted to saturate the molecules absorbing at the center of the band without complete saturation of those absorbing at frequencies on either side, the result is a "hole-burning" at the exact center of the band. The "burned hole," or Lamb dip, can be used for precise measurement of the resonant frequency ν_0. This effect is illustrated in Fig. 5.5 by the Lamb dip in a Doppler-broadened OCS line. Several rotational frequencies of OCS as measured with the method are listed in Table 5.5, together with the rotational constants derived from them. These measurements were made with a klystron-driven harmonic generator, a source

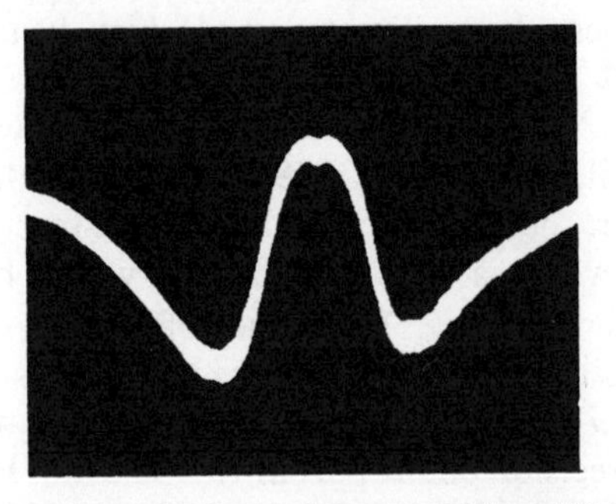

Fig. 5.5 Lamb dip in the $J=15\rightarrow16$ rotational line of OCS at 194,586.4333 MHz. The broad downward curve is the Fabry-Perot cavity resonance; the broad upward curve is the Doppler-broadened line (width about 310 kHz); the small, sharp depression at the center of the line is the Lamb dip (width about 10 kHz). From Winton and Gordy [31].

Table 5.5 Ground-state Rotational Frequencies and Constants of $^{16}O^{12}C^{32}S$ from Measurements of Lamb Dips[a]

	Frequencies in MHz		
J^b	*Measured*	*Calculated*	*Difference*
8	109,463.063	109,463.064	−0.001
9	121,624.638	121,624.638	0.000
10	133,785.900	133,785.900	0.000
11	145,946.821	145,946.818	+0.003
12	158,107.360	158,107.361	−0.001
13	170,267.494	170,267.498	−0.004
14	182,427.198	182,427.198	0.000
15	194,586.433	194,586.433	0.000
16	206,745.161	206,745.161	0.000
19	243,218.040	243,218.042	−0.002
20	255,374.461	255,374.460	+0.001

Molecular Constants in MHz

$B_0=6081.49205\pm0.0002$
$D_J=(1.3008\pm0.0006)\times10^{-3}$
$H_J=(-0.85\pm0.8)\times10^{-9}$

[a]From Winton and Gordy [31].
[b]Rotational quantum number for lower level.

of relatively low power; the saturation power was achieved through amplification with a high-Q, Fabry-Perot cell.

Cazzoli et al. [32] have used Lamb-dip spectroscopy to measure the first two rotational transitions of $D^{12}C^{15}N$. The Lamb dips of the D hyperfine structure are well resolved. The value they obtained for $(eQq)_D$, 0.207 MHz, compares favorably with 0.194 MHz, that from the beam maser experiments [22].

Microwave–microwave double resonance is becoming an increasingly important method, not only for ultra-high resolution, but also for special types of measurements, particularly for study of line shapes and molecular collision processes. Notable studies by Oka [33] and by Lees [34] have been made on the linear molecules HCN and OCS. These studies, the experimental techniques, and their potentialities are described in an excellent review by Baker [35].

4 OBSERVATIONS ON LINEAR TRIATOMIC MOLECULES

Alkali Hydroxides

Through measurements of rotational transitions of KOH and CsOH vapors in a high-temperature microwave spectrometer, Kuczkowski et al. [36] gained the first evidence of the linear configuration of these molecules. This preliminary finding was confirmed by more complete measurements on RbOH and CsOH by Lide and his associates [37, 38], who obtained, with isotopic substitution, accurate values for the internuclear distances. The ground-state rotational constants and structures are given in Table 5.6. The dipole moments of these

Table 5.6 Molecular Constants of Alkali Hydroxides

Hydroxide			r_0 *and* r_e (Å)	
MOH	B_0 (*MHz*)	D_0 (*kHz*)	MO	OH
NaOH	12,567.054(10)[a]	28.72(5)[b]	r_0 1.95[b]	(0.96)[b]
^{39}KOH	8,208.679(10)[c]	12.19(6)[c]	r_0 2.212[c]	0.912[c]
^{39}KOD	7,494.827(10)[c]	9.46(6)[c]	r_e 2.200[d]	0.968[d]
^{85}RbOH	6,290.10[e]		r_0 2.316[e]	0.913[e]
^{85}RbOD	5,720.77[e]		r_e 2.301(2)[f]	0.957(10)[f]
CsOH	5,501.08[g]		r_0 2.403[g]	0.920[g]
CsOD	4,996.83[g]		r_e 2.391(2)[f]	0.960(10)[f]

[a]Pearson and Trueblood [39].
[b]Kuijpers et al. [42].
[c]Pearson and Trueblood [40].
[d]Pearson et al. [43].
[e]Matsumura and Lide [38].
[f]Lide and Matsumura [41].
[g]Lide and Kuczkowski [37].

structures are very high, 7.1 D for CsOH for example, indicating a completely ionic structure of the form Cs^+O^-H, as would be expected for the electronegativity differences of Cs and OH. Pearson and Trueblood [39, 40] made similarly thorough measurements on NaOH, KOH, and KOD. The accurate values they obtained for the ground-state constants are listed in Table 5.6. Other measurements on alkali hydroxides in several vibrational states have led to accurate values for the equilibrium structures and rotation–vibration interaction constants [41–43]. The r_e values are listed with the r_0 values in Table 5.6 for all except NaOH. Generally, the r_e distances are found to be shorter than r_0 for MO and longer than that for OH. A preliminary study with molecular-beam electric resonance [44] has been made on LiOH, but details apparently are not yet published. Its structure may not be linear.

Boron Compounds

The millimeter wave spectra of several isotopic species of the unstable thioborine molecule HBS have been measured by Pearson and McCormick [45], who proved that the molecule has a linear structure. They obtained the accurate substitutional structure, $r_s(\text{HB}) = 1.1692$ Å and $r_s(\text{SB}) = 1.5994$ Å. They produced the HBS by causing the H_2S (or D_2S) to flow over solid boron at 1100°C in a "free-space" hot cell. Afterward, microwave spectra were measured for four isotopic species of ClBS by Kirby and Kroto [46], who showed that it also has a linear structure, with $r_s(\text{ClB}) = 1.681$ Å and $r_s(\text{BS}) = 1.606$ Å.

Cyanides and Phosphides

Table 5.7 displays the bond lengths and ground-state rotational constants for the most abundant isotopic species of some related cyanides and phosphides. Since the first observation [47] in 1948, hydrogen cyanide has been observed many times with millimeter wave spectroscopy; its ground-state rotational constants for the various isotopic combinations [48], including the tritium species [49], are accurately known. Those for its most abundant species, listed in Table 5.7, were measured precisely with a molecular beam maser. It was one of the earlier molecules to be detected with microwave spectroscopy in interstellar space [50], and it has proved to be very useful in various types of interstellar observations. It is responsible for the strong, much used, submillimeter maser source [51, 52] at 0.33 mm.

It is interesting to compare the lengths of the same bonds in the cyanides and phosphides of Table 5.7. The CF bond is significantly shorter in FCN than in FCP, primarily because of the greater conjugation of the single and triple bonds formed with the first-row elements. This conjugation, which may be represented as contribution of valence structures of the form $F^+{=}C{=}N^-$, has much more influence in shortening the single bond than in lengthening the triple bond. The CN length of FCN is only slightly longer than that of HCN. Note that the lengths of the CN bonds in all the cyanides listed are quite close, as are the CP lengths also. There is, of course, considerable conjugation of the bonds in FCP, although less than in FCN. With no such conjugation, the FC

Table 5.7 Rotational Constants and Bond Lengths of Some Linear Cyanides and Phosphides for the Ground Vibrational State and the Most Abundant Isotopic Species

Bond Lengths (Å)	B_0 (*MHz*)	D_0 (*kHz*)	*Ref.*
$H\overset{1.063}{—}C\overset{1.155}{\equiv}N$	44,315.9757(4)	87.24(6)	*a*
$H\overset{0.987}{—}N\overset{1.171}{\equiv}C$	45,332.005(40)	101.9(50)	*b*
$H\overset{1.067}{—}C\overset{1.542}{\equiv}P$	19,976.005(9)	21.23(10)	*c,d*
$F\overset{1.262}{—}C\overset{1.159}{\equiv}N$	10,554.20(2)	5.3(5)	*e*
$F\overset{1.285}{—}C\overset{1.541}{\equiv}P$	5,257.80(3)	1.0	*f*
$N\overset{1.159}{\equiv}C\overset{1.378}{—}C\overset{1.544}{\equiv}P$	2,704.4803(19)	0.216(13)	*g*

[a]De Lucia and Gordy [22].
[b]Creswell et al. [53]; also, Pearson et al. [48].
[c]J. W. C. Johns, J. M. R. Stone, and G. Winnewisser, *J. Mol. Spectrosc.*, **38**, 437 (1971).
[d]J. K. Tyler, *J. Chem. Phys.*, **40**, 1170 (1964).
[e]J. K. Tyler and J. Sheridan, *Trans. Faraday Soc.*, **59**, 2661 (1963).
[f]Kroto et al. [9].
[g]T. A. Cooper, H. W. Kroto, J. F. Nixon, and O. Ohashi, *J. Chem. Soc., Chem. Commun.*, No. 8, 333 (1980).

bond would be much longer than 1.285 Å; the CF single-bond length in CH_3F is 1.38 Å. Without reduction by contributions from structures of the form $F^+{=}C{=}P^-$, the dipole moment of FCP would be expected to be appreciably larger than the observed value [9] of 0.279 D.

Hydrogen Isocyanide HNC

Not until 1976 was the microwave rotational spectrum of HNC, hydrogen isocyanide observed in the laboratory [53–55], and that only after this elusive species had been discovered six years earlier in interstellar space through observation of its $J=0\rightarrow1$ transition [56, 57] with a rest frequency of 90,663.9 MHz and tentative identification of the line by use of a theoretically calculated structure [58]. Although the series CH_3NC, C_2H_5NC, and so on, are commonly known chemicals studied in the gas phase by microwave spectroscopy, the first evidence for the existence of HNC was the infrared spectral observation by Milligan and Jacox [59] made in 1963 on photolized methylazide in an argon matrix at 4°K. Creswell et al. [53] measured the $J=0\rightarrow1$ and $1\rightarrow2$ rotational transitions of HNC, DNC, and $HN^{13}C$; they obtained precise values of the ground-state rotational constants and structural parameters. The $J=0\rightarrow1$ line was measured at 90,663.602 MHz, thus providing experimental confirmation of the correct assignment of 90,663.9(5) MHz as the rest frequency of HNC in the earlier radio-astronomical observation. Pearson et al. [48] extended the laboratory observations to the $J=2\rightarrow3$ transitions and included species containing the ^{13}C isotope. The constants of HNC are recorded in Table 5.7.

5 LONG LINEAR MOLECULES ON EARTH AND IN OUTER SPACE

Much interest in long linear molecules has been generated by the discovery with microwave spectroscopy of carbon-chain cyanides in interstellar space. The ground-state rotational constants and bond lengths for these cyanides, as derived from spectral measurements in the laboratory, are listed in Table 5.8. The constants of HCN are repeated for comparison.

All molecules listed in Table 5.8 have been observed in interstellar space. The longest of them, HC_9N and $HC_{11}N$, have been observed only in interstellar space. Detection of their microwave rotational lines and their assignment were achieved by means of a theoretical model based on the projections of the measured constants of the preceding members of the HC_nN series recorded in Table 5.8. Because of the orderliness of this sequence, the projections proved reliable and provided a close fitting of the interstellar lines observed. The lines of HC_9N and $HC_{11}N$ from outer space are very weak. To this time, it is the longest molecule to be observed in outer space. Whether either of these mole-

Table 5.8 Rotational Constants and Bond Lengths of Some Linear HC_nN Molecules for their Ground Vibrational States and Most Abundant Isotopic Species

HC_nN	B_0 (*MHz*)	D_0 (*kHz*)	*Ref.*
HCN	44,315.9757(4)	87.24(6)	*a*
HC_3N	4,549.0579(4)	0.54311(45)	*b*
HC_5N	1,331.332714(46)	0.0301016(58)	*c,d*
HC_7N	564.00074(16)	0.003821(87)	*e*
HC_9N	(Detected only in interstellar space)		*f*
$HC_{11}N$	(Detected only in interstellar space)		*g*

Bond Lengths in Å

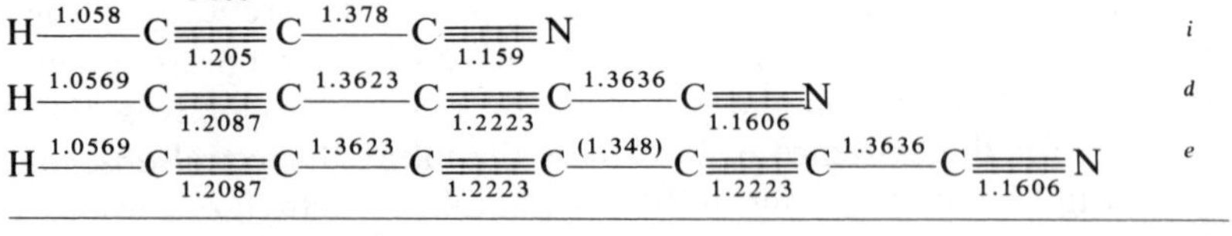

Bond lengths	Ref.
$H\overset{1.0631}{-}C\underset{1.155}{\equiv}N$	*h*
$H\overset{1.058}{-}C\underset{1.205}{\equiv}C\overset{1.378}{-}C\underset{1.159}{\equiv}N$	*i*
$H\overset{1.0569}{-}C\underset{1.2087}{\equiv}C\overset{1.3623}{-}C\underset{1.2223}{\equiv}C\overset{1.3636}{-}C\underset{1.1606}{\equiv}N$	*d*
$H\overset{1.0569}{-}C\underset{1.2087}{\equiv}C\overset{1.3623}{-}C\underset{1.2223}{\equiv}C\overset{(1.348)}{-}C\underset{1.2223}{\equiv}C\overset{1.3636}{-}C\underset{1.1606}{\equiv}N$	*e*

[a]De Lucia and Gordy [22].
[b]R. A. Creswell, G. Winnewisser, and M. C. L. Gerry, *J. Mol. Spectrosc.*, **65**, 4201 (1977).
[c]M. Winnewisser, *J. Chem. Soc., Faraday Discussions* **71**, 1 (1981).
[d]Alexander et al. [63].
[e]Kirby et al. [64].
[f]Broton et al. [60].
[g]M. B. Bell, P. A. Feldman, S. Kwok, and H. E. Matthews, *Nature*, **295**, 399 (1982).
[h]G. Winnewisser, A. H. Maki, and D. R. Johnson, *J. Mol. Spectrosc.*, **39**, 149 (1971).
[i]J. K. Tyler and J. Sheridan, *Trans. Faraday Society*, **59**, 266 (1963).

cules is sufficiently stable under achievable earth conditions to be observed in an absorption cell in a laboratory seems problematical.

Broten et al. [60], who detected the HC_9N, speculate on the possibility of detection in interstellar space of still longer members of the chain, to $HC_{13}N$, and conclude that this would be extremely difficult with the equipment presently available. The highly significant discoveries of the long molecules, HC_5N, HC_7N, and HC_9N, in outer space [60–62] have resulted from a fortunate collaboration between Oka et al., of the Herzberg Institute in Ottawa, and Kroto et al., at the University of Sussex. The laboratory synthesis and measurement of HC_5N and HC_7N by the Kroto group greatly aided the astronomical observations. The shorter members of this series, HCN and HC_3N, were among the first molecules to be discovered in interstellar space [50, 65]. Their discovery was aided by the frequencies that were accurately known from previous measurements in the laboratory.

Although rotational transitions in highly excited bending vibrational states of the long, linear molecules HC_nN are observed in the laboratory, only transitions of the ground vibrational state were detected for molecules in the cold interstellar spaces. Like most other lines of interstellar molecules, those of the long-chain cyanides were observed through maser emission. The mechanisms of the required population inversion of the rotational states are not certain, but collisional interactions and radiative transfer are possible causes [66]. The mechanisms of formation of such long-chain molecules in outer space and their role in the history of the evolving universe are challenging problems.

The microwave rotational spectrum of the rather long molecule O=C=C=C=S (tricarbon oxide sulfide) has been measured for the first time by Winnewisser and Christiansen [67]. Their measurements prove that its structure is strictly linear, in contrast to the similar molecule OCCCO, which is quasilinear (see Section 7). Rotational constants for the ground vibrational state and most abundant isotopic combination are $B_0 = 1{,}413.898$ MHz and $D_0 = 0.046$ kHz. A later study [68] of the molecule in excited bending states yielded the equilibrium constants: $B_e = 1{,}407.230$ MHz and $D_e = 0.0347$ kHz. Values for q, γ, and x_{ll} were also obtained. The dipole moment for the ground state was found to be 0.662 D. Knowledge of these constants should be helpful in the search for this long molecule in interstellar space.

6 EQUILIBRIUM STRUCTURES

An important objective for measurement of the rotational structure of excited vibrational states is to obtain the equilibrium structures of the molecule. Descriptions of the various types of structures of polyatomic molecules—equilibrium, r_e, ground state, r_0, substitutional, r_s, and so on—and methods for their calculation from the spectral measurements are described in Chapter XIII. Although it is not particularly difficult to obtain equilibrium structures for diatomic molecules, the measurements required for even the simplest polyatomic molecules are an order of magnitude greater. Nevertheless, many accurate equilibrium structures have now been obtained for linear triatomic molecules, most of them since the earlier edition of this book in 1970. The progress is accelerating partly because of the improved sensitivity and the increased frequency range of microwave spectrometers and partly because of new techniques for excitation of the higher vibrational states.

Since there are two independent structural parameters of nonsymmetrical linear triatomic molecules, measurements on two isotopic species are required for determination of their structures. For calculation of r_e structures, the B_e value for each of these species is required. To obtain each B_e value, one must measure rotational transitions in excited levels of each of the three fundamental vibrational modes and over a sufficient frequency range for evaluation of the centrifugal distortions. Theoretically, only the lowest excited levels must be measured, but for greater accuracy, especially where there are Fermi resonances or other complications, it is desirable to make more extended measurements. Also, it is desirable for a consistency test to include more than one isotopic pair.

The minimum of data for giving accurate B_e values of the two isotopic species necessary for evaluation of the equilibrium structures of a linear triatomic molecule is illustrated by the constants of ICN in Table 5.9. Although accurate measurements on excited v_1 and v_2 states of ICN were measured in 1972 by Simpson et al. [69], the necessary measurements on an excited level of v_3, which corresponds to the CN vibration, were accomplished much later, in 1978, by Cazzoli et al. [70], who excited the v_3 states by energy transfer from activated N_2.

Equilibrium structures for ICN and a few other linear triatomic molecules are given in Table 5.10 as an illustration of the accuracy obtainable. Even though considerable microwave data were available, some data from infrared rotation–vibration spectra were included in the solutions for HCN and OCS. The degree of consistency of the derivations for different isotopic combinations of OCSe is illustrated in Table 5.11.

Table 5.9 Equilibrium Constants for ICN[a]

Constants	$I^{12}C^{14}N$	$I^{13}C^{14}N$
B_e (MHz)	3229.159(6)	3180.674(6)
α_1	11.895(4)[b]	11.634(5)
α_2	−9.497(2)[b]	−8.861(3)
α_3	14.332(7)	13.353(5)
D_e (kHz)	0.587(27)	0.570(9)
β_1	0.022[b]	0.028
β_2	0.011[b]	0.008
β_3	−0.040	−0.021

[a]Cazzoli et al. [70].
[b]Simpson et al. [69].

Table 5.10 Examples of Equilibrium Structures of Linear Triatomic Molecules

Molecule XYZ	Equilibrium Distance $r_e(XY)$	Equilibrium Distance $r_e(YZ)$	Ref.
HCN	1.06549(24)	1.15321(5)	a
ICN	1.99209(22)	1.16044(33)	b
OCS	1.15446	1.56295	c
OCSe	1.1535(1)	1.7098(1)	d

[a] G. Winnewisser, A. G. Maki, and R. D. Johnson, *J. Mol. Spectrosc.*, **39**, 149 (1971).
[b] *Cazzoli et al.* [70].
[c] Y. Morino and T. Nakagawa, *J. Mol. Spectrosc.*, **26**, 496 (1968).
[d] A. G. Maki, R. L. Sams, and R. Pearson, Jr., *J. Mol. Spectrosc.*, **64**, 452 (1977).

Table 5.11 Equilibrium Structures for Carbonyl Selenide as Derived from Various Isotopic Pairs[a]

Isotopic Pairs	r_e(C—O) (Å)	r_e(C—Se) (Å)
$r_e(^{18}O^{12}C^{80}Se—^{16}O^{13}C^{80}Se)$[b]	1.15347(4)	1.70978(3)
$r_e(^{18}O^{12}C^{80}Se—^{16}O^{12}C^{80}Se)$[b]	1.15333(21)	1.70989(16)
$r_e(^{16}O^{12}C^{80}Se—^{16}O^{13}C^{80}Se)$[b]	1.15372(36)	1.70960(26)
$r_e(^{18}O^{12}C^{80}Se—^{16}O^{13}C^{80}Se)$[c]	1.15341(4)	1.70971(3)
$r_e(^{18}O^{12}C^{80}Se—^{16}O^{12}C^{80}Se)$[c]	1.15356(23)	1.70959(18)
$r_e(^{16}O^{12}C^{80}Se—^{16}O^{13}C^{80}Se)$[c]	1.15313(40)	1.70991(29)

[a] From A. G. Maki, R. L. Sams, and R. Pearson, *J. Mol. Spectrosc.*, **64**, 452 (1977).
[b] Excited states used: $(10^0 0, 01^1 0, 00^0 1)$.
[c] Excited states used: $(10^0 0, 02^0 0, 00^0 1)$.

Equilibrium structures have now been obtained for most of the alkali hydroxides. Their r_e values are given in Table 5.6, where they are compared with the ground state structures, r_0.

7 QUASI-LINEAR MOLECULES

The quasi-linear characterization of certain molecules apparently was made first by Thorson and Nakagawa [71] in 1960. They based their criteria for quasi linearity on a theoretical model of a two-dimensional harmonic oscillator with a superimposed barrier at the minimum of the harmonic potential. Specifically, the model was designed to simulate deviation in the behavior of a triatomic linear molecule when barriers of different heights are superimposed at the minimum of the bending potential. A computer program was made for calculation of the energy levels, eigenfunctions, dipole transition moments, and other significant properties as a function of the imposed barrier heights. Not surprisingly, Thorson and Nakagawa found that in the high barrier limit the calculated properties are those of a nonlinear, or bent, triatomic molecule. However, the calculations for the low-to-moderate barriers and those for anharmonic vibrations in a well with no barrier, revealed properties that could not be assigned either to bent or to strictly linear molecules. They concluded that certain molecules cannot be correctly classified as either bent or linear and are best classified as quasi-linear molecules. Their theory has been applied by Johns [72] to HCN in an excited bending state and to certain bent triatomic molecules such as H_2O in their lowest bending state. He found that the effects of quasi linearity give rise to large anharmonicities in the bending vibrations.

Yamada and M. Winnewisser [73] have proposed a relatively simple test for quasi linearity which is based on the degree of separation of the coordinates of the rotational and vibrational motions in the Born-Oppenheimer approximation [74]. The degree of separation required by the approximation is based on the ratio of the differences of the rotational energy levels ΔE_r and those of the vibrational levels ΔE_v and is approximately

$$\frac{\Delta E_r}{\Delta E_v} \sim \sqrt{\frac{m_e}{m_p}} \approx 10^{-2} \tag{5.43}$$

where m_e and m_p signify electron and proton masses. Yamada and Winnewisser label the energy ratio as

$$\gamma = \frac{\Delta E_r}{\Delta E_v} \tag{5.44}$$

and show that the parameter γ_0 as defined by

$$\gamma_0 = 1 - 4\gamma \tag{5.45}$$

is a measure of quasi linearity when the ΔE_r and ΔE_v are properly chosen.

The proposed relationship is suggested by consideration of the limiting cases of the ideal linear molecule and the ideal bent molecule. In an ideal linear molecule the coordinates of one of the three rotational degrees of freedom is not separable from those of the bending vibrations, with the result that a completely linear molecule has only two rotational degrees of freedom. In contrast, an ideal bent molecule has three rotational degrees of freedom; that is, the rotational coordinates are completely separable from the coordinates for vibration. The parameter γ_0, defined by (5.45), gives a measure of the allowed departure of "well-behaved" molecules from these ideal cases.

An ideal bent molecule has $A \ll \nu_{\text{bend}}$ and $\gamma \approx 0$; consequently, from (5.46), $\gamma_0 \approx 1$. An ideal linear molecule, which conforms to the harmonic approximation of a two-dimensional isotropic oscillator, has $E(02^00) = 2E(01^10)$; thus from (5.47) $\gamma = \frac{1}{2}$. Likewise, (5.46) gives $\gamma = \frac{1}{2}$ for this case. With this γ value, (5.45) gives $\gamma_0 = -1$ for the idealized linear molecule.

Upon the assumption that a well-behaved bent molecule conforms to the Born-Oppenheimer approximations, a deviation of 0.01 from $\gamma=0$ is allowed by (5.43). If a comparable deviation of 1% is allowed in the value $\gamma=\frac{1}{2}$ for linear molecules, the well-behaved linear molecule could have $\gamma=0.495$ to 0.505, and hence $\gamma_0=-1.02$ to 0.98. With these considerations, linear and bent molecules are classified by Yamada and Winnewisser [73] in terms of the γ_0 parameter, as shown in Table 5.12.

Yamada and Winnewisser listed numerous molecules on an ordered scale according to their γ_0 values. Most of those listed fall within, or close to, the limits of the well-behaved linear or bent molecules. Certain ones, however, fall far from the limits of either. Those falling outside the limits of the well-behaved linear and well-behaved bent molecules are classified as quasi-linear. Figure 5.6 shows an abbreviated version of this scale as given by Winnewisser and Christiansen [67]. This plot includes the newly observed C_3OS (see Section 5) which, according to its position on the scale is a well-behaved linear molecule. Note, however, that the closely related C_3O_2 molecule is definitely quasi-linear, as is the interesting HCNO molecule discussed in Section 1.

For evaluation of the bent triatomic molecule, Yamada and Winnewisser [73] chose the fundamental bending frequency ν_{bend} for ΔE_v and the rotational constant A for ΔE_r and set

$$\gamma=\frac{A}{\nu_{\text{bend}}} \tag{5.46}$$

Conventionally, the spectral constant A is a measure of the energy of rotation about the axis of least moment of inertia, or the a axis. In slightly, or moderately, bent molecules the a axis would correspond to the figure axis in the limiting case of a completely linear molecule. Thus, in tests for quasi-linearity, the constant A should be usable for ΔE_r in all cases, as indicated by Yamada and Winnewisser.

For a linear triatomic molecule the level $E(0\,1^1\,0)$ represents the energy of the lowest excited, degenerate bending state for which the energy is rotational in form, and $E(02^00)$ represents the energy of the lowest excited bending state for which the energy is vibrational in form. Both levels are measured relative to that of the ground state, $E(00^00)$. Therefore, Yamada and Winnewisser set

Table 5.12 Limiting γ_0 Values for Well-Behaved Linear Molecules[a]

Molecule	*Ideal Molecule*	*Well-Behaved Molecule*
Linear	$\gamma_0=-1$	-1.02 to -0.98
Bent	$\gamma_0=+1$	$+1.00$ to $+0.96$

[a]From Yamada and Winnewisser [73].

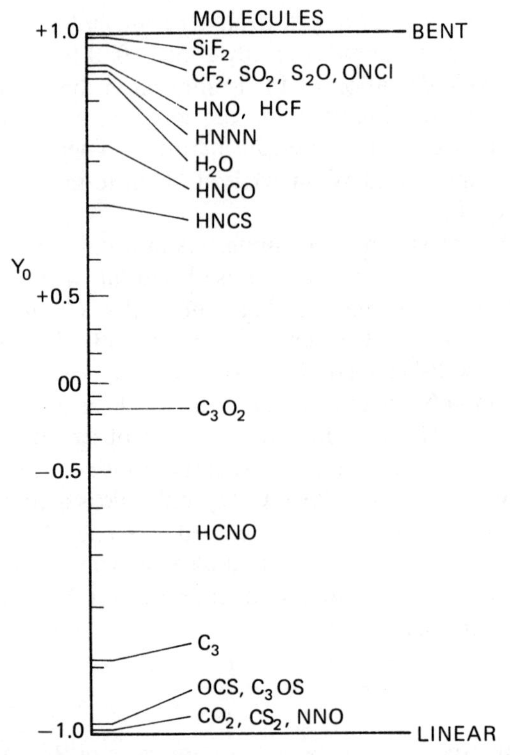

Fig. 5.6 Ordering of some chain form molecules in the sequence of quasi linearity according to $\gamma_0 = 1 - 4\gamma$, where γ is defined by (5.48). The quadratic scale is chosen to expand around +1 and −1. From Winnewisser and Christiansen [67].

for this case

$$\gamma = \frac{E(01^10)}{E(02^00)} \tag{5.47}$$

or, more generally,

$$\gamma = \frac{E(\text{lowest state with } K, \text{ or } l, =1)}{E(\text{lowest excited state with } K, \text{ or } l, =0} \tag{5.48}$$

where E represents the energy of the indicated state above that of the ground state. The molecules HNCO and HNCS are also classified as quasi-linear although they fall nearer to the limits of the bent rather than the linear molecules on the γ_0 scale.

The quasi-linear carbon suboxide molecule O=C=C=C=O has no observed rotational spectrum, but transitions of its vibration–rotation spectrum have been observed in the 545 to 595 GHz (0.55 to 0.50 mm) submillimeter wave region by Burenin et al. [75]. The part of the spectrum observed is the Q-branch, with some R- and P-branch lines of the v_7 bending vibrational transition, $(0000^00^00^0) \rightarrow (0000^00^01^1)$. Except for the inversion-type transitions of NH_3, this appears to be the first observation of a vibration–rotation spectrum with microwave spectroscopy. Other observations of this type will no doubt follow as further expansion of the submillimeter wave coverage with coherent microwave techniques occurs.

8 HYDROGEN-BONDED DIMERS

Since the early thirties hydrogen-bonded systems have been studied with infrared vibration spectroscopy. Most of these studies were made on liquids or solids, but the vibration bands of a few dimers in the gaseous state were observed. Notable among them are dimers of hydrogen fluoride, hydrogen cyanide, and acetic acid. Structural information has also been obtained about hydrogen-bonded dimers from studies with gas-phase electron diffraction. As is well known, hydrogen bonding has great significance in biological and chemical processes, and the quantitative information about this bonding now being gained from microwave spectroscopy is likely to be of much importance.

The first radiofrequency and microwave investigation of gas-phase, hydrogen-bonded complexes appears to be that on the dimers of hydrogen fluoride, $(HF)_2$, $(DF)_2$, and HFDF, by Dyke et al. [76] with the molecular-beam, electric resonance method. They concluded that these dimers are not linear, that the HF units are 60 to 70° from the FF axis. Although this interesting system is not a suitable subject for elaboration in this chapter, it is relevant to the linear dimers studied with microwave spectroscopy. The dipole moment was found to be 2.987(3) D for $(HF)_2$ and the FF distance, to be 2.79(5) Å. Perhaps the most interesting discovery in this study is a hydrogen-tunneling motion which involves the breaking and reforming of the hydrogen bond in $(HF)_2$ and $(DF)_2$, but not in HFDF. This tunneling was revealed by a splitting of the rotational energy levels of the symmetrical dimers. This MBER study was followed by conventional microwave spectral investigation of the nonlinear dimers $(H_2O_2)_2$, by Dyke and Muenter [77] in 1974, and $H_2O \cdots HF$ by Bevan et al. [78] in 1975.

The first microwave spectral observation of a linear, hydrogen-bonded dimer was made by Legon et al. [79] on $HCN \cdots HF$ in 1976. They measured three ground-state rotational transitions, $J=1\rightarrow 2$, $3\rightarrow 4$, $4\rightarrow 5$, and established the linear structure of the dimer. Later, they carried out more thorough investigations [80] which included dimers of four different isotopic combinations and measurements on several excited vibrational states. The additional vibrational modes resulting from the dimer formation were labeled v_σ, v_β, and v_B, as indicated in Fig. 5.7. Excited states were observed for the parallel vibrational mode v_σ and the lower frequency degenerate bending mode v_β. For each of the four isotopic species they obtained the rotational constants B_0 and D_v, the rotation–vibration constants α_σ and α_β, and the l-type doubling constants q_β. They also derived the fundamental vibrational frequencies, $\nu_\sigma=197\ \text{cm}^{-1}$ and $\nu_\beta=91\ \text{cm}^{-1}$, for the most abundant species. The ν_σ represents the vibrational frequency of one molecular unit against the other and corresponds to

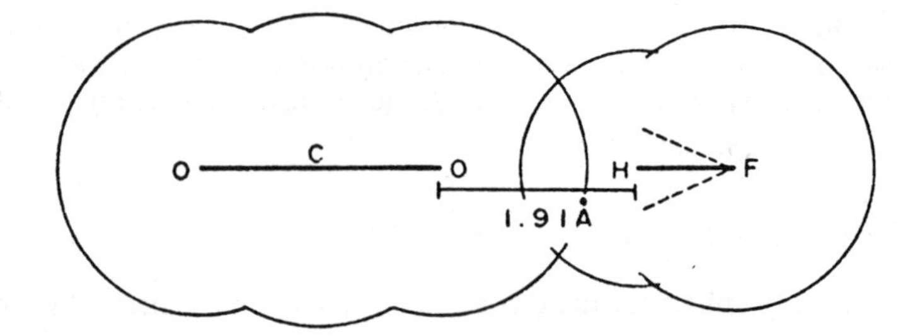

Fig. 5.7 Estimated equilibrium structure of CO_2—HF. The broken lines represent the average HF position. From Baiocchi, Dixon, Joyner, and Klemperer [89].

the hydrogen-bond stretching vibration. The v_β mode, corresponding to hydrogen-bond bending, is doubly degenerate. They also calculated the force constants for the normal vibrations. Values for the other force constants may be found in the original paper [80]. The rotational constants and lengths of the hydrogen bridge, $r_0(\mathrm{N}\cdots\mathrm{F})$ found for the H and D species are given in Table 5.13. It is surprising that no significant difference in length is found for the hydrogen and the deuterium bridges.

From Stark-effect measurements [81] on two rotational transitions of the HCN···HF, an electric dipole moment of 5.612 D was obtained for the dimer. It is of interest that this moment is greater by 0.80 D than the sum of the separate HCN and HF moments, thus indicating that the dipole moment is significantly enhanced by the formation of the hydrogen bond. A comparison of this dipole enhancement with that in other hydrogen-bonded complexes is given in Table 5.14.

From intensity measurements of the rotational lines, Legon et al. [82] derived the dissociation energy of the hydrogen bond for the ground vibrational state and for the equilibrium configuration of the HCN··· HF dimer. The final values obtained [80] are $\mathscr{D}_0=18.9$ and $\mathscr{D}_e=26.1$ in kJ/mole. The method described is applicable for measurement of dissociation energy for other hydrogen-bonded complexes. It gives a measure of the hydrogen-bond strength for the particular complex.

The hydrogen cyanide dimer HCN···HCN has been observed with microwave spectroscopy by Legon et al. [83]. Four rotational transitions, $J=7\rightarrow8$ to $J=10\rightarrow11$, were measured for $(HCN)_2$ and for $(DCN)_2$. The regularity of the spectrum and the changes in spacing caused by the isotopic substitutions verify the structure as linear. However, rather large differences appear between the rotational constants for $(HC^{14}N)_2$ and $(HC^{15}N)_2$ and the corresponding ones measured later by Buxton et al. [84], who concluded that the ground-state constants reported earlier for these species were likely those for an excited vibrational state. The constants reproduced for $(HCN)_2$ in Table 5.13 are the later ones, whereas those for $(DCN)_2$ are from the earlier work. Note that the $r_0(\mathrm{N}\cdots\mathrm{C})$ distance for $(DCN)_2$ is 0.055 Å less than that for

Table 5.13 Rotational Constants and Structural Parameters of Some Linear, Hydrogen-bonded Complexes

Complex	B_0 *(MHz)*	D_J *(kHz)*	$r_0(Y \cdots X)$ (Å)	θ*(deg.)*[a]	*Ref.*
			$r_0(N \cdots Hal)$		
$HCN \cdots HF$	3591.11	5.2	2.796	18.7	b,c
$DCN \cdots DF$	3351.87	5.5	2.792		b
$NCCN \cdots HF$	1195.0616	0.34	2.862	20.1	c
$HCN \cdots H^{79}Br$	1396.5958	1.89	3.610	21.36	d
$NN \cdots HF$	3195.353	17.2	3.082	25.5	e
$NN \cdots H^{35}Cl$	1745.664	8.56	3.701	26.0	f
			$r_0(C \cdots Hal)$		
$OC \cdots HF$	3063.899	9.77	3.068	21.9	g,h
$OC \cdots H^{35}Cl$	1671.723	4.802	3.696	23.0	g,i
$OC \cdots H^{79}Br$	1150.512	2.703	3.917	23.2	g,j
			$r_0(N \cdots C)$		
$HCN \cdots HCN$	1745.8097	2.133	3.287		k
$DCN \cdots DCN$	1698.37		3.232		l
			$r_0(O \cdots Hal)$		
$OCO \cdots HF$	1951.170	10.69	2.84	25.1	m
$OCO \cdots H^{35}Cl$	1109.886	4.83	3.42	25.0	n
$SCO \cdots HF$	1302.462	2.46	2.87	25.0	m

[a] θ is the vibrationally averaged angle of HX with the inertial axis a of the complex, approximately with the $Y \cdots X$ bridge. Some workers designate this angle as γ and use θ to designate the angle between the averaged HX direction and the line between the mass centers of the two components of the complex. In complexes with linear equilibrium structures like those in this table, the two angles are equal, or nearly so.

[b] Legon et al. [80].

[c] Legon et al. [91].

[d] E. J. Campbell, A. C. Legon, and W. H. Flygare, *J. Chem. Phys.*, **78**, 3494 (1983).

[e] P. D. Soper, A. C. Legon, W. G. Read, and W. H. Flygare, *J. Chem. Phys.*, **76**, 292 (1982).

[f] R. S. Altman, M. D. Marshall, and W. Klemperer, *J. Chem. Phys.*, **79**, 57 (1983).

[g] Legon et al. [85].

[h] Legon et al. [88].

[i] Soper et al. [87].

[j] Keenan et al. [86].

[k] Buxton et al. [84].

[l] Legon et al. [83].

[m] Baiocchi et al. [89].

[n] R. S. Altman, M. D. Marshall, and W. Klemperer, *J. Chem. Phys.*, **77**, 4344 (1982).

Table 5.14 Dipole Moment Enhancement due to Hydrogen Bonding

$M_1 \cdots M_2$	$\mu(D)$	$(\mu_1+\mu_2)^a$	$\Delta\mu(D)$
$HCN \cdots HF$	5.612^b	4.812	0.800^b
$OCO \cdots HF$	2.2465^c	1.827	0.420
$OCO \cdots DF$	2.3024^c	1.827	0.475
$SCO \cdots HF$	3.2085^c	2.542	0.667
$NN \cdots H^{35}Cl$	1.244^d	1.109	0.14
$OCO \cdots H^{35}Cl$	1.451^e	1.109	0.34

[a] Sum of moments of separated monomers.
[b] Legon et al. [80].
[c] Baiocchi et al. [89].
[d] R. S. Altman, M. D. Marshall, and W. Klemperer, *J. Chem. Phys.*, **79**, 57 (1983).
[e] R. S. Altman, M. D. Marshall, and W. Klemperer, *J. Chem. Phys.*, **77**, 4344 (1982).

$(HCN)_2$. This is consistent with the difference recorded for $OC \cdots DF$ and $OC \cdots HF$. If real, these differences indicate a slightly stronger bonding by D than by H, resulting from the lower zero-point vibrational energy of D.

Weakly bonded dimers comprised of carbon monoxide hydrogen-bonded to the hydrogen halides, HX (X=F, Cl, or Br), have been discovered with microwave spectroscopy by Legon et al. [85–88]. Detectable dimers were produced by pulsed injection of mixtures of CO (4%), the halide (4%), and argon into a Fabry-Perot absorption cavity of a highly sensitive spectrometer. From its rotational spectrum, each dimer was shown to have a linear, or nearly linear, structure, with the hydrogen bonding to the carbon, $OC \cdots HX$. For the $OC \cdots HCl$ dimer, this structure was verified by measurements on dimers of four different isotopic combinations. Additional evidence for collinearity of the nuclei at equilibrium was obtained from the observed nuclear quadrupole couplings of the Cl and Br.

The ground-state rotational constants for the $OC \cdots HX$ dimers and the hydrogen bond lengths $r(C \cdots H)$ obtained from them with known monomer structures, assumed unchanged, are listed in Table 5.13. By subtraction of the HX length, one can obtain the $r(C \cdots H)$ distance. With r_e values for the hydrogen halides, this gives $r(C \cdots H) = 2.132$ Å, 2.430 Å, and 2.503 Å for X=F, Cl, and Br, respectively. This suggests a decreasing strength of the bonding from F to Br. It is of interest that $r(C \cdots D) = 2.410$ Å for the $OC \cdots DCl$ complex, 0.01 Å less than that for the $OC \cdots HCl$ complex.

The later, more complete, study of $OC \cdots HBr$ includes four isotopic combinations [86]. Although the vibrationally averaged structure is that of a prolate symmetric top with the averaged Br—H bond axis 23° off the linear $O—C \cdots Br$ axis, the data, like those for the other $OC \cdots HX$ complexes in Table 5.13, are consistent with a linear equilibrium structure.

Hydrogen-bonded complexes formed by CO_2 and by OCS with the hydrogen fluoride have been studied by Klemperer's group [89] with microwave and radiofrequency spectroscopy. Their rotational spectra show that these molecules have linear equilibrium structure with the chemical form OCO ··· HF. The geometrical structure is like that indicated for OCO ··· HF in Fig. 5.8. Although the equilibrium structure is linear, the zero-point vibrational energy causes the vibrationally averaged HF angle to deviate from the molecular axis by approximately 25°, as indicated by the dotted lines in Fig. 5.8. This angular deviation, as for the other H-bonded structures described in this section, is revealed and measured by the nuclear hyperfine structure. This rotational spectrum is like that of a linear molecule or of a symmetric-top rotor with $K=0$. Rotational constants B_0 and D_0 obtained from the measurements are given in Table 5.13. The O ··· H or O ··· F distances are derived from B_0 with the assumption that the structures of the component molecules are the same as those for the separated monomers. The O ··· H distance for SCO ··· HF is 1.94 Å, somewhat larger than the 1.91 Å for OCO ··· HF. Distances of O ··· F are given in Table 5.13.

It is surprising that the NNO ··· HF complex, which is isoelectric with the linear OCO ··· HF, was found *not* to have a linear structure [90]. Instead, the NNO axis makes an angle of about 47° with the center-of-mass line. The complexes Ar ··· CO_2 and Ar ··· N_2O have almost identical T-shaped structures [90].

The rotational spectrum of the linear hydrogen-bonded dimer NCCC ··· HF has been observed by Legon et al. [91]. Their measurements included three isotopic species with ^{15}N substituted in different positions. Rotational constants obtained for the most abundant isotopic species are reproduced in Table 5.13 with the length of the N ··· F hydrogen bridge derived from B_0. No significant difference from this length was observed for the ^{15}N-substituted species. The averaged angle, θ in Table 5.13, derived from the nuclear magnetic complex of the HF, is 5° less than that for OCO ··· HF.

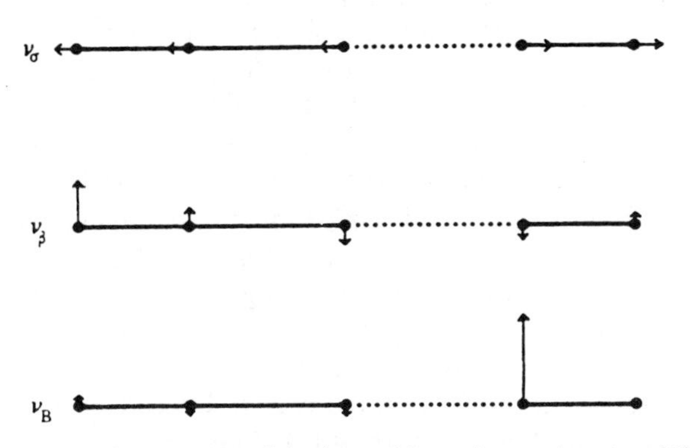

Fig. 5.8 Diagrammatic representation of the three additional normal modes of HCN ··· HF that result on dimer formation. The low and high frequency bending modes ν_β and ν_B are doubly degenerate. From Legon, Millen, and Rogers [80].

The angle θ, listed for a number of dimers in Table 5.13, is the vibrationally averaged angle of the HX bond with the principal inertial axis a of the dimer. The high θ values result from the large amplitudes of the perpendicular vibrations of the light, weakly constrained hydrogen, even in the ground vibrational state. Since the amplitudes of zero-point vibrations of the heavier atoms are small as compared with those for H or D, these atoms remain near their equilibrium positions in the ground vibrational state and are thus approximately along the inertial axis a, as indicated in Fig. 5.8. The perpendicular vibrations of the H are damped by attraction to the atom X, which tends to align the H with the heavier atoms. The angle θ thus provides one measure of the strength of the hydrogen bond.

The rotational spectra for the ground state do not directly provide a measure of θ since the rotational transitions about the symmetry axis, like those for the symmetric top, are not detectable. Rather accurate values of θ can, however, be derived from hyperfine structure of the rotational spectra, as were the values of θ listed in Table 5.13. The hyperfine structure may be due to nuclear magnetic interactions, as in HF, or to nuclear quadrupole interactions, as in HCl. It may also be due to spin–spin interactions, as between H and F, but other magnetic interactions in HF are stronger. All these interactions are symmetric about the molecular axis of a diatomic molecule. Any coupling tensor that has axial symmetry about a bond axis can be transformed to a principal rotational axis, as explained in Chapter IX, Section 4. For a nuclear quadrupole interaction [see (9.112)], this may be explained as

$$\chi_a = \tfrac{1}{2}\chi_z(3\langle\cos^2\theta_{z,a}\rangle - 1) \tag{5.49}$$

where χ_z is the coupling value along the H X bond axis and $\theta_{z,a}$ is the instantaneous value of the angle between the HX bond axis z and the principal inertial axis a. The average is taken over all the vibrational motions. The ratio of the observed coupling for the dimer to that observed for the free HX monomer, χ_a/χ_z, gives, with (5.49), the averaged value $\langle\cos^2\theta_{z,a}\rangle$ and hence the vibrationally averaged value for θ. The magnetic coupling constant C_z of H or of X measured in the free HX monomer has the same formal relationship as that of (5.49), and the ratio of C_a/C_z can in a similar manner be used to determine the value for the vibrationally averaged θ. Accurate values of the magnetic coupling constants of the HF monomer are given by Muenter [92]. Quadrupole coupling constants for the hydrogen halides are given in Chapter XIV.

In Table 5.15 we have recorded for several dimers fundamental hydrogen-bond stretching frequencies ν_s and related force constants k_s corresponding to the parallel mode v_σ indicated at the top of Fig. 5.7. The stretching force constants k_s or k_σ depend directly on the strength of the hydrogen bonds and provide a comparison of the relative energies of these bonds for the molecules in Table 5.15. The frequencies were not directly measured because they fall outside the microwave range, in the far-infrared region. Instead, they were calculated from the centrifugal stretching constants listed in Table 5.13 by simple approximate methods.

Table 5.15 Vibrational Frequencies and Hydrogen-bond Force Constants

Complex	Frequency (cm^{-1}) $v_s(v_\sigma)$	Force Constant (mdyne/Å) $k_s(k_\sigma)$	Ref.
HCN ··· HF	197	0.24	a
HCN ··· HCN	120	0.11	b
OC ··· HF	125	0.108	c
OC ··· DF	127	0.118	c
OC ··· $H^{35}Cl$	69	0.0446	d
OC ··· $D^{35}Cl$	72	0.0491	d
OC ··· $H^{79}Br$	52	0.0330	e
OCO ··· HF	51	0.021	f
OCO ··· DF	49	0.020	f
SCO ··· HF	56	0.028	f

[a]Legon et al. [80].
[b]Buxton et al. [84].
[c]Legon et al. [88].
[d]Soper et al. [87].
[e]Keenan et al. [86].
[f]Baiocchi et al. [89].

The v_σ mode corresponds closely to the vibration of one molecular unit of the dimer against the other. For calculation of v_s and k_s one can neglect the distortions of the molecular units and treat the dimer as a diatomic molecule with the masses of the two weakly bonded molecular units concentrated at their respective mass centers. The fundamental frequency $v_s(=v_\sigma)$ is then calculated from (4.16) with the measured B_0 and D_J values in Table 5.13 used for B_e and D_e. Thus,

$$v_s^2 = \frac{4B_0^3}{D_J} \tag{5.50}$$

The force constants k_s are calculated from the harmonic oscillator formula

$$v_s = \frac{1}{2\pi}\sqrt{\frac{k_s}{\mu_{\text{dimer}}}} \tag{5.51}$$

where the reduced mass of the dimer is

$$\mu_{\text{dimer}} = \frac{M_1 M_2}{M_1 + M_2} \tag{5.52}$$

in which M_1 and M_2 are the masses of the two molecules linked by the hydrogen bond.

Values for the fundamental bending frequencies v_b and force constants k_b corresponding to the v_B mode of Fig. 5.7 have been estimated [86, 87, 89] from the vibrationally averaged θ values listed in Table 5.13. The theory used is that described in Section 9, (5.53)–(5.59). However, the simplifying assumptions required are more uncertain than those for the stretching vibration ν_σ. Later, more complete measurements on the rotational structure of excited vibrational states, or direct measurements of the bending frequencies in the infrared region, should provide accurate values for the bending constants.

9 RARE GAS ATOM–MOLECULE COMPLEXES

Rare gas atoms—neon, argon, krypton, and xenon—form weakly bonded van der Waals complexes with a variety of molecules. Microwave and radio-frequency spectra of a number of them have been measured with molecular beam, electric resonance techniques by Klemperer and his group at Harvard [93–100] and with special microwave spectral techniques by Flygare and his group at Illinois [101–106]. Many of the complexes in which the molecular component is diatomic are found to have linear equilibrium structures although the wide-angle bending vibrational motions they undergo, even in the ground state, produce an averaged structure that conforms to that of a slender prolate symmetric top. Similar features are found for the hydrogen-bonded dimers described in Section 8. In fact some of these complexes may be classified as weak, hydrogen-bonded complexes.

These atom–molecule complexes are usually produced by the expansion of a highly pressurized mixture of the constituents when the mixture is passed through a restricting supersonic nozzle into the evacuated cavity of the spectrometer. The molecular constituent is the more dilute, only a few percent of the mixture. The Illinois group use a pulsed microwave Fourier-transform spectrometer and pulse the supersonic nozzle to admit the gas at a frequency synchronized with the pulse of the spectrometer. Descriptions of their highly sensitive system are given in some of their papers [104, 107].

The rapid expansion of the pressurized gas as it escapes into the evacuated cavity cools the molecular complexes to a very low temperature so that thermal motions do not dissociate the very weak van der Waals bonds. Furthermore, at this low temperature, only the lower rotational states are populated. Although the number of molecular complexes in the escaping beam is very small compared to the molecules in a conventional microwave absorption cell, they are detectable with the exceptionally sensitive, pulsed Fourier-transform microwave spectrometer just mentioned [104, 107] and with the comparably sensitive molecular-beam, electric-resonance method. Favoring the detection is the high population of the lowest rotational state.

The first observations of the microwave spectrum of a linear complex of this kind appears to be those on ArHCl by Novick et al. [93, 97], who obtained the rotational constants, nuclear couplings, and electric dipole moments for the ground vibrational states of $ArH^{35}Cl$ and $ArD^{35}Cl$. They found that the complex has the atomic arrangement Ar—HCl with the averaged Ar $\cdots$ Cl distance of 4.006 Å. The large, vibrationally averaged bending angle θ of 45° for ArHCl and 34° for ArDCl shows that the complex is highly flexible.

The vibrationally averaged angle θ, sometimes designated γ, is obtained from the experimentally measured nuclear quadrupole coupling or the nuclear magnetic couplings, as described in Section 8. See (5.49). Rotational constants and derived structural constants r_0 and θ are recorded in Table 5.16 for a number of rare gas atom complexes with diatomic halides. These complexes include argon, krypton, and xenon, but not neon. Evidences have been obtained [99, 108] that similar complexes are formed with Ne but that they are nonpolar, probably because of a very large vibrationally averaged angle θ which leads to restricted internal rotation of the molecular components [99]. The references cited in Table 5.16 are those which give the latest, or most complete, information about these van der Waals complexes. Further information and references to earlier or preliminary work on them may be found in the cited papers.

Bond-stretching and bond-bending frequencies and force constants which provide a comparison of the relative strength of the van der Waals bonds in some

Table 5.16 Rotational and Structural Constants for Some Rare Gas Atom–Molecule Complexes

Complex	B_0 *(MHz)*	D_J *(kHz)*	r_0(X ··· Hal) (Å)	θ*(deg)*	*Ref.*
Ar ··· HF	3065.7099(4)	70.90(6)	3.5445(2)	41	a,b
Ar ··· DF	3039.8529(4)	59.53(6)	3.5349(2)	32.7(2)	a,b
Ar ··· $H^{35}Cl$	1678.511	20.0	4.006	45	c,d
Ar ··· $D^{35}Cl$	1657.627	17.1	4.025	34	c,d
Ar ··· $H^{79}Br$	1106.6695(2)	12.397(4)	4.146	42.1	e
Ar ··· $D^{79}Br$	1087.5089(3)	8.497(6)	4.182	34.4	e
^{84}Kr ··· HF	2392.4152(7)	31.94(10)	3.647	39	f
^{84}Kr ··· DF	2353.1407(7)	27.31(10)	3.640	30.8	f
^{84}Kr ··· $H^{35}Cl$	1200.6255(7)	7.389(16)	4.111	37.8	g
^{84}Kr ··· $D^{35}Cl$	1183.7735(5)	6.530(12)	4.125	30.5	g
^{84}Kr ··· $H^{79}Br$	684.2295(2)	2.597(1)	4.257	38.0	e
^{84}Kr ··· $D^{79}Br$	675.8486(2)	2.190(1)	4.281	31.1	e
^{129}Xe ··· $H^{35}Cl$	994.1448(1)	3.813(2)	4.275	34.5	h
^{129}Xe ··· $D^{35}Cl$	979.4096(1)	3.452(1)	4.287	27.9	h
Ar ··· ^{35}ClF	1327.113(5)	4.66(20)	3.330 (Ar ··· Cl)	8.69	i
^{84}Kr ··· ^{35}ClF	925.186(5)	1.86(20)	3.388 (Ar ··· Cl)	8.11	j

[a] Keenan et al. [106].
[b] Dixon et al. [100].
[c] Novick et al. [97].
[d] Novick et al. [93].
[e] Keenan et al. [103].
[f] Buxton et al. [105].
[g] Balle et al. [102].
[h] Keenan et al. [104].
[i] Harris et al. [95].
[j] Novick et al. [96].

of the complexes are given in Table 5.17. The bond-stretching frequencies are calculated from the bond-stretching constants D_J in Table 5.16 with (5.50). The bond-bending frequencies are calculated from the θ values in Table 5.16 with the assumption of a harmonic bending potential

$$V(\theta)=\tfrac{1}{2}k_b\theta^2 \tag{5.53}$$

which leads to the bending frequency [103]

$$\nu_b=\frac{\hbar}{2\pi\mu_b\langle\theta^2\rangle} \tag{5.54}$$

For the triatomic complexes in Table 5.16, the reduced mass of the bending vibration may be obtained from the general formula for the binding mode ν_2 of a linear XYZ molecule for which [109]

$$\frac{1}{\mu_b}=\frac{1}{m_X l_{XY}^2}+\frac{1}{m_Z l_{YZ}^2}+\frac{1}{m_Y}\left(\frac{1}{l_{XY}}+\frac{1}{l_{HZ}}\right)^2 \tag{5.55}$$

Table 5.17 Derived Vibrational Frequencies and Force Constants for Rare Gas Atom–Molecule Complexes

Complex	ν_s *(cm^{-1})*	k_s *(mdyne/Å)*	ν_b*(cm^{-1})*	k_b *(mdyne/Å)*	*Ref.*
Ar···HF	42.5	0.0142	79.8	0.0031	a
Ar···DF	45.6	0.0168	65.5	0.0039	a
Ar···$H^{35}Cl$	32.4	0.0117		0.0015	b,a
Ar···$D^{35}Cl$	34.4	0.0134		0.0018	b,a
Ar···$H^{79}Br$	22.1	0.0076	30.8	0.00114	c
Ar···$D^{79}Br$	26.0	0.0106	23.4	0.00129	c
^{84}Kr···HF	43.7	0.0187	87.5	0.00378	d
^{84}Kr···DF	43.1	0.0210	74.7	0.00533	d
^{84}Kr···$H^{35}Cl$	32.3	0.0155	52.9	0.00128	e
^{84}Kr···$D^{35}Cl$	33.6	0.0171	39.8	0.00142	e
^{84}Kr···$H^{79}Br$	23.4	0.0132	37.7	0.00170	c
^{84}Kr···$D^{79}Br$	25.0	0.0152	28.7	0.00195	c
^{129}Xe···$H^{35}Cl$	33.9	0.0190	56.8	0.00311	f
^{129}Xe···$D^{35}Cl$	34.8	0.0205	44.3	0.00370	f
Ar···^{35}ClF	47.2	0.0301	41.0	0.0215	g
^{84}Kr···^{35}ClF	43.6	0.0369	48.4	0.0311	g

[a] Dixon et al. [100].
[b] Novick et al. [97].
[c] Keenan et al. [103].
[d] Buxton et al. [105].
[e] Balle et al. [102].
[f] Keenan et al. [104].
[g] Novick et al. [96].

in which m_X, m_Y, and m_Z are the atomic masses of X, Y, and Z, and the l's are the indicated atomic distances. For the hydride or deuteride complexes, $m_Y = m_H$ or m_D, $m_Z = m_{Hal}$, and m_X is the mass of the rare gas atom. In these, $m_Y \ll m_X \cong m_Z$, and approximately,

$$\frac{1}{\mu_b} \cong \frac{1}{m_H}\left(\frac{1}{l_{XH}} + \frac{1}{l_{HZ}}\right)^2 \tag{5.56}$$

The l_{HZ} may be taken as the $r_0(\text{H—Hal})$ of the free hydrogen halide and

$$l_{XH} = r_0(X\cdots\text{Hal}) - r_0(\text{H—Hal}) \tag{5.57}$$

where the $r_0(X\cdots\text{Hal})$ values are those recorded in Table 5.16. Because $r_0(\text{H—Hal}) < l_{XH}$, this expression is sometimes reduced further

$$\mu_b \approx m_H r_{HZ}^2 \tag{5.58}$$

which is simply the moment of inertia of the free diatomic molecule. With the calculated μ_b, the bending frequency can be obtained from (5.54), and the bending force constant, from the harmonic oscillator formula, (5.51),

$$k_b = 4\pi^2 v_b^2 \mu_b \tag{5.59}$$

In Table 5.17 we have recorded vibrational frequencies and force constants derived from experimental D_J and θ values. The constants provide a comparison of the relative bonding forces in these van der Waals complexes. Note that the bond-stretching constants k_s are generally smaller than those for the hydrogen-bonded dimers listed in Table 5.15.

10 LINEAR MOLECULAR IONS: INTERSTELLAR AND LABORATORY

Observation of molecular ions is relatively new in microwave spectroscopy. The first laboratory observation of a microwave rotational transition of an ion was that of CO^+, reported in 1975 by Dixon and Woods [110]. Before that time, only fine-structure transitions in the hydrogen molecule ion had been observed [111], in the low-frequency range of 4 to 1276 MHz. Microwave astronomers have led in the detection of microwave lines of linear polyatomic ions. Rotational constants of all three linear triatomic ions observed to date are listed in Table 5.18. Lines of each of these ions were first detected from interstellar space and were later identified by laboratory measurements.

The most extensive studies have been made on HCO^+, both in the laboratory and in outer space. It was the first molecular ion to be detected with microwave spectroscopy in interstellar space. In 1970, Buhl and Snyder [112] observed from several interstellar sources an unknown line at 89,190 MHz. Because they were unable to assign it to any known species, they tentatively labeled the unknown species X-ogen. Soon after this discovery, Klemperer [113] on theoretical grounds suggested HCO^+ as the source of the unknown line and predicted that its $J = 0 \rightarrow 1$ transition would occur at 89,246 MHz. Later, Herbst

Table 5.18 Rotational Constants of Some Linear Triatomic Ions

Ion	$B_0(MHz)$	$D_0(kHz)$	*Ref.*
$(H^{12}C^{16}O)^+$	44,594.420(2)	82.39(7)	a,b
$(H^{13}C^{16}O)^+$	43,377.32(4)	78(8)	b
$(H^{12}C^{18}O)^+$	42,581.21(4)	65(8)	b
$(D^{12}C^{16}O)^+$	36,019.776(3)	55.875(5)	a,b
$(D^{13}C^{16}O)^+$	35,366.71(1)	53.2(5)	b
$(D^{12}C^{18}O)^+$	34,413.798(1)	50.29(4)	b
$(H^{16}O^{12}C)^+$	44,743.924(14)	114.9(5)	c
$(H^{14}N^{14}N)^+$	46,586.863(15)	87.5(5)	d
$(D^{14}N^{14}N)^+$	38,554.717(14)	60.8(4)	d
$(H^{12}C^{32}S)^+$	$\nu(J=1\rightarrow2)=85{,}347.90(3)$ MHz		e

[a] K. V. L. N. Sastry, E. Herbst, and F. C. De Lucia, *J. Chem. Phys.*, **75**, 4169 (1981).
[b] Bogey et al. [122].
[c] Blake et al. [131].
[d] Sastry et al. [126].
[e] Gudeman et al. [128].

and Klemperer [114] proposed a theory for formation of molecules in interstellar clouds in which HCO^+ has a significant function as a reactive intermediary. It is believed that the failure to observe CO^+ in interstellar molecular clouds where CO is relatively abundant is due to conversion of these ions to HCO^+ by reaction with H atoms [114, 115]. Support for HCO^+ as the unknown X-ogen came from observation by Snyder et al. [116] of an interstellar line at 86,754 MHz which could be assigned to $H^{13}CO^+$ and by more detailed theoretical predictions [117, 118] of the structure of HCO^+. Experimental proof of the identification came with the laboratory detection of HCO^+ by Woods et al. [119], who measured the rest frequency of the $J=0\rightarrow1$ transition at 89,188.545(20) MHz, quite close to some of the more recent astronomical measurements [120, 121].

The laboratory measurements of the important HCO^+ ion have been extended to include several isotopic species and several other rotational transitions. The accurate ground-state rotational constants listed in Table 5.18 should be useful to microwave astronomers in detection of other lines and other isotopic species. The references cited in Table 5.18 give the rest frequencies of the measured transitions.

Bogey et al. [122] have used their observed constants of several isotopic species of HCO^+ to calculate the interatomic distances of the ion, as listed in Table 5.19. The equilibrium structure predicted by Wahlgren et al. [117] and by Kraemer and Dierksen [118] were found to agree closely with the experimental r_0 and r_s values. Generally, r_s should be somewhat closer than r_0 to r_e (see Chapter XIII, Section 8).

Table 5.19 Observed and Theoretical Structures of the Formyl Ion, HCO^+

HCO^+	*Experimental*[a]		*Theoretical*[b]	
	r_0	r_s	r_e	
r(H—C) Å	1.0913(4)	1.0930(1)	1.095[c]	1.091[d]
r(C—O) Å	1.1093(1)	1.1070(2)	1.1045[c]	1.103[d]

[a]Bogey et al. [122].
[b]Ab initio C.I. calculation of equilibrium structure.
[c]Wahlgren et al. [117].
[d]Kraemer and Diercksen [118].

The beautiful hyperfine pattern shown in Fig. 1.4 is that for the $J=0\rightarrow1$ transition of interstellar HNN^+. Nuclear quadrupole coupling by the terminal ^{14}N gives the gross triplet with total spread of about 4 MHz. The substructure partially resolved in the center line is due to quadrupole coupling by the central ^{14}N. This interstellar transition was first observed by Turner [123] in 1974, but its identification, like that for HCO^+, remained uncertain until later calculations and observations could be made. The proposal that the observed triplet comes from HNN^+ was quickly made by Green et al. [124]. The close agreement of the theoretical and observed hyperfine patterns in Fig. 1.4 confirms their assignment. The first laboratory observation of this ion was made in 1976 by Saykally et al. [125], who resolved the gross triplet of the $J=0\rightarrow1$ transition but could not detect splitting by the central ^{14}N because of the weakness and broadness of the lines. They mention that the HNN^+ spectrum proved to be considerably more difficult to study than that of HCO^+. In 1981, Sastry et al. [126] measured several higher millimeter wave transitions for both HN_2^+ and DN_2^+ and from them derived the accurate values for ground-state rotational constants listed in Table 5.18.

In 1981 Thaddeus et al. [127] observed from several astronomical sources a line at 85,348 MHz which they assigned to the $J=1\rightarrow2$ transition of the linear ion HCS^+. Their assignment was confirmed shortly afterward by Gudeman et al. [128] with a laboratory detection of a line at 85,347.90 MHz in a glow discharge under conditions expected to produce HCS^+.

A laboratory observation of the $J=0\rightarrow1$ transition of HOC^+ has been made by Gudeman and Woods [129] and used by Woods et al. [130] for a tentative identification of interstellar HOC^+ in the constellation Sgr B2. In the Duke laboratory, Blake et al. [131] measured the higher transitions, $J=1\rightarrow2$, $2\rightarrow3$, and $3\rightarrow4$, and derived the accurate rotational constants for this ion listed in Table 5.18. Note the significantly higher stretching constant D_0 of HOC^+ over that of its isomer HCO^+.

The principal cause for the long delay in the development of microwave spectroscopy of molecular ions is, of course, their instability relative to the neutral singlet Σ molecules commonly studied in the past. The "free-space" absorption cell usable with semioptical techniques at the higher frequencies of the millimeter wave region made possible the observation of spectra of molecular ions directly as they are produced in a glow discharge enclosed in glass. Nevertheless, many spectroscopists were probably reluctant to attempt spectral detection and identification of short-lived ions in a discharge column composed of mixed constituents. One group evidently not reluctant is that led by R. C. Woods at Wisconsin University. They made the first successful laboratory observations on all the molecular ions discussed in this section. A strong impetus to the development of techniques for microwave spectroscopy of molecular ions within the laboratory has been the detection of molecular ions in interstellar space.

One factor contributing to the success of the Woods group was their lowering of the temperature of the discharge column, usually to that of liquid air. In this way they both increased the stability of the ions and decreased the collision broadening. Line widths observed for these molecular ions in a glow discharge mixture at liquid air temperature proved to be surprisingly small, a fortunate circumstance aiding their detection. Anderson et al. [132] made careful measurement of the line-breadth parameter for the $J=0\rightarrow1$ transition of HCO^+ broadened by H_2 at a temperature near 100°K. Measurements were taken over a pressure range of 4 to 40 mTorr. At that temperature and pressure range, the line-breadth parameter $(\Delta\nu)_1$ was found to be 29.3 MHz/Torr. This parameter is comparable in magnitude to the self-broadening parameters of many polar molecules at room temperature (see Table 3.1).

11 LINEAR FREE RADICALS: LABORATORY AND INTERSTELLAR

Since the detection of spectral lines of molecular free radicals in interstellar space, microwave spectroscopists have become increasingly interested in the study of these unstable species in the laboratory. As already mentioned in Chapter I, the first molecular species observed in interstellar space was the OH radical. The microwave lines of this light radical as well as those of CH observed in interstellar space are Λ doublet transitions of their $^2\Pi$ ground states (see Chapter IV, Section 2). Rotational transitions of linear polyatomic free radicals are now observed from interstellar space and from molecules in the laboratory. Microwave magnetic resonance spectra of such free radicals are treated in Chapter XI, Section 9.

Discovery of the interesting linear free radical CCH in the interstellar medium was announced by Tucker et al. [133] from detection of its lowest rotational transition, $N=0\leftarrow1$. Several years later the assignment was confirmed by detection of the $N=2\leftarrow3$ transition of the interstellar radicals by Ziurys et al. [134] and by laboratory measurement of the $N=1\rightarrow2$, $2\rightarrow3$, and $3\rightarrow4$ transitions by Sastry et al. [135]. The CCH radical has the $^2\Sigma$ ground state, and its spectrum is analyzed by the theory described in Chapter IV, Section 2 for diatomic molecules. The only hyperfine splitting is that due to the magnetic interaction of the H nucleus, and this is smaller than the electron spin–rotation interaction. Therefore the coupling scheme used in the analysis is

$$\mathbf{N}+\mathbf{S}=\mathbf{J} \quad \text{and} \quad \mathbf{J}+\mathbf{I}=\mathbf{F}$$

There is no nuclear quadrupole interaction, and the $C_I\mathbf{N}\cdot\mathbf{I}$ term is negligible. The observed spectra from outer space [134] and in the laboratory [135] were analyzed satisfactorily with the Hamiltonian

$$\mathscr{H}=[B_0-D_0N(N+1)]\mathbf{N}^2+\gamma_0(\mathbf{N}\cdot\mathbf{S})+b(\mathbf{I}\cdot\mathbf{S})+cI_zS_z \tag{5.60}$$

The energy level diagram of Fig. 5.9 is that for the $N=2$ and $N=3$ states of CCH calculated with the $\mathscr{H}$ of (5.60) by Ziurys et al. [134]. The allowed emission transitions are indicated by arrows, with those observed in bold face. The signals observed from the interstellar source, OMC-1, are shown in Fig. 5.10. These consist of the $J=\frac{5}{2}\rightarrow\frac{3}{2}$, $\frac{7}{2}\rightarrow\frac{5}{2}$ fine structure doublet with a partially resolved proton hyperfine structure. Table 5.20 shows the spectral constants of the Hamiltonian of (5.60) as derived from the laboratory measurement compared with those derived from the interstellar spectra. The agreement between the two sets proves, without doubt, the existence of the CCH radical in outer space.

Microwave lines observed from interstellar sources have been tentatively assigned to the radicals CCCCH and CCCN on the basis of predicted frequencies of theoretical models. From structural considerations both radicals are expected to be linear and to have $^2\Sigma$ electronic ground states. No laboratory

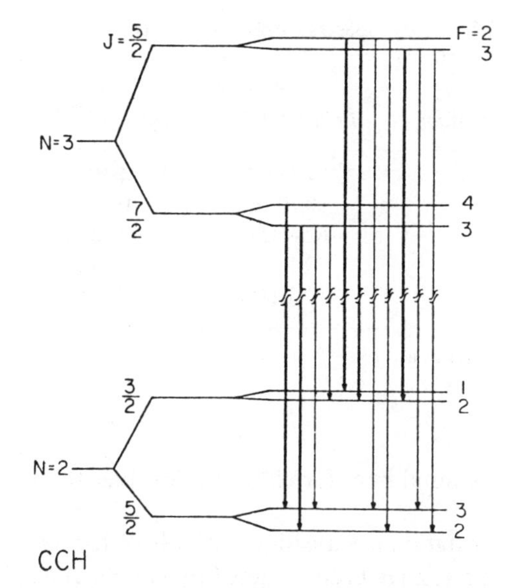

Fig. 5.9 Energy level diagram of the $N=2$ and 3 levels of the $^2\Sigma$ state of CCH including the fine and hyperfine splittings, which are drawn to scale. The N levels are separated by approximately 262 MHz. Allowed emission transitions are indicated by arrows. Arrows in bold face indicate observed interstellar transitions shown in Fig. 5.10. From Ziurys, Saykally, Plambeck, and Erickson [134].

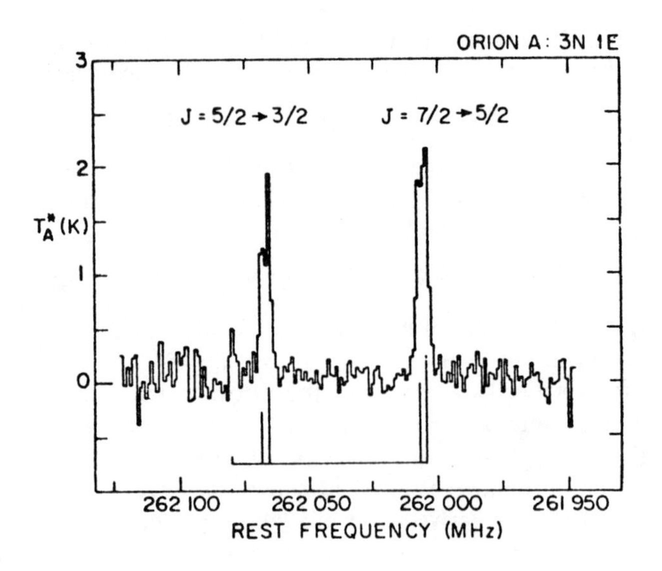

Fig. 5.10 Interstellar lines of the $N=2\leftarrow3$ rotational transition of CCH, observed in OMC-1. The bars indicate the theoretically predicted lines indicated in Fig. 5.9 by arrows in bold face. From Ziurys, Saykally, Plambeck, and Erickson [134].

Table 5.20 Spectroscopic Constants of CCH in MHz

Constant	*Laboratory*[a]	*Interstellar*[b]
B_0	43,674.514(5)	43,674.7(10)
D_0	0.1052(3)	0.1137(6)
γ_0	−62.66(2)	−62.62(5)
b	40.40(19)	40.39(14)
c	12.33(11)	12.32(9)

[a]Sastry et al. [135].
[b]Ziurys et al. [134].

measurement of rotational lines for either radical has, to our knowledge, been reported.

Four successive rotational transitions, $N=8\leftarrow9$ to $11\leftarrow12$, observed from an astronomical source, have been assigned by Guélin et al. [136] to the linear C_4H radical in the $^2\Sigma$ state. The measured frequencies fit closely those predicted from the theoretical model of Wilson and Green [137]. The observed B_0 is 4758.48(10) MHz compared with the theoretically predicted value of 4753 MHz.

The evidences for identification of C_3N in interstellar space, although strong, are somewhat less than those for C_4H. Only two observed rotational transitions, the $N=8\leftarrow9$ and $9\leftarrow10$, each with a doublet structure, are reported in the original paper by Guélin and Thaddeus [138], but Guélin et al. [136] mention later observations of a third, the $N=10\leftarrow11$. The frequencies of the observed lines fit closely those predicted from the Hartree-Fock theoretical model calculated by Wilson and Green [137]. The observed B_0 is 4947.66(10) MHz compared with the theoretical value of 4955 MHz, and the observed magnitude of the spin–rotation constant, $|\gamma_0|=18.7(5)$ MHz, is within the range of that expected for the linear CCCN radical. So far this molecule has not been detected on earth by any method.

Rotational spectra of linear polyatomic molecules having electronic orbital momentum, Π, Δ, or higher states are, except for bending vibrational effects, similar to those for diatomic molecules in the same electronic states as described in Chapter IV, Section 2, Theory for the Π and Δ states is reviewed in Chapter IV, Section 2 and Chapter IX, Section 9. Formulation of the theory and the nature of the solution for the energy eigenvalues for either the diatomic or the linear polyatomic molecule having both electronic spin and orbital angular momentum depend sensitively on the relative magnitudes of the couplings of the various spin and angular momentum vectors, including the nuclear spins. No generally applicable formulation can be given.

Hougen [139] has given a theoretical treatment for linear triatomic molecules in $^2\Pi$ states which takes into account the effects of bending vibrations. An instructive application of Hougen's theory is made by Saito and Amano [140], who measured and calculated rotational transitions for both the $^2\Pi_{1/2}$ and $^2\Pi_{3/2}$ states of the linear radical NCO.

References

1. A. H. Nielsen, *J. Chem. Phys.*, **11**, 160 (1943).
2. R. S. Mulliken, *J. Phys. Chem.*, **41**, 159 (1937).
3. F. C. De Lucia and P. A. Helminger, *J. Chem. Phys.*, **67**, 4262 (1977).
4. H. H. Nielsen and W. H. Shaffer, *J. Chem. Phys.*, **11**, 140 (1943).
5. H. H. Nielsen, *Rev. Mod. Phys.*, **23**, 90 (1951).
6. P. Kuijpers, T. Törring, and A. Dymanus, *Z. Naturforsch.*, **32a**, 930 (1977).
7. M. Bogey and A. Bauer, *J. Mol. Spectrosc.*, **84**, 170 (1980).
8. G. Amat, H. H. Nielsen, and G. Tarrago, *Rotation Vibration of Polyatomic Molecules*, Dekker, New York, 1971.
9. H. W. Kroto, J. F. Nixon, and N. P. C. Simmons, *J. Mol. Spectrosc.*, **82**, 185 (1980).
10. R. G. Shulman and C. H. Townes, *Phys. Rev.*, **77**, 421 (1950).
11. J. F. Westerkamp, *Phys. Rev.*, **93**, 716 (1954).
12. T. Törring, *Z. Physik*, **161**, 179 (1961).
13. H. H. Nielsen and G. Amat, *J. Phys. Radium*, **15**, 601 (1954).
14. M. Winnewisser and H. K. Bodenseh, *Z. Naturforsch.*, **22a**, 1724 (1967).
15. J. Ramadier and G. Amat, *J. Phys. Radium*, **19**, 915 (1958).
16. M. Winnewisser and B. P. Winnewisser, *J. Mol. Spectrosc.*, **41**, 143 (1972).

17. B. P. Winnewisser, M. Winnewisser, and F. Winther, *J. Mol. Spectrosc.*, **51**, 65 (1974).
18. K. Yamada, B. P. Winnewisser, and M. Winnewisser, *J. Mol. Spectrosc.*, **56**, 449 (1975).
19. B. P. Winnewisser and M. Winnewisser, *J. Mol. Spectrosc.*, **56**, 471 (1975).
20. P. R. Bunker, B. M. Landsberg, and B. P. Winnewisser, *J. Mol. Spectrosc.*, **74**, 9 (1979).
21. E. Fermi, *Z. Physik*, **71**, 250 (1931).
22. F. C. De Lucia and W. Gordy, *Phys. Rev.*, **187**, 58 (1969).
23. D. Marcuse, *Proc. IRE*, **49**, 1706 (1961).
24. A. K. Garrison and W. Gordy, *Phys. Rev.*, **108**, 899 (1957).
25. J. R. Rusk and W. Gordy, *Phys. Rev.*, **127**, 817 (1962).
26. C. Huiszoon and A. Dymanus, *Phys. Lett.*, **21**, 164 (1966).
27. F. A. Van Dijk and A. Dymanus, *Chem. Phys. Lett.*, **2**, 235 (1968).
28. F. A. Van Dijk and A. Dymanus, *Chem. Phys. Lett.*, **5**, 387 (1970).
29. W. E. Lamb, *Phys. Rev.*, **134**, A1429 (1964).
30. C. C. Costain, *Can. J. Phys.*, **47**, 2431 (1969).
31. R. S. Winton and W. Gordy, *Phys. Lett.*, **32A**, 219 (1970).
32. G. Cazzoli, C. D. Esposti, and P. G. Favero, *J. Phys. Chem.*, **84**, 1756 (1980).
33. T. Oka, *Adv. Atom. Mol. Phys.*, **9**, 127 (1973).
34. R. M. Lees, *J. Mol. Spectrosc.*, **69**, 225 (1978).
35. J. G. Baker, "Microwave–Microwave Double Resonance," in *Modern Aspects of Microwave Spectroscopy*, G. W. Chantry, Ed., Academic, London, 1979, pp. 65–122.
36. R. L. Kuczkowski, D. R. Lide, and L. C. Krisher, *J. Chem. Phys.*, **44**, 3131 (1966).
37. D. R. Lide and R. L. Kuczkowski, *J. Chem. Phys.*, **46**, 4768 (1967).
38. C. Matsumura and D. R. Lide, *J. Chem. Phys.*, **50**, 71 (1969).
39. E. F. Pearson and M. B. Trueblood, *Astrophys. J.*, **179**, L145 (1973).
40. E. F. Pearson and M. B. Trueblood, *J. Chem. Phys.*, **58**, 826 (1973).
41. D. R. Lide and C. Matsumura, *J. Chem. Phys.*, **50**, 3080 (1969).
42. P. Kuijpers, T. Törring, and A. Dymanus, *Chem. Phys.*, **15**, 457 (1976).
43. E. F. Pearson, B. P. Winnewisser, and M. B. Trueblood, *Z. Naturforsch.*, **31a**, 1259 (1976).
44. S. M. Freund, P. D. Godfrey, and W. Klemperer, *25th Symposium on Molecular Structure*, The Ohio State University, 1970. Paper E8.
45. E. F. Pearson and R. V. McCormick, *J. Chem. Phys.*, **58**, 1619 (1973).
46. C. Kirby and H. W. Kroto, *J. Mol. Spectrosc.*, **83**, 130 (1980).
47. A. G. Smith, W. Gordy, J. W. Simmons, and W. V. Smith, *Phys. Rev.*, **75**, 260 (1949).
48. E. F. Pearson, R. A. Creswell, M. Winnewisser, and G. Winnewisser, *Z. Naturforsch.*, **31a**, 1394 (1976).
49. F. C. De Lucia, *J. Mol. Spectrosc.*, **55**, 271 (1975).
50. L. E. Snyder and D. Buhl, *Astrophys. J.*, **163**, L47 (1971).
51. H. A. Gebbie, N. W. B. Stone, and F. D. Findly, *Nature*, **202**, 685 (1966).
52. D. R. Lide and A. G. Maki, *Appl. Phys. Lett.*, **11**, 52 (1967).
53. R. A. Creswell, E. F. Pearson, M. Winnewisser, and G. Winnewisser, *Z. Naturforsch.*, **31a**, 221 (1976).
54. R. J. Saykally, P. G. Szanto, T. G. Anderson, and R. C. Woods, *Astrophys. J.*, **204**, 143 (1976).
55. G. L. Blackman, R. D. Brown, P. D. Godfrey, and H. I. Gunn, *Nature*, **261**, 395 (1976).
56. L. E. Snyder and D. Buhl, *Bull. Am. Astron. Soc.*, **3**, 388 (1971).
57. B. Zukerman, M. Morris, P. Palmer, and B. E. Turner, *Astrophys. J.*, **173**, L125 (1972).
58. J. Barsuhn, *Astrophys. Lett.*, **12**, 169 (1972).

59. D. E. Milligan and M. E. Jacox, *J. Chem. Phys.*, **39**, 712 (1963).
60. N. W. Broten, T. Oka, L. W. Avery, J. M. MacLeod, and H. W. Kroto, *Astrophys. J.*, **223**, L105 (1978).
61. L. W. Avery, N. W. Broten, J. M. MacLeod, T. Oka, and H. W. Kroto, *Astrophys. J.*, **205**, L173 (1976).
62. H. W. Kroto, C. Kirby, D. R. M. Walton, L. W. Avery, N. W. Broten, J. M. MacLeod, and T. Oka, *Astrophys. J.*, **219**, L133 (1978).
63. A. J. Alexander, H. W. Kroto, and D. R. M. Walton, *J. Mol. Spectrosc.* **62**, 175 (1976).
64. C. Kirby, H. W. Kroto, and D. R. M. Walton, *J. Mol. Spectrosc.*, **83**, 261 (1980).
65. E. B. Turner, *Astrophys. J.*, **163**, L38 (1971).
66. L. E. Snyder, "Molecules in Space," *MTP International Review of Science, Physical Chemistry*, Vol. 3, *Spectroscopy*, D. A. Ramsey, Ed., Butterworths, London, 1972, pp. 193–240.
67. M. Winnewisser and J. J. Christiansen, *Chem. Phys. Lett.*, **37**, 270 (1976).
68. M. Winnewisser, E. W. Peau, K. Yamada, and J. J. Christiansen, *Z. Naturforsch.*, **36a**, 819 (1981).
69. J. B. Simpson, J. G. Smith, and D. H. Whiffen, *J. Mol. Spectrosc.*, **44**, 558 (1972).
70. G. Cazzoli, C. D. Esposti, and P. G. Favero, *J. Mol. Struct.*, **48**, 1 (1978).
71. W. R. Thorson and I. Nakagawa, *J. Chem. Phys.*, **33**, 994 (1960).
72. J. W. C. Johns, *Can. J. Phys.*, **45**, 2639 (1967).
73. K. Yamada and M. Winnewisser, *Z. Naturforsch.*, **31a**, 139 (1976).
74. M. Born and J. R. Oppenheimer, *Ann. Physik*, **84**, 157 (1927).
75. A. V. Burenin, E. N. Karyakin, A. F. Krupnov, and S. M. Shapin, *J. Mol. Spectrosc.*, **78**, 181 (1979).
76. T. R. Dyke, B. J. Howard, and W. Klemperer, *J. Chem. Phys.*, **56**, 2442 (1972).
77. T. R. Dyke and J. S. Muenter, *J. Chem. Phys.*, **60**, 2929 (1974).
78. J. W. Bevan, A. C. Legon, D. J. Millen, and S. C. Rogers, *J. Chem. Soc. Commun.*, 341 (1975).
79. A. C. Legon, D. J. Millen, and S. C. Rogers, *Chem. Phys. Lett.*, **41**, 137 (1976).
80. A. C. Legon, D. J. Millen, and S. C. Rogers, *Proc. R. Soc.* (*London*) **A370**, 213 (1980).
81. A. C. Legon, D. J. Millen, and S. C. Rogers, *J. Mol. Spectrosc.*, **70**, 209 (1978).
82. A. C. Legon, D. J. Millen, P. J. Mjoberg, and S. C. Rogers, *Chem. Phys. Lett.*, **55**, 157 (1978).
83. A. C. Legon, D. J. Millen, and P. J. Mjoberg, *Chem. Phys. Lett.*, **47**, 589 (1977).
84. L. W. Buxton, E. J. Campbell, and W. H. Flygare, *Chem. Phys.*, **56**, 399 (1981).
85. A. C. Legon, P. D. Soper, M. R. Keenan, T. K. Minton, T. J. Balle, and W. H. Flygare, *J. Chem. Phys.*, **73**, 583 (1980).
86. M. R. Keenan, T. K. Minton, A. C. Legon, T. J. Balle, and W. H. Flygare, *Proc. Natl. Acad. Sci. USA*, **77**, 5583 (1980).
87. P. D. Soper, A. C. Legon, and W. H. Flygare, *J. Chem. Phys.*, **74**, 2138 (1981).
88. A. C. Legon, P. D. Soper, and W. H. Flygare, *J. Chem. Phys.*, **74**, 4944 (1981).
89. F. A. Baiocchi, T. A. Dixon, C. H. Joyner, and W. Klemperer, *J. Chem. Phys.*, **74**, 6544 (1981).
90. C. H. Joyner, T. A. Dixon, F. A. Baiocchi, and W. Klemperer, *J. Chem. Phys.*, **74**, 6550 (1981).
91. A. C. Legon, P. D. Soper, and W. H. Flygare, *J. Chem. Phys.*, **74**, 4936 (1981).
92. J. S. Muenter, *J. Chem. Phys.*, **56**, 5409 (1972).
93. S. E. Novick, P. Davies, S. J. Harris, and W. Klemperer, *J. Chem. Phys.*, **59**, 2273 (1973).
94. S. J. Harris, S. E. Novick, and W. Klemperer, *J. Chem. Phys.*, **60**, 3208 (1974).
95. S. J. Harris, S. E. Novick, and W. Klemperer, *J. Chem. Phys.*, **61**, 193 (1974).
96. S. E. Novick, S. J. Harris, K. C. Janda, and W. Klemperer, *Can. J. Phys.*, **53**, 2007 (1975).

97. S. E. Novick, K. C. Janda, S. L. Holmgren, M. Walden, and W. Klemperer, *J. Chem. Phys.*, **65**, 1114 (1976).

98. F. A. Baiocchi, T. A. Dixon, C. H. Joyner, and W. Klemperer, *J. Chem. Phys.*, 75, 2041 (1981).

99. K. V. Chance, K. H. Bowen, J. S. Winn, and W. Klemperer, *J. Chem. Phys.*, **70**, 5157 (1979).

100. T. A. Dixon, C. H. Joyner, F. A. Baiocchi, and W. Klemperer, *J. Chem. Phys.*, **74**, 6539 (1981).

101. T. J. Balle, E. J. Campbell, M. R. Keenan, and W. H. Flygare, *J. Chem. Phys.*, **71**, 2723 (1979).

102. T. E. Balle, E. J. Campbell, M. R. Keenan, and W. H. Flygare, *J. Chem. Phys.*, **72**, 922 (1980).

103. M. R. Keenan, E. J. Campbell, T. J. Balle, L. W. Buxton, T. K. Minton, P. D. Soper, and W. H. Flygare, *J. Chem. Phys.*, **72**, 3070 (1980).

104. M. R. Keenan, L. W. Buxton, E. J. Campbell, T. J. Balle, and W. H. Flygare, *J. Chem. Phys.*, **73**, 3523 (1980).

105. L. W. Buxton, E. J. Campbell, T. J. Balle, and W. H. Flygare, *Chem. Phys.*, **54**, 173 (1981).

106. M. R. Keenan, L. W. Buxton, E. J. Campbell, A. C. Legon, and W. H. Flygare, *J. Chem. Phys.*, **74**, 2133 (1981).

107. E. J. Campbell, L. W. Buxton, and W. H. Flygare, *J. Chem. Phys.*, **74**, 813 (1981).

108. S. E. Novick, P. Davies, T. R. Dyke, and W. Klemperer, *J. Am. Chem. Soc.*, **95**, 8547 (1973).

109. G. Herzberg, *Infrared and Raman Spectra of Polyatomic Molecules*, D. Van Nostrand Company, New York, 1945, p. 173.

110. T. A. Dixon and R. C. Woods, *Phys. Rev. Lett.*, **34**, 61 (1975).

111. K. B. Jefferts, *Phys. Rev. Lett.*, **23**, 1476 (1969).

112. D. Buhl and L. E. Snyder, *Nature*, **228**, 267 (1970).

113. W. Klemperer, *Nature*, **227**, 1230 (1970).

114. E. Herbst and W. Klemperer, *Astrophys. J.*, **185**, 505 (1973).

115. L. M. Hobbs, *Astrophys. J.*, **181**, 795 (1973).

116. L. E. Snyder, J. M. Hollis, B. L. Ulich, F. J. Lovas, and D. Buhl, *Bull. Am. Astron. Soc.*, **7**, 497 (1975).

117. U. Wahlgren, B. Liu, P. K. Pearson, and H. F. Schaefer, *Nature*, **246**, 4 (1973).

118. W. P. Kraemer and G. H. F. Diercksen, *Astrophys. J.*, **205**, L97 (1976).

119. R. C. Woods, T. A. Dixon, R. J. Saykally, and P. G. Szanto, *Phys. Rev. Lett.*, **35**, 1269 (1975).

120. M. Morris, B. Zukerman, B. E. Turner, and P. Palmer, *Astrophys. J.*, **192**, L27 (1974).

121. J. M. Hollis, L. E. Snyder, D. Buhl, and P. T. Giguere, *Astrophys. J.*, **200**, 584 (1975).

122. M. Bogey; C. Demuynck, and J. L. Destombes, *Mol. Phys.*, **46**, 679 (1982).

123. B. E. Turner, *Astrophys. J.*, **193**, L83 (1974).

124. S. Green, J. A. Montgomery, and P. Thaddeus. *Astrophys. J.*, **193**, L89 (1974).

125. R. J. Saykally, T. A. Dixon, T. G. Anderson, P. G. Szanto, and R. C. Woods, *Astrophys. J.*, **205**, L101 (1976).

126. K. V. L. N. Sastry, P. Helminger, E. Herbst, and F. C. De Lucia, *Chem. Phys. Lett.*, **84**, 286 (1981).

127. P. Thaddeus, M. Guélin, and R. A. Linke, *Astrophys. J.*, **246**, L41 (1981).

128. C. S. Gudeman, N. N. Haese, N. D. Piltch, and R. C. Woods, *Astrophys. J.*, **246**, L47 (1981).

129. C. S. Gudeman and R. C. Woods, *Phys. Rev. Lett.*, **48**, 1344 (1982).

130. R. C. Woods, C. S. Gudeman, R. L. Dickman, P. F. Goldsmith, G. R. Huguenin, W. M. Irvine, Å. Hjalmarson, L.-Å. Nyman, and H. Olofsson, *Astrophys. J.*, **270**, 583 (1983).

131. G. A. Blake, P. Helminger, E. Herbst, and F. C. De Lucia, *Astrophys. J.*, **264**, L69 (1983).

132. T. G. Anderson, C. S. Gudeman, T. A. Dixon, and R. C. Woods, *J. Chem. Phys.*, **72**, 1332 (1980).

133. K. D. Tucker, M. L. Kutner, and P. Thaddeus, *Astrophys. J.*, **193**, L115 (1974).
134. L. M. Ziurys, R. J. Saykally, R. L. Plambeck, and N. R. Erickson, *Astrophys. J.*, **254**, 94 (1982).
135. K. V. L. N. Sastry, P. Helminger, A. Charo, E. Herbst, and F. C. De Lucia, *Astrophys. J.*, **251**, L119 (1981).
136. M. Guélin, S. Green, and P. Thaddeus, *Astrophys. J.*, **224**, L27 (1978).
137. S. Wilson and S. Green, *Astrophys. J.*, **212**, L87 (1977).
138. M. Guélin and P. Thaddeus, *Astrophys. J.*, **212**, L81 (1977).
139. J. T. Hougen, *J. Chem. Phys.*, **36**, 519 (1962).
140. S. Saito and T. Amano, *J. Mol. Spectrosc.*, **34**, 353 (1970).

Chapter **VI**

SYMMETRIC-TOP MOLECULES

A molecule in which two of the principal moments of inertia are equal is a symmetric-top rotor. This condition is generally met when the molecule has an axis of symmetry that is trigonal or greater. The molecules PH_3, CH_3F, CH_3CCH, CH_3SiH_3, and CF_3SF_5 are examples of symmetric-top molecules. A linear molecule can be treated as a special case of a symmetric top in which the angular momentum about the symmetry axis is zero. The rotational energies of the symmetric-top molecule were calculated first by Dennison [1] with matrix mechanical methods and later by Reiche and Rademacher [2] and by Kronig and Rabi [3] with the Schrödinger wave equation.

1 MOLECULES IN THE GROUND VIBRATIONAL STATE

The Rigid Rotor Approximation

The classical mechanics of the symmetric top is discussed in Chapter II, Section 1. Classically, the molecule rotates about the symmetry axis while this axis in turn precesses about a fixed direction in space corresponding to the direction of the total angular moment **P**. This motion is illustrated by Fig. 6.1.

In a symmetric top, one of the principal axes of inertia must lie along the molecular axis of symmetry. The principal moments of inertia which have their axes perpendicular to this axis are equal. If a, the axis of least moment of inertia ($I_a < I_b = I_c$), lies along the symmetry axis, the molecule is a prolate symmetric top (CH_3CCH, for example). If c, the axis of the greatest moment of inertia ($I_a = I_b < I_c$), lies along the symmetry axis, the molecule is an oblate symmetric top (BCl_3, for example). Most of the symmetric-top molecules observed in the microwave region are prolate. With the a axis chosen along the symmetry axis ($I_c = I_b$) and with $P^2 = P_a^2 + P_b^2 + P_c^2$, the Hamiltonian operator of (2.17) may be expressed as

$$\mathscr{H}_r = \frac{P^2}{2I_b} + \frac{1}{2}\left(\frac{1}{I_a} - \frac{1}{I_b}\right) P_a^2 \tag{6.1}$$

Eigenvalues for the angular momentum operators P^2, P_z, and P_Z for the symmetric-top rotor (Chapter II, Section 2) are

$$(J, K, M|P^2|J, K, M) = \hbar^2 J(J+1) \tag{6.2}$$

$$(J, K, M|P_z|J, K, M) = \hbar K \tag{6.3}$$

$$(J, K, M|P_Z|J, K, M) = \hbar M \tag{6.4}$$

where

$$J = 0, 1, 2, 3, \ldots$$
$$K = 0, \pm 1, \pm 2, \pm 3, \ldots, \pm J$$
$$M = 0, \pm 1, \pm 2, \pm 3, \ldots, \pm J$$

In the x, y, z system, z is chosen as the symmetry axis of the top. In the a, b, c system used here for designation of the principal axes of inertia, z becomes a

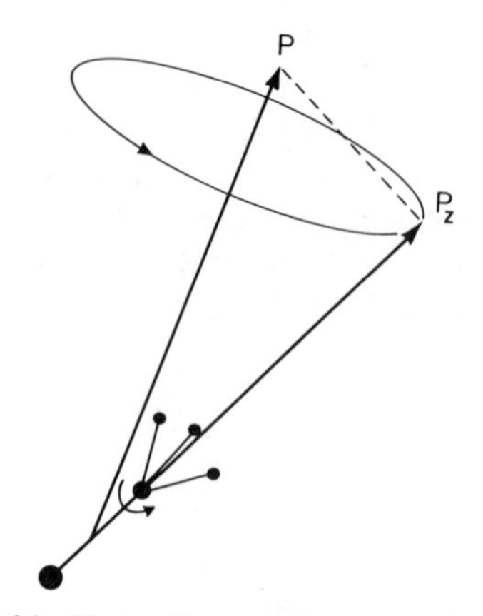

Fig. 6.1 Vector diagram of a symmetric rotor.

for the prolate top, and z becomes c for the oblate top. In the field-free rotor, the rotational energies do not depend on M. It is evident that the Hamiltonian of (6.1) commutes with P^2 and P_α and is therefore diagonal in the J, K representation. Its eigenvalues, which are the quantized rotational energies of the rigid prolate symmetric top, are therefore

$$E_{J,K}=(J, K|\mathscr{H}_r|J, K)=\frac{1}{2I_b}(J, K|P^2|J, K)+\frac{1}{2}\left(\frac{1}{I_\alpha}-\frac{1}{I_b}\right)(J, K|P_\alpha^2|J, K)$$

$$=\left(\frac{h^2}{8\pi^2 I_b}\right)J(J+1)+\left(\frac{h^2}{8\pi^2}\right)\left(\frac{1}{I_a}-\frac{1}{I_b}\right)K^2 \tag{6.5}$$

With the designation $A=h/(8\pi^2 I_a)$ and $B=h/(8\pi^2 I_b)$, $E_{J,K}$ can be written

$$E_{J,K}=h[BJ(J+1)+(A-B)K^2] \tag{6.6}$$

Since P_a is a component of P, the values of K cannot exceed those of J in magnitude. Although K can have both negative and positive values, the $+$ and $-$ values do not lead to separate sets of energy levels because K appears only as a squared term in (6.6). Thus all K levels except those for $K=0$ are doubly degenerate. This K degeneracy cannot be removed by either external or internal fields. In addition to the K degeneracy, there is a $(2J+1)$, M degeneracy in the field-free symmetric rotor as in the linear molecule. Unlike the K degeneracy, the M degeneracy can be lifted completely by the application of an external electric or magnetic field (see Chapters X and XI).

In a true symmetric top, any permanent dipole moment must of necessity lie along the symmetry axis. All matrix elements of this dipole moment resolved along a space-fixed axis vanish except those between states corresponding to $J\rightarrow J$ or $J\pm 1$, $K\rightarrow K$. See Chapter II, Section 6. The selection rules for the field-free rotor are therefore

$$\Delta J=0,\ \pm 1 \qquad \Delta K=0 \tag{6.7}$$

The rule corresponding to absorption of radiation is $J\rightarrow J+1$ and $K\rightarrow K$. Application of these rules to (6.6) gives the formula for the absorption frequencies for the rigid symmetric top

$$\nu=2B(J+1) \tag{6.8}$$

which is exactly that for the linear molecule. However, as we shall see below, centrifugal stretching separates the lines corresponding to different $|K|$ by small frequency differences which are usually sufficient to make these lines resolvable in the microwave region.

The Distortable Rotor

Centrifugal stretching is treated as a perturbation on the eigenstates of the rigid rotor. If $\mathscr{H}_r$ represents the Hamiltonian of the rigid rotor and $\mathscr{H}_d$ represents that of the distortional energy, the rotational Hamiltonian is

$$\mathscr{H}=\mathscr{H}_r+\mathscr{H}_d \tag{6.9}$$

In Chapter VIII it is shown that the centrifugal distortional Hamiltonian has the form

$$\mathscr{H}_d = \frac{\hbar^4}{4} \sum_{\alpha\beta\gamma\delta} \tau_{\alpha\beta\gamma\delta} P_\alpha P_\beta P_\gamma P_\delta \tag{6.10}$$

where α, β, γ, δ represent the principal coordinate axes of the moments of inertia and where each must be summed over all three coordinate axes. However, in the first-order perturbation treatment, the terms with odd powers in any angular momentum which occur in the sum average to zero. Furthermore, by use of the angular momenta commutation rules and the fact that many of the τ's are equal, the Hamiltonian can be further simplified. This is discussed in Chapter VIII. For the first-order treatment, which is adequate for the usual microwave measurements, the Hamiltonian can be written as

$$\begin{aligned}\mathscr{H}_d^{(1)} &= \frac{1}{4} \sum \tau'_{\alpha\alpha\beta\beta} P_\alpha^2 P_\beta^2 \\ &= \frac{1}{4} \left[\tau'_{xxxx} P_x^4 + \tau'_{yyyy} P_y^4 + \tau'_{zzzz} P_z^4 + \tau'_{xxyy}(P_x^2 P_y^2 + P_y^2 P_x^2)\right. \\ &\quad \left. + \tau'_{xxzz}(P_x^2 P_z^2 + P_z^2 P_x^2) + \tau'_{yyzz}(P_y^2 P_z^2 + P_z^2 P_y^2)\right]\end{aligned} \tag{6.11}$$

in which x, y, z represent the principal axes and z represents the symmetry axis of the top. The τ''s are defined in Table 8.4. To first order, the distortion energy is

$$E_d^{(1)} = (J, K| \mathscr{H}_d^{(1)} |J, K) \tag{6.12}$$

With the angular momentum matrix elements given in Table 8.19, this expression can be evaluated readily. The resulting expression can be condensed because of the fact that for the symmetric top certain of the coefficients τ' are equal, that is, $\tau'_{xxxx} = \tau'_{xxyy} = \tau'_{yyyy}$. Therefore, the terms which contain contributions from $\langle P_x^4 \rangle$ and $\langle P_y^4 \rangle$ which are not simple functions of P^2 and K^2 cancel with those from $\langle P_x^2 P_y^2 + P_y^2 P_x^2 \rangle$. The nonvanishing terms can be combined in an expression of the form

$$E_d^{(1)} = -h[D_J J^2(J+1)^2 + D_{JK} J(J+1) K^2 + D_K K^4] \tag{6.13}$$

in which the D_J, D_{JK}, and D_K are the usual first-order centrifugal stretching constants of the symmetric-top molecule expressed in frequency units. The D's obviously represent a combination of the τ's. For molecules of $\mathscr{C}_{3v}$ symmetry, expressions of the centrifugal stretching constants in terms of τ's are given in Table 8.11 and 8.12.

Addition of $E_d^{(1)}$ to the rigid rotor values of (6.6) gives the usual expression for the rotational energy of the nonrigid prolate symmetric-top molecule

$$E_{J,K} = h[B_0 J(J+1) + (A_0 - B_0) K^2 - D_J J^2(J+1)^2 - D_{JK} J(J+1) K^2 + D_K K^4] \tag{6.14}$$

For the oblate symmetric top, the expression in the parentheses of the second term becomes $(C-B)$ instead of $(A-B)$. With the selection rules $J \to J+1$,

$K \rightarrow K$, this equation gives the rotational frequencies as

$$\nu = 2B_0(J+1) - 4D_J(J+1)^3 - 2D_{JK}(J+1)K^2 \tag{6.15}$$

This frequency equation holds for both the prolate and the oblate tops. Note that neither the second nor the last term in the energy expression of (6.14) influences the pure rotational frequencies. These terms involve only rotation about the symmetry axis. The last term of (6.15) splits the rotational transition into $J+1$ closely spaced components whose separation increases as K^2.

The next higher order correction for centrifugal stretching leads to the expression

$$\begin{aligned}\nu = h\{&2B_0(J+1) - 4D_J(J+1)^3 - 2D_{JK}(J+1)K^2 \\ &+ H_J(J+1)^3[(J+2)^3 - J^3] + 4H_{JK}(J+1)^3K^2 + 2H_{KJ}(J+1)K^4\}\end{aligned} \tag{6.16}$$

for the ground-state rotational frequencies of symmetric-top molecules. The first-order expression, (6.15), has been found entirely adequate for most measurements. Small, higher-order terms have been detected in a few cases [4, 5].

Figure 6.2 indicates the appearance of the first few rotational transitions of a symmetric-top molecule. The K splitting is greatly exaggerated. Figure 6.3 shows the K lines of the $J=8 \rightarrow 9$ transition of CF_3H. Generally, pyramidal molecules such as NH_3 and PH_3 have D_{JK} negative so that the lines of a given J transition occur at higher frequencies as K increases, whereas those for which the symmetrical group occurs on the end of a linear group, such as CH_3CN and CH_3CCH, have D_{JK} positive and have the lines of increasing K falling at lower frequencies. Those such as CF_3H (shown in Fig. 6.3) and CCl_3H, which have a heavy symmetrical group and the light H atom attached to a central carbon, have negative D_{JK} like the pyramidal molecules, whereas those with the lighter H atoms forming the symmetrical group of CH_3F, CH_3Cl, and so on, have D_{JK} positive. The values of B_0, of D_J, and of D_{JK} for selected molecules of different types are compared in Table 6.1. For all types, D_J is positive; therefore the D_J term shifts the different J transitions to lower frequencies.

The centrifugal stretching constants of symmetric-top molecules are related to the bond force constants and fundamental vibrational frequencies, but the

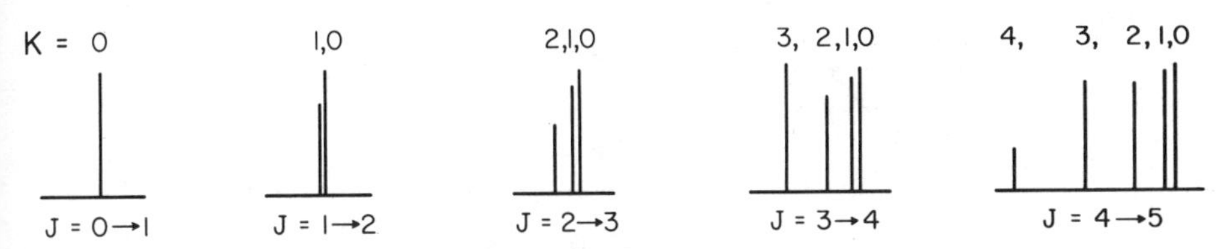

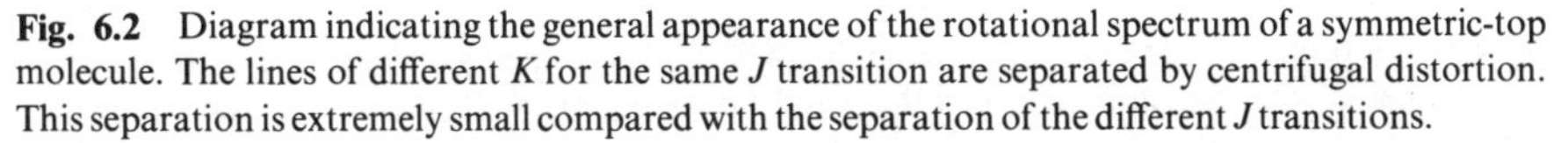

Fig. 6.2 Diagram indicating the general appearance of the rotational spectrum of a symmetric-top molecule. The lines of different K for the same J transition are separated by centrifugal distortion. This separation is extremely small compared with the separation of the different J transitions.

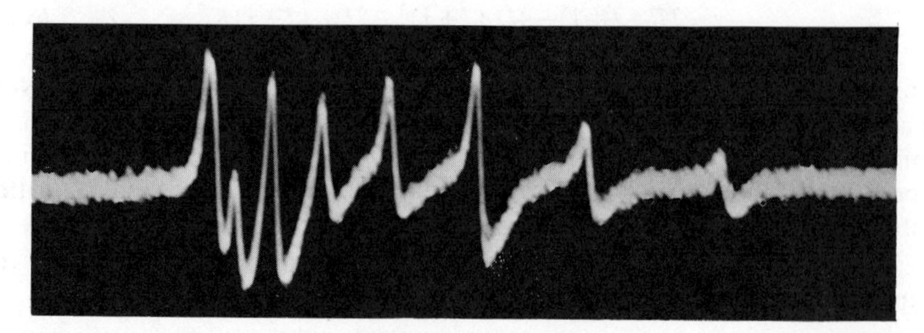

Fig. 6.3 Cathode ray display of the eight K lines of the $J=8\rightarrow 9$ transition of CF_3H occurring at $\lambda=1.61$ mm. The centrifugal distortion constant D_{JK} is negative (-18 kHz) so that the higher K components fall at higher frequencies to the right. The total frequency spread of the multiplet is 21 MHz. From C. A. Burrus and W. Gordy, *J. Chem. Phys.* **26**, 391 (1957).

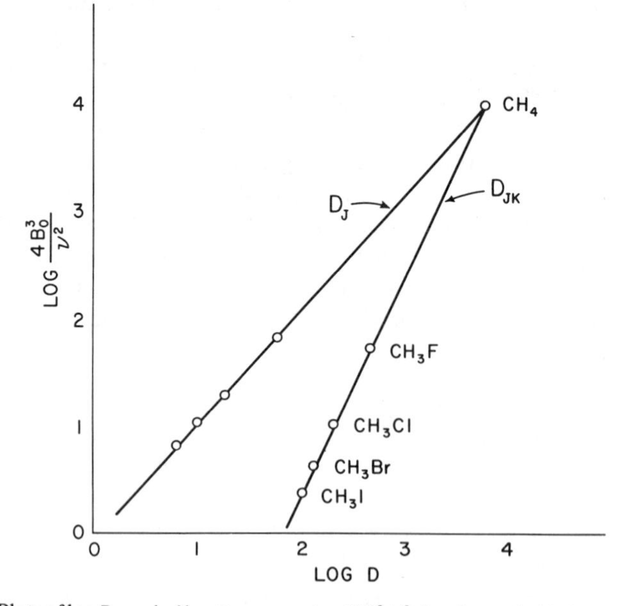

Fig. 6.4 Plots of log D_J and of log D_{JK} versus log $(4B_0^3/\nu^2)$ for the methyl halides where ν represents the fundamental vibrational frequency of the parallel mode υ_3 for the D_J plot and the fundamental frequency of the perpendicular mode υ_6 for the D_{JK} plot. From Orville-Thomas, Cox, and Gordy [7].

Table 6.1 Selected Rotational Constants of Some Symmetric-top Molecules in the Ground Vibrational State

Molecule	B_0 *(MHz)*	D_J *(kHz)*	D_{JK} *(kHz)*	*Ref.*
PH_3	133,480.15	3950	−5180	a
AsH_3	112,470.59(3)	2925(3)	−3718(4)	b
$^{121}SbH_3$	88,038.99(3)	1884(4)	−2394(15)	b
NF_3	10,681.02(1)	14.53(7)	−22.69(10)	c,d
$N^{35}Cl$	3,468.603(3)	1.863(2)	−3.015(6)	e
CH_3F	25,536.147(2)	59.87(20)	420.3(4)	f,g
$CH_3{}^{35}Cl$	13,292.869(10)	18.0	198(3)	g
$CH_3{}^{79}Br$	9,568.19(5)	9.87	128(1)	g
CH_3I	7,501.276(5)	6.31	98.7(1)	g
CH_3CN	9,198.897(6)	3.81(8)	176.9(2)	h
SiH_3CN	4,973.009(15)	1.48(10)	63(2)	i
CH_3CCH	8,545.869	2.875	163.0	j,k
SiH_3CCH	4,828.687	2.1	63	l
CH_3CNO	3,914.796(4)	0.42(2)	153.1(3)	m
SiH_3NCO	2,517.888(14)	0.9(2)	641(13)	n
SiH_3NCS	1,516.040(1)	0.037(4)	41.958(6)	o

[a]Helminger and Gordy [77].
[b]Helminger et al. [78].
[c]A. M. Mirri and G. Cazzoli, *J. Chem. Phys.*, **47**, 1197 (1967).
[d]Otake et al. [117].
[e]G. Cazzoli, P. G. Favero, and A. Dal Borgo, *J. Mol. Spectrosc.*, **50**, 82 (1974).
[f]R. S. Winton and W. Gordy, *Phys. Lett.*, **A 32**, 219 (1970).
[g]T. E. Sullivan and L. Frenkel, *J. Mol. Spectrosc.*, **39**, 185 (1971).
[h]A. Bauer and S. Maes, *J. Phys. (Paris)*, **30**, 169 (1968).
[i]Careless and Kroto [31].
[j]C. A. Burrus and W. Gordy, *J. Chem. Phys.*, **26**, 391 (1957).
[k]A. Bauer and J. Burie, *Compt. Rend.*, **268 B**, 1569 (1969).
[l]Gerry and Sugden [30].
[m]H. K. Bodenseh and K. Morgenstern, *Z. Naturforsch.*, **25a**, 150 (1970).
[n]M. C. L. Gerry, J. C. Thompson, and T. M. Sugden, *Nature*, **211**, 846 (1966).
[o]K. F. Dössel and A. G. Robiette, *Z. Naturforsch.*, **32a**, 462 (1977).

relationship is generally quite complicated because of the many vibrational degrees of freedom of the polyatomic molecules (see Chapter VIII). The methyl halides provide perhaps the simplest group of organic symmetric-top molecules for which attempts at relating the vibrational frequencies with the stretching constants have been made. From the infrared vibrational spectra Chang and Dennison [6] have calculated values of $D_J = 18.4$ kHz and $D_{JK} = 189$ kHz for $CH_3{}^{35}Cl$ which agree very well with the observed microwave values [7], 18.1 and 198 kHz, respectively.

One can, to a useful approximation, treat the methyl halides as diatomic molecules in which the CH_3 group as a unit vibrates against the halogen. To this simplified model one can then apply (4.16) for diatomic molecules

$$D_J = \frac{4B^3}{\omega_i^2} \tag{6.17}$$

where ω_i represents the fundamental parallel vibrational frequency corresponding to the stretching of the C—X bond. Figure 6.4 shows a plot [7] of log D_J versus log (B^3/ω_i^2) for the methyl halides with ω_i taken as the parallel frequency ω_3. The relationship is linear as expected from (6.17). In the same figure is given a plot of log D_{JK} versus log (B^3/ω_i^2) with ω_i taken as the frequency of the mode ω_6 in which the atoms vibrate in planes perpendicular to the symmetry axis. This relationship is also found to be linear. As the mass of X is reduced to that of H, both the frequencies ω_3 and ω_6 go over into the deformation frequency ω_4 in methane [8]. Likewise, the parameters D_J and D_{JK} go over into a single distortion parameter D for the spherical-top molecule. It is therefore interesting to note that the two plots of Fig. 6.4 intersect at the point for methane if ω_i is taken as its deformation frequency ω_4 and if $D_J = D_{JK} = D$ represents the distortion constant of CH_4 from optical spectroscopy.

2 MOLECULES IN EXCITED VIBRATIONAL STATES

For symmetric-top molecules in excited vibrational states the effective rotational constants differ somewhat from those of the ground state. The effective values of the constants in excited states are expected to have, in the first approximation, variation with the vibrational quantum number similar to that for linear molecules. To the approximation usually required,

$$B_v = B_e - \sum \alpha_i^b \left(v_i + \frac{d_i}{2} \right) \tag{6.18}$$

$$A_v = A_e - \sum \alpha_i^a \left(v_i + \frac{d_i}{2} \right) \tag{6.19}$$

The higher-order expression B_v is similar to that for the linear molecule, (5.26). In these relations B_e and A_e are the equilibrium values, v_i is the vibrational quantum number for the ith mode, α_i^a and α_i^b are small anharmonic constants, and d_i represents the degeneracy of the ith mode. Slight differences in the stretching constants from those of the ground vibrational state are also expected. These differences, which are usually negligible, may be represented by a formula similar to that of (4.12). Silver and Shaffer [9, 10] have shown that (6.18) and (6.19) hold for pyramidal molecules; Shaffer [11], for axially symmetric molecules. Nielsen [12] has given a general treatment of the subject.

For molecules in nondegenerate vibrational modes, the rotational frequencies of either the prolate or the oblate symmetric top can be expressed by

$$\nu = 2B_v(J+1) - 4D_J^{(v)}(J+1)^3 - 2D_{JK}^{(v)}(J+1)K^2 \tag{6.20}$$

where B_v is given by (6.18), with $d_i = 1$. For molecules in degenerate vibrational modes, $d_i > 1$, the change in B_v can also be expressed by (6.18), but (6.20) does not adequately predict the rotational spectrum. In addition to the change in A, B, or C for the degenerate vibrational modes, there is a splitting of the rotational lines, which is similar to the l-type doubling for linear molecules.

The displacement of rotational lines of symmetric-top molecules by vibrational excitation is illustrated in Fig. 6.5. This recording shows $J=11\rightarrow 12$ rotational transitions in the 21 GHz region of $(CH_3)_3C—C\equiv C^{35}Cl$ in excited vibrational states up to $v=5$ of the fundamental mode believed to be that corresponding to a bending mode between the tertiary C atom and the acetylene group [13]. The labeling on the figure indicates the main peaks of the rotational lines in the vibrational states $v=0$ to $v=5$ which have unresolved l-type doubling as described later, also unresolved nuclear quadrupole structure and K structure. From a fitting of the relative intensities of the lines with those expected from the Boltzmann relation (Chapter III), the fundamental frequency of the vibration was found to be 120 cm^{-1}, approximately.

For symmetric-top molecules such as CH_3CCH which have a linear group along the symmetry axis, there are degenerate bending vibrational modes which give rise to l-type doubling of the rotational levels similar to that described for linear molecules in Chapter V. Also, degenerate vibrational modes involving perpendicular motions of the atoms which are off the axis can give rise to components of vibrational angular momentum along the symmetry axis. Thus degenerate modes leading to vibrational angular momentum occur in pyramidal XY_3 molecules and even in planar XY_3 molecules. Because of Coriolis interaction between vibration and rotation, a splitting of the levels occurs which is known as l-type doubling. There is a component of this vibrational angular momentum along the figure axis which adds to, or subtracts from, the pure rotational angular momentum about the symmetry axis. This component has the value $\zeta l\hbar$, where ζ is the Coriolis coupling constant and

$$l=v_d,\ v_d-2,\ldots,\ -v_d \tag{6.21}$$

where v_d is the vibrational quantum number of the degenerate mode. For the most important case, $v_d=1$ and $l=\pm 1$. When the vibrating motions are perpendicular to the symmetry axis, $\zeta=1$ and the vibrational component along z is $l\hbar$, just as in the linear molecules. In the symmetric top, however, the

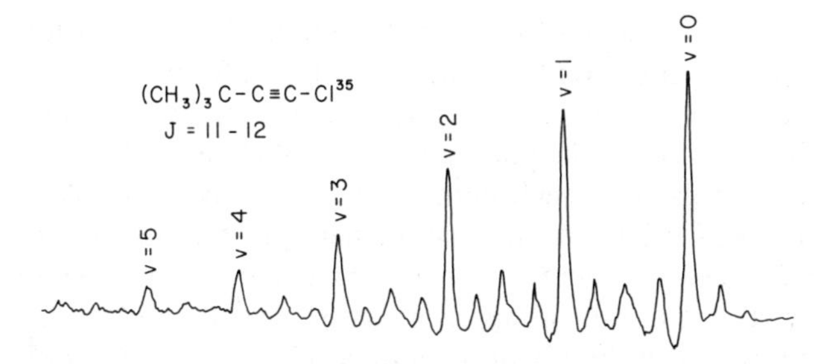

Fig. 6.5 Recording of the $J=11\rightarrow 12$ rotational transition of $(CH_3)_3C—C\equiv C^{35}Cl$ in excited states up to $v=5$ of a bending vibrational mode. The substructure of the lines caused by l-type doubling, centrifugal distortion, and nuclear coupling is not resolved. From Bodenseh, Gegenheimer, Mennicke, and Zeil [13].

component along the symmetry axis can, in general, be less than $l\hbar$, or $0 \leqslant |\zeta| \leqslant 1$. In any event, the quantum mechanics of the symmetric top requires that the overall angular momentum along the symmetry axis, that is, the sum of that caused by pure rotation and vibration, must be an integral multiple of $\hbar$ instead of that caused by vibration or rotation alone. Likewise, the square of the overall angular momentum including that arising from the degenerate vibrational modes is quantized with values $J(J+1)\hbar^2$.

If P_x, P_y, and P_z represent the overall angular momentum about the principal axis including that caused by pure rotation of the molecule and that caused by vibration, the pure rotational Hamiltonian can be expressed as

$$\mathscr{H}_r = \frac{(P_x - p_x)^2}{2I_x} + \frac{(P_y - p_y)^2}{2I_y} + \frac{(P_z - p_z)^2}{2I_z} \tag{6.22}$$

where p_x, p_y, and p_z represent the components of the angular momentum which arises from the degenerate vibratory motions. Let us assume a prolate symmetric top with $I_x = I_y = I_b$, $I_z = I_a$, and $p_x = p_y = 0$. The $\mathscr{H}_r$ can then be expressed as

$$\begin{aligned} \mathscr{H}_r &= \frac{P^2}{2I_b} - \frac{P_z^2}{2I_b} + \frac{(P_z - p_z)^2}{2I_a} \\ &= \frac{P^2}{2I_b} + \frac{1}{2}\left(\frac{1}{I_a} - \frac{1}{I_b}\right) P_z^2 - \frac{P_z p_z}{I_a} \end{aligned} \tag{6.23}$$

In the last expression we neglect the term $p_z^2/(2I_a)$, which represents pure vibrational energy and which does not change with the rotational state. Since

$$P^2 = \hbar^2 J(J+1), \qquad P_z = K\hbar, \qquad \text{and} \qquad p_z = \zeta l \hbar \tag{6.24}$$

$$\frac{E_r}{h} = B_v J(J+1) + (A_v - B_v) K^2 - 2\zeta l K A_v \tag{6.25}$$

where

$$\begin{aligned} J &= 0, 1, 2, 3, \ldots \\ K &= 0, \pm 1, \pm 2, \ldots, \pm J \end{aligned} \tag{6.26}$$

and where l values are given by (6.21). To obtain the formula for the oblate symmetric top one simply replaces the A_v by C_v. A formula like that of (6.25) was originally derived by Teller [14] and by Johnston and Dennison [15].

Since both K and l can take plus and minus values, (6.25) indicates a doublet splitting of the levels with each component retaining a double degeneracy. However, certain of the levels may be further split by interactions of higher order. When $|K| = 3n+1$ where n is an integer including zero, the term with Kl positive represents two coinciding levels of species A whereas the term with Kl negative represents a doubly degenerate level of species E. For $|K| = 3n+2$, the reverse is true. When $|K| = 3n$ but not including zero, the terms with Kl positive or negative represent separately doubly degenerate levels of species E.

The double degeneracy of the levels of species E cannot be removed [16], but the coinciding levels of species A may be split by interactions of higher order. H. H. Nielsen [17] has shown that the further splitting of the levels of species A can be appreciable for $K=1$, but that it is negligible for other K values.

More complete treatments of the interaction of degenerate vibrations with rotation in symmetric-top molecules have been given by Nielsen [17, 18], by Grenier-Beeson and Amat [19], and by others [20–22]. For singly excited degenerate bending, with $v_d=1$, $l=\pm 1$, Nielsen's formula for the rotational energy which includes stretching as well as higher-order interaction with the degenerate vibration is

$$\begin{aligned} E_{JK}=h\{&B_vJ(J+1)+(A_v-B_v)K^2-2A_vKl\zeta-D_JJ^2(J+1)^2\\ &-D_{JK}J(J+1)K^2-D_KK^4+2\\ &\times[(2D_J+D_{JK})J(J+1)+(2D_K+D_{JK})K^2]Kl\zeta+P(J,\ K,\ l)\} \end{aligned} \tag{6.27}$$

where for $K=l=\pm 1$

$$P=\pm\tfrac{1}{2}J(J+1)q \tag{6.28}$$

where $q\approx 2B^2/\omega$, and ω is the fundamental bending vibrational frequency. For $K\neq l=\pm 1$,

$$P=\pm\frac{[J(J+1)-K(K\mp 1)][J(J+1)-(K\mp 1)(K\mp 2)]}{8(K\mp 1)[(1-\zeta)A_v-B_v]}\,q^2 \tag{6.29}$$

The upper signs are taken when K and l have the same sign; the lower signs, when K and l are of different sign. With the selection rules,

$$J\rightarrow J+1, \qquad K\rightarrow K, \qquad l\rightarrow l \tag{6.30}$$

the predicted frequencies are

$$\begin{aligned} \nu=&2B_v(J+1)-4D_J(J+1)^3-2D_{JK}(J+1)K^2\\ &+4(2D_J+D_{JK})(J+1)Kl\zeta+\Delta P(J,\ K,\ l) \end{aligned} \tag{6.31}$$

where

$$\Delta P=\pm q(J+1) \tag{6.32}$$

for $K=l=\pm 1$, and

$$\Delta P=\pm\frac{(J+1)[(J+1)^2-(K\mp 1)^2]}{4(K\mp 1)[(1-\zeta)A_v-B_v]}\,q^2 \tag{6.33}$$

for $K\neq l=\pm 1$. Upper signs are taken when K and l have the same sign; the lower signs hold when K and l have different signs.

The qualitative features predicted by Nielsen's theory, including the rather wide separation of the $K=l=\pm 1$ lines by $2q(J+1)$, were verified [23] by early measurements on CH_3CCH. The theory has been applied to a number of other symmetric-top molecules [24], including SiH_3NCS which, surprisingly, is a symmetric top, and to several different rotational transitions of CH_3CN

[25]. Table 6.2 illustrates the extent of agreement of the calculated with the observed frequencies of the $J=10 \rightarrow 11$ lines of the $v_8=1$ bending vibrational state of CH_3CN. Although it was found that the positions of the various doublets for CH_3CN were given adequately by Nielsen's theory, the separations of the individual doublets were not satisfactorily predicted. Venkateswarlu et al. [25] found that these separations could be closely predicted, as in Table 6.2, if the term $4\varepsilon K\zeta(J+1)$ where $\varepsilon \approx 0.034$ is inserted into (6.31). Values of interaction constants were found to be $\zeta=0.878$ and $q=17.775$ MHz. Measurements and analysis of CH_3CN have been extended by Bauer [26] to include several rotational transitions of the second excited state, $v_8=2$.

Nielsen's theoretical work on l-type doubling has been extended and improved by others. Formulas for $J \rightarrow J+1$ rotational frequencies including l-type doubling are provided by Grenier-Beeson and Amat [19] for first excited states, $v_t=1$, of molecules having C_{3v} symmetry. The formulation is extended by Tarrago [27] to second excited vibrations having $l=1$, that is, $v_t=2^1$ states.

Table 6.2 The Observed and Calculated Frequencies and the Assignments of the $J=10 \rightarrow 11$ Lines Corresponding to the Molecules in the Excited State $v_8=1$ of CH_3CN[a]

J	Transition K	Transition l	*Observed Frequency (MHz)*	*Calculated Frequency (MHz)*
$10 \rightarrow 11$	± 1	± 1	203,161.23	203,161.53
			202,769.94	202,770.08
	± 2	± 1	202,972.63	202,972.53
	0	∓ 1	202,950.97	202,950.87
	± 3	± 1	202,956.31	202,956.28
	± 1	∓ 1	202,943.39	202,943.68
	± 4	± 1	202,935.67	202,935.61
	± 2	∓ 1	202,924.94	202,925.23
	± 5	± 1	202,907.98	202,907.97
	± 3	∓ 1	202,897.68	202,898.15
	± 6	± 1	202,872.91	202,872.87
	± 4	∓ 1	202,862.38	202,862.88
	± 7	± 1	202,830.05	202,830.11
	± 5	∓ 1	202,819.06	202,819.63
	± 8	± 1	202,779.70	202,779.64
	± 6	∓ 1	202,768.06	202,768.47
	± 9	± 1	202,721.62	202,721.40
	± 7	∓ 1	202,709.07	202,709.43
	± 10	± 1	202,655.71	202,655.39
	± 8	∓ 1	202,642.27	202,642.51

[a]From Venkateswarlu et al. [25].

The formulas by Grenier-Beeson and Amat were applied by them [19] to the available data on $F_3CC{\equiv}CH$ and found to provide consistent agreement. Others have successfully applied their theory: Otake et al. [28], to measurements on the degenerate v_3 and v_4 modes of NF_3; Whittle et al. [29], to the degenerate v_7 and v_8 modes of CF_3CN; Gerry and Sugden [30], to the degenerate bending mode v_{10} of SiH_3CCH, as examples. The formulation of Tarrago [27] has been applied with consistent fitting to rotational transitions of a number of C_{3v} molecules in $v=2^1$ states: to CH_3CN, by Bauer [26]; to SiH_3CN, by Careless and Kroto [31]. The latter work extends the formulation to include $v_8=3$ and 4 states of SiH_3CN.

Direct transitions between the l-type doublets in the $J=17$ to 24 rotational levels of the $v_4=1$ state of PF_3 have been observed and analyzed by Hirota [32]. The l-type doubling constant, $q_4=29.49270(8)$ MHz, and the 2-1 interaction constant, $r_4=3.013(5)$ MHz, were obtained. Direct l-type doublet transitions have also been measured and analyzed in rotational levels of the $v_6=1$ state of CF_3H and CF_3D by Kawashima and Cox [33]. The values, $|q_6|=36.27917(13)$ MHz and $|r_6|=0.97(2)$ MHz, were obtained. These papers provide instructive examples of the analysis of l-type doublet spectra in symmetric-top molecules.

An example of Fermi resonance between different vibrational states is provided by the work of Morino and Hirose [34] in the measurement of shifts in the lines due to interaction of the v_5 and v_3+v_6 vibrational states of CH_3I. Accidentally strong resonances between certain rotational levels of the $v=2$ vibrational states of both $CH_3{}^{12}CN$ and $CH_3{}^{13}CN$ have been observed by Bauer and Maes [35].

3 MOLECULAR INVERSIONS

According to quantum mechanics, pyramidal XY_3 symmetric-top molecules can execute inversion motion in which the X atom moves through the Y_3 plane to achieve an identical but inverted pyramidal configuration. The inverted configuration is obtained mathematically by a change of sign of the coordinates of all the particles measured from the center of mass. Theoretically, such inversion is possible in any nonplanar molecule (Chapter III, Section 4), but, practically, it is of significance for only a very few. The inversion potential curve of the NH_3 molecule is illustrated by Fig. 6.6. The potential energy of the molecule plotted as a function of the distance between the N atom and the H_3 plane has double minima corresponding to the two equivalent equilibrium configurations.

The wave functions of the molecule can be expressed as linear combinations of ϕ_L and ϕ_R of those of the molecule in the equivalent right and left configurations. These functions,

$$\psi_+=\frac{1}{\sqrt{2}}(\phi_L+\phi_R) \tag{6.34}$$

$$\psi_-=\frac{1}{\sqrt{2}}(\phi_L-\phi_R) \tag{6.35}$$

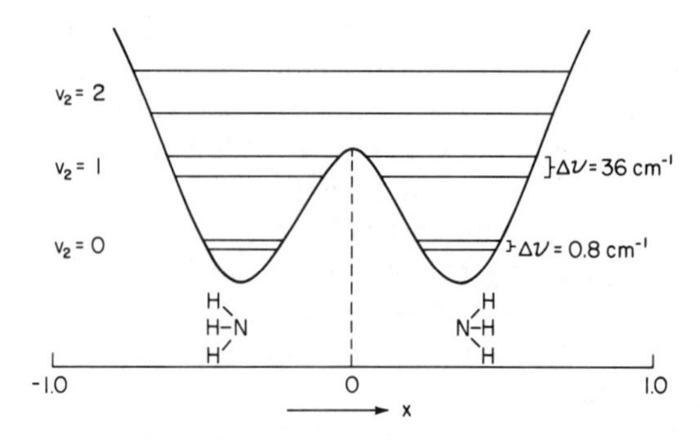

Fig. 6.6 Potential curve of the NH_3 inversion.

have opposite symmetry. For an infinitely high barrier they represent degenerate states; but when the barrier is sufficiently low, as in NH_3, the two states are separated by a measurable interaction energy. Selection rules (Chapter III, Section 4) allow transitions $+\leftrightarrow-$ between the states giving rise to inversion spectra. The transition moment is given by (2.125).

In addition to the inversion motion, the molecule has a symmetrical vibrational mode in which the X atom moves back and forth along the symmetry axis in opposition to the motions of the Y_3 atoms. In the first and second excited symmetrical modes this vibrational energy for NH_3 is lower than the potential energy barrier between the two minima. In all other XY_3 molecules the potential hill between the vibrational and potential energy is much greater relative to the vibrational energy. Thus, in lower symmetrical vibrational modes the XY_3 molecule executes vibrational motions in the potential valley on either side of the potential barrier. Classically, it is not possible for the molecule to invert when in these vibrational states which lie below the potential maximum. Quantum mechanically, though, it is possible for the molecule to achieve inversion transition through "barrier tunneling." In NH_3, the N atom tunnels through the H_3 plane at approximately 23 GHz to give rise to the well-known inversion spectrum in the 1.3-cm wave region, the first spectrum to be observed in the microwave region [36–38]. The corresponding inversion frequency of ND_3 is approximately 1.7 GHz, and for NT_3 it is 0.306 GHz. No other symmetric-top molecules have inversion frequencies that fall in the microwave region. The next highest inversion frequency [39, 40], that of PH_3, is predicted to be very small. So far, the inversion splitting of the rotational lines of PH_3 has proved undetectable with the highest resolution available. Other related molecules such as AsH_3 are predicted to have such long inversion time, of the order of days [40], that no observable effects on microwave spectra are expected. However, related barrier-tunneling effects associated with internal rotation and torsional oscillation of groups within symmetric-top and asymmetric-top molecules are observable in microwave spectra. These effects are treated in Chapter XII.

Theoretically, one can obtain the inversion frequencies from a solution of the wave equation for the vibrating rotor, as is done for diatomic molecules with the assumption of a double minima potential of the appropriate form. This is a very difficult problem, however. In a first-order approximation one can neglect the interaction between vibration and rotation and can obtain separate solutions of the wave equation for the vibration and rotation and can then, with the perturbation theory, correct for the interaction. The simplest model that accounts for the inversion is one which treats the three hydrogens in a plane as a unit vibrating against the N atom, somewhat as a diatomic molecule. The reduced mass of the vibration is $\mu=3mM/(3m+M)$. The wave equation for this simple model is

$$\frac{d^2\psi}{dx^2}+\frac{8\pi^2\mu}{h^2}[E_v-V(x)]\psi=0 \tag{6.36}$$

With a double minima potential function, the solution yields the vibrational levels split by inversion. Solutions have been obtained by Manning [41], also by Dennison and Uhlenbeck [42], whose solutions for inversion energies and frequencies are

$$\frac{\Delta E_{\text{inv.}}}{\Delta E_v}=\frac{1}{\pi A^2} \quad \text{or} \quad \nu_{\text{inv}}=\nu_{\text{vib}}\left(\frac{1}{\pi A^2}\right) \tag{6.37}$$

where

$$A=\exp\left\{\frac{2\pi}{h}\int_0^{x_1}[2\mu(V(x)-E_v)]^{1/2}\,dx\right\} \tag{6.38}$$

in which

ΔE_v = separation of the vibrational levels
ΔE_{inv} = inversion splitting of the vibrational levels
ν_{inv} = the inversion frequency
ν_{vib} = the fundamental vibrational frequency
E_v = the total vibrational energy
$x=x_1$ when $V=E_v$

The quantity A measures the area under the potential curve between the potential minima and is not particularly sensitive to the exact shape of the curve. Dennison and Uhlenbeck used the *W-K-B* method and assumed the curves in the region of the minima to be parabolic. Manning assumed a potential function of the form

$$V\ (\text{in cm}^{-1})=66{,}551\ \text{sech}^4\frac{x}{2\rho}-109{,}619\ \text{sech}^2\frac{x}{2\rho} \tag{6.39}$$

where $\rho=6.98\times10^{-8}/\mu^{1/2}$. This formula was found to predict the frequencies quite well. It allows, through the effects of μ, the prediction of the isotopic shifts of the inversion frequency.

The effects of rotation on the inversion splitting have been taken into account in various semiempirical formulas, the nature of which is suggested by the centrifugal stretching terms of the noninverting symmetric-top rotor or by the nature of the function A of (6.38). The various constants in these expressions are then evaluated by a fitting of them to the accurately measured rotational fine structure of the inversion. The formula that fits most closely all the accurately measured fine structure is an exponential function suggested by Costain [43] from a consideration of the nature of the function of (6.37) as well as the nature of the centrifugal stretching terms in symmetric-top rotors. It is evident that centrifugal distortion would alter slightly both V and E_v of (6.38). Costain's formula was extended to higher order by Schnabel et al. [44] and its various constants determined by a fitting of the formula to the most complete and accurate measurements on the NH_3 available to 1965. Their formula and values for the constants are given in the earlier edition of this book.

Because the lowest rotational frequency of NH_3 and ND_3 falls in the submillimeter wave region, the many early microwave measurements on these symmetric molecules were limited to their inversion spectra, although combined rotation–inversion transitions were observed for the mixed species NHD_2, NH_2D which are asymmetric rotors. In 1957, observation of the $J=0\rightarrow 1$ rotational transition of ND_3, which occurs at $\lambda=0.97$ mm, was reported [45], but the corresponding rotational transitions of $^{14}NH_3$ and $^{15}NH_3$, which occur in the 0.52-mm wave region, were not observed until 1967 [46]. In 1974, the rotation–inversion spectrum of NT_3 in the $v_2=0$ states was measured [47].

Figure 6.7 shows the energy level diagram of the first few rotation–inversion levels of NH_3 and ND_3 with the observable transitions indicated. Note that for NH_3, only one of the inversion levels occurs when $K=0$, whereas for ND_3

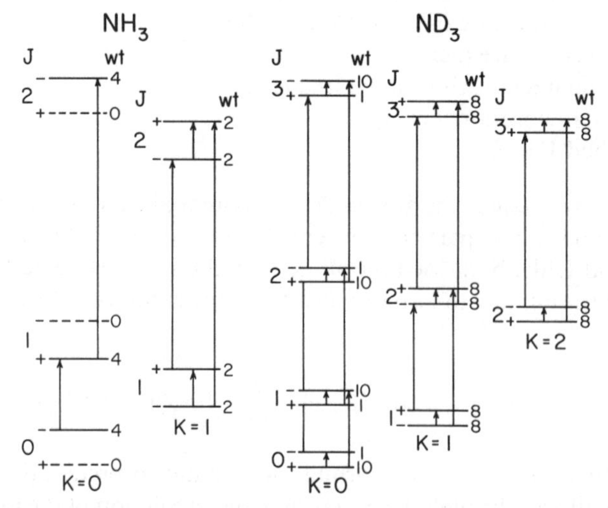

Fig. 6.7 Diagram of the lower rotation-inversion levels and transitions of NH_3 and ND_3 with the allowed transitions indicated.

both the + and − levels occur, but with differing statistical weights. This difference is due to the effects of nuclear spin statistics discussed in Chapter III, Section 4 and conforms to the statistical weight formula given in Table 3.3.

Figure 6.8 indicates how the measured centimeter wave inversion frequencies for the upper and lower levels were combined with the $J=0\rightarrow 1$ submillimeter wave rotation–inversion frequencies [48] to give accurate values for the rotational constants B_0 for $^{14}NH_3$ and $^{15}NH_3$. After correction of the rotational frequencies for inversion effects, they were used in (6.15) with infrared D_J values for calculation of the B_0 values listed on Fig. 6.8. Measured inversion frequencies of ND_3 in the 2-GHz microwave region [49, 50] have also been combined with the observed submillimeter wave rotation–inversion transitions for derivation of the rotational constants [48] B_0, D_J, and D_{JK} for the $^{14}ND_3$ and $^{15}ND_3$, entirely from microwave data.

The inversion splitting of NT_3, together with its rotational constants, has been obtained by Helminger et al. [47], entirely from measurements of millimeter and submillimeter wave rotation–inversion transitions. They used the frequency formula

$$\nu=2B_0(J+1)-4D_J(J+1)^3-2D_{JK}(J+1)K^2\pm[(\nu_i)_0+C_1(J+1)^2+C_2K^2] \tag{6.40}$$

and adjusted the constants by the least-squares method to reproduce the observed frequencies as shown in Table 6.3. The resulting constants are given at the bottom of Table 6.3. The $J=0\rightarrow 1$ frequency is corrected for ^{14}N nuclear quadrupole splitting. The bracketed term in (6.40) represents the inversion

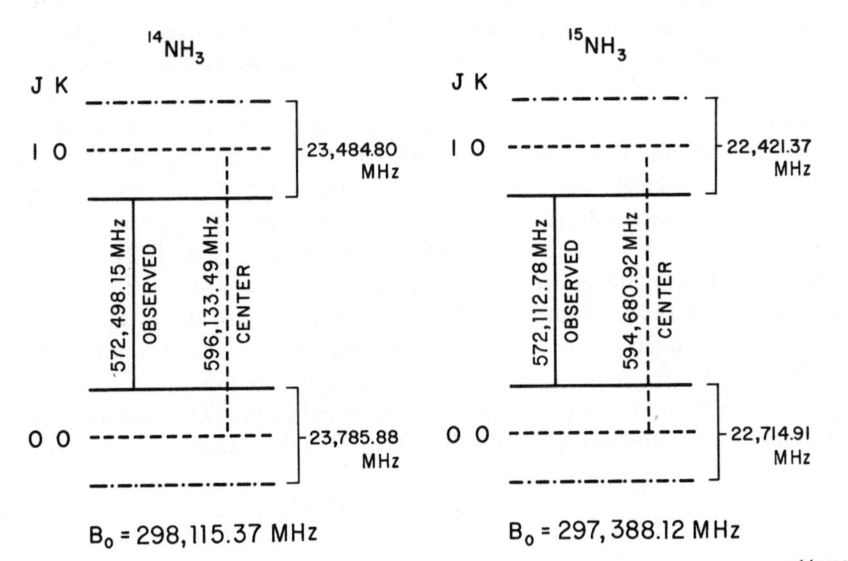

Fig. 6.8 Illustration of the derivation of B_0 from the observed rotational frequencies of $^{14}NH_3$ and $^{15}NH_3$ as measured. From Helminger et al. [48].

Table 6.3 Rotation–Inversion Frequencies and Derived Spectral Constants of $^{14}NT_3$ [a]

Transition	*Inversion Component*	*Observed Frequency (MHz)*	*Calculated Frequency (MHz)*	*Difference (MHz)*
$J=0\to1, K=0$	lower	210,814.885	210,815.023	−0.138
$J=1\to2, K=1$	lower	421,891.743	421,891.608	0.135
$J=1\to2, K=0$	upper	422,482.040	422,482.014	0.026
$J=1\to2, K=1$	upper	422,500.800	422,500.871	−0.071
$J=2\to3, K=0$	lower	632,810.839	632,810.768	0.071
$J=2\to3, K=1$	lower	632,836.513	632,836.626	−0.113
$J=2\to3, K=2$	lower	632,914.199	632,914.202	−0.003
$J=2\to3, K=1$	upper	633,440.324	633,440.321	0.003
$J=2\to3, K=2$	upper	633,523.744	633,523.724	0.020

Inversion Constants (MHz)	*Rotation Constants (MHz)*
$(\nu_i)_0=305.89\pm0.11$	$B_0=105{,}565.373\pm0.034$
$C_1=-0.557\pm0.020$	$D_J=2.5981\pm0.0024$
$C_2=0.971\pm0.038$	$D_{JK}=-4.472\pm0.006$

[a] From Helminger et al. [47].

splitting; $(\nu_i)_0$ is the inversion frequency; the C_1 and C_2 terms correct the inversion frequency for centrifugal distortion effects.

The effective and substitution structures for the ground state of ammonia, as derived from the B_0 values of various isotopic species, are given in Section 6, Table 6.7.

The inversion doubling frequencies $(\nu_i)_0$ for the ground state are now known accurately from microwave measurements for all three species, NH_3, ND_3, and NT_3. The much larger inversion frequencies for the $v_2=1$ excited state have been measured with infrared vibration–rotation spectroscopy: for NH_3 and ND_3 by Benedict and Plyler [51] and for NT_3 by Rao et al. [52]. Figure 6.9 shows the relationship between $\log_{10}(\nu_i/\nu_0)$ and $\mu^{1/2}$, where ν_i is the inversion frequency, ν_0 is the fundamental vibrational frequency of the v_2 mode, and μ is the reduced mass of the particular isotopic species of the ammonia. The relationship in each state, $v_2=0$ or 1, is approximately linear.

The relationship revealed in Fig. 6.9 is suggested by the Dennison-Uhlenbeck formula, (6.37). Substitution of (6.38) into (6.37) shows that

$$\frac{\nu_{\text{inv}}}{\nu_{\text{vib}}}=\frac{1}{\pi}\exp\left(-\frac{2}{\hbar}\int_0^{x_1}[2\mu(V(x)-E_v)]^{1/2}\,dx\right) \tag{6.41}$$

and thus

$$\ln\left(\frac{\nu_{\text{inv}}}{\nu_{\text{vib}}}\right)\sim\sqrt{\mu}\left(\int_0^{x_1}[V(x)-E_v]^{1/2}\,dx\right) \tag{6.42}$$

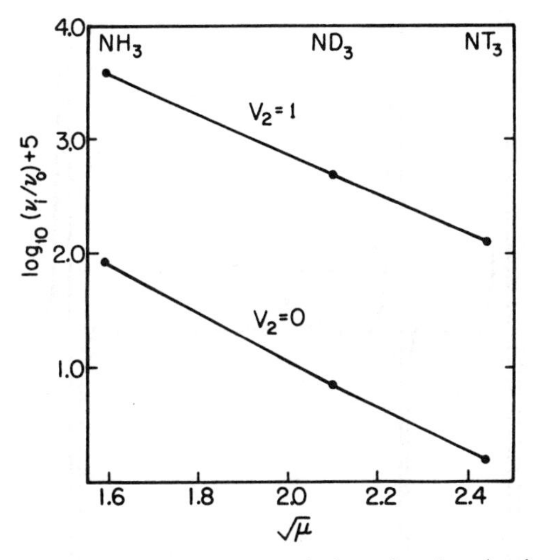

Fig. 6.9 Correlation of the inversion splitting with the reduced molecular mass for the three isotopic species of ammonia in different v_2 vibrational states. From Helminger, De Lucia, Gordy, Morgan, and Staats [47].

where, in Fig. 6.9, $\nu_i \equiv \nu_{\text{inv}}$ and $\nu_0 \equiv \nu_{\text{vib}}$. The fact that the experimental relationship for each vibrational state, $v^2=0$ or $v^2=1$, is approximately linear indicates that the quantity expressed by the integral varies little with the change of isotopic species, that is, approximately $\ln(\nu_i/\nu_0) \sim \mu^{1/2}$ for a given v_2 vibrational state.

The simple, one-dimensional, double-minimum potential model of Dennison and Uhlenbeck [42] and Manning [41], (6.37)–(6.39), has been extended and refined by Swalen and Ibers [53] and by Damburg and Propin [54]. A more comprehensive, vibration–inversion–rotation Hamiltonian that includes centrifugal distortion and Coriolis interactions in the ground and excited inversion states has been developed by Papoušek et al. [55]. The Hamiltonian is a logical extension of one developed by Hougen et al. [56] for triatomic molecules. Further development of this Hamiltonian has been carried out by Špirko et al. [57], who give procedures for a least-squares fitting of its solution to the vibration–inversion–rotation energy levels of NH_3, ND_3, and NT_3. This later form of the ammonia Hamiltonian is related to the one for the inversion–rotation of triatomic molecules with large-amplitude bending motions, developed earlier by Hoy and Bunker [58].

The effective vibration–inversion–rotation Hamiltonian developed by Papoušek et al. [55, 57] has been used in successful analyses of extensive submillimeter wave measurements on the inversion and inversion–rotation spectrum of NH_3 in the $v_2=1$ vibration state by Belov et al. [59]. A section of the submillimeter wave spectrum of $^{14}NH_3$ in the $v_2=1$ vibrational state, obtained with the Krupnov spectrometer, is shown in Fig. 6.10. Scappini and Guarnieri

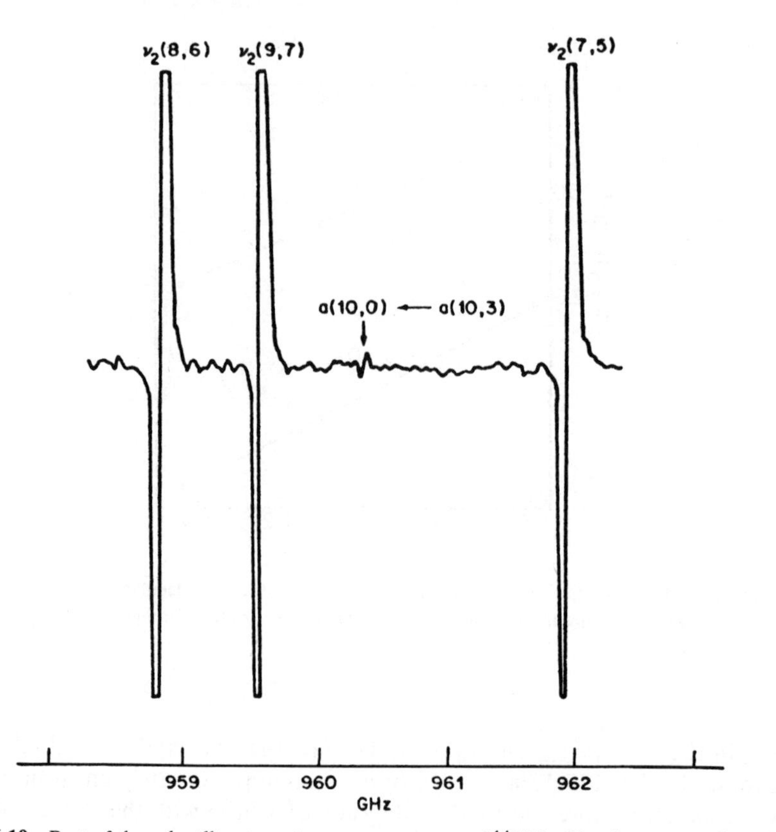

Fig. 6.10 Part of the submillimeter microwave spectrum of $^{14}NH_3$. The three strong lines result from inversion transitions, $s(J, K) \rightarrow a(J, K)$, of the excited $v_2 = 1$ state. The weak line, denoted $a(10, 0) \leftarrow a(10, 3)$, is a perturbation-allowed transition of the ground state, $v_2 = 0$. From S. Urban, V. Špirko, D. Papoušek, J. Kauppinen, S. P. Belov, L. I. Gershtein, and A. F. Krupnov, *J. Mol. Spectrosc.*, **88**, 274 (1981).

[60] have measured the millimeter wave spectrum of ND_3 in the $v_i = 1$ vibrational state and obtained the accurate value of 106,354.35(3) MHz for $(\nu_i)_1$ in that state.

4 "FORBIDDEN" ROTATIONAL TRANSITIONS

Because there is no permanent dipole component perpendicular to the symmetry axis, resonance radiation does not normally induce changes in the K quantum number of symmetric-top rotors; that is, the selection rule $\Delta K = 0$ applies in first order. Thus, from normal rotational spectra, only one inertial constant, I_b or B, can be obtained for symmetric-top rotors, whereas three independent inertial constants can be obtained for most asymmetric rotors. Because of theoretical developments and considerable improvement in experimental techniques and spectrometer sensitivity, it has become possible to detect in polar, symmetric-top rotors the "forbidden" rotational transitions of the ground vibrational state corresponding to changes in the K quantum number. Likewise, it has become possible to detect $\Delta J=1$, $\Delta K\neq 0$ rotational transitions of the ground vibrational states for certain nonpolar, spherical-top molecules. These advances, which came after publication of the earlier edition of this book (1970), have led to the precise microwave measurement of the spectral constant C_0 (or A_0 for prolate-top rotors) of several symmetric-top molecules and B_0 values for some spherical tops such as CH_4. For pyramidal XY_3 molecules, such as PH_3, that have only two structural parameters, these developments have made possible complete determination of the molecular structure of the ground state without isotopic substitution. Examples are given in Section 6, Table 6.8.

The possibility of observing forbidden rotational transitions in the ground state of symmetric-top molecules was apparently reported first by Hansen [61]. In the same year, Oka [62] considered collision-induced $\Delta K=\pm 3$ transitions of NH_3 in mixtures with rare gases. Pure rotational spectra of spherical-top molecules in their ground vibrational states were theoretically predicted by Fox [63] and Watson [64] in 1971. Mizushima and Venkateswarlu [65] had predicted earlier that nonpolar, symmetric-top molecules having T_d symmetry (CH_4, CF_4, SiH_3, etc.) can have vibrationally induced dipole moments and possibly observable pure rotational spectra when in excited, degenerate vibrational states. Their theory was extended to other molecular types by Mills et al. [66]. Pure rotational spectra of nonpolar molecules in excited vibrational states would be very difficult to observe because of the relatively low populations of the excited states. So far as we know, none have been observed.

The first detection of the forbidden $\Delta J=0$, $\Delta K=\pm 3$ rotational transitions in symmetric-top molecules was achieved by Chu and Oka [67], who observed $\Delta J=0$, $\Delta K=\pm 3\,(K=\pm 1\rightarrow \pm 2)$ transitions in PH_3, PD_3, and AsH_3. The first observation of the forbidden rotational transitions in a nonpolar molecule of T_d symmetry, CH_4, was achieved by Rosenberg et al. [68] in far infrared spectroscopy. The first microwave detection of pure rotational of CH_4 was accomplished by Holt et al. [69]. An excellent review of the earlier work on forbidden rotational transitional is given by Oka [70].

A general theory of ground-state, forbidden rotational transitions in polyatomic molecules, with predictions of the selection rules and transition probabilities, has been developed by Watson [64]. The theory for ground-state transitions of tetrahedral molecules like CH_4 was developed independently by Fox [63, 71] and was extended to other molecular types by Aliev and Mikhaylov [72]. Generally, the ground-state transition moment results from centrifugal distortions of the molecules. For those with C_{3v} symmetry, the rotation of the molecule about the b axis produces a small distortion moment perpendicular to the symmetry axis which can give rise to observable $\Delta J=\pm 1, \Delta K=\pm 3$ transitions. In molecules having T_d symmetry and no permanent dipole moment (CH_4, SiH_4, etc.), rotation about a bond axis distorts the molecule so as to produce a weak dipole moment along that axis which can give rise to observable $\Delta J = \pm 1$, $\Delta K = 0$ transitions. In addition to the ground-state, centrifugal-distortion mechanism, Watson [64] has shown that vibration–rotation interaction of the ground state with an excited, degenerate vibrational mode can also contribute to the probability of a forbidden transition.

Watson [64] defined an effective dipole-moment operator which includes the induced distortion moments as well as any fixed dipole components. The operator for the effective dipole moment is expressed as a power series in the vibrational and rotational operators in a manner similar to that used in Chapter VIII, Section 2 for development of the Wilson-Howard Hamiltonian, (8.10). In the molecule-fixed axes, Watson's effective dipole operator is expressed by

$$\mu_\alpha = \mu_\alpha^{(e)} + \sum_{\beta\gamma} \theta_\alpha^{\beta\gamma} P_\beta P_\gamma \tag{6.43}$$

where $\mu_\alpha^{(e)}$ signifies the components of the fixed dipole operator, and the last term signifies the components of the induced dipole operator in which P_β, P_γ are molecule-fixed components of the total angular-momentum operator. The symmetrized, space-fixed components of the operator of (6.43) are

$$\mu_F = \frac{1}{2} \sum (\Phi_{f\alpha}\mu_\alpha + \mu_\alpha \Phi_{f\alpha}) \tag{6.44}$$

where $\Phi_{F\alpha}$ are the direction cosines of the molecule-fixed axes with the space-fixed axes. In calculation of the intensities of the spectra lines, Watson used the matrix elements of the effective dipole operator μ_f of (6.44) in the representation that diagonalizes the Wilson-Howard rotational Hamiltonian, (8.10).

Molecules with C_{3v} Symmetry

Molecules having C_{3v} symmetry comprise the largest class of symmetric tops investigated with microwave spectroscopy. Forbidden $\Delta K = \pm 3$ transitions have been observed in a number of the simpler ones. The $\Delta K = 3$ selection rule for these not-strictly-forbidden transitions arises from the nature of the molecular wave function for the C_{3v} symmetry and an intramolecular interaction $\mathscr{H}_d$ such as rotational distortion that mixes the K and $K' = K \pm 3$ states. Exclusive of the M dependency, which we omit for simplicity, the undistorted symmetric-top rotational function may be expressed, (2.103), by

$$\psi_{JK} = e^{i\phi K} e^{i\phi K} \Theta_{JK} \tag{6.45}$$

where K has both $+$ and $-$ values and ϕ is the angle about the symmetry axis. For a molecule of C_{3v} symmetry, a rotation of 120° exchanges two pairs of identical nuclei. The two configurations are indistinguishable. Thus, $\psi_{JK}\psi_{JK}^*$ is left unchanged for an operation that simply rotates the molecules through $\pm 120°$ or $\pm 2\pi/3$ about the symmetry axis. Under the C_s operation

$$\psi_{JK} = e^{(\pm 2\pi i/3)K} \Theta_{JK} \tag{6.46}$$

If J, the total angular quantum number exclusive of nuclear spin, is assumed to remain a good quantum number, the degree of admixture of the K and K'

states is measured by $\langle J, K|\mathscr{H}_d|J, K'\rangle$, where $\mathscr{H}_d$ is the Hamiltonian for the admixing interaction. With the restricted $\psi_{JK}=|JK\rangle$ of (6.46), this can be expressed [70]

$$\langle J, K|\mathscr{H}_d|J, K'\rangle = e^{(2\pi i/3)(K'-K)}\langle J, K|\mathscr{H}_d|J, K'\rangle \tag{6.47}$$

Only when $K-K'=3n$ where n is an integer including zero or when the interaction $\langle J, K|\mathscr{H}_d|J, K'\rangle$ is zero, are the two sides of (6.47) equal. Thus, with a finite interaction, the selection rules $\Delta K=\pm 3$, ± 6, and so on, may occur in addition to the normal $\Delta K=0$ rule.

The interaction $\mathscr{H}_d$, responsible for admixing of the $|J, K)$ and $|J, K\pm 3)$ levels of the ground vibrational state of symmetric-top rotors having C_{3v} symmetry, is the centrifugal distortion term

$$\mathscr{H}_d = \left(\frac{\hbar^4 \tau_{xxxz}}{4}\right)[(P_+^3+P_-^3)P_z + P_z(P_+^3+P_-^3)] \tag{6.48}$$

where τ_{xxxz} is the centrifugal distortion constant and $P_\pm = P_x \pm iP_y$ are the rotational ladder operators. The $|J, K)$ levels of the ground state can also intermix slightly with $|J, K\pm 2, l\mp 1)$ levels of excited degenerate vibrational states by means of the vibration–rotation interaction term of the form [70]

$$\mathscr{H}_{v-r} = -2\pi B a_{t_1}^{(xx)}(q_+P_-^2 + q_-P_+^2)\left(\frac{ch}{\nu_t}\right)^{1/2} \tag{6.49}$$

where ν_t is the frequency of the degenerate vibration, B is the rotational constant, $a_{t_1}^{(xx)} = (\partial I_{xx}/\partial Q_{t_1})$, Q_{t_1}, Q_{t_2} are the pair of degenerate coordinates, and $q_\pm$ and $P_\pm$ are the vibrational and rotational operators.

Watson [64] shows that the effective dipole-moment operator of (6.44) may be expressed for molecules of D_{3h} or C_{3v} symmetry by

$$\mu_F = \tfrac{1}{2}(\theta_x^{xx})_{\text{eff}}[(\Phi_{Fx}+i\Phi_{Fy})P_+^2 + (\Phi_{Fx}-i\Phi_{Fy})P^2] \tag{5.50}$$

where

$$(\theta_x^{xx})_{\text{eff}} = \theta_x^{xx} + \frac{\hbar^4 \tau_{xxxz}\mu_z^{(e)}}{2hc(B_x - B_z)} \tag{6.51}$$

In (6.51) $\mu_z^{(e)}$ is the permanent dipole moment, τ_{xxxz} is the centrifugal distortion constant, and B_x and B_z are rotational constants. The z coordinate is along the symmetry axis; hence, $B_x=B$ and $B_z=C$ for the oblate top and A for the prolate top. It should be noted that for molecules of D_{3h} symmetry, which have no permanent dipole moment, the last term of (6.51) vanishes and $(\theta_x^{xx})_{\text{eff}}=\theta_x^{xx}$.

The line strength S is proportional to the squared transition matrix elements of μ_F. These squared matrix elements of the μ_F of (6.50) were derived by Watson [64] for molecules of D_{3h} symmetry, planar XY_3-type. The line-strength factors of the allowed transitions (those for which the matrix elements are nonvanishing) are given in Table I of Watson's paper. These nonvanishing matrix elements prove the $\Delta K=\pm 3$ selection rules for molecules of D_{3h} and C_{3v} symmetry. By a unitary transformation of the $\mathscr{H}_d$ of (6.48) and of the μ_F expressed by (6.50), he derived the (θ_x^{xx}) for molecules of C_{ev} symmetry given by (6.51). Thus, by substitution of $(\theta_x^{xx})_{\text{eff}}$ from (6.51) for θ_x^{xx} in the formulas of his Table I, the corresponding line strengths for symmetric-top molecules of C_{3v} symmetry are obtained. For example, the line strengths for the $\Delta J=0$, $\Delta K=\pm 3$ transitions of phosphine or arsine are

$$\Phi(J, K\leftrightarrow J, K\pm 3)=\tfrac{1}{4}(\theta_x^{xx})_{\text{eff}}^2(J\mp K)(J\mp K-1)(J\mp K-2)$$
$$\times(J\pm K+1)(J\pm K+2)(J\pm K+3)(2J+1)/J(J+1) \qquad (6.52)$$

where

$$(\theta_x^{xx})_{\text{eff}}=\theta_x^{xx}+\frac{\hbar^4\tau_{xxxz}\mu_z}{2h(B-C)} \qquad (6.53)$$

and

$$\theta_x^{xx}=2B^2\left[\frac{a_3^{xx}}{v_3^2}\left(\frac{\partial\mu_x}{\partial Q_3}\right)+\frac{a_4^{xx}}{v_4^2}\left(\frac{\partial\mu_x}{\partial Q_4}\right)\right] \qquad (6.54)$$

is the induced dipole component resulting from the admixture of the excited degenerate vibrational modes v_3 and v_4 with the ground vibrational states. Note that Watson [64], also Oka [70], uses k for + and − values of K and defines $K=|k|$, whereas we allow K to have + or − values defined by (2.62).

Chu and Oka [67] used (6.52) in calculations of the intensities of forbidden $(J, K)=(J, \pm 1)\leftarrow(J, \mp 2)$ rotational transitions of PH_3 for J values ranging from 6 to 17 which they observed in the frequency range of 47.4 to 43.7 GHz. The intensities ranged in strength from 2.1×10^{-9} cm^{-1} for $J=6$ to 5.5×10^{-9} cm^{-1} for $J=10$ and dropped to 1.1×10^{-9} cm^{-1} for $J=17$. Helms and Gordy [73] observed $(J, 0)\leftarrow(J, 3)$ transitions of PH_3 for J values from 3 to 14 which range in frequency from 143.7 to 134.6 GHz. The intensity distribution of the observed lines, as calculated with (6.52), is shown in Fig. 6.11. The maximum intensity, 48.5×10^{-9} cm^{-1}, occurs for $J=10$. Intensities for the same transitions in PD_3 are an order of magnitude lower than those for PH_3. For AsH_3, they are still lower [74], with the maximum intensity for the $J\rightarrow J$, $K=\pm 2\rightarrow\mp 1$ transitions only 2×10^{-10} cm^{-1}.

It is obvious that millimeter wave spectrometers of high sensitivity are required for observation of these forbidden spectra. The spectrometer used at Duke for measurement of the $\Delta K=\pm 3$ transition in phosphine [73] and arsine [74] employed a high-Q, Stark-modulated Fabry-Perot absorption cavity and had an estimated sensitivity of 2×10^{-10} cm^{-1} at 144 GHz for a time constant of 1 sec. Figure 6.12 shows four ^{75}As hyperfine components of a $\Delta J+0$, $\Delta K=\pm 3$ transition of AsH_3 recorded with this spectrometer.

For C_{3v} molecules such as phosphine and arsine having a relatively large μ_z and high degenerate vibrational frequencies, v_3 and v_4, the intensities of the $\Delta K=\pm 3$ transitions are due primarily to centrifugal distortion in the ground vibrational state. For example, Chu and Oka [67] calculated $\theta_x^{xx}=1.6\times 10^{-5}$ D and the distortion moment, last term of (6.53), to be 8.3×10^{-5} D. Thus the

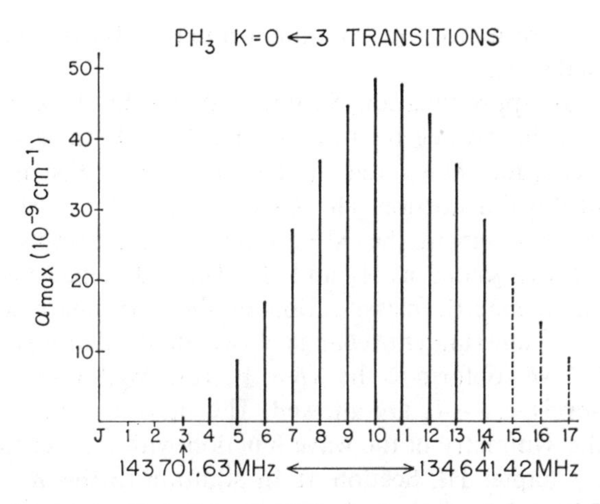

Fig. 6.11 Calculated intensities of the $J \leftarrow J$, $K=0 \leftarrow 3$ rotational lines of PH_3 in the ground vibrational state. The solid lines indicate observed transitions. From Helms and Gordy [73].

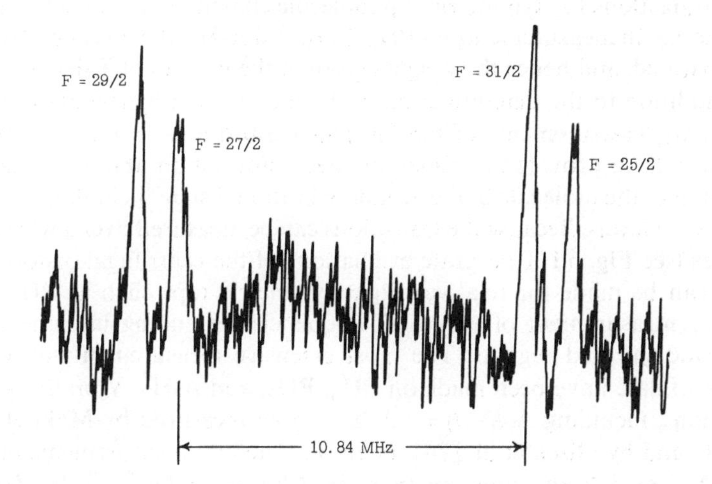

Fig. 6.12 Single scan of the "forbidden" $J=14$, $K=\pm 5 \rightarrow \pm 2$ rotational transition of AsH_3 showing $F \rightarrow F$ components of the ^{75}As hyperfine structure. From Helms and Gordy [74].

contribution of the vibration–rotation interaction to the line intensities is only $(1.6)^2/[(1.6)^2+(8.3)^2]=3.6\%$. Oka [70] describes the contribution from centrifugal distortion, which mixes the (J, K) and $(J, K \pm 3)$ rotational levels of the ground vibrational states, as "rotational intensity borrowing" from the normal $\Delta J=1$, $\Delta K=0$ transitions. The contribution of the vibration–rotation interaction, (6.49), which mixes the ground-state rotational levels with those of excited, degenerate vibrational states, he describes as "vibrational intensity borrowing." In this description, the observed $\Delta K = \pm 3$ transitions in phosphine and arsine are due primarily to "rotational intensity borrowing" within the ground vibrational state.

In the rigid-rotor approximation, Section 1, the $|K|$ levels of the symmetric-top molecule are all doubly degenerate except for $K=0$. In the distortable rotor, the $|K|$ levels, except for $|K|=3$, remain degenerate with E symmetry but are displaced by centrifugal distortion. The $K=0$ and $|K|=3, 6, \ldots$ have symmetry A. In addition to displacement, the $|K|=3$ levels are split into two components with symmetry A, designated as A_1 and A_2. These designations indicate the symmetric or antisymmetric classification of the coordinate wave function (exclusive of nuclear spin) under exchange of two identical atoms. The selection rules $A \leftrightarrow A$ and $E \leftrightarrow E$ conform to the $\Delta K = \pm 3$ rule. Within the transitions for the split $K=3$ levels, $A_1 \leftrightarrow A_2$ are allowed. The signs $+$ and $-$ are used for designation of the symmetry of the wave function with respect to inversion of the coordinates (Chapter III, Section 4). In addition to the $K=\pm 3$ rule, the transitions must comply with the parity selection rules, sym$\leftrightarrow$sym and antisym $\leftrightarrow$antisym, for the overall wave function including nuclear spin, as described in Chapter III, Section 4. The effects of nuclear-spin statistics on the intensities of the transitions for symmetric-top molecules having C_{3v} symmetry are given in Table 3.3. In measurements on PH_3, PD_3, and AsH_3, the inversion levels were not separated, and hence the weights given at the bottom of Table 3.3 apply.

In addition to the determination of the moments of inertia about the symmetry axis, measurements of the forbidden transitions in C_{3v} symmetric-top molecules have provided considerable new information about the centrifugal distortion of the molecule in the ground vibrational state including the splitting of the $K=3$ levels. Because the transitions can be measured over a wide range of J values (see Fig. 6.11), accurate evaluations of the centrifugal distortion constants can be made for relatively light symmetric tops such as PH_3 without need for measurement of very high frequencies extending into the submillimeter and infrared regions. The most extensive measurements of forbidden spectra to date have been made on PH_3, PD_3, and AsH_3. Vibration–rotation transitions, including $\Delta(K-l)=\pm 3$, have been measured by Maki et al. [75] for PH_3 and by Olsen et al. [76] for AsH_3. The first measurements of ground state, $\Delta K=\pm 3$, transitions are those by Chu and Oka [67] the $J \rightarrow J$, $K=\pm 2 \rightarrow \mp 1$ transitions of PH_3, PD_3, and AsH_3. Helms and Gordy [73, 74] extended these measurements to include: $J \rightarrow J$, $K=3 \rightarrow 0$ in PH_3; $J \rightarrow J$, $K=3 \rightarrow 0$, and $K=\pm 4 \rightarrow \pm 1$ in PD_3; $J \rightarrow J$, $K=\pm 4 \rightarrow \pm 1$, and $K=\pm 5 \rightarrow \pm 2$ in AsH_3. Each K transition was measured for several J values. Helms and Gordy combined the measured frequencies of all these forbidden transitions with those measured earlier [77, 78] for the normal $\Delta J=1$, $\Delta K=0$ of these molecules in a computer analysis to obtain the values of their rotational constants, listed in Table 6.4. The molecular structures obtained from this analysis are given in Table 6.8.

The analysis of the phosphine and arsine data consists of a least-squares computer fitting of the involved spectral parameters to the observed frequencies. The computer program was made to include all frequencies of the different types of transitions measured. It was designed to compute the differences in the

Table 6.4 Ground-state Spectral Constants[a] of PH_3, PD_3, and AsH_3

Constant	PH_3	PD_3	AsH_3
B_0 (MHz)	133,480.15(12)	69,471.10(3)	112,470.597(30)
C_0 (MHz)	117,488.85(16)	58,974.37(5)	104,884.665(43)
D_J (MHz)	3.947(16)	1.021(2)	2.9257(39)
D_{JK} (MHz)	−5.182(1)	−1.312(1)	−3.7164(4)
D_K (MHz)	4.177(11)	1.023(2)	3.4126(21)
$\hbar^4\tau_{xxxz}/4h$ (MHz)	0.62(3)	0.221(5)	0.679(3)
H_K (kHz)			1.990(47)
H_{JK} (kHz)	−1.09(4)	−0.091(4)	−0.239(9)
h_0 (Hz)	67.6(22)	9.41(22)	
L_{JK} (Hz)	4.7(7)		−1.729(67)
L_{JJK} (Hz)	0.52(3)	0.044(5)	0.701(11)

[a]These constants were calculated by Helms and Gordy [73, 74] from $J \to J$, $\Delta K = 3$ transitions measured by them and by Chu and Oka [67] and from normal $\Delta J = 1$, $\Delta K = 0$ transitions measured by Helminger et al. [77, 78].

energy levels involved in these transitions with assumed values of the spectral parameters in the appropriate energy equations. Transition frequencies were thus automatically computed and compared with the measured frequencies to generate new estimates of the spectral constants. This iterative process was repeated by the computer until convergence was obtained. The degree of fitting of the composite set of frequencies determined the estimated errors (standard deviations) indicated in Table 6.4. The energy equations for the $K=3$ levels involved in the observed $K=0 \leftrightarrow 3$ transitions of PH_3 and PD_3 include the term $\pm h_0 J(J+1)[J(J+1)-2][J(J+1)-6]$, first derived by Nielsen and Dennison [79] in their calculation of the anomalous effects of the $K=3$ lines in the ammonia inversion spectrum observed by early microwave spectroscopists. The AsH_3 energy equations include the nuclear hyperfine interactions described in Chapter IX. The energy expressions for all transitions included the pure rotational energy with the centrifugal distortion terms expressed in the form

$$\begin{aligned} E(J,K)/h = {} & B_0[J(J+1)-K^2]+C_0K^2-D_JJ^2(J+1)^2 \\ & -D_{JK}J(J+1)K^2-D_KK^4+H_{JJK}J^2(J+1)^2K^2 \\ & +H_{KJ}J(J+1)K^4+L_{JK}J^2(J+1)^2K^4 \\ & +H_KK^6+L_{JJK}J^3(J+1)^3K^2 \end{aligned} \tag{6.55}$$

Complete expressions for the energy terms for both $K \pm 3$ and $K=3$ levels are given in the papers cited. See also Amat et al. [22].

Recently (1981) Belov et al. [80] measured the PH_3 rotational spectrum in the region of 300 to 1070 GHz. In addition to the normal rotational transitions, $\Delta J=1$, $\Delta K=0$, they measured the $J \leftrightarrow J$, $|K|=2 \to 5$ transitions which occur in the region of 310 to 334 GHz for J values ranging from 6 to 16. They

Table 6.5 Ground-state Spectral Constants[a] of $^{16}O^{31}P^{19}F_3$

Parameter	*Value*	*Parameter*	*Value*
A_0 (MHz)	4811.7579(18)	H_J (10^{-3} Hz)	−6.7(10)
B_0 (MHz)	4594.2624(5)	H_{JK} (10^{-3} Hz)	9.10(27)
D_J (kHz)	1.0119(12)	H_{KJ} (10^{-3} Hz)	−6.8(8)
D_{JK} (kHz)	1.2971(7)	H_K (10^{-3} Hz)	−73(30)
D_K (kHz)	−1.114(12)		

[a]From Kagann et al. [82].

reanalyzed the previously measured millimeter wave frequencies together with their extended submillimeter measurements to obtain a new set of parameter values. These higher-frequency data made possible an improvement in the centrifugal distortion constants, particularly those of higher order.

The symmetry selection rules permit direct transitions between the two components of the split $K=3$ levels. The transition frequencies between the two components in PH_3 have been measured with molecular beam resonance techniques for $|K|=3$, $J=3$ to 9 by Davies et al. [81]. These frequencies were included in the analyses of Belov et al. [80].

Belov et al. [59] have also used the Krupnov spectrometer to measure some $\Delta K=\pm 3$ transitions in the rotation–inversion spectrum of NH_3 in the excited v_2 vibrational state as well as the $v_2=0$ ground state. One of the ground-state lines, the $J=10$, $K=0\leftarrow 3$ is shown in the center of Fig. 6.10. The theory used for analysis of these transitions, designated by them as "perturbation allowed," is that developed by Papoušek et al. [59].

Ground-state, forbidden $\Delta J=0$, $\Delta K=\pm 3$ transitions of OPF_3, an XYZ_3 type of prolate, symmetric-top molecule, have been observed by Kagann et al. [82]. A total of 152 lines of $^{16}O^{31}P^{19}F_3$ were measured in the frequency range of 9 to 18 GHz. These included six K transitions in the range from $K=9\rightarrow 6$ to $15\leftarrow 12$, each for a wide range of high J values. For example, the $K=11\leftarrow 8$ transitions were measured for 31 J values ranging from 52 to 82. Thus the high-order distortion constants could be accurately obtained from the data analysis. As an aid to the assignment of the J values for the different K transitions, they included in their computer program a reanalysis of the normal $J\rightarrow J+1$ transition (R-branch) measured earlier [83]. The various rotational and distortion constants obtained in the analysis are recorded in Table 6.5. The molecular structural parameters are listed in Table 6.9. Section 6.

Avoided-crossing, Molecular Beam Resonance Method for Measurement of K-level Separations

Ozier and Meerts [84–87] have developed an effective method for measurement of the separation of K levels between which normal transitions are forbidden. The method is capable of measurement of symmetric-top energy-level separations corresponding to $\Delta K=\pm 1$ and ± 2 transitions as well as to the

$\Delta K = \pm 3$ transitions discussed in Chapter II, Section 3. In molecular-beam, electric resonance experiments a strong, uniform electric $\mathscr{E}$ field is applied to bring selected Stark components of the upper and lower K levels into collision. The Stark shift for the upper-level component must, of course, be negative and that for the lower be positive. In the vicinity of the crossing, or anticrossing, region, the wave functions of the upper and lower levels are strongly mixed. This admixing is similar to that described in Chapter X, Section 5 for the mixing by an imposed Stark field of the wave functions of the l-doublet states of a linear molecule in a degenerate bending vibrational state, or those of the inversion doublet states of NH_3. It is also similar to the admixing of near-degenerate pairs of levels in slightly asymmetric rotors described in Chapter X, Section 3. As in other cases of near-degeneracy, the two colliding levels appear to repel, or to avoid, crossing. This effect is demonstrated by Fig. 6.13, where the $M_J = \pm 1$ Stark components of $J=1$, $K=\pm 1$ and $J=1$, $K=0$ states of CF_3H are brought

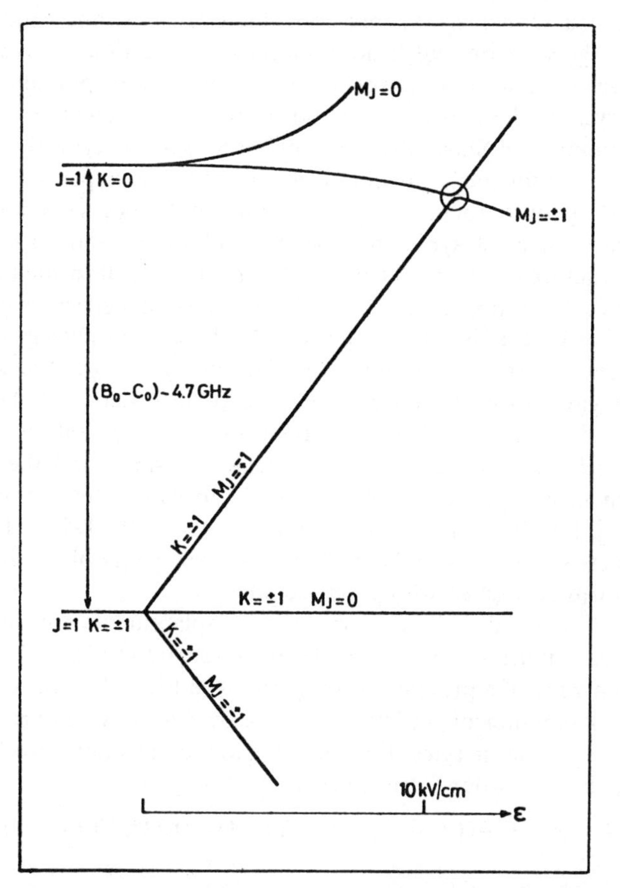

Fig. 6.13 Schematic plot against the electric field of the energy levels in the $J=1$ state of fluoroform. The nuclear magnetic quantum numbers and the hyperfine effects are not shown. The circle indicates the anticrossing region. For clarity, the curvature of the levels in this region and the quadratic Stark effect of the $K=0$ state have been greatly exaggerated. From Meerts and Ozier [86].

into proximity by an increasing Stark field. In the encircled "anticrossing" region, the $|J, K, M_J) = |1, \pm 1, \mp 1)$ and $|1, 0, \pm 1)$ states are strongly mixed. As a result, M_J components are reversed, giving rise to a detectable change in the strength of the molecular beam resonance signal. The admixing causes changes in the symmetry of the rotational wave functions and, in some instances, changes to ortho-para conversions in the nuclear spin states.

If the effects of centrifugal distortion are neglected, the separation of the $|J, K=0)$ and $|J, K=\pm 1)$ states is $(B_0 - C_0)$ for the oblate symmetric top CF_3H, as indicated on the diagram of Fig. 6.13. From the diagram it is evident that the sum of the Stark energies required to bring the two M_J components to the critical anticrossing region is equal to $(B_0 - C_0)$. The critical field $\mathscr{E}_c$ for the anticrossing signal can be accurately measured; from the known dipole moment μ_9, these Stark displacement energies can be calculated. Thus the value of $(B_0 - C_0)$ is obtained. By inclusion of the centrifugal distortion terms and the measurement of additional transitions, the K-dependent distortion constants can be obtained. Obviously, the presence of nuclear hyperfine structure further complicates the solution, but it does not make the method unworkable. One then observes the critical anticrossing regions of Stark components of certain hyperfine levels of the two K states and proceeds to calculate the sum of the upper and lower level Stark displacement energies at the critical field $\mathscr{E}_c$. One must also correct for the zero-field displacement of the hyperfine components employed. Hyperfine structure in symmetric-top molecules is treated in Chapter IX. The Stark effect of symmetric-top molecules with and without nuclear hyperfine structure is treated in Chapter X. For the case illustrated in Fig. 6.13, the Stark effect in the upper $(K=0)$ level is a second-order effect whereas that for the lower $(K=1)$ level is a first-order effect. There is no nuclear quadrupole hyperfine structure for CF_3H, but the nuclear magnetic hyperfine structure of the F and H caused a broadening of the signal of the order of 20 kHz.

The avoided-crossing method is simplest and most accurate for symmetric tops for which $(A_0 - B_0)$ or $(B_0 - C_0)$ is relatively small and the permanent dipole moment is large. The molecule OPF_3, to which the method was first applied [84], fits these criteria well with $(A_0 - B_0) = 217.495$ MHz and $\mu_0 = 1.868$ D. Because the crossing fields $\mathscr{E}_c$ are not excessively high, the first-order Stark effect can be applied with good accuracy.

The $J \leftrightarrow J$ first-order, Stark-component displacements for molecules in which nuclear splittings are negligible are given by (10.3), and the $E(J, K)$ rotational energies of a prolate top are given by (6.14). If $\mathscr{E}_c$ is the crossing field for two Stark components of levels differing by 3 in K which are undergoing avoided crossing, the energies corresponding to the transitions $J \leftrightarrow J$, $K \rightarrow K+3$, as given by the first-order Stark effect, are

$$\Delta E_r(J \rightarrow J, K \rightarrow K+3) = [(A_0 - B_0) - D_{JK} J(J+1) - D_K(2K^2 + 6K + 9)](2K+3)$$
$$= \frac{\mu_0 \mathscr{E}_c M_J}{J(J+1)} \qquad (6.56)$$

where K is algebraically the smaller of the K values for the levels involved.

Table 6.6 Rotational Constants A_0 and C_0 from Avoided-crossing, MBER Measurements of "Forbidden" K Transitions

Molecule	B_0[a] *(MHz)*	A_0[b] *(MHz)*	C_0[b] *(MHz)*
CF_3H	10,348.867(2)[c]		5,673.46(10)[d]
CF_3D	9,921.126(2)		5,673.21(10)[d]
CF_3CH_3	5,185.1387(24)[e]	5,498.85(7)[e]	

[a]From normal, $J \rightarrow J+1$ rotational spectra.
[b]Obtained from (A_0-B_0) and (B_0-C_0) measured with avoided-crossing, MBER method and B_0 values from normal spectroscopy.
[c]From T. E. Sullivan and L. Frankel, *J. Mol. Spectrosc.*, **39**, 185 (1971).
[d]From Meerts and Ozier [86].
[e]From Meerts and Ozier [85].

Derivations of similar first-order equations for $\Delta K=1$, or 2, will be obvious. If $\mathscr{E}_c$ for sufficient J, K levels are measured, the (A_0-B_0), D_{JK}, and D_K can be obtained from (6.56). However, D_{JK} can be measured more accurately from the normal, $\Delta J=1$, $\Delta K=0$ transition. With B_0 and D_{JK} from the normal spectrum, the constants D_K and A_0 (or C_0 for the oblate top) are easily evaluated from the $\mathscr{E}_c$ measurements. Meerts and Ozier [86] estimated the absolute accuracy of (A_0-B_0) or (B_0-C_0), as determined mainly by the long-range stability and reproduction of the source voltage, to be 0.002%.

For determination of the value of (A_0-B_0) for OPF_3, Ozier and Meerts [84, 86] measured $\mathscr{E}_c$ for the $K=\mp1\rightarrow\pm2$ crossing with $M_J=J$ for levels of $J=2$ to 6. They employed D_{JK} and D_K values from the distortion-moment spectra observed by Kagann et al. [82] and obtained $(A_0-B_0)=217.494(4)$ MHz, in good agreement with the value of 217.495(2) MHz derived from the distortion moment spectroscopy. The resulting A_0 value is 4,811.756(5) MHz. Values of A_0 or C_0 obtained for other molecules by this method are listed in Table 6.6.

In the avoided-crossing experiments [85] on CH_3CF_3, internal rotational splittings were also observed for the levels corresponding to ±1 in the torsional quantum number σ, as well as for those corresponding to changes in K. From measurements of these splittings, the height of the barrier to internal rotation of the CH_3 group relative to the CF_3 was found to be $V_3=3.16(11)$ kcal/mole. The value of 3.480 kcal/mole (Table 12.7) was found from microwave spectroscopy. In addition, the moment of inertia of the CH_3 group relative to the symmetry axis, $I_\alpha=3.17(11)$ amu Å^2, was derived. This experiment demonstrates a new method of obtaining these constants.

Tetrahedral, Spherical-top Molecules

Although spherical-top molecules, strictly speaking, are not symmetric tops, they have many properties in common with symmetric tops. They are the limiting case of a symmetric top as the moment of inertia about the symmetry axis approaches I_b. Because they have no permanent dipole moment, spherical-top molecules have no normal rotational spectra. However, as predicted by Fox [63, 71], Watson [64], and others [72], in the ground vibrational state they can have weak, but observable, pure rotational spectra resulting from distortion dipole moments. The moments, which are similar to the θ_x^{xx} moment of (6.54), arise from the vibration–rotation interaction of the ground-state rotational levels with those of excited degenerate vibrational states. Terms like the last one of (6.53) make no contribution to the distortion moment because $\mu_z^{(e)}=0$. Consequently, the line strength results entirely from the "vibrational intensity borrowing" [70].

A distortion dipole moment is induced when the tetrahedral molecule rotates about a C_3 axis; it is zero for rotation about an S_4 axis. In contrast, the centrifugal distortion energy is greatest when the rotation is about the S_4 axis and least when it is about the C_3 axis [70]. Thus the distortion moment spectrum results from rotation about a C_3 axis and resembles that of a C_{3v} symmetric top.

As for other molecules, the selection rules depend on the overall symmetry of the wave functions, including nuclear spin functions (Chapter III, Section 4). The tetrahedral molecules, such as CH_4, SiH_4, and so on, belong to the point group T_d with symmetry A, E, F. There are two types of nondegenerate A wave functions, A_1 and A_2, one doubly degenerate species E, and two triply degenerate species, F_1 and F_2. The selection rules for the distortion moment spectra are

$$\Delta J=0, \pm 1, \qquad \Delta M=0, \pm 1$$

$$A_1 \leftrightarrow A_2, \qquad E \leftrightarrow E, \qquad F_1 \leftrightarrow F_2$$

The nuclear spin functions have symmetries determined by the magnitude of the resultant spin: $T_I=I_1+I_2+I_3+I_4=2, 1, 0$, for spins of $\frac{1}{2}$ as in CH_4. Those for $T_I=2$ have symmetry A_1; for $T_I=0$, symmetry E; and for $T_I=1$, symmetry F_2.

The line-strength factor for rotational distortion spectra corresponding to the $J \rightarrow J+1$ transitions of spherical tops as calculated by Watson [64] (neglecting effects of nuclear statistical weights), is

$$\lambda(J \leftrightarrow J+1)=\frac{2}{35}(\theta_z^{xy})^2 J(J+2)(2J-1)(2J+1)(2J+3)(2J+5) \qquad (6.57)$$

For CH_4, Watson calculated $\theta_x^{xy}=(2.6 \text{ or } 1.2)\times 10^{-5}$ D from the vibration frequencies ν_3 and ν_4 and vibrational intensities measured by Mills [88]. The distortion moment for the ground vibrational state of CH_4 has been measured by Ozier [89] with molecular beam resonance techniques. The value obtained for $\mu_\varepsilon=5.38\times 10^{-6}$ D$=(20)^{-1/2}\theta_z^{xy}$ gives $\theta_z^{xy}=2.41\times 10^{-5}$ D, in good agreement with the upper value estimated by Watson [64].

A rigid spherical-top molecule has only one value for the moment of inertia, and its pure rotational spectrum—if it could be observed—would be like that for a rigid linear molecule (Fig. 4.1) with rotational frequencies, $\nu_r=2B(J+1)$.

Centrifugal distortion is necessary, however, to produce the dipole moment making possible the detection of the lines. Thus the detected lines must include distortion effects in their frequencies. These effects can be taken into account by the inclusion of scalar distortion terms similar to those of a linear molecule. The frequency equation is

$$\nu_{J\to J+1}=2B_0(J+1)-4D_s(J+1)^3+H_s(J+1)^3[(J+2)^3-J^3] \quad (6.58)$$

Rosenberg et al. [68] found that they could fit the lines they observed for $^{12}CH_4$ to (6.58) with the constants $B_0=5.245(4)$ cm^{-1} and $D_s=1.19(9)\times10^{-4}$ cm^{-1}, and $H_s=0$. Ten $J\to J+1$ transitions were observed in the region of 20 to 200 cm^{-1} for $J=7$ to 16. Equation 6.57 was also used in the analysis of the observed $J\to J+1$ frequency of SiH_4 [90] and GeH_4 [91].

Despite the good fitting of the R-branch transitions achievable with (6.58), the scalar distortion constants thus obtained do not give a measure of the complete molecular distortion of the tetrahedral molecules, as it does for linear molecules. The distortion of the spherical top is more nearly like that of the symmetric top, for which the centrifugal distortion is not completely determined by the normal $J\to J+1$, $\Delta K=0$ transition, but requires also the measurement of the $\Delta K\neq 0$ transitions for determination of the centrifugal distortion constant D_K. The spherical top is like the symmetric top in having, under centrifugal distortion, an observable Q-branch, $\Delta J=0$, spectrum. These Q-branch transitions are, however, more complicated than the "forbidden" Q-branch transitions of a symmetric top, described in Chapter II, Section 3. They result not only from distortion dipole moments but also from centrifugal distortion effects on the inertial moments. For example, if the effects of distortion on the moments of inertia are omitted, the Q-branch, $\Delta J=0$, $\Delta K=3$, frequency of the prolate symmetric top would be $(6K+9)(A_0-B_0)$. In the limiting case of a rigid spherical top, $A_0=B_0(=C_0)$, these frequencies vanish. Centrifugal distortion caused by rotation about the b axis can, however, destroy the equality, causing A_0 (or C_0) to differ slightly from B_0 and thus producing a "forbidden" Q-branch spectrum analogous to that of a symmetric top. Because the inequality of the inertial moments caused by centrifugal distortion is not large, the Q-branch frequencies fall in the lower microwave or radiofrequency region. They have been measured directly with microwave spectroscopy by Ozier, Gerry, and their associates for CH_4 [69, 92, 93], for SiH_4 [94], and for GeH_4 [95]; they have been measured with microwave–infrared, double-resonance techniques by Curl et al. [96, 97] for CH_4, and by Kreiner and Oka [98] for SiH_4.

Because the $\Delta J=0$ (Q-branch) spectra depend entirely on differences in moments of inertia and on transition dipole moments caused by centrifugal distortion, the spectra are rather complex. As shown in the early theory by Hecht [99], and in later developments by Fox [71], the rotational energy of a tetrahedral molecule consists of a scalar term E_s and a tensor term E_T

$$E_{\rm rot}=E_s+E_T \quad (6.59)$$

The scalar term is like that for linear molecules, (5.13). The tensor term results entirely from distortion. Expressions for both E_s and E_T are given by Kirschner and Watson [100], Ozier [101], and others [94]. Values for the tensor distortion constants are obtained only from measurements of the Q-branch transitions, whereas the scalar energies as well as the scalar distortion constants are obtained from the $\Delta J \rightarrow J+1$ transitions. However, it is evident from (6.59) that a precise evaluation of the rotational energies E_{rot}, of B_0, and the scalar distortion constants depends on a knowledge of the tensor distortion energies E_T. Thus accurate measurements of both R-branch and Q-branch transitions, with a combined analysis of the frequencies of both, are required for precise values of the rotational constants. Such an analysis has been made for SiH_4 by Pierre et al. [102] and improved with more accurate Q-branch frequencies by Ozier et al. [94]. Their final values for $^{28}SiH_4$ are: $B_0 = 2.859065(10)$ cm^{-1}, $D_s = 3.690(13) \times 10^{-5}$ cm^{-1}, and $H_s = 2.14(48) \times 10^{-9}$ cm^{-1}. Tabulation of the measured tensor distortion constants and B_0 values for $^{12}CH_4$, $^{28}SiH_4$, and GeH_4 is given by Kagann et al. [95] (see their Table 7). For $^{12}CH_4$, $B_0 = 5.241036(10)$ cm^{-1} was obtained by Tarrago et al. [103].

5 LINE INTENSITIES

The absorption coefficients for microwave rotational lines of symmetric-top molecules can be found from (3.24), or the peak absorption $\alpha_{\max}$ from (3.27) by substitution of the appropriate values of F_m and of the dipole moment matrix elements. The values of $F_m \equiv F_{J,K}$ may be obtained from (3.66) by substitution of Q_r from (3.68). The dipole moment matrix elements corresponding to the rotational absorption line transitions $J \rightarrow J+1$, $K \rightarrow K$ are given by (2.124). When these expressions, along with the values of N from (3.25), are substituted in (3.24) with numerical values for the various constants, the resulting expressions for the absorption coefficient becomes

$$\alpha_\nu = \frac{11.58 \times 10^{-12} F_v i_c p_{\text{mm}} \mu^2 \sigma g_I g_K B \sqrt{A}\, \nu^2}{T^{7/2}} \left(1 - \frac{24 \times 10^{-6} \nu}{T}\right) \times \left[\frac{\Delta\nu}{(\nu-\nu_0)^2 + (\Delta\nu)^2}\right] \left[\frac{(J+1)^2 - K^2}{(J+1)}\right] \times \exp\left\{\frac{-48 \times 10^{-6}[BJ(J+1) + (A-B)K^2]}{T}\right\} \tag{6.60}$$

In this expression the frequencies ν, $\Delta\nu$ and spectral constants A and B are expressed in MHz; α_ν, in cm^{-1}. Here and in later expressions it is assumed that the effects of centrifugal distortion cause the different K lines to be resolved, but these effects are otherwise neglected.

Microwave spectroscopists are usually interested in the absorption coefficient $\alpha_{\max}$ at the center of the resonance corresponding to the frequency $\nu = \nu_0$, for which the maximum absorption occurs. The lines of gases are most often observed under conditions such that the peak intensity is independent of the pressure (Chapter III, see Section 2). Under these conditions we can set $\Delta\nu = 300\, p_{mm} (\Delta\nu)_1/T$, (3.26). In the numerical calculations of α_{max} it is more convenient to express the resonant frequency ν_0 and the spectral constants A, B, or C in GHz units (1 GHz = 1000 MHz) rather than in MHz. Because of its smaller value, we shall leave $(\Delta\nu)_1$ in MHz units. With $\nu_0 = 2B(J+1)$, the formula for the peak absorption can be expressed

$$\alpha_{max} = X\nu_0^3\left(1-\frac{0.024\nu_0}{T}\right)e^{-0.024J\nu_0/T}\left[1-\frac{K^2}{(J+1)^2}\right]e^{-0.048(A-B)K^2/T} \tag{6.61}$$

$$\approx X\nu_0^3\left(1-\frac{0.024\nu_0}{T}\right)\left(1-\frac{0.024J\nu_0}{T}\right)\left[1-\frac{K^2}{(J+1)^2}\right]\times\left[1-\frac{0.048(A-B)K^2}{T}\right] \tag{6.62}$$

where

$$X = \frac{6.11\times 10^{-4} F_v i_c \mu^2 \sigma g_I g_K \sqrt{A}^{1/2}}{(\Delta\nu)_1 T^{5/2}} \tag{6.63}$$

and where ν_0, A and B are in GHz and $(\Delta\nu)_1$ is in MHz. When $J\nu_0 \ll T$ and $(A-B)K^2 \ll T$, as is true for relatively low J transitions, the simpler expression

$$\alpha_{max} \approx X\nu_0^3\left[1-\frac{K^2}{(J+1)^2}\right] \tag{6.64}$$

is often adequate. In these expressions

F_v = fraction of molecules in the particular vibrational state observed; see (3.61).
i_c = fractional concentration of the isotopic species observed
T = temperature of observation in absolute scale
μ = molecular dipole moment in debye units
ν_0 = resonant frequency in GHz units
$(\Delta\nu)_1$ = line breadth in MHz for $p = 1$ mm of Hg pressure when $T = 300°K$
A = spectral constant in GHz units for rotation about symmetry axis. This constant becomes C for a prolate top
B = spectral constant in GHz units for rotation about axis perpendicular to symmetry axis
g_K = 1 for $K = 0$
= 2 for $K > 0$
σ = symmetry number described in Chapter III, Section 4
g_I = reduced nuclear statistical weight factor described in Chapter III, Section 4

For molecules in symmetric ground electronic states having C_{3v} symmetry with no resolvable inversion splitting, such as CH_3F or CH_3CCH, which have one set of identical off-axis nuclei with nuclear spin I,

$$\sigma=3$$

$$g_I=\frac{1}{3}\left[1+\frac{2}{(2I+1)^2}\right] \quad \text{for } K=0, 3, 6, \ldots \tag{6.65}$$

and

$$g_I=\frac{1}{3}\left[1-\frac{1}{(2I+1)^2}\right] \quad \text{for } K \text{ not divisible by 3} \tag{6.66}$$

For molecules of C_3 symmetry in symmetric ground electronic states having two sets of three identical off-axis nuclei with spin I_1 and I_2, such as CH_3SiF_3,

$$\sigma=3$$

$$g_I=\frac{1}{3}\left[1+\frac{2}{(2I_1+1)^2(2I_2+1)^2}\right] \quad \text{for } K=0, 3, 6, \ldots \tag{6.67}$$

and

$$g_I=\frac{1}{3}\left[1-\frac{1}{(2I_1+1)^2(2I_2+1)^2}\right] \quad \text{for } K \text{ not divisible by 3} \tag{6.68}$$

The use of these values for g_I assumes that there is no resolvable inversion doubling. Such doubling has been found for ammonia only.

From examination of the foregoing expressions it is evident that $\sigma g_I \approx 1$ when $I>\frac{1}{2}$ and for two sets of identical nuclei even when $I_1=I_2=\frac{1}{2}$. For the latter case, with K divisible by 3, $\sigma g_I=\frac{9}{8}\approx 1$; for K not divisible by 3, $\sigma g_I=\frac{15}{16}\approx 1$. Thus, to the approximation normally desired in intensity calculations, one can set $\sigma g_I=1$, or simply drop the factor σg_I from the equations except for the important class of molecules with one set of three identical nuclei with $I=\frac{1}{2}$. For these, $\sigma g_I=\frac{3}{2}$ when $K=0, 3, 9, \ldots$, and $\sigma g_I=\frac{3}{4}$ when K is not divisible by 3. For these, the weights of the lines for the first group are twice those for the latter group. Note that when $I=0$, $g_I=0$ for K not divisible by 3. Hence, only the lines for $K=0, 3, 6, \ldots$ occur.

In computations of the absorption coefficients it is convenient to calculate those for the $K=0$ lines first and multiply the coefficient for the $K=0$ line by the K-dependent factor to obtain the α_{max} for the other K components. Thus the formula can be expressed, from (6.61) as

$$\alpha_{max}(K=0)=X\nu_0^3\left(1-\frac{0.024\nu_0}{T}\right)e^{-0.024J\nu_0/T} \tag{6.69}$$

$$\alpha_{max}(K\neq 0)=2\alpha_{max}(K=0)\left[\frac{g_I(K\neq 0)}{g_I(K=0)}\right]\left[1-\frac{K^2}{(J+1)^2}\right]e^{-0.048(A-B)K^2/T} \tag{6.70}$$

It is evident from comparison of (6.69) with (4.82) that the intensities of the $K=0$ lines of the symmetric top vary with frequency in the same manner as do those for diatomic and linear polyatomic molecules. For J not too large, the α_{max} for the $K=0$ lines increases approximately with the cube of the frequency.

The K lines for a given J transition decrease with increasing K if effects of g_I and g_K are neglected.

As an illustration, we shall calculate the peak absorption coefficients for the microwave rotational lines of CH_3CCH, which is a rather typical organic, symmetric-top molecule. For $^{12}CH_3{}^{12}C^{12}CH$, the natural isotopic concentration $i_c=0.966$, $\mu=0.75$ debye, $B_0=8.545$ GHz, $A=153$ GHz, $\sigma=3$, $g_I=\frac{1}{2}$ for $K=0, 3, 6, \ldots$, and $g_I=\frac{1}{4}$ for $K=1, 2, 4, 5, 7, 8, \ldots$. Although the line breadth parameter is not accurately known and is not exactly independent of J and K, we assume for an approximate calculation that $(\Delta\nu)_1=10$ MHz. We choose the temperature of dry ice, $T=195°$K, the one at which measurements on this molecule are commonly made. The molecule has five nondegenerate, symmetrical modes of vibration and five doubly degenerate modes. However, in calculating F_v we need to consider only the two lowest modes. Other modes have negligible populations at $T=195°$K. The C—C≡C bending mode [104] ω_{10} equals 336 cm^{-1}, and the C≡C—H bending mode [104] ω_9 equals 643 cm^{-1}. Both are doubly degenerate. With these frequencies substituted into (3.61) with $v=0$ for the ground vibrational state,

$$F_{v=0}=\prod_i (1-e^{-h\omega_i/kT})^{d_i} \tag{6.71}$$

$$F_{v=0}(CH_3CCH)=(1-e^{-1.44(336)/195})^2(1-e^{-1.44(643)/195})^2\cdots=0.83 \tag{6.72}$$

With these various constants, $X=9.68\times10^{-10}$. When centrifugal distortion effects are negligible, $\nu_0=2B(J+1)$ for the frequency of the $K=0$ lines and the absorption coefficient for the $K=0$ lines expressed in terms of J, the lower rotational quantum number for the transition, is

$$\alpha_{max}(K=0)=4.83\times10^{-6}(J+1)^3[1-2.10\times10^{-3}(J+1)]\,e^{-2.10\times10^{-3}J(J+1)} \tag{6.73}$$

Figure 6.14 shows a plot of $\alpha_{max}(K=0)$ as a function of J for all the significantly populated $K=0$ lines of $^{12}CH_3{}^{12}C^{12}CH$. The coefficients for the lines with $K\neq0$ can be obtained from these with the relation

$$\alpha_{max}(K\neq0)=4g_I\alpha_{max}(K=0)\left[1-\frac{K^2}{(J+1)^2}\right]e^{-0.036K^2} \tag{6.74}$$

where $g_I=\frac{1}{2}$ for K values divisible by 3 and $g_I=\frac{1}{4}$ for K values not divisible by 3. In the inset of Fig. 6.14 we have plotted the K lines for the $J=10\rightarrow11$ transition.

At 195°K the strongest rotational lines of methyl acetylene occur in the wavelength region of 0.65 mm. The absorption coefficient of the $K=0$ line of the $J=26\rightarrow27$ transition which occurs in the optimum region is 0.02 cm^{-1}. This indicates that a one-meter cell filled with the gas under the assumed pressure range would absorb $(1-e^{-3})\equiv86\%$ of the energy at the peak or resonant frequency. The strongest component of this transition is the $K=3$ lines which has $\alpha_{max}=0.03$ cm^{-1}; it would absorb 95% of the energy with a cell one meter in length.

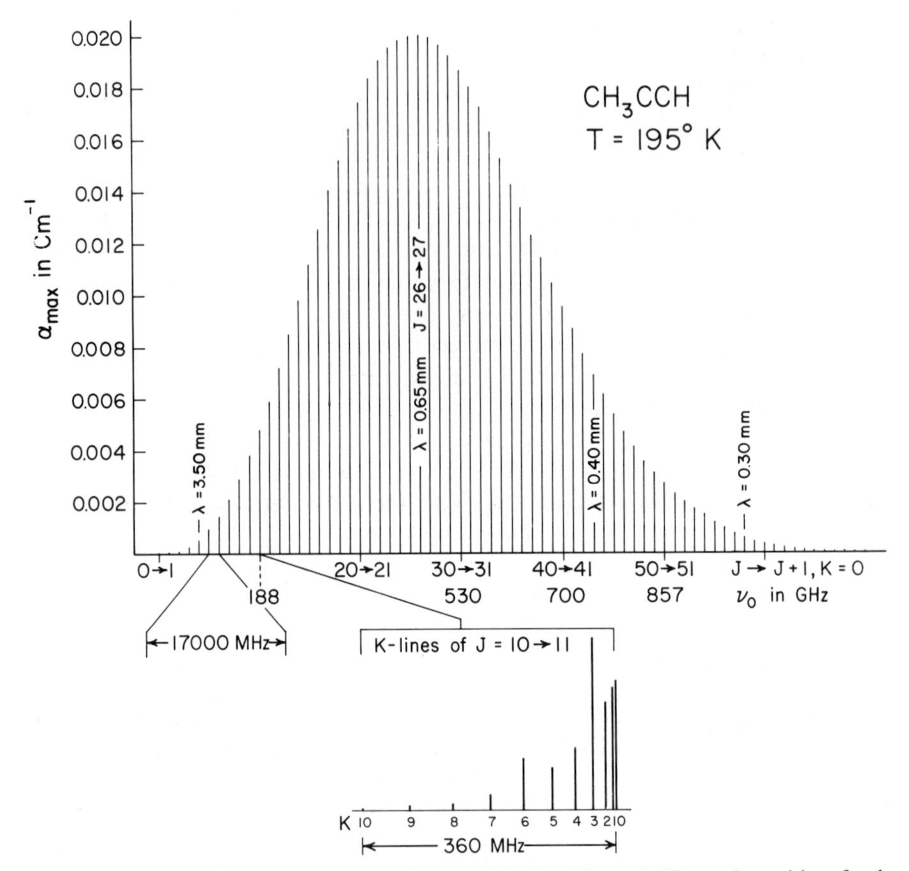

Fig. 6.14 Plots of the peak absorption coefficients of the $K=0$ lines of different J transitions for the symmetric-top molecule CH_3CCH. The inset shows the variation of intensity among the lines of different K for a particular J transition. The lines are assumed to be pressure broadened within the range in which the line width is directly proportional to pressure.

Optimum Regions for Detection of Symmetric-top Rotational Spectra

One can predict the spectral region where the rotational absorption lines of symmetric-top molecules are strongest by equating the derivative of α_{max} with respect to J equal to zero and by solving for J_{opt}, just as for linear molecules. Since we are interested only in locating the region where the strongest lines occur, we can simplify the problem by setting $K=0$. The problem thus reduces to that for the linear molecule, and we can employ (5.43) or (5.44), already derived in Chapter V, Section 4. By substitution of J_{opt} from (5.43) into (5.44) one obtains

$$\nu_{opt}\,(\text{in GHz})=2B+11(BT)^{1/2} \tag{6.75}$$

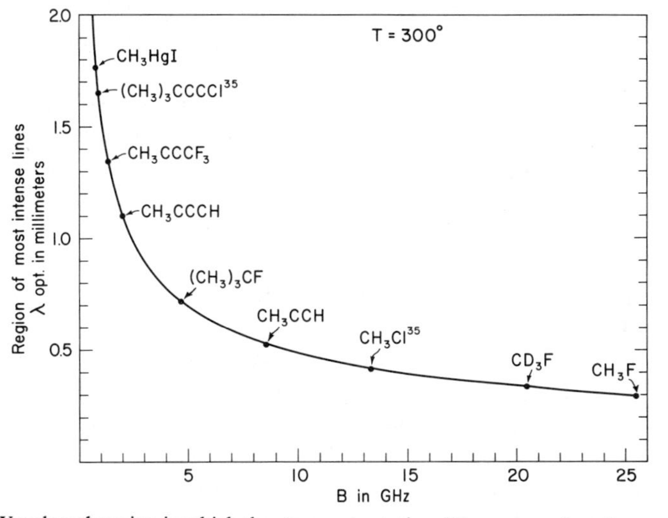

Fig. 6.15 Wavelength region in which the strongest rotational lines occur, λ_{opt}, for symmetric-top molecules of different B. The temperature is assumed to be 300° K.

where B is in GHz and T is the absolute temperature on the Kelvin scale. The optimum wavelength region in millimeters is

$$\lambda_{\text{opt}}(\text{in mm}) = \frac{300}{2B + 11(BT)^{1/2}} \tag{6.76}$$

For convenience we give in Figure 6.15 a plot of the wavelength in millimeters for optimum absorption as a function of B in GHz at a temperature of 300°K. The optimum regions for a few selected molecules are indicated on the chart. It is evident from this chart that the region of strongest absorption for practically all symmetric-top molecules falls between 0.4 and 2 mm wavelength. This is also true for diatomic or linear polyatomic molecules.

6 MOLECULAR STRUCTURES

Ground-state Structures

Ground-state structures for a very large number of symmetric-top molecules have been found from microwave spectroscopy. Listings of the structural parameters of those reported in the literature up to about 1969 may be found in Tables A.VIII.3 and A.VIII.4 of the 1970 edition of this book. A complete listing of the structures derived to a later date, 1982, may be found in the pertinent volumes of the Landolt-Börnstein tables [105]. Here, we give only a few examples of the many symmetric-top molecular structures found from microwave spectroscopy. Methods for their calculation from the observed spectral constants are described in Chapter XIII.

Probably the most complete information on ground-state structure is available for the group V hydrides: NH_3, PH_3, AsH_3, and SbH_3. These pyramidal molecules have only two independent structural parameters; hence only two independent rotational constants are required for their solution. Extensive measurements of rotational transitions of several isotopic species of ammonia and stibene in their ground vibrational states have made possible different kinds of solutions for the various structural parameters listed in Table 6.7. The effective ground-state structures, r_0, are obtainable from only two isotopic forms whereas substitution structures, r_s, require three isotopic forms. Descriptions of both r_0 and r_s structures with methods for their calculation are given in Chapter XIII.

Because of the relatively large zero-point vibrations of the light hydrogen, a substitution of D for H is expected to cause a much greater degree of nonconformity of the structures than does isotopic substitution of the heavier central atom. Thus the effective structures for NH_3 and ND_3 and for SbH_3 and SbD_3 listed in Table 6.7 provide an indication of the degree of nonconformity between the normal and the completely deuterated structures. Note that for both molecules the effective bond lengths in the deuterated species are slightly shorter than those in the corresponding hydrogen species. In Table 6.7 are also given substitution structures r_s for six different isotopic combinations of ammonia, which include tritiated as well as deuterated species. These structures

Table 6.7 Ground-state Structures of Ammonia and of Stibene

Isotopic Combinations	*Bond Length* (Å)	*Bond Angle*	*Ref.*
	Effective Structures of Ammonia		
$^{14}NH_3$, $^{15}NH_3$	1.0156	107° 17′	a
$^{14}ND_3$, $^{15}ND_3$	1.0143	107° 04′	a
	Substitution Structures of Ammonia		
$^{14}NH_3$: $^{15}NH_3$, $^{14}ND_3$	1.0138	107° 14′	a
$^{15}NH_3$: $^{14}NH_3$, $^{15}ND_3$	1.0138	107° 14′	a
$^{14}ND_3$: $^{15}ND_3$, $^{14}NH_3$	1.0136	107° 04′	a
$^{15}ND_3$: $^{14}ND_3$, $^{15}NH_3$	1.0137	107° 04′	a
$^{14}NH_3$: $^{15}NH_3$, $^{14}NT_3$	1.0132	107° 13′	b
$^{14}ND_3$: $^{15}ND_3$, $^{14}NT_3$	1.0128	107° 02′	b
	Effective Structures of Stibene		
$^{121}SbH_3$, $^{123}SbH_3$	1.7102	91° 42′	c
$^{121}SbD_3$, $^{123}SbD_3$	1.7075	91° 42′	c

[a] P. Helminger, F. C. De Lucia, and W. Gordy, *J. Mol. Spectrosc.*, **39**, 94 (1971).
[b] P. Helminger, F. C. De Lucia, W. Gordy, H. W. Morgan, and P. A. Staats, *Phys. Rev.*, **A9**, 12 (1974).
[c] Helminger et al. [78].

are remarkably consistent, and the bond lengths are slightly shorter than those for the r_0 structures. Note that the bond lengths decrease slightly as the combination used includes more of the heavier hydrogen species.

The structures of phosphine and arsine, shown in Table 6.8, are among the few symmetric-top structures that have been derived without the aid of isotopic substitution. These derivations were made possible by observation of the moment of inertia about the symmetry axis from "forbidden" $\Delta K = \pm 3$ transitions, as described in Section 4. With only two structural parameters, a complete structural solution is possible when C_0 (or A_0 for a prolate top) as well as B_0 is known.

The effective ground-state structures derived from C_0 and B_0 for PH_3, PD_3, and AsH_3 are recorded in Table 6.8. As for ammonia, the bond length of the deuterated species is slightly shorter than that of the normal species. This slight shortening, 0.0024 Å for r_0(PD), is expected from the lower vibrational energy of the deuterium. The effect of the lower vibrational energy on the bond angle is more difficult to assess, but the measured bond angle is slightly greater in PD_3 than in PH_3. Equilibrium structures have also been derived for PH_3 and PD_3. As expected, no measurable differences could be detected in the r_e structures of the PH_3 and PD_3. These r_e structures are given for comparison at the bottom of Table 6.8. Note that the equilibrium dimensions are slightly smaller than those of the effective structures.

Table 6.8 Structures of Phosphine and Arsine Derived from B_0 and C_0 without Isotopic Substitution

Species	*Bond Length* (Å)	*Bond Angle* (°)	*Ref.*
	Effective Ground-state Structure		
PH_3	1.4200	93.345	a
PD_3	1.4176	93.359	a
AsH_3	1.52014	91.9758	b
	Average Ground-state Structure		
PH_3	1.4270(2)	93.229(5)	a,c
PD_3	1.4227(1)	93.257(4)	a,c
AsH_3	1.5275(3)	91.93(5)	c
	Equilibrium Structure		
PH_3	1.4116(6)	93.33(2)	a,c
PD_3	1.4116(6)	93.33(2)	a,c
AsH_3	1.511(2)	91.94(20)	c

[a] Helms and Gordy [73].
[b] Helms and Gordy [74].
[c] Chu and Oka [67].

Table 6.9 Effective Ground-state Structures of OPF_3, $OPCl_3$, and $OVCl_3$

OXY_3	$r_0(OX)$ (Å)	$r_0(OY)$ (Å)	$\angle(YXY)$ (*deg.*)	*Ref.*
OPF_3	1.437(4)	1.5222(4)	101.14(10)	a
$OPCl_3$	1.455(5)	1.989(2)	103.7(2)	b
$OVCl_3$	1.595(5)	2.131(1)	111.8(2)	c

[a]Kagann et al. [82].
[b]Y. S. Li, M. M. Chen, and J. R. Durag, *J. Mol. Struct.*, **14**, 261 (1972).
[c]K. Karakida, K. Kuchitau, and M. Matsumura, *Chem. Lett.*, 273 (1972).

The very interesting organometallic molecule C_5H_5BeCl (cyclopentabenzylberyllium) has been investigated with both microwave spectroscopy and electron diffraction. The microwave spectrum, observed by Bjørseth et al. [106] conforms to that of a symmetric top having $\mathscr{C}_{5v}$ symmetry. Through isotopic substitution these observers were able to obtain a complete r_0 structure and a partial r_s structure. The probable nature of the bonding of the Be to the cyclopental ring is described by Drew and Haaland [107].

Although the ground-state structures of the methyl, silyl, and germanyl halides were obtained in the early years of microwave spectroscopy, spectral measurements on the chemically unstable stannic halides, SnH_3X, were not achieved until 1971. These measurements, by Wolf, Krisher, Gsell, et al. [108–110], led to accurate values of the substitutional structures and nuclear quadrupole coupling, which provide information about the nature of the Sn-Hal bonds. Remeasurement of some of the germanyl halides [111–113] has provided more accurate values of the ground-state structures of these halides.

The complete ground-state structure r_0 of vanadyl chloride, $OVCl_3$, has been obtained by Karakida et al. [114] from their microwave measurements of its three isotopic forms. This structure is compared with that obtained by Li et al. [115] for the similar molecule $OPCl_3$ in Table 6.9. The newly prepared, reactive molecule CH_3CP has been studied by Kroto et al. [116] with isotopic substitution. The r_s structure of CH_3CP was found to be: $r_s(CH)=1.108(1)$, $r_s(CC)=1.465(3)$, $r_s(CP)=1.544(4)$, $\angle_s(HCC)=110.30(9)^{\circ}$.

Equilibrium Structures

The determination of the equilibrium constant $B_e=h/(8\pi^2 I_e)$ requires measurements on rotational lines for molecules in all the different fundamental vibrational modes. Symmetric-top molecules with n atoms have $3n-6$ fundamental vibrational modes, some of which are doubly degenerate. The pyramidal AB_3 molecules, the polar symmetric-top molecules with the fewest number of atoms, have six fundamental vibrational modes, two of which are doubly degenerate. Therefore, to obtain the equilibrium structure one must measure B_v values for at least four different vibrational states even for this simplest symmetric top The doubly degenerate modes may also be complicated by l-type doubling, as described in Chapter VI, Section 2.

Otake et al. [117] have measured two rotational transitions in all the fundamental vibrational states of two isotopic species of the molecule, $^{14}NF_3$ and $^{15}NF_3$. From the results they were able to derive the complete equilibrium structure of this molecule. The results are summarized in Table 6.10. From (6.18) for the ground state and the first excited state of the ith mode, $v_i=1$, all others zero, it is evident that

$$\alpha_i^b = B_0 - B_v \tag{6.77}$$

Of the fundamental modes, v_1, v_2, v_3, and v_4, the last two are degenerate. Thus the equilibrium value of B from (6.17) is

$$B_e = B_0 + \tfrac{1}{2}\alpha_1^b + \tfrac{1}{2}\alpha_2^b + \alpha_3^b + \alpha_4^b \tag{6.78}$$

From the B_e values of the two isotopic species the two equilibrium parameters can be found by methods described in Chapter XIII. As would be expected from the anharmonicity in the potential functions, the r_e(NF) length is slightly less than the r_0(NF) length. However, the equilibrium value of the bond angle

Table 6.10 Anharmonic Constants and the Equilibrium Structure of NF_3 from Microwave Spectroscopy (from Otake et al. [117])

	$^{14}NF_3$		$^{15}NF_3$	
State	B_v *(MHz)*	α_i^b *(MHz)*	B_v *(MHz)*	α_i^b *(MHz)*
v_0	10,681.02 ± 0.01	—	10,629.44 ± 0.03	—
v_4	10,676.54 ± 0.01	4.48 ± 0.02	10,624.80 ± 0.02	4.64 ± 0.05
v_2	10,642.38 ± 0.02	38.65 ± 0.03	10,589.30 ± 0.02	40.14 ± 0.05
v_3	10,602.22 ± 0.03	78.81 ± 0.04	10,553.77 ± 0.02	75.67 ± 0.05
v_1	10,724.47 ± 0.03	−43.45 ± 0.04	10,667.85 ± 0.04	−38.41 ± 0.07
$2v_4$	10,671.90 ± 0.05	9.12 ± 0.06	10,620.12 ± 0.03	9.32 ± 0.06
	B_e *(MHz)*		B_e *(MHz)*	
	10761.91 ± 0.2		10710.63 ± 0.2	

Molecular Structure

r_0(NF) = 1.371 ± 0.002 Å ($\angle$FNF)$_0$ = 102°10′ ± 2′
r_e(NF) = 1.365 ± 0.002 Å ($\angle$FNF)$_e$ = 102°22′ ± 2′

was found to be slightly less than the effective value. Both these structures are given at the bottom of Table 6.10.

Duncan [118] has combined the A_e values derived from infrared vibration–rotational spectra with B_e values from microwave rotational spectra for calculation of the equilibrium structures of the methyl halides. Because the moment of inertia I_a about the symmetry axes is small, arising only from the inertial moments of the light, off-axis hydrogens, the A_v values are sufficiently large for measurement with good accuracy from vibration–rotation spectra. The I_b values are an order of magnitude larger, and the small B_v values are measured with significant accuracy only with microwave rotational spectroscopy. Since the A_v values are not obtained by microwave spectroscopy, except in special cases of second-order transitions as described in Section 4, the combination of infrared and microwave data as employed by Duncan is a useful procedure.

Next to the AB_3 molecules, the methyl halides belong to the simplest class of symmetric-top molecules measurable with microwave spectroscopy. Even so, the complete determination of their equilibrium structures required an enormous amount of infrared and microwave data. Many spectroscopists contributed to these data. The numerous papers, microwave and infrared, from which Duncan selected these data are referenced by him and will not be repeated here. Table 6.11 gives the α_i^b (or α_i^B) values that he used in calculation of the B_e values of the chloride, bromide, and iodide. The corresponding infrared α_i^a and A_e constants may be found in his paper.

The methyl halides have nine vibrational modes, three of which are degenerate, yielding six which are distinguishable. Thus, there are only six α_i^b constants, as listed in Table 6.11. The three symmetric modes, v_1, v_2, v_3 (species A_1), are nondegenerate whereas the perpendicular modes, v_4, v_5, v_6 (species E), are each

Table 6.11 Equilibrium Constants B_e and α_i^b for Some Methyl Halides[a]

Halide	B_e	α_1^b	α_2^b	α_3^b	α_4^b	α_5^b	α_6^b
$CH_3{}^{35}Cl$	13,407	1.8	121	115	−5.4	−47.4	49.2
$CD_3{}^{35}Cl$	10,925	5.2	179	84	0	−90	39
$CH_3{}^{Av}Br$	9,635	9	63.9	72.5	−4.2	−16	34
$CD_3{}^{Av}Br$	7,753	6.6	60.6	49.8	0	−27	24
CH_3I	7,564	5.1	41.1	54.3	−3.6	−7.5	24
CD_3I	6,077	4.05	30.9	37.8	0	−13	15

[a]From Duncan [118]. The α_i values are those chosen by Duncan [118] from various sources, to which references may be found in his tabulations. Here the equilibrium constants given by Duncan in cm^{-1} units are converted to MHz, and α_i^b is used for his α_i^B. A corresponding set of A_e, α_i^A values obtained from infrared spectroscopy may be found in Duncan's paper. The $CH_3{}^{Av}Br$ is the average of the ^{79}Br and ^{81}Br isotopic forms.

doubly degenerate. It is generally simpler to analyze the rotational spectra of the nondegenerate, parallel modes than the degenerate perpendicular ones, because of the l-type doubling in the latter. However, perpendicular modes usually have lower vibrational frequencies and hence higher excited-state populations which make their rotational lines easier to detect than those of excited parallel modes. In addition to l-type doubling, the coincidence, or near coincidence, of certain excited states leading to Fermi resonance between fundamental or combination bands can complicate the analysis of the rotational structure of excited vibrational states and the evaluation of α_i for symmetric-top molecules. Other complications in the evaluation of α_i arise from effects of large Coriolis interactions between fundamental modes that are nearly coincident [118]. These effects were found to be especially large between the v_2 and v_5 modes of CH_3F and of CD_3Cl.

The equilibrium structures for the methyl halides as obtained by Duncan [118] are given in Table 6.12. The structure for methyl fluoride was considered as least reliable of the four, primarily because of uncertainties in some of the α_i values resulting from the strong overlapping of the v_2 and v_5 fundamental bands already mentioned. Because of these uncertainties, the probable errors in the structural parameters were not estimated for the fluoride.

With infrared, tunable-laser spectroscopy or with microwave–infrared double resonance it should be possible in the future to improve the accuracy of the A_e values of methyl halides to the level of that already achieved for the B_e values.

Table 6.12 Equilibrium Structures of the Methyl Halides[a]

	Bond Distances (Å)		*Bond Angle*
Halide	r_e(CH)	r_e(C-Hal)	$\angle_e$(HCH)
CH_3F[b]	1.095	1.382	110°20′
CH_3Cl[c]	1.086(4)	1.778(2)	110°40′±40′
CH_3Br[c]	1.086(3)	1.933(2)	110°10′±25′
CH_3I[c]	1.085(3)	2.133(2)	111°17′±25′

[a]From Duncan [118].

[b]Derived from the equilibrium constants A_e^{H}, B_e^{H}, B_e^{D} for CH_3F, CH_3F, and CD_3F, respectively. The values from A_e^{D}, B_e^{H}, and B_e^{D} are 1.094 Å, 1.382 Å, and 110°26′. The limits of error were not determined for the fluoride. See discussion in text.

[c]These values are determined from two sets: those derived from the combinations (I) A_e^{H}, B_e^{H}, B_e^{D}, and those from (II) A_e^{D}, B_e^{H}, B_e^{D}. The sets are essentially indistinguishable for the bromide and iodide. For methyl chloride the sets are (I) 1.087 Å, 1.778 Å, 110°47′, and (II) 1.084 Å, 1.777 Å, 110°27′; the chosen structure is weighted to favor set (I).

However, this improvement in accuracy may not bring about a corresponding improvement in accuracy of the structures for the following reasons. There are three independent structural parameters in the methyl halide molecules, and one must resort to use of isotopic substitution for a third observable. The degree of conformability of the normal and substituted structures will probably limit the accuracy of the structures to approximately that already obtained, regardless of the accuracy of the A_e and B_e values.

The equilibrium structure of trifluorosilane ($HSiF_3$) has been derived by Hoy et al. [119] from B_e values of different isotopic species that were obtained principally from microwave spectroscopy. Because the molecule has three independent structural parameters, B_e values for a minimum of three isotopic species are required for determination of its equilibrium structure. Six isotopic species, those listed in Table 6.13, were observed. Like the methyl halides, $HSiF_3$ has six fundamental modes, three of which are doubly degenerate. The constants, α_2^b, α_3^b, α_4^b, α_5^b, α_6^b, were measured for $H^{28}SiF_3$ and $D^{28}SiF_3$; the α_1^b, which could not be measured, was theoretically deduced from the cubic anharmonic force field derived partly from vibrational force constants obtained from infrared spectroscopy. Also, some of the values of α_i^b for the less abundant ^{29}Si and ^{30}Si species were derived from the cubic anharmonic force field. The B_e values were then obtained from the measured B_0 values with the expression

$$B_e = B_0 + \sum_{i=1}^{6} \alpha_i^b \left(\frac{d_i}{2}\right) \tag{6.79}$$

Table 6.13 Rotational Constants and Equilibrium Structure of Trifluorosilane[a]

	Rotational Constants (MHz)			
Isotopic Species	B_0 (obs.)	$\Sigma\alpha_i(d_i/2)$	B_e (obs.)	B_e (obs.-calc.)[b]
$H^{28}SiF_3$	7208.05	21.97	7230.02	+0.02
$H^{29}SiF_3$	7195.75	21.89	7217.64	−0.00
$H^{30}SiF_3$	7183.79	21.82	7205.61	−0.01
$D^{28}SiF_3$	6890.10	22.97	6913.07	−0.02
$D^{29}SiF_3$	6880.20	22.90	6903.10	+0.01
$D^{30}SiF_3$	6870.58	22.80	6893.38	+0.01

Equilibrium Structure

r_e(SiH)	r_e(SiF)	$\angle_e$(HSiF)
1.4468(5) Å	1.5624(1) Å	110.64(3)°

[a]From Hoy et al. [119].
[b]Difference between the observed values of B_e and those calculated from the equilibrium structure.

The resulting B_e values with the B_0 and the summed α_i^b contributions for each isotopic species are given in Table 6.13. The equilibrium structure derived from these B_0 values is shown at the bottom of Table 6.13. This paper also gives derived l-type doubling constants for the three degenerate modes and makes corrections for perturbations of Fermi resonance between the v_2 and $2v_3$ vibrations.

Hydrogen-bonded Symmetric-top Complexes

The first example of a hydrogen-bonded complex shown by microwave spectroscopy to have a symmetric-top structure was the hetero-dimer $CH_3CN \cdots HF$. In 1980 Bevan et al. [120] measured pure rotational transitions of the complex and correctly assigned the transitions to a characteristic symmetric-top spectrum. Shortly thereafter, they observed [121] a symmetric-top-like rotational spectrum attributable to the large complex $(CH_3)_3CCN \cdots HF$. From analysis of the spectrum the $C—C{\equiv}N \cdots HF$ fragment of the complex was shown to be linear. The hydrogen bridges in both these complexes are similar to that in the linear hetero-dimer $HCN \cdots HF$ described in Chapter V, Section 8.

The rotational constants and hydrogen-bridge lengths, $r_0(N \cdots F)$ found for the cyanide complexes are listed in Table 6.14. Those for the linear $HCN \cdots HF$ complex are included for comparison. Although the bridge lengths are approximately the same for all three complexes, there is a definite trend to a shorter bridge length with increasing size of the R group. This trend, which

Table 6.14 Rotational Constants and Bridge Lengths for Some Hydrogen-bonded Complexes

Complex	B_0 *(MHz)*	D_J *(kHz)*	D_{JK} *(kHz)*	$r_0(N \cdots F)$ *or* $r_0(Bz \cdots Cl)^a$	*Ref.*
$HCN \cdots HF$	3591.11	5.2		2.796	b
$CH_3CN \cdots HF$	1853.37	0.82	67	2.759	c
$CD_3CN \cdots HF$	1686.21	1.0	50	2.768	c
$CH_3CN \cdots DF$	1832.42	1.0	62	2.753	c
$(CH_3)_3CCN \cdots HF$	876.25			2.725	d
$(CH_3)_3CCN \cdots DF$	862.10			2.725	d
$C_6H_6 \cdots H^{35}Cl$	1237.6836	1.22	13.35	3.594^a	e
$C_6H_6 \cdots D^{35}Cl$	1228.2440	1.19	14.57	3.570^a	e
$C_6D_6 \cdots H^{35}Cl$	1165.1542	1.08	10.93	3.591^a	e

[a] Distance from center of gravity of benzene ring to the Cl.
[b] A. C. Legon, D. W. Millen, and S. C. Rogers, *Proc. R. Soc. London*, **A370**, 213 (1980).
[c] Bevan et al. [120].
[d] Georgiou et al. [121].
[e] Read et al. [122].

indicates an increasing hydrogen-bond strength with R=H to $(CH_3)_3C$, is borne out by hydrogen-bond stretching force constant $k_s(k_\sigma)=0.24$ mdyne/Å for R=H and 0.27 mdyne/Å for $R=(CH_3)_3C$.

A very interesting complex formed by HCl and benzene was shown from measurements on its rotational spectrum to have a C_{6v} symmetric-top equilibrium structure [122]. The vibrationally averaged direction of the HCl is along the perpendicular symmetry axis of the benzene with its H directed toward the center of the ring. Evidently the positive H interacts with the π-electron cloud to form the hydrogen-bonded dimer $HCl\cdots C_6H_6$. From the rotational constants (Table 6.14), the averaged distance of the Cl above the plane of the ring is found to be 3.59 Å. An approximate hydrogen bond-stretching constant, $k_s=0.080$ mdyne/Å, and bonding energy (well depth), $\varepsilon=720\ \text{cm}^{-1}$, are obtained. Other information with discussion of the various results are given in the paper.

From microwave measurements, HCN and HCl are found to form T-type hydrogen-bonded structures [123, 124] with HC≡C—H in which the HCN, or HCl, is along the perpendicular bisector of the acetylene so that the hydrogen forming the bond is nearest to the center of the negative lobes of the π bonds. Hydrogen cyanide forms a similar T-type complex with ethylene [125] in which the H of the HCN interacts with the π-bond lobe density of the $H_2C{=}CH_2$. Although these T-structures are asymmetric rotors, they have approximate (accidental) prolate, symmetric-top structures with $B\approx C\ll A$.

The cyclopropane–HCN complex, found by Kukolich [126] from microwave spectral measurements, is also very close to an accidental prolate symmetric-top with $B\approx C\ll A$. The HCN was found to lie in the plane of the carbon ring and to be directed from outside along a perpendicular to the center of one of the C–C bonds with the H toward the bond. The distance between the center of this C–C bond and the carbon of the HCN, the hydrogen-bridge length, was found to be 3.474 Å, the hydrogen bond stretching force constant to be $k_s=0.062$ mdyne/Å, and the bonding energy of the complex to be $860\ \text{cm}^{-1}$, approximately. The "outward bending," caused by the strain in the bonds of the saturated, three-membered, cyclopropane ring evidently gives rise to the interaction of the bond with the proton of the HCN. Hydrogen bonds are not known to occur through direct interaction with a normal σ-bond pair. The most commonly occurring hydrogen bonds are formed by interaction of the hydrogen with a nonbonding electron pair like those of $CH_3CN\cdots HF$ or the linear complexes described in Chapter V, Section 8.

Hydrogen-bonded complexes, like those described, in which the bonding hydrogen interacts with π-electron lobes or with "strained" σ bonds should be useful for locating reactive sites in molecules where reactions with other electrophilic agents are most likely to begin.

The methods for derivation of various properties of the hydrogen-bonded complexes from microwave spectral measurements such as hydrogen-bond force constants and bond energies, bridge length, and dipole moment enhancement are like those described for linear hydrogen-bonded complexes in Chapter

V, Sections 8 and 9. Also, the spectrometers used for detection and measurement of the symmetric-top complexes are similar to those described or referenced in those sections. For both types of structures the complex is produced in very cold beams by expansion of a pressurized mixture through a restricting nozzle into the evacuated observation cavity.

References

1. D. M. Dennison, *Phys. Rev.*, **28**, 318 (1926); *Rev. Mod. Phys.*, **3**, 280 (1931).
2. F. Reiche and H. Rademacher, *Z. Physik*, **39**, 444 (1927).
3. R. de L. Kronig and I. I. Rabi, *Phys. Rev.*, **29**, 262 (1927).
4. T. E. Sullivan and L. Frenkel, *J. Mol. Spectrosc.*, **39**, 185 (1971).
5. J. G. Smith and I. Thompson, *Mol. Phys.*, **32**, 1247 (1976).
6. Tsu-Shen Chang and D. M. Dennison, *J. Chem. Phys.*, **21**, 1293 (1953).
7. W. J. Orville-Thomas, J. T. Cox, and W. Gordy, *J. Chem. Phys.*, **22**, 1718 (1954).
8. J. Wagner, *Z. Physik. Chem.*, **B40**, 36 (1938).
9. S. Silver and W. H. Shaffer, *J. Chem. Phys.*, **9**, 599 (1941).
10. W. H. Shaffer, *J. Chem. Phys.*, **9**, 607 (1941).
11. W. H. Shaffer, *J. Chem. Phys.*, **10**, 1 (1942).
12. H. H. Nielsen, *Phys. Rev.*, **60**, 764 (1941).
13. H. K. Bodenseh, R. Gegenheimer, J. Mennicke, and W. Zeil, *Z. Naturforsch.*, **22a**, 523 (1967).
14. E. Teller, *Hand-u. Jahrb. Chem. Phys.*, **9**, II, 43 (1934).
15. M. Johnston and D. M. Dennison, *Phys. Rev.*, **48**, 868 (1935).
16. E. B. Wilson, *J. Chem. Phys.*, **3**, 818 (1935).
17. H. H. Nielsen, *Phys. Rev.*, **77**, 130 (1950).
18. H. H. Nielsen, *Rev. Mod. Phys.*, **23**, 90 (1951).
19. M. L. Grenier-Beeson and G. Amat, *J. Mol. Spectrosc.*, **8**, 22 (1962).
20. S. Maes and G. Amat, *Can. J. Phys.*, **43**, 321 (1964).
21. S. Maes, *Cah. Phys.*, **14**, 125 (1960).
22. G. Amat, H. H. Nielsen, and G. Tarrago, *Rotation Vibration of Polyatomic Molecules*, Dekker, New York, 1971.
23. R. Trambarulo and W. Gordy, *J. Chem. Phys.*, **18**, 1613 (1950).
24. D. R. Jenkins, R. Kewley, and T. M. Sugden, *Trans. Faraday Soc.*, **58**, 1284 (1962).
25. P. Venkateswarlu, J. G. Baker, and W. Gordy, *J. Mol. Spectrosc.*, **6**, 215 (1961).
26. A. Bauer, *J. Mol. Spectrosc.*, **40**, 183 (1971).
27. G. Tarrago, *J. Mol. Spectrosc.*, **34**, 23 (1970).
28. M. Otake, F. Hirota, and Y. Morino, *J. Mol. Spectrosc.*, **28**, 325 (1968).
29. M. J. Whittle, J. G. Baker, and G. Corbelli, *J. Mol. Spectrosc.*, **40**, 388 (1971).
30. M. C. L. Gerry and T. M. Sugden, *Trans. Faraday Soc.*, **61**, 209 (1965).
31. A. J. Careless and H. W. Kroto, *J. Mol. Spectrosc.*, **57**, 198 (1975).
32. E. Hirota, *J. Mol. Spectrosc.*, **37**, 20 (1971); **38**, 195 (1971).
33. Y. Kawashima and A. P. Cox, *J. Mol. Spectrosc.*, **61**, 435 (1976).
34. Y. Morino and C. Hirose, *J. Mol. Spectrosc.*, **22**, 99 (1967).
35. A. Bauer and S. Maes, *J. Mol. Spectrosc.*, **40**, 207 (1971).
36. C. E. Cleeton and N. H. Williams, *Phys. Rev.*, **45**, 234 (1934).
37. B. Bleaney and R. P. Penrose, *Nature*, 157, 339 (1946).

38. W. E. Good, *Phys. Rev.*, **70**, 213 (1946).
39. C. C. Costain and G. B. B. M. Sutherland, *J. Phys. Chem.*, **56**, 321 (1952).
40. C. H. Townes and A. L. Schawlow, *Microwave Spectroscopy*, McGraw-Hill, New York, 1955, p. 307.
41. M. F. Manning, *J. Chem. Phys.*, **3**, 136 (1935).
42. D. M. Dennison and G. E. Uhlenbeck, *Phys. Rev.*, **41**, 313 (1932).
43. C. C. Costain, *Phys. Rev.*, **82**, 108 (1951).
44. E. Schnabel, T. Törring, and W. Wilke, *Z. Physik.*, **188**, 167 (1965).
45. G. E. Erlandsson and W. Gordy, *Phys. Rev.*, **106**, 513 (1957).
46. P. Helminger and W. Gordy, *Bull. Am. Phys. Soc.*, **12**, 543 (1967).
47. P. Helminger, F. C. De Lucia, W. Gordy, H. W. Morgan, and P. A. Staats, *Phys. Rev.*, **A 9**, 12 (1974).
48. P. Helminger, F. C. De Lucia, and W. Gordy, *J. Mol. Spectrosc.*, **39**, 94 (1971).
49. R. G. Nuckolls, L. J. Reuger, and H. Lyons, *Phys. Rev.*, **89**, 1101 (1953).
50. G. Hermann, *J. Chem. Phys.*, **29**, 875 (1958).
51. W. S. Benedict and E. K. Plyler, *Can. J. Phys.*, **35**, 1235 (1957).
52. K. N. Rao, W. W. Brim, J. M. Hoffman, L. H. Jones, and R. S. McDowell, *J. Mol. Spectrosc.*, **7**, 362 (1961).
53. J. D. Swalen and J. A. Ibers, *J. Chem. Phys.*, **36**, 1914 (1962).
54. R. J. Damburg and R. Kh. Propin, *J. Phys. B: Atom. Mol. Phys.*, **5**, 1861 (1972).
55. D. Papoušek, J. M. R. Stone, and V. Špirko, *J. Mol. Spectrosc.*, **48**, 17 (1973).
56. J. T. Hougan, R. R. Bunker, and J. W. C. Johns, *J. Mol. Spectrosc.*, **34**, 136 (1970).
57. V. Špirko, J. M. R. Stone, and D. Papoušek, *J. Mol. Spectrosc.*, **60**, 151 (1980).
58. A. R. Hoy and P. R. Bunker, *J. Mol. Spectrosc.*, **52**, 439 (1974).
59. S. P. Belov, L. I. Gershtein, A. F. Krupnov, A. V. Maslovskij, S. Urban, V. Špirko, and D. Papoušek, *J. Mol. Spectrosc.*, **84**, 288 (1980).
60. F. Scappini and A. Guarnieri, *J. Mol. Spectrosc.*, **95**, 20 (1982).
61. H. M. Hanson, *J. Mol. Spectrosc.*, **23**, 287 (1967).
62. T. Oka, *J. Chem. Phys.*, **47**, 5410 (1967).
63. K. Fox, *Phys. Rev. Lett.*, **27**, 233 (1971).
64. J. K. G. Watson, *J. Mol. Spectrosc.*, **40**, 536 (1971).
65. M. Mizushima and P. Venkateswarlu, *J. Chem. Phys.*, **21**, 705 (1953).
66. I. M. Mills, J. K. G. Watson, and W. L. Smith, *Mol. Phys.*, **16**, 329 (1969).
67. F. Y. Chu and T. Oka, *J. Chem. Phys.*, **60**, 4612 (1974).
68. A. Rosenberg, I. Ozier, and A. K. Kudian, *J. Chem. Phys.*, **57**, 568 (1972).
69. C. W. Holt, M. C. L. Gerry, and I. Ozier, *Phys. Rev. Lett.*, **31**, 1033 (1973).
70. T. Oka, "'Forbidden' Rotational Transitions," in *Molecular Spectroscopy: Modern Research*, Vol. II, K. N. Rao, Ed. Academic, New York, 1976, pp. 229–253.
71. K. Fox, *Phys. Rev.*, **A 6**, 907 (1972).
72. M. R. Aliev and V. M. Mikhaylov, *J. Mol. Spectrosc.*, **49**, 18 (1974).
73. D. A. Helms and W. Gordy, *J. Mol. Spectrosc.*, **66**, 206 (1977).
74. D. A. Helms and W. Gordy, *J. Mol. Spectrosc.*, **69**, 473 (1978).
75. A. H. Maki, R. L. Sams, and W. B. Olson, *J. Chem. Phys.*, **58**, 4502 (1973).
76. W. B. Olson, A. G. Maki, and R. L. Sams, *J. Mol. Spectrosc.*, **55**, 252 (1975).
77. P. Helminger and W. Gordy, *Phys. Rev.*, **188**, 100 (1969).
78. P. Helminger, E. Beeson, and W. Gordy, *Phys. Rev.*, **A3**, 122 (1971).
79. H. H. Nielsen and D. M. Dennison, *Phys. Rev.*, **72**, 1011 (1947).

80. S. P. Belov, A. V. Burenin, L. I. Gershtein, A. F. Krupnov, V. N. Markov, A. V. Maslovsky, and S. M. Shapin, *J. Mol. Spectrosc.*, **86**, 184 (1981).
81. P. B. Davies, R. M. Neumann, S. C. Wofsky, and W. Klemperer, *J. Chem. Phys.*, **55**, 3564 (1971).
82. R. H. Kagann, I. Ozier, and M. C. L. Gerry, *J. Mol. Spectrosc.*, **71**, 281 (1978).
83. J. G. Smith, *Mol. Phys.*, **32**, 621 (1976).
84. I. Ozier and W. L. Meerts, *Phys. Rev. Lett.*, **40**, 226 (1978).
85. W. L. Meerts and I. Ozier, *Phys. Rev. Lett.*, **41**, 1109 (1978).
86. W. L. Meerts and I. Ozier, *J. Chem. Phys.*, **75**, 596 (1981).
87. I. Ozier and W. L. Meerts, *Can. J. Phys.*, **59**, 150 (1981).
88. I. M. Mills, *Mol. Phys.* **1**, 107 (1958); *Spectrochim. Acta.*, **16**, 35 (1960).
89. I. Ozier, *Phys. Rev. Lett.*, **27**, 1329 (1971).
90. A. Rosenberg and I. Ozier, *Can. J. Phys.*, **52**, 575 (1974).
91. I. Ozier and A. Rosenberg, *Can. J. Phys.*, **51**, 1882 (1973).
92. C. W. Holt, M. C. L. Gerry, and I. Ozier, *Can. J. Phys.*, **53**, 1791 (1975).
93. I. Ozier, R. M. Lees, and M. C. L. Gerry, *J. Chem. Phys.*, **65**, 1795 (1976).
94. I. Ozier, R. M. Lees, and M. C. L. Gerry, *Can. J. Phys.*, **54**, 1094 (1976).
95. R. H. Kagann, I. Ozier, G. A. McRae, and M.C.L. Gerry, *Can. J. Phys.*, **57**, 593 (1979).
96. R. F. Curl, T. Oka, and D. S. Smith, *J. Mol. Spectrosc.*, **46**, 518 (1973).
97. R. F. Curl, *J. Mol. Spectrosc.*, **48**, 165 (1973).
98. W. A. Kreiner and T. Oka, *Can. J. Phys.*, **53**, 2000 (1975).
99. K. T. Hecht, *J. Mol. Spectrosc.*, **5**, 355 (1960); **5**, 390 (1960).
100. S. M. Kirschner and J. K. G. Watson, *J. Mol. Spectrosc.*, **47**, 347 (1973).
101. I. Ozier, *J. Mol. Spectrosc.*, **53**, 336 (1974).
102. G. Pierre, G. Guelachvili, and C. Amiot, *J. Phys. Paris.*, **36**, 487 (1975).
103. G. Tarrago, M. Dang-Nhu, G. Poussigue, G. Guelachvili, and C. Amiot, *J. Mol. Spectrosc.*, **57**, 246 (1975).
104. G. Herzberg, *Infrared and Raman Spectra of Polyatomic Molecules.*, Van Nostrand, New York, 1945, p. 339.
105. This collection of data initiated by B. Starck as Vol. II/4 is now extended to Vols. II/6, II/7, and II/14. A complete reference is given in Chapter I, Ref. 65.
106. A. Bjørseth, D. A. Drew, K. M. Marstokk, and H. Møllendal, *J. Mol. Struct.* **13**, 233 (1972).
107. D. A. Drew and A. Haland, *Chem. Commun.*, 1551 (1971).
108. L. C. Krisher, R. A. Gsell, and J. M. Bellama, *J. Chem. Phys.*, **54**, 2287 (1971).
109. S. N. Wolf, L. C. Krisher, and R. A. Gsell, *J. Chem. Phys.*, **54**, 4605 (1971).
110. S. N. Wolf, L. C. Krisher, and R. A. Gsell, *J. Chem. Phys.*, **55**, 2106 (1971).
111. J. R. Durig, A. B. Mohamad, P. L. Trowell, and Y. S. Li, *J. Chem. Phys.*, **75**, 2147 (1981).
112. L. C. Krisher, J. A. Morrison, and W. A. Watson, *J. Chem. Phys.*, **57**, 1357 (1972).
113. S. N. Wolf and L. C. Krisher, *J. Chem. Phys.*, **56**, 1040 (1972).
114. K. Karakida, K. Kuchitsu, and C. Matamura, *Chem. Lett.*, 293 (1972).
115. Y. S. Li, M. M. Chen, and J. R. Durag, *J. Mol. Struct.*, **14**, 26 (1972).
116. H. W. Kroto, J. F. Nixon, and N. P. C. Simmons, *J. Mol. Spectrosc.*, **77**, 270 (1979).
117. M. Otake, C. Matsumura, and Y. Morino, *J. Mol. Spectrosc.*, **28**, 316 (1968).
118. J. L. Duncan, *J. Mol. Struct.*, **6**, 477 (1970).
119. A. R. Hoy, M. Bertram, and I. M. Mills, *J. Mol. Spectrosc.*, **46**, 429 (1973).
120. J. W. Bevan, A. C. Legon, D. J. Millen, and S. C. Rogers, *Proc. R. Soc. London* **A370**, 239 (1980).

121. A. S. Georgiou, A. C. Legon, and D. J. Millen, *Proc. R. Soc., London* **A370**, 257 (1980)
122. W. G. Read, E. J. Campbell, and G. Henderson, *J. Chem. Phys.*, **78**, 3501 (1983).
123. P. D. Aldrich, S. G. Kukolich, and E. J. Campbell, *J. Chem. Phys.*, **78**, 3521 (1983).
124. A. C. Legon, P. D. Aldrich, and W. H. Flygare, *J. Chem. Phys.*, **75**, 625 (1981).
125. S. G. Kukolich, W. G. Read, and P. D. Aldrich, *J. Chem. Phys.*, **78**, 3553 (1983).
126. S. G. Kukolich, *J. Chem. Phys.* **78**, 4832 (1983).

Chapter **VII**

ASYMMETRIC-TOP MOLECULES

1 QUALITATIVE DESCRIPTION

When none of the three principal moments of inertia of a molecule is zero and if no two are equal, considerable complexity is encountered in its pure rotational spectrum. The rotational frequencies can no longer be expressed in convenient equations, as can be done for linear or symmetric-top molecules. Only for certain low J values can the energy levels of the asymmetric rotor be expressed in closed form, even if centrifugal distortion effects are neglected. The increased complexity of the pure rotational spectrum over that of the symmetric-top rotor is illustrated in Fig. 7.1. The various methods that have been used to give the energy levels and wave functions for the asymmetric rotor are considerably more involved than are those for symmetric rotors. The general procedure is to assume that the wave functions can be expanded in terms of an orthogonal set of functions (a natural basis would be the symmetric-top functions) and to set up the secular equations for the unknown coefficients and energies. The resulting secular determinant can be broken down into a number

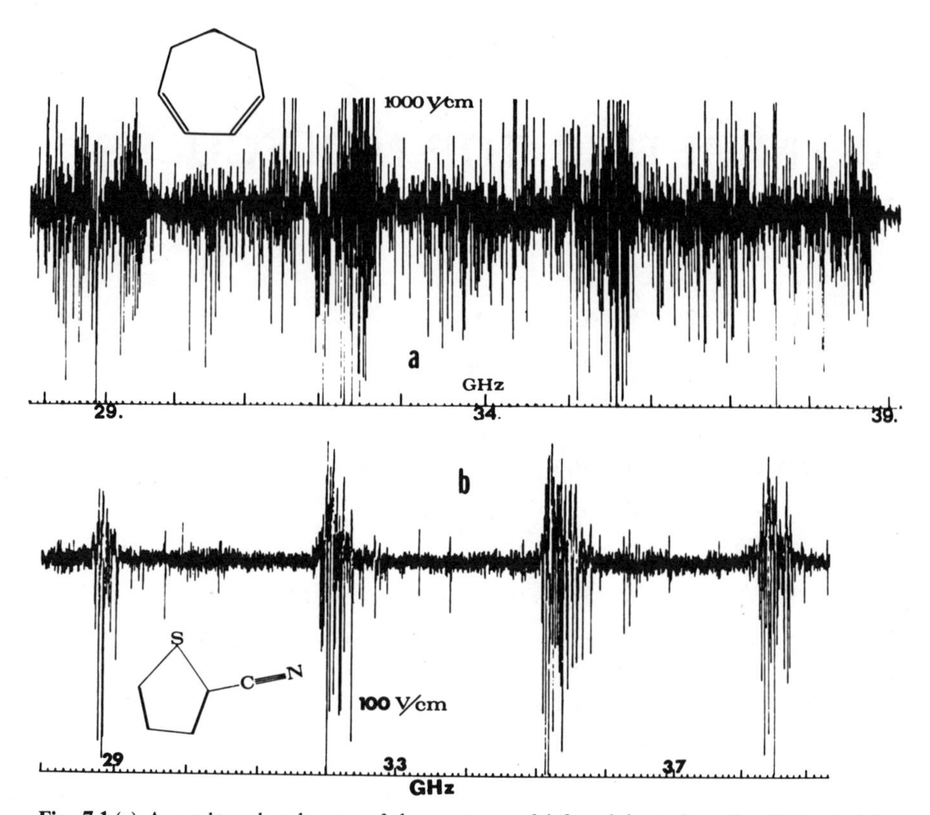

Fig. 7.1 (*a*) A condensed-scale scan of the spectrum of 1, 3-cycloheptadiene ($\kappa \cong 0.85$) which has "*b*"-type transitions. Frequency markers are spaced 100 MHz apart and the sweep rate was about 5 MHz/sec. The zero field lines are down, Stark lobes are up. The separation between the ${}^{b}Q_{1-1}$ band heads is approximately 3110 MHz. $A=3419.424$, $B=3297.290$, and $C=1799.961$ MHz. From Avirah et al. [48]. (*b*) Condensed-scale, *R*-band spectrum of 2-cyanothiophene. The sample pressure was about 30 μm and the scan was about 3 MHz/sec. The Stark lobes are up and the zero field lines are down. Groups of closely spaced absorption lines ($J \rightarrow J+1$, $K_{-1} \rightarrow K_{-1}$) are observed separated by approximately 3175 MHz ($\simeq B+C$). From T. K. Avirah, T. B. Malloy Jr., and R. L. Cook, *J. Mol. Struct.*, **29**, 47 (1975).

of subdeterminants, the order of which increases with *J*. The solution of these subdeterminants yields the required energy levels and expansion coefficients. Details of these energy level calculations will be discussed in the next section.

Despite this complexity, much information useful to the chemist can be obtained from the microwave spectrum of an asymmetric-top molecule, in many instances without long labor or advanced mathematical skill. This is a fortunate circumstance since most molecules are of the asymmetric-top class. In the first place, both qualitative and quantitative spectrochemical analysis can be accomplished by measurements of microwave rotational spectra without an interpretation of their patterns. Second, somewhere in the wide span of the now workable microwave region there are usually low *J* transitions whose frequency can be expressed with closed algebraic equations, from which the principal moments of inertia can be immediately evaluated. Furthermore, except for a few very light molecules such as H_2O, the centrifugal distortion effects on very low J transitions can usually be neglected. These lines cannot be identified from the simple pattern of the rotational spectrum as they can for the symmetric-top rotor, but the low J transitions can be readily identified by their Stark patterns from which the electric dipole moment can also be obtained (see Chapter X). In some cases, the transition can be identified by nuclear quadrupole hyperfine structure. Third, extensive tables are available from which energy levels for various degrees of asymmetry can be obtained to a useful degree of approximation for high J values.

To apply the equations and selection rules developed in the following sections, one must first become familiar with the notation used to designate the levels. Here we shall outline briefly the qualitative characteristics of the asymmetric rotor energy levels. In an asymmetric rotor there is no internal component of the angular momentum that is a constant of the motion; that is P_z no longer commutes with $\mathscr{H}$ and only J and M are "good" quantum numbers. As for the symmetric rotor, we can write $P^2=(h/2\pi)^2J(J+1)$ and $P_Z=hM/2\pi$, where J and M are integers as previously defined. Pseudo-quantum numbers, customarily designated by subscripts on J, are employed in the designation of the levels. The double subscript system of King et al. [1] is perhaps the most descriptive and useful; it will be employed extensively. Their system is best understood by a comparison of the limiting prolate and oblate symmetric tops. In the conventional order, $I_a<I_b<I_c$. Thus, when $I_b\rightarrow I_c$, the prolate symmetric top is approached; and when $I_b\rightarrow I_a$, the oblate symmetric top is approached. We can describe the behavior of the asymmetric rotor in terms of the parameter:

$$\kappa=\frac{2B-A-C}{A-C} \tag{7.1}$$

which is a measure of its asymmetry, with A, B, C, the rotational constants with respect to the a, b, c axes. The limiting values for κ, -1 and $+1$, correspond to the prolate and oblate symmetric tops, respectively. The most asymmetric top has $\kappa=0$. The energy levels of asymmetric rotors ($\kappa\approx-1$ or $\approx+1$) differ from the limiting symmetric-top ones essentially in that the levels corresponding to $-K$ and $+K$, which are always degenerate in the symmetric rotor, are separated in the asymmetric rotor. Thus, an asymmetric rotor has $(2J+1)$ distinct rotational sublevels for each value of J, whereas the symmetric rotor has only $(J+1)$ distinct sublevels for each value of J. With an increase in asymmetry, the "K splitting" increases until there is no longer any close correspondence between the two levels and the degenerate K levels of the symmetric top. Nevertheless, by connecting the K levels for a given J of the limiting prolate symmetric top with those of the limiting oblate symmetric top in the ordered sequence—highest to highest, next highest to next highest, and so on, as indicated in Fig. 7.2—one may obtain a qualitative indication of the levels of the asymmetric rotor. This chart also reveals the significance of the King-Hainer-Cross notation,

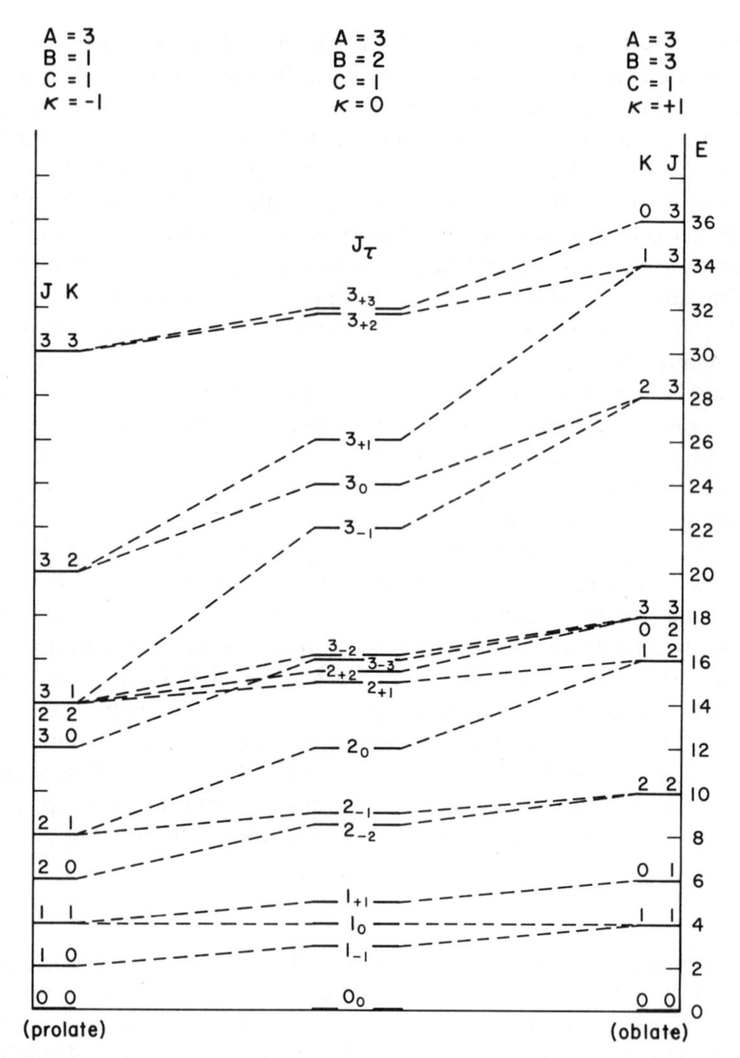

Fig. 7.2 Relation of the asymmetric rotor energy levels to those of the limiting prolate and oblate symmetric top. Note there is no crossing of sublevels of a given J although those of different J may cross. Also, the straight-line representation of the variation of the energy levels with k is only approximate.

$J_{K_{-1}K_1}$. The first subscript, K_{-1}, represents the K value of the limiting prolate top (left side in Fig. 7.2) with which the asymmetric-top level connects as κ approaches -1. The second subscript, K_1, represents the K value of the limiting oblate top with which the particular level connects as κ approaches 1. Note that the highest sublevels of the prolate symmetric top have the highest K values, whereas the highest sublevels for the oblate symmetric top have the lowest K values. Another important point evident from Fig. 7.2 is that the asymmetry

Table 7.1 Effective Rigid-Rotor Constants for the Ground Vibrational State of Some Asymmetric-top Molecules

Molecule	$A_0(MHz)$	$B_0(MHz)$	$C_0(MHz)$	*Ref.*
CH_3OCl	42,064.35	6,296.88	5,670.62	[b]
CH_2F_2	49,138.4	10,603.89	9,249.20	[c]
$CH_2(CN)_2$	20,882.137	2,942.477	2,616.774	[d]
CH_3CH_2F	36,070.30	9,364.54	8,199.74	[e]
CH_3CHO[a]	56,920.5	10,165.1	9,100.0	[f]
CH_3COF	11,039.28	9,685.65	5,322.05	[g]
$CH_2{=}CHF$	64,582.7	10,636.83	9,118.18	[h]
$CH_3N{=}CH_2$[a]	52,523.75	10,666.13	9,377.19	[i]
$CH_3CH_2CH_3$	29,207.36	8,446.07	7,458.98	[j]
$CH_3CH_2CH_2Cl$ (*gauche*)	11,829.22	3,322.58	2,853.06	[k]
$CH_3CH{=}CHF$ (*cis*)	17,826.09	5,656.57	4,406.91	[l]
$CH_3CH{=}C{=}O$	38,920	4,507.349	4,136.983	[m]
$(CH_3)_2O$	38,788.5	10,056.6	8,886.9	[n]
$(CH_3)_2CO$	10,165.60	8,514.95	4,910.17	[o]
C_4H_4O (furan)	9,446.96	9,246.61	4,670.88	[p]
C_6H_5F	5,663.54	2,570.64	1,767.94	[q]
C_6H_5OH	5,650.46	2,619.20	1,789.84	[r]
p-$CH_3C_6H_4F$	5,702.722	1,430.322	1,143.551	[s]

[a]Rotational constants of the A lines which have not been corrected for internal rotation effects.

[b]J. S. Rigden and S. S. Butcher, *J. Chem. Phys.*, **40**, 2109 (1964).

[c]D. R. Lide, Jr., *J. Am. Chem. Soc.*, **74**, 3548 (1952).

[d]E. Hirota and Y. Morino, *Bull. Chem. Soc. Japan*, **33**, 158 (1960).

[e]J. Kraitchman and B. P. Dailey, *J. Chem. Phys.*, **23**, 184 (1955).

[f]R. W. Kilb, C. C. Lin, and E. B. Wilson, Jr., *J. Chem. Phys.*, **26**, 1695 (1957).

[g]L. Pierce and L. C. Krisher, *J. Chem. Phys.*, **31**, 875 (1959).

[h]A. M. Mirri, A. Guarnieri, and P. Favero, *Nuovo Cim.*, **19**, 1189 (1961).

[i]K. V. L. N. Sastry and R. F. Curl, Jr., *J. Chem. Phys.*, **41**, 77 (1964).

[j]D. R. Lide, Jr., *J. Chem. Phys.*, **33**, 1514 (1960).

[k]T. N. Sarachman, *J. Chem. Phys.*, **39**, 469 (1963).

[l]R. A. Beaudet and E. B. Wilson, Jr., *J. Chem. Phys.*, **37**, 1133 (1962).

[m]B. Bak, J. J. Christiansen, K. Kunstmann, L. Nygaard, and J. Rastrup-Andersen, *J. Chem. Phys.*, **45**, 883 (1966).

[n]U. Blukis, P. H. Kasai, and R. J. Myers, *J. Chem. Phys.*, **38**, 2753 (1963).

[o]R. Nelson and L. Pierce, *J. Mol. Spectrosc.*, **18**, 344 (1965).

[p]B. Bak, D. Christensen, W. B. Dixon, L. Hansen-Nygaard, J. Rastrup-Andersen and M. Schottlander, *J. Mol. Spectrosc.*, **9**, 124 (1962).

[q]B. Bak, D. Christensen, L. Hansen-Nygaard, and E. Tannenbaum, *J. Chem. Phys.*, **26**, 134 (1957).

[r]T. Kojima, *J. Phys. Soc. Japan.*, **15**, 284 (1960).

[s]H. D. Rudolph and H. Seiler, *Z. Naturforsch.*, **20a**, 1682 (1965).

splitting of the K levels decreases as K increases, and for a given K, increases as J increases. The double subscript notation also gives the symmetry of the wave functions of the level, which is useful information for the application of the symmetry selection rules to be given later.

In the older notation still used widely in the literature, a single subscript on J, designated τ, is employed. For the highest sublevels of a given J, the subscript τ is given the value J; for the next highest, $J-1$, and so on to the lowest level for which the subscript τ is assigned the value of $-J$. Since there are $(2J+1)$ discrete rotational sublevels for a given value of J, there are $(2J+1)$ values of τ ranging from $\tau=J$ for the highest, to $\tau=-J$ for the lowest level. From Fig. 7.2 it is clear that

$$\tau=K_{-1}-K_1 \tag{7.2}$$

It should be emphasized that τ is not a quantum number (nor is K_{-1} or K_1); it is simply a number used to designate the sequence of the sublevels. It is often called a pseudo-quantum number.

Characteristic rotational energies of the rigid asymmetric rotor for J values up to 15 have been obtained [2] in the form of algebraic equations which, in general, increase in power as J increases. For very low J values, only linear and quadratic equations are involved, and these can be readily solved to give the desired energies. Expressions for the energies for such cases have been given in terms of the rotational constants [3] A, B, and C, and in terms of the asymmetry parameter [1] κ. These are collected in Tables 7.6 and 7.7. The values of A, B, and C for some selected molecules are listed in Table 7.1. One can obtain the microwave rotational frequencies, as before, by finding the difference in the energies of the two levels between which a transition occurs. Various expansions can be employed for finding the energies of special classes of asymmetric rotors, such as those near the prolate or near the oblate limits. These methods will be discussed in Section 3 after a more complete description of the asymmetric rotor energy levels is given.

2 ENERGIES OF THE RIGID ASYMMETRIC ROTOR

The qualitative behavior of the energy levels of an asymmetric rotor has been described; however, quantitative calculations of the rotational energies require a discussion of the quantum mechanics of the system.

In general, the rotational problem is treated in terms of a Cartesian axis system (x, y, z) tied, so to speak to the molecule so that it rotates with the molecule and has its origin located at the center of mass of the system. Furthermore, the molecule-fixed axis system is usually oriented so that its axes coincide with the principal axes of inertia, designated as a, b, and c. The quantum mechanical Hamiltonian describing the rotation of a rigid asymmetric body is then

$$\mathscr{H}=AP_a^2+BP_b^2+CP_c^2 \tag{7.3}$$

where $A=h^2/(8\pi^2 I_a)$, $B=h^2/(8\pi^2 I_b)$, and $C=h^2/(8\pi^2 I_c)$. This Hamiltonian is analogous to the classical Hamiltonian except that the components of the angular momentum of rotation P_a, P_b, P_c are replaced in quantum mechanics by their corresponding operators. For convenience, the $\hbar$ factor has been included in the definition of the rotational constants so that the angular momenta are now expressed in units of $\hbar$. The calculation of the energy levels is facilitated by rearrangement of (7.3) as proposed by Ray [4]:

$$\mathscr{H}=\tfrac{1}{2}(A+C)P^2+\tfrac{1}{2}(A-C)\mathscr{H}(\kappa) \tag{7.4}$$

$$\mathscr{H}(\kappa)=P_a^2+\kappa P_b^2-P_c^2 \qquad \text{(Reduced Hamiltonian)} \tag{7.5}$$

with $P^2=P_a^2+P_b^2+P_c^2$ and κ a dimensionless number measuring the degree of asymmetry. That this expression is equivalent to (7.3) is readily seen by substitution of the definition of Ray's asymmetry parameter, (7.1) and collection of terms. The advantage of this formulation is that the reduced energies, which are the eigenvalues of $\mathscr{H}(\kappa)$, depend only on the inertial asymmetry parameter κ and not on the individual rotational constants, and are hence easily tabulated. It may be noted that the reduced energies are simply eigenvalues of (7.3) for a hypothetical rotor with $A=+1$, $B=\kappa$, $C=-1$.

Unlike the symmetric rotor Hamiltonian, this Hamiltonian is such that the Schrödinger wave equation cannot be solved directly; thus, a closed general expression for the asymmetric rotor wave functions is not possible. However, they may be represented by a linear combination of symmetric rotor functions, that is,

$$\psi_{J_\tau M}=\sum_{J,K,M} a_{JKM}\psi_{JKM} \tag{7.6}$$

where the a_{JKM}'s are numerical constants and where ψ_{JKM} is given by (2.103). The symmetric rotor wave functions are orthonormal:

$$(J', K', M'|J, K, M)=\begin{cases}0 & \text{if } J'K'M'\neq JKM\\ 1 & \text{if } J'K'M'=JKM\end{cases} \tag{7.7}$$

and characterize a representation in which P^2, P_z, and P_Z are simultaneously diagonal. Since $\mathscr{H}$ is a function of P_x^2, P_y^2, and P_z^2, the matrix elements of these operators will be required. In Table 7.2 we summarize the angular momenta matrix elements, with angular momentum measured in units of $\hbar$. These have been discussed in Chapter II, Section 2.

Following the procedure outlined in Chapter II, Section 4, and noting that the Hamiltonian matrix elements vanish unless $J', M'=J, M$ (see Table 7.2), we obtain the following set of homogeneous linear equations in the coefficients a_{JKM}:

$$\sum_{K=-J}^{+J}(\mathscr{H}_{K'K}-\delta_{K'K}\lambda)a_{JKM}=0 \qquad K'=-J,\ldots,+J \tag{7.8}$$

where $\mathscr{H}_{K'K}=(J, K', M|\mathscr{H}|J, K, M)$. The square array $[\mathscr{H}_{K'K}]$ is in particular

Table 7.2 Angular Momentum Matrix Elements in a Symmetric Rotor Representation[a]

Molecule-fixed Axis System
$(J, K, M\|P_z\|J, K, M) = K$
$(J, K, M\|P_y\|J, K\pm 1, M) = \mp i(J, K, M\|P_x\|J, K\pm 1, M) = \frac{1}{2}[J(J+1) - K(K\pm 1)]^{1/2}$
$(J, K, M\|P^2\|J, K, M) = J(J+1)$
$(J, K, M\|P_z^2\|J, K, M) = K^2$
$(J, K, M\|P_y^2\|J, K, M) = (J, K, M\|P_x^2\|J, K, M) = \frac{1}{2}[J(J+1) - K^2]$
$(J, K, M\|P_y^2\|J, K\pm 2, M) = -(J, K, M\|P_x^2\|J, K\pm 2, M) = \frac{1}{4}\{[J(J+1) - K(K\pm 1)][J(J+1) - (K\pm 1)(K\pm 2)]\}^{1/2}$

[a]Phase choice is that of King et al. [1] and the angular momentum is measured in units of $\hbar$. Note the matrices are Hermitian, i.e., the elements are related by $(R|P_\alpha|R') = (R'|P_\alpha|R)^*$ where R stands for the totality of quantum numbers JKM and the asterisk stands for the complex conjugate. The P_α^2 matrix elements are obtained by the ordinary matrix multiplication rule: $(R|P_\alpha^2|R') = \sum_{R''}(R|P_\alpha|R'') \times (R''|P_\alpha|R')$.

a matrix representation of the Hamiltonian operator (energy matrix) with respect to the symmetric rotor basis functions. The foregoing set of equations has nontrivial solutions for the expansion coefficients only for certain values of λ. The special values of λ (the allowed energy levels for an asymmetric rotor) are those for which the secular determinant vanishes $|\mathscr{H}-\mathbf{I}\lambda|=0$ where $\mathbf{I}$ is a unit matrix. Because there are no off-diagonal matrix elements in J, the matrices for each value of J are independent and may be treated separately (see Fig. 7.3). The fact that the energy matrix is diagonal in J is to be expected since we noted previously that the total angular momentum is a constant of motion. Moreover, the matrix elements are independent of the value of M (the energy independent of the spacial orientation of $\mathbf{P}$) so that its value need not concern us further. Consequently, we will have a secular determinant to solve for each value of J; and since K takes on all integral values from $-J$ to $+J$, each determinant will have $2J+1$ rows and columns. The energy eigenvalues or characteristic values are obtained by solution of the secular determinant.

To set up the energy matrix explicitly for a given J, it is necessary to know the nonvanishing matrix elements of the Hamiltonian, (7.4). These may be obtained from Table 7.2. Since $\frac{1}{2}(A+C)P^2$ contributes only a constant diagonal term, being independent of K, and since the factor $\frac{1}{2}(A-C)$ multiplies each matrix element of $\mathscr{H}(\kappa)$, a reduced energy matrix $\mathbf{E}(\kappa)$ may be defined, involving only the operator $\mathscr{H}(\kappa)$, with the secular determinant

$$|\mathbf{E}(\kappa)-\mathbf{I}\lambda|=0 \tag{7.9}$$

The total rotational energy for a particular level is given by

$$E=\tfrac{1}{2}(A+C)J(J+1)+\tfrac{1}{2}(A-C)E_{J_\tau}(\kappa) \tag{7.10}$$

Here the $2J+1$ solutions of (7.9) are labeled by $E_{J_\tau}(\kappa)$ where the magnitude of the subscript τ increases as the magnitude of the reduced energy increases. The nonvanishing matrix elements of $\mathbf{E}(\kappa)$ are

$$E_{K,K}=(J,K,M|\mathscr{H}(\kappa)J,K,M)=F[J(J+1)-K^2]+GK^2 \tag{7.11}$$

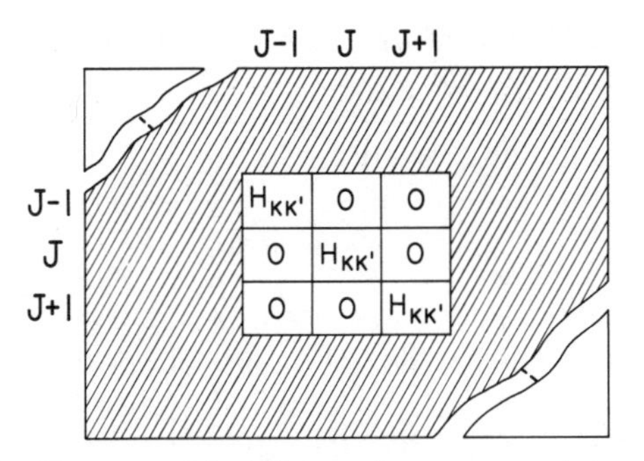

Fig. 7.3 Factoring of the infinite order energy matrix into J-blocks.

$$E_{K,K\pm2}=(J,K,M|\mathscr{H}(\kappa)|J,K\pm2,M)=H[f(J,K\pm1)]^{1/2} \tag{7.12}$$

where

$$f(J,K\pm1)=\tfrac{1}{4}[J(J+1)-K(K\pm1)][J(J+1)-(K\pm1)(K\pm2)] \tag{7.13}$$

with the fact that

$$E_{K,K}=E_{-K,-K}, \qquad E_{K,K+2}=E_{K+2,K}=E_{-K,-K-2}=E_{-K-2,-K} \tag{7.14}$$

The values of $f(J, K+1)$ have been tabulated [1] up to $J=30$. The constants F, G, and H depend on the particular way in which the a, b, and c axes are identified with the axes x, y, and z. There are six possible ways to make this identification, which are summarized in Table 7.3, along with the possible values of F, G, and H. The I^r and III^l type representation are sufficient to handle conveniently most asymmetric tops. Both identifications, of course, give the same energy levels; however, each identification has a particular advantage for certain ranges of κ. For example, if we consider a prolate rotor where $\kappa=-1$ and where the unique axis, the symmetry axis z, is associated with the a axis, that is, type I^r, the energy matrix becomes diagonal, $H=0$. The diagonal matrix elements are simply

$$-J(J+1)+2K_{-1}^2$$

and from (7.10) the rotational energies of a prolate symmetric top are

$$E=CJ(J+1)+(A-C)K_{-1}^2 \tag{7.15}$$

Thus for an asymmetric rotor with κ near -1, type I^r is the most convenient form to use in determining the energy levels since the matrix has small off-diagonal elements and thus is nearly diagonal. For an oblate rotor where $\kappa=+1$ with $z=c$, that is, type III^l, the energy matrix is again diagonal with elements

$$J(J+1)-2K_1^2$$

and the oblate top energy levels are given by

$$E=AJ(J+1)-(A-C)K_1^2 \tag{7.16}$$

Table 7.3 Identification of a, b, c with x, y, z and the Coefficients for the Matrix Elements of $E(\kappa)$

	I^r	II^r	III^r	I^l	II^l	III^l
x	b	c	a	c	a	b
y	c	a	b	b	c	a
z	a	b	c	a	b	c
F	$\frac{1}{2}(\kappa-1)$	0	$\frac{1}{2}(\kappa+1)$	$\frac{1}{2}(\kappa-1)$	0	$\frac{1}{2}(\kappa+1)$
G	1	κ	-1	1	κ	-1
H	$-\frac{1}{2}(\kappa+1)$	1	$\frac{1}{2}(\kappa-1)$	$\frac{1}{2}(\kappa+1)$	-1	$-\frac{1}{2}(\kappa-1)$

Therefore for an asymmetric oblate type top with κ close to $+1$ type III^l is the most useful form to employ.

The explicit form of the secular determinant, (7.9), for $J=3$ is

$$
\begin{array}{c|ccccccc|}
K'|K & -3 & -2 & -1 & 0 & 1 & 2 & 3 \\
\hline
-3 & E_{-3-3}-\lambda & 0 & E_{-3-1} & 0 & 0 & 0 & 0 \\
-2 & 0 & E_{-2-2}-\lambda & 0 & E_{-20} & 0 & 0 & 0 \\
-1 & E_{-1-3} & 0 & E_{-1-1}-\lambda & 0 & E_{-11} & 0 & 0 \\
0 & 0 & E_{0-2} & 0 & E_{00}-\lambda & 0 & E_{02} & 0 \\
1 & 0 & 0 & E_{1-1} & 0 & E_{11}-\lambda & 0 & E_{13} \\
2 & 0 & 0 & 0 & E_{20} & 0 & E_{22}-\lambda & 0 \\
3 & 0 & 0 & 0 & 0 & E_{31} & 0 & E_{33}-\lambda
\end{array} = 0 \tag{7.17}
$$

where, for clarity, the rows and columns are labeled in terms of the possible values of K. The numerical values of the matrix elements $E_{KK'}$ are obtained directly from (7.11) and (7.12) once a value of κ is specified. It may be pointed out that this determinant can be factored into two subdeterminants made up, respectively, of even and odd K indices, because there are no matrix elements connecting even and odd K. However, a further factoring is possible, which is suggested by the additional symmetry of the energy matrix about the secondary diagonal, and this will be discussed in the next section.

Factorization of the Energy Matrix from Symmetry Properties

Although the energy levels can be found from a solution of (7.9) and thus, in principle, the energy level calculation for an asymmetric rotor has been solved, a further simplification can be obtained by consideration of the symmetry properties of the Hamiltonian. The symmetry properties of the rotational problem may be deduced from its ellipsoid of inertia. The ellipsoid of inertia is symmetric not only to an identity operation E but also to a rotation of 180°, C_2, about any one of its principal axes of inertia. This set of symmetry operations form a group which in the language of group theory is known as the Four-group designated by $V(a, b, c)$. These symmetry operations cause the angular momentum to transform in the following manner:

$$
\begin{aligned}
&E\colon P_a \to P_a,\ P_b \to P_b,\ P_c \to P_c \\
&C_2^a\colon P_a \to P_a,\ P_b \to -P_b,\ P_c \to -P_c \\
&C_2^b\colon P_a \to -P_a,\ P_b \to P_b,\ P_c \to -P_c \\
&C_2^c\colon P_a \to -P_a,\ P_b \to -P_b,\ P_c \to P_c
\end{aligned}
$$

The Hamiltonian of a rigid asymmetric rotor is invariant under these operations and, therefore, has the symmetry of the Four-group. The symmetry group of the Hamiltonian is extremely important in quantum mechanics because a knowledge of it allows the classification of quantum states, simplification thereby of the energy matrix, and the determination of selection rules with relative ease. In the present case, each asymmetric rotor wave function may be classified according to its behavior under $V(a, b, c)$, the symmetry group of the asymmetric rotor Hamiltonian. The character table for the Four-group is given in Table 7.4. The notation for the symmetry species of V indicates the axis about which the rotation has a character of $+1$. A wave function which is, for example, symmetric (multiplied by $+1$) for a twofold rotation about axis a and antisymmetric (multiplied by -1) for a twofold rotation about the other two axes may be classified as belonging to species B_a of the group. A function that is invariant with respect to all symmetry operations of the group obviously belongs to species A.

It would be advantageous to use a set of basis functions in the calculations of the rotational energies which also belong to this group. Each wave function could then be classified according to one of the symmetry species A, B_a, B_b, B_c of V, and hence the matrix elements of the Hamiltonian $(\psi_i|\mathscr{H}|\psi_j)$ would be nonzero only between states of the same symmetry. This follows since the matrix elements are just numerical quantities and hence must be invariant under a transformation of coordinates such as the group symmetry operations that carry the system into an equivalent configuration, that is, one which is indistinguishable from the original. Therefore, since $\mathscr{H}$ belongs to the species A being unchanged by any symmetry operations of the group, the wave functions must both have the same symmetry; otherwise, the matrix elements will change sign for two of the symmetry operations and must then vanish. (See also the discussion at the end of Appendix A.) As a consequence, the secular determinant for any value of J will factor into four independent subdeterminants, one for each of the symmetry species of the Four-group (see Fig. 7.4). This, of course, considerably simplifies the diagonalization problem and has the further advantage that pairs of degenerate or nearly degenerate K levels are separated into different submatrices. Without this separation of near degeneracies, the numerical evaluation of the roots by continued fraction techniques become more difficult.

This factoring of the secular determinant is a typical example of the simplification that can result if basis functions are chosen which may be classified according to the symmetry species of the group which reflects the symmetry

Table 7.4 Character Table of the Four-Group $V(a, b, c)$

Symmetry Species	E	C_2^a	C_2^b	C_2^c
A	1	1	1	1
B_a	1	1	-1	-1
B_b	1	-1	1	-1
B_c	1	-1	-1	1

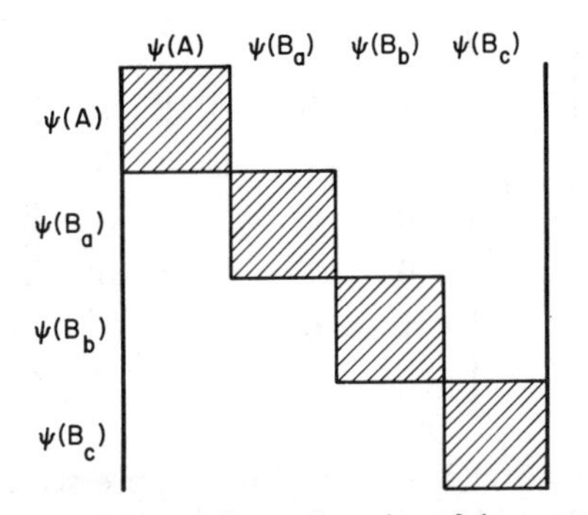

Fig.7.4 Schematic illustration of the maximum factoring of the energy matrix by proper choice of basis functions. $\psi(A)$, for example, signifies a set of basis functions with symmetry A. Nonzero matrix elements of the Hamiltonian are present only in the shaded blocks. Each of the smaller submatrices may be solved separately since there are no connecting elements present.

of the problem. Molecular orbital calculations, which are of particular interest to the chemist, are greatly aided by exploitation of the symmetry of the molecule.

The symmetric rotor wave functions which have been used as the basis functions do not belong to V but to the continuous two-dimensional rotation group D_∞, which is comprised of C_∞^z, indicating full rotational symmetry about the z axis and an infinite number of twofold axes of symmetry, C_2's, at right angles to the principal symmetry axis z. This group has an infinite number of symmetry species. However, as pointed out by Mulliken [5], the Wang [6] linear combinations of symmetric rotor functions do belong to the Four-group. The appropriate linear combinations are constructed from functions of the doubly degenerate K-states which differ in the sign of K and are defined for a given J and M as follows

$$
\begin{aligned}
S_{JKM\gamma} &= \frac{1}{\sqrt{2}}\left[\psi_{JKM}^x + (-1)^\gamma \psi_{J-KM}^x\right], \qquad \text{for } K \neq 0 \\
S_{J0M0} &= \psi_{J0M}^x, \qquad \text{for } K = 0
\end{aligned}
\tag{7.18}
$$

where γ is 0 or 1 and K now takes on only positive values. The $\gamma=0$ functions are the symmetric Wang functions whereas those with $\gamma=1$ are the antisymmetric Wang functions. Here $\psi_{JKM}^x = (-1)^\beta \psi_{JKM}$ with $\beta = K$ if $K \geqslant M$ or $\beta = M$ if $K \leqslant M$. This latter modification was suggested by Van Vleck [7] in discussing the phase of the wave functions. The phases of the symmetric rotor wave functions, of course, have no effect on the energies. However, they are important when the symmetry classification of Wang's functions are considered [5].

The foregoing equations relate the old basis (the symmetric rotor functions) to the new basis (the Wang functions or symmetrized functions) and may be written in matrix form, $\mathbf{S} = \tilde{\mathbf{X}}\psi^x$, in terms of the Wang symmetrizing transformation matrix, $\mathbf{X}$, which defines the change of basis. For a given J we have explicitly:

$$\begin{bmatrix} \vdots \\ S_{21} \\ S_{11} \\ S_{00} \\ S_{10} \\ S_{20} \\ \vdots \end{bmatrix} = \frac{1}{\sqrt{2}} \begin{bmatrix} \ddots & & & & & ⋰ \\ & -1 & 0 & 0 & 0 & 1 & \\ & 0 & -1 & 0 & 1 & 0 & \\ & 0 & 0 & \sqrt{2} & 0 & 0 & \\ & 0 & 1 & 0 & 1 & 0 & \\ & 1 & 0 & 0 & 0 & 1 & \\ ⋰ & & & & & \ddots \end{bmatrix} \begin{bmatrix} \vdots \\ \psi^x_{-2} \\ \psi^x_{-1} \\ \psi^x_0 \\ \psi^x_1 \\ \psi^x_2 \\ \vdots \end{bmatrix} \tag{7.19}$$

where all the subscripts except K and γ are suppressed and where the square matrix $\tilde{\mathbf{X}} = \mathbf{X}^{-1} = \mathbf{X}$ (the tilde indicates the transpose).

Each asymmetric rotor wave function may now be expressed as a linear combination of the symmetrized wave functions

$$A_{J_\tau M} = \sum_K a_K^{J_\tau M} S_{JKM\gamma} \tag{7.20}$$

Alternately, we might use the notation $A_{JK_{-1}K_1M}$ to designate the asymmetric rotor wave function. The index M is not necessary since, as we have seen, the energy does not depend on the spatial orientation, at least in the absence of external fields, and is included only for completeness. Since the symmetrized functions can be classified to particular symmetry species of the asymmetric rotor group, the sum over K is to be carried out only over these $S_{JKM\gamma}$'s that belong to the same symmetry species of V as does $A_{J_\tau M}$. Thus, the number of terms in (7.20) is much less than in (7.6). The change from an expansion in terms of the ψ_{JKM} to an expansion in terms of the $S_{JKM\gamma}$ will affect the form of the energy matrix. The new energy matrix $\mathbf{E}'(\kappa)$ may be readily obtained by the following transformation on the original energy matrix $\mathbf{E}(\kappa)$ (see Appendix A):

$$\mathbf{E}'(\kappa) = \tilde{\mathbf{X}}\mathbf{E}(\kappa)\mathbf{X} = \mathbf{E}^+(\kappa) + \mathbf{O}^+(\kappa) + \mathbf{E}^-(\kappa) + \mathbf{O}^-(\kappa) \tag{7.21}$$

This is the usual rule for relating the matrix representation of an operator (here the Hamiltonian) in an old basis to that in a new basis. The result of the matrix multiplication [use being made of (7.14)] is that $\mathbf{E}(\kappa)$ splits into two submatrices corresponding to $\gamma=0$ and $\gamma=1$. Each of these can be further factored as noted previously by arrangement of rows and columns so that even values of K are grouped together and odd values of K are grouped together. Thus in the notation used in (7.21) E and O refer to the evenness or oddness of the K values involved in the matrix elements and + and − to the evenness or oddness of γ. Portions of the four submatrices are:

$$\mathbf{E}^+(\kappa) = \begin{bmatrix} E_{00} & \sqrt{2}E_{02} & 0 & \cdot & \cdot \\ \sqrt{2}E_{02} & E_{22} & E_{24} & 0 & \cdot \\ 0 & E_{24} & E_{44} & E_{46} & \cdot \\ \cdot & 0 & E_{46} & E_{66} & \cdot \\ \cdot & \cdot & \cdot & \cdot & \cdot \end{bmatrix},$$

$$\mathbf{E}^-(\kappa)=\begin{bmatrix} E_{22} & E_{24} & 0 & \cdot & \cdot \\ E_{24} & E_{44} & E_{46} & 0 & \cdot \\ 0 & E_{46} & E_{66} & E_{68} & \cdot \\ \cdot & 0 & E_{68} & E_{88} & \cdot \\ \cdot & \cdot & \cdot & \cdot & \cdot \end{bmatrix},$$

$$\mathbf{O}^+(\kappa)=\begin{bmatrix} E_{11}+E_{-11} & E_{13} & 0 & \cdot & \cdot \\ E_{13} & E_{33} & E_{35} & 0 & \cdot \\ 0 & E_{35} & E_{55} & E_{57} & \cdot \\ \cdot & 0 & E_{57} & E_{77} & \cdot \\ \cdot & \cdot & \cdot & \cdot & \cdot \end{bmatrix}, \tag{7.22}$$

$$\mathbf{O}^-(\kappa)=\begin{bmatrix} E_{11}-E_{-11} & E_{13} & 0 & \cdot & \cdot \\ E_{13} & E_{33} & E_{35} & 0 & \cdot \\ 0 & E_{35} & E_{55} & E_{57} & \cdot \\ \cdot & 0 & E_{57} & E_{77} & \cdot \\ \cdot & \cdot & \cdot & \cdot & \cdot \end{bmatrix}$$

where the elements are calculable from (7.11)–(7.14). Each of these submatrices may now be diagonalized independently to give the reduced energies, $E_{J_\tau}(\kappa)$. Therefore, by symmetry considerations, the size of the secular determinant has been reduced; the order of the submatrices being approximately $J/2$.

Symmetry Classification of the Energy Levels

We have seen that the energy levels are conveniently labeled in terms of the limiting prolate and oblate K values, that is, J_{K_{-1},K_1}. This particular notation has the added advantage that the symmetry classification of a level to the species of $V(a, b, c)$ is given uniquely in terms of the parities of these limiting K indices. Consider the limiting case of a symmetric prolate rotor. If K_{-1} is an even integer $(0, \pm 2, \ldots)$, the rotational wave function ψ_{JKM} is symmetric with respect to a rotation of π degrees about a; that is, it is an even function, not changing in sign with the rotation. The even characteristic is indicated by e. This is easily seen because such a rotation about the molecule-fixed symmetry axis (the a axis) changes the angle ϕ into $\phi=\phi+\pi$, while leaving the other angles unaffected (see Chapter II, Section 5, Fig. 2.3). Since the angle ϕ enters ψ_{JKM} through the factor $e^{iK\phi}$, we obtain

$$C_2^a\colon \psi_{JKM} \rightarrow e^{iK\pi}\psi_{JKM}=(-1)^K\psi_{JKM}$$

If $K_{-1}=\pm 1, \pm 3, \ldots$ (odd integer), ψ_{JKM} is antisymmetric [changes sign, odd (o) function] with respect to this operation. Similarly, for an oblate rotor, if K_1 is an even integer, ψ_{JKM} is symmetric (e function) with respect to a rotation of π degrees about the symmetry axis, that is, the c axis; and if K_1 is an odd integer, ψ_{JKM} is antisymmetric with respect to this operation. Now the symmetry of an asymmetric-top level must be invariant to changes in the moments of inertia, that is, κ. It must then possess the behavior of the two limiting cases with which it correlates. Therefore, if K_{-1} is *e* and K_1 is *o*, then the asymmetric wave function is symmetric with respect to a rotation about *a* and antisymmetric with respect to a rotation about *c*. Since a rotation about any two of the axes in succession is equivalent to a rotation about the third, a rotation about the *b* axis in the present case must yield an antisymmetric result. From Table 7.4 this implies that the function belongs to species B_a. Similarly, if $K_{-1}K_1$ is *ee*, *oo*, or *oe*, the wave function must belong to A, B_b, and B_c, respectively. Thus a knowledge of the evenness or oddness of K_{-1}, K_1 gives us the symmetry classification of the level directly.

Another important point is the symmetry classification of each submatrix which has been discussed by King et al. [1]. The symmetrized functions $S_{JKM\gamma}$ that occur in any particular submatrix must all belong to the same symmetry species of the Four-group. This fact may be used for classification of the submatrices. However, the symmetrized functions are constructed relative to the axes x, y, z and not relative to the axes a, b, c, that is, the z axis is always taken as the symmetry axis which may coincide with a if the rotor is a prolate top or with c if it is an oblate top. Therefore, the functions are characterized by the representation A, B_x, B_y, B_z of the Four-group $V(x, y, z)$ (here as before the species notation shows directly the axis of rotation for which the character is $+1$). To classify the symmetrized functions one must find their behavior with respect to the group operations E, C_2^x, C_2^y, C_2^z, of $V(x, y, z)$, as Mulliken [5] has done, with the result

$$\begin{aligned}
&E: S_{JKM\gamma} \to S_{JKM\gamma}\\
&C_2^x: S_{JKM\gamma} \to (-1)^{J+K+\gamma} S_{JKM\gamma}\\
&C_2^y: S_{JKM\gamma} \to (-1)^{J+\gamma} S_{JKM\gamma}\\
&C_2^z: S_{JKM\gamma} \to (-1)^{K} S_{JKM\gamma}
\end{aligned} \tag{7.23}$$

The parities are seen to depend only on the even or odd character of $J+\gamma$ and K, and this character is readily determined from the designation of each submatrix. Consider, for example, the $\mathbf{E}^+$ submatrix where K and γ are even; therefore, for J_{even}, all of the $S_{JKM\gamma}$'s of $\mathbf{E}^+$ are symmetric with respect to all operations of the group and must belong to symmetric species A. The symmetry classification of the various submatrices in $V(x, y, z)$ are collected in Table 7.5.

The symmetry classification of the submatrices with respect to $V(a, b, c)$ will depend on the correlation of x, y, z with a, b, c; in particular, A corresponds to A, and B_a, B_b, B_c correspond to B_x, B_y, B_z in the same way as a, b, c are related to x, y, z. The symmetry classification in $V(a, b, c)$ is also given in Table 7.5 along with the parities of K_{-1}, K_1 which are included in parenthesis for the assignments considered. Thus a knowledge of the symmetry of a particular level also tells us in which of the four submatrices the level may be found.

Table 7.5 Symmetry Classification of the Submatrices in $V(x, y, z)$ and $V(a, b, c)$

Symmetry Species in $V(x, y, z)$

Submatrix	K	γ	$J+\gamma$			
			J_{even}	J_{odd}	J_{even}	J_{odd}
$\mathbf{E}^+$	e	e	e	o	A	B_z
$\mathbf{E}^-$	e	o	o	e	B_z	A
$\mathbf{O}^+$	o	e	e	o	B_y	B_x
$\mathbf{O}^-$	o	o	o	e	B_x	B_y

Symmetry Species in $V(a, b, c)$[a]

Type:	I^r		II^r		III^r		I^l		II^l		III^l	
Submatrix	J_{even}	J_{odd}	J_{even}	J_{odd}	J_{even}	J_{odd}	J_{even}	J_{odd}	J_{even}	J_{odd}	J_{even}	J_{odd}
$\mathbf{E}^+$	$A(ee)$	$B_a(eo)$	$A(ee)$	$B_b(oo)$	$A(ee)$	$B_c(oe)$	$A(ee)$	$B_a(eo)$	$A(ee)$	$B_b(oo)$	$A(ee)$	$B_c(oe)$
$\mathbf{E}^-$	$B_a(eo)$	$A(ee)$	$B_b(oo)$	$A(ee)$	$B_c(oe)$	$A(ee)$	$B_a(eo)$	$A(ee)$	$B_b(oo)$	$A(ee)$	$B_c(oe)$	$A(ee)$
$\mathbf{O}^+$	$B_c(oe)$	$B_b(oo)$	$B_a(eo)$	$B_c(oe)$	$B_b(oo)$	$B_a(eo)$	$B_b(oo)$	$B_c(oe)$	$B_c(oe)$	$B_a(eo)$	$B_a(eo)$	$B_b(oo)$
$\mathbf{O}^-$	$B_b(oo)$	$B_c(oe)$	$B_c(oe)$	$B_a(eo)$	$B_a(eo)$	$B_b(oo)$	$B_c(oe)$	$B_b(oo)$	$B_a(eo)$	$B_c(oe)$	$B_b(oo)$	$B_a(eo)$

[a]Symmetry classification of the levels are indicated with the parities of $K_{-1}K_1$ indicated in parenthesis for various representations.

Evaluation of the Energy Levels, Wave Functions, and Average Values

For relatively low values of J one can find the energy levels by expanding the determinant of each submatrix of (7.22) and finding the roots of the resulting polynomial. Explicit expressions for $E(\kappa)$, for certain low J values, are given in Table 7.6. Explicit expressions for the total rotational energy in terms of the rotational constants are given in Table 7.7. From a knowledge of the energy eigenvalues, the eigenvectors of the energy matrix can then be obtained by solution of a set of simultaneous equations, similar to (7.8), for each submatrix.

Table 7.6 Explicit Expressions for the Reduced Energy $E(\kappa)$[a]

$J_{K_{-1}K_1}$	$E(\kappa)$
0_{00}	0
1_{10}	$\kappa+1$
1_{11}	0
1_{01}	$\kappa-1$
2_{20}	$2[\kappa+(\kappa^2+3)^{1/2}]$
2_{21}	$\kappa+3$
2_{11}	4κ
2_{12}	$\kappa-3$
2_{02}	$2[\kappa-(\kappa^2+3)^{1/2}]$
3_{30}	$5\kappa+3+2(4\kappa^2-6\kappa+6)^{1/2}$
3_{31}	$2[\kappa+(\kappa^2+15)^{1/2}]$
3_{21}	$5\kappa-3+2(4\kappa^2+6\kappa+6)^{1/2}$
3_{22}	4κ
3_{12}	$5\kappa+3-2(4\kappa^2-6\kappa+6)^{1/2}$
3_{13}	$2[\kappa-(\kappa^2+15)^{1/2}]$
3_{03}	$5\kappa-3-2(4\kappa^2+6\kappa+6)^{1/2}$
4_{40}	—
4_{41}	$5\kappa+5+2(4\kappa^2-10\kappa+22)^{1/2}$
4_{31}	$10\kappa+2(9\kappa^2+7)^{1/2}$
4_{32}	$5\kappa-5+2(4\kappa^2+10\kappa+22)^{1/2}$
4_{22}	—
4_{23}	$5\kappa+5-2(4\kappa^2-10\kappa+22)^{1/2}$
4_{13}	$10\kappa-2(9\kappa^2+7)^{1/2}$
4_{14}	$5\kappa-5-2(4\kappa^2+10\kappa+22)^{1/2}$
4_{04}	—
5_{42}	$10\kappa+6(\kappa^2+3)^{1/2}$
5_{24}	$10\kappa-6(\kappa^2+3)^{1/2}$

[a]In general, the sum rule for the reduced energy levels of a given J is $\sum_\tau E_{J_\tau}(\kappa)=\frac{1}{3}J(J+1)(2J+1)\kappa$.

***Table* 7.7** Explicit Expressions for the Total Rotational Energy in Terms of the Rotational Constants[a]

$J_{K_{-1}K_1}$	$E(A, B, C)$
0_{00}	0
1_{10}	$A+B$
1_{11}	$A+C$
1_{01}	$B+C$
2_{20}	$2A+2B+2C+2[(B-C)^2+(A-C)(A-B)]^{1/2}$
2_{21}	$4A+B+C$
2_{11}	$A+4B+C$
2_{12}	$A+B+4C$
2_{02}	$2A+2B+2C-2[(B-C)^2+(A-C)(A-B)]^{1/2}$
3_{30}	$5A+5B+2C+2[4(A-B)^2+(A-C)(B-C)]^{1/2}$
3_{31}	$5A+2B+5C+2[4(A-C)^2-(A-B)(B-C)]^{1/2}$
3_{21}	$2A+5B+5C+2[4(B-C)^2+(A-B)(A-C)]^{1/2}$
3_{22}	$4A+4B+4C$
3_{12}	$5A+5B+2C-2[4(A-B)^2+(A-C)(B-C)]^{1/2}$
3_{13}	$5A+2B+5C-2[4(A-C)^2-(A-B)(B-C)]^{1/2}$
3_{03}	$2A+5B+5C-2[4(B-C)^2+(A-B)(A-C)]^{1/2}$
4_{40}	—
4_{41}	$10A+5B+5C+2[4(B-C)^2+9(A-C)(A-B)]^{1/2}$
4_{31}	$5A+10B+5C+2[4(A-C)^2-9(A-B)(B-C)]^{1/2}$
4_{32}	$5A+5B+10C+2[4(A-B)^2+9(A-C)(B-C)]^{1/2}$
4_{22}	—
4_{23}	$10A+5B+5C-2[4(B-C)^2+9(A-C)(A-B)]^{1/2}$
4_{13}	$5A+10B+5C-2[4(A-C)^2-9(A-B)(B-C)]^{1/2}$
4_{14}	$5A+5B+10C-2[4(A-B)^2+9(A-C)(B-C)]^{1/2}$
4_{04}	—
5_{42}	$10A+10B+10C+6[(B-C)^2+(A-B)(A-C)]^{1/2}$
5_{24}	$10A+10B+10C-6[(B-C)^2+(A-B)(A-C)]^{1/2}$

[a]In general, the sum rule for the energy levels of a given J is $\sum_\tau E_{J_\tau}(A, B, C)=\frac{1}{3}J(J+1)(2J+1)(A+B+C)$.

As pointed out in Appendix A, the matrix formed from the normalized eigenvectors (each column corresponding to an eigenvector associated with a particular energy level) constitutes a transformation matrix $\mathbf{T}$ which diagonalizes the reduced energy matrix, that is,

$$\tilde{\mathbf{T}}\mathbf{E}\mathbf{T}=\mathbf{\Lambda} \tag{7.24}$$

where $\mathbf{\Lambda}$ is a diagonal matrix formed by the eigenvalues of $\mathbf{E}$. A knowledge of $\mathbf{T}$ is required whenever matrix elements of operators in the asymmetric rotor basis are needed.

Obviously, this technique of direct expansion becomes impractical for all but small J values, and iterative matrix diagonalization procedures must be used. The rapid development of high-speed digital computers has, however, made possible the relatively easy solution of such mathematical problems by the use of iterative procedures. A very useful diagonalization procedure due to Jacobi [8] determines the eigenvalues and eigenvectors simultaneously. This method is applicable to symmetric matrices and is unaffected by the presence of degenerate eigenvalues. The method consists in the application of a series of plane rotations (in a space of n-dimensions) which are chosen to reduce the size of the off-diagonal elements, the process being continued until the off-diagonal elements are small enough for the matrix to be considered diagonal. The product of the rotation matrices thus constitutes a transformation matrix which diagonalizes the energy matrix and hence is the matrix of eigenvectors.

Alternately, since the four submatrices are of tridiagonal form, that is, with nonzero elements along the diagonal and immediately above and below the principal diagonal, the secular determinant may be cast into a continued fraction form. The continued fraction expansion is particularly efficient for numerical evaluation of the eigenvalues and allows the roots to be determined very accurately. Both first- [1, 9] and second-order [10] iterative techniques for the continued fraction method have been described, as well as efficient procedures for evaluation of eigenvectors. This method is discussed in Appendix B.

A number of tables are available which give numerical values of $E_{J_\tau}(\kappa)$ for values of κ in steps of 0.1 [1, 11] and 0.01 [12] with values of J up to 12. Also available are tables for values of κ in steps of 0.1 [13] and 0.001 [14] up to $J=40$ and $J=18$, respectively. These tables provide a convenient means by which the energies of any asymmetric rotor may be approximated. To calculate the characteristic rotational energy for a particular level, one merely selects from the table the reduced energy, $E_{J_\tau}(\kappa)$, corresponding to the desired J_τ and κ values, and then makes use of (7.10). Interpolation between tabulated values of κ may be used, but in most tables the intervals in κ are not small enough to give the required accuracy for microwave work. However, the tabulation for intervals of 0.001 in κ are sufficiently close to allow accurate interpolation. Usually only positive or negative values are tabulated since the reduced energy matrix has a form such that

$$E_{J_\tau}(\kappa) = -E_{J_{-\tau}}(-\kappa)$$

or (7.25)

$$E_{J_{K_{-1}K_1}}(\kappa) = -E_{J_{K_1K_{-1}}}(-\kappa)$$

The quantities of interest in the calculation are, of course, the rotational transition frequencies, obtained by division of the energy difference by Planck's constant. It is common practice to express the rotational spectroscopic constants in units of megahertz so that

$$A = \frac{h}{8\pi^2 I_a}, \text{ etc.}$$

or (7.26)

$$A\ (\text{MHz}) = \frac{5.05376 \times 10^5}{I_a(\text{amu-Å}^2)}, \text{ etc.}$$

This conversion factor is based on the ^{12}C atomic mass scale and constants of Appendix D.

In (7.20) the asymmetric rotor wave functions were synthesized from the symmetrized symmetric rotor wave functions by means of the expansion coefficients $a_K^{J\tau}$, which give the relative contributions of the various symmetric rotor states to the asymmetric rotor state with their squares giving the probability for rotation with a particular angular momentum J and $\pm K$. For a very slightly asymmetric rotor, one of the $a_K^{J\tau}$'s approaches unity while the others approach zero and each asymmetric rotor wave function A_{J_τ} will be closely approximated by one $S_{JK\gamma}$. The different sets of coefficients, the eigenvectors of the energy matrix, are discriminated by means of the index τ. The eigenvectors are required for the calculation of line strengths of asymmetric rotors. They are also useful for calculation of the average values of P_z^2 and P_z^4, that is, the values of P_z^2 and P_z^4 averaged over the asymmetric rotor wave functions. Such quantities find frequent use particularly in the analysis of centrifugal distortion, quadrupole and internal rotation effects. Consider, for example, the average value of P_z^2, where z represents the unique molecule-fixed axis of quantization, we have

$$\langle P_z^2 \rangle = (J, \tau | P_z^2 | J, \tau) \tag{7.27}$$

where $\langle P_z^2 \rangle$ is the average value of P_z^2 associated with the level J, τ. From (7.20) which gives A_{J_τ} in terms of $S_{JK\gamma}$'s, recalling the definition of $S_{JK\gamma}$ and the fact that P_z^2 is diagonal in the ψ_{JK}^x basis, we obtain

$$\langle P_z^2 \rangle = \sum_K (a_K^{J\tau})^2 K^2 \tag{7.28}$$

The summation over K will be only over those K's that occur along the main diagonal of the submatrix to which the level J, τ belongs. A similar expression would result for $\langle P_z^4 \rangle$ except that K^2 would be replaced by K^4. Usually only a limited number of terms in the sum are required to give results of sufficient accuracy. For a slightly asymmetric rotor, where one of the $a_K^{J\tau}$'s approaches unity and the others approach zero, the $\langle P_z^2 \rangle$ approaches the value of K^2 for the limiting symmetric rotor level. Tables of $\langle P_z^2 \rangle$ and $\langle P_z^4 \rangle$ tabulated in increments of 0.1 [15] and 0.002 [16] in κ are available. Formulas for $\langle P_z^2 \rangle$ and $\langle P_z^4 \rangle$ based on continued fractions have been given [17], which are very convenient when the rigid rotor energy levels are calculated by the continued fraction method. It is also possible to express the $\langle P_z^4 \rangle$ simply in terms of P^2, E, and $\langle P_g^2 \rangle$, $g = x, y, z$ (see Chapter VIII, Section 3).

The average values of P_x^2, P_y^2, and P_z^2 are also directly related to the derivatives of the rotational energy with respect to the rotational constants. Specifically, it has been shown by Bragg and Golden [18, 19] that

$$\langle P_a^2 \rangle = \frac{\partial E}{\partial A} \tag{7.29}$$

$$\langle P_b^2 \rangle = \frac{\partial E}{\partial B} \tag{7.30}$$

$$\langle P_c^2 \rangle = \frac{\partial E}{\partial C} \tag{7.31}$$

Therefore, on differentiating (7.10) with respect to A, B, and C, we find the following convenient expressions

$$\langle P_a^2 \rangle = \frac{\partial E}{\partial A} = \frac{1}{2}\left[J(J+1) + E(\kappa) - (\kappa+1)\frac{\partial E(\kappa)}{\partial \kappa} \right] \tag{7.32}$$

$$\langle P_b^2 \rangle = \frac{\partial E}{\partial B} = \frac{\partial E(\kappa)}{\partial \kappa} \tag{7.33}$$

$$\langle P_c^2 \rangle = \frac{\partial E}{\partial C} = \frac{1}{2}\left[J(J+1) - E(\kappa) + (\kappa-1)\frac{\partial E(\kappa)}{\partial \kappa} \right] \tag{7.34}$$

The $\langle P_x^2 \rangle$, $\langle P_y^2 \rangle$, $\langle P_z^2 \rangle$ are identified with the $\langle P_a^2 \rangle$, $\langle P_b^2 \rangle$, $\langle P_c^2 \rangle$ according to the way a, b, c are related to x, y, z. The values of $\partial E/\partial \kappa$ may be obtained from Tables of $E(\kappa)$ if differences of $E(\kappa)$ between two appropriate values of κ are taken. Tables with small increments in κ should be employed.

To indicate the use of Table 7.5 and to illustrate some of the previous concepts, we look at a simple example. Consider the case of a prolate asymmetric rotor in a type I^r representation for $J=2$. The allowed energy levels are 2_{02}, 2_{12}, 2_{11}, 2_{21}, and 2_{20}. We see from the Table 7.5 that the levels will be found in the following submatrices: 2_{02}, 2_{20} in $\mathbf{E}^+$; 2_{21} in $\mathbf{E}^-$; 2_{12} in $\mathbf{O}^+$, and 2_{11} in $\mathbf{O}^-$. $\mathbf{E}^+$ gives a 2×2 secular determinant, and the others are all one-dimensional. The wave functions associated with the levels in the various submatrices have the symmetry: $\mathbf{E}^+ \leftrightarrow A$, $\mathbf{E}^- \leftrightarrow B_a$, $\mathbf{O}^+ \leftrightarrow B_c$, and $\mathbf{O}^- \leftrightarrow B_b$. The largest eigenvalue in $\mathbf{E}^+$ will be associated with 2_{20} level and the other with the 2_{02} level. The $\mathbf{E}^+$ matrix has the following secular equation:

$$\begin{vmatrix} E_{00}-\lambda & \sqrt{2}E_{02} \\ \sqrt{2}E_{02} & E_{22}-\lambda \end{vmatrix} = \begin{vmatrix} 3(\kappa-1)-\lambda & -\sqrt{3}(\kappa+1) \\ -\sqrt{3}(\kappa+1) & (\kappa+3)-\lambda \end{vmatrix} = 0$$

Upon expansion, the determinant yields a quadratic equation in λ with the roots:

$$E_{2_{20}} = E_{2_2} = 2[\kappa + (\kappa^2+3)^{1/2}]$$
$$E_{2_{02}} = E_{2\,_{-2}} = X[\kappa - (\kappa^2+3)^{1/2}]$$

The asymmetric rotor wave functions for the two states have the form:

$$A_{2_2} = a_0^{2\,2_2} S_{200} + a_2^{2\,2_2} S_{220}$$
$$A_{2\,_{-2}} = a_0^{2\,_{-2}} S_{200} + a_2^{2\,_{-2}} S_{220}$$

where the M index is suppressed. The coefficients $a_K^{J_\tau}$ in the above expressions are determined from the simultaneous equations:

$$(E_{00}-\lambda)a_0^{2\tau}+\sqrt{2}E_{02}a_2^{2\tau}=0$$
$$\sqrt{2}E_{02}a_0^{2\tau}+(E_{22}-\lambda)a_2^{2\tau}=0$$

A system of equations such as these do not determine the $a_K^{J_\tau}$'s uniquely, but give only their ratios. We can evaluate an arbitrary or relative set from either of the foregoing equations by setting one $a_K^{J_\tau}=1$ and solving for the remaining one. Using the first of the foregoing equations and setting $\lambda=E_{2_2}$ we find the relative eigenvectors to be

$$a_0^{2_2}=1 \qquad \text{and} \qquad a_2^{2_2}=\frac{(\kappa-3)-2(\kappa^2+3)^{1/2}}{\sqrt{3}(\kappa+1)}$$

The values are fixed by means of the relation,

$$a_K^{J_\tau}=\frac{a_K^{J_\tau}\text{ (relative)}}{\{\Sigma_K\,[a_K^{J_\tau}\text{ (relative)}]^2\}^{1/2}}$$

which ensures the eigenvectors will be normalized, that is, $\sum_K (a_K^{J_\tau})^2=1$. The normalized set of eigenvectors are:

$$a_0^{2_2}=\frac{(\sqrt{3}/2)(\kappa+1)}{\{(\kappa^2+3)^{1/2}[2(\kappa^2+3)^{1/2}+(3-\kappa)]\}^{1/2}}$$

$$a_2^{2_2}=-\tfrac{1}{2}\left[2+\frac{(3-\kappa)}{(\kappa^2+3)^{1/2}}\right]^{1/2}$$

Likewise, we find for the 2_{02} level that

$$a_2^{2_{-2}}=a_0^{2_2} \qquad \text{and} \qquad a_0^{2_{-2}}=-a_2^{2_2}$$

If for example, $\kappa=-0.968$, we find that

$$E_{2_2}=2.0324, \qquad a_0^{2_2}=0.00698, \qquad a_2^{2_2}=-0.99998$$
$$E_{2_{-2}}=-5.9044, \qquad a_0^{2_{-2}}=0.99998, \qquad a_2^{2_{-2}}=0.00698$$

By means of the eigenvectors we calculate

$$\langle P_z^2\rangle_{2_2}=3.9998 \qquad \text{and} \qquad \langle P_z^2\rangle_{2_{-2}}=0.0002$$

As mentioned previously, the original energy matrix can be brought to diagonal form by a similarity transformation, (7.23), with the matrix of eigenvectors. In the present case, we have explicitly:

$$\overset{\tilde{\mathbf{T}}}{\begin{bmatrix}0.99998 & 0.00698\\ 0.00698 & -0.99998\end{bmatrix}}\overset{\mathbf{E}^+}{\begin{bmatrix}-5.9040 & -0.0554\\ -0.0554 & 2.0320\end{bmatrix}}\overset{\mathbf{T}}{\begin{bmatrix}0.99998 & 0.00698\\ 0.00698 & -0.99998\end{bmatrix}}$$
$$=\begin{bmatrix}-5.9044 & 0.0000\\ 0.0000 & 2.0324\end{bmatrix}$$

which provides a check of our calculations.

3 SLIGHTLY ASYMMETRIC ROTORS

It would be convenient to have expressions that give the explicit dependence of the energies on the rotational quantum numbers. This is not possible for molecules with large asymmetry; however, there are a large number of molecules that may be classified as slightly asymmetric tops, for example, HN_3, HNCO, CH_2DBr, NOBr, CH_3SH, H_2S_2, $P^{35}Cl_2{}^{37}Cl$, etc. The spectrum of a slightly asymmetric rotor is illustrated in Fig. 7.5. For this type, the energies can be expressed with sufficient accuracy as a series expansion in powers of an asymmetry parameter. Various asymmetry parameters have been employed for finding the energies of this special class of asymmetric rotors. An asymmetry parameter introduced by Wang [6] is particularly useful for such expansions. The rotational Hamiltonian for a near-prolate rotor type I^r, in terms of Wang's asymmetry parameter is written in the form

$$\mathscr{H}=AP_a^2+BP_b^2+CP_c^2=\tfrac{1}{2}(B+C)P^2+[A-\tfrac{1}{2}(B+C)]\mathscr{H}(b_p) \tag{7.35}$$

with

$$\mathscr{H}(b_p)=P_a^2+b_p(P_c^2-P_b^2) \tag{7.36}$$

where the a axis is the unique axis of quantization and the asymmetry parameter is defined as

$$b_p=\frac{C-B}{2A-B-C}=\frac{\kappa+1}{\kappa-3} \tag{7.37}$$

in which $-1\leqslant b_p\leqslant 0$ with $b_p=-\frac{1}{3}$ corresponding to the maximum degree of asymmetry. In a symmetric rotor representation the nonvanishing matrix elements are:

$$(J,K,M|P^2|J,K,M)=J(J+1) \tag{7.38}$$

$$(J,K,M|\mathscr{H}(b_p)|J,K,M)=K^2 \tag{7.39}$$

$$(J,K,M|\mathscr{H}(b_p)|J,K\pm 2,M)=b_p[f(J,K\pm 1)]^{1/2} \tag{7.40}$$

As with Ray's formulation, the determination of the eigenvalues of $\mathscr{H}$ is reduced to finding those of $\mathscr{H}(b_p)$. If the matrix of the Wang operator, (7.39) and (7.40),

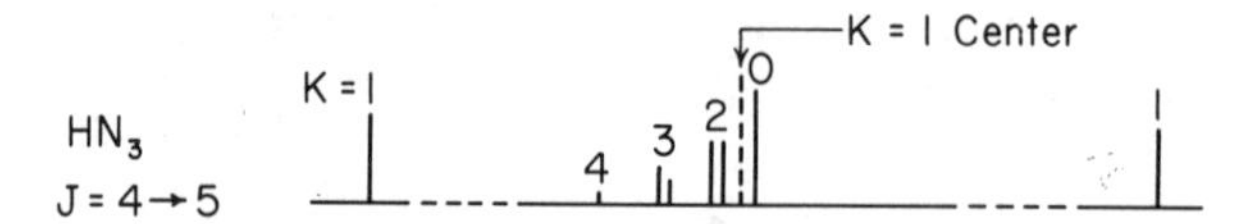

Fig. 7.5 Illustration of the rotational spectrum of the slightly asymmetric prolate rotor HN_3 ($b_p=-20.9\times10^{-5}$) observed in the millimeter region. The frequencies of the absorption lines increase from left to right. Separation of the lines for different values of K is due to centrifugal distortion effects. The $K=1$ and $K=2$ lines are split (1263.1 and 1.5 MHz, respectively) because of the small asymmetry of the molecule. The $K=3$ line shows quadrupole splitting which is unresolved in the other lines. After R. Kewley, K. V. L. N. Sastry, and M. Winnewisser, *J. Mol. Spectrosc.* **12**, 387 (1964).

is subjected to the Wang transformation, four matrices will result similarly to the $\mathbf{E}^{\pm}$, $\mathbf{O}^{\pm}$ matrices, discussed previously. The only difference is that now the matrix elements are given by (7.39) and (7.40). The roots of these secular determinants are the reduced energies $W_{J_\tau}(b_p)$ with the characteristic rotational energy being given by:

$$E=\tfrac{1}{2}(B+C)J(J+1)+[A-\tfrac{1}{2}(B+C)]W_{J_\tau}(b_p) \tag{7.41}$$

All of the symmetry properties given previously are still valid. For the prolate symmetric limit, $B=C$, we have $b_p=0$ and $W_{J_\tau}=K_{-1}^2$ so that the energy expression reduces to that given for a symmetric top. Alternately, near the symmetric rotor limit, that is, $b_p \approx 0$, the reduced energies may be expressed as:

$$W_{J_\tau}(b_p)=K_{-1}^2+C_1 b_p+C_2 b_p^2+C_3 b_p^3+\cdots \tag{7.42}$$

where K_{-1} is the limiting prolate index of the level. The coefficients C_i for the powers of b_p can be evaluated by standard perturbation techniques, with the term in b_p of $\mathscr{H}(b_p)$ being treated as the perturbation operator. The Wang linear combinations of symmetric rotor wave functions are used as the basis functions, with the coefficient of b_p^n being provided by nth order perturbation. The expressions for the coefficients [13, 20] up to C_5 are given in Table 7.8. They depend only on the rotational quantum numbers. Numerically tabulated coefficients for various energy levels have also been reported [12, 20].

We may, for example, by use of Table 7.8 immediately write down the asymmetry splitting between the two $K=1$ levels of a slightly asymmetric rotor. We find (retaining terms up to b_p^2)

$$\Delta W(b_p)=b_p J(J+1)$$

which indicates directly that the degenerate level splitting increases with J. An approximate general expression for the asymmetry splitting of the K levels has been given by Wang [6]

$$\Delta W(b)=\frac{b^K(J+K)!}{8^{K-1}(J-K)![(K-1)!]^2} \tag{7.43}$$

in which ΔW is the reduced energy difference with b and K, respectively, b_p and K_{-1} for a near prolate top, or b_o and K_1 for a near oblate top. Equation 7.43 reveals that the degenerate level splitting decreases with increasing K (see Fig. 7.2). A more accurate expression, correct to $(K+2)$-power of b, is also available [21].

The energy expression for a near-oblate top, type III^l, is obtained by a simple interchange of A and C in the preceding relations. The total rotational energy for an asymmetric oblate top is then:

$$E=\tfrac{1}{2}(A+B)J(J+1)+[C-\tfrac{1}{2}(A+B)]W_{J_\tau}(b_o) \tag{7.44}$$

with

$$b_o=\frac{A-B}{2C-B-A}=\frac{\kappa-1}{\kappa+3} \tag{7.45}$$

Table 7.8 General Expressions for the Coefficients C_l for a Near Prolate Top[a]

$$C_1 = \pm\frac{J(J+1)}{2} \qquad \text{for } K=1, \mathbf{O}^{\pm}$$

$$C_1 = 0 \text{ for all other } K \text{ values}$$

$$C_2 = \left[\frac{f(J, K-1)}{4(K-1)} - \frac{f(J, K+1)}{4(K+1)}\right] \qquad \text{for all } K \text{ values}$$

$$C_3 = \pm\frac{J(J+1)f(J, 2)}{128} \qquad \text{for } K=1, \mathbf{O}^{\mp}$$

$$C_3 = \pm\frac{J(J+1)f(J, 2)}{128} \qquad \text{for } K=3, \mathbf{O}^{\pm}$$

$$C_3 = 0 \qquad \text{for all other } K \text{ values}$$

$$C_4 = C_4' - \frac{J^2(J+1)^2 f(J, 2)}{2048} \qquad \text{for } K=1, C_4' \text{ given below}$$

$$C_4 = C_4' + \frac{J^2(J+1)^2 + f(J, 2)}{2048} \qquad \text{for } K=3, C_4' \text{ given below}$$

$$C_4 = C_4' = \frac{f(J, K-1)}{128(K-1)^2}\left[\frac{2f(J, K+1)}{(K+1)} - \frac{2f(J, K-1)}{(K-1)} + \frac{f(J, K-3)}{(K-2)}\right]$$
$$- \frac{f(J, K+1)}{128(K+1)^2}\left[\frac{2f(J, K-1)}{(K-1)} - \frac{2f(J, K+1)}{(K+1)} + \frac{f(J, K+3)}{(K+2)}\right] \qquad \text{for all other } K \text{ values}$$

$$C_5 = \frac{\pm J(J+1)f(J, 2)}{294{,}912}[108f(J, 2) - 9J^2(J+1)^2 - 28f(J, 4)] \qquad \text{for } K=1, \mathbf{O}^{\pm}$$

$$C_5 = \frac{\pm J(J+1)f(J, 2)}{294{,}912}[108f(J, 2) - 9J^2(J+1)^2 - 27f(J, 4)] \qquad \text{for } K=3, \mathbf{O}^{\mp}$$

$$C_5 = \frac{\pm J(J+1)f(J, 2)f(J, 4)}{294{,}912} \qquad \text{for } K=5, \mathbf{O}^{\pm}$$

$$C_5 = 0 \qquad \text{for all other } K \text{ values}$$

[a] Here K refers to the limiting prolate index of a level. The symbol $\mathbf{O}^{\pm}$ identifies the submatrix to which a level belongs, see Table 7.5, and indicates which sign to use; for example, if level in $\mathbf{O}^{+}$ for $K=1$, minus sign is to be used for calculation of C_3. The functions $f(J, K\pm1)$ are given by (7.13) of text, except that $f(J, 1) = \frac{1}{2}J(J+1)[J(J+1)-2]$. Furthermore, $f(J, 1)$ must be taken to be zero in computation of C_2 or C_4 for the $\mathbf{E}^{-}$ levels. Also $f(J, 0)$ and $f(J, -1)$ must always be set equal to zero. After Allen and Cross [13].

The range is the same as b_p, that is, $-1 \leqslant b_o \leqslant 0$ and the most symmetric case is again represented by the value $-\frac{1}{3}$. For an oblate limit $A = B$, we have $b_0 = 0$, so that an expansion in terms of b_o is the most appropriate for expressing the reduced energy of a slightly asymmetric oblate rotor

$$W_{J_\tau}(b_o) = K_1^2 + C_1 b_o + C_2 b_o^2 + C_3 b_o^3 + \cdots \tag{7.46}$$

where K_1 is the appropriate limiting oblate index. The C's for the oblate case may be obtained from Table 7.8 for the prolate case as follows: If the oblate level of interest is J_{K_{-1},K_1}, then one computes the coefficients for the prolate rotor level $J_{K_1,K_{-1}}$. The use of such expansions for energy level calculations are, of course, dictated by the size of the asymmetry parameter b and the accuracy that is desired. If the asymmetry is too large, the series expansion, (7.42) or (7.46), will not converge rapidly enough. The expansions are, in general, most accurate for very small asymmetries and become progressively worse with increasing asymmetry.

For asymmetric rotors where the power series expansion is not appropriate, the Wang formulation suffers from the fact that extensive tables of reduced energies are available in terms of κ rather than Wang's asymmetry parameter. However, if necessary, Wang's reduced energy can be calculated from the tabulated values of Ray's reduced energy by means of the relation

$$E_{J_\tau}(\kappa)=FJ(J+1)+(G-F)W_{J_\tau}(b) \tag{7.47}$$

where b is either b_p or b_o, depending on whether one is considering a prolate or oblate asymmetric rotor. The values of F and G are given in Table 7.2 for the two possible cases. This relation can be obtained simply by equating the different expressions for the total energy.

In cases where the expansion can be used satisfactorily, we can write the average values of P_z^2 in an asymmetric rotor basis as an expansion in terms of the asymmetry parameter, thereby circumventing the more laborious calculations using the eigenvectors. As an example, consider the derivation of $\langle P_z^2\rangle$. For the a axis identified with the z axis, type I^r, we have from (7.41), on differentiating with respect to A,

$$\left(\frac{\partial E}{\partial A}\right)=W(b_p)+[A-\tfrac{1}{2}(B+C)]\left(\frac{\partial W(b_p)}{\partial A}\right) \tag{7.48}$$

Now

$$\left(\frac{\partial W(b_p)}{\partial A}\right)=\left(\frac{\partial W(b_p)}{\partial b_p}\right)\left(\frac{\partial b_p}{\partial A}\right)=\frac{-b_p}{A-\tfrac{1}{2}(B+C)}\left(\frac{\partial W(b_p)}{\partial b_p}\right)$$

Therefore

$$\langle P_z^2\rangle=W(b_p)-b_p\left(\frac{\partial W(b_p)}{\partial b_p}\right) \tag{7.49}$$

This is a general expression for $\langle P_z^2\rangle$. If, however, we assume $W(b_p)$ is given by (7.42), we obtain the following expression

$$\langle P_z^2\rangle=K_{-1}^2+\sum_{n=1}(1-n)C_nb_p^n \tag{7.50}$$

From a knowledge of $\langle P_z^2\rangle$, we can calculate $\langle P_x^2\rangle$ and $\langle P_y^2\rangle$ from the relations

$$\frac{\partial E}{\partial B}=\langle P_x^2\rangle=\tfrac{1}{2}[J(J+1)-(1+\sigma)\langle P_z^2\rangle+\sigma W(b_p)] \tag{7.51}$$

$$\frac{\partial E}{\partial C} = \langle P_y^2 \rangle = \tfrac{1}{2}[J(J+1) + (\sigma - 1)\langle P_z^2 \rangle - \sigma W(b_p)] \tag{7.52}$$

with $\sigma = -1/b_p$. The method of calculation of $\langle P_z^2 \rangle$ and $W(b_p)$ in the foregoing expressions is, of course, dictated by the size of the asymmetry. For slightly asymmetric rotors, the $\langle P_z^4 \rangle$ can be roughly approximated by

$$\langle P_z^4 \rangle \simeq \langle P_z^2 \rangle^2 \tag{7.53}$$

A more accurate approximation has been given by Kivelson and Wilson [17] for slightly asymmetric rotors.

4 SELECTION RULES AND INTENSITIES

The allowed changes in J for dipole absorption of radiation are:

$$\Delta J = 0, \pm 1 \tag{7.54}$$

As in previous instances, these "permitted" transitions result from the nonvanishing property of the dipole matrix elements, $\int A_{J_\tau} \mu A_{J'_{\tau'}} \, d\tau$, when $J' = J$ or $J' = J \pm 1$. For other values of J', all matrix elements of the dipole components along fixed axes in space are found to be zero. This is to be expected since the asymmetric rotor wave functions are expressed as linear combinations of symmetric rotor functions, all having the same value of J. Therefore, only those matrix elements will be nonvanishing for change of J which were nonvanishing for the symmetric rotor. In an asymmetric rotor, each of the three changes might give rise to an absorption line, whereas, for a symmetric rotor, the level with higher J always lies highest, and only $\Delta J = +1$ gives rise to rotational absorption. The $\Delta J = -1$ transitions are designated as P-branch; the $\Delta J = 0$, as Q-branch; and the $\Delta J = +1$, as R-branch transitions.

In addition to these selection rules for J, there are also restrictions on the changes that can occur in the subscripts of J, in the pseudo-quantum numbers. These restrictions result from the symmetry properties of the ellipsoid of inertia. To discuss these selection rules, we must again inquire into the nonvanishing property of the matrix elements of the dipole components along axes fixed in space. The component of the electric moment along a space-fixed axis F, can be written as

$$\mu_F = \sum_g \cos(Fg) \mu_g \qquad F = X, Y, Z; \; g = a, b, c \tag{7.55}$$

where μ_g are the components of the permanent molecular dipole moment resolved along the principal axes of inertia. The $\cos(Fg)$ are the cosines of the angles between the nonrotating F and rotating g axes. These quantities may be expressed as explicit functions of the Eulerian angles θ, ϕ, χ which specify the orientation of the molecule-fixed axis system with respect to the space-fixed axis system. The direction cosines are commonly indicated by the symbol Φ_{Fg}. They find extensive use whenever it is desirable to refer a vector whose components are known in the rotating coordinate system to the nonrotating system, or vice versa. For the case where the electric moment lies only along one of the principal axes, say the a axis, since the direction cosines are functions of the rotational coordinates, the dipole matrix element is

$$(J, K_{-1}, K_1|\mu_F|J', K'_{-1}, K'_1)=\mu_a(J, K_{-1}, K_1|\cos(Fa)|J', K'_{-1}, K'_1) \quad (7.56)$$

If this integral is not to vanish, it must be unchanged (no change in sign) for any operation that carries the system into a configuration indistinguishable from the original. This is tantamount to saying that the integrand $A^*_{JK_{-1},K_1}$ cos $(Fa)A_{J'K'_{-1},K'_1}$ transforms according to the totally symmetric representation A. The symmetry in which we are interested is not the molecular symmetry but the symmetry of the inertia ellipsoid, which for the asymmetric rotor is characterized by the Four-group operations. If the integrand is to belong to species A, then for each symmetry operation of V either each term of the integrand is even, or one term is even and the other two are odd. Under the Four-group operations, the direction cosines have B symmetries. For a rotation of 180° about a, C_2^a, the cos (Fa) does not change, but it does change sign for a rotation about b or c since the angle changes from Fa to $\pi - Fa$. It transforms, therefore, according to the $B_a(eo)$ representation. Likewise, cos (Fb), does not change sign for the C_2^b symmetry operation but does change sign for the C_2^a and C_2^c operations and thus transforms as the species $B_b(oo)$. For similar reasons, cos (Fc) has $B_c(oe)$ symmetry. For the case under consideration ($\mu_a \neq 0$ with $\mu_b=\mu_c=0$) where the direction cosine transforms as B_a, it is readily apparent from Table 7.3 that if the integrand is to have symmetry A, the functions $A^*_{JK_{-1},K_1}$ and $A_{J'K'_{-1},K'_1}$ must be of symmetry A and B_a, or of symmetry B_b and B_c. The allowed transitions are then $A \leftrightarrow B_a$, and $B_b \leftrightarrow B_c$. The other selection rules for electric dipole components along the b and c axes are similarly established. For $\mu_b \neq 0$ with $\mu_a=\mu_c=0$, the allowed transitions are $A \leftrightarrow B_b$ and $B_a \leftrightarrow B_c$, whereas for $\mu_c \neq 0$ and $\mu_a=\mu_b=0$ they are $A \leftrightarrow B_c$ and $B_a \leftrightarrow B_b$. Often, the symmetry selection rules are stated in terms of the evenness or oddness of the K_{-1}, K_1 subscripts. In this notation they are:

Dipole Component	Permitted Transitions
$\mu_a \neq 0$	$ee \leftarrow \rightarrow eo$
(along axis of least moment of inertia)	$oe \leftarrow \rightarrow oo$
$\mu_b \neq 0$	$ee \leftarrow \rightarrow oo$
(along axis of intermediate moment of inertia)	$oe \leftarrow \rightarrow eo$
$\mu_c \neq 0$	$ee \leftarrow \rightarrow oe$
(along axis of greatest moment of inertia)	$eo \leftarrow \rightarrow oo$

In terms of permitted changes in the K_{-1} and K_1 subscripts, the selection rules are:

Dipole Component	ΔK_{-1}	ΔK_1
$\mu_a \neq 0$ (along axis of least moment of inertia)	$0, \pm 2, \ldots$	$\pm 1, \pm 3, \ldots$
$\mu_b \neq 0$ (along axis of intermediate moment of inertia)	$\pm 1, \pm 3, \ldots$	$\pm 1, \pm 3, \ldots$
$\mu_c \neq 0$ (along axis of greatest moment of inertia)	$\pm 1, \pm 3, \ldots$	$0, \pm 2, \ldots$

If the dipole moment lies wholly along one of the principal inertial axes, only those changes in the subscript notation listed in the preceding tables for that component are allowed. If there are dipole components along each of the principal inertial axes, all changes listed in the tables are allowed. Any given transition will be due to only one component of the molecular dipole, for example, μ_a, μ_b, or μ_c. The transitions due to the μ_a component are designated as "*a*"-type transitions, those due to μ_b as "*b*"-type, and those due to μ_c as "*c*"-type transitions.

The presence of different types of transitions for asymmetric rotors, although at times creating additional complexity, can also be an advantage. For instance, with moderately large organic molecules one or more rotational isomers may exist, and one conformation might, from preliminary information, be expected to exhibit "*a*"-type transitions whereas another both "*a*"- and "*b*"-type transitions. This information can be of great aid in distinguishing the isomers and in the assignment of the spectrum. The dependence of the type of transition on the structure in the asymmetric rotor [22] is illustrated with H_2S_2 in Fig. 7.6 (see also Fig. 12.19).

While any changes indicated in the previous stable are permitted by the selection rules, it is found that absorption lines corresponding to large changes in either subscript (K_{-1} or K_1) are weak. Also, if the dipole along a particular principal axis is small, the transitions arising from this component will be weak. Furthermore, if the rotor is near the limiting prolate symmetric-top case ($\kappa \approx -1$), the changes in K_{-1} which correspond to the symmetric-top selection rules, $\Delta K_{-1} = 0, \pm 1$, will be the only ones of significant strength, but those for larger changes in ΔK_1 may have observable strength. Conversely, if the molecule approximates the oblate symmetric rotor ($\kappa \approx 1$), only the lines corresponding to $\Delta K_1 = 0, \pm 1$ will have significant strength, but relatively large changes in K_{-1} can give rise to measurable lines. When neither symmetric-top case is approached ($\kappa \approx 0$, or asymmetry large), the strongest lines will correspond to 0 and ± 1 change in both K_{-1} and K_1, but lines of significant strength may occur which correspond to larger changes in either or both subscripts. It should also be pointed out that when the dipole moment lies wholly along the symmetry axis, the symmetric-top selection rule is simply $\Delta K = 0$ (commonly referred to

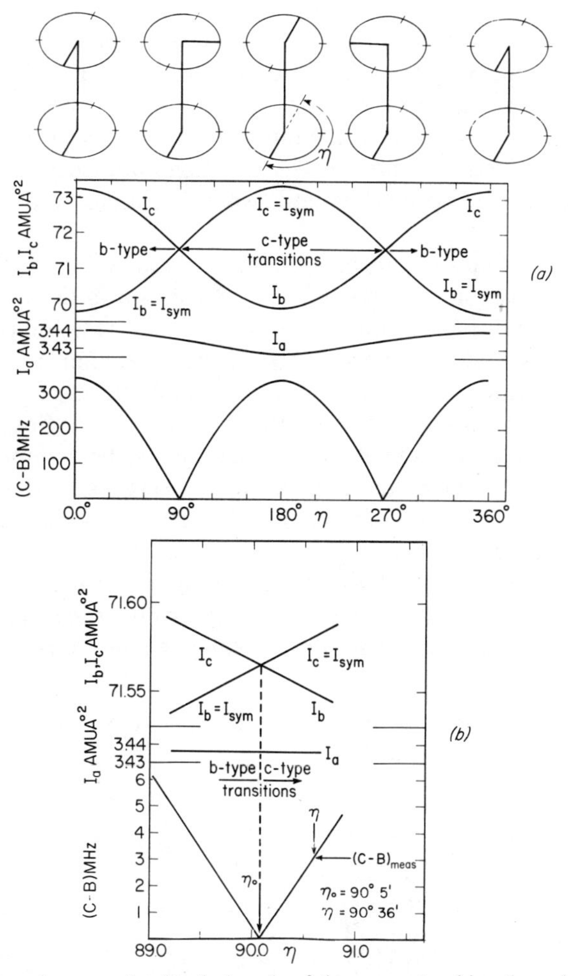

Fig. 7.6 The dependence on the dihedral angle of the moments of inertia and of the difference of the rotational constants $(C-B)$ for HSSH. (*a*) The dihedral angle varies from 0 to 360°. (*b*) The dihedral angle varies from 89.0 to 91.0°. After Winnewisser et al. [22].

as parallel-type transitions), and that when it is entirely perpendicular to the symmetry axis, $\Delta K = \pm 1$ (perpendicular-type transitions). Hence, we can obtain further information about the probable intensity of a transition by noting the direction of the dipole moment with reference to that principal inertial axis which would in the limit become the symmetry axis.

It is often convenient in the analysis of spectra to group transitions into series or branches which are characterized by the changes in J, K_{-1} and K_1. The notation used to specify each branch, which is frequently found in the literature, is

$$^{g}Q_{\Delta K_{-1}\Delta K_1},\ ^{g}R_{\Delta K_{-1}\Delta K_1},\ ^{g}P_{\Delta K_{-1}\Delta K_1}$$

Here $Q(\Delta J=0)$ and $R(\Delta J=+1)$ indicate the changes in $J(J\rightarrow J$ and $J\rightarrow J+1)$ while the change in K_{-1} is indicated by the first subscript. The second subscript gives ΔK_1. The dipole component responsible for the transition is given as a superscript, $g=a$, b, and c. A summary of the stronger asymmetric rotor transitions is provided in Table 7.9. These stronger branches can often provide more easily recognized spectral patterns which are useful in the assignment of rotational spectra.

The intensities of absorption lines are proportional to the squares of absolute values of direction cosine matrix elements. As shown in (2.108), the direction cosine matrix elements in a symmetric rotor basis may be written as a product of three factors. Each factor depends only on the rotational quantum numbers indicated. The various factors have been given in Table 2.1. They are also seen to depend on the internal axes g and the external axes F. For the linear and symmetric-top molecules, the electric moment could be assumed to be along the axis z so that only the matrix elements of Φ_{F_z} were required. In contrast, the asymmetric rotor often has components of its dipole moment along all three principal axes x, y, z.

For an asymmetric rotor basis the direction cosine matrix elements may also be written in a form similar to (2.108). Each asymmetric rotor wave function is expressed as a linear combination of Wang symmetric rotor functions all

Table 7.9 Stronger Asymmetric Rotor Transitions[a]

Near-Oblate Rotor
${}^aQ_{01}$, ${}^aQ_{0-1}$, ${}^aQ_{2-1}$, ${}^aQ_{-21}$
${}^aR_{01}$, ${}^aP_{0-1}$,[b] ${}^aR_{2-1}$, ${}^aP_{-21}$
${}^bQ_{1-1}$, ${}^bQ_{-11}$
${}^bR_{11}$, ${}^bP_{-1-1}$, ${}^bR_{3-1}$, ${}^bP_{-31}$
${}^cQ_{10}$, ${}^cQ_{-10}$
${}^cR_{10}$, ${}^cP_{-10}$
Near-Prolate Rotor
${}^aQ_{01}$, ${}^aQ_{0-1}$
${}^aR_{01}$, ${}^aP_{0-1}$
${}^bQ_{1-1}$, ${}^bQ_{-11}$
${}^bR_{11}$, ${}^bP_{-1-1}$, ${}^bR_{-13}$, ${}^bP_{1-3}$
${}^cQ_{10}$, ${}^cQ_{-10}$, ${}^cQ_{-12}$, ${}^cQ_{1-2}$
${}^cR_{10}$, ${}^cP_{-10}$, ${}^cP_{1-2}$, ${}^cR_{-12}$

[a]The stronger branches are indicated. Here the "g"-type, $\Delta J=0$ and $\Delta J=+1$ branches are specified with the notation ${}^gQ_{\Delta K_{-1}\Delta K_1}$ and ${}^gR_{\Delta K_{-1}\Delta K_1}$ respectively.

[b]Such reverse transitions are designated by $P(\Delta J=-1)$.

having the same value of J and M but different values of K. The J and M dependence of the matrix elements in an asymmetric rotor basis will thus be the same as that for a symmetric rotor basis, and we may write

$$(J, \tau, M|\Phi_{Fg}|J', \tau', M')=(J|\Phi_{Fg}|J')\cdot(J, \tau|\Phi_{Fg}|J'\tau')\cdot(J, M|\Phi_{Fg}|J', M') \quad (7.57)$$

The calculation of the $(J, \tau|\Phi_{Fg}|J', \tau')$ factor requires a knowledge of the eigenvectors associated with the two states J_τ and $J_{\tau'}$. Evaluation of this direction cosine matrix has been discussed by Cross et al. [23] and by Schwendeman [24]. The latter author has given the matrix elements of the direction cosines in the Wang symmetric rotor basis and has discussed the properties of the matrices, symmetry, and so on. Also the forms chosen for the matrix elements are convenient for transformation to the asymmetric rotor basis by means of a digital computer. In the case of slightly asymmetric rotors, perturbation theory may be used to give the asymmetric rotor wave functions by use of the Wang functions as basis functions. Simple expressions may then be obtained for the direction cosine matrix elements in terms of the symmetric rotor matrix elements and correction terms dependent on the asymmetry. This has been carried out by Lide [25] to second-order, and the results applicable to molecules with $|\kappa|\geqslant 0.9$ are collected in Table 7.10.

For a transition $J, \tau \rightarrow J', \tau'$ caused by a component of the electric moment μ_g, a convenient quantity called the line strength is defined as

$$\lambda_g(J, \tau; J', \tau')=\sum_{FMM'} |(J, \tau, M|\Phi_{Fg}|J', \tau', M')|^2 \quad (7.58)$$

where the sum extends over the three directions of the space-fixed system and over all values of M and M'. The latter sum over M and M' takes into account all possible transitions which in the absence of external fields will be degenerate. The line strength depends only on the inertial asymmetry parameter κ and not on the individual rotational constants; thus it is easily tabulated. Extensive tables of line strengths for various values of κ are available [23, 26–28]. The summation over F in the absence of external fields can be accomplished by multiplying the results for a given F by 3, that is,

$$\lambda_g(J, \tau; J'\tau')=3\sum_{MM'} |(J, \tau, M|\Phi_{Fg}|J', \tau', M')|^2 \quad (7.59)$$

The dipole moment matrix element appearing in (3.24) may be expressed in terms of the line strength. We have for a field-free rotor

$$|(J, \tau|\mu|J', \tau')|^2=\sum_{M'F} |(J, \tau, M|\mu_F|J', \tau', M')|^2$$

$$=\mu_g^2 \sum_{M'F} |(J, \tau, M|\Phi_{Fg}|J', \tau', M')|^2 \quad (7.60)$$

since $\mu_F=\sum_g \mu_g\Phi_{Fg}$ and for a given transition only one term in the sum over g

Table 7.10 Coefficients Required to Evaluate the Direction Cosine Matrix Elements of a Slightly Asymmetric Prolate Type Rotator[a,b]

Coefficients		α	β
$K=0$		0	$\frac{\sqrt{2}}{8}[f(J,1)]^{1/2}\left\{\delta+\frac{\delta^2}{2}\right\}$
$K=1,$	$\gamma=0$	0	$\frac{1}{16}[f(J,2)]^{1/2}\left\{\delta+\frac{[16+J(J+1)]\delta^2}{32}\right\}$
$K=1,$	$\gamma=1$	0	$\frac{1}{16}[f(J,2)]^{1/2}\left\{\delta+\frac{[16-J(J+1)]\delta^2}{32}\right\}$
$K=2,$	$\gamma=0$	$-\frac{\sqrt{2}}{8}[f(J,1)]^{1/2}\left\{\delta+\frac{\delta^2}{2}\right\}$	$\frac{1}{24}[f(J,3)]^{1/2}\left\{\delta+\frac{\delta^2}{2}\right\}$
$K=2,$	$\gamma=1$	0	$\frac{1}{24}[f(J,3)]^{1/2}\left\{\delta+\frac{\delta^2}{2}\right\}$
$K=3,$	$\gamma=0$	$-\frac{1}{16}[f(J,2)]^{1/2}\left\{\delta+\frac{[16+J(J+1)]\delta^2}{32}\right\}$	$\frac{1}{32}[f(J,4)]^{1/2}\left\{\delta+\frac{\delta^2}{2}\right\}$
$K=3,$	$\gamma=1$	$-\frac{1}{16}[f(J,2)]^{1/2}\left\{\delta+\frac{[16-J(J+1)]\delta^2}{32}\right\}$	$\frac{1}{32}[f(J,4)]^{1/2}\left\{\delta+\frac{\delta^2}{2}\right\}$
$K>3$		$-\frac{[f(J,K-1)]^{1/2}}{8(K-1)}\left\{\delta+\frac{\delta^2}{2}\right\}$	$\frac{[f(J,K+1)]^{1/2}}{8(K+1)}\left\{\delta+\frac{\delta^2}{2}\right\}$

[a]From Lide [25].

[b]Here K stands for K_{-1} and $\gamma=0$ if $J+K+K_1$ is even and $\gamma=1$ if $J+K+K_1$ is odd. The parameter $\delta=(B-C)/(A-C)$. The direction cosine matrix elements are evaluated from the relations:

$$(J, K, K_1|\Phi_{Fg}|J', K, K_1')=[1-(\alpha^2+\beta^2+\alpha'^2+\beta'^2)/2](J, K|\Phi_{Fg}^0|J', K)+\alpha\alpha'(J, K-2|\Phi_{Fg}^0|J', K-2) + \beta\beta'(J, K+2|\Phi_{Fg}^0|J', K+2)$$

and

$$(J, K, K_1|\Phi_{Fg}|J', K+1, K_1')=[1-(\alpha^2+\beta^2+\alpha'^2+\beta'^2)/2](J, K|\Phi_{Fg}^0|J', K+1)+\alpha'(J, K|\Phi_{Fg}^0|J', K-1) + \beta(J, K+2|\Phi_{Fg}^0|J', K+1)+\alpha\alpha'(J, K-2|\Phi_{Fg}^0|J', K-1) + \beta\beta'(J, K+2|\Phi_{Fg}^0|J', K+3)$$

The $(J, K|\Phi_{Fg}^0|J', K')$ are the symmetric rotor direction cosine matrix elements and α, β, calculable from the above table, refer to the level J, K, K_1, and α', β' to the level J', K', K_1'. For $(J, 0|\Phi_{Fg}^0|J', 1)$ the expression given in Table 2.1 must, in this particular case, be multiplied by $2^{1/2}$.

remains from symmetry arguments. The results are

$$\lambda_g(J, \tau; J', \tau') = (2J+1) \sum_{FM'} |(J, \tau, M|\Phi_{Fg}|J', \tau', M')|^2$$

$$= (2J+1)|(J, \tau|\mu|J', \tau')|^2/\mu_g^2 \tag{7.61}$$

$$= (2J'+1) \sum_{FM} |(J', \tau', M'|\Phi_{Fg}|J, \tau, M)|^2$$

$$= (2J'+1)|(J', \tau'|\mu|J, \tau)|^2/\mu_g^2 \tag{7.62}$$

The direction cosine matrix elements can also be conveniently expressed in terms of the line strengths. For $F = Z$ and any g, the line strength can be expressed in terms of the matrix elements $|(J, \tau|\Phi_{Zg}|J', \tau')|^2$ by use of (7.57), Table 2.1, and carrying out the summation. Note that the sum over M' reduces to one term since Φ_{Zg} is diagonal in M [see (10.71)–(10.73)]. We find

$$\lambda_g(J, \tau; J', \tau') = \tfrac{1}{4}R|(J, \tau|\Phi_{Zg}|J', \tau')|^2 \tag{7.63}$$

where R is defined as $1/J$, $(2J+1)/J(J+1)$, or $1/(J+1)$ for $J' = J-1$, J, or $J+1$, respectively. Since from (7.57)

$$|(J, \tau|\Phi_{Zg}|J', \tau'|^2 = \frac{|(J, \tau, M|\Phi_{Zg}|J', \tau', M)|^2}{|(J|\Phi_{Zg}|J')|^2 \cdot |(J, M|\Phi_{Zg}|J', M)|^2} \tag{7.64}$$

one obtains from (7.63) and Table 2.1

$$|(J, \tau, M|\Phi_{Zg}|J-1, \tau', M)|^2 = \frac{J^2 - M^2}{J(2J-1)(2J+1)} \lambda_g(J, \tau; J-1, \tau') \tag{7.65}$$

$$|J, \tau, M|\Phi_{Zg}|J, \tau', M)|^2 = \frac{M^2}{J(J+1)(2J+1)} \lambda_g(J, \tau; J, \tau') \tag{7.66}$$

$$|(J, \tau, M|\Phi_{Zg}|J+1, \tau', M)|^2 = \frac{(J+1)^2 - M^2}{(J+1)(2J+1)(2J+3)} \lambda_g(J, \tau; J+1, \tau') \tag{7.67}$$

These expressions multiplied by μ_g^2 represent the dipole matrix elements for particular Stark or Zeeman components with the field along Z analogous to (2.109) and (2.115) for a symmetric top

$$|(J, \tau, M|\mu_Z|J', \tau', M)|^2 = \mu_g^2|(J, \tau, M|\Phi_{Zg}|J', \tau', M)|^2 \tag{7.68}$$

It may also be readily shown that

$$\sum_{M=-J}^{+J} |(J, \tau, M|\Phi_{Zg}|J', \tau', M)|^2 = \tfrac{1}{3}\lambda_g(J, \tau; J', \tau') \tag{7.69}$$

since $\Sigma M^2 = \frac{1}{3}J(J+1)(2J+1)$. A number of useful sum rules exist because of the orthogonal properties of the direction cosine matrices [23]. Rudolph [29, 30] has also derived a number of additional relations involving the line strengths, reduced energy, and average of P_g^2.

A knowledge of the line strength is important in accessing the intensity of a given line and how the intensity can be expected to change with κ or variations in J or K. The larger $|\Delta\tau|$ in the transition the smaller the value of the corresponding λ. The line strength is independent of the direction of the transition. For a given κ

$$\lambda_g(J, \tau; J', \tau') = \lambda_g(J', \tau'; J, \tau) \tag{7.70}$$

Also the line strength for an oblate asymmetric rotor transition $J_{K_{-1}K_1} \rightarrow J'_{K'_{-1}K'_1}$ can be obtained from the prolate asymmetric rotor transition $J_{K_1K_{-1}} \rightarrow J'_{K'_1K'_{-1}}$ with the subscripts reversed by the means of the relation

$$\lambda_g(J, \tau; J', \tau'; \kappa) = \lambda_{g'}(J, -\tau; J', -\tau'; -\kappa) \tag{7.71}$$

Here the explicit dependence on κ is indicated, and the axes change $g \rightarrow g'$ as follows: $a \rightarrow c$, $b \rightarrow b$, and $c \rightarrow a$.

For small asymmetry the direction cosine matrix elements may be evaluated with the aid of Table 7.10 and hence the line strengths [see (7.63)]. Explicit expressions for the line strengths of low J rotational transitions have been given by Gora [31]. The line strengths for the symmetric top limit, which are useful simple approximations to the line strengths for near asymmetric tops, follow from (7.63) and Table 7.10 with $\delta = 0$ and Table 2.1.

The dipole matrix element factor appearing in (3.27) for the absorption coefficient is related to the line strength from (7.61) as follows:

$$|(J, \tau|\mu|J', \tau')|^2 = \frac{\mu_g^2 \lambda_g(J, \tau; J', \tau')}{(2J+1)} \tag{7.72}$$

If (7.72) and (3.25), (3.26), and (3.66) (the latter equation modified for an asymmetric rotor as indicated in Chapter III, Section 3) are inserted into (3.27), the peak absorption coefficient for the transition $J, \tau \rightarrow J', \tau'$ may be written as

$$\alpha_{\max} = 3.85 \times 10^{-14} F_v \sigma (ABC)^{1/2} \mu_g^2 \lambda_g(J, \tau; J', \tau') \left(\frac{\nu_0^2}{(\Delta\nu)_1 T^{5/2}} \right) g_I \, e^{-E_{J\tau}/kT} \tag{7.73}$$

where it is assumed that $h\nu_0 \ll kT$. The rotational constants A, B, C, the resonant frequency ν_0, and the line breadth $(\Delta\nu)_1$ are in megahertz, with μ_g in debye units and $\alpha_{\max}$ in cm^{-1}. If the lowest vibrational frequency is much greater than kT (say 1000 cm^{-1} for room temperature), the vibrational partition function is nearly equal to unity, and, for the ground vibrational state, F_v may hence be taken to be unity. Other things being equal (such as the Boltzmann factor), the strongest lines of a spectrum will be those which have the largest line strength. For asymmetric rotors with no symmetry σ, $g_I = 1$. When equivalent nuclei may be exchanged, the rotational level populations are affected and hence the relative intensities. The effects of nuclear spin statistics and the evaluation of g_I have been discussed in Chapter III, Section 4.

5 IDENTIFICATION AND ANALYSIS OF ROTATIONAL SPECTRA

Before molecular information can be derived from the rotational spectrum, the major task of assignment of the rotational quantum numbers to the observed spectral transitions must be accomplished. Therefore some remarks need to be made on the assignment of the transitions, application of selection rules, and evaluation of the rotational constants.

Common Aids for Identification of Observed Transitions

The assignment of spectra can be relatively easy for linear and symmetric-top molecules. Because of the greater complexity of the spectra of asymmetric-tops, assignment of their rotational transitions can be very difficult. In the investigation of any molecule it is possible to obtain an initial set of structural parameters from similar molecules, or other studies, for example, infrared, X-ray diffraction, or simply an intelligent guess, and thus the rotational constants with which the rigid-rotor spectrum can be tentatively predicted. Some idea of the dipole components μ_g, or of their relative magnitudes, is also required, since the strength of a particular transition will be proportional not only to the line strength but also to the square of the dipole component. Thus the transitions that dominate the spectrum are dependent on the size of μ_g. Bond dipole moments or dipole moments of similar molecules can be used for approximation of these components. Often the molecular symmetry can be used to argue for, or against, the presence of a particular component. Such calculations can usually indicate via the selection rules at least the type of transitions and the frequency region in which one can expect to find certain useful transitions, for example, particular low J transitions.

Spectral patterns, the Stark effect (see Chapter X, Section 8), quadrupole hyperfine splittings (see Chapter IX, Section 4), nuclear spin statistics (see Chapter III, Section 4), and the rigid rotor fit can all be helpful in establishing and confirming assignments. If the conditions permit, the Stark effect can be a useful technique. By observation of the number of components (Stark lobes) into which a given line splits in the presence of an externally applied electric field, the smaller of the two J's involved in a transition may be determined. For Q-branch lines, the intensity of the Stark lobes increases with M_J. The reverse is true for R-branch lines. If nuclear quadrupole hyperfine structure is present, it may be used as a guide, and transitions may be identified on the basis of their characteristic quadrupole splitting pattern, as well as the self-consistency of the spectral fit. Often a characteristic spectral pattern is an aid to assignment. Both the line patterns and intensities provide useful information. Finally, the technique of microwave double resonance can be employed (see later in this section).

One of the more useful patterns is obtained for molecules with "a"-type R-branch transitions ($^aR_{01}$) for a prolate asymmetric rotor and "c"-type for an oblate rotor ($^cR_{10}$). Consider a prolate rotor. There are two transitions

$(J \to J+1, K_{-1} \to K_{-1})$ for each value of K_{-1} except for $K_{-1}=0$. The structure hence consists of $2J+1$ lines for a given $J \to J+1$ transition. Figure 7.1 shows a condensed-scale spectrum of 2-cyanothiophene which is characteristic of a near-prolate asymmetric rotor with an "a"-type spectrum. (See also Fig. 7.5). The grouping of the lines $J_{K_{-1}K_1} \to (J+1)_{K_{-1}K_1+1}$ into clusters for a given $J \to J+1$ transition is clearly apparent. The intensity of the transitions within a given group fall off with increasing K_{-1} because of the Boltzmann factor. As J increases, the number of lines in each group increases, and the lines are spread out over a larger frequency range. The lower K_{-1}-doublet transitions are more widely split by the asymmetry than the higher K_{-1} lines. Hence as K_{-1} increases, the splitting between a given K_{-1} doublet decreases. If the asymmetry is not too large, the pair of higher K_{-1} lines will coalesce.

For very slightly asymmetric rotors, the frequencies of the $K_{-1}=1$ and 2 lines can be readily used for evaluation of the rotational constants. These relations collected in Table 7.11 are also useful for more asymmetric rotors as rough approximations to the rotational constants.

Another identifying characteristic for this kind of transition is the Stark effect. If the dominant contribution to the Stark effect arises from μ_a for a prolate top, the degenerate transitions at high K_{-1} will exhibit a very fast Stark effect and give rise to a first-order Stark pattern, that is, Stark components on both sides of the line which move away very rapidly with increasing electric field. This is due to the degenerate levels $J_{K_{-1}K_1}$, $J_{K_{-1}}K'_1$ interacting via μ_a (see Chapter X). The $K_{-1}=0$ line will show a characteristic second-order effect, whereas the lower K_{-1} lines with larger splitting will also show a second-order pattern, although the Stark effect can be quite fast because of the near degeneracies of the interacting levels. The higher K_{-1} lines are hence more easily modulated and will often show up even at low modulation voltages. The Stark patterns of low K_{-1} lines can be useful even if not resolved. The Stark lobes for the higher frequency K_{-1} line are on the low-frequency side of the line; whereas, the Stark lobes for the low-frequency K_{-1} line will be found on the high-frequency side. This pattern, along with an approximate frequency separation of a given pair of low K_{-1} lines, can be very helpful in limiting assignment possibilities.

With the development of broadbanded detectors and BWO tubes as microwave sources, rotational spectra can be conveniently recorded over wide frequency ranges. Under fast-sweep conditions, rotational spectra with significantly reduced resolution are obtained, and a whole microwave band (e.g., R-band 26 to 40 GHz) can be displayed on a recording similar in size to an IR spectral trace (see, e.g., Fig. 7.1). They are useful as an aid in the assignment of high resolution spectra. Such spectra were first reported by Scharpen [32]. Illustrative applications to structure studies are given in Chapter XII (see also [33] and [34]).

The ${}^aR_{01}$ and ${}^cR_{10}$ transitions discussed previously can often, under low resolution, give rise to a symmetric-top, bandlike spectrum. Useful band spectra arising from overlapping of high K lines have been obtained for molecules with $|\kappa| \geqslant 0.70$ and $\mu_{a \text{ or } c} > 0.50$ D [35]. The frequency of the band center in terms

Table 7.11 Useful Relations for Characterization of Rotational Spectra of Asymmetric Rotors

Relation	*Comments*
$\left.\begin{array}{l}\nu_h+\nu_l=2(B+C)(J+1)\\ \nu_h-\nu_l=(B-C)(J+1)\end{array}\right\}\begin{array}{l}J\to J+1\\ K_{-1}=1\to 1\end{array}$	Approximate location of the high (ν_h) and low (ν_l) frequency components of the $K_{-1}=1$ doublet of the ${}^aR_{01}$ series for a prolate rotor.
$\nu_h-\nu_l=\dfrac{(B-C)^2/4}{2A-(B+C)}\,J(J+1)(J+2)\left\{\begin{array}{l}J\to J+1\\ K_{-1}=2\to 2\end{array}\right.$	Approximate splitting of the $K_{-1}=2$ lines of the ${}^aR_{01}$ series for a prolate rotor. For an oblate rotor (${}^cR_{10}$) interchange A and C here and above expression.
$\nu=(B+C)(J+1)$ $J\to J+1$, $K_{-1}\to K_{-1}$ $\Delta\nu=(B+C)$ Band separation	Frequency of the band center for the ${}^aR_{01}$ transitions of a near prolate rotor. For transitions (${}^cR_{10}$) of a near oblate rotor replace C by A.
$(B+C)/(B_0+C_0)=1+(0.025\pm0.005)(1+\kappa)$	Empirical relation [37] to correct low resolution observed $(B+C)$ to ground-state quantity (B_0+C_0). κ may be estimated from assumed structure.
$\nu=(A+B)(J+1)$ $(J\to J+1,\ K_1\to K_1+1)$ $+(2C-A-B)(K_{+1}+\frac{1}{2})$	Approximate location (their mean) of the high K_1 R-branch transitions of a highly asymmetric rotor. The separation between the different pairs of transitions is $2C$.
$\nu=(2J+1)C+\left(\dfrac{A+B}{2}\right)(K_1=J\to J+1)$	The $K_1=J\to J+1$ transitions can give rise to band spectra separated by $2C$ characterized by this relation.
$\nu=[A-\frac{1}{2}(B+C)]\Delta W_{JK}$, $K_{-1}\to K_{-1}$ $\cong\left(\dfrac{B-C}{2}\right)J(J+1)$ $K=1$	Location of the Q-branch transitions between K doublets where ΔW_{JK} is the energy level splitting between a given K doublet.
$\nu=[A-\frac{1}{2}(B+C)](2K_{-1}+1)$ $K_{-1}\to K_{-1}+1$	Q-branch series beginning $(J=K_{-1}+1)$ for a near-prolate rotor with "b"-type transitions. Separation between band heads is $2A-B-C$.
$\nu=[\frac{1}{2}(A+B)-C](2K_1-1)$ $K_1\to K_1-1$	Q-branch series beginning $(J=K_1)$ for a near-oblate rotor with "b"-type transitions. Separation between band heads is $A+B-2C$.

of the rotational constants is given in Table 7.11; a typical band spectrum [36] is shown in Fig. 7.7. Corrections to the band center to extract the ground state constant $(B_0 + C_0)$ given in Table 7.11, have been discussed by Farge and Bohn [37].

The width of the band varies from about 50 to 200 MHz because of the spreading effect of the asymmetry; in fact, the half width of the band can be used as a qualitative measure of κ [35]. This can be useful in comparing different conformers of a compound if the κ's differ by at least 0.05. The broadening of the band with application of a static field [38] can also be used to approximate the magnitude of μ_a. With large molecules there are often more than one configuration of the molecule that can give distinct spectra, and more than one band series will be found. The ratio of the integrated band intensities of two conformers of a molecule can be related to the ratio of the number of molecules, N, of each conformer [39].

For "a"-type R-branch transitions of a highly asymmetric toror ($|\kappa| < 0.3$), the pair of transitions $J_{0,J} \rightarrow (J+1)_{0,J+1}$ and $J_{1,J} \rightarrow (J+1)_{1,J+1}$ turn out to be coincident, or nearly coincident, in frequency, and adjacent pairs are separated by $2C$. Because of this, and the fact they have large line strengths, they tend to overlap and stand out in the spectrum. In addition, the Stark effect is particularly fast when μ_c is nonzero due to the near degeneracy of the $J_{0,J}$ and $J_{1,J}$ levels; mirror-image Stark patterns can be obtained. The lines are then easily observable even at low modulation voltages. Such transitions have been found [40–44] to be particularly useful in the assignment of rotation spectra of highly asymmetric rotors. A similar situation holds for "b"-type transitions $J_{1,J} \rightarrow (J+1)_{0,J+1}$ and $J_{0,J} \rightarrow (J+1)_{1,J+1}$. The characteristics of these transitions are most easily explained by noting that for a given κ at high J many of the higher K_{-1} energy

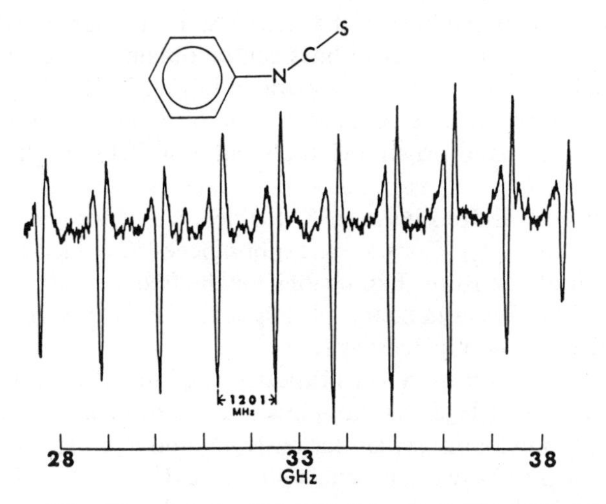

Fig. 7.7 Low-resolution $^aR_{01}$ spectrum of phenylisothiocyanate. The constant separation $(B+C)$ in the bands is apparent. From Higgins, et al. [36].

levels are prolate in character, whereas the lower K_{-1} levels are quite oblate in character [45] (see Table 7.11).

These high K_1-lines, when accompanied by vibrational satellite lines, can also give rise to band spectra under appropriate experimental conditions [46], but these band spectra are not as common as those described previously. Little application has been made of them so far. An analogous situation holds for "c"-type transitions of a highly asymmetric rotor from levels of high K_{-1}, for example, $J_{J,0} \to (J+1)_{J+1,0}$ and $J_{J,1} \to (J+1)_{J+1,1}$. The separation between different adjacent pairs of such transitions is now $2A$.

Q-branch transitions which often give rise to recognizable spectral patterns can provide the initial key to an assignment (see, e.g., [45], [47]–[50]). The stronger subbranches have been listed in Table 7.9. Even when no special spectral pattern is present there may be certain prominent Q-branch transitions which can be used as an aid in the assignment [51]. For the ${}^bQ_{1\,-1}$ series, a given $K \to K'$ transition consists of two transitions (except $K=0$) of the same J; hence two series can be formed characterized by K and various values of J. These two series diverge from each other as J increases. Only one part of the series may be observed in a given frequency region. If κ is large or a series with large K_{-1} (or K_1) is being studied, the two series can overlap and a series of strong lines relatively closely spaced can be obtained with each line consisting of two superimposed transitions. However, for a given κ, the pair of superimposed transitions will split at sufficiently high J because of asymmetry effects, with their separation increasing with increasing J. The $K=0 \to 1$ lines form a single series moving to higher or lower frequency depending on the type of transition involved. The details of the series structure depend on K, J, κ and the type of transition. These details will be clear from calculated spectra, but the less the inertial asymmetry, the less the splitting and the more easily recognizable will be the series.

If the splittings between lines of the series are not too large and if a high K series is studied, a broad cluster of lines can be obtained if recording is done under low resolution conditions, as illustrated in Fig. 7.1. Each cluster of lines or band is characterized by a particular value of K and various values of J. Expressions for the band origin and the separation between the band heads are given in Table 7.11.

The Q-branch transitions between K doublets $K \to K$ form a series for each value of K (see, e.g., [52]). These are the subbranches ${}^aQ_{0\,-1}$ for a prolate rotor and ${}^cQ_{10}$ for an oblate rotor. Expressions for the frequencies in the series are given in Table 7.11. The asymmetry splitting of the levels ($K>0$), and hence the frequency, decreases with increasing K.

Often the type of transition to be expected is quite sensitive to a particular structural parameter(s). H_2S_2 has been investigated by Winnewisser et al. [22]. By analogy with previous observations on H_2O_2 and F_2S_2, the molecule H_2S_2 would be expected to have a skew-chain structure (H—S—S—H) and hence a C_2 axis of rotational symmetry. Furthermore, the dihedral angle (η), the angle defining the relative position of the two S—H bonds, might be expected to be close to that found in F_2S_2, that is, somewhere around 90°. Because of the C_2 axis of symmetry, a variation in the line intensities is expected due to nuclear spin statistics (see Chapter III, Section 4); preliminary calculation indicates that a dominant feature of the spectrum will be a series of closely spaced Q-branch-type transitions, namely, $J \rightarrow J$, $K_{-1} = 0 \rightarrow 1$. However, the asymmetry of the molecule and the type of transition which will be observed depend critically on the dihedral angle. Figure 7.6a gives the dependence of I_a, I_b, I_c and $(C-B)$ on the dihedral angle. It is apparent that I_b and I_c vary significantly as η is varied, although I_a remains essentially constant. For values of η near 90° (or 270°) the molecule is a very slightly asymmetric rotor. At η_0 (90°5′) $I_b = I_c$, the molecule becomes an accidentally symmetric top, and the C_2 symmetry axis changes from the b axis to the c axis. For values of $\eta < 90°$ the molecule becomes more asymmetric, and "b"-type transitions are expected, while for $90° < \eta < 270°$ "c"-type transitions are expected. A detailed analysis of the spectrum confirms the chainlike structure and reveals that the molecule is an almost perfect prolate rotor with "c"-type transitions. The sensitivity of the observed value of $C - B = -3.08$ MHz to the dihedral angle is illustrated in the expanded plot of Fig. 7.6, where η is found to be 90°36′. A portion of the Q-branch spectrum of D_2S_2 is shown in Fig. 3.5.

Computer Assignment and Analysis

We have been discussing the use of various spectral characteristics in selecting likely candidates for analysis. An alternate approach to the problem of assigning spectra has been described [53, 54] which makes use of only observed frequencies. This is a computer-based assignment technique. It is possible, in principle, to write a computer program to assign spectra from input line frequencies and use the closeness of fit as the criterion for an assignment. Some constraints can be placed on the process to reduce the computer time [54, 55], but the results will be no better than the input data.

In the absence of any similarity between the observed and calculated spectrum it will be necessary to assume a new molecular model and repeat the calculations until some similarity appears between the two. Once an assignment of three or more transitions has been made, the rotational constants for a tentative model can be evaluated. At least one of the three transitions must, however, be a P- or R-branch transition, since the Q-branch transitions depend only on κ and $(A-C)/2$. This new set of A, B, and C values should yield predictions of new transitions that can be found where predicted. Only in this way can one confirm the assignment. The new measured lines are added to the analysis and the procedure repeated until the agreement between observation and calculation is satisfactory and the uncertainties in the rotational constants are small. This general "bootstrap" assignment procedure is typical of the analysis of any rotational spectra. Where a rigid rotor Hamiltonian does not provide an adequate basis for analysis, the effects of centrifugal distortion, and so on, may have to be included. The fit of the spectrum should reflect an experimental accuracy typically 0.1 MHz or better.

If very low J transitions have been observed, their frequencies may be related directly to the spectroscopic constants by application of Tables 7.6 and 7.7. Transitions are frequently observed for which simple closed-form expressions for the energy levels are not possible. For them, it is convenient to write the rigid rotor energy as

$$E=E_0+\left(\frac{\partial E}{\partial A}\right)\delta A+\left(\frac{\partial E}{\partial B}\right)\delta B+\left(\frac{\partial E}{\partial C}\right)\delta C \tag{7.74}$$

where E_0 is the rigid rotor energy evaluated from an initial estimate of the rotational constants; $(\partial E/\partial A)$, and so on, are evaluated as indicated previously; and δA, δB, δC are the corrections to the initial set of rotational constants. By use of (7.74) the observed frequencies may be expressed as a function of δA, δB, δC; hence the corrections to the rotational constants are readily evaluated. If the initial estimates of A, B, and C are poor, the analysis will have to be iterated. With several transitions a least squares analysis based on (7.74) should be employed.

In particular, in the calculation of an improved set of rotational constants from the set of observed frequencies, ν_{obs}, one uses a linear variation of the parameters in the form

$$\nu_{\text{obs}}-\nu_{\text{cal}}=\left[\left(\frac{\partial E}{\partial \alpha}\right)_U-\left(\frac{\partial E}{\partial \alpha}\right)_L\right]\delta\alpha+\left[\left(\frac{\partial E}{\partial \beta}\right)_U-\left(\frac{\partial E}{\partial \beta}\right)_L\right]\delta\beta$$
$$+\left[\left(\frac{\partial E}{\partial \gamma}\right)_U-\left(\frac{\partial E}{\partial \gamma}\right)_L\right]\delta\gamma \tag{7.75}$$

Here the initial set of rotational constants A_0, B_0, C_0, is employed to calculate the transition frequencies ν_{cal}. The α, β, γ are, in general, linear combinations of the rotational constants. The subscripts U and L refer to the upper and lower levels of the transition. A least-squares analysis based on this equation provides the corrections $\delta\alpha$, $\delta\beta$, and $\delta\gamma$. If the Wang representation is used, we have for a prolate asymmetric top (I^r)

$$E=\tfrac{1}{2}(B+C)P^2+[A-\tfrac{1}{2}(B+C)]\langle P_a^2\rangle+\tfrac{1}{2}(B-C)\langle P_b^2-P_c^2\rangle \tag{7.76}$$

where the average values are to be evaluated in the asymmetric top basis with the initial rotational constants. The linear combinations of the rotational constants are conveniently defined as

$$\begin{aligned}\alpha&=A-\tfrac{1}{2}(B+C)\\ \beta&=\tfrac{1}{2}(B+C)\\ \gamma&=\tfrac{1}{2}(B-C)\end{aligned} \tag{7.77}$$

with

$$\left(\frac{\partial E}{\partial \beta}\right) = P^2 = \langle P_a^2 \rangle + \langle P_b^2 \rangle + \langle P_c^2 \rangle = J(J+1)$$
$$\left(\frac{\partial E}{\partial \alpha}\right) = \langle P_a^2 \rangle \tag{7.78}$$
$$\left(\frac{\partial E}{\partial \gamma}\right) = \langle P_b^2 - P_c^2 \rangle = \frac{1}{b_P}[\langle P_a^2 \rangle - W(b_P)]$$

The foregoing are to be substituted into (7.75). For an oblate asymmetric top, A and C and a and c may be simply interchanged in the foregoing expressions. The least-squares analysis gives the corrections $\delta\alpha$, $\delta\beta$, and $\delta\gamma$ and an improved set of rotational constants are calculated from the initial set A_0, B_0, and C_0; for example,

$$A = A_0 + \delta\alpha + \delta\beta$$
$$B = B_0 + \delta\beta + \delta\gamma \tag{7.79}$$
$$C = C_0 + \delta\beta - \delta\gamma$$

The best set of A, B, and C values are calculated by iteration until the corrections are negligibly small.

A set of Q-branch transitions can be analyzed as outlined previously, but only two linear combinations of the three rotational constants can be evaluated. It is often convenient in checking assignments, however, to employ a graphical technique, the so called "Q-branch plot." If Ray's representation of the energy is employed we have for the frequency of a Q-branch transition

$$\nu = \left(\frac{A-C}{2}\right)\Delta E(\kappa) \tag{7.80}$$

where $\Delta E(\kappa)$ is the difference in the reduced energy $E(\kappa)$ between the upper and lower states of the transition. For a given κ, the reduced energy difference may be obtained, for example, from a table of $E(\kappa)$, and $(A-C)/2$ may be calculated from the observed ν. For the observed Q-branch series, if κ is plotted against $(A-C)/2$, the curves will approximate straight lines over a small range of κ; they should intersect at a given point which specifies the values of $(A-C)/2$ and κ. Lines that do not intersect at the same point reflect an incorrect assignment of the quantum numbers. Some small spread in the intersection point will have to be expected if the effects of centrifugal distortion are important. An assignment verified by a Q-branch plot from Durig et al. [56] is illustrated in Fig. 7.8. If the Wang representation of the energy is used, then, for example, $A - \frac{1}{2}(B+C)$ and b_p would be evaluated from the Q-branch data.

If only Q-branch transitions have been measured, as we have seen, only two rotational parameters can be evaluated. However, if one is dealing with a planar

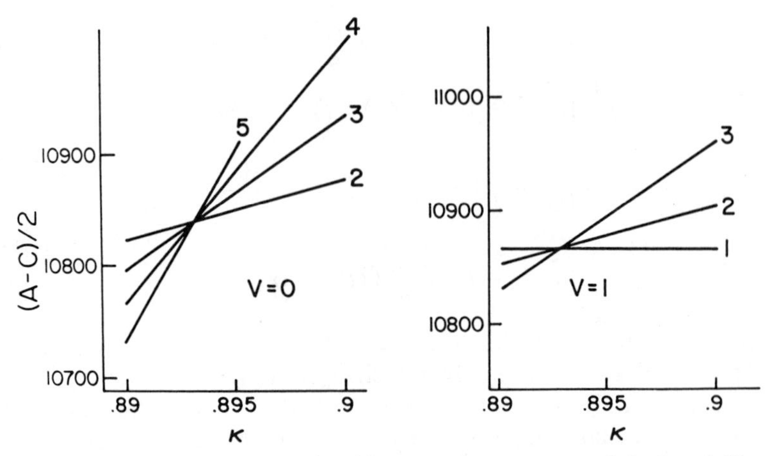

Fig. 7.8 Q-branch plot for the ground and first excited torsion states of *cis*-glyoxal. The numbers on the lines indicate the J levels. For the ground state $\kappa=-0.893$ and $(A-C)/2=10{,}840$ MHz. From Durig, Tong and Li [56].

asymmetric rotor, the three rotational constants can be obtained to a first approximation from an analysis of only the Q-branch data [57]. The well-known planarity condition ($I_c=I_a+I_b$) provides the relation

$$\frac{1}{C}=\frac{1}{A}+\frac{1}{B} \tag{7.81}$$

Here the vibrational contributions to the rotational constants are assumed negligible (see Chapter XIII). Making use of the foregoing, the following relation by Doraiswamy and Sharma [57] has been derived expressing $(A+C)$ in terms of κ and $(A-C)$

$$(A+C)=(A-C)\{1+[2(\kappa+1)]^{1/2}\} \tag{7.82}$$

The linear combinations $(A+C)$ and $(A-C)$ readily yield A and C, and B may be evaluated from

$$B=\frac{A+C}{2}+\kappa\left(\frac{A-C}{2}\right) \tag{7.83}$$

The results can be expected to be more accurate for planar molecules with small inertial defects.

With availability of high-speed computers, observed spectra are almost exclusively analyzed with the aid of computer programs that evaluate the various asymmetric rotor energy levels and eigenvectors and compute the line frequencies, intensities, Stark coefficients, quadrupole splitting, and so forth. Various computer programs have been discussed in the literature [58–61].

Double Resonance as an Aid to Spectral Assignment

In this section we explain briefly the application of microwave–microwave double resonance (MMDR) for the identification of transitions in spectra of asymmetric rotors. Double resonance is generally useful in a variety of experiments to obtain information on such properties as energy level structure, relaxation phenomena, rotational energy transfer, pure rotational transitions in nonpolar molecules, selection rules in collision-induced transitions, excited-state microwave transitions, and the enhancement of weak or forbidden transitions. The technique has also found application as a method of modulation. Reviews and summaries on the subject are available [62–68].

In a typical double resonance experiment, two radiation fields are applied simultaneously to the molecular sample. One is a high-power source of "pump" radiation which is resonant with a particular dipole-allowed transition; the other is a low-powered source termed the "signal" or "probe" radiation, which is also resonant with a dipole-allowed transition at a different frequency. The frequency of the pump or signal can be in the microwave, radio frequency, infrared, or visible/ultraviolet region. We consider here only microwave detection techniques, that is, microwave-(MMDR) and radio frequency–microwave double resonance (RFMDR) experiments and limit the discussion to its use as an assignment aid. If the pump radiation is of sufficient intensity and the two transitions share a common energy level, the pump radiation will affect the microwave absorption signal. Double resonance effects are hence characteristic of the energy level structure since they depend on the coupling of two transitions through a common energy level. Herein lies its usefulness as an aid to the assignment of complex spectra. The double resonance effects result because of population changes induced by the high-powered pump radiation and by coherence effects due to the time-dependent mixing by the coherent radiation of the wave functions characterizing the energy levels [62]. This latter effect is particularly important in radio frequency–microwave double resonance since the population changes are negligible. Earlier work on both the experimental [69–78] and theoretical [79–81] aspects of this technique have been carried out by a number of investigators.

For a three-level system (lower portion of Fig. 7.9), Javan [79] has discussed the effect of the double resonance on the line shape of a transition. To a first approximation ($|y|/\nu_p \ll 1$), the absorption signal is split into two absorption peaks, at frequencies [79, 82]

$$\nu_{s1} = \nu_s^0 + \tfrac{1}{2}[\pm\Delta\nu_p - (\Delta\nu_p^2 + |y|^2)^{1/2}] \tag{7.84}$$

$$\nu_{s2} = \nu_s^0 + \tfrac{1}{2}[\pm\Delta\nu_p + (\Delta\nu_p^2 + |y|^2)^{1/2}] \tag{7.85}$$

where ν_s^0 and ν_p^0 are the resonance frequencies of the signal and pump transitions, and where the amount by which the pump is off resonance is specified by

$$\Delta\nu_p = (\nu_p^0 - \nu_p) \tag{7.86}$$

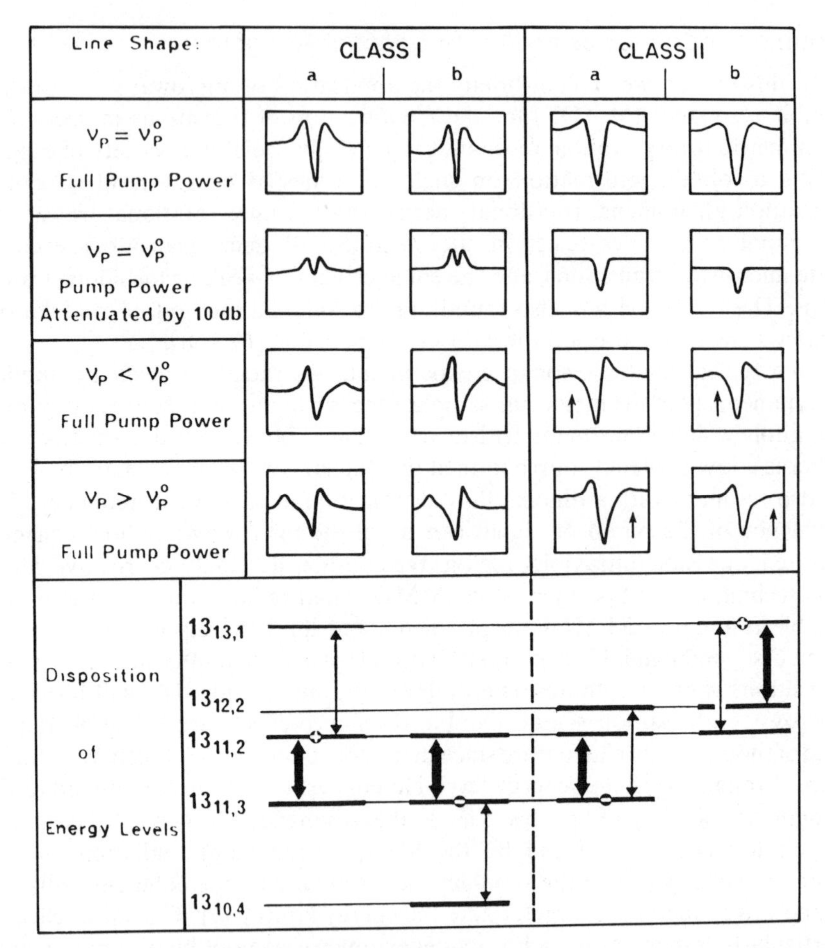

Fig. 7.9 Frequency-modulated double resonance signals for various pumping conditions, and the three-level energy level arrangements in isoazole. The pump transition is indicated with a bold face arrow and the signal transition with a light arrow. Class I: pump transition lies outside the signal energy levels; Class II: pump transition lies inside the signal energy levels. The positions of the weak double-photon transitions of Class II are indicated for clarity in the upper portion of the figure with an arrow. The sign displayed on the energy levels specifies whether the pump decreases or increases the population of the level. The zero-pump lines are pointing downward. Note for Class I, for example, with $\nu_p < \nu_p^0$, a derivative profile is obtained with a double-photon transition at higher frequency. For $\nu_p = \nu_p^0$, a symmetrical profile is obtained. Here and in other traces, the zero-pump line is directed downward opposite to the coherence doublet of Fig. 7.10. For $\nu_p > \nu_p^0$ a derivative profile with the double-photon transition is at lower frequency. From Stiefvater [83].

and

$$y = |\mu_p| \frac{E_p^0}{h} \tag{7.87}$$

where E_p^0 is the electric field amplitude of the pump radiation and $|\mu_p|$ is the dipole moment matrix element for the pump transition. The choice of sign in front of $\Delta\nu_p$ in (7.84) and (7.85) depends on the relative order of the energy levels: plus sign for class I, negative sign for class II. If the pump frequency is on resonance $\nu_p = \nu_p^0$, a symmetrical doublet is obtained that is centered about the ν_s^0 position and split by $|y|$

$$\nu_{s1} = \nu_s^0 - \frac{|y|}{2} \tag{7.88}$$

$$\nu_{s2} = \nu_s^0 + \frac{|y|}{2} \tag{7.89}$$

This coherence splitting is illustrated in Fig. 7.10. On the other hand, when the pump radiation is above or below resonance, an asymmetrical doublet is obtained. One of the components, the "main line," moves toward the unsplit position ν_s^0 and becomes stronger while the other component moves away from ν_s^0 and becomes weaker, as shown in Figure 7.10. This component, which is called the "creeper," corresponds to a double photon transition between, for example, the lower and upper state of the three-level system. For $\Delta\nu_p \gg |y|$

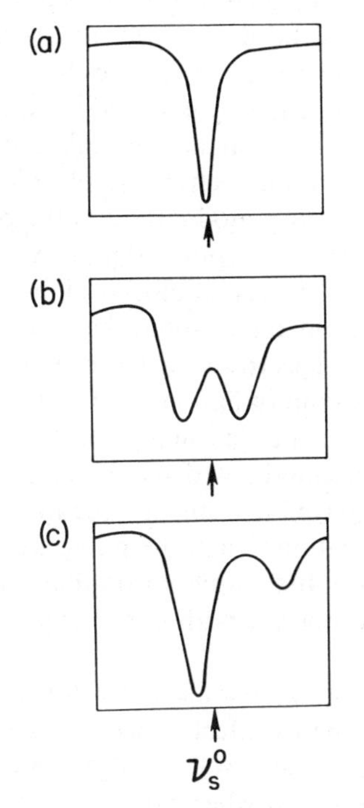

Fig. 7.10 The effect on an absorption line for the pump frequency on- and near resonance. (*a*) Absorption line pump radiation off. (*b*) Symmetrical doublet obtained when the pump radiation is on resonance ($\nu_p = \nu_p^0$). (*c*) Asymmetrical doublet obtained when the pump radiation is off resonance ($\nu_p \neq \nu_p^0$). From Stiefvater [83].

we have

$$\nu_{s1} \simeq \nu_s^0 \mp \frac{|y|^2}{4\Delta\nu_p}$$
$$\simeq \nu_s^0 \tag{7.90}$$

$$\nu_{s2} \simeq \nu_s^0 \pm \Delta\nu_p \pm \frac{|y|^2}{4\Delta\nu_p}$$
$$\simeq \nu_s^0 \pm \Delta\nu_p \tag{7.91}$$

It is clear that the maximum signal absorption in the presence of the pump radiation does not occur at exactly ν_s^0 but is shifted slightly. Likewise, the creeper is displaced from ν_s^0 by essentially the amount that the pump frequency is off resonance but in opposite directions [taking the upper sign in (7.91)]. If, for example, $\nu_p > \nu_p^0$, then one signal peak for class I is shifted to higher frequency by a small amount; the other is shifted to lower frequency by essentially $\Delta\nu_p$. More accurate relations are available to higher order in $|y|$ [69, 82].

Various experimental configurations have been described for observation of microwave–microwave double resonance [74, 77, 83–92]. In a conventional Stark-modulated spectrometer, the usual spectrum is displayed and double resonances are detected by observation of changes in the line shape and intensity of a given line as the pump frequency is varied [74]. One of the more useful techniques makes use of double resonance modulation [77, 83] where only certain transitions that have a common energy level will be observed. Here one makes use of square-wave frequency modulation of the pump to modulate, via the double resonance effect, the absorption signal. A narrow-band amplifier tuned to the modulation frequency is employed along with the usual phase-sensitive detector. The spectrometer is somewhat more complicated than a conventional Stark-modulated spectrometer because it requires two radiation sources. Both the signal and pump radiation propagate through the waveguide absorption cell. The lower frequency is usually chosen as the pump transition, since it is easy to prevent the modulated pump radiation from reaching the signal detector by use of a waveguide transition section with a cutoff frequency beyond that of the pump radiation. Often the pump frequency is selected, and the signal frequency is swept in the search for double resonant signals. A spectrometer may be designed to operate with either Stark modulation or double resonance modulation [83].

Figure 7.9 shows various energy level schemes for a three-level system that can give rise to a double resonance effect. Since the signal ν_s and the pump ν_p transitions share a common energy level, the signal transition will be affected by the high-power pump radiation when both are applied simultaneously to the sample. Since this double resonance effect allows the selective modulation of ν_s via ν_p, only the modulated ν_s reaches the detector. Thus, essentially only the transitions that share an energy level with the pump transition will be observed.

This virtually eliminates double resonance signals involving different rotational energy levels of a given molecule and from different molecules, for example, isotopic species. This is in contradistinction to Stark effect modulation where all possible absorptions are essentially modulated and hence dense spectra are possible. The appearance of the double resonance modulated signal for various pump frequencies is also illustrated in Fig. 7.9.

It should be noted that the "zero-pump" line is slightly shifted from ν_s^0 by an amount that depends on the size of pump power modulation. Under the usual conditions of the experiment, however, ν_s^0 is obtained to better than ± 0.2 MHz, and ν_p^0 should be better than ± 0.5 MHz by adjustment of ν_p to give the symmetrical pattern [83] of Fig. 7.9. The attainment of the spectral patterns of Fig. 7.9 provides unambiguous evidence that the signal and pump transitions are coupled through a common energy level. This knowledge of the energy level connections, as pointed out previously, provides the basis for spectral assignments.

A double resonance effect of reasonable intensity can be observed [93, 94] even when the pump frequency is away from the resonant frequency ν_p^0. This detectability, can make it difficult to identify the desired transition. On the other hand, if the pump transitions are not too dense, this effect can be very useful. If more than one pump transition lies within say 50 MHz of another, the appropriately connected signal transitions will be simultaneously modulated and a scan of the signal region will reveal these transitions. Subsequent adjustment of ν_p to each resonance point allows measurement of the various ν_s^0 and ν_p^0. Even when the pump frequency is far off resonance, a signal can be observed that arises from the so-called high-frequency Stark effect [95, 96].

It is also possible to have a pump transition share an energy level with more than one signal transition. In isobutyryl fluoride [83], for example, if the pump frequency is $(4_{23} \rightarrow 5_{14})_p = 24{,}249$ MHz, a signal can be observed at 25,103, 25,860, 31,416, and at 32,199 MHz corresponding, respectively, to the signal transitions $(3_{12} \rightarrow 4_{23})_s$, $(4_{23} \rightarrow 5_{24})_s$, $(4_{13} \rightarrow 5_{14})_s$, $(5_{14} \rightarrow 6_{15})_s$, and $(5_{14} \rightarrow 6_{25})_s$.

A further useful guide in the assignment of spectra, as pointed out by Woods et al. [77] and by Stiefvater [97], is the two different line profiles that are obtained, depending on the position of the common energy level with respect to the other two levels of the three-level system. In the Class I line shape category, the common energy level is the intermediate energy level; for Class II, the common level is the highest or lowest energy level. For Class I, which is the most common, Fig. 7.9 shows that the double resonance lobes are stronger, the signal and lobes tend to cancel each other with a reduction of pump power in such a way that the lobes extinguish the "zero-pump" line, and the creeper moves to high (low) frequency if the pump radiation is tuned below (above) exact resonance by a few megahertz [upper sign in (7.91)]. On the other hand, for Class II, the double resonance lobes are much weaker and virtually disappear when the pump power is reduced (leaving essentially the "zero-pump" line), and the creeper moves in the same direction as the pump frequency relative to the exact resonance position ν_p^0 [lower sign in (7.91)].

Clearly, a knowledge of the relative position of the common energy level further restricts the number of possible assignments that may be made for the pump and signal transitions, and hence provides additional information toward the assignment of an observed spectrum. Furthermore, these distinguishable line shapes are also useful in the assignment of isotopic species and vibrational excited states. For a particular double resonance signal of the ground state of the parent species, the line shape for the same three-level system for different isotopic species and vibrational states must be the same.

A number of applications to the assignment of rotational spectra may be found in the previous references; see also [98]. The simplification possible by use of microwave–microwave double resonance is illustrated in Fig. 7.11 from the work of Pearson et al. [91]. The upper trace taken with Stark modulation shows a number of absorptions. The lower trace, a double resonance scan of the same region, shows only two features—an on-resonance line shape and an off-resonance profile.

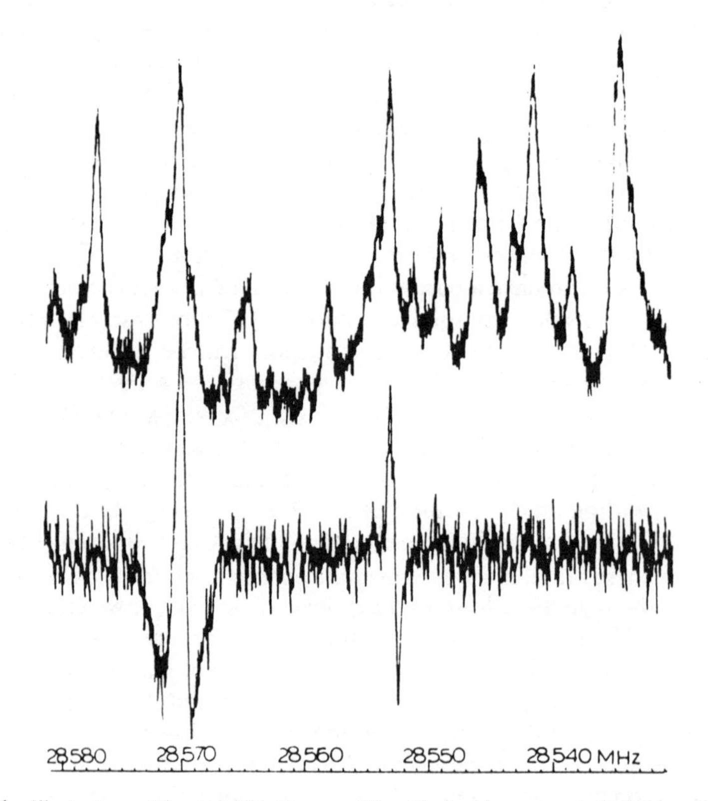

Fig. 7.11 Illustration of the simplification possible with double resonance techniques. The upper trace is taken with the Stark modulation. The lower trace is taken with frequency modulated microwave–microwave double resonance, and only two signals are apparent. From Pearson et al. [91].

The power of the technique is illustrated by the assignment of 2,6-difluoropyridine by Stiefvater [99], who has assigned some 10 isotopic forms of the molecule, most of which were studied in their natural abundances. This can be difficult with use of Stark modulation only. Initial calculations indicated some key double resonance connections. For a detailed description of the process of the assignments, the reader should consult the original work [99].

A particularly convenient double resonance technique for identifying microwave transitions makes use of radiofrequency pump radiation. Transitions in this region arise from accidental near degeneracies between levels of asymmetric tops and transitions between K-type doublets, as well as l-type doublets of linear molecules. Double resonance experiments with radio frequencies were first carried out by Autler and Townes [69]. More recently Wodarczyk and Wilson [82] have discussed the use of the technique for the assignment of complex asymmetric top spectra and have described details of the technique. The strong radiofrequency source is amplitude modulated with square-wave modulation. The modulated radiofrequency radiation can be introduced into the sample cell by its application across a conventional Stark waveguide cell as is done with a Stark square wave generator. Hence, both Stark spectroscopy and radiofrequency–microwave spectroscopy can be carried out with the same waveguide cell. Radiofrequency fields on the order of 3 to 15 V/cm are sufficient. Without paying special attention to impedance matching, the upper limit for the RF signal is about 100 MHz, and a RF power amplifier is usually required. As before, only those signal transitions will be observed which are modulated via the pump radiation; therefore, the pump and signal transitions must share a common energy level. Since the radiofrequency field is "off" for half of the modulation cycle and "on" for the other half, the zero-field absorption signal and the double resonance signal can be displayed by use of phase-sensitive detection with opposite polarity.

In the three-level systems of Fig. 7.9, when the energy levels of the pump transitions are close enough, the pump frequency can be in the RF region and RF pumping can be employed. Closely spaced levels are frequently encountered for slightly asymmetric tops. A typical energy level diagram for K-type doublets of a slightly asymmetric rotor is shown in Fig. 7.12. Two radiofrequency transitions (ν_{p1}^0, ν_{p2}^0) and two microwave transitions (ν_{s1}^0, ν_{s2}^0) are possible. Note that both signal transitions have energy levels in common with either pump transition. When ν_p is near ν_{p1}^0 or ν_{p2}^0, the three-level situation applies. If the pump frequencies are close enough, say $(\nu_{p2}^0 - \nu_{p1}^0) \lesssim 15$ MHz, the pump radiation over a range of pump frequencies will be near resonance to both transitions. For such a four-level system (Fig. 7.12) with ν_p near ν_{p1}^0, or ν_{p2}^0, a creeper and a main line appear about both zero-field signal frequencies ν_{s1}^0 and ν_{s2}^0. Figure 7.13 shows the line shape changes of two adjacent microwave double resonance signals arising from a pair of K-doublet transitions as the pump frequency is varied from slightly above ν_{p1}^0 to slightly below ν_{p2}^0. The two creepers that face each other, clearly approach each other, merge at $\bar{\nu}_p$, and again approach the zero-field positions as ν_p nears ν_{p2}^0. This distinctive mirror-image pattern can be

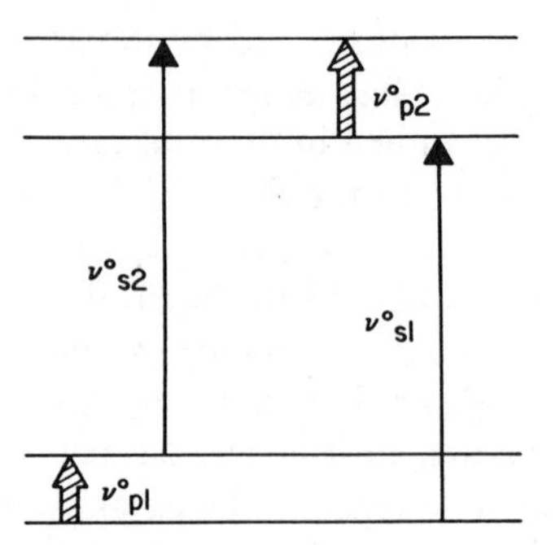

Fig. 7.12 A four-level double resonance system. A K-doublet energy level arrangement of a typical slightly asymmetric top is shown.

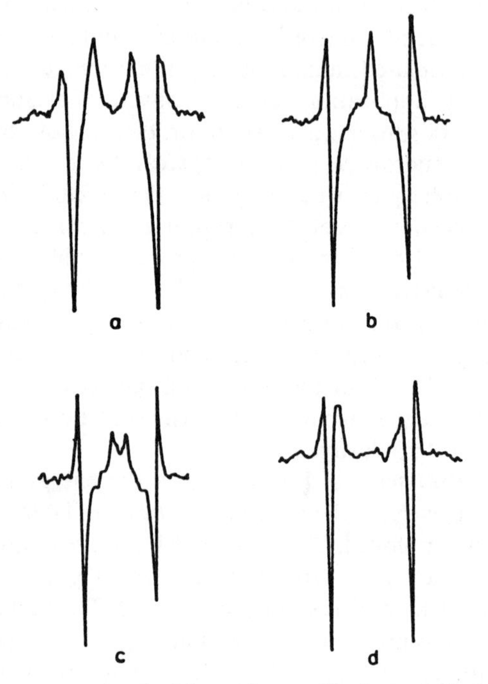

Fig. 7.13 The double resonance signals of skew-1-butene. The $J=3\rightarrow4$, $K=2$ transitions are shown at different fixed pump frequencies, where $\nu_{p1}^0=2.04$ MHz and $\nu_{p2}^0=6.10$ MHz. (*a*) $\nu_p=3.00$ MHz; (*b*) $\nu_p=4.05$ MHz; (*c*) $\nu_p=4.50$ MHz; (*d*) $\nu_p=6.00$ MHz. From Wodarczyk and Wilson [82].

very useful in assignment of these transitions. The frequency at which the double resonance creepers from the adjacent double resonance signals of the four-level system merge is given by [82]

$$\bar{\nu}_p \simeq \frac{\nu_{p1}^0+\nu_{p2}^0}{2} \tag{7.92}$$

This four-level effect can be used to measure ν_{p1}^0, and ν_{p2}^0 when direct measurement by tuning ν_p for a symmetrical line profile is not satisfactory (as with a usual three-level system). Here $\bar{\nu}_p$, ν_{s1}^0, ν_{s2}^0 are measured and (7.92) and the relation

$$\nu_{s1}^0 + \nu_{p2}^0 = \nu_{p1}^0 + \nu_{s2}^0, \tag{7.93}$$

which follows from Fig. 7.12, are solved simultaneously for ν_{p1}^0 and ν_{p2}^0. These frequencies provide additional data for the determination of the rotational constants.

In the radiofrequency region, higher-order terms affecting the intensities and positions of the double resonance components can be important although they are negligible in the microwave region. This is discussed elsewhere [82]. As before, the pump frequency does not have to be on resonance, and double resonance signals can be observed over a range of pump frequencies.

Radiofrequency pumping is also particularly useful as an aid in the assignment of vibrational, excited-state transitions. Often the shift of the frequency of the pump transition with vibrational state is very small, and by pumping at the appropriate frequency for the ground state a whole series of excited-state transitions can be detected as the signal frequency is scanned [100]. Other illustrative application of RFMDR may also be cited [101, 102].

6 EFFECTS OF VIBRATION

The vibrational motions of a molecule, which are ignored in the rigid-rotor approximation, affect the moments of inertia and result in a dependence of the moments of inertia on the vibrational state. The spectroscopic constants ascertained from an analysis of the rotational spectra are hence effective constants. The dependence of the effective rotational constants on the vibrational quantum numbers may be expressed by

$$A_v = A_e - \sum_i \alpha_i^a (v_i + \tfrac{1}{2}) \tag{7.94}$$

$$B_v = B_e - \sum_i \alpha_i^b (v_i + \tfrac{1}{2}) \tag{7.95}$$

$$C_v = C_e - \sum_i \alpha_i^c (v_i + \tfrac{1}{2}) \tag{7.96}$$

where A_e, etc., are the equilibrium constants and where the vibrational state v is specified by the quantum numbers $(v_1, v_2, \ldots, v_i, \ldots)$ with v_i the vibrational quantum number of the ith vibration, and where the sum is over the various vibrations.

The vibration–rotation constants depend on both the potential constants and the Coriolis coupling constants. To account for various effects of vibration in the rotational spectrum, and to obtain explicit expressions for the interaction constants such as the α_i's in terms of the fundamental molecular constants, a general rotation–vibration Hamiltonian is required. In this section, the appropriate Hamiltonian is described and a perturbation technique for obtaining the energies to the desired order is outlined. Nonrigidity effects arising from centrifugal distortion also shift the rigid rotor energy levels. These corrections are discussed further in Chapter VIII. Large amplitude motions are treated in Chapter XII and in Chapter VIII, Section 9.

Rotation–Vibration Hamiltonian

If in addition to rotational motion vibrational motion is included, then an internal angular momentum π_α relative to the molecule-fixed axis system, will arise from the rotation–vibration interaction. If P_α signifies the overall angular momentum components, then the rotational Hamiltonian may be generalized as

$$\mathscr{H}=\tfrac{1}{2}\hbar^2 \sum_{\alpha,\beta} \mu_{\alpha\beta}(P_\alpha-\pi_\alpha)(P_\beta-\pi_\beta)+\tfrac{1}{2}\sum_k P_k^2+V \tag{7.97}$$

where $\mu_{\alpha\beta}$ (α, $\beta=x$, y, z) is an element of the inverse moment of inertia tensor and is here a function of the vibrational coordinates, which may be taken as the $3N-6$ normal coordinates Q_k. The P_k is the vibrational momentum conjugate to Q_k ($P_k=-i\hbar\partial/\partial Q_k$), and V a vibrational potential function.

A rigorous derivation of this Hamiltonian for a nonlinear molecule starts with construction of the classical Hamiltonian, employing a center-of-mass, rotating, coordinate system and a set of coordinates that specify the positions of the atoms relative to each other in the rotating system. This is followed by transformation to a quantum mechanical Hamiltonian operator. The derivation of the general rotation–vibration Hamiltonian has been discussed by Wilson and Howard [103], Darling and Dennison [104], Nielsen [105], and more recently further simplified by Watson [106] to the form given in (7.97). This is a consequence of the commutator relation

$$\sum_\alpha [\pi_\alpha, \mu_{\alpha\beta}]=0 \tag{7.98}$$

A derivation of (7.97) starting with the Schrödinger equation has been given by Louck [107]. For further references discussing aspects of the rotation–vibration Hamiltonian, see Chapter VIII and [108] through [113].

The general Hamiltonian contains only terms quadratic in P_α, whereas we have seen in previous chapters that the effective rotational Hamiltonian contains higher terms in the angular momentum. The usual analysis procedure entails expanding $\mu_{\alpha\beta}$ and V in terms of the vibrational coordinates followed by a perturbation treatment to obtain the energies correct to a given order.

The zeroth-order vibrational Hamiltonian involves only the quadratic terms from V

$$\mathscr{H}^0_{\text{vib}}=\frac{1}{2}\sum_k (P_k^2+\lambda_k Q_k^2) \tag{7.99}$$

with $\lambda_k = 4\pi^2 c^2 \omega_k^2$ and ω_k the harmonic vibrational frequency in cm^{-1} units of the kth normal mode Q_k. It is convenient to employ the dimensionless normal coordinates q_k and momentum p_k defined as follows

$$q_k = \gamma_k^{1/2} Q_k,\ p_k = \frac{P_k}{\gamma_k^{1/2}\hbar},\ \gamma_k = \frac{\lambda_k^{1/2}}{\hbar} \tag{7.100}$$

thus

$$\mathscr{H}^0_{\text{vib}} = \frac{1}{2}\sum_k \omega_k (p_k^2 + q_k^2) \tag{7.101}$$

These and other terms in the Hamiltonian (7.97) will be divided by hc to express all terms in consistent wavenumber units (cm^{-1}). The eigenfunctions of $\mathscr{H}^0_{\text{vib}}$ are product functions of harmonic oscillator eigenfunctions

$$\Psi_v = \prod_{k=1}^{3N-6} \psi_{v_k}(q_k) \tag{7.102}$$

where v represents the set of vibrational quantum numbers v_k specifying the vibrational state and eigenvalues

$$\mathscr{H}^0_{\text{vib}}\Psi_v = E_v^0 \Psi_v = \sum_k \omega_k (v_k + \tfrac{1}{2})\psi_v \tag{7.103}$$

For a perturbation treatment, matrix elements of q_k and p_k will be required

$$(v_k|q_k|v_k+1) = [(v_k+1)/2]^{1/2},\ (v_k|q_k|v_k-1) = [v_k/2]^{1/2} \tag{7.104}$$

$$(v_k|p_k|v_k+1) = -i[(v_k+1)/2]^{1/2},\ (v_k|p_k|v_k-1) = i[v_k/2]^{1/2} \tag{7.105}$$

Other matrix elements may be readily obtained from tabulations given elsewhere [111].

The general vibrational potential function is expressed as

$$V = \frac{1}{2}\sum_k \omega_k q_k^2 + \frac{1}{6}\sum_{lmn} k_{lmn} q_l q_m q_n + \cdots + U \tag{7.106}$$

where k_{lmn} are the cubic anharmonic potential constants. The next terms in the expansion would involve the quartic constants k_{rstu}. The term U is a pseudopotential term of quantum mechanical origin [106]

$$U = -\frac{\hbar^2}{8}\sum_\alpha \mu_{\alpha\alpha} \tag{7.107}$$

which depends only on the vibrational coordinates and represents a small mass-dependent correction to the vibrational potential energy. This term may be neglected except for very precise work.

The effective vibrational angular momentum in (7.97) is defined by (in units of $\hbar$)

$$\pi_\alpha = \sum_{kl} \zeta_{kl}^{(\alpha)} \frac{Q_k P_l}{\hbar} = \sum_{kl} \zeta_{kl}^{(\alpha)} q_k p_l \left(\frac{\omega_l}{\omega_k}\right)^{1/2} \tag{7.108}$$

where $\zeta_{kl}^{(\alpha)}$ ($\zeta_{kl}^{(\alpha)}=-\zeta_{lk}^{(\alpha)}$ and $\zeta_{kl}^{(\alpha)}=0$ if $k=l$) are the Coriolis coupling constants, coupling Q_k and Q_l via rotation about α (see Chapter XIII for further discussion).

The $\mu_{\alpha\beta}$ may be expanded about the equilibrium configuration as

$$\mu_{\alpha\beta}=\mu_{\alpha\beta}^{(e)}+\sum_{k}\mu_{\alpha\beta}^{(k)}q_k+\sum_{kl}\mu_{\alpha\beta}^{(kl)}q_kq_l+\cdots \tag{7.109}$$

which is assumed to converge rapidly (small oscillations approximation). Here

$$\mu_{\alpha\beta}^{(e)}=\frac{I_{\alpha\beta}^{e}}{I_{\alpha}^{e}I_{\beta}^{e}} \tag{7.110}$$

and the derivatives are given by [106, 108]

$$\mu_{\alpha\beta}^{(k)}=-\frac{a_k^{(\alpha\beta)}}{I_\alpha^e I_\beta^e \gamma_k^{1/2}} \tag{7.111}$$

$$\mu_{\alpha\beta}^{(kl)}=\frac{3}{8}\sum_{\gamma}\frac{a_k^{(\alpha\gamma)}a_l^{(\beta\gamma)}+a_k^{(\beta\gamma)}a_l^{(\alpha\gamma)}}{I_\alpha^e I_\beta^e I_\gamma^e(\gamma_k\gamma_l)^{1/2}} \tag{7.112}$$

with $\gamma_k=\lambda_k^{1/2}/\hbar$ and the inertial derivatives are

$$\alpha_k^{(\alpha\beta)}=a_k^{(\beta\alpha)}=\left(\frac{\partial I_{\alpha\beta}}{\partial Q_k}\right)_e \tag{7.113}$$

Since the internal coordinates R_i are related to the normal coordinates by the transformation $R_i=\sum L_{ik}Q_k$, (7.113) can be written in the form

$$a_k^{(\alpha\beta)}=\sum_{i}J_{\alpha\beta}^{(i)}L_{ik} \tag{7.114}$$

$$J_{\alpha\beta}^{(i)}=\left(\frac{\partial I_{\alpha\beta}}{\partial R_i}\right)_e \tag{7.115}$$

The L_{ik} are obtained from a solution of the vibrational eigenvalue secular equation (see Appendix H). Derivation of expressions for $J_{\alpha\beta}^{(i)}$ are given in Chapter VIII.

Insertion of (7.109) and (7.106) into (7.97) gives for the expanded rotation–vibration Hamiltonian

$$\begin{aligned}\mathscr{H}=&\frac{\hbar^2}{2}\sum_{\alpha,\beta}\left\{\mu_{\alpha\beta}^{(e)}+\sum_{k}\mu_{\alpha\beta}^{(k)}q_k+\sum_{k,l}\mu_{\alpha\beta}^{(kl)}q_kq_l+\cdots\right\}\\&\times\{P_\alpha P_\beta-(P_\alpha\pi_\beta+\pi_\alpha P_\beta)+\pi_\alpha\pi_\beta\}\\&+\frac{1}{2}\sum_{k}\omega_k(p_k^2+q_k^2)+\frac{1}{6}\sum_{lmn}k_{lmn}q_lq_mq_n+\cdots\end{aligned} \tag{7.116}$$

Clearly the expansion can be carried to higher order if required. The various terms resulting in the expanded Hamiltonian may be classified with the notation $\mathscr{H}_{nm}$, where n is the degree in the vibrational operators q^n (q and/or p for π_α, $n=2$)

and m is the degree in the rotational operators P^m,

$$\mathscr{H} = \mathscr{H}_{02}(\kappa^2) + \mathscr{H}_{20}(\kappa^0) + \mathscr{H}_{21}(\kappa^2) + \mathscr{H}_{12}(\kappa^3) + \mathscr{H}_{22}(\kappa^4) + \mathscr{H}_{30}(\kappa) + \mathscr{H}_{31}(\kappa^3) + \cdots \quad (7.117)$$

The order of magnitude of the coefficients is specified in terms of the Born-Oppenheimer expansion parameter $\kappa = (m_e/m_n)^{1/4} \cong \frac{1}{10}$ as suggested by Oka [114] and further clarified by Pedersen [115]. This parameter measures the size of a term relative to the vibrational energy $\kappa^{n+2m-2}E_v$. The κ-order is also indicated in (7.117). The rigid rotor rotational constants are thus smaller than E_v by a factor κ^2, and smaller than a typical electronic energy E_e by κ^4. This classification allows one to choose the particular terms required to obtain the energies by means of a perturbation treatment to a particular κ^P. Terms such as $\mu_{\alpha\beta}^{(e)}\pi_\alpha\pi_\beta$ of order κ^2, which represent a purely vibrational correction, are often not required. Furthermore, since the Hamiltonian must be invariant to the symmetry operations of the particular molecular point group, many terms will be found to vanish in a specific application. For a term such as $\mathscr{H}_{12} \sim a_k^{(\alpha\beta)} q_k P_\alpha P_\beta$, a given $a_k^{(\alpha\beta)}$ can be nonvanishing only if the product of the symmetry species of q_k, P_α and P_β is totally symmetric. Similarly, a Coriolis coupling constant $\zeta_{kl}^{(\alpha)}$ is nonvanishing only when $q_k p_l P_\alpha$ belongs to the totally symmetric species. An excellent discussion of the molecular symmetry group for both rigid and nonrigid molecules has been given by Bunker [116] (see also references cited therein). In addition, a number of sum rules exist for the inertial derivatives $a_k^{(\alpha\beta)}$ (e.g., $\sum_k (a_k^{(\alpha\alpha)})^2 = 4I_\alpha^e$) and the Coriolis coupling constants $\zeta_{kl}^{(\alpha)}$ which can be used to simplify the general interactions of (7.119)–(7.121). These relations are discussed elsewhere [117–122]. Similarly, isotopic relations between such parameters are very useful and have been reported [123].

We now consider some of the more important terms of (7.116). The zeroth-order vibrational Hamiltonian $\mathscr{H}_{20}$ of order κ^0 has already been defined in (7.101). The zeroth-order or rigid-rotor rotational Hamiltonian is given by the leading term of $\mu_{\alpha\beta}$

$$\mathscr{H}_{02}(\kappa^2) = \frac{\hbar^2}{2} \sum_{\alpha\beta} \mu_{\alpha\beta}^{(e)} P_\alpha P_\beta = \sum_\alpha B_\alpha P_\alpha^2 \quad (7.118)$$

where the rotational constant $B_\alpha = \hbar^2/2hcI_\alpha^e$ is in wavenumber units. Since the principal axes system is chosen for the equilibrium configuration, $I_{\alpha\beta}^e(a \neq \beta)$ will vanish. In a similar way we find, for example,

$$\begin{aligned}\mathscr{H}_{21}(\kappa^2) &= -2 \sum_\alpha B_\alpha \pi_\alpha P_\alpha \\ &= \sum_{k,l} \left[-2 \left(\frac{\omega_l}{\omega_k}\right)^{1/2} \sum_\alpha B_\alpha \zeta_{kl}^{(a)} P_\alpha \right] q_k p_l \\ &= \sum_{k,l} R_k^l q_k p_l \end{aligned} \quad (7.119)$$

$$\mathscr{H}_{12}(\kappa^3)=\sum_k\left[-\sum_{\alpha\beta}C_k^{(\alpha\beta)}\omega_k P_\alpha P_\beta\right]q_k$$

$$=\sum_k R_k q_k \tag{7.120}$$

$$\mathscr{H}_{22}(\kappa^4)=\sum_{kl}\left[\frac{3}{8}\sum_{\alpha\beta\gamma}\omega_k\omega_l(C_k^{(\alpha\gamma)}C_l^{(\beta\gamma)}+C_k^{(\beta\gamma)}C_l^{(\alpha\gamma)})P_\alpha P_\beta/B_\gamma\right]q_k q_l$$

$$=\sum_{k,l}R_{kl}q_k q_l \tag{7.121}$$

$$\mathscr{H}_{30}(\kappa)=\frac{1}{6}\sum_{lmn}k_{lmn}q_l q_m q_n \tag{7.122}$$

The definitions of the R-operators [124] are enclosed in brackets. The forms of $\mathscr{H}_{nm}$ in terms of the R's are convenient for the perturbation treatment to be discussed. The dimensionless parameter $C_k^{(\alpha\beta)}$ is defined by

$$C_k^{(\alpha\beta)}=C_k^{(\beta\alpha)}=\frac{a_k^{(\alpha\beta)}}{2I_\alpha^e I_\beta^e \gamma_k^{3/2}} \tag{7.123}$$

Recall that the operators q_k, p_k, P_α, π_α are dimensionless and that the $\mathscr{H}_{nm}$ are in wavenumber units. The $\mathscr{H}_{21}$ term represents a Coriolis interaction, $\mathscr{H}_{12}$ and $\mathscr{H}_{22}$ represent correction terms to the rigid rotor which arise from the linear and quadratic dependence of μ on the vibrational coordinates, respectively, and $\mathscr{H}_{30}$ represents an anharmonicity correction to the potential energy which can give rise to anharmonic resonance effects.

For a perturbation calculation of the rotation–vibration energies, this Hamiltonian may be subjected to successive contact transformations chosen to effectively diagonalize the Hamiltonian in the vibrational quantum numbers, thereby giving an effective rotational Hamiltonian for each vibrational state. This procedure has been extensively discussed in the literature [108, 124–126]. Detailed expressions (although somewhat complicated) of the transformed Hamiltonian and appropriate matrix elements for various types of rotors have been compiled in a monograph by Amat et al. [108] to a high order of approximation. The Van Vleck transformation has also been applied [103, 127]. An alternate approach uses standard perturbation theory, to the appropriate order, with the harmonic oscillator functions as basis functions to effectively diagonalize $\mathscr{H}$ in v and thus provide the effective rotational Hamiltonian $\tilde{\mathscr{H}}$ for a given vibrational state. The κ-order and the type of rotational operator dependence sought are used as a guide for choice of the appropriate terms to treat in the expanded Hamiltonian. A few examples are given here to illustrate the procedure. The first- and second-order perturbation correction are expressed by

$$(v|\tilde{\mathscr{H}}|v)=(v|\mathscr{H}'|v) \tag{7.124}$$

$$(v|\tilde{\mathscr{H}}|v)=\sum_{v'}\frac{(v|\mathscr{H}'|v')(v'|\mathscr{H}'|v)}{E_v^0-E_{v'}^0} \tag{7.125}$$

where $\mathscr{H}'$ represents the perturbing Hamiltonian and, in general, contains all the terms of (7.117) except the zeroth-order vibrational Hamiltonian $\mathscr{H}_{20}$. The transformed Hamiltonian operator $\tilde{\mathscr{H}}_{nm}$ also labeled by n and m are given by the diagonal elements $(v|\tilde{\mathscr{H}}|v)$ via the foregoing equations. This type of treatment requires that the spacing of the rotational energy levels be small compared to the vibrational spacings and that no near-degenerate vibrational levels are connected by the perturbation. This approach is analogous to the Van Vleck perturbation technique (Appendix C) under similar assumptions. The rotational operators in $\mathscr{H}'$ are treated as noncommuting parameters independent of v. Third- and fourth-order perturbation formulas appropriate to this treatment, which take into account that the order of the $\mathscr{H}'_{vv'}$ elements in the perturbation formulae are important in view of the noncommuting character of the rotational operators, have been given by Georghiou [128].

In a particular rotation–vibration interaction, only those terms in $\mathscr{H}'$ that contribute to the interaction need be retained. The leading term, for example, representing the effects of centrifugal distortion (considered further in Chapter VIII) involves purely rotational terms of P^4. Thus the equilibrium quartic distortion Hamiltonian is specified by the transformed Hamiltonian $\tilde{\mathscr{H}}_{04}$ of order κ^6. A second-order treatment of $\mathscr{H}_{nm}(\kappa^r)$ and $\mathscr{H}_{n'm'}(\kappa^s)$ gives a contribution of order κ^{r+s} and reduces the power of the vibrational operators by two. Thus the desired form of $\tilde{\mathscr{H}}_{04}$ arises from a second-order treatment of $\mathscr{H}_{12}(\kappa^3)$. A discussion regarding the selection of the terms to be considered to obtain a particular $\tilde{\mathscr{H}}_{nm}$ has been given by Mills [129]. Setting $\mathscr{H}'=\mathscr{H}_{12}$ in (7.125) we have, because of (7.104),

$$(v|\tilde{\mathscr{H}}_{04}|v)=\sum_k R_k R_k\left[\frac{|(v_k|q_k|v_k+1)|^2}{-\omega_k}+\frac{|(v_k|q_k|v_k-1)|^2}{\omega_k}\right]$$

$$=-\frac{1}{2}\sum_k\sum_{\alpha\beta\gamma\delta} C_k^{(\alpha\beta)}C_k^{(\gamma\delta)}\omega_k P_\alpha P_\beta P_\gamma P_\delta$$

$$=\frac{1}{4}\sum_{\alpha\beta\gamma\delta}\tau_{\alpha\beta\gamma\delta}P_\alpha P_\beta P_\gamma P_\delta \tag{7.126}$$

which gives the correct rotational operator dependence; the distortion coefficient is defined by

$$\tau_{\alpha\beta\gamma\delta}=-2\sum_k \omega_k C_k^{(\alpha\beta)}C_k^{(\gamma\delta)} \tag{7.127}$$

This may be expressed in terms of $a_k^{(\alpha\beta)}$ from (7.123).

When the vibrational and rotational energy differences are not widely separated, this perturbation treatment will be inappropriate. In this case, a more appropriate basis for the perturbation treatment are the product functions of $\mathscr{H}_{20}$ and $\mathscr{H}_{02}$, that is, $\Psi_v\Psi_{J\tau}$. The denominators in (7.126) will then also involve explicitly the rotational energy difference. An example of this type of effect is provided by HNCO [130] where the $K_{-1}=7$ level of the ground state is very close in energy to the $K_{-1}=6$ level of the lowest bending vibrational state $v_5=1$.

The vibrational correction to the rotational constants $\alpha_k^{(\beta)}$ are coefficients of $(v_k+\frac{1}{2})P_\beta^2$ which appears in the transformed Hamiltonian $\tilde{\mathscr{H}}_{22}(\kappa^4)$ since $\frac{1}{2}(v_k|q_k^2+p_k^2|v_k)=v_k+\frac{1}{2}$. This type of term results [129] from a first-order treatment of $\mathscr{H}_{22}(\kappa^4)$, and second-order treatments of $\mathscr{H}_{21}(\kappa^2)\times\mathscr{H}_{21}(\kappa^2)$ and $\mathscr{H}_{30}(\kappa)\times\mathscr{H}_{12}(\kappa^3)$. The origin of these terms is clear from inspection of the individual $\mathscr{H}_{nm}$, and the three contributions may be classified, respectively, as a harmonic, Coriolis, and anharmonic contribution. The resulting expression is [131]

$$\alpha_k^{(\beta)}=\alpha_k^{(\beta)}(\text{har})+\alpha_k^{(\beta)}(\text{Cor})+\alpha_k^{(\beta)}(\text{anhar})$$

$$=-\frac{3}{4}\omega_k^2\sum_\gamma\frac{(C_k^{(\beta\gamma)})^2}{B_\gamma}-\frac{2B_\beta^2}{\omega_k}\sum_l{}'(\zeta_{kl}^{(\beta)})^2\frac{3\omega_k^2+\omega_l^2}{\omega_k^2-\omega_l^2}-\frac{1}{2}\sum_l k_{kkl}C_l^{(\beta\beta)} \tag{7.128}$$

A first-order treatment of $\mathscr{H}_{22}(\mathscr{H}'=\mathscr{H}_{22})$, for example, readily yields the first term of (7.128).

If it is desirable to obtain the definition of the equilibrium coefficients of the sextic distortion constants, which are coefficients of P^6 of order κ^{10}, five terms [124, 128] are found to contribute to $\tilde{\mathscr{H}}_{06}$ which require a laborious calculation. The appropriate terms arise from a third-order treatment of $\mathscr{H}_{12}(\kappa^3)\times\mathscr{H}_{12}(\kappa^3)\times\mathscr{H}_{22}(\kappa^4)$ and a fourth-order treatment of $\mathscr{H}_{30}(\kappa^1)\times\mathscr{H}_{12}(\kappa^3)\times\mathscr{H}_{12}(\kappa^3)\times\mathscr{H}_{12}(\kappa^3)$, $\mathscr{H}_{21}(\kappa^2)\times\mathscr{H}_{21}(\kappa^2)\times\mathscr{H}_{12}(\kappa^3)\times\mathscr{H}_{12}(\kappa^3)$, $\mathscr{H}_{02}(\kappa^2)\times\mathscr{H}_{02}(\kappa^2)\times\mathscr{H}_{12}(\kappa^3)\times\mathscr{H}_{12}(\kappa^3)$, and $\mathscr{H}_{21}(\kappa^2)\times\mathscr{H}_{02}(\kappa^2)\times\mathscr{H}_{12}(\kappa^3)\times\mathscr{H}_{12}(\kappa^3)$. Each of these contributions is of order of magnitude κ^{10}. The presence of n products of $\mathscr{H}_{nm}$ indicates the use of nth-order perturbation theory. If $\mathscr{H}'=\mathscr{H}_{12}+\mathscr{H}_{22}$ is set in the appropriate third-order perturbation formula, which will give terms of the type $(\mathscr{H}_{12}\times\mathscr{H}_{12}\times\mathscr{H}_{22})/(\Delta E_{\text{vib}}^0)^2$, the contribution of the first-indicated term may be evaluated [128]. Other product terms which might appear to contribute can be shown either to vanish or to have inappropriate rotational operator dependence. Also contributions of apparent degree greater than six involving $\mathscr{H}_{02}$, which is diagonal in v, can be reduced to sixth degree, one power for each $\mathscr{H}_{02}$ [128]. In fact, it is possible to drop either two powers of q or p, or one power of P for each ΔE_v^0 in the denominator of a perturbation formula [129]. Hence, the terms to include are not always obvious and caution must be exercised that all the appropriate contributions have been included.

For symmetric tops, complications arise from the presence of degenerate vibrations, but the analysis is similar to that just described. Discussions of vibration–rotation effects are given in Chapter VI, and further treatments [110, 129, 132–134] and illustrative applications may be cited [135–138]. Coriolis interactions in asymmetric tops are treated in the next section.

Coriolis Perturbations

Because of the lower rotational symmetry of asymmetric rotors there are no inherent degenerate modes of vibration, and hence there are no first-order Coriolis effects such as were discussed previously for symmetric rotors in a degenerate vibration state. However, higher-order Coriolis perturbations of the rotational energy level structure are possible and can be particularly significant when there is an accidental vibrational degeneracy. This condition has been discussed by Wilson [139], and the treatment may also be applied with appropriate modification to symmetric rotors having two nondegenerate nearby vibrational states.

The Coriolis interaction is specified by $\mathscr{H}_{21}$ given in (7.119), which we write here in the form

$$\mathscr{H}' = -\hbar^2 \sum_{\alpha} \frac{\pi_\alpha P_\alpha}{I_\alpha} \tag{7.129}$$

The $\pi_\alpha(\alpha = x, y, z)$, which are components of the vibrational angular momentum along the molecule fixed axes, are defined as

$$\pi_\alpha = \sum_{kl} \zeta_{kl}^{(\alpha)} q_k p_l \left(\frac{\omega_l}{\omega_k}\right)^{1/2} \tag{7.130}$$

in terms of the dimensionless q_k and p_l. The previous perturbation treatment is not adequate when near-degenerate vibrational levels are connected by the perturbation term; rather, the energy matrix of the interacting levels is diagonalized directly.

The Hamiltonian including coupling between rotation and vibration may be written as

$$\mathscr{H} = \frac{\hbar^2}{2}\left(\frac{P_x^2}{I_x} + \frac{P_y^2}{I_y} + \frac{P_z^2}{I_z}\right) - \hbar^2\left(\frac{\pi_x P_x}{I_x} + \frac{\pi_y P_y}{I_y} + \frac{\pi_z P_z}{I_z}\right) + \mathscr{H}_v \tag{7.131}$$

where the first term is the rigid rotor Hamiltonian, $\mathscr{H}_v$ is the vibrational Hamiltonian, and the second term represents the coupling between rotation and vibration.

The energy matrix may be constructed with basis functions which are products of symmetric rotor functions and vibrational functions. The vibrational functions are solutions of the vibrational Hamiltonian and may, to a good approximation, be taken as harmonic oscillator functions. The energy matrix will have both diagonal and off-diagonal elements in v. The off-diagonal elements in v come from the operator π_α in the coupling terms. For nondegenerate states the diagonal elements of π_α vanish, and hence the Coriolis interaction does not contribute elements diagonal in v. Part of the infinite energy matrix is illustrated in Figure 7.14. When the two vibrational states v and v' have nearly the same energy $E_v \simeq E_{v'}$ the rotation–vibration coupling can be large. If the other vibrational states have energies significantly different

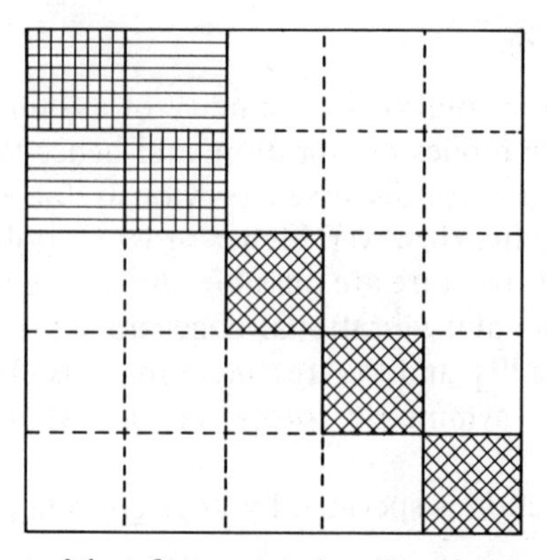

Fig. 7.14 Illustration of part of the infinite matrix for $\mathscr{H}$ in the basis $\psi_{JK}\psi_v$. The doubly shaded blocks contain the matrix elements connecting the two vibrational states of nearly the same energy. The unshaded blocks, which represent the connections between states of widely different energy, may be neglected. The 2×2 block submatrix may hence be factored from the remaining matrix. This submatrix is diagonal in J and has $2(2J+1)$ rows and columns labeled by the values of $K(0, \pm1, \ldots, \pm J)$ and $v(v, v')$.

from E_v and $E_{v'}$, the connections between these states and v or v' may be neglected, and only the elements connecting the nearly degenerate states v and v' need be retained. The submatrix composed of the diagonal and off-diagonal elements of the two interacting vibrational states v, v' can thus be separated from the rest of the matrix (see Fig. 7.14). Diagonalization of this 2×2 block submatrix gives the energy levels. Employing Table 7.2 we find that the nonvanishing matrix elements of the secular equation are:

$$(J, K, v|J, K, v)=K^2+\sigma_v-\sigma \tag{7.132}$$

$$(J, K, v'|J, K, v')=K^2+\sigma_{v'}-\sigma \tag{7.133}$$

$$\begin{aligned}(J, K, v|J, K\pm2, v)&=(J, K, v'|J, K\pm2, v')\\&=-\tfrac{1}{2}b\{[J^2-(K\pm1)^2][(J+1)^2-(K\pm1)^2]\}^{1/2}\end{aligned} \tag{7.134}$$

$$(J, K, v|J, K, v')=-(J, K, v'|J, K, v)=iG_zK \tag{7.135}$$

$$\begin{aligned}(J, K, v|J, K\pm1, v')&=-(J, K, v'|J, K\pm1, v)\\&=\tfrac{1}{2}(iG_y\mp G_x)[(J\mp K)(J\pm K+1)]^{1/2}\end{aligned} \tag{7.136}$$

where here

$$\begin{aligned}a&=\left(\frac{1}{2c}\right)\left(\frac{1}{I_x}+\frac{1}{I_y}\right)\\b&=\left(\frac{1}{2c}\right)\left(\frac{1}{I_x}-\frac{1}{I_y}\right)\\c&=\frac{1}{I_z}-\frac{1}{2}\left(\frac{1}{I_x}+\frac{1}{I_y}\right)\end{aligned} \tag{7.137}$$

$$\sigma = \left(\frac{2}{\hbar^2 c}\right)E - J(J+1)a \tag{7.138}$$

$$\sigma_v = \left(\frac{2}{\hbar^2 c}\right)E_v$$

$$G_x = \left(\frac{2}{cI_x}\right)\sum_{k,l} \zeta_{kl}^{(x)} I_{kl}$$

$$G_y = \left(\frac{2}{cI_y}\right)\sum_{k,l} \zeta_{kl}^{(y)} I_{kl} \tag{7.139}$$

$$G_z = \left(\frac{2}{cI_z}\right)\sum_{k,l} \zeta_{kl}^{(z)} I_{kl}$$

$$I_{kl} = i \int_i \psi_{v'}^* q_k p_l \left(\frac{\omega_l}{\omega_k}\right)^{1/2} \psi_v \, d\tau \tag{7.140}$$

In the above matrix elements of the secular equation it has been assumed that the effective moments of inertia are the same in the two vibrational states.

The G_α may be evaluated explicitly if sufficient information on the molecular force field is known, or they may be treated as empirical parameters. For G_α to be nonvanishing, $\psi_v\psi_{v'}$ must have the same symmetry as π_α under the molecular point group. With asymmetric rotors of symmetry V, V_h, and C_{2v}, no more than one of the G_α will be nonvanishing for a given pair of states v, v'. Furthermore, by use of a new set of basis functions $S_{JK\gamma}\psi_v$ and $S_{JK\gamma}\psi_{v'}$, where the rotational functions are the Wang functions, the matrix may be factored into four smaller submatrices [139].

For I_{kl} to be of reasonable magnitude, the two states v and v' are required to differ in only two of their vibrational quantum numbers v_k, and these two quantum numbers must change by one unit from v to v'. For the levels to have nearly the same energy, this essentially limits the state v' to one in which $v_k' = v_k + 1$ and $v_l' = v_l - 1$ for a pair of normal modes k and l. Evaluating I_{kl} with harmonic oscillator functions [See (7.104) and (7.105)] yields

$$I_{kl} = \frac{1}{2}\left[(v_k+1)v_l\left(\frac{\omega_l}{\omega_k}\right)\right]^{1/2} \tag{7.141}$$

where ω_k is the frequency of the kth vibration.

Some examples of Coriolis interactions studied in asymmetric rotors are SO_2F_2 [140], $CH_2(CN)_2$ [141], F_2CO [142], H_2CO [143], FNO_2 [144], O_3 [145], OF_2 [146], HNCS [147], HNCO [148], and $(CH_3)_2CCCH_2$ [149]. Such studies give information on Coriolis coupling constants and on the separation of the interacting vibrational states.

7 ASYMMETRIC ROTOR STRUCTURES

Most applications must be postponed until treatments of centrifugal distortion, hyperfine structure, internal rotation, Stark and Zeeman effects have been given and until methods for calculation of molecular parameters from spectral constants have been described. However, a few remarks will be made here regarding the use of rotational constants in the derivation of molecular structures. Illustrative values of spectral constants A, B, and C are given in Table 7.1.

Analysis of the rotational lines of an asymmetric top gives the three principal moments of inertia from which, with known atomic masses, the internuclear distances and bond angles can be evaluated. However, except for the simplest asymmetric-top molecules there are more structural parameters to be determined than moments of inertia from one isotopic species; therefore a number of isotopic species must be investigated. The investigation of a sufficient number of isotopic species can be tedious, especially when the line intensities are not large enough for observation of the isotopic species in natural abundance and chemical synthesis of a number of isotopic species becomes necessary. Nevertheless, the structures of numerous molecules have been evaluated by this procedure. Various methods of calculating molecular structures of asymmetric rotors from rotational constants are described in Chapter XIII.

In principle, very precise determinations of bond lengths and bond angles can be made from measurements of pure rotational spectra. Unfortunately, because the moments of inertia generally obtained are not equilibrium moments but effective ground-state moments of inertia, which are contaminated with zero-point vibrational effects, ambiguities are transmitted to the derived structural parameters. The methods of evaluation of molecular structures from pure rotational spectra and techniques for minimizing vibrational effects will be discussed in Chapter XIII. To obtain the equilibrium moments of inertia one must measure rotational transitions in excited vibrational states to enable the evaluation of the α_i's in (7.94)–(7.96). Usually, some or all of the vibrational frequencies are greater than 500 cm^{-1}, and the rotational transitions in the excited vibrational states are weak because of the small population. The population F_v in the excited vibrational state can be increased by heating the gas under observation. However, the line intensity cannot always be increased sufficiently because of the offsetting effect due to the inverse dependence of the intensity on the temperature [see (7.73)]. So far only a few equilibrium structures of asymmetric rotors have been evaluated (see Chapter XIII, Section 6).

In a previous edition of this book [150] we gave a reasonably complete listing of structures through about 1969. Since most molecules are asymmetric tops and since the output of molecular structures by microwave spectroscopists has been enormous during the last ten years, such a listing is no longer feasible. However, some illustrative structures of asymmetric tops are given in Chapter XIII. For more complete compilations the reader is referred to those referenced in Chapter XIII and the Landolt-Börnstein tables [151].

References

1. G. W. King, R. M. Hainer, and P. C. Cross, *J. Chem. Phys.*, **11**, 27 (1943).
2. See, for example, H. H. Nielsen, *Rev. Mod. Phys.*, **23**, 90 (1951) and the references quoted therein.

3. W. Gordy, W. V. Smith, and R. F. Trambarulo, *Microwave Spectroscopy*, Wiley, New York, 1953.
4. B. S. Ray, *Z. Physik.*, **78**, 74 (1932).
5. R. S. Mulliken, *Phys. Rev.*, **59**, 873 (1941).
6. S. C. Wang, *Phys. Rev.*, **34**, 243 (1929).
7. J. H. Van Vleck, *Phys. Rev.*, **33**, 467 (1929).
8. See, for example, A. Ralston and H. S. Wilf, *Mathematical Methods for Digital Computers*, Wiley, New York, 1960.
9. M. W. P. Strandberg, *Microwave Spectroscopy*, Wiley, New York, 1954.
10. J. D. Swalen and L. Pierce, *J. Math. Phys.*, **2**, 736 (1961).
11. R. M. Hainer, P. C. Cross, and G. W. King, *J. Chem. Phys.*, **17**, 826 (1949).
12. C. H. Townes and A. L. Schawlow, *Microwave Spectroscopy*, McGraw-Hill, New York, 1955.
13. H. C. Allen, Jr. and P. C. Cross, *Molecular Vib-Rotors*, Wiley, New York, 1963.
14. J. W. Blaker, M. Sidran, and A. Kaercher, *J. Chem. Phys.*, **37**, 684 (1962); M. Sidran, F. J. Nolan, and J. W. Blaker, *J. Mol. Spectrosc.*, **11**, 79 (1963); F. J. Nolan, M. Sidran, and J. W. Blaker, *J. Chem. Phys.*, **41**, 588 (1964).
15. R. H. Schwendeman, "A Table of $\langle P_z^2 \rangle$ and $\langle P_z^4 \rangle$ for Asymmetric Rotor Molecules," Michigan State University, East Lansing, Michigan.
16. H. Dreizler and R. Peter, *J. Chem. Phys.*, **39**, 1132 (1963). See also H. Dreizler, R. Peter, and H. D. Rudolph, *Z. Naturforsch.*, **21a**, 2058 (1966).
17. D. Kivelson and E. B. Wilson, Jr., *J. Chem. Phys.*, **20**, 1575 (1952).
18. J. K. Bragg and S. Golden, *J. Chem. Phys.*, **75**, 735 (1949).
19. This may be easily seen as follows: the energy of an asymmetric rotor may be written in the form $E = A\langle P_a^2 \rangle + B\langle P_b^2 \rangle + C\langle P_c^2 \rangle$, where the average values are evaluated in the asymmetric-rotor basis. This expression readily yields to first-order, (7.29)–(7.31).
20. R. H. Schwendeman, *J. Chem. Phys.*, **27**, 986 (1957). R. H. Schwendeman, "A Table of Coefficients for the Energy Levels of a Near Symmetric Top," Department of Chemistry, Harvard University, Cambridge, Massachusetts.
21. D. Kivelson, *J. Chem. Phys.*, **21**, 536 (1953).
22. G. Winnewisser, M. Winnewisser, and W. Gordy, *J. Chem. Phys.*, **49**, 3465 (1968).
23. P. C. Cross, R. M. Hainer, and G. W. King, *J. Chem. Phys.*, **12**, 210 (1944).
24. R. H. Schwendeman, *J. Mol. Spectrosc.*, **7**, 280 (1961).
25. D. R. Lide, Jr., *J. Chem. Phys.*, **20**, 1761 (1952).
26. R. H. Schwendeman and V. W. Laurie, *Table of Line Strengths for Rotational Transitions of Asymmetric Rotor Molecules*, Pergamon, London, 1958.
27. P. F. Wacker and M. R. Pratto, *Microwave Spectral Tables*, Vol. II. *Line Strengths of Asymmetric Rotors* (Natl. Bur. Std. Monograph 70), 1964.
28. See [12], Appendix V.
29. H. D. Rudolph, *Z. Phys.*, **211**, 419 (1968).
30. H. D. Rudolph, *Z. Naturforsch.*, **23a**, 1020 (1968).
31. E. K. Gora, *J. Mol. Spectrosc.*, **2**, 259 (1958).
32. L. H. Scharpen, *Symposium on Molecular Structure and Spectroscopy*, Paper 09, Ohio State University, Columbus, Ohio 1969.
33. L. H. Scharpen and V. W. Laurie, *Anal. Chem.*, **44**, 378R (1972).
34. G. E. Jones and R. L. Cook, '*CRC Crit. Rev. in Anal. Chem.*,' **3**, 455 (1974).
35. W. E. Steinmetz, *J. Am. Chem. Soc.*, **96**, 685 (1974).
36. R. J. Higgins, L. L. Combs, T. B. Malloy, Jr., and R. L. Cook, *J. Mol. Struct.*, **28**, 121 (1975).
37. M. S. Farag and R. K. Bohn, *J. Chem. Phys.*, **62**, 3946 (1975).

38. E. M. Bellott, Jr., and E. B. Wilson, Jr., *Tetrahedron*, **31**, 2896 (1975).
39. E. M. Bellott, Jr., and E. B. Wilson, Jr., *J. Mol. Spectrosc.*, **66**, 41 (1977).
40. W. J. Lafferty, *J. Mol. Spectrosc.*, **38**, 84 (1970).
41. T. Ikeda, R. Kewley, and R. F. Curl, Jr., *J. Mol. Spectrosc.*, **44**, 459 (1972).
42. H. M. Pickett, *J. Am. Chem. Soc.*, **95**, 1770 (1973).
43. R. W. Kitchin, T. K. Avirah, T. B. Malloy, Jr., and R. L. Cook, *J. Mol. Struct.*, **24**, 337 (1975).
44. R. W. Kitchin, T. B. Malloy, Jr., and R. L. Cook, *J. Mol. Spectrosc.*, **57**, 179 (1975).
45. D. R. Lide, Jr., "Microwave Spectroscopy" in *Methods of Experimental Physics*, Vol. 3, *Molecular Physics*, Part A, D. Williams, Ed., Chapter 2.1 Academic, New York, 1974.
46. S. J. Borchert, *J. Mol. Spectrosc.*, **57**, 312 (1975).
47. J. A. Wells and T. B. Malloy, Jr., *J. Chem. Phys.*, **60**, 3987 (1974).
48. T. K. Avirah, T. B. Malloy, Jr., and R. L. Cook, *J. Chem. Phys.*, **71**, 2194 (1979).
49. T. Ikeda, K. V. L. N. Sastry, and R. F. Curl, Jr., *J. Mol. Spectrosc.*, **56**, 411 (1975).
50. R. A. Creswell and W. J. Lafferty, *J. Mol. Spectrosc.*, **46**, 371 (1973).
51. T. K. Avirah, T. B. Malloy, Jr., and R. L. Cook, *J. Mol. Struct.*, **26**, 267 (1975).
52. M. Winnewisser, G. Winnewisser, T. Honda, and E. Hirota, *Z. Naturforsch.*, **30a**, 1001 (1975).
53. B. P. van Eijck, *J. Mol. Spectrosc.*, **38**, 149 (1971).
54. A. B. Delfino and K. R. Ramaprasad, *J. Mol. Struct.*, **25**, 293 (1975).
55. A. Bouchy and G. Roussy, *J. Mol. Spectrosc.*, **78**, 395 (1979).
56. J. R. Durig, C. C. Tong, and Y. S. Li, *J. Chem. Phys.*, **57**, 4425 (1972).
57. S. Doraiswamy and S. D. Sharma, *J. Mol. Spectrosc.*, **58**, 323 (1975).
58. F. Kneubuhl, T. Gaumann, and H. Günthard, *J. Mol. Spectrosc.*, **3**, 349 (1959).
59. R. A. Beaudet, Ph.D. Dissertation, Harvard University, 1962.
60. G. F. Pollnow and A. J. Hopfinger, *J. Chem. Educ.*, **45**, 528 (1968).
61. K. M. Marstokk and H. Møllendal, *J. Mol. Struct.*, **4**, 470 (1969).
62. J. G. Baker, "Microwave–Microwave Double Resonance" in *Modern Aspects of Microwave Spectroscopy*, G. W. Chantry, Ed., Academic, New York, 1979, pp. 65–121.
63. H. Jones, "Infrared–Microwave Double Resonance Techniques" in *Modern Aspects of Microwave Spectroscopy*, G. W. Chantry, Ed., Academic, New York, 1979, pp. 123–216.
64. J. I. Steinfeld and P. L. Houston, "Double Resonance Spectroscopy" in *Laser and Coherence Spectroscopy*, J. I. Steinfeld, Ed., Plenum Press, New York, 1978, pp. 1–123.
65. K. Shimoda, "Infrared–Microwave Double Resonance," pp. 29–44; T. Oka, "Infrared–Microwave (Radiofrequency) Two-Photon Spectroscopy," pp. 413–431; J. C. McGurk, C. L. Norris, T. G. Schmalz, E. F. Pearson, and W. H. Flygare, "Infrared–Microwave Double Resonance Measurements of T_1 in Methyl Fluoride and Methyl Chloride," pp. 541–554, in *Laser Spectroscopy*, R. G. Brewer and A. Mooradian, eds., Plenum Press, New York, 1974.
66. K. Shimoda and T. Shimizer, "Non-Linear Spectroscopy of Molecules" in *Progress in Quantum Electronics*, J. H. Sanders and S. Stenholm, Eds., Pergamon, Oxford, 1972, Vol. 2, pp. 43–139.
67. T. Oka, *Ad. At. Mol. Phys.*, **9**, 127 (1973).
68. T. Oka, "Forbidden Rotational Transitions" in *Molecular Spectroscopy: Modern Research*, K. N. Rao, Ed., Academic, 1976, Vol. II, pp. 229–253.
69. S. H. Autler and C. H. Townes, *Phys. Rev.*, **100**, 703 (1955).
70. K. S. Shimoda, *J. Phys. Soc. Japan* **14**, 954 (1959).
71. A. Battaglia, A. Gozzini, and E. Polacco, *Nuovo Cimento* **14**, 1076 (1959).
72. T. Yajima and K. Shimoda, *J. Phys. Soc. Japan*, **15**, 1668 (1960).
73. T. Yajima, *J. Phys. Soc. Japan*, **16**, 1709 (1961).

74. A. P. Cox, G. W. Flynn, and E. B. Wilson, Jr., *J. Chem. Phys.*, **42**, 3094 (1965).
75. M. L. Unland, V. Weiss, and W. H. Flygare, *J. Chem. Phys.*, **42**, 2138 (1965).
76. M. L. Unland and W. H. Flygare, *J. Chem. Phys.*, **45**, 2421 (1966).
77. R. C. Woods, III, A. M. Ronn, E. B. Wilson, Jr., *Rev. Sci. Instrum.*, **37**, 927 (1966).
78. P. G. Favero, F. Scappini, and A. M. Mirri, *Boll. Sci. della Facolta' di Chim. Indust.* (*Bologna*), **24**, 93 (1966).
79. A. Javan, *Phys. Rev.*, **107**, 1579 (1957).
80. A. DiGaicomo, *Nuovo Cimento* **14**, 1082 (1959).
81. T. Yajima, *J. Phys. Soc. Japan*, **16**, 1594 (1961).
82. F. J. Wodarczyk and E. B. Wilson, Jr., *J. Mol. Spectrosc.*, **37**, 445 (1971).
83. O. L. Stiefvater, *Z. Naturforsch.*, **30a**, 1742 (1975).
84. H. Mäder, H. Dreizler, and A. Guarnieri, *Z. Naturforsch.*, **30a**, 693 (1975).
85. H. D. Rudolph, H. Dreizler, and U. Andresen, *Z. Naturforsch.*, **26a**, 233 (1970).
86. J. Ekkers, A. Bauder, and Hs. H. Günthard, *J. Phys. E. Sci. Instrum.*, **8**, 819 (1975).
87. J. Ekkers, A. Bauder, and Hs. H. Günthard, *Rev. Sci. Instrum.*, **45**, 311 (1974).
88. M. C. Lee and W. F. White, *Rev. Sci. Instrum.*, **43**, 638 (1972).
89. G. K. Pandey and H. Dreizler, *Z. Naturforsch.*, **31a**, 357 (1976).
90. T. Oka, *Can. J. Phys.*, **47**, 2343 (1969).
91. R. Pearson, Jr., A. Choplin, V. Laurie, and J. Schwartz, *J. Chem. Phys.*, **62**, 2949 (1975).
92. R. G. Ford, *J. Chem. Phys.*, **65**, 354 (1976).
93. G. W. Flynn, *J. Mol. Spectrosc.*, **43**, 353 (1972).
94. P. Glorieux, J. Legrand, B. Macke, and J. Messelyn, *C. R. Acad. Sci. Paris* **270B**, 1412 (1970).
95. P. Glorieux, J. Legrand, B. Macke, and J. Messelyn, *J. Quant. Spectrosc. Radiat. Transfer*, **12**, 731 (1972).
96. B. Macke and P. Glorieux, *J. Mol. Spectrosc.*, **45**, 302 (1973).
97. O. L. Stiefvater, *Z. Naturforsch.*, **30a**, 1756 (1975).
98. See for example: O. L. Stiefvater, *J. Chem. Phys.*, **63**, 2580 (1975); O. L. Stiefvater, H. Jones, and J. Sheridan, *Spectrochim. Acta*, **26a**, 825 (1970); ref. 55.
99. O. L. Stiefvater, *Z. Naturforsch.*, **30a**, 1756 (1975).
100. R. G. Ford, *J. Mol. Spectrosc.*, **49**, 117 (1974).
101. R. D. Suenram, *J. Mol. Struct.*, **33**, 1 (1976).
102. G. N. Mathur and M. D. Harmony, *J. Mol. Spectrosc.*, **69**, 37 (1978).
103. E. B. Wilson and J. B. Howard, *J. Chem. Phys.*, **4**, 260 (1936).
104. B. T. Darling and D. M. Dennison, *Phys. Rev.*, **57**, 128 (1940).
105. H. H. Nielsen, *Rev. Mod. Phys.*, **23**, 90 (1951).
106. J. K. G. Watson, *Mol. Phys.*, **15**, 479 (1968).
107. J. D. Louck, *J. Mol. Spectrosc.*, **61**, 107 (1976).
108. G. Amat, H. H. Nielsen, and G. Tarrago, *Rotation-Vibration of Polyatomic Molecules*, Marcel Dekker, New York, 1971.
109. J. E. Wollrab, *Rotational Spectra and Molecular Structure*, Academic, New York, 1967.
110. H. W. Kroto, *Molecular Rotation Spectra*, Wiley, New York, 1975.
111. E. B. Wilson, J. C. Decius, and P. C. Cross, *Molecular Vibrations*, McGraw-Hill, New York, 1955.
112. H. M. Pickett, *J. Chem. Phys.*, **56**, 1715 (1972).
113. A. Attanasio, A. Bauder, Hs. H. Günthard, and H. J. Keller, *Mol. Phys.*, **23**, 35 (1971).
114. T. Oka, *J. Chem. Phys.*, **47**, 5410 (1967).

115. T. Pedersen, *J. Mol. Spectrosc.*, **80**, 229 (1980).
116. P. R. Bunker, *Molecular Symmetry and Spectroscopy*, Academic, New York, 1979.
117. J. H. Meal and S. R. Polo, *J. Chem. Phys.*, **24**, 1119, 1126 (1956).
118. T. Oka and Y. Morino, *J. Mol. Spectrosc.*, **6**, 472 (1961); **11**, 349 (1963).
119. L. Nemes, *J. Mol. Spectrosc.*, **28**, 59 (1968).
120. T. Oka, *J. Mol. Spectrosc.*, **29**, 84 (1969).
121. J. K. G. Watson, *J. Mol. Spectrosc.*, **39**, 364 (1971).
122. M. R. Aliev, *J. Mol. Struct.*, **23**, 411 (1974).
123. A. D. Bykov, Y. S. Makushkin, and O. N. Ulenikov, *J. Mol. Spectrosc.*, **85**, 462 (1981).
124. M. R. Aliev and J. K. G. Watson, *J. Mol. Spectrosc.*, **61**, 29 (1976).
125. A. Niroomand-Rad and P. M. Parker, *J. Mol. Spectrosc.*, **85**, 40 (1981); **75**, 454 (1979).
126. F. W. Birss, *Mol. Phys.*, **30**, 111 (1975).
127. T. Pedersen, *J. Mol. Spectrosc.*, **73**, 360 (1978).
128. C. Georghiou, *Mol. Phys.*, **32**, 1279 (1976).
129. I. M. Mills, "Vibration–Rotation Structure in Asymmetric- and Symmetric-top Molecules," in *Molecular Spectroscopy: Modern Research*, K. N. Rao and C. W. Mathews, Eds., Academic, New York, 1972.
130. K. Yamada, *J. Mol. Spectrosc.*, **81**, 139 (1980).
131. J. K. G. Watson, "Aspects of Quartic and Sextic Centrifugal Effects on Rotational Energy Levels," in *Vibrational Spectra and Structure*, Vol. 6, J. R. Durig, Ed., Marcel Dekker, New York, 1977.
132. E. Hirota, *J. Mol. Spectrosc.*, **43**, 36 (1972).
133. A. J. Careless and H. W. Kroto, *J. Mol. Spectrosc.*, **57**, 189 (1975).
134. M. R. Aliev and J. K. G. Watson, *J. Mol. Spectrosc.*, **75**, 150 (1979).
135. Y. Y. Kwan and E. A. Cohen, *J. Mol. Spectrosc.*, **58**, 54 (1975).
136. P. N. Brier, S. R. Jones, J. G. Baker, and C. Georghiou, *J. Mol. Spectrosc.*, **64**, 415 (1977).
137. J. H. Carpenter, J. D. Muse, C. E. Small, and J. G. Smith, *J. Mol. Spectrosc.*, **93**, 286 (1982).
138. M. Winnewisser, E. F. Pearson, J. Galica, and B. P. Winnewisser, *J. Mol. Spectrosc.*, **91**, 255 (1982).
139. E. B. Wilson, Jr., *J. Chem. Phys.*, **4**, 313 (1936).
140. D. R. Lide, Jr., D. E. Mann, and R. E. Fristrom, *J. Chem. Phys.*, **26**, 734 (1957).
141. E. Hirota, *J. Mol. Spectrosc.*, **7**, 242 (1961).
142. V. W. Laurie and D. T. Pence, *J. Mol. Spectrosc.*, **10**, 155 (1963).
143. T. Oka, K. Takagi, and Y. Morino, *J. Mol. Spectrosc.*, **14**, 27 (1964).
144. T. Tanaka and Y. Morino, *J. Mol. Spectrosc.*, **5**, 436 (1969).
145. T. Tanaka and Y. Morino, *J. Mol. Spectrosc.*, **33**, 538 (1970).
146. Y. Morino and S. Saito, *J. Mol. Spectrosc.*, **19**, 435 (1966).
147. K. Yamada and M. Winnewisser, *J. Mol. Spectrosc.*, **72**, 484 (1978).
148. K. Yamada and M. Winnewisser, *J. Mol. Spectrosc.*, **68**, 307 (1977).
149. J. Demaison, D. Schwoch, B. T. Tan, and H. D. Rudolph, *J. Mol. Spectrosc.*, **68**, 97 (1977).
150. W. Gordy and R. L. Cook, *Microwave Molecular Spectra*, Wiley-Interscience, New York, (1970).
151. Landolt-Börnstein, *Numerical Data and Functional Relationships in Science and Technology*. For a complete reference, see Chapter I, [65].

Chapter **VIII**

THE DISTORTABLE ROTOR

1 INTRODUCTION

In the initial treatment of the asymmetric rotor we regarded the nuclear framework as rigid. This has proved to be a very useful approximation in the interpretation of rotational spectra. However, in reality, the nuclei are held together by finite restoring forces. Thus, bond distances and angles will vary because of the centrifugal force produced by rotation, which gives rise to a centrifugal distortion. The rotating molecule will no longer be in its equilibrium configuration but in a distorted configuration. Rotation of CH_3F, for example, about the symmetry axis would lead to a spreading of the hydrogens and a lengthening of the C—H bond, while end-over-end rotation would produce a closing of the methyl group umbrella and a stretching of the C—F bond. The centrifugal distortion of the molecular parameters of a few molecules for various rotational states is illustrated in Table 8.1. A general expression for the centrifugally induced changes in molecular geometry has been given [1]. Specific application has been made [2] to various isotopic forms of water and hydrogen sulfide which allows calculation of the geometric changes for any rotational level. The effects are usually large for light molecules because of the small moments of inertia. As a consequence of this distortion, the moments of inertia can no longer be considered constant and independent of the rotational state. Therefore, the rotational spectrum will not be simply that of a rigid rotor characterized by a set of equilibrium moments of inertia. Even for low-lying levels with relatively small rotational energies, the precision of microwave measurements is such that the effects of centrifugal distortion can be observed.

Table 8.1 Centrifugal Distortion of Molecular Parameters[a]

Molecule	*Equilibrium Configuration*		*Change due to Centrifugal Distortion*		
			$J=$ 10	30	
OH^b	r(Å)	0.9706	0.0109	0.0922	
HCl^b	r(Å)	1.2746	0.0070	0.0595	
$CO_2{}^b$	r(Å)	1.1615	0.0000	0.0004	
			$J, K=$ 10, 10	10, 5	10, 0
$NH_3{}^c$	r(Å)	1.0124	0.00093	0.00246	0.00297
	α (deg)	106.67	0.237	−0.337	−0.528
	β (deg)		0.217	−0.308	−0.483
			$J_{K_{-1},K_1}=$ $10_{0,10}$	$10_{3,7}$	$10_{10,0}$
H_2O^d	r(Å)	0.9572	0.00283	0.00352	0.00983
	α (deg)	104.52	0.016	0.781	−5.741

[a]From M. Toyama et al. [1].
[b]The rotational state is indicated by the value of J.
[c]The rotational state is indicated by the value of J, K; r denotes N—H bond distance; α, the H—N—H bond angle; and β, the angle between the N—H bond and the symmetry axis.
[d]The rotational state is indicated by the value of J_{K_{-1},K_1}; r denotes O—H bond distance; and α, the H—O—H bond angle.

The theory of centrifugal distortion is considerably more complex for asymmetric rotors, and the effects are in general larger for them than for linear or symmetric-top molecules. In asymmetric rotors numerous transitions between levels of large rotational energies can occur in the microwave region, especially for perpendicular-type transitions. On the other hand, for symmetric tops and linear molecules, transitions involving high J values usually are observed for molecules with large moments of inertia and consequently small rotational energies. In Table 8.2 we show for the asymmetric rotor NSF the shift of the observed frequencies because of centrifugal distortion for various rotational transitions. It is apparent from the table that the spectrum varies significantly from that of a rigid rotor.

Although we must consider the influence of centrifugal distortion in order to accurately account for the positions of rotational transitions, its effects still represent only a small fraction of the rotational energy which is accounted for mainly by the rigid rotor term. Therefore in many cases it can be treated as a perturbation of the rigid rotor Hamiltonian.

Besides introducing a complication in the rotational spectrum, the study of centrifugal distortion provides useful kinds of information. First of all, such a study provides very accurate ground-state spectroscopic constants. For small molecules these data can be profitably combined with infrared vibration–rotation measurements to yield even more accurate values for the rotational energy levels of the excited vibrational states. Such results are important in the study of the fundamental processes of complex molecular laser systems; see, for example, De Lucia et al. [3, 4]. Second, a study of distortion effects allows

Table 8.2 Comparison of Observed and Rigid Rotor Frequencies for the Asymmetric Rotor NSF (MHz)[a]

Transition	*Observed Frequency*	*Calculated Rigid Rotor Frequency*	*Distortion Shift*[b]
$1_{0,1} \leftarrow 0_{0,0}$	16,105.42	16,105.45	−0.03
$2_{1,2} \leftarrow 1_{0,1}$	71,897.10	71,898.95	−1.85
$4_{2,3} \leftarrow 5_{1,4}$	34,757.97	34,780.70	−22.73
$9_{3,7} \leftarrow 10_{2,8}$	36,366.59	36,418.53	−51.94
$14_{2,12} \leftarrow 15_{1,15}$	13,904.62	14,014.98	−110.36
$15_{2,14} \leftarrow 14_{3,11}$	11,267.85	11,267.46	0.39
$21_{2,20} \leftarrow 20_{3,17}$	39,528.46	39,281.63	246.83
$24_{4,20} \leftarrow 23_{5,19}$	34,948.10	35,440.36	−492.26
$25_{2,24} \leftarrow 24_{3,21}$	16,312.53	15,373.17	939.36
$28_{3,25} \leftarrow 29_{2,28}$	39,307.27	41,451.29	−2,144.02
$30_{7,24} \leftarrow 31_{6,25}$	27,957.29	27,938.85	18.44

[a]After Cook and Kirchhoff [163].
[b]Observed frequency minus rigid rotor frequency.

one to predict unmeasured transition frequencies, in many cases, with a high degree of confidence. The more varied the sample data set, the lower will be the calculated uncertainty limits in the predicted frequencies. This is particularly useful for the transitions of molecules of potential astrophysical interest. Radio astronomers have discovered to date some 50 molecules, and it can be expected that many more will be found as the sensitivity of radio telescopes increases. Contrary to previous expectations, which considered only diatomic molecules as likely to be found in interstellar space, most of the molecules found have been considerably more complex. The importance of laboratory measurements of accurate spectroscopic constants to aid in the detection of molecules in interstellar space through observation of rotational transitions is well documented in reviews on the subject by Winnewisser et al. [5] and Snyder [6]. Third, lines of most interest in a distortion analysis are also particularly useful in the application of microwave spectroscopy to chemical analysis. The location of absorption lines provides the basis for qualitative analysis. The transitions of most use for analytical identification purposes are naturally the most intense transitions. These strong transitions will usually be high J transitions, the ones assigned and employed in the analysis of distortion effects, but not the ones most likely used to evaluate just the rotational constants. However, if a distortion analysis is performed, high J lines can be assigned, and one can be sure that these lines belong to the molecule in question and not to a possible impurity. This is illustrated in Fig. 8.1, which shows a small portion of the R-band spectrum of methylene cyanide [7]. The spectrum extends some 800 MHz with the major frequency markers separated by 100 MHz. The three strongest lines correspond to transitions with $J > 30$. The two lower J transitions are shifted around 300 MHz from their rigid rotor positions while the higher J transition is shifted around 1900 MHz. These transitions have been measured and assigned on the basis of the distortion analysis. It perhaps should be noted that the rigid rotor positions of these lines cannot even be indicated in Fig. 8.1 since the limited frequency range shown does not encompass these frequency positions. Such

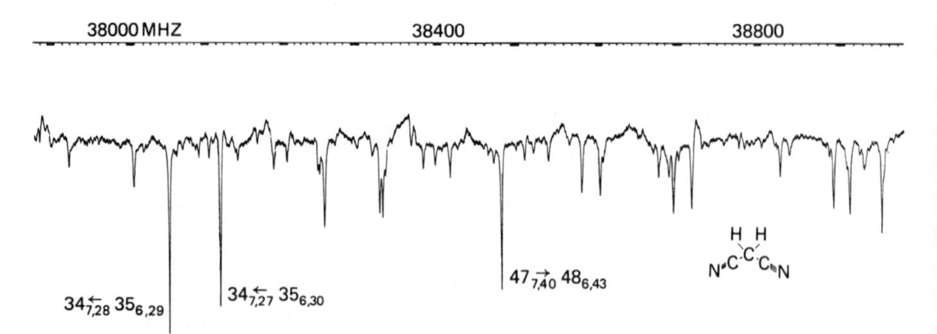

Fig. 8.1 A small portion of the microwave spectrum of methylene cyanide. The Stark modulation voltage is 800 V. The three stronger transitions are high J lines. The $J = 47 \rightarrow 48$ transition is shifted about 1900 MHz from its rigid-rotor position. From Cook et al. [7].

strong lines would naturally be most useful for compilation in a microwave atlas. With the continued interest in microwave spectroscopy as an analytical tool, it might prove useful for molecules where the distortion effects have been studied, to specifically note some of the most intense lines of the spectrum. Finally, the distortion constants themselves provide direct information on the vibrational potential function. This has proved particularly useful in the case of small molecules. A knowledge of the molecular force constants provides fundamental information for the elucidation of interatomic forces and bonding. Examples of this application are reviewed in Section 8.

2 HAMILTONIAN FOR THE DISTORTABLE ROTOR

Before we can discuss the perturbation treatment of rotational distortion for a semirigid rotor, a knowledge of the quantum mechanical Hamiltonian is required. The general theory of centrifugal distortion of asymmetric rotors has been formulated by Wilson and Howard [8] and by Nielsen [9]. A simpler derivation, similar to that given earlier by Wilson [10], is described here.

In general, three coordinates are required to describe the translational motion of the center of mass of a molecule. In addition, three coordinates are needed for a nonlinear molecule (two for a linear molecule) to describe the rotational motion about the center of mass. Therefore, for an N-atom molecule a total of $3N-6$ (or $3N-5$ for a linear molecule) coordinates are necessary to describe the vibrations of the molecule. The molecule-fixed Cartesian coordinate system rotating with the molecule provides the reference framework for description of the vibrational motions. Various coordinates can be employed which measure the displacements of the atoms from their equilibrium positions, and since the potential energy is a function of the distortion of the molecule from its equilibrium configuration it can be expressed as a power series expansion in terms of the displacement coordinates. For our purposes it is convenient to set up the vibrational problem in terms of a set of $3N-6$ independent internal displacement coordinates.

Consider a set of $3N-6$ internal parameters (e.g., bond distances r_i and angles α_i) required to describe the molecular configuration of a polyatomic molecule. Let $\{R_i\}$ be a set of $3N-6$ internal displacement coordinates that describe the displacements of the internuclear distances and bond angles of the molecule from their equilibrium values (e.g., $R_1=\delta r_1=r_1-r_1^e$, $R_2=\delta r_2=r_2-r_2^e$, $R_3=\delta\alpha_1=\alpha_1-\alpha_1^e,\ldots$). If harmonic forces are assumed, that is, only terms quadratic in the displacements are retained, the potential energy can be expressed as

$$V=\frac{1}{2}\sum_{i,j} f_{ij}R_iR_j \tag{8.1}$$

with f_{ij} the harmonic force constants and $f_{ij}=f_{ji}$. The nature of the internal coordinates R_i and R_j determines whether the force constant f_{ij} will represent a bond-stretching or angle-bending force constant, or an interaction force constant, for example, a stretch–bend interaction constant. For small displacements this potential function is a good approximation to the actual potential of the molecule. Thus the classical Hamiltonian for a vibrating rotating polyatomic molecule may be written approximately as

$$\mathscr{H}=\frac{1}{2}\sum_{\alpha,\beta}\mu_{\alpha\beta}P_\alpha P_\beta+\frac{1}{2}\sum_{i,j}G_{ij}p_ip_j+V \tag{8.2}$$

where α and β take on the values x, y, or z of the molecule-fixed Cartesian coordinate system, where i and j enumerate the various internal coordinates, and where V is given by (8.1). The first term represents the rotational energy of the system with $\mu_{\alpha\beta}$ as an element of the inverse moment of inertia tensor and P_α as the α component of the total angular momentum. At this point the $\mu_{\alpha\beta}$ coefficients must be considered to be functions of the vibrational coordinates, since the requirement that the molecule be rigid has been dropped. The last two terms made up of kinetic and potential energy terms represent the vibrational energy of the molecular system. Here p_i is the momentum conjugate to the internal coordinate R_i, and G_{ij} are the elements of the well-known G-matrix that arises when internal coordinates are used in molecular vibration problems [11]. The G-matrix elements are simply functions of the masses and molecular geometry.

In the present discussion we are interested in the situation where the molecular system is not vibrating but only rotating. This rotational motion will result in a distortion of the molecule because of the stretching effects of the centrifugal forces and the nonrigidity of the molecular framework. One of Hamilton's equations of motion, $\dot{p}_i=-\partial\mathscr{H}/\partial R_i$, when applied to (8.2), with $\mu_{\alpha\beta}$ depending on the R_i's, yields

$$\frac{1}{2}\sum_{\alpha,\beta}\frac{\partial\mu_{\alpha\beta}}{\partial R_i}P_\alpha P_\beta+\frac{\partial V}{\partial R_i}=0, \qquad i=1,2,\ldots,3N-6 \tag{8.3}$$

since both p_i and $\dot{p}_i$ are zero under the assumption of no vibration. This relation expresses the equilibrium condition between the restoring potential forces and the centrifugal forces due to rotation.

The coefficient $\mu_{\alpha\beta}$ can be approximated by a series expansion about the equilibrium value $\mu_{\alpha\beta}^e$ (i.e., the value when all displacement coordinates R_i are zero, which corresponds physically here to no rotation), as follows

$$\mu_{\alpha\beta}=\mu_{\alpha\beta}^e+\sum_i\mu_{\alpha\beta}^{(i)}R_i+\ldots \tag{8.4}$$

with

$$\mu_{\alpha\beta}^{(i)}=\left(\frac{\partial\mu_{\alpha\beta}}{\partial R_i}\right)_e \tag{8.5}$$

This latter quantity, which is the partial derivative of the $\alpha\beta$ component of the reciprocal moment of inertia tensor with respect to the internal coordinate R_i, is to be evaluated at the equilibrium position. In the expansion, only linear terms are retained since the displacements are assumed to be small. From this expansion in conjunction with the potential energy expression, (8.1), one immediately obtains from (8.3)

$$\frac{1}{2}\sum_{\alpha,\beta} \mu_{\alpha\beta}^{(i)} P_\alpha P_\beta + \sum_j f_{ij} R_j = 0, \qquad i = 1, 2, \ldots, 3N-6 \tag{8.6}$$

This set of linear equations may be solved for R_j giving

$$R_j = -\frac{1}{2}\sum_{i,\alpha,\beta} (f^{-1})_{ji} \mu_{\alpha\beta}^{(i)} P_\alpha P_\beta \tag{8.7}$$

where $(f^{-1})_{ji}$ is an element of the matrix inverse to the matrix of force constants f_{ji}. Inserting this in (8.4) allows one to express the instantaneous values of $\mu_{\alpha\beta}$ in terms of the components of the total angular momentum as follows

$$\mu_{\alpha\beta} = \mu_{\alpha\beta}^{e} - \frac{1}{2}\sum_{i,j,\gamma,\delta} \mu_{\alpha\beta}^{(i)} (f^{-1})_{ij} \mu_{\gamma\delta}^{(j)} P_\gamma P_\delta \tag{8.8}$$

If (8.7) is inserted into (8.1), the potential energy becomes

$$V = \frac{1}{8}\sum_{i,j,\alpha,\beta,\gamma,\delta} \mu_{\alpha\beta}^{(i)} (f^{-1})_{ij} \mu_{\gamma\delta}^{(j)} P_\alpha P_\beta P_\gamma P_\delta \tag{8.9}$$

Note $\sum_l f_{il}(f^{-1})_{lj} = \delta_{ij}$. When these expressions for $\mu_{\alpha\beta}$ and V are substituted in (8.2) one obtains

$$\mathscr{H} = \frac{1}{2}\sum_{\alpha,\beta} \mu_{\alpha\beta}^{e} P_\alpha P_\beta + \frac{1}{4}\sum_{\alpha,\beta,\gamma,\delta} \tau_{\alpha\beta\gamma\delta} P_\alpha P_\beta P_\gamma P_\delta \tag{8.10}$$

with

$$\tau_{\alpha\beta\gamma\delta} = -\frac{1}{2}\sum_{ij} \mu_{\alpha\beta}^{(i)} (f^{-1})_{ij} \mu_{\gamma\delta}^{(j)} \tag{8.11}$$

This is the classical rotational Hamiltonian for a semirigid nonvibrating molecule; it represents the quantum mechanical Hamiltonian when the angular momentum components are taken as the corresponding angular momentum operators.

The first term on the right in (8.10) represents the usual rigid rotor energy while the second term represents the energy of distortion. The $\tau_{\alpha\beta\gamma\delta}$ are the distortion constants, and the distortion contribution will involve angular momentum about all three axes. The dependence of the distortion constants on the inverse elements of the molecular force constant matrix is physically reasonable, since, all things being equal, one would expect that weaker bonds with small force constants should show larger distortion effects than the strong bonds with large force constants.

The equilibrium rotational and centrifugal distortion constants in $\mathscr{H}$ which are independent of the rotational coordinates are also seen to be independent of the vibrational coordinates. Within the approximation considered here, we have a rigid rotor with a correction term for centrifugal distortion. However, in an actual molecule, vibrational motion is unavoidable, and the rotational and distortion constants must be considered as functions of the vibrational coordinates and thus dependent on the vibrational state in which the rotational spectrum is observed. In practice then, the principal moments of inertia obtained from the ground state spectrum are moments averaged over the ground vibrational state, that is, effective moments of inertia. The general vibration–rotation quantum mechanical Hamiltonian operator of a polyatomic molecule has been discussed by Wilson and Howard [8] and by Darling and Dennison [12]. Perturbation treatments to various order have been given for the evaluation of the vibration–rotation energy levels [8, 9, 13, 14]. These general formulations admit vibration–rotation interaction terms in the Hamiltonian from which the vibrational dependence of the rotational constants and centrifugal distortion constants arises as well as other vibration–rotation interactions such as Coriolis coupling. We may also cite some of the more recent discussions [15–20] on the rotation–vibration Hamiltonian (see also Chapter VII).

As a simple application of (8.10), let us consider the case of a semirigid diatomic rotor. The rotation of a diatomic molecule is mathematically equivalent to the rotation of a single particle of reduced mass m at a distance r from the axis of rotation and we have

$$\mu=\frac{1}{I}; \qquad I=mr^2=\frac{m_1 m_2}{m_1+m_2}r^2 \tag{8.12}$$

The distortion term of (8.10) with the aid of (8.11) reduces to

$$-\frac{1}{8}\left(\frac{\partial\mu}{\partial R}\right)_e^2\left(\frac{1}{f}\right)P^4 \tag{8.13}$$

where f is the force constant appearing in the potential energy which is simply $V=\frac{1}{2}fR^2$ where $R=\delta r=r-r_e$, r_e is the equilibrium bond distance, and r is the instantaneous value. Now the partial derivative, evaluated at equilibrium, of the reciprocal moment of inertia is

$$\left(\frac{\partial\mu}{\partial R}\right)_e=\left(\frac{\partial\mu}{\partial r}\right)_e=\frac{-2}{mr_e^3}=\frac{-2}{m}\left(\frac{m}{I^e}\right)^{3/2} \tag{8.14}$$

In the case of polyatomic molecules the evaluation of such quantities is more complicated. If we now impose the quantum mechanical restriction that the angular momentum must be quantized in units of $\hbar$, that is, $P^2=\hbar^2 J(J+1)$, then (8.13) may be written in the familiar form $-hD_J J^2(J+1)^2$ with the distortion constant defined as

$$D_J=\frac{\hbar^4}{2h}\frac{m}{f(I^e)^3}=\frac{4B_e^3}{\omega_e^2} \tag{8.15}$$

where (8.14) and the relations $B_e=h/(8\pi^2 I^e)$ and $\omega_e=(1/2\pi)(f/m)^{1/2}$ have been utilized. This is the usual definition of D_J given for a diatomic molecule.

Calculation of the τ's

The distortion constants can either be evaluated from the rotational spectrum (this will be discussed in the next section) or if the force constants are known they may be obtained by means of (8.11) once the derivatives, which depend on the molecular geometry, are evaluated. For a general molecule the partial derivatives appearing in (8.11) are most easily determined by the method proposed by Kivelson and Wilson [21]. They first express the partial derivatives of the inverse inertia tensor components in terms of the partial derivatives of the components of the inertia tensor. In the matrix notation we have, since $\boldsymbol{\mu}=\mathbf{I}^{-1}$, the equation $\boldsymbol{\mu}\mathbf{I}=\mathbf{E}$ with $\mathbf{E}$ the unit matrix, and taking partial derivatives we obtain

$$\frac{\partial\boldsymbol{\mu}}{\partial R_i}\mathbf{I}+\boldsymbol{\mu}\frac{\partial\mathbf{I}}{\partial R_i}=0 \tag{8.16}$$

thus

$$\frac{\partial\boldsymbol{\mu}}{\partial R_i}=-\boldsymbol{\mu}\frac{\partial\mathbf{I}}{\partial R_i}\boldsymbol{\mu} \tag{8.17}$$

The coordinate axes are now so chosen that the undistorted molecule will be in its principal axis system with origin at the center of mass. Then the inertia matrix is diagonal with equilibrium elements I_{xx}^e, I_{yy}^e, I_{zz}^e, and we find for the components of (8.17) evaluated at equilibrium

$$\left(\frac{\partial\mu_{\alpha\beta}}{\partial R_i}\right)_e=\frac{-[J_{\alpha\beta}^{(i)}]_e}{I_{\alpha\alpha}^e I_{\beta\beta}^e} \tag{8.18}$$

since $\mu_{\alpha\alpha}^e=1/I_{\alpha\alpha}^e$, etc. Here

$$J_{\alpha\beta}^{(i)}=\frac{\partial I_{\alpha\beta}}{\partial R_i} \tag{8.19}$$

If (8.18) is inserted into (8.11), the distortion constants may be re-expressed as

$$\tau_{\alpha\beta\gamma\delta}=-\tfrac{1}{2}(I_{\alpha\alpha}^e I_{\beta\beta}^e I_{\gamma\gamma}^e I_{\delta\delta}^e)^{-1}\sum_{i,j}[J_{\alpha\beta}^{(i)}]_e(f^{-1})_{ij}[J_{\gamma\delta}^{(j)}]_e \tag{8.20}$$

An alternate definition in terms of partial derivatives with respect to the normal coordinates has been given in Chapter VII, Section 6.

Evaluation of the partial derivatives of components of the inertia matrix requires special consideration. The components of the inertia matrix are given by

$$I_{\alpha\alpha}=\sum_l m_l(\beta_l^2+\gamma_l^2) \tag{8.21}$$

$$I_{\alpha\beta}=I_{\beta\alpha}=-\sum_l m_l\alpha_l\beta_l \tag{8.22}$$

where m_l and α_l are, respectively, the mass and the α-coordinate of the lth atom as measured in the body-fixed axis system. Here, and in the following equations,

α, β, and γ are taken in cyclic order; that is, if $\alpha_l = x_l$, then β_l and γ_l stand for y_l and z_l, respectively. What is required are the variations of the components of the inertia tensor when the molecule is distorted from its equilibrium configuration. Now the variations in the $3N$ Cartesian coordinates, $\delta\alpha_l$, $\delta\beta_l$, $\delta\gamma_l$, cannot be made arbitrarily but must satisfy the six Eckart conditions [22, 23] which serve to define the rotating coordinate system so as to provide maximum separation of rotation and vibration. Physically, we may view these as restrictions on the distortions that insure that the center of mass of the molecule does not change and that no angular momentum is imparted to the molecule. With the Eckart conditions it can be shown that [21]

$$J_{\alpha\alpha}^{(i)} = \frac{2}{\delta R_i} \sum_l m_l(\beta_l \delta\beta_l + \gamma_l \delta\gamma_l) \tag{8.23}$$

$$J_{\alpha\beta}^{(i)} = -\frac{2}{\delta R_i I_{\gamma\gamma}} \left\{ I_\alpha \sum_l m_l \beta_l \delta\alpha_l + I_\beta \sum_l m_l \alpha_l \delta\beta_l \right\} \tag{8.24}$$

with $I_\alpha = \sum_k m_k \alpha_k^2$, and so on, and $I_{\gamma\gamma}$ given by (8.21). These relations allow the displacements $\delta\alpha_l$, and so on, to be entirely arbitrary while still satisfying the Eckart conditions. The expressions are to be evaluated at the equilibrium configuration of the molecule ($R_i = 0$, $i = 1, \ldots, 3N-6$) and are referred to the rotating principal axis system. The Cartesian coordinates α_l are thus measured from the center of mass in the principal axis system.

For the finding of a derivative, the principal axis system may be translated to a convenient position in the molecule. Small displacements are then given to the nuclei so that the set of increments $\delta\alpha_l$ will change the angle or internuclear distance that is involved in a given internal coordinate R_i, in such a way that the internal coordinate will change by an increment δR_i while leaving all other internal coordinates with their equilibrium values (see Fig. 8.2). The coordinates α_l are related to those of the displaced coordinate system (x', y', z') by $\alpha_l = \alpha_l' - \bar{\alpha}$, where $\bar{\alpha}$ is the α component of the center of mass in the displaced system.

When there are more internal displacement coordinates than vibrational degrees of freedom, that is, redundancies are present, additional considerations are necessary. These have been discussed by Gold et al. [24]. If the molecule has symmetry, calculation of the distortion constants can be simplified by use of a set of internal symmetry coordinates $\{S_i\}$ which are particular orthonormal linear combinations of the internal displacement coordinates. These symmetry coordinates are chosen so that, under the symmetry operations of the molecular point group, each coordinate may be classified as belonging to one of the symmetry species of the group. Equation 8.20 is still applicable when both the partial derivatives and the force constant matrix are expressed in terms of the symmetry coordinates. To obtain the $J_{\alpha\beta}^{(i)}$, we choose increments $\delta\alpha_l$ so as to produce an increment δS_i in a particular symmetry coordinate while the other symmetry coordinates are left unchanged. When symmetry coordinates are used, the matrix of force constants, and its inverse, will factor into submatrices

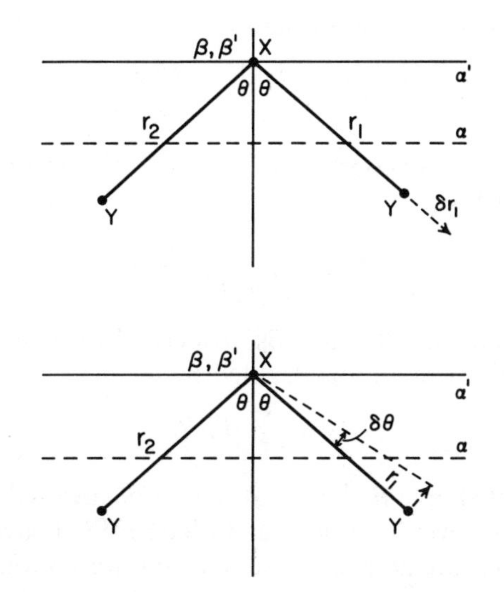

Fig. 8.2 Principal and translated axes for bent XY_2-type molecule. Displacement of r_1 bond length (top figure). Displacement of valence bond angle (bottom figure).

corresponding to different symmetry species of the group; there will be no coupling between $J_{\alpha\beta}^{(i)}$ and $J_{\alpha\beta}^{(j)}$ if the ith and jth symmetry coordinates belong to different symmetry species [21]. In Table 8.3 (on pages 308–311) explicit expressions for the derivatives of some common types of molecules are given. Expressions for other types such as axially symmetry ZX_3Y molecules [25], pyramidal XYZ_2 molecules [26], and axially symmetric ZX_3YW molecules [27] have also been reported. General expressions for a number of commonly used internal displacement coordinates may be found in [21]. Corrections for non-tetrahedral angles in the symmetric tops ZX_3Y and XY_3 have been described [28].

Since calculation of the $J_{\alpha\beta}^{(i)}$ elements is somewhat tedious, alternate methods of evaluation that are more convenient have been formulated. The Cartesian displacements in the principal inertial axis system can be conveniently expressed, as pointed out by Parent and Gerry [29], in terms of the ρ vectors $(\rho_i = \rho_{i1}^{(x)}, \rho_{i1}^{(y)}, \rho_{i1}^{(z)}, \rho_{i2}^{(x)}, \ldots, \rho_{iN}^{(z)})$

$$\delta\alpha_l = \rho_{il}^{(\alpha)} \tag{8.25}$$

with $\rho_{il}^{(\alpha)}$ the α component of the ρ vector for the lth atom associated with the ith internal coordinate. These ρ vectors, introduced by Polo [30], can also be used to evaluate the inverse kinetic energy matrix $\mathbf{G}^{-1}$, which is useful in the solution of the vibrational problem. An alternate formulation which makes use of the Wilson **s** vectors has been discussed by a number of authors (see Aliev and Aleksanyan [31], Cyvin et al. [32–34], Pulay and Sawodny [35], and Klauss and Strey [36].

Let R_i denote an internal displacement coordinate and $\delta\alpha_l = \alpha_l - \alpha_l^e$, the Cartesian displacement coordinate of the lth atom ($\alpha = x$, y, or z). These are related by the well-known transformation

$$R_i=\sum_{l,\alpha} B_{il}^{(\alpha)}\delta\alpha_l \begin{cases} l=1,\ldots N;\ \alpha=x,y,z \\ i=1,\ldots 3N-6 \end{cases} \tag{8.26}$$

The **B** matrix elements are constructed from the Wilson **s** vectors and give the kinetic energy matrix **G**,

$$G_{ij}=\sum_{\alpha,l}\left(\frac{1}{m_l}\right)B_{il}^{(\alpha)}B_{jl}^{(\alpha)} \tag{8.27}$$

with the three components $B_{il}^{(x)}$, $B_{il}^{(y)}$, $B_{il}^{(z)}$ associated with atom l and coordinate R_i. The inverse transformation is denoted by

$$\delta\alpha_l=\sum_i A_{li}^{(\alpha)}R_i \tag{8.28}$$

The matrix **A** is not simple $\mathbf{B}^{-1}$ since **B** cannot be inverted because it is a rectangular matrix. However, Crawford and Fletcher [37] have shown that if the six Eckart conditions are added, the desired transformation is given by

$$A_{li}^{(\alpha)}=\sum_k \frac{1}{m_l} B_{kl}^{(\alpha)}G_{ki}^{-1} \tag{8.29}$$

This relation forms the basis for evaluating the derivatives of $I_{\alpha\beta}$ satisfying the Eckart conditions. Taking the partial derivatives of the components of the inertia tensor, with respect to the internal coordinates, gives

$$\begin{aligned}\frac{\partial I_{\alpha\alpha}}{\partial R_i}&=2\sum_l m_l\left(B_l\frac{\partial\beta_l}{\partial R_i}+\gamma_l\frac{\partial\gamma_l}{\partial R_i}\right)\\ \frac{\partial I_{\alpha\beta}}{\partial R_i}&=-\sum_l m_l\left(\alpha_l\frac{\partial\beta_l}{\partial R_i}+\beta_l\frac{\partial\alpha_l}{\partial R_i}\right)\end{aligned} \tag{8.30}$$

From (8.28)

$$\frac{\partial\alpha_l}{\partial R_i}=A_{li}^{(\alpha)} \tag{8.31}$$

and substitution in (8.30), with cognizance of (8.29), gives

$$\begin{aligned}\frac{\partial I_{\alpha\alpha}}{\partial R_i}&=2\sum_k\sum_l(\beta_l B_{kl}^{(\beta)}+\gamma_l B_{kl}^{(\gamma)})G_{ki}^{-1}=\sum_k G_{ik}^{-1}T_{\alpha\alpha}^{(k)}\\ \frac{\partial I_{\alpha\beta}}{\partial R_i}&=-\sum_k\sum_l(\alpha_l B_{kl}^{(\beta)}+\beta_l B_{kl}^{(\alpha)})G_{ki}^{-1}=\sum_k G_{ik}^{-1}T_{\alpha\beta}^{(k)}\end{aligned} \tag{8.32}$$

The definitions of $T_{\alpha\alpha}^{(k)}$ and $T_{\alpha\beta}^{(k)}$ are apparent from the foregoing equations. Evaluation of these in the equilibrium configuration yields the desired quantities

$$J_{\alpha\beta}^{(i)}=\left(\frac{\partial I_{\alpha\beta}}{\partial R_i}\right)_e \tag{8.33}$$

Table 8.3 The $[J_{\alpha\beta}^{(i)}]_e$ Expressions for Some Particular Cases

	$[J_{xx}^{(i)}]_e$	$[J_{zz}^{(i)}]_e$	$[J_{xz}^{(i)}]_e$
Internal Coordinate			
	For Nonlinear XY_2 Molecules[a]		
δr_1	$2m_Y rs^2$	$\dfrac{2m_X m_Y rc^2}{M}$	$\dfrac{2m_Y rsc}{1+2m_Y m_X^{-1} s^2}$
δr_2	$2m_Y rs^2$	$\dfrac{2m_X m_Y rc^2}{M}$	$\dfrac{-2m_Y rsc}{1+2m_Y m_X^{-1} s^2}$
$\delta\alpha$	$2m_Y r^2 sc$	$\dfrac{-2m_X m_Y r^2 sc}{M}$	0
Symmetry Coordinate			
S_1	$2\sqrt{2} m_Y rs^2$	$\dfrac{2\sqrt{2} m_X m_Y rc^2}{M}$	0
S_2	$2m_Y r^2 sc$	$\dfrac{-2m_X m_Y r^2 sc}{M}$	0
S_3	0	0	$\dfrac{2\sqrt{2} m_Y rsc}{1+2m_Y m_X^{-1} s^2}$

Definition of Symmetry Coordinates for XY_2

A_1 species	B_1 species
$S_1 = \dfrac{\delta r_1 + \delta r_2}{\sqrt{2}}$	$S_3 = \dfrac{\delta r_1 - \delta r_2}{\sqrt{2}}$
$S_2 = \delta\alpha$	

For Planar XYZ_2 Molecules[b]

	$[J_{xx}^{(i)}]_e$	$[J_{zz}^{(i)}]_e$	$[J_{xz}^{(i)}]_e$
S_1	$\sqrt{8} m_Z (l + r_1 c) c$	$\sqrt{8} m_Z r_1 s^2$	0
S_2	$2m_Y (r_3 - l)$	0	0
S_3	$\dfrac{-4m_Z r_1 s (l + r_1 c)}{\sqrt{6}}$	$\dfrac{4m_Z r_1^2 sc}{\sqrt{6}}$	0
S_4	0	0	$\dfrac{-\sqrt{8} m_Z [I_x s (l + r_1 c) + I_z r_1 sc]}{I_{yy}}$
S_5	0	0	$\dfrac{-\sqrt{2} m_Y I_x r_3 (r_3 - l)}{I_{yy}}$

Table 8.3 Continued

	$[J_{xx}^{(i)}]_e$	$[J_{zz}^{(i)}]_e$	$[J_{xz}^{(i)}]_e$

Definition of Symmetry Coordinates for XYZ_2

A_1 species	B_1 species	B_2 species
$S_1=\dfrac{\delta r_1+\delta r_2}{\sqrt{2}}$	$S_4=\dfrac{\delta r_1-\delta r_2}{\sqrt{2}}$	$S_6=\delta\tau$
$S_2=\delta r_3$	$S_5=\dfrac{\delta\beta_1-\delta\beta_2}{\sqrt{2}}$	
$S_3=\dfrac{2\delta\alpha-\delta\beta_1-\delta\beta_2}{\sqrt{6}}$		

For Pyramidal XY_3 Molecules[c]

	$[J_{xx}^{(i)}]_e$	$[J_{zz}^{(i)}]_e$	$[J_{xz}^{(i)}]_e$
S_1	$\dfrac{m_Y r^2 t(3m_Y-m_X)}{\sqrt{3}M}$	$\dfrac{2}{\sqrt{3}}m_Y r^2 t$	0
S_2	$\dfrac{\sqrt{3}rm_Y}{M}[M+(m_X-3m_Y)c^2]$	$2\sqrt{3}m_Y rs^2$	0
S_{1x}	$\dfrac{\sqrt{2}}{\sqrt{3}}m_Y r^2 t$	0	$\dfrac{-4m_X m_Y r^2 ct I_x}{\sqrt{6}MsI_{yy}}$
S_{2x}	$\dfrac{\sqrt{6}}{2}m_Y rs^2$	0	$\dfrac{\sqrt{6}m_Y rsc(m_X I_x+MI_z)}{MI_{yy}}$
S_{1y}, S_{2y}	0	0	0

Definition of Symmetry Coordinates for XY_3

A_1 species	E species
$S_1=\dfrac{\delta\alpha_{12}+\delta\alpha_{13}+\delta\alpha_{23}}{\sqrt{3}}$	$S_{1x}=\dfrac{2\delta\alpha_{12}-\delta\alpha_{13}-\delta\alpha_{23}}{\sqrt{6}}$
$S_2=\dfrac{\delta r_1+\delta r_2+\delta r_3}{\sqrt{3}}$	$S_{2x}=\dfrac{-2\delta r_3+\delta r_1+\delta r_2}{\sqrt{6}}$
	$S_{1y}=\dfrac{\delta\alpha_{13}-\delta\alpha_{23}}{\sqrt{2}}$
	$S_{2y}=\dfrac{\delta r_1-\delta r_2}{\sqrt{2}}$

The Eckart conditions are satisfied automatically, and the G-matrix elements and the $J_{\alpha\beta}^{(i)}$ can be calculated together with a set of s vectors evaluated in the principal inertial axis system of the equilibrium configuration. The equilibrium coordinates α_i^e, however, need not refer to the center of gravity of the molecule as their origin [33].

If desired, (8.20) may be written in an alternate form in terms of the $T_{\alpha\beta}^{(k)}$. The vibrational eigenvalue problem may be expressed as $\mathbf{GFL}=\mathbf{L\Lambda}$, where $\mathbf{\Lambda}$ is a diagonal matrix of eigenvalues $\lambda_i=4\pi^2c^2\omega_i^2$ and $\mathbf{L}$ relates the internal coordinates R_i to the normal coordinates Q_i, $\mathbf{R}=\mathbf{LQ}$. Since in matrix notation $\tilde{\mathbf{L}}\mathbf{FL}=\mathbf{\Lambda}$ and $\tilde{\mathbf{L}}\mathbf{G}^{-1}\mathbf{L}=\mathbf{E}$, with $\mathbf{E}$ a unit matrix, it follows that $\mathbf{G}^{-1}\mathbf{F}^{-1}\mathbf{G}^{-1}=\tilde{\mathbf{L}}^{-1}\mathbf{\Lambda}^{-1}\mathbf{L}^{-1}$. By use of (8.32), (8.20) may be rewritten as

$$\tau_{\alpha\beta\gamma\delta}=-\tfrac{1}{2}(I_{\alpha\alpha}^e I_{\beta\beta}^e I_{\gamma\gamma}^e I_{\delta\delta}^e)^{-1}\sum_{k,k'} T_{\alpha\beta}^{(k)}T_{\gamma\delta}^{(k')}\theta_{kk'} \qquad (8.34)$$

with

$$\theta_{kk'}=\sum_{ij} G_{ki}^{-1}F_{ij}^{-1}G_{jk'}^{-1}=\sum_i \frac{L_{ik}^{-1}L_{ik'}^{-1}}{\lambda_i} \qquad (8.35)$$

The elements of $\theta_{kk'}$ are obtained from the usual normal-coordinate analysis of the molecular vibrations. (See Appendix H).

Application of Table 8.3 may be illustrated by considering a symmetrical bent triatomic molecule, for example, SO_2, for which there are four force constants and four independent τ's. From (8.20) and the $[J_{\alpha\beta}^{(i)}]_e$'s of Table 8.3, with introduction of the factor $\hbar^4/h$, the explicit relations between the distortion constants (equilibrium values) and the force constants are found to be [38]

$$\frac{-\tau_{aaaa}R}{A^2}=2(F^{-1})_{11}+\tan^2\theta(F^{-1})_{22}-2\sqrt{2}\tan\theta(F^{-1})_{12} \qquad (8.36)$$

---

[a] After Kivelson and Wilson [21]. The xz plane is taken as the molecular plane with the x axis parallel to the C_2 axis, and $M=2m_Y+m_X$, $c=\cos\theta$, $s=\sin\theta$, where θ is one-half the $Y-X-Y$ angle (α) and r is the $Y-X$ equilibrium bond distance. Derivatives of I_{yy} with respect to S_1 and S_2 can be obtained from the table since $I_{yy}=I_{xx}+I_{zz}$ for a planar molecule.

[b] From M. G. K. Pillai and F. F. Cleveland, *J. Mol. Spectrosc.*, **6**, 465 (1961). The symbols are defined as follows: m_Z = mass of Z atom, m_Y = mass of Y atom, r_1 is the X—Z bond distance and r_3 is the X—Y bond distance, $c=\cos(\alpha/2)$, $s=\sin(\alpha/2)$ with α the Z—X—Z angle, l is the distance between the center of mass and the X atom, measured along the X—Y bond, which is taken as the z axis. For symmetry coordinates: r_1 and r_2 are the X—Z_i bond distances; r_3, the X—Y bond distance; α, the Z—X—Z angle; β_i, the Y—X—Z_i angle; and τ, out-of-plane angle. Derivatives of I_{yy} with respect to S_1, S_2, and S_3 are easily obtained from the table since the molecule is planar. All other derivatives vanish.

[c] From P. N. Schatz, *J. Chem. Phys.*, **29**, 481 (1958). Here the C_3 axis has been chosen as the z axis. The symbols are defined as follows: $c=\cos\theta$, $s=\sin\theta$, with θ the angle between the $C_3(z)$ axis and an X—Y bond ($\theta>90°$), m_X = mass of X atom, m_Y = mass of Y atom, M = total mass, $r=X$—Y bond distance, $t=\sin\alpha$, where $\alpha=Y$—X—Y angle.

$$\frac{-\tau_{bbbb}R}{B^2}=2(F^{-1})_{11}+\cot^2\theta(F^{-1})_{22}+2\sqrt{2}\cot\theta(F^{-1})_{12} \tag{8.37}$$

$$\frac{-\tau_{aabb}R}{AB}=2(F^{-1})_{11}-(F^{-1})_{22}+\sqrt{2}(\cot\theta-\tan\theta)(F^{-1})_{12} \tag{8.38}$$

$$\frac{-\tau_{abab}R}{AB}=2Mm_X^{-1}(1+2m_Ym_X^{-1}\sin^2\theta)^{-2}(F^{-1})_{33} \tag{8.39}$$

Here h is in erg-sec, masses are in amu and

$$R=\frac{r^2\times 10^{-22}}{2h}$$

with θ as one-half the Y—X—Y angle, $M=2m_Y+m_X$, r as the X—Y bond distance expressed in angstrom units, and with the τ's and rotational constants in megahertz. The $(F^{-1})_{ij}$ are the elements of the matrix inverse to the potential constant matrix $\mathbf{F}$ (dynes/cm) written in terms of the symmetry coordinates given in Table 8.3 [see (8.199)]. It is clear that the observed τ's provide a basis for evaluation of the force constant matrix. Since τ_{abab} depends only on $\omega_3(B_1)$, an even simpler expression may be written [39] in terms of the vibrational frequency

$$\tau_{abab}=-\frac{16cABC}{\omega_3^2} \tag{8.40}$$

with τ_{abab} in Hz and A, B, C, ω_3 in cm^{-1}, and c the speed of light in cm/sec.

From these relations it is apparent that for triatomic planar molecules, where the C_{2v} symmetry is unaltered by isotopic substitution, the distortion constants for the isotopic species can be calculated from the distortion constants of the parent species. In particular, for Y^*-X^*-Y^* we have

$$\tau^*_{aaaa}=\left(\frac{A^*}{A}\right)^2\tau_{aaaa} \tag{8.41}$$

$$\tau^*_{bbbb}=\left(\frac{B^*}{B}\right)^2\tau_{bbbb} \tag{8.42}$$

$$\tau^*_{aabb}=\left(\frac{A^*B^*}{AB}\right)\tau_{aabb} \tag{8.43}$$

$$\tau^*_{abab}=\left(\frac{A^*B^*}{AB}\right)\left(\frac{M^*m_X}{Mm_X^*}\right)\times\left[\frac{1+(2m_Y/m_X)\sin^2\theta}{1+(2m_Y^*/m_X^*)\sin^2\theta}\right]^2\tau_{abab} \tag{8.44}$$

where A, B, and C are the rotational constants; the asterisk refers to the isotopically substituted molecule. It can be noted that for the isotope X^*Y_2, where X is on the C_2 symmetry axis (α), the corresponding moment of inertia is unchanged, and hence the corresponding $\tau_{\alpha\alpha\alpha\alpha}$ is unchanged. An even simpler isotopic relation follows from (8.40) for τ_{abab}. Isotopic relations for the τ constants of various type molecules have been given [40, 41], as well as relations for the P^6 constants of tetrahedral XY_4 molecules [42].

From an examination of the extremal properties of $\tau_{\alpha\alpha\alpha\alpha}$ and $\tau_{\alpha\beta\alpha\beta}$, it has been shown that these constants are always negative [43]. Simple empirical [44] and theoretical [45] relationships between the distortion constants $\tau_{\alpha\alpha\alpha\alpha}$ and the rotational constants B_α have been discussed to obtain order-of-magnitude estimates of the distortion constants. Relations for the upper and lower limits of $\tau_{\alpha\alpha\alpha\alpha}$, $\tau_{\alpha\alpha\beta\beta}$, and $\tau_{\alpha\beta\alpha\beta}$ have been developed [43, 46, 47] that are determined from the values of the rotational constants and the lowest and highest frequency vibration.

The relation for $(-\tau_{\alpha\alpha\alpha\alpha})_{\max}=16B_\alpha^3/\omega_{\min}^2$ is particularly simple, and interchanging max↔min gives the lower bound, $(-\tau_{\alpha\alpha\alpha\alpha})_{\min}$. These relations may be employed to calculate an approximate value of a constant which can help in deciding whether such a constant should be included in the distortion Hamiltonian or whether the observed constant is consistent with the observed vibrational frequencies. Bounds for the spectroscopic constants D_J, D_{JK}, D_K for symmetric tops [46, 47] and Δ_J, Δ_{JK}, and so on, for asymmetric tops [47] have been discussed. From these results it may be concluded that D_J is positive and D_{JK}, D_K may be positive or negative. For a planar symmetric top D_{JK} and D_K are negative and positive, respectively. The linear combinations $(\Delta_J+2\delta_J)$, $(\Delta_J-2\delta_J)$, and $(\Delta_J+\Delta_{JK}+\Delta_K)$ are necessarily positive. For linear triatomic molecules, the experimental values of D_J are very close to their upper limits and thus are not very useful as supplemental data for force constant evaluations. In this regard, a very simple formula for D_J, applicable to a triatomic molecule with a very low stretching frequency, has been obtained [48]. This is particularly useful for linear van der Waals molecules. For nonlinear triatomic molecules, however, the experimental values are not close to their upper limits and hence are useful in force field evaluations, as has been found. The fact that the lower frequency vibrational modes are primarily responsible for the centrifugal distortion effects has been used to give information on the skeletal motions of some molecules of moderate size [49].

3 FIRST-ORDER PERTURBATION TREATMENT OF ROTATIONAL DISTORTION

Taking the quantum mechanical Hamiltonian for a semirigid rotor in the form of (8.10) and choosing the principal axes system for the rigid rotor problem, we may express the Hamiltonian as

$$\mathscr{H}' = \mathscr{H}'_r + \mathscr{H}'_d \tag{8.45}$$

$$\mathscr{H}'_r = A'P_z^2 + B'P_x^2 + C'P_y^2 \tag{8.46}$$

$$\mathscr{H}'_d = \frac{\hbar^4}{4} \sum_{\alpha,\beta,\gamma,\delta} \tau_{\alpha\beta\gamma\delta} P_\alpha P_\beta P_\gamma P_\delta \tag{8.47}$$

with the angular momentum in units of $\hbar$, with α, β, γ, $\delta = x$, y, or z, and with $A' = \hbar^2/2I_z$, and so on. As mentioned previously, a rigorous treatment of the problem indicates that the rotation and distortion constants in the foregoing Hamiltonian must be regarded as effective constants and thus dependent on the vibrational state. The effects of centrifugal distortion on the rotational spectra will depend on both the constants $\tau_{\alpha\beta\gamma\delta}$ and the angular momentum operators. A simple closed-form expression for the effect of these operators, like that obtained for a symmetric rotor, cannot be obtained for an asymmetric rotor. However, the distortion effects for a large number of asymmetric rotors may be treated to a good approximation by first-order perturbation theory. Furthermore, the first-order energy expression is particularly convenient for analysis of the rotational spectrum for the distortion constants, since these constants enter linearly into the energy expression. A first-order treatment of centrifugal distortion has been discussed by Kivelson and Wilson [50] and more recently by Watson [51]. The first-order perturbation treatment will involve averaging the perturbing operator $\mathscr{H}'_d$ over the wave functions associated with $\mathscr{H}'_r$, that is, the asymmetric rigid rotor wave functions. The first-order distortion energy for a particular level J_τ is then

$$E_d = \langle \mathscr{H}'_d \rangle = (J, \tau | \mathscr{H}'_d | J, \tau) \tag{8.48}$$

In the following discussion it will be seen that by consideration of only a first-order approximation, or molecules with particular symmetry, the general distortion Hamiltonian, (8.47), may be greatly simplified. In Section 4, order-of-magnitude arguments are used to provide a more general simplified Hamiltonian.

As noted before, an important quantum mechanical property of angular momentum components is that they do not commute. When the total angular momentum is referred to axes fixed in the molecule (x, y, z), the following commutation rules are found to exist:

$$\begin{aligned} P_xP_y - P_yP_x &= -iP_z \\ P_yP_z - P_zP_y &= -iP_x \\ P_zP_x - P_xP_z &= -iP_y \end{aligned} \tag{8.49}$$

where P_α is in units of $\hbar$. Except for the sign change on the right-hand side of the equations, these are similar to the set of commutation rules that apply for the components of the angular momentum resolved along the space-fixed (nonrotating) axes (X, Y, Z). Because of the noncommuting character of the angular momentum components P_α, there is a total of 81 terms in the sum of (8.47). However, from (8.20) it is seen that many distortion constants are equivalent, and we have, in fact,

$$\tau_{\alpha\beta\gamma\delta} = \tau_{\alpha\beta\delta\gamma} = \tau_{\beta\alpha\delta\gamma} = \tau_{\beta\alpha\gamma\delta} = \tau_{\gamma\delta\alpha\beta} = \tau_{\gamma\delta\beta\alpha} = \tau_{\delta\gamma\beta\alpha} = \tau_{\delta\gamma\alpha\beta} \tag{8.50}$$

Furthermore, many of the terms will not contribute to first-order. The rigid asymmetric rotor wave functions are classified according to the Four-group V, and hence if the integral (8.48) is to be nonvanishing, the product of the angular momentum components must be invariant to the symmetry operations of the group, that is, belong to the totally symmetric species A. The transformation properties of P_α, P_b, P_c have already been given (see Chapter VII, Section 2), and we may summarize the results by saying that under the group V, P_a transforms according to B_a, P_b according to B_b, and P_c according to B_c. The symmetry of a given angular momentum term will be given by the direct product of the symmetries of the individual members of the term. It follows, therefore, that all terms involving odd powers of any angular momentum component will not contribute to first-order because such terms will belong to one of the B symmetry species and their average values will vanish. Thus, for any asymmetric top only 21 constants (many of which are equal to each other) will contribute to a first-order approximation. Moreover, when the symmetry of the molecule is taken into account ($\mathscr{H}'$ and thus $\mathscr{H}'_d$ must be invariant under the symmetry operations of the molecular point group), it is found that the distortion constants of these odd-power terms must vanish identically for molecules of orthorhombic symmetry [52, 53]. Orthorhombic symmetry includes molecules belonging to either the C_{2v}, V_h, or V point groups. The eight asymmetric rotor point groups are summarized in Table 8.4. Noting that P_α transforms as the corresponding rotation R_α of the point group, and carrying out the symmetry operations of the group on $P_\alpha P_\beta P_\gamma P_\delta$, Parker [53] found that the nonvanishing $\tau_{\alpha\beta\gamma\delta}$ are as given in Table 8.4. All other terms are antisymmetric under one or more symmetry operations of the particular group. As a consequence, for molecules of orthorhombic symmetry there are only 21 nonvanishing coefficients in $\mathscr{H}'_d$ to any approximation; since many of these constants are equal, one is left finally with only nine different quartic distortion constants:

$$\tau_{\alpha\alpha\alpha\alpha},\ \tau_{\alpha\alpha\beta\beta}=\tau_{\beta\beta\alpha\alpha},\ \tau_{\alpha\beta\alpha\beta}=\tau_{\alpha\beta\beta\alpha}=\tau_{\beta\alpha\beta\alpha}=\tau_{\beta\alpha\alpha\beta} \qquad (\alpha, \beta=x, y, \text{ or } z) \tag{8.51}$$

with $\alpha \neq \beta$. Therefore, the various surviving $P_\alpha P_\beta P_\gamma P_\delta$ terms of $\mathscr{H}'_d$ may be arranged into nine groups, each of which has the same τ coefficient.

A further reduction of the distortion Hamiltonian to six groups of terms is still possible by an appeal to the commutation rules. By means of the commutation rules, (8.49), the following relations may be derived [48]:

$$(P_\alpha P_\beta + P_\beta P_\alpha)^2 = 2(P_\alpha^2 P_\beta^2 + P_\beta^2 P_\alpha^2) + 3P_\gamma^2 - 2P_\alpha^2 - 2P_\beta^2 \tag{8.52}$$

here $\alpha \neq \beta \neq \gamma$ and with α, β, γ to be taken in cyclic order. Note that since $P_\alpha P_\beta$ do not commute, there will be four different P^4 terms when the left-hand side is multiplied out. From these three relations, further simplification results, since one can eliminate the $\tau_{\alpha\beta\alpha\beta}(P_\alpha P_\beta + P_\beta P_\alpha)^2$ terms of $\mathscr{H}'_d$. As a result of this simplification, the coefficients $\tau_{\alpha\beta\alpha\beta}$ are folded into those of $(P_\alpha^2 P_\beta^2 + P_\beta^2 P_\alpha^2)$, and terms in P_α^2, P_β^2, and P_γ^2 are introduced which can be absorbed into the rigid rotor part of the Hamiltonian $\mathscr{H}'_r$, thus giving a new definition to the rotational constants. This procedure leads to the following form of the Hamiltonian

$$\mathscr{H} = \mathscr{H}_r + \mathscr{H}_d \tag{8.53}$$

Table 8.4 Asymmetric-Rotator Point Groups and Nonvanishing τ Constants[a]

Crystallographic Nomenclature	*Group Symbol*	*Group Operations Other Than Identity Operation*	*Sets with Nonzero Distortion Constants*[b]
Triclinic	C_1	None	I_1, I_2, II_1, II_2
	$C_i=S_2$	i	I_1, I_2, II_1, II_2
Monoclinic	$C_S=C_{1h}$	σ, (a) $\sigma(xy)$	I_1, I_2
		(b) $\sigma(yz)$	I_1, II_1
		(c) $\sigma(zx)$	I_1, II_2
	C_2	C_2, (a) $C_2(z)$	I_1, I_2
		(b) $C_2(x)$	I_1, II_1
		(c) $C_2(y)$	I_1, II_2
	C_{2h}	C_2, σ_h, i, (a) $C_2(z), \sigma(xy)$	I_1, I_2
		(b) $C_2(x), \sigma(yz)$	$I_1 II_1$
		(c) $C_2(y), \sigma(zx)$	$I_1 II_2$
Orthorhombic	C_{2v}	C_2, two σ_v,	I_1
	$V=D_2$	three mutually $\perp C_2$	I_1
	$V_h=D_{2h}$	three mutually $\perp C_2$, i, three mutually $\perp \sigma$,	I_1

	Nonvanishing $\tau_{\alpha\beta\gamma\delta}$	*Set*
1	τ_{xxxx}	
2	τ_{yyyy}	
3	τ_{zzzz}	
4	$\tau_{yyzz}=\tau_{zzyy}$	
5	$\tau_{zzxx}=\tau_{xxzz}$	I_1
6	$\tau_{xxyy}=\tau_{yyxx}$	
7	$\tau_{yzyz}=\tau_{zyzy}=\tau_{yzzy}=\tau_{zyyz}$	
8	$\tau_{zxzx}=\tau_{xzxz}=\tau_{zxxz}=\tau_{xzzx}$	
9	$\tau_{xyxy}=\tau_{yxyx}=\tau_{xyyx}=\tau_{yxxy}$	
10	$\tau_{xxxy}=\tau_{xxyx}=\tau_{xyxx}=\tau_{yxxx}$	
11	$\tau_{yyyx}=\tau_{yyxy}=\tau_{yxyy}=\tau_{xyyy}$	I_2
12	$\tau_{xyzz}=\tau_{yxzz}=\tau_{zzxy}=\tau_{zzyx}$	
13	$\tau_{xzzy}=\tau_{yzzx}=\tau_{zxyz}=\tau_{zyxz}=\tau_{zxzy}=\tau_{zyzx}=\tau_{xzyz}=\tau_{yzxz}$	
14	$\tau_{yyyz}=\tau_{yyzy}=\tau_{yzyy}=\tau_{zyyy}$	
15	$\tau_{zzzy}=\tau_{zzyz}=\tau_{zyzz}=\tau_{yzzz}$	
16	$\tau_{yzxx}=\tau_{zyxx}=\tau_{xxyz}=\tau_{xxzy}$	II_1
17	$\tau_{yxxz}=\tau_{zxxy}=\tau_{xyzx}=\tau_{xzyx}=\tau_{xyxz}=\tau_{xzxy}=\tau_{yxzx}=\tau_{zxyx}$	
18	$\tau_{xxxz}=\tau_{xxzx}=\tau_{xzxx}=\tau_{zxxx}$	
19	$\tau_{zzzx}=\tau_{zzxz}=\tau_{zxzz}=\tau_{xzzz}$	
20	$\tau_{zxyy}=\tau_{xzyy}=\tau_{yyzx}=\tau_{yyxz}$	II_2
21	$\tau_{zyyx}=\tau_{xyyz}=\tau_{yzxy}=\tau_{yxzy}=\tau_{yzyx}=\tau_{yxyz}=\tau_{zyxy}=\tau_{xyzy}$	

[a]From Parker [53].

[b]Corresponding angular momentum operators of Set I may have nonvanishing matrix elements in a symmetric rotor basis of the type $(K|K)$, $(K|K\pm 2)$, $(K|K\pm 4)$; whereas Set II operators may have nonvanishing elements of the type $(K|K\pm 1)$ and $(K|K\pm 3)$ only.

$$\mathscr{H}_r = AP_z^2 + BP_x^2 + CP_y^2 \tag{8.54}$$

$$\mathscr{H}_d = \frac{1}{4}\sum_{\alpha,\beta} \tau'_{\alpha\alpha\beta\beta} P_\alpha^2 P_\beta^2 \tag{8.55}$$

This Hamiltonian is applicable for molecules with orthorhombic symmetry or for any molecule to a first order of approximation. The relations between the old and new coefficients are given in Table 8.5. From the table we see that the corrected rotational constants A, B, and C depend on the distortion constants. This introduces another ambiguity into the definition of the moments of inertia in addition to that arising from the vibrational effects. If the distortion constants are not known, with the result that A', B', C' cannot be obtained, additional uncertainty is transmitted to the derived structural parameters although this effect is very small.

The first-order distortion energy may now be written from (8.55) in the following form

$$E_d = \frac{1}{8}\sum_{\alpha,\beta} \tau'_{\alpha\alpha\beta\beta} \langle P_\alpha^2 P_\beta^2 + P_\beta^2 P_\alpha^2 \rangle \qquad (\alpha, \beta = x, y, \text{ or } z) \tag{8.56}$$

with $\tau'_{\alpha\alpha\beta\beta} = \tau'_{\beta\beta\alpha\alpha}$ and with $\langle P_\alpha^2 P_\beta^2 + P_\beta^2 P_\alpha^2 \rangle$ as the average value of the appropriate operators. To proceed further, one needs explicit expressions for the various average values appearing in the above equation. These are usually expressed in terms of P^2, $\langle P_z^n \rangle$ ($n = 2$ and 4), and the rigid rotor energy E_r, since the latter quantities are easily determined from the diagonalization of the rigid rotor problem. Employing the expressions for $\mathscr{H}_r$ and P^2, one may write

$$P_x^2 = \frac{[\mathscr{H}_r - (A-C)P_z^2 - CP^2]}{(B-C)} \tag{8.57}$$

$$P_y^2 = \frac{[\mathscr{H}_r - (A-B)P_z^2 - BP^2]}{(C-B)} \tag{8.58}$$

Table 8.5 Definition of the Coefficients in the Distortion Hamiltonian

$\tau'_{xxxx} = \hbar^4 \tau_{xxxx}$	$\tau'_{xxzz} = \hbar^4(\tau_{xxzz} + 2\tau_{xzxz})$
$\tau'_{yyyy} = \hbar^4 \tau_{yyyy}$	$\tau'_{xxyy} = \hbar^4(\tau_{xxyy} + 2\tau_{xyxy})$
$\tau'_{zzzz} = \hbar^4 \tau_{zzzz}$	$\tau'_{yyzz} = \hbar^4(\tau_{yyzz}\ ''2\tau_{yzyz})$

$$A = A' + \frac{\hbar^4}{4}(3\tau_{xyxy} - 2\tau_{xzxz} - 2\tau_{yzyz})$$

$$B = B' + \frac{\hbar^4}{4}(3\tau_{yzyz} - 2\tau_{xyxy} - 2\tau_{xzxz})$$

$$C = C' + \frac{\hbar^4}{4}(3\tau_{xzxz} - 2\tau_{xyxy} - 2\tau_{yzyz})$$

from which one may construct the desired average value equations noting that with P in units of $\hbar$

$$
\begin{aligned}
\langle P^4\rangle &= J^2(J+1)^2 \\
\langle \mathscr{H}_r^2\rangle &= E_r^2 \\
\langle \mathscr{H}_r P^2 + P^2 \mathscr{H}_r\rangle &= 2E_r J(J+1) \\
\langle \mathscr{H}_r P_z^2 + P_z^2 \mathscr{H}_r\rangle &= 2E_r \langle P_z^2\rangle \\
\langle P^2 P_z^2 + P_z^2 P^2\rangle &= 2J(J+1)\langle P_z^2\rangle
\end{aligned}
\tag{8.59}
$$

The average values of these operators are evaluated in a rigid asymmetric rotor basis, the basis in which the matrix representation of $\mathscr{H}_r$ is diagonal with diagonal elements $E_r = \langle \mathscr{H}_r\rangle = A\langle P_z^2\rangle + B\langle P_x^2\rangle + C\langle P_y^2\rangle$. The appropriate average value equations of the quartic angular momentum terms are listed in Table 8.6. It may be pointed out that the average value equations can also be expressed as desired in terms of Ray's $E(\kappa)$ or Wang's $W(b_p)$ reduced energy by simple definition of A, B, C in the appropriate manner, that is, $A \to 1$, $B \to \kappa$, $C \to -1$, $E_r \to E(\kappa)$ or $A \to 1$, $B \to -b_p$, $C \to b_p$, $E_r \to W(b_p)$.

Watson [51] has shown that, in addition to the five equations of (8.59), another relation may be obtained from the average over the commutation relation

$$
\begin{aligned}
i[\mathscr{H}_r, (P_xP_yP_z + P_zP_yP_x)] = {} & 2(C-B)(P_x^2P_y^2 + P_y^2P_x^2 + 2P_z^2) \\
& + 2(A-C)(P_y^2P_z^2 + P_z^2P_y^2 + 2P_x^2) \\
& + 2(B-A)(P_x^2P_z^2 + P_z^2P_x^2 + 2P_y^2)
\end{aligned}
\tag{8.60}
$$

Here the following relations,

$$
\begin{aligned}
[P_\alpha^2, (P_xP_yP_z + P_zP_yP_x)] = {} & 2i(P_\alpha^2P_\beta^2 + P_\beta^2P_\alpha^2) - 2i(P_\alpha^2P_\gamma^2 + P_\gamma^2P_\alpha^2) \\
& + 4i(P_\gamma^2 - P_\beta^2)
\end{aligned}
\tag{8.61}
$$

(with α, β, $\gamma = x$, y, or z; $\alpha \neq \beta \neq \gamma$, and α, β, and γ taken in cyclic order) derived from the angular momentum commutation rules, have been used to reduce the right-hand side of (8.60). Now the average value of the commutator $<[\mathscr{H}, O_p]>$ vanishes for any operator O_p averaged over eigenfunctions of $\mathscr{H}$. Therefore, the average value of the commutator on the left-hand side of (8.60) vanishes, and one obtains a relation between the average values of the quartic and quadratic angular momentum terms in the rigid rotor basis. The vanishing of such commutators has also found application to internal rotation problems [54].

By insertion of the appropriate average value equations given in Table 8.6, the following expression is attained [51]

$$
\begin{aligned}
E_r^2 = {} & (B-C)^2\langle P_z^2\rangle + (A-C)(C-B)\langle P_x^2\rangle + (A-B)(B-C)\langle P_y^2\rangle \\
& + (B+C)E_rJ(J+1) - BCJ^2(J+1)^2 - 2(AB+AC-2BC) \\
& \times J(J+1)\langle P_z^2\rangle - 3(A-B)(A-C)\langle P_z^4\rangle \\
& + 2(2A-B-C)E_r\langle P_z^2\rangle
\end{aligned}
\tag{8.62}
$$

This relation may now be used to eliminate E_r^2 from the average value equations of Table 8.6 so that any given average value will at most be a function of five different terms, P^4, $P^2\langle P_z^2\rangle$, $\langle P_z^4\rangle$, P^2E_r, and $E_r\langle P_z^2\rangle$. As a result, the distortion

Table 8.6 The Average Values of the Distortion Operators in the Asymmetric Rigid Rotor Basis[a]

$$\langle P_x^4\rangle=\left(\frac{1}{B-C}\right)^2[C^2P^4+2C(A-C)P^2\langle P_z^2\rangle+(A-C)^2\langle P_z^4\rangle-2CP^2E_r-2(A-C)E_r\langle P_z^2\rangle+E_r^2]$$

$$\langle P_y^4\rangle=\left(\frac{1}{B-C}\right)^2[B^2P^4+2B(A-B)P^2\langle P_z^2\rangle+(A-B)^2\langle P_z^4\rangle-2BP^2E_r$$
$$-2(A-B)E_r\langle P_z^2\rangle+E_r^2]$$

$$\langle P_x^2P_y^2+P_y^2P_x^2\rangle=-2\left(\frac{1}{B-C}\right)^2[BCP^4+(AB+AC-2BC)P^2\langle P_z^2\rangle$$
$$+(A-B)(A-C)\langle P_z^4\rangle-(B+C)P^2E_r-(2A-B-C)E_r\langle P_z^2\rangle+E_r^2]$$

$$\langle P_x^2P_z^2+P_z^2P_x^2\rangle=-2\left(\frac{1}{B-C}\right)[CP^2\langle P_z^2\rangle+(A-C)\langle P_z^4\rangle-E_r\langle P_z^2\rangle]$$

$$\langle P_y^2P_z^2+P_z^2P_y^2\rangle=2\left(\frac{1}{B-C}\right)[BP^2\langle P_z^2\rangle+(A-B)\langle P_z^4\rangle-E_r\langle P_z^2\rangle]$$

[a] $P^2=J(J+1)$, E_r is the rigid rotor energy and $\langle P_z^n\rangle$ is the average value of P_z^n in the rigid asymmetric rotor basis.

energy to first order will depend on only five combinations of the $\tau'_{\alpha\alpha\beta\beta}$. The terms in $\langle P_x^2\rangle$, $\langle P_y^2\rangle$, and $\langle P_z^2\rangle$ introduced by the elimination of E_r^2 can be absorbed into E_r, thus resulting in a redefinition of the effective rotational constants A, B, and C. This will not significantly affect the average values of the quartic terms, and the average values in (8.64) may be identified with the asymmetric rotor basis specified by (8.65). This produces, however, another small ambiguity in the definition of the moments of inertia, since the quantities A', B', and C' should define the effective moments of inertia, whereas those of (8.65) are determined in practice.

By use then of (8.62) and Table 8.6 and by insertion of the average values in (8.56) followed by collection of terms and considerable manipulation, one arrives at the following first-order expression for the energy of a semirigid asymmetric rotor [50, 51]

$$E=E_r+E_d \tag{8.63}$$

$$E_d=-d_JJ^2(J+1)^2-d_{JK}J(J+1)\langle P_z^2\rangle-d_K\langle P_z^4\rangle$$
$$-d_{EJ}E_rJ(J+1)-d_{EK}E_r\langle P_z^2\rangle \tag{8.64}$$

Here E_r is the energy of a rigid rotor with the following rotational constants

$$\mathscr{A}=A+16R_6$$
$$\mathscr{B}=B-\frac{16R_6(A-C)}{B-C}$$
$$\mathscr{C}=C+\frac{16R_6(A-B)}{B-C} \tag{8.65}$$

The d coefficients of Watson [51] are defined in Table 8.7 in terms of the D's, R's, and δ_J used by Kivelson and Wilson [50]. These latter constants are linear combinations of the τ's and are also defined in Table 8.7. Equation (8.61) could equally well have been used for elimination of $\langle P_z^4 \rangle$ rather than E_r^2. The calculation of $\langle P_z^4 \rangle$ could therefore be avoided, but the analogy with the symmetric top would be lost. The energy expression obtained by elimination of $\langle P_z^4 \rangle$ may be found elsewhere [55]. By application of (8.60) it is possible to express the average values of the quartic terms, $\langle P_\alpha^2 P_\beta^2 \rangle (= \langle P_\beta^2 P_\alpha^2 \rangle)$, in terms of the average values of the quadratic terms, $\langle P_\alpha^2 \rangle$. The detailed expressions are given elsewhere [55]. Results for the d coefficients of some asymmetric rotors are collected in Table 8.8.

It has been found [56–58] in studies of a number of nonplanar molecules with the first-order expression, (8.63), that there is high correlation between d_K and d_{EK} and between d_{JK} and d_{EK}. Also, for $\kappa < -0.97$, the normal equations become highly ill-conditioned (see Section 4). Equation 8.64 can be written in a more suitable form involving the Δ coefficients to be introduced later. In particular, elimination of E_r from (8.64) by means of the relations

$$\mathscr{H}_r = \tfrac{1}{2}(\mathscr{B}+\mathscr{C})J(J+1) + [\mathscr{A} - \tfrac{1}{2}(\mathscr{B}+\mathscr{C})]P_z^2 + \tfrac{1}{2}(\mathscr{B}-\mathscr{C})(P_x^2 - P_y^2) \quad (8.66)$$

and

$$\langle \mathscr{H}_r P_z^2 + P_z^2 \mathscr{H}_r \rangle = 2E_r \langle P_z^2 \rangle, \ \langle \mathscr{H}_r \rangle = E_r \quad (8.67)$$

Table 8.7 Distortion Coefficients of the First-Order Energy Expression

$$d_J = D_J - \frac{2\delta_J(B+C)}{B-C} - 2R_6$$

$$d_{JK} = D_{JK} - 2\sigma\delta_J + 4(R_5 + 2\sigma R_6)\frac{B+C}{B-C} + 12R_6$$

$$d_K = D_K + 4\sigma(R_5 + 2\sigma R_6) - 10R_6$$

$$d_{EJ} = \frac{4\delta_J}{B-C}$$

$$d_{EK} = \frac{-8(R_5 + 2\sigma R_6)}{B-C}$$

$$\sigma = \frac{2A-B-C}{B-C}$$

$$D_J = -\tfrac{1}{32}\{3\tau_{xxxx} + 3\tau_{yyyy} + 2(\tau_{xxyy} + 2\tau_{xyxy})\}\hbar^4$$
$$D_K = D_J - \tfrac{1}{4}\{\tau_{zzzz} - (\tau_{xxzz} + 2\tau_{xzxz}) - (\tau_{yyzz} + 2\tau_{yzyz})\}\hbar^4$$
$$D_{JK} = -D_J - D_K - \tfrac{1}{4}\tau_{zzzz}\hbar^4$$
$$R_5 = -\tfrac{1}{32}\{\tau_{xxxx} - \tau_{yyyy} - 2(\tau_{xxzz} + 2\tau_{xzxz}) + 2(\tau_{yyzz} + 2\tau_{yzyz})\}\hbar^4$$
$$R_6 = \tfrac{1}{64}\{\tau_{xxxx} + \tau_{yyyy} - 2(\tau_{xxyy} + 2\tau_{xyxy})\}\hbar^4$$
$$\delta_J = -\tfrac{1}{16}\{\tau_{xxxx} - \tau_{yyyy}\}\hbar^4$$

Table 8.8 Rotational and Centrifugal Distortion Constants (*d* Coefficients) of Some Asymmetric Rotors (MHz)[a]

Molecule	$\mathscr{A}$	$\mathscr{B}$	$\mathscr{C}$	d_J	d_{JK}	d_K	$d_{EJ}(\times 10^6)$	$d_{EK}(\times 10^6)$	*Ref.*
$(CH_3)_2SO$	7,036.49	6,910.93	4,218.78	0.043	10.550	−4.172	−5.20	−1517	*d*
$(CH_3)_2S$[c]	17,809.73	7,621.10	5,717.77	−0.03158	−0.1503	0.058	5.943	7.280	*d*
$CH_2(OH)CHO$[b]	18,446.41	6,525.04	4,969.27	−0.02714	−0.0732	0.0883	5.648	−3.40	*e*
CH_2FCONH_2[c]	9,884.36	4,059.67	2,932.47	−0.00206	−0.0312	−0.0507	0.801	8.41	*f*
1, 3 Dioxane[b]	4,999.93	4,807.61	2,757.11	0.00198	−0.2088	0.09118	−0.1839	42.21	*g*
Pyridine	6,029.28	5,804.95	2,959.25	0.00367	0.1738	−0.0868	−0.22	−29.87	*h*
3-Fluorophenol[c]	3,748.49	1,797.71	1,215.05	0.00023	−0.0110	−0.0150	−0.20	7.2	*i*
trans.CH_3CHCH_2NH[c]	16,892.44	6,533.61	5,761.30	−0.00457	0.020	0.020	1.228	−5.0	*j*
CH_3SCl	17,341.83	4,603.78	3,719.10	−0.01167	−0.2087	−0.3692	3.468	37.23	*k*
HBF_2[c]	52,896.09	10,498.29	8,740.46	−0.0386	−1.134	−2.96	4.91	89.5	*l*
N_2F_4	5,576.19	3,189.42	2,813.16	−0.01389	−0.2566	−0.2087	4.7984	81.164	*m*

[a]Note constants d_{EJ} and d_{EK} are dimensionless.

[b]III^r representation.

[c]I^r representation.

[d]H. Dreizler, *Z. Naturforsch*, **21a**, 1719 (1966).

[e]K. M. Marstokk and H. Møllendal, *J. Mol. Struct.*, **5**, 205 (1970).

[f]K. M. Marstokk and H. Møllendal, *J. Mol. Struct.*, **22**, 287 (1974).

[g]R. S. Lowe and R. Kewley, *J. Mol. Spectrosc.*, **60**, 312 (1976).

[h]R. R. Filgueira, A. C. Fantoni, and L. M. Boggia, *J. Mol. Spectrosc.*, **78**, 175 (1979).

[i]A. I. Jaman, R. N. Nandi, and D. K. Ghosh, *J. Mol. Spectrosc.*, **86**, 269 (1981).

[j]C. F. Su and E. B. Beason, *J. Chem. Phys.*, **59**, 759 (1973).

[k]A. Guarnieri, *Z. Naturforsch.*, **25a**, 18 (1970).

[l]T. Kasuya, W. J. Lafferty, and D. R. Lide, *J. Chem. Phys.*, **48**, 1 (1968).

[m]V. K. Kaushik and P. Venkateswarlu, *Chem. Phys. Lett.*, **46**, 426 (1977).

Table 8.9 Relations Between the d- and Δ Distortion Coefficients[a]

$$d_J = \Delta_J - \frac{2\delta_J(\mathscr{B}+\mathscr{C})}{\mathscr{B}-\mathscr{C}}$$

$$d_{JK} = \Delta_{JK} - \frac{2\delta_K(\mathscr{B}+\mathscr{C})}{\mathscr{B}-\mathscr{C}} - \frac{2\delta_J(2\mathscr{A}-\mathscr{B}-\mathscr{C})}{\mathscr{B}-\mathscr{C}}$$

$$d_K = \Delta_K - \frac{2\delta_K(2\mathscr{A}-\mathscr{B}-\mathscr{C})}{\mathscr{B}-\mathscr{C}}$$

$$d_{EJ} = \frac{4\delta_J}{\mathscr{B}-\mathscr{C}}$$

$$d_{EK} = \frac{4\delta_K}{\mathscr{B}-\mathscr{C}}$$

[a] Here $(x, y, z) \equiv (b, c, a)$, alternate choices made by reidentification of $\mathscr{A}$, $\mathscr{B}$, $\mathscr{C}$.

yields

$$E_d = -\Delta_J J^2(J+1)^2 - \Delta_{JK} J(J+1)\langle P_z^2\rangle - \Delta_K\langle P_z^4\rangle - 2\delta_J J(J+1)\langle P_x^2 - P_y^2\rangle - \delta_K\langle P_z^2(P_x^2-P_y^2)+(P_x^2-P_y^2)P_z^2\rangle \quad (8.68)$$

The relations between the d- and Δ coefficients are given in Table 8.9. In the present discussion we have used an I^r representation ($x\to b, y\to c, z\to a$) for $\mathscr{H}_r$. Alternate choices are readily obtained by reidentification of $\mathscr{A}$, $\mathscr{B}$, $\mathscr{C}$.

Simplifications for Planar Asymmetric Rotors and Symmetric Tops

For planar asymmetric rotor molecules, significant simplification results, as is shown by Dowling [59]. For a planar, prolate-type rotor, the unique symmetric-top axis z will lie in the plane of the molecule, whereas for an oblate top the limiting symmetric-top axis will be perpendicular to this plane. For purposes of discussion, assume that the molecule lies in the xz plane. Then if α or β is y in (8.24), all the $J_{\alpha\beta}^{(i)}$'s will be zero since all the y_l coordinates of the atoms are zero at equilibrium. Therefore it follows from (8.20) that $\tau_{xyxy} = \tau_{yzyz} = 0$. Similarly, (8.23) yields $J_{xx}^{(i)} + J_{zz}^{(i)} = J_{yy}^{(i)}$ when all y_l coordinates are set equal to zero. This relation in conjunction with (8.20) provides three relations among the τ's. The relations summarized in Table 8.10 are appropriate for both prolate, type I^r, and oblate, type III^l, cases. These relations can be used to reduce the number of independent τ's to only four, for example, τ_{xxxx}, τ_{zzzz}, τ_{xxzz}, and τ_{xzxz}. In this particular case the first-order energy expression can be written in terms of four τ's. The coefficients of the four τ's in the energy expression can be easily found by use of the average values of Table 8.6 and the relations of Table 8.10 in conjunction with (8.56). Evaluation of the four τ's from the

Table 8.10 Distortion Constant Relationships for Planar Asymmetric Rotor Molecules[a]

$$\tau_{bcbc}=\tau_{acac}=0$$

$$\tau_{cccc}=\left(\frac{C}{A}\right)^4\tau_{aaaa}+2\frac{C^4}{A^2B^2}\tau_{aabb}+\left(\frac{C}{B}\right)^4\tau_{bbbb}$$

$$\tau_{bbcc}=\left(\frac{C}{B}\right)^2\tau_{bbbb}+\left(\frac{C}{A}\right)^2\tau_{aabb}$$

$$\tau_{aacc}=\left(\frac{C}{A}\right)^2\tau_{aaaa}+\left(\frac{C}{B}\right)^2\tau_{aabb}$$

[a]The coordinate axes are to be assigned $a\leftrightarrow z$, $b\leftrightarrow x$, $c\leftrightarrow y$ for the prolate case and $a\leftrightarrow y$, $b\leftrightarrow x$, and $c\leftrightarrow z$ in the oblate case; in each case the molecule lies in the ab plane.

Table 8.11 Rotational Distortion Constants for Molecules of C_{3v} Symmetry[a]

Nonvanishing τ's

$\tau_{xxxx}=\tau_{yyyy}$	τ_{zzzz}
$\tau_{xxzz}=\tau_{yyzz}$	$\tau_{xxyy}=\tau_{xxxx}-2\tau_{xyxy}$
$\tau_{xzxz}=\tau_{yzyz}$	$\tau_{yyyz}=-\tau_{xyzx}=-\tau_{xxyz}$[b]

Distortion Coefficients

$$D_J=-\frac{\hbar^4}{4}\tau_{xxxx}\qquad D_{JK}=-2D_J-\frac{\hbar^4}{2}\{\tau_{xxzz}+2\tau_{xzxz}\}$$

$$D_K=-D_J-D_{JK}-\frac{\hbar^4}{4}\tau_{zzzz}$$

[a]y axis lies in a symmetry plane with $I^c_{xx}=I^c_{yy}$, $R_5=R_6=\delta_J=0$. The symmetry axis is z.
[b]Do not contribute in a first-order approximation.

spectrum and use of the relations allows direct evaluation of the seven individual τ's. The rotational constants to be used in the relations of Table 8.10 should be the equilibrium rotational constants, but in practice these usually must be replaced by the effective rotational constants. Also, if the harmonic approximation is not adequate, then the relations are not strictly correct.

For a $C_{pv}(p\neq 4)$ symmetric rotor, R_5, R_6, and δ_J are identically zero while the factors multiplying these constants remain finite in this limit. Furthermore, $\langle P_z^2\rangle=K^2$ and $\langle P_z^4\rangle=K^4$ and hence the first-order (8.64) will reduce to that given previously for a symmetric rotor. For molecules with C_{3v} symmetry,

Table 8.12 Distortion Coefficients for Planar Symmetric Rotors[a]

$D_J = -\frac{\hbar^4}{4}\tau_{xxxx}$	$D_K = -\frac{\hbar^4}{4}\{\tau_{xxxx} - 3\tau_{zzzz}\}$
$D_{JK} = \frac{\hbar^4}{2}\{\tau_{xxxx} - 2\tau_{zzzz}\}$	$D_{JK} = -\frac{2}{3}(D_J + 2D_K)$

[a]The z axis is taken as the symmetry axis that is perpendicular to the plane of the molecule.

symmetric rotors, direct calculation [60] or group theoretical considerations [21] lead to the nonvanishing τ's listed in Table 8.11, and the explicit definitions of D_J, D_K, and D_{JK} in terms of the τ's can be simplified [25]. These results are also included in Table 8.11. For planar symmetric rotors (oblate tops) additional relations exist between the τ's, with further simplification of the distortion coefficients [59]. These are given in Table 8.12. For this case it is evident that if any two distortion constants are determined, the remaining one may be immediately evaluated.

4 REDUCED HAMILTONIAN

From (8.64) it is apparent that no more than five linear combinations of the nine possible τ's can be obtained from an analysis of the rotational spectrum of an asymmetric rotor. It might appear from (8.56) and Table 8.6 that six linear combinations of τ's could be determined from the observed spectrum; because of the linear relation (8.62), however, the resulting energy level equation would have an inherent indeterminacy and would thus not be satisfactory for evaluation of unique values of the distortion constants, although it would be applicable to calculation of the distortion effects with known constants. In (8.64) this indeterminacy has been removed by application of (8.62), leaving five determinable distortion coefficients. Such a problem does not arise for planar molecules when the planarity conditions (see the previous section) are applied, since this effectively removes the indeterminacy resulting from the linear relation between the average values. Most early analysis of centrifugal distortion employed the planar relations and thus avoided the problem. This indeterminacy was first reported by Dreizler, Dendl, and Rudolph [61, 62] in their analysis of dimethylsulfoxide and dimethylsulfide for six quartic distortion constants. They found the least-squares equations to be quite ill-conditioned, and determinate values for the constants could not be obtained.

This is an example of a more general problem that can arise when one attempts to evaluate parameters in any Hamiltonian from the corresponding eigenvalues. The eigenvalues may depend on only certain linear combinations of the parameters. To properly fit the experimental data, one must employ a

Hamiltonian that has any experimentally indeterminable parameters removed from it. This Hamiltonian, which may be obtained by a contact transformation, is termed a "reduced Hamiltonian." The reduced Hamiltonian is to be used to fit experimental data. On the other hand, to interpret the quantities in terms of fundamental molecular parameters, the experimentally determined coefficients must be related to the constants in the original "first-principles Hamiltonian." These relations are, of course, specified by the contact transformation which relates the two Hamiltonians.

The nature of the problem can be readily appreciated by consideration of the rigid rotor Hamiltonian. For a general asymmetric top only three principal moments of inertia can be obtained from the eigenvalues of $\mathscr{H}_r$. For an arbitrary orientation of the molecule-fixed axis, however, the rotational Hamiltonian contains six terms corresponding to six elements of the inertia tensor; see (2.6). But, by application of a unitary transformation a particular axis orientation can be selected for which the inertia tensor has a diagonal form; $\tilde{\mathscr{H}}_r$ then depends on only three independent constants, the three principal moments of inertia, and it is these parameters that can be evaluated from the observed spectrum.

A unitary operator ($U^{-1}=U\dagger$) can be conveniently written as

$$U=e^{iS} \tag{8.69}$$

where S is Hermitian. The transformed Hamiltonian is given by

$$\tilde{\mathscr{H}}=U^{-1}\mathscr{H}U \tag{8.70}$$

The result of applying the appropriate transformation is that the form and eigenvalues of $\tilde{\mathscr{H}}$ are unchanged from $\mathscr{H}$, but the individual coefficients are altered. This fact, along with order of magnitude arguments, have been used by Watson in a series of papers [55, 63, 64] to obtain the reduced Hamiltonian. Since the eigenvalues must remain the same, only those linear combinations of coefficients which remain invariant to the transformation can be evaluated from the eigenvalues, and hence are determinable coefficients.

The most general form of the rotational Hamiltonian, correct to terms in P^6, is given by

$$\mathscr{H}=\sum_\alpha B_\alpha'^2 P_\alpha^2+\frac{\hbar^4}{4}\sum \tau_{\alpha\beta\gamma\delta}P_\alpha P_\beta P_\gamma P_\delta+\hbar^6\sum \tau_{\alpha\beta\gamma\delta\varepsilon\eta}P_\alpha P_\beta P_\gamma P_\delta P_\varepsilon P_\eta \tag{8.71}$$

where we denote the rotational constants by $B_\alpha'=\hbar^2/2I_\alpha$. From our previous discussion of the P^4 terms we have shown that for orthorhombic symmetry there are nine independent $\tau_{\alpha\beta\gamma\delta}$ (Table 8.4) and that reduction of the fourth-power angular momentum terms $P_\alpha P_\beta P_\gamma P_\delta$ of $\mathscr{H}_d'$ by means of the commutation relations yields a Hamiltonian with six terms, (8.55). Similar simplifications are possible for the P^6 terms. Initially, there are a total of $3^6=729$ coefficients in the P^6 sum. $\mathscr{H}$ must, however, be invariant under all symmetry operations of the point group of the molecule, and a number of coefficients in the P^6 sum must vanish. Application of the group operations of the asymmetric-top point groups to the Hamiltonian reveals that for the orthorhombic point groups (C_{2V}, D_2, D_{2h}) a total of 183 coefficients are nonzero, for the monoclinic point groups (C_{1h}, C_2, C_{2h}) 365 coefficients are nonzero, and for the triclinic point groups (C_1, C_i) all 729 coefficients are nonzero. Enumeration of the P^6 distortion constants, $\tau_{\alpha\beta\gamma\delta\varepsilon\eta}$, for the orthorhombic point groups has been discussed by Chung and Parker [65]. They show that, in general

$$\tau_{\alpha\beta\gamma\delta\varepsilon\eta} = \tau_{\eta\varepsilon\delta\gamma\beta\alpha} \tag{8.72}$$

The 183 coefficients are hence reduced to 105 distinct ones as follows:

1. Three coefficients of the type $\tau_{\alpha\alpha\alpha\alpha\alpha\alpha}$.
2. Fifty-four coefficients of the type $\tau_{\alpha\alpha\alpha\alpha\beta\beta}(\alpha \neq \beta)$, with the subscripts occurring in any order, taking (8.72) into account.
3. Forty-eight coefficients of the type $\tau_{\alpha\alpha\beta\beta\gamma\gamma}(\alpha \neq \beta \neq \gamma)$, with the subscripts occurring in any order, with (8.72) taken into account.

Because of (8.72), coefficients such as τ_{xyyyxy} and τ_{yxyyyx} occurring in (2) are equal, and likewise coefficients such as τ_{xyyxzz} and τ_{zzxyyx} occurring in (3) are equal. Compilation of all 105 distinct coefficients may be found elsewhere [65].

Reduction of the sixth-power angular momentum terms of the rotational Hamiltonian appropriate to an orthorhomic molecule by means of the commutation relations has been discussed by Kneizys et al. [66]. This reduction results in various P^6 coefficients being folded into other coefficients. It is found that the Hamiltonian can be written in the form [63]

$$\begin{aligned}\mathscr{H} = &\sum_{\alpha} B_\alpha P_\alpha^2 + \sum_{\alpha,\beta} T_{\alpha\beta} P_\alpha^2 P_\beta^2 + \sum_{\alpha} \Phi_{\alpha\alpha\alpha} P_\alpha^6 \\ &+ \sum_{\alpha \neq \beta} \Phi_{\alpha\alpha\beta}(P_\alpha^4 P_\beta^2 + P_\beta^2 P_\alpha^4) \\ &+ \Phi_{xyz}(P_x^2 P_y^2 P_z^2 + P_z^2 P_y^2 P_x^2)\end{aligned} \tag{8.73}$$

where the coefficients are all real. (See also Table 8.13 for an alternate form.) The rotational constants B_α are defined in Table 8.5, as

$$B_\alpha = (\hbar^2/2I_\alpha) + (\hbar^2/4)(3\tau_{\beta\gamma\beta\gamma} - 2\tau_{\alpha\beta\alpha\beta} - 2\tau_{\alpha\gamma\alpha\gamma}),\ \alpha \neq \beta \neq \gamma \equiv (x, y, \text{or } z) \tag{8.74}$$

Once a particular choice of representation has been made, the various B_x, B_y, B_z can be identified with the A, B, C in a particular order. Here an alternate notation has been adopted for the effective quartic distortion coefficients

$$T_{\alpha\alpha} = \frac{1}{4}\tau'_{\alpha\alpha\alpha\alpha} = \frac{\hbar^4}{4}\tau_{\alpha\alpha\alpha\alpha} \tag{8.75}$$

$$T_{\alpha\beta} = \frac{1}{4}\tau'_{\alpha\alpha\beta\beta} = \frac{\hbar^4}{4}(\tau_{\alpha\alpha\beta\beta} + 2\tau_{\alpha\beta\alpha\beta}),\ \alpha \neq \beta \tag{8.76}$$

with $T_{\alpha\beta} = T_{\beta\alpha}$. The 10 $\Phi_{\alpha\beta\gamma}$ coefficients are the effective sextic distortion coefficients. They are linear combinations of the various 105 P^6 distortion constants mentioned previously, and explicit expressions are available [66]. The subscripts on the distortion coefficients are indicative of the angular momentum operator associated with the coefficient.

An alternate and more convenient form of the general power series of (8.71) has been introduced by Watson [55], the so-called standard form, that is,

$$\mathscr{H}_{st} = \sum_{p,q,r=0}^{\infty} h_{pqr}(P_x^p P_y^q P_z^r + P_z^r P_y^q P_x^p) \tag{8.77}$$

Any term in the general Hamiltonian, (8.71), can, by means of the commutation relations, be cast into this form at the expense of introducting terms of lower degree. There is one term for each combination of powers of P_x, P_y, P_z. Equation 8.55 is hence in standard form, and the fact that $\tau_{\alpha\beta\alpha\beta}$ cannot be separated from $\tau_{\alpha\alpha\beta\beta}$ is taken care of automatically by writing the Hamiltonian in the standard form. (Note, e.g., $T_{xy}=h_{220}$.)

If one exploits the requirement that the Hamiltonian is invariant to the operation of Hermitian conjugation and time reversal [55], only terms with even values of $n=p+q+r$ are allowed, with the corresponding coefficients h_{pqr} real. The number of terms of total degree $n=p+q+r$ is $\frac{1}{2}(n+1)(n+2)$. Thus, for $n=2$, 4, and 6, there are 6, 15, and 28 terms, respectively. Of these terms for a given n, there are $\frac{1}{8}(n+2)(n+4)$ terms with $pqr=eee$ (e even) and $\frac{1}{8}n(n+2)$ terms each with $pqr=eoo$, oeo, and ooe (o odd). Since the Hamiltonian must belong to the totally symmetric symmetry species of the molecular point group, one finds [55] for molecules of orthorhombic symmetry that p, q, and r are, in addition, each required to be even in (8.77). Thus, it is clear that for $n=2$, 4, and 6, there are in the standard form $\frac{1}{8}(n+2)(n+4)=3$, 6, and 10 terms, respectively, in agreement with (8.73).

It should be noted that the Hamiltonian (8.73) holds even for molecules of non-orthorhombic symmetry. For such molecules, additional terms such as ooe appear in the standard form, for example, h_{310}, h_{112}, and these contribute in higher order. Elements of the unitary transformation S_3 (S_5 for $\mathscr{H}_d^{(6)}$) can, however, be chosen to eliminate such terms, and thereby to give a Hamiltonian of orthorhombic form with coefficients modified from their original values. The important changes occur in the higher-order P^6 coefficients, and the modified definitions of the $\Phi_{\alpha\beta\gamma}$ are given elsewhere [64].

The orthorhombic form of the Hamiltonian can be further reduced by means of a unitary transformation as indicated previously. The transformation is conveniently represented as two successively applied transformations

$$U=e^{iS_3}e^{iS_5} \tag{8.78}$$

where terms to P^6 are to be considered. The invariant properties discussed previously show that S has a form similar to (8.77) with $p+q+r=n$, and n odd, viz.

$$S=\sum s_{pqr}(P_x^p P_y^q P_z^r + P_z^r P_y^q P_x^p) \tag{8.79}$$

Terms of the same degree are denoted by S_n. For molecules of orthorhombic symmetry, only the terms [55]

$$S_3=s_{111}(P_xP_yP_z+P_zP_yP_x) \tag{8.80}$$

and

$$S_5 = s_{311}(P_x^3 P_y P_z + P_z P_y P_x^3) + s_{131}(P_x P_y^3 P_z + P_z P_y^3 P_x) + s_{113}(P_x P_y P_z^3 + P_z^3 P_y P_x) \tag{8.81}$$

are required. The s_{pqr} are real coefficients and are the parameters of the unitary transformation that are to be chosen to achieve the reduction. Other terms of S_3 and S_5 come into play when one wishes to remove nonorthorhombic terms in (8.77) as mentioned previously. For such molecules, the terms in $\mathscr{H}$ that are not totally symmetric in the Four-group are removed by non-totally symmetric terms in S and they appear in terms of higher degree in the Hamiltonian.

Now the order of magnitude of the terms in (8.77) may be taken approximately as

$$h_{pqr} \sim \kappa^{2(p+q+r)} E_e \tag{8.82}$$

where the rotational, vibrational, and electronic energy are related approximately by $E_r \sim \kappa^2 E_v \sim \kappa^4 E_e (\kappa \ll 1)$. The rigid rotor Hamiltonian is thus of order $\kappa^4 E_e$. Likewise, the quartic distortion Hamiltonian is of order $\kappa^8 E_e$, and the sextic distortion Hamiltonian is of order $\kappa^{12} E_e$. The s_{pqr} satisfy the order-of-magnitude relation

$$s_{pqr} \sim \kappa^{2(p+q+r-1)} \tag{8.83}$$

Thus, for example, s_{111} is of order of magnitude κ^4, which is on the order of the ratio of a quartic distortion coefficient to a rotational constant (τ/B). Expressing the exponentials in (8.70) as a power series expansion (see Appendix C) and arranging by order of magnitude, one finds

$$\tilde{\mathscr{H}} = U^{-1} \mathscr{H} U = \tilde{\mathscr{H}}_r + \tilde{\mathscr{H}}_d^{(4)} + \tilde{\mathscr{H}}_d^{(6)} \tag{8.84}$$

where

$$\tilde{\mathscr{H}}_r = \mathscr{H}_r \tag{8.85}$$

$$\tilde{\mathscr{H}}_d^{(4)} = \mathscr{H}_d^{(4)} + i[\mathscr{H}_r, S_3] \tag{8.86}$$

$$\tilde{\mathscr{H}}_d^{(6)} = \mathscr{H}_d^{(6)} + i[\mathscr{H}_r, S_5] + i[\mathscr{H}_d^{(4)}, S_3] - \tfrac{1}{2}[[\mathscr{H}_r, S_3], S_3] \tag{8.87}$$

which is correct to order $\kappa^{12} E_e$, where $[A, B]$ is the commutator $AB - BA$. Note, for example, that $\mathscr{H}_d^{(4)}$ and $[\mathscr{H}_r, S_3]$ are both of order $\kappa^8 E_e$. Here s_{111} specifies the reduction of $\mathscr{H}_d^{(4)}$, whereas s_{311}, s_{131}, and s_{113} specify the reduction of $\tilde{\mathscr{H}}_d^{(6)}$. Once a choice of the s_{pqr} parameters has been made to eliminate the maximum number of terms from $\mathscr{H}$, the transformed Hamiltonian becomes a reduced Hamiltonian. In principle, an infinite number of reductions are possible, and a number have been discussed in the literature [55, 63, 64, 67–71]. In the next section two specific reductions will be considered.

The most general Hamiltonian for a symmetric-top and the procedure for obtaining the corresponding reduced Hamiltonian via (8.84), which depends on the molecular symmetry, are discussed elsewhere [64], though the particular results will be given here.

Asymmetric Top Reduction (*A*)

For simplicity we consider only the reduction of $\tilde{\mathscr{H}}_d^{(4)}$ in demonstrating the procedure. The results for the reduction of $\tilde{\mathscr{H}}_d^{(6)}$ will be simply quoted. Consult [63] for further details. Evaluation of the commutator in (8.86) has been given in (8.60), and with a change in notation we have

$$
\begin{aligned}
i[\mathscr{H}_r, S_3] = 2s_{111}\{&(B_y - B_x)(P_x^2P_y^2 + P_y^2P_x^2 + 2P_z^2) \\
&+ (B_z - B_y)(P_y^2P_z^2 + P_z^2P_y^2 + 2P_x^2) \\
&+ (B_x - B_z)(P_z^2P_x^2 + P_x^2P_z^2 + 2P_y^2)\}
\end{aligned} \tag{8.88}
$$

and hence, from (8.73) and (8.84),

$$
\tilde{\mathscr{H}} = \sum_{\alpha} \tilde{B}_\alpha P_\alpha^2 + \sum_{\alpha,\beta} \tilde{T}_{\alpha\beta} P_\alpha^2 P_\beta^2 \tag{8.89}
$$

This has the same form as $\mathscr{H}$; however, the coefficients are modified as follows:

$$
\begin{aligned}
\tilde{B}_x &= B_x + 4(B_z - B_y)s_{111} \\
\tilde{B}_y &= B_y + 4(B_x - B_z)s_{111} \\
\tilde{B}_z &= B_z + 4(B_y - B_x)s_{111}
\end{aligned} \tag{8.90}
$$

and

$$
\begin{aligned}
\tilde{T}_{\alpha\alpha} &= T_{\alpha\alpha} \\
\tilde{T}_{xy} &= T_{xy} + 2(B_y - B_x)s_{111} \\
\tilde{T}_{yz} &= T_{yz} + 2(B_z - B_y)s_{111} \\
\tilde{T}_{xz} &= T_{xz} + 2(B_x - B_z)s_{111}
\end{aligned} \tag{8.91}
$$

It will be convenient to recast (8.73) into a form involving the constants D_J, D_{JK}, D_K, δ_J, R_5, and R_6 defined in Table 8.7, which is more suitable for numerical calculations. With considerable manipulations one finds [64]

$$
\begin{aligned}
\mathscr{H} = &(B_x - 4R_6)P_x^2 + (B_y - 4R_6)P_y^2 + (B_z + 6R_6)P_z^2 \\
&- D_J P^4 - D_{JK} P^2 P_z^2 - D_K P_z^4 - \delta_J P^2 (P_+^2 + P_-^2) \\
&+ R_5 \{P_z^2 (P_+^2 + P_-^2) + (P_+^2 + P_-^2) P_z^2\} \\
&+ R_6 (P_+^4 + P_-^4)
\end{aligned} \tag{8.92}
$$

with $P_\pm = (P_x \pm iP_y)$ and $P^2 = P_x^2 + P_y^2 + P_z^2$. A more compact notation [72] for this Hamiltonian is found in Table 8.13. The relations between the coefficients of the cylindrical tensor form and that of (8.73) are also given in Table 8.13. The form of $\tilde{\mathscr{H}}$ of (8.92) will also be the same, with the coefficients identified by a tilde, however; with use of (8.90), (8.91), and Table 8.7, they are defined as

$$
\begin{aligned}
\tilde{D}_J &= D_J + \tfrac{1}{2}(B_x - B_y)s_{111} \\
\tilde{D}_{JK} &= D_{JK} - 3(B_x - B_y)s_{111} \\
\tilde{D}_K &= D_K + \tfrac{5}{2}(B_x - B_y)s_{111} \\
\tilde{\delta}_J &= \delta_J \\
\tilde{R}_5 &= R_5 + \tfrac{1}{2}(B_x + B_y - 2B_z)s_{111} \\
\tilde{R}_6 &= R_6 + \tfrac{1}{4}(B_x - B_y)s_{111}
\end{aligned} \tag{8.93}
$$

Table 8.13 Relations Between the Cylindrical and Cartesian Coefficients

Cylindrical Tensor Form of $\mathscr{H}$[a]

$$\mathscr{H}=\sum_{l,m}\Phi_{2l,2m0}+\tfrac{1}{2}\sum_{\substack{l,n,m\\ n>0}}\Phi_{2l,2m,2n}$$

$$\Phi_{2l2m,2n}\begin{cases}=B_{2l,2m,2n} & \text{for } 2l+2m+2n=2\\ =T_{2l,2m,2n} & \text{for } 2l+2m+2n=4\end{cases}$$

P^2 *Terms*[b]	P^6 *Terms*[b]
$B_{200}=\frac{1}{2}(B_x+B_y)-4T_{004}$	$\Phi_{600}=\frac{5}{16}(\Phi_{xxx}+\Phi_{yyy})+\frac{1}{8}(\Phi_{xxy}+\Phi_{yyx})$
$B_{020}=B_z-B_{200}+6T_{004}$	$\Phi_{420}=\frac{3}{4}(\Phi_{xxz}+\Phi_{yyz})+\frac{1}{4}\Phi_{xyz}-3\Phi_{600}$
$B_{002}=\frac{1}{4}(B_x-B_y)$	$\Phi_{240}=(\Phi_{zzx}+\Phi_{zzy})-2\Phi_{420}-3\Phi_{600}$
P^4 *Terms*[b]	$\Phi_{060}=\Phi_{zzz}-\Phi_{240}-\Phi_{420}-\Phi_{600}$
$T_{400}=\frac{1}{8}(3T_{xx}+3T_{yy}+2T_{xy})=-D_J$	$\Phi_{402}=\frac{15}{64}(\Phi_{xxx}-\Phi_{yyy})+\frac{1}{32}(\Phi_{xxy}-\Phi_{yyx})$
$T_{220}=(T_{xz}+T_{yz})-2T_{400}=-D_{JK}$	$\Phi_{222}=\frac{1}{2}(\Phi_{xxz}-\Phi_{yyz})-2\Phi_{402}$
$T_{040}=T_{zz}-T_{220}-T_{400}=-D_K$	$\Phi_{042}=\frac{1}{2}(\Phi_{zzx}-\Phi_{zzy})-\Phi_{222}-\Phi_{402}$
$T_{202}=\frac{1}{4}(T_{xx}-T_{yy})=-\delta_J$	$\Phi_{204}=\frac{3}{32}(\Phi_{xxx}+\Phi_{yyy})-\frac{1}{16}(\Phi_{xxy}+\Phi_{yyx})$
$T_{022}=\frac{1}{2}(T_{xz}-T_{yz})-T_{202}=2R_5$	$\Phi_{024}=\frac{1}{8}(\Phi_{xxz}+\Phi_{yyz}-\Phi_{xyz})-\Phi_{204}$
$T_{004}=\frac{1}{16}(T_{xx}+T_{yy}-2T_{xy})=R_6$	$\Phi_{006}=\frac{1}{64}(\Phi_{xxx}-\Phi_{yyy})-\frac{1}{32}(\Phi_{xxy}-\Phi_{yyx})$

[a] $P_\pm=(P_x\pm iP_y)$, $P^2=P_x^2+P_y^2+P_z^2$. Note the subscripts $2l$, $2m$, and $2n$ refer, respectively, to the powers of P, P_z, and $P\pm$.
[b] From Watson [64].

with the effective rotational constants

$$\tilde{B}_x-4\tilde{R}_6,\ \tilde{B}_y-4\tilde{R}_6,\ \tilde{B}_z+6\tilde{R}_6 \tag{8.94}$$

The most advantageous reduction, initially introduced by Watson, involves chosing s_{111} so that $R_6=0$. In this case, the $(K/K\pm4)$ matrix elements of $\mathscr{H}$ in a symmetric rotor basis are eliminated. $\tilde{\mathscr{H}}_d^{(4)}$ then contains only five terms, and the matrix of $\tilde{\mathscr{H}}_d^{(4)}$ has the same tridiagonal form as that for a rigid rotor. The condition $\tilde{R}_6=0$ requires

$$s_{111}=-\frac{4R_6}{B_x-B_y} \tag{8.95}$$

and the coefficients in the reduced form are

$$\begin{aligned}\Delta_J&=\tilde{D}_J=D_J-2R_6\\ \Delta_{JK}&=\tilde{D}_{JK}=D_{JK}+12R_6\\ \Delta_K&=\tilde{D}_K=D_K-10R_6\\ \delta_J&=\tilde{\delta}_J\\ \delta_K&=-2\tilde{R}_5=-2R_5-\frac{4(2B_z-B_x-B_y)R_6}{B_x-B_y}\end{aligned} \tag{8.96}$$

where the determinable constants $\tilde{D}_J$ and so on, are given an alternate notation.

Note that Δ_K is invariant to a permutation of axes, whereas Δ_J, Δ_{JK}, δ_J, and δ_K change on permuting axes. Hence these latter constants depend on the axis representation used in the analysis. From (8.92) with $P_+^2 + P_-^2 = 2(P_x^2 - P_y^2)$, the reduced Hamiltonian for the A reduction including the P^6 terms is, therefore,

$$\tilde{\mathscr{H}}^{(A)} = \mathscr{H}_r + \mathscr{H}_d^{(4)} + \mathscr{H}_d^{(6)} \tag{8.97}$$

$$\begin{aligned}\mathscr{H}_r &= B_x^{(A)}P_x^2 + B_y^{(A)}P_y^2 + B_z^{(A)}P_z^2\\ &= \tfrac{1}{2}(B_x^{(A)} + B_y^{(A)})P^2 + [B_z^{(A)} - \tfrac{1}{2}(B_x^{(A)} + B_y^{(A)})]P_z^2\\ &\quad + \tfrac{1}{2}(B_x^{(A)} - B_y^{(A)})(P_x^2 - P_y^2)\end{aligned} \tag{8.98}$$

$$\begin{aligned}\mathscr{H}_d^{(4)} &= \Delta_J P^4 - \Delta_{JK} P^2 P_z^2 - \Delta_K P_z^4 - 2\delta_J P^2(P_x^2 - P_y^2)\\ &\quad - \delta_K[P_z^2(P_x^2 - P_y^2) + (P_x^2 - P_y^2)P_z^2]\end{aligned} \tag{8.99}$$

$$\begin{aligned}\mathscr{H}_d^{(6)} &= \Phi_J P^6 + \Phi_{JK} P^4 P_z^2 + \Phi_{KJ} P^2 P_z^4 + \Phi_K P_z^6 + 2\phi_J P^4(P_x^2 - P_y^2)\\ &\quad + \phi_{JK} P^2[P_z^2(P_x^2 - P_y^2) + (P_x^2 - P_y^2)P_z^2]\\ &\quad + \phi_K[P_z^4(P_x^2 - P_y^2) + (P_x^2 - P_y^2)P_z^4]\end{aligned} \tag{8.100}$$

where Δ_J, Δ_{JK}, Δ_K, δ_J, and δ_K are the quartic distortion coefficients defined in terms of various notations in Table 8.14, and where Φ_J, Φ_{JK}, Φ_{KJ}, Φ_K, ϕ_J, ϕ_{JK}, and ϕ_K are the sextic distortion coefficients. Note P_z^2 and $(P_x^2 - P_y^2)$ do not commute. The notation H_J, H_{JK}, H_{KJ}, H_K, h_J, h_{JK}, h_K has been used in the past for the P^6 constants, and many constants have been reported with this notation using the A reduction. However, as has been suggested [64], we have used the

Table 8.14 Quartic Distortion Coefficients of the Reduced Distortion Hamiltonian[a]

$$\Delta_J = -\tfrac{1}{8}\{\tau'_{xxxx} + \tau'_{yyyy}\} = D_J - 2R_6$$

$$\Delta_{JK} = \tfrac{3}{8}\{\tau'_{xxxx} + \tau'_{yyyy}\} - \tfrac{1}{4}\{\tau'_{yyzz} + \tau'_{xxzz} + \tau'_{xxyy}\} = D_{JK} + 12R_6$$

$$\Delta_K = -\tfrac{1}{4}\{\tau'_{xxxx} + \tau'_{yyyy} + \tau'_{zzzz}\} + \tfrac{1}{4}\{\tau'_{yyzz} + \tau'_{xxzz} + \tau'_{xxyy}\} = D_K - 10R_6$$

$$\delta_J = -\tfrac{1}{16}\{\tau'_{xxxx} - \tau'_{yyyy}\}$$

$$\delta_K = \frac{\tfrac{1}{8}\tau'_{xxxx}(B_x - B_z)}{B_x - B_y} + \frac{\tfrac{1}{8}\tau'_{yyyy}(B_y - B_z)}{B_x - B_y} + \tfrac{1}{8}\left\{\tau'_{yyzz} - \tau'_{xxzz} + \frac{\tau'_{xxyy}(2B_z - B_x - B_y)}{B_x - B_y}\right\}$$

$$= -2\left[R_5 + \frac{2R_6(2B_z - B_x - B_y)}{B_x - B_y}\right]$$

$$\Delta_J = -\tfrac{1}{2}(T_{xx} + T_{yy})$$

$$\Delta_{JK} = \tfrac{3}{2}(T_{xx} + T_{yy}) - (T_{yz} + T_{xz} + T_{xy})$$

$$\Delta_K = -(T_{xx} + T_{yy} + T_{zz}) + (T_{yz} + T_{xz} + T_{xy})$$

$$\delta_J = -\tfrac{1}{4}(T_{xx} - T_{yy})$$

$$\delta_K = -\tfrac{1}{2}(T_{xz} - T_{yz}) + \tfrac{1}{4}(T_{xx} - T_{yy}) - \frac{(2B_z - B_x - B_y)}{4(B_x - B_y)}(T_{xx} + T_{yy} - 2T_{xy})$$

[a] $\tau'_{\alpha\alpha\alpha\alpha} = \hbar^4\tau_{\alpha\alpha\alpha\alpha}$, $T_{\alpha\alpha} = \tfrac{1}{4}\tau'_{\alpha\alpha\alpha\alpha}$, $\tau'_{\alpha\alpha\beta\beta} = \hbar^4(\tau_{\alpha\alpha\beta\beta} + 2\tau_{\alpha\beta\alpha\beta})$, $T_{\alpha\beta} = \tfrac{1}{4}\tau'_{\alpha\alpha\beta\beta}$ with $(\alpha \neq \beta)$. B_α are the rotational constants.

alternate notation Φ_J, Φ_{JK}, Φ_{KJ}, Φ_K, ϕ_J, ϕ_{JK}, and ϕ_K in order to retain the $H_J, \ldots, H_K$ notation for the symmetric-top reduced Hamiltonian of the next section, since the diagonal P^6 constants for the A reduction do not correlate appropriately with the symmetric-top constants. The P^6 spectroscopic constants are defined in Table 8.15 in terms of the $\Phi_{\alpha\beta\gamma}$. The effective rotational constants are defined as

$$\begin{aligned} B_x^{(A)} &= B_x - \frac{16R_6(B_z - B_y)}{B_x - B_y} \\ &= \mathcal{B}_x - 2\Delta_J - \Delta_{JK} + 2\delta_J + 2\delta_K \end{aligned} \tag{8.101}$$

$$\begin{aligned} B_y^{(A)} &= B_y + \frac{16R_6(B_z - B_x)}{B_x - B_y} \\ &= \mathcal{B}_y - 2\Delta_J - \Delta_{JK} - 2\delta_J - 2\delta_K \end{aligned} \tag{8.102}$$

$$\begin{aligned} B_z^{(A)} &= B_z + 16R_6 \\ &= \mathcal{B}_z - 2\Delta_J \end{aligned} \tag{8.103}$$

where $\mathcal{B}_x$, $\mathcal{B}_y$, and $\mathcal{B}_z$ are given in (8.118) and are readily derivable from the effective rotational constants using the Δ-constants obtained from the spectrum. This Hamiltonian has been found to be applicable to a large range of asymmetric tops. It is apparent that the effective Hamiltonian has been reduced to one that contains only determinable constants and is appropriate for fitting purposes. The reduced Hamiltonian in general contains $(n+1)$ distortion coefficients for each degree n in the angular momentum. There are $(n/2+1)$ purely diagonal angular momentum operators, and the remaining ones are off-diagonal, involving $(P_x^2 - P_y^2)$.

For light molecules higher-order effects are important, and numerous distortion coefficients are required to adequately fit the spectrum. The reduced

Table 8.15 Sextic Distortion Coefficients of the Reduced Distortion Hamiltonian in Terms of the Cylindrical Tensor Coefficients[a]

$$\Phi_J = \Phi_{600} + 2\Phi_{204}$$

$$\Phi_{JK} = \Phi_{420} - 12\Phi_{204} + 2\Phi_{024} + 16\sigma\Phi_{006} + 8T_{022}T_{004}/B_{002}$$

$$\Phi_{KJ} = \Phi_{240} + \tfrac{10}{3}\Phi_{420} - 30\Phi_{204} - \tfrac{10}{3}\Phi_{JK}$$

$$\Phi_K = \Phi_{060} - \tfrac{7}{3}\Phi_{420} + 28\Phi_{204} + \tfrac{7}{3}\Phi_{JK}$$

$$\phi_J = \Phi_{402} + \Phi_{006}$$

$$\phi_{JK} = \Phi_{222} + 4\sigma\Phi_{204} - 10\Phi_{006} + 2(T_{220} - 2\sigma T_{202} - 4T_{004})T_{004}/B_{002}$$

$$\phi_K = \Phi_{042} + \frac{4\sigma}{3}\Phi_{024} + \left(\frac{32}{3}\sigma^2 + 9\right)\Phi_{006} + 4\left[T_{040} + \frac{\sigma}{3}T_{022} - 2(\sigma^2 - 2)T_{004}\right]T_{004}/B_{002}$$

[a]See Table 8.13 for definition of the coefficients in the cylindrical tensor form of $\mathcal{H}$ in terms of $\Phi_{\alpha\beta\gamma}$ of (8.73). From Watson [64].

Hamiltonian can be readily generalized to include P^8 and P^{10} effects, and the following terms are added to the Hamiltonian

$$\begin{aligned}\mathscr{H}_d^{(8)}=&L_JP^8+L_{JJK}P^6P_z^2+L_{JK}P^4P_z^4+L_{KKJ}P^2P_z^6+L_KP_z^8+2l_JP^6(P_x^2-P_y^2)\\&+l_{JK}P^4[P_z^2(P_x^2-P_y^2)+(P_x^2-P_y^2)P_z^2]+l_{KJ}P^2[P_z^4(P_x^2-P_y^2)+(P_x^2-P_y^2)P_z^4]\\&+l_K[P_z^6(P_x^2-P_y^2)+(P_x^2-P_y^2)P_z^6],\end{aligned} \tag{8.104}$$

$$\begin{aligned}\mathscr{H}_d^{(10)}=&P_JP^{10}+P_{JJK}P^8P_z^2+P_{JK}P^6P_z^4+P_{KJ}P^4P_z^6+P_{KKJ}P^2P_z^8+P_KP_z^{10}\\&+2p_JP^8(P_x^2-P_y^2)+p_{JJK}P^6[P_z^2(P_x^2-P_y^2)+(P_x^2-P_y^2)P_z^2]\\&+P_{JK}P^4[P_z^4(P_x^2-P_y^2)+(P_x^2-P_y^2)P_z^4]\\&+p_{KKJ}P^2[P_z^6(P_x^2-P_y^2)+(P_x^2-P_y^2)P_z^6]+p_K[P_z^8(P_x^2-P_y^2)+(P_x^2-P_y^2)P_z^8]\end{aligned} \tag{8.105}$$

The L, l are octic centrifugal distortion constants, and the P, p are the corresponding dectic constants. These coefficients must be regarded at this point as empirical constants in the sense that they have not as yet been theoretically related to the vibrational potential constants. On the other hand, the quartic and sextic distortion constants have been related to the potential constants, in particular to the quadratic and cubic force constants, respectively.

Symmetric Top Reduction (S)

For very slightly asymmetric tops, the previous reduction can present problems. It is clear from the definition of s_{111} that if $B_x \simeq B_y$, s_{111} gets very large and the order-of-magnitude arguments employed previously no longer apply. Specifically, the breakdown depends on the ratio of a P^4 distortion constant to (B_x-B_y). Different representations can be chosen to minimize this effect by making (B_x-B_y) larger and R_6 as small as possible. However, it is preferable to employ an alternate reduction that reduces in a well-defined way to the symmetric top limit. In this case it is convenient to take $\tilde{R}_5=0$ which requires

$$s_{111}=\frac{2R_5}{2B_z-B_x-B_y} \tag{8.106}$$

This reduction has been proposed by Winnewisser [69] and Van Eijck [70]. It has been demonstrated by Van Eijck [70], who discussed $\mathscr{H}_d^{(4)}$, and by Typke [71] who extended the results to $\mathscr{H}_d^{(6)}$, that a more satisfactory analysis can be obtained for slightly asymmetric tops with this reduction. Watson [64] also was instrumental in the introduction of this form of the reduction.

The coefficients in the reduced form of (8.92) are now

$$\begin{aligned}\tilde{D}_J&=D_J+\frac{R_5}{\sigma}\\&=\Delta_J-\frac{\delta_K}{2\sigma}\end{aligned}$$

$$\begin{aligned}
\tilde{D}_{JK} &= D_{JK} - \frac{6R_5}{\sigma} \\
&= \Delta_{JK} + \frac{3\delta_K}{\sigma} \\
\tilde{D}_K &= D_K + \frac{5R_5}{\sigma} \\
&= \Delta_K - \frac{5\delta_K}{2\sigma} \\
\tilde{\delta}_J &= \delta_J \\
\tilde{R}_6 &= R_6 + \frac{R_5}{2\sigma} \\
&= -\frac{\delta_K}{4\sigma}
\end{aligned} \tag{8.107}$$

where

$$\sigma = \frac{2B_z - B_x - B_y}{B_x - B_y} \tag{8.108}$$

for an I^r and III^l representation, $\sigma = -1/b_p$ and $-1/b_0$, respectively. The notation adopted to retain the similarity of the reduced Hamiltonian to that of a symmetric top is

$$D_J = \tilde{D}_J,\ D_{JK} = \tilde{D}_{JK},\ D_K = \tilde{D}_K,\ d_1 = -\delta_J,\ d_2 = \tilde{R}_6 \tag{8.109}$$

The reduced Hamiltonian for the S reduction is therefore

$$\tilde{\mathscr{H}}^{(S)} = \mathscr{H}_r + \mathscr{H}_d^{(4)} + \mathscr{H}_d^{(6)} \tag{8.110}$$

$$\begin{aligned}
\mathscr{H}_r &= B_x^{(S)} P_x^2 + B_y^{(S)} P_y^2 + B_z^{(S)} P_z^2 \\
&= \tfrac{1}{2}(B_x^{(S)} + B_y^{(S)})P^2 + [B_z^{(S)} - \tfrac{1}{2}(B_x^{(S)} + B_y^{(S)})]P_z^2 + \tfrac{1}{4}(B_x^{(S)} - B_y^{(S)})(P_+^2 + P_-^2)
\end{aligned} \tag{8.111}$$

$$\mathscr{H}_d^{(4)} = -D_J P^4 - D_{JK} P^2 P_z^2 - D_K P_z^4 + d_1 P^2(P_+^2 + P_-^2) + d_2(P_+^4 + P_-^4) \tag{8.112}$$

$$\begin{aligned}
\mathscr{H}_d^{(6)} = {} & H_J P^6 + H_{JK} P^4 P_z^2 + H_{KJ} P^2 P_z^4 + H_K P_z^6 \\
& + h_1 P^4(P_+^2 + P_-^2) + h_2 P^2(P_+^4 + P_-^4) + h_3(P_+^6 + P_-^6)
\end{aligned} \tag{8.113}$$

where the P^6 terms have also been included. Note, however, that the constants D_J, D_{JK}, and D_K used in the foregoing Hamiltonian are not the same constants as those defined in Table 8.7. The effective rotational constants, with cognizance of (8.90), are

$$\begin{aligned}
B_x^{(S)} &= \tilde{B}_x - 4\tilde{R}_6 = B_x - 4R_6 + 4R_5 + \frac{2R_5}{\sigma} \\
&= B_x - 2D_J - D_{JK} - 2d_1 - 4d_2
\end{aligned} \tag{8.114}$$

$$B_y^{(S)}=\tilde{B}_y-4\tilde{R}_6=B_y-4R_6-4R_5+\frac{2R_5}{\sigma}$$

$$=\mathcal{B}_y-2D_J-D_{JK}+2d_1-4d_2 \tag{8.115}$$

$$B_z^{(S)}=\tilde{B}_z+6\tilde{R}_6=B_z+6R_6-\frac{5R_5}{\sigma}$$

$$=\mathcal{B}_z-2D_J-6d_2 \tag{8.116}$$

where $\mathcal{B}_x$, $\mathcal{B}_y$, $\mathcal{B}_z$ are defined in (8.118). The coefficients in $\mathscr{H}^{(A)}$ and $\mathscr{H}^{(S)}$ can be related to each other (see Table 8.16), and the coefficients of the two reduced Hamiltonians provide the same information.

It is apparent that this Hamiltonian can be readily generalized to include higher-order constants. If we extend the Hamiltonian to P^8 terms, the following terms must be added

$$\tilde{\mathscr{H}}^{(8)}=L_J P^8+L_{JJK}P^6P_z^2+L_{JK}P^4P_z^4+L_{KKJ}P^2P_z^6+L_K P_z^8$$
$$+l_1P^6(P_+^2+P_-^2)+l_2P^4(P_+^4+P_-^4)+l_3P^2(P_+^6+P_-^6)+l_4(P_+^8+P_-^8) \tag{8.117}$$

Determinable Combinations of Coefficients

From (8.90) and (8.91) s_{111} can be eliminated from the nine equations to give eight linear combinations of quadratic and quartic coefficients. These coefficients are the determinable combinations of coefficients since the energies cannot depend on s_{111}. The experimentally determinable coefficients can be conveniently taken as

$$\mathcal{B}_x=B_x-2T_{yz}$$
$$\mathcal{B}_y=B_y-2T_{zx}$$
$$\mathcal{B}_z=B_z-2T_{xy} \tag{8.118}$$

Table 8.16 Relations Between the Distortion Coefficients in the A and S Reduced Hamiltonians[a]

$B_x^{(A)}=B_x^{(S)}-4(2\sigma+1)d_2$		$B_y^{(A)}=B_y^{(S)}+4(2\sigma-1)d_2$
	$B_z^{(A)}=B_z^{(S)}+10d_2$	
$\Delta_J=D_J-2d_2$	$\Delta_{JK}=D_{JK}+12d_2$	$\Delta_K=D_K-10d_2$
$\delta_J=-d_1$		$\delta_K=-4\sigma d_2$
$\Phi_J=H_J+2h_2$		$\Phi_{JK}=H_{JK}-12h_2+16\sigma h_3$
$\Phi_{KJ}=H_{KJ}+10h_2-(160\sigma/3)h_3$		$\Phi_K=H_K+(112\sigma/3)h_3$
	$\phi_J=h_1+h_3$	
	$\phi_{JK}=4\sigma h_2-10h_3-8d_2\{D_{JK}+2\sigma d_1+4d_2\}/(B_x-B_y)$	
	$\phi_K=(32\sigma^2/3+9)h_3-16d_2\{D_K+2(\sigma^2-2)d_2\}/(B_x-B_y)$	

[a] $\sigma=(2B_z-B_x-B_y)/(B_x-B_y)$. From Watson [140].

$$\begin{aligned}&T_{xx},\ T_{yy},\ T_{zz},\\&T_1=T_{yz}+T_{xz}+T_{xy}\\&T_2=B_xT_{yz}+B_yT_{xz}+B_zT_{xy}\end{aligned} \tag{8.119}$$

In terms of the τ notation, the determinable coefficients are

$$\begin{aligned}\mathcal{B}_x&=B_x-\tfrac{1}{2}\tau'_{yyzz}\\\mathcal{B}_y&=B_y-\tfrac{1}{2}\tau'_{xxzz}\\\mathcal{B}_z&=B_z-\tfrac{1}{2}\tau'_{xxyy}\end{aligned} \tag{8.120}$$

$$\begin{aligned}&\tau'_{xxxx},\ \tau'_{yyyy},\ \tau'_{zzzz},\\&\tau_1=\tau'_{yyzz}+\tau'_{xxzz}+\tau'_{xxyy}\\&\tau_2=B_x\tau'_{yyzz}+B_y\tau'_{xxzz}+B_z\tau'_{xxyy}\end{aligned} \tag{8.121}$$

The T_1 (or τ_1) and T_2 (or τ_2) have the same value for any permutation of the axes. The constant T_2 (or τ_2) is often scaled so that it has same units as the other constants, that is,

$$T_2\rightarrow\frac{T_2}{S}\quad\text{or}\quad\tau_2\rightarrow\frac{\tau_2}{S} \tag{8.122}$$

with $S=B_x+B_y+B_z$. Likewise the P^6 determinable coefficients can be taken as [64]

$$\begin{aligned}&\Phi_{xxx},\ \Phi_{yyy},\ \Phi_{zzz},\\&\Phi_1=3\sum_{\alpha\neq\beta}\Phi_{\alpha\alpha\beta}+\Phi_{xyz}\\&\Phi_2=(B_x-B_z)\Phi_{xxy}+(B_x-B_y)\Phi_{xxz}-2(T_{xx}-T_{xy})(T_{xx}-T_{xz})\\&\Phi_3=(B_y-B_x)\Phi_{yyz}+(B_y-B_z)\Phi_{yyx}-2(T_{yy}-T_{yz})(T_{yy}-T_{yx})\\&\Phi_4=(B_z-B_y)\Phi_{zzx}+(B_z-B_x)\Phi_{zzy}-2(T_{zz}-T_{zx})(T_{zz}-T_{zy})\end{aligned} \tag{8.123}$$

These determinable combinations of constants are hence directly related to the constants of the first-principles Hamiltonian and can thus be directly related to the molecular force field.

The foregoing determinable combinations may be expressed in terms of the constants evaluated from the rotational spectrum via the Hamiltonians of (8.97) and (8.110). These relations are given in Tables 8.17 and 8.18. Once an axis representation is chosen, those relations yield the constants T_{aa}, $T_{bb}, \ldots,$ referred to the principal inertial axis system.

In reporting the rotation and distortion constants, the effective rotational constants $\mathscr{A}$, $\mathscr{B}$, $\mathscr{C}$ and A, B, C should be listed. The type of reduction (A or S) used and the representation I^r, II^r, and so on, that is employed should be indicated in the table of constants. In the cases where large numbers of constants are required in the fit, the addition of an A or S superscript to the constants would avoid the need for two sets of symbols for the higher-order constants, for example, P^8 constants ($L_J^{(S)}$ or $L_J^{(A)}$).

Table 8.17 Determinable Combinations in Terms of the Spectroscopic Constants of the A Reduction[a]

$$B_x = B_x^{(A)} + 2\Delta_J + \Delta_{JK} - 2\delta_J - 2\delta_K$$
$$B_y = B_y^{(A)} + 2\Delta_J + \Delta_{JK} + 2\delta_J + 2\delta_K$$
$$B_z = B_z^{(A)} + 2\Delta_J$$
$$T_{xx} = -\Delta_J - 2\delta_J,\ T_{yy} = -\Delta_J + 2\delta_J,\ T_{zz} = -\Delta_J - \Delta_{JK} - \Delta_K$$
$$T_1 = -3\Delta_J - \Delta_{JK}$$
$$T_2 = -(B_x + B_y + B_z)\Delta_J - \tfrac{1}{2}(B_x + B_y)\Delta_{JK} + (B_x - B_y)(\delta_J + \delta_K)$$
$$\Phi_{xxx} = \Phi_J + 2\phi_J,\ \Phi_{yyy} = \Phi_J - 2\phi_J$$
$$\Phi_{zzz} = \Phi_J + \Phi_{JK} + \Phi_{KJ} + \Phi_K$$
$$\Phi_1 = 30\Phi_J + 10\Phi_{JK} + 3\Phi_{KJ}$$
$$\Phi_2 + \Phi_3 = -3[B_z - \tfrac{1}{2}(B_x + B_y) + 10R_6]\Phi_J + (B_x - B_y)(5\phi_J + 2\phi_{JK}) - 8\delta_J(\delta_J - \delta_K)$$
$$\Phi_2 - \Phi_3 = -2[B_z - \tfrac{1}{2}(B_x + B_y) + 10R_6]\phi_J + \tfrac{1}{2}(B_x - B_y)(9\Phi_J + 2\Phi_{JK}) + 4\delta_J\Delta_{JK}$$
$$\Phi_4 = [B_z - \tfrac{1}{2}(B_x + B_y) + 10R_6](3\Phi_J + 2\Phi_{JK} + \Phi_{KJ}) + (B_x - B_y)(\phi_J + \phi_{JK} + \phi_K) - \tfrac{1}{2}(\Delta_{JK} + 2\Delta_K)^2 + 2(\delta_J + \delta_K)^2$$

[a] $R_6 = \frac{1}{16}(T_{xx} + T_{yy} - 2T_{xy})$. From Watson [64].

Table 8.18 Determinable Combinations in Terms of the Spectroscopic Constants of the S Reduction[a]

$$B_x = B_x^{(S)} + 2D_J + D_{JK} + 2d_1 + 4d_2$$
$$B_y = B_y^{(S)} + 2D_J + D_{JK} - 2d_1 + 4d_2$$
$$B_z = B_z^{(S)} + 2D_J + 6d_2$$
$$T_{xx} = -D_J\ ''2d_1 + 2d_2, \quad T_{yy} = -D_J - 2d_1 + 2d_2$$
$$T_{zz} = -D_J - D_{JK} - D_K, \quad T_1 = -3D_J - D_{JK} - 6d_2$$
$$T_2 = -(B_x + B_y + B_z)D_J - \tfrac{1}{2}(B_x + B_y)D_{JK} - (B_x - B_y)d_1 - 6B_z d_2$$
$$\Phi_{xxx} = H_J + 2h_1 + 2h_2 + 2h_3$$
$$\Phi_{yyy} = H_J - 2h_1 + 2h_2 - 2h_3$$
$$\Phi_{zzz} = H_J + H_{JK} + H_{KJ} + H_K$$
$$\Phi_1 = 30H_J + 10H_{JK} + 3H_{KJ} - 30h_2$$
$$\Phi_2 + \Phi_3 = -[B_z - \tfrac{1}{2}(B_x + B_y) + 10R_6](3H_J - 10h_2) + 5(B_x - B_y)(h_1 - 3h_3) - 8d_1^2 - 16d_2(D_{JK} + 4d_2)$$
$$\Phi_2 - \Phi_3 = -2[B_z - \tfrac{1}{2}(B_x + B_y) + 10R_6](h_1 - 15h_3) + \tfrac{1}{2}(B_x - B_y)(9H_J + 2H_{JK} - 6h_2) - 4d_1(D_{JK} + 12d_2)$$
$$\Phi_4 = [B_z - \tfrac{1}{2}(B_x + B_y) + 10R_6](3H_J + 2H_{JK} + H_{KJ}) + (B_x - B_y)h_1 - \tfrac{1}{2}(D_{JK} + 2D_K)^2 + 2d_1^2$$

[a] $R_6 = \frac{1}{16}(T_{xx} + T_{yy} - 2T_{xy})$. From Watson [64].

Symmetric and Spherical Tops

For a symmetric rotor $B_x^{(S)} = B_y^{(S)}$, and the reduced Hamiltonian has the form

$$\tilde{\mathscr{H}} = B_x P^2 + (B_z - B_x)P_z^2 - D_J P^4 - D_{JK} P^2 P_z^2 - D_K P_z^4 + H_J P^6 + H_{JK} P^4 P_z^2 + H_{KJ} P^2 P_z^4 + H_K P_z^6 + \tilde{\mathscr{H}}_{\text{split}} \qquad (8.124)$$

The D's have their usual meaning for a symmetric top. The $\tilde{\mathscr{H}}_{\text{split}}$ term arises from the off-diagonal terms in $\mathscr{H}^{(S)}$ and leads to splitting of the K-degenerate levels of a symmetric top. The form of the reduced splitting term depends on the molecular point group. In particular, correct to P^6 terms

$$\tilde{\mathscr{H}}_{\text{split}} = (d_2 + h_2 P^2)(P_+^4 + P_-^4) + h_3(P_+^6 + P_-^6) \tag{8.125}$$

where, for a p-fold axis of rotation [61]

$$d_2 = h_2 = h_3 = 0, \qquad \text{if } p = 5 \text{ or } \geqslant 7 \tag{8.126}$$

$$d_2 = h_2 = 0, \qquad \text{if } p = 3 \text{ or } 6 \tag{8.127}$$

$$h_3 = 0, \qquad \text{if } p = 4 \tag{8.128}$$

The constants d_2, h_2, and h_3 are particularly important in the splitting of the $K = \pm 2$ and $K = \pm 3$ levels, respectively. For a C_{3V} molecule [72]

$$\begin{aligned}
H_J &= \tfrac{1}{2}(\Phi_{xxx} + \Phi_{yyy}) \\
H_{JK} &= 2\Phi_{zxx} - 3H_J \\
H_{KJ} &= 2\Phi_{zzx} - 2H_{JK} - 3H_J \\
H_K &= \Phi_{zzz} - H_{KJ} - H_{JK} - H_J \\
2h_3 &= \tfrac{1}{2}(\Phi_{xxx} - \Phi_{yyy})
\end{aligned} \tag{8.129}$$

The Hamiltonian for a spherical rotor has been discussed by a number of authors [73–80]. It may be conveniently written in terms of spherical tensor operators Ω_l of rank $l(l = 0, 2, 4, 6, \ldots)$ [78]

$$\mathscr{H} = BP^2 - D_S P^4 + D_{4T}\Omega_4 + H_S P^6 + H_{4T} P^2 \Omega_4 + H_{6T}\Omega_6 \tag{8.130}$$

where the D's and H's are, respectively, the effective quartic and sextic distortion constants. The subscript T indicates a coefficient of a tensor operator, $l \neq 0$, and S indicates coefficients of the scalar terms, $l = 0$. For a T_d molecule [72]

$$\begin{aligned}
H_S &= \frac{1}{105}(45\Phi_{xxx} + 36\Phi_{xxy} + 2\Phi_{xyz}) \\
H_{4T} &= \frac{1}{110}(-15\Phi_{xxx} + 8\Phi_{xxy} + \Phi_{xyz}) \\
H_{6T} &= \frac{1}{231}(6\Phi_{xxx} - 12\Phi_{xxy} + 4\Phi_{xyz})
\end{aligned} \tag{8.131}$$

The energy levels are conveniently labeled by J, M_J, C, t, where J, M_J have their usual meaning, C specifies the irreducible symmetry representation of the level, and t distinguishes different levels of the same representation. Precise measurements [81] have indicated the need for higher-order terms in the Hamiltonian. The P^8 effects add the following terms

$$\mathscr{H}^{(8)} = L_S P^8 + L_{4T} P^4 \Omega_4 + L_{6T} P^2 \Omega_6 + L_{8T}\Omega_8 \tag{8.132}$$

It has been shown that this can be written in an alternate form [80]

$$\mathscr{H}^{(8)} = L_S P^8 + L_{4T} P^4 \Omega_4 + L_{6T} P^2 \Omega_6 + L_{44T} \Omega_4^2 \quad (8.133)$$

which replaces the evaluation of the matrix of Ω_8 with that for Ω_4^2. The latter matrix is readily constructed from that for Ω_4. The coefficients of lower degree are also modified slightly by this formulation. In a similar way one obtains the following additional terms to higher-order for the reduced Hamiltonian to twelfth degree [80]

$$\mathscr{H}^{(10)} = P_S P^{10} + P_{4T} P^6 \Omega_4 + P_{6T} P^4 \Omega_6 + P_{44T} P^2 \Omega_4^2 + P_{46T}(\Omega_4 \Omega_6 + \Omega_6 \Omega_4) \quad (8.134)$$

$$\mathscr{H}^{(12)} = R_S P^{12} + R_{4T} P^8 \Omega_4 + R_{6T} P^6 \Omega_6 + R_{44T} P^4 \Omega_4^2 + R_{46T} P^2(\Omega_4 \Omega_6 + \Omega_6 \Omega_4) + R_{444T} \Omega_4^3 \quad (8.135)$$

where the P's and R's are additional distortion constants.

A discussion of the evaluation of the energy levels is found in Section 5 (see also Chapter VI). The tensor operators in the foregoing expressions are constructed from the angular momentum components P_α so as to be totally symmetric under the operations of the molecular point group, for example, [77]

$$\Omega_4 = 6P^4 - 10(P_x^4 + P_y^4 + P_z^4) - 2P^2 \quad (8.136)$$

and

$$\Omega_6 = 15P^6 + \tfrac{77}{2}(P_x^6 + P_y^6 + P_z^6) - \tfrac{35}{2}(P_x^4 + P_y^4 + P_z^4)(3P^2 - 7) - \tfrac{135}{2}P^4 + 19P^2 \quad (8.137)$$

5 EVALUATION OF THE DISTORTION CONSTANTS FROM OBSERVED SPECTRA

For molecules with very small moments of inertia (e.g., H_2O, H_2S) or levels with high rotational quantum numbers, a first-order treatment of centrifugal distortion may not be adequate. A detailed study of the distortion effects in OF_2 by Pierce et al. [82] has shown that at high J values, $20 \leqslant J \leqslant 40$, second- and higher-order distortion effects, though small compared to the first-order effects, are nevertheless significant. For these cases, as with the rigid rotor problem, the complete energy matrix can be set up and diagonalized directly to give the total energy of the molecule, including distortion corrections. Earlier distortion studies employed the Hamiltonian of (8.53) and were carried out on planar molecules; the planarity relations were used to reduce the number of τ constants from six to four, for example, τ_{aaaa}, τ_{bbbb}, τ_{aabb}, and τ_{abab}. The indeterminacy mentioned previously was thus removed. For the case of molecules with orthorhombic symmetry, (8.53), the Hamiltonian matrix in a symmetric rotor basis is diagonal in J and M, with nonvanishing matrix elements in K of the type $(K|K)$, $(K|K \pm 2)$, and $(K|K \pm 4)$. For molecules with non-orthorhombic symmetry, additional distortion constants (e.g., $\tau_{\alpha\alpha\alpha\beta}$) are introduced which also contribute in higher order [50, 64, 72, 83]. In the reduced

Hamiltonian, these constants do not appear explicitly but rather modify the definitions of the higher-order reduced constants. The matrix elements for the distortion Hamiltonian (8.55) may be obtained, as before, by matrix multiplication and use of the matrix elements given previously. Explicit expressions are given in Table 8.19. The matrix will still factor into four submatrices upon application of the Wang symmetrizing transformation [84]. Portions of the submatrices in terms of the original matrix elements are:

$$\mathbf{E}^{\pm}=\left[\begin{array}{c|cccccc} E_{00} & \sqrt{2}E_{02} & \sqrt{2}E_{04} & 0 & \cdot & \cdot & \cdot \\ \hline \sqrt{2}E_{02} & E_{22}\pm E_{-2,2} & E_{24} & E_{26} & 0 & \cdot & \cdot \\ 2E_{04} & E_{24} & E_{44} & E_{46} & E_{48} & 0 & \cdot \\ 0 & E_{26} & E_{46} & E_{66} & E_{68} & E_{6,10} & \cdot \\ \cdot & \cdot & \cdot & \cdot & \cdot & \cdot & \cdot \end{array}\right]$$

$$\mathbf{O}^{\pm}=\left[\begin{array}{ccccccc} E_{11}\pm E_{-1,1} & E_{13}\pm E_{-1,3} & E_{15} & 0 & \cdot & \cdot & \cdot \\ E_{13}\pm E_{-1,3} & E_{33} & E_{35} & E_{37} & 0 & \cdot & \cdot \\ E_{15} & E_{35} & E_{55} & E_{57} & E_{59} & 0 & \cdot \\ 0 & E_{37} & E_{57} & E_{77} & E_{79} & E_{7.11} & \cdot \\ \cdot & \cdot & \cdot & \cdot & \cdot & \cdot & \cdot \end{array}\right] \tag{8.138}$$

Table 8.19 Matrix Elements of $P_\alpha^2 P_\beta^2$ in a Symmetric Rotor Representation[a]

$(K|P_z^4|K)=K^4$

$(K|P_y^4|K)=\frac{1}{4}[(P^2-K^2)^2+\frac{1}{4}\{f_+(0)f_+(1)+f_-(0)f_-(1)\}]$

$(K|P_x^4|K)=(K|P_y^4|K)$

$(K|P_x^2P_y^2+P_y^2P_x^2|K)=\frac{1}{2}[(P^2-K^2)^2-\frac{1}{4}\{f_+(0)f_+(1)+f_-(0)f_-(1)\}]$

$(K|P_y^2P_z^2+P_z^2P_y^2|K)=K^2[P^2-K^2]$

$(K|P_x^2P_z^2+P_z^2P_x^2|K)=(K|P_y^2P_z^2+P_z^2P_y^2|K)$

$(K|P_y^4|K\pm 2)=\frac{1}{8}\{2P^2-K^2-(K\pm 2)^2\}\{f_\pm(0)f_\pm(1)\}^{1/2}$

$(K|P_x^4|K\pm 2)=-(K|P_y^4|K\pm 2)$

$(K|P_y^2P_z^2+P_z^2P_y^2|K\pm 2)=\frac{1}{4}\{K^2+(K\pm 2)^2\}\{f_\pm(0)f_\pm(1)\}^{1/2}$

$(K|P_x^2P_z^2+P_z^2P_x^2|K\pm 2)=-(K|P_y^2P_z^2+P_z^2P_y^2|K\pm 2)$

$(K|P_y^4|K\pm 4)=\frac{1}{16}\{f_\pm(0)f_\pm(1)f_\pm(2)f_\pm(3)\}^{1/2}$

$(K|P_x^4|K\pm 4)=(K|P_y^4|K\pm 4)$

$(K|P_x^2P_y^2+P_y^2P_x^2|K\pm 4)=-2(K|P_y^4|K\pm 4)$

$P^2=J(J+1);\ f_\pm(l)=\{P^2-(K\pm l)(K\pm l\pm 1)\}$

[a]The phase choice used here is that of G. W. King, R. M. Hainer, and P. C. Cross, *J. Chem. Phys.*, **11**, 27 (1943). For simplicity we do not display the J and M labeling, it being understood the matrix elements are diagonal in these quantum numbers.

For $\mathbf{E}^-$, the first row and column must be deleted as indicated by the dashed lines. In these derivations, use has been made of the relations

$$E_{K,K}=E_{-K,-K}; \qquad E_{K,K+l}=E_{K+l,K}=E_{-K,-K-l}=E_{-K-l,-K} \quad (8.139)$$

with $l=2$ or 4. Table 7.5 is still applicable for identification of the explicit energy levels contained in each submatrix. Each matrix element of (8.138) will be made up of contributions from $\mathscr{H}_r$ and $\mathscr{H}_d$. Diagonalization of the complete energy matrix gives the rotational energy, including distortion corrections. For sufficiently low values of J, the subdeterminants of (8.138) may be solved directly to give expressions for the energy levels, including the effects of centrifugal distortion. Oka and Morino [85] and Chung and Parker [19] have reported expressions for the rotational energies of particular low J levels which are in the form of linear or quadratic expressions. In Table 8.20 we give explicit expressions for the energy levels that can be expressed in linear forms. By use of this table, the frequency for certain low J transitions can be easily evaluated if the τ's are known.

Sum rules involving the rotational and distortion constants have also been obtained [86–90]. These are derived by use of the invariance of the trace of a matrix to a similarity transformation. Thus, for each J, the trace of a submatrix of, for example, (8.138) equals the sum of the corresponding energy levels.

Table 8.20 Linear Expressions for the Rotational Energy Levels Including Centrifugal Distortion Effects[a]

J_{K_{-1},K_1}	E
$0_{0,0}$	0
$1_{0,1}$	$B+C+\frac{\hbar^4}{4}\{\tau_{bbbb}+\tau_{cccc}+2(\tau_{bbcc}+2\tau_{bcbc})\}$
$1_{1,1}$	$A+C+\frac{\hbar^4}{4}\{\tau_{aaaa}+\tau_{cccc}+2(\tau_{aacc}+2\tau_{acac})\}$
$1_{1,0}$	$A+B+\frac{\hbar^4}{4}\{\tau_{aaaa}+\tau_{bbbb}+2(\tau_{aabb}+2\tau_{abab})\}$
$2_{1,2}$	$A+B+4C+\frac{\hbar^4}{4}\{\tau_{aaaa}+\tau_{bbbb}+16\tau_{cccc}+8(\tau_{bbcc}+2\tau_{bcbc})+8(\tau_{aacc}+2\tau_{acac})+2(\tau_{aabb}+2\tau_{abab})\}$
$2_{1,1}$	$A+4B+C+\frac{\hbar^4}{4}\{\tau_{aaaa}+16\tau_{bbbb}+\tau_{cccc}+8(\tau_{bbcc}+2\tau_{bcbc})+2(\tau_{aacc}+2\tau_{acac})+8(\tau_{aabb}+2\tau_{abab})\}$
$2_{2,1}$	$4A+B+C+\frac{\hbar^4}{4}\{16\tau_{aaaa}+\tau_{bbbb}+\tau_{cccc}+2(\tau_{bbcc}+2\tau_{bcbc})+8(\tau_{aacc}+2\tau_{acac})+8(\tau_{aabb}+2\tau_{abab})\}$
$3_{2,2}$	$4(A+B+C)+4\hbar^4\{\tau_{aaaa}+\tau_{bbbb}+\tau_{cccc}+2(\tau_{bbcc}+2\tau_{bcbc})+2(\tau_{aacc}+2\tau_{acac})+2(\tau_{aabb}+2\tau_{abab})\}$

[a]From Oka and Morino [85]. Note that for planar molecules $\tau_{acac}=\tau_{bcbc}=0$.

Because the latter information is often unavailable, however, the sum rules are hence of limited value. Sum rules for the reduced Hamiltonian are also available [64].

For higher values of J, the order of the secular determinant requires numerical diagonalization techniques. To set up the energy matrix, (8.138), the numerical values of the τ's are required. If the force constants and structural parameters are available, the distortion constants may be calculated directly, and the corresponding energy matrix diagonalized to give the effects of centrifugal distortion. The centrifugal distortion constants are, however, best evaluated from the experimental data of the rotational spectrum. An approximate set of rotation and distortion constants may be obtained by use of the first-order expression (8.56) for a planar molecule; all seven individual τ's may be obtained from the analysis (if the relations of Table 8.10 are assumed). These constants along with the rotational constants are used for setting up the complete energy matrix, (8.138), and the higher effects may be taken into account by use of the rigid-rotor basis distortion analysis to be discussed later. Distortion constants of some earlier studies, as well as more recent ones employing the planar relations in the analysis, are listed in Table 8.21.

Although the form of $\mathscr{H}_d$, (8.55), is applicable to planar molecules with use of the planarity conditions, it is not suitable for the general case. The maximum number of quartic distortion coefficients that can be obtained from the experimental data is five, whereas $\mathscr{H}_d$ of (8.55) contains six coefficients. Furthermore, as we shall see, use of the planarity relations usually introduces significant model errors. To remove this problem, the Hamiltonian was transformed by means of a unitary operator to a form that contains only five independent terms in the angular momenta. The coefficients of each term are combinations of the original coefficients. This reduced Hamiltonian, which has been discussed in Section 4, has been given correct to P^{10} terms in the Hamiltonian of (8.97)–(8.105)

$$\mathscr{H}^{(A)} = \mathscr{H}_r + \mathscr{H}_d \tag{8.140}$$

$$\mathscr{H}_d = \mathscr{H}_d^{(4)} + \mathscr{H}_d^{(6)} + \mathscr{H}_d^{(8)} + \mathscr{H}_d^{(10)} \tag{8.141}$$

This form is particularly suited to the empirical determination of the centrifugal distortion constants if higher-order effects are to be considered in nonplanar molecules or in planar molecules where the planarity conditions are not employed. The energy matrix, like the rigid rotor energy matrix, has only $(K|K)$ and $(K|K\pm 2)$ matrix elements in a symmetric rotor basis, and the Wang transformation gives four separate submatrices $\mathbf{E}^+$, $\mathbf{E}^-$, $\mathbf{0}^+$, $\mathbf{0}^-$ for each J, the form of which is specified by (7.22). Furthermore, the continued-fraction technique, which is particularly efficient, may be used for diagonalization of the matrix (see Appendix B). Thus, if a program applicable to calculation of the rigid rotor energy levels is available, it can be readily applied to the analysis of distortion effects by modification of the matrix elements. In particular, the matrix elements of (7.22) contain terms arising from the various distortion constants

$$E_{KK'} = (J, K|\mathscr{H}|J, K') = (J, K|\mathscr{H}_r|J, K') + (J, K|\mathscr{H}_d|J, K') \tag{8.142}$$

Table 8.21 Rotational Distortion Constants for Some Planar Asymmetric Rotors[a]

Molecule	$\tau_{aaaa}(MHz)$	$\tau_{bbbb}(MHz)$	$\tau_{aabb}(MHz)$	$\tau_{abab}(MHz)$	*Ref.*
SO_2	−10.1557	−0.0356	0.4644	−0.0485	c
O_3	−23.2588	−0.0762	0.4538	−0.3009	d
OF_2	−6.6154	−0.0950	0.3996	−0.1625	e
ClO_2	−7.8422	−0.0433	0.4748	−0.0416	f
Cl_2O	−5.3889	−0.0091	0.1146	−0.0188	g
$S(CN)_2$	−0.5838	−0.0144	0.0833	−0.0023	h
NF_2	−7.75	−0.081	0.297	−0.126	i
NOF	−15.5869	−0.0996	0.4186	−0.2896	j
NSF	−8.8343	−0.0564	0.4480	−0.0896	k
NOCl	−17.322	−0.0292	0.2555	−0.1101	l
NO_2Cl	−0.0427	−0.0173	0.0038	−0.0253	m
$H^{11}BF_2$	−9.04	−0.0563	0.409	−0.099	n
ClNCO	−59.138	−0.0107	0.6923	−0.0101	o
CHO-COOH	−0.0397	−0.0033	0.0018	−0.0368	p
$^{11}BH_2OH$	−25.64	−0.362	0.54	−1.676	q
$ArCO_2$	−0.0160[b]	−0.0983	0[b]	−0.907	r

[a]The planarity conditions have been invoked in these analyses.
[b]Fixed.
[c]Kivelson [38].
[d]Pierce [153].
[e]Pierce et al. [82].
[f]M. G. K. Pillai and R. F. Curl, Jr., *J. Chem. Phys.*, **37**, 2921 (1962).
[g]Herberich et al. [138].
[h]L. Pierce, R. Nelson, and C. Thomas, *J. Chem. Phys.*, **43**, 3423 (1965).
[i]R. D. Brown, F. R. Burden, P. D. Godfrey, and I. R. Gillard, *J. Mol. Spectrosc.*, **52**, 301 (1974).
[j]Cook [157].
[k]Cook and Kirchhoff [163].
[l]G. Cazzoli, R. Cervellati, and A. M. Mirri, *J. Mol. Spectrosc.*, **56**, 422 (1975).
[m]R. R. Filgueira, P. Forti, and G. Corbelli, *J. Mol. Spectrosc.*, **57**, 97 (1975).
[n]Robiette and Gerry [178].
[o]W. H. Hocking and M. C. L. Gerry, *J. Mol. Spectrosc.*, **42**, 547 (1972).
[p]K. M. Marstokk and H. Møllendal, *J. Mol. Struct.*, **15**, 137 (1973).
[q]Y. Kawashima, H. Takeo, and C. Matsumura, *J. Chem. Phys.*, **74**, 5430 (1981).
[r]J. M. Steed, T. A. Dixon, and W. Klemperer, *J. Chem. Phys.*, **70**, 4095 (1979).

The matrix elements required for $\mathscr{H}_d$ in a symmetric rotor basis are summarized in Table 8.22. Those for the rigid rotor part, $\mathscr{H}_r$, have been given in Chapter VII. The terms independent of $(P_x^2 - P_y^2)$ are purely diagonal in the symmetric top basis, while those involving $(P_x^2 - P_y^2)$ are off-diagonal with $\Delta K = \pm 2$.

For evaluating the centrifugal distortion constants from the rotational spectrum, two procedures are particularly convenient—a rigid rotor basis [91, 92] and a semirigid rotor basis [93] distortion analysis. The former procedure makes use of the rigid rotor basis, that is, the basis in which $\mathscr{H}_r$ is diagonal, to derive a first-order energy expression from which the

Table 8.22 Matrix Elements for the Angular Momentum Operators in a Symmetric Rotor Basis[a]

$$(J, K|P^{2n}|J, K)=J^n(J+1)^n \qquad n, m=0, 1, 2, \ldots$$
$$(J, K|P_z^{2n}|J, K)=K^{2n}$$
$$(J, K|P^{2n}P_z^{2m}|J, K)=J^n(J+1)^nK^{2m}$$
$$(J, K|P^{2n}(P_x^2-P_y^2)|J, K\pm 2)=-\tfrac{1}{2}J^n(J+1)^n f_\pm(J, K)$$
$$(J, K|P^{2n}[P_z^{2m}(P_x^2-P_y^2)+(P_x^2-P_y^2)P_z^{2m}]|J, K\pm 2)$$
$$=-\tfrac{1}{2}J^n(J+1)^n[K^{2m}+(K\pm 2)^{2m}]f_\pm(J, K)$$
$$f_\pm(J, K)=\{[J(J+1)-K(K\pm 1)][J(J+1)-(K\pm 1)(K\pm 2]\}^{1/2}$$

[a]Phase choice that of G. W. King, R. M. Hainer, and P. C. Cross, *J. Chem. Phys.*, **11**, 27 (1943). Angular momentum is in units of $\hbar$.

rotation-distortion constants can be obtained with least-squares techniques.

If the distortion effects are treated as a perturbation on the rigid rotor, the first-order distortion energy may be found by evaluating (8.97) in the rigid asymmetric rotor basis. If the Wang representation of $\mathscr{H}_r$ is employed, (7.35) with ($x\to b$, $y\to c$, $z\to a$, that is, I^r), and if it is recognized that P^2 and $\mathscr{H}(b_p)$ are diagonal in the asymmetric rotor basis, the following average value equations for the angular momentum operators may be derived from (7.35)

$$\langle P_x^2-P_y^2\rangle=\sigma[W(b_p)-\langle P_z^2\rangle] \tag{8.143}$$

$$\langle P_z^2(P_x^2-P_y^2)+(P_x^2-P_y^2)P_z^2\rangle=2\sigma[W(b_p)\langle P_z^2\rangle-\langle P_z^4\rangle] \tag{8.144}$$

$$\langle P_z^4(P_x^2-P_y^2)+(P_x^2-P_y^2)P_z^4\rangle=2\sigma[W(b_p)\langle P_z^4\rangle-P_z^6\rangle] \tag{8.145}$$

The first-order expression for the semirigid prolate asymmetric rotor for the A reduction becomes [91]

$$E^{(A)}=E_r+E_d^{(4)}+E_d^{(6)} \tag{8.146}$$

$$E_r=\tfrac{1}{2}(\mathscr{B}+\mathscr{C})J(J+1)+[\mathscr{A}-\tfrac{1}{2}(\mathscr{B}+\mathscr{C})]W(b_p) \tag{8.147}$$

$$\begin{aligned}E_d^{(4)}=&-\Delta_J J^2(J+1)^2-\Delta_{JK}J(J+1)\langle P_z^2\rangle-\Delta_K\langle P_z^4\rangle\\&-2\delta_J\sigma J(J+1)[W(b_p)-\langle P_z^2\rangle]\\&-2\delta_K\sigma[W(b_p)\langle P_z^2\rangle-\langle P_z^4\rangle]\end{aligned} \tag{8.148}$$

$$\begin{aligned}E_d^{(6)}=&\Phi_J J^3(J+1)^3+\Phi_{JK}J^2(J+1)^2\langle P_z^2\rangle+\Phi_{KJ}J(J+1)\langle P_z^4\rangle\\&+\Phi_K\langle P_z^6\rangle+2\phi_J\sigma J^2(J+1)^2[W(b_p)-\langle P_z^2\rangle]\\&+2\sigma\phi_{JK}J(J+1)[W(b_p)\langle P_z^2\rangle-\langle P_z^4\rangle]\\&+2\sigma\phi_K[W(b_p)\langle P_z^4\rangle-\langle P_z^6\rangle]\end{aligned} \tag{8.149}$$

where $\sigma=-1/b_p=(2\mathscr{A}-\mathscr{B}-\mathscr{C})/(\mathscr{B}-\mathscr{C})$, $W(b_p)$ is the Wang reduced energy, and $\langle P_z^n\rangle$ is the average of P_z^n in the rigid asymmetric rotor basis. The latter quantities are obtained in conjunction with the diagonalization of the rigid rotor energy matrix. The $\mathscr{A}(=B_z^{(A)})$, $\mathscr{B}(=B_x^{(A)})$, $\mathscr{C}(=B_y^{(A)})$ are the effective rotational constants

$$\begin{aligned}
\mathscr{A} &= A + 16R_6 \\
&= A - 2\Delta_J \\
\mathscr{B} &= B - \frac{16R_6(A-C)}{(B-C)} \\
&= B - 2\Delta_J - \Delta_{JK} + 2\delta_J + 2\delta_K \\
\mathscr{C} &= C + \frac{16R_6(A-B)}{(B-C)} \\
&= C - 2\Delta_J - \Delta_{JK} + 2\delta_J + 2\delta_K
\end{aligned} \tag{8.150}$$

If these results are generalized to include P^8 and P^{10} effects, the following terms are added to the first-order energy expression [92]

$$\begin{aligned}
E_d^{(8)} = {} & L_J J^4(J+1)^4 + L_{JJK} J^3(J+1)^3 \langle P_z^2 \rangle \\
& + L_{JK} J^2(J+1)^2 \langle P_z^4 \rangle + L_{KKJ} J(J+1) \langle P_z^6 \rangle \\
& + L_K \langle P_z^8 \rangle + 2\sigma l_J J^3(J+1)^3 [W(b_p) - \langle P_z^2 \rangle] \\
& + 2\sigma l_{JK} J^2(J+1)^2 [W(b_p)\langle P_z^2 \rangle - \langle P_z^4 \rangle] \\
& + 2\sigma l_{KJ} J(J+1) [W(b_p)\langle P_z^4 \rangle - \langle P_z^6 \rangle] \\
& + 2\sigma l_K [W(b_p)\langle P_z^6 \rangle - \langle P_z^8 \rangle]
\end{aligned} \tag{8.151}$$

$$\begin{aligned}
E_d^{(10)} = {} & P_J J^5(J+1)^5 + P_{JJK} J^4(J+1)^4 \langle P_z^2 \rangle \\
& + P_{JK} J^3(J+1)^3 \langle P_z^4 \rangle + P_{KJ} J^2(J+1)^2 \langle P_z^6 \rangle \\
& + P_{KKJ} J(J+1) \langle P_z^8 \rangle + P_K \langle P_z^{10} \rangle \\
& + 2\sigma p_J J^4(J+1)^4 [W(b_p) - \langle P_z^2 \rangle] \\
& + 2\sigma p_{JJK} J^3(J+1)^3 [W(b_p)\langle P_z^2 \rangle - \langle P_z^4 \rangle] \\
& + 2\sigma p_{JK} J^2(J+1)^2 [W(b_p)\langle P_z^4 \rangle - \langle P_z^6 \rangle] \\
& + 2\sigma p_{KKJ} J(J+1) [W(b_p)\langle P_z^6 \rangle - \langle P_z^8 \rangle] \\
& + 2\sigma p_K [W(b_p)\langle P_z^8 \rangle - \langle P_z^{10} \rangle]
\end{aligned} \tag{8.152}$$

One may apply these equations to an oblate top ($x \to b$, $y \to a$, $z \to c$; i.e., III^l) by interchanging $\mathscr{A}$ and $\mathscr{C}$ and setting $\sigma = -1/b_0$ in the foregoing energy expression. For a III^r representation, $\sigma = 1/b_0$. The effective rotational constants for the III^l representation are now defined by

$$\begin{aligned}
\mathscr{A} &= A + \frac{16R_6(C-B)}{(B-A)} \\
&= A - 2\Delta_J - \Delta_{JK} - 2\delta_J - 2\delta_K \\
\mathscr{B} &= B - \frac{16R_6(C-A)}{(B-A)} \\
&= B - 2\Delta_J - \Delta_{JK} + 2\delta_J + 2\delta_K, \\
\mathscr{C} &= C + 16R_6 \\
&= C - 2\Delta_J
\end{aligned} \tag{8.153}$$

In order to obtain the rotational and distortion constants, a least-squares analysis of the observed frequencies is carried out based on (8.146). In particular, the difference between the observed frequency and the calculated rigid rotor frequency is taken as the distortion effect, and the distortion constants are fitted to these distortion shifts ($\nu_{obs}-\nu_r$) by means of the least-squares technique. Usually, linear variations of the rigid rotor energy with respect to the three rigid rotor parameters are also considered, that is,

$$\frac{\partial E_r}{\partial \mathscr{A}}\delta\mathscr{A}+\frac{\partial E_r}{\partial \mathscr{B}}\delta\mathscr{B}+\frac{\partial E_r}{\partial \mathscr{C}}\delta\mathscr{C} \tag{8.154}$$

The three variables $\delta\mathscr{A}$, $\delta\mathscr{B}$, $\delta\mathscr{C}$ are thus added to the analysis. For (8.147), variations in $(\mathscr{B}+\mathscr{C})/2$, $(\mathscr{B}-\mathscr{C})/2$, and $\mathscr{A}-(\mathscr{B}+\mathscr{C})/2$ would be more appropriate (see Chapter VII, Section 5). The observed frequency for a transition can thus be expressed in the form

$$\nu_{obs}=\nu_r+\sum\Delta\langle\Pi_i\rangle T_i \tag{8.155}$$

where ν_r is the rigid rotor frequency, $\langle\Pi_i\rangle$ are the appropriate operators averaged in the rigid rotor basis, $\Delta\langle\Pi_i\rangle$ is the average value difference between the upper and lower energy level of the transition, and the T_i represent the distortion constants and corrections to the rotational constants which are to be evaluated by use of the least-squares criterion.

If higher-order distortion effects, such as P^6 distortion terms, have to be considered in the analysis, then a first-order treatment will be be adequate, and higher-order perturbation effects must be considered. When the higher-order effects are not extremely large, a convenient iterative procedure may be used which was first employed by Pierce et al. [82] and subsequently extended to light asymmetric tops by Helminger, et al. [91]. This procedure takes into account higher-order perturbation effects while allowing the actual analysis for the distortion constants to be linear. The technique entails first analyzing the spectrum by means of (8.146) to obtain an approximate set of rotational and distortion constants. These constants are then employed to construct the complete energy matrix of (8.140). The energy for a given J value may be written, correct to P^6 terms, for example, as

$$E=\tfrac{1}{2}(\mathscr{B}+\mathscr{C})J(J+1)-\Delta_J J^2(J+1)^2+H_J J^3(J+1)^3+[\mathscr{A}-\tfrac{1}{2}(\mathscr{B}+\mathscr{C})]W, \tag{8.156}$$

where W is an energy eigenvalue of the reduced energy submatrix $\mathbf{W}$, which is composed of K-dependent matrix elements of $\mathscr{H}_r$, $\mathscr{H}_d^{(4)}$, and $\mathscr{H}_d^{(6)}$, in which the original distortion constants have been divided by the factor $[\mathscr{A}-\tfrac{1}{2}(\mathscr{B}+\mathscr{C})]$. This scaling is convenient for the diagonalization problem. Diagonalization of $\mathbf{W}$ gives the energy levels of the complete Hamiltonian; by taking energy differences, the transition frequencies are obtained which include the corrections due to centrifugal distortion. These frequencies are "exact" in that the complete energy matrix, (8.142), has been considered. The difference in the transition frequency found from (8.156) via diagonalization of the complete energy matrix and the frequency computed from the first-order energy expression, (8.146), gives the second- and higher-order centrifugal distortion effect to the given transition (the so-called "second-order" effects). The observed frequencies are then corrected for these effects and reanalyzed with first-order theory.

$$\nu_{\text{"second-order"}} = \nu_{\text{exact}} - \nu_{\text{first-order}} \tag{8.157}$$

$$\nu_{\text{corr-obs}} = \nu_{\text{obs}} - \nu_{\text{"second-order"}} \tag{8.158}$$

The observed frequencies, when corrected for these higher-order effects, essentially represent a set of first-order observed frequencies, that is, those which would be observed if there were no higher-order distortion effects present. If the corrected-observed frequencies, (8.158), are then reanalyzed with the first-order theory, a slightly different set of rotation and distortion constants is obtained. The whole procedure is then repeated with these new constants, and another set of exact frequencies is thus obtained, which are again compared with first-order frequencies (computed with the new constants), to give the second- and higher-order corrections and subsequently a new set of corrected-observed frequencies. This iterative procedure may be continued until the higher effects are stable. In this way accurate distortion constants can be obtained while the higher-order distortion effects are still taken into account. This procedure may be applied in similar problems where a first-order treatment is not sufficient and higher-order effects necessitate diagonalization of a matrix in which the constants to be evaluated appear explicitly. Usually only a few iterations are required to stabilize the higher-order effects.

This mode of analysis is very convenient, since only the average values of the operator, P_z^n in the rigid rotor basis are required, and is applicable to all heavy molecules; it has also been applied successfully to the light molecules HDS [91] and HDO [92]. However, when the "second-order" effects become very large, the convergence of this procedure becomes slow. It is then convenient to employ an alternative procedure which involves linearization of the problem from the start and is very similar to that used in obtaining the rotational constants from a rigid rotor spectrum.

We follow the discussion of Cook et al. [93] and for convenience, denote the Hamiltonian of (8.140) as

$$\mathscr{H} = \sum_i T_i \Pi_i \tag{8.159}$$

where T_i correspond to the various rotation and distortion constants of (8.140) which we are interested in evaluating, and the Π_i represent the corresponding quantum mechanical operators. If (8.159) is evaluated in the semirigid rotor basis, that is, the basis in which $\mathscr{H}$ is diagonal, we have

$$\langle \mathscr{H} \rangle = E(T_i) = \sum_i T_i \langle \Pi_i \rangle \tag{8.160}$$

The $\langle \Pi_i \rangle$ are averages such as $\langle P_x^2 - P_y^2 \rangle$, $\langle P_z^4 \rangle$, and so on. If the T_i are the correct set of constants, then (8.160) represents the true energy of the level.

In general, the energy may be expanded in a Taylor series about the approximate parameters ($T_i{}^0$), and to first order we have

$$E(T_i)=E(T_i^0)+\sum_i\left(\frac{\partial E}{\partial T_i}\right)_0 \delta T_i \tag{8.161}$$

This expression, which is linear in the parameter corrections δT_i, may be used as the basis for a least-squares analysis in which the corrections to the constants are calculated. The derivatives are given by

$$\left(\frac{\partial E}{\partial T_i}\right)_0=\langle \Pi_i \rangle_0 \tag{8.162}$$

where the average value on the right is calculated in the semirigid rotor basis employing the T_i^0. These require the matrix elements of Table 8.22 and the eigenvectors for each energy level that should be obtained as part of the diagonalization procedure.

The procedure for the analysis in the semirigid rotor basis is as follows. Approximate values of the spectral constants T_i^0 are used for the construction of the energy matrix of (8.140), and subsequent diagonalization of this matrix yields the $E(T_i^0)$ and the $\langle \Pi_i \rangle_0$. From (8.161) we can write for an observed transition

$$\nu_{\text{obs}}-\nu_{\text{calc}}=\sum \Delta\langle \Pi_i \rangle_0 \delta T_i \tag{8.163}$$

where $\nu_{\text{calc}}=\Delta E(T_i^0)$. Here Δ indicates the difference between the energy (or average value) of the upper and lower level of the transition. Equation 8.163 and the least-squares criterion are used in an analysis of the observed transitions for the calculation of the δT_i's. These corrections yield an improved approximation to the constants given by

$$T_i=T_i^0+\delta T_i \tag{8.164}$$

This procedure is then repeated with the improved approximations to the rotation and distortion constants. This numerical iteration procedure is continued until the δT_i's are negligibly small.

Both the rigid rotor basis distortion analysis and the semirigid rotor basis distortion analysis will produce the same parameters. In general, when the "second-order" effects are large, the rigid rotor basis tends to be inconvenient because of the number of iterations that are necessary, whereas the semirigid rotor basis has been found to converge very rapidly [93]. Reasonably good approximations to the constants can usually be obtained from a rigid rotor basis first-order analysis. As additional higher-order distortion constants are subsequently added to the analysis, these constants may be initially started at zero with success. In fact, with an initial choice of zero for all the distortion constants, no difficulty is usually experienced in the convergence of the analysis. In all the calculations described, double precision (16-digit) arithmetic should be employed.

Examples of distortion constants obtained from rotational spectra with use of the A reduction are given in Table 8.23.

The recent advances in submillimeter wave rotational spectroscopy and the accompanying theoretical advances in centrifugal distortion theory have provided significant impetus to the study of light asymmetric rotors. In the past, such molecules have eluded an accurate microwave study because of the occurrence of many, if not most, of their rotational transitions at frequencies inaccessible to conventional microwave spectrometers. Reviews of millimeter and submillimeter spectroscopy have been given by Winnewisser et al. [94], by De Lucia [95], and by Krupnov and Burenin [96]. In addition, inadequacies in the previous formulations of centrifugal distortion theory have handicapped the empirical determination of the distortion constants from observed spectra. Present theoretical formulations provide an adequate solution, but require on the order of 20 parameters to properly characterize the rotational spectrum of this class of molecules. Results of a typical analysis requiring a complete set of P^4 and P^6 terms, six P^8 terms, and one P^{10} term are given in Table 8.24 for D_2S [93]. In addition, studies of HDS [91], H_2S [97], $H_2{}^{16}O$ [98], $H_2{}^{17}O$ [99, 101], $H_2{}^{18}O$ [100, 101], $D_2{}^{16}O$ [102, 103], $D_2{}^{17,18}O$ [104], HDO [92, 103], T_2O [105, 106], HTO [105, 107], DTO [107], H_2Se [108], NH_2D and ND_2H [109, 110] have been reported. Table 8.25 summarizes some of the results for this class of molecules.

For the symmetric-top reduction, the analysis procedure discussed previously can be applied; however the energy matrix of

$$\mathscr{H}^{(S)} = \mathscr{H}_r + \mathscr{H}_d^{(4)} + \mathscr{H}_d^{(6)} \tag{8.165}$$

has matrix elements of the type $(K|K)$, $(K|K\pm 2)$, $(K|K\pm 4)$, $(K|K\pm 6)$, and the four submatrices given in (8.138) contain more elements, as indicated in (8.166).

$$\mathbf{E}^{\pm} = \begin{bmatrix} E_{0,0} & \sqrt{2}E_{0,2} & \sqrt{2}E_{0,4} & \sqrt{2}E_{0,6} & 0 & \cdot \\ \sqrt{2}E_{0,2} & E_{2,2}\pm E_{-2,2} & E_{2,4}\pm E_{-2,4} & E_{2,6} & E_{2,8} & \cdot \\ \sqrt{2}E_{0,4} & E_{2,4}\pm E_{-2,4} & E_{4,4} & E_{4,6} & E_{4,8} & \cdot \\ \sqrt{2}E_{0,6} & E_{2,6} & E_{4,6} & E_{6,6} & E_{6,8} & \cdot \\ 0 & E_{2,8} & E_{4,8} & E_{6,8} & E_{8,8} & \cdot \\ \cdot & \cdot & \cdot & \cdot & \cdot & \cdot \end{bmatrix}$$

$$\mathbf{O}^{\pm} = \begin{bmatrix} E_{1,1}\pm E_{-1,1} & E_{1,3}\pm E_{-1,3} & E_{1,5}\pm E_{-1,5} & E_{1,7} & 0 & \cdot \\ E_{1,3}\pm E_{-1,3} & E_{3,3}\pm E_{-3,3} & E_{3,5} & E_{3,7} & E_{3,9} & \cdot \\ E_{1,5}\pm E_{-1,5} & E_{3,5} & E_{5,5} & E_{5,7} & E_{5,9} & \cdot \\ E_{1,7} & E_{3,7} & E_{5,7} & E_{7,7} & E_{7,9} & \cdot \\ 0 & E_{3,9} & E_{5,9} & E_{7,9} & E_{9,9} & \cdot \\ \cdot & \cdot & \cdot & \cdot & \cdot & \cdot \end{bmatrix} \tag{8.166}$$

Table 8.23 Rotational and Centrifugal Distortion Constants of Some Asymmetric Rotors (A Reduction) (MHz)[a]

Molecule	$\mathscr{A}$	$\mathscr{B}$	$\mathscr{C}$	$\Delta_J(10^3)$	$\Delta_{JK}(10^2)$	$\Delta_K(10^0)$	$\delta_J(10^4)$	$\delta_K(10^2)$	*Ref.*
HOF	585,631.20	26,760.31	25,510.19	94.14	241.46	82.520	43.2	112.0	*b*
SO_2	60,778.52	10,318.07	8,799.70	6.5875	−11.724	2.5899	17.008	2.5259	*c*
S_2O	41,915.44	5,059.10	4,507.16	1.895	−3.192	1.197	3.453	1.223	*d*
SCL_2	14,613.60	2,920.87	2,430.69	1.3249	−1.4604	0.1380	3.397	0.3927	*e*
F_2SO	8,614.80	8,356.95	4,952.94	4.586	−0.2475	0.01001	15.50	0.2225	*f*
DN_3	344,746.59	11,350.98	10,964.75	4.281	44.451	92.242	1.864	36.51	*g*
HN_3†	610,996.2	12,034.15	11,781.45	4.673	79.1186	230.0	0.888	0	*h*
HNO_3	13,011.03	12,099.86	6,260.64	14.038	−2.0178	0.00742	11.828	−2.0565	*i*
$HC{\equiv}CCOOH$	12,110.02	4,146.94	3,084.49	0.5376	2.132	−0.007715	1.605	1.219	*j*
$CH_2(CN)_2$	20,882.75	2,942.30	2,616.72	1.8554	−6.7922	0.8622	4.8926	0.6750	*k*
HCOOH	77,512.23	12,055.10	10,416.15	9.989	−8.625	1.7023	19.492	4.260	*l*
$H_2C{=}CHCN$	49,850.70	4,971.21	4,513.83	2.2448	−8.5442	2.7183	4.5716	2.4575	*m*
$CH_2{=}CHF$	64,584.69	10,636.88	9,118.03	8.4352	−7.593	1.3294	17.6591	3.5654	*n*
H_2CO	281,970.52	38,836.04	34,002.20	75.2979	129.038	19.4125	104.4571	102.5757	*o*
trans.-HCOSH	62,036.09	6,125.30	5,569.64	3.4287	−4.3167	1.2507	4.3473	1.6516	*p*
cis-HCOSH	62,927.71	6,134.26	5,584.75	3.6786	−4.7962	1.3057	4.6312	3.7653	*p*
COF_2	11,813.54	11,753.06	5,880.90	6.130	−0.3073	0.0133	25.775	0.4293	*q*
$COCL_2$	7,918.79	3,474.96	2,412.21	0.9712	−0.1533	0.01041	3.481	0.1793	*r*
CH_2DF	119,675.05	24,043.44	22,959.37	49.371	34.268	1.3774	23.29	6.87	*s*
$H_2C{=}C{=}CHCN$	25,981.00	2,689.29	2,474.81	1.28	−8.55	1.94	3.10	0.6	*t*
cis-CH_3CH_2COF	10,042.54	3,762.21	2,832.69	0.776	0.3474	0.0042	1.832	0.1774	*u*

†Δ_K fixed and δ_K set to zero.

[a] I^r representation employed. Constant times 10^k is tabulated.

[b] Pearson and Kim [120].

[c] Carpenter [115].

[d] Cook et al. [155].

[e] Davis and Gerry [152].

[f] Lucas and Smith [170].

[g] J. Bendtsen and M. Winnewisser, *Chem. Phys.*, **40**, 359 (1979).

[h] J. Bendtsen and M. Winnewisser, *Chem. Phys. Lett.*, **33**, 141 (1975).

[i] Cazzoli and De Lucia [3].

[j] R. W. Davis and M. C. L. Gerry, *J. Mol. Spectrosc.*, **59**, 407 (1976).

[k] Cook et al. [7].

[l] Willemot et al. [142].

[m] M. C. L. Gerry and G. Winnewisser, *J. Mol. Spectrosc.*, **48**, 1 (1973).

[n] Gerry [174].

[o] R. Cornet and G. Winnewisser, *J. Mol. Spectrosc.*, **80**, 438 (1980).

[p] W. H. Hocking, G. Winnewisser, *Z. Naturforsch.*, **31a**, 422 (1976).

[q] Carpenter [115].

[r] J. H. Carpenter and D. F. Rimmer, *J.C.S. Faraday II*, **74**, 466 (1978).

[s] W. W. Clark and F. C. De Lucia, *J. Mol. Struct.*, **32**, 29 (1976).

[t] Bouchy et al. [58].

[u] F. Scappini and H. Dreizler, *Z. Naturforsch.*, **36a**, 1327 (1981).

Table 8.24 Rotational and Distortion Constants of D_2S (MHz)[a]

$\mathscr{A}=164{,}571.118\pm0.045$	
$\mathscr{B}=135{,}380.313\pm0.045$	
$\mathscr{C}=73{,}244.068\pm0.071$	
$\Delta_J=13.0763\pm0.0026$	$\Delta_{JK}=-41.7800\pm0.0066$
$\Delta_K=29.2170\pm0.0113$	$\delta_J=-1.95725\pm0.00068$
$\delta_k=47.2516\pm0.0037$	
$\Phi_J=3.783\times10^{-3}\pm0.041\times10^{-3}$	$\Phi_{JK}=-3.206\times10^{-2}\pm0.020\times10^{-2}$
$\Phi_{KJ}=6.373\times10^{-2}\pm0.044\times10^{-2}$	$\Phi_K=-3.620\times10^{-2}\pm0.075\times10^{-2}$
$\phi_J=-9.84\times10^{-4}\pm0.16\times10^{-4}$	$\phi_{JK}=2.443\times10^{-2}\pm0.019\times10^{-2}$
$\phi_K=8.804\times10^{-2}\pm0.065\times10^{-2}$	
$L_{JJK}=2.747\times10^{-5}\pm0.101\times10^{-5}$	$L_{JK}=-8.85\times10^{-5}\pm1.07\times10^{-5}$
$L_{KKJ}=7.8\times10^{-5}\pm2.1\times10^{-5}$	$l_J=7.72\times10^{-7}\pm1.02\times10^{-7}$
$l_{JK}=-1.750\times10^{-5}\pm0.131\times10^{-5}$	$l_K=-5.76\times10^{-4}\pm0.25\times10^{-4}$
$p_K=1.95\times10^{-6}\pm0.32\times10^{-6}$	

[a]The errors quoted are for 95% confidence limits. There is a high positive correlation ($|\rho|\geqslant0.90$) between (Φ_J, Δ_J), (ϕ_J, δ_J), (L_{JK}, Φ_K), and (l_J, L_{JJK}); and a high negative correlation between (L_{KKJ}, Φ_K), (L_{KKJ}, L_{JK}), (l_J, ϕ_J) and (l_K, ϕ_K). *A*-reduction and III^l representation employed in analysis. From Cook et al. [93].

It is not usually necessary to consider these higher-order distortion effects to obtain satisfactory distortion constants if high J transitions are excluded from the analysis, because only for high J (except for light molecules such as H_2O with large rotational energies even at low J) do these higher-order effects become appreciably large. Therefore, a first-order treatment that is particularly simple can be profitably employed. It may be remarked that even for high J certain ${}^bQ_{1-1}$ transitions exhibit very small centrifugal distortion effects. The procedure for finding the values of J that correspond to these transitions is discussed elsewhere [111].

Since the matrices no longer have tridiagonal form, the continued fraction diagonalization procedure is no longer directly applicable, and a slower diagonalization procedure is required, for example, the Jacobi method [112] which employs a series of rotation matrices to bring the energy matrix to diagonal form. Alternately, the Givens [112] method can be used which first reduces the matrix to tridiagonal form and then proceeds with a continued fraction procedure to complete the diagonalization. The off-diagonal terms $(K|K\pm4)$ arise from the constants d_2 and h_2, whereas the $(K|K\pm6)$ elements arise from the h_3 term in the Hamiltonian. If these terms are not too large, they may be neglected; the submatrices then have the same form as that for the rigid rotor problem. The neglected terms could then be treated by perturbation theory

$$E_d^{(1)}=d_2\langle P_+^4+P_-^4\rangle+h_2J(J+1)\langle P_+^4+P_-^4\rangle+h_3\langle P_+^6+P^6\rangle \quad (8.167)$$

The approximation should be quite good for the P^6 terms, but may not be satisfactory for the d_2 term.

Table 8.25 Rotational, Quartic, and Sextic Distortion Constants of Some Light Molecules (MHz)[a]

	H_2O[b]	HDO[b]	T_2O[b]	H_2S[c]	HDS[b]	$H_2{}^{80}Se$[c]
$\mathscr{A}$	835,840.29	701,931.50	338,810.92	310,182.24	292,351.30	244,095.53
$\mathscr{B}$	435,351.72	272,912.60	145,665.42	270,884.05	147,861.80	232,566.13
$\mathscr{C}$	278,138.70	192,055.25	100,259.42	141,705.88	96,704.12	116,874.62
Δ_J	37.594	10.838	4.1456	49.851	2.6134	34.822
Δ_{JK}	−172.91	34.208	−22.039	−159.696	28.6933	−118.293
Δ_K	973.29	377.078	144.138	111.851	−11.297	88.965
δ_J	15.210	3.647	1.6098	−6.0191	0.8554	−2.0788
δ_K	41.05	63.087	5.441	262.17	19.4078	463.32
$\Phi_J \cdot 10^3$	15.66	1.128	0.557	28.1	—	—
$\Phi_{JK} \cdot 10^2$	−4.21	7.344	−0.245	−22.83	1.326	−117.2
$\Phi_{KJ} \cdot 10^1$	−5.10	−2.740	−0.273	4.59	−0.2027	259.4
$\Phi_K \cdot 10^0$	3.733	1.4651	0.2036	−0.277	0.0130	—
$\phi_J \cdot 10^4$	78.0	6.55	2.68	−58.4	1.07	−58.44
$\phi_{JK} \cdot 10^2$	−2.52	3.096	0.0277	24.3	0.53	22.34
$\phi_K \cdot 10^1$	10.97	5.549	0.3637	28.70	0.283	85.11
Reference	d	e	f	g	h	i

[a] *A*-reduction.
[b] I^r representation.
[c] III^l representation.
[d] De Lucia et al. [98].
[e] De Lucia et al. [92].
[f] De Lucia et al. [106].
[g] Helminger et al. [97].
[h] Helminger et al. [91].
[i] Helminger and De Lucia [108].

The important new kinds of matrix elements required to set up the energy matrix are given in Table 8.26, and the remaining ones may be found in Table 8.22. Indications are that for a given set of data, the correlations between the constants are somewhat less for the S-reduced Hamiltonian than for the A-reduced Hamiltonian, but the quality of the fit for the same number of constants is the same [64]. Most of the results reported have been obtained with the A-reduced Hamiltonian. Examples employing the S reduction, which is particularly useful for slightly asymmetric rotors, are given in Table 8.27. To date, only a few molecules have been analyzed by means of the S-reduced Hamiltonian. A limited number of molecules have also been analyzed by the K reduction [68, 113]. In this reduction, the value of s_{111} is constrained to zero, and a sixth parameter τ_3 is fixed by means of the planarity relations. Details of this reduction may be found elsewhere [68].

Different representations affect the size of s_{111} and the distribution of the angular momentum operators P_a, P_b, P_c appearing in the distortion Hamiltonian. The importance of the choice of axis representation (choice of a, b, c to x, y, z) has been studied by Typke [114] and Carpenter [115] for the A reduction. If only the quartic distortion effects are considered, it is found that the standard deviation of the fit and the value of the distortion constants depend on the representation. However, when the sextic constants are added to the analysis, these effects are essentially eliminated. Different representations result in a different distribution of terms in the Hamiltonian between quartic and sextic terms, since s_{111} is different, as indicated by (8.86) and (8.87). Truncation of the Hamiltonian (S or A) can produce a significant dependence on the representation because a substantial quartic effect in the standard Hamiltonian can wind up in the sextic terms of the reduced Hamiltonian. The addition of higher-order constants ensures there will be no higher-order contributions absorbed in the lower-order constants. On the other hand, for molecules where a large number of higher-order terms are required, a representation that minimizes the size of s_{111} and hence improves the convergence of the Hamiltonian would be more appropriate [116]. Since Δ_K should be invariant to the choice of representation, it has been useful as a check on the representation chosen and whether sufficient terms have been retained in the Hamiltonian. The T_{aa}, T_{bb},

Table 8.26 Matrix Elements of $(P_+^{2n}+P_-^{2n})$ in a Symmetric Rotor Basis[a]

$$(J, K|P_+|J, K+1)=i\{f_+(0)\}^{1/2};\ (J, K|P_-|J, K-1)=-i\{f_-(0)\}^{1/2}$$
$$(J, K|P^{2n}(P_+^{2m}+P_-^{2m})|J, K')=J^n(J+1)^n(J, K|P_+^{2m}+P_-^{2m}|J, K')$$
$$(J, K|P_+^2+P_-^2|J, K\pm 2)=-\{f_\pm(0)f_\pm(1)\}^{1/2}$$
$$(J, K|P_+^4+P_-^4|J, K\pm 4)=\{f_\pm(0)f_\pm(1)f_\pm(2)f_\pm(3)\}^{1/2}$$
$$(J, K|P_+^6+P_-^6|J, K\pm 6)=-\{f_\pm(0)f_\pm(1)f_\pm(2)f_\pm(3)f_\pm(4)f_\pm(5)\}^{1/2}$$
$$f_\pm(l)=[J(J+1)-(K\pm l)(K\pm l\pm 1)]$$

[a]Phase choice that of G. W. King, R. M. Hainer, and P. C. Cross, *J. Chem. Phys.*, **11**, 27 (1943). Angular momentum in units of $\hbar$. $P_\pm = P_x \pm iP_y$, $n, m=1, 2, \ldots$.

Table 8.27 Rotational and Centrifugal Distortion Constants of Some Asymmetric Rotors (S Reduction) (MHz)[a]

Molecule	$\mathscr{A}$	$\mathscr{B}$	$\mathscr{C}$	$D_J(10^3)$	$D_{JK}(10^2)$	$D_K(10^0)$	$d_1(10^6)$	$d_2(10^6)$	*Ref.*
$H_2C{=}C{=}S$[b]	286,655	5,659.48	5,544.51	1.0857	16.8269	23.5	−25.46	−5.21	e
HNCS	1,357,250.0	5,883.46	5,845.61	1.1939	−102.537	51,570.0	−13.781	−4.59	f
HNCO	918,504.4	11,071.01	10,910.57	3.4863	93.170	6,065.6	−72.995	−36.5	g
H_2CNN[b]	272,979.1	11,305.43	10,845.22	4.1462	39.682	22.0	−192,550.	−57.27	h
$H_2N\text{-}NC(O^+)$[c]	282,757	10,757.28	10,525.34	5.2720	42.265	214	−176.75	−34.34	i
$H_2N\text{-}NC(O^-)$[c]	282,616	10,765.77	10,525.25	2.2669	42.092	214	175.33	−34.34	i
NH_2D	290,125.38	192,194.18	140,795.26	15,820.7	−2394.5	10.954	4.182×10^6	1.324×10^5	j
CD_2S[b]	146,399.87	14,904.28	13,495.86	12.579[d]	29.61	5.208	−1409.	−287.2	k

[a] I^r representation employed except for NH_2D. Constant times 10^k is tabulated.

[b] D_K fixed.

[c] 0^+, 0^- inversion states.

[d] Fixed by planarity relation.

[e] M. Winnewisser, E. Schäfer, *Z. Naturforsch.*, **35a**, 483 (1980).

[f] K. Yamada, M. Winnewisser, G. Winnewisser, L. B. Szalanski, and M. C. L. Gerry, *J. Mol. Spectrosc.*, **78**, 189 (1979).

[g] K. Yamada, *J. Mol. Spectrosc.*, **79**, 323 (1980).

[h] E. Schäfer and M. Winnewisser, *J. Mol. Spectrosc.*, **97**, 154 (1983).

[i] E. Schäfer and M. Winnewisser, *Ber. Bunsenges, Phys. Chem.*, **86**, 780 (1982).

[j] Cohen and Pickett [110].

[k] A. P. Cox, S. D. Hubbard, and H. Kato, *J. Mol. Spectrosc.*, **93**, 196 (1982).

T_{cc}, T_1, T_2 (or τ_{aaaa}, τ_{bbbb}, τ_{cccc}, τ_1, τ_2) referred to the principal inertial axis should also be invariant to the choice of representation or the reduction chosen for the Hamiltonian. It should also be noted that the signs of the spectroscopic constants can change from one representation to another and this must be considered in comparing constants (P^4 and P^6, etc.) evaluated with different representations. It is apparent from Table 8.14, for example, that the quartic constants $\delta_J(\mathrm{III}^r) = -\delta_J(\mathrm{III}^l)$ and $\delta_K(\mathrm{III}^r) = -\delta_K(\mathrm{III}^l)$ change sign, whereas Δ_J, Δ_{JK}, and Δ_K are invariant to this change in representation. Likewise, the P^6 constants ϕ_J, ϕ_{JK}, and ϕ_K are reversed in sign in going from a III^l to a III^r representation.

Relations between the quartic distortion constants for various representations and reductions have been given by Yamada and Winnewisser [117]. Some of the pertinent results are summarized in Table 8.28.

Studies by G. Winnewisser et al. [69, 118, 119] of HSSH ($\kappa = -0.9996$) and DSSD ($\kappa = -0.9999993$), which are very slightly asymmetric tops, revealed that the asymmetry splitting of the K levels can be smaller than that arising from centrifugal distortion effects; usually the reverse is true. The K doubling arises from the inertial asymmetry operator $(P_+^2 + P_-^2)$ and the centrifugal-distortion operator $(P_+^4 + P_-^4)$. The coefficient of $(P_+^2 + P_-^2)$ is dependent on

Table 8.28 Relations Between the Spectroscopic Constants in Different Axis Representations[a]

Transformations between representations:

$D_j = Q_j^{-1} Q_i D_i$, ($i, j = \mathrm{I}^r$, II^r, and III^r)

$$D = \begin{bmatrix} \Delta_J \\ \Delta_{JK} \\ \Delta_K \\ \delta_J \\ \delta_K \end{bmatrix}$$

$\mathrm{II}^r \to \mathrm{III}^r$

$$Q_{\mathrm{III}}^{-1} Q_{\mathrm{II}} = \begin{bmatrix} 1 & \frac{1}{2} & \frac{1}{2} & -1 & 0 \\ 0 & -\frac{1}{2} & -\frac{3}{2} & 3 & 0 \\ 0 & 0 & 1 & 0 & 0 \\ 0 & -\frac{1}{4} & -\frac{1}{4} & -\frac{1}{2} & 0 \\ 0 & 0 & -\dfrac{B-C}{2(A-B)} & 0 & -\dfrac{A-C}{A-B} \end{bmatrix}$$

$\mathrm{I}^r \to \mathrm{III}^r$

$$Q_{\mathrm{III}}^{-1} Q_{\mathrm{I}} = \begin{bmatrix} 1 & \frac{1}{2} & \frac{1}{2} & 1 & 0 \\ 0 & -\frac{1}{2} & -\frac{3}{2} & -3 & 0 \\ 0 & 0 & 1 & 0 & 0 \\ 0 & \frac{1}{4} & \frac{1}{4}. & -\frac{1}{2} & 0 \\ 0 & 0 & -\dfrac{A-C}{2(A-B)} & 0 & \dfrac{B-C}{A-B} \end{bmatrix}$$

$\mathrm{I}^r \to \mathrm{II}^r$

$$Q_{\mathrm{II}}^{-1} Q_{\mathrm{I}} = \begin{bmatrix} 1 & \frac{1}{2} & \frac{1}{2} & -1 & 0 \\ 0 & -\frac{1}{2} & -\frac{3}{2} & 3 & 0 \\ 0 & 0 & 1 & 0 & 0 \\ 0 & -\frac{1}{4} & -\frac{1}{4} & -\frac{1}{2} & 0 \\ 0 & 0 & \dfrac{A-B}{2(A-C)} & 0 & -\dfrac{B-C}{A-C} \end{bmatrix}$$

For $Q_{\mathrm{I}}^{-1} Q_{\mathrm{III}}$, replace last row of $Q_{\mathrm{II}}^{-1} Q_{\mathrm{I}}$ by $(0, 0, (A-C)/\{2(B-C)\}, 0, (A-B)/(B-C)]$
For $Q_{\mathrm{I}}^{-1} Q_{\mathrm{II}}$, replace last row of $Q_{\mathrm{III}}^{-1} Q_{\mathrm{I}}$ by $[0, 0 (A-B)/\{2(B-C)\}, 0, -(A-C)/(B-C)]$
For $Q_{\mathrm{II}}^{-1} Q_{\mathrm{III}}$, replace last row of $Q_{\mathrm{III}}^{-1} Q_{\mathrm{I}}$ by $[0, 0-(B-C)/\{2(A-C)\}, 0, -(A-B)/(A-C)]$

[a] The vector D of spectroscopic constants depends on the choice of axis representation. $\mathrm{I}^r(xyz \equiv bca)$, $\mathrm{II}^r(xyz \equiv cab)$, and $\mathrm{III}^r(xyz \equiv abc)$. From Yamada and Winnewisser [117].

the rotational constants; the coefficient of $(P_+^4 + P_-^4)$ depends on the distortion constants and hence on the molecular force field. For the $K=2$ levels the distortion splitting contribution is so large that the order of the levels $(J_{2,J-2} > J_{2,J-1})$ is reversed from that expected for a rigid rotor. This could produce problems in labeling the energy levels properly [120]. This large centrifugal-distortion splitting of the K levels is illustrated in Fig. 8.3 along with that expected from the asymmetry splitting only.

Symmetric and Spherical Tops

The diagonal terms of the symmetric-top Hamiltonian, (8.124), give the usual energy level expression to sixth degree

$$E_{JK} = B_x J(J+1) + (B_z - B_x)K^2 - D_J J^2(J+1)^2 - D_{JK} J(J+1)K^2 - D_K K^4 \\ + H_J J^3(J+1)^3 + H_{JK} J^2(J+1)^2 K^2 + H_{KJ} J(J+1)K^4 + H_K K^6 \tag{8.168}$$

For cases where $\tilde{\mathscr{H}}_{\text{split}}$ is important, off-diagonal matrix elements are also present in the energy matrix. Additional symmetry factoring of the energy matrix, (8.166), in the Wang basis is possible [64]. For $p=3$ or 6, the reduced splitting Hamiltonian is

$$\tilde{\mathscr{H}}_{\text{split}} = h_3(P_+^6 + P_-^6) \tag{8.169}$$

and matrix elements of the type $(K|K \pm 6)$ are introduced. A first-order treatment is given by the diagonal matrix elements of (8.166). Thus, the usual expression, (8.168), applies except for the $K = \pm 3$ levels. Matrix elements of (8.169) directly connect the pair of originally degenerate $K = \pm 3$ states, and the $E_{-3,3}$ element gives rise to a splitting of the $|K| = 3$ level

$$\Delta E = \pm E_{-3,3} = \mp h_3 J(J+1)[J(J+1)-2][J(J+1)-6] \tag{8.170}$$

The $K=3$ splitting constant $2h_3$ depends, for example, on τ_{xxxz} and the cubic potential constants. A detailed expression in terms of molecular parameters is available [72]. The coefficient has also been denoted by the symbol A and Δ_3 in the literature. This type of splitting has been observed [121–125]. This splitting, for example, in the ground state of PH_3 has been studied in the radiofrequency region by Davies et al. [123]. The observed splitting as a function of J was found to fit (8.170). The energy level diagram for the $J=7$, $K=3$ state is given in Fig. 8.4. The observed K-doubling amounts to 12,969.0 kHz. Additional hyperfine splitting is also shown in the figure.

The splitting for higher $K=6, 9, \ldots$ is caused by higher-order interactions and is usually very small. The next first-order effect would arise from the higher-order distortion term $(P_+^{12} + P_-^{12})$ in the reduced Hamiltonian which leads to a splitting of the $K=6$ level. In general, the splitting for $K=3n (n=1, 2, \ldots)$ arising from such terms may be written to first-order as

$$\Delta \nu_{3n} = \Delta_{3n} \frac{(J+3n)!}{(J-3n)!} \tag{8.171}$$

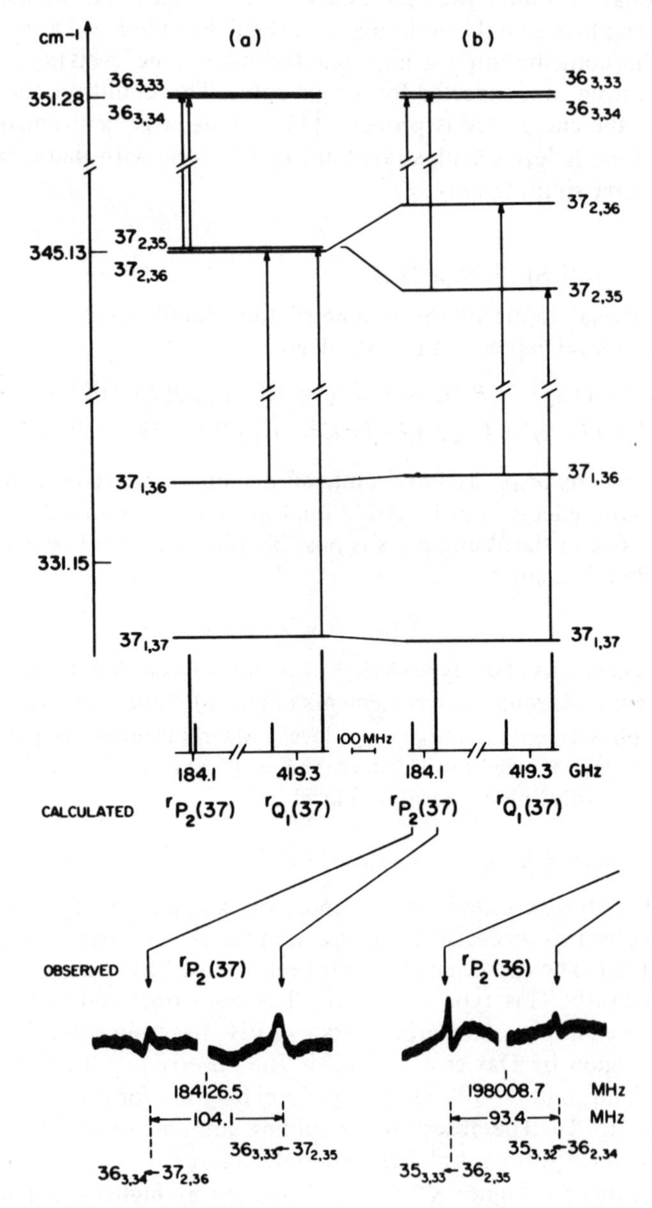

Fig. 8.3 Energy level diagram for HSSH. (*a*) *K* doubling due to inertial asymmetry. (*b*) *K* doubling due to asymmetry and centrifugal distortion splitting. The pattern of the $P_2(37)$ doublet is reversed in going from (*a*) to (*b*). The calculated intensity pattern of the $P_2(37)$ doublet may be compared with the observed transition $P_2(37)$. The effects of nuclear statistics are also illustrated for the $P_2(37)$ and $P_2(36)$ doublets, where the intensity pattern is weak, strong; strong, weak. From G. Winnewisser [119].

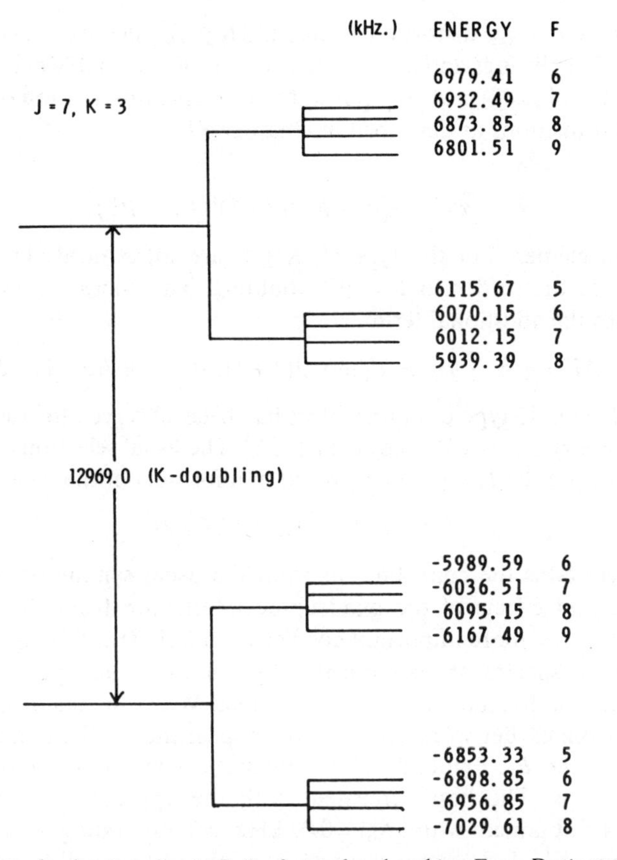

Fig. 8.4 Energy level pattern for $J=7$, $K=3$ state for phosphine. From Davis et al. [123].

Small $K=6$ splittings have been observed for NH_3 [122]. On the other hand, large splittings have been observed for $K=3$, 6, and 9 in excited vibrational states of PH_3 by Bernard and Oka [125]. To account for these splittings, the perturbation treatment described above was found inadequate. Here the dominant term in the reduced constant Δ_3 arises from τ_{xxxz}. Instead of the transformed or reduced Hamiltonian, the original general Hamiltonian for a symmetric top was used in the analyses. The terms off-diagonal in K by ± 3 arise from the three second-order τ's of Table 8.11. Because of (8.50) (See also Table 8.4), there are 16 angular momentum operators associated with these τ's, all of which (because of the relations in Table 8.11) may be associated with the same τ coefficient usually taken as τ_{xxxz} (x in symmetry plane). These angular momentum operators may be reduced by the commutation relations (8.49) and written in the compact form given in (6.48). To account for the observed splittings, the distortion term (6.48) was added directly to the usual symmetric rotor Hamiltonian, and the resulting secular equation was solved directly. The off-diagonal elements of the τ_{xxxz} distortion terms, $(K|K \pm 3)$, effectively explain the observed $K=3n$ splittings and provide τ_{xxxz} values of reasonable magnitude [125]. This operator of (6.48) is also important in allowing observation of "forbidden" rotational transitions as described in Chapter VI.

For $p=4$

$$\tilde{\mathscr{H}}_{\text{split}} = \{d_2 + h_2 J(J+1)\}(P_+^4 + P_-^4) \tag{8.172}$$

and matrix elements of the type $(K|K \pm 4)$ are introduced. Thus d_2 and h_2 contribute to first-order to $K = \pm 2$ doubling. The energy expression (8.168) applies with the additional term

$$\Delta E = \pm E_{-2,2} = \pm \{d_2 + h_2 J(J+1)\} J(J+1)[J(J+1)-2] \tag{8.173}$$

for $|K|=2$. This K-type doubling effect has been observed for the first time in BrF_5 (symmetry C_{4V}) by Bradley et al. [126]. The usual selection rules $\Delta J = +1$, $\Delta K = 0$, indicate the $K = \pm 2$ lines would fit a formula of the type

$$\nu = \nu_0 \pm 4 d_2 J(J+1)(J+2) \tag{8.174}$$

where ν_0 contains the contributions from the usual symmetric-top frequency expression and effects of the quadrupole interaction from Br. The splitting constant d_2 is the more important coefficient in (8.173) and is equivalent to R_6. The observed spectra are complicated by nuclear quadrupole structure and contain many individual and overlapped lines. When the quadrupole coupling constants from earlier work are used, the appearance of the spectrum depends on only D_{JK} and R_6. Figure 8.5 shows theoretical spectra for various values of these constants. The best agreement with the appearance of the observed spectrum is obtained with $|R_6| = 0.07$ kHz corresponding to a splitting of 2.8 MHz for $|K|=2$. Direct measurement of the $|K|=2$ splitting for a lower J transition ($J=8 \to 9$) has also been obtained [127]. Both the quadrupole structure and the $|K|=2$ splitting are resolved, and the value of the splitting constant obtained from fitting the spectral line contours is confirmed. The $|K|=2$ splittings in SF_5Cl [128], ClF_5 [129], and IF_5 [130] have also been analyzed. Additional discussions are available [131, 132].

For spherical tops, the calculation of the energy levels has been considered by a number of authors [73–80]. The eigenvalues $f(J, \kappa)$ of the fourth-rank tensor Ω_4 have been discussed by Hecht [73]. Kirschner and Watson [77] have treated the sixth-rank tensor Ω_6 by first-order perturbation theory employing the eigenfunctions of the operator Ω_4

$$g(J, \kappa) = (J, \kappa|\Omega_6|J, \kappa) \tag{8.175}$$

These quantities are independent of the distortion constants and are readily tabulated; tables for the $f(J, \kappa)$ and $g(J, \kappa)$ for $J \leq 20$ are available [77]. To sixth degree, the rotational energy is given approximately by

$$\begin{aligned} E(J, \kappa) = {} & BJ(J+1) - D_S J^2(J+1)^2 + H_S J^3(J+1)^3 \\ & + [D_{4T} + H_{4T} J(J+1)] f(J, \kappa) + H_{6T} g(J, \kappa) \end{aligned} \tag{8.176}$$

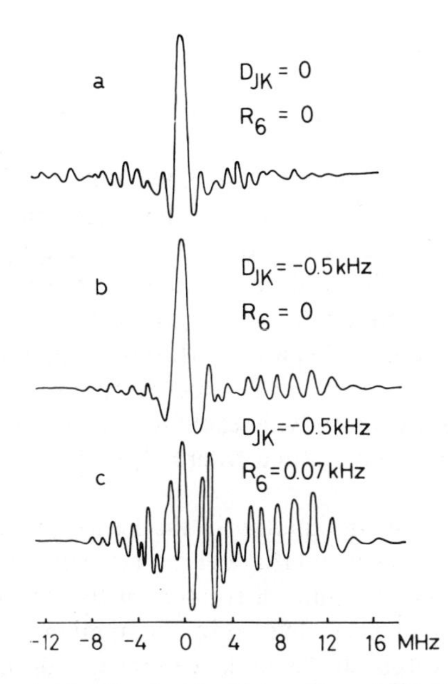

Fig. 8.5 Calculated spectra for the $J=16\rightarrow17$ transition of $^{79}BrF_5$ for the values of D_{JK} and R_6 shown. The bottom pattern is similar to the observed spectrum. From Bradley et al. [126].

Perturbation treatment of Ω_8 has been discussed by Ozier [78]. This adds the following terms to the energy expression

$$E^{(8)}(J,\kappa)=L_S J^4(J+1)^4+L_{4T}J^2(J+1)^2 f(J,\kappa)+L_{6T}J(J+1)g(J,\kappa) + L_{8T}h(J,\kappa)+\left(\frac{H_{6T}^2}{D_{4T}}\right)\tilde{g}(J,\kappa) \tag{8.177}$$

Tables to include the effect of the octic terms $(h, \tilde{g})$ have also been given $(J\leqslant 20)$ [78]. Here the $\kappa=C^{(t)}$ label is used to distinguish the energy levels for each J. Matrix elements for the Hamiltonian are available [64, 76] for numerical diagonalization if higher-order effects are to be considered. An idea of the order-of-magnitude of the constants may be given by quoting the results obtained for SiH_4:$D_{4T}=74{,}751.4$; $H_{4T}=-6.044$; $H_{6T}=2.598$; $L_{4T}=46.5\times10^{-5}$; $L_{6T}=-37.9-10^{-5}$; and $L_{8T}=-76.6\times10^{-5}$ Hz [133].

Choice of Hamiltonian

Assignment procedures and applications of statistical techniques for model testing have been discussed by Kirchhoff [68] and others [7, 91–93, 102, 134, 135]. A detailed treatment of least-squares fitting of spectroscopic data has been given by Albritton et al. [136].

The distortion analyses techniques discussed here give rise to a set of observational equations, such as (8.155), appropriate to a linear least-squares analysis.

The least-squares criterion leads to the usual normal equations for estimates of the parameters, $T_j (j=1, 2, \ldots, p)$ employing the n observed transition frequencies, where $p \leqslant n$. The so-called square matrix of normal equations is designated here by $\mathbf{A}$

Usually, unit weight ($W_i=1$) will be associated with the microwave measurements unless there is a known difference in the measurement accuracy of the transitions in the fit. However, when combining, for example, microwave and infrared data, different weights will be required. The weights, in general, are taken as inversely proportional to the variances, $W_i=1/\sigma_i^2$. A typical set of weights could be constructed [136] for the IR data with $\sigma_{\mathrm{IR}} \simeq 10^{-3}\ \mathrm{cm}^{-1}$, and for the microwave data subset, $\sigma_{\mathrm{M}} \simeq 10^{-6}\ \mathrm{cm}^{-1}$. The weighting factors can be conveniently introduced by multiplying each observational equation by $(W_i)^{1/2}$ and treating the system as unweighted where W_i is the weight assigned to the ith transition.

An important measure of the quality of the least-squares fit and precision of the data is provided by the standard deviation σ, which is the square root of the variance σ^2, defined as the sum of squares of the deviations between observed and calculated frequencies divided by the number of degrees of freedom $(n-p)$. It is possible to establish a confidence interval for σ, $(\sigma_1 \leqslant \sigma \leqslant \sigma_2)$ since σ^2 follows a chi-square (χ^2) distribution [135].

If, for example, the data set is divided into two groups (one with predominately high J transitions), and analyzed, two 95% confidence intervals can be obtained from the two standard deviations of the fits. If these do not overlap, then some evidence exists for the need of higher-order distortion terms. An example of the use of confidence intervals on σ to indicate the need for higher-order constants in SO_2 has been given [68].

An estimate of the standard error in the ith parameter is given by

$$\sigma_i=\sigma\sqrt{\mathbf{A}_{ii}^{-1}} \tag{8.178}$$

with $\mathbf{A}^{-1}$ the inverse of the normal equation matrix and σ the standard deviation of the fit. Usually, the uncertainties in the parameters are quoted as $T_i \pm 2\sigma_i$, which corresponds to essentially a 95% confidence level; that is, one can be 95% confident that the true values lie within the interval. The factor of 2 is satisfactory [136] as long as $(n-p) \gtrsim 30$. For ease in constructing other confidence limits, one might report the number of degrees of freedom $(n-p)$ explicitly.

The uncertainty in the derived parameters gives rise to an uncertainty in the calculated frequency. The standard deviation of the calculated frequency σ_ν is given by

$$\sigma_\nu^2=\sum_{i,j}^{p} \sigma_i \sigma_j \rho_{ij} C_i C_j=\sigma^2 \sum_{i,j}^{p} A_{ij}^{-1} C_i C_j \tag{8.179}$$

where σ_i is the standard error in the ith spectral constant and C_i the corresponding coefficient in the frequency expression. The ρ_{ij} are the correlation coefficients [see (8.180)]. The σ_ν are useful for detection of bad transition data and as a guide to selection of transitions for measurement.

Table 8.29 shows results of an analysis of the data of HDS used to confirm the assignment of the lines. The transitions listed in the table have been omitted from the data set and the frequency of each line calculated, along with its corresponding standard deviation. It is apparent that the underlined transition is not satisfactorily predicted when compared to the expected σ_ν. When this transition was remeasured, it agreed with that calculated well within the predicted uncertainty. On the other hand, when this transition was included in the analysis, it was not apparent that the transition was mismeasured. This often happens with least-squares analysis where the error is often distributed over other transitions or particular constants.

Another useful statistical test for location of a bad transition is the standardized residuals [68] $t(\Delta\nu)=\Delta\nu/\sigma(\Delta\nu)$, where $\Delta\nu=\nu_{\text{obs}}-\nu_{\text{calc}}$ and $\sigma(\Delta\nu)$ is the standard deviation of $\Delta\nu$. The $t(\Delta\nu)$ measures the number of standard deviations by which $\Delta\nu$ differs from the expected value of zero; a large value indicates the transition is suspect. The values of this ratio should follow a student's t_{n-p} distribution. Details of application of this test and expressions for $\sigma(\Delta\nu)$ may be found elsewhere [68, 137].

The correlation coefficients ρ_{ij} between the ith and jth constants may be evaluated from the inverse normal equation matrix, that is,

$$\rho_{ij}=A_{ij}^{-1}/\sqrt{A_{ii}^{-1}}\,\sqrt{A_{jj}^{-1}} \qquad (-1\leqslant\rho_{ij}\leqslant 1) \tag{8.180}$$

These are of interest in themselves since they give information on the linear dependence of the constants [138] and associated errors [136]. The diagonal elements are always 1. A value of $\rho_{ij}=+1$ indicates a perfect positive correlation or interdependence, while a $\rho_{ij}=0$ indicates no correlation. Note that the ρ_{ij} can be calculated for a given Hamiltonian and choice of transition data and does not involve the actual measurements. The variation of ρ_{ij} with the data set can thus be studied, and transitions which might lead to a decrease in a given correlation can be determined.

When the spectroscopic constants are highly correlated, a number of significant figures beyond that indicated by the standard error are required to re-

Table 8.29 Confirmation of HDS Assignment

Transition	*Observed Frequency (MHz)*	*Calculated Frequency (MHz)*	*Calculated Error[a] (MHz)*
$1_{1,0}-1_{1,1}$	51,073.27	51,073.36	± 0.08
$2_{1,1}-2_{1,2}$	153,179.16	153,179.32	± 0.18
$5_{3,2}-4_{4,1}$	<u>112,689.59[b]</u>	<u>112,684.90</u>	<u>± 2.32</u>
$6_{2,4}-6_{1,5}$	598,805.17	598,804.92	± 0.74
$7_{3,4}-7_{2,5}$	628,018.53	628,016.90	± 3.80

[a]95% confidence limits.
[b]Remeasurement 112,685.45.

produce the calculated quantities. The number of significant figures required can be determined by trial and error, if additional digits in each spectral constant are kept until the rounded numbers actually reproduce the observations. Three or four digits beyond the one standard error digit should be a reasonable rule of thumb. More specific criteria have also been reported [137, 139].

If the confidence interval $T_i \pm 2\sigma_i$ includes zero, then to a 95% confidence limit, T_i is not significant and can be omitted from the fit. A more sophisticated statistical criterion of whether an additional constant is needed is given by the F test. This is essentially a measure of the improvement in the sum of squares of the deviations, S, obtained by introduction of another fitting parameter. In particular,

$$F=(n-p)\frac{(S_{p-1}-S_p)}{S_p} \tag{8.181}$$

with n the number of transition frequencies and S_p the sum of squares of the deviations for the p-parameter fit. For example, consider a p versus a $p-1$ parameter P^6 fit. If comparison of the calculated F with the tabulated $F_{1,n-p}$ value (5% critical value) indicates that $F<F_{1,n-p}$, the fit with p constants does not significantly improve the fit over that with $p-1$ constants, and the constant with 95% confidence may be omitted. An example of use of the F-test to decide on the significance of certain terms in $\mathscr{H}_d^{(6)}$ has been given for $CH_2(CN)_2$ [7].

In evaluating the distortion constants from the rotational spectrum one usually selects a set of low J lines, with small distortion effects, to obtain the three rigid rotor parameters. Next, additional lower J transitions are assigned which have sufficient distortion effects to provide credible predictions. Some difficulty may be experienced in this first stage, if an incorrectly measured or misassigned transition is present in the data set. Once enough data are available for a reasonably reliable analysis, the frequency predictions and their standard deviations become meaningful. Those predicted transitions with small estimated uncertainties can then be readily found. Attempts to assign transitions with large uncertainties (e.g., >100 MHz) will usually be unsuccessful. These new transitions may then be added to the distortion analysis to improve the spectroscopic constants and thus predict other higher J transitions with greater reliability. This technique is repeated as one observes and adds to the analysis higher and higher J transitions, until an adequate data set is obtained that yields accurate spectroscopic constants and predicts the remainder of the spectrum to a given J value with uncertainties near the measurement error. One, of course, must be careful about employing predictions for high J transitions or certain subbranches, if such data are not in the data set. The determination of accurate distortion constants requires a number of rotational transitions with distortion effects of reasonable size. Also, it is important to select transitions from as many different subbranches as possible. (A subbranch is designated by the changes in J, K_{-1}, and K_1, i.e., ΔJ, ΔK_{-1}, and ΔK_1.)

To guard against possible misassignments or incorrect measurements, the following procedure can be followed. Separate distortion analyses are carried out by use of the final Hamiltonian with one line (or a small arbitrary group of lines) omitted from the data set. The spectral constants, line-frequency predictions, and the standard deviation σ_ν in each line-frequency prediction are evaluated for each analysis. By use of an analysis that does not include the line itself, each line is compared with its predicted frequency and required to fall within, or very close to, the range expected considering $2\sigma_\nu$ and the experimental uncertainty. A useful measure for this comparison is also given by the values of $t(\Delta\nu)$, discussed previously. No attempt at this stage should be made to improve the predictions by including more constants in the Hamiltonian. The standard deviation of the least-squares fit should also be virtually the same for the various analyses and, in addition, the various analyses should produce essentially the same set of spectral constants.

As additional transitions are measured and used in the analysis, higher-order, for example, P^6 constants may have to be included to adequately fit the experimental observations. This will become apparent from a significant increase in the standard deviation of the least-squares fit as higher J transitions are added, with the assumption that the measurements and assignments of the higher J transitions are correct.

However the observed data may not be sensitive to all the P^6 constants. The P^6 constants selected to describe the data may be obtained as follows. The data are analyzed with a full complement of sextic terms, and the most statistically undetermined constant (standard deviation greater than the constant) is removed and the analysis repeated. It will usually be clear that the constant omitted did not improve the fit. For the six-parameter fit, another constant may now be statistically undetermined; this constant can be removed and the analysis repeated with one less fitting parameter. This process is continued until all the remaining constants are judged to be statistically determined. The selection process is complicated if there are high correlations between the constants. Because of the possible high correlation among some of the constants, it is not always clear in which order the terms should be eliminated. In such situations, inspection of the results of the several alternatives will usually indicate the proper choice. In certain cases, the uncertainty in the constant may be such that it is not clear whether the constant should be retained or rejected. In fact, as judged from the standard deviation of the least-squares, the constant could possibly be omitted without significantly affecting the fit. For such borderline cases a more sophisticated statistical criterion, the F-test, might be used to test for the significance of a particular constant as indicated previously. Instead of constraining one or more P^6 constants at zero, as judged from the previous discussion, all the constants could, of course, be included in the analysis. However, their inclusion would add nothing to the fit, and there is little reason to expect that their inclusion would improve the prediction of unmeasured lines, particularly those which might be sensitive to these constants. Since a sextic planar relation is now available, it would be preferable to use the planarity condition to eliminate a particular constant rather than to fix the constant to zero. If the data are still not sensitive to the remaining constants, then the above procedure can be used to remove additional constants from the fit. Such a statistical treatment essentially allows the available data to decide which of the distortion terms are important in the characterization of the spectrum.

For light asymmetric rotors such as H_2O, D_2O, H_2S, D_2S, and so on, the distortion analysis is further complicated [91–93, 102]. For these light molecules, the series converges very slowly, which necessitates the inclusion of a large number of terms in the Hamiltonian to adequately characterize the observed spectra. This slow convergence also makes difficult the accurate prediction of transitions too far outside the particular data set used. Even the relatively low J lines have significant effects due to the higher-order distortion coefficients.

Of critical importance to the proper analysis of these molecules is the selection of distortion terms to be retained in the Hamiltonian. This selection is complicated by the rather slow convergence of the power series in the angular momentum as well as by the high statistical correlation among many of the terms. The analysis procedure is similar to that described previously and is considered in more detail elsewhere [92, 93].

One error that should be guarded against is the truncation of $\mathscr{H}$ too early when the data contain higher-order effects. That the importance of a term is not simply characterized by its degree, but depends on both the coefficient and the corresponding operator, has been pointed out by Steenbeckeliers and Bellet [102]. For small K_{-1}, both diagonal and off-diagonal terms are of similar importance, whereas for large K_{-1}, the diagonal terms are the more important.

In general, for both light and heavy molecules, the final selection of terms (choice of model) should produce a Hamiltonian that [92, 93]

1. Fits the data to within experimental uncertainty.
2. Adequately predicts lines removed from the fit, to within expected theoretical and experimental error.
3. Has the minimum number of terms consistent with (1) and (2).
4. Contains constants that are statistically meaningful.
5. Is consistent with a converging power series.

With regard to (5), this implies, for example, that L_K should be better determined than P_K. It is furthermore important that the Hamiltonian, as has been found, results in essentially the same values of rotation and lower-order distortion constants independent of the details of either the data set or the higher-order terms retained in the Hamiltonian.

6 PLANARITY RELATIONS FOR CONSTANTS OF THE REDUCED HAMILTONIAN

For a planar molecule, the number of coefficients in the reduced Hamiltonian can be reduced. The planarity relations discussed previously can be used to write the following planar relations

$$\tau_{aacc}=\frac{C^2}{2A^2}\tau_{aaaa}-\frac{C^2A^2}{2B^4}\tau_{bbbb}+\frac{A^2}{2C^2}\tau_{cccc} \tag{8.182}$$

$$\tau_{bbcc}=\frac{C^2}{2B^2}\tau_{bbbb}-\frac{B^2C^2}{2A^4}\tau_{aaaa}+\frac{B_2}{2C^2}\tau_{cccc} \tag{8.183}$$

Recall that only $\tau_{abab}\neq 0$ and hence $\tau'_{aabb}=\tau_{aabb}+2\tau_{abab}$, $\tau'_{aacc}=\tau_{aacc}$, and $\tau'_{bbcc}=\tau_{bbcc}$. These relations and the planarity condition $AB=(A+B)C$ can be used to give a single relation between the five determinable coefficients. In particular

$$\tau_2=C\tau_1+(A+B)\tau_{cccc} \tag{8.184}$$

or in terms of the T's

$$T_2=CT_1+(A+B)T_{cc} \tag{8.185}$$

These relations are strictly correct only in the small oscillations approximation. In principle, equilibrium rotational constants and distortion constants should be used in these relations, although one is usually forced to use effective constants which contain vibrational contributions. If the planarity relations are to be used to eliminate one of the P^4 coefficients in the reduced Hamiltonian leaving only four quartic coefficients, it is convenient to express the planarity relation in terms of the parameters of the A- or S-reduced Hamiltonians [140]. If the previous equations are employed, the relations are given in Table 8.30 for two particular representations $\mathrm{I}^r(x\to b,\ y\to c,\ z\to a)$ and $\mathrm{III}^r(x\to a,\ y\to b,\ z\to c)$, appropriate for a prolate and oblate top, respectively.

A relation for the sextic coefficients has also been derived by Watson [140]

$$\begin{aligned}3(A+B)\Phi_{ccc}+2\{-(B-C)\Phi_{cca}-(A-C)\Phi_{ccb}\\ -2(T_{cc}-T_{ac})(T_{cc}-T_{bc})\}+4T_{cc}^2=0\end{aligned} \tag{8.186}$$

The expression in braces corresponds to Φ_3, Φ_2, or Φ_4 for I^r, II^r, or III^r representations, respectively. (Note for II^r: $x\to a, y\to b, z\to c$). This relation leads to an expression between the parameters of the A- or S-reduced Hamiltonian also given in Table 8.30. Application of this planarity constraint can be used to reduce by one the number of sextic coefficients in the reduced Hamiltonian, leaving only six terms.

For a planar symmetric top, the molecule is an oblate top $A=B=2C$ and $d_1=0$, $T_{ac}=T_{bc}=2T_{cc}$. Using these conditions for the S reduction in a III^r representation gives the two relations for a symmetric top, also included in Table 8.30.

Planarity relations for the P^6 constants of XYX and XYZ asymmetric tops have also been discussed by Niroomand-Rad et al. [141]. Here, four planarity relations among the 10 original sextic coefficients of (8.73) are given. These could be employed to eliminate four of the ten sextic constants and thus reduce the general "first principles" Hamiltonian so that it would contain only determinable coefficients and thus remove the inherent indeterminacy in $\mathscr{H}^{(6)}$.

Table 8.30 Quartic and Sextic Planarity Relations for the Spectroscopic Constants of the Reduced Hamiltonian[a]

Asymmetric Top

Reduction A, Representation I^r

$$4C\Delta_J-(B-C)\Delta_{JK}-2(2A+B+C)\delta_J+2(B-C)\delta_K=0$$
$$6C\Phi_J-(B-C)\Phi_{JK}-2(2A+B+3C)\phi_J+2(B-C)\phi_{JK}$$
$$+4\Delta_J^2-4\delta_J(4\Delta_J+\Delta_{JK}-2\delta_J-2\delta_K)=0$$

Reduction S, Representation III^r

$$4C\Delta_J+(A+B+2C)\Delta_{JK}+2(A+B)\Delta_K+2(A-B)(\delta_J+\delta_K)=0$$
$$6C\Phi_J+(A+B+4C)\Phi_{JK}+2(A+B+C)\Phi_{KJ}+3(A+B)\Phi_K+2(A-B)(\phi_J+\phi_{JK}+\phi_K)$$
$$+(2\Delta_J+\Delta_{JK})(2\Delta_J+3\Delta_{JK}+4\Delta_K)+4(\delta_J+\delta_K)^2=0$$

Reduction A, Representation I^r

$$4CD_J-(B-C)D_{JK}+2(2A+B+C)d_1-4(4A+B-3C)d_2=0$$
$$6CH_J-(B-C)H_{JK}-2(2A+B+3C)h_1+4(4A+B-2C)h_2-6(6A+B-5C)h_3$$
$$+4(D_J+2d_1-2d_2)^2+4(D_{JK}-2d_1+4d_2)(d_1-4d_2)=0$$

Reduction S, Representation III^r

$$4CD_J+(A+B+2C)D_{JK}+2(A+B)D_K-2(A-B)d_1=0$$
$$6CH_J+(A+B+4C)H_{JK}+2(A+B+C)H_{KJ}+3(A+B)H_K+2(A-B)h_1$$
$$+(2D_J+D_{JK})(2D_J+3D_{JK}+4D_K)+4d_1^2=0$$

Symmetric Top

$$2D_J+3D_{JK}+4D_K=0$$
$$3H_J+4H_{JK}+5H_{KJ}+6H_K=0$$

[a]Here, A, B, C are the equilibrium rotational constants. From Watson [140].

Unless the data are such that least-squares equations are ill-conditioned, the quartic planarity constraint should not be applied (see next section). On the other hand, the sextic constraint can, in many cases, be profitably employed to improve the determination of the constants when the uncertainties in the sextic constants are large.

7 EFFECTS OF VIBRATION

The derived distortion coefficients obtained from an analysis of the observed ground state spectrum are effective values containing contributions from the vibrational averaging effects. The vibrational dependence can be expressed, for example, as

$$\Delta_J^{[v]}=\Delta_J^e-\sum \beta_s^J\left(v_s+\frac{d_s}{2}\right) \tag{8.187}$$

$$\Delta_{JK}^{[v]}=\Delta_{JK}^e-\sum \beta_s^{JK}\left(v_s+\frac{d_s}{2}\right) \tag{8.188}$$

and so on, and the dependence in the T constants might be expressed by

$$T_{\alpha\beta}^{[v]} = T_{\alpha\beta}^{e} + \sum \rho_s^{\alpha\beta}\left(v_s + \frac{d_s}{2}\right) \tag{8.189}$$

where the superscript e indicates equilibrium values, $[v]$ represents the set of vibrational quantum number, v_s, specifying the vibrational state, with the sum over the various vibrations of degeneracy d_s. For asymmetric rotors $d_s=1$. Very general definitions of $\rho_s^{\alpha\beta}$ are available [19]. Just as with the rotational constants, in order to obtain the distortion coefficients associated with the equilibrium vibrationless state, measurements must be made in all $(3N-6)$ vibrational states of the molecule. Because of intensity considerations, such an analysis has been made in only a limited number of cases. Examples of distortion constants obtained in the ground and excited vibrational states are available [142]. The difference in these values gives one some idea of the size of the vibrational effects and hence the possible error introduced into force field calculations when effective rather than equilibrium values are employed in the analysis. Differences between the equilibrium and ground state distortion constants on the order of 5% or less are common. Observed variations in the distortion constants with vibrational state which are particularly large, can often be explained in terms of a Coriolis interaction [143] or a large amplitude motion.

A detailed study of trimethylene oxide (TMO) with C_{2v} symmetry by Creswell and Mills [144] shows for the distortion constants an anomalous zig-zag dependence on the ring puckering vibrational quantum number. Similar dependence has been exhibited by the rotational constants [145] (see Chapter XII). The distortion constants were determined for the ground state and in the excited puckering vibrational states through $v_p=5$. The vibrational dependence is illustrated in Fig. 8.6. All states could be analyzed with the same effective Hamiltonian so the vibrational dependence in the quartic constants is not affected by the choice of model. The constant δ_J is essentially independent of the puckering vibration, while Δ_K, $-\Delta_{JK}$, and $-\delta_K$ show qualitatively similar patterns in the vibrational dependence. The experimental uncertainty in Δ_J is believed to be too large to draw significant conclusions.

The low-frequency anharmonic ring puckering vibration of TMO can to a good approximation be separated from the remaining vibrational modes which are essentially harmonic in nature. The contribution of this mode to $\tau_{\alpha\beta\gamma\delta}$ may hence be separated from the contributions of the other modes which are independent of v_p. The contribution from the puckering mode which depends on v_p may be evaluated from the known ring puckering potential constants (see Chapter XII, Section 14). From the symmetry of the puckering coordinate Q_p, one finds the only nonvanishing $a_p^{(\alpha\beta)}$ is $a_p^{(ac)}$. Thus only τ_{acac} will carry a vibrational dependence which implies only Δ_{JK}, Δ_K, and Δ_K (III^r representation) are affected. Explicit calculation shows that the vibrational dependence for δ_J, Δ_{JK}, Δ_K, and δ_K given in Fig. 8.6 is accounted for.

Similar results have been obtained for 1-pyrazoline [146]; however, for $v_p=0$ and 1, because of the close separation of these vibrational levels

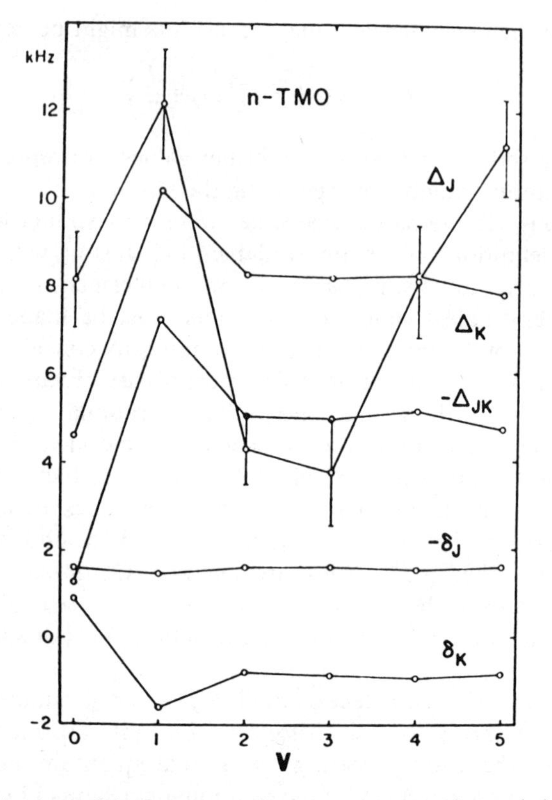

Fig. 8.6 Observed values of the five quartic distortion constants of trimethylene oxide (TMO) as a function of the quantum number in the puckering vibration. Uncertainty in Δ_J is shown as an error bar equal to the standard error in each case; the uncertainty in the remaining four distortion constants is too small to show on these graphs. From Creswell and Mills [144].

(7.61 cm^{-1}), it was necessary for a vibration–rotation coupling term to be included in the effective $\mathscr{H}$ and for these two states to be analyzed together.

Centrifugal Defect

A number of examples are available of the failure of the planarity relations when applied to effective ground state constants rather than equilibrium quantities. This can be well demonstrated if an unconstrained fit of a planar molecule is carried out to five quartic constants that are then used to test how well (8.184) is obeyed. This has been demonstrated by Steenbeckeliers [134] for SO_2 and for a number of molecules by Kirchhoff [68]. In terms of the scaled value of τ_2, one can define a centrifugal defect analogous to the inertial defect

$$\delta_{cd} = \tau_{cccc} - \tau'_2 - \frac{C}{A+B}(\tau'_2 - \tau_1) \tag{8.190}$$

with $\tau'_2 = \tau_2/(A+B+C)$. Values of δ_{cd} calculated from the observed constants are collected in Table 8.31. If the relations were exactly obeyed, $\delta_{cd} = 0$. The

Table 8.31 Values of the Centrifugal Defect for Several Planar Molecules[a]

Molecule	$\delta_{cd}(kHz)$	*Ref.*
SO_2	-0.248 ± 0.009	b
F_2O	-0.56 ± 0.06	b
NSF	-0.323 ± 0.015	b
SiF_2	-1.19 ± 0.37	b
NOF	-0.08 ± 0.18	b
COF_2	0 ± 0.12	b
CF_2	-0.22 ± 0.07	c
SF_2	-0.42 ± 0.14	d
NH_2CHO	-0.12 ± 0.07	e
HNCS	4.363 ± 0.003	f
HClCO	-0.017 ± 0.016	g
SCl_2	-0.045 ± 0.032	h
KCN	-0.86 ± 0.02	i
$CH_2{=}CHNC$	-0.133 ± 0.001	j

[a]Uncertainties represent one standard deviation.
[b]Kirchhoff [68].
[c]Kirchhoff et al. [137].
[d]Kirchhoff et al. [154].
[e]W. H. Kirchhoff and D. R. Johnson, *J. Mol. Spectrosc.*, **45**, 159 (1973).
[f]K. Yamada, M. Winnewisser, G. Winnewisser, L. B. Szalanski, and M. C. L. Gerry, *J. Mol. Spectrosc.*, **79**, 295 (1980).
[g]M. Suzuki, K. Yamada, and M. Takami, *J. Mol. Spectrosc.*, **88**, 207 (1981).
[h]Davis and Gerry [152].
[i]T. Törring, J. P. Bekooy, W. L. Meerts, J. Hoeft, E. Tiemann, and A. Dymanus, *J. Chem. Phys.*, **73**, 4875 (1980).
[j]K. Yamada and M. Winnewisser, *Z. Naturforsch.*, **30a**, 672 (1975).

results indicate that planarity condition should not be used unless the data are clearly insufficient to evaluate all five determinable coefficients. Examples of the quartic defect employing the planarity relations for the parameters of the reduced Hamiltonian are given in Table 8.32 for isotopes of water. The effects of vibration for water are readily apparent from the rather large nonzero values obtained. It is further apparent that there is a systematic decrease in δ with increasing mass or effectively decreasing amplitude of vibration.

A study of the sextic planarity relations for several triatomic molecules has also been reported [140]. From Table 8.32 it is apparent that the sextic defect is significant for H_2O and for most of the isotopic forms of water. For H_2S and D_2S the sextic defect is only barely significant. For O_3 it has been found that there is good cancellation of terms and that the sextic defect is insignificant.

Table 8.32 Terms in Planarity Relations for the Reduced Hamiltonian for Isotopes of Water[a,b]

	Isotopes					
	$H_2{}^{16}O$	$HD^{16}O$	$D_2{}^{16}O$	$HT^{16}O$	$DT^{16}O$	$T_2{}^{16}O$
	Quartic Planarity Relation $(GHz)^2$					
$4C\Delta_J$	41.83	8.326	5.397	3.137	2.477	1.663
$-(B-C)\Delta_{JK}$	27.18	−2.766	3.328	−2.316	0.821	1.001
$-2(2A+B+C)\delta_J$	−72.56	−13.632	−9.499	−4.820	−4.316	−2.974
$2(B-C)\delta_K$	12.91	10.202	1.517	4.900	1.366	0.494
Sum	9.36	2.130	0.743	0.901	0.348	0.184
	±0.15	±0.009	±0.003	±0.007	±0.003	±0.002
	Sextic Planarity Relation $[10^2(MHz)^2]$					
$6C\Phi_J$	261	13.0	15.2	4.7	4.3	3.35
$-(B-C)\Phi_{JK}$	66	−59.4	5.6	−18.8	−2.2	1.11
$-2(2A+B+3C)\phi_J$	−459	−29.5	−28.6	−7.2	−8.0	−6.03
$2(B-C)\phi_{JK}$	−79	50.1	−3.9	16.4	3.1	0.25
$4\Delta_J^2$	57	4.7	3.5	1.1	1.1	0.69
$-4\delta_J(4\Delta_J+\Delta_{JK}-2\delta_J-2\delta_K)$	82	8.2	5.4	2.0	1.9	1.26
Sum	−72	−12.9	−2.8	−1.8	+0.2	+0.63
	±22	±1.6	±0.5	±0.9	±0.9	±0.34
Refs. to Data	[98]	[92]	[102]	[107]	[107]	[106]

[a]The uncertainties of the sums are calculated from the uncertainties of the individual terms without correlation. All constants refer to the zero-point vibrational state. From Watson [140].
[b]The relations for a I^rA Hamiltonian are employed.

For SO_2 a rather large sextic defect is obtained, which is probably due to truncation of the Hamiltonian at the sextic terms in analysis of the data.

8 CENTRIFUGAL DISTORTION CONSTANTS AND MOLECULAR FORCE FIELDS

A considerable amount of effort has been expended in the calculation of force constants from the fundamental vibrational frequencies. The fundamental vibrational frequencies ν of a molecule are related to the $3N-6$ (or $3N-5$ for linear case) roots λ of the secular determinant [11]

$$|\mathbf{FG}-\lambda\mathbf{E}|=0 \tag{8.191}$$

where $\mathbf{E}$ is a unit matrix, $\mathbf{F}$ the force-constant matrix with elements f_{ij}, and $\lambda=4\pi^2\nu^2$. However, in most cases the number of quadratic force constants exceeds the number of normal vibrations. This difficulty is avoided by either the use of isotopic information or the assumption that the interaction force constants may be neglected. Unfortunately, the latter assumption is not always justifiable. Furthermore, in many cases it is found that the isotopic vibrational frequency data, especially when hydrogen is not substituted, are not sufficient to uniquely determine the force field, and additional data are necessary [147]. The experimentally determined distortion constants provide one such valuable source of force-constant information. These data can be used to supplement the vibrational data and thus uniquely fix the force field or enable the determination of additional force constants. In special cases, for diatomic or bent symmetric triatomic molecules, they can be used directly in calculation of the force constants. In addition to the distortion constants, the inertial defect [148] and Coriolis coupling coefficients [149], as well as mean-square amplitudes [150, 151], are also useful sources of information, since they depend on the quadratic potential constants.

The force constant information available from distortion constants is, of course, limited by the number of independent combinations of τ's that can be derived from the rotational spectrum. As we have seen, in the general case, no more than five independent linear combinations can be determined. For planar asymmetric molecules, only four distortion constants are independent; for a symmetric rotor, although no more than three independent linear combinations exist, only two can be determined from pure rotational spectra.

All of the earlier work to obtain information about molecular force constants, utilizing distortion data of asymmetric rotors determined from microwave spectra, was centered around planar molecules. Several calculations utilizing only distortion information have been carried out for symmetrical bent triatomic molecules, for example, SO_2, for which there are four force constants and four independent τ's. The inverse relationships to those of (8.36)–(8.39) are [138]:

$$(F^{-1})_{11} = -\frac{1}{2} R \sin^2\theta \cos^2\theta \left(\frac{\cot^2\theta \tau_{aaaa}}{A^2} + \frac{\tan^2\theta \tau_{bbb}}{B^2} + \frac{2\tau_{aabb}}{AB} \right) \tag{8.192}$$

$$(F^{-1})_{22} = -R \sin^2\theta \cos^2\theta \left(\frac{\tau_{aaaa}}{A^2} + \frac{\tau_{bbbb}}{B^2} - \frac{2\tau_{aabb}}{AB} \right) \tag{8.193}$$

$$(F^{-1})_{12} = -\frac{R}{\sqrt{2}} \sin^2\theta \cos^2\theta \left[\frac{-\cot\theta \tau_{aaaa}}{A^2} + \frac{\tan\theta \tau_{bbbb}}{B^2} + \frac{(\cot\theta - \tan\theta)\tau_{aabb}}{AB} \right] \tag{8.194}$$

$$(F^{-1})_{33} = -\frac{Rm_X}{2M} \frac{(1 + 2m_Y m_X^{-1} \sin^2\theta)^2 \tau_{abab}}{AB} \tag{8.195}$$

Or alternately in terms of τ_{cccc} rather than τ_{aabb} [152]:

$$(F^{-1})_{11} = -\frac{R \sin^2\theta \cos^2\theta}{2} \left[\frac{A \cot^2\theta - B}{A^3} \tau_{aaaa} + \frac{B \tan^2\theta - A}{B^3} \tau_{bbbb} + \frac{AB}{C^4} \tau_{cccc} \right] \tag{8.196}$$

$$(F^{-1})_{22} = -R\sin^2\theta\cos^2\theta\left[(A+B)\left(\frac{\tau_{aaaa}}{A^3}+\frac{\tau_{bbbb}}{B^3}\right)-\frac{AB}{C^4}\tau_{cccc}\right] \tag{8.197}$$

$$(F^{-1})_{12} = -\frac{R\sin^2\theta\cos^2\theta}{2\sqrt{2}}\left[\frac{B(\tan\theta-\cot\theta)-2A\cot\theta}{A^3}\tau_{aaaa}\right.$$

$$\left.+\frac{A(\tan\theta-\cot\theta)+2B\tan\theta}{B^3}\tau_{bbbb}-\frac{AB(\tan\theta-\cot\theta)}{C^4}\tau_{cccc}\right] \tag{8.198}$$

The expression for $(F^{-1})_{33}$ is still given by (8.195). Here, as before,

$$R = \frac{r^2 \times 10^{-22}}{2h}$$

with h in erg-sec, θ as one-half the Y—X—Y angle, $M = 2m_Y + m_X$, r as the X—Y bond distance expressed in angstrom units, and with the τ's and rotational constants in megahertz. The **F** matrix has the following form (elements in dynes/cm)

$$\begin{bmatrix} f_r + f_{rr} & \dfrac{\sqrt{2}f_{r\alpha}}{r} & 0 \\ \dfrac{\sqrt{2}f_{r\alpha}}{r} & \dfrac{f_\alpha}{r^2} & 0 \\ 0 & 0 & f_r - f_{rr} \end{bmatrix} \tag{8.199}$$

These relations are sufficient to determine the complete quadratic-valence force field

$$2V = f_r(\delta r_1^2 + \delta r_2^2) + f_\alpha \delta\alpha^2 + 2f_{r\alpha}(\delta r_1 + \delta r_2)\delta\alpha + 2f_{rr}\delta r_1 \delta r_2 \tag{8.200}$$

from the distortion constants obtained by an analysis of the rotational spectrum. The force constants can, of course, then be used for calculation of the vibrational frequencies of the molecule. In certain cases, force constants determined in this manner have yielded vibrational frequencies which were in fairly good agreement ($\sim 4\%$) with observed values [82, 138, 153]. Such calculations, as well as yielding force constants, provide an independent confirmation of the assignment of the vibrational frequency.

If the planarity relations are employed and the spectrum analyzed for 4 τ's, the model errors are averaged in a complicated way. The use of these constants to evaluate a force field for XY_2 can, of course, be expected to introduce an unspecified error in the derived constants, although the results have, in fact, been quite useful. Alternately, if the five quartic constants have been evaluated, then the planarity relations can be used to extract, for example, τ_{aaaa}, τ_{bbbb}, τ_{aabb}, and τ_{abab}, and subsequently the force constants from (8.192)–(8.195). However, the evaluation of τ_{aabb} and τ_{abab} is overdetermined and different values can be obtained depending on which combination of constants is employed because the planarity relations are not satisfied exactly. This ambiguity has been pointed out by Kirchhoff [68, 154] and by Cook et al. [93, 155, 156], and a detailed consideration has been given by Yamada and Winnewisser [117]. In Table 8.33 the results for OF_2 are given for three sets of derived τ's. In particular, τ_{aabb} and τ_{abab} are evaluated from τ_{aaaa}, τ_{bbbb}, and (τ_1, τ_2), (τ_1, τ_{cccc}), or (τ_2, τ_{cccc}). The value of τ_{aabb} will have the same value for the latter two combinations but not τ_{abab}. This gives three sets of force constants. Note that F_{33} is evaluated solely from τ_{abab}. The spread of the force constants and calculated vibrational frequencies thus obtained, gives some idea of the effects of using effective rather than equilibrium parameters in the planarity relations. The use of effective rotational constants can also amplify the model errors. It is clear that the vibrational effects can be significant. It would be better to minimize the use of the planarity relations. Since τ_{aaaa}, τ_{bbbb}, τ_{cccc} can be obtained directly from the analysis of the rotational spectrum without the use of the planar relations it is more appropriate to use these constants to extract information on the force constants F_{11}, F_{12}, F_{22} (A_1 vibrations) [155, 156], for example, from (8.196)–(8.198). The correct calculation procedure for obtaining a value of τ_{abab} from the observed spectroscopic constants that minimizes the vibrational effects remains uncertain.

The most fruitful approach for obtaining information about the molecular potential function is the combination of both microwave and infrared data. A technique for combining vibrational and rotational distortion data for symmetric bent triatomic molecules has been discussed by Kivelson [38] and Pierce et al. [82, 153]. The criterion applied is that the force constants reproduce the observed vibrational frequencies exactly and fit as consistently as possible the frequency shifts in the rotational spectrum which are due to centrifugal distor-

Table 8.33 Force Field Calculations from Microwave Data for OF_2[a]

			Force Constants (mdyne/Å)			
Set[b]	τ_{aabb}(kHz)	τ_{abab}(kHz)	F_{11}	F_{12}	F_{22}	F_{33}
I	403.4±2.4	−158.2±0.9	4.65±0.04	0.202±0.004	0.7051±0.0009	2.969±0.016
II	385.3±2.7	−142.5±2.2	4.45±0.04	0.206±0.004	0.7161±0.0014	3.297±0.050
III	385.3±2.7	−140.0±2.4	4.45±0.04	0.206±0.004	0.7161±0.0014	3.354±0.057

	Infrared Spectrum Predictions (cm^{-1})[c]		
	ω_1	ω_2	ω_3
I	916.5±3.2	455.9±0.3	807.5±2.2
II	897.9±3.0	458.3±0.5	851.0±6.5
III	897.9±3.0	458.3±0.5	858.3±7.3

[a]The uncertainties represent one standard deviation as estimated by the least-squares fit. From Kirchhoff [68].
[b]Set I: Calculated from τ_{aaaa}, τ_{bbbb}, τ_1, τ_2; Set II: Calculated from τ_{aaaa}, τ_{bbbb}, τ_{cccc}, and τ_1; Set III: Calculated from τ_{aaaa}, τ_{bbbb}, τ_{cccc}, and τ_2.
[c]Observed: 928, 461, and 831.

tion. Extension of this method to bent unsymmetrical triatomic molecules has been discussed by Cook [157].

For larger molecules the number of force constants increases rapidly. For an unsymmetrical molecule, the number of harmonic and anharmonic potential constants is $\frac{1}{2}n(n+1)$ and $\frac{1}{6}n(n+2)$, respectively with $n=3N-6$ (or $3N-5$). For molecules of higher symmetry there are fewer constants and the number of independent constants can be obtained [158] by group-theoretical arguments and the representations of Γ_{vib}, Γ_{vib}^2 and Γ_{vib}^4. The most general and convenient approach for combining distortion data, and other spectroscopic quantities dependent on the force field, with the vibrational isotopic frequency data makes use of a least-squares analysis. The analysis procedure is very similar to that discussed previously for evaluation of spectroscopic constants from rotational spectra. Linear equations are constructed describing the changes in the observed quantities due to small changes in a particular force constant. From an initial trial force field, the least-squares criterion is employed to evaluate the changes in the force constants required to obtain agreement between observed and calculated quantities.

The vibrational problem is usually linearized by means of a Taylor series expansion

$$\omega_k=\omega_k^0+\sum_{i,j}\left(\frac{\partial\omega_k}{\partial F_{ij}}\right)_0 F_{ij}$$

$$=\omega_k^0+\left(\frac{\omega_k^0}{2\lambda_k^0}\right)\sum_i L_{ik}^0 L_{ik}^0 \delta F_{ii}+\left(\frac{\omega_k^0}{\lambda_k^0}\right)\sum_{i<j} L_{ik}^0 L_{jk}^0 \delta F_{ij} \tag{8.201}$$

where $\lambda_k=4\pi^2c^2\omega_k^2$, with c the speed of light, and ω_k a fundamental vibrational frequency in cm^{-1} units. The L_{ij} are elements of the L-matrix, where $\lambda_k=\sum_{i,j}L_{ik}L_{jk}F_{ij}$. The ω_k^0 and L_{ij}^0 quantities are calculated from the approximate set of force constants $\mathbf{F}^0$ by solution of the vibrational eigenvalue problem (see Appendix H). The foregoing expression and the following equations are used as the basis for a least-squares analysis in which the corrections, δF_{ij}, to the constants, F_{ij}^0, are calculated. The quantities ω_k^0 and L_{ij}^0 are calculated after each iteration with the improved force constants, $F_{ij}^0+\delta F_{ij}$. The iterations are continued until all the δF_{ij} become sufficiently small. It is assumed in applying such a procedure that the initial $\mathbf{F}^0$ represents a good approximation to the force field and that the data are reasonably sensitive to the force constants so that convergence of successive iterations is assured.

Likewise, for the distortion constants

$$\tau_{\alpha\beta\gamma\delta}=\tau_{\alpha\beta\gamma\delta}^0+\sum_{i,j}\left(\frac{\partial\tau_{\alpha\beta\gamma\delta}}{\partial F_{ij}}\right)_0 \delta F_{ij} \tag{8.202}$$

with

$$\left(\frac{\partial\tau_{\alpha\beta\gamma\delta}}{\partial F_{ij}}\right)=\sum_{k,l}\left(\frac{\partial\tau_{\alpha\beta\gamma\delta}}{\partial F_{kl}^{-1}}\right)\left(\frac{\partial F_{kl}^{-1}}{\partial F_{ij}}\right) \tag{8.203}$$

where the derivations of the distortion constants with respect to the F_{kl}^{-1}, follow directly from (8.20). By taking the partial derivations of the matrix equation $\mathbf{F}^{-1}\mathbf{F}=\mathbf{E}$, we find

$$\frac{\partial \mathbf{F}^{-1}}{\partial F_{ij}} = -\mathbf{F}^{-1}\left(\frac{\partial \mathbf{F}}{\partial F_{ij}}\right)\mathbf{F}^{-1} \tag{8.204}$$

which gives

$$\frac{\partial F_{kl}^{-1}}{\partial F_{ij}} = -F_{ki}^{-1}F_{jl}^{-1} \tag{8.205}$$

Note $F_{ij}^{-1}=F_{ji}^{-1}$. These relations provide the basis for combining the vibrational frequency data and distortion constant data. The relations between the τ's and the observed spectroscopic constants have been given previously.

In addition to the distortion constants, the Coriolis constants $\zeta_{kl}^{(\alpha)}=\sum_{i,j} L_{ki}^{-1}C_{ij}^{(\alpha)}L_{lj}^{-1}$ depend on the force constants (through $\mathbf{L}$) and measure the coupling of the two vibrations Q_k and Q_l via rotation about the α axis. The $C_{ij}^{(\alpha)}$ elements depend on the masses and geometry of the molecules. The derivatives $(\partial\zeta_{kl}^{(\alpha)}/\partial F_{ij})$ have been given by Mills [159] and others [160].

For planar molecules, the inertial defect for the vibrational state $[v]$, $\Delta_{[v]}=\sum_s \Delta_s(v_s+\frac{1}{2})+\Delta_{\text{cent}}+\Delta_{\text{elec}}$ defined in (13.134) is a function of the harmonic vibrational frequencies and the Coriolis constants. The differences between the inertial defect in successive vibrational states are particularly useful because they are virtually independent of electronic and centrifugal distortion contributions. The derivatives $(\partial\Delta_s/\partial F_{ij})$ are derivable from (13.135). Mean square amplitudes [151] also depend on the force field and are obtained in gas-phase electron diffraction structure studies. The sensitivity of all these quantities to the force field can vary from molecule to molecule, and usually as much information as possible is required to uniquely specify the force constants. Some illustrative examples of force fields determined from both microwave (MW) distortion data and infrared (IR) data for some simple molecules are given in Table 8.34.

In the combination of IR and MW data a weighted least-squares analysis is usually employed. However, weighting factors are complicated by the use of ground state rather than equilibrium values. Simply specifying the weights $W_i=1/\sigma_i^2$ in terms of the estimated experimental uncertainties does not take zero-point vibrational effects into consideration. The weighting scheme should allow for the effects of anharmonicity in the vibrational frequencies, which are usually larger for the highest frequencies. Furthermore, the distortion data should be weighted with the realization that the effective ground state constants may be contaminated with vibrational effects of 2 to 5%.

Water and hydrogen sulfide are two of the molecules for which the anharmonic corrections to the observed vibrational frequencies are known. In these rotors the corrections are particularly large. Hence, it is possible to judge directly the merits of the potential function obtained by combination of observed IR and MW data. Least-squares calculations employing only IR data,

Table 8.34 Potential Constants Determined by Combination of Infrared and Microwave Date (mdyne/Å)

Molecule	f_r	f_{rr}	f_{α/r^2}	$f_{r\alpha/r}$		*Ref.*
SO_2	10.006	0.024	0.793	0.189		e
O_3	5.70	1.52	1.28	0.332		f
OF_2	3.950	0.806	0.724	0.137		g
CL_2O[a]	2.88	0.31	0.423	0.17		h
CLO_2	7.018	−0.170	0.651	0.006		i
NO_2	11.043	2.140	1.109	0.481		j
GeF_2	4.08	0.26	0.316	−0.01		k
SeO_2	6.91	0.03	0.488	0.009		l
SiF_2	5.03	0.31	0.439	0.138		m
CF_2	6.19	1.56	1.373	0.545		n
SCL_2	2.27	0.25	0.262	0.065		o
NF_2[a]	5.14	0.18	1.103	0.335		p
	f_{r1}	f_{r2}	f_{α/d^2}	$f_{r1\alpha/d}$	$f_{r2\alpha/d}$	f_{r1r2}	
NOF[b]	15.08	2.09	1.08	0[d]	0.17	1.85	q
NOCL[b]	15.29	1.30	0.56	0.09	0.07	1.45	r
NOBr[b]	14.3	1.42	0.43	−0.19	0.02	−1.5	s
NSF[c]	10.703	2.872	0.411	0[d]	0.014	0[d]	t
NSCL[c]	10.031	1.465	0.244	0[d]	−0.038	0[d]	u

[a] Determined from microwave distortion data only
[b] For the NOX molecules $r_1 = N\text{–}O$, $r_2 = N\text{–}X$, and $d = (r_1 r_2)^{1/2}$.
[c] Here $r_1 = S\text{–}N$, $r_2 = S\text{–}X$.
[d] Assumed value.
[e] Kivelson [38].
[f] Pierce [153].
[g] Pierce et al. [82].
[h] M. G. K. Pillai and R. F. Curl, Jr., *J. Chem. Phys.*, **37**, 2921 (1962).
[i] Herberich et al. [138].
[j] G. R. Bird, J. C. Baird, A. W. Jache, J. A. Hodgeson, R. F. Curl, Jr., A. C. Kunkle, J. W. Bransford, J. Rastrup-Anderson, and J. Rosenthal, *J. Chem. Phys.*, **40**, 3378 (1964).
[k] H. Takeo, R. F. Curl, and P. W. Wilson, *J. Mol. Spectrosc.*, **38**, 464 (1971).
[l] H. Takeo, E. Hirota, and Y. Morino, *J. Mol. Spectrosc.*, **34**, 370 (1970).
[m] H. Shoji, T. Tanaka, and E. Hirota, *J. Mol. Spectrosc.*, **47**, 268 (1973).
[n] Kirchhoff et al. [137].
[o] Davis and Gerry [152].
[p] R. D. Brown, F. R. Burden, P. D. Godfrey, and I. R. Gillard, *J. Mol. Spectrosc.*, **52**, 301 (1974).
[q] Cook [157].
[r] Mirri and Mazzariol [147]; and G. Cazzoli, R. Cirvellali, and A. M. Mirri, *J. Mol. Spectrosc.*, **56**, 422 (1975).
[s] Mirri and Mazzoriol [147].
[t] Cook and Kirchhoff [163].
[u] T. Beppu, E. Hirota, and Y. Morino, *J. Mol. Spectrosc.*, **36**, 386 (1970).

MW data (τ_{aaaa}, τ_{bbbb}, τ_{cccc}), and IR and MW data have been reported [156, 161], and various weighting schemes have been used. A comparison of the force fields for H_2O obtained in various ways is summarized in Table 8.35. For the IR and MW entry, F_{33} comes from the ω_3 data only, because of the limitations in calculating an accurate value of τ_{abab}, as discussed previously. Even for these light molecules, the vibrational frequencies (ω_1, ω_2) predicted by use of the microwave data only are quite satisfactory. The largest discrepancy is about 5%. Clearly, one advantage of the IR and MW force field for the ground state is that it fits both spectral regions, and the interaction constant is positive and close to the value found for the harmonic force field. The valence bond force constants obtained from the symmetrized force constants are also compared in Table 8.35. These results indicate that if corrections for the effects of vibration are to be ignored, the "best" or most representative force field is obtained from a combination of both IR and MW data, even when the vibrational frequency data are rather extensive. Furthermore, even though the anharmonicity effects are particularly large, the vibration–rotation data, when included, lead to a meaningful potential function.

There is, in general, more than one possible solution to the force field that will reproduce the vibrational frequency data. For an $n \times n$ secular equation, there are $n!$ sets of force constants corresponding to the $n!$ possible set of frequency assignments, that is, the correlation of a vibrational frequency with a

Table 8.35 Force Constant Calculation for $H_2O^{a,f}$

Symmetry Force Constants	$MW(\omega_e)^b$	$MW(\omega_0)^b$	MW^c	$IR(\omega_0)$ & MW^d
F_{11}(m dyne/Å)	8.355	7.396	7.481	7.653
F_{22}(m dyne/Å)	0.761	0.771	0.675	0.700
F_{12}(m dyne/Å)	0.345	−0.517	0.443	0.536
F_{33}(m dyne/Å)	8.555	7.838		7.838[e]

Valence bond force constants (m dyne/Å)

	f_r	f_θ	f_{rr}	$f_{r\theta}$
$IR(\omega_e)$	8.455	0.761	−0.100	0.244
$IR(\omega_0)$	7.617	0.771	−0.221	−0.366
$IR(\omega_0)$ & MW	7.746	0.700	−0.093	0.379

[a] Weighting $\sim 1/\omega$ and $1/\tau$. $IR(\omega_e)$: vibrational frequencies corrected for effects of anharmonicity. $IR(\omega_0)$: observed uncorrected vibrational frequencies.

[b] H_2O, D_2O, HDO data.

[c] H_2O, D_2O, T_2O data.

[d] H_2O, D_2O data.

[e] From $IR(\omega_0)$ calculation.

[f] From [156].

certain force constant or internal displacement coordinate. This results because the force constants are derived from the solution of a set of polynomial equations. If the frequency assignments are in doubt or not largely dependent on a given normal mode, other data such as isotopic frequency shifts, distortion constants, and so on, are needed to resolve the ambiguity. A general method of arriving at the various possible solutions has been discussed by Toman and Pliva [162]. An example of the problem is provided by the nonlinear NSF molecule [163] where the centrifugal-distortion information is employed to decide between the possible assignments.

Even with a rather varied data set, the data may not be sufficient to discriminate between two or more physically realistic force fields. For example, for NF_3, two harmonic force fields [164] can be found for the E-species (2×2) which reproduces not only the vibrational frequencies but also the distortion constants, Coriolis coupling constants, and nitrogen isotopic frequency shifts. This ambiguity associated with a 2×2 degenerate species has been considered in a general way by Hoy et al. [165]. In favorable cases, one of the solutions can be rejected on physical grounds. In principle, for XY_3 molecules resolution of the ambiguity can be obtained by use of vibrational frequencies of an unsymmetrical substituted species XY_2Y^*. However, this is not possible for trifluorides. Fortunately, the mean square amplitudes do provide a means for discriminating between the two force fields.

Some force-constant investigations have already been mentioned in the previous tables. Over the last 10 years a considerable number of studies have employed MW data to aid in the determination of molecular force fields and one cannot cite all of them here. Recent reviews on force constant calculations have been given by Duncan [166] and by Mills [167]. Harmonic force field studies have been reported for Cl_2CO [168], NOX [169], (X=F, Cl, Br), S_2O [155], SOF_2 [170], HCOOH [171], HCOSH [172], HONO [173], H_2=CHF [174], $FClO_2$ [29], CH_3X(X=F, Cl, Br, I) [175], BF_3 [165], PF_3 [165], NCl_3 [176], PCl_3 [176], $SbCl_3$ [177], HBF_2 [178], NO_2Cl [179], H_2CO [180], and F_2CO [181], to mention a few.

In general, limitations on the accuracy of the potential constants arise from neglect of certain interaction constants and from use of moments of inertia averaged over the ground vibrational state rather than equilibrium moments of inertia, as well as from the fact that the distortion constants and the vibrational frequencies contain contributions from anharmonic terms in the potential energy which are usually neglected. Some feeling for the significance of these latter effects on the derived constants has been obtained. However, the evaluation of distortion constants for the equilibrium state for a few asymmetric tops will allow a better assessment of the errors introduced when present calculation procedures are employed.

In addition to work on the harmonic part of the potential function, work has been done on small molecules to extract information on the anharmonic part of the potential function. Evaluation of the rotation–vibration interaction constants from the rotational spectrum provides information on the cubic force

Table 8.36 Comparison of Calculated and Observed Sextic Centrifugal Distortion Constants

	H_2S		SO_2	
Constants	Calc.[a]	Obs.[b]	Calc.[a]	Obs.[c]
Φ_J	25.2	28.1 ± 1.1	0.159	0.148 ± 0.014
ϕ_J	5.7	5.8 ± 0.1	-0.080	-0.074 ± 0.007
Φ_{JK}	-218.6	-228.3 ± 2.7	-0.863	-0.818 ± 0.073
ϕ_{JK}	-201.8	-242.8 ± 5.2	0.349	0.332 ± 0.029
Φ_{KJ}	447.3	459.0 ± 16.0	1.271	1.220 ± 0.100
Φ_K	-254.0	-276.5 ± 14.1	-0.568	-0.553 ± 0.041
ϕ_K	-2767.3	-2870.3 ± 39	-0.282	-0.275 ± 0.020
	NH_3			
	Calc.[d,e]	Obs.[f]		
H_J	1.944	2.38 ± 0.34		
H_{JK}	-6.693	-8.78 ± 0.91		
H_{KJ}	7.786	10.5 ± 1.2		
H_K	-2.962	—		
h_3	0.064	0.1203		
	CH_4			
	Calc.[d,g]	Obs.[h]		
H	137.1	—		
H_{4T}	-12.0	-16.984 ± 0.023		
H_{6T}	2.9	11.034 ± 0.019		

[a]Calculated values from Georghiou [201]. Units: kHz.
[b]III^r axis representation. From Helminger et al. [97].
[c]II^r axis representation. From Carpenter [115].
[d]Calculated values from Aliev and Watson [72].
[e]Units: 10^{-7} cm^{-1}.
[f]J. M. Dowling, *J. Mol. Spectrosc.*, **27**, 527 (1968) and E. Schnabel, T. Törring, and W. Wilke, *Z. Physik*, **188**, 167 (1965).
[g]Units: Hz.
[h]C. W. Holt, M. C. L. Gerry, and I. Ozier, *Can. J. Phys.*, **53**, 1791 (1975).

constants F_{ijk}. An accurate harmonic force field is needed in such calculations. A useful review has been given by Morino [182]. Relations involving the rotation–vibration information and the anharmonic potential constants has been given by Mills [183]. Anharmonic potential constants have been reported for HCN [184], HNC [185], HCP [186], ClCN [187], OCS [188], SO_2 [189], O_3 [190], OF_2 [191], SeO_2 [192], SF_2 [193], SiF_2 [194], GeF_2 [195], and

H_2O [196], to list a few. A summary of cubic force fields obtained from microwave data may be found in a review on anharmonic force fields by Pliva [197].

Calculation of the Sextic Distortion Constants

The recent theoretical elucidation of the explicit relations between the sextic distortion constants and the cubic potential constants

$$V_{\text{anhar}} = \frac{1}{6} \sum_{l,m,n} k_{lmn} q_k q_l q_n \tag{8.206}$$

($q_k = \gamma_k^{1/2} Q_k$ dimensionless normal coordinate) now makes possible the use of these constants to provide information on the anharmonic potential constants. Derivation of these relations requires application of fourth-order perturbation (see Chapter VII) and presents a particularly grueling computation. Theoretical treatments have been given by Aliev [198] for axial symmetric tops, by Sumberg and Parker [199] for nonlinear triatomic molecules; by Chung and Parker [200], Georghiou [201], and Aliev and Watson [72] for a general polyatomic molecule. The latter two formulations are particularly convenient for practical calculations, and relatively compact expressions of them have been formulated.

Comparison between observed and calculated sextic coefficients are given in Table 8.36 for H_2S, SO_2, NH_3, and CH_4. The rotational parameters and the cubic potential constants employed in the calculations may be found in the references cited. The results, though quite satisfactory, could undoubtedly be improved by adjustment of the force constants, particularly for CH_4. However, it should be realized that the observed sextic constants have undetermined zero-point vibrational contributions and that some discrepancy is to be expected, as with the quartic constants. Similiar calculations have been reported for O_3 [200].

9 ALTERNATE FORMULATIONS

In the usual treatment, based on the small oscillations approximation, the vibrational potential function and elements of the inertia tensor are expanded in terms of the vibrational coordinates, and a perturbation treatment is applied to give an effective rotational Hamiltonian for a given vibrational state (see Chapter VII). However, the Hamiltonian, represented by a Taylor series expansion in components of the angular momentum, converges very slowly for certain molecules. These extreme distortion effects can arise because of the strong dependence of a moment inertia on a bending coordinate such as in water, or the presence of a large amplitude coordinate such as the inversion coordinate in ammonia and related molecules. In such cases, the previous treatment may not be the most appropriate and alternate formulations have been described.

Large Amplitude Motion–Rotation Hamiltonian

The general rotation–vibration Hamiltonian allowing for a large amplitude bending motion in triatomic molecules has been derived by Hougen et al. [18].

Further developments and applications of this Hamiltonian have been given by Bunker and Stone [202] (rigid-bender Hamiltonian), Hoy and Bunker [203] (nonrigid bender Hamiltonian), and Bunker and Landsberg [204] (semirigid bender Hamiltonian). Extension to tetraatomic molecules has been discussed by Sarka [205] and Stone [206]. Treatments for a large amplitude inversion motion have also been given [207–209].

To calculate the energy levels in the nonrigid bender formulation requires for H_2O a knowledge of the structure and potential constants to quartic terms—21 parameters [203]. The particular advantage of this approach is not in a better fit to the rotational data but in the direct way information on the potential energy surface is obtained. In HCNO, with a large amplitude H-bending motion, the vibrational and rotational energy levels have been fit with the semirigid bender formulation by use of geometrical parameters and a simple quadratic-quartic bending potential function [210]. The results indicate a linear equilibrium structure but a bent form in the ground vibrational state and a zero-point HCN bending amplitude of $\pm 34°$. A similar treatment has been applied to carbon suboxide [211].

The Padé Hamiltonian

An alternate treatment to overcome the slow convergence of the angular momentum power series is to express the energy as a Padé approximant. The success of Padé approximants to approximate the value of a function is well known [212, 213], and its application has been growing in various areas of physics although little use of the method has been made to problems in spectroscopy [214]. The usual Padé approximant to a function $F(x)$ is the ratio of two polynomials $P_n(x)/Q_m(x)$ of degree n and m.

The first application of this technique was made by Young and Young [215] to fit the inversion spectrum of NH_3. The Padé approximant with 19 terms, for example, gave a much better fit than the usual power series in $J(J+1)$ and K^2 with 21 terms. Based on a representation of the effective Hamiltonian in the form of a Padé operator, Belov et al. [216] have obtained an improved fit of the rotational spectrum of PH_3 in the ground vibrational state. The Padé Hamiltonian has the form

$$\mathscr{H}_{\text{Padé}} = \frac{1}{2}\left[H\left(\frac{1}{1+h}\right)+\left(\frac{1}{1+h}\right)H\right] \tag{8.207}$$

where the operators H and h both have the form of the usual Hamiltonian operator and in general do not commute. Both H and h contain n constants and hence $\mathscr{H}_{\text{Padé}}$ consists of $2n$ constants. The constants associated with h are dimensionless. It is convenient [216] to use basis functions which are eigenfunctions of h

$$(R|h|R) = \lambda_R \tag{8.208}$$

where R denotes the collection of quantum numbers required to specify the particular eigenfunction and energy level. The eigenfunction Φ_R is actually a linear combination of appropriate basis functions. The energy matrix of the Padé Hamiltonian then has the form

$$(R|\mathscr{H}_{\text{Padé}}|R') = \frac{1}{2}\left(\frac{1}{1+\lambda_R} + \frac{1}{1+\lambda_{R'}}\right)(R|H|R') \tag{8.209}$$

For a symmetric top

$$H = BP^2 + (C-B)P_z^2 - D_J P^4 - D_{JK} P^2 P_z^2 - D_K P_z^4 + \cdots \tag{8.210}$$

and

$$h = bP^2 + (c-b)P_z^2 - d_J P^4 - d_{JK} P^2 P_z^2 - d_K P_z^4 + \cdots \tag{8.211}$$

By expansion of the Padé operator, the Padé constants of (8.210) and (8.211) may be related to the spectroscopic constants of the usual Hamiltonian. For example [216],

$$D_J^{sp} = D_J + Bb \tag{8.212}$$

$$D_{JK}^{sp} = D_{JK} + (C-B)b + B(c-b) \tag{8.213}$$

$$D_K^{sp} = D_K + (C-B)(c-b) \tag{8.214}$$

These spectroscopic constants may be interpreted with existing theories. This analysis technique should be useful for molecules with large rotational constants to provide a better fit of the observed rotational frequencies than is possible with the usual Hamiltonian containing $2n$ constants.

References

1. M. Toyama, T. Oka, and U. Morino, *J. Mol. Spectrosc.*, **13**, 193 (1964).
2. R. M. Garvey, *J. Mol. Spectrosc.*, **65**, 330 (1977).
3. G. Cazzoli and F. C. De Lucia, *J. Mol. Spectrosc.*, **76**, 131 (1979).
4. F. C. De Lucia and P. Helminger, *J. Chem. Phys.*, **67**, 4262 (1977).
5. G. Winnewisser, E. Churchwell, and C. M. Walmsley, "Astrophysics of Interstellar Molecules," in *Modern Aspects of Microwave Spectroscopy*, G. W. Chantry, Ed., Academic, New York, 1979.
6. L. E. Snyder, "Molecules in Space," in *MTP International Review of Science*, Vol. 3, *Spectroscopy*, D. A. Ramsey, Ed., Butterworth, London, 1972.
7. R. L. Cook, R. T. Walden, and G. E. Jones, *J. Mol. Spectrosc.*, **53**, 370 (1974).
8. E. B. Wilson, Jr., and J. B. Howard, *J. Chem. Phys.*, **4**, 260 (1936).
9. H. H. Nielsen, *Rev. Mod. Phys.*, **23**, 90 (1951).
10. E. B. Wilson, Jr., *J. Chem. Phys.*, **4**, 526 (1936).
11. E. B. Wilson, Jr., J. C. Decius, and P. C. Cross, *Molecular Vibrations*, McGraw-Hill, New York, 1955.
12. B. T. Darling and D. M. Dennison, *Phys. Rev.*, **57**, 128 (1940).
13. M. Goldsmith, G. Amat, and H. H. Nielsen, *J. Chem. Phys.*, **24**, 1178 (1956); **27**, 838 (1957).
14. G. Amat and H. H. Nielsen, *J. Chem. Phys.*, **27**, 845 (1957); **29**, 665 (1958); **36**, 1859 (1962). See also M. L. Grenier-Besson, G. Amat, and H. H. Nielsen, *J. Chem. Phys.*, **36**, 3454 (1962); G. Amat, H. H. Nielsen, and G. Tarrago, *Rotation-Vibration of Polyatomic Molecules*, Dekker, New York, 1971.

15. J. K. G. Watson, *Mol. Phys.*, **15**, 479 (1968).
16. L. S. Rothman and S. A. Clough, *J. Chem. Phys.*, **54**, 3246 (1971).
17. A. M. Walsh-Bakke, *J. Mol. Spectrosc.*, **40**, 1 (1971).
18. J. T. Hougen, P. R. Bunker, and J. W. C. Johns, *J. Mol. Spectrosc.*, **34**, 136 (1970).
19. K. T. Chung and P. M. Parker, *J. Chem. Phys.*, **38**, 8 (1963); M. Y. Chan, L. Wilardjo, and P. M. Parker, *J. Mol. Spectrosc.*. **40**, 473 (1971).
20. J. D. Louck, *J. Mol. Spectrosc.*, **61**, 107 (1976).
21. D. Kivelson and E. B. Wilson, Jr., *J. Chem. Phys.*, **21**, 1229 (1953).
22. C. Eckart, *Phys. Rev.*, **47**, 552 (1935).
23. S. M. Ferigle and A. Weber, *Am. J. Phys.*, **21**, 102 (1953).
24. R. Gold, J. M. Dowling, and A. G. Meister, *J. Mol. Spectrosc.*, **2**, 9 (1958).
25. J. M. Dowling, R. Gold, and A. G. Meister, *J. Mol. Spectrosc.*, **1**, 265 (1957); **2**, 411 (1958).
26. G. Thyagarajan, S. Sundaram, and F. F. Cleveland, *J. Mol. Spectrosc.*, **5**, 307 (1960).
27. S. Sundaram and F. F. Cleveland, *J. Chem. Phys.*, **32**, 166 (1960).
28. I. McNaught, *J. Mol. Spectrosc.*, **91**, 492 (1982).
29. C. R. Parent and M. C. L. Gerry, *J. Mol. Spectrosc.*, **49**, 343 (1974).
30. S. R. Polo, *J. Chem. Phys.*, **24**, 1133 (1956).
31. M. R. Aliev and Aleksanyan, *Dokl. Akad. Nauk SSSR, Ser. Khim.* **169**, 1229 (1966). See also [40] and [41].
32. S. J. Cyvin and G. Hagen, *Chem. Phys. Lett.*, **1**, 645 (1968).
33. S. J. Cyvin, B. N. Cyvin, and G. Hagen, *Z. Naturforsch.*, **23A**, 1649 (1968).
34. G. O. Sørensen, G. Hagen, and S. J. Cyvin, *J. Mol. Spectrosc.*, **35**, 489 (1970).
35. P. Pulay and W. Sawodny, *J. Mol. Spectrosc.*, **26**, 150 (1968).
36. K. Klauss and G. Strey, *Z. Naturforsch.*, **23A**, 1308 (1968).
37. B. L. Crawford, Jr. and W. H. Fletcher, *J. Chem. Phys.*, **19**, 141 (1951).
38. D. Kivelson, *J. Chem. Phys.*, **22**, 904 (1954).
39. I. Gamo, *J. Mol. Spectrosc.*, **30**, 216 (1969).
40. A. P. Aleksandrov, M. R. Aliev, and V. T. Aleksanyan, *Opt. Spectrosc.*, **29**, 568 (1970).
41. A. P. Aleksandrov and M. R. Aliev, *J. Mol. Spectrosc.*, **47**, 1 (1973).
42. J. K. G. Watson, *J. Mol. Spectrosc.*, **74**, 483 (1979).
43. A. Müller, N. Mohan, and A. Alix, *J. Chem. Phys.*, **59**, 6112 (1973).
44. J. Demaison, *J. Mol. Struct.*, **31**, 233 (1976).
45. L. Nemes, *J. Mol. Struct.*, **46**, 67 (1978).
46. M. R. Aliev, *Opt. Spectrosc.*, **31**, 568 (1971); **33**, 858 (1972).
47. M. R. Aliev and J. K. G. Watson, *J. Mol. Spectrosc.*, **74**, 282 (1979).
48. S. E. Novick, *J. Mol. Spectrosc.*, **68**, 77 (1977).
49. R. L. Cook, *J. Mol. Struct.*, **26**, 126 (1975).
50. D. Kivelson and E. B. Wilson, Jr., *J. Chem. Phys.*, **20**, 1575 (1952).
51. J. K. G. Watson, *J. Chem. Phys.*, **45**, 1360 (1966); **48**, 181 (1968).
52. E. B. Wilson, Jr., *J. Chem. Phys.*, **5**, 617 (1937).
53. P. M. Parker, *J. Chem. Phys.*, **37**, 1596 (1962).
54. R. M. Lees, *J. Mol. Spectrosc.*, **33**, 124 (1970).
55. J. K. G. Watson, *J. Chem. Phys.*, **46**, 1935 (1967).
56. K. M. Marstokk and H. Møllendal, *J. Mol. Struct.*, **8**, 234 (1971).
57. V. Typke, *Z. Naturforsch.*, **26A**, 1775 (1971).
58. A. Bouchy, J. Demaison, G. Roussy, and J. Barriol, *J. Mol. Struct.*, **18**, 211 (1973).
59. J. M. Dowling, *J. Mol. Spectrosc.*, **6**, 550 (1961).

60. Z. I. Slawsky and D. M. Dennison, *J. Chem. Phys.*, **7**, 509 (1939).
61. H. Dreizler, G. Dendl, *Z. Naturforsch.*, **20A**, 30 (1965).
62. H. Dreizler and H. D. Rudolph, *Z. Naturforsch.*, **20A**, 749 (1965).
63. J. K. G. Watson, *J. Chem. Phys.*, **48**, 4517 (1968).
64. J. K. G. Watson, "Aspects of Quartic and Sextic Centrifugal Effects on Rotational Energy Levels," In *Vibrational Spectra and Structure*, J. R. Durig, Ed., Vol. 6, Marcel Dekker, New York (1977).
65. K. T. Chung and P. M. Parker, *J. Chem. Phys.*, **43**, 3865, 3869 (1965).
66. F. X. Kneizys, J. N. Freedman, and S. A. Clough, *J. Chem. Phys.*, **44**, 2552 (1966).
67. K. K. Yallabandi and P. M. Parker, *J. Chem. Phys.*, **49**, 410 (1968).
68. W. H. Kirchhoff, *J. Mol. Spectrosc.*, **41**, 333 (1972).
69. G. Winnewisser, *J. Chem. Phys.*, **56**, 2944 (1972).
70. B. P. Van Eijck, *J. Mol. Spectrosc.*, **53**, 246 (1974).
71. V. Typke, *J. Mol. Spectrosc.*, **63**, 170 (1976).
72. M. R. Aliev and J. K. G. Watson, *J. Mol. Spectrosc.*, **61**, 29 (1976).
73. K. T. Hecht, *J. Mol. Spectrosc.*, **5**, 355, 390 (1960).
74. J. Moret-Bailly, *J. Mol. Spectrosc.*, **15**, 344 (1965).
75. J. Moret-Bailly, L. Gauthier, and J. Montagutelli, *J. Mol. Spectrosc.*, **15**, 355 (1965).
76. A. J. Dorney and J. K. G. Watson, *J. Mol. Spectrosc.*, **42**, 135 (1972).
77. S. M. Kirschner and J. K. G. Watson, *J. Mol. Spectrosc.*, **47**, 347 (1973).
78. I. Ozier, *J. Mol. Spectrosc.*, **53**, 336 (1974).
79. F. Michelot, J. Moret-Bailly, and K. Fox, *J. Chem. Phys.*, **60**, 2606 (1974).
80. J. K. G. Watson, *J. Mol. Spectrosc.*, **55**, 498 (1975).
81. C. W. Holt, M. C. L. Gerry, and I. Ozier, *Phys. Rev. Lett.*, **31**, 1033 (1973).
82. L. Pierce, N. Di Cianni, and R. H. Jackson, *J. Chem. Phys.*, **38**, 730 (1963).
83. D. W. Posener and M. W. P. Strandberg, *Phys. Rev.*, **95**, 374 (1954).
84. R. E. Hillger and M. W. P. Strandberg, *Phys. Rev.*, **83**, 575 (1951).
85. T. Oka and Y. Morino, *J. Mol. Spectrosc.*, **8**, 300 (1962).
86. W. S. Benedict and E. K. Plyer, *J. Res. Nat. Bur. Stand.*, **46**, 246, (1951).
87. H. C. Allen, Jr. and W. B. Olson, *J. Chem. Phys.*, **37**, 212 (1962).
88. P. E. Fraley and K. Narahari Rao, *J. Mol. Spectrosc.*, **19**, 133 (1966).
89. L. H. Ford, K. K. Yallabandi, and P. M. Parker, *J. Mol. Spectrosc.*, **30**, 241 (1969).
90. R. H. Hill and T. H. Edwards, *J. Mol. Spectrosc.*, **11**, 433 (1963).
91. P. Helminger, R. L. Cook, and F. C. De Lucia, *J. Mol. Spectrosc.*, **40**, 125 (1971).
92. F. C. De Lucia, R. L. Cook, P. Helminger, and W. Gordy, *J. Chem. Phys.*, **55**, 5334 (1971).
93. R. L. Cook, F. C. De Lucia, and P. Helminger, *J. Mol. Spectrosc.*, **41**, 123 (1972).
94. G. Winnewisser, M. Winnewisser, and B. P. Winnewisser, "Millimeter Wave Spectroscopy," in *MTP International Review of Science*, Vol. 3, *Spectroscopy*, D. A. Ramsay, Ed., Butterworths, London, 1972.
95. F. C. De Lucia, "Millimeter- and Submillimeter-Wave Spectroscopy," in *Molecular Spectroscopy: Modern Research*, Vol. II, K. N. Rao, Ed., Academic, New York, 1972.
96. A. F. Krupnov and A. V. Burenin, "New Methods in Subillimeter Microwave Spectroscopy," in *Molecular Spectroscopy: Modern Research*, Vol. II, K. N. Rao, Ed., Academic, New York, 1972.
97. P. Helminger, R. L. Cook, and F. C. De Lucia, *J. Chem. Phys.*, **56**, 4581 (1972).
98. F. C. De Lucia, P. Helminger, R. L. Cook, and W. Gordy, *Phys. Rev.*, **A5**, 487 (1972).

99. F. C. De Lucia and P. Helminger, *J. Mol. Spectrosc.*, **56**, 138 (1975).
100. F. C. DeLucia, P. Helminger, R. L. Cook, and W. Gordy, *Phys. Rev.*, **A6**, 1324 (1972).
101. P. Helminger and F. C. DeLucia, *J. Mol. Spectrosc.*, **70**, 263 (1978).
102. G. Steenbeckeliers and J. Bellet, *J. Mol. Spectrosc.*, **45**, 10 (1973).
103. J. Bellet and G. Steenbeckeliers, *Compt. Rend.*, **271B**, 1208 (1971). See also G. Steenbeckeliers and J. Bellet, *Compt. Rend.*, **273B**, 471 (1971) and W. Lafferty, J. Bellet, and G. Steenbeckeliers, *Compt. Rend.*, **273B**, 388 (1971).
104. J. Bellet, W. J. Lafferty, and G. Steenbeckeliers, *J. Mol. Spectrosc.*, **47**, 388 (1973).
105. J. Bellet, G. Steenbeckeliers, and P. Stouffs, *Compt. Rend.*, **275B**, 501 (1972).
106. F. C. DeLucia, P. Helminger, W. Gordy, H. W. Morgan, and P. A. Staats, *Phys. Rev.*, **A8**, 2785 (1973).
107. P. Helminger, F. C. DeLucia, and W. Gordy, P. A. Staats, and H. W. Morgan, *Phys. Rev.*, **A10**, 1072 (1974).
108. P. Helminger and F. C. DeLucia, *J. Mol. Spectrosc.*, **58**, 375 (1975).
109. F. C. DeLucia and P. Helminger, *J. Mol. Spectrosc.*, **54**, 200 (1975).
110. E. A. Cohen and H. M. Pickett, *J. Mol. Spectrosc.*, **93**, 83 (1982).
111. A. Bouchy and G. Roussy, *J. Mol. Struct.*, **84**, 1 (1982).
112. See for example, M. Newman, "Matrix Computations," in *A Survey of Numerical Analysis*, J. Todd, Ed., McGraw-Hill, New York, 1962.
113. See also, e.g., W. H. Kirchhoff and D. R. Johnson, *J. Mol. Spectrosc.*, **45**, 159 (1973); D. R. Johnson, F. J. Lovas, and W. H. Kirchhoff, *J. Chem. Phys. Ref. Data*, **1**, 1011 (1972); Y. Beers, G. P. Klein, W. H. Kirchhoff, and D. R. Johnson, *J. Mol. Spectrosc.*, **44**, 553 (1972).
114. V. Typke, *Z. Naturforsch.*, **26A**, 175 (1971).
115. J. H. Carpenter, *J. Mol. Spectrosc.*, **46**, 348 (1973).
116. L. L. Strow, *J. Mol. Spectrosc.*, **97**, 9 (1983).
117. K. Yamada and M. Winnewisser, *Z. Naturforsch.*, **31A**, 131 (1976).
118. G. Winnewisser and P. Helminger, *J. Chem. Phys.*, **56**, 2967; 2954 (1972).
119. G. Winnewisser, *J. Chem. Phys.*, **57**, 1803 (1972).
120. E. F. Pearson and H. Kim, *J. Chem. Phys.*, **57**, 4230 (1972).
121. C. C. Costain, *Phys. Rev.*, **82**, 108 (1951).
122. E. Schnabel, T. Törring, and W. Wilke, *Z. Physik*, **188**, 167 (1965).
123. P. B. Davies, R. M. Neumann, S. C. Wofsy, and W. Klemperer, *J. Chem. Phys.*, **55**, 3564 (1971).
124. W. B. Olson, A. G. Maki and R. L. Sams, *J. Mol. Spectrosc.*, **55**, 252 (1975).
125. P. Bernard and T. Oka, *J. Mol. Spectrosc.*, **75**, 181 (1979).
126. R. H. Bradley, P. N. Brier, and M. J. Whittle, *J. Mol. Spectrosc.*, **44**, 536 (1972).
127. S. R. Jones, P. N. Brier, D. M. Brookbanks, and J. G. Baker, *J. Mol. Spectrosc.*, **47**, 351 (1973).
128. R. Jurek, J. Chanussot, and J. Bellet, *Compt. Rend.*, **277B**, 53 (1973); J. Chanussot, R. Jurek, and P. Suzean, *Compt. Rend.*, **276B**, 729 (1973).
129. R. Jurek, P. Suzean, J. Chanussot, and J. P. Champion, *J. Physique*, **35**, 533 (1974); P. Goulet, R. Jurek, and J. Chanussot, *J. Physique*, **37**, 495 (1976).
130. B. Balikci and P. N. Brier, *J. Mol. Spectrosc.*, **85**, 109 (1981); **89**, 254 (1981).
131. G. Amat and L. Henry, *J. Phys. Radium*, **21**, 728 (1960).
132. P. Kupecek, *J. Phys.*, **25**, 831 (1964).
133. I. Ozier, R. M. Lees, and M. C. L. Gerry, *Cand. J. Phys.*, **54**, 1094 (1976).
134. G. Steenbeckeliers, *Ann. Soc. Sci. Brux.*, **82**, 331 (1968).

135. W. C. Hamilton, *Statistics in Physical Science*, Ronald, New York, 1964.

136. D. L. Albritton, A. L. Schmeltekopf, and R. N. Zare, "An Introduction to the Least-Squares Fitting of Spectroscopic Data, in *Molecular Spectroscopy: Modern Research*, Vol. 2, K. N. Rao, Ed., Academic, New York, 1976.

137. W. H. Kirchhoff and D. R. Lide, Jr., and F. X. Powell, *J. Mol. Spectrosc.*, **47**, 491 (1973).

138. G. E. Herberich, R. H. Jackson, and D. J. Millen, *J. Chem. Soc.*, (**A**), 336 (1966).

139. J. K. G. Watson, *J. Mol. Spectrosc.*, **66**, 500 (1977).

140. J. K. G. Watson, *J. Mol. Spectrosc.*, **65**, 123 (1977).

141. A. Niroomand-Rad, M. Y. Chan, and P. M. Parker, *J. Mol. Spectrosc.*, **69**, 450 (1978).

142. See, e.g., J. G. Smith, *Mol. Phys.*, **35**, 461 (1978); K. M. Marstokk and H. Møllendal, *J. Mol. Struct.*, **32**, 191 (1976); K. M. Marstokk and H. Møllendal, *J. Mol. Struct.*, **40**, 1 (1977); A. Dubrulle and J. L. Destombes, *J. Mol. Struct.*, **13**, 461 (1972); E. Willemot, D. Dangoisse, and J. Bellet, *J. Mol. Spectrosc.*, **73**, 96 (1978).

143. J. Demaison, D. Schwoch, B. Tan, and H. D. Rudolph, *J. Mol. Spectrosc.*, **59**, 226 (1976).

144. R. A. Creswell and I. M. Mills, *J. Mol. Spectrosc.*, **52**, 392 (1974).

145. S. I. Chan, T. R. Borgers, J. W. Russell, H. L. Strauss, and W. D. Gwinn, *J. Chem. Phys.*, **44**, 1103 (1966).

146. L. Halonen, E. Friz, A. G. Robiette, and I. M. Mills, *J. Mol. Spectrosc.*, **79**, 432 (1980).

147. See, for example, J. L. Duncan and I. M. Mills, *Spectrochim. Acta.*, **20**, 523 (1964); A. M. Mirri and E. Mazzariol, *Spectrochim. Acta.*, **22**, 785 (1966).

148. T. Oka and Y. Morino, *J. Mol. Spectrosc.*, **6**, 472 (1961).

149. J. H. Meal and S. R. Polo, *J. Chem. Phys.*, **24**, 1119, 1126 (1956).

150. S. J. Cyvin, *Molecular Structures and Vibrations*, Elsevier, Amsterdam, 1972.

151. S. J. Cyvin, "Molecular Vibrations and Mean Square Amplitudes," Elsevier, Amsterdam, 1968.

152. R. W. Davis and M. C. L. Gerry, *J. Mol. Spectrosc.*, **65**, 455 (1977).

153. L. Pierce, *J. Chem. Phys.*, **24**, 139 (1956).

154. W. H. Kirchhoff, D. R. Johnson, and F. X. Powell, *J. Mol. Spectrosc.*, **48**, 157 (1973).

155. R. L. Cook, G. Winnewisser, and D. C. Lindsey, *J. Mol. Spectrosc.*, **46**, 276 (1973).

156. R. L. Cook, F. C. DeLucia, and P. Helminger, *J. Mol. Spectrosc.*, **53**, 62 (1974).

157. R. L. Cook, *J. Chem. Phys.*, **42**, 2927 (1965).

158. J. K. G. Watson, *J. Mol. Spectrosc.*, **41**, 229 (1972).

159. I. M. Mills, *J. Mol. Spectrosc.*, **5**, 334 (1960); **17**, 164 (1965).

160. See also G. Strey, *J. Mol. Spectrosc.*, **17**, 265 (1965) and CH. V. S. R. Rao, *J. Mol. Spectrosc.*, **41**, 105 (1972).

161. R. L. Cook, F. C. DeLucia, and P. Helminger, *J. Mol. Spectrosc.*, **28**, 237 (1975).

162. S. Toman and J. Pliva, *J. Mol. Spectrosc.*, **21**, 362 (1966).

163. R. L. Cook and W. H. Kirchhoff, *J. Chem. Phys.*, **47**, 4521 (1967).

164. A. Allan, J. L. Duncan, J. H. Holloway, and D. C. McKean, *J. Mol. Spectrosc.*, **31**, 368 (1969).

165. A. R. Hoy, J. M. R. Stone, and J. K. G. Watson, *J. Mol. Spectrosc.*, **42**, 393 (1972).

166. J. D. Duncan, "Force Constant Calculations in Molecules," in *Molecular Spectroscopy*, Vol. 3, The Chemical Society of London (1973).

167. I. M. Mills, "Harmonic Force Field Calculations," in *Critical Evaluation of Chemical and Physical Structural Information*, D. R. Lide and J. A. Paul, Eds., National Academy of Science, Washington, 1974.

168. J. H. Carpenter and D. F. Rimmer, *J. C. S. Faraday II*, **74**, 466 (1978).

169. L. H. Jones, L. B. Asprey, and R. R. Ryan, *J. Chem. Phys.*, **47**, 3371 (1967); L. H. Jones, R. R. Ryan, and L. B. Asprey, *J. Chem. Phys.*, **49**, 581 (1968); J. Laane, L. H. Jones, R. R. Ryan, and L. B. Asprey, *J. Mol. Spectrosc.*, **30**, 489 (1969).
170. N. J. D. Lucas and J. G. Smith, *J. Mol. Spectrosc.*, **43**, 327 (1972).
171. R. W. Davis, A. G. Robiette, M. C. L. Gerry, E. Bjarnov, and G. Winnewisser, *J. Mol. Spectrosc.*, **81**, 93 (1980).
172. B. P. Winnewisser and W. H. Hocking, *J. Phys. Chem.*, **84**, 1771 (1980).
173. D. J. Finnigan, A. P. Cox, A. H. Brittain, and J. G. Smith, *J. C. S. Faraday II*, **68**, 548 (1972).
174. M. C. L. Gerry, *J. Mol. Spectrosc.*, **45**, 71 (1973).
175. J. L. Duncan, D. C. McKean, and G. K. Speirs, *Mol. Phys.*, **24**, 353 (1972).
176. G. Cazzoli, *J. Mol. Spectrosc.*, **53**, 37 (1974).
177. G. Cazzoli and W. Caminati, *J. Mol. Spectrosc.*, **62**, 1 (1976).
178. A. G. Robiette and M. C. L. Gerry, *J. Mol. Spectrosc.*, **80**, 403 (1980).
179. R. R. Filgueira, P. Forti, and G. Corbelli, *J. Mol. Spectrosc.*, **57**, 97 (1975).
180. J. L. Duncan and P. D. Mallinson, *Chem. Phys. Lett.*, **23**, 597 (1973).
181. P. D. Mallinson, D. C. McKean, J. H. Holloway, and I. A. Oxton, *Spectrochim. Acta*, **31A**, 143 (1975).
182. Y. Morino, *Pure Appl. Chem.*, **18**, 323 (1969).
183. I. M. Mills, "Vibration–Rotation Structure in Asymmetric- and Symmetric-Top Molecules," in *Molecular Spectroscopy: Modern Research*, K. N. Rao and C. W. Mathews, Eds., Academic, New York, 1972.
184. T. Nakagawa and Y. Morino, *Bull. Chem. Soc. Japan*, **42**, 2212 (1969).
185. R. A. Creswell and A. G. Robiette, *Mol. Phys.*, **36**, 869 (1978).
186. G. Strey and I. M. Mills, *Mol. Phys.*, **26**, 129 (1973).
187. C. B. Murchison and J. Overend, *Spectrochim. Acta.* **27A**, 1801 (1971).
188. Y. Morino and T. Nakagawa, *J. Mol. Spectrosc.*, **26**, 496 (1968).
189. Y. Morino, Y. Kikuchi, S. Saito, and E. Hirota, *J. Mol. Spectrosc.*, **13**, 95 (1964).
190. T. Tanaka and Y. Morino, *J. Mol. Spectrosc.*, **33**, 538 (1970).
191. Y. Morino and S. Saito, *J. Mol. Spectrosc.*, **19**, 435 (1966).
192. H. Takeo, E. Hirota, and Y. Morino, *J. Mol. Spectrosc.*, **41**, 420 (1972); **34**, 370 (1970).
193. Y. Endo, S. Saito, E. Hirota, and T. Chikaraishi, *J. Mol. Spectrosc.*, **77**, 222 (1979).
194. H. Shoji, T. Tanaka, and E. Hirota, *J. Mol. Spectrosc.*, **47**, 268 (1973).
195. H. Takeo, R. F. Curl, Jr., and P. W. Wilson, *J. Mol. Spectrosc.*, **38**, 464 (1971).
196. D. F. Smith and J. Overend, *Spectrochim. Acta.*, **28A**, 471 (1972; A. R. Hoy, I. M. Mills, and G. Strey, *Mol. Phys.*, **24**, 1265 (1972).
197. J. Pliva, "Anharmonic Force Fields," in *Critical Evaluation of Chemical and Physical Structural Information*, D. R. Lide and M. A. Paul, Eds., National Academy of Science, Washington, 1974.
198. M. R. Aliev, *J. Mol. Spectrosc.*, **52**, 171 (1974).
199. D. A. Sumberg and P. M. Parker, *J. Mol. Spectrosc.*, **48**, 459 (1973).
200. M. Y. Chan and P. M. Parker, *J. Mol. Spectrosc.*, **65**, 190 (1977). See also A. Niroomand-Rad and P. M. Parker, *J. Mol. Spectrosc.*, **75**, 177 (1979).
201. C. Georghiou, *Mol. Phys.*, **32**, 1279 (1976).
202. P. R. Bunker and J. M. R. Stone, *J. Mol. Spectrosc.*, **41**, 310 (1972).
203. A. R. Hoy and P. R. Bunker, *J. Mol. Spectrosc.*, **52**, 439 (1974); **74**. 1 (1979).
204. P. R. Bunker and B. M. Landsberg, *J. Mol. Spectrosc.*, **67**, 374 (1977).
205. K. Sarka, *J. Mol. Spectrosc.*, **38**, 545 (1971).

206. J. M. R. Stone, *J. Mol. Spectrosc.*, **54**, 1 (1975).
207. D. Papousek, J. M. R. Stone, and V. Spirko, *J. Mol. Spectrosc.*, **48**, 17 (1973).
208. V. Spirko, J. M. R. Stone, and D. Papoušek, *J. Mol. Spectrosc.*, **60**, 159 (1976).
209. V. Danielis, D. Papoušek, V. Spirko, and M. Horak, *J. Mol. Spectrosc.*, **54**, 339 (1975).
210. P. R. Bunker, B. M. Landsberg, and B. P. Winnewisser, *J. Mol. Spectrosc.* **74**, 9 (1979).
211. P. R. Bunker, *J. Mol. Spectrosc.*, **80**, 422 (1980).
212. G. A. Baker, Jr., *Essentials of Padé Approximants*, Academic, New York, 1975.
213. E. B. Saff and R. S. Varga, Eds., *Padé and Rational Approximation*, Academic, New York, 1977.
214. M. Mizushima, *The Theory of Rotating Diatomic Molecules*, Wiley, New York, 1975.
215. L. D. G. Young and A. T. Young, *J. Quant. Spectrosc. Radiat. Transfer*, **20**, 533 (1978).
216. S. P. Belov, A. V. Burenin, O. L. Polyansky, and S. M. Shapin, *J. Mol. Spectrosc.*, **90**, 579 (1981).

Chapter **IX**

NUCLEAR HYPERFINE STRUCTURE IN MOLECULAR ROTATIONAL SPECTRA

1 INTRODUCTION

Nuclear hyperfine structure in molecular rotational spectra may arise from either magnetic or electric interactions of the molecular fields with the nuclear moments, or from a combination of the two. The most important of these interactions is that of the molecular field gradient with the electric quadrupole moments of the nuclei. Such an interaction is not possible, however, for isotopes with nuclear spins of 0 or $\frac{1}{2}$ since such nuclei are spherically symmetric and hence have no quadrupole moments. The nuclear quadrupole hyperfine structure provides a measure of the molecular field gradients from which much information about the electronic structure and chemical bonds can be obtained. Nuclear quadrupole interaction in molecules was first detected in molecular beam experiments on diatomic molecules in the radiofrequency region [1, 2]. The basic theory worked out by Casimer[3] was extended to diatomic molecules by Nordsieck [4] and by Feld and Lamb [5]. The first detection of quadrupole hyperfine structure in the microwave region was that of the ammonia inversion spectrum by Good [6]. The theory for the interaction was extended to symmetric tops by Coles and Good [7] and by Van Vleck [8], to asymmetric rotors by Bragg and Golden [9, 10].

Practically all stable molecules, particularly organic molecules, have singlet Σ electronic ground states. All their electrons are paired, either in the atomic subshells or in the molecular orbital valence shells, so that in the first-order approximation their electronic magnetism whether from electronic spin or from orbital motion is canceled. When the molecules are not rotating, the molecular magnetism is canceled even in the higher orders of approximation. The end-over-end rotation of the molecule, however, generates weak magnetic fields which can interact with the nuclear magnetic moments to produce a slight magnetic splitting or displacement of the lines. Although the complexity of this slight magnetic field generated by rotation usually prevents measurements sufficiently accurate to give reliable information about chemical bonding like that given by nuclear quadrupole interactions, one must evaluate these small magnetic displacements when they are evident if the evaluation of the nuclear quadrupole coupling and rotational constants is to be the most precise. A general theory for nuclear magnetic interactions of rotating molecules in Σ singlet states has been worked out by Gunther-Mohr et al. [11].

For convenience in the prediction of hyperfine structure we have provided a table of nuclear moments (Appendix E).

2 CLASSICAL HAMILTONIAN FOR NUCLEAR QUADRUPOLE COUPLING

The nuclear quadrupole interaction results from a nonspherical distribution of nuclear charge which gives rise to a nuclear quadrupole moment and a nonspherical distribution of electronic charge about the nucleus which gives rise to an electric field gradient at the nucleus. If either the nuclear charge or the electronic charge about the nucleus is spherically symmetric, no such interaction is observed. The interaction puts a twisting torque on the nucleus, tending to align its spin moment in the direction of the field gradient. As a result of this torque, the spin axis will precess about the direction of the resultant field gradient, giving rise to precessional frequencies and nuclear quadrupole spectra. In solids, the field gradients are fixed in direction, and pure nuclear quadrupole spectra analogous to nuclear magnetic resonance are observable. In gases, the field gradients at the nucleus depend on the rotational state of the molecule; the nuclear quadrupole interaction differs for each rotational state and leads to a hyperfine structure of the rotational levels.

Let us first consider the classical interaction energy of a nuclear charge with a static potential V arising from extranuclear charges. Suppose $V(XYZ)$ to be the potential at a point with coordinates X, Y, Z in a cartesian system fixed in space and having its origin at the center of the nucleus, as indicated in Fig. 9.1. The value of V at the origin is V_0. To a first approximation, the electrical energy is ZeV_0, the usual energy that holds the electrons in their orbits. Because of the finite size of the nucleus, however, the electrical energy is correctly expressed by

$$E=\int \rho_n V\, d\tau_n \tag{9.1}$$

where $\rho_n=\rho(X, Y, Z)$ represents the density of the nuclear charge in the elemental volume $d\tau_n=dX\, dY\, dZ$ and where the integration is taken over the

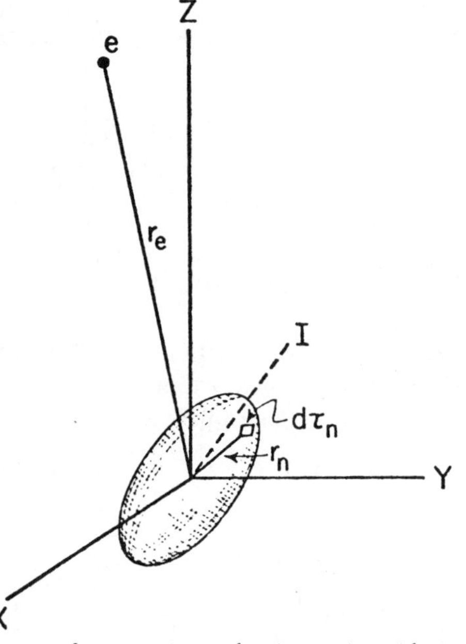

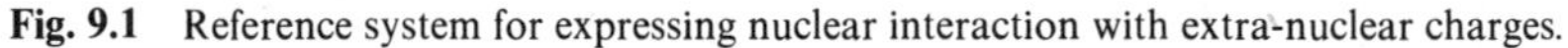
Fig. 9.1 Reference system for expressing nuclear interaction with extra-nuclear charges.

nuclear volume. The charge density ρ_n can be taken as uniform throughout the volume. In evaluation of the integral it is convenient to express the potential as a Taylor's expansion. In this expansion,

$$V=V_0+\left(\frac{\partial V}{\partial X}\right)_0 X_n+\left(\frac{\partial V}{\partial Y}\right)_0 Y_n+\left(\frac{\partial V}{\partial Z}\right)_0 Z_n+\frac{1}{2}\left(\frac{\partial^2 V}{\partial X^2}\right)_0 X_n^2$$
$$+\frac{1}{2}\left(\frac{\partial^2 V}{\partial Y^2}\right)_0 Y_n^2+\frac{1}{2}\left(\frac{\partial^2 V}{\partial Z^2}\right)_0 Z_n^2+\left(\frac{\partial^2 V}{\partial X\,\partial Y}\right)_0 X_nY_n \tag{9.2}$$
$$+\left(\frac{\partial^2 V}{\partial X\,\partial Z}\right)_0 X_nZ_n+\left(\frac{\partial^2 V}{\partial Y\,\partial Z}\right)_0 Y_nZ_n\cdots$$

where $X_n=\Delta X$, $Y_n=\Delta Y$, and $Z_n=\Delta Z$ represent the nuclear coordinates at the point of the elemental volume $d\tau_n$ in the nucleus. Substitution of V from (9.2) into (9.1) shows the first term in the expansion

$$E_0=V_0\int \rho_n\,d\tau_n \tag{9.3}$$

to be the usual monopole interaction which is independent of nuclear orientation. The second or dipole term,

$$E_d=\left(\frac{\partial V}{\partial X}\right)_0\int X_n\rho_n\,d\tau_n+\left(\frac{\partial V}{\partial Y}\right)_0\int Y_n\rho_n\,d\tau_n+\left(\frac{\partial V}{\partial Z}\right)_0\int Z_n\rho_n\,d\tau_n=0 \tag{9.4}$$

vanishes because $\rho(X, Y, Z)$ is a symmetric function, and the integral with $+$ and $-$ values of the coordinates cancels. This is in agreement with the fact that no nuclear electric dipole moment has ever been observed experimentally. The term of next higher order represents the quadrupole interaction,

$$E_Q=\frac{1}{2}\left(\frac{\partial^2 V}{\partial X^2}\right)_0\int \rho_n X_n^2\,d\tau_n+\frac{1}{2}\left(\frac{\partial^2 V}{\partial Y^2}\right)_0\int \rho_n Y_n^2\,d\tau_n$$
$$+\frac{1}{2}\left(\frac{\partial^2 V}{\partial Z^2}\right)_0\int \rho_n Z_n^2\,d\tau_n+\left(\frac{\partial^2 V}{\partial X\,\partial Y}\right)_0\int \rho_n X_nY_n\,d\tau_n \tag{9.5}$$
$$+\left(\frac{\partial^2 V}{\partial X\,\partial Z}\right)_0\int \rho_n X_nZ_n\,d\tau_n+\left(\frac{\partial^2 V}{\partial Y\,\partial Z}\right)_0\int \rho_n Y_nZ_n\,d\tau_n$$

Let us now choose a body-fixed system x, y, z with z along the spin axis $\mathbf{I}$. As a consequence of rotational symmetry about the spin axis; x, y, z are principal axes; the integrals of the cross terms vanish; and

$$\int \rho_n x_n^2\,d\tau_n=\int \rho_n y_n^2\,d\tau_n=\frac{1}{2}\int \rho_n(x_n^2+y_n^2)\,d\tau_n=\frac{1}{2}\int \rho_n(r_n^2-z_n^2)\,d\tau_n \tag{9.6}$$

Consequently, (9.5) can be expressed as

$$E_Q=\frac{1}{4}\left[\left(\frac{\partial^2 V}{\partial x^2}\right)_0+\left(\frac{\partial^2 V}{\partial y^2}\right)_0\right]\int \rho_n(r_n^2-z_n^2)\,d\tau_n+\frac{1}{2}\left(\frac{\partial^2 V}{\partial z^2}\right)_0\int \rho_n z_n^2\,d\tau_n \tag{9.7}$$

Because the charge giving rise to the field gradient can be considered to be zero over the nuclear volume, Laplace's equation $\nabla^2 V = 0$ holds, and

$$\left(\frac{\partial^2 V}{\partial x^2}\right)_0 + \left(\frac{\partial^2 V}{\partial y^2}\right)_0 = -\left(\frac{\partial^2 V}{\partial z^2}\right)_0 \tag{9.8}$$

With this relation, the classical interaction can be expressed as

$$E_Q = \frac{1}{4}\left(\frac{\partial^2 V}{\partial z^2}\right)_0 \int \rho_n(3z_n^2 - r_n^2)\, d\tau_n \tag{9.9}$$

The quantity

$$Q^* = \frac{1}{e}\int \rho_n(3z_n^2 - r_n^2)\, d\tau_n \tag{9.10}$$

is defined as the intrinsic nuclear quadrupole moment. Hence

$$E_Q = \frac{1}{4}\left(\frac{\partial^2 V}{\partial z^2}\right)_0 eQ^* \tag{9.11}$$

In this classical expression z is along the spin axis. Quantum mechanically, however, the spin $\mathbf{I}$ and Q^* are not observable quantities since only $\mathbf{I}^2$ and its components along a fixed direction in space have eigenvalues. However, the value of Q^* can be obtained from the observable quantity Q with the relationship, (9.20), derived later. Q^* is a measure of the deviation of the nuclear shape from spherical symmetry. A positive Q^* indicates that the nucleus is elongated along the spin axis, is prolate; a negative Q^* indicates that the nucleus is flattened along this axis, is oblate. For a spherical nucleus Q^* vanishes.

3 NUCLEAR QUADRUPOLE INTERACTIONS IN FIXED MOLECULES OF SOLIDS

Let us now transform the field gradient of (9.11) to a space-fixed system X, Y, Z, with its origin at the center of the nucleus and chosen so that X, Y, and Z are principal axes of the field gradient. In the principal system the cross terms vanish, and the transformation is expressed by

$$\begin{aligned}\frac{\partial^2 V}{\partial z^2} &= \frac{\partial^2 V}{\partial X^2}\left(\frac{\partial X}{\partial z}\right)^2 + \frac{\partial^2 V}{\partial Y^2}\left(\frac{\partial Y}{\partial z}\right)^2 + \frac{\partial^2 V}{\partial Z^2}\left(\frac{\partial Z}{\partial z}\right)^2 \\ &= \frac{\partial^2 V}{\partial X^2}\left(\frac{I_X}{|\mathbf{I}|}\right)^2 + \frac{\partial^2 V}{\partial Y^2}\left(\frac{I_Y}{|\mathbf{I}|}\right)^2 + \frac{\partial^2 V}{\partial Z^2}\left(\frac{I_Z}{|\mathbf{I}|}\right)^2\end{aligned} \tag{9.12}$$

The last form follows from the fact that $\mathbf{I}$ is along z and I_X, I_Y, and I_Z are components of $\mathbf{I}$; hence the direction cosines are

$$\frac{\partial X}{\partial z} = \cos(X, z) = \frac{I_X}{|\mathbf{I}|}, \ldots \tag{9.13}$$

By treating the spin components as quantum mechanical operators and substituting for $\mathbf{I}^2$ its eigenvalue $I(I+1)$, one can obtain a spin Hamiltonian operator by substitution of (9.12) into (9.11). Thus

$$\mathscr{H}_Q=\frac{eQ^*}{4I(I+1)}\left[\left(\frac{\partial^2 V}{\partial X^2}\right)_0 I_X^2+\left(\frac{\partial^2 V}{\partial Y^2}\right)_0 I_Y^2+\left(\frac{\partial^2 V}{\partial Z^2}\right)_0 I_Z^2\right] \tag{9.14}$$

When the field gradient is axially symmetric about Z,

$$\left(\frac{\partial^2 V}{\partial X^2}\right)_0=\left(\frac{\partial^2 V}{\partial Y^2}\right)_0=-\frac{1}{2}\left(\frac{\partial^2 V}{\partial Z^2}\right)_0 \tag{9.15}$$

where use is made of Laplace's equation $\nabla^2 V=0$. With (9.15) and $I^2=I_X^2+I_Y^2+I_Z^2$, (9.14) becomes, for the axially symmetric coupling,

$$\mathscr{H}_Q=\frac{eQ^*}{8I(I+1)}\left(\frac{\partial^2 V}{\partial Z^2}\right)_0 (3I_Z^2-\mathbf{I}^2) \tag{9.16}$$

and

$$(E_Q)_{M_I}=\frac{eQ^*(\partial^2 V/\partial Z^2)_0}{8I(I+1)}\left[3M_I^2-I(I+1)\right] \tag{9.17}$$

where M_I is the eigenvalue of I_Z in units of $\hbar$. Except for the use of the intrinsic quadrupole moment Q^*, (9.14) and (9.16) are the usual Hamiltonians for pure quadrupole resonance in the solid state.

A simple, if not rigorous, derivation of the relationship between Q^* and the observable Q can be found by comparison of (9.17) with the classical energy, (9.11), derived for axially symmetric coupling about z, the spin axis. Conventionally, Q is defined as the effective component for the most complete alignment along Z, that is, the component for $M_I=I$. From (9.17),

$$(E_Q)_{M_I=I}=\frac{eQ^*(\partial^2 V/\partial Z^2)_0(2I-1)}{8(I+1)} \tag{9.18}$$

Since the effective quadrupole moment for coupling with $(\partial^2 V/\partial Z^2)$ is Q, the coupling energy found by analogy with (9.11) is

$$E_Q=\frac{1}{4}\left(\frac{\partial^2 V}{\partial Z^2}\right)_0 eQ \tag{9.19}$$

Equation of this energy to $(E_Q)_{M_I=I}$ from (9.18) yields

$$Q^*=\frac{2(I+1)}{(2I-1)}Q \tag{9.20}$$

Substitution of this value into (9.14) gives the solid state Hamiltonian operator in the form

$$\mathscr{H}_Q=\frac{1}{2I(2I-1)}(\chi_{XX}I_X^2+\chi_{YY}I_Y^2+\chi_{ZZ}I_Z^2) \tag{9.21}$$

where the customary designations for the principal values of the coupling constants have been made:

$$\chi_{XX} \equiv eQq_{XX}, \qquad \chi_{YY} \equiv eQq_{YY}, \qquad \chi_{ZZ} \equiv eQq_{ZZ} \tag{9.22}$$

where

$$q_{XX} \equiv \left(\frac{\partial^2 V}{\partial X^2}\right)_0, \qquad q_{YY} \equiv \left(\frac{\partial^2 V}{\partial Y^2}\right)_0, \qquad q_{ZZ} \equiv \left(\frac{\partial^2 V}{\partial Z^2}\right)_0 \tag{9.23}$$

With axial symmetry about Z, it is evident from (9.17) that

$$E_Q = \frac{\chi_{ZZ}}{4I(2I-1)}\left[3M_I^2 - I(I+1)\right] \tag{9.24}$$

where

$$M_I = I,\ I-1,\ I-2, \ldots, -I \tag{9.25}$$

In the general case when there is no axial symmetry, the secular equation must be set up and solved as described in Chapter II, Section 4, or perturbation methods must be used. The required matrix elements of the squared operators of (9.21) are

$$\begin{aligned}
(I, M_I|I_Z^2|I, M_I) &= M_I^2 \\
(I, M_I|I_X^2|I, M_I) &= (I, M_I|I_Y^2|I, M_I) = \tfrac{1}{2}[I(I+1) - M_I^2] \\
(I, M_I|I_X^2|I, M_I \pm 2) &= -(I, M_I|I_Y^2|I, M_I \pm 2) \\
&= -\tfrac{1}{4}[I(I+1) - M_I(M_I \pm 1)]^{1/2} \\
&\quad \times [I(I+1) - (M_I \pm 1)(M_I \pm 2)]^{1/2}
\end{aligned} \tag{9.26}$$

These matrix elements have the same form as those for P_x^2, P_y^2, and P_z^2 in the symmetric-top representation if I is identified with J and M_I with K. However, I and M_I may have half-integral as well as integral values, whereas J and K have only integral values. Note that (9.21) for $\mathscr{H}_Q$ has the same form as the Hamiltonian for the rigid asymmetric rotor if the couplings are related to the moments of inertia by

$$\frac{\chi_{XX}}{2I(2I-1)} \to \frac{1}{2I_x}, \qquad \frac{\chi_{YY}}{2I(2I-1)} \to \frac{1}{2I_y}, \qquad \frac{\chi_{ZZ}}{2I(2I-1)} \to \frac{1}{2I_z} \tag{9.27}$$

With these relations the various solutions already obtained for the rigid asymmetric rotor can be used for the finding of quadrupole coupling energies for integral spin values. This relationship was first pointed out by Bersohn [12]. Because the values of I are small, exceeding $\frac{9}{2}$ for only a very few isotopes, the problem of finding the quadrupole coupling energies in solid state asymmetric field gradients is simpler than that of calculating the energies of the asymmetric rotor for which populated states with high values are possible. Solutions for various spin values are given by Cohen and Reif [13] and by Das and Hahn [14].

Frequently, the field gradient is almost axially symmetric, if not completely so, about a bond to the coupling atom. Such cases are most conveniently treated by perturbation theory in a manner similar to that for treatment of the slightly asymmetric rotor. The most nearly symmetric axis is defined as the Z axis, and the asymmetry parameter is defined by

$$\eta = \frac{\chi_{XX} - \chi_{YY}}{\chi_{ZZ}} \tag{9.28}$$

Because Laplace's equation holds, $\chi_{XX}+\chi_{YY}+\chi_{ZZ}=0$, there are only two independent coupling parameters. These are chosen as χ_{ZZ} and η. The Hamiltonian can then be expressed so that the off-diagonal terms contain η as a coefficient. The off-diagonal terms can then be evaluated with perturbation theory. The asymmetry parameter η has been related by Das and Hahn [14] to the asymmetry parameter b of the slightly asymmetric rotor so that the energy formula for the slightly asymmetric rotor (Chapter VII) can be used to give the quadrupole energies when $\eta \ll 1$.

Transitions between the quadrupole hyperfine levels in solids can be observed through coupling of the magnetic dipole moment of the nucleus to the radiation field. The selection rule for pure quadrupole absorption spectra is $|\Delta M_I|=1$. For an axially symmetric field, the frequencies from (9.24) are

$$\nu = |\chi_{ZZ}| \left[\frac{3}{4I(2I-1)}\right](2|M_I|-1) \tag{9.29}$$

where M_I is the larger of the two quantum numbers involved. Figure 9.2 shows an energy level diagram indicating the transitions and pure quadrupole spectrum for a nucleus with spin $I=\frac{7}{2}$ which is in an axially symmetric field.

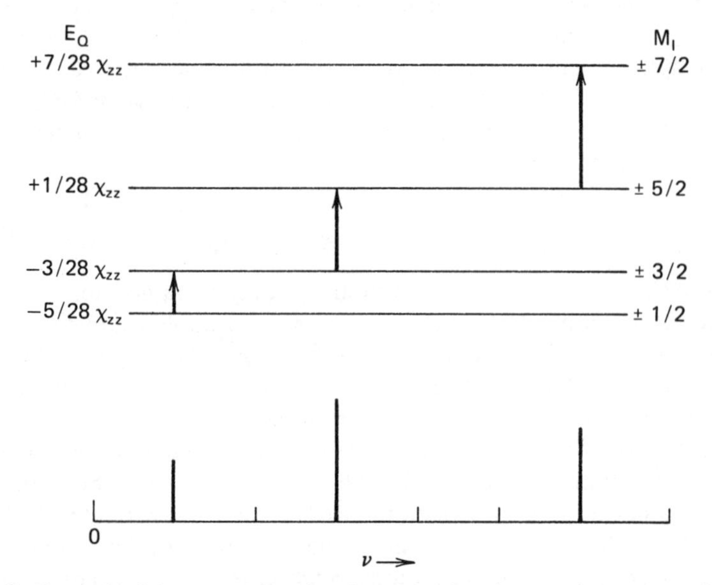

Fig. 9.2 Energy level diagram and predicted pure quadrupole resonance spectrum for a nucleus with spin $I=\frac{7}{2}$.

Pure quadrupole resonance frequencies of solids, first detected by Dehmelt and Krüger [15], generally occur at radiofrequencies below those of the microwave region. For this reason we shall not discuss them further. Comprehensive treatments are given by Cohen and Reif [13], and by Das and Hahn [14]. The coupling values obtained are closely related to those measured by microwave spectroscopy of gases, and in Chapter XIV we shall discuss some of the solid-state results. Pure quadrupole resonance in solids is a simpler phenomenon than quadrupole coupling in gases, where the coupling gradient is a function of the rotational state. This is particularly true when there is more than one coupling nucleus in the molecule. In solids, the interaction of one nucleus does not perturb the field gradient at the other coupling nuclei, and the theory derived for a single coupling nucleus applies when there are other coupling nuclei in the same molecule. In contrast, plural coupling greatly complicates the problem in rotating molecules.

4 QUADRUPOLE INTERACTIONS BY A SINGLE COUPLING NUCLEUS IN A ROTATING MOLECULE

General Theory of the Interaction

The quadrupole coupling of a nucleus in a rotating molecule of a gas is more complicated than that in frozen molecules of a solid because the field gradients depend upon the rotation. For an external-field-free molecule, the nuclear spin **I** is coupled to the molecular rotational angular momentum **J** to form a resultant **F**. In the vector model, **J** and **I** can be considered as precessing about **F**, as indicated in Fig. 9.3. The total angular momentum of the molecule with nuclear coupling is thus represented by **F** rather than **J**, which designates the total angular momentum exclusive of nuclear spin. However, in the first-order treatment, which is adequate for most coupling cases, $\mathbf{J}^2$ is still a constant of the motion. The good quantum numbers are thus F, M_F, J, and I. The new angular momentum quantum numbers are

$$F = J+I,\ J+I-1,\ J+I-2, \ldots, |J-I| \tag{9.30}$$

and

$$M_F = F,\ F-1,\ F-2, \ldots, -F \tag{9.31}$$

where J and I have the values assigned previously in Chapter II. In units of $\hbar$ the eigenvalues of the square of the total angular momentum and its components along an axis in space are

$$(F, M_F|\mathbf{F}^2|F, M_F) = F(F+1) \tag{9.32}$$

$$(F, M_F|F_Z|F, M_F) = M_F \tag{9.33}$$

Let us consider again the general expansion of (9.5) for the classical quadrupole interaction. By subtraction and addition of $\frac{1}{6}\nabla^2 V \int \rho_n R_n^2\, d\tau_n$ to the right

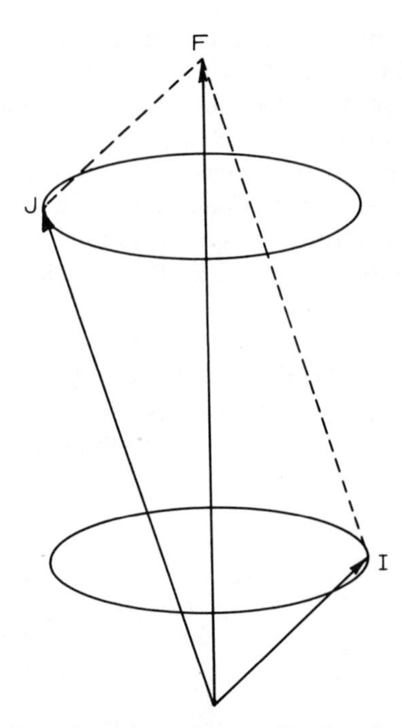

Fig. 9.3 Vector diagram of coupling of a nuclear spin **I** with molecular rotational momentum **J**.

side, this expression can be put into the form:

$$E_Q=\frac{1}{6}\left[\left(\frac{\partial^2 V}{\partial X^2}\right)_0 \int (3X_n^2-R_n^2)\rho_n\,d\tau_n+\left(\frac{\partial^2 V}{\partial Y^2}\right)_0 \int (3Y_n^2-R_n^2)\rho_n\,d\tau_n\right.$$
$$+\left(\frac{\partial^2 V}{\partial Z^2}\right)_0 \int (3Z_n^2-R_n^2)\rho_n\,d\tau_n+6\left(\frac{\partial^2 V}{\partial X\,\partial Y}\right)_0 \int X_nY_n\rho_n\,d\tau_n$$
$$\left.+6\left(\frac{\partial^2 V}{\partial X\,\partial Z}\right)_0 \int X_nZ_n\rho_n\,d\tau_n+6\left(\frac{\partial^2 V}{\partial Y\,\partial Z}\right)_0 \int Y_nZ_n\rho_n\,d\tau_n\right]$$
$$+\frac{1}{6}\left[\left(\frac{\partial^2 V}{\partial X^2}\right)_0+\left(\frac{\partial^2 V}{\partial Y^2}\right)_0+\left(\frac{\partial^2 V}{\partial Z^2}\right)_0\right]\int R_n^2\rho_n\,d\tau_n \tag{9.34}$$

For reasons already given, Laplace's equation $\nabla^2 V=0$ holds over the nuclear volume; hence the last term is vanishingly small and can be omitted. The remaining expression can be written as the scalar product (double-dot product) of two symmetric dyadics

$$E_Q=-\tfrac{1}{6}\mathbf{Q}:\nabla\mathbf{E}=\frac{1}{6}\sum_{i,j=X,Y,Z} Q_{ij}V_{ij} \tag{9.35}$$

where $V_{ij}=-\nabla E_{ij}$ and where the dyadic $\nabla\mathbf{E}$ is the gradient of the electric field of the extra nuclear charges

$$\nabla \mathbf{E} = \boldsymbol{e}_X \frac{\partial \mathbf{E}}{\partial X} + \boldsymbol{e}_Y \frac{\partial \mathbf{E}}{\partial Y} + \boldsymbol{e}_Z \frac{\partial \mathbf{E}}{\partial Z} \tag{9.36}$$

and $\mathbf{Q}$ is the quadrupole moment dyadic

$$\mathbf{Q} = \int \rho_n [3\mathbf{R}_n \mathbf{R}_n - \mathbb{I} R_n^2] \, d\tau_n \tag{9.37}$$

Here $\mathbf{R}_n$ represents the vector locating points of the nuclear volume in the space-fixed X, Y, Z reference system. $\mathbb{I} = \boldsymbol{e}_X \boldsymbol{e}_X + \boldsymbol{e}_Y \boldsymbol{e}_Y + \boldsymbol{e}_Z \boldsymbol{e}_Z$ is the unit dyadic and $\boldsymbol{e}_X$, $\boldsymbol{e}_Y$, and $\boldsymbol{e}_Z$ are unit vectors along X, Y, and Z.

We now develop the quantum mechanical operator for the quadrupole interaction energy. The potential V at the nucleus arises from all the various extra nuclear charges of the molecule, but we need consider only the extra nuclear electrons since the charges of other nuclei are well screened. We can therefore express the potential as

$$V = \sum_k e \left(\frac{1}{R_e} \right)_k \tag{9.38}$$

where $(1/R_e)_k$ represents the distance from the kth electron to the nucleus and where the summation is taken over all electrons contributiong to V. Let us substitute this value of V into the tensor V_{ij}. We consider only one component because the transformations of other components of the tensor are similar. Since

$$\frac{\partial^2}{\partial X^2} \left(\frac{1}{R} \right) = \left(\frac{\partial^2}{\partial X^2} \right) (X^2 + Y^2 + Z^2)^{-1/2} = \left(\frac{3X^2 - R^2}{R^5} \right) \tag{9.39}$$

Thus it is evident that

$$\begin{aligned} V_{XX} = \left(\frac{\partial^2 V}{\partial X^2} \right)_0 &= e \sum_k \left\langle \left(\frac{1}{R_e^3} \right)_k \right\rangle \left\langle \left[3 \left(\frac{X_e}{R_e} \right)^2 - 1 \right]_k \right\rangle \\ &= e \sum_k \left\langle \left(\frac{1}{R_e^3} \right)_k \right\rangle \langle (3 \cos^2 \theta_X - 1)_k \rangle \end{aligned} \tag{9.40}$$

where θ_X is the angle between R_e and the X axis fixed in space (see Fig. 9.1). The average is taken over the electronic orbitals and over the vibrational and rotational states of the molecule. However, $\langle (1/R_e^3)_k \rangle$ is independent of the rotation and in a particular vibrational state may be treated as a constant. We can then write

$$V_{XX} = \sum_k C_k \left\langle \left(\frac{3X_e^2}{R_e^2} - 1 \right)_k \right\rangle \tag{9.41}$$

Because the orbitals of all the electrons are fixed in the molecule and rotate with the molecular frame about the axis of $\mathbf{J}$, the angular dependence of their averaged sum will be the same as that of $\mathbf{J}$, and their resultant field gradient will resolve along space-fixed axes in the same proportions as the components of **J**; the rapid rotation about **J** will effectively average out the components of the field gradient perpendicular to **J** and make **J** an axis of symmetry for the gradient. From (9.41) one thus obtains

$$(V_{XX})_{op} = C\left(3\frac{J_X^2}{\mathbf{J}^2} - 1\right) \tag{9.42}$$

where C is a proportionality constant to be evaluated, a constant which depends on the electronic distribution in the whole molecule but primarily in the atom of the particular coupling nucleus. In units of $\hbar$, the eigenvalues of $\mathbf{J}^2$ are $J(J+1)$, and hence (9.42) can be written

$$(V_{XX})_{op} = \frac{C}{J(J+1)}[3J_X^2 - J(J+1)] \tag{9.43}$$

The expressions for V_{YY} and V_{ZZ} are similar. Because the component operators do not commute, however, the analogous operators for the cross-product terms must be symmetrized. For example,

$$V_{XY} = \sum_k C_k \left\langle \left(3\frac{X_e Y_e}{R_e^2}\right)_k \right\rangle \tag{9.44}$$

and the conjugate operator is

$$(V_{XY})_{op} = \frac{C}{J(J+1)}\left(3\frac{J_X J_Y + J_Y J_X}{2}\right) \tag{9.45}$$

It is customary to evaluate the constant C by defining the coupling constant q_J as that observed for the maximum projection of **J** along a space-fixed axis, that is, for the state $M_J = J$. Hence from the ZZ component

$$q_J = (J, J|V_{ZZ}|J, J) = \frac{C}{J(J+1)}[3J^2 - J(J+1)] \tag{9.46}$$

and

$$C = \frac{J+1}{2J-1} q_J \tag{9.47}$$

Hence

$$(V_{ZZ})_{op} = \frac{q_J}{J(2J-1)}[3J_Z^2 - J(J+1)] \tag{9.48}$$

The generalized expression for the elements of the tensor operator of the field gradient is consequently

$$-(\nabla E_{ij})_{op} = (V_{ij})_{op} = \frac{q_J}{J(2J-1)}\left(3\frac{J_i J_j + J_j J_i}{2} - \delta_{ij}\mathbf{J}^2\right) \tag{9.49}$$

where $i, j = X, Y,$ or Z and $\delta_{ij} = 1$ when $i = j$ and 0 when $i \neq j$.

Since the nuclear quadrupole tensor Q_{ij}, (9,37), has rotational symmetry about the spin axis, it can be treated in a manner similar to V_{ij}. Accordingly, the nuclear quadrupole operator can be expressed as

$$(Q_{ij})_{\text{op}}=C_n\left(3\,\frac{I_iI_j+I_jI_i}{2}-\delta_{ij}\mathbf{I}^2\right) \tag{9.50}$$

where C_n is the proportionality constant evaluated by definition of Q as the value for maximum resolution along a space-fixed axis, that is, when $M_I=I$. Thus

$$eQ\equiv(I,\,I|Q_{ZZ}|I,\,I)=C_n(I,\,I|3I_Z^2-I(I+1)|I,\,I)=C_n[3I^2-I(I+1)] \tag{9.51}$$

Therefore

$$C_n=\frac{eQ}{I(2I-1)} \tag{9.52}$$

and

$$(Q_{ij})_{\text{op}}=\frac{eQ}{I(2I-1)}\left(3\,\frac{I_iI_j+I_jI_i}{2}-\delta_{ij}\mathbf{I}^2\right) \tag{9.53}$$

Substitution of the operators of (9.49) and (9.53) into (9.35) yields the Hamiltonian operator for the quadrupole interaction

$$\mathcal{H}_Q=\frac{1}{6}\,\frac{eq_JQ}{J(2J-1)I(2I-1)}\sum_{i,j=X,Y,Z}\left(3\,\frac{I_iI_j+I_jI_i}{2}-\delta_{ij}\mathbf{I}^2\right)\left(3\,\frac{J_iJ_j+J_jJ_i}{2}-\delta_{ij}\mathbf{J}^2\right) \tag{9.54}$$

If the indicated multiplication is performed and the results expressed in terms of vector products, this Hamiltonian can be put into a more convenient form. Because the components of **I** commute with those of **J**, the term

$$\begin{aligned}\sum_{i,j} I_iI_jJ_iJ_j&=\sum_i (I_iJ_i)\sum_j (I_jJ_j)\\&=(I_XJ_X+I_YJ_Y+I_ZJ_Z)(I_XJ_X+I_YJ_Y+I_ZJ_Z)=(\mathbf{I}\cdot\mathbf{J})^2\end{aligned} \tag{9.55}$$

The sum of all such terms is $\frac{9}{2}(\mathbf{I}\cdot\mathbf{J})^2$. Because $\delta_{ij}=0$ when $i\neq j$, the terms

$$-\sum_{i,j}\tfrac{3}{2}(J_iJ_j+J_jJ_i)\delta_{ij}\mathbf{I}^2=-3\mathbf{J}^2\mathbf{I}^2$$

$$-\sum_{i,j}\tfrac{3}{2}(I_iI_j+I_jI_i)\delta_{ij}\mathbf{J}^2=-3\mathbf{J}^2\mathbf{I}^2$$

and

$$\sum_{i,j}\delta_{ij}\mathbf{I}^2\delta_{ij}\mathbf{J}^2=3\mathbf{J}^2\mathbf{I}^2 \tag{9.56}$$

Because the components of **J** and of **I** do not commute among themselves, that is, $J_i J_j \neq J_j J_i$ and $I_i I_j \neq I_j I_i$ when $i \neq j$, the commutation rules given in Chapter II must be applied for evaluation of the remaining terms which are of the form $I_i I_j J_j J_i$. The transformations, which are rather lengthy, are given by Ramsey [16]. The results are

$$\sum_{i,j} I_i I_j J_j J_i = (\mathbf{I} \cdot \mathbf{J})^2 + \mathbf{I} \cdot \mathbf{J} \tag{9.57}$$

and the total of such terms is $\frac{9}{2}[(\mathbf{I} \cdot \mathbf{J})^2 + (\mathbf{I} \cdot \mathbf{J})]$. Collection of all these terms, with slight simplification, yields the quadrupole Hamiltonian in the compact form,

$$\mathcal{H}_Q = \frac{eQq_J}{2J(2J-1)I(2I-1)} [3(\mathbf{I} \cdot \mathbf{J})^2 + \tfrac{3}{2}\mathbf{I} \cdot \mathbf{J} - \mathbf{I}^2\mathbf{J}^2] \tag{9.58}$$

This Hamiltonian, originally derived by Casimer [3], is applicable to a coupling nucleus in any type of molecule, or to a free atom if **J** is treated as the total electronic angular momentum exclusive of nuclear spin. The quantity q_J depends on the particular type of molecule, whether linear, symmetric rotor, or asymmetric rotor.

The advantage of the general Hamiltonian of (9.58) is that its eigenvalues for the terms in the bracket can be obtained easily for the field-free rotor in which **J** and **I** are coupled to form a resultant **F**, as indicated in Fig. 9.3. Since **F** is the vector sum of **J** and **I**,

$$\mathbf{F}^2 = (\mathbf{J} + \mathbf{I})^2 = \mathbf{J}^2 + 2\mathbf{I} \cdot \mathbf{J} + \mathbf{I}^2 \tag{9.59}$$

and

$$\mathbf{I} \cdot \mathbf{J} = \tfrac{1}{2}(\mathbf{F}^2 - \mathbf{J}^2 - \mathbf{I}^2) \tag{9.60}$$

The eigenvalues of $\mathbf{I} \cdot \mathbf{J}$ are therefore

$$(F, J, I|\mathbf{I} \cdot \mathbf{J}|F, J, I) = \tfrac{1}{2}[F(F+1) - J(J+1) - I(I+1)] = \tfrac{1}{2}C \tag{9.61}$$

and

$$(F, J, I|(\mathbf{I} \cdot \mathbf{J})^2|F, J, I) = \tfrac{1}{4}C^2 \tag{9.62}$$

Substitution of these values with $\mathbf{J}^2\mathbf{I}^2 = J(J+1)I(I+1)$ into (9.58) yields

$$E_Q = \frac{eQq_J}{2J(2J-1)I(2I-1)} [\tfrac{3}{4}C(C+1) - J(J+1)I(I+1)] \tag{9.63}$$

where

$$C = F(F+1) - J(J+1) - I(I+1) \tag{9.64}$$

Let us now evaluate the quantity q_J. To do this we must first transform the field gradient to a reference system fixed in the molecule. Although later transformations will be made to special axes chosen to coincide with the bond axis to the coupling nucleus, the principal inertial axes a, b, c at this stage in the development have the advantage of applying to molecules of all classes. The cross terms in the field gradient, such as $(\partial^2 V/\partial a\,\partial b)$, vanish when averaged over the rotational state, as will be explained later.

As defined previously,

$$q_J = \left\langle \left(\frac{\partial^2 V}{\partial Z^2}\right)_0 \right\rangle_{M_J=J} = \int \psi_{J,i}^{*M_J=J} \frac{\partial^2 V}{\partial Z^2} \psi_{J,i}^{M_J=J}\, d\tau = \left(J, i, J \left| \frac{\partial^2 V}{\partial Z^2} \right| J, i, J \right) \tag{9.65}$$

where Z is fixed in space and where i represents any internal rotational quantum numbers such as K. In the inertial axes a, b, c,

$$\begin{aligned}\frac{\partial^2 V}{\partial Z^2} = {} & \frac{\partial^2 V}{\partial a^2}\left(\frac{\partial a}{\partial Z}\right)^2 + \frac{\partial^2 V}{\partial b^2}\left(\frac{\partial b}{\partial Z}\right)^2 + \frac{\partial^2 V}{\partial c^2}\left(\frac{\partial c}{\partial Z}\right)^2 \\ & + 2\frac{\partial^2 V}{\partial a\,\partial b}\left(\frac{\partial a}{\partial Z}\right)\left(\frac{\partial b}{\partial Z}\right) + 2\frac{\partial^2 V}{\partial a\,\partial c}\left(\frac{\partial a}{\partial Z}\right)\frac{\partial c}{\partial Z} \\ & + 2\frac{\partial^2 V}{\partial b\,\partial c}\left(\frac{\partial b}{\partial Z}\right)\left(\frac{\partial c}{\partial Z}\right)\end{aligned} \tag{9.66}$$

The quantities

$$\frac{\partial a}{\partial Z} = \cos\theta_{Z,a} = \Phi_{Z,a}, \ldots \tag{9.67}$$

are the direction cosines of the principal axes with the space-fixed Z axis. The quantities

$$\frac{\partial^2 V}{\partial a^2} = q_{aa}, \qquad \frac{\partial^2 V}{\partial a\,\partial b} = q_{ab}, \ldots \tag{9.68}$$

are the field gradients with reference to the principal inertial axes. They depend on the electronic state and to a slight extent on the vibrational state of the molecule, but to a high order of approximation they are independent of the molecular rotational state and can in the present evaluation be treated as constants. Thus

$$\begin{aligned}\frac{\partial^2 V}{\partial Z^2} = {} & q_{aa}\Phi_{Za}^2 + q_{bb}\Phi_{Zb}^2 + q_{cc}\Phi_{Zc}^2 + 2q_{ab}\Phi_{Za}\Phi_{Zb} \\ & + 2q_{ac}\Phi_{Za}\Phi_{Zc} + 2q_{bc}\Phi_{Zb}\Phi_{Zc}\end{aligned} \tag{9.69}$$

To obtain q_J we must average the terms on the right side of (9.69) over the rotational wave function $\Psi_r^{M_J=J}$. The factors q_{aa}, q_{ab}, . . . , are constant and can be removed from the integral. When averaged over the rotational state, the off-diagonal terms for an asymmetric rotor vanish, for example,

$$q_{ab} \int \Psi_r^{*M_J=J} \Phi_{Za}\Phi_{Zb} \Psi_r^{M_J=J}\, d\tau = 0, \ldots \tag{9.70}$$

This results from the symmetry properties of the momental ellipsoid (see Chapter III). Either Ψ_r must be symmetric (+) or antisymmetric (−) with respect to a rotation of π degrees about a principal axis. In either case, however, the product $\Psi_r\Psi_r^*$ will be positive for the same rotational function. It is evident that for a rotation of π about the a axis, Φ_{Za} will be unchanged while Φ_{Zb} will change sign only. Similar relationships apply for b and c. Thus the integrals of (9.70) must change in sign for at least two of the operations C_2^a, C_2^b, C_2^c. However, these definite integrals cannot change sign for such operations as change the system into an indistinguishable one. They are therefore identically zero. The off-diagonal terms, however, are not zero in the second-order approximation. For linear and symmetric-top molecules the principal inertial axes are principal axes of the quadrupole coupling tensor, and hence the cross terms vanish automatically since $q_{ab}, \ldots$ vanish.

Expressed in the principal inertial axes, q_J becomes

$$
\begin{aligned}
q_J = {} & q_{aa}(J, i, M_J=J|\Phi_{Za}^2|J, i, M_J=J)+q_{bb}(J, i, M_J=J|\Phi_{Zb}^2|J, i, M_J=J) \\
& +q_{cc}(J, i, M_J=J|\Phi_{Zc}^2|J, i, M_J=J)
\end{aligned} \tag{9.71}
$$

where q_{aa}, q_{bb}, and q_{cc} are the molecular field gradients at the coupling nucleus with reference to the inertial axes.

For brevity, the coupling constant eQq is often designated by χ. With this designation the general expression for the quadrupole coupling energy is obtained by substitution of (9.71) into (9.63)

$$
E_Q = \sum_{g=a,b,c} \chi_{gg}(J, i, M_J=J|\Phi_{Zg}^2|J, i, M_J=J)\left[\frac{\frac{3}{4}C(C+1)-J(J+1)I(I+1)}{2J(2J-1)I(2I-1)}\right] \tag{9.72}
$$

where

$$
\chi_{aa}=eQq_{aa}, \qquad \chi_{bb}=eQq_{bb}, \qquad \chi_{cc}=eQq_{cc} \tag{9.73}
$$

Because of the relationship

$$
(J, i, M_J=J|\Phi_{Zg}^2|J, i, M_J=J)=\frac{2}{(J+1)(2J+3)}(J, i|J_g^2|J, i)+\frac{1}{2J+3} \tag{9.74}
$$

which exists between the diagonal matrix elements of Φ_{Zg}^2 when $M_J=J$ and the diagonal matrix elements of J_g^2 in the J, i, M_J representation of the unperturbed rotor, (9.72) can be put in the alternative form

$$
E_Q=\frac{2}{(J+1)(2J+3)}\sum_{g=a,b,c} \chi_{gg}(J, i|J_g^2|J, i)\left[\frac{\frac{3}{4}C(C+1)-J(J+1)I(I+1)}{2J(2J-1)I(2I-1)}\right] \tag{9.75}
$$

The term $(2J+3)^{-1}$ of (9.74) does not enter in this expression since we have used Laplace's relation, (9.76). This expression of the quadrupole coupling energies, which was first obtained by Bragg and Golden [10], is particularly useful in the calculation of the nuclear quadrupole perturbation in the asymmetric rotor. Equation 9.74 can be readily proved for the symmetric-top wave functions and hence for the asymmetric-top functions by comparison of the matrix elements derived from Table 2.1 $(J, K, M=J|\Phi_{Zg}^2|J, K, M_J=J)$ with those for the symmetric top $(J, K|J_g^2|J, K)$, which are given in Chapter II.

Because Laplace's equation holds,

$$\chi_{aa}+\chi_{bb}+\chi_{cc}=0 \tag{9.76}$$

and there are only two independent coupling constants in the most general case. These are usually expressed in terms of the coupling constant with reference to one of the axes and an asymmetric parameter η. If the reference axis is chosen as the c axis, the two coupling constants would be χ_{cc} and

$$\eta=\frac{\chi_{aa}-\chi_{bb}}{\chi_{cc}} \tag{9.77}$$

The reference axis is usually chosen as the one for which the coupling is the most nearly symmetric, that is, for which η is smallest.

The matrix elements of the squared direction cosines differ for the different classes of molecules. For diatomic or linear polyatomic molecules and for symmetric-top molecules they can be evaluated explicitly, as will be done below. For asymmetric rotors a closed-form evaluation can be made only for low J states, but values can be obtained by numerical techniques (especially with computers) for higher J values.

It should be appreciated that the values of q_J as expressed in the preceding formulas are not exact, but are first-order perturbation values obtained when the field gradient is averaged over the wave functions of the unperturbed rotor. However, the off-diagonal elements that are omitted in the average are entirely negligible except for the coupling by the few nuclei that have large quadrupole moments. For them, the second-order corrections described in Section 6 are adequate.

Diatomic and Linear Polyatomic Molecules

In a linear molecule, diatomic or polyatomic, the molecular field gradient is symmetric about the bond axis. Let us designate a molecule-fixed coordinate system x, y, z with z along the bond axis. Obviously, this system coincides with the principal axes of inertia with $a\rightarrow z$. From the symmetry and from Laplace's equation

$$q_{xx}=q_{yy}=-\tfrac{1}{2}q_{zz} \tag{9.78}$$

Because there is only one coupling axis it is customary to drop the subscript and to set $q_{zz}=q$. Substitution of these relations into (9.71) gives

$$\begin{aligned} q_J&=q(J, M_J=J|\Phi_{Zz}^2-\tfrac{1}{2}[\Phi_{Zx}^2+\Phi_{Zy}^2]|J, M_J=J) \\ &=q\left(J, M_J=J\left|\frac{3\Phi_{Zz}^2-1}{2}\right|J, M_J=J\right) \end{aligned} \tag{9.79}$$

The matrix elements of Φ_{Zz} for the linear molecule correspond to those of the symmetric top with $K=0$. Although the diagonal matrix elements of Φ for a linear molecule are zero, those of Φ^2 are not. From the matrix product rule,

$$(J, M_J=J|\Phi_{Zz}^2|J, M_J=J)=\sum_{J'}(J, J|\Phi_{Zz}|J', J)(J', J|\Phi_{Zz}|J, J)=\frac{1}{2J+3} \tag{9.80}$$

where the elements of the product are evaluated from Table 2.1 with $K=0$ and $M_J=J$. Substitution of (9.80) into (9.79) yields

$$q_J = -\frac{qJ}{2J+3} \tag{9.81}$$

The first-order quadrupole coupling energies for a single coupling nucleus in a linear molecule are found from combination of (9.63) and (9.81) or (9.72) and (9.80) to be

$$E_Q = -eQqY(J, I, F) = -\chi Y(J, I, F) \tag{9.82}$$

where

$$Y(J, I, F) = \frac{\frac{3}{4}C(C+1) - I(I+1)J(J+1)}{2(2J-1)(2J+3)I(2I-1)} \tag{9.83}$$

For convenience in the calculation of hyperfine structure, numerical tabulations of $Y(J, I, F)$ and of the relative intensities of the hyperfine components for $J \to J+1$ transitions are given in Appendix I.

Selection rules for hyperfine transitions in rotational absorption spectra are

$$J \to J+1, \qquad F \to F, \qquad F \to F \pm 1, \qquad I \to I$$

The rotational frequencies perturbed by quadrupole coupling are

$$\nu = \frac{E_r(J+1) - E_r(J)}{h} + \frac{E_Q(J+1, I, F') - E_Q(J, I, F)}{h} \tag{9.84}$$

where $F' = F,\ F \pm 1$. It is customary to express the coupling constants in frequency units so that

$$\nu = \nu_0 - eQq[Y(J+1, I, F') - Y(J, I, F)] \tag{9.85}$$

where ν_0 is the "unperturbed" rotational frequency that would be observed if there were no quadrupole coupling. By measurement of only two frequencies of the hyperfine multiplet, $\chi \equiv eQq$ and ν_0 may be obtained. For best accuracy one should measure the most widely spaced components provided that they are sufficiently strong for precise measurement. As a test of the accuracy, other components are usually measured. It is evident that wherever the rotational transitions have resolved hyperfine structure one must measure eQq in order to obtain ν_0 and hence the rotational constant B_0 described in Chapters IV and V.

Figure 9.4 illustrates the calculation of the frequencies of the hyperfine components caused by ^{127}I in the $J=7 \to 8$ rotational transition of ^{127}ICN. The spin of ^{127}I is $\frac{5}{2}$, and the total number of components for all transitions for which $J > I$ is 15. However, it will be noted that only the six $F \to F+1$ components are of significant strength. Figure 9.5 shows these six $F \to F+1$ components as they appear on the cathode ray scope. The splitting caused by ^{14}N ($I=1$) is not resolvable for this J transition.

For transitions where $J > I$, the variation in the hyperfine pattern is rather uniform with increasing J. The $F \to F-1$ components become weaker and

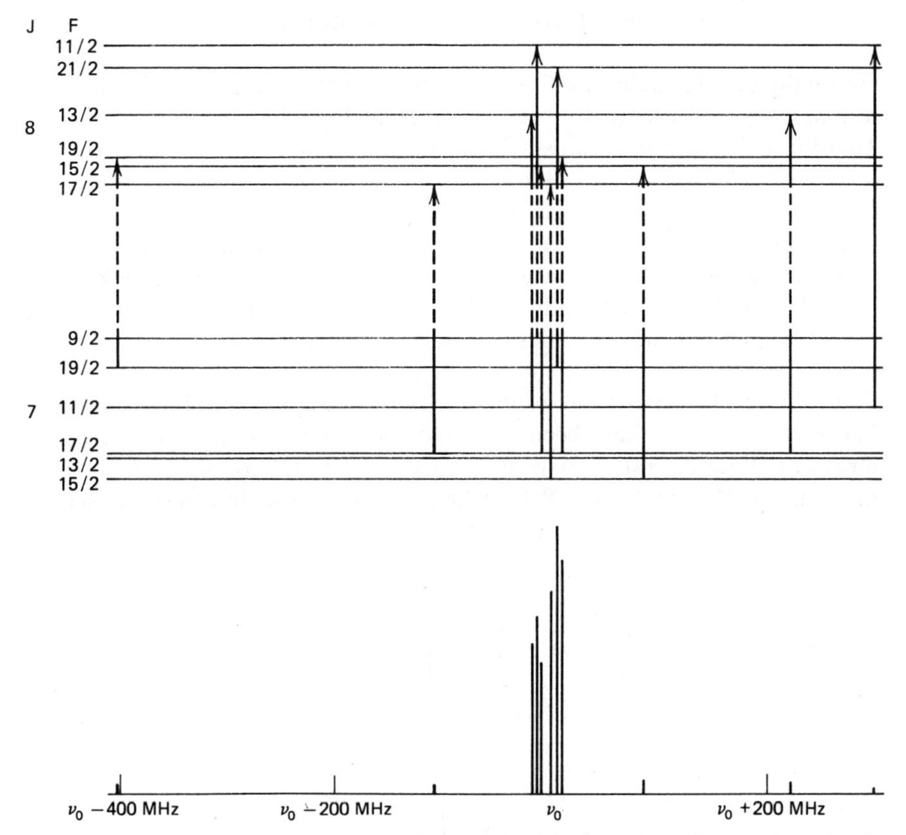

Fig. 9.4 Energy level diagram and calculated nuclear quadrupole hyperfine structure for the $J=7 \rightarrow 8$ rotational transition of a linear molecule with a single coupling nucleus having a spin of $\frac{5}{2}$. All $F \rightarrow F-1$ components are less than one-thousandth the strength of the strongest component and hence are omitted. Here eQq is assumed to be -2400 MHz, as it is for ICN in Fig. 9.5.

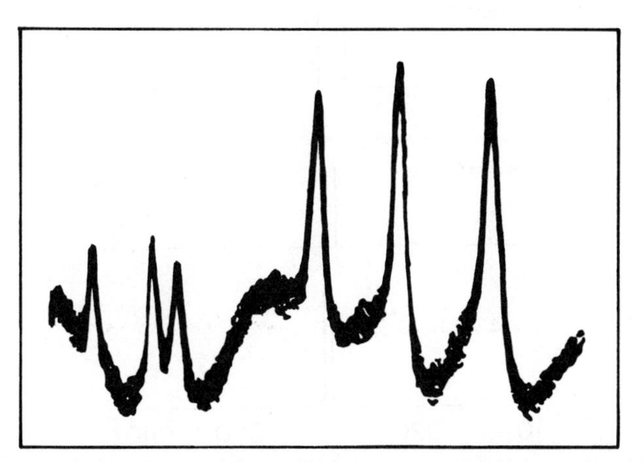

Fig. 9.5 The $F \rightarrow F+1$ components of the $J=7 \rightarrow 8$ transition of ICN observed at 5.87 mm wavelength (compare with the rheoretical components of Fig. 9.4). The observed components are due to ^{127}I (spin $\frac{5}{2}$, $eQq = -2400$ MHz). The substructure caused by ^{14}N (spin 1, $eQq = -3.8$ MHz) is not resolved.

eventually undetectable; the $F \to F+1$ components remain strong but converge in frequency and eventually become unresolvable as J continues to increase. Thus for high J values only a single line is observable except when the coupling is very large. When $J > I$, there are $2I+1$ components for $F \to F+1$, $2I$ components for $F \to F$, and $2I-1$ components for $F \to F-1$. However, for the $J=0 \to 1$ transition, there are only three hyperfine components whatever the value of I.

These variations are illustrated in Fig. 9.6 for a coupling nucleus with $I=\frac{3}{2}$. Although there are four $F \to F+1$ transitions expected for $I=\frac{3}{2}$ when $J > I$, two pairs of the four lines coincide so that only two $F \to F+1$ lines are observable.

Table 9.1 illustrates some of the nuclear quadrupole coupling constants of linear molecules that have been measured from analysis of rotational hyperfine structure. Note the rather large differences in the ^{14}N coupling of the related molecules FCN, ClCN, and BrCN. Also note the large difference between the

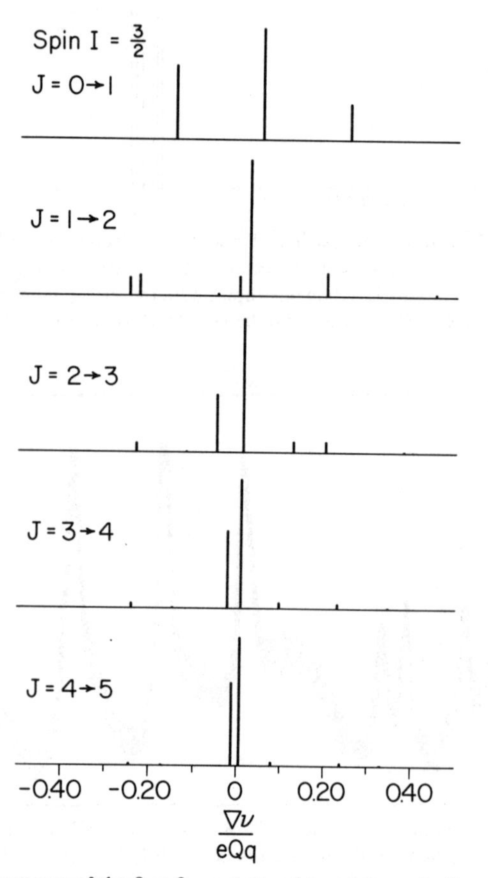

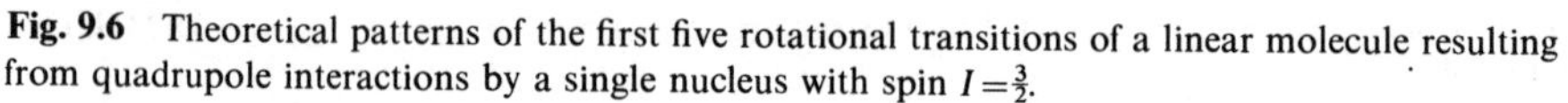

Fig. 9.6 Theoretical patterns of the first five rotational transitions of a linear molecule resulting from quadrupole interactions by a single nucleus with spin $I=\frac{3}{2}$.

Table 9.1 Nuclear Quadrupole Coupling in Some Diatomic and Linear Polyatomic Molecules in the Ground Vibrational State

Molecule	*eQq (MHz)*	*Ref.*	*Molecule*	*eQq (MHz)*	*Ref.*
D coupling			*^{33}S coupling*		
DBr	0.1469(14)	a	CS	12.83(3)	m
DCN	0.1944(25)	b	SiS	10.90(20)	n
DCP	0.233(40)	c	GeS	6.96(30)	n
DCCF	0.212(10)	d	OCS	−29.07(1)	o
^{11}B coupling			*^{35}Cl coupling*		
BF	4.5(4)	e	HCl	−67.80(10)	p
HBS	−3.71(3)	f	ClCn	−83.2752(4)	q
			HCCCl	−79.67	r
^{14}N coupling					
			^{79}Br coupling		
PN	−5.2031(5)	g			
HCN	−4.7091(15)	b	DBr	530.65(7)	p
FCN	−2.67(5)	h	BrCN	685.6(4)	q
HCNO	0.245(5)	i	HCCBr	648.10(2)	s
HCCCN	−4.28(5)	j			
			^{127}I coupling		
^{17}O coupling			HI	−1828.42(20)	p
CO	4.48(10)	k	ICN	−2418.8(5)	t
OCS	−1.32(7)	l	HCCI	−2250.6(55)	u

[a]B. P. Van Eijck, *J. Mol. Spectrosc.*, **53**, 246 (1974).
[b]F. C. De Lucia and W. Gordy, *Phys. Rev.*, **187**, 58 (1969).
[c]S. L. Harford, W. C. Allen, C. L. Allen, C. L. Harris, E. F. Pearson, and W. H. Flygare, *Chem. Phys. Lett.*, **18**, 153 (1973).
[d]V. W. Weiss and W. H. Flygare, *J. Chem. Phys.*, **45**, 8 (1966).
[e]F. J. Lovas and D. R. Johnson, *J. Chem. Phys.*, **55**, 41 (1971).
[f]E. F. Pearson, C. L. Norris, and W. H. Flygare, *J. Chem. Phys.*, **60**, 1761 (1974).
[g]J. Raymonda and W. Klemperer, *J. Chem. Phys.*, **55**, 232 (1971).
[h]J. Sheridan, J. K. Tyler, E. E. Aynsley, R. E. Dodd, and R. Little, *Nature*, **185**, 96 (1960).
[i]W. Hüttner, H. K. Bodenseh, P. Nowicki, and K. Morgenstern, *J. Mol. Spectrosc.*, **71**, 246 (1978).
[j]J. K. Tyler and J. Sheridan, *Trans. Faraday Soc.*, **59**, 2661 (1963).
[k]W. H. Flygare and V. W. Weiss, *J. Chem. Phys.*, **45**, 2785 (1966).
[l]S. Geschwind, R. Gunther-Mohr, and J. Silvey, *Phys. Rev.*, **85**, 474 (1952).
[m]M. Bogey, C. Demuynck, and J. L. Destombes, *Chem. Phys. Lett.*, **81**, 256 (1981).
[n]J. Hoeft, F. J. Lovas, E. Tiemann, and T. Törring, *J. Chem. Phys.*, **53**, 2736 (1970).
[o]C. H. Townes and S. Geschwind, *Phys. Rev.*, **74**, 626 (1948).
[p]F. C. De Lucia, P. Helminger, and W. Gordy, *Phys. Rev.*, **A3**, 1849 (1971).
[q]J. M. L. J. Reinartz, W. L. Meerts, A. Dymanus, *Chem. Phys.*, **45**, 387 (1980).
[r]H. Jones, M. Takami, and J. Sheridan, *Z. Naturforsch.*, **33a**, 156 (1978).
[s]A. P. Porter and P. D. Godfrey, *J. Mol. Spectrosc.*, **68**, 492 (1977).
[t]T. Oka, H. Hirakawa and A. Miyahara, *J. Phys. Soc. Japan*, **12**, 39 (1957).
[u]E. Schafer and J. J. Christiansen, *J. Mol. Struct.*, **97**, 101 (1983).

coupling of the middle nitrogen and that of the end nitrogen in NNO, as well as that between the ^{33}S coupling in CS and in OCS. It is evident that the nuclear quadrupole coupling depends upon the nature of the chemical bonding to the coupling atom. This relationship is treated in Chapter XIV.

Symmetric-top Molecules

The most commonly studied symmetric-top hyperfine structure arises from a single coupling nucleus on the symmetry axis. A single coupling isotope, or two such isotopes which are off the axis, would destroy the symmetry and convert the molecule into an asymmetric rotor. It is possible, however, to have three off-axis coupling isotopes, for example, ^{35}Cl in $^{35}Cl_3CH$, in a symmetric top. At this point we consider the simpler case of a single coupling nucleus that is on the symmetry axis.

The symmetry axis will, of course, be a principal axis of inertia, either a (prolate top) or c (oblate top). To include both cases we designate this axis as z and the other two body-fixed axes as x and y. With (9.78), which also holds for symmetric tops, (9.71) becomes

$$q_J = q\left(J, K, M_J = J\left|\frac{3\Phi_{Zz}^2 - 1}{2}\right| J, K, M_J = J\right) \tag{9.86}$$

From the matrix elements given in Table 2.1

$$\begin{aligned}(J, K, M_J = J|\Phi_{Zz}^2|\, J, K, M_J = J) &= (J, K, J|\Phi_{Zz}|\, J, K, J)^2 \\ &\quad + (J, K, J|\Phi_{Zz}|\, J+1, K, J)^2 + (J, K, J|\Phi_{Zz}|\, J-1, K, J)^2 \\ &= \frac{K^2}{(J+1)^2} + \frac{(J+1)^2 - K^2}{(J+1)^2(2J+3)}\end{aligned} \tag{9.87}$$

Substitution of these values into (9.86) gives

$$q_J = q\,\frac{J}{2J+3}\left[\frac{3K^2}{J(J+1)} - 1\right] \tag{9.88}$$

The resulting formula for the quadrupole energy of the symmetric top is, from (9.63),

$$E_Q = eQq\left[\frac{3K^2}{J(J+1)} - 1\right] Y(J, I, F) \tag{9.89}$$

where $Y(J, I, F)$ is the function of Eq. (9.83), which is numerically tabulated in Appendix I, and

$$eQq = eQ\left\langle\frac{\partial^2 V}{\partial z^2}\right\rangle = \chi \tag{9.90}$$

is the coupling constant with reference to the molecular axis of symmetry. Selection rules are $\Delta J = \pm 1$; $\Delta K = 0$; $\Delta F = 0 \pm 1$; $\Delta I = 0$.

It is evident from examination of (9.89) that each K component of a $J \rightarrow J+1$ rotational transition of a symmetric top will have its own hyperfine multiplet.

For $K=0$ the equation reduces to that for the linear molecule, and the pattern for this component alone will be similar to those illustrated in Figs. 9.4–9.6. The splitting of other K components differs by a scale factor only and hence individually resembles the patterns for the linear molecule. When, as is often true, the separation of the K components by centrifugal stretching is less than the quadrupole splitting, the multiplets for the different K components are scrambled so that they produce a rather complicated spectrum, such as is illustrated in Fig. 9.7. Comparison of these various patterns was used for proof that the nuclear spin of radioactive ^{129}I is $\frac{7}{2}$. The slight disagreement between the observed and calculated patterns for $I=\frac{7}{2}$ are due to second-order effects, described in Section 6, which were not included in the calculations. Since the spin of the stable isotope ^{127}I is $\frac{5}{2}$, the hyperfine structure of the $J=2\rightarrow 3$ transition of normal methyl iodide is like that for the theoretical pattern for $I=\frac{5}{2}$.

Selected values of quadrupole couplings in symmetric-top molecules as obtained from microwave spectroscopy are given in Table 9.2.

Asymmetric Rotors

The general expression for nuclear quadrupole interaction, (9.72), applies to all classes of molecules. By multiplication and division of this equation by $J/(2J+3)$ it can be expressed as

$$E_Q = eQq_J \frac{2J+3}{J} Y(J, I, F) \tag{9.91}$$

In the asymmetric-top representation,

$$q_J = \sum_{g=a,b,c} q_{gg}(J_{K_{-1}K_1} M_J = J|\Phi_{Zg}^2| J_{K_{-1}K_1}, M_J = J) \tag{9.92}$$

The function $Y(J, I, F)$ is expressed by (9.83) and is tabulated in Appendix I. Except for low J values, the matrix elements of the squared direction cosines in (9.92) cannot be expressed in closed form; hence q_J cannot be expressed in closed form. Because the line intensities depend on the squared matrix elements, the problem of calculation of q_J can, however, be reduced to that already encountered in the calculation of line intensities. For example, from the matrix product rule

$$(J_{K_{-1}K_1}, J)\Phi_{Za}^2| J_{K_{-1}K_1}, J) = \sum_{J'_{K'_{-1}K'_1}} |(J_{K_{-1}K_1}, M_J = J|\Phi_{Za}| J'_{K'_{-1}K'_1}, M_J = J)|^2 \tag{9.93}$$

If the dipole moment is along the principal axis a, the line strength of the transitions $(J_{K_{-1}K_1}, M_J=J)\rightarrow(J'_{K'_{-1}K'_1}, M_J=J)$ is proportional to the squared matrix elements of this expression. By thus relating the diagonal matrix elements of

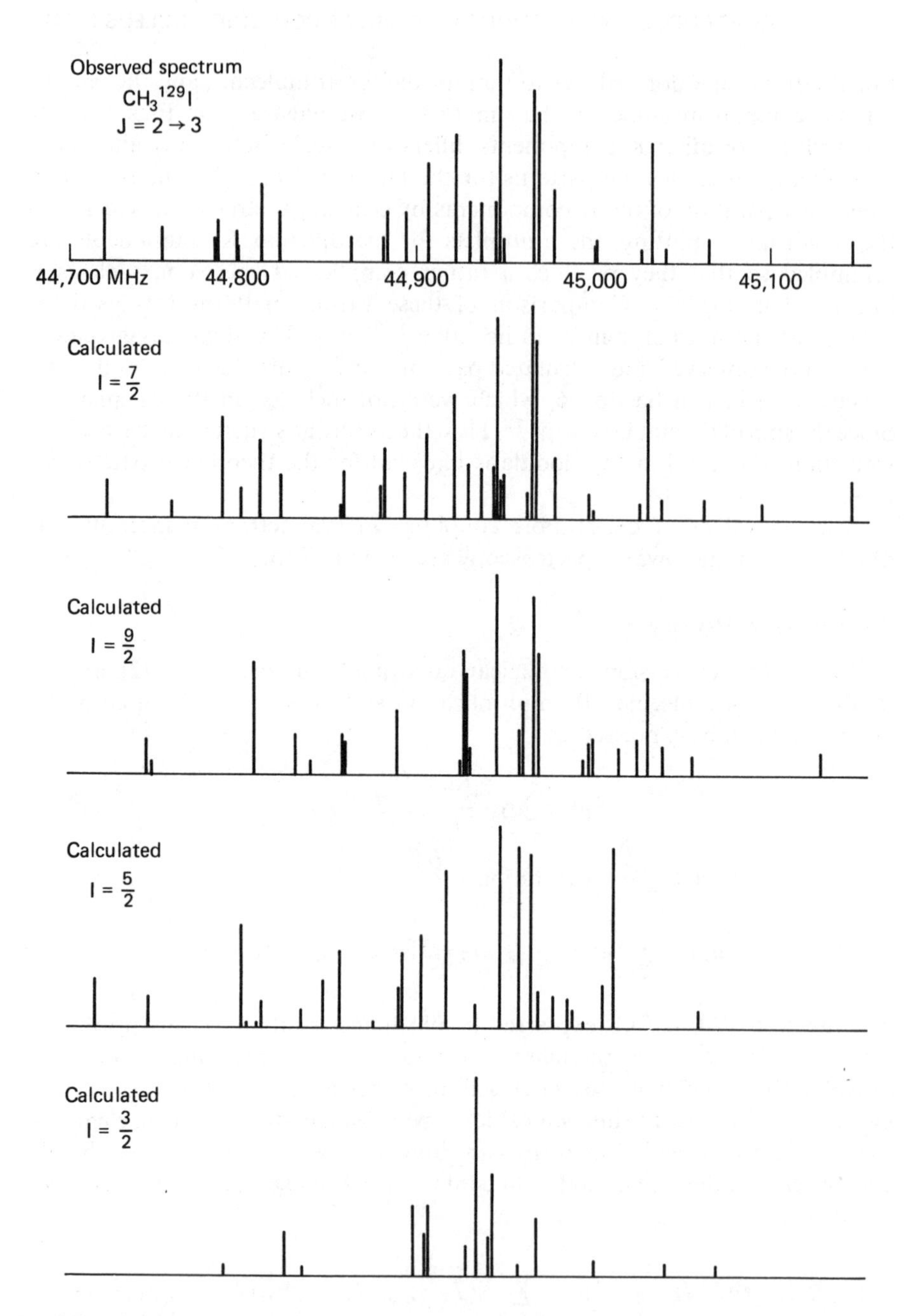

Fig. 9.7 Observed nuclear quadrupole hyperfine pattern for the $J=2\rightarrow3$ rotational transition of $CH_3\,{}^{129}I$ (top figure) compared with the patterns predicted with first-order theory for various nuclear spin values assumed for the iodine (lower figures). The close similarity of the two top patterns proved that the spin of ${}^{129}I$ is $\frac{7}{2}$. From R. Livingston, O. R. Gilliam, and W. Gordy, *Phys. Rev.*, **76**, 149 (1949).

Table 9.2 Nuclear Quadrupole Coupling in Some Symmetric-top Molecules in the Ground Vibrational State

Molecule	*eQq (MHz)*	*Ref.*	*Molecule*	*eQq (MHz)*	*Ref.*
D coupling			*^{35}Cl coupling*		
CH_3D	0.192	a	NCl_3	−108(3)	l
CF_3D	0.171(2)	b	CH_3Cl	−74.77(10)	m
CD_3CN	0.168(4)	c	SiH_3Cl	−40.0	n
			GeH_3Cl	−46.95(2)	o
^{11}B coupling			SnH_3Cl	−41.6(3)	p
BH_3CO	1.60(10)	d			
CH_3BS	−3.714(20)	e	*^{79}Br coupling*		
^{14}N coupling			CH_3Br	577.15(10)	m
NH_3	−4.0842(3)	f	SiH_3Br	334.970(6)	q
NF_3	−7.07(10)	g	GeH_3Br	384(2)	r
CH_3CN	−4.2244(15)	c	SnH_3Br	350(6)	s
CH_3NC	−0.4885(10)	h			
SiH_3CN	−4.77(3)	i	*^{127}I coupling*		
			CH_3I	−1940.41(7)	t
^{73}Ge coupling			SiH_3I	−1240(30)	u
GeH_3F	−93.0(1)	j	GeH_3I	−1381(4)	r
GeH_3CCH	+32.5	k	SnH_3I	−1273(8)	v

[a]S. C. Wofsy, J. S. Muenter, and W. Klemperer, *J. Chem. Phys.*, **53**, 4005 (1970).
[b]S. G. Kukolich, A. C. Nelson, and D. J. Ruben, *J. Mol. Spectrosc.*, **40**, 33 (1971).
[c]S. G. Kukolich, D. J. Ruben, J. H. S. Wang, and J. R. Williams, *J. Chem. Phys.*, **58**, 3155 (1973).
[d]A. C. Venkatachar, R. C. Taylor, and R. L. Kuczkowski, *J. Mol. Struct.*, **38**, 17 (1977).
[e]C. Kirby and H. W. Kroto, *J. Mol. Spectrosc.*, **83**, 1 (1980).
[f]Gunther-Mohr et al. [41].
[g]J. Sheridan and W. Gordy, *Phys. Rev.*, **79**, 513 (1950).
[h]S. G. Kukolich, *Chem. Phys. Lett.*, **10**, 52 (1971).
[i]A. J. Careless and H. W. Kroto, *J. Mol. Spectrosc.*, **57**, 198 (1975).
[j]L. C. Krisher, J. A. Morrison, and W. A. Watson, *J. Chem. Phys.*, **57**, 1357 (1972).
[k]E. C. Thomas and V. C. Laurie, *J. Chem. Phys.*, **44**, 2602 (1966).
[l]G. Cazzoli, P. G. Favero, and A. Dal Borgo, *J. Mol. Spectrosc.*, **50**, 82 (1974).
[m]J. Kraitchman and B. P. Dailey, *J. Chem. Phys.*, **22**, 1477 (1954).
[n]A. H. Sharbaugh, *Phys. Rev.*, **74**, 1870 (1948).
[o]S. Geshwind, R. Gunther-Mohr, and C. H. Townes, *Phys. Rev.*, **81**, 288 (1951).
[p]L. C. Krisher, R. A. Gsell, and J. M. Bellama, *J. Chem. Phys.*, **54**, 2287 (1971).
[q]K. F. Dössel and D. H. Sutter, *Z. Naturforsch.*, **32a**, 1444 (1977).
[r]S. N. Wolf and L. C. Krisher, *J. Chem. Phys.*, **56**, 1040 (1971).
[s]S. N. Wolf, L. C. Krisher, and R. A. Gsell, *J. Chem. Phys.*, **54**, 4605 (1971).
[t]A. Dubrulle, J. Burie, D. Boucher, F. Herlemont, and J. Demaison, *J. Mol. Spectrosc.*, **88**, 394 (1981).
[u]A. H. Sharbaugh, G. A. Heath, L. F. Thomas, and J. Sheridan, *Nature*, **171**, 87 (1953).
[v]S. N. Wolf, L. C. Krisher, and R. A. Gsell, *J. Chem. Phys.*, **55**, 2106 (1971).

Φ_{Zg}^2 to those of the transition dipole moments, Bragg [9] has derived the expression

$$q_J = \frac{2J}{(2J+1)(2J+3)} \sum_{K'_{-1}K'_1} [q_{aa}\lambda_a(J_{K_{-1}K_1}; J_{K'_{-1}K'_1}) + q_{bb}\lambda_b(J_{K_{-1}K_1}; J_{K'_{-1}K'_1}) + q_{cc}\lambda_c(J_{K_{-1}K_1}; J_{K'_{-1}K'_1})] \tag{9.94}$$

in which the λ's are the line strength factors for the transitions $J_{K_{-1}K_1} \rightarrow J_{K'_{-1}K'_1}$ tabulated by Cross et al. [17]. Although these tables include only values for line strengths of molecules having an asymmetry parameter κ of ∓ 1, ∓ 0.5, 0, ± 0.5, and ± 1, these can be interpolated to give reasonably approximate values of q_J for other degrees of asymmetry. More extensive tabulations are referenced in Chapter VII.

A second, and perhaps more useful, expression of q_J in terms of the reduced energies and Ray's asymmetry parameter κ has been derived by Bragg and Golden [10]. This derivation is based upon the relationship between the matrix elements of the direction cosines and the angular momentum expressed by (9.74), which allows q_J to be expressed by

$$q_J = \frac{2}{(J+1)(2J+3)} \sum_{g=a,b,c} q_{gg} \langle J_g^2 \rangle \tag{9.95}$$

where the asymmetric-top representation is used for the average values

$$\langle J_g^2 \rangle = (J, K_{-1}, K_1 | J_g^2 | J, K_{-1}, K_1) \tag{9.96}$$

As discussed in Chapter VII,

$$\langle J_a^2 \rangle = \frac{1}{2}\left[J(J+1) + E(\kappa) - (\kappa+1)\frac{\partial E(\kappa)}{\partial \kappa} \right] \tag{9.97}$$

$$\langle J_b^2 \rangle = \frac{\partial E(\kappa)}{\partial \kappa} \tag{9.98}$$

$$\langle J_c^2 \rangle = \frac{1}{2}\left[J(J+1) - E(\kappa) + (\kappa-1)\frac{\partial E(\kappa)}{\partial \kappa} \right] \tag{9.99}$$

which lead to an expression for the quadrupole energy in the form

$$E_Q = \frac{1}{J(J+1)} \left\{ \chi_{aa}\left[J(J+1) + E(\kappa) - (\kappa+1)\frac{\partial E(\kappa)}{\partial \kappa} \right] + 2\chi_{bb}\frac{\partial E(\kappa)}{\partial \kappa} + \chi_{cc}\left[J(J+1) - E(\kappa) + (\kappa-1)\frac{\partial E(\kappa)}{\partial \kappa} \right] \right\} Y(J, I, F) \tag{9.100}$$

in which $E(\kappa)$ is the reduced energy of the asymmetric rotor described in Chapter VII and Ray's parameter is

$$\kappa = \frac{2B - A - C}{A - C} \tag{9.101}$$

where A, B, and C are the rotational constants with regard to the respective a, b, and c axes. The reduced energies for various values of κ have been tabulated (see Chapter VII). For low J values, for which nuclear quadrupole coupling is best resolved, explicit expressions for $E(\kappa)$ have been obtained. Selection rules $\Delta F=0, \pm 1$ and $\Delta I=0$ apply with those given for $J_{K_{-1}K_1}$ in Chapter VII.

We illustrate the application of (9.100) with a simple example, the $1_{01}\rightarrow 1_{10}$ transition of $H_2{}^{32}S$. For the 1_{01} level:

$$E(\kappa)=\kappa-1, \qquad \frac{\partial E(\kappa)}{\partial \kappa}=1 \tag{9.102}$$

and

$$E_Q(1_{01})=(\chi_{bb}+\chi_{cc})Y(F)=-\chi_{aa}Y(F) \tag{9.103}$$

For the 1_{10} level:

$$E(\kappa)=\kappa+1, \qquad \frac{\partial E(\kappa)}{\partial \kappa}=1 \tag{9.104}$$

and

$$E_Q(1_{10})=(\chi_{aa}+\chi_{bb})Y(F)=-\chi_{cc}Y(F) \tag{9.105}$$

For both levels:

$$J=1, \qquad I=\tfrac{3}{2}, \qquad F=\tfrac{5}{2}, \tfrac{3}{2}, \tfrac{1}{2}$$

$$Y(1, \tfrac{3}{2}, \tfrac{5}{2})=\tfrac{1}{20}, \qquad Y(1, \tfrac{3}{2}, \tfrac{3}{2})=-\tfrac{1}{5}, \qquad Y(1, \tfrac{3}{2}, \tfrac{1}{2})=\tfrac{1}{4}$$

The resulting energy levels, allowed transitions, and predicted lines are indicated in Fig. 9.8. Comparison of theoretical with observed frequencies [18] leads to the values

$$\chi_{aa}=-32\text{ MHz}, \qquad \chi_{bb}=-8\text{ MHz}, \qquad \chi_{cc}=40\text{ MHz},$$
$$\nu_0=168{,}322.63\text{ MHz}$$

and hence the asymmetry parameter of the quadrupole coupling is

$$\eta=\frac{\chi_{aa}-\chi_{bb}}{\chi_{cc}}=-0.60$$

For slightly asymmetric rotors, (9.100) can be expressed in a more convenient form by the use of expansions for $W(b)$ given in Chapter VII. The resulting expression for the nearly prolate or oblate symmetric top is

$$E_Q=\frac{\chi}{J(J+1)}[3K^2-J(J+1)-3b^2(C_2+2C_3b)$$
$$+\eta(C_1+2C_2b+3C_3b^2+4C_4b^3)]Y(J, I, F) \tag{9.106}$$

where the b's are the inertial parameters defined in Chapter VII, Section 3 and where the C's are parameters given in Table 7.8. For a prolate rotor $b=b_p$,

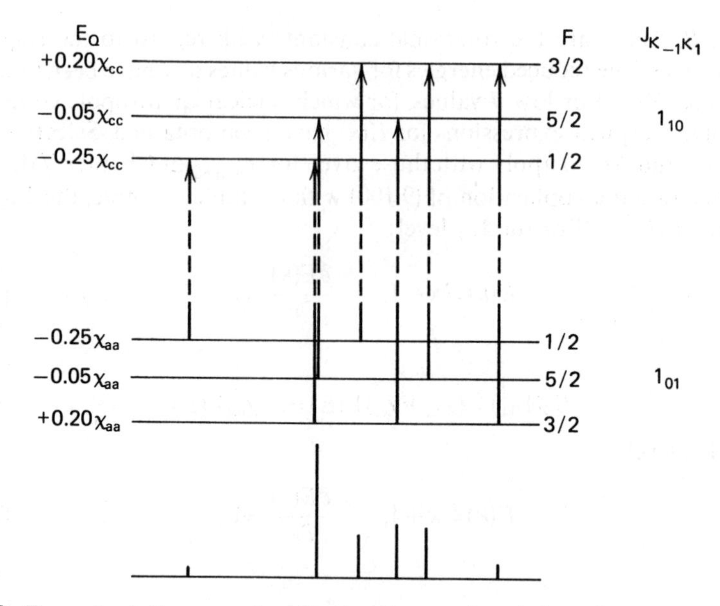

Fig. 9.8 Energy level diagram and predicted nuclear quadrupole hyperfine structure caused by ^{33}S ($I=\frac{3}{2}$) in the $1_{01}\rightarrow 1_{10}$ rotational transition of $H_2{}^{33}S$.

$\chi=\chi_{aa}$ and $\eta=(\chi_{cc}-\chi_{bb})/\chi_{aa}$, whereas for an oblate rotor $b=b_o$, $\chi=\chi_{cc}$, and $\eta=(\chi_{aa}-\chi_{bb})/\chi_{cc}$.

Although the diagonal elements χ_{aa}, χ_{bb}, and χ_{cc} of the coupling with reference to the principal inertial axes are the coupling constants directly observable from the rotational hyperfine structure, they are not necessarily the most convenient for interpretation of the quadrupole coupling in terms of the properties of chemical bonds. While the cross terms $\chi_{ab}, \ldots$, are averaged out by the relation [see (9.70)], the principal inertial axes are not generally principal axes of the nuclear coupling in the tensor, even though they may be in special cases, such as linear molecules. Let us therefore transform χ_{aa}, χ_{bb}, and χ_{cc} to the principal axes of the quadrupole coupling tensor which we shall designate by x, y, z. Since eQ is constant, we need only to transform q.

$$q_{aa}=\left(\frac{\partial^2 V}{\partial a^2}\right)=\left(\frac{\partial^2 V}{\partial x^2}\right)\left(\frac{\partial x}{\partial a}\right)^2+\left(\frac{\partial^2 V}{\partial y^2}\right)\left(\frac{\partial y}{\partial a}\right)^2+\left(\frac{\partial^2 V}{\partial z^2}\right)\left(\frac{\partial z}{\partial a}\right)^2 \tag{9.107}$$

with similar expressions for q_{bb} and q_{cc}. Cross terms such as $(\partial^2 V/\partial x\,\partial y)$ are zero because x, y, z are principal axes of the field gradient.

Now

$$\frac{\partial x}{\partial a}=\cos\theta_{xa}, \ldots$$

$$\frac{\partial^2 V}{\partial x^2}=q_{xx}, \ldots$$

and

$$\chi_x = eQq_{xx}, \ldots \tag{9.108}$$

Therefore

$$\chi_{aa} = \chi_x \cos^2 \theta_{xa} + \chi_y \cos^2 \theta_{ya} + \chi_z \cos^2 \theta_{za} \tag{9.109}$$

$$\chi_{bb} = \chi_x \cos^2 \theta_{xb} + \chi_y \cos^2 \theta_{yb} + \chi_z \cos^2 \theta_{zb} \tag{9.110}$$

$$\chi_{cc} = \chi_x \cos^2 \theta_{xc} + \chi_y \cos^2 \theta_{yc} + \chi_z \cos^2 \theta_{zc} \tag{9.111}$$

The difficulty in the solution for the principal elements is apparent from these equations. In the general case where the angles as well as the magnitudes χ_x, χ_y, and χ_z are unknown, there are too few equations linking these unknowns with observable quantities. To diagonalize the tensor in the usual way we must also know the off-diagonal elements χ_{ab}, χ_{ac}, and so on. However, these off-diagonal elements are not directly observable in the first-order quadrupole hyperfine structure usually measured. For nuclei with large quadrupole coupling it is sometimes possible to obtain very approximate values of off-diagonal elements from measurement of the second-order displacements in the hyperfine component (Section 6), but only diagonal elements in the inertial system are usually measured. Devious methods for use of these diagonal elements to give information about the principal elements are described later.

In some molecules the directions of one or more of the principal axes in the coupling can be ascertained from a consideration of the symmetry of the electronic structure around the coupling nucleus. Then, with the molecular structure and the inertial axes *a*, *b*, and *c* known from analysis of the rotational spectra, the angles θ can be found; hence the coupling constants χ_x, χ_y, and χ_z can be obtained from the measured quantities χ_{aa}, χ_{bb}, and χ_{cc} by solution of (9.109)–(9.111). In some cases the coupling has axial symmetry about a bond axis or other axis. If z is chosen as such an axis of symmetry, $\chi_x = \chi_y = -\frac{1}{2}\chi_z$. There is then only one independent coupling constant, and (9.109)–(9.111) can be put in the form

$$\chi_{aa} = \tfrac{1}{2}\chi_z(3 \cos^2 \theta_{za} - 1) \tag{9.112}$$

$$\chi_{bb} = \tfrac{1}{2}\chi_z(3 \cos^2 \theta_{zb} - 1) \tag{9.113}$$

$$\chi_{cc} = \tfrac{1}{2}\chi_z(3 \cos^2 \theta_{zc} - 1) \tag{9.114}$$

where χ_z represents the coupling constant with reference to the symmetry axis.

In a molecule having a plane of symmetry, the axis perpendicular to the plane will be a principal axis of inertia and also a principal axis of the field gradient or coupling. We assume that this is the *c* axis of the inertial moment and that it coincides with the principal axis *y* of the field gradient. Then (9.109)–(9.111) reduce to the form

$$\chi_{aa} = \chi_x \sin^2 \theta_{za} + \chi_z \cos^2 \theta_{za} \tag{9.115}$$

$$\chi_{bb} = \chi_x \cos^2 \theta_{za} + \chi_z \sin^2 \theta_{za} \tag{9.116}$$

Table 9.3 Nuclear Quadrupole Coupling in Some Asymmetric-top Molecules in the Ground Vibrational State

	Coupling Constants (MHz)			
	Inertial Axes	*Bond Axis System*		
Molecule	χ_{aa} χ_{bb} χ_{cc}	χ_{xx} χ_{yy} χ_{zz}	*Orientation of x, y, z Axes*	*Ref.*
		D Coupling		
HOD	0.2765(9) −0.1110(15) −0.1658(11)	−0.1658 −0.1477 0.3132	z along OD bond	a
CH_3OD		χ_{zz}=0.303(12)	z along OD bond	b
		^{14}N *Coupling*		
$S(CN)_2$	−1.51 0.30 1.21	2.27 1.21 −3.48		c
$CH_3C(=O)CN$	−4.34(7) 2.03(31) 2.31(31)	2.06 2.31 −4.37		d
Pyridine	−4.88(4) 1.43(3) 3.45(2)	1.43(3) 3.45(2) −4.88(4)	See Fig. 14.4 for coordinate directions	e,f
		^{17}O *Coupling*		
H_2CO	$\chi_{aa}=-1.90(12)$ $\eta=-12.0(8)$			g
H_2C—CH_2 (with O bridging, epoxide)		12.6(1) −5.2(1) −7.4(1)	$x\perp$ to plane $z\parallel$ to CC	h
		^{35}Cl *Coupling*		
CH_2Cl_2	−41.07(1) 1.24(2) 39.83(1)	−76.92(2) 37.09(2) 39.83(1)	x along C–Cl bond y in Cl–C–Cl plane $\perp$ to x	i
CH_2FCl		31.6(10) 38.8(10) −70.5(10)	z along C–Cl bond $y\perp$ to molecular plane	j
CH_3CH_2Cl	$\chi_{aa}=-49.20(10)$ $\eta=0.4479(34)$	$\chi_{zz}=-68.80(15)$ $\eta=0.035(3)$	z along C–Cl bond $\theta_{az}=26°$	k
$CH_3C(=O)$—Cl	−58.0(10) 36.4 21.6	−59.2 37.6 21.6	x along C–Cl bond $z\perp$ to C–O–Cl plane $\theta_{ax}=6°28'$	l

Table 9.3 Continued

		^{35}Cl *Coupling* (continued)		
$CH_2{=}CCl_2$		$\chi_{zz}=-78.7(10)$ $\eta=0.12(1)$	z along C–Cl bond $x \perp$ to molecular plane	[m]
CH_2——CHCl \\ / CH_2	$-56.64(40)$ $36.72(22)$ $19.92(22)$	 $\chi_{zz}=-71.40(51)$ $\eta=0.029(11)$	 z along C–Cl bond $\theta_{az}=21.9°$	[n]
		^{79}Br *Coupling*		
Br_2CO	$316.9(21)$ $-83.5(17)$ $-233.4(11)$	-229.4 -397.1 626.5		[o]
CH_2—CHBr	472.1 -220.7 -251.4	-251.4 -307.5 558.9	$\theta_{az}=18.45°$	[p]
C_6H_5Br	$\chi_{aa}=567(4)$	$\chi_{zz}=567(4)$ $\eta=-0.049(20)$	z along C–Br bond	[q]
		^{127}I *Coupling*		
CH_3CH_2I	$-1478.11(7)$ $564.46(8)$ $913.65(7)$ $\chi_{ab}=896.4(5)$	$893(10)$ $878(50)$ $-1771(10)$	z along C–I bond	[r,s]
$CH_2{=}CHI$	$-1656(10)$ $770(10)$ $-654(10)$ $\chi_{ab}=-765$	886 991 -1877	z along C–I bond $x \perp$ to molecular plane	[t,u]

[a]H. A. Fry and S. G. Kukolich, *J. Chem. Phys.*, **76**, 4387 (1982).
[b]K. H. Casleton and S. G. Kukolich, *Chem. Phys. Lett.*, **22**, 331 (1973).
[c]L. Pierce, R. Nelson, and C. Thomas, *J. Chem. Phys.*, **43**, 3423 (1965).
[d]F. Scappini and H. Dreizler, *Z. Naturforsch.*, **31a**, 840 (1976).
[e]G. O. Sorensen, *J. Mol. Spectrosc.*, **22**, 325 (1967).
[f]G. O. Sorensen, L. Mahler, N. Rastrup-Andersen, *J. Mol. Struct.*, **20**, 119 (1974).
[g]R. Cornet, B. M. Landsberg, and G. Winnewisser, *J. Mol. Spectrosc.*, **82**, 253 (1980).
[h]R. A. Creswell and R. H. Schwendeman, *Chem. Phys. Lett.*, **27**, 521 (1974).
[i]W. H. Flygare and W. D. Gwinn, *J. Chem. Phys.*, **36**, 787 (1962).
[j]N. Muller, *J. Am. Chem. Soc.*, **75**, 860 (1953).
[k]R. H. Schwendeman and G. D. Jacobs, *J. Chem. Phys.*, **36**, 1245 (1962).
[l]K. M. Sinnott, *J. Chem. Phys.*, **34**, 851 (1961).
[m]S. Sekino and T. Nishikawa, *J. Phys. Soc. Japan*, **12**, 43 (1957).
[n]R. H. Schwendeman, G. D. Jacobs, and T. M. Krigas, *J. Chem. Phys.*, **40**, 1022 (1964).
[o]J. H. Carpenter, J. G. Smith, I. Thompson, and D. H. Whiffen, *Chem. Soc. Faraday Trans.*, 2, **73**, 384 (1977).
[p]J. Maroor and M. de Hemptinne, *Bull. Acad. R. Belg.*, **58**, 956 (1972).
[q]E. Rosenthal and B. A. Dailey, *J. Chem. Phys.*, **43**, 2093 (1965).
[r]T. Kasuya and T. Oka, *J. Phys. Soc. Japan*, **15**, 296 (1960).
[s]D. Boucher, A. Dubrulle, and J. Damaison, *J. Mol. Spectrosc.*, **84**, 375 (1980).
[t]H. W. Morgan and J. H. Goldstein, *J. Chem. Phys.*, **22**, 1427 (1954).
[u]C. D. Cornwell and R. L. Poynter, *J. Chem. Phys.*, **22**, 1257 (1954).

$$\chi_{cc}=\chi_y \tag{9.117}$$

where θ_{za} is the angle between the a and z axes. From the first two of these equations,

$$\chi_x=\frac{\chi_{aa}\sin^2\theta_{za}-\chi_{bb}\cos^2\theta_{za}}{\sin^2\theta_{za}-\cos^2\theta_{za}} \tag{9.118}$$

$$\chi_z=\frac{\chi_{aa}\cos^2\theta_{za}-\chi_{bb}\sin^2\theta_{za}}{\cos^2\theta_{za}-\sin^2\theta_{za}} \tag{9.119}$$

When the direction of one principal axis of the field gradient is known, even if the molecule does not have a symmetry plane, the coupling equations can be reduced by rotation of the coordinates a, b, and c to forms which have only one unknown angle. Nevertheless, there will remain four unknown quantities and only three independent equations.

An important way of providing the additional equations needed in solutions for the principal values of the quadrupole coupling tensor is isotopic substitution, which shifts the axes a, b, and c to new ones a', b', and c' that produce different values of the measurable couplings $\chi_{a'a'}$, $\chi_{b'b'}$, and $\chi_{c'c'}$. If the isotopic substitution is made in some nucleus other than the coupling one, there will be no change in the principal values χ_x, χ_y, and χ_z. If the principal quadrupole axes system is transformed (by a rotation of coordinates) to either the parent or isotopic principal inertial axis system, the coupling tensor will have off-diagonal elements. If the coupling nucleus is substituted, no new unknown couplings will be introduced when the ratios ρ of the quadrupole moments of the substituted isotopes are known.

As an illustration, consider an isotopic substitution of an atom which is in the plane of symmetry, where (9.115)–(9.117) apply. This substitution will change neither the coupling axes nor the inertial axis c which is perpendicular to the symmetry plane, but will alter the direction of the inertial axes a and b relative to the coupling axes x and z and hence $\chi_{a'a'}$, $\chi_{b'b'}$, and $\theta_{za'}$. We indicate the new values by primes and set $\chi'_x=\rho\chi_x$, $\chi'_z=\rho\chi_z$ when the substitution is in the coupling nucleus. Thus

$$\chi_{a'a'}=\rho\chi_x\sin^2\theta_{za'}+\rho\chi_z\cos^2\theta_{za'} \tag{9.120}$$

$$\chi_{b'b'}=\rho\chi_x\cos^2\theta_{za'}+\rho\chi_z\sin^2\theta_{za'} \tag{9.121}$$

Since $\chi_{a'a'}$ and $\chi_{b'b'}$ are measurable quantities and since the quadrupole ratio ρ is presumed to be known, only one new unknown quantity, $\theta_{za'}$, is introduced, whereas two new equations are provided. Thus the four equations (9.115), (9.116), (9.120), and (9.121) have only four unknowns and are therefore solvable. Also, $\Delta\theta=\theta_{za'}-\theta_{za}$ can be obtained from the change in the principal moments of inertia ΔI_a and ΔI_b which was caused by the isotopic substitution.

Diagonal elements of nuclear quadrupole coupling constants with reference to the inertial axes a, b, and c have been measured for numerous asymmetric-top molecules. Selected ones are given in Table 9.3. A complete tabulation of measurements of these elements up to 1982 may be found in the Landolt-Börnstein tables [19]. For certain of these, principal elements of the coupling tensors have also been found.

5 QUADRUPOLE COUPLING BY MORE THAN ONE NUCLEUS IN A ROTATING MOLECULE

If there are two or more nuclei with quadrupole coupling in a molecule, its rotational hyperfine structure is generally quite complicated. The reason for this complexity is easy to see. The coupling of each nucleus perturbs the rotational axes and alters the averaged field gradient which interacts with all the other coupling nuclei. One cannot use the wave function Ψ_r of the rigid rotor to average the quantity $(\partial^2 V/\partial Z^2)$, as was done for a single coupling nucleus, but must use the functions for the perturbed state to find the correct average of the field gradient along Z.

Since the complexity in the quadrupole couplings by plural nuclei arises through their perturbations of Ψ_r, it is obvious that no such complexity arises for solid state resonances. In frozen molecules the quadrupole interaction of one nucleus is virtually independent of that of all others. Thus when there are two or more coupling nuclei in the same molecule, the coupling constants χ are most simply obtained through measurement of pure quadrupole resonance in the solid state. However, the coupling constants of gaseous molecules and of frozen molecules are not exactly the same, and it is often an advantage to have measurements of both.

When the coupling of one nucleus is large as compared with that of a second, the problem of a dual nuclear coupling in a rotating molecule is not very formidable, as we shall explain later. Also, when an experimentalist is better at chemistry than at mathematics, he or she may often circumvent the complexity in the gaseous state problem by substitution of noncoupling isotopes for all except one of the nuclei in a given molecule. Fortunately, nature has already done this in many isotopic species of common molecules.

Coupling by Two Nuclei

The theory of quadrupole interaction by two nuclei in a rotating molecule has been worked out by Bardeen and Townes [20], who give formulas for analysis of the hyperfine structure for certain spin combinations.

The simplest case occurs when the coupling by one nucleus is large as compared with that of the other. When $\chi_1 \gtrsim 10\chi_2$, the spin $\mathbf{I}_1$ will in the first-order approximation couple to $\mathbf{J}$ to form a resultant $\mathbf{F}_1$ about which they both precess. The spin of the second nucleus $\mathbf{I}_2$ will then couple with $\mathbf{F}_1$ to form a resultant $\mathbf{F}$ about which they both precess. This coupling case is indicated by the vector diagram of Fig. 9.9. For this vector model to hold, the precession of $\mathbf{J}$ about $\mathbf{F}_1$ must be sufficiently rapid as compared with the precession of $\mathbf{I}_2$ that the components of the gradient at nucleus 2 which are normal to $\mathbf{F}_1$ average out, leaving an effective component only along $\mathbf{F}_1$. In this model

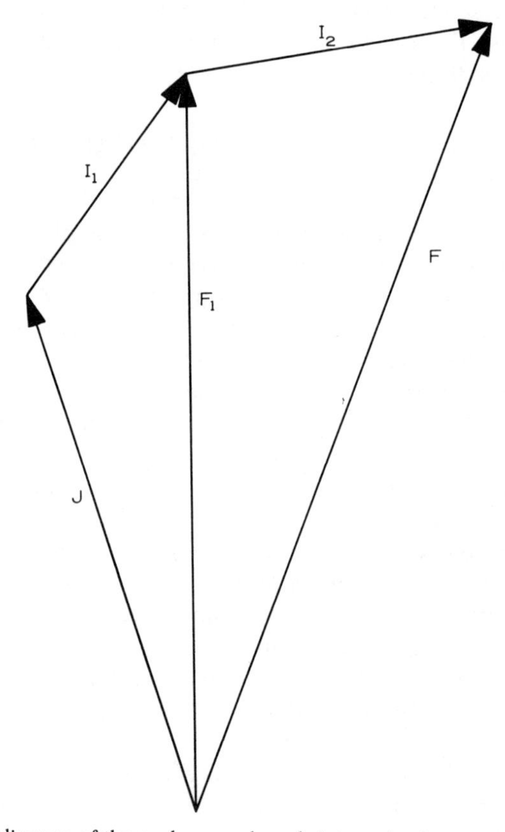

Fig. 9.9 Vector diagram of the nuclear quadrupole interaction by two nuclei in a molecule with the coupling of $\mathbf{I}_1$ assumed to be stronger than that of $\mathbf{I}_2$.

$\mathbf{F}$, $\mathbf{F}_1$, $\mathbf{J}$, $\mathbf{I}_1$, and $\mathbf{I}_2$ are constants of the motion. The associated quantum numbers are

$$F_1 = J + I_1,\ J + I_1 - 1,\ J + I_1 - 2, \ldots, |J - I_1|$$

and

$$F = F_1 + I_2,\ F_1 + I_2 - 1,\ F_1 + I_1 - 2, \ldots, |F_1 - I_2| \tag{9.122}$$

Suppose that $\mathscr{H}_{Q_1}$ represents the Hamiltonian operator for the interaction energy of $\mathbf{I}_1$ and $\mathbf{J}$, and that $\mathscr{H}_{Q_2}$ represents the operator for the interaction of $\mathbf{I}_2$ and $\mathbf{J}$. The coupling scheme $\mathbf{F}_1 = \mathbf{J} + \mathbf{I}_1$ and $\mathbf{F} = \mathbf{F}_1 + \mathbf{I}_2$ generates a set of functions $\Psi_{F_1,F}$ for the combined system $\mathscr{H}_{Q_1} + \mathscr{H}_{Q_2}$. These are eigenfunctions of $\mathscr{H}_{Q_1}$; if the effect of $\mathscr{H}_{Q_2}$ is negligible, states specified by a given F_1 and different values of F will be degenerate. This degeneracy will, however, be removed if $\mathscr{H}_{Q_2}$ is not negligible. Likewise, the coupling scheme $\mathbf{F}_2 = \mathbf{J} + \mathbf{I}_2$ and $\mathbf{F} = \mathbf{F}_2 + \mathbf{I}_1$ generates a set of functions $\Psi_{F_2,F}$ for the combined system which are eigenfunctions of $\mathscr{H}_{Q_2}$. These two sets of basis functions are related

$$\Psi_{F_2,F}=\sum_{F_1} C(F_2, F_1)\Psi_{F_1,F} \qquad \text{or} \qquad \Psi_{F_1,F}=\sum_{F_2} C(F_2, F_1)\Psi_{F_2,F} \tag{9.123}$$

where the transformation coefficients are chosen to be real. With the assumed coupling scheme of Fig. 9.9, the eigenvalues of $\mathscr{H}_{Q_1}$ are $(F_1|\mathscr{H}_{Q_1}|F_1)=E_{Q_1}(F_1)$, the same as those already obtained in Section 4 for the coupling by a single nucleus. The allowed values of $\mathscr{H}_{Q_2}$ treated as a perturbation on $\mathscr{H}_{Q_1}$ are

$$(F_1|\mathscr{H}_{Q_2}|F_1)=\sum [C(F_2, F_1)]^2(F_2|\mathscr{H}_{Q_2}|F_2)=\sum [C(F_2, F_1)]^2 E_{Q_2}(F_2) \tag{9.124}$$

The combined quadrupole coupling energy values to first-order are

$$E_Q=E_{Q_1}(F_1)+\sum_{F_2} [C(F_2, F_1)]^2 E_{Q_2}(F_2) \tag{9.125}$$

The values of $E_{Q_1}(F_1)$ and of $E_{Q_2}(F_2)$ are obtained by substitution of F_1 and F_2, respectively, for F in the formulas for a single nucleus (Section 4). The selection rules for F_1 are $\Delta F_1=0, \pm 1$ and the selection rules for F are $\Delta F=0, \pm 1$. The weighting coefficients $C(F_2, F_1)$ for a few spin combinations ($I_1=I$, $I_2=1$ and $\frac{3}{2}$) are given by Bardeen and Townes [20] and reproduced by Townes and Schawlow [21].

Transformation coefficients relating the two schemes of coupling three angular momenta find extensive use in quantum mechanical problems. In terms of the so-called $6j$ symbol, which have been tabulated for spin values up to 8 (discussed next), the foregoing coefficients are expressed as

$$C(F_2, F_1)=(-1)^{J+F+I_1+I_2}[(2F_1+1)(2F_2+1)]^{1/2}\begin{Bmatrix} I_1 & J & F_1 \\ I_2 & F & F_2 \end{Bmatrix} \tag{9.126}$$

To obtain the energy values for the intermediate case where the couplings of the two nuclei are of comparable magnitude and for accidental degeneracies in the strong-weak coupling described previously, one must set up and solve the secular equation. Treatment of these more involved cases is given in the original paper by Bardeen and Townes [20]. The theory has been extended and applied to asymmetric rotors by Myers and Gwinn [22], also by Robinson and Cornwell [23].

Coupling by Three or More Nuclei

Quadrupole coupling by three nuclei in a molecule can produce a very complex hyperfine structure of rotational spectra. Matrix elements of the combined Hamiltonian $\mathscr{H}_{Q_1}+\mathscr{H}_{Q_2}$ "$\mathscr{H}_{Q_3}$ have been worked out by Bersohn [24]. Using his results, Mizushima and Ito [25] have calculated the expected theoretical pattern for the $J=0\rightarrow 1$ transition for spin values of 1, $\frac{3}{2}$, 2, and $\frac{5}{2}$. In only a few instances of symmetric-top molecules having three identical symmetrically placed coupling nuclei, including $CHCl_3$ and $CFCl_3$, have attempts been made at application of the theory to observed hyperfine patterns

[26–28]. In these efforts some inconsistencies have been encountered with transitions involving $K=0$ because of failure to take into account off-diagonal elements caused by contributions from $K=+1$ and $K=-1$ states in the averaging of $(\partial^2 V/\partial Z^2)$. This difficulty was overcome by Wolf et al. [28] who made use of a procedure developed by Svidzinskii [29] from group theory for calculation of the hyperfine interactions in symmetrical molecules. A description of the rather complex theory involved in the calculations is given by Wolf et al. [28]. They also provide specific formulas and tables to aid in the calculation of the hyperfine structure for the case of three identical nuclei with spin $I=\frac{3}{2}$ in symmetric-top molecules. They compared their theoretical calculation with the observed hyperfine patterns for the $J=1\rightarrow 2$ and $2\rightarrow 3$ transitions for $CH^{35}Cl_3$ and $CFCl_3$ and reported good agreement. The derived bond axial coupling constants of ^{35}Cl, -77.9 ± 0.05 MHz in $CHCl_3$ and -79.9 ± 0.6 MHz in $CFCl_3$, are close to the solid-state values, -76.98 and -79.63 MHz, observed by Livingston [30]. The case of three identical nuclei with spin $I=1$ has been treated theoretically by Hadley [31] and observed experimentally for ND_3 by Hermann [32].

The general case for any number of coupling nuclei has been treated by Thaddeus et al. [33], who expressed the matrix elements of $\mathscr{H}_Q$ in terms of the $6j$ Wigner coefficients [34, 35], which are tabulated [36]. This treatment is similar to that used for magnetic structures of plurally coupling nuclei described in Section 8 and can be applied when the plural coupling is both magnetic and quadrupole. The matrix elements of $\mathscr{H}_{QL}$ for the Lth coupling nucleus are expressed in the representation $(J, i, I_L, F_1 \cdots F_{N-1}, F)$ chosen as described in Section 8. The diagonal elements are obtained from the assumed vector model of the coupled system; the off-diagonal elements, from the tabulated $6j$ coefficients [36] with (9.185) of Section 8, from Thaddeus et al. [33]. A general discussion employing irreducible tensor methods is given in Chapter XV.

6 SECOND-ORDER EFFECTS IN NUCLEAR QUADRUPOLE INTERACTIONS

The formulas for E_Q in the previous sections represent the diagonal elements of $\mathscr{H}_Q$ in the representation J, i, F, I, and M_F, where i represents any inner quantum numbers such as K. However, the matrix of $\mathscr{H}_Q$ is not exactly diagonal in this representation because the rotational part of the molecular wave function indicated by J, i is perturbed by the nuclear interactions, and hence J is not a perfectly good quantum number. Thus

$$E_Q=(J, i, I, F, M_F|\mathscr{H}_Q|J, i, I, F, M_F) \tag{9.127}$$

earlier evaluated in Section 4, is a calculation of first-order perturbation only. The off-diagonal elements are negligible, however, except for coupling by a few nuclei, the most notable of which is ^{127}I, which in some molecules give large couplings. Even then the off-diagonal elements are small, and adequate correction can be made by inclusion of the second-order term in the perturbation theory, except perhaps in a few instances of near degeneracy in the levels of asymmetric rotors. The second-order term is

$$E_Q^{(2)}=\sum_{J',i'} \frac{|(J, i, I, F, M_F|\mathscr{H}_Q|J', i', I, F, M_F)|^2}{E_{J,i}-E_{J',i'}} \tag{9.128}$$

where $J', i' \neq J, i$. In this expression i represents the inner quantum number K for the symmetric top and the pseudo-quantum numbers τ or $K_{-1}K_1$ for the asymmetric top. For the linear molecule, $i=0$. For the asymmetric rotors, the off-diagonal elements, like the diagonal elements, cannot be expressed in closed formulas except for every low J values. They can be evaluated numerically by high-speed computers when the degree of asymmetry of the molecule is known.

Bardeen and Townes [37] have evaluated the off-diagonal elements of $\mathscr{H}_Q$ for symmetric tops, and Bragg [9] has given a general expression for the matrix elements of an asymmetric top. The nonvanishing off-diagonal matrix elements of $\mathscr{H}_Q$ in the $JiIFM_F$ representation are of the type $(J, i|J\pm 1, i')$ and $(J, i|J\pm 2, i')$. In particular

$$(J, i, I, F, M_F|\mathscr{H}_Q|J+1, i', I, F, M_F)=\frac{eQ(J, i, M_J=J|\partial^2 V|\partial Z^2|J+1, i', M_J=J)}{8I(2I-1)J(2J+1)^{1/2}}$$
$$\times[F(F+1)-I(I+1)-J(J+2)]$$
$$\times[(I+J+F+2)(I-J+F)(J-I+F+1)(J+I-F+1)]^{1/2} \tag{9.129}$$

$$(J, i, I, F, M_F|\mathscr{H}_Q|J+2, i', I, F, M_F)=\frac{eQ(J, i, M_J=J|\partial^2 V/\partial Z^2|J+2, i', M_J=J)}{16I(2I-1)[(2J+1)(J+1)]^{1/2}}$$
$$\times[(I+J+F+2)(I+J+F+3)(I-J+F-1)(I-J+F)$$
$$\times(J-I+F+1)(J-I+F+2)(I+J-F+1)(I+J-F+2)]^{1/2} \tag{9.130}$$

Since the matrix is Hermitian, the matrix elements for $J'=J-1$ and $J'=J-2$ may also be obtained from these expressions.

For symmetric tops the matrix elements of the electric field gradient appearing in the foregoing equations can be readily evaluated from Table 2.1 by expressing the electric field gradient along the space-fixed Z axis in terms of the field gradients along the principal axes of inertia; see (9.69). Since for a symmetric top with a single coupling nucleus the principal inertial axes coincide with the principal axes of the coupling tensor and $q_{xx}=q_{yy}=-\frac{1}{2}q_{zz}=-\frac{1}{2}q$, we have

$$\frac{\partial^2 V}{\partial Z^2}=q\left(\frac{3\Phi_{Zz}^2-1}{2}\right) \tag{9.131}$$

The matrix elements will hence be diagonal in the internal quantum number K.

Table 9.4 Illustration of Second-Order Effects in Nuclear Quadrupole Hyperfine Structure[a]

	Theoretical Displacement from the $F=\frac{7}{2}\to\frac{9}{2}$, $K=0$ Line			
F Transitions	*First Order* $(\Delta\nu)_1$ *(MHz)*	*Second Order* $(\Delta\nu)_2$ *(MHz)*	$(\Delta\nu)_1+(\Delta\nu)_2$	*Observed Displacement* $(\Delta\nu)_{obs}$
		For $K=0$		
$\frac{3}{2}\to\frac{1}{2}$	+74.59	−0.27	+74.32	+74.33
$\frac{3}{2}\to\frac{3}{2}$	−174.06	−0.37	−174.43	−174.47
$\frac{3}{2}\to\frac{5}{2}$	−450.35	+2.19	−448.16	−448.04
$\frac{5}{2}\to\frac{3}{2}$	+406.14	+0.40	+406.54	+406.47
$\frac{5}{2}\to\frac{5}{2}$	+129.85	+2.95	+132.70	+132.72
$\frac{5}{2}\to\frac{7}{2}$	+33.15	−0.39	+32.76	+32.73
$\frac{7}{2}\to\frac{5}{2}$	−276.29	+3.35	−272.94	−273.04
$\frac{7}{2}\to\frac{7}{2}$	−372.99	0	−372.99	−373.04
$\frac{7}{2}\to\frac{9}{2}$	—	—	—	0
		For $K=1$[b]		
$\frac{3}{2}\to\frac{1}{2}$	+281.73	+2.76	+284.49	
$\frac{3}{2}\to\frac{3}{2}$	+163.02	+6.00	+169.02	+168.96
$\frac{3}{2}\to\frac{5}{2}$	+24.85	+3.20	+28.05	+28.09
$\frac{5}{2}\to\frac{3}{2}$	−127.08	+3.72	−123.36	−123.49
$\frac{5}{2}\to\frac{5}{2}$	−265.25	+0.92	−264.33	−264.28
$\frac{5}{2}\to\frac{7}{2}$	−313.60	+2.22	−311.38	−311.28
$\frac{7}{2}\to\frac{5}{2}$	−62.18	+2.17	−60.01	−60.15
$\frac{7}{2}\to\frac{7}{2}$	−110.53	+3.47	−107.06	−107.12
$\frac{7}{2}\to\frac{9}{2}$	+75.99	+0.57	+76.56	+76.65

[a]The observed and calculated displacements are from the $\frac{7}{2}\to\frac{9}{2}$, $K=0$ line in the $J=1\to2$ transition of methyl iodide (CH_3I). The splitting arises from ^{127}I with $eQq=-1934$ MHz. Data taken from W. Gordy, J. W. Simmons, and A. G. Smith, *Phys. Rev.*, **74**, 243 (1948).
[b]The $K=1$ lines have an additional correction of −0.38 MHz for effects of centrifugal stretching.

The $(J|J+1)$ matrix element is evaluated as follows

$$\left(J, K, J\left|\frac{\partial^2 V}{\partial Z^2}\right|J+2, K, J\right)=q\left(J, K, J\left|\frac{3\Phi_{Zz}^2-1}{2}\right|J+1, K, J\right)$$

$$=\tfrac{3}{2}q(J, K, J|\Phi_{Zz}^2|J+1, K, J)$$

$$=\frac{3qK[(J+1)^2-K^2]^{1/2}}{(J+1)(J+2)[2J+3]^{1/2}} \tag{9.132}$$

Likewise

$$\left(J, K, J\left|\frac{\partial^2 V}{\partial Z^2}\right|J+2, K, J\right)=\frac{3q\{[(J+1)^2-K^2][(J+2)^2-K^2]\}^{1/2}}{(J+2)(2J+3)[(J+1)(2J+5)]^{1/2}} \tag{9.133}$$

These expressions along with (9.129) and (9.130) yield the off-diagonal elements of $\mathscr{H}_Q$ for symmetric tops. They apply also to linear molecules when K is set equal to zero. It is evident that for linear molecules $\mathscr{H}_Q$ has only off-diagonal elements of the type $(J|J\pm 2)$.

The rigid rotor approximation of $E_{J,K}$ is adequate for evaluation of the energy difference of the denominator of (9.128). By substitution of the non-vanishing matrix elements and the energy differences $E_{J+1,K}-E_{J,K}=2B(J+1)$, $E_{J+2,K}-E_{J,K}=2B(2J+3)$ into (9.128) followed by summation, the second-order correction energies are found. Because these calculations are rather tedious, numerical evaluations are given in Appendix J for most of the spins and levels for which they are needed.

In Table 9.4 the second-order corrections are shown for the ^{127}I nuclear quadrupole hyperfine structure of the $J=1\rightarrow 2$ transition in CH_3I, for which the coupling $eQq=\chi_z=1934$ MHz. Although these corrections are not large, they are well beyond the experimental error. For $CH_3{}^{79}Br$, $\chi_z=577$ MHz; the second-order corrections are still measurable, but very much smaller than those for CH_3I.

For asymmetric rotors off-diagonal elements in τ will appear. These must be evaluated numerically for each individual case. Nonvanishing contributions to $(J,\tau|\partial^2 V/\partial Z^2|J',\tau')$ will appear whenever the direct product of the symmetries of J_τ, $J'_{\tau'}$, Φ_{Zg}, and $\Phi_{Zg'}$ belong to the representation A of the Four-group (see Table 10.1 for the symmetries of the direction cosines and their products).

7 MAGNETIC HYPERFINE STRUCTURE OF MOLECULES IN SINGLET Σ STATES

Theory of Magnetic Interaction

The classical Hamiltonian for the interaction of a magnetic dipole with a magnetic field is

$$\mathscr{H}_M=-\boldsymbol{\mu}\cdot\mathbf{H} \tag{9.134}$$

The dipole moments we are considering here are the nuclear spin magnetic moments, and the field $\mathbf{H}$ is that generated by the molecular rotation.

Let us first consider the magnetic interaction with a single nucleus having spin $\mathbf{I}$. Its magnetic moment will be

$$\boldsymbol{\mu}_I=g_I\beta_I\mathbf{I} \tag{9.135}$$

where β_I is the nuclear magneton and g_I is the dimensionless gyromagnetic ratio or g factor for the particular nucleus. Because the interaction of $\boldsymbol{\mu}_I$ with the molecular field is weak, J can be considered a good quantum number, and the vector model of Fig. 9.3 applies. The molecular rotation is so rapid compared with the precession of $\mathbf{I}$ that the component fields normal to $\mathbf{J}$ are effectively averaged out, and $\boldsymbol{\mu}_I$ interacts only with the averaged field which is in the direction of $\mathbf{J}$. Thus we can set

$$\mathbf{H}_{\text{eff}}=\langle H_J\rangle\frac{\mathbf{J}}{|\mathbf{J}|}=\frac{\langle H_J\rangle\mathbf{J}}{[J(J+1)]^{1/2}} \tag{9.136}$$

Substitution of (9.135) and (9.136) in (9.134) yields the first-order expression

$$\mathscr{H}_M = -\boldsymbol{\mu}_I \cdot \mathbf{H}_{\text{eff}} = -\frac{g_I \beta_I \langle H_J \rangle}{[J(J+1)]^{1/2}} \mathbf{I} \cdot \mathbf{J} \tag{9.137}$$

When the principal magnetic field axes x, y, z are also principal inertial axes, as is true for linear and symmetric-top molecules and for certain asymmetric-top molecules which have a plane of symmetry, we can set

$$\begin{aligned} H_J &= H_x \cos \theta_{xJ} + H_y \cos \theta_{yJ} + H_z \cos \theta_{zJ} \\ &= H_x \frac{J_x}{|J|} + H_y \frac{J_y}{|J|} + H_z \frac{J_z}{|J|} \\ &= [J(J+1)]^{-1/2}(h_{xx} J_x^2 + h_{yy} J_y^2 + h_{zz} J_z^2) \end{aligned} \tag{9.138}$$

This last step follows from the assumption that $H_x = h_{xx} J_x$, and so on. Since the magnetic field is generated by the molecular rotation, it is reasonable to assume that the field components are proportional to the components of angular momentum. This assumption is verified by the treatment in Chapter XI, Section 4. In the general case where the principal axes of magnetic susceptibilities do not coincide with the principal inertial axes, off-diagonal elements will be generated, and

$$\begin{aligned} H_x &= h_{xx} J_x + h_{xy} J_y + h_{xz} J_z \\ H_y &= h_{yy} J_y + h_{yx} J_x + h_{yz} J_z \\ H_z &= h_{zz} J_z + h_{zx} J_x + h_{zy} J_y \end{aligned} \tag{9.139}$$

Substitution of these values into (9.138) yields

$$H_J = [J(J+1)]^{-1/2} \sum_{gg' = x,y,z} h_{gg'} J_g J_{g'} \tag{9.140}$$

where x, y, z are the principal axes of inertia. For reasons similar to those given for cancellation of the field gradients when they are averaged over the rotational state (Section 4), the off-diagonal elements in H_J will drop out, leaving

$$\langle H_J \rangle = [J(J+1)]^{-1/2}[h_{xx}\langle J_x^2 \rangle + h_{yy}\langle J_y^2 \rangle + h_{zz}\langle J_z^2 \rangle] \tag{9.141}$$

where the average is taken over the rotational state. Therefore the Hamiltonian operator can be expressed as

$$\mathscr{H}_M = C_{J,i} \mathbf{I} \cdot \mathbf{J} \tag{9.142}$$

where

$$C_{J,i} = \frac{1}{J(J+1)} \sum_{g = x,y,z} C_{gg}(J, i | J_g^2 | J, i) \tag{9.143}$$

and

$$C_{gg} = -g_I \beta_I h_{gg} \tag{9.144}$$

The term C_{gg} represents coupling constants that are independent of the rotational state, whereas $C_{J,i}$ is constant only for the particular rotational state J, i over which H_J is averaged. Here i represents any inner quantum numbers such as K of the symmetric top or pseudo-quantum numbers τ or $K_{-1}K_1$ in the asymmetric rotor. The $\mathbf{I}\cdot\mathbf{J}$ term can be expressed (Section 4) as

$$\mathbf{I}\cdot\mathbf{J}=\tfrac{1}{2}(\mathbf{F}^2-\mathbf{I}^2-\mathbf{J}^2) \tag{9.145}$$

where $\mathbf{F}$ is the vector sum of $\mathbf{I}$ and $\mathbf{J}$, as indicated in Fig. 9.3. Since J is a good quantum number in the assumed conditions, as are I and F also, the evaluation can be reduced to

$$E_M=\frac{C_{J,i}}{2}[F(F+1)-I(I+1)-J(J+1)] \tag{9.146}$$

where

$$C_{J,i}=\frac{1}{J(J+1)}(J,i|C_{xx}J_x^2+C_{yy}J_y^2+C_{zz}J_z^2|J,i) \tag{9.147}$$

$$F=J+I,\ J+I-1,\ldots,|J-I|$$

and where the C_{gg}'s are the diagonal elements of the nuclear magnetic coupling tensor. When the principal magnetic axes coincide with the principal inertial axes, these C_{gg}'s represent principal elements of the nuclear coupling tensor. This is true for linear and symmetric-top molecules, also for certain asymmetric rotors which have a plane of symmetry.

Although the parameter $C_{J,i}$ depends on the rotational state, it is a constant for a particular rotational state. In spectral analysis it is convenient to measure first the $C_{J,i}$'s or their differences for connecting rotational states and from these values to derive the constants C_{gg}. For an R-branch $J\rightarrow J+1$ transition, the frequencies of the strong $F\rightarrow F+1$ components are displaced by the magnetic interaction thus

$$(\Delta\nu)_{F\rightarrow F+1}^{J\rightarrow J+1}=\frac{(C_{J',i'}-C_{J,i})}{2h}[F(F+1)-J(J+1)-I(I+1)]+\frac{C_{J',i'}}{h}(F-J) \tag{9.148}$$

For a Q-branch transition $J\rightarrow J$, the $F\rightarrow F$ components are the stronger. Their frequencies are

$$(\Delta\nu)_{F\rightarrow F}^{J\rightarrow J}=\frac{C_{J,i'}-C_{J,i}}{2h}[F(F+1)-J(J+1)-I(I+1)] \tag{9.149}$$

where the primed values are for the upper state. These magnetic displacements are to be added to any quadrupole displacement by the same nucleus for the same transitions.

To simplify the derivation, we have assumed that the molecular magnetic field generated by the rotation is proportional to the angular momentum. A Hamiltonian of the same form is obtained if more rigorous theoretical treatments are applied. The magnetic field arises from two sources, the more obvious of which is the rotation of formal charges with the molecular frame. The second, not so obvious, source is the induction by the rotation of a slight admixture of excited electronic states with angular momentum L into the ground electronic state of the molecule. A molecule in a $^1\Sigma$ state when at rest can have no electronic angular momentum because of the strict quantization of the total angular momentum. When the molecule is rotating, the total angular momentum is still strictly quantized, but the total can now represent an admixture of end-over-end rotation with a component of orbital angular momentum. The interaction between the electronic and end-over-end rotation is a second-order effect which results in nuclear coupling constants proportional to $\Sigma_n|(0|L_g|n)|^2/(E_n-E_0)$ where L_g represents the components of the electronic orbital moment along the principal coupling axes and where the summation is taken over all excited electronic states having orbital angular momentum. The quantity E_n-E_0 represents the difference in energy between the excited electronic states n and the ground state. The nuclear coupling has the same origin as does the rotational g factor and is derived in a similar manner. See Chapter XI, Section 4.

Attempts to relate the molecular magnetic coupling constants of $^1\Sigma$ state molecules to the electronic structure of the molecule have been made by Townes et al. [38]. Although their treatment gives an insight into the nature of the coupling, it does not provide a means for obtaining reliable information about the chemical bond because of the very small values of the observed coupling combined with the complex nature of the coupling mechanism.

Diatomic and Linear Polyatomic Molecules

Linear molecules in the ground vibrational state and in $^1\Sigma$ electronic states have no resultant angular momentum about the molecular axis, chosen as the z inertial axis. Because of symmetry, $C_{xx}=C_{yy}=C_I$. With these conditions, the matrix of (9.147) transforms to

$$C_J=\frac{1}{J(J+1)}C_I(J|J^2|J)=C_I\frac{J(J+1)}{J(J+1)}=C_I \tag{9.150}$$

and from (9.146) the magnetic hyperfine energies are found to be

$$E_M=\frac{C_I}{2}[F(F+1)-I(I+1)-J(J+1)] \tag{9.151}$$

where C_I is the nuclear magnetic coupling constant. Selection rules are the same as those for quadrupole hyperfine structure. For a rotational absorption transition

$$J\to J+1, \qquad F\to F\pm 1 \text{ and } F\to F, \qquad I\to I \tag{9.152}$$

Equation 9.151 gives the frequency of the $J\to J+1$ transition to be

$$\nu_{F\to F+1}=\nu_0-\left(\frac{C_I}{h}\right)(J+1)+\left(\frac{C_I}{h}\right)(F+1) \tag{9.153}$$

$$\nu_{F\to F}=\nu_0-\left(\frac{C_I}{h}\right)(J+1) \tag{9.154}$$

$$\nu_{F\to F-1}=\nu_0-\left(\frac{C_I}{h}\right)(J+1)-\left(\frac{C_I}{h}\right)F \tag{9.155}$$

where ν_0 is the frequency of the line unperturbed by the magnetic interaction. As will be seen from an examination of the intensity tables in Appendix I, only the $F\to F+1$ lines have significant intensity except for very low J values. Therefore, except for $J\lesssim 3$, the noticeable hyperfine components are the $F\to F+1$ lines. When $J\geqslant I$, the values of F range in integral values from $J-I$ to $J+I$. Thus there are $2I+1$, $F\to F+1$ components with equal spacing of C_I/h. Usually the value of C_I/h for molecules in $^1\Sigma$ states is only of the order of 10 kHz and hence cannot be resolved in most experiments because of Doppler broadening (Chapter III, Section 2). When, however, the $F\to F+1$ components are already separated by nuclear quadrupole coupling, as is usually true for spins greater than $\frac{1}{2}$, the extra displacement caused by the magnetic hyperfine interaction is often measurable. Note that the outside components are the most displaced in frequency, one up and the other down.

The small splitting of the magnetic hyperfine structure can often be resolved with molecular beam masers or molecular beam resonance techniques which avoid both collision broadening and Doppler broadening. Some of the magnetic coupling constants given in Table 9.5 were measured by molecular beam resonance techniques.

From the observed nuclear coupling constants C_I, it is possible to calculate the molecular magnetic field at the respective coupling nucleus for the different rotational states. Since $g_I=\mu_I\,(\mathrm{nm})/I$, the field values from (9.141) and (9.144) can be expressed as

$$\langle H_J\rangle=\frac{C_I}{g_I\beta_I}[J(J+1)]^{1/2}=\frac{C_I\,(\mathrm{Hz})hI}{\mu_I(\mathrm{nm})\beta_I}[J(J+1)]^{1/2} \tag{9.156}$$

or

$$\langle H_J\rangle\,(\mathrm{gauss})=1.311\,\frac{C_I(\mathrm{kHz})\,I}{\mu_I(\mathrm{nm})}[J(J+1)]^{1/2} \tag{9.157}$$

As an illustration, let us calculate the field at the nitrogen nucleus in HCN, for which $C_I(^{14}\mathrm{N})$ is 10 kHz (see Table 9.5). For ^{14}N, $I=1$ and $\mu_I=0.407$. Substitution of these values into (9.157) shows that

$$\langle H_J\rangle=32.5[J(J+1)]^{1/2}\ \mathrm{gauss}$$

Thus for the rotational state $J=1$, the magnetic field at the N nucleus is approximately 46 gauss; for the $J=10$ rotational state it is about 340 gauss.

Symmetric-top Molecules

In the symmetric-top molecule two of the coupling coefficients of (9.147) are equal because of axial symmetry. We choose z as the symmetry axis and set

Table 9.5 Illustrative Values of Nuclear Magnetic Coupling Constants in Linear Molecules

Molecule	*Nucleus*	C_I *(kHz)*	*Ref.*
DCl	^{35}Cl	0.026(11)	a
DBr	^{79}Br	0.148(9)	a
DI	^{127}I	0.165(6)	a
HI	^{127}I	0.349(10)	a
LiF	^{7}Li	2.2(6)	b
LiF	^{19}F	32.9(1)	b
PN	^{14}N	10.4(5)	c
PN	^{31}P	−78.2(5)	c
HCN	^{14}N	10.4(3)	d
DCN	D	−0.6(3)	d
ClCN	^{14}N	2.5(8)	b
OCS	^{17}O	−4.0(15)	b

[a]F. C. De Lucia, P. Helminger, and W. Gordy, *Phys. Rev.*, **A3**, 1849 (1971).
[b]White [45].
[c]J. Raymonda and W. Klemperer, *J. Chem. Phys.*, **55**, 232 (1971).
[d]F. C. De Lucia and W. Gordy, *Phys. Rev.*, **187**, 58 (1969).

$C_{xx}=C_{yy}=C_N$ and $C_{zz}=C_K$. In the symmetric-top representation the matrix elements of (9.147) are then

$$\begin{aligned} C_{J,K} &= \frac{1}{J(J+1)}(J, K|C_N(J_x^2+J_y^2)+C_K J_z^2|J, K) \\ &= \frac{1}{J(J+1)}(J, K|C_N(J^2-J_z^2)+C_K J_z^2|J, K) \\ &= C_N+(C_K-C_N)\frac{K^2}{J(J+1)} \end{aligned} \tag{9.158}$$

Substitution of these values into (9.146) yields the formula for the nuclear magnetic splitting of the symmetric-top rotational levels by a single coupling nucleus

$$E_M=\frac{1}{2}\left[C_N+(C_K-C_N)\frac{K^2}{J(J+1)}\right][F(F+1)-I(I+1)-J(J+1)] \tag{9.159}$$

where C_K is the principal value of the nuclear coupling along the symmetry axis and where C_N is the principal value normal to this axis.

Selection rules for F are the same as those for linear molecules, (9.152). Only the $F \to F+1$ components of the rotational absorption transition $J \to J+1$, $K \to K$ have significant strength except for very low J values. From

(9.159) the magnetic displacements of these transitions are found to be

$$\Delta\nu_{F\to F+1} = C_N(F-J) + \frac{(C_N - C_K)K^2}{J(J+1)(J+2)} \times [J(J-F) + F(F+1) - I(I+1) - J(J+1)] \qquad (9.160)$$

Small deviations in the positions of the hyperfine components of the NH_3 inversion spectrum from those predicted from nuclear quadrupole theory were first measured by Simmons and Gordy [39] and threated theoretically by Henderson [40]. The observed displacement can be fitted to a formula [40] like (9.159) with $C_N = 11.4$ kHz and $C_K = 13.6$ kHz. From a relationship like that of (9.157)

$$\langle H_K \rangle\,(\text{gauss}) = 1.311\,\frac{C_K\,(\text{kHz})\,IK}{\mu_I\,(\text{nm})} \qquad (9.161)$$

it is found that the magnetic field at the nitrogen nucleus in the direction of the symmetry axis is approximately 44 gauss when $K=1$. For large K values it is a few hundred gauss. The NH_3 hyperfine structure is further complicated by magnetic coupling by the hydrogen nuclei (see Section 8).

Asymmetric Rotors

The magnetic coupling by a single nucleus in an asymmetric rotor is given by (9.147), but the constants C_{xx}, C_{yy}, and C_{zz} with reference to the principal inertial axes are not necessarily the principal elements of the magnetic coupling in the molecular system: they are simply the averaged values with reference to the inertial axes a, b, c. We designate these constants as C_a, C_b, and C_c. The $C_{J,i}$ of (9.143) can then be expressed as

$$C_{JK_{-1},K_1} = \frac{1}{J(J+1)}(J, K_{-1}, K_1|C_aJ_a^2 + C_bJ_b^2 + C_cJ_c^2|J, K_{-1}, K_1) = \frac{1}{J(J+1)}\sum_{g=a,b,c} C_g\langle J_g^2\rangle \qquad (9.162)$$

where the average is over the wave function for the rigid asymmetric rotor. It is obvious that the quantity in vertical bars has the same form as the Hamiltonian of the rigid asymmetric rotor (see Chapter VII). Hence the energies already found for the rigid rotor can be used for its evaluation. The values of $\langle J_g^2\rangle$, already used in the evaluation of quadrupole coupling energies, are given by (9.97)–(9.99). Substitution of these expressions into (9.162) and combination with (9.146) yields the first-order values

$$E_H = \frac{1}{4J(J+1)}\left\{C_a\left[J(J+1) + E(\kappa) - (\kappa+1)\frac{\partial E(\kappa)}{\partial\kappa}\right] + 2C_b\frac{\partial E(\kappa)}{\partial\kappa} + C_c\left[J(J+1) - E(\kappa) + (\kappa-1)\frac{\partial E(\kappa)}{\partial\kappa}\right]\right\} \times [F(F+1) - I(I+1) - J(J+1)] \qquad (9.163)$$

for the magnetic coupling energies.

8 PLURAL NUCLEAR COUPLING—QUADRUPOLE AND MAGNETIC

Effects of plural nuclear magnetic coupling were noted in microwave spectra when distortions of certain NH_3 inversion lines by the hydrogen nuclei were observed with a conventional gaseous spectrometer by Gunther-Mohr et al. [41]. The invention of the molecular beam maser by Townes and associates, which first operated on the NH_3 inversion spectrum, made possible further resolution of this closely spaced hyperfine structure as well as the resolution of composite hyperfine structure in other molecules. The many lines arising from the four coupled nuclei in NH_3 were partially resolved by Gordon [42] with the first molecular beam spectrometer, and later, by Shimoda and Kondo [43]. An exceptionally high resolution of this hyperfine structure has since been achieved by Kukolich [44] with a two-cavity molecular beam maser.

Stimulated by these initial observations on NH_3, Gunther-Mohr et al. [11] developed a comprehensive theory of plural nuclear coupling in NH_3 which has been extended and adapted to molecules of different types by others. Reviews of the early treatments are given by Townes and Schawlow [21] and by White [45]. Theory for the combined nuclear quadrupole and magnetic hyperfine interactions by plural nuclei has been treated by Posener [46], and a convenient formulation with adaptation to a symmetric rotor is provided by Thaddeus et al. [33]. We give here only an outline of the essential elements. A more detailed discussion is given in Chapter XV.

Hybrid Coupling by Two Nuclei: Strong Quadrupole and Weak Magnetic Coupling

When there is coupling by two nuclei, one of which has both a quadrupole moment and a magnetic moment and the other only a magnetic moment, the quadrupole coupling is usually much the stronger for molecules in Σ singlet states. The outstanding exception is the combination of D coupling with H coupling which must be treated by the more general methods discussed later. The more typical case of strong nuclear quadrupole coupling combined with weak magnetic coupling can be treated by use of the vector model, as here described.

The quadrupolar nucleus with spin $\mathbf{I}_1$ forms a resultant

$$\mathbf{F}_1 = \mathbf{J} + \mathbf{I}_1 \tag{9.164}$$

and the magnetic nucleus with spin $\mathbf{I}_2$ couples to $\mathbf{F}_1$ to form the resultant

$$\mathbf{F} = \mathbf{F}_1 + \mathbf{I}_2 \tag{9.165}$$

as in the vector model of Fig. 9.9. The quadrupole interaction of the first nucleus is calculated in the usual manner with the neglect of the weak magnetic coupling of I_2. This energy can be expressed as

$$E_{Q_1} = eQ_1 q_J \left(\frac{2J+3}{J}\right) Y(J, I_1, F_1) \tag{9.166}$$

where q_J depends on the nuclear type and where $Y(J, I_1, F_1)$ is defined by (9.83). Nucleus one will also have an $\mathbf{I}\cdot\mathbf{J}$ interaction caused by its magnetic moment which can be expressed from (9.146) by

$$E_{M_1}=\frac{(C_{J,i})_1}{2}[F_1(F_1+1)-I_1(I_1+1)-J(J+1)] \tag{9.167}$$

In the assumed vector model, **J** precesses so rapidly about $\mathbf{F}_1$ that only the components of $\langle H_J\rangle$ of (9.137) that are parallel to $\mathbf{F}_1$ effectively interact with $\mathbf{I}_2$ which precesses much more slowly about **F**. The Hamiltonian for the magnetic interaction is therefore

$$\mathscr{H}_{M_2}=C_{F_1}\mathbf{I}_2\cdot\mathbf{F}_1=\left(\frac{C_{F_1}}{2}\right)(\mathbf{F}^2-\mathbf{I}_2^2-\mathbf{F}_1^2) \tag{9.168}$$

under the conditions that F_1, I_2, and F are all good quantum numbers. Hence

$$E_{M_2}=\left[\frac{C_{F1}}{2}\right][F(F+1)-I_2(I_2+1)-F_1(F_1+1)] \tag{9.169}$$

where

$$C_{F_1}=-\frac{g_{I_2}\beta_I\langle H_J\rangle\cos\theta_{J,F_1}}{[F_1(F_1+1)]^{1/2}}=(C_{J,i})_2\frac{[J(J+1)]^{1/2}}{[F_1(F_1+1)]^{1/2}}\cos\theta_{J,F_1} \tag{9.170}$$

with $\cos\theta_{J,F_1}$ evaluated from the vector model and the law of the cosine

$$C_{F_1}=(C_{J,i})_2\frac{F_1(F_1+1)-I_1(I_1+1)+J(J+1)}{2F_1(F_1+1)} \tag{9.171}$$

The total hyperfine energy is

$$E_{1,2}=E_{Q_1}+E_{M_1}+E_{M_2} \tag{9.172}$$

The hyperfine quantum numbers are

$$F_1=J+I_1,\ J+I_1-1,\ldots,|J-I_1| \tag{9.173}$$

$$F=F_1+I_2,\ F_1+I_2-1,\ldots,|F_1-I_2| \tag{9.174}$$

The selection rules for F_1 and F are the same as those for the F of a single coupling nucleus. If $F_1\geqslant I_2$ each level of F_1 which is due to nucleus one is split into $2I_2+1$ components by the interaction of nucleus two. For $F_1\leqslant I_2$ the level splits into $2F_1+1$ components. Because nuclei with spin $I=\frac{1}{2}$ have magnetic moments only, with no quadrupole displacements, their splitting can seldom be resolved except in molecules with unbalanced electronic angular momentum (Section 9).

Couplings of Comparable Strength

When two or more of the nuclei have couplings of comparable strength, either quadrupole or magnetic, the vector model treatment given in the previous section breaks down. Because the off-diagonal elements of the coupling matrix are large, one must solve a secular equation to find the correct level splitting. A general method for finding the required matrix elements for N-coupling nuclei, which may be a combination of quadrupole and magnetic coupling, has been described by Thaddeus et al. [33]. Their treatment will be described briefly with the omission of the spin–spin interaction, which is quite complicated and usually negligibly small.

The Hamiltonian for the $\mathbf{I}\cdot\mathbf{J}$ interaction of any one of the coupling nuclei is like that for a single coupling nucleus just described. Thus the Hamiltonian for the combined $\mathbf{I}\cdot\mathbf{J}$ interactions can be expressed as

$$\mathscr{H}_M=\sum_L (C_{J,i})_L \mathbf{I}_L\cdot\mathbf{J} \tag{9.175}$$

where $\mathbf{I}_L$ is the spin of the Lth coupling nucleus, where the summation is taken over all coupling nuclei, and where

$$(C_{J,i})_L=\frac{1}{J(J+1)}\sum_{g=x,y,z}(C_{gg})_L\langle J_g^2\rangle \tag{9.176}$$

In this form of the Hamiltonian J is assumed to be a good quantum number. The problem is to express the matrix elements of $\mathscr{H}_M$ in a suitable representation so that its quantized values can be found either by perturbation theory or from solution of the secular equation. It is desirable to express the matrix of $\mathscr{H}_M$ in the representation in which off-diagonal elements are smallest. Since all the magnetic perturbations are small, J can be assumed to be a good quantum number. The matrix $\mathscr{H}_M$ is therefore diagonal in J as well as in all the individual spin numbers $I_1 \cdots I_N$. It is diagonal also in F, the number representing the total angular momentum, which is always a constant of the motion. Thus $\mathscr{H}_M$ commutes with the operators $\mathbf{J}, \mathbf{I}_1, \ldots, \mathbf{I}_N, \mathbf{F}$ but not necessarily with the component operators $\mathbf{F}_1, \mathbf{F}_2, \ldots, \mathbf{F}_N$ defined next. The representation which makes the off-diagonal elements smallest when the degree of coupling is substantially different, is [33]

$$|I_1 \cdots I_N, J, F_1 \cdots F_{N-1}, F) \tag{9.177}$$

where the $F_1 \cdots F_{N-1}$ indices are chosen by the following coupling scheme. Of the distinguishable nuclei the strongest coupling one with spin $\mathbf{I}_1$ is assumed to be coupled first to $\mathbf{J}$ to form a resultant $\mathbf{F}_1$; then the nucleus having coupling next in strength with spin indicated by $\mathbf{I}_2$ is assumed to be coupled with $\mathbf{F}_1$ to form the resultant $\mathbf{F}_2$, and so on

$$\mathbf{F}_1=\mathbf{J}+\mathbf{I}_1, \quad \mathbf{F}_2=\mathbf{F}_1+\mathbf{I}_2, \quad \ldots, \quad \mathbf{F}=\mathbf{F}_{N-1}+\mathbf{I}_N \tag{9.178}$$

The evaluation of the interaction for each nucleus can be separated. The total energy in first-order is the sum

$$E_M=E_1+E_2+\cdots+E_N \tag{9.179}$$

where E_L is the diagonal matrix element of the interaction of the particular nucleus in the representation of (9.177). In this representation the matrix of the Hamiltonian for the Lth nucleus diagonal in J and i is

$$(C_{J,i})_L(I_L, J, F'_1 \cdots F'_{L-1}, F|\mathbf{I}_L\cdot\mathbf{J}|I_L, J, F_1 \cdots F_{L-1}, F) \tag{9.180}$$

The diagonal elements of the matrix of $\mathbf{I}_L\cdot\mathbf{J}$ can be obtained from the assumed vector model of the coupled systems. The off-diagonal elements can be calculated [33] most easily from the Wigner $6j$ coefficients tabulated by Rotenberg et al. [36] with the formula [33–35]

$$\begin{aligned}\langle F'_1 \cdots F'_{L-1}|\mathbf{I}_L\cdot\mathbf{J}|F_1 \cdots F_{L-1}\rangle &= (-1)^r\{J(J+1)(2J+1)[(2F'_1+1)(2F_1+1)]\cdots \\ &\quad\times[(2F'_{L-1}+1)(2F_{L-1}+1)]I_L(I_L+1)(2I_L+1)\}^{1/2} \\ &\quad\times\begin{Bmatrix}F'_0 & F'_1 & I_1\\ F_1 & F_0 & 1\end{Bmatrix}\cdots\begin{Bmatrix}F'_{L-2} & F'_{L-1} & I_{L-1}\\ F_{L-1} & F_{L-2} & 1\end{Bmatrix}\begin{Bmatrix}F_L & I_L & F'_{L-1}\\ 1 & F_{L-1} & I_L\end{Bmatrix}\end{aligned} \tag{9.181}$$

where

$$r=(L-1)+\sum_{i=1}^{L-1}(F'_{i-1}+I_i+F_i)+(F_{L-1}+I_L+F_L) \tag{9.182}$$

The derivation of (9.181) and (9.184) is given in Chapter XV, Section 3. The quantities in the large braces are the $6j$ coefficients. The operator $\mathbf{I}_L\cdot\mathbf{J}$ is diagonal in all the F_i not lying in the range $F_1 \cdots F_{L-1}$, and does not depend on the quantum numbers $F_{L+1} \cdots F$. In these coefficients $F_0=J=F'_0$ and $F_N=F$. When $L=2$ the product of the $6j$ symbols immediately on either side of the leaders in (9.181) reduces to one symbol. When $L=1$, only the symbol on the far right of the chain remains, and the product $[(2F'_1+1)(2F_1+1)]\cdots[(2F'_{L-1}+1)(2F_{L-1}+1)]$ reduces to unity. For two coupling nuclei the off-diagonal elements of $(\mathbf{I}_2\cdot\mathbf{J})$ are of the type $(F_1\pm1|F_1)$.

Some of the N-coupling nuclei may also have quadrupole interactions. The composite Hamiltonian of the quadrupole interaction can likewise be represented as the sum of the interactions of the separate nuclei, as described in Section 4,

$$\mathscr{H}_Q=\sum_L \frac{(eQq_J)_L}{2J(2J-1)I_L(2I_L-1)}[3(\mathbf{I}_L\cdot\mathbf{J})^2+\tfrac{3}{2}(\mathbf{I}_L\cdot\mathbf{J})+\mathbf{I}_L^2\mathbf{J}^2] \tag{9.183}$$

In terms of the $6j$ Wigner coefficients the matrix elements of the quadrupolar Hamiltonian for the Lth coupling nucleus in the representation just described are [33, 34]

$$\langle F_1' \cdots F_{L-1}' | \mathcal{H}_{QL} | F_1 \cdots F_{L-1} \rangle = (-1)^t (eQq_J)_L \times \left\{ \frac{(2J+1)(2J+2)(2J+3)}{8J(2J-1)} [(2F_1'+1)(2F_1+1)] \cdots \times [(2F_{L-1}'+1)(2F_{L-1}+1)] \frac{(2I_L+1)(2I_L+2)(2I_L+3)}{8I_L(2I_L-1)} \right\}^{1/2} \times \begin{Bmatrix} F_0' & F_1' & I_1 \\ F_1 & F_0 & 2 \end{Bmatrix} \cdots \begin{Bmatrix} F_{L-2}' & F_{L-1}' & I_{L-1} \\ F_{L-1} & F_{L-2} & 2 \end{Bmatrix} \begin{Bmatrix} F_L & I_L & F_{L-1}' \\ 2 & F_{L-1} & I_L \end{Bmatrix} \quad (9.184)$$

where

$$t = \sum_{i=1}^{L-1} (F_{i-1}' + I_i + F_i) + (F_{L-1} + I_L + F_L) \quad (9.185)$$

As before, when $L=1$, only the $6j$ symbol on the far right of the chain remains, and the product $[(2F_1'+1)(2F_1+1)] \ldots [(2F_{L-1}'+1)(2F_{L-1}+1)]$ reduces to unity; for $L=2$ the product reduces to $[(2F_1'+1)(2F_1+1)]$, and only two $6j$ symbols remain. See also Chapter XV, Section 3.

Coupling by Identical Nuclei

When there are identical nuclei in the molecule, linear combinations of their identical spin functions must be chosen to give independent functions. In this case, however, the plural hyperfine coupling is sometimes simplified. Consider the coupling by two identical nuclei in a molecule such as $H_2{}^{16}O$ or CH_2O. Because the over-all symmetry of the wave function must be antisymmetric for an operator which exchanges these identical nuclei (see Chapter III, Section 4), only spin functions that are antisymmetric can combine with the even rotational states (if the electronic and vibrational functions are assumed to be even), and only even nuclear spin functions can combine with the odd rotational levels. For $I_1 = I_2 = \frac{1}{2}$ the combinations $(\frac{1}{2}, \frac{1}{2})$ and $(-\frac{1}{2}, -\frac{1}{2})$ are symmetric since an exchange of the two nuclei produces an indistinguishable combination. For the degenerate combinations $(-\frac{1}{2}, \frac{1}{2})$ and $(\frac{1}{2}, -\frac{1}{2})$, the independent functions are the linear combinations

$$\frac{1}{\sqrt{2}} [(\tfrac{1}{2}, -\tfrac{1}{2}) + (-\tfrac{1}{2}, +\tfrac{1}{2})] \quad (9.186)$$

$$\frac{1}{\sqrt{2}} [(\tfrac{1}{2}, -\tfrac{1}{2}) - (-\tfrac{1}{2}, +\tfrac{1}{2})] \quad (9.187)$$

The first of these is symmetric; the second, antisymmetric. Thus there are three symmetric combinations. These are equivalent to an effective spin $I'=1$ with M_I values of 1, 0, and -1. Therefore the hyperfine structure of the odd rotational levels corresponds to that of a single nucleus with spin $I=1$, and the calculation is the same as that for a single nucleus with $I=1$. There is only one spin combination that is antisymmetric with effective spin $I'=0$, and thus the even rotational levels are singlets and cannot be split by the nuclear interaction. For methods of finding the symmetry of rotational levels in asymmetric rotors, see Chapter III, Section 4.

The hyperfine interaction by three identical nuclei is likewise reduced by the symmetry of the rotational wave function, but its calculation is not so simple as that for two nuclei, described previously. These effects are treated for the three identical coupling H nuclei in NH_3 by Gunther-Mohr et al. [11].

Spin–Spin Interactions

The direct dipole–dipole or spin–spin interaction energy between the different nuclei is of the order of only a few kilohertz and is usually not resolvable even with the beam masers. For this reason we shall not treat it here. The Hamiltonian of this interaction is derived by Gunther-Mohr et al. [11], and a generalized method for finding its matrix elements which employ the tabulated $6j$ Wigner coefficients is given by Thaddeus et al. [33]. A brief discussion is given in Chapter XV, Section 5.

Observed Spectra in Asymmetric Rotors—Maser Resolutions

Nuclear magnetic hyperfine structure has been measured in the rotational spectra of only a few, relatively simple asymmetric-top molecules. Most of the coupling nuclei are protons or protons and deuterons. The splittings expected from $(\mathbf{I}\cdot\mathbf{J})$ interactions of the various nuclei are of the order of 10 kHz or less. Such splittings are usually obscured by Doppler broadening in the conventional microwave spectrometer used for the study of gases, but further displacements of this order might be observed when the $2I+1$ degeneracy caused by a given nucleus is already lifted by quadrupole splitting. Since the Doppler broadening decreases with frequency, the best possibilities for measurement of the shifts for gaseous molecules in thermal equilibrium are on low frequency transitions.

Posener [47] as well as Treacy and Beers [48] have measured the D and H couplings in HDO with sensitive gaseous absorption spectrometers designed to operate on low frequency transitions. Similarly, Flygare [49] has measured these constants in HDCO. However, the instrument best suited for observation of magnetic hyperfine structure is the molecular beam maser which avoids both Doppler and pressure broadening. Because of difficulties in experimental techniques and low sensitivity, it can, unfortunately, be used only for highly selected molecules and transitions.

One of the simplest organic molecules for which magnetic hyperfine structure has been studied with the beam maser is $^{12}CH_2{}^{16}O$. In this molecule there is no nucleus having a quadrupole moment, and the hyperfine structure arises solely from the magnetic moments of the two protons. The spins of the two protons combine to form a resultant spin $T=1$ or 0, which gives a coupling like that of a single nucleus with $I=1$, or with no coupling $I=0$, according to the symmetry of the rotational levels. The splitting of the rotational levels which combine with the symmetric spin functions can be calculated by use of (9.163) with $I=1$. The rotational levels are split into a triplet corresponding to $F=J+I$, J, $J-1$. For Q-branch transitions $J \rightarrow J$, there are three strong lines that correspond to the $F \rightarrow F$ transition. Resolution of this hyperfine structure was first achieved by Okaya [50], who detected two of the strong $F \rightarrow F$ components in the $4_{14} \rightarrow 4_{13}$ transition. Later Takuma et al. [51] observed the three expected $F \rightarrow F$ components of the $3_{12} \rightarrow 3_{13}$ transition. More recently, Thaddeus et al. [33] resolved beautifully all three of the $F \rightarrow F$ components of the $2_{11} \rightarrow 2_{12}$ transition, and, in addition, detected one of the weaker $F \rightarrow F-1$ components. This spectrum, which is shown in Fig. 9.10 provides an illustration of the resolution which can be achieved with the molecular beam maser. The total spread of the spectrum is about 40 kHz. The observed proton magnetic coupling constant for the 2_{11} level $C_H(2_{11})$ is only 0.65 kHz, and $C_H(2_{11})-C_H(2_{12})=$ 2.26 kHz. When a deuterium atom, $I=1$, is substituted for one of the hydrogens, the symmetry is destroyed, and the splitting results from the quadrupole coupling of the D atom, for which $eQq=170$ kHz, as well as the additional, slight magnetic displacement by D and the magnetic splitting by the remaining H. Figure 9.11 shows the spectrum for the same $2_{11} \rightarrow 2_{12}$ transition of CHDO, which, when compared with Fig. 9.10, illustrates the marked changes produced by the D substitution.

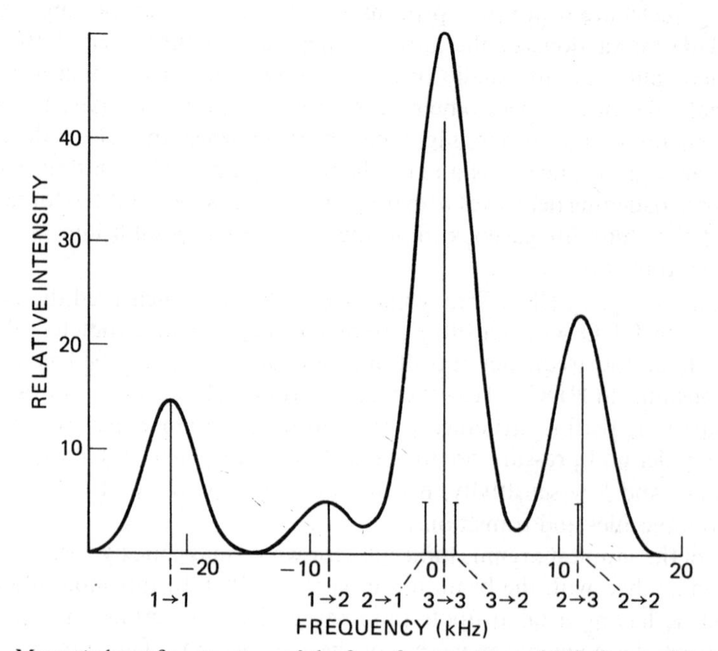

Fig. 9.10 Magnetic hyperfine structure of the $2_{11} \rightarrow 2_{12}$ transition of CH_2O caused by the two protons. Bars represent calculated components. The curve simulates the observed maser pattern. From Thaddeus et al. [33].

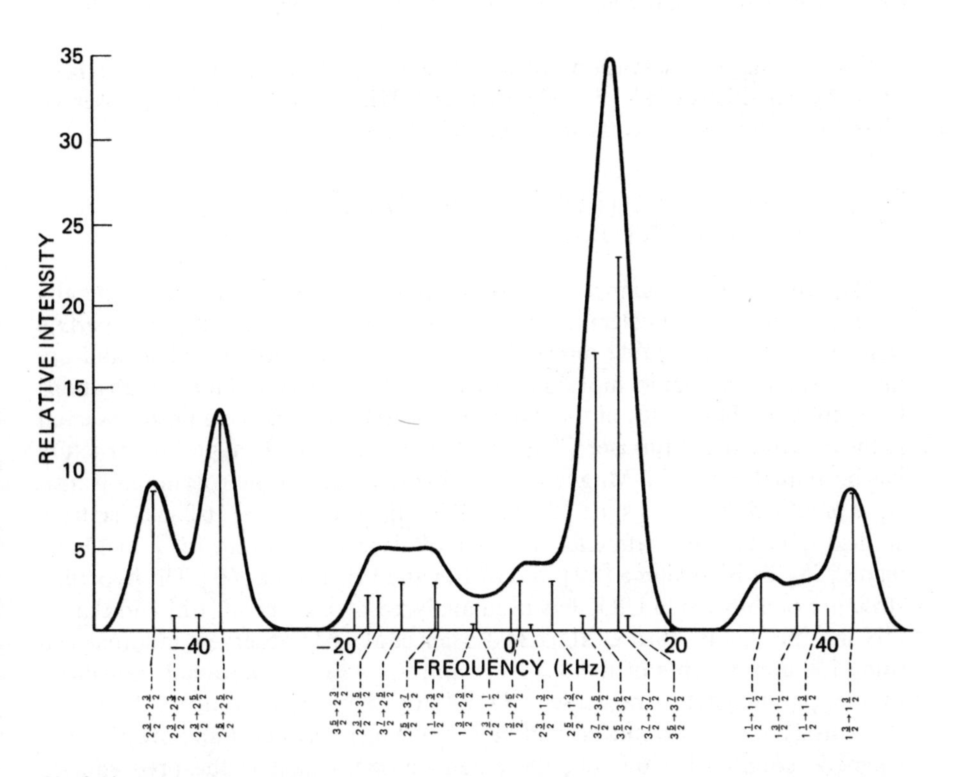

Fig. 9.11 Composite quadrupole and magnetic hyperfine structure in the $2_{11} \rightarrow 2_{12}$ transition of CHDO caused by the proton and deuteron. Bars represent calculated components. The curve simulates the observed maser pattern. From Thaddeus et al. [33].

Table 9.6 Hyperfine Splitting Parameters of D and H for Particular Rotational States, as Measured with the Molecular Beam Maser

Molecule	*Rotational State*	$(eq_JQ)_D$ (*kHz*)	$(C_H)_J$ (*kHz*)	$(C_D)_J$ (*kHz*)	*Ref.*
HDO	2_{20}	79.3 ± 0.3	-43.47 ± 0.11	-2.33 ± 0.02	a
	2_{21}	79.6 ± 0.3	-43.63 ± 0.13	-2.20 ± 0.02	a
HDS	2_{20}	42.9 ± 0.4	-25.03 ± 0.13	-0.47 ± 0.02	a
	2_{21}	43.3 ± 0.4	-27.45 ± 0.13	-0.22 ± 0.02	a
NH_2D	3_{13}	-62.5 ± 1.0	-13.6 ± 0.5	-2.7 ± 0.3	b
	3_{03}	-73.9 ± 1.0	-12.3 ± 0.5	-2.7 ± 0.3	b

[a] Thaddeus et al. [33].
[b] Thaddeus et al. [52].

Similar maser observations of composite hyperfine structures have been made by Thaddeus et al. [33, 52] for HDO, HDS, and NH_2D. The observed H and D coupling constants are given in Table 9.6.

9 MOLECULES WITH UNBALANCED ELECTRONIC ANGULAR MOMENTUM

Molecules with unbalanced electronic spin or orbital momentum generally have large magnetic nuclear coupling and widely spaced magnetic hyperfine structure. However, stable molecules of this kind are not prevalent and are mostly small, inorganic molecules such as O_2, NO, and NO_2. Transitions between Λ doublet states of the free radicals OH and OD have been observed in the microwave regions (see Chapter IV), and their H or D hyperfine structure has been analyzed [53]. Magnetic hyperfine structure of the rotational transitions of NO in the $^2\Pi_{1/2}$ state [54, 55] and in the $^2\Pi_{3/2}$ state [56] has also been studied. Theoretical treatments are given by Frosch and Foley [57], by Dousmanis [58], by Mizushima [59], and by Lin and Mizushima [60]. The hyperfine structure of chlorine in ClO_2 has been analyzed by Curl et al. [61] and that of ^{14}N in $^{14}N^{16}O_2$ by Bird et al. [62]. Foster et al. [63] observed and analyzed some 150 components of the $8_{08}\rightarrow 7_{17}$ and $9_{19}\rightarrow 10_{0,10}$ rotational transitions of the paramagnetic molecule $^{17}O^{14}N^{16}O$. This very complex hyperfine structure is due to two nuclei, ^{14}N and ^{17}O, which have both quadrupole and magnetic couplings. However, the magnetic component is the predominant one for these high J values. Magnetic hyperfine structure for such molecules depends upon the nature and degree of the coupling between the various angular momenta and spin vectors. Because of its complexity and limited applications in rotational spectra we shall not treat the subject here. The original literature cited on these various types of molecules provides the necessary theory and illustrations for those who are concerned with analysis of hyperfine structure in paramagnetic molecules.

10 MOLECULES IN EXCITED VIBRATIONAL STATES

Although the analysis of nuclear quadrupole transitions of molecules in excited vibrational states is generally similar to that for the ground vibrational state, the spacing of the hyperfine structure and the nuclear coupling constants depend significantly on the vibrational state. Most of the microwave measurements of hyperfine structures in excited states have been made on the heavier diatomic molecules. Examples of the observed quadrupole couplings in them are given in Table 9.7. The coupling constant eQq_v for the parallel vibrational states may be fitted to the power series expansion

$$eQq_v = eQq_e + eQq_I(v+\tfrac{1}{2}) + eQq_{II}(v+\tfrac{1}{2})^2 + \cdots \qquad (9.188)$$

where v is the vibrational quantum number and eQq_e, eQq_I, and so on, are constants. For example, the observed eQq_v values of ^{127}I in $^{69}Ga^{127}I$ listed in

Table 9.7 Nuclear Quadrupole Coupling of Selected Diatomic Molecules in Excited Vibrational States

Molecule	*Coupling Atom*	*Coupling in MHz*				*Ref.*
		$v=0$	$v=1$	$v=2$	$v=4$	
BiF	^{209}Bi	−1148.2(1)	−1144.0(2)	−1139.7(3)		*a*
RbBr	^{85}Rb	−47.1(2)	−46.7(2)	−46.6(3)		*b*
	^{79}Br	3.5(3)	4.1(3)	5.0(3)		*b*
IBr	^{127}I	−2753.5(1)	−2752.9(3)	−2752.4(3)		*c*
	^{79}Br	696.8(2)	697.7(2)	698.5		*c*
GaCl	^{69}Ga	−92.1(2)	−91.4(2)	−90.8(2)	−90.0(2)	*d*
	^{35}Cl	−13.3(1)	−13.5(1)	−13.7(1)	−13.9(2)	*d*
GaI	^{69}Ga	−81.1(8)	−80.8(1)	−80.0(2)	−79.8(2)	*e*
	^{127}I	−370.6(1)	−373.3(2)	−375.9(2)	−378.3(2)	*e*

[a] P. Kuijpers and A. Dymanus, *Chem. Phys.*, **24**, 97 (1977).
[b] E. Tiemann, B. Hölzer, and J. Hoeft, *Z. Naturforsch.*, **32a**, 123 (1977).
[c] E. Tiemann and T. Möller, *Z. Naturforsch.*, **30a**, 986 (1975).
[d] E. Tiemann, M. Grashoff, and J. Hoeft, *Z. Naturforsch.*, **27a**, 753 (1972).
[e] K. P. R. Nair and J. Hoeft, *J. Mol. Spectrosc.*, **85**, 301 (1981).

Table 9.7 can be fitted to (9.188) with $eQq_e = -369.4(1)$ MHz, $eQq_I = -2.5(2)$ MHz, and $eQq_{II} = 0$.

An examination of the couplings listed in Table 9.7 will show that the magnitude of the coupling constant increases with the vibrational quantum number v for the more electronegative atom of the bond (i.e., the atom forming the negative pole of a polar bond) and decreases with v for the less electronegative atom (i.e., the one forming the positive pole of the polar bond). This trend is evident for other single-bonded, diatomic molecules listed in the Landolt-Börnstein tables [19]. In contrast, the coupling of the more electronegative ^{14}N atom in the triply bonded PN molecule decreases with increase in the vibrational quantum number, as it does in the triply bonded, nonpolar N_2. The ^{79}Br coupling in the nonpolar Br_2 decreases with increase of v, whereas in the polar IBr the ^{79}Br coupling increases with v. Thus the nuclear quadrupole coupling may increase or decrease with an increase of the vibrational state, depending on the nature of the bonding to the coupling atom. The relationship of quadrupole coupling and chemical bonding is discussed in Chapter XIV.

11 SELECTION RULES AND RELATIVE INTENSITIES OF HYPERFINE COMPONENTS

Selection rules for hyperfine structure, magnetic or quadrupole, depend on the nonvanishing matrix elements of the molecular dipole moment along space-fixed axes. The relative intensities are proportional to the square of these matrix elements expressed in the appropriate wave functions describing the states. When there is only one coupling nucleus and **J** forms with **I** the resultant **F**, the proper representation is J, i, I, F, M_F, and the line intensities I are proportional to

$$I \sim \sum_{M_F M'_F} |(J', i', I', F', M'_F|\mu|J, i, I, F, M_F)|^2 \tag{9.189}$$

The wave functions representing the hyperfine states can be expanded in terms of the uncoupled basis functions $|J, i, M_J, I, M_I)$ of the molecule as follows

$$\psi(J, i, I, F, M_F) = \sum_{M_J} C(JIF; M_J M_I M_F)\psi(J, i, M_J)\phi(I, M_I) \tag{9.190}$$

where the expansion coefficients are the Clebsch-Gordon coefficients and where $\psi(J, i, M_J)$ and $\phi(I, M_I)$ are the unperturbed rotational and nuclear spin functions, respectively. The molecular dipole moment is independent of the nuclear spin functions $\phi(I, M_I)$ which can be factored out of (9.189) and normalized to unity. Equation 9.189 then becomes

$$I \sim \sum_{M_F M'_F M_J M'_J} |C(J'IF'; M'_J M_I M'_F)|^2 |C(JIF; M_J M_I M_F)|^2 \times |(J', i', M'_J|\mu|J, i, M_J)|^2 \tag{9.191}$$

The squared matrix elements $|(J', i', M'_J|\mu|J, i, M_J|^2$ are those already given in Chapter II, Section 6 for the rigid rotor. Although the foregoing discussion brings out the basic procedure for evaluating the intensities, (9.191) is not in a form very convenient for calculation. Evaluation of (9.189) is most easily accomplished by the use of the calculus of tensor operators which is discussed further in Chapter XV, Section 6. It is found [64] that the relative intensities of the components of a hyperfine multiplet of a particular rotational transition are given by

For $J-1 \to J$

$F-1 \to F$:

$$\frac{B(J+F+I+1)(J+F+I)(J+F-I)(J+F-I-1)}{F} \tag{9.192}$$

$F \to F$:

$$-\frac{B(J+F+I+1)(J+F-I)(J-F+I)(J-F-I-1)(2F+1)}{F(F+1)} \tag{9.193}$$

$F+1 \to F$:

$$\frac{B(J-F+I)(J-F+I-1)(J-F-I-1)(J-F-I-2)}{F+1} \tag{9.194}$$

For $J \to J$

$F-1 \to F$:

$$-\frac{A(J+F+I+1)(J+F-I)(J-F+I+1)(J-F-I)}{F} \tag{9.195}$$

$F \to F$:

$$\frac{A[J(J+1)+F(F+1)-I(I+1)]^2(2F+1)}{F(F+1)} \tag{9.196}$$

$F+1 \to F$:

$$-\frac{A(J+F+I+2)(J+F-I+1)(J-F+I)(J-F-I-1)}{F+1} \tag{9.197}$$

where A and B are constants that depend on the strength of the unsplit rotational line. For convenience, the relative intensities calculated from these expressions are given in Appendix I. They are the same as those tabulated for atoms in various books on atomic spectra where F, J, I are replaced, respectively, by J, L, S.

When there is plural nuclear coupling, the foregoing formula no longer applies. The relative intensities must be found by evaluation of the squared matrix elements of the dipole moment in the representation of the coupled system. Understandably, the evaluations are more involved than those for a single coupling nucleus. For the case of two coupling nuclei, in which the coupling of I_1 is much stronger than that of I_2, the relative intensities of the components arising from I_1 can be obtained from the foregoing formulas, and the relative intensities of the additional hyperfine structure on the I_1 components due to I_2 may likewise be computed from (9.192)–(9.197) by replacement of J with F_1 and I with I_2. The general case for N-coupled nuclei is treated by Thaddeus et al. [33] and is discussed further in Chapter XV, Section 6. The relative intensities evaluated in the representation specified by (9.178) may be written

$$I(\alpha', F' \to \alpha, F) \sim |C(\alpha', F'; \alpha F)|^2 \tag{9.198}$$

where

$$C(\alpha', F'; \alpha F) = (-1)^u \{[(2F'_1+1)(2F_1+1)] \cdots [(2F'_N+1)(2F_N+1)]\}^{1/2} \times \begin{Bmatrix} F'_0 & F'_1 & I_1 \\ F_1 & F_0 & 1 \end{Bmatrix} \cdots \begin{Bmatrix} F'_{N-1} & F'_N & I_N \\ F_N & F_{N-1} & 1 \end{Bmatrix} \tag{9.199}$$

and

$$u = \sum_{i=1}^{N} (I_i + F'_{i-1} + F_i) + N \tag{9.200}$$

Here α stands for $F_0, F_1, \cdots F_{N-1}$ and $J' = F'_0$, $J = F_0$; $F = F_N$. Numerical values of the $6j$ symbols may be obtained from tables [36]. Simple formulas for some special cases are given by Edmonds [34]. If the intermediate F_i are not good quantum numbers, the intensities given by (9.198) may not be adequate. The matrix of **C** must then be transformed by means of a similarity transformation to the basis which diagonalizes the hyperfine Hamiltonian, i.e.,

$$\mathbf{C}' = \tilde{\mathbf{T}}_i \mathbf{C} \mathbf{T}_f \tag{9.201}$$

where $\mathbf{T}_i$ diagonalizes the energy matrix of the initial state of the transition and $\mathbf{T}_f$ diagonalizes the energy matrix of the final state. The squares of the elements of $\mathbf{C}'$ yield the relative intensities.

References

1. J. M. B. Kellogg, I. I. Rabi, N. F. Ramsey, and J. R. Zacharias, *Phys. Rev.*, **57**, 677 (1940).
2. N. F. Ramsey, *Molecular Beams*, Oxford Univ. Press, London, 1955.
3. H. B. G. Casimer, *On the Interaction between Atomic Nuclei and Electronics*, Teyler's Tweede Genootschap, E. F. Bohn, Haarlem, 1936.
4. A. Nordsieck, *Phys. Rev.*, **58**, 310 (1940).
5. B. T. Feld and W. E. Lamb, *Phys. Rev.*, **67**, 15 (1945).
6. W. E. Good, *Phys. Rev.*, **70**, 213 (1946).
7. D. K. Coles and W. E. Good, *Phys. Rev.*, **70**, 979 (1946).
8. J. H. Van Vleck, *Phys. Rev.*, **71**, 468 (1947).
9. J. K. Bragg, *Phys. Rev.*, **74**, 533 (1948).
10. J. K. Bragg and S. Golden, *Phys. Rev.*, **75**, 735 (1949).
11. G. R. Gunther-Mohr, C. H. Townes, and J. H. Van Vleck, *Phys. Rev.*, **94**, 1191 (1954).
12. R. Bersohn, *J. Chem. Phys.*, **20**, 1505 (1952).
13. M. H. Cohen and F. Reif, "Nuclear Quadrupole Effects in Solids," in *Solid State Physics*, Vol. 5, F. Seitz and D. Turnbull, Eds. Academic, New York, 1957, pp. 321–438.
14. T. P. Das and E. L. Hahn, *Nuclear Quadrupole Resonance Spectroscopy*, Academic, New York, 1958.
15. H. G. Dehmelt and H. Krüger, *Naturwiss.*, **37**, 111 (1950).
16. N. F. Ramsey, *Nuclear Moments*, Wiley, New York, 1953.
17. P. C. Cross, R. M. Hainer, and G. W. King, *J. Chem. Phys.*, **12**, 210 (1944).
18. C. A. Burrus and W. Gordy, *Phys. Rev.*, **92**, 274 (1953).
19. Landolt-Börnstein, *Numerical Data and Functional Relations in Science and Technology*. For a complete reference, see Chapter 1, [65].
20. J. Bardeen and C. H. Townes, *Phys. Rev.*, **73**, 97 (1948).
21. C. H. Townes and A. L. Schawlow, *Microwave Spectroscopy*, McGraw-Hill, New York, 1955.
22. R. J. Myers and W. D. Gwinn, *J. Chem. Phys.*, **20**, 1420 (1952).
23. G. W. Robinson and C. D. Cornwell, *J. Chem. Phys.*, **21**, 1436 (1953).
24. R. Bersohn, *J. Chem. Phys.*, **18**, 1124 (1950).
25. M. Mizushima and T. Ito, *J. Chem. Phys.*, **19**, 739 (1951).
26. P. N. Wolfe, *J. Chem. Phys.*, **25**, 976 (1956).
27. M. W. Long, Q. Williams, and T. L. Weatherly, *J. Chem. Phys.*, **33**, 508 (1960).
28. A. A. Wolf, Q. Williams, and T. L. Weatherly, *J. Chem. Phys.*, **47**, 5101 (1967).
29. K. K. Svidzinskii, "Theory of the Hyperfine Structure in the Rotational Spectra of Molecules," in *Soviet Maser Research*, D. V. Skobel'tsyn, Ed., Consultants Bureau, New York, 1964. (Translated from *Proc. P. N. Lebedev Phys. Inst.*, **21**, 88 (1963).

30. R. Livingston, *Phys. Rev.*, **82**, 289 (1951).
31. G. F. Hadley, *J. Chem. Phys.*, **26**, 1482 (1957).
32. G. Herrmann, *J. Chem. Phys.*, **29**, 875 (1958).
33. P. Thaddeus, L. C. Krisher, and J. H. N. Loubser, *J. Chem. Phys.*, **40**, 257 (1964).
34. A. R. Edmonds, *Angular Momentum in Quantum Mechanics*, 2nd ed., Princeton Univ. Press, Princeton, 1960.
35. B. R. Judd, *Operator Techniques in Atomic Spectroscopy*, McGraw-Hill, New York, 1963.
36. M. Rotenberg, R. Bivens, N. Metropolis, and J. K. Wooten, *The 3j and 6j Symbols*, The Technology Press, Massachusetts Inst. Tech., Cambridge, 1959.
37. J. Bardeen and C. H. Townes, *Phys. Rev.*, **73**, 647, 1204 (1948).
38. C. H. Townes, G. C. Dousmanis, R. L. White, and R. F. Schwarz, *Discussion Faraday Soc.*, **19**, 56 (1955).
39. J. W. Simmons and W. Gordy, *Phys. Rev.*, **73**, 713 (1948).
40. R. S. Henderson, *Phys. Rev.*, **74**, 107 (1948).
41. G. R. Gunther-Mohr, R. L. White, A. L. Schawlow, W. E. Good, and D. K. Coles, *Phys. Rev.*, **94**, 1184 (1954).
42. J. P. Gordon, *Phys. Rev.*, **99**, 1253 (1955).
43. K. Shimoda and K. Kondo, *J. Phys. Soc. Japan.*, **15**, 1125 (1960); **20**, 437 (1965).
44. S. G. Kukolich, *Phys. Rev.*, **156**, 83 (1967).
45. R. L. White, *Rev. Mod. Phys.*, **27**, 276 (1955).
46. D. W. Posener, *Australian J. Phys.*, **11**, 1 (1958).
47. D. W. Posener, *Australian J. Phys.*, **10**, 276 (1957); **13**, 168 (1960).
48. E. B. Treacy and Y. Beers, *J. Chem. Phys.*, **36**, 1473 (1962).
49. W. H. Flygare, *J. Chem. Phys.*, **41**, 206 (1964).
50. A. Okaya, *J. Phys. Soc. Japan.*, **11**, 258 (1956).
51. H. Takuma, T. Shimizu, and K. Shimoda, *J. Phys. Soc. Japan.*, **14**, 1595 (1959).
52. P. Thaddeus, L. C. Krisher, and P. Cahill, *J. Chem. Phys.*, **41**, 1542 (1964).
53. G. C. Dousmanis, T. M. Sanders, and C. H. Townes, *Phys. Rev.*, **100**, 1735 (1955).
54. C. A. Burrus and W. Gordy, *Phys. Rev.*, **92**, 1437 (1953).
55. J. J. Gallagher and C. M. Johnson, *Phys. Rev.*, **103**, 1727 (1956).
56. P. G. Favero, A. M. Mirri, and W. Gordy, *Phys. Rev.*, **114**, 1534 (1959).
57. R. A. Frosch and H. M. Foley, *Phys. Rev.*, **88**, 1337 (1952).
58. G. C. Dousmanis, *Phys. Rev.*, **97**, 967 (1955).
59. M. Mizushima, *Phys. Rev.*, **94**, 569 (1954).
60. C. C. Lin and M. Mizushima, *Phys. Rev.*, **100**, 1726 (1955).
61. R. F. Curl, J. L. Kinsey, J. G. Baker, J. C. Baird, G. R. Bird, R. F. Heidelberg, T. M. Sugden, D. R. Jenkins, and C. N. Kenney, *Phys. Rev.*, **121**, 1119 (1961).
62. G. R. Bird, J. C. Baird, A. W. Jache, J. A. Hodgeson, R. F. Curl, A. C. Kundle, J. W. Bransford, J. Rastrup-Anderson, and J. Rosenthal, *J. Chem. Phys.*, **40**, 3378 (1964).
63. P. D. Foster, J. A. Hodgeson, and R. F. Curl, Jr., *J. Chem. Phys.*, **45**, 3760 (1966).
64. E. U. Condon and G. H. Shortley, *The Theory of Atomic Spectra*, Cambridge Univ. Press, Cambridge, England, 1959.

31. R. Livingston, [illegible] 31 [illegible]
32. [illegible]
33. [illegible] Chem. Phys. [illegible]
34. [illegible] and [illegible], Chem. Phys. [illegible]
35. [illegible]
36. [illegible] McGraw-Hill, New York, 1966.
37. [illegible] Technology, Cambridge, 1952.
38. [illegible] Phys. Rev. [illegible]
39. [illegible] Chem. Phys. [illegible]
40. [illegible]
41. [illegible]
42. [illegible] Phys. Rev. 99, 1124 [illegible]
43. [illegible]
44. [illegible]
45. [illegible]
46. [illegible]
47. [illegible]
48. [illegible] Chem. Phys. [illegible]
49. [illegible]
50. [illegible]
51. [illegible]
52. [illegible] J. Chem. Phys. [illegible]
53. [illegible]
54. [illegible]
55. [illegible]
56. [illegible] Phys. Rev. [illegible]
57. [illegible]
58. [illegible] Phys. Rev. [illegible]
59. [illegible]
60. [illegible]
61. [illegible] Phys. Rev. 121 [illegible]
62. [illegible]
63. [illegible]
64. [illegible]

Chapter **X**

EFFECTS OF APPLIED ELECTRIC FIELDS

The Stark effect is a particularly useful auxiliary in microwave spectroscopy. From it, the most accurate evaluation of electric dipole moments of gaseous molecules can be made. It is also useful in the identification of pure rotational lines, particularly of asymmetric-top molecules, and it is widely employed as an aid to the detection of spectral lines [1].

1 LINEAR AND SYMMETRIC-TOP MOLECULES WITHOUT NUCLEAR COUPLING

Classically, the interaction energy of a dipole moment μ in a field $\mathscr{E}$ is $-\boldsymbol{\mu}\cdot\boldsymbol{\mathscr{E}}$. In the Stark effect of rotational spectra, $\mathscr{E}$ is an electric field fixed in space, and

μ is an electric dipole moment fixed in the molecule. Here $\mathscr{E}$ is assumed to be constant in magnitude and to have the fixed direction Z in space, and μ is assumed to be constant in the molecule-fixed reference system chosen as the principal inertial axes x, y, z. With these conditions, the Stark effect Hamiltonian operator can then be expressed as

$$\mathscr{H}_{\mathscr{E}} = -\mathscr{E} \sum_{g=x,y,z} \mu_g \Phi_{Zg} \tag{10.1}$$

where Φ_{Zg} are the direction cosines of the x, y, z axes with reference to the space-fixed Z axis. A small term $\alpha\mathscr{E}^2\Phi_{Zg}$ caused by anisotropic polarizability of the molecule which is neglected in (10.1) is treated in Section 6. Linear and symmetric-top molecules have a dipole moment component only along the symmetry axis z. For them, $\mu_z = \mu$, $\mu_x = \mu_y = 0$; the Hamiltonian then becomes simply

$$\mathscr{H}_{\mathscr{E}} = -\mu\mathscr{E}\Phi_{Zz} \tag{10.2}$$

where μ and $\mathscr{E}$ are constants and where Φ_{Zz} is the direction cosine of the axis of molecular symmetry with reference to the direction of the applied field. Although the magnitude of the field is periodically changed in a Stark modulation spectrometer, a square-wave modulation of low frequency is generally employed so that $\mathscr{E}$ can still be considered constant in value during the time interval when the Stark components are displaced. In comparison to the effects of the permanent dipole moment which must be present for detection of rotational spectra, those of the much small polarization moments induced through electronic displacements by the applied field are entirely negligible. We can consider μ as having a constant value in the molecule-fixed reference system even though centrifugal distortion causes it to vary slightly with vibrational state. In the present treatment we shall neglect these small effects and consider only Φ as varying with rotation.

For rotational lines observed in the microwave region with the field values usually applicable, the Stark energies can be evaluated with significant accuracy from perturbation theory. The first-order energy is simply the average of $\mathscr{H}_{\mathscr{E}}$ over the unperturbed rotational state. Since the linear molecule can be treated as a special case of the symmetric top with $K=0$, we shall express the average first in the unperturbed symmetric-top wave functions. In this representation the first-order Stark energies are

$$\begin{aligned} E_{\mathscr{E}}^{(1)} &= (J, K, M_J|\mathscr{H}_{\mathscr{E}}|J, K, M_J) \\ &= -\mu\mathscr{E}(J, K, M_J|\Phi_{Zz}|J, K, M_J) \\ &= -\frac{\mu\mathscr{E}KM_J}{J(J+1)} \end{aligned} \tag{10.3}$$

where the direction cosine matrix elements are evaluated from Table 2.1. Note that M_J is designated by M in Table 2.1.

When $K=0$, it is seen from (10.3) that $E_{\mathscr{E}}^{(1)}=0$. Thus there is no first-order Stark effect for a linear molecule nor for the $K=0$ levels of a symmetric-top molecule. We can also prove this by averaging Φ_{Zz} over the wave function ψ_{J,M_J} of the linear molecule.

The first-order energy can be obtained from the vector model of Fig. 10.1 which we shall describe because of the insight it gives into the Stark effect. In this model, **K** represents the direction of the symmetry axis and hence the direction of **μ**. In the normal motion of a symmetric top, **K**, which is a component of **J**, precesses about the direction of **J** while **J** precesses about the direction of the applied field, as indicated in the diagram. When the rotational energy is large as compared with the Stark energy, the precession of **K** about **J** is so rapid compared with that of **J** about $\mathscr{E}$ that the components of μ normal to **J** are averaged out, leaving only the component μ_J along **J** which effectively interacts with the field. Therefore the first-order Stark energy can be expressed as

$$E_{\mathscr{E}}^{(1)} = -\mathscr{E}\mu_J \cos\theta_{Z,J} \tag{10.4}$$

where

$$\cos\theta_{Z,J} = \frac{J_Z}{|\mathbf{J}|} = \frac{M_J}{[J(J+1)]^{1/2}} \tag{10.5}$$

and

$$\mu_J = \mu\cos\theta_{z,J} = \mu\frac{J_z}{|\mathbf{J}|} = \frac{\mu K}{[J(J+1)]^{1/2}} \tag{10.6}$$

Substitution of these values into (10.4) gives the expression already obtained, (10.3). It is seen that the vanishing of the first-order effect when $K=0$ can be attributed to the fact that $\cos\theta_{z,J}=0$ and hence that the molecular axis z is normal to **J**.

The second-order term of the Stark perturbation for the symmetric-top molecule is

$$\begin{aligned} E_{\mathscr{E}}^{(2)} &= \sum_{J'}{}' \frac{|(J,K,M_J|\mathscr{H}_{\mathscr{E}}|J',K,M_J)|^2}{E_{J,K}-E_{J',K}} \\ &= \mu^2\mathscr{E}^2\left[\frac{(J,K,M_J|\Phi_{Zz}|J+1,K,M_J)^2}{E_{J,K}-E_{J+1,K}} + \frac{(J,K,M_J|\Phi_{Zz}|J-1,K,M_J)^2}{E_{J,K}-E_{J-1,K}}\right] \end{aligned} \tag{10.7}$$

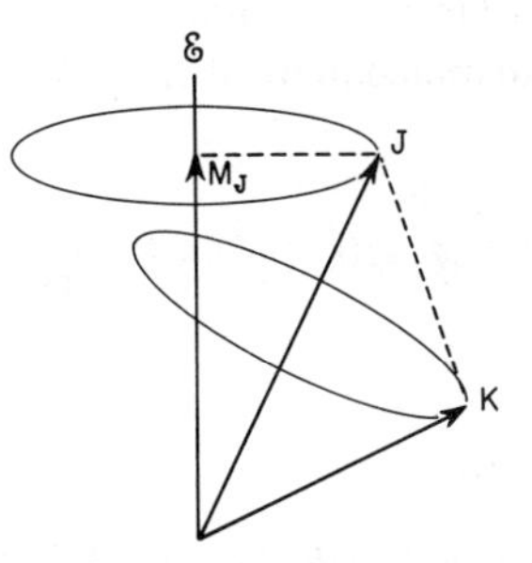

Fig. 10.1 Vector model of a symmetric-top molecule in an electric field.

The direction cosine matrix elements are found from Table 2.1 and the energy differences

$$E_{J,K}-E_{J+1,K}=-2hB(J+1) \tag{10.8}$$

$$E_{J,K}-E_{J-1,K}=2hBJ \tag{10.9}$$

are evaluated from (6.6). Substitution of the values into (10.7) shows that the second-order Stark energy of the symmetric-top molecule is

$$E_{\mathscr{E}}^{(2)}=\frac{\mu^2\mathscr{E}^2}{2hB}\left\{\frac{(J^2-K^2)(J^2-M_J^2)}{J^3(2J-1)(2J+1)}-\frac{[(J+1)^2-K^2][(J+1)^2-M_J^2]}{(J+1)^3(2J+1)(2J+3)}\right\} \tag{10.10}$$

The Stark energy for the linear molecule, obtained by setting $K=0$ in (10.10) is

$$E_{\mathscr{E}}^{(2)}=\frac{\mu^2\mathscr{E}^2}{2hB}\frac{[J(J+1)-3M_J^2]}{J(J+1)(2J-1)(2J+3)} \tag{10.11}$$

Although the $J=0$ level cannot be split by the field since for it $M_J=0$, it has a second-order Stark displacement which is given by

$$E_{\mathscr{E}}^{(2)}(J=0)=-\frac{\mu^2\mathscr{E}^2}{6hB} \tag{10.12}$$

as can be seen by substitution of $J=0$, $K=0$, and $M_J=0$ into (10.10).

The M_J selection rules for rotational lines are

$$M_J\rightarrow M_J \qquad \text{and} \qquad M_J\rightarrow M_J\pm 1 \tag{10.13}$$

The first of these transitions $\Delta M_J=0$, sometimes designated as π components, are observed when the Stark field is parallel to the electric vector of the microwave radiation. The $\Delta M_J=\pm 1$ transitions, designated as σ components, are observed when the dc field is perpendicular to the electric vector of the radiation. The most commonly observed are the $\Delta M_J=0$ components. The first-order frequency displacement of the $K\neq 0$ lines of these components are seen from (10.3) to be

$$\Delta\nu^{(1)}(\Delta M_J=0)=2\left(\frac{\mu\mathscr{E}}{h}\right)\frac{KM_J}{J(J+1)(J+2)} \tag{10.14}$$

To this must be added the small second-order corrections when the splitting becomes large (when $\mu\mathscr{E}$ is large). For linear molecules or for symmetric-top molecules with $K=0$, the displacement of the $\Delta M_J=0$ components caused by the second-order effect are

$$\Delta\nu^{(2)}(\Delta M_J=0,\ J=0\rightarrow 1)=\frac{8}{15}\frac{\mu^2\mathscr{E}^2}{h^2\nu_0} \tag{10.15}$$

and

$$\Delta\nu^{(2)}(\Delta M_J=0,\ J\rightarrow J+1,\ J\neq 0)$$

$$=\frac{2\mu^2\mathscr{E}^2}{h^2\nu_0}\left[\frac{3M_J^2(8J^2+16J+5)-4J(J+1)^2(J+2)}{J(J+2)(2J-1)(2J+1)(2J+3)(2J+5)}\right] \tag{10.16}$$

In these equations ν_0 represents the frequency of the unsplit rotational line observed when no field is applied; $\Delta\nu$ represents the displacements caused by the applied field $\mathscr{E}$. If the factor $2/h^2$ in (10.16) is replaced by 0.5069 and the factor $8/15h^2$ in (10.15) is replaced by 0.1352, the equations give $\Delta\nu$ in MHz when ν_0 is in MHz units, μ is in debye units, and $\mathscr{E}$ is in volts per centimeter.

It is evident from the foregoing equations that the magnitude of the Stark splitting of the rotational levels decreases with increase in J. The largest displacement occurs for $M_J = J$. For high J values the maximum second-order displacement varies inversely with J^2 approximately. It also varies inversely with B so that the second-order Stark effect becomes increasingly difficult to observe as the submillimeter wave region is approached. If one chooses a light molecule so that high-frequency transitions occur at low J values, the large B value then makes low the second-order Stark sensitivity. Note, however, that the first-order effect, (10.3), which is very much more sensitive than the second-order effect, is independent of B. The maximum first-order $\Delta\nu$, which occurs for $K = J$, $M_J = J$, varies approximately as $1/J$; but, because of the great sensitivity of the first-order effect, the splittings of symmetric-top lines for $K \neq 0$ can be observed with ease for very high J values of transitions occurring in the submillimeter wave region. Figure 10.2 illustrates the first-order Stark effect

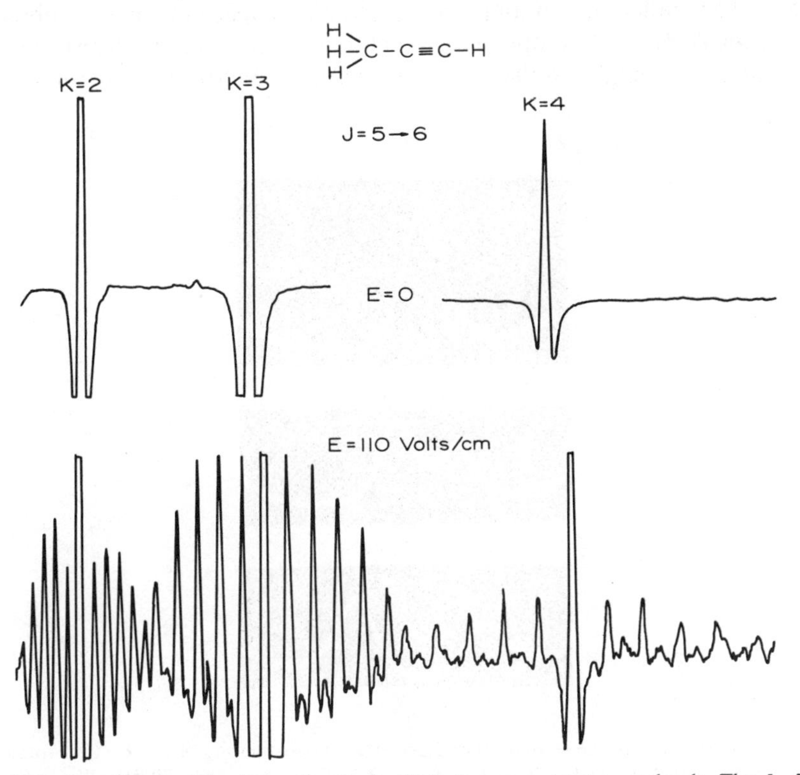

Fig. 10.2 Illustration of the first-order Stark effect of a symmetric-top molecule. The $J = 5 \rightarrow 6$ transition of CH_3CCH. From P. A. Steiner, Ph.D. dissertation, Duke University, 1964.

of a symmetric-top molecule observed in the 2.9-mm wave region with a field of 110 v/cm. Figure 10.3 illustrates the second-order effect for a linear molecule as observed in the centimeter wave region with a field up to 1070 v/cm.

Figure 10.4 shows an energy level diagram of the Stark splitting of the $J=1 \rightarrow 2$ transition of a symmetric-top molecule with the π and σ absorption transitions indicated by solid arrows. Such a diagram cannot be drawn to scale because the separations of the rotational levels are generally of the order of 100 times greater than the Stark splitting; and for the same field value, the first-order splitting of the $K=1$ lines is of the order of 10 to 100 times that of the second-order splitting of the $K=0$ lines. Nevertheless, the diagram gives a qualitative indication of the expected Stark effect. The $K=0$, $\Delta M_J=0$ transitions correspond to the observed components shown for a linear molecule in Fig. 10.3.

Because the field orientations required for observations of the π and σ components are different, these components are generally not observed at the same time. Nearly all measurements in the past have been made on $\Delta M_J=0$ components, which require the dc electric field to be parallel to the electric vector of the microwave radiation. These are the components observed with a rectangular waveguide Stark cell having for the Stark electrode a metal strip held by dielectric supports in the center of the waveguide. The development of the parallel-plate, millimeter-wave Stark cell made possible the observation of the $\Delta M_J=\pm 1$ components, which require that the dc electric field be imposed at right angles to the electric vector of the microwave radiation.

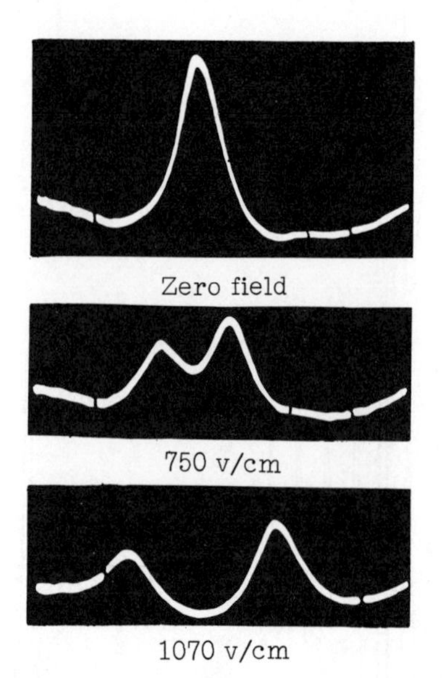

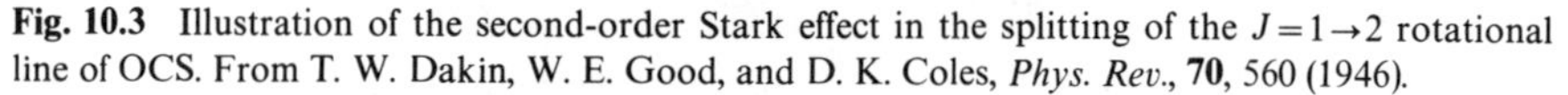

Fig. 10.3 Illustration of the second-order Stark effect in the splitting of the $J=1 \rightarrow 2$ rotational line of OCS. From T. W. Dakin, W. E. Good, and D. K. Coles, *Phys. Rev.*, **70**, 560 (1946).

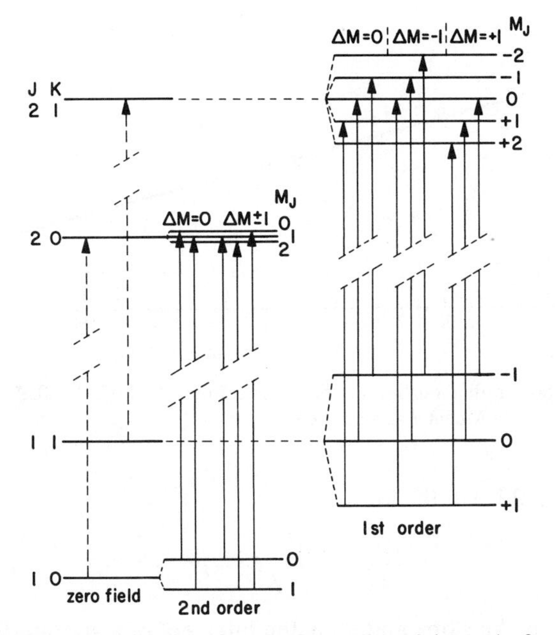

Fig. 10.4 Diagram (not drawn to scale) of the Stark effect of the $J=1\rightarrow 2$ transitions of a symmetric-top molecule.

Perturbation theory to second order is sufficient to account for the Stark splitting within the accuracy of most measurements which are made with an imposed field of a few thousand volts/centimeter. When there is a first-order effect, one needs to apply only a few hundred volts/centimeter to obtain a very wide separation of the Stark components ($\sim$100 MHz or more), and first-order plus second-order terms are completely adequate. When there is no first-order effect, one often finds it advantageous to apply several thousand volts/centimeter to obtain sufficiently large displacements for precise measurement of $\Delta\nu$. With voltage of the order of 5000 v/cm, Muenter and Laurie [2] detected deviations of as much as 1% from the second-order theory in $K=0$ transitions of some symmetric-top molecules. For linear molecules or symmetric tops with $K=0$, all odd perturbation terms are zero. Hence the next term of significance is of fourth order. To correct for the fourth-order perturbation in the precise measurement of a dipole moment it is fortunately not necessary to employ explicitly the complicated fourth-order formula. The displacement of a particular Stark component for which there is no odd-order effect can be expressed by the simple equation

$$\Delta\nu = a\mu^2\mathscr{E}^2 + b\mu^4\mathscr{E}^4 + \cdots \tag{10.17}$$

where a and b are constants. The constant a can be obtained from the second-order formula, (10.11), or for the $\Delta M_J=0$ component more simply from (10.16).

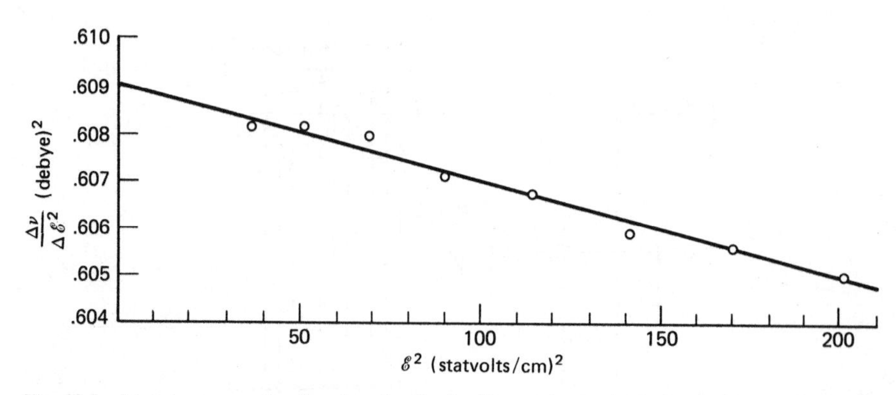

Fig. 10.5 Plot demonstrating fourth-order Stark effect in the $J=1\rightarrow2$, $K=0$, $\Delta M_J=0$ transition of methyl acetylene. From Muenter and Laurie [2].

If one expresses (10.17) in the form

$$\frac{\Delta\nu}{a\mathscr{E}^2}=\left(\frac{b\mu^4}{a}\right)\mathscr{E}^2+\mu^2 \tag{10.18}$$

it is seen $(b\mu^4/a)$ is the slope and μ^2 is the intercept of a straight line obtained by a plotting of $(\Delta\nu/a\mathscr{E}^2)$ versus $\mathscr{E}^2$. Figure 10.5 shows such a plot for the $J=1\rightarrow2$, $K=0$ transition of CH_3CCH by Muenter and Laurie. From projection of the straight line one obtains μ^2 as the intercept. With this value one can then obtain b from the slope if this is desirable.

Effects of anisotropic polarization, detectable in some molecules at very high field values, are treated in Section 6.

The relative intensities of the Stark components of molecules without hyperfine structure depend only on J and M_J and hence are the same for all classes of molecules. The relative intensity formulas given in Section 7 are simply the M_J-dependent terms of the squared direction-cosine matrix elements given in Chapter II.

2 LINEAR AND SYMMETRIC-TOP MOLECULES WITH NUCLEAR QUADRUPOLE COUPLING

The theory of the Stark effect in the rotational spectra of linear molecules with nuclear quadrupole coupling was first treated by Fano [3]; that of symmetric-top molecules, by Low and Townes [4]. Buckingham and Stephens [5] developed this theory further to take into account such cross-product terms as $\mathscr{H}_{\mathscr{E}}\mathscr{H}_Q$ and $\mathscr{H}^2_{\mathscr{E}}\mathscr{H}_Q$ which have measurable effects for molecules having large nuclear quadrupole coupling. Coester [6] has treated the combined Stark-Zeeman effect for symmetric-top molecules with nuclear quadrupole coupling.

The complete Hamiltonian for a rotating molecule having nuclear quadrupole coupling with an imposed electric field can be expressed as

$$\mathscr{H}=\mathscr{H}_r+\mathscr{H}_Q+\mathscr{H}_{\mathscr{E}} \tag{10.19}$$

where $\mathscr{H}_r$ is the Hamiltonian operator for the pure rotational energy described in Chapter II, $\mathscr{H}_Q$ is that for the nuclear quadrupole interaction described in Chapter IX, and $\mathscr{H}_\mathscr{E}$ is that just described for the Stark interaction of molecules without hyperfine structure. In finding the characteristic energies it is convenient to consider three cases separately. First is the weak-field case, in which $\mathscr{H}_\mathscr{E} \ll \mathscr{H}_Q$; second, the intermediate-field case, in which $\mathscr{H}_\mathscr{E} \sim \mathscr{H}_Q$; and third, the strong-field case, in which $\mathscr{H}_\mathscr{E} \gg \mathscr{H}_Q$. The weak- and strong-field cases are simpler than the intermediate one, and the greater number of measurements of Stark effect and dipole moments have been made with applied fields such that one or the other of these simple cases holds. When the nuclear coupling is relatively strong, as that for Br or I in organic molecules, the weak-field case can be used; when the coupling is relatively weak, as that for B or N, the strong-field case can be easily achieved. Perturbation theory allows the derivation of a closed formula for weak- and strong-field cases; for the intermediate field-cases a secular equation must be solved. Plural nuclear coupling will be treated in the strong-field case only, for which the effect is particularly simple.

The Weak-field Case

When $\mathscr{H}_\mathscr{E} \ll \mathscr{H}_Q$, the Stark interaction can be treated as a perturbation on the hyperfine state. The first-order Stark energy of a symmetric-top is then the average of the $\mathscr{H}_\mathscr{E}$ over the wave function $\psi(J, K, I, F, M_F)$ of the nuclear-coupled molecule unperturbed by the Stark field, that is, the wave function of the operator $\mathscr{H}_r + \mathscr{H}_Q$. This wave function can be expanded in terms of the function for the molecule without nuclear coupling:

$$\psi(J, K, I, F, M_F) = \sum_{M_J} C(JIF; M_J M_I M_F)\psi(J, K, M_J)\phi(I, M_I) \quad (10.20)$$

where the C's are known as Clebsch-Gordon coefficients, values of which are given by Condon and Shortley [7]. (See also Rose [8].) These coefficients arise when, as here, the problem of combining two commuting angular momenta to form a resultant is considered, in particular $\mathbf{F} = \mathbf{J} + \mathbf{I}$. These coefficients are independent of K and vanish unless $M_F = M_J + M_I$ and unless F is one of the following values (see Chapter XV for further discussion).

$$F = J + I,\ J + I - 1, \ldots, |J - I|$$

The first-order Stark term is

$$\begin{aligned} E_\mathscr{E}^{(1)} &= (J, K, I, F, M_F|\mathscr{H}_\mathscr{E}|J, K, I, F, M_F) \\ &= -\mu\mathscr{E} \sum_{M_J} |C(JIF; M_J M_I M_F)|^2 (J, K, M_J|\Phi_{Zz}|J, K, M_J) \quad (10.21) \end{aligned}$$

Since the direction cosine operator Φ_{Zz} is independent of the nuclear spin function $\phi(I, M_I)$, these spin functions are factored out and normalized to unity $(\phi|\phi) = 1$. The quantity $(J, K, M_J|\Phi_{Zz}|J, K, M_J)$ has been evaluated as

$KM_J/[J(J+1)]$ in (10.3). Therefore

$$E_{\mathscr{E}}^{(1)} = -\mu\mathscr{E} \sum_{M_J} |C(JIF;\, M_J M_I M_F)|^2 \frac{KM_J}{J(J+1)}$$

$$= -\frac{\mu\mathscr{E}K}{J(J+1)} \sum_{M_J} |C(JIF;\, M_J M_I M_F)|^2 M_J \qquad (10.22)$$

The quantity in the summation is the average of M_J over the function $\psi(J, K, I, F, M_F)$ which is simply the component of J along the space-fixed axis in the $\mathbf{I}\cdot\mathbf{J}$-coupled vector model of Fig. 10.6. This value can easily be found from the vector model. For example, the component of $\mathbf{J}$ along $\mathbf{F}$ is $|\mathbf{J}| \cos(\mathbf{F}, \mathbf{J})$, and the averaged component along Z is

$$\langle M_J \rangle = |\mathbf{J}| \cos(\mathbf{F}, \mathbf{J}) \cos(\mathbf{F}, \mathbf{Z}) \qquad (10.23)$$

From application of the law of the cosine with the vector model

$$\cos(\mathbf{F}, \mathbf{J}) = \frac{\mathbf{F}^2 + \mathbf{J}^2 - \mathbf{I}^2}{2|\mathbf{F}|\,|\mathbf{J}|} = \frac{F(F+1) + J(J+1) - I(I+1)}{2[F(F+1)]^{1/2}[J(J+1)]^{1/2}} \qquad (10.24)$$

also

$$\cos(\mathbf{F}, \mathbf{Z}) = \frac{M_F}{|\mathbf{F}|} = \frac{M_F}{[F(F+1)]^{1/2}} \qquad (10.25)$$

and hence

$$\langle M_J \rangle = M_F \frac{F(F+1) + J(J+1) - I(I+1)}{2F(F+1)} \qquad (10.26)$$

Substitution of this value for $\langle M_J \rangle$ for the summation into (10.22) shows the first-order Stark energy to be

$$E_{\mathscr{E}}^{(1)} = -\frac{\mu\mathscr{E}KM_F[F(F+1) + J(J+1) - I(I+1)]}{2J(J+1)F(F+1)} \qquad (10.27)$$

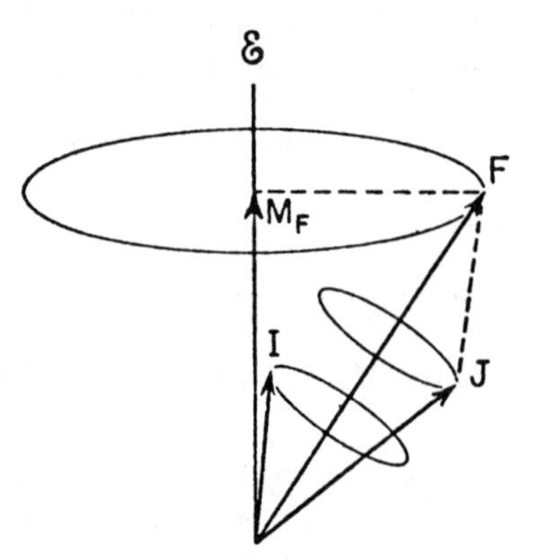

Fig. 10.6 Vector model of the weak-field Stark effect of a symmetric-top molecule with nuclear coupling.

One can derive (10.27) entirely from the vector model of Fig. 10.6 by first resolving the component μ, which is wholly along the symmetry axis **K**, from **K** to **J**. Thus,

$$\mu_J = \mu \frac{K}{[J(J+1)]^{1/2}} \tag{10.28}$$

$$\mu_F = \mu_J \cos(\mathbf{F}, \mathbf{J}) \tag{10.29}$$

and

$$\mu_{\mathscr{E}} = \mu_F \cos(\mathbf{F}, \mathbf{Z}) = \mu_F \frac{M_F}{[F(F+1)]^{1/2}} \tag{10.30}$$

The Stark energy is just $-\mathscr{E}$ times the average dipole moment in the direction of the field. Therefore

$$E_{\mathscr{E}} = -\mu_{\mathscr{E}}\mathscr{E} = -\frac{\mu\mathscr{E}KM_F}{[J(J+1)]^{1/2}[F(F+1)]^{1/2}} \cos(\mathbf{F}, \mathbf{J}) \tag{10.31}$$

Substitution of the value of cos (**F**, **J**) from (10.24) into (10.31) yields (10.27).

It should be noted that the first-order Stark effect vanishes when $K=0$, as it does for molecules without nuclear coupling. This condition also applies for linear molecules. When $M_F = F = J + I$, the first-order formula reduces to the simple form,

$$E_{\mathscr{E}}^{(1)} = -\frac{\mu\mathscr{E}K}{J+1} \qquad \text{when } M_F = F = J + I \tag{10.32}$$

which is equivalent to (10.3) for molecules without quadrupole coupling when $M_J = J$.

The second-order term in the weak-field case is

$$E_{\mathscr{E}}^{(2)} = \sum_{F',J'}{}' \frac{|(J, K, I, F, M_F|\mathscr{H}_{\mathscr{E}}|J', K, I, F', M_F)|^2}{E_{J,K,I,F,M_F} - E_{J',K,I,F',M_F}} \tag{10.33}$$

Note that the term $F=F'$, $J=J'$ is to be excluded from the sum. This expression can be separated into two sums. By substitution of the $\mathscr{H}_{\mathscr{E}}$ from (10.2) it becomes

$$\begin{aligned} E_{\mathscr{E}}^{(2)} &= \mu^2\mathscr{E}^2 \sum_{F' \neq F} \frac{|(J, K, I, F, M_F|\Phi_{Zz}|J, K, I, F', M_F)|^2}{E_Q(F) - E_Q(F')} \\ &+ \mu^2\mathscr{E}^2 \sum_{F', J' \neq J} \frac{|(J, K, I, F, M_F|\Phi_{Zz}|J', K, I, F', M_F)|^2}{E_r(J) - E_r(J')} \end{aligned} \tag{10.34}$$

where the approximation $E_{J,K,F} - F_{J',K,F'} \approx E_{J,K} - E_{J',K}$ has been made. The first of these expressions is the larger because of the smaller energy difference in the denominator. To evaluate the terms on the right, one expresses the functions of $|J, K, I, F, M_F)$ in terms of those of the field-free rotor with the nucleus decoupled, as in the expansion of (10.20). When evaluated, the first term on the right contains the factor K and hence vanishes when $K=0$. The second term gives the second-order correction for symmetric-top levels when $K=0$, or for linear molecules. It has the values [3, 4]

$$E_{\mathscr{E}}^{(2)}(K=0)=-\frac{\mu^2\mathscr{E}^2[3M_F^2-F(F+1)][3D(D-1)-4F(F+1)J(J+1)]}{hB2J(J+1)(2J-1)(2J+3)2F(F+1)(2F-1)(2F+3)} \tag{10.35}$$

in which

$$D=F(F+1)-I(I+1)+J(J+1) \tag{10.36}$$

As before, the case where $M_F=F=J+I$ is particularly simple since the Stark effect is independent of the effects of nuclear quadrupole coupling. As noted previously, because of the smaller energy difference in its denominator, the first term on the right of (10.34) is larger than the second term, but it is absent unless the much larger, first-order effect of (10.27) is present. Since the field values must be sufficiently small to insure that the Stark splitting is small compared with the quadrupole splitting for the weak field case to apply, the second-order terms are not of much value except when $K=0$. Hence we shall not reproduce here the rather complicated second-order term which vanishes when $K=0$. An explicit formula for the term is given by Coester [6]. In addition to the above terms, Buckingham and Stephens [5] show that for strong nuclear quadrupole coupling, which is necessary for practical application of the weak-field case, cross terms of the form

$$2\sum_{J'\neq J}\frac{(J,K,I,F,M_F|\mathscr{H}_{\mathscr{E}}|J',K,I,F,M_F)(J',K,I,F,M_F|\mathscr{H}_Q|J,K,I,F,M_F)}{E(J)-E(J')}$$

are important. With this term included, the Stark displacement of a particular component in the weak-field case can be expressed [5] as

$$\Delta\nu=a\mu\mathscr{E}+b\frac{(\mu\mathscr{E})(eQq)}{B}+c\frac{\mu^2\mathscr{E}^2}{eQq} \tag{10.37}$$

where a, b, and c depend only on the quantum numbers for the transition.

The M_F selection rules are

$$M_F\rightarrow M_F \qquad \text{and} \qquad M_F\rightarrow M_F\pm 1 \tag{10.38}$$

The $\Delta M_F=0$ transitions are observed when the dc Stark field is parallel to the radiofrequency electric field, while $\Delta M_F=\pm 1$ transitions are observed when these fields are perpendicular.

The Strong-field Case: Single or Plural Nuclear Coupling

The simplest and probably the most useful case is that in which the applied field $\mathscr{E}$ is sufficiently strong to break down the nuclear coupling between **I** and **J** but not large enough to perturb significantly the rotational state. This case occurs when $\mathscr{H}_Q \ll \mathscr{H}_{\mathscr{E}} \ll \mathscr{H}_r$. It is represented by the vector model of Fig. 10.7 in which both **J** and **I** precess about the direction of the field. This model is similar to that of the familiar Paschen-Back effect observed with strong magnetic fields in atomic spectra.

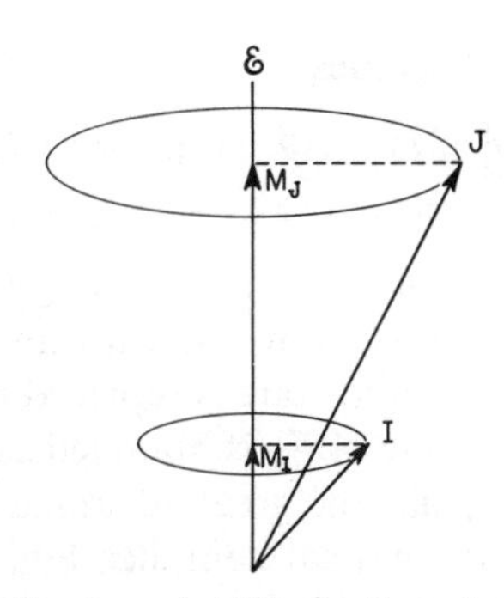

Fig. 10.7 Vector model of the strong-field Stark effect of a molecule with nuclear coupling.

Since **J** and **I** are decoupled, the quantum number F is destroyed as a good quantum number and the J, I, F scheme is no longer appropriate. The representation appropriate to the strong-field case is J, K, I, M_J, M_I, and in this representation the required matrix elements of $\mathscr{H}_r$, $\mathscr{H}_Q$, and $\mathscr{H}_{\mathscr{E}}$ are easily found. The rotational Hamiltonian $\mathscr{H}_r$ is independent of I and M_I, and the values are those already obtained in Chapter VI. Since the matrix elements of $\mathscr{H}_{\mathscr{E}}$ are likewise independent of the nuclear spin functions $|I, M_I)$ in this decoupled case, the values of the Stark energy are the same as those for molecules without nuclear coupling described in Section 1. Hence we have only to evaluate $\mathscr{H}_Q$.

Under the assumed conditions, only the Z component of the molecular field component, or $V_{ZZ}=(\partial^2 V/\partial Z^2)$, interacts with the nuclear quadrupole moment, and only with the Z component of eQ_{ZZ}. The precession of **J** about **Z** is so rapid relative to that of **I** that the X and Y components are averaged out. The quadrupole Hamiltonian can be expressed as

$$\mathscr{H}_Q=\tfrac{1}{4}Q_{ZZ}V_{ZZ} \tag{10.39}$$

With the value of $(Q_{ZZ})_{\rm op}$ obtained from (9.53) with $i=j=Z$ and with V_{ZZ} given by (9.48), it is seen that

$$(\mathscr{H}_Q)_{ZZ}=\frac{eQq_J}{4J(2J-1)I(2I-1)}[3J_Z^2-J(J+1)][3I_Z^2-I(I+1)] \tag{10.40}$$

This Hamiltonian is diagonal in the specified representation with J_Z and I_Z having the eigenvalues M_J and M_I, respectively. The quantity q_J, which is independent of M_J and M_I, has already been evaluated in Chapter IX, Section 7. For the symmetric top, its values are given to first order by (9.88). Thus for the symmetric top in the strong-field case the quadrupole energies are

$$E_Q=\frac{eQq}{4(2J-1)(2J+3)I(2I-1)}\left[\frac{3K^2}{J(J+1)}-1\right] \\ \times[3M_J^2-J(J+1)][3M_I^2-I(I+1)] \tag{10.41}$$

which for the linear molecule becomes

$$E_Q = -\frac{eQq[3M_J^2 - J(J+1)][3M_I^2 - I(I+1)]}{4(2J-1)(2J+3)I(2I-1)} \tag{10.42}$$

The nuclear quadrupole interaction in the strong-field case is very similar to the interaction of a quadrupole moment with an axially symmetric field in a solid. If q_{ZZ} of (9.24) for the solid state is replaced by the value of V_{ZZ} from (9.48), it is seen that the same value of E_Q is obtained as is given by (10.41).

The strong-field Stark effect has the great advantage that coupling by more than one nucleus causes no theoretical complication, as it does for field-free molecules or for the weak-field or intermediate-field Stark effect. The value for the E_Q of each coupling nucleus is calculated from (10.41), and the results are added:

$$E_Q = E_{Q_1}(I_1) + E_{Q_2}(I_2) + \cdots$$

The presence of magnetic hyperfine structure also causes no particular complication in the strong-field case, for which all nuclear coupling is broken down. For molecules in $^1\Sigma$ states the magnetic coupling is weak and easily decoupled. The strong-field energies are given by

$$E_m = (J, K, M_J, M_I|\mathscr{H}_m|J, K, M_J, M_I) = C_{J,K}M_JM_I \tag{10.43}$$

where $C_{J,K}$ depends only on the quantum numbers J and K and is given by (9.158). The magnetic energy for the different nuclei are additive, like those for the quadrupole energy. This simplification is possible because the precession of **J** is so dominated by the Stark field that coupling by a given nucleus does not significantly alter the value of the electric field gradient or the magnetic field value at the other nuclei. This is similar to the situation in solids where the nuclear coupling does not alter the field gradient because the directions of the field gradients are "frozen in." Hence the quadrupole couplings by the different nuclei of the solid state are simply additive, as in the strong-field case. In addition to the usual selection rules for M_J, we have $\Delta M_{I_i} = 0$.

The resultant energy for the strong-field case is the sum

$$E = E_r + E_{\mathscr{E}}^{(1)} + E_{\mathscr{E}}^{(2)} + \sum_i E_{Q_i} + \sum_i E_{m_i} \tag{10.44}$$

where the values of $E_{\mathscr{E}}^{(1)}$ and $E_{\mathscr{E}}^{(2)}$ are those given in Section 1 and where E_{Q_i} and E_{m_i} are those given above.

Although the strong-field case can generally be achieved when eQq is not large, there is one important exception for $K=0$ which occurs for integral spin values. When $M_J = \pm 1$ and also $M_I = \pm 1$, there is an admixture of the $M_J = \pm 1$ and the $M_I = \mp 1$ states which causes a breakdown in the $M_J = \pm 1$ degeneracy and hence a failure of the strong-field formulas. To obtain the energy for this case one must solve the secular equation

$$\begin{vmatrix} E_{\mathscr{E}}(M_J=1) + (1, -1|\mathscr{H}_Q|1, -1) - E & (1, -1|\mathscr{H}_Q|-1, 1) \\ (-1, 1|\mathscr{H}_Q|1, -1) & E_{\mathscr{E}}(M_J=-1) + (-1, 1|\mathscr{H}_Q|-1, 1) - E \end{vmatrix} = 0 \tag{10.45}$$

for the intermediate-field case described in the next section. The abbreviation $|M_J, M_I)$ is used for $|J, I, M_J, M_I)$. It is apparent from (10.45) that there are off-diagonal matrix elements that directly connect $M_J=+1$ and $M_J=-1$. For $E_\mathscr{E}(|M_J|=1)$, the values from (10.10) can be used. The matrix elements of $\mathscr{H}_Q$ are obtained from (10.53). Solution of this equation is like that for (10.55).

The Intermediate-field Case

When the applied field is such as to make $\mathscr{H}_\mathscr{E}$ of comparable magnitude to $\mathscr{H}_Q$, one must treat $\mathscr{H}_Q+\mathscr{H}_\mathscr{E}$ as a common perturbation on the pure rotational states instead of treating $\mathscr{H}_\mathscr{E}$ as a perturbation on the functions of $\mathscr{H}_r+\mathscr{H}_Q$, as was done for weak fields. This can be achieved by solution of a secular equation which is obtained from

$$\mathscr{H}\psi=E\psi \tag{10.46}$$

where

$$\mathscr{H}=\mathscr{H}_r+\mathscr{H}_Q+\mathscr{H}_\mathscr{E} \tag{10.47}$$

To find E, it is convenient to express ψ in the eigenfunctions of the operator $\mathscr{H}_r$ of the unperturbed rotor

$$\psi=\sum C_i\psi_i(J, K, I, M_J, M_I) \tag{10.48}$$

Subsitution of these functions into (10.46), multiplication by ψ_j, followed by integration and transformation, yield

$$\sum_i C_i[(\psi_j|\mathscr{H}_r+\mathscr{H}_\mathscr{E}+\mathscr{H}_Q|\psi_i)-E\delta_{ij}]=0 \tag{10.49}$$

Since $\mathscr{H}_r$ is diagonal in all the quantum numbers J, K, I, M_J, M_I, and since $\mathscr{H}_Q$ is diagonal in J, K, I, (10.49) can be expressed as

$$\sum_i C_i[(J, K, I, M_J, M_I|\mathscr{H}_\mathscr{E}|J', K, I, M_J, M_I) \\ +(J, K, I, M_J, M_I|\mathscr{H}_Q|J, K, M_J', M_I')+(E_{J,K}-E)\delta_{ij}]=0 \tag{10.50}$$

where $E_{J,K}$ is the unperturbed rotational energy. Because $\mathscr{H}_\mathscr{E}$ has off-diagonal elements only in J, the matrix can be conveniently reduced by a procedure that begins with evaluation of $\mathscr{H}_\mathscr{E}$ for the rotor without nuclear coupling, as was made in Section 1. Let the resulting values be

$$E_\mathscr{E}(M_J)=E_\mathscr{E}^{(1)}(M_J)+E_\mathscr{E}^{(2)}(M_J) \tag{10.51}$$

where $E_\mathscr{E}^{(1)}(M_J)$ and $E_\mathscr{E}^{(2)}(M_J)$ are given by (10.3) and (10.10). The secular equation can then be expressed as

$$|(J, K, I, M_J, M_I|\mathscr{H}_Q|J, K, I, M_J', M_I')+[E_{J,K}+E_\mathscr{E}(M_J)-E\delta_{M_JM_J'}\delta_{M_IM_I'}|=0 \tag{10.52}$$

This simplification is possible because the second-order perturbation in J by the Stark effect can be neglected in the evaluation of the matrix elements of

$\mathscr{H}_Q$. The required matrix elements [9, 6] are

$$
\begin{aligned}
&(J, K, I, M_J, M_I|\mathscr{H}_Q|J, K, I, M'_J, M'_I) \\
&\quad = P(J, K, I)\{\delta_{M'_J, M_J}\delta_{M'_I, M_I}[3M_I^2 - I(I+1)] \\
&\qquad \times [3M_J^2 - J(J+1)] + (\delta_{M'_J, M_J \pm 1}\delta_{M'_I, M_I \mp 1}) \\
&\qquad \times [\tfrac{3}{2} + 3M_I M_J + 3(M_I \mp 1)(M_J \pm 1)] \\
&\qquad \times ([I(I+1) - M_I(M_I \mp 1)][J(J+1) - M_J(M_J \pm 1)])^{1/2} + (\delta_{M'_J M_J \pm 2}\delta_{M'_I M_I \mp 2}) \\
&\qquad \times \tfrac{3}{2}([J(J+1) - M_J(M_J \pm 1)] \times [J(J+1) - (M_J \pm 1)(M_J \pm 2)] \\
&\qquad \times [I(I+1) - M_I(M_I \mp 1)] \times [I(I+1) - (M_I \mp 1)(M_I \mp 2)])^{1/2}\}
\end{aligned}
\tag{10.53}
$$

where

$$P = \frac{eqQ}{4(2J-1)(2J+3)I(2I-1)}\left(\frac{3K^2}{J(J+1)} - 1\right)$$

and where the δ's are all zero except when the indicated subscript values are equal. Because $\mathscr{H}_Q$ is diagonal in J, K, I, an independent set of equations is obtained for each J, K, I combination. Furthermore, $\mathscr{H}_Q$ is diagonal in $M \equiv M_J + M_I$, and hence the matrix can be factored into submatrices with common values of $M_J + M_I$. In writing out the matrix it is convenient to group together the common values $M = M_J + M_I$.

For the maximum values $M_J = J$, $M_I = I$, and $M = I + J$, the secular equation is of first degree and gives

$$E = E_{J,K} + E_{\mathscr{E}}(M_J = J) + (M_J = J, M_I = I|\mathscr{H}_Q|M_J = J, M_I = I) \tag{10.54}$$

when $M = I + J - 1$, the secular equation is the quadratic

M_J, M_I	$M_J = J, M_I = I-1$	$M_J = J-1, M_I = I$	
$\begin{cases} M_J = J \\ M_I = I-1 \end{cases}$	$E_{J,K} + E_{\mathscr{E}}(M_J = J)$ $+ (J, I-1|\mathscr{H}_Q|J, I-1) - E$	$(J-1, I|\mathscr{H}_Q|J, I-1)$	$=0$
$\begin{cases} M_J = J-1 \\ M_I = I \end{cases}$	$(J-1, I|\mathscr{H}_Q|J, I-1)$	$E_{J,K} + E_{\mathscr{E}}(M_J = J-1)$ $+ (J-1, I|\mathscr{H}_Q|J-1, I) - E$	

(10.55)

In these equations we have used the abbreviation $|M_J, M_I)$ for $|J, K, I, M_J, M_I)$. The equation of next higher order, that for $M = M_J + M_I = J + I - 2$, is a cubic; the next, for $M = J + I - 3$, is a quartic. High-speed computers can be used for solution of these equations of higher power when it is desirable. However, one can often employ transitions involving only the energy values obtained from equations of first and second power for accurate evaluation of dipole moments. The relevant selection rules for M are $\Delta M = 0, \pm 1$; M_J and M_I are no longer good quantum numbers.

Explicit values of E for symmetric rotors ($K \neq 0$) for which $M = M_J + M_I = J + I$ are derived from (10.54).

$$E_{J,K,M_J=J,M_I=I}=E_{J,K}-\frac{\mu\mathscr{E}K}{J+1}-\frac{\mu^2\mathscr{E}^2[(J+1)^2-K^2]}{2hB(J+1)^3(2J+3)}+\frac{eQq[3K^2-J(J+1)]}{4(J+1)(2J+3)} \tag{10.56}$$

For linear molecules this equation reduces to

$$E_{J,M_J=J,M_I=I}=E_J-\frac{\mu^2\mathscr{E}^2}{2hB(J+1)(2J+3)}-\frac{eQqJ}{4(2J+3)} \tag{10.57}$$

With a Stark spectrometer which permits measurement of the σ component $M\rightarrow M'(=M+1)$ of the $J\rightarrow J'(=J+1)$ transition, one can measure frequency shifts determined entirely by (10.56) or (10.57). From these shifts one can obtain μ and eQq.

The quadratic equation (10.55) is readily solvable and yields energies of the levels for $M=M_J+M_I=J+I-1$ which are

$$E_{\pm}=E_{J,K}+\tfrac{1}{2}(\alpha+\beta+a+b)\pm\tfrac{1}{2}[(\alpha-\beta+a-b)^2+4c^2]^{1/2} \tag{10.58}$$

where

$$\alpha=E_{\mathscr{E}}(M_J=J)=-\frac{\mu\mathscr{E}K}{J+1}-\frac{(\mu\mathscr{E})^2}{2hB}\frac{[(J+1)^2-K^2]}{(J+1)^3(2J+3)} \tag{10.59}$$

$$\beta=E_{\mathscr{E}}(M_J=J-1)$$
$$=-\frac{\mu\mathscr{E}K(J-1)}{J(J+1)}+\frac{(\mu\mathscr{E})^2}{2hB}\left\{\frac{(J^2-K^2)}{J^3(2J+1)}-\frac{4J[(J+1)^2-K^2]}{(J+1)^3(2J+1)(2J+3)}\right\} \tag{10.60}$$

$$a=(M_J=J,\ M_I=I-1|\mathscr{H}_Q|M_J=J,\ M_I=I-1)$$
$$=PJ(2J-1)(2I^2-7I+3) \tag{10.61}$$

$$b=(M_J=J-1,\ M_I=I|\mathscr{H}_Q|M_J=J-1,\ M_I=I)$$
$$=PI(2I-1)(2J^2-7J+3) \tag{10.62}$$

$$c=(M_J=J-1,\ M_I=I|\mathscr{H}_Q|M_J=J,\ M_I=I-1)$$
$$=3P[(2J-1)(2I-1)](IJ)^{1/2} \tag{10.63}$$

$$P=\frac{eQq}{4(2J-1)(2J+3)I(2I-1)}\left[\frac{3K^2}{J(J+1)}-1\right] \tag{10.64}$$

The levels corresponding to E_+ and E_- in (10.58) represent a mixture of $M_J=J$ and $M_J=J-1$ states. The π-type transitions ($\Delta M=0$) are possible from, or to, either the plus or minus levels to other levels having the same M values, provided that the selection rules for J and K are not violated. A commonly observed transition for measurement of dipole moments is a π-type one from the $M=M_J+M_I=J+I$ (the energy of which can be calculated with 10.56) to the next highest rotational level, represented by $J'=J+1$, $M'=J'+I-1=J+I$. Since $M=M'$, this is a $\Delta M=0$ transition which is observed with the dc electric field parallel to the electric vector of the microwave radiation. However, a doublet will be observed that corresponds to transitions to the plus and minus substates having the same M value. The relative intensities of the doublet will depend on the relative weights of the $M'_J = J'$ and $M'_J = J' - 1$ components in the admixed states. These weights can be found by substitution of the $E_\pm$ values into (10.50) followed by solution for the C_i^2 values. The relative intensities will be proportional to the C_i^2 values of the two admixed functions.

In deriving the secular equation for the intermediate-field values Low and Townes [4] used the representation J, I, F, M in which $\mathscr{H}_Q$ is diagonal and $\mathscr{H}_\mathscr{E}$ is therefore not diagonal. This representation is a bit complicated since one must then express the basis function in terms of the set $|J, K, I, M_J, M_I)$ in which the matrix elements of $\mathscr{H}_\mathscr{E}$ are known. This transformation can be achieved by use of (10.20) with the Clebsch-Gordon [7, 8] coefficients.

3 ASYMMETRIC-TOP MOLECULES WITHOUT NUCLEAR COUPLING

The theory of the Stark effect for asymmetric rotors has been developed by Penney [10] and by Golden and Wilson [11]. The Stark effect for asymmetric-top molecules is calculated with perturbation theory in a manner similar to that for linear or symmetric rotors. However, the evaluation of the direction cosine matrix elements becomes more involved since the wave functions for asymmetric rotors are considerably more complicated. With symmetric or linear molecules the dipole moment was associated with one of the axes, whereas for asymmetric rotors we have now the possibility of permanent molecular dipole components along each of the three principal axes of inertia. The Stark effect Hamiltonian operator will thus have the form given in (10.1). The application of an electric field will, via the interaction of the dipole moment and the field, perturb the rotational motion and thus the rotational energy levels. In particular, the levels will no longer be $(2J+1)$-fold degenerate in the space orientation quantum number M_J. Each energy level of the asymmetric rotor consequently splits up into a number of sublevels corresponding to various values of $|M_J|$.

In a perturbation treatment of $\mathscr{H}_\mathscr{E}$, the asymmetric rotor functions $|J, \tau, M_J)$ for the unperturbed or field-free rotor are used to give the matrix elements of the direction cosines. The relevant direction cosine matrix elements in an asymmetric rotor basis may be expressed as

$$(J, \tau, M_J|\Phi_{Zg}|J', \tau', M_J) = (J|\Phi_{Zg}|J')(J, \tau|\Phi_{Zg}|J', \tau')(J, M_J|\Phi_{Zg}|J', M_J) \qquad (10.65)$$

The J and J, M_J dependent factors are the same as for a symmetric rotor basis and are found in Table 2.1. The direction cosine matrix elements will be nonvanishing only if the product of the symmetries of $\psi_{J_\tau M_J}$, Φ_{Zg}, and $\psi_{J'_{\tau'} M_J}$ belongs to species A. The symmetries of the direction cosines (see Chapter VII, Section 4) and their products, under the rotation group $V(x, y, z)$, are summarized in Table 10.1. The correlation of the species of $V(x, y, z)$ with those of $V(a, b, c)$ may be readily found by use of Table 7.5. With the aid of Table 10.1 the possible

Table 10.1 The Direction Cosine Symmetries in $V(x, y, z)$

Direction Cosine	*Symmetry*
$\Phi_{Zx}, \Phi_{Zy}\Phi_{Zz}$	B_x
$\Phi_{Zy}, \Phi_{Zx}\Phi_{Zz}$	B_y
$\Phi_{Zz}, \Phi_{Zx}\Phi_{Zy}$	B_z
$\Phi_{Zx}^2, \Phi_{Zy}^2, \Phi_{Zz}^2$	A

nonvanishing matrix elements of $\mathscr{H}_{\mathscr{E}}$ may be found. These are listed in Table 10.2. It is apparent that the perturbation connects the various diagonal blocks of $\mathscr{H}_r$. From (10.71)–(10.73), which give the direction cosine matrix elements in terms of the line strengths, we see that the connections are of the type $(J|J)$ and $(J|J\pm1)$. The complete energy matrix with $\mathscr{H}_{\mathscr{E}}$ included can, of course, be constructed and subsequently diagonalized to give the energy levels [12]. Usually, however, perturbation techniques are sufficiently accurate. The diagonal elements of $\mathscr{H}_{\mathscr{E}}$, in the basis which diagonalizes $\mathscr{H}_r$, are the first-order perturbation corrections. However, since $\mathscr{H}_{\mathscr{E}}$ has (from Table 10.2) no diagonal matrix elements, there should be no first-order effects. Usually the Stark effect for asymmetric rotors will be of second-order, that is, proportional to $\mathscr{E}^2$. It frequently happens, however, that a pair of interacting levels for the unperturbed rotor are degenerate or nearly degenerate. Under these conditions a first-order effect can arise. This situation will be deferred until consideration is given to the case where the unperturbed energy levels are nondegenerate, that is, widely separated.

The Nondegenerate Case

From Table 2.1 it is seen that the Φ_{Zg} matrix elements are diagonal in M_J; hence, the total rotational energy with a Stark field perturbation can be expressed for a specific value of M_J as:

$$E_{J_\tau M_J}=E^0_{J_\tau}+\sum_g [E_g^{(2)}]_{J_\tau M_J} \tag{10.66}$$

Table 10.2 Structure of the $\mathscr{H}_{\mathscr{E}}$ Matrix

$\psi_{J_\tau}\vert\psi_{J'_{\tau'}}$	A	B_x	B_y	B_z
A	...	$\mu_x\Phi_{Zx}$	$\mu_y\Phi_{Zy}$	$\mu_z\Phi_{Zz}$
B_x	$\mu_x\Phi_{Zx}$	...	$\mu_z\Phi_{Zz}$	$\mu_y\Phi_{Zy}$
B_y	$\mu_y\Phi_{Zy}$	$\mu_z\Phi_{Zz}$	...	$\mu_x\Phi_{Zx}$
B_z	$\mu_z\Phi_{Zz}$	$\mu_y\Phi_{Zy}$	$\mu_x\Phi_{Zx}$	...

where $E^0_{J_\tau}$ is the rotational energy of the unperturbed rotor, and the second-order Stark energy is given by the conventional nondegenerate perturbation expression as

$$[E_g^{(2)}]_{J_\tau M_J} = \mu_g^2 \mathscr{E}^2 \sum_{J',\tau'}{}' \frac{|(J, \tau, M_J|\Phi_{Zg}|J', \tau', M_J)|^2}{E^0_{J_\tau} - E^0_{J'_{\tau'}}} \tag{10.67}$$

The summation over J', τ' includes all states that interact with the state J_τ, that is to say, all states which are connected by a nonvanishing direction cosine matrix element. The level J_τ, however, is to be excluded as indicated by the prime on the summation. There are no cross products of dipole components in (10.67) because of symmetry restrictions (see Table 10.1). The total Stark energy of a level characterized by J, τ, M_J is the sum of the contributions arising from each dipole component.

As mentioned previously, the only nonvanishing direction cosine matrix elements are those for which $J' = J-1$, J, and $J+1$. If (10.65) is used to separate out the J and M_J factors, the energy shift arising from the gth component of the permanent dipole is [11]

$$\begin{aligned}[E_g^{(2)}]_{J_\tau M_J} = \mu_g^2 \mathscr{E}^2 \Bigg[&\frac{(J^2 - M_J^2)}{4J^2(4J^2-1)} \sum_{\tau'} \frac{|(J, \tau|\Phi_{Zg}|J-1, \tau')|^2}{E^0_{J_\tau} - E^0_{J-1_{\tau'}}} \\ &+ \frac{M_J^2}{4J^2(J+1)^2} \sum_{\tau'}{}' \frac{|(J, \tau|\Phi_{Zg}|J, \tau')|^2}{E^0_{J_\tau} - E^0_{J_{\tau'}}} \\ &+ \frac{(J+1)^2 - M_J^2}{4(J+1)^2(2J+1)(2J+3)} \sum_{\tau'} \frac{|(J, \tau|\Phi_{Zg}|J+1, \tau')|^2}{E^0_{J_\tau} - E^0_{J+1_{\tau'}}} \Bigg]\end{aligned} \tag{10.68}$$

This expression in conjunction with (10.66) yields to second order the rotational energies in the presence of an applied electric field. The Stark energies are seen to depend on M_J^2, and hence the $(2J+1)$-fold degeneracy in M_J is partially removed, a given rotational level being split into $(J+1)$ distinct sublevels. To calculate the Stark shifts one must know the dipole moment, the direction cosine matrix elements, and the energy level differences. In many instances, some of the energy differences will not correspond to observed frequencies and will have to be evaluated by known rotational constants. Conversely, when the energy levels and Stark shifts are known, the dipole moment may be calculated. The evaluation of $|(J, \tau|\Phi_{Zg}|J', \tau')|^2$ in an asymmetric rotor basis is rather tedious for any except the lowest J values. Separating the M_J^2 dependence, Golden and Wilson [11] write (10.68) in the equivalent form,

$$[E_g^{(2)}]_{J_\tau M_J} = \frac{2\mu_g^2 \mathscr{E}^2}{A+C} [A_{J_\tau}(\kappa, \alpha) + M_J^2 B_{J_\tau}(\kappa, \alpha)] \tag{10.69}$$

where $\alpha = (A-C)/(A+C)$ and the $(A+C)/2$ dependence of the E^0's has been factored out. They have tabulated these reduced Stark coefficients A_{J_τ} and B_{J_τ} for various values of α and κ up to $J=2$. By means of these tabulated quantities, the second-order Stark energies may be calculated with a minimum of effort.

For higher values of J, the calculations can be further simplified if it is noted that the matrix elements of interest are related to the line strengths, (7.58), for which extensive tables are available. In the absence of external fields, X, Y, and Z are equivalent, and the line strengths may be written as

$$\lambda_g(J, \tau; J', \tau') = 3 \sum_{M} |(J, \tau, M_J|\Phi_{Zg}|J', \tau', M_J)|^2$$

$$= \left[3|(J|\Phi_{Zg}|J')|^2 \cdot \sum_{M_J} |(J, M_J|\Phi_{Zg}|J', M_J)|^2 \right] |(J, \tau|\Phi_{Zg}|J', \tau')|^2 \tag{10.70}$$

In writing this expression we have taken cognizance of the fact that Φ_{Zg} has nonvanishing elements only for $M'_J = M_J$. The term in brackets may be readily evaluated. Using Table 2.1 and carrying out the sum over M_J, noting that $\sum_{M_J=-J}^{+J} M_J^2 = (\frac{1}{3})J(J+1)(2J+1)$, one obtains:

$$\lambda_g(J, \tau; J+1, \tau') = \left[\frac{1}{4(J+1)}\right] |(J, \tau|\Phi_{Zg}|J+1, \tau')|^2 \tag{10.71}$$

$$\lambda_g(J, \tau; J, \tau') = \left[\frac{2J+1}{4J(J+1)}\right] |(J, \tau|\Phi_{Zg}|J, \tau')|^2 \tag{10.72}$$

$$\lambda_g(J, \tau; J-1, \tau') = \left(\frac{1}{4J}\right) |(J, \tau|\Phi_{Zg}|J-1, \tau')|^2 \tag{10.73}$$

Thus the required matrix elements $|(J, \tau|\Phi_{Zg}|J', \tau')|^2$ may be evaluated from the foregoing equation and the tabulated line strengths [13]. Interpolation of the line strength tables may be used when necessary. As J increases, the number of possible interacting levels increases rapidly, especially when there is more than one dipole component. However, in many cases certain of the matrix elements or line strengths connecting two levels will be vanishingly small and can be neglected.

Usually the electric field applied is such that $\Delta M_J = 0$ transitions are observed. A completely resolved spectral line with $\Delta J = \pm 1$ will have $(J+1)$ Stark components, while for $\Delta J = 0$ there will be J components, where J is the smaller of the two J's involved in the transition. These cases are illustrated in Fig. 10.8. The loss of one component for $\Delta J = 0$ transition is due to the vanishing intensity of the $M_J = 0$ component which is apparent from the intensity expressions given later. According to the magnitude and sign of the Stark coefficients for the two levels involved in the transition, the Stark components may be on one or both sides of the zero-field line. When the Stark coefficient differences, that is, ΔA and ΔB, have the same sign, the largest displacement will occur for the Stark component associated with the largest possible M_J value.

The evaluation of the $|M_J|$ associated with a particular Stark component can be made if necessary from measurements on the frequency displacements [11]. Let the frequency shift for a given M_J component be expressed as

$$\Delta\nu_{M_J} = \mathscr{E}^2(A' + B' M_J^2) \tag{10.74}$$

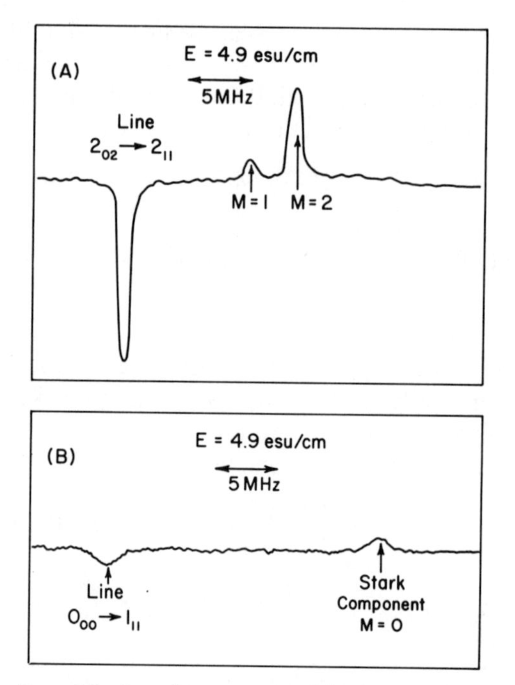

Fig. 10.8 (*A*) Recording of the $2_{02} \rightarrow 2_{11}$ transition of SO_2 at 53,529 MHz. (*B*) Recording of the $0_{00} \rightarrow 1_{11}$ transition of SO_2 at 69,576 MHz. Obtained with a Stark-modulation spectrometer, employing phase-sensitive detection. After Crable [15].

The M_J value for successive components will differ by unity and the following relations for three consecutive components may be written

$$\Delta\nu_{M_J} - \Delta\nu_{M_J \pm 1} = \mp \mathscr{E}^2 B'(2M_J \pm 1) \tag{10.75}$$

$$\Delta\nu_{M_J \pm 1} - \Delta\nu_{M_J \pm 2} = \mp \mathscr{E}^2 B'(2M_J \pm 3 \tag{10.76}$$

Taking the ratio of the component separations gives

$$r = \frac{\Delta\nu_{M_J} - \Delta\nu_{M_J \pm 1}}{\Delta\nu_{M_J \pm 1} - \Delta\nu_{M_J \pm 2}} = \frac{2M_J \pm 1}{2M_J \pm 3} \tag{10.77}$$

which in turn yields

$$M_J = \pm \left(\frac{3r - 1}{2 - 2r} \right) \tag{10.78}$$

The values of M_J can also be related to the relative intensities of the components [11]; however, frequency measurements can be made with greater accuracy, and hence the $|M_J|$ can be most satisfactorily estimated from the frequency displacements.

With very large electric fields, higher-order Stark corrections may need to be considered. For nondegenerate levels the odd-order corrections vanish [11], and hence the next correction will be proportional to $\mathscr{E}^4$. If fourth-order Stark corrections are important, the procedure for handling the data discussed in Section 1 can be employed which, although explicitly including fourth-order effects, does not require a knowledge of the fourth-order Stark coefficients. The second-order Stark coefficients in (10.18) are obtained from (10.68). This method may be generalized for the case where the Stark effect depends on more than one dipole component μ_g, but additional intercepts from different transitions (e.g., different M_J values) will be required in the solution for the dipole components.

As an illustration of the application of (10.68), consider the evaluation of the dipole moment of SO_2 from the Stark splitting of the $0_{00} \rightarrow 1_{11}$ transition for which only one Stark component is observed, viz., $M_J = 0$ (see Fig. 10.8), and no near-degeneracies occur. The symmetry of SO_2 is such that the dipole moment will be entirely along the twofold axis of symmetry which is the b axis. Therefore, the levels which perturb the 0_{00} level by means of the dipole component μ_b must have symmetry B_b. Thus only the 1_{11} level perturbs the 0_{00} level. On the other hand, the 1_{11} level interacts via the Stark perturbation with the 0_{00}, 2_{02}, and 2_{20} levels. The spectroscopic constants of SO_2 are [14]: $A = 60{,}778.79$ MHz, $B = 10{,}318.10$ MHz, and $C = 8{,}799.96$ MHz. Keeping only three decimal places we find $\kappa = -0.942$. The reduced energies of the various levels may be obtained from the expressions of Tables 7.6 and 7.7 or existing reduced energy tables. The total rigid-rotor energy of the various levels is found to be (MHz):

$$E_{0_{00}} = 0$$

$$E_{1_{11}} = A + C = 69{,}578.75$$

$$E_{2_{02}} = 3(A + C) + \frac{(A - C)}{2}(-5.8273) = 57{,}288.13$$

$$E_{2_{20}} = 3(A + C) + \frac{(A - C)}{2}(2.0593) = 262{,}256.25$$

The appropriate matrix elements of the direction cosines are obtainable from the tabulated line strengths and (10.71)–(10.73). The necessary quantities may be summarized as follows:

$J_{K_{-1},K_1} - J'_{K'_{-1},K'_1}$	$\lambda_b(J_{K_{-1},K_1};\, J'_{K'_{-1},K'_1})$	$\lvert(J_{K_{-1},K_1}\lvert\Phi_{Zb}\rvert J'_{K'_{-1},K'_1})\rvert^2$
$0_{00} - 1_{11}$	1.0000	4.0000
$1_{11} - 2_{02}$	0.5191	4.1528
$1_{11} - 2_{20}$	1.4809	11.8472

These are for a κ of -0.95 from [13] which is sufficiently accurate for our purpose. In practice, however, an interpolation would be carried out to give the values corresponding to the observed κ. With these results, the Stark energy for the 0_{00} and 1_{11} levels may be readily calculated from (10.68). The separation $\Delta\nu$ of the Stark component ($M_J=0$) from the zero-field rotational transition is given by the following expression

$$\Delta\nu=[E_b^{(2)}]_{1_{11}}-[E_b^{(2)}]_{0_{00}}=(0.50344)^2\mu_b^2\mathscr{E}^2[9.3973-(-4.7907]\times 10^{-6}$$
$$=\mu_b^2\mathscr{E}^2(0.50344)^2(14.1880\times 10^{-6})$$

Here the conversion factor 0.50344 (MHz) (debye volt/cm)$^{-1}$ is included so that the Stark splittings may be expressed in megahertz with μ_g in debye units, and $\mathscr{E}$ in v/cm. For $\mathscr{E}=2005.1$ v/cm, the observed Stark splitting is found to be $\Delta\nu=36.2$ MHz [15], and one obtains from the above expression $\mu=\mu_b=1.58$ D for the dipole moment. The dipole moment obtained with interpolation and measurement of different splittings is 1.59 ± 0.01 D [16].

The Degenerate Case

In asymmetric rotors degeneracies frequently occur; both approximate symmetric-rotor degeneracy and various types of accidental degeneracies are possible. Table 10.3 indicates for $J\leqslant 3$ which pairs of levels can become accidentally degenerate. When degeneracies are present, (10.68) obviously fails, since it will contain terms with vanishing denominators. Thus, if rotational degeneracies or near-degeneracies are encountered, a simple second-order perturbation treatment is not applicable. Van Vleck [17], however, has developed a convenient perturbation technique which is very effective for dealing

Table 10.3 Possible Degeneracies in the Asymmetric Rotor[a]

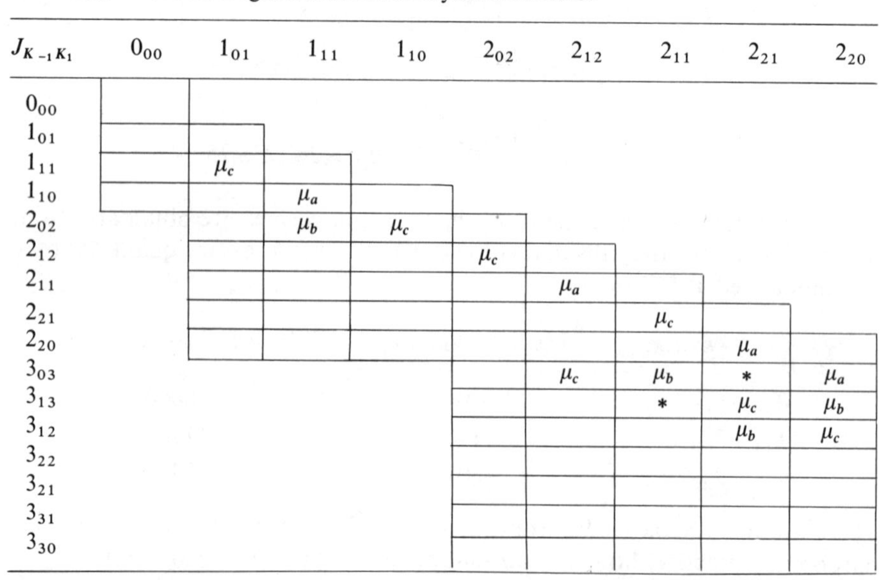

$J_{K_{-1}K_1}$	0_{00}	1_{01}	1_{11}	1_{10}	2_{02}	2_{12}	2_{11}	2_{21}	2_{20}
0_{00}									
1_{01}									
1_{11}		μ_c							
1_{10}			μ_a						
2_{02}			μ_b	μ_c					
2_{12}					μ_c				
2_{11}						μ_a			
2_{21}							μ_c		
2_{20}								μ_a	
3_{03}						μ_c	μ_b	*	μ_a
3_{13}							*	μ_c	μ_b
3_{12}								μ_b	μ_c
3_{22}									
3_{21}									
3_{31}									
3_{30}									

[a]From Golden and Wilson [11]. Entries indicate which component of the dipole moment becomes important in the accidental degeneracy. The asterisk denotes no coupling term. Blanks indicate no accidental degeneracy possible.

with the problem of degeneracies or near-degeneracies. It consists of application of a transformation to the energy matrix, that is, the matrix of $\mathscr{H}_r + \mathscr{H}_\mathscr{E}$ evaluated in an asymmetric-rotor basis. The transformation is so constructed that the elements connecting degenerate blocks after the transformation are reduced to second order and can hence be neglected for results correct up to fourth order in the energy. This perturbation technique is discussed in more detail in Appendix C and in the section on internal rotation. The net result of the transformation is that one obtains a number of small submatrices, each associated with a group of degenerate or nearly degenerate levels. The diagonal elements of these submatrices contain correction terms from the Van Vleck transformation and are given by (10.66) and (10.68), except that in the case of degeneracies or near-degeneracies the summation is only over those levels for which $E^0_{J'_{\tau'}}$ is not near $E^0_{J_\tau}$. The perturbed energies are then found from solution of the usual secular determinant. Usually the degeneracy occurs only between a pair of levels; the secular equation then has the form

$$\begin{vmatrix} E_{J_\tau M_J} - E & \mathscr{E}\xi \\ \mathscr{E}\xi^* & E_{J'_{\tau'} M_J} - E \end{vmatrix} = 0 \tag{10.79}$$

where ξ is the appropriate off-diagonal matrix element. The two possible solutions are

$$E_\pm = \frac{E_{J_\tau M_J} + E_{J'_{\tau'} M_J}}{2} \pm \left[\left(\frac{E_{J_\tau M_J} - E_{J'_{\tau'} M_J}}{2} \right)^2 + \mathscr{E}^2 |\xi|^2 \right]^{1/2} \tag{10.80}$$

where for $E_{J_\tau M_J} > E_{J'_{\tau'} M_J}$, the plus sign is associated with the level $J_\tau M_J$ and the minus sign is associated with the level $J'_{\tau'} M_J$, since two such interacting states repel one another. The off-diagonal elements are made up of the Stark coupling term for the levels and correction terms from the transformation

$$\mathscr{E}\xi = (J, \tau, M_J | \mathscr{H}_\mathscr{E} | J', \tau', M_J) + \text{Correction terms} \tag{10.81}$$

These off-diagonal elements depend on the two levels under consideration, and Golden and Wilson [11] have given expressions for various types of near-degeneracies.

We shall discuss only the case of near-degeneracies between asymmetry doublet states, which is often encountered even in rather asymmetric rotors. These levels will appear in pairs having different symmetry, except for the $K=0$ level which is nondegenerate in K. For a near-prolate rotor the pairs of K levels will have either symmetries A, B_a or symmetries B_b, B_c, whereas for a near-oblate rotor the symmetries will be B_a, B_b or A, B_c. For two members of an asymmetry doublet, J_τ and $J_{\tau'}$, which interact in the presence of an electric field by means of a dipole component μ_g, the off-diagonal matrix element connecting the two states will have the simple form

$$\mathscr{E}^2|\xi|^2 = |(J, \tau, M_J | \mathscr{H}_\mathscr{E} | J, \tau', M_J)|^2 = \mu_g^2 \mathscr{E}^2 \frac{M_J^2}{4J^2(J+1)^2} |(J, \tau | \Phi_{Zg} | J, \tau')|^2 \tag{10.82}$$

where g is a or c depending on the symmetry of the two near-degenerate K levels. The secular determinant, (10.79), with the off-diagonal elements given by (10.82), is just what is expected from conventional perturbation theory for two such nearby levels, except that the diagonal elements here contain, in addition, the second-order Stark contributions of the remaining nondegenerate levels. Each of the two levels, J_τ and $J_{\tau'}$, splits into $(J+1)$ distinct sublevels given by (10.80) and (10.82), the $+M_J$ and $-M_J$ levels coinciding. If $|E_{J_\tau M_J}-E_{J_{\tau'} M_J}|\ll|\mathscr{E}\xi|$, then (10.80) may be expanded to give an energy that is a linear function of the electric field

$$E_\pm=\tfrac{1}{2}(E^0_{J_\tau}+E^0_{J_{\tau'}})\pm\mathscr{E}|\xi| \tag{10.83}$$

or explicitly for the two levels

$$E(J_\tau M_J)=\tfrac{1}{2}(E^0_{J_\tau}+E^0_{J_{\tau'}})+\mu_g\mathscr{E}\,\frac{|M_J|}{2J(J+1)}(J,\tau|\Phi_{Zg}|J,\tau') \tag{10.84}$$

$$E(J_{\tau'}\,M_J)=\tfrac{1}{2}(E^0_{J_\tau}+E^0_{J_{\tau'}})-\mu_g\mathscr{E}\,\frac{|M_J|}{2J(J+1)}(J,\tau|\Phi_{Zg}|J,\tau') \tag{10.85}$$

In this approximation, the asymmetry splitting has been ignored along with possible second-order Stark effects. If, on the other hand, $|E_{J_\tau M_J}-E_{J_{\tau'} M_J}|\gg|\mathscr{E}\xi|$, the expansion will give the conventional second-order result, although possibly a large second-order effect because of the presence of a small energy denominator. If the separation of the levels is such that for a given field neither expansion is sufficiently accurate, that is, an intermediate case is attained, then the complete energy expression, (10.80), must be employed. Under the appropriate conditions it is possible with increasing electric field to observe a transition from a Stark-effect quadratic in the electric field to one linear in $\mathscr{E}$.

As with the nondegenerate case, the M_J value can be expressed in terms of the Stark-component separations. For the first-order Stark effect in asymmetry doublets one finds

$$M_J=\pm\frac{\Delta\nu_{M_J}}{\Delta\nu_{M_J}-\Delta\nu_{M_J\pm1}} \tag{10.86}$$

For slightly asymmetric rotors, the direction cosine matrix elements are given to a good approximation by those for symmetric tops. Hence the off-diagonal matrix element would reduce to

$$\mathscr{E}^2|\xi|^2=\left[\frac{\mu_g\mathscr{E}KM_J}{J(J+1)}\right]^2 \tag{10.87}$$

with K the appropriate quantum number for the limiting symmetric top. When the asymmetry splitting vanishes, $E^0_{J_\tau}=E^0_{J_{\tau'}}$ (levels degenerate), it is apparent that (10.83), with the aid of (10.87), reduces to the first-order expression given for a symmetric rotor.

Figure 10.9 illustrates both a first- and second-order Stark pattern for the nearly prolate asymmetric rotor HNCO. The asymmetry splitting is very small, and the $3_{22}\rightarrow4_{23}$ and $3_{21}\rightarrow4_{22}$ $(K=2)$ transitions appear as a single line in the absence of an electric field. This K degeneracy leads to a first-order Stark pattern like that of a symmetric top, and $(2J+1)$ Stark components are

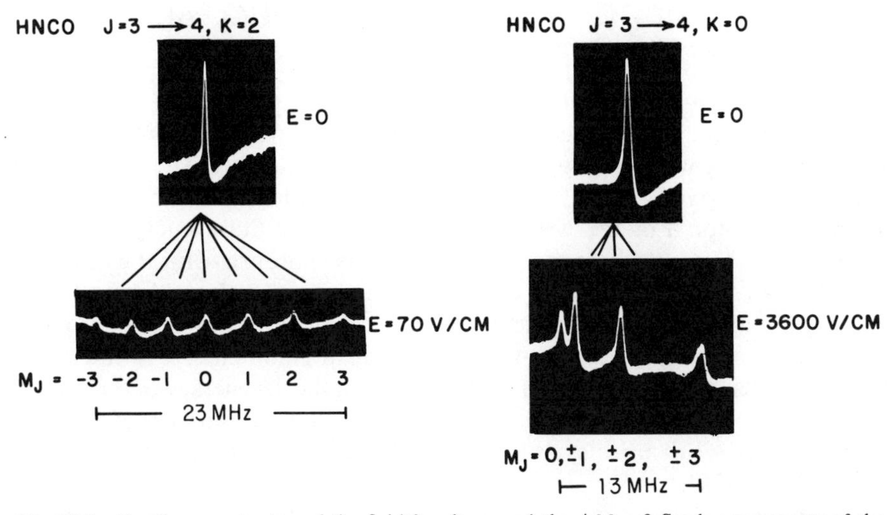

Fig. 10.9 Oscilloscope tracing of the field-free lines and the $\Delta M_J=0$ Stark components of the $J=3\rightarrow4$, $K=2$ transition (first-order) and the $J=3\rightarrow4$, $K=0$ transition (second-order) for the slightly asymmetric rotor HNCO. The frequency increases from left to right. Obtained with a parallel-plate Stark cell operating in the millimeter wave region. From K. White, Ph.D. dissertation, Duke University, 1965.

obtained. For the $3_{03}\rightarrow4_{04}$ ($K=0$) transition, which is a nondegenerate case, a typical second-order pattern is observed. For the $J=3\rightarrow4$, $K=1$ lines (not pictured) an intermediate case is found. A Stark pattern similar to the $K=0$ transition is obtained for each $K=1$ line, except that the Stark components for the high-frequency $K=1$ line are on the low-frequency side whereas for the low-frequency $K=1$ line they are on the high-frequency side.

A case of a highly degenerate Stark effect has been observed in the near-oblate rotor CH_3CHF_2 by Kwei and Herschbach [18]. Here the Stark effect exhibits strong perturbations arising from the two components of the dipole moment (μ_a, μ_c) and all three of the $J=1$ rotational levels and all five of the $J=2$ levels had to be treated as degenerate. The Van Vleck transformation is used to separate the blocks of different J, giving a 3×3 secular equation for $J=1$ and a 5×5 secular equation for $J=2$.

For molecules with internal rotation it is possible for E levels with $K>0$ to exhibit a first-order Stark effect that is dependent on the barrier height. This is discussed more fully in Chapter XII. It suffices to say here that because of the presence in the Hamiltonian of a linear term in the angular momentum, the $\pm M_J$ degeneracy can be removed and a first-order Stark effect obtained. This splitting between the $+M_J$ and the $-M_J$ Stark components is sensitive to the barrier height and thus allows a means of its evaluation.

The degree of planarity of molecules and the nature of the potential function which governs the out-of-plane bending vibration are questions in which the Stark effect can yield valuable information. A discussion of this may be found elsewhere [19, 20].

4 ASYMMETRIC-TOP MOLECULES WITH NUCLEAR QUADRUPOLE COUPLING

As with linear and symmetric-top molecules, the presence of nuclei with quadrupole coupling complicates the Stark effect of asymmetric rotors. Plural nuclear coupling will be treated for the strong-field case only. The appropriate Hamiltonian is given by (10.19) with $\mathscr{H}_r$, now the rotational Hamiltonian for an asymmetric rotor. For the weak-field case where the Stark energy is much smaller than the quadrupole energy, the functions appropriate for the field-free Hamiltonian $\mathscr{H}_r + \mathscr{H}_Q$ are the $|J, \tau, I, F, M_F)$ basis functions for which $\mathbf{J}^2$, $\mathbf{I}^2$, $\mathbf{F}^2$, and $\mathbf{M}_F$ are all diagonal. A particular level is characterized by the set of good quantum numbers J, I, F, M_F and the pseudo-quantum number τ. In the weak-field case when an external electric field is applied, the degeneracy of the spatial orientations of F will be removed. Each F level is split into a number of different Stark levels ($F+1$ distinct levels for a second-order Stark effect) corresponding to the values of M_F.

The basis functions required for the perturbation treatment of $\mathscr{H}_{\mathscr{E}}$ may be constructed from product functions of the orthonormal asymmetric-rotor wave functions $\psi(J_\tau M_J)$ and the orthonormal nuclear spin functions $\phi(IM_I)$ as follows

$$\psi(J_\tau IFM_F) = \sum_{M_J} C(JIF;\, M_J M_I M_F)\psi(J_\tau M_J)\cdot\phi(IM_I) \tag{10.88}$$

The expansion coefficients, C's, are the Clebsch-Gordon coefficients [7, 8] noted previously. These coefficients are independent of τ. For asymmetric rotors without degenerate or near-degenerate rotational energy levels, the second-order Stark perturbation energies of the hyperfine levels are given by

$$E^{(2)}_{J_\tau IFM_F} = \sum_g [E^{(2)}_g]_{J_\tau IFM_F} \tag{10.89}$$

with

$$[E^{(2)}_g]_{J_\tau IFM_F} = \mu_g^2 \mathscr{E}^2 \sum_{J',\tau',F'}{}' \frac{|(J, \tau, I, F, M_F|\Phi_{Zg}|J', \tau', I, F', M_F)|^2}{E_{J_\tau, I, F} - E_{J'_{\tau'}, I, F'}} \tag{10.90}$$

Usually the separation of the rotational energy levels coupled by the Stark perturbation are large compared to the hyperfine splitting of these levels and $E_{J_\tau,I,F} - E_{J'_{\tau'},I,F'} \approx E_{J_\tau} - E_{J'_{\tau'}}$. With this approximation and the substitution of (10.88) into (10.90) one obtains, upon integration [4, 21, 22],

$$[E^{(2)}_g]_{J_\tau IFM_F} = \sum_{M_J} |C(JIF;\, M_J M_I M_F)|^2 [E^{(2)}_g]_{J_\tau M_J} \tag{10.91}$$

Note that

$$\sum_{F'} C^*(J'IF';\, M_J M_I M_F) C(J'IF';\, M_J M_I M_F) = 1 \tag{10.92}$$

Expressing $[E^{(2)}_g]_{J_\tau, M_J}$ as in (10.69) gives

$$[E^{(2)}_g]_{J_\tau IFM_F} = \frac{2\mu_g^2 \mathscr{E}^2}{A+C} [A_{J_\tau}(\kappa, \alpha) + \langle M_J^2 \rangle B_{J_\tau}(\kappa, \alpha)] \tag{10.93}$$

since $\sum_{M_J} |C|^2=1$ and $\sum_{M_J} |C|^2 M_J^2=\langle M_J^2\rangle$. This latter quantity is simply the quantum mechanical average of M_J^2 in the J, I, F, M_F representation and when evaluated can be expressed as [3]

$$\langle M_J^2\rangle=\frac{[3M_F^2-F(F+1)][3D(D-1)-4F(F+1)J(J+1)]}{6F(F+1)(2F-1)(2F+3)}+\frac{J(J+1)}{3} \tag{10.94}$$

where $D=F(F+1)+J(J+1)-I(I+1)$. The Stark coefficients A_{J_τ} and B_{J_τ} are those discussed previously where quadrupole coupling was absent. When $I=0$, it is easily seen that $\langle M_J^2\rangle=M_F^2=M_J^2$ and hence, (10.93) reduces to the Stark-effect expression for asymmetric rotors without quadrupole hyperfine structure. When a nucleus with a quadrupole moment is present, (10.93) indicates how each hyperfine level characterized by J, τ, F is further split by the Stark effect into $(F+1)$ sublevels corresponding to the values of M_F, each M_F level being doubly degenerate except for $M_F=0$. A given hyperfine line of a rotational transition will thus exhibit a number of Stark components. In addition to the selection rules for the unperturbed rotational transition we have $\Delta F=0, \pm 1$; and for the usual experimental set up have $\Delta M_F=0$.

When the hyperfine splitting as compared with the rotational splitting cannot be ignored, that is, when $E_{J_\tau,I,F}-E_{J'_{\tau'},I,F'} \not\approx E_{J_\tau}-E_{J'_{\tau'}}$, then the expression becomes more complicated. The evaluation of the direction cosine matrix elements in the $|J_\tau, I, F, M_F)$ basis has been carried out by Mizushima [22], using Racah's [23] method. The Stark energy for the weak-field case is now given by

$$\begin{aligned}
E^{(2)}_{J_\tau IFM_F}=\sum_g \mu_g^2\mathscr{E}^2\Bigg[& f_1(J, I, F, M_F)\sum_{\tau'}\frac{\lambda_g(J,\tau;J+1,\tau')}{E_{J_\tau,F}-E_{J+1_{\tau'},F}} \\
& +f_2(JIFM_F)\sum_{\tau'}\frac{\lambda_g(J,\tau;J+1,\tau')}{E_{J_\tau,F}-E_{J+1_{\tau'},F+1}} \\
& +f_3(JIFM_F)\sum_{\tau'}\frac{\lambda_g(J,\tau;J+1,\tau')}{E_{J_\tau,F}-E_{J+1_{\tau'},F-1}} \\
& +f_4(JIFM_F)\sum_{\tau'}\frac{\lambda_g(J,\tau;J,\tau')}{E_{J_\tau,F}-E_{J_{\tau'},F+1}} \\
& +f_5(JIFM_F)\sum_{\tau'\neq\tau}\frac{\lambda_g(J,\tau;J,\tau')}{E_{J_\tau,F}-E_{J_{\tau'},F}} \\
& +f_4(JI, F-1, M_F)\sum_{\tau'}\frac{\lambda_g(J,\tau;J,\tau')}{E_{J_\tau,F}-E_{J_{\tau'},F-1}} \\
& +f_1(J-1, IFM_F)\sum_{\tau'}\frac{\lambda_g(J,\tau;J-1,\tau')}{E_{J_\tau,F}-E_{J-1_{\tau'},F}} \\
& +f_2(J-1, I, F-1, M_F)\sum_{\tau'}\frac{\lambda_g(J,\tau;J-1,\tau')}{E_{J_\tau,F}-E_{J-1_{\tau'},F-1}} \\
& +f_3(J-1, I, F+1, M_F)\sum_{\tau'}\frac{\lambda_g(J,\tau;J-1,\tau')}{E_{J_\tau,F}-E_{J-1_{\tau'},F+1}}\Bigg]
\end{aligned} \tag{10.95}$$

in which the λ_g's are the line strengths and the f coefficients are given in Table 10.4.

For the strong-field case where the Stark energies are much larger than the quadrupole coupling energies, F is no longer a good quantum number. This case may be obtained if the quadrupole coupling constants are small, or if the electric fields are sufficiently large. The $|J, \tau, M_J, I, M_I)$ functions are suitable basis functions for treating the strong-field case. In this representation $\mathbf{J}^2$, $\mathbf{I}^2$, $\mathbf{M}_J$, and $\mathbf{M}_I$ are diagonal. A particular level in the strong-field case will be characterized by the set of quantum numbers J, I, M_J, M_I, and the label τ. The Stark levels are those determined previously for asymmetric rotors without quadrupole coupling effects.

If only diagonal elements need be considered, that is, if $\mathscr{H}_{\mathscr{E}}$ is large enough that the effect of the off-diagonal elements of $\mathscr{H}_Q$ can be ignored, the quadrupole perturbation energies of the Stark levels [see (10.53) and (10.98)] are given by

$$E_Q^{(1)}=(J, \tau, I, M_J, M_I|\mathscr{H}_Q|J, \tau, I, M_J, M_I)$$
$$=eQq_J\frac{[3M_J^2-J(J+1)][3M_I^2-I(I+1)]}{4J(2J-1)I(2I-1)} \tag{10.96}$$

where q_J is given by (9.88). The foregoing expression indicates how each Stark level characterized by J, τ, M_J is split into a number of hyperfine levels identified by different values of M_I. Note that the $+M_I$ and the $-M_I$ hyperfine levels have the same energy. A given Stark line of a rotational transition will thus have an accompanying quadrupole hyperfine structure. The additional selection rule $\Delta M_I=0$ is applicable. If two or more coupling nuclei are present, their respective quadrupole energies may be simply added because of the nuclear decoupling in the strong-field case.

Table 10.4 f Coefficients for the Stark Energy in the Weak-field Case[a]

$$f_1(JIFM_F)=\frac{M_F^2(J+I+F+2)(I+F-J)(I+J+1-F)(F+J+1-I)}{4F^2(F+1)^2(2J+3)(2J+1)(J+1)}$$

$$f_2(JIFM_F)=\frac{(J+I+F+3)(J+I+F+2)(J-I+F+2)(J-I+F+1)\{(F+1)^2-M_F^2\}}{4(F+1)^2(2F+3)(2F+1)(2J+3)(2J+1)(J+1)}$$

$$f_3(JIFM_F)=\frac{(I+F-J)(I+F-J-1)(I+J-F+2)(I+J-F+1)(F^2-M_F^2)}{4F^2(2F-1)(2F+1)(2J+3)(2J+1)(J+1)}$$

$$f_4(JIFM_F)=\frac{(I+J+F+2)(I+J-F)(I+F-J+1)(J+F-I+1)\{(F+1)^2-M_F^2\}}{4J(J+1)(2J+1)(F+1)^2(2F+3)(2F+1)}$$

$$f_5(JIFM_F)=\frac{M_F^2\{J(J+1)+F(F+1)-I(I+1)\}^2}{4J(J+1)(2J+1)F^2(F+1)^2}$$

[a]From Mizushima [22].

For the intermediate-field case, whether the weak-field or the strong-field quantization scheme is chosen, both diagonal and off-diagonal matrix elements are important, and it is necessary to solve secular equations to obtain the desired energy levels. Mizushima [22] has given the following secular equation for the intermediate-field case in the $|J, I, M_J, M_I)$ basis

$$|(J, \tau, I, M_J, M_I|\mathcal{H}_Q|J, \tau, I, M'_J, M'_I)+(E_{J_\tau,I=0,M_J}-E)\delta_{M_J,M'_J}\delta_{M_I,M'_I}|=0 \quad (10.97)$$

Where $E_{J_\tau,I=0,M_J}$ is obtained from (10.66). The matrix elements of $\mathcal{H}_Q$ are given by (10.53) where

$$P=(\tfrac{1}{2})eQq_J\left[\frac{1}{2J(2J-1)I(2I-1)}\right] \quad (10.98)$$

with q_J given by (9.95). The secular equation can be factored into blocks for each value of $M=M_J+M_I$, since matrix elements of $\mathcal{H}_Q$ between different M's are zero, as noted previously. For the usual experimental arrangement the selection rule for M is $\Delta M=0$. In the transition from a weak-field case to a strong-field case, M remains a good quantum number, becoming M_F in the weak-field limit and M_J+M_I in the strong-field limit.

The strong-field Stark effect for molecules with two quadrupole nuclei of spin $\frac{3}{2}$ has been discussed by Howe and Flygare [24]. They note that the $J=0\rightarrow1$, $M_J=0$ component will, in the strong-field limit, appear as a symmetric triplet with the center component the most intense and independent of the quadrupole coupling. This obviously simplifies the determination of the dipole moment.

Eagle et al. [25] have discussed the Stark effect when $\mathcal{H}_\mathcal{E}$, $\mathcal{H}_Q$ and rotational near-degeneracy are comparable. Off-diagonal elements of $\mathcal{H}_Q$ and $\mathcal{H}_\mathcal{E}$ coupling near-degenerate rotational levels must be considered since they can make significant contributions to the energy. Approximations are discussed, and the secular equation applicable to planar near-prolate rotors such as NOCl is given in the $|J, \tau, M_J, I, M_I)$ basis.

5 MOLECULES IN EXCITED VIBRATIONAL STATES—INVERSION DOUBLETS AND *l* DOUBLETS

Although the theory described in other sections for the Stark effect in rotational spectra of ground vibrational states is also applicable to molecules in excited, nondegenerate vibrational modes, it is not applicable to *l*-type doublets of degenerate bending modes or to vibrational modes with inversion doubling. The Stark effect of *l*-type doublets and inversion doublets can be treated with the same theory. The treatment given here is very similar to that in Section 3, where the Stark effect of near-degenerate levels in asymmetric rotors is considered. The two near-degenerate levels of these doublets are assumed to be close in energy as compared with the energy separation from the doublets of all other levels of the molecule with which intermixing induced by the applied field is possible, that is, with which the off-diagonal elements of $\mathcal{H}_\mathcal{E}$ are not zero.

This approximation holds for inversion doublets of symmetric tops like NH_3 and for the l-type doublets commonly observed in linear molecules. In this approximation the problem can be treated as a two-level system with eigenfunctions $\psi_1^{(0)}$ and $\psi_2^{(0)}$ corresponding to the doublet energies $E_1^{(0)}$ and $E_2^{(0)}$. The Hamiltonian operator of the perturbed system is

$$\mathscr{H} = \mathscr{H}^{(0)} + \mathscr{H}_{\mathscr{E}} \tag{10.99}$$

Since there are only two states, the wave functions of the perturbed system can be expressed as the linear combination of the functions $\psi_1^{(0)}$ and $\psi_2^{(0)}$. Since these functions are eigenfunctions of $\mathscr{H}^{(0)}$ but not of $\mathscr{H}_{\mathscr{E}}$, there are no off-diagonal elements of $\mathscr{H}^{(0)}$ and no diagonal elements of $\mathscr{H}_{\mathscr{E}}$. Hence the secular equation is

$$\begin{vmatrix} (1|\mathscr{H}^{(0)}|1) - E & (1|\mathscr{H}_{\mathscr{E}}|2) \\ (2|\mathscr{H}_{\mathscr{E}}|1) & (2|\mathscr{H}^{(0)}|2) - E \end{vmatrix} = 0 \tag{10.100}$$

where $(1|\mathscr{H}^{(0)}|1) = E_1^{(0)}$, $(2|\mathscr{H}^{(0)}|2) = E_2^{(0)}$, and $(1|\mathscr{H}_{\mathscr{E}}|2) = (2|\mathscr{H}_{\mathscr{E}}|1) = \mathscr{E}(1|\mu|2)$. Solution of the equation yields the perturbed energies

$$E_\pm = \tfrac{1}{2}[E_1^{(0)} + E_2^{(0)}] \pm \tfrac{1}{2}[(E_1^{(0)} - E_2^{(0)})^2 + 4\mathscr{E}^2|(1|\mu|2)|^2]^{1/2} \tag{10.101}$$

For the symmetric-top inversion levels, the matrix elements are

$$(1|\mu|2) = \mu(J, K, M_J(1)|\Phi_{Zz}|J, K, M_J(2)) = \mu\frac{KM_J}{J(J+1)} \tag{10.102}$$

and the Stark-perturbed levels can be obtained by substitution of these elements into (10.101). The average of the two unperturbed levels $(E_1^{(0)} + E_2^{(0)})/2 = E_{J,K}$ is simply the rotational energy the molecule would have in the absence of inversion doubling; it is given by (6.14). The energy difference $E_1^{(0)} - E_2^{(0)} = h\nu_{\text{inv}}$ where ν_{inv} is the inversion frequency for the particular rotational and vibrational state. The energies in the presence of a field $\mathscr{E}$ can thus be expressed as

$$E_\pm = E_{J,K} \pm \left[\left(\frac{h\nu_{\text{inv}}}{2}\right)^2 + \left(\frac{\mu\mathscr{E}KM_J}{J(J+1)}\right)^2\right]^{1/2} \tag{10.103}$$

When the inversion splitting is zero or negligible, as it is for all symmetric tops except ammonia, the formula becomes that of the ordinary first-order effect for a symmetric-top.

In Chapter V, Section 2 it is pointed out that the linear molecule in an excited bending mode has angular momentum $l\hbar$ about the figure axis. Its motion simulates that of a symmetric top with $l \to K$. As was first shown by Penney [10], the off-diagonal elements of μ connecting the l-type doublet states are like those of the symmetric top with l replacing K. For the l doublets, therefore

$$|(1|\mu|2)|^2 = \frac{\mu^2 l^2 M_J^2}{J^2(J+1)^2} \tag{10.104}$$

The Stark-perturbed l-doublet levels where $E_1^{(0)} > E_2^{(0)}$ are then

$$E_1 = \tfrac{1}{2}(E_1^{(0)} + E_2^{(0)}) + \tfrac{1}{2}\left[(E_1^{(0)} - E_2^{(0)})^2 + \left(2\frac{\mu\mathscr{E}lM_J}{J(J+1)}\right)^2\right]^{1/2} \tag{10.105}$$

and

$$E_2=\tfrac{1}{2}(E_1^{(0)}+E_2^{(0)})-\tfrac{1}{2}\left[(E_1^{(0)}-E_2^{(0)})^2+\left(2\frac{\mu\mathscr{E}lM_J}{J(J+1)}\right)^2\right]^{1/2} \tag{10.106}$$

It is evident from these equations that when the Stark perturbation is large as compared with the l-doublet separation, the Stark displacement is a first-order effect, like that of the symmetric top. When the Stark splitting is small as compared with the doublet separation, a quadratic Stark effect is observed. This can be seen if $(E_1^{(0)}-E_2^{(0)})^2$ is factored from the radical and if the term under the radical is expanded by the binomial theorem. When $(E_1^{(0)}-E_2^{(0)})^2 \gg [(2\mu\mathscr{E}lM_J)/J(J+1)]^2$, the resulting second-order formulas can be expressed as

$$E_1=E_1^{(0)}+\frac{(\mu\mathscr{E})^2l^2M_J^2}{J^2(J+1)^2(E_1^{(0)}-E_2^{(0)})} \tag{10.107}$$

and

$$E_2=E_2^{(0)}-\frac{(\mu\mathscr{E})^2l^2M_J^2}{J^2(J+1)^2(E_1^{(0)}-E_2^{(0)})} \tag{10.108}$$

The Stark components of the higher doublet level (1) are shifted upward, and those of the lower level (2) are shifted downward.

In the region where $[(2\mu\mathscr{E}lM_J)/J(J+1)]^2 \gg (E_1^{(0)}-E_2^{(0)})^2$, the first-order effect holds, and the $M_J \to M_J$ Stark components of the rotational transitions $J \to J+1$ have the frequencies

$$\nu_1 \approx \nu_r^{(0)}-\frac{2\mu\mathscr{E}l|M_J|}{hJ(J+1)(J+2)} \tag{10.109}$$

$$\nu_2 \approx \nu_r^{(0)}+\frac{2\mu\mathscr{E}l|M_J|}{hJ(J+1)(J+2)} \tag{10.110}$$

where $\nu_r^{(0)}=(\nu_1^{(0)}+\nu_2^{(0)})/2$ is the rotational frequency that would be obtained if the l doubling were not present. This approximation, which neglects the l doubling $(E_1^{(0)}-E_2^{(0)})$, shows that the linear molecule can have a first-order Stark effect when in excited bending vibrational modes.

Although the doublet energies E_1 and E_2 for the same J and M_J are pushed apart, the $M_J \to M_J$ components of the rotational frequencies ν_1 and ν_2 are pushed together by the Stark field. Thus for a particular field value the Stark components of ν_1 and ν_2 for the same M_J values will coincide. When this coincidence occurs in the region where the first-order effect holds, the field value for the intersection is

$$\mathscr{E}=\frac{hJ(J+1)(J+2)}{4\mu l|M_J|}(\Delta\nu)^{(0)} \tag{10.111}$$

where $(\Delta\nu)^{(0)}=\nu_1^{(0)}-\nu_2^{(0)}$ is the separation of the frequencies of the rotational doublets in the absence of a field.

Figure 10.10 from Strandberg et al. [26] shows the Stark frequencies as a function of $\mathscr{E}$ for the $|M_J|=1$ and 2 Stark components of the $J=2 \to 3$ transition

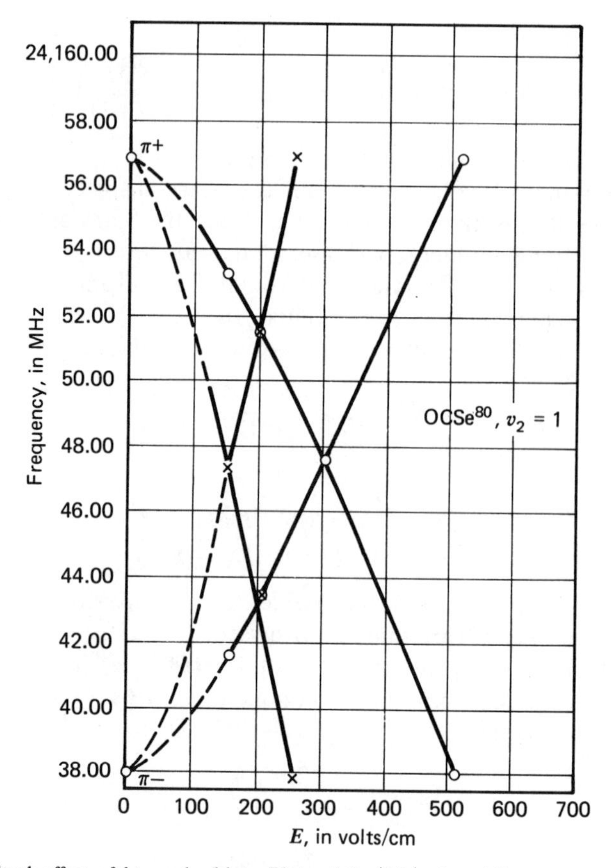

Fig. 10.10 Stark effect of l-type doublets. Plots of the $|M_J|=1$ and 2, π components versus field strength for the $J=2\rightarrow 3$ transition of OCSe in the $v_2=1, l=1$ vibrational state. From Strandberg et al. [26].

of OCSe in the $v_2=1$, $l=1$ bending vibrational state. The components for the same $|M_J|$ cross at the field values given by (10.111). This crossing point provides a convenient method for evaluation of μ since the only frequency difference which must be measured is the field-free $(\Delta\nu)^{(0)}$ which is greater than the Stark splitting except at very high fields.

6 ANISOTROPIC POLARIZABILITIES FROM THE STARK EFFECT

In the preceding discussions we have not considered the Stark effects of induced dipole moments caused by polarizability of the molecule. This omission is justified because the small effects of polarizability are seldom detected and indeed for most molecules are probably not measurable with meaningful accuracy. However, with favorably selected molecules and with accurately ground parallel plate Stark cells which will permit measurements up to 50,000 v/cm it is possible to measure polarizability anisotropy with useful accuracy. As early as 1957, Marshall and Weber [27] detected polarizability effects in their measurements of the Stark effect of OCS, from which they derived a rough value of polarizability anisotropy. They made measurements with fields up to 24,000 v/cm on Stark components of the $J=1\rightarrow 2$ transitions. More recently, in 1967, Scharpen et al. [28] extended the measurements to higher J transitions and to higher fields (up to 50,000 v/cm) and thus achieved better discrimination between the effects of the permanent and induced moments and thereby obtained a more reliable measurement of the polarizability anisotropy in the molecule. They obtained for OCS the value $(\alpha_{\parallel}-\alpha_{\perp})=(5.34\pm 0.06)\times 10^{-24}\ \text{cm}^3$ where $\alpha_{\parallel}$ and $\alpha_{\perp}$ are the polarizabilities parallel and perpendicular to the molecular axis. A reevaluation has led to correction [29] of this value to $4.63\times 10^{-24}\ \text{cm}^3$.

Mizushima [30] has predicted polarizability effects for several diatomic molecules and for HCN. Most of these molecules have nuclear quadrupole coupling which complicates the measurement and evaluation because of the changing degrees of coupling as the field is increased. Mizushima derived the formulas for the weak-field case, but these are not very useful because of the high fields that must be applied to produce measurable polarizability effects. Here we shall consider only molecules without nuclear coupling.

The Hamiltonian for the interaction of a field with an induced moment μ_a is $-\boldsymbol{\mu}_a\cdot\mathscr{E}$, but here, in contrast to the dipole interaction considered earlier, μ_a is itself a function of the field strength. Because the polarizability is anisotropic, the induced moment is also a function of the orientation of the molecule in the field. Generally, the polarizability is a tensor quantity, and the principal axes of the polarizability do not necessarily coincide with the principal axes of inertia. However, in linear or symmetric-top molecules considered here, and in many asymmetric tops, these axes are coincident. If $g=x, y, z$ are the principal axes of the polarizability tensor in the molecule, the classical energy which is due to the polarization by the field is

$$\mathscr{H}_{\alpha}=-\frac{1}{2}\sum_{g=x,y,z}\alpha_{gg}\mathscr{E}_g^2 \tag{10.112}$$

where α_{gg} represents the principal values of the polarizability and $\mathscr{E}_g$ represents the components of the field with reference to the molecule-fixed axes. Since $\mathscr{E}$ is imposed along the space-fixed axis chosen as Z,

$$\mathscr{E}_g=\mathscr{E}\Phi_{Zg} \tag{10.113}$$

where Φ_{Zg} represents the direction cosine of the axis g with Z. Therefore, the polarizability Hamiltonian can be expressed as

$$\mathscr{H}_{\alpha}=-\tfrac{1}{2}\mathscr{E}^2(\alpha_{xx}\Phi_{Zx}^2+\alpha_{yy}\Phi_{Zy}^2+\alpha_{zz}\Phi_{Zz}^2) \tag{10.114}$$

For symmetric-top or linear molecules, the polarizability axes are the same as the inertial axes, and only the diagonal elements of the polarizability tensor,

α, are nonzero. Although there are no permanent components of the dipole moment normal to the symmetry axis, there are components of the induced moment perpendicular to the molecular axis even in linear molecules. Nevertheless, the polarizability is axially symmetric about the figure axis z, and we can set $\alpha_{xx}=\alpha_{yy}=\alpha_\perp$ and $\alpha_{zz}=\alpha_\parallel$. With these conditions, with the omission of $\alpha_\perp\mathscr{E}^2$ which does not depend on the orientation, and with the relation $\sum\Phi_{Zg}^2=1$, (10.114) transforms to

$$\mathscr{H}_\alpha=-\tfrac{1}{2}(\alpha_\parallel-\alpha_\perp)\mathscr{E}^2\Phi_{Zz}^2=-\tfrac{1}{2}\alpha\mathscr{E}^2\Phi_{Zz}^2 \tag{10.115}$$

where for simplicity we designate

$$\alpha=\alpha_\parallel-\alpha_\perp \tag{10.116}$$

The omitted term shifts all levels by the same amount and hence has no effect on the observed spectra. The first term in the contribution of the polarizability to the Stark energy is the average of $\mathscr{H}_\alpha$ over the unperturbed wave function $|J, K, M_J)$. Actually, this represents a second-order contribution to the total Stark effect. Since Φ_{Zz} is diagonal in K and M_J,

$$\begin{aligned}E_{J,K,M_J}(\alpha)&=(J, K, M_J|\mathscr{H}_a|J, K, M_J)\\&=-\tfrac{1}{2}\alpha\mathscr{E}^2(J, K, M_J|\Phi_{Zz}^2|J, K, M_J)\\&=-\tfrac{1}{2}\alpha\mathscr{E}^2\sum_{J'}(J, K, M_J|\Phi_{Zz}|J', K, M_J)^2\end{aligned} \tag{10.117}$$

The direction cosine matrix elements, which may be obtained from Table 2.1, are the same as those used in calculation of the second-order term of the permanent moment. In practice it is unlikely that the polarizability term will be observable for levels with $K\neq0$ when the large first-order contribution from the permanent moment is present. Hence we shall write the specific formula only for symmetric-top levels of $K=0$, which applies for linear molecules also. When the direction cosine matrix elements for $K=0$ are substituted into (10.117) it is seen that

$$E_{J,M_J}(\alpha)=-\tfrac{1}{2}\alpha\mathscr{E}^2\left[\frac{(J+1)^2-M_J^2}{(2J+1)(2J+3)}+\frac{J^2-M_J^2}{(2J+1)(2J-1)}\right] \tag{10.118}$$

This term is added to that caused by the permanent dipole moment, given by (10.11), for a total Stark perturbation to second order:

$$E_{J,M_J}=E_{J,M_J}(\mu)+E_{J,M_J}(\alpha) \tag{10.119}$$

The displacement of a particular Stark component can be expressed as

$$\Delta\nu=(a_1\mu^2+a_2\alpha)\mathscr{E}^2=a\mathscr{E}^2 \tag{10.120}$$

where a_1 and a_2 depend on the quantum numbers but are constant for the particular Stark component. These constants are calculable: a_1 from (10.11) and a_2 from (10.118). By measurement on two or more different Stark components with different values of a_1 and a_2, both μ and $\alpha=(\alpha_\parallel-\alpha_\perp)$ can be obtained. Notice that a_1 decreases more rapidly with increasing J than does a_2.

Thus for measurement of the small term in α, it is advantageous to use relatively high J values with very high field values to achieve splittings which are accurately measurable.

When carried to higher order, the Stark shift of a particular component can be expressed as

$$\Delta\nu = a\mathscr{E}^2 + b\mathscr{E}^4 \tag{10.121}$$

where

$$a = a_1\mu^2 + a_2\alpha \tag{10.122}$$

and

$$b = b_1\alpha^2 + b_2\alpha\mu^2 + b_3\mu^4 \tag{10.123}$$

If (10.121) is divided by $\mathscr{E}^2$, it is seen that a plot of values of $\Delta\nu/\mathscr{E}^2$ versus $\mathscr{E}^2$ gives a straight line with b as the slope and a as the intercept. Thus a values, from which μ and α are obtained, can be evaluated from (10.121) without specific knowledge of the coefficients b_1, b_2, and b_3. However, Scharpen et al. [28] give the higher-order formulas required for calculation of these constants if they are needed. Only the $b_3\mu^4$ term is likely to have measurable effects.

It is interesting to compare the anisotropic polarizability obtained by the Stark effect for OCS with the values of 2.2×10^{-24} cm^3 for CO_2 and 9.6×10^{-24} cm^3 for CS_2 evaluated by other methods [31]. The average of these two values 5.9×10^{-24} cm^3 is not far from that of 4.63×10^{-24} cm^3 obtained for OCS [29]. Probably many other polarizability anisotropies will be measured by the microwave Stark effect as experimental techniques are further improved.

The electric polarizability anisotropy of the nonpolar oxygen molecule has been measured through observation of the splitting of one of its millimeter-wave, fine structure lines by a strong external electric field [32]. The observed splitting of the $J=0\rightarrow1$, $N=1\rightarrow1$ line that occurs at 118 GHz is shown in Fig. 10.11. Since O_2 has no permanent dipole moment, the splitting is due entirely to anisotropic polarization induced by the applied field. From the Hamiltonian of (10.115), the anisotropic polarization energy for O_2 can be obtained in the form [32]

$$\begin{aligned} E_{\mathscr{E}} = &-(\alpha_\parallel-\alpha_\perp)\mathscr{E}^2\frac{2J+1}{3(2N-1)(2N+3)} \\ &\times\sum_{M_S=-1}^{1}\begin{pmatrix}1 & N & J\\ M_S & M_J-M_S & -M_J\end{pmatrix}^2[N(N+1)-3(M_J-M_S)^2] \end{aligned}\tag{10.124}$$

where the first quantity in the summation is a squared $3j$ symbol explained in Chapter XV. This expression leads to the splitting for the $J=0\rightarrow1$, $N=1\rightarrow1$ oxygen line of

$$(\Delta\nu)_{\Delta M_J=0} = \frac{1}{15h}(\alpha_\parallel-\alpha_\perp)\mathscr{E}^2 \tag{10.125}$$

$$(\Delta\nu)_{\Delta M_J=\pm1} = -\frac{1}{30h}(\alpha_\parallel-\alpha_\perp)\mathscr{E}^2 \tag{10.126}$$

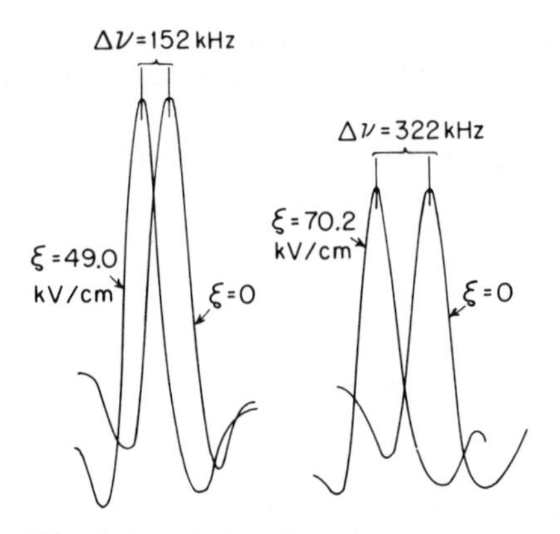

Fig. 10.11 The Stark shift of the 118 GHz line of oxygen ($\Delta M_J = \pm 1$). From Gustafson and Gordy [32].

Equation 10.126, when applied to the observed splittings (see Fig. 10.11), gives the anisotropic polarizability of O_2 as

$$\alpha_{\parallel} - \alpha_{\perp} = +1.12 \pm 0.07 \text{ cm}^3 \times 10^{-24}$$

A comparable magnitude of $1.09 \text{ cm}^3 \times 10^{-24}$ has been estimated from laser depolarization scattering and bulk polarizability measurements [33, 34].

Perturbation corrections to the Stark energy arising from polarization for asymmetric rotors have been discussed by Golden and Wilson [11]. Although the off-diagonal elements of the polarizability tensor do not vanish, in general, for asymmetric rotors, only the diagonal elements of the tensor will contribute in a perturbation treatment correct to second-order in the field. The polarization contribution is obtained if $\mathscr{H}_\alpha$, (10.114) is averaged over the functions $|J, \tau, M_J)$

$$\begin{aligned} E_{J_\tau, M_J}(\alpha) = &-\frac{1}{2} \sum_g \alpha_{gg} \mathscr{E}^2 \\ &\times \left[\frac{J^2 - M_J^2}{4J^2(4J^2-1)} \sum_{\tau'} |(J, \tau|\Phi_{Zg}|J-1, \tau')|^2 \right. \\ &+ \frac{M_J^2}{4J^2(J+1)^2} \sum_{\tau'} |(J, \tau|\Phi_{Zg}|J, \tau')|^2 \\ &\left. + \frac{(J+1)^2 - M_J^2}{4(J+1)^2(2J+1)(2J+3)} \sum_{\tau'} |(J, \tau|\Phi_{Zg}|J+1, \tau')|^2 \right] \end{aligned} \tag{10.127}$$

This may be expressed in terms of the line strengths by use of (10.71)–(10.73) or more conveniently in terms of the average values of the angular momenta, that is, $\langle J_a^2 \rangle$, $\langle J_b^2 \rangle$, and $\langle J_c^2 \rangle$ (see Chapter VII, Section 2).

7 RELATIVE INTENSITIES

The relative intensities are important for the interpretation of the observed spectra. When the M_J degeneracy is lifted by an external field, the intensities of the Stark components will depend upon M_J. Since the M_J dependence of the direction cosine matrix elements is the same for asymmetric rotors as for symmetric rotors, the M_J dependence of the intensities for a particular line will be proportional to $|(J, M_J|\Phi_{Fg}|J', M_J')|^2$. Thus the relative intensities of the $M_J \rightarrow M_J$ Stark components are from Table 2.1 seen to be

$$I_{M_J} = PM_J^2 \qquad \Delta J = 0 \quad \Big\} M_J = 0 \tag{10.128}$$

$$I_{M_J} = Q[(J+1)^2 - M_J^2] \qquad \Delta J = \pm 1 \tag{10.129}$$

where J is the smaller of the two J's involved in the transition. For a second-order Stark effect a completely resolved Stark pattern will have $(J+1)$ components for $\Delta J = \pm 1$ and J components for $\Delta J = 0$.

The relative intensities of the $M_J \rightarrow M_J \pm 1$ Stark components, also derivable from Table 2.1, are given by

$$I_{M_J} = \frac{Q}{4}(J \pm M_J + 1)(J \pm M_J + 2) \qquad \Delta J = +1 \tag{10.130}$$

$$I_{M_J} = \frac{P}{4}(J \mp M_J)(J \pm M_J + 1) \qquad \Delta J = 0 \quad \Big\} \Delta M_J = \pm 1 \tag{10.131}$$

$$I_{M_J} = \frac{Q}{4}(J \mp M_J)(J \mp M_J + 1) \qquad \Delta J = -1 \tag{10.132}$$

where J is the smaller of the two J's involved in the transition and where the upper sign is taken throughout for the $M_J \rightarrow M_J + 1$ transition and the lower sign for the $M_J \rightarrow M_J - 1$ transition. For a Stark effect dependent on M_J^2, a completely resolved Stark pattern will have $(2J+1)$ components. Figure 10.12 illustrates some typical Stark patterns.

The coefficients P and Q are independent of M_J but do depend on the strength of the unsplit line which is a function of the other quantum numbers involved in the transition. For asymmetric rotors the coefficients are also dependent on the asymmetry of the molecule. When the Stark energy is proportional to M_J^2, the two states $+M_J$ and $-M_J$ are degenerate, except for $M_J = 0$, and have a statistical weight factor of 2. Therefore, for transitions with $M_J = 0$ a factor of $\frac{1}{2}$ must be included in the intensity expressions. Furthermore, it is noted that for $\Delta J = 0$, $\Delta M_J = 0$ the $M_J = 0$ component is forbidden, that is, it has a vanishing intensity.

In derivation of the foregoing expressions it is assumed that perturbation of the wave functions by the electric field is small and that zero-field functions are adequate. However, the applied electric field, if sufficiently strong, can measurably alter the relative intensities. In particular, the Stark perturbation

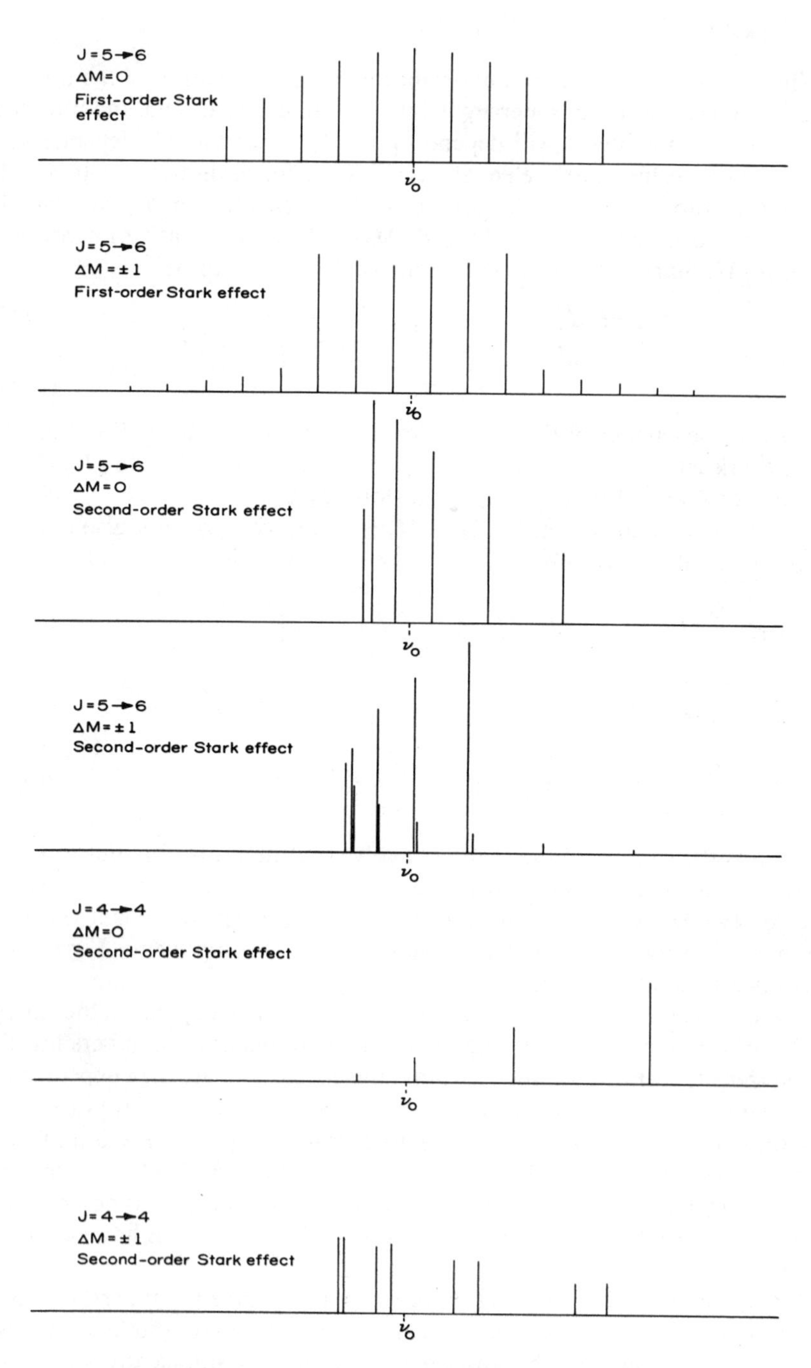

Fig. 10.12 Some typical first- and second-order Stark patterns.

causes a mixing of the zero-field wave functions with the degree of mixing being dependent on the field strength. Even though these effects may be small, an originally forbidden transition, for example, $|\Delta J|>1$, can, when an electric field is applied, have a nonvanishing transition probability, and observation of such forbidden transitions becomes possible [11, 21]. The presence of near-degeneracies can cause rather large perturbations, as in the case of an l-type doublet in linear rotors, and of a K-type doublet in asymmetric rotors.

The relative intensities, when there is quadrupole hyperfine structure can be readily found if the limiting weak- or strong-field approximation is valid.

For strong-field case the JIM_JM_I representation is appropriate, and the relative intensities of the various Stark components are the same as for molecules without hyperfine structure. For the usual $\Delta M_J=0$ transitions the intensities are hence given by (10.128) and (10.129). However, a given Stark line is further split into $I+1$ hyperfine lines, all of which have the same intensity except for $M_I=0$, which is half as strong because of the loss of the twofold degeneracy, viz., $\pm M_I$.

For the weak-field case the $JIFM_F$ representation is appropriate. In this representation the possible projections of the total angular momentum $\mathbf{F}$ along the field direction are specified by the values of M_F. The intensities are hence obtained by replacing J by F and M_F in the preceding equations. For the $M_F \rightarrow M_F$ transitions, the Stark component intensities are

$$I_{M_F}=RM_F^2 \qquad \Delta F=0 \qquad \Delta M_F=0 \tag{10.133}$$

$$I_{M_F}=S[(F+1)^2-M_F^2] \qquad \Delta F=\pm 1 \qquad \Delta M_F=0 \tag{10.134}$$

where R and S are parameters independent of the quantum number M_F and where F is the smaller of the two values involved in the particular hyperfine transition.

The intensities for the intermediate-field case can be approximated by an interpolation between the strong- and weak-field limits. The exact intensities are proportional to the squares of the dipole-moment matrix elements evaluated with the intermediate-field wave functions. Evaluation of the intensities hence requires the solution of the secular equation and evaluation of the wave function for each level. The intermediate-field wave functions are linear combinations of the original basis functions chosen to construct the intermediate-field secular equation. They are defined once the expansion coefficients, the C_i's, are specified [see (10.48)].

8 IDENTIFICATION AND MODULATION OF ROTATIONAL LINES WITH THE STARK EFFECT

A knowledge of the intensities, spacing, and number of Stark components is frequently helpful in the assignment of rotational transitions, especially for asymmetric rotors. For the usual $\Delta M_J=0$ transitions the relative intensities are useful for distinction between the $\Delta J=0$ type transitions and the $\Delta J=\pm 1$ type. In the former case, the components with the largest values of M_J will have the greater intensity, while in the latter case just the reverse is true. The measured spacing between successive M_J components gives information regarding the value of the quantum number M_J corresponding to a particular component of a Stark pattern as well as to the smaller J involved in the transition. This latter information is obtained by determination of the maximum value of $|M_J|$ which must be equal to the smaller J. For relatively low J lines, where it is usually possible to obtain a completely resolved Stark pattern, a direct count of the number of M_J components leads to the determination of the smaller of the two J's involved in a transition. Note that for a second-order Stark effect there will be a maximum of $(J+1)$ Stark components when $\Delta J = \pm 1$ and J components when $\Delta J = 0$, while for a first-order Stark effect there will be $(2J+1)$ components when $\Delta J = \pm 1$ and $2J$ components when $\Delta J = 0$. Once the value of J and the type of transition are known, the assignment of the τ and τ' involved in a transition may often be found by a trial-and-error fitting of observed transition frequencies to calculated rigid rotor frequencies.

A valuable application of the Stark effect is its use for modulation of rotational absorption lines [35] in the Stark modulation spectrometers introduced by Hughes and Wilson [1]. The modulation is accomplished by insertion of an electrode in the waveguide absorption cell and application of a low radio-frequency (e.g., ~ 100 kHz) square wave voltage. The electrode consists of a flat strip of brass placed parallel to the broad side of the rectangular waveguide and mounted on a grooved dielectric such as Teflon. This arrangement results in a modulation of the microwave energy in the vicinity of an absorption line. If a narrow bandwidth detector is tuned to the modulating frequency, a significant increase in sensitivity is obtained. If one side of the square wave is clamped to zero voltage, the undisplaced absorption line is observed during the zero field part of the cycle; during the "field-on" portion, a series of absorption lines which are displaced because of the Stark effect will appear. When phase-sensitive detection is used, the zero-field line and its Stark components will be displayed with opposite sense; see, for example, Fig. 10.8. For the millimeter wave region, source-frequency modulation is more useful because of the high losses of power in conventional Stark cells. If Stark modulation is desired, a parallel plate cell can best be employed. Rather thorough treatments of the Stark effect for the varying fields often used in Stark modulation spectrometers are given by Autler and Townes [36] and in the text by Townes and Schawlow [37].

A method for approximate measurement of dipole moments which depends on Stark modulation is described by Lindfors and Cornwell [38]. The method consists in measurement of the peak-to-peak intensity of an absorption line for various field strengths (modulation amplitude). It requires the observation of an absorption line with a first- or second-derivative line shape over the range of modulation amplitudes. This is usually possible at low modulation amplitudes when one uses a spectrometer employing zero-based square wave Stark modulation and phase-sensitive detection. In general, a range of field strengths must be chosen such that the Stark shifts are smaller than the linewidth. Although this method, "the rate of growth of line intensity," is not as accurate as the Stark splitting method for determination of the dipole moment, it has the advantage of not requiring resolution of individual Stark components and hence can be applied to high J transitions for which the Stark pattern can be modulated but not completely resolved. The necessary working equations and a discussion of experimental procedures are given in the original paper [38].

9 DIPOLE MOMENTS FROM THE STARK EFFECT

Although the Stark effect is a valuable aid in the detection and identification of rotational lines, its most important chemical application is in the measurement of electric dipole moments. In contrast to that of the methods based on measurement of the dielectric constant, its accuracy is not influenced by impurities in the sample. With the Stark effect, the dipole moment is measured for a particular isotopic species and for a particular vibrational state and rotational transition. Hence it is possible under favorable conditions to measure the small differences which are due to isotopic substitution or change in vibrational or rotational state.

In the earlier period of microwave spectroscopy, up to 1957, the Stark field was applied to an electrode in the center of a rectangular waveguide. The uncertainty in the effective spacing of the electrode in such a cell and the tendency for potential breakdowns at high voltage limited the accuracy of these measurements to a percent or so. Improvement by an order of magnitude in the accuracy of dipole moment measurement was made possible by the introduction of the parallel plate Stark cell [27, 39, 40] with optically ground quartz spacers. Such a cell is highly practical in the shorter millimeter wave region where the radiation is easily focused with horns and lenses of convenient size and where the rotational absorption lines are so strong that a cell of short length can be used. With such a cell it is possible to apply a higher voltage without breakdown than is possible with the conventional waveguide Stark cell. Furthermore, the parallel-plate cell can easily be designed to transmit modes which allow measurement [40, 41] of the σ, or $\Delta M_J = \pm 1$, components.

In measurements of Stark splitting it is convenient to use as secondary standards the dipole moments or Stark splitting values of molecules which have been calibrated previously by careful, high-precision measurement. This practice eliminates the necessity for making precise voltage measurements with a potentiometer and precise electrode spacing with an interferometer every time a dipole moment is measured. Dipole moments of the linear molecules OCS and HCN, also the symmetric-top molecules CH_3F, CD_3F, CH_3CN, and CD_3CN, have been measured to high precision with parallel-plate cells. These are absolute measurements in the sense that the plate separation, the potential, and the frequency were measured directly with precision instruments without the use of a secondary molecular standard. These moments can be used as secondary standards for calibration of Stark spectrometers. Errors in the OCS values mentioned later are, hopefully, now cleared up. Those of carbonyl sulfide and methyl fluoride are more satisfactory to use because they offer no complications from nuclear quadrupole hyperfine structure.

Chemically, OCS is simple to prepare and easy to handle. Its microwave spectrum is simple. All three nuclei of the most abundant isotopic species $^{16}O^{12}C^{32}S$ have zero spins, and hence the rotational spectrum of this species has no hyperfine structure. The Stark pattern of the $J=1 \rightarrow 2$ transition is the simple, strong doublet shown in Fig. 10.3. Unfortunately, there is a rather large discrepancy in the value of the dipole moment of this molecule originally measured from the $M_J \rightarrow M_J$ Stark components of the $J=1 \rightarrow 2$ transition with a parallel-plate cell by Marshall and Weber [27] in 1957 and the recent measurement (1968) with molecular beam electric resonance on the $J=1$, $M_J \rightarrow M_J \pm 1$ transition, also with a parallel-plate cell, by Muenter [29]. The value obtained Marshall and Weber is 0.7124 D with a stated error limit of ± 0.0002, whereas the value obtained by Muenter is 0.71521 ± 0.00020 D. Although the two values have the same estimated error limit, they differ by a factor of more than ten times this limit. The reason for this discrepancy is not clear, but we accept the new Muenter value because it, when used as a secondary standard for measurement of the CH_3F moment, gives a result that is consistent with the absolute value of the CH_3F moment measured in our own laboratory, as described later.

It is of interest that the measurement of the OCS moment which was made with a waveguide cell and widely used as a secondary standard until replaced by the measurement of Marshall and Weber is 0.709 ± 0.003 D, a value lower by 0.0062 D (twice the estimated error limits) than the most recent OCS value. Therefore it appears that the many dipole moment values reported in the literature up to 1968 for which OCS was used as a secondary standard have absolute errors greater than the error limits quoted for their standard.

Because of the decreasing sensitivity of the Stark effect with increasing J, the OCS moment is not satisfactory for calibration of Stark spectrometers operating in the shorter millimeter range. For that region of the spectrum, the dipole moments of HCN, CH_3F, or CD_3F provide accurate secondary standards. The lines of HCN are the strongest, but the most abundant isotopic species of HCN is complicated by ^{14}N quadrupole coupling. Absolute measurements of the moments of CH_3F and CD_3F have been made [41, 42] with both $K=0$ and $K=1$ lines and with π and σ components of the $J=1 \rightarrow 2$ transition, which falls at 102 GHz for CH_3F and at 81.7 GHz for CD_3F. A summary of the results is given in Table 10.5. There is no measurable difference in the values obtained for $K=0$ and $K=1$ transitions although there is a difference in the moments of the two isotopic species. Figure 10.4 shows the energy level diagram (not drawn to scale) and indicates the different transitions employed for these measurements.

In some symmetric-top molecules with $K=0$ at high fields, Muenter and Laurie [2] found measurable contributions of fourth-order Stark terms. Though the fourth-order perturbation term is calculable, the resulting expression is complicated and inconvenient to apply. They therefore used the method described in Section 1 which bypassed the calculations of the fourth-order co-

Table 10.5 Summary of Dipole Moments of CH_3F and CD_3F Measured with σ and π Components of the $J=1\rightarrow2$ Transition[a]

Transition			Dipole Moment μ (D)	
			CH_3F	CD_3F
$K=1$	$\Delta M=0$		1.858 ± 0.0010	1.868 ± 0.0010
$K=1$	$\Delta M=\pm1$		1.857 ± 0.0011	1.868 ± 0.0010
$K=0$	$\Delta M=0$		1.857 ± 0.0010	1.868 ± 0.0013
$K=0$	$\Delta M=\pm1$		1.857 ± 0.0011	1.868 ± 0.0011
		Average	1.8572 ± 0.0010	1.8682 ± 0.0010

[a]From Steiner and Gordy [42].

efficient but allowed the fourth-order effect to be included in the evaluation of the dipole moment.

Using a short waveguide Stark cell calibrated with the HCN moment, Burrus [43, 44] observed the Stark effect on a number of molecules in the wavelength range from 2.5 to 0.93 mm and derived the dipole moments of the simple molecules DCl, DBr, DI, CO, and PH_3, which have no microwave transitions in the lower microwave frequencies. The moments of the deuterium halides which have quadrupole coupling by the halogen were measured with the weak-field formula, (10.35). The weak-field Stark displacement of each component is a quadratic function of the field strength,

$$\Delta\nu=k(J, I, F, M_F)\mu^2\mathscr{E}^2 \tag{10.135}$$

where k is constant for a particular component and is calculable from (10.35). Thus if one plots $\Delta\nu$ as a function of $\mathscr{E}^2$, a straight line is expected where the weak-field case holds. Such plots of the observed data by Burrus for $D^{81}Br$ are shown in Fig. 10.13. It is seen that the straight-line function holds well up to about 8000 v/cm. The slope of these plots is $k\mu^2$ from which the dipole moment is obtained with the k's calculated from the weak-field formula. Burrus similarly observed that the weak-field case holds for $D^{35}Cl$ to voltages of approximately 4000 v/cm, for which eQq is only 67 MHz. Because CO has a very small moment it was necessary to use fields up to 24,000 v/cm to produce accurately measurable shifts. With such fields, the polarizability corrections (Section 6) become important. Mizushima [30] has calculated this effect which, he indicates, would add about 0.002 D to the moment as given by Burrus [43]. With this correction the Burrus value of the permanent CO dipole moment would be 0,114 D.

The dipole moment of CH_3CN of Table 10.10 was measured with the strong-field Stark effect which is readily achieved with the first-order effect on the $K\neq0$ lines having the relatively weak quadrupole coupling $eQq\approx4$ MHz. Shulman et al. [45] used the intermediate-field theory to derive the dipole moments of the methyl halides from their Stark measurements. The value that

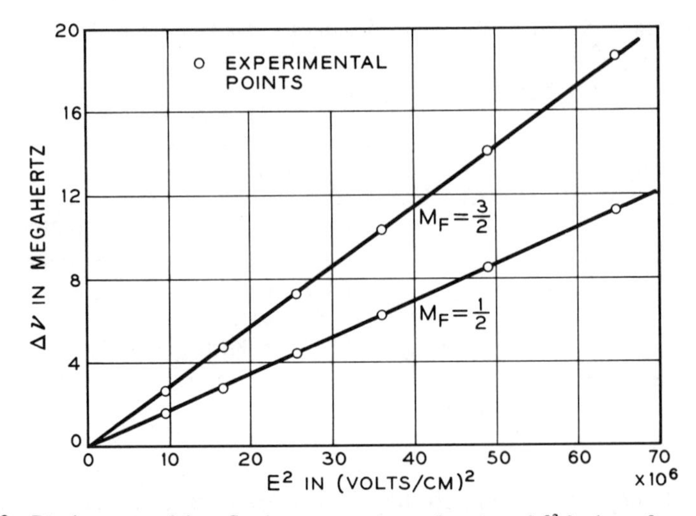

Fig. 10.13 Displacement of the π Stark components as a function of $\mathscr{E}^2$ for hyperfine components of the $J=0\rightarrow 1$ transition of $D^{81}Br$. From Burrus [44].

they obtained from CH_3I has been recalculated to higher order by Buckingham and Stephens [5], who reduced the value from 1.647 to 1.618 D. Likewise, SiH_3Cl, SiH_3Br, and GeH_3Cl were treated with intermediate-field theory by Mays and Dailey [46].

Since the effective bond length r_v is generally longer for excited bond-stretching vibrational states, one might expect that, other things being equal, there would be an accompanying increase in the dipole moment for these states. In LiCl, where the bonding is simple and largely ionic, this is found to be true; but in CS, for which the bonding is largely covalent with a π-bond component, the dipole moment actually decreases with vibration (see Table 10.6). Unfortunately, not many examples are available for comparisons. It has not been feasible to measure such light molecules as HBr and CO in excited vibrational states.

As a general rule, the dipole moments of diatomic molecules tend to increase with the bond-stretching vibrational quantum numbers. This may be expected from the greater effective bond lengths of the higher vibrational states. The trend is followed by the highly ionic molecules LiCl and KCl, shown in Table 10.6, but not by CS, for which the bonding is more complex. The dipole moments of linear polyatomic molecules tend to be slightly lower for excited bending states than for the ground state. This trend is illustrated by a few triatomic molecules in Table 10.7. Qualitatively, this effect may be attributed to the effective shortening of the overall molecular lengths of the bending vibrations. Since there is also some redistribution of the molecular charges resulting from the vibrations, no such simple rule can be expected to hold for all molecules.

Deuterium substitution would be expected to lower the dipole moment of a particular bond because of the lowered zero-point vibrational energy com-

Table 10.6 Illustrations of the Effect of Bond-stretching Vibration on Dipole Moment

	Dipole Moment (D)					
Molecule	$v=0$	$v=1$	$v=2$	$v=3$	$v=4$	*Ref.*
$^{12}C^{32}S$	1.966	1.944				a
$^{6}Li^{35}Cl$	7.119_5	7.206_9	7.296_4	7.386_5		b
$^{39}K^{35}Cl$	10.269_0	10.329_0	10.389_4	10.450_2	10.511_4	c

[a]G. Winnewisser and R. L. Cook, *J. Mol. Spectrosc.*, **28**, 266 (1968). Values are here corrected to conform to new value of OCS moment used as a standard.
[b]D. R. Lide, Jr., P. Cahill, and L. P. Gold, *J. Chem. Phys.*, **40**, 156 (1964).
[c]R. van Wachem and A. Dymanus, *J. Chem. Phys.*, **46**, 3749 (1967).

Table 10.7 Illustrations of Effect of Bending Vibration on Dipole Moment

		Dipole Moment μ *(D)*		
Molecule	$J \rightarrow J+1$	$v=0$	$v_2=1$	$v_2=3$
HCN	$0 \rightarrow 1$	2.985[a]	2.957[b]	
OCS	$1 \rightarrow 2$	0.71521[c]	0.700[d]	
OCSe	$2 \rightarrow 3$	0.754[e]	0.728[e]	0.730[e]

[a]B. N. Bhattacharya and W. Gordy [40].
[b]R. G. Shulman and C. H. Townes, *Phys. Rev.*, **77**, 421 (1948).
[c]Muenter [29].
[d]R. G. Shulman and C. H. Townes, *Phys. Rev.*, **77**, 500 (1950).
[e]Strandberg et al. [26].

bined with a slightly anharmonic, Morse-type potential function. However, the electronic structure of the molecule is also perturbed by this change in vibrational energy so that it is not possible to predict in a simple way the effect of isotopic substitution on a dipole moment. For example, deuterium substitution decreases the moment in CH_3CCH from 0.7809 to 0.7668 D for the same transition in CH_3CCD, but increases the moment from 0.390 D in HCP to 0.397 D in DCP. The dipole moment is also raised by deuterium substitution in methyl fluoride (see Table 10.5), but is lowered in methyl silane and methyl germane.

The effects of centrifugal distortion on the dipole moments of HNCO, DNCO and HN_3 have been observed by White and Cook [47] from measurements of the rotational Stark spectra in the ground vibrational state. Analysis of the Stark effect for the $J=3 \rightarrow 4$ transitions of these molecules yields the dipole moment component along the axis of smallest moment of inertia. For these very slightly asymmetric prolate rotors, with A rotational constants between half a million and a million megahertz, the dipole moment is observed to decrease with increasing K, the changes being surprisingly large. Recent measurements of the $\mu_b(=1.35\text{ D})$ component of HNCO by Hocking et al. [48] further clarified the variation of μ_a with K and the results are collected in Table 10.8. A simple model [47] can explain the observed K dependence of μ_a which is essentially quadratic. K can be approximately interpreted as the magnitude of the spin about the NCO molecular chain. Thus as K increases, the HNC angle decreases due to centrifugal distortion and the projection of the N–H bond moment on the a axis decreases. The calculated [48] change in the HNC angle and corresponding changes in μ_a are also shown in Table 10.8.

Although only the absolute magnitudes of the dipole components along the three principal axes of an asymmetric-top molecule are determined from the rotational Stark effect, the direction of these components can usually be decided by an appeal to electronegativities, bond moments, or other chemical information. Therefore, the orientation of the dipole moment vector in the molecule can be determined. However, when a clear-cut decision as to the direction of the dipole components cannot be made, the effect of isotopic substitution can sometimes be used to give the relative direction of the components. Numerous determinations of the dipole moment direction employing isotopic substitution have been reported in the literature. We give here only two examples to demonstrate the method. Consider, for example, formyl fluoride (HCOF) studied by Favero and Baker [49] for which $\mu_a=0.58\pm0.02$ and $\mu_b=1.91\pm0.03$ (debye units). The location of the principal axes is shown in Fig. 10.14. It is most probable that the oxygen and fluorine atoms because of their greater electronegativity are negative with respect to the hydrogen atom and μ_b is oriented as

Table 10.8 Illustration of the Effect of Centrifugal Distortion on the Dipole Moment of Isocyanic acid (HNCO)

Transition $J=3\rightarrow4$	$\mu_a^{(a)}$	$\Delta\mu_a^{(b)}$	$\Delta\theta^{(c)}$	$\Delta\mu_a^{(c)}$
$K=0$	1.577	—	—	—
$K=1$	1.572	0.005	0.32	0.009
$K=2$	1.540	0.037	1.30	0.036
$K=3$	1.500	0.077	2.92	0.081

[a] Dipole component along the axis of smallest moment of inertia in debye units. From White and Cook [47], and Hocking et al. [48].
[b] Observed decrease in μ_a relative to $K=0$.
[c] Calculated decrease in HNC angle (θ in degrees) and in μ_a relative to $K=0$ from Hocking et al. [48].

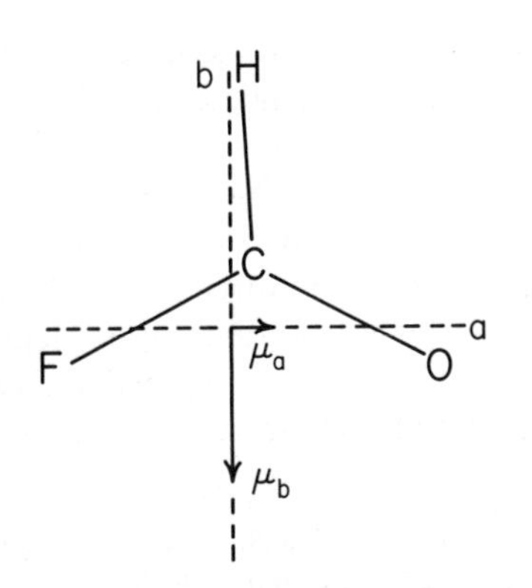

Fig. 10.14 Principal inertial axes of HCOF. The direction of μ_b may be selected from chemical arguments but not that of μ_a. However, for μ_a to increase on deuteration, which results in a rotation of the b axis towards the D atom, the direction must be as shown. The arrowhead indicates the negative end of the dipole moment components. After Favero and Baker [49].

shown. The arrowheads indicate the negative end of the dipole moment. Substitution of deuterium for the hydrogen atom will result in a rotation ($\sim 1°$) of the principal axes of inertia, the b axis moving towards the D atom. For DCOF the dipole moment components are found to be $\mu_a = 0.61 \pm 0.02$ and $\mu_b = 1.96 \pm 0.03$. In order for μ_a to increase, its orientation must be as shown in Fig. 10.14 if it is assumed that the increase arises from the rotation of the axes and that the orientation of the molecular dipole relative to the molecule is essentially unchanged. A method for finding the sign of the dipole moment from the Zeeman effect is described in Section 6 of Chapter XI.

For formic acid, HCOOH, which has been discussed by Kim et al. [50], the direction of μ_a (see Fig. 10.15) can be similarly selected, but not that of μ_b because of its small size. The measured dipole components for HCOOH are (in debye units): $\mu_a = 1.391 \pm 0.005$, $\mu_b = 0.26 \pm 0.04$, and $\mu = 1.415 \pm 0.01$ and for HCOOD they are $\mu_a = 1.377 \pm 0.005$, $\mu_b = 0.22 \pm 0.02$, and $\mu = 1.39 \pm 0.01$. From the experimental value of μ_b^2 (HCOOH)$=0.07$, the angle θ between the total dipole moment and the a inertial axis is found to be $+10.8°$ or $-10.8°$, depending

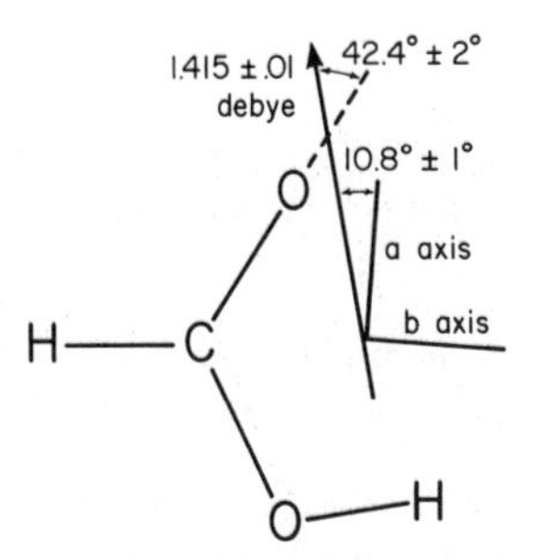

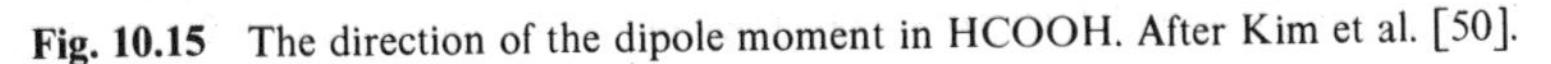
Fig. 10.15 The direction of the dipole moment in HCOOH. After Kim et al. [50].

on the direction of μ_b. Deuteration of the acid hydrogen will produce a rotation of the inertial axes which can be calculated from the known structure given by Kwei and Curl [51]. If the positive direction of θ is taken as clockwise in Fig. 10.15, then the inertial axes are rotated by -1.8° upon deuteration. Therefore, θ in HCOOD should be either $+12.6^\circ$ or -9.0°. From μ_b^2 for HCOOD one finds $\theta = \pm 9.0^\circ$; and if we assume that the direction of the dipole moment relative to the molecule remains basically the same upon deuteration, then it follows that θ is -10.8° in HCOOH. Therefore μ_b in HCOOH is directed counterclockwise from the μ_a direction. These results are summarized in Fig. 10.15.

Table 10.9 Dipole Moments of Some Diatomic and Linear Polyatomic Molecules in the Ground Vibrational State[a]

Molecule	μ *(D)*	*Ref.*	*Molecule*	μ *(D)*	*Ref.*
DCl	1.12(4)	b	HCP	0.390(5)	j
DBr	0.83(2)	b	DCP	0.397(5)	j
DI	0.445(20)	b	FCP	0.279(1)	k
LiCl	7.119(8)	c	NNO	0.166(2)	l
−CO+	0.112(5)	d	HCNO	3.06(13)	m
	0.10980(3)	e	HCCD	0.01001(15)	n
−CS+	1.966(5)	f	HCCF	0.7207	o
CSe	1.99(4)	g	DCCF	0.7329	p
−OCS+	0.7152(2)	h	HCCCl	0.44(1)	m
HCN	2.985(4)	i	FCCCl	0.25(10)	q

[a]All measurements are made from the $J=0\rightarrow1$ transition except those of OCS($1\rightarrow2$), FCP($2\rightarrow3$), and the acetylene derivatives. Signs of certain of the moments, as indicated were determined by the Zeeman method as described in Chapter XI, Section 6.

[b]Burrus [44].

[c]D. R. Lide, P. Cahill, and L. P. Gold, *J. Chem. Phys.*, **40**, 156 (1964).

[d]Burrus [43].

[e]From molecular beam resonance, by J. S. Muenter, *J. Mol. Spectrosc.*, **55**, 490 (1975).

[f]G. Winnewisser and R. L. Cook, *J. Mol. Spectrosc.*, **28**, 266 (1968).

[g]J. McGurk, H. L. Tigelear, S. L. Rock, C. L. Norris, and W. H. Flygare, *J. Chem. Phys.*, **58**, 1420 (1973).

[h]Muenter [29].

[i]Bhattacharya and Gordy [40].

[j]J. K. Tyler, *J. Chem. Phys.*, **40**, 1170 (1964).

[k]W. H. Kroto, J. F. Nixon, and N. C. P. Simmons, *J. Mol. Spectrosc.*, **82**, 185 (1980).

[l]Shulman et al. [45].

[m]M. Winnewisser and H. K. Bodenseh, *Z. Naturforsch.*, **22a**, 1724 (1967).

[n]K. Matsumara, T. Tanaka, Y. Endo, S. Saita, and E. Hirota, *J. Chem. Phys.*, **84**, 1793 (1980).

[o]T. Tanaka, C. Yamada, and E. Hirota, *J. Mol. Spectrosc.*, **63**, 335 (1978).

[p]K. Matsumura, K. Tanaka, C. Yamada, and T. Tanaka, *J. Mol. Spectrosc.*, **80**, 209 (1980).

[q]A. Guarnieri and M. Andolfatto, *Z. Naturforsch.*, **36a**, 899 (1981).

The total dipole moment lies between the C—H and C=O bonds, making an angle of 42.4° ± 2° with the C=O bond. A method for finding the orientation of the resultant dipole moment from the observed dipole components of the isomers for molecules exhibiting rotational isomerism has been described by Botskor [52].

Illustrative values of dipole moments of some diatomic and linear polyatomic molecules are given in Table 10.9. Similar listings for symmetric tops are given in Table 10.10 and for asymmetric tops in Table 10.11. Interpretations of dipole moments in terms of chemical bond properties may be found in Chapter XIV, Section 16.

Dipole moments of hundreds of molecules—diatomic, linear, polyatomic, symmetric and asymmetric tops—have now been measured with the Stark

Table 10.10 Dipole Moments of Symmetric-top Molecules from the Stark Effect in Microwave Rotational Spectra[a]

Molecule	μ *(D)*	*Ref.*	*Molecule*	μ *(D)*	*Ref.*
BH_3CO	1.795(10)	b	SiH_3F	1.298(4)	j
BH_3NH_2	5.216(17)	c	SiH_3Cl	1.303(10)	k
CH_3F	1.8572(10)	d	SiH_3Br	1.318(8)	l
CH_3Cl	1.869(10)	e	GeH_3Cl	2.124(20)	k
CH_3Br	1.797(15)	e	SiH_3CH_3	0.7380	d
CH_3I	1.618	f	GeH_3CH_3	0.6450	d
CH_3CN	3.913(2)	d	SnH_3CH_3	0.68(3)	m
CH_3CP	1.499(1)	g	CH_3CCH	0.7835	d
CF_3H	1.6526	h	SiH_3CCH	0.317(3)	n
CF_3Cl	0.500(10)	i	GeH_3CCH	0.136(2)	o

[a]All values are for the ground vibrational state.
[b]M. W. P. Strandberg, C. S. Pearsall, and M. T. Weiss, *J. Chem. Phys.*, **17**, 429 (1949).
[c]L. R. Thorne, R. D. Suenram, and F. J. Lovas, *J. Chem. Phys.*, **78**, 167 (1983); see also R. D. Suenram and L. R. Thorne, *Chem. Phys. Lett.*, **78**, 157 (1981).
[d]Steiner and Gordy [42].
[e]Shulman et al. [45].
[f]Buckingham and Stephens [5].
[g]H. W. Kroto, F. J. Nixon, and N. P. C. Simmons, *J. Mol. Spectrosc.*, **77**, 270 (1979).
[h]Muenter and Laurie [2]. Values are here corrected to conform to new value for OCS moment used as a standard.
[i]R. C. Johnson, T. L. Weatherly, and Q. Williams, *J. Chem. Phys.*, **35**, 2261 (1961).
[j]C. Georghiou, J. G. Baker, and S. R. Jones, *J. Mol. Spectrosc.*, **63**, 89 (1976).
[k]J. M. Mays and B. P. Dailey, *J. Chem. Phys.*, **20**, 1695 (1952).
[l]K. F. Dossel and D. H. Sutter, *Z. Naturforsch.*, **32a**, 1444 (1977).
[m]D. R. Lide, *J. Chem. Phys.*, **19**, 1605 (1951).
[n]J. S. Muenter and V. C. Laurie, *J. Chem. Phys.*, **39**, 1181 (1963). See footnote *h*.
[o]E. C. Thomas and V. C. Laurie, *J. Chem. Phys.*, **44**, 2602 (1966).

Table 10.11 Dipole Moments of Some Asymmetric Rotors from the Stark Effect of Microwave Spectra[a]

	Magnitude of Dipole Moment (debye units)				
Molecule	μ_a	μ_b	μ_c	μ	*Ref.*
HN_3	0.837[b]	1.48(5)		1.70(5)	c
NSCl	0.57(3)	1.77(2)		1.87(2)	d
H_2CO				2.339	e
HCOOH	1.391(5)	0.26(4)		1.415(10)	f
H_3NH_2	0.304		1.247	1.238	g
CH_3OCl	1.373(4)	1.146(7)		1.788	h
CD_3OCl	1.409(4)	1.106(8)		1.791(8)	h
$(CH_3)_2SO$		3.94(4)	0.40(1)	3.96(4)	i
H_3CH_2CN	3.85(2)	0.123(2)		4.05(3)	j
CH_2CHCN	3.68	1.25		3.89(8)	k
CH_3CHCO	1.755(10)	0.35(3)		1.79(2)	l
CH_3CHCH_2O (ring)	0.95(1)	1.67(2)	0.56(1)	2.00(2)	m
$CH_2CH_2CH_2S$ (ring)	1.85(3)		0.00(3)	1.85(3)	n
CH_2CH_2CO–O (ring)	3.67(3)	1.99(2)		4.17(3)	o

[a]All values are for the ground vibrational state and, with the exception of CD_3OCl, are for the most abundant isotopic species.
[b]K. G. White and R. L. Cook, *J. Chem. Phys.*, **46**, 143 (1967).
[c]J. Bendtsen and M. Winnewisser, *Chem. Phys. Lett.*, **33**, 141 (1975).
[d]A. Guarnieri, Z. *Naturforsch.*, **26a**, 1246 (1971).
[e]J. N. Shoolery and A. H. Sharbaugh, *Phys. Rev.*, **82**, 95 (1951).
[f]H. Kim, R. Keller, and W. D. Gwinn, *J. Chem. Phys.*, **37**, 2748 (1962).
[g]D. R. Lide, Jr., *J. Chem. Phys.*, **27**, 343 (1957).
[h]M. Suzuki and A. Guarnieri, Z. *Naturforsch.*, **31a**, 1242 (1976).
[i]H. Dreizler and G. Dendl, *Z. Naturforsch.*, **19a**, 512 (1964).
[j]H. M. Heise, H. Lutz, and H. Dreizler, *Z. Naturforsch.*, **29a**, 1345 (1974).
[k]W. S. Wilcox, J. H. Goldstein, and J. W. Simmons, *J. Chem. Phys.*, **22**, 516 (1954).
[l]B. Bak, J. J. Christiansen, K. Kunstmann, L. Nygaard, and J. Rastrup-Andersen, *J. Chem. Phys.*, **45**, 883 (1966).
[m]J. D. Swalen and D. R. Herschbach, *J. Chem. Phys.*, **27**, 100 (1957).
[n]M. S. White and E. L. Beeson, Jr., *J. Chem. Phys.*, **43**, 1839 (1965).
[o]D. W. Boone, C. O. Britt, and J. E. Boggs, *J. Chem. Phys.*, **43**, 1190 (1965).

effect in microwave rotational spectra, many of them to five significant figures. It is feasible to record only a small fraction of them in Tables 10.9 through 10.11. Comprehensive tables of measured dipole moments are available in the Landolt-Börnstein tables [53]. These tabulations are frequently revised to bring them up to date.

References

1. R. H. Hughes and E. B. Wilson, *Phys. Rev.*, **71**, 562 (1947).
2. J. S. Muenter and V. W. Laurie, *J. Chem. Phys.*, **45**, 855 (1966).

3. U. Fano, *J. Res. Natl. Bur. Stand.*, **40**, 215 (1948).
4. W. Low and C. H. Townes, *Phys. Rev.* **76**, 1295 (1949).
5. A. D. Buckingham and P. J. Stephens, *Mol. Phys.*, **7**, 481 (1963/64).
6. F. Coester, *Phys. Rev.*, **77**, 454 (1950).
7. E. U. Condon and G. H. Shortley, *The Theory of Atomic Spectra*, Macmillan, New York, 1935, pp. 76–77.
8. M. E. Rose, *Elementary Theory of Angular Momentum*, Wiley, New York, 1957.
9. J. B. M. Kellogg, I. I. Rabi, N. F. Ramsey, Jr., and J. R. Zacharias, *Phys. Rev.*, **57**, 677 (1940).
10. W. G. Penney, *Phil. Mag.*, **11**, 602 (1931).
11. S. Golden and E. B. Wilson, Jr., *J. Chem. Phys.*, **16**, 669 (1948).
12. See, for example, B. G. West and M. Mizushima, *J. Chem. Phys.*, **38**, 251 (1963); C. C. Lin, *Bull. Am. Phys. Soc.*, **1**, 13 (1956).
13. P. E. Wacher and M. R. Pratto, *Microwave Spectral Tables*, Vol. II. *Line Strengths of Asymmetric Rotors* (Natl. Bur. Stand. Monograph *70*), 1964. See Chapter VII for further references.
14. D. Kivelson, *J. Chem. Phys.*, **22**, 904 (1954).
15. G. F. Crable, Ph.D. dissertation, Duke University, 1951.
16. G. F. Crable and W. V. Smith, *J. Chem. Phys.*, **19**, 502 (1951).
17. J. H. Van Vleck, *Phys. Rev.*, **33**, 467 (1929). See also O. M. Jordahl. *Phys. Rev.*, **45**, 87 (1934); E. C. Kemble, *Fundamental Principles of Quantum Mechanics*, Dover, New York, 1958.
18. G. H. Kwei and D. R. Herschbach, *J. Chem. Phys.*, **32**, 1270 (1960).
19. S. I. Chan, J. Zinn, J. Fernandez, and W. D. Gwinn, *J. Chem. Phys.*, **33**, 1643 (1960).
20. S. S. Butcher and C. C. Costain, *J. Mol. Spectrosc.*, **15**, 40 (1965).
21. C. H. Townes and A. L. Schawlow, *Microwave Spectroscopy*, McGraw-Hill, New York, 1955.
22. M. Mizushima, *J. Chem. Phys.*, **21**, 539 (1953).
23. G. Racah, *Phys. Rev.*, **62**, 438 (1942). See also [8].
24. J. A. Howe and W. H. Flygare, *J. Chem. Phys.*, **36**, 650 (1962).
25. D. F. Eagle, T. L. Weatherly, and Q. Williams, *J. Chem. Phys.*, **44**, 847 (1966).
26. M. W. P. Strandberg, T. Wentink, Jr., and A. G. Hill, *Phys. Rev.*, **75**, 827 (1949).
27. S. A. Marshall and J. Weber, *Phys. Rev.*, **105**, 1502 (1957).
28. L. H. Scharpen, J. S. Muenter, and V. Laurie, *J. Chem. Phys.*, **46**, 2431 (1967).
29. J. S. Muenter, *J. Chem. Phys.*, **48**, 4544 (1968).
30. M. Mizushima, in *Advances in Molecular Spectroscopy*, Vol. 3 (Proc. IVth International Meeting on Molecular Spectroscopy), A. Magnini, Ed., Macmillan, New York, 1962, pp. 1167–1182.
31. J. O. Hirschfelder, C. F. Curtiss, and R. B. Bird, *Molecular Theory of Gases and Liquids*, Wiley, New York, 1954, p. 950.
32. S. Gustafson and W. Gordy, *Phys. Lett.*, **49A**, 161 (1974).
33. N. J. Bridge and A. D. Buckingham, *Proc. R. Soc.* (London), **A295**, 334 (1966).
34. R. L. Rowell and G. M. Aval, *J. Chem. Phys.*, **54**, 1960 (1971).
35. K. B. McAfee, Jr., R. H. Hughes; and E. B. Wilson, Jr., *Rev. Sci. Instrum.*, **20**, 821 (1949).
36. S. H. Autler and C. H. Townes, *Phys. Rev.*, **100**, 703 (1955).
37. Townes and Schawlow, Reference 21, pp. 273–283.
38. K. L. Lindfors and C. D. Cornwell, *J. Chem. Phys.*, **42**, 149 (1965).
39. B. N. Bhattacharya, W. Gordy, and O. Fujii, *Bull. Am. Phys. Soc.*, **2**, 213 (1957).
40. B. N. Bhattacharya and W. Gordy, *Phys. Rev.*, **119**, 144 (1960).
41. D. M. Larkin and W. Gordy, *J. Chem. Phys.*, **38**, 2329 (1963).
42. P. A. Steiner and W. Gordy, *J. Mol. Spectrosc.*, **21**, 291 (1966).
43. C. A. Burrus, *J. Chem. Phys.*, **28**, 427 (1958).

44. C. A. Burrus, *J. Chem. Phys.*, **31**, 1270 (1959).
45. R. G. Shulman, B. P. Dailey, and C. H. Townes, *Phys. Rev.*, **78**, 145 (1950).
46. J. M. Mays and B. P. Dailey, *J. Chem. Phys.*, **20**, 1695 (1952).
47. K. J. White and R. L. Cook, *J. Chem. Phys.*, **46**, 143 (1967).
48. W. H. Hocking, M. C. L. Gerry, and G. Winnewisser, *Can. J. Phys.*, **53**, 1869 (1975).
49. P. Favero and J. G. Baker, *Nuovo Cimento*, **17**, 734 (1960).
50. H. Kim, R. Keller, and W. D. Gwinn, *J. Chem. Phys.*, **37**, 2748 (1962).
51. G. H. Kwei and R. F. Curl, Jr., *J. Chem. Phys.*, **32**, 1592 (1960).
52. I. Botskor, *Z. Naturforsch.*, **35a**, 748 (1980).
53. Landolt-Börnstein, *Numerical Data and Functional Relations in Science and Technology*. For a complete reference, see Chapter I, [65].

Chapter **XI**

EFFECTS OF APPLIED MAGNETIC FIELDS

Most stable molecules have Σ singlet ground electronic states and hence when not rotating are nonmagnetic in the sense that they have no permanent magnetic dipole moment exclusive of nuclear moments. However, magnetic moments are generated by molecular rotation which gives rise to Zeeman splittings [1–3], usually resolvable with fields of several kilogauss. Magnetic moments are also induced by the applied field because of the magnetic susceptibility of the molecules, but these induced moments are very much smaller than those caused by rotation, and only the anisotropic components can affect the observed spectra [4]. The nuclear magnetic moments influence the rotational Zeeman effect only when the moments are coupled to the molecular axes so that an observable hyperfine structure results. First we shall consider molecules in $^1\Sigma$ states without nuclear coupling.

1 ZEEMAN EFFECT IN MOLECULES WITHOUT NUCLEAR COUPLING

Interaction Energies

The classical Hamiltonian for the interaction of a magnetic dipole with a field **H** is

$$\mathscr{H}_H = -\boldsymbol{\mu}\cdot\mathbf{H} \tag{11.1}$$

In the molecular rotational Zeeman effect the magnetic moment is fixed in the molecular frame, and **H** is fixed in space. The component of the molecular magnetic moment along the total angular momentum vector **J** (or along **F** when there is nuclear coupling) interacts with **H**, causing **J** to precess about the direction of **H**, as indicated by the vector diagram of Fig. 11.1. With the

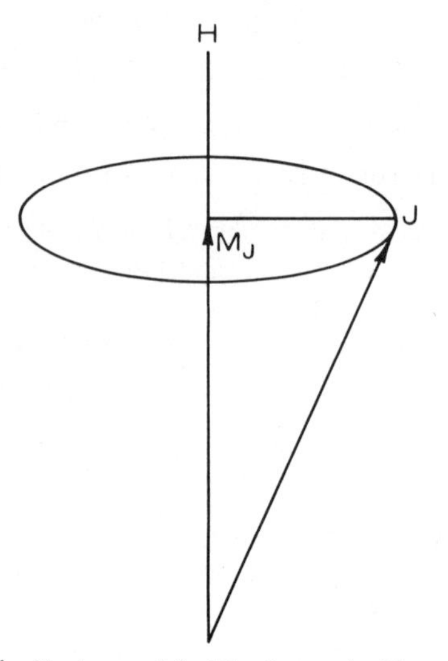

Fig. 11.1 Vector model of the first-order Zeeman effect.

weak magnetic moments of molecules in Σ states, the precession of **J** about **H** is very much slower than the molecular rotation, and only the components of $\boldsymbol{\mu}$ along **J** effectively interact with **H**. Those normal to **H** are averaged out by the rotation. Even in molecules with unbalanced electronic moments this approximation holds for ordinary field strengths. If $\langle \mu_J \rangle$ represents the averaged value of the molecular magnetic moment along **J**, the Zeeman Hamiltonian can be expressed as

$$\mathscr{H}_H = -\langle \mu_J \rangle \frac{\mathbf{J}\cdot\mathbf{H}}{|\mathbf{J}|} = -\langle \mu_J \rangle H \cos(\mathbf{J}, \mathbf{H}) \tag{11.2}$$

From the vector diagram of Fig. 11.1 it is seen that

$$\cos(\mathbf{J}, \mathbf{H}) = \frac{J_Z}{|\mathbf{J}|} = \frac{M_J}{[J(J+1)]^{1/2}} \tag{11.3}$$

Therefore (11.2) can be expressed as

$$E_H = -\langle \mu_J \rangle \frac{HM_J}{[J(J+1)]^{1/2}} \tag{11.4}$$

Although (11.4) is a rather general first-order expression which applies to molecules of different types, symmetric tops or asymmetric rotors, the quantity $\langle \mu_J \rangle$ varies with molecular type and rotational state. Therefore, if we are to find the specific value of E_H in (11.4), we must find the corresponding operator for μ_J and average it over the molecular wave function.

When the principal magnetic axes are the same as the inertial axes x, y, z, the components of $\boldsymbol{\mu}$ generated along the axes are proportional to the respective components of rotational momentum, as verified in Section 3. They can be expressed as

$$\mu_x = \beta_I g_{xx} J_x, \qquad \mu_y = \beta_I g_{yy} J_y, \qquad \mu_z = \beta_I g_{zz} J_z \tag{11.5}$$

where $\beta_I g_{xx}$, and so on, are proportionality constants. The nuclear magneton β_I is introduced so that g will represent the ratio of μ in nuclear magneton units to the angular momentum J in units of $\hbar$, the customary designation used in the definition of the dimensionless gyromagnetic ratio of the nuclei. It gives the g factors in numbers of convenient magnitude for molecules in Σ states. The magnitude of μ_J is found by resolution of the components of (11.5) along **J**:

$$\begin{aligned} \mu_J &= \mu_x \cos(x, J) + \mu_y \cos(y, J) + \mu_z \cos(z, J) \\ &= \mu_x \frac{J_x}{|\mathbf{J}|} + \mu_y \frac{J_y}{|\mathbf{J}|} + \mu_z \frac{J_z}{|\mathbf{J}|} \\ &= \frac{\beta_I}{[J(J+1)]^{1/2}} (g_{xx} J_x^2 + g_{yy} J_y^2 + g_{zz} J_z^2) \end{aligned} \tag{11.6}$$

When the principal magnetic axes are not the same as the inertial axes, off-diagonal elements in the magnetic moment are generated so that

$$\begin{aligned}\mu_x &= \beta_I(g_{xx}J_x + g_{xy}J_y + g_{xz}J_z)\\ \mu_y &= \beta_I(g_{yx}J_x + g_{yy}J_y + g_{yz}J_z)\\ \mu_z &= \beta_I(g_{zx}J_x + g_{zy}J_y + g_{zz}J_z)\end{aligned} \tag{11.7}$$

Resolution of these components along **J** gives

$$\mu_J = \frac{\beta_I}{[J(J+1)]^{1/2}} \sum_{g,g'=x,y,z} g_{gg'} J_g J_{g'} \tag{11.8}$$

When, however, μ_J is averaged over the rotational wave functions, the off-diagonal elements will vanish, leaving only the diagonal elements of the **g** tensor. Thus,

$$\langle \mu_J \rangle = \frac{\beta_I}{[J(J+1)]^{1/2}} \sum_{g=x,y,z} g_{gg} \langle J_g^2 \rangle \tag{11.9}$$

Because $\langle \mu_J \rangle$ is independent of M_J, it is possible to define an effective g factor for the particular rotational state $|J, \tau)$ by

$$g_{J,\tau} = \frac{\langle \mu_J \rangle / \beta_I}{[J(J+1)]^{1/2}} = \frac{1}{J(J+1)} \sum_g g_{gg}(J, \tau | J_g^2 | J, \tau) \tag{11.10}$$

and hence

$$\langle \mu_J \rangle = g_{J,\tau} \beta_I [J(J+1)]^{1/2} \tag{11.11}$$

where τ becomes K for symmetric tops, is zero for linear molecules, and represets the pseudo-quantum numbers $K_{-1} - K_1$ for the asymmetric rotor. Substitution of this value of $\langle \mu_J \rangle$ from (11.11) into (11.4) gives the Zeeman energies as

$$E_H(\mu_J) = -g_{J,\tau} \beta_I H M_J \tag{11.12}$$

in which

$$g_{J,\tau} = \frac{1}{J(J+1)} (g_{xx}\langle J_x^2 \rangle + g_{yy}\langle J_y^2 \rangle + g_{zz}\langle J_z^2 \rangle) \tag{11.13}$$

where g_{xx}, g_{yy}, and g_{zz} are the diagonal elements of the **g** tensor with reference to the inertial axes and where the averages are taken over the wave functions of the unperturbed rotors. The quantity $g_{J,\tau}$ is called the rotational g factor or the spectroscopic splitting factor. As shown in Section 3, its origin in $^1\Sigma$ molecules is due partly to the rotation of formal charges and partly to a slight admixture of electronic states having orbital angular momentum with the Σ singlet ground electronic state.

We did not include in (11.1) a small term arising from the moments induced in the molecule by the applied field. At very high fields this term can have detectable effects on the Zeeman displacements when the magnetic susceptibility of the molecules is anisotropic. These effects are discussed in Section 4.

Strictly speaking, we should include in (11.1) the Hamiltonian for the nuclear magnetic interaction with the applied field

$$\mathscr{H}_n = -\beta_I \sum_i g_{I_i} \mathbf{I}_i \cdot \mathbf{H} \tag{11.14}$$

where the sum is taken over all nuclei having nonzero spins. On the other hand, when the nuclear coupling is negligible, as for nuclei with no quadrupole interaction, the perturbation of the molecular rotation can be neglected, and the allowed values of $\mathscr{H}_n$ are

$$E_H(\mu_I) = -\beta_I H \sum_i g_{I_i} M_{I_i} \tag{11.15}$$

Because of the selection rule $\Delta M_{I_i} = 0$, there is no observable effect of $E_H(\mu_I)$ on the observed microwave spectra.

Linear Molecules

The remaining problem in the evaluation of the Zeeman splitting is the specialization of (11.13) to the particular molecular type under consideration. For linear molecules $J_z = 0$, $g_{zz} = 0$, and from symmetry $g_{xx} = g_{yy} = g$. Since $J_x^2 + J_y^2 = J^2$ and, in units of $\hbar$, $\langle J^2 \rangle = J(J+1)$,

$$g_J = \frac{1}{J(J+1)} g \langle J^2 \rangle = g \tag{11.16}$$

Therefore, in the first-order treatment for linear molecules g_J is a constant, independent of the rotational state. The associated magnetic moment $\langle \mu_J \rangle$ is perpendicular to the molecular axis and gives rise to a first-order Zeeman effect. The level splitting is given by (11.12). In contrast, there is no first-order Stark effect for linear molecules because the electric dipole moment is parallel to the molecular axis.

The M_J selection rules for the Zeeman effect are the same as those for the Stark effect

$$M_J \to M_J \qquad \text{and} \qquad M_J \to M_J \pm 1 \tag{11.17}$$

For transitions which are induced by electric dipole coupling with the radiation, the magnetic field must be parallel to the electric vector of the radiation for observation of the $M_J \to M_J$ transitions, called π components; for observation of the $M_J \to M_J \pm 1$ transitions, called σ components, it must be imposed perpendicular to the electric vector of the radiation.

With the foregoing selection rules and with (11.12), the frequencies of the Zeeman components of the $J \to J+1$ rotational transition are

$$\nu(\pi) = \nu_0 \tag{11.18}$$

$$\nu(\sigma) = \nu_0 \pm \frac{g_J \beta_I H}{h} \tag{11.19}$$

where ν_0 is the frequency of the line when no field is applied. Since from (11.16) g_J is a constant, these first-order effects of a linear molecule are independent of J. One therefore expects the same Zeeman pattern for all the lines of an unsplit π component and for a doublet σ component, as shown in Fig. 11.2. Together they form the normal Zeeman triplet. Since it does not depend on J, the σ splitting can be observed for very high J transitions. Only the undisplaced rotational line, the π component, is observable when the magnetic field is parallel to the microwave **E** vector.

In the higher-order approximations, g_J is not exactly the same for all values of J. The frequency expressions for this situation may be obtained from (11.23) and (11.24) by replacement of $g_{J,K}$ and $g_{J+1,K}$ with g_J and g_{J+1}, respectively. When $g_J \neq g_{J+1}$, there are $(2J+1)$ π components corresponding to the possible values of M_J. Although g_J and g_{J+1} are not likely to differ sufficiently to make this structure resolvable, they may differ enough to cause a detectable broadening of the line in some molecules. Likewise, a difference in g_J and g_{J+1} would cause each component of the σ doublet to be split into $2J+1$ components, but such a splitting has not been resolved. Nevertheless, differences in g_J and g_{J+1} might cause a distortion of this line shape, as indicated by the theoretical pattern of Fig. 11.3. The difference between g_J and g_{J+1} assumed in these patterns is much greater than that likely to be encountered for linear molecules. Note that the center of the sub-multiplet structure does not occur at the peak absorption and that the asymmetry reverses when g_{J+1} changes from greater than g_J to smaller than g_J. If $2\Delta\nu$ represents the frequency difference between the centers of the σ doublet multiplets, g_{J+1} can be found from

$$g_{J+1} = \frac{h\Delta\nu}{\beta_I H} \tag{11.20}$$

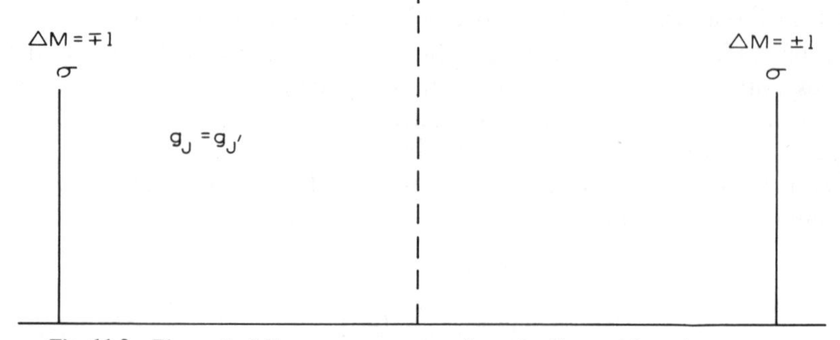

Fig. 11.2 Theoretical Zeeman components for a $J \to J'$ transition when $g_J = g_{J'}$.

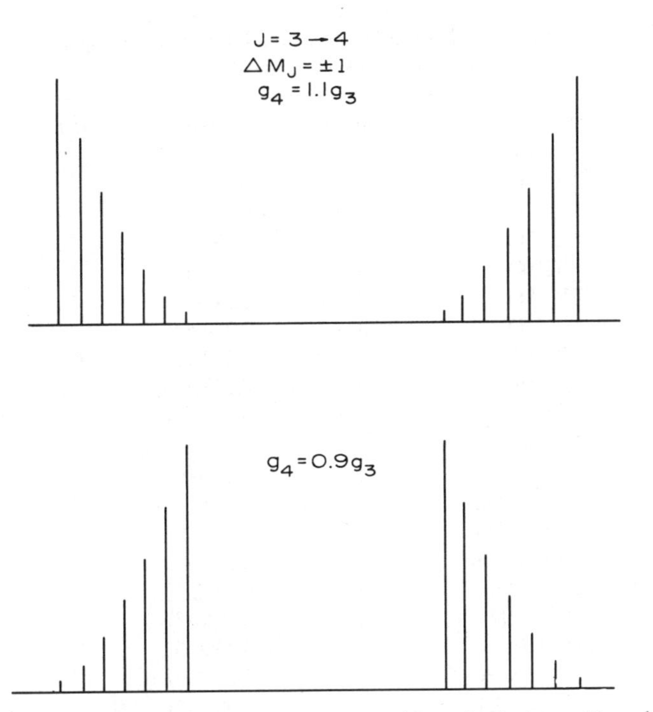

Fig. 11.3 The $\Delta M = \pm 1$ Zeeman pattern (σ components) for a $J \rightarrow J+1$ transition when $g_J \neq g_{J+1}$.

The most complete data available are for OCS. The $\Delta M = \pm 1$ components of the Zeeman splitting have been observed for several OCS transitions, and the simple doublet pattern expected from (11.19) is found for each. That g_J does not vary significantly with the rotational state is demonstrated more quantitatively in Table 11.1, where the effective g_J values observed for different transitions are shown. Not surprisingly, the greatest deviation is observed for the $J=0 \rightarrow 1$ transition. This doublet splitting was first observed by Jen [5], who found the g value from $J=1 \rightarrow 2$ transition to be 0.029 ± 0.006. By use of circularly polarized radiation, Eshbach and Strandberg [3] later established the sign of g_J as negative and found for the $J=1 \rightarrow 2$ transition the average $g_J = -0.0251 \pm 0.002$. The magnitude of their value appears too small in comparison to other values, and its deviation is well outside the estimated limits of error of the highly accurate value later measured by Flygare et al. which is given in Table 11.1.

The g_J factor of CO is an order of magnitude larger than that of OCS. It has been measured from the $J=0 \rightarrow 1$ transition by three different groups of workers [6–8]. All the values are in good agreement. Rosenblum et al. [7], who made measurements on different isotopic species, obtained measurably different g_J values for them. Burrus [8] extended the measurements on $^{12}C^{16}O$ to $J=1 \rightarrow 2$ transition and found g_J to be independent of J within the accuracy of measurement. The g_J measured for the $J=0 \rightarrow 1$ transition is strictly that for

Table 11.1 Molecular Rotational g Factors for Selected Diatomic and Linear Molecules[a]

Molecule	*J State*	$g_J = g_\perp$	*Ref.*
$^{12}C^{16}O$	1	−0.2691(5)	*b*
$^{12}C^{17}O$	1	−0.2623(5)	*b*
$^{13}C^{16}O$	1	−0.2570(5)	*b*
$^{12}C^{18}O$	1	−0.2562(5)	*b*
$^{14}C^{16}O$	1	−0.2466(5)	*b*
$^{13}C^{18}O$	1	−0.2442(5)	*b*
$^{12}C^{32}S$	1	−0.272(2)	*c,d*
$^{12}C^{80}Se$	1	−0.2431(16)	*e*
$D^{79}Br$	1	+0.181(15)	*f*
$D^{81}Br$	1	+0.184(15)	*f*
$D^{127}I$	1	+0.096(10)	*f*
$^{16}O^{12}C^{32}S$	1	−0.02871(4)	*g,h*
$^{16}O^{12}C^{34}S$	1	−0.02813(4)	*h*
$^{16}O^{12}C^{79}Se$	1	−0.0195(2)	*i*
$H^{12}C^{14}N$	1	−0.0962	*j*
$H^{12}C^{15}N$	1	−0.0917(7)	*k*
$H^{12}CP$	1	−0.430(10)	*e*
$^{15}N^{15}N^{16}O$	1	+0.07606(10)	*l*
$F^{12}C^{12}CH$	1	−0.0077(2)	*m*
$^{35}Cl^{12}C^{12}CH$	1	−0.00630(14)	*n*

[a]The g factors are consistent with the magnetic moment expressed in nuclear magneton units.
[b]Rosenblum et al. [7].
[c]Gustafson and Gordy [31].
[d]H. F. Bates, J. J. Gallagher, and V. E. Derr, *J. Appl. Phys.*, **39**, 3218 (1968).
[e]McGurk et al. [40].
[f]Burrus [8].
[g]Hüttner et al. [17].
[h]Flygare et al. [20].
[i]Shoemaker and Flygare [30].
[j]B. N. Bhattacharya, Ph.D. dissertation, Duke University, 1958.
[k]Gustafson and Gordy [41].
[l]Flygare et al. [22].
[m]Shoemaker and Flygare [18].
[n]Allen and Flygare [36].

the $J=1$ state since M_J equals zero in the lower state. The g_J measured for the $J=1\rightarrow 2$ transition is essentially that for the $J=2$ state, as indicated by (11.20).

Symmetric-top Molecules

For the symmetric top, as for the linear molecule, z represents the symmetry axis, and we can set $g_{xx}=g_{yy}=g_\perp$, $g_{zz}=g_K$, and $\tau=K$. Since $J_x^2+J_y^2=J^2-J_z^2$, the average is $(J, K, M_J|J_x^2+J_y^2|J, K, M_J)=J(J+1)-K^2$. Equation 11.10 specialized to the symmetric top becomes

$$g_{J,K}=g_\perp+(g_K-g_\perp)\frac{K^2}{J(J+1)} \tag{11.21}$$

and from (11.12) we get

$$E_{J,K,M_J}=-g_{J,K}\beta_I HM_J \tag{11.22}$$

Equation 11.21 can also be derived from the vector model of the symmetric top in an applied field, Fig. 10.1.

The selection rules governing M_J are the same as those for a linear molecule, (11.17), and for a rotational transition $J\to J+1$ and $K\to K$. Since K does not change, g_K is the same for both upper and lower transitions. Furthermore, the assumption that $g_\perp(J, K)=g_\perp(J+1, K)=g_\perp$ seems justified from the data on linear molecules, Table 11.1. From (11.22) the Zeeman frequencies are found to be

$$\nu_\pi(M_J\to M_J)=\nu_0+\frac{\beta_I H}{h}(g_{J,K}-g_{J+1,K})M_J \tag{11.23}$$

$$\nu_\sigma(M_J\to M_J\pm 1)=\nu_0+\frac{\beta_I H}{h}(g_{J,K}-g_{J+1,K})M_J\mp\frac{\beta_I Hg_{J+1,K}}{h} \tag{11.24}$$

where ν_0 is the rotational frequency when no field is applied. When g_K and $g_\perp$ are constants

$$g_{J,K}-g_{J+1,K}=\frac{2(g_K-g_\perp)K^2}{J(J+1)(J+2)} \tag{11.25}$$

Because $g_K-g_\perp$ is not negligible for many molecules, the π splitting is observable for low J values. It vanishes for $K=0$ and for a given K decreases approximately as $1/J^3$, but for $K=J$ it decreases only as $1/J$, approximately.

We illustrate the application of the foregoing theory with CH_3CCH. The upper curve of Fig. 11.4 shows the zero field and the components of the $J=3\to 4$ transition. The different K lines are separated by centrifugal distortion. The bottom curve shows the same transition with a field of 10 kG imposed at right angles to the electric vector of the microwave radiation polarized in such a way that only the $M_J\to M_J\pm 1$ components are observable. Note that the $K=0$ line is unsplit and not even measurably broadened by the field. When $K=0$, the splitting is due entirely to $g_\perp$, as may be readily seen from (11.21). The expected splitting is like that for a linear molecule with $g_J=g_\perp$. Thus the failure to observe splitting or broadening of the $K=0$ line proves that $g_\perp$ is small. For the other components we can set $g_\perp\approx 0$ and obtain

$$g_{J,K}=\frac{g_K K^2}{J(J+1)} \qquad gg_{J+1,K}=\frac{g_K K^2}{(J+1)(J+2)} \tag{11.26}$$

The splitting is best resolved for the $K=3$ line for which

$$g_{33}=\frac{3g_K}{4} \qquad \text{and} \qquad g_{43}=\frac{9g_K}{20}$$

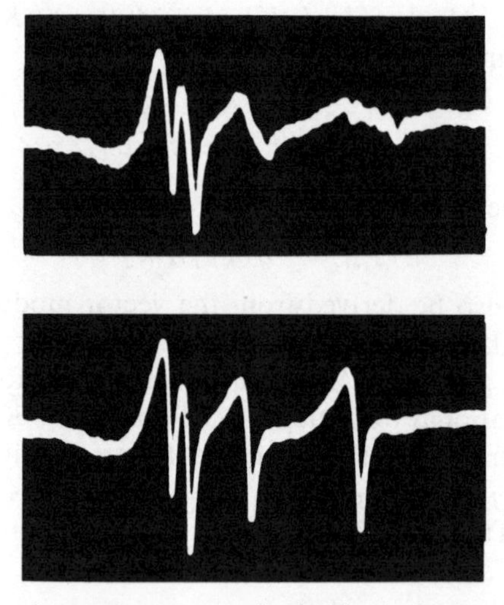

Fig. 11.4 Zeeman effect of a symmetric-top molecule. The $J=3\rightarrow4$ transition ($K=0$, 1, 2, 3 components from right to left) of CH_3CCH with the field arranged for observation of the $\Delta M_J=\pm1$ components. The upper figure is for zero field; the lower, for a field of 10 kgauss. From Cox and Gordy [6].

The splittings of the two energy levels are given by

$$E_{33}=-\frac{3g_K\beta_I HM_J}{4h} \qquad \text{and} \qquad E_{43}=-\frac{9g_K\beta_I HM_J}{20h}$$

in frequency units. The energy level diagram (not drawn to scale) and the indicated $M_J\rightarrow M_J\pm1$ transitions are shown in Fig. 11.5. The spacing of the upper multiplet level is only $\frac{3}{5}$ that of the lower multiplet. The relative intensities of the $M_J\rightarrow M_J-1$ components with increasing frequencies are 2, 6, 12, 20, 42, and 56 as calculated from (10.127). Those of the $M_J\rightarrow M_J+1$ components are the same, but in reverse order. The superimposed $M_J\rightarrow M_J-1$ and $M_J\rightarrow M_J+1$ components obtained when $g_{43}=(\frac{3}{5})g_{33}$ are shown in Fig. 11.6. The four observed Zeeman components of the $K=3$ line shown in Fig. 11.4 correspond to the four strong lines of the theoretical pattern. The others are too weak to be observable. From the separation of the outer of the four observed lines, g_K is found [6] to be 0.298 ± 0.006. The corresponding $g_{J,K}$ factor for the $K=2$ line is only $\frac{4}{9}$ that for the $K=3$ line, and the spread of the unresolved multiplet is about one-half that of the $K=3$ line. The g factor for the $K=1$ line, only $\frac{1}{9}$ that of the $K=3$ line, is too small for any resolution of the splitting.

Shoemaker and Flygare [9] have remeasured the Zeeman effect of methyl acetylene with a magnetic field of 30,000 G and obtained the more accurate value $g_\parallel=+0.312\pm0.002$. They were also able to measure the small $g_\perp$ for CH_3CCH. These and g factors for some other symmetric-top molecules are listed in Table 11.2. The g value caused by rotation of the methyl group in different symmetric tops appears to have the approximately constant value of 0.310.

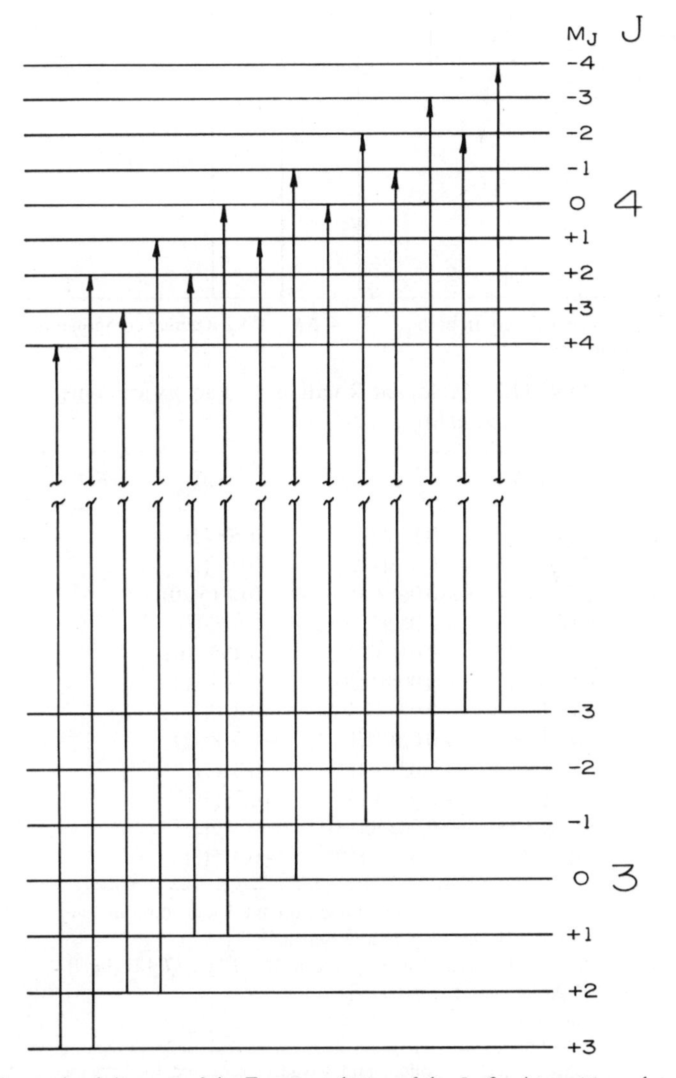

Fig. 11.5 Energy level diagram of the Zeeman splitting of the $J=3\rightarrow4$ transition when $g_4=(\frac{3}{5})g_3$ with indicated $\Delta M_J=\pm1$ transition. This diagram corresponds to that for the $K=3$ line shown for CH_3CCH in Fig. 11.4 and to the theoretical spectrum shown in Fig. 11.6.

Asymmetric Rotors

The first-order Zeeman perturbation energies of asymmetric-top molecules are given by (11.12) with the $g_{J,\tau}$ given by (11.13). The problem of obtaining the average $\langle J_g^2\rangle$ over the wave functions of the asymmetric rotor is the same as that already encountered in the evaluation of the rigid rotor energies. The values given in terms of the reduced energies, (9.97)–(9.99), can be employed.

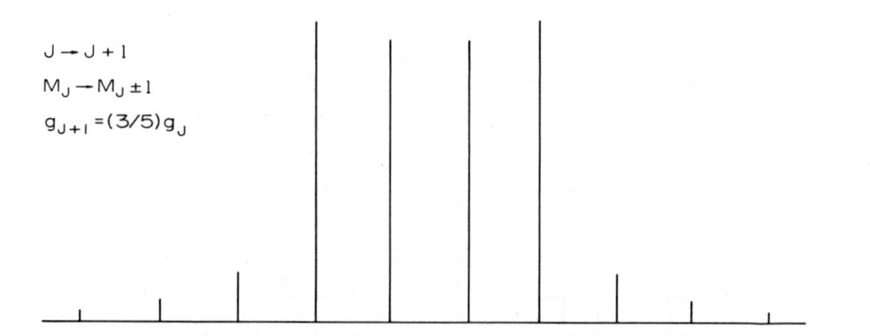

Fig. 11.6 The theoretical pattern of the $J=3 \to 4$, $\Delta M_J = \pm 1$ Zeeman components when $g_4 = (\frac{3}{5}) g_3$.

Table 11.2 Molecular Rotational g Factors for Some Symmetric-top Molecules

Molecule	$g_\perp$	$g_\parallel = g_K$	*Ref.*
$^{15}NH_3$	0.563(2)	0.500(2)	b
PH_3	−0.033(1)	0.017(1)	c
PF_3	−0.0659(3)	−0.0815(20)	d
CH_3F	−0.062(5)	[0.31(3)][a]	e,f
$CH_3{}^{35}Cl$	−0.01653(28)	[0.305(20)]	g
$CH_3{}^{79}Br$	−0.00569(30)	0.294(16)	g,h
CH_3I	−0.00677(40)	0.310(16)	g
$SiH_3{}^{81}Br$	−0.02185(13)	−0.3185(5)	h
CH_3CN	−0.0338(8)	[0.31(3)]	i
CH_3NC	−0.0546(15)	[0.31(3)]	i
CH_3CCH	0.00350(15)	0.312(2)	j
SiH_3NCS	−0.01521(10)	−0.315(5)	k

[a]Values in brackets are assumed from methyl group contribution of 0.31. See Flygare and Benson [57].
[b]S. G. Kukolich and W. H. Flygare, *Mol. Phys.*, **17**, 127 (1969).
[c]Kukolich and Flygare [28].
[d]Stone et al. [23].
[e]Cox and Gordy [6].
[f]Shoemaker, Kukolich, and Flygare, reported by Flygare and Benson [57].
[g]VanderHart and Flygare [21].
[h]Dösel and Sutter [51].
[i]Pochan et al. [32].
[j]Shoemaker and Flygare [24].
[k]Dössel and Sutter [52].

With these expressions we have

$$g_{J,\tau} = \frac{1}{2J(J+1)} \left\{ g_{aa} \left[J(J+1) + E_{J,\tau}(\kappa) - (\kappa+1) \frac{\partial E_{J,\tau}(\kappa)}{\partial \kappa} \right] + 2g_{bb} \left[\frac{\partial E_{J,\tau}(\kappa)}{\partial \kappa} \right] + g_{cc} \left[J(J+1) - E_{J,\tau}(\kappa) + (\kappa-1) \frac{E_{J,\tau}(\kappa)}{\partial \kappa} \right] \right\} \tag{11.27}$$

where a, b, and c are the principal axes of inertia, $E(\kappa)$ is the reduced rotational energy, $\kappa=(2B-A-C)/(A-C)$ is the asymmetry parameter, and A, B, and C are the usual spectral constants (see Chapter VII). In analysis of the spectra and evaluation of g_{aa}, g_{bb}, and g_{cc}, it is convenient to use first the frequency equations

$$\nu_\pi(M_J\to M_J)=\nu_0+\frac{\beta_I H}{h}(g_{J,\tau}-g_{J',\tau'})M_J \tag{11.28}$$

$$\nu_\sigma(M_J\to M_J\pm1)=\nu_0+\frac{\beta_I H}{h}(g_{J,\tau}-g_{J'\tau'})M_J\mp\frac{\beta_I H}{h}g_{J'\tau'} \tag{11.29}$$

to find the $g_{J,\tau}$ values for different states from the observed Zeeman patterns and then to employ these values with (11.27) to solve for g_{aa}, g_{bb}, and g_{cc}. Obviously, $g_{J,\tau}$ must be measured for three different states for evaluation of these elements.

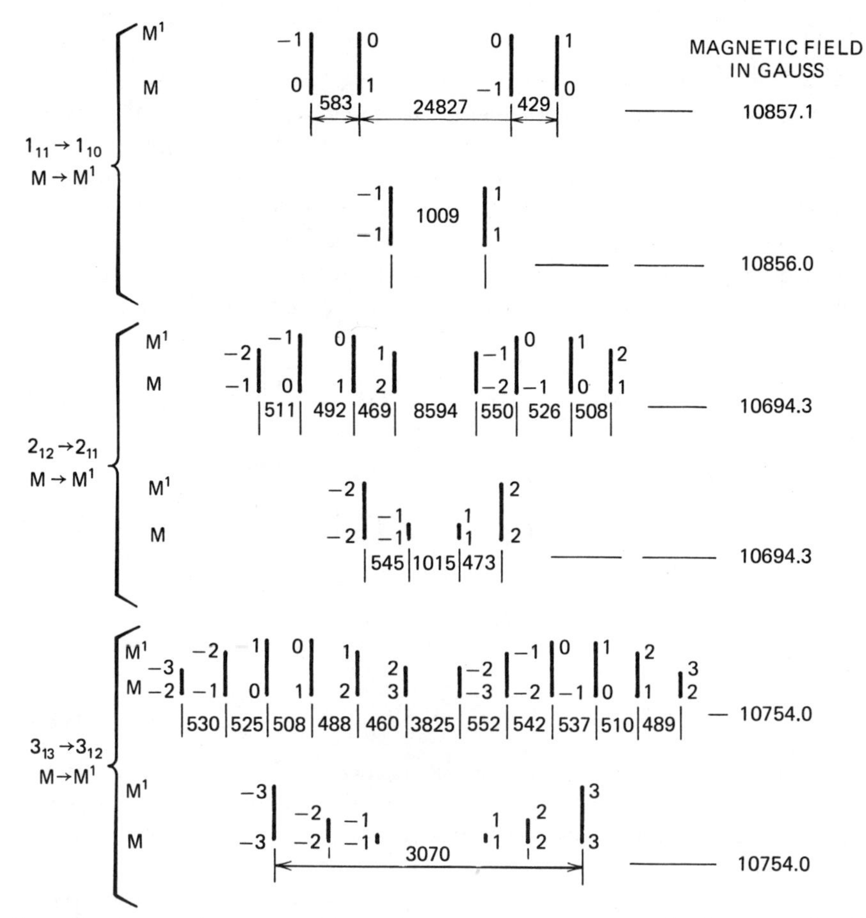

Fig. 11.7 Theoretical Q-branch ($\Delta J=0$) Zeeman pattern with observed splittings for H_2CO. From Hüttner et al. [17].

Because J does not appear explicitly in the energy formula (11.12), the foregoing frequency formulas are applicable to $J \to J$ transitions as well as the $J \to J \pm 1$ transitions, all of which may occur in asymmetric rotors. Since the intensity formulas are different, (10.125)–(10.129), the appearance of the Zeeman patterns for the Q, P, and R branches are generally different. Figure 11.7 illustrates the types of patterns which occur for the $\Delta J = 0$, Q-branch transition. The small differences in the observed component spacings of a particular set are due to effects of anisotropic susceptibility, described in Section 4.

Table 11.5, in Section 4, gives examples of principal g values of asymmetric rotors measured with the microwave Zeeman effect. For these molecules that have a plane of symmetry normal to the molecular plane, the magnetic axes, x, y, and z, coincide with the inertial axes, and the diagonal elements g_{aa}, g_{bb}, g_{cc} can be taken as principal values of the **g** tensor.

2 ZEEMAN EFFECT OF NUCLEAR HYPERFINE STRUCTURE

Weak-field Case

When the nuclear coupling energy is much stronger than the Zeeman interaction, the rotational angular momentum **J** will form with the various coupled nuclear spin vectors a resultant angular momentum **F** which will precess about the direction of the field, as indicated in Fig. 10.6. In this weak-field case only the averaged values of the magnetic moment along **F** will effectively interact with **H** since the components of $\boldsymbol{\mu}_J$ or of $\boldsymbol{\mu}_I$ normal to **F** will be averaged out by the precession about **F**. The Zeeman interaction energy will then be

$$E_H(M_F) = -\langle \mu_F \rangle H \cos(\mathbf{F}, \mathbf{H}) = -\frac{\langle \mu_F \rangle H M_F}{[F(F+1)]^{1/2}} \tag{11.30}$$

By comparison with (11.11) g_F is so defined that

$$\langle \mu_F \rangle = g_F \beta_I [F(F+1)]^{1/2} \tag{11.31}$$

Hence

$$E_H(M_F) = -g_F \beta_I H M_F \tag{11.32}$$

where

$$M_F = F, F-1, F-2, \ldots, -F \tag{11.33}$$

where F is the hyperfine quantum number defined in Chapter IX. The selection rules for M_F are the same as those for M_J, that is, $\Delta M_F = 0, \pm 1$. The frequencies of the Zeeman component can be expressed as

$$\nu = \nu_0 + \frac{\beta_I H}{h}[(g_F - g_{F'})M_F - g_{F'}\Delta M_F] \tag{11.34}$$

where ν_0 represents the frequency of the $F \to F'$ component when no field is applied and where M_F and g_F are for the lower rotational level. When the magnetic field is parallel to the electric vector of the microwave radiation, only the π components corresponding to $\Delta M_F = 0$ are observable. When the field is perpendicular to this vector, the $\Delta M_F = \pm 1$ or σ components are observable.

The relative intensity formulas are the same as those for the Stark effect (10.130) and (10.131) and are like those for molecules without hyperfine structure if J is replaced by F and M_J by M_F.

From analysis of the observed Zeeman patterns with (11.34) values of g_F for the different hyperfine states can be obtained. From them it is possible to derive the nuclear g factor and the diagonal elements of the molecular g factor with reference to the inertial axes. To do this we must resolve both μ_I and μ_J along **F**. When **I** and **J** are coupled, the nuclear magnetic moment μ_I which is along **I** contributes to the Zeeman splitting of the lines, whereas it does not when the coupling is negligible. In fact, the contribution of μ_I to the Zeeman splitting of the hyperfine components is of the same order of magnitude as that of μ_J.

We now express g_F in terms of g_J and g_I. First let us consider a molecule with a single coupling nucleus. The derivation is independent of the type of coupling (magnetic or quadrupole), but only quadrupole coupling is sufficiently strong to prevent decoupling of **J** and **I** for fields sufficiently strong for resolution of the Zeeman patterns. For the weak-field case the vector model of Fig. 10.6 applies. The angles between **J** and **F** and between **I** and **F** are constant, and with the relation $\langle \mu_F \rangle = g_J \beta_I [J(J+1)]^{1/2} \cos(\mathbf{J}, \mathbf{F}) + g_I \beta_I [I(I+1)]^{1/2} \cos(\mathbf{I}, \mathbf{F})$ we find

$$g_F = g_J \alpha_J + g_I \alpha_I \tag{11.35}$$

where

$$\alpha_J = \frac{[J(J+1)]^{1/2}}{[F(F+1)]^{1/2}} \cos(\mathbf{J}, \mathbf{F}) \qquad \text{and} \qquad \alpha_I = \frac{[I(I+1)]^{1/2}}{[F(F+1)]^{1/2}} \cos(\mathbf{I}, \mathbf{F})$$

From the law of the cosine applied to Fig. 10.6,

$$\alpha_J = \frac{F(F+1) + J(J+1) - I(I+1)}{2F(F+1)} \tag{11.36}$$

$$\alpha_I = \frac{F(F+1) + I(I+1) - J(J+1)}{2F(F+1)} \tag{11.37}$$

where

$$F = J+I, J+I-1, \ldots, |J-I| \tag{11.38}$$

Equation 11.32 for the Zeeman energies can now be expressed in the more explicit form

$$E_H(M_F) = -(g_J \alpha_J + g_I \alpha_I) \beta_I H M_F \tag{11.39}$$

For a linear molecule, $g_J \equiv g_\perp$ is approximately constant, independent of the rotational state (Section 1). For symmetric-top molecules, the value of $g_J \equiv g_{J,K}$ is given by Eq. 11.21; for asymmetric rotors, $g_J \equiv g_{J,\tau}$ is given by (11.27).

With the g_I measured from the Zeeman effect one can obtain values for the nuclear magnetic moment from

$$\mu_I = g_I \beta_I I \tag{11.40}$$

The value of μ_I for radioactive ^{131}I was first measured with this method [10], also that of ^{33}S [11]. However, the Zeeman effect cannot compete with the more accurate nuclear resonance method, which has already been used for measurement of μ_I for essentially all known isotopes in their ground states. It is common practice to use these known nuclear moments (tabulated in Appendix E) to simplify the determination of g_J from (11.39).

When there is a second nucleus with coupling weaker than the first, one first finds

$$g_{F_1} = g_J \alpha_J + g_{I_1} \alpha_{I_1} \tag{11.41}$$

by resolution of μ_J and μ_{I_1}, along $\mathbf{F}_1$ the resultant of $\mathbf{J}$ and $\mathbf{I}_1$, as described above; one then resolves μ_{F_1} and μ_{I_2} along $\mathbf{F}$, which is the resultant of $\mathbf{F}_1$ and $\mathbf{I}_2$ (see vector diagram of Fig. 9.9). The result is

$$g_F = g_{F_1} \alpha_{F_1} + g_{I_2} \alpha_{I_2} \tag{11.42}$$

where

$$\alpha_{F_1} = \frac{F(F+1) + F_1(F_1+1) - I_2(I_2+1)}{2F(F+1)} \tag{11.43}$$

$$\alpha_{I_2} = \frac{F(F+1) + I_2(I_2+1) - F_1(F_1+1)}{2F(F+1)} \tag{11.44}$$

The Zeeman splitting of the hyperfine level designated by F is given by (11.32) with the g_F of (11.42). Note that g_F is a function of g_J, g_{I_1}, and g_{I_2}. Definition of the quantum numbers is given by (9.122) and (9.123). When there is a third nucleus with still weaker coupling, derivation of g_F is accomplished by a further compounding of the vectors in the same manner.

Jen [2] first observed the Zeeman effect of rotational hyperfine structure in the $0 \to 1$ transition of CH_3Cl. Burrus [8] has observed the effect in hyperfine components of DBr and DI. Figure 11.8 shows the σ and π components which he observed for the $F = \frac{5}{2} \to \frac{7}{2}$ transitions of $D^{132}I$. The measured values of g_J are listed in Table 11.1. Note that the sign as well as the magnitude of g_J is given. The known sign of g_I makes possible the determination of the sign of the molecular g factor from application of (11.35). For molecules without nuclear coupling, determination of the sign of g_J requires the use of circularly polarized radiation [3].

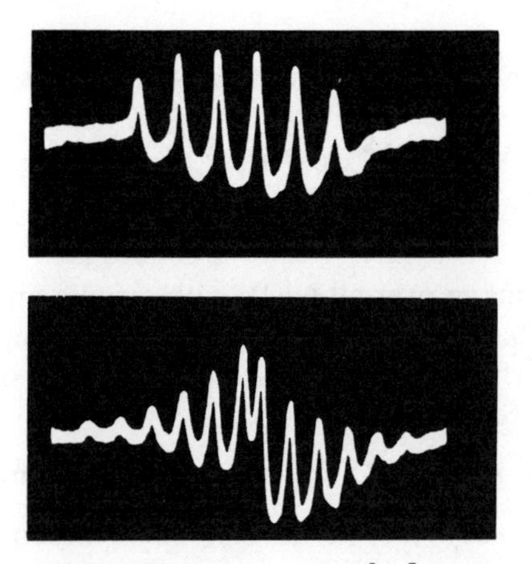

Fig. 11.8 Zeeman pattern observed for the $J=0\rightarrow1$, $F=\frac{5}{2}\rightarrow\frac{7}{2}$ transition of DI at 195 GHz. The upper trace shows the π components ($\Delta M_F=0$). The lower trace shows the σ components ($\Delta M_F=\pm1$). From Burrus [8].

Strong-field Case

When the strength of the magnetic field is such as to make the Zeeman energy large as compared with the nuclear hyperfine energy, the nuclear coupling is broken down, and **J** and **I** precess separately about the field direction, as indicated in Fig. 10.7. The resultant **F** is not formed, and F and M_F are no longer good quantum numbers. Instead, the quantum numbers which describe the state are J, M_J, I, and M_I. The strong-field case is often called the Paschen-Back effect from analogy with the breaking down of the spin–orbit coupling in atoms by strong magnetic fields or the Back-Goudsmit effect, which applies to the similar breaking of the nuclear and the electron coupling in atomic spectra. It is very similar to the strong-field Stark case described in this section.

The Zeeman energy for the decoupled, strong-field case *per se* is the same as that for molecule without nuclear coupling, Section 1. However, the nuclear coupling energies are not canceled but simply altered by the applied field, and they must still be taken into account. The nuclear magnetic coupling energy in strong fields is very simple:

$$E_\mu^{(1)}(M_J M_I)=C_{J,\tau}M_J M_I \tag{11.45}$$

where $C_{J,\tau}$ is the coupling constant for the particular rotational state J, τ derived for various molecular types in Chapter IX, Section 7. The nuclear quadrupole coupling energy for the strong-field case, which is the average $(J, M_J, I, M_I|\mathscr{H}_Q|J, M_J, I, M_I)$, already evaluated in Chapter X, Section 2, is

$$E_Q^{(1)}(M_J M_I)=\frac{eQq_J}{4J(2J-1)I(2I-1)}[3M_J^2-J(J+1)][3M_I^2-I(I+1)] \tag{11.46}$$

where q_J for different molecular types is given in Chapter IX. If there is more than one coupling nucleus, the E_Q for each one is calculated with the same formula, (11.46), and the results are added. The total first-order perturbation energy of the rotating molecule in the strong-field case is

$$E^{(1)} = -g_{J,\tau}\beta_I H M_J - \beta_I H \sum_i g_{I_i} M_{I_i} + \sum_i (C_{J,\tau})_i M_J M_{I_i} + \sum_i E_{Q_i}(M_J M_{I_i}) \tag{11.47}$$

where the sums are taken over all nuclei with coupling moments. The strong-field Zeeman effect has the same advantage as the strong-field Stark effect in simplifying the hyperfine structure caused by more than one coupling nucleus. The selection rules are

$$M_J \rightarrow M_J \qquad M_I \rightarrow M_I \qquad \pi \text{ components} \tag{11.48}$$

$$M_J \rightarrow M_J \pm 1 \qquad M_I \rightarrow M_I \qquad \sigma \text{ components} \tag{11.49}$$

Because g_{I_i} is the same for all states and because the nuclear spin does not change orientation; $\Delta M_{I_i}=0$, the second term on the right, has no influence on the spectrum and can be dropped. For molecules in $^1\Sigma$ states the term in $C_{J,\tau}$ is usually negligible. Since the first term corresponding to $E_H(M_J)$ is larger than $E_Q(M_J M_I)$, it is convenient to calculate first the splitting caused by this term and then to superimpose the hyperfine splitting on each of the Zeeman states.

The relative intensities of the gross components of the Zeeman pattern are the same as those for molecules without hyperfine structure. Each gross Zeeman component is split into a hyperfine substructure. Because of the selection rules $\Delta M_I=0$, there is no reorientation of the nuclei during a transition, and the nuclear orientation states are equally populated or very nearly so. Thus the hyperfine components corresponding to different values of M_I are all of equal intensity. However, for nuclear quadrupole interaction the levels corresponding to $+M_I$ and $-M_I$ are degenerate. For this reason the components for M_I not zero are twice as strong as those for $M_I=0$. When there is a single nucleus with quadrupole coupling having $I=\frac{3}{2}$, each Zeeman component will be split into a doublet of equal intensity corresponding to $M_I=\pm\frac{1}{2}$ and $M_I=\pm\frac{3}{2}$. For $I=1$, the substructure would still be a doublet, but the components would have intensity ratios of 1 to 2 corresponding to $M_I=0$ and $M_I=\pm 1$. If the molecule has two nuclei with quadrupole coupling, it is convenient to calculate first the hyperfine pattern which is due to the stronger coupling nucleus and then to calculate the substructure of each of its components caused by the coupling of the second nucleus. Each calculation is independent of the other. Suppose, for example, that the stronger coupling nucleus has $I=\frac{5}{2}$. Each Zeeman component would have a calculated triplet substructure of equal intensity caused by this nucleus. If the second coupling nucleus has $I=1$, each component of this triplet is further split into doublets with intensity ratios of 1:2.

Figure 11.9 illustrates the strong-field spectrum [12] for N_2O in which there are two ^{14}N nuclei with weak but unequal quadrupole coupling. The

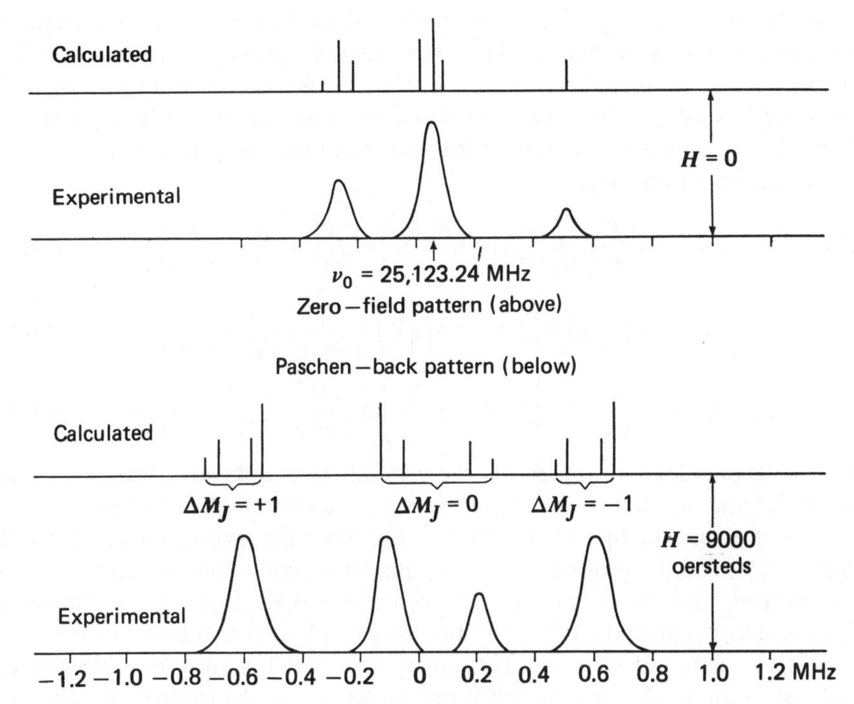

Fig. 11.9 Theoretical and observed strong-field Zeeman effect of the hyperfine structure of the $J=0\rightarrow1$ transition of $^{14}N^{14}NO$. From Jen [12].

nuclear magnetic coupling is negligible. The gross triplet is due to the π and σ components arising from the term $-g_J\beta_I HM_J$, the substructure to the nuclear quadrupole coupling. The coupling by one ^{14}N splits each Zeeman component into a doublet with intensity ratios 1:2; the second ^{14}N further splits each of these into doublets with intensity ratios 1:2, thus giving each Zeeman component a four-line substructure. The splitting by the end nitrogen is the greater.

Because of its infrequent application we shall not treat the intermediate-field case in which the Zeeman and the nuclear coupling energies are of comparable magnitude. The treatment is similar to that described for the intermediate-field Stark effect in Chapter X, Section 2. Usually the external field may be adjusted so that either the weak- or strong-field case applies.

3 THEORY OF MAGNETISM IN CLOSED-SHELL MOLECULES ($^1\Sigma$ STATES)

The most common molecules have closed electronic shells ($^1\Sigma$ states) and have no magnetic moments when field-free and not rotating. Molecular rotation, however, generates a small magnetic moment which is proportional to the rotational angular momentum, as is assumed in Section 1. Also, an externally imposed magnetic field may generate a magnetic moment caused by the molecular susceptibility which is proportional to the strength of the magnetic field, as is assumed in Section 11.4. The magnetic susceptibility is a familiar quantity which is treated thoroughly in Van Vleck's classic work on magnetic and electric susceptibilities [13]. The susceptibility per molecule consists of a diamagnetic as well as a paramagnetic contribution, the principal elements of which can be expressed as

$$\chi_{xx} = -\frac{e^2}{4mc^2}\left(0\left|\sum_i (y_i^2+z_i^2)\right|0\right) + \frac{e^2}{2m^2c^2}\sum_{n\neq 0}\frac{|(n|L_x|0)|^2}{E_n-E_0} \tag{11.50}$$

$$\chi_{yy} = -\frac{e^2}{4mc^2}\left(0\left|\sum_i (z_i^2+x_i^2)\right|0\right) + \frac{e^2}{2m^2c^2}\sum_{n\neq 0}\frac{|(n|L_y|0)|^2}{E_n-E_0} \tag{11.51}$$

$$\chi_{zz} = -\frac{e^2}{4mc^2}\left(0\left|\sum_i (x_i^2+y_i^2)\right|0\right) + \frac{e^2}{2m^2c^2}\sum_{n\neq 0}\frac{|(n|L_z|0)|^2}{E_n-E_0} \tag{11.52}$$

In these expressions $|0)$ represents the ground electronic wave function of the molecule; and $|n)$, that of the excited state; L_g represents the electronic orbital angular momentum operator; and $E_n - E_0$, the difference in energy between the states $|n)$ and the ground state. The subscripts signify the coordinates of the ith electron, and the summation is taken over all electrons in the molecule. The constants e, m, and c represent the charge and mass of the electron and the velocity of light, respectively. The molar susceptibility may be obtained by multiplication of the susceptibility per molecule by Avogadro's number N. The quantity measured in the usual bulk susceptibility experiments is the average

$$\chi = \tfrac{1}{3}(\chi_{xx}+\chi_{yy}+\chi_{zz}) \tag{11.53}$$

The first term on the right of Eqs. (11.50)–(11.52) arises from the Larmor precession of the electronic charge about the applied field and represents the diamagnetic component. The last term is the paramagnetic component that arises from an induced admixture of excited paramagnetic states with the nonmagnetic ground states. Although these excited states have orbital angular momentum, this momentum is counterbalanced by the opposite circulation of charges giving the diamagnetic term so that the total, overall angular momentum remains zero. The electronic spins remain paired and thus make no contribution to the magnetism.

Condon [14] originally treated rotational magnetic moments as caused by the rotation of charges fixed in the molecule. However, the early measurements of rotational magnetism by the molecular beam resonance method showed that the concept of fixed charges rotating with the molecular frame was inadequate. Wick [15] considered the separate contribution of the electron and nuclei and showed that the measured moment of H_2 could be explained if the electrons were considered as having angular momentum apart from their rotation with the molecular frame. Jen [5] extended this theory to polyatomic molecules and assumed that the inner-shell electrons rotate with the nuclei and that only the valence electrons have angular momentum relative to the molecular frame.

With the Wick theory one can consider the rotation of the charges as generating a magnetic field which then induces a magnetic moment in the electronic cloud because of its magnetic susceptibility. In the Hamiltonian for the rotating molecule Eshbach and Strandberg [3] approached the problem in a more basic manner by considering separately the angular momentum due to the nuclear system from that due to the electrons. Their treatment, which does not explicitly employ the concept of magnetic susceptibility, is followed here.

Rotating molecules in $^1\Sigma$ states can have a slight component of orbital electronic angular momentum which is due to a rotation-induced admixture of electronic states having orbital angular momentum with the ground electronic state. Because of the rigorous requirement that the total angular momentum must be a constant of the motion, no such components are possible in the nonrotating molecule, for which the electronic orbital angular momentum must either be exactly zero, as for $^1\Sigma$ states, or must have the magnitude $\hbar[L(L+1)]^{1/2}$ where L is an integer, as for Π states.

In the rotating molecule let us represent the total angular momentum by

$$\mathbf{J}=\mathbf{N}+\mathbf{L} \tag{11.54}$$

where **N** represents the angular momentum caused by rotation of the nuclei alone and **L**, that due to the electrons. Only **J**, the total angular momentum, must be precisely quantized. The rotational Hamiltonian for the nuclear system plus the Hamiltonian for the unperturbed electronic energies is

$$\begin{aligned}\mathscr{H}&=\frac{1}{2}\sum_g \frac{N_g^2}{I_g}+\mathscr{H}_e=\frac{1}{2}\sum_g \frac{(J_g-L_g)^2}{I_g}+\mathscr{H}_e\\ &=\frac{1}{2}\sum_g \frac{J_g^2}{I_g}-\sum_g \frac{J_gL_g}{I_g}+\frac{1}{2}\sum_g \frac{L_g^2}{I_g}+\mathscr{H}_e \end{aligned}\tag{11.55}$$

where $g=x, y, z$ represents the principal inertial axes and where I_g represents the principal moments of inertia due to the nuclear masses alone. We have assumed the nuclear frame to be rigid, that is, have neglected the effects of vibration and centrifugal distortion. Since L_g is very small, the third term in (11.55) can be neglected, and the second term can be treated as a perturbation. Therefore

$$\mathscr{H}=\mathscr{H}^0+\mathscr{H}' \tag{11.56}$$

where

$$\mathscr{H}^0=\frac{1}{2}\sum \frac{J_g^2}{I_g}+\mathscr{H}_e \tag{11.57}$$

and

$$\mathscr{H}'=-\sum_g \frac{J_gL_g}{I_g} \tag{11.58}$$

Here the first term in $\mathscr{H}^0$ is like the Hamiltonian of the rigid rotor considered in earlier chapters except that I_g represents the moment of inertia resulting from the nuclear masses alone, whereas in the earlier chapters the electronic masses were lumped with the nuclear masses in determination of I_g. See the discussion in Section 7. The electronic Hamiltonian is a function of the kinetic and potential energy of the electrons; however, the exact form of $\mathscr{H}_e$ is not required for our discussion. The angular momentum of the electrons will have an associated magnetic moment

$$\boldsymbol{\mu}_L = -\frac{e}{2mc}\mathbf{L} \tag{11.59}$$

Averaged over the unperturbed ground electronic state (pure Σ state), this will be zero.

$$(0|\boldsymbol{\mu}_L|0) = -\frac{e}{2mc}(0|\mathbf{L}|0) = 0 \tag{11.60}$$

However, the term $\mathscr{H}'$ perturbs the ground electronic state when the molecule is rotating. To obtain the electronic contribution to the rotational magnetic moment one may correct the electronic basis functions for the perturbation term. According to perturbation theory the first-order perturbed wave function is

$$\psi_0^{(1)} = \psi_0^{(0)} - \sum_{n \neq 0} \frac{(n|\mathscr{H}'|0)}{E_n - E_0}\psi_n^0 \tag{11.61}$$

where $\psi_0^{(0)}$ signifies the wave function of the unperturbed state and $\psi_n^{(0)}$ signifies those of other states for which the matrix elements of $\mathscr{H}'$ are not zero. The E_n are the unperturbed electronic energies. Since the unperturbed electronic functions are independent of the J_g operators, the first-order perturbed electronic function is found from (11.58) and (11.61) to be

$$\psi_0^{(1)} = \psi_0^{(0)} + \sum_{n \neq 0}\sum_g \frac{J_g}{I_g}\frac{(n|L_g|0)}{E_n - E_0}\psi_n^{(0)} \tag{11.62}$$

with I_g assumed to be constant. The averaged component $\langle \mu \rangle_{\text{electronic}}$ along the g axis caused by the electronic angular momentum is

$$\begin{aligned}\langle \mu_g \rangle_{\text{electronic}} &= -\frac{e}{2mc}(\psi_0^{(1)}|L_g|\psi_0^{(1)}) \\ &= -\frac{e}{2mc}(\psi_0^{(0)}|L_g|\psi_0^{(0)}) - \frac{e}{2mc}\sum_{n \neq 0}\sum_{g'}\frac{J_{g'}}{I_{g'}} \\ &\quad \times \frac{(0|L_g|n)(n|L_{g'}|0) + (0|L_{g'}|n)(n|L_g|0)}{E_n - E_0}\end{aligned} \tag{11.63}$$

where higher-order terms are omitted. The first term on the right is zero, from (11.60). The second term is a tensor quantity, the diagonal elements of which give the observable quantities along the principal axes of inertia.

These are

$$\langle \mu_g \rangle_{\text{electronic}} = -2\beta \frac{J_g}{I_g} \sum_{n \neq 0} \frac{|(n|L_g|0)|^2}{E_n - E_0} = \beta_I g_{gg}^e J_g \tag{11.64}$$

where β ($=e\hbar/2mc$) is the Bohr magneton, β_I is the nuclear magneton, and J_g is now in units of $\hbar$. The electronic contribution is seen to arise purely from an admixture with paramagnetic states. There is no diagmagnetic contribution to the rotational magnetic moment from the precessional motions of the electronic cloud.

The contribution to μ_g by the rotating nuclei is easy to calculate because it need not be averaged over the electronic functions. It is simply the classical expression for the magnetic moment produced by the fixed nuclear charges $Z_k e$ rotating about the principal axes with angular velocity $\omega_g = \hbar J_g/I_g$. The component about the principal axis x is, for example,

$$(\mu_x)_{\text{nuclear}} = M_p \beta_I \frac{J_x}{I_x} \sum_k Z_k(y_k^2 + z_k^2) = \beta_I g_{xx}^n J_x \tag{11.65}$$

where M_p is the proton mass, β_I is the nuclear magneton $e\hbar/2M_p c$, Z_k is the atomic number, and y_k and z_k are the coordinates of the kth nucleus in the principal inertial axes, and the summation is taken over all nuclei in the molecule. The y and z components are similar. The foregoing equation defines the diagonal elements of the $\mathbf{g}^n$ tensor; $\mathbf{g}^n$, like $\mathbf{g}^e$, is, in general, a tensor quantity, but in first-order only the diagonal elements enter, as observed previously. When the inertial axes and the magnetic axes are the same, as in symmetric-top molecules, the diagonal elements of $\mathbf{g}$ represent the principal values. The total rotational moment along the principal inertial axis g is

$$\mu_g = (\mu_g)_{\text{nuclear}} + (\mu_g)_{\text{electronic}} \tag{11.66}$$

where the values on the right are obtained from (11.64) and (11.65). It is seen that μ_g is proportional to J_g, as assumed in Section 1. The diagonal elements of the molecular rotational g factor in the principal inertial axes are readily obtainable from these values of μ_g and from (11.5) which defines $g_{gg} = g_{gg}^e + g_{gg}^n$. These elements are

$$g_{xx} = \frac{M_p}{I_x} \sum_k Z_k(y_k^2 + z_k^2) - \frac{2M_p}{mI_x} \sum_{n \neq 0} \frac{|(n|L_x|0)|^2}{E_n - E_0} \tag{11.67}$$

$$g_{yy} = \frac{M_p}{I_y} \sum_k Z_k(z_k^2 + x_k^2) - \frac{2M_p}{mI_y} \sum_{n \neq 0} \frac{|(n|L_y|0)|^2}{E_n - E_0} \tag{11.68}$$

$$g_{zz} = \frac{M_p}{I_z} \sum_k Z_k(x_k^2 + y_k^2) - \frac{2M_p}{mI_z} \sum_{n \neq 0} \frac{|(n|L_z|0)|^2}{E_n - E_0} \tag{11.69}$$

The first term on the right in these expressions is easily calculable from the moments of inertia and the molecular structures which are found from analysis of the rotational spectra. Since $I_x \simeq \sum M_p Z_k(y_k^2 + z_k^2)$, and so on, these quantities are very closely equal to unity. The last, the paramagnetic term, can then be measured from the principal elements of the g factors. These values of the paramagnetic terms can then be substituted into (11.50)–(11.52). When the corresponding elements of paramagnetic susceptibility are measured, the summed averages of all the electrons $\sum_i \langle 0|y_i^2+z_i^2|0\rangle$ over the electronic ground state function can be obtained. From the sum of these expressions $\sum_i \langle 0|r_i^2|0\rangle$ can be evaluated. For this, it is convenient to eliminate the paramagnetic term between these expressions and to solve for the summed averages. Thus

$$\sum_i \langle 0|y_i^2+z_i^2|0\rangle=\sum_k Z_k(y_k^2+z_k^2)-\frac{I_x}{M_p}g_{xx}-\frac{4mc^2}{e^2}\chi_{xx} \tag{11.70}$$

$$\sum_i \langle 0|z_i^2+x_i^2|0\rangle=\sum_k Z_k(z_k^2+x_k^2)-\frac{I_y}{M_p}g_{yy}-\frac{4mc^2}{e^2}\chi_{yy} \tag{11.71}$$

$$\sum_i \langle 0|x_i^2+y_i^2|0\rangle=\sum_k Z_k(x_k^2+y_k^2)-\frac{I_z}{M_p}g_{zz}-\frac{4mc^2}{e^2}\chi_{zz} \tag{11.72}$$

For a linear molecule, $g_{xx}=g_{yy}=g$ and (11.67)–(11.69) reduce to the simpler expression

$$g=\frac{M_p}{I}\sum_k Z_k z_k^2-\frac{2M_p}{mI}\sum_{n\neq 0}\frac{|(n|L_x|0)|^2}{E_n-E_0} \tag{11.73}$$

where z_k is the distance of the kth nucleus from the center of gravity. For a linear molecule the summed average from (11.70) to (11.72) is

$$\sum_i \langle 0|r_i^2|0\rangle=\sum_k Z_k z_k^2-\frac{gI}{M_p}-\frac{6mc^2}{e^2}\chi \tag{11.74}$$

where $\chi=\frac{1}{3}(\chi_{\parallel}+2\chi_{\perp})$, z_k is the distance of the kth nucleus from the center of gravity, I is the moment of inertia due to the nuclear masses alone, and $r_i^2=x_i^2+y_i^2+z_i^2$.

4 ANISOTROPIC MAGNETIC SUSCEPTIBILITIES

Measurement of Molecular Anisotropies

In addition to the molecular magnetic moments generated by rotation, which were considered previously, there are much smaller magnetic moments introduced in the molecule by the applied Zeeman field. When the magnetic susceptibility of the molecule is completely isotropic the induced moment is independent of the rotational state, and there is no detectable effect on the spectra. An anisotropic component can, in sufficiently strong fields, produce a measurable effect on the Zeeman pattern. Essentially all molecules, except the spherically symmetric ones which generally have no observable rotational spectra, do have anisotropic susceptibility.

The effects of anisotropic susceptibility on the Zeeman effect are very similar to those of anisotropic polarizability on the Stark effect considered in

Chapter X, Section 6. It is convenient to express the induced moments with reference to the inertial axes. Although these axes are not necessarily the principal axes of magnetic susceptibility, the off-diagonal elements will average out in the first-order treatment so that we need to consider only the diagonal elements. The induced value of the moment along the principal inertial axis x is

$$\begin{aligned}\mu_x(\chi) &= \tfrac{1}{2}(\chi_{xx}H_x + \chi_{xy}H_y + \chi_{xz}H_z) \\ &= \tfrac{1}{2}H(\chi_{xx}\Phi_{Zx} + \chi_{xy}\Phi_{Zy} + \chi_{xz}\Phi_{Zz}) \end{aligned} \tag{11.75}$$

with H taken along the space-fixed Z axis. The interaction of this component with the field will be

$$\begin{aligned}-\boldsymbol{\mu}_x \cdot \mathbf{H} &= -\mu_x H \Phi_{Zx} \\ &= -\tfrac{1}{2}H^2(\chi_{xx}\Phi_{Zx}^2 + \chi_{xy}\Phi_{Zx}\Phi_{Zy} + \chi_{xz}\Phi_{Zx}\Phi_{Zz}) \end{aligned} \tag{11.76}$$

The interactions with μ_y and μ_z are similar, and the total interaction is

$$\mathscr{H}(\chi) = -\tfrac{1}{2}H^2 \sum_{g,g'=x,y,z} \chi_{gg'}\Phi_{Zg}\Phi_{Zg'} \tag{11.77}$$

where $\chi_{gg'}$ are elements of the magnetic-susceptibility tensor referred to the principal inertial axis system. The first-order energy is the average of this Hamiltonian over the wave function of the rotor. In this average the off-diagonal terms do not contribute; only the diagonal terms are involved. Thus

$$E_H(\chi) = -\tfrac{1}{2}H^2[\chi_{xx}\langle\Phi_{Zx}^2\rangle + \chi_{yy}\langle\Phi_{Zy}^2\rangle + \chi_{zz}\langle\Phi_{Zz}^2\rangle] \tag{11.78}$$

in a symmetric-top or linear molecule the principal inertial axes are also principal axes of χ, and $\chi_{xx} = \chi_{yy} = \chi_{\perp}$, $\chi_{zz} = \chi_{\parallel}$, where z is the symmetry axis. Therefore we write

$$E_H(\chi) = -\tfrac{1}{2}H^2[\chi_{\perp}\langle\Phi_{Zx}^2 + \Phi_{Zy}^2\rangle + \chi_{\parallel}\langle\Phi_{Zz}^2\rangle] \tag{11.79}$$

With $\Phi_{Zx}^2 + \Phi_{Zy}^2 = 1 - \Phi_{Zz}^2$ and with neglect of the term $-\frac{1}{2}\chi_{\perp}H^2$ which does not depend on the rotational quantum numbers, we obtain

$$E_H(\chi) = -\tfrac{1}{2}H^2(\chi_{\parallel} - \chi_{\perp})\langle\Phi_{Zz}^2\rangle \tag{11.80}$$

where for a symmetric rotor basis

$$\langle\Phi_{Zz}^2\rangle = (J, K, M_J|\Phi_{Zz}^2|J, K, M_J) = \sum_{J'} |(J, K, M_J|\Phi_{Zz}|J', K, M_J)|^2 \tag{11.81}$$

The required cosine matrix elements are given in Table 2.1. Substitution of these matrix elements gives the perturbation energy for the symmetric top,

$$\begin{aligned}E_{J,K,M_J}(\chi) = -\tfrac{1}{2}(\chi_{\parallel} - \chi_{\perp})H^2 &\left\{\frac{[(J+1)^2 - K^2][(J+1)^2 - M_J^2]}{(J+1)^2(2J+1)(2J+3)}\right. \\ &\left. + \frac{K^2M_J^2}{J^2(J+1)^2} + \frac{(J^2 - K^2)(J^2 - M_J^2)}{J^2(4J^2 - 1)}\right\} \end{aligned} \tag{11.82}$$

The formula for a linear molecule is obtained when $K=0$ is set in (11.82), which then becomes

$$E_{J,M_J}(\chi)=-\tfrac{1}{2}(\chi_{\parallel}-\chi_{\perp})H^2\left[\frac{(J+1)^2-M_J^2}{(2J+1)(2J+3)}+\frac{J^2-M_J^2}{4J^2-1}\right] \tag{11.83}$$

One must add $E_H(\chi)$ to $E_H(\mu)$ to obtain the total perturbation of a magnetic field. The total energy of the rotating molecule in a magnetic field is

$$E_H=E_r^0+E_H(\mu)+E_H(\chi) \tag{11.84}$$

where E_r^0 is the energy of the unperturbed rotor. With the magnetic fields commonly available in the laboratory, one is not justified in extending the calculations to higher order. Here $E_H(\mu)$ represents the first-order magnetic interaction derived in Section 1.

The anisotropic susceptibility energies for asymmetric-top molecules are obtained from the average of the squared direction cosines of (11.78) over the functions of the asymmetric rotor. These averages can be expressed in terms of the line strengths. However, a more usable formula is obtained by expression of the average values of the squared direction cosines in terms of the average values of the squared angular momentum components. The resulting formula can be expressed in the form [16]

$$E_{J,\tau}(\chi)=-H^2\left[\frac{3M_J^2-J(J+1)}{(2J-1)(2J+3)J(J+1)}\right]\sum_{g=a,b,c}(\chi_{gg}-\chi)\langle J_g^2\rangle \tag{11.85}$$

where

$$\chi=\tfrac{1}{3}(\chi_{aa}+\chi_{bb}+\chi_{cc}) \tag{11.86}$$

is the isotropic component and where a, b, and c represent the principal axes of inertia. Values of $\langle J_g^2\rangle$ in terms of the reduced rotational energies and their derivatives are given by (9.97)–(9.99). See also Chapter VII.

The total energy of the asymmetric rotor in a strong magnetic field, exclusive of nuclear interactions, may be expressed by (11.84) with $E_H(\mu)$ from (11.12) and $E_H(\chi)$ from (11.85). Thus

$$E_H(J,\tau,M_J)=E_0(J,\tau,M_J)-\frac{H\beta_I M_J}{J(J+1)}\sum_{g=a,b,c}g_{gg}\langle J_g^2\rangle$$
$$-H^2\left[\frac{3M_J^2-J(J+1)}{(2J-1)(2J+3)J(J+1)}\right]\sum_{g=a,b,c}(\chi_{gg}-\chi)\langle J_g^2\rangle \tag{11.87}$$

where the first term on the right, $E_0(J,\tau,M_J)$, is the field-free rotational energy which may be found as described in Chapter IX. A term, $-\frac{1}{2}H^2$, is omitted because it displaces all levels equally and has no influence on the spectra. The Zeeman splitting of the rotational lines caused by the second term on the right is described in Section 1. Further shifts of these components resulting from the last term are measured as a function of the field H. From these measurements are found the magnitudes of the three components, g_{aa}, g_{bb}, and g_{cc}, along the principal inertial axes, a, b, and c, of the asymmetric rotor, together with the anisotropies.

$$3(\chi_{aa}-\chi)=2\chi_{aa}-\chi_{bb}-\chi_{cc} \tag{11.88}$$

$$3(\chi_{bb}-\chi)=2\chi_{bb}-\chi_{aa}-\chi_{cc} \tag{11.89}$$

$$3(\chi_{cc}-\chi)=2\chi_{cc}-\chi_{aa}-\chi_{bb} \tag{11.90}$$

These expressions are easily derived by addition of χ_{aa}, χ_{bb}, or χ_{cc} to both sides of (11.86) and rearrangement of the terms. The third $3(\chi_{cc}-\chi)$ is seen to be the negative sum of the first two. Because of the relationship of (11.86), there are only two independent molecular anisotropies. These are usually expressed by the right-hand sides of (11.88) and (11.89). When χ can be reliably calculated or derived from bulk susceptibility measurements, the diagonal elements, χ_{aa}, χ_{bb}, and χ_{cc}, can be obtained from (11.88)–(11.90).

As indicated in Section 3, the molecular susceptibilities defined by (11.50)–(11.52) have diamagnetic as well as paramagnetic components.

$$\chi_{gg}=\chi_{gg}^{d}+\chi_{gg}^{p} \tag{11.91}$$

The diamagnetic components correspond to the first terms on the right in (11.50)–(11.52), and in the asymmetric rotor, inertial axes, a, b, and c, are

$$\chi_{aa}^{d}=-\frac{e^2}{4mc^2}\sum_{i}^{\text{electrons}}\langle 0|b_i^2+c_i^2|0\rangle \tag{11.92}$$

$$\chi_{bb}^{d}=-\frac{e^2}{4mc^2}\sum_{i}^{\text{electrons}}\langle 0|c_i^2+a_i^2|0\rangle \tag{11.93}$$

$$\chi_{cc}^{d}=-\frac{e^2}{4mc^2}\sum_{i}^{\text{electrons}}\langle 0|a_i^2+b_i^2|0\rangle \tag{11.94}$$

If these values are substituted in (11.70)–(11.72) with the coordinate change $x\rightarrow a$, $y\rightarrow b$, $z\rightarrow c$, solutions for $\chi_{aa}-\chi_{aa}^{d}$, and so on, yield the paramagnetic components

$$\chi_{aa}^{p}=-\frac{e^2}{4mc^2}\left[\frac{hg_{aa}}{8\pi^2AM_p}-\sum_{k}^{\text{nuclei}}Z_k(b_k^2+c_k^2)\right] \tag{11.95}$$

$$\chi_{bb}^{p}=-\frac{e^2}{4mc^2}\left[\frac{hg_{bb}}{8\pi^2BM_p}-\sum_{k}^{\text{nuclei}}Z_k(c_k^2+a_k^2)\right] \tag{11.96}$$

$$\chi_{cc}^{p}=-\frac{e^2}{4mc^2}\left[\frac{hg_{cc}}{8\pi^2CM_p}-\sum_{k}^{\text{nuclei}}Z_k(a_k^2+b_k^2)\right] \tag{11.97}$$

where we have substituted $I_a=h/8\pi^2A$, and so on. In these expressions Z_k is the atomic number of the kth nucleus, and the summation is taken over all nuclei in the molecule. To express the susceptibility components in mole units, the terms on the right are multiplied by Avagadro's number, N.

Because of their symmetry, the principal axes of magnetic susceptibility of linear and symmetric-top molecules coincide with their principal inertial axes, indicated simply as the parallel and perpendicular axes in (11.82) and (11.83) and in the tabulated values for these quantities. See Tables 11.3 and 11.4. The axes, a, b, and c, in which g_{gg} or χ_{gg} are expressed for the asymmetric rotors, as in (11.87), are principal axes of inertia and do not necessarily coincide with the principal magnetic axes. In general, the derived magnetic quantities g_{gg} and χ_{gg} are diagonal elements of magnetic tensors that may have nonzero, off-diagonal elements. However, any such off-diagonal elements of χ are averaged out by the molecular rotation and hence are not detected in these experiments. In planar molecules having $\mathscr{C}_{2v}$ symmetry, such as those listed in Table 11.5, the principal magnetic axes designated as x, y, z, coincide with the principal inertial axes a, b, c; for these, the principal values of the magnetic quantities are obtained directly from the analysis of the spectra.

Table 11.3 Anisotropy in Magnetic Susceptibility and Molecular Quadrupole Moment of Some Linear Molecules

Molecule	*Susceptibility Anisotropy,* $\chi_\perp - \chi_\parallel$ (10^{-6} *erg/G*^{2} *·mole*)	*Quadrupole Moment,* $Q_\parallel$ (10^{-26} *esu·cm*^{2})	*Ref.*
CO	8.2(9)	−2.0(10)	a
CS	24(3)	0.8(14)	a,b
CSe	27.8(14)	−2.6(16)	b
SiO	11.1(9)	−4.6(11)	c
AlF	5.2(5)	−5.9(5)	d
CuF	6.5(7)	−6.1(8)	e
OCS	9.27(10)	−0.88(15)	f
OCSe	10.06(18)	−0.32(24)	g
$HC^{15}N$	7.6(8)	3.5(9)	h
HCP	8.4(9)	4.4(9)	b
NNO	10.15(15)	−3.65(25)	i
FCCH	5.19(15)	3.96(14)	j
^{35}ClCCH	9.3(5)	8.8(4)	k

[a] Gustafson and Gordy [31].
[b] McGurk et al. [40].
[c] Hornerjäger and Tischer [38].
[d] Hornerjäger and Tischer [42].
[e] Hornerjäger and Tischer [45].
[f] Flygare et al. [20].
[g] Shoemaker and Flygare [30].
[h] Gustafson and Gordy [41].
[i] Flygare et al. [22].
[j] Shoemaker and Flygare [18].
[k] Allen and Flygare [36].

Table 11.4 Anisotropy in Magnetic Susceptibility and Molecular Quadrupole Moment of Some Symmetric-top Molecules

Molecule	*Susceptibility Anisotropy,* $\chi_\perp - \chi_\parallel$ (10^6 *erg/G*^{2}*·mole*)	*Molecular Quadrupole Moment,* $Q_\parallel$ (10^{-26} *esu·cm*^{2})	*Ref.*
NH_3	−0.37(4)	−2.32(7)	a
PH_3	2.7(8)	−2.1(10)	b
PF_3	1.3(5)	24(3)	c
CH_3Cl	8.0(5)	1.2(8)	d
CH_3Br	8.5(4)	3.4(8)	d
CH_3I	11.0(5)	5.4(9)	d
CHCN	10.2(10)	−1.8(12)	e
CH_3NC	13.5(17)	−2.7(16)	e
CH_3CCH	7.70(14)	4.82(23)	f
CH_3CCCCH	13.1(2)	9.9(8)	g
SiH_3Br	2.7(3)	−0.1(4)	h
SiH_3NCS	28.0(30)	16.6(37)	i

[a]Kukolich [29].
[b]Kukolich and Flygare [28].
[c]Stone et al. [23].
[d]VanderHart and Flygare [21].
[e]Pochan et al. [32].
[f]Shoemaker and Flygare [24].
[g]Shoemaker and Flygare, reported by Sutter and Flygare [58].
[h]Dössel and Sutter [51].
[i]Dössel and Sutter [52].

Taft and Dailey [4] have measured the anisotropic susceptibility for OCS from the $J=1\rightarrow 2$, $M_J\rightarrow M_J$ transition with a magnitude field of 23 kG. The $M_J\rightarrow M_J$ or π transitions are observed with the field imposed parallel to the microwave electric vector. With this arrangement no shift or broadening of the rotational lines is caused by the rotational magnetic moment when $g_J=g_{J+1}$. For linear molecules the g value for the different rotational states is expected to be the same except for very slight effects caused by centrifugal distortion. In OCS the possible difference in the g value observed for different rotational states can give rise to a splitting of the π components of only a few hertz, even at the 23 kG fields employed by Taft and Dailey. However, it is seen from Eq. (11.82) that the magnetic states have a slight J dependence caused by the anisotropic susceptibility. As a result, the $J=1\rightarrow 2$, $M_J=0\rightarrow 0$ component is separated from the $J=1\rightarrow 2$, $M_J=\pm 1\rightarrow \pm 1$ component. Taft and Dailey observed that the $\pm 1\rightarrow \pm 1$ component was shifted from the position of the zero-field line by 132 kHz. Using this observation they obtained the anisotropic susceptibility from (11.83) as $\chi_\perp - \chi_\parallel = 8.35\times 10^{-6}$ erg/(G^2·mole). A remeasurement by Flygare et al. with a field of 30 kG gives the more accurate value listed in Table 11.3.

In an asymmetric rotor the molecular g factor differs according to the rotational state, and therefore the $\Delta M_J=0$ transitions are split, even in the absence of anisotropic susceptibility. Without anisotropic effects, however, all the Zeeman components of a given multiplet (whether π or σ) are equally spaced. Addition of the quadratic term $E_H(\chi)$ of (11.85) cause a difference in the spacing which allows measurement of the $(\chi_{gg}-\chi)$. Hüttner et al. [17] were the first to measure the components of the magnetic susceptibilities of an asymmetric rotor, H_2CO, from the differences in the spacing of the Zeeman components, as shown in Fig. 11.7. Later values of the magnetic constants of H_2CO obtained with magnetic fields of 30 kG are listed in Table 11.5.

Following the experiments just discussed, there have been a large number of measurements of anisotropies in magnetic susceptibilities, some of which are cited here [18–56]. Description of the method and review of the essential results to 1970 are given by Flygare and Benson [57] and to 1976, by Sutter and Flygare [58]. Selected experimental values are given in Tables 11.3, 11.4, and

Table 11.5 Zeeman Parameters and Molecular Quadrupole Moments for Selected Asymmetric-top Molecules[a]

Molecule (axes: y vertical, x horizontal)	g_{xx} g_{yy} g_{zz}	$2\chi_{xx}-\chi_{yy}-\chi_{zz}$ $2\chi_{yy}-\chi_{xx}-\chi_{zz}$ (10^{-6} *erg/G*^{2}·*mole*)	Q_{xx} Q_{yy} Q_{zz}	*Ref.*
H_2O (H╲O╱H)	0.718(7) 0.657(1) 0.645(6)	−0.199(48) 0.464(24)	−0.13(3) 2.63(2) −2.50(2)	b,c
$H_2C{=}O$ (H╲C=O╱H)	−2.9017(8) −0.2243(1) −0.0994(1)	25.5(5) −3.9(3)	−0.1(3) 0.2(2) −0.1(5)	d
$H_2C{=}C{=}O$ (H╲C=C=O╱H)	−0.4182(9) −0.0356(13) −0.0238(6)	−5.0 −0.2(6)	−0.7(3) 3.8(4) −3.1(4)	e,f
$HFC{=}CHF$ (H, F above; C=C; H, F below)	−0.0421(5) −0.0466(4) −0.0119(4)	−2.3(6) 7.7(5)	2.4(5) −0.9(4) −1.5(8)	d

Table 11.5 (Continued)

y └x Molecule	g_{xx} g_{yy} g_{zz}	$2\chi_{xx}-\chi_{yy}-\chi_{zz}$ $2\chi_{yy}-\chi_{xx}-\chi_{zz}$ (10^{-6} *erg/G*^{2}·*mole*)	Q_{xx} Q_{yy} Q_{zz}	*Ref.*
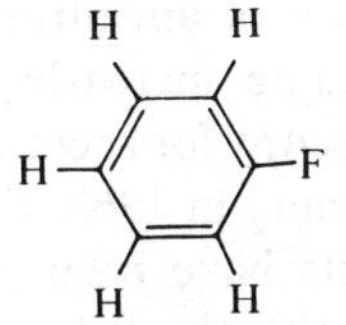 Fluorobenzene (H, H, H, F, H, H)	−0.06892(22) −0.04161(13) −0.02627(9)	51.68(26) 61.69(45)	−1.52(36) 7.34(48) −5.82(62)	*g*
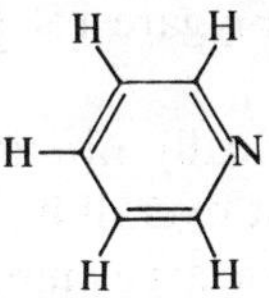 Pyridine (H, H, H, N, H, H)	−0.0770(5) −0.1010(8) 0.0428(4)	54.3(6) 60.5(8)	−3.5(9) 9.7(11) −6.2(15)	*h*
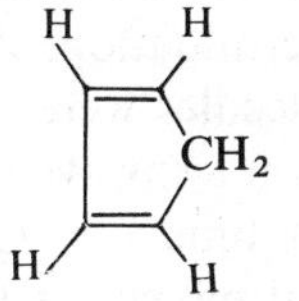 Cyclopentadiene (H, H, CH_2, H, H)	−0.0827(3) −0.0700(3) 0.0385(2)	37.8(3) 30.7(3)	3.7(4) 1.4(4) −5.1(5)	*i*
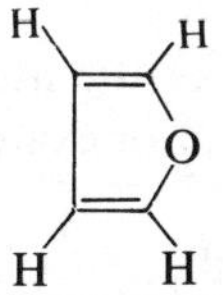 Furan (H, H, O, H, H)	−0.0911(7) −0.0913(2) 0.0511(1)	43.0(2) 34.4(2)	0.2(4) 5.9(3) −6.1(4)	*j*

[a]In these molecules the principal axes of susceptibility, *x*, *y*, and *z*, coincide with the inertial axes, *a*, *b*, and *c*.
[b]J. Verhoeven and A. Dynamus, *J. Chem. Phys.*, **52**, 3222 (1970).
[c]S. G. Kukolich, *J. Chim. Phys.*, **50**, 3751 (1969).
[d]Blickensderfer et al. [26].
[e]Lo et al. [19].
[f]Hüttner et al. [25].
[g]Stolze et al. [56].
[h]Wang and Flygare [33].
[i]Benson and Flygare [34].
[j]Sutter and Flygare [27].

11.5. In these tables are also given molecular quadrupole moments derived from the anisotropies and measured *g* values, described in Section 5. Certain of the molecules listed have nuclear quadrupole coupling which can complicate the interpretation of the Zeeman patterns, as described in Section 2. Treatment of these complications for particular molecules may be found in the references cited. Diamagnetic and paramagnetic susceptibility components from (11.92)–(11.97) are illustrated in Table 11.7 in Section 5.

Calculation of Molecular Susceptibilities from Local Atomic Components

Early in the century (1910) Pascal [59] proposed the concept of additive, local susceptibilities for calculation of bulk susceptibilities from atomic contributions. In 1970 Benson and Flygare [60] broadened the concept to include tensor elements of local atomic susceptibilities for calculation of anisotropic susceptibility components of molecules. This concept became increasingly useful with the widespread applications of microwave spectroscopy for accurate measurement of anisotropic molecular susceptibilities, beginning in 1968. For the calculations, useful sets of atomic susceptibility elements have been developed and published in convenient tables. The first such table, by Schmalz et al. [61] in 1973, was revised and extended by Sutter and Flygare [58] in 1976.

The tabulated local-atom susceptibility elements are empirically obtained by adjustment of assumed local parameter elements for the best overall fitting of calculated with observed susceptibility elements χ_{aa}, χ_{bb}, χ_{cc}, of a large number of open-chain or nonaromatic molecules for which contributions from delocalized π-type bonds can be assumed negligible. For determination of the original set of local elements by Schmalz et al. [61], 14 molecules were used, for which good values of χ_{aa}, χ_{bb}, and χ_{cc} were available from microwave spectroscopy and bulk susceptibility measurements. Since that time the values have been improved and extended by inclusion of additional molecules from later microwave Zeeman measurements [54, 55, 58].

The molecular susceptibilities are calculated by summation of the atomic components resolved along the molecular inertial axes, a, b, and c. For example,

$$\chi_{aa}^{\text{mol}}=\sum_i \{\chi_{xx}^i \cos^2 (a, x^i)+\chi_{yy}^i \cos^2 (a, y^i)+\chi_{zz}^i \cos^2 (a, z^i)\} \tag{11.98}$$

where x^i, y^i, z^i are the principal axes of the local atomic tensor χ^i of the ith atom and where the summation is taken over all the local tensors contribution to the χ_{aa}^{mol}. The expressions for χ_{bb}^{mol} and χ_{cc}^{mol} are similar. For obtaining the cosines in (11.98), the directions of the diagonal elements of the local tensors relative to the molecular frame must be known. Their directions are related to the bond directions formed by the atoms within the molecule. Generally, the complete structure of the molecule must be known for resolution of all the local tensor elements along a, b, and c. Atoms forming different types of localized bonds, such as sp, sp_2, and sp_3 bonds, are assigned different atomic elements for each type of bonding. Some of the more commonly used elements are reproduced in Table 11.6.

Two examples of calculated (local atomic) and experimental χ_{gg} values for molecules having negligible delocalized contributions are methyl formate [61] for which

$$\chi_{aa}^{\text{calc}}=-28.6 \qquad \chi_{bb}^{\text{calc}}=-29.8 \qquad \chi_{cc}^{\text{calc}}=-36.7$$
$$\chi_{aa}^{\text{exp}}=-28.3(5) \qquad \chi_{bb}^{\text{exp}}=-30.9(6) \qquad \chi_{cc}^{\text{exp}}=-36.7(8)$$

Table 11.6 Some Local Atomic Susceptibilities Useful for Calculation of Molecular Susceptibilities[a,b]

(y, x axes)	χ_{xx}	χ_{yy}	χ_{zz}
H—	−1.17(53)	−2.08(38)	−2.08(38)
F—	−8.22(116)	−6.87(111)	−5.42(52)
$C_{sp3}(H_3C—)$	−9.92(181)	−8.27(149)	−8.27(149)
C_{sp3} (H_2C<)	−7.45(118)	−7.19(125)	−8.26(107)
C_{sp2} (>C=)	−3.64(51)	−3.75(60)	−7.33(55)
O_{ether} (O<)	−8.73(101)	−10.35(104)	−8.23(95)
$O_{carbonyl}$(O=)	+1.90(120)	−1.29(120)	−5.70(107)
N_{sp2} (>N—)	−13.82	−10.35	−6.13
S<	−17.03	−17.07	−15.62

[a]Insufficient tests were available for estimation of probable error in the last two entries. The units are 10^{-6} erg/G^2·mole.
[b]From Sutter and Flygare [58].

and maleic anhydride [61] for which

$$\chi_{aa}^{\text{calc}} = -25.1 \qquad \chi_{bb}^{\text{calc}} = -27.5 \qquad \chi_{cc}^{\text{calc}} = -54.3$$
$$\chi_{aa}^{\text{exp}} = -25.7(15) \qquad \chi_{bb}^{\text{exp}} = -28.2(15) \qquad \chi_{cc}^{\text{exp}} = -53.5(17)$$

all in units of 10^{-6} erg/G^2·mole. These two molecules are among the original fourteen used to test the assumed local atomic increments [61].

Delocalized, π-bond contributions to the molecular susceptibilities cannot be estimated by the atomic-increment method. The technique, however, is quite useful for obtaining the nonlocal contributions indirectly. One first calculates the contribution of the local elements to the particular molecular component and subtracts this from the overall measured molecular component to obtain the nonlocalized contribution

$$\chi_{gg}^{\text{nonlocal}} = \chi_{gg}^{\text{exp}} - \chi_{gg}^{\text{local(calc)}} \tag{11.99}$$

This indirect method is particularly useful for estimation of the π-bonding components in planar aromatic ringed compounds where all the localized σ-bond contributions are in the molecular plane and the π lobes are perpendicular to the plane and make large contributions to $\chi_{\perp}$. In such molecules the inertial c

axis is generally perpendicular to the molecular plane, and the a and b axes are in the ring. Thus, for planar rings one can usually assign $\chi_{\parallel}=(\chi_{aa}+\chi_{bb})/2$ and $\chi_{\perp}=\chi_{cc}$. As an example, consider fluorobenzene [55], for which the experimental values in units of 10^{-6} erg/G^2·mole are

$$\chi_{\parallel}^{\text{exp}}=-35.9 \qquad \text{and} \qquad \chi_{\perp}=-97.2$$

and the corresponding values calculated from the local atomic tensors are

$$\chi_{\parallel}^{\text{local(calc)}}=-37 \qquad \text{and} \qquad \chi_{\perp}^{\text{local(calc)}}=-59$$

Thus,

$$\chi_{\parallel}^{\text{nonlocal}}=\chi_{\parallel}^{\text{exp}}-\chi_{\parallel}^{\text{local(calc)}}=1.1$$

and

$$\chi_{\perp}^{\text{nonlocal}}=\chi_{\perp}^{\text{exp}}-\chi_{\perp}^{\text{local(calc)}}=-38.2$$

The nonlocal contribution in the plane of the ring is insignificant, whereas the magnitude of the nonlocal contribution perpendicular to the ring is comparatively large, -38. The $\chi_{\perp}^{\text{nonlocal}}$ is evidently due to the π electron lobes above and below the plane of the rings. For furan, H_4C_4O, similar analysis gives

$$\chi_{\parallel}^{\text{nonlocal}}=-0.8 \qquad \text{and} \qquad \chi_{\perp}^{\text{nonlocal}}=-25.3$$

and for thiofuran, H_4C_4S,

$$\chi_{\parallel}^{\text{nonlocal}}=-2.8 \qquad \text{and} \qquad \chi_{\perp}^{\text{nonlocal}}=-39.1$$

A similar trend is found in other aromatic ringed compounds. The $\chi_{\perp}^{\text{nonlocal}}$ values give a measure of electron delocalization or an indication of the aromaticity of the planar ringed compounds [61].

5 SECOND MOMENTS AND MOLECULAR QUADRUPOLE MOMENTS FROM THE ZEEMAN EFFECT

Molecular Quadrupole Moments

The molecular quadrupole moment is a measure of the asymmetry of the charge cloud of a molecule. Its definition is similar to that of the nuclear quadrupole moment. By comparison with (9.37) the molecular quadrupole tensor can be expressed in dyadic notation as

$$\mathbf{Q}_{\text{mol}}=\int \rho(3\mathbf{rr}-\mathbb{I}r^2)\,d\tau \tag{11.100}$$

where ρ is the molecular charge density at the point x, y, z in the molecule-fixed reference system at the position of the elemental volume $d\tau$ and where the integral is taken over the charge cloud of the molecule. This definition is like that for the nuclear quadrupole moment defined by (9.37) and expresses the quadrupole moment with reference to the center of mass of the molecule.

It is more convenient to express $Q_{\rm mol}$ in terms of the summation of the contribution from the charges of all different particles, rather than in terms of the integral over the molecular charge cloud. In terms of these sums, the diagonal elements of the $Q_{\rm mol}$ tensor are

$$Q_{xx} = -\frac{e}{2}\sum_i (0|3x_i^2 - r_i^2|0) + \frac{e}{2}\sum_k Z_k(3x_k^2 - r_k^2) \qquad (11.101)$$

$$Q_{yy} = -\frac{e}{2}\sum_i (0|3y_i^2 - r_i^2|0) + \frac{e}{2}\sum_k Z_k(3y_k^2 - r_k^2) \qquad (11.102)$$

$$Q_{zz} = -\frac{e}{2}\sum_i (0|3z_i^2 - r_i^2|0) + \frac{e}{2}\sum_k Z_k(3z_k^2 - r_k^2) \qquad (11.103)$$

where e represents the magnitude of the electronic charge. The first term on the right in these equations represents the electronic contributions; the indicated average is taken over the orbital of each electron; and the summation is taken over all electrons of the molecule. The last term arises from the nuclei, and the specified coordinates x_k, y_k, z_k and r_k are the fixed distances of the kth nucleus relative to the molecule-fixed reference system x, y, z. The summation is taken over all nuclei of the molecule.

By simultaneous solution of (11.70)–(11.72) one can obtain the summed averages $\sum_i (0|x_i^2|0)$, $\sum_i (0|y_i^2|0)$, and $\sum_i (0|z_i^2|0)$ in terms of the diagonal elements of the rotational g factors and susceptibilities χ. Addition of these terms gives $\sum_i (0|r_i^2|0)$. By substitution of these values into (11.101)–(11.103) the diagonal elements of the molecular quadrupole moment can be obtained in terms of the measurable magnetic constants and moments of inertia. The results are

$$Q_{xx} = -\frac{e}{2M_p}(2g_{xx}I_x - g_{yy}I_y - g_{zz}I_z) - \frac{2mc^2}{e}(2\chi_{xx} - \chi_{yy} - \chi_{zz}) \qquad (11.104)$$

$$Q_{yy} = -\frac{e}{2M_p}(2g_{yy}I_y - g_{zz}I_z - g_{xx}I_x) - \frac{2mc^2}{e}(2\chi_{yy} - \chi_{zz} - \chi_{xx}) \qquad (11.105)$$

$$Q_{zz} = -\frac{e}{2M_p}(2g_{zz}I_z - g_{xx}I_x - g_{yy}I_y) - \frac{2mc^2}{e}(2\chi_{zz} - \chi_{xx} - \chi_{yy}) \qquad (11.106)$$

in which e is the magnitude of the electronic charge. For linear and symmetric-top molecules and for asymmetric tops of C_{2v} symmetry, the diagonal elements are also principal elements. For a linear molecule, $Q_{xx} = Q_{yy} = -\frac{1}{2}Q_{zz}$, and there is only one independent element of Q. This is designated by Buckingham [62] as $Q_{\rm mol} = Q_{zz}$ and by earlier workers [63–65] as $Q_{\rm mol} \equiv 2Q_{zz}$. Since for linear molecules $I_z = 0$, $I_x = I_y = I$, $g_{xx} = g_{yy} = g$, $g_{zz} = 0$, and $\chi_{yy} = \chi_{xx} = \chi_\perp$, $\chi_{zz} = \chi_\parallel$, the quadrupole moment in the Buckingham units is, from (11.106),

$$Q_{\rm mol} = Q_{zz} = Q_\parallel = \frac{eIg_\perp}{M_p} - \frac{4mc^2}{e}(\chi_\parallel - \chi_\perp) \qquad (11.107)$$

for linear molecules. If χ is measured in units of erg/G$^2\cdot$mole, the last terms of (11.104)–(11.106) must be divided by Avagadro's number, N.

The experimental elements of the g and χ tensors for asymmetric-top molecules are measured with reference to the inertial axes, a, b, and c. See (11.87). It is thus convenient to express the molecular quadrupole moment with reference to those axes. The diagonal elements of Q_{mol} in the inertial axes are, however, averaged ones for the rotating molecules and not necessarily the principal elements of the molecule at rest. It is also desirable to express the moments of inertia in terms of the spectral constants, A, B, and C. With these transformations, (11.104)–(11.106) take the form

$$Q_{aa}=-\frac{eh}{16\pi^2 M_p}\left\{\frac{2g_{aa}}{A}-\frac{g_{bb}}{B}-\frac{g_{cc}}{C}\right\}-\frac{2mc^2}{e}(2\chi_{aa}-\chi_{bb}-\chi_{cc}) \quad (11.108)$$

The corresponding expression for Q_{bb} and Q_{cc} may be written from cyclic permutation of the coordinates and spectral constants. Note that the term in parentheses is the directly measured susceptibility anisotropy, (11.88).

There is only one susceptibility anisotropy, $\chi_\perp-\chi_\parallel$, that is observable for a symmetric-top molecule and only one principal molecular quadrupole element, $Q_\parallel$. A formula for $Q_\parallel$ may be easily derived from (11.108). With symmetry axes chosen as the a inertial axis, $g_{aa}=g_\parallel$, $g_{bb}=g_{cc}=g_\perp$, $B=C$, and (11.108) reduces to

$$Q_\parallel=-\frac{eh}{8\pi^2 M_p}\left\{\frac{g_\parallel}{A}-\frac{g_\perp}{B}\right\}-\frac{4mc^2}{e}(\chi_\parallel-\chi_\perp) \quad (11.109)$$

for the symmetric-top molecule.

In the early fifties derivations of $Q_{\text{mol}}$from Zeeman parameters were achieved by N. F. Ramsey and his group [66–68] on the hydrogen molecule from the parameters measured with molecular beam resonance. Following the 1968 microwave spectral publications of Taft and Dailey [4] on the linear molecule OCS and Hüttner et al. [17] on the asymmetric rotor H_2CO, quadrupole moments of many molecules have been evaluated from Zeeman parameters obtained from microwave spectral measurements. Selected examples of the results for linear molecules are given in Table 11.3, for symmetric-top molecules in Table 11.4, and for asymmetric rotors in Tables 11.5 and 11.7.

The magnetic constants and the quadrupole moment of benzene cannot be measured directly with microwave spectroscopy because benzene has no electric dipole moment and hence no detectable rotational spectra. However, Shoemaker and Flygare [66], with some reasonable approximations, indirectly derived the molecular quadrupole elements of this basically important molecule from the measured moments of monofluorobenzene. The principal assumptions are that the substitution of the fluorine does not appreciably distort the ring structure nor the electronic charge distribution difference, $\langle b^2\rangle-\langle c^2\rangle$. They concluded that the limiting factor in the accuracy of the derivation was the inaccuracy in the measured constants of the fluorobenzene. Stolze et al. [56] later remeasured the magnetic parameters of fluorobenzene with a spectrometer having significantly improved resolution and have obtained the more accurate values for the parameters as listed in Table 11.5. Using the same assumptions

Table 11.7 Some Molecular Parameters from the Zeeman Effect: Dia- and Paramagnetic Susceptibilities, Second Moments of Charge Distributions, and Quadrupole Moments

(axes: b up, a right)	(*a*) H_2C—CH_2 bridged by O (ethylene oxide)	(*b*) F, F substituted benzene ring
	Units: 10^{-16} *erg/G*^{2}·*mole*	
χ^p_{aa}	60.7(3)	330(9)
χ^p_{bb}	67.1(2)	646(9)
χ^p_{cc}	87.4(2)	890(9)
χ^d_{aa}	−85.3(13)	−378(11)
χ^d_{bb}	−97.8(15)	−691(11)
χ^d_{cc}	−124.3(16)	−987(11)
	Units: 10^{-16} *cm*^{2}	
Nuclei		
$\sum_k Z_k a_k^2$	12.81(2)	143(2)
$\sum_k Z_k b_k^2$	8.99(5)	71(2)
$\sum_k Z_k c_k^2$	3.41(2)	0.0
Electrons		
$\langle 0\|\sum_i a_i^2\|0\rangle$	16.1(6)	153(5)
$\langle 0\|\sum_i b_i^2\|0\rangle$	13.2(6)	79(5)
$\langle 0\|\sum_i c_i^2\|0\rangle$	6.96(6)	10(5)
	Units: 10^{-26} *esu·cm*^{2}	
Q_{aa}	2.60(5)	−5.0(9)
Q_{bb}	−3.69(7)	7.6(10)
Q_{cc}	1.10(11)	−2.6(13)

[a]Parameter values, from Sutter and Flygare [58].
[b]Parameter values, from Czieslik et al. [48].

as those of Shoemaker and Flygare, they recalculated Q_{mol} for benzene and obtained the higher value

$$Q_{cc}^{\text{benzene}} = Q_\perp = -8.5(14) \times 10^{-26}\ \text{esu}\cdot\text{cm}^2$$

as compared with the value [69] $-5.6(28) \times 10^{-26}$ esu·cm^2 reproduced in the earlier edition of this book.

The molecular quadrupole moments now being measured, with prospects of improved accuracies in future values made possible by newer microwave

Zeeman spectrometers [56] of higher resolving power, offer an incentive for molecular theorists to predict the observed moments with various assumed sets of molecular orbitals. The tailoring of the assumed functions to yield the best fitting of accurately measured moments should provide a better knowledge of the electronic structure of molecules. Such calculations are already being made, but it is outside the scope of this volume to describe them.

Certain qualitative deductions from the moments in Tables 11.5 and 11.7 are noteworthy. For example, the small values of the Q_{mol} elements of H_2CO are due to the fact that the electronic and nuclear contributions which have opposite signs are nearly equal. All Q_{zz} moments are negative for the aromatic planar-ring molecules shown, whereas the Q_{yy} values are all positive. The negative, out-of-plane Q_{zz} values result from the fact that the negative, π-electron clouds are above and below the molecular plane, whereas the positive nuclear charges are in the plane normal to the z axis. The positive Q_{zz} for ethylene oxide (Table 11.7) is due to the semishielded hydrogen nuclei that project above or below the plane of the ring. Note that the Q_{yy} values are positive for all molecules in Table 11.5 except H_2CCF_2. These positive values of Q_{yy} result from the semishielded proton charges which extend in the y direction. In H_2CCF_2 these proton changes are more than counterbalanced by the negatively charged F atoms which also have projections along y.

The molecular quadrupole moments derived from the Zeeman method described here are defined with reference to the center of mass of the molecule (Eq. 11.100), whereas those obtained from electro-optical bulk properties are defined [65] with reference to an "effective quadrupolar center." The two moments [65] differ by $2\mu z_0$ where μ is the electrical dipole moment of the molecule and z_0 is the distance between the two centers.

The line-breadth method for measurement of molecular quadrupole moments, which was developed in the early years of microwave spectroscopy mainly by W. V. Smith and his co-workers [63, 64], gives only the approximate magnitude of Q_{mol}, whereas the Zeeman method provides both its sign and magnitude. With magnets having very high fields and high homogeneity over the rather large volume of a gaseous absorption cell, now becoming available, it seems that the accuracy of the Zeeman method of measurement of Q_{mol} can be further improved and that it will become more widely employed. Already it appears superior to all other known methods for measurement of this important molecular property.

Second Moments of the Electron Charge Distributions

The summed averages, $\sum_i \langle 0|x_i^2|0\rangle$, $\sum_i \langle 0|y_i^2|0\rangle$, $\sum_i \langle 0|z_i^2|0\rangle$, which make up the terms on the left in (11.70)–(11.72) and which are equivalent to the averaged sums, $\langle 0|\sum_i x_i^2|0\rangle$, $\langle 0|\sum_i y_i^2|0\rangle$, $\langle 0|\sum_i z_i^2|0\rangle$, are sometimes written simply as $\langle x^2\rangle$, $\langle y^2\rangle$, and $\langle z^2\rangle$. These second moments of the electron charge distributions give a measure of the averaged extension of the electron clouds in the direction of the principal axes. For example, in a planar, aromatic ringed compound in which the z axis is perpendicular to the plane, $\langle z^2 \rangle$ gives a measure of the extension of the electron cloud above and below the molecular plane.

Equations 11.70–11.72 are easily solved for the second moments of electric charge distributions. For $\langle x^2 \rangle$, one obtains

$$\left\langle 0 \left| \sum_i^{\text{electrons}} x_i^2 \right| 0 \right\rangle = \sum_k^{\text{nuclei}} Z_k x_k^2 + \frac{1}{2M_p}(I_x g_{xx} - I_y g_{yy} - I_z g_{zz}) + \frac{2mc^2}{e^2}(\chi_{xx} - \chi_{yy} - \chi_{zz}) \tag{11.110}$$

from which $\langle y^2 \rangle$ and $\langle z^2 \rangle$ can be obtained by cyclic permutation of the coordinates. In the inertial axes of the asymmetric rotor with a, b, c identified with x, y, z, respectively, and with the moment of inertia expressed in terms of the spectral constants A, B, C, (11.110) transforms to

$$\left\langle 0 \left| \sum_i a_i^2 \right| 0 \right\rangle = \sum_k Z_k a_k^2 + \frac{h}{16\pi^2 M_p}\left(\frac{g_{aa}}{A} - \frac{g_{bb}}{B} - \frac{g_{cc}}{C}\right) + \frac{2mc^2}{e^2}(\chi_{aa} - \chi_{bb} - \chi_{cc}) \tag{11.111}$$

The corresponding formulas for $\langle b^2 \rangle$ and $\langle c^2 \rangle$ are easily obtained by cyclic permutation of the coordinates.

Examination of (11.110) or (11.111) reveals that numerical solutions for the second moments require values for all three susceptibility components. Without a knowledge of χ_{bulk}, the two independent anisotropic susceptibilities measured with the Zeeman effect are insufficient for evaluation of the individual elements, χ_{aa}, χ_{bb}, and χ_{cc}. However, the second-moment differences $\langle b^2 \rangle - \langle a^2 \rangle$, and so on, can be obtained from microwave spectral data alone. By subtraction of (11.71) from (11.70), one obtains

$$\left\langle 0 \left| \sum_i y_i^2 \right| 0 \right\rangle - \left\langle 0 \left| \sum_i x_i^2 \right| 0 \right\rangle = \sum_k Z_k(y_k^2 - x_k^2) + \frac{1}{M_p}(I_y g_{yy} - I_x g_{xx}) + \frac{4mc^2}{e^2}(\chi_{yy} - \chi_{xx}) \tag{11.112}$$

Note that the susceptibility difference in the last term can be expressed in terms of the two measurable anisotropies by

$$(\chi_{yy} - \chi_{xx}) = \tfrac{1}{3}[(2\chi_{yy} - \chi_{xx} - \chi_{zz}) - (2\chi_{xx} - \chi_{yy} - \chi_{zz})] \tag{11.113}$$

With this equivalence and a transformation to the inertial coordinates, (11.112) becomes

$$\left\langle 0 \left| \sum_i^{\text{electrons}} b_i^2 \right| 0 \right\rangle - \left\langle 0 \left| \sum_i^{\text{electrons}} a_i^2 \right| 0 \right\rangle = \sum_k^{\text{nuclei}} Z_k(b_k^2 - a_k^2) + \frac{h}{8\pi^2 M_p}\left(\frac{g_{bb}}{B} - \frac{g_{aa}}{A}\right) + \frac{4mc^2}{3e^2}[(2\chi_{bb} - \chi_{aa} - \chi_{cc}) - (2\chi_{aa} - \chi_{bb} - \chi_{cc})] \tag{11.114}$$

The g values, spectral constants, and susceptibility anisotropies are measurable quantities. Calculation of the nuclear sums (first term on the right) requires a knowledge of the coordinates of all the nuclei in the inertial axis system and hence the complete molecular structure.

Illustrative values of nuclear and electron second moments are given in Table 11.7, together with other related parameters.

6 SIGN OF THE ELECTRIC DIPOLE MOMENT FROM THE ZEEMAN EFFECT

Townes et al. [67] have shown how the rotational g factor, when measured for two isotopic species of a molecule, can be used to give the sign of, and the approximate magnitude of, the electric dipole moment. Although the magnitude can be most accurately measured with the Stark effect, this method does not give the sign. The Zeeman method depends on the fact that the magnetic susceptibility is independent of the origin of the coordinates [13]. One can derive the required formulas most simply by starting with (11.70)–(11.72), which express the coordinates of the electrons and nuclei in the principal inertial system in terms of the diagonal elements of g and χ. Equation (11.70) can be expressed as

$$I_x g_{xx} = M_p \left[\sum_k Z_k (y_k^2 + z_k^2) - \sum_i (0|\, y_i^2 + z_i^2 |0) \right] - \frac{4mc^2 M_p}{e^2} \chi_{xx} \qquad (11.115)$$

Now assume that there is an isotopic substitution which shifts the center of gravity by Δx, Δy, Δz so as to change all the coordinates $y_k \rightarrow y_k - \Delta y$, and so on, and to change I_x and g_{xx} to I'_x and g'_{xx}. If the inertial axes are also principal magnetic axes, χ_{xx} will not change. With the higher-order terms $(\Delta y)^2$ and $(\Delta x)^2$ neglected, the equation for the new isotopic species can be written as

$$I'_x g'_{xx} = M_p \left[\sum_k Z_k (y_k^2 - 2y_k \Delta y + z_k^2 - 2z_k \Delta z) \right.$$
$$\left. - \sum_i (0|\, y_i^2 - 2y_i \Delta y + z_i^2 - 2z_i \Delta z |0) \right] - \frac{4mc^2 M_p}{e^2} \chi_{xx} \qquad (11.116)$$

By subtraction of (11.115) from (11.116) one obtains

$$I'_x g'_{xx} - I_x g_{xx} = -2M_p \left[\sum_k Z_k (y_k \Delta y + z_k \Delta z) - \sum_i (0|\, y_i \Delta y + z_i \Delta z |0) \right] \qquad (11.117)$$

Now

$$\sum_k Z_k e y_k - \sum_i e(0|\, y_i |0) = (\mu_e)_y \qquad (11.118)$$

$$\sum_k Z_k e z_k - \sum_i e(0|\, z_i |0) = (\mu_e)_z \qquad (11.119)$$

where e is the magnitude of the charge on the electron and μ_e signifies the electric dipole moment. Thus (11.117) can be expressed as

$$I'_x g'_{xx} - I_x g_{xx} = -\frac{2M_p}{e}\left[(\mu_e)_y \Delta y + (\mu_e)_z \Delta z\right] \tag{11.120}$$

Similarly,

$$I'_y g'_{yy} - I_y g_{yy} = -\frac{2M_p}{e}\left[(\mu_e)_z \Delta z + (\mu_e)_x \Delta x\right] \tag{11.121}$$

and

$$I'_z g'_{zz} - I_z g_{zz} = -\frac{2M_p}{e}\left[(\mu_e)_x \Delta x + (\mu_e)_y \Delta y\right] \tag{11.122}$$

When the structure of the molecule is known, the moments of inertia and Δx, Δy, and Δz are also known or can be calculated. If the elements of g' are measured, the components of the electric dipole moments can thus be found.

For a linear molecule, $I_z=0$, $I_x=I_y=I$, $(\mu_e)_x=0$, $(\mu_e)_y=0$, $(\mu_e)_z=\mu_e$, $g_{xx}=g_{yy}=g$, and Eqs. (11.120)–(11.122) reduce to the simpler form

$$\mu_e = \frac{-(I'g' - Ig)e}{2M_p \Delta z} \tag{11.123}$$

Table 11.8 Polarity of the Electric Dipole Moment of Some Molecules as Determined from the Zeeman Method

Direction of Moment	*Ref.*	*Magnitude of Moment*[a] *(debye)*
⊖←⊕ CO	[b]	0.112
⊖←⊕ OCS	[c]	0.7152
⊕→⊖ H_2CO	[d]	2.339
⊕→⊖ CH_3CN	[e]	3.913
⊕→⊖ CH_3CCH	[e]	0.7835

[a]For literature source, see Tables 10.9–10.11.
[b]Rosenblum, Nethercot, and Townes [7].
[c]Flygare et al. [20].
[d]Hüttner et al. [17].
[e]Shoemaker and Flygare [24].

With the method described and with Zeeman measurements on the isotopic species $^{12}C^{16}O$, $^{13}C^{16}O$, and $^{14}C^{16}O$, Rosenblum et al. [7] were able to show that the polarity of the electric dipole moment of carbon monoxide is $^{-}C^{+}O$, in opposite direction to that which would be expected from the difference in the electronegativity of the two atoms. Hüttner et al. [17], with the same method, showed that the moment of the CO bond in formaldehyde has the opposite sign, or $H_2{}^{+}C^{-}O$ (H and H bonded to C), as would be expected from the greater electronegativity of oxygen over carbon. Table 11.8 shows the sign of the dipole moments of some molecules that are found in this way. Polarities of many others have now been found with the method [36, 40, 51, 52].

7 ELECTRONIC EFFECTS ON INERTIAL CONSTANTS: RELATION TO g FACTOR

In the calculation of molecular structures from the experimentally observed moments of inertia, the masses of the neutral atoms are employed rather than the fully ionized nuclear masses. Generally, it is assumed that the masses of the electrons associated with a particular nucleus can be considered as concentrated at the nucleus. This is a very good approximation for most molecules, and for most polyatomic molecules is about the only feasible approximation. Especially in the lighter diatomic molecules, higher-order corrections for electronic motions, for L uncoupling, for unequal sharing of the electrons by the atoms (ionic character), and for nonspherical distribution of the electronic clouds about the atoms are sometimes desirable and feasible. The Zeeman effect is very useful in evaluation of these higher-order corrections. If the assumption that the electrons can be lumped with the nuclei were to hold strictly, the magnetic moment caused by the rotation of the electrons would be exactly canceled by the equal rotation of the oppositely charged nuclei, and there would be no rotational magnetic moment. Deviations from this assumption are evidenced through the rotational magnetic moment, or g factor, that is observed.

In making corrections for the electronic effects on the molecular rotation we might add the masses of the inner-shell electrons to the nuclear masses in the usual way and treat separately the effects of the valence-shell electrons, as was done by Jen [5] in treating rotational magnetic moments. However, it is simpler to consider angular momentum of all the electrons separately from that of the nuclei, as was done by Eshbach and Strandberg [3] in treating the rotational magnetic moment and by Rosenblum et al. [7] in their analysis of the microwave spectrum of CO. The rotational angular momentum of the electrons which is due to the end-over-end rotation of the nuclear frame is included in the total L rather than with N. The net result is the same. The changes in the rotational energy are taken into account through the modification of the effective moments of inertia by the electronic interactions.

We start with the rotational Hamiltonian of Eq. (11.56) and consider the molecule to be not in the pure $^1\Sigma$ state $\psi_0^{(0)}$ but in the perturbed state $\psi_0^{(1)}$ which has some electronic momentum. We obtain the correct form for the effective rotational Hamiltonian to second order by averaging $\mathscr{H}$ over the electronic wave function $\psi_0^{(1)}$.

$$\mathscr{H}_{\text{eff}} = (\psi_0^{(1)} | \mathscr{H}_r + \mathscr{H}' | \psi_0^{(1)}) \tag{11.124}$$

Since $\mathscr{H}_r = \frac{1}{2} \sum J_g^2 / I_g$ does not contain the L operator, one has only to evaluate $\mathscr{H}' = -J_g L_g / I_g$. The first-order evaluation of $\mathscr{H}'$ is zero, since

$$(0|\mathscr{H}'|0) = -\frac{J_g}{I_g} \sum_g (0|L_g|0) = 0 \tag{11.125}$$

and the second-order evaluation is

$$\sum_{n \neq 0} \frac{|(n|\mathscr{H}'|0)|^2}{E_n - E_0} = -\sum_{n \neq 0} \sum_g \frac{J_g^2}{I_g^2} \frac{|(n|L_g|0)|^2}{E_n - E_0} \tag{11.126}$$

where cross terms $J_g J_{g'}$ are ignored. Hence the effective rotational Hamiltonian in the perturbed electronic state is, to second order,

$$\mathscr{H}_{\text{eff}} = \frac{1}{2} \sum_g J_g^2 \left(\frac{1}{I_g} - \frac{2}{I_g^2} \sum_{n \neq 0} \frac{|(n|L_g|0)|^2}{E_n - E_0} \right) \tag{11.127}$$

This is equivalent to a rotational Hamiltonian with an effective moment of inertia defined by

$$\frac{1}{(I_g)_{\text{eff}}} = \frac{1}{I_g} - \frac{2}{I_g^2} \sum_{n \neq 0} \frac{|(n|L_g|0)|^2}{E_n - E_0} \tag{11.128}$$

where the I_g values on the right are determined from the nuclear masses only. The quantities in the summation on the right can be obtained from the measured g values by use of (11.67)–(11.69). For diatomic and linear polyatomic molecules, $I_g = I_x = I_y = I$. One finds by combination of (11.73) and (11.128) that

$$\frac{1}{I_{\text{eff}}} = \frac{1}{I} - \frac{m}{I^2} \sum_k Z_k z_k^2 + \frac{mg}{M_p I} \tag{11.129}$$

The effective spectral constant is

$$B_{\text{eff}} = B_n - B_n \frac{m}{I} \sum_k Z_k z_k^2 + B_n \frac{m}{M_p} g \tag{11.130}$$

where B_n represents the spectral constant resulting from the nuclear masses and where g signifies the rotational g factor. Since z_k represents the distances of the different nuclei from the center of gravity and since $M_p Z_k$ represents closely the mass of the kth nucleus, I_n very closely equals $M_p \sum Z_k z_k^2$. With this substitution, the middle term on the right reduces to $(m/M_p)B_n$, and

$$B_{\text{eff}}=B_n-\frac{m}{M_p}B_n+\frac{mg}{M_p}B_n \tag{11.131}$$

It should be remembered that $B_n \equiv h/(8\pi^2 I_n)$ where I_n is the moment of inertia determined from the nuclear masses alone. Addition of the electronic masses to the nuclear masses would give the usual moment of inertia $I_b = \sum_k Z_k z_k^2(M_p+m) = I_n(1+m/M_p)$. The usual B value calculated with atomic masses is therefore

$$B=\frac{h}{8\pi^2 I_n(1+m/M_p)}=B_n\left[1-\frac{m}{M_p}+\left(\frac{m}{M_p}\right)^2-\cdots\right] \tag{11.132}$$

Since $(m/M_p)^2 = 3\times 10^{-7}$, we neglect it and the higher-order terms. Thus

$$B_{\text{eff}}=B+\frac{m}{M_p}gB_n \tag{11.133}$$

or, in general, for a polyatomic molecule

$$B_{\text{eff}}=B+\frac{m}{M_p}g_{bb}B_n \tag{11.134}$$

where the correction for the various rotational constants depends on the corresponding diagonal element of the g tensor. For convenience we can set $B_n = B$ in the last term since the error caused would be less than that in the usual g measurement. The B for the linear molecule can then be expressed as

$$B=\frac{B}{1+(m/M_p)g}\approx B_{\text{eff}}\left(1-\frac{g}{1836}\right) \tag{11.135}$$

where g is the molecular rotational g factor consistent with the magnetic moment (in nuclear magneton units) and where B is the spectral constant corresponding to the moment of inertia of the rigid molecule derived in the usual way from atomic masses (electrons plus nucleus) separated by the internuclear distances. We see from (11.135) that the B value calculated in the usual way from atomic masses deviates from the observed B_{eff} by the factor $[1-(g/1836)]$. This deviation is in addition to that due to centrifugal stretching or the interaction of rotation and vibration considered in earlier chapters. In Dunham's notation one can set $B_{\text{eff}} \approx Y_{01}$. See Chapter IV, Section 1.

For OCS, $g = -0.02871$ (see Table 11.1) and $B = B_{\text{eff}}(1+1.56\times 10^{-5})$. Since $B_{\text{eff}} = 6081.49$ MHz, one must add a correction of only $+0.095$ MHz to obtain B as usually defined. However, for $^{12}C^{16}O$, $g = -0.2691$, $(B_e)_{\text{eff}} = 57{,}898.57$ MHz, and the corresponding correction [7] $B_e = (B_e)_{\text{eff}}(1+1.47\times 10^{-4})$ is $+8.48$ MHz. The very light molecule ^{7}LiD has $g = -0.272$ and a large $B_{\text{eff}} \approx Y_{01} = 126{,}905$ MHz, and the corresponding correction [68] is $+18.8$ MHz. The alkali fluorides have positive g values, and the corresponding dirrection [69] of B for ^{6}LiF is -2.015 MHz; that for $^6Li^{35}Cl$ is -1.369 MHz. It is evident from these examples that this effect on B caused by admixtures of the $^1\Sigma$

ground electronic state with excited Π states, called L uncoupling, is significant for light diatomic molecules. In the determination of precise r_e values or mass ratios it must be taken into account. For most polyatomic molecules it is not likely to be of consequence because of larger uncertainties due to vibrational effects.

8 MOLECULES WITH ELECTRONIC ANGULAR MOMENTUM

The general formula for analysis of the Zeeman effect in molecules with electronic angular momentum is the same as that already described in Section 1 for molecules in Σ states. However, there is now a large magnetic moment associated with the orbital angular momentum $\boldsymbol{\mu}_L = -\beta\mathbf{L}$ and with the spin angular momentum $\boldsymbol{\mu}_s = -g_s\beta\mathbf{S}$. The minus sign indicates that the magnetic moment and the angular momentum vectors are oppositely directed (β positive). When there is no nuclear coupling, $\mathbf{J}$ represents the overall angular momentum, including the electronic (orbital and spin) angular momentum. The resultant magnetic moment resolved along $\mathbf{J}$ or $\langle\mu_J\rangle$, which will be defined here as equal to $-g_J\beta[J(J+1)]^{1/2}$, will be the only component which interacts with the applied field in the first-order, vector-model treatment which applies in the weak-field case. As in Section 1, the allowed Zeeman energies are

$$E_H = -\boldsymbol{\mu}\cdot\mathbf{H} = g_J\beta H M_J \tag{11.136}$$

Here, however, it is more convenient, because of the larger value of the magnetic moments associated with the unpaired electrons, to express $\langle\mu_J\rangle$ in Bohr magnetons rather than nuclear magnetons. Thus β in Eq. (11.136) is the Bohr magneton and g_J is consistent with these units. The expressions for the σ and π components are of the same form as for $^1\Sigma$ molecules, or

$$\nu(\pi) = \nu_0 + \frac{\beta H}{h}(g_{J'} - g_J)M_J \qquad M_J \to M_J \tag{11.137}$$

$$\nu(\sigma) = \nu_0 + \frac{\beta H}{h}(g_{J'} - g_J)M_J \pm \frac{\beta H g_{J'}}{h} \qquad M_J \to M_J \pm 1 \tag{11.138}$$

where ν_0 represents the frequency of the unsplit line when no field is applied and g_J represents the effective g factor for the lower state. The relative intensities of the π and σ components are described in Chapter X, Section 7. Thus these spectra are similar in appearance to those already described for molecules in $^1\Sigma$ states.

The principal differences in the Zeeman effects for molecules with unbalanced electronic moments and those with closed electron shells are the greater sensitivity of the former to the magnetic field and the change in expression of the g factor. The greater sensitivity to the magnetic field, of the order of 10^4 greater, results in a breakdown in the coupling vectors and a failure of the first-order effect with moderate field values. One can usually achieve measurable splitting of the lines with only a few gauss. Expressions for g_J for different coupling cases are given in this section.

Molecules with Orbital Anular Momentum (Π or Δ States)

The form of the g factor depends upon the nature of the coupling between the various component vectors of the angular momentum. The most common coupling cases are described in Chapter IV, Section 2. For diatomic or linear polyatomic molecules having unbalanced orbital momentum (Π or Δ states) Hund's case (*a*) applies, and the electronic magnetic moments (both spin and orbital) are resolved along the molecular axes, as shown in Fig. 4.4. The combined orbital and spin moment along the molecular axis is

$$\mu_\Omega = -\beta(\Lambda + g_s\Sigma) \tag{11.139}$$

where Λ and Σ are orbital and spin quantum numbers defined in Section 4.2 and where $g_s = 2.0023$ is the g factor of the free electron spin. In addition, there will be a very small rotational magnetic moment $\mu_r = g_r\beta[O(O+1)]^{1/2}$ generated along the rotational axis $\mathbf{O}$ like that for Σ states. The components of these moments along $\mathbf{J}$ are easily found from the vector model of Figure 4.4. Accordingly,

$$\begin{aligned} \langle \mu_J \rangle &= \mu_\Omega \cos(\mathbf{\Omega}, \mathbf{J}) + \mu_r \cos(\mathbf{O}, \mathbf{J}) \\ &= -\beta \frac{(\Lambda + g_s\Sigma)|\Lambda + \Sigma|}{[J(J+1)]^{1/2}} + \frac{g_r\beta O(O+1)}{[J(J+1)]^{1/2}} \end{aligned} \tag{11.140}$$

The value $g_J = (-\langle \mu_J \rangle/\beta)/[J(J+1)]^{1/2}$ for weak fields that applies for Π, $\Delta, \ldots,$ states is

$$g_J = \frac{|\Lambda + \Sigma|(\Lambda + g_s\Sigma)}{J(J+1)} - \frac{g_r O(O+1)}{J(J+1)} \tag{11.141}$$

All g's in this expression are defined via (11.136) where the magnetic energy is expressed in terms of the Bohr magneton. Usually the last term is negligible. Note that this g_J factor with (11.136) indicates that the Zeeman splitting of the energy levels decreases rapidly as J increases.

For a $^2\Pi_{1/2}$ state $\Lambda = 1$ and $\Sigma = -\frac{1}{2}$, or $\Lambda = -1$, $\Sigma = \frac{1}{2}$. The orbital moments and spin moments are opposed, and since $g_s = 2.0023$ they cancel to a high order of approximation, leaving the molecule "nonmagnetic" although it has unbalanced angular momentum $\Omega = \frac{1}{2}$. An interesting example of this is NO, which has a $2\Pi_{1/2}$ ground state. Equation (11.141) indicates a g_J value of zero and hence no significant Zeeman effect except at fields of a few thousand gauss for which the small rotational moment is sufficient to give a splitting. Nevertheless, NO in the $^2\Pi_{1/2}$ state has been found [70] to have a resolvable Zeeman splitting for fields of the order of 100 G. This pronounced Zeeman effect results primarily from admixtures of the $^2\Pi_{1/2}$ ground state with the $^2\Pi_{3/2}$ magnetic state which is only 123 cm^{-1} above the ground state. It is further complicated by ^{14}N hyperfine structure and by near-degeneracy in the Λ doubling. The theory developed by Mizushima et al. [70] for this case is applicable to other molecules in $^2\Pi_{1/2}$ states.

Molecules with Electronic Spin but Not Orbital Angular Momentum ($^2\Sigma$ or $^3\Sigma$ States)

Molecules that have electronic spin angular momentum with no orbital momentum, that is, those in $^2\Sigma$ or $^3\Sigma$ states, for which $\Lambda=0$ but $S\neq0$, generally qualify for Hund's case (*b*), Chapter IV, Section 2. When Λ is zero, O equals N, and the end-over-end rotation is customarily designated by N. Since there is no orbital moment to couple the spin moment to the internuclear axis, **S** becomes coupled to **N** through the weak moment μ_N generated by the molecular rotation. The vector model for this case is shown in Fig. 4.6. To find the effective g_J one must resolve the spin moment $\mu_s=-g_s\beta[S(S+1)]^{1/2}$ and, to be exact, must also resolve the weaker rotational moment $\mu_N=g_r\beta[N(N+1)]^{1/2}$ along **J**. This can be done simply by application of the law of the cosines to the vector model of Fig. 4.6. With the relation $\langle\mu_J\rangle=\mu_s \cos(\mathbf{S},\mathbf{J})+\mu_N\cos(\mathbf{N},\mathbf{J})=-g_J\beta[J(J+1)]^{1/2}$, the resultant g factor is found to be

$$g_J=\left(\frac{g_s}{2}\right)\frac{J(J+1)+S(S+1)-N(N+1)}{J(J+1)}-\left(\frac{g_r}{2}\right)\frac{J(J+1)+N(N+1)-S(S+1)}{J(J+1)} \tag{11.142}$$

in which both g_r and g_s are consistent with μ_J expressed in Bohr magneton units. Since $g_s=2.0023$ and $g_r\sim10^{-4}$, this factor is, to a good approximation,

$$g_J=\frac{J(J+1)+S(S+1)-N(N+1)}{J(J+1)} \tag{11.143}$$

Although μ_N can usually be neglected in calculation of the Zeeman splitting, indirectly it is an essential factor in the coupling of **S** to **J**.

Zeeman measurements with weak fields, of less than 100 G, for a number of millimeter wave lines of O_2 ($^3\Sigma$ state) have been made [71]. Similar measurements have been made on the $N=3\rightarrow4$, $J=2\rightarrow3$ transition [72] of SO ($^3\Sigma$ state) at 159 GHz. In all instances the observed splittings were found to be in agreement with those predicted from (11.136) with the g factor from (11.143).

Although the simple, first-order treatment presented is adequate for very weak fields, moderate or strong fields cause a decoupling of the vector and a failure of the theory. A complete treatment that applies for all field strengths to molecules in Σ states is given by Tinkham and Strandberg [73].

Hyperfine Structure in Molecules with Electronic Angular Momentum

Zeeman effect of hyperfine structure in molecules with electronic angular momentum is treated in the same way as for molecules in $^1\Sigma$ states, Section 2. As before for the weak-field case, **J** forms with the nuclear spin **I** a resultant **F**.

For a single coupling nucleus, the energies are given by (11.39) with α_J and α_I given by (11.36) and (11.37). Molecules with two coupling nuclei are treated similarly with (11.32) and (11.42). The only difference is that one must use for g_J the expressions appropriate to the molecule with electronic angular momentum, (11.141) or (11.142), and a slight modification must be made for the change in the magnetic energy expression. Compare (11.136) and (11.32) with respect to sign and units. Molecules with electronic angular momentum will, of course, have strong magnetic coupling to nuclei with nonzero spins. This will not be easily broken down by the applied field.

When the strength of the magnetic field is increased to the point where the weak-field theory begins to fail, the problem becomes more complex and must be treated with intermediate-field theory. The degree of difficulty will depend on the relative magnitude of the couplings of the different vectors. Only those which are partially decoupled need be treated with intermediate-field theory. Because of the complexity of the problem and its rather limited applications we shall not treat the intermediate-field case. The simple Paschen-Back case can frequently be achieved in paramagnetic resonance experiments such as those described later.

9 GAS-PHASE ELECTRON PARAMAGNETIC RESONANCE SPECTRA

As was first demonstrated in the pioneering work of Beringer and his associates [74–79], transitions between Zeeman components of a particular rotational level—electron paramagnetic resonance—can be detected for gaseous molecules having electronic angular momentum. Although there are only a few stable molecules having ground states with electronic moments, the most notable of which is oxygen, potentially there are numerous unstable free radicals which might be studied to advantage through gas-phase paramagnetic resonance. The first observation of the paramagnetic resonance of an unstable free radical in the gas phase was accomplished by Radford [78] in 1961 on the OH radical previously studied in the microwave region through Λ doublet transitions by Dousmanis et al. [79]. Since that time several other unstable radicals have been similarly investigated, as will be described. Experimental difficulties which delayed the development of this branch of microwave spectroscopy have been mostly overcome. The remaining difficulties are largely chemical, involving the preparation and handling of the unstable radical species. The method appears very promising not only for the finding of structural properties of unstable radicals but as an aid in the investigation of certain chemical reactions in the gas phase.

Since for the gas phase one observes only the molecules that populate a particular rotational or fine-structure level, the fraction of molecules of a sample that contributes to a particular resonance signal is small. This contrasts with electron spin resonance of the solid state where there is often little or no fine structure and where essentially all the free radicals in the sample may contribute to the signals. To offset this disadvantage, paramagnetic resonance transitions in the gas phase can be induced through electric dipole moments [77] due to the coupling of the electronic angular momentum to the molecular axis. Magnetic dipole transitions can also be used for detection of paramagnetic resonance of the gas phase, as was achieved [74] for O_2, but these transitions are weaker than those induced by the electric dipole; a magnetic moment of 1 Bohr magneton is about 100 times smaller than an electric moment of 1 debye. For more complex molecules where there are many gas-phase levels to be populated, detection of the resonance through magnetic dipole coupling is impractical, especially in unstable free radicals. Probably this explains why there has been no observation of paramagnetic resonance in the gas phase of the important free radical CH_3 which is planar and has no electric dipole moment. Most gaseous free radicals are electrically polar, however. As will be seen from the discussion to follow, electric dipole transitions not only enhance detectability of the resonance but also give more information about the radical than do magnetic dipole transitions.

Molecules with Orbital Angular Momentum

When there is orbital angular momentum, the coupling of the electronic moments to the internuclear axis is so strong, Hund's case (*a*), that one can usually achieve the strong-field, Paschen-Back case for the nuclear coupling without significant decoupling of the electronic vectors. Then **J** and **I** precess separately about the direction Z of the imposed field. One can represent the Hamiltonian of the interaction of **J** and **I** with the field H by

$$\mathscr{H}_H = -\langle \mu_J \rangle \frac{\mathbf{J}\cdot\mathbf{H}}{|\mathbf{J}|} - \frac{|\boldsymbol{\mu}_I|\mathbf{I}\cdot\mathbf{H}}{|\mathbf{I}|} \tag{11.144}$$

where the average of μ_J is taken over the electronic state. For case (*a*) coupling, the value of $\langle \mu_J \rangle$ can be found from the vector model of Fig. 4.4 as was done in Section 8. The value of the electronic moment μ_Ω along the molecular axis is, however, not exactly $-\beta(\Lambda + g_s\Sigma)$. The components of μ along the molecular axis, and consequently along J, are slightly different for the two Λ doublet states described by the zero-order functions ψ_0^+ and ψ_0^- which are linear combinations of basis functions of the type

$$\frac{1}{\sqrt{2}}[\psi(\Lambda, \Sigma, \Omega, J, M_J, I, M_I) \pm \psi(-\Lambda, -\Sigma, -\Omega, J, M_J, I, M_I)]$$

We signify the moments of these two states by $\langle \mu_J \rangle^{\pm} = -g_J^{\pm}\beta J(J+1)$, where g_J^+ and g_J^- differ slightly, with their average value being given approximately by (11.141). From the vector model of Fig. 10.7, the first-order magnetic energy is given by

$$E_H = g_J^{\pm}\beta H M_J - g_I \beta_I H M_I \tag{11.145}$$

Because of the selection rules $\Delta M_I = 0$, the last term has no effect on the observed spectra and will be omitted from further discussion. Higher-order evaluations of the $\mu_J \cdot \mathbf{H}$ term must be included, however, for an interpretation of the observed spectra. These higher-order perturbation terms, although not large, produce a fine structure in the observed spectra by causing the Zeeman levels which have equal separation in first order to have slightly different spacing in the higher-order approximations. The higher-order terms have been calculated for NO by Margenau and Henry [80] and for OH by Radford [78]. The corrections as expressed by Radford are

$$\Delta E_H = (K_0 + K_2 M_J^2)\frac{(\beta H)^2}{hc} + (K_1 M_J + K_3 M_J^3)\frac{(\beta H)^3}{(hc)^2} \tag{11.146}$$

where the K's (dimensionless) are constant for a particular electronic and rotational state. They can be expressed in terms of the molecular quantum numbers and rotational constants, but the expressions are complicated and not always useful for finding molecular properties because the perturbations are generally so small that the K's cannot be evaluated with accuracy. The K_0 term is independent of M_J and has no effect on the spectra. The terms in H^3 and higher are generally negligible. Hence, in the following development we shall set $\Delta E_H = K_2 M_J^2 (\beta H)^2/hc$ and shall treat K_2 simply as an empirical constant to be evaluated from the spectra. Hence, the important magnetic terms are

$$E_H = g_J^{\pm} \beta H M_J + \frac{K_2 \beta^2 M_J^2 H^2}{hc} \tag{11.147}$$

The nuclear interactions can be calculated with sufficient accuracy from first-order theory of the strong-field case (decoupled representation) in which I is resolved along the space-fixed Z axis. The nuclear magnetic interactions involve the interaction of the nuclear moments with the magnetic fields associated with the electronic orbital angular momentum and the spin angular momentum. The expressions for the magnetic hyperfine energy are similar to those discussed for the case of the strong-field Stark effect, (10.44). The magnetic interaction has the simple form $A_I M_J M_I$, where A_I represents the nuclear magnetic coupling constant. However, the magnetic coupling constant is measurably different for the two Λ doublet levels of a given J level. This can be shown from an average of the nuclear coupling Hamiltonian over the electronic wave functions ψ_0^+ and ψ_0^- which represent the two Λ doublet states. The difference can be taken into account most simply by replacement of A_I in the strong-field formula by $A_I^+ = A_I + \Delta A_I$ for the ψ_0^+ state and by $A_I^- = A_I - \Delta A_I$ for the ψ_0^- state. The quantity ΔA_I can be considered as the hyperfine doubling constant. It is generally very small in comparison with A_I, which represents the mean coupling constant for the two Λ doublet states. We shall not describe the relation of these coupling constants to the electronic orbitals and nuclear moments. This relationship is developed in the basic paper by Frosch and Foley [81]. Further elucidation of the theory of hyperfine structure in free radicals is given by Dousmanis et al. [79].

As described in Chapter IV, Section 2, the diatomic molecule in coupling case (*a*) simulates the motion of a symmetric top with an angular momentum along the symmetry axis of $\Omega=|\Lambda+\Sigma|$. Consequently, the first-order, strong-field quadrupole coupling energy can be obtained by a simple replacement of K by $(\Lambda+\Sigma)$ in the strong-field formula, (10.41), already developed for the symmetric top. Thus

$$E_Q(M_J M_I)=eQq^{\pm}\left[\frac{3(\Lambda+\Sigma)^2}{J(J+1)}-1\right]\frac{[3M_J^2-J(J+1)][3M_I^2-I(I+1)]}{4(2J-1)(2J+3)I(2I-1)} \tag{11.148}$$

where the $\pm$ sign is an indication that the two doublet states may have a slightly different coupling constant.

The total energy of the molecule can now be expressed as

$$E^{\pm}=E_0\pm\tfrac{1}{2}E_{\Lambda}+g_J^{\pm}\beta HM_J+\frac{K_2(\beta HM_J)^2}{hc}+\sum_i A_{I_i}^{\pm}M_J M_{I_i}+\sum_i E_{Q_i}(M_J M_{I_i}) \tag{11.149}$$

where $E_0\pm\frac{1}{2}E_{\Lambda}$ represent the energy of the molecule exclusive of nuclear and magnetic field perturbations, E_{Λ} being the Λ doubling energy. The $\pm$ superscripts refer to the symmetry of the overall wave functions for the two doublet states. The summations are taken over all coupling nuclei. When the Λ doubling is small, as in most molecules in Π and Σ states (the light OH is an exception), one can neglect the small difference of g_J and A_I, and for the two Λ doublet states and can set $g_J^{\pm}=g_J$, $A_I^{\pm}=A_I$, and $q^{\pm}=q$ in (11.149).

The selection rules for magnetic resonance spectra are

For magnetic dipole transitions (observed when the applied field H is perpendicular to the magnetic vector of the radiation)

$$\Delta J=0, \qquad +\leftrightarrow+, \qquad -\leftrightarrow-, \qquad \Delta M_J=1, \qquad \Delta M_I=0 \tag{11.150}$$

For electric dipole transitions (observed when H is perpendicular to the electric vector of the radiation)

$$\Delta J=0, \qquad +\leftrightarrow-, \qquad \Delta M_J=1, \qquad \Delta M_I=0 \tag{11.151}$$

The electric dipole transitions thus take place between the Zeeman levels of the two Λ doublet components, whereas the magnetic dipole transitions take place between the Zeeman levels in the two Λ doublet states. For the magnetic dipole transitions it is evident from Eq. (11.149) that neither E_0 nor E_{Λ} affect the observed frequencies, whereas for the electric dipole transitions the Λ doublet frequency is superimposed upon the magnetic resonance frequencies. The Λ doublet components of the resonance are separated by $2\nu_{\Lambda}$, where ν_{Λ} is the transition frequency between the Λ doublet levels.

The parity selection rules for magnetic dipole transitions given in (11.150) follow from the discussion in Chapter III, Section 4 if it is noted that the magnetic dipole operator is invariant under inversion (a pseudovector) in contradistinction to the electric dipole operator.

In the original experiments by Beringer and Castle [76] on the $^2\Pi_{3/2}$ state of the $J=\frac{3}{2}$ level of NO, only the magnetic dipole transitions were observed. The observed g_J value was found to be approximately $\frac{4}{5}$ as predicted from (11.141). Later the experiment was repeated with the cell arranged for observation of the electric dipole transitions, and the Λ doublet frequency for the $J=\frac{3}{2}$ level was found to be 1.7 MHz [77]. The small value of the Λ doubling constant accounts for the fact that g_J agrees well with the value predicted from Eq. (11.141), or that Hund's case (*a*) applies rather closely for this state.

Radford [78, 82, 83] has observed electric-dipole-induced paramagnetic resonance spectra of the $J=\frac{3}{2}$ level in the $^2\Pi_{3/2}$ ground electronic state of the unstable hydride free radicals OH and OD, SH, SeH, and TeH. The SH and also SD resonance have likewise been observed by McDonald [84]. Figure 11.10 shows the energy level diagram for OH and indicates the electric dipole transitions which were observed. The parity of the levels is also shown. The dotted arrows indicate the magnetic dipole transitions which were not observed in Radford's experiment. The energy levels and transitions are similar for the other hydride radicals except that the L uncoupling and hence the Λ doublet separation is much the greater for OH. Figure 11.11 shows the observed spectra for SH. The wider doublet splitting is the Λ doublet splitting (equivalent to $2\nu_\Lambda$). The triplet substructure arises from the second-order term in the Zeeman energy which leads to slightly unequal separations of the four Zeeman levels. Each component of the triplet is split into a doublet by the proton magnetic hyperfine interaction.

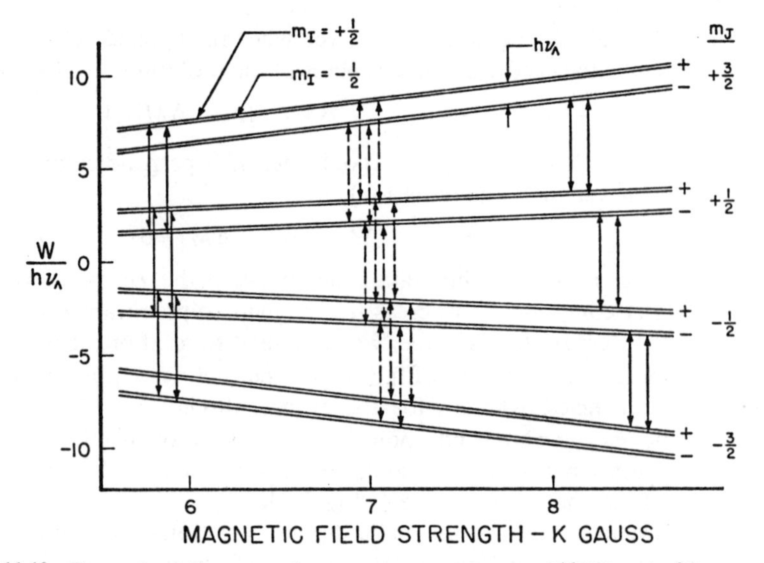

Fig. 11.10 Energy level diagram and paramagnetic transitions of ^{16}OH in the $^2\Pi_{3/2}$, $J=\frac{3}{2}$ state. Electric dipole transitions are indicated by solid arrows; magnetic dipole transitions, by dotted lines. From Radford [78].

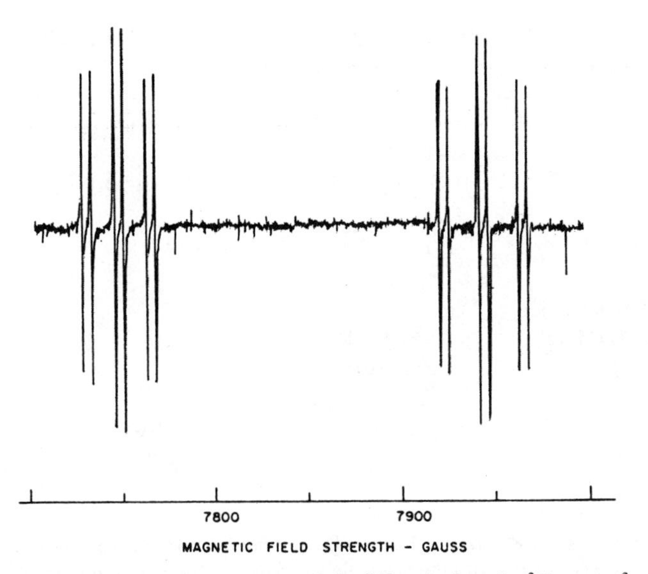

Fig. 11.11 Paramagnetic resonance spectrum of the SH radical in the $^2\Pi_{3/2}$, $J=\frac{3}{2}$ ground state at a frequency of 9216 MHz. From Radford and Linzer [82].

Tables 11.9 and 11.10 summarize the observed and derived constants for the hydrides. The rotational constants B_0 and the spin–orbit coupling constant A are derived from the second-order perturbation terms. Details of the analysis are given in the paper on the OH radical [78]. The OH radical is the only one for which significant differences occur for the two Λ states. Note that the g_J factor for OH also deviates most widely from the theoretical value of $\frac{4}{5}$ predicted from (11.141).

In addition to the early work on NO already mentioned [76], microwave magnetic resonance spectra have been observed for several heavier (non-hydride) diatomic free radicals in $^2\Pi_{3/2}$ states, including CF, NS, ClO, BrO,

Table 11.9 Molecular Constants of the $^2\Pi_{3/2}$, $J=\frac{3}{2}$ Ground States of OH and OD as Derived from Paramagnetic Resonance Spectra[a]

Constant		OH	OD
Molecular g factor[b]	g_J^+	0.93493 ± 0.00003	0.88920 ± 0.00003
	g_J^-	0.93622 ± 0.00003	0.88971 ± 0.00003
Λ Doubling frequency	Λ (MHz)	$1{,}666.34 \pm 0.10$	310.12 ± 0.08
Second-order Zeeman	K_2	$(3.13 \pm 0.03) \times 10^{-3}$	$(3.78 \pm 0.02) \times 10^{-3}$
Magnetic coupling	A_I (MHz)	27.01 ± 0.05	4.84 ± 0.03
	ΔA_I (MHz)	0.51 ± 0.05	0.03 ± 0.03

[a]From Radford [78].
[b]The g factors are consistent with the magnetic moment expressed in Bohr magneton units.

Table 11.10 Molecular Constants of the $^2\Pi_{3/2}$, $J=\frac{3}{2}$ Ground States of SH, SeH, and TeH as Derived from Paramagnetic Resonance Spectra[a]

Constant	SH	SeH	TeH
Molecular g factor[b] $\bar{g}_J$	0.83792 ± 0.00002	0.80800 ± 0.00010	0.80366 ± 0.00020
Λ doubling frequency ν_Λ (MHz)	111.42 ± 0.04	14.4 ± 0.1	6.6 ± 0.2
Magnetic coupling constant A_I (MHz)	5.61 ± 0.04	1.9 ± 0.1	1.8 ± 0.2
ΔA_I (MHz)	0.08 ± 0.04		
Rotational constant B_0 (cm^{-1})		7.98 ± 0.08	5.56 ± 0.15
Spin-orbit constant A_0 (cm^{-1})		$-1{,}600 \pm 50$	$-2{,}250 \pm 200$

[a]Data for SH come from Radford and Linzer [82]; those for SeH and TeH, from Radford [83].
[b]The g factors are consistent with the magnetic moment expressed in Bohr magneton units.

IO, SF, SeF, and LiF. Journal references and a description of some of these spectra may be found in Carrington's monograph on free radicals [85]. The LiO was measured [86] with the molecular beam, electric resonance method. Their spectra are similar to those of the hydrides mentioned previously except that the nuclear quadrupole interaction is present in some and the Λ doubling is generally smaller. The spectrum for ClO has been analyzed in detail [87]. Figure 11.12 shows the resonance for ClO for which the Λ doubling is too small to be resolved. The gross triplet splitting is due to the unequal separation of the four Zeeman levels of the $J=\frac{3}{2}$ state; the substructure is due to the ^{35}Cl and ^{37}Cl nuclear interactions. The four stronger components of each submultiplet are due to ^{35}Cl, with abundance of 75%; the weaker lines, due to ^{37}Cl, are not well resolved. Both isotopes have spins $I=\frac{3}{2}$, and thus both have a quartet hyperfine

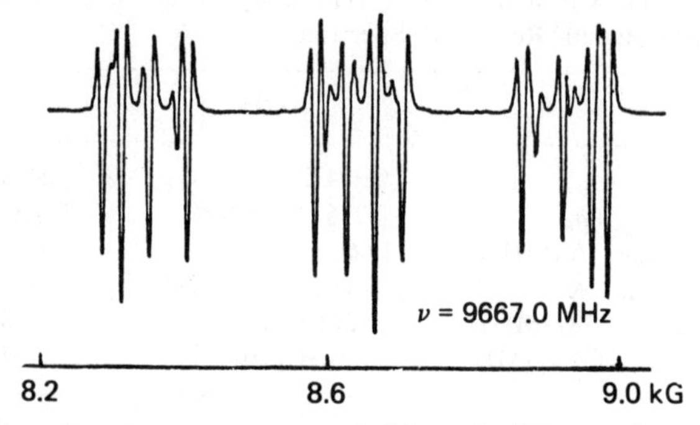

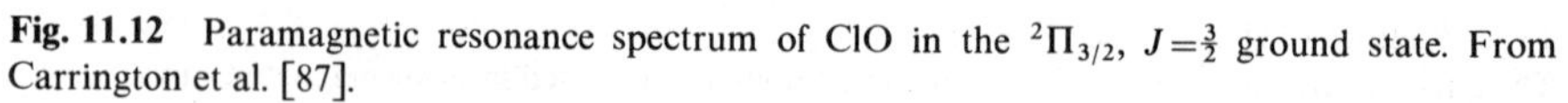

Fig. 11.12 Paramagnetic resonance spectrum of ClO in the $^2\Pi_{3/2}$, $J=\frac{3}{2}$ ground state. From Carrington et al. [87].

structure. The nuclear splitting is due partly to magnetic and partly to electric quadrupole interaction. The inequality of the hyperfine spacing of the two outer multiplets results from the quadrupole interaction. The derived constants for this molecule are listed in Table 11.11. The Λ doubling is not evident in this radical. Carrington et al. [87] give the basic Hamiltonians for the different interactions and the matrix elements in the coupled representation ($\mathbf{F}=\mathbf{I}+\mathbf{J}$) required for derivation of the secular equations for free radicals of this type. Solution of these equations gives the energies in terms of functions which explicitly involve B_0 and the spin-orbit constant A. However, the general solution cannot be represented in such a compact form as (11.149). Solutions of the secular equation are most simply obtained for particular numerical values of the various quantum numbers of the basis functions.

One free radical, SO, has been observed [88] in the excited $^1\Delta$ state with $J=2$. Analysis of the $^1\Delta$ state is simpler than that for the $^2\Pi_{3/2}$ state since there is no spin moment and hence no spin–orbit coupling to include. Furthermore, the Λ doubling of a Δ state is much smaller than that of a Π state, and its effects on the constants are generally negligible. The energy can be represented in the strong-field case by (11.149) with $g_J^{\pm}=g_J=4/J(J+1)$, $A_I^{\pm}=A_I$, and with $E_\Lambda \approx 0$. Matrix elements for solution of the secular equations are given by Carrington et al. [88]. From the analysis of the magnetic resonance spectra they derive the internuclear distance of SO in the $^1\Delta$ state to be $r_0=1.493$ Å compared with $r_0=1.484$ Å for the ground $^3\Sigma^-$ state.

Paramagnetic resonances of a few unstable triatomic free radicals in the gaseous state have been reported. Those for the linear species NCO and NCS were observed by Carrington et al. [89, 90] in their $^2\Pi_{3/2}$ ($v_2=0$) ground states, in the $^2\Delta_{5/2}$ ($v_2=1$) excited states, and in their lowest rotational states.

Table 11.11 Molecular Constants of the $^2\Pi_{3/2}$, $J=\frac{3}{2}$ Ground States of $^{35}Cl^{16}O$ and $^{37}Cl^{16}O$ as Derived from Their Paramagnetic Resonance Spectra[a]

Constant	$^{35}Cl^{16}O$	$^{37}Cl^{16}O$	*Ref.*
Molecular g factor[b]	0.798 ± 0.01	0.798 ± 0.01	[c]
Rotational B_0 (cm^{-1})	0.622 ± 0.001	0.611 ± 0.001	[d]
Λ doubling frequency ν_Λ	~ 0	~ 0	[d]
Spin-orbit A (cm^{-1})	-282 ± 9	-282 ± 9	[d]
Magnetic coupling A_I (MHz)	111 ± 2	93 ± 2	[d]
Quadrupole coupling eQq (MHz)	-88 ± 6	-69 ± 6	[d]

[a]For more extensive results on ClO, see Amano et al. [108]. For results on BrO, see Powell and Johnson [109].

[b]The g factors are consistent with the magnetic moment expressed in Bohr magneton units.

[c]From A. Carrington and D. H. Levy, *J. Chem. Phys.*, **44**, 7500 (1966).

[d]From Carrington, Dyer, and Levy et al. [87].

Resonances for the nonlinear HCO and DCO radicals have been detected by Bowater et al. [91, 92]. The spectra of these triatomic radicals are described in some detail in Carrington's monograph [85]. Rotational transitions of a number of triatomic radicals, including NCO, NCS, and HSO, have been measured by Saito et al. [93–95].

Microwave EPR as well as the laser magnetic resonance of the slightly asymmetric rotor HO_2 has been investigated by Barnes et al. [96], following earlier infrared laser magnetic resonance studies of the radical by Radford et al. [97, 98]. With the aid of their results, the $1_{00} \rightarrow 0_{00}$ pure rotational transition was measured at the 65 GHz microwave resonance by Beers and Howard [99]. Other rotational transitions were measured in the 30 to 140 GHz region by Saito [100]. More recently, Charo and De Lucia [101] carried out extensive measurements on the rotational spectra of H_2O in the 150 to 550 GHz millimeter and submillimeter range. As a result of these various studies, the spectral constants of this important peroxide radical are accurately known. A tabulation of the various constants may be found in the paper by Charo and De Lucia [101].

Three millimeter wave transitions of the triatomic free radical PH_2 in the ground vibrational state have been measured and analyzed by Endo et al. [102]. Accurate values of the rotational constants, the spin–rotation constants, and the P and H nuclear coupling constants were obtained. Cook et al. [103] have derived the spin–rotation constants and the ^{14}N and ^{1}H nuclear couplings of the related NH_2 radical from microwave–optical double resonance (MODR).

Molecules in Multiplet Σ States

For molecules in multiplet Σ states, $\Lambda = 0$ but $S \neq 0$, Hund's case (*b*) applies. **S** is coupled to **N** through the rotational magnetic moment. For $N = 0$ states, this coupling vanishes, and one can observe essentially a free electron spin resonance. For low-J states of heavy free radicals such as NO_2, the coupling is so weak that it is broken down by the magnetic fields required for observation of paramagnetic resonance in the centimeter wave region. In this case, **N**, **S**, and **I** all precess independently about the direction of the field, and the energies are given to first order by

$$E = E_0 + g_s \beta H M_S + \gamma M_N M_S + \sum A_{I_i} M_S M_{I_i} + \sum E_{Qi}(M_N, M_{I_i}) \qquad (11.152)$$

where E_0 is the energy of the molecule in the absence of a magnetic field, of electron spin coupling, and of nuclear coupling. In the above formula the contribution of the molecular rotational moment is neglected. The selection rules for paramagnetic resonance are $\Delta M_S = 1$, $\Delta M_N = 0$, and $\Delta M_I = 0$. The term $g_s \beta H M_S$ represents the principal electron resonance energy where $g_s = 2.0023$. The term $\gamma M_N M_S$ represents a fine structure arising from the interaction between the Z components of the molecular rotational magnetic moment and the electron spin moment. The last terms represent the nuclear magnetic coupling and the nuclear quadrupole coupling energies in the strong-field representation.

This equation has been found to account adequately for the paramagnetic resonance of $^{14}NO_2$ observed by Castle and Beringer [75] with a magnetic field of 15,000 G. The ground state of $^{14}NO_2$ is $^2\Sigma$. For this state $S=\frac{1}{2}$, and for ^{14}N, $I=1$.

Paramagnetic resonance spectra have been observed for O_2 [74] and for SO [84, 104] and for SeO [105] in the $^3\Sigma$ ground state. For these diatomic molecules the coupling between N and S is so strong that the decoupling between N and S is not achieved completely by fields required for observation of the resonance at microwave frequencies, and the simple theory described above is not adequate. A rigorous, intermediate-field theory for the paramagnetic resonance of molecules in multiplet Σ states has been developed by Tinkham and Strandberg [73] and applied by Daniels and Dorain [104] in their analysis of SO. We shall not describe this rather complex theory here. Most of the information derived from the magnetic resonance of O_2 and SO has been obtained more accurately from measurement of fine structure transitions of the field-free molecules.

10 STARK EFFECT OF GAS-PHASE ELECTRON PARAMAGNETIC RESONANCE

Carrington et al. [105] have observed the Stark effect of electron magnetic resonance and have thus derived values of the electric dipole moments of a number of unstable gaseous free radicals. Their experiments are of significance not only because they provide values of dipole moments of unstable species which were not previously known, but also because the method appears to be a promising one for other such measurements and for Stark modulation of gas-phase paramagnetic resonance lines as an aid to detection.

For case (*a*) coupling, the weak-field Stark effect of strong-field magnetic resonance is relatively simple. This case applies to most molecules in Π and Δ states and hence to most free radicals so far observed with gas-phase electron magnetic resonance. A brief treatment of this case will be given.

With the weak electric field superimposed on a strong magnetic field, the Stark effect can be treated as a perturbation upon the Zeeman states, but the near-degeneracy of the Λ doubling prevents treatment of the problem by the simple, first-order perturbation theory. The problem is not difficult, however, and is like that already solved for two nearly degenerate inversion doublets of symmetric tops (Chapter X, Section 5). There, the Stark-perturbed energy levels of two nearly degenerate levels of energy E_1 and E_2 were shown to be

$$E=\tfrac{1}{2}(E_1^0+E_2^0)\pm\tfrac{1}{2}[(E_1^0-E_2^0)^2+4\mathscr{E}^2|(1|\mu|2)|^2]^{1/2} \qquad (11.153)$$

where $\mathscr{E}$ is the electric field and $(1|\mu|2)$ represents the dipole moment matrix elements connecting the nearly degenerate states (1) and (2) having zero-field energies E_1^0 and E_2^0. The required matrix elements can be obtained from those of (10.101) by identification of Ω with K of the symmetric-top rotor. This identification holds since both Ω and K represent the quantized rotational momentum about the symmetry axis, the axis of the permanent electric dipole moment μ. Thus

$$(1|\mu|2)=\mu(J, \Omega, M_J, 1|\Phi_{Zz}|J, \Omega, M_j, 2)=\frac{\mu\Omega M_J}{J(J+1)} \tag{11.154}$$

To simplify the discussion we shall omit the hyperfine structure, but any hyperfine components will be superimposed on each level, as previously described in Section 9. Now we can identify

$$E_1^0=E_0+E_M+\tfrac{1}{2}E_\Lambda \tag{11.155}$$

$$E_2^0=E_0+E_M-\tfrac{1}{2}E_\Lambda \tag{11.156}$$

where E_0 is the energy of the molecule exclusive of Λ doubling and without applied fields either magnetic or electric and where E_M is the magnetic resonance energy. As before, E_Λ represents the energy separation of the Λ doublet levels of the particular state being considered. Obviously,

$$\tfrac{1}{2}(E_1^0+E_2^0)=E_0+E_M \tag{11.157}$$

and

$$E_1^0-E_2^0=E_\Lambda \tag{11.158}$$

Therefore (11.153), for the levels in the presence of the Stark field, becomes

$$\begin{aligned} E&=E_0+E_M\pm\left\{\tfrac{1}{2}E_\Lambda^2+\left[\frac{2\mu\mathscr{E}\Omega M_J}{J(J+1)}\right]^2\right\}^{1/2} \\ &=E_0+E_M\pm\left\{\left[\frac{h\nu_\Lambda}{2}\right]^2+\left[\frac{\mu\mathscr{E}\Omega M_J}{J(J+1)}\right]^2\right\}^{1/2} \end{aligned} \tag{11.159}$$

where ν_Λ represents the Λ doublet frequency. Since E_0 is independent of M_J, it has no effect on the magnetic resonance frequencies to first order. The value of E_M can be obtained from Eq. (11.149). When the Λ doublet splitting is negligible, as is true for ClO, BrO, and for most radicals in Δ states, a simple, first-order Stark effect is observed, like that for a symmetric top. This can be seen if E_Λ is set equal to zero into Eq. (11.159), which then becomes

$$E=E_0+E_M\pm\frac{\mu\mathscr{E}\Omega|M_J|}{J(J+1)} \tag{11.160}$$

Most of the free radicals studied by Carrington et al. were in $^2\Pi_{3/2}$ states with no end-over-end rotation, $\Omega=J=\frac{3}{2}$. Figure 11.13 shows the energy levels and indicates the transitions for this important case when the strong magnetic field is constant and the electric field is varied. Compare this with Fig. 11.10 for the same state when there is no electric field applied and when the strong magnetic field is varied.

In addition to the electric dipole transitions, $+\leftrightarrow-$, normally observed, magnetic-dipole type transitions, $-\leftrightarrow-$ and $+\leftrightarrow+$, are induced when a Stark field is applied, even with the magnetic field parallel to the magnetic component

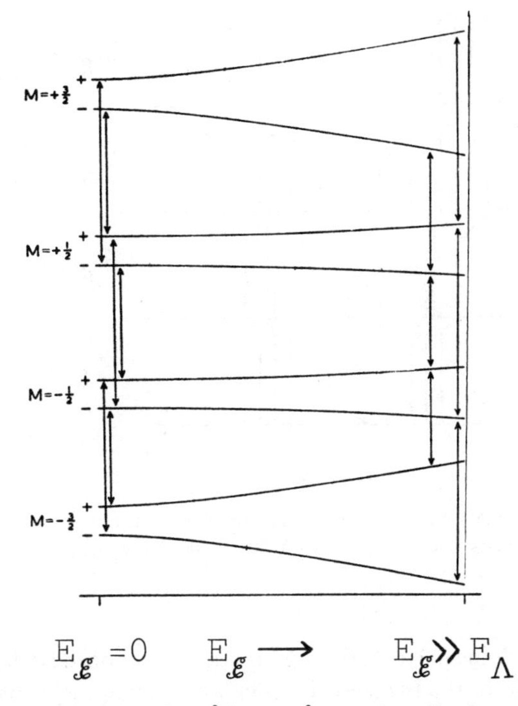

Fig. 11.13 Energy level diagram for a $^2\Pi_{3/2}$, $J=\frac{3}{2}$ state of a molecule in a strong, fixed magnetic field and a variable electric field. The arrows indicate transitions observed with fixed magnetic field perpendicular to the electric vector of the radiation. From Carrington et al. [106].

of the microwave radiation. These transitions are caused by an induced admixture of the + and − doublet states by the Stark field. Such transitions are weak for very weak Stark fields but increase as the Stark field increases. In the absence of an electric field, the magnetic dipole transitions would show no Λ doublet splitting, but because of the unequal displacement of the levels by the Stark field they, for $J=\frac{3}{2}$ states, give rise to a pair of weak doublets at low electric fields which transforms into a strong, more widely spaced pair of doublets at high electric fields. The transitions indicated on the right in Fig. 11.13 are the transitions which are the predominant ones observed at high electric field values; the arrows to the left indicate the electric-dipole type transitions observed for zero electric fields which are the stronger transitions for very weak electric fields when the static magnetic field is perpendicular to the electric component of the observed radiation, as assumed in this discussion. Figure 11.14 indicates the type of patterns observed for different values of the electric field when transitions of both types occur, but with unequal strength. In these figures nuclear hyperfine splitting is not shown. Carrington et al. [106] developed formulas for calculation of the relative intensities of the two sets of components for different values of the Stark energy $E_{\mathscr{E}}$ relative to E_Λ. A detailed analysis of the transition probabilities, which entails evaluation of the dipole

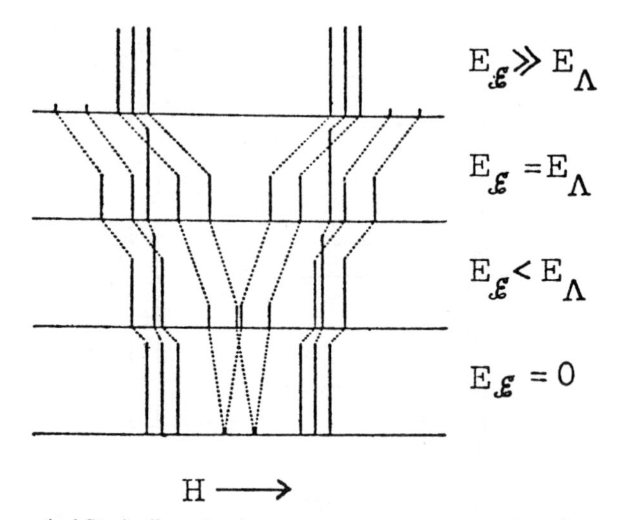

Fig. 11.14 Theoretical Stark effect of paramagnetic resonance patterns for a $^2\Pi_{3/2}$, $J=\frac{3}{2}$ state for different values of the Stark energy $E_{\mathcal{E}}$ relative to the Λ doubling energy E_Λ. From Carrington et al. [106].

matrix elements with the perturbed wave functions (composed of field-dependent linear combinations of the two zeroth-order wave functions) shows that for the $M=-\frac{1}{2}\leftrightarrow+\frac{1}{2}$ transitions the intensities are unaffected by the Stark field (see Fig. 11.14).

Table 11.12 lists dipole moments of free radicals measured by Carrington et al. [106] for a few gaseous free radicals. Note that SO was measured in the excited $^1\Delta$ electronic state. Compare this value of 1.47 D with that of 1.55 D observed for its $^3\Sigma$ ground state [107]. The other three were measured in the $^2\Pi_{3/2}$ ground state.

In a more recent paper, Amano et al. [108] have reported measurements on both the $^2\Pi_{1/2}$ and the $^2\Pi_{3/2}$ states of ClO and have given more accurate values than those listed in Tables 11.11 and 11.12. Powell and Johnson [109] have likewise obtained more extensive results on BrO. Further work on the NS radical has been reported by Uehara and Morino [110], also by Amano et al. [111].

Table 11.12 Electric Dipole Moments from the Stark Effect of Paramagnetic Resonance[a]

ClO	BrO	$^1\Delta$SO	SH
1.18	1.55 D	1.47 D	0.62 D

[a]From Carrington et al. [106]. For additional parameters of BrO, see Powell and Johnson [109].

References

1. D. K. Coles and W. E. Good, *Phys. Rev.*, **70**, 979 (1946).
2. C. K. Jen, *Phys. Rev.*, **74**, 1396 (1948).
3. J. R. Eshbach and M. W. P. Strandberg, *Phys. Rev.*, **85**, 24 (1952).
4. H. Taft and B. P. Dailey, *J. Chem. Phys.*, **48**, 597 (1968).
5. C. K. Jen, *Phys. Rev.*, **81**, 197 (1951); *Physica*, **17**, 379 (1951).
6. J. T. Cox and W. Gordy, *Phys. Rev.*, **101**, 1298 (1956).
7. B. Rosenblum, A. H. Nethercot, and C. H. Townes, *Phys. Rev.*, **109**, 400 (1958).
8. C. A. Burrus, *J. Chem. Phys.*, **30**, 976 (1959).
9. R. L. Shoemaker and W. H. Flygare, private communication.
10. W. Gordy, O. R. Gilliam, and R. Livingston, *Phys. Rev.*, **76**, 443 (1949).
11. J. R. Eshbach, R. E. Hillger, and C. K. Jen, *Phys. Rev.*, **80**, 1106 (1950).
12. C. K. Jen, *Phys. Rev.*, **76**, 1494 (1949).
13. J. H. Van Vleck, *Theory of Electric and Magnetic Susceptibilities*, Clarendon Press, Oxford, 1932.
14. E. U. Condon, *Phys. Rev.*, **30**, 781 (1927).
15. G. C. Wick, *Phys. Rev.*, **73**, 51 (1948).
16. W. Hüttner and W. H. Flygare, *J. Chem. Phys.*, **47**, 4137 (1967).
17. W. Hüttner, M. -K. Lo, and W. H. Flygare, *J. Chem. Phys.*, **48**, 1206 (1968).
18. R. L. Shoemaker and W. H. Flygare, *Chem. Phys. Lett.*, **2**, 610 (1968).
19. M. -K. Lo, P. D. Foster, and W. H. Flygare, *J. Chem. Phys.*, **48**, 948 (1968).
20. W. H. Flygare, W. Hüttner, R. L. Shoemaker, and P. D. Foster, *J. Chem. Phys.*, **50**, 1714 (1969).
21. D. VanderHart and W. H. Flygare, *Mol. Phys.*, **18**, 77 (1969).
22. W. H. Flygare, R. L. Shoemaker, and W. Hüttner, *J. Chem. Phys.*, **50**, 2414 (1969).
23. R. G. Stone, J. M. Pochan, and W. H. Flygare, *Inorg. Chem.*, **8**, 2647 (1969).
24. R. L. Shoemaker and W. H. Flygare, *J. Am. Chem. Soc.*, **91**, 5417 (1969).
25. W. Hüttner, P. D. Foster, and W. H. Flygare, *J. Chem. Phys.*, **50**, 1710 (1969).
26. R. Blickensderfer, J. H. S. Wang, and W. H. Flygare, *J. Chem. Phys.*, **51**, 3196 (1969).
27. D. H. Sutter and W. H. Flygare, *J. Am. Chem. Soc.*, **91**, 4063 (1969).
28. S. G. Kukolich and W. H. Flygare, *Chem. Phys. Lett.*, **6**, 45 (1970).
29. S. G. Kukolich, *Chem. Phys. Lett.*, **5**, 401 (1970).
30. R. L. Shoemaker and W. H. Flygare, *Chem. Phys. Lett.*, **6**, 576 (1970).
31. S. Gustafson and W. Gordy, *J. Chem. Phys.*, **52**, 579 (1970).
32. J. M. Pochan, R. L. Shoemaker, R. G. Stone, and W. H. Flygare, *J. Chem. Phys.*, **52**, 2478 (1970).
33. J. H. S. Wang and W. H. Flygare, *J. Chem. Phys.*, **52**, 5636 (1970).
34. R. C. Benson and W. H. Flygare, *J. Am. Chem. Soc.*, **92**, 7523 (1970).
35. R. L. Shoemaker and W. H. Flygare, reported in refs. [57, 58], Flygare and Benson [57].
36. W. C. Allen and W. H. Flygare, *Chem. Phys. Lett.*, **15**, 461 (1972).
37. B. Bak, E. Hamer, D. H. Sutter, and H. Dreizler, *Z. Naturforsch.*, **27a**, 705 (1972).
38. R. Honerjäger and R. Tischer, *Z. Naturforsch.*, **28a**, 1372 (1973).
39. W. Czieslik, U. Andersen, and H. Dreizler, *Z. Naturforsch.*, **28a**, 1906 (1973).
40. J. McGurk, H. L. Tigelear, S. L. Rock, C. L. Norris, and W. H. Flygare, *J. Chem. Phys.*, **58**, 1420(1973).

41. S. Gustafson and W. Gordy, *J. Chem. Phys.*, **58**, 5181 (1973).
42. R. Hornerjäger and R. Tischer, *Z. Naturforsch.*, **29a**, 342 (1974).
43. K. V. L. N. Sastry and A. Guarnieri, *Z. Naturforsch.*, **29a**, 1495 (1974).
44. W. Czieslik and D. H. Sutter, *Z. Naturforsch.*, **29a**, 1820 (1974).
45. R. Hornerjäger and R. Tischer, *Z. Naturforsch.*, **29a**, 1919 (1974).
46. M. Suzuki and A. Guarnieri, *Z. Naturforsch.*, **30a**, 497 (1975).
47. F. Scappini and A. Guarnieri, *Z. Naturforsch.*, **31a**, 369 (1976).
48. W. Czieslik, J. Wiese, and D. H. Sutter, *Z. Naturforsch.*, **31a**, 1210 (1976).
49. K. -F. Dössel, J. Wiese, and D. H. Sutter, *Z. Naturforsch.*, **33a**, 21 (1978).
50. G. K. Pandy and D. H. Sutter, *Z. Naturforsch.*, **33a**, 29 (1978).
51. K. -F. Dössel and D. H. Sutter, *Z. Naturforsch.*, **34a**, 469 (1979).
52. K. -F. Dössel and D. H. Sutter, *Z. Naturforsch.*, **34a**, 482 (1979).
53. F. Rohwer and A. Guarnieri, *Z. Naturforsch.*, **35a**, 336 (1980).
54. J. Wiese and D. H. Sutter, *Z. Naturforsch.*, **35a**, 712 (1980).
55. D. Hübner, M. Stolze, and D. H. Sutter, *Z. Naturforsch.*, **36a**, 332 (1981).
56. W. H. Stolze, M. Stolze, D. Hübner, and D. H. Sutter, *Z. Naturforsch.*, **37a**, 1165 (1982).
57. W. H. Flygare and R. C. Benson, *Mol. Phys.*, **20**, 225 (1971).
58. S. H. Sutter and W. H. Flygare, "The Molecular Zeeman Effect," in *Topics in Current Chemistry*, 63: *Bonding and Structure*, Springer-Verlag, Berlin, 1976, pp. 89–196.
59. P. Pascal, *Ann. Chim. Phys.*, **19**, 5 (1910).
60. R. C. Benson and W. H. Flygare, *J. Chem. Phys.*, **53**, 4470 (1970).
61. T. G. Schmalz, C. L. Norris, and W. H. Flygare, *J. Am. Chem. Soc.*, **95**, 7961 (1973).
62. A. D. Buckingham, *Q. Rev. (London)*, **13**, 183 (1959).
63. W. V. Smith and R. R. Howard, *Phys. Rev.*, **79**, 128 (1950): **79**, 132 (1950).
64. R. M. Hill and W. V. Smith, *Phys. Rev.*, **82**, 451 (1951).
65. A. D. Buckingham and H. C. Longuet-Higgins, *Mol. Phys.*, **14**, 63 (1968). See also A. D. Buckingham, R. L. Disch, and D. A. Dunmur, *J. Am. Chem. Soc.*, **90**, 3104 (1968).
66. R. L. Shoemaker and W. H. Flygare, *J. Chem. Phys.*, **51**, 2988 (1969).
67. C. H. Townes, G. C. Dousmanis, R. L. White, and R. F. Schwarz, *Discussions Faraday Soc.*, **19**, 56 (1955).
68. E. F. Pearson and W. Gordy, *Phys. Rev.*, **177**, 52 (1969).
69. E. F. Pearson and W. Gordy, *Phys. Rev.*, **177**, 59 (1969).
70. M. Mizushima, J. T. Cox, and W. Gordy, *Phys. Rev.*, **98**, 1034 (1955).
71. R. M. Hill and W. Gordy, *Phys. Rev.*, **93**, 1019 (1954).
72. M. Winnewisser, K. V. L. N. Sastry, R. L. Cook, and W. Gordy, *J. Chem. Phys.*, **41**, 1687 (1964).
78. M. Tinkham and M. W. P. Strandberg, *Phys. Rev.*, **97**, 951 (1955).
74. R. Beringer and J. G. Castle, Jr., *Phys. Rev.*, **75**, 1963 (1949).
75. J. G. Castle, Jr., and R. Beringer, *Phys. Rev.*, **80**, 114 (1950).
76. R. Beringer and J. G. Castle, Jr., *Phys. Rev.*, **78**, 581 (1950).
77. R. Beringer, E. B. Rawson, and A. F. Henry, *Phys. Rev.*, **94**, 343 (1954).
78. H. E. Radford, *Phys. Rev.*, **122**, 114 (1961).
79. G. C. Dousmanis, T. M. Sanders, and C. H. Townes, *Phys. Rev.*, **100**, 1735 (1955).
80. H. Margenau and A. Henry, *Phys. Rev.*, **78**, 587 (1950).
81. R. A. Frosch and H. M. Foley, *Phys. Rev.*, **88**, 1337 (1952).

82. H. E. Radford and M. Linzer, *Phys. Rev. Lett.*, **10**, 443 (1963).
83. H. E. Radford, *J. Chem. Phys.*, **40**, 2732 (1964).
84. C. C. McDonald, *J. Chem. Phys.*, **39**, 2587 (1963).
85. A. Carrington, *Microwave Spectroscopy of Free Radicals*, Academic, London, 1974.
86. S. M. Freund, E. Herbst, R. P. Mariella, and W. A. Klemperer, *J. Chem. Phys.*, **56**, 1467 (1972).
87. A. Carrington, P. N. Dyer, and D. H. Levy, *J. Chem. Phys.*, **47**, 1756 (1967).
88. A. Carrington, D. H. Levy, and T. A. Miller, *Proc. R. Soc.* (London), **A293**, 108 (1966).
89. A. Carrington, A. R. Fabris, and N. J. D. Lucas, *J. Chem. Phys.*, **49**, 5545 (1968).
90. A. Carrington, A. R. Fabris, and N. J. D. Lucas, *Mol. Phys.*, **16**, 195 (1969).
91. I. C. Bowater, J. M. Brown, and A. Carrington, *J. Chem. Phys.*, **54**, 4957 (1971).
92. I. C. Bowater, J. M. Brown, and A. Carrington, *Proc. R. Soc.* (*London*), **A333**, 265 (1973).
93. S. Saito, *J. Chem. Phys.*, **53**, 2544 (1970).
94. S. Saito and T. Amano, *J. Mol. Spectrosc.*, **34**, 383 (1970).
95. Y. Endo, S. Saito, and E. Hirota, *J. Chem. Phys.*, **75**, 4379 (1981).
96. C. E. Barnes, J. M. Brown, A. Carrington, J. Pinkstone, and T. J. Sears, *J. Mol. Spectrosc.*, **72**, 86 (1978).
97. H. E. Radford, K. M. Evenson, and C. J. Howard, *J. Chem. Phys.*, **60**, 3178 (1974).
98. J. T. Hougen, H. E. Radford, K. M. Evenson, and C. J. Howard, *J. Mol. Spectrosc.*, **56**, 210 (1975).
99. Y. Beers and C. J. Howard, *J. Chem. Phys.*, **63**, 4212 (1975).
100. S. Saito, *J. Mol. Spectrosc.*, **65**, 229 (1977).
101. A. Charo and F. C. De Lucia, *J. Mol. Spectrosc.*, **94**, 426 (1982).
102. Y. Endo, S. Saito, and E. Hirota, *J. Mol. Spectrosc.*, **97**, 204 (1983).
103. J. M. Cook, G. W. Hills, and R. F. Curl, Jr., *J. Chem. Phys.*, **67**, 1450 (1977).
104. J. M. Daniels and P. B. Dorain, *J. Chem. Phys.*, **45**, 26 (1966).
105. A. Carrington, G. N. Currie, D. H. Levy, and T. A. Miller, *Mol. Phys.*, **17**, 535 (1979).
106. A. Carrington, D. H. Levy, and T. A. Miller, *J. Chem. Phys.*, **47**, 3801 (1967).
107. F. X. Powell and D. R. Lide, *J. Chem. Phys.*, **41**, 1413 (1964).
108. T. Amano, S. Saito, E. Hirota, Y. Morino, D. R. Johnson, and F. X. Powell, *J. Mol. Spectrosc.*, **30**, 275 (1969).
109. F. X. Powell and D. R. Johnson, *J. Chem. Phys.*, **50**, 4596 (1969).
110. H. Uehara and Y. Morino, *Mod. Phys.*, **17**, 239 (1969).
111. T. Amano, S. Saito, E. Hirota, and Y. Morino, *J. Mol. Spectrosc.*, **32**, 97 (1969).

Chapter **XII**

INTERNAL MOTIONS

1 INTRODUCTION

In this chapter we consider hindered internal motions essentially of large amplitude, such as internal rotation and ring puckering, which may be effectively separated from the other vibrational modes. Inversion motion considered in Chapter VI is another example of large amplitude motion. Numerous threefold barriers to internal rotation for molecules with methyl groups have been evaluated from a study of the microwave spectrum. For molecules with asymmetric tops, the most stable conformations have been obtained, and in certain cases, the potential function governing interconversion. Similarly for ring compounds, a microwave study provides the preferred conformations. In the case of certain small rings, the conformation and the ring puckering potential function have been evaluated. Studies of potential functions, barrier heights, stability of rotational isomers and ring conformations all provide basic information for testing and improving predictive methods and for understanding the origin of barriers and the forces responsible for conformational preference.

The most active of these areas in microwave spectroscopy is the study of internal rotation. The phenomenon of internal rotation has for many years been a subject of considerable interest to both chemists and physicists. For molecules which consist of two groups connected by a single bond, the groups can rotate with respect to one another about the single bond. In ethane, $H_3C—CH_3$, one CH_3 group rotates with respect to the other about the C—C bond. If the barrier were very high, the internal motion of the methyl groups about the C–C bond would correspond to simple harmonic torsional oscillation, whereas if the barrier were very low the internal motion would be essentially free rotation about the C–C bond. For a long time, this rotation about a single bond was believed to be completely free because it had been impossible to separate any isomers which could arise from different orientations of one part of the molecule relative to the other. In contrast, for two groups connected by a double bond, the relative rotation of the two groups requires uncoupling of the π-electrons and is expected to be very difficult. This is reflected in the fact that stable isomers can be isolated, for example, *cis*- and *trans*-1, 2 dichloroethylene.

In 1936 Kemp and Pitzer [1] concluded, on the basis of thermodynamic evidence, that the relative rotation of the two methyl groups in ethane is not entirely free but is restricted by a potential barrier. In Fig. 12.1 are shown the

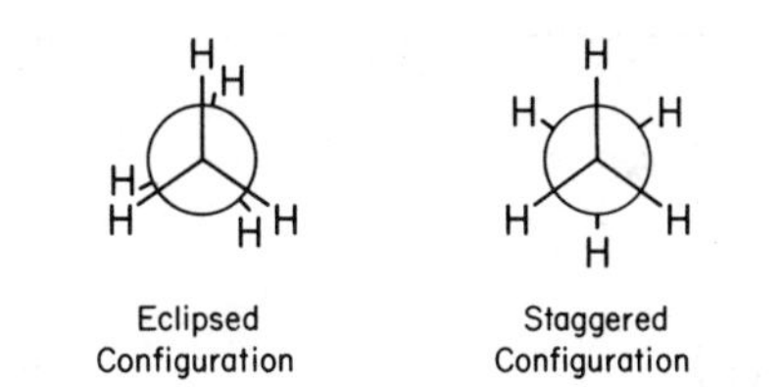

Fig. 12.1 Newman projection formulas for ethane.

configurations of ethane where the two methyl groups are in an eclipsed and staggered position relative to each other. Within one complete revolution there are three equivalent staggered and eclipsed configurations. The staggered configuration turns out to be the one of minimum energy [2] while the eclipsed configuration corresponds to an energy maximum. In the transition from a staggered to an eclipsed conformation an energy barrier of about 3000 cal/mole [3] has to be surmounted. Soon after the work of Kemp and Pitzer it became apparent that rotation about single bonds is restricted in many molecules, although the barriers (on the order of a few thousand calories) are not large enough to permit chemical isolation.

For over 30 years the method of microwave spectroscopy has been employed to study the problem of hindered internal rotation. It has the great advantage that potential barriers can be obtained with an accuracy of 5% or better and that other important information is obtainable, such as the structural parameters and the equilibrium configuration of the methyl group. It is applicable to polar molecules with barriers ranging from several calories/mole to about 4000 cal/mole. In most of the remaining part of this chapter a quantitative study of the effects of internal rotation on the rotational spectrum of a molecule will be considered. In short, the effect of internal rotation on the rotational spectrum is that each rotational transition will exhibit a fine structure caused by the interaction of internal and overall rotation. This fine structure depends on the height of the potential barrier hindering internal rotation. Analysis of this fine structure with the high resolution and accurate measurements of microwave spectroscopy leads to the evaluation of the potential barrier.

A typical spectrum is illustrated in Fig. 12.2 which shows the main features of the $J=2\rightarrow3$ spectrum of thioacetaldehyde, CH_3CHS [4]. Thioacetaldehyde is a near-prolate asymmetric rotor. The lines are doublets made up of A and E components. The ground state $K_{-1}=0$, 1, and 2 lines are split by approximately 0.7, 4, and 7 MHz, respectively. Strong satellite lines with larger splittings are also observed. The A torsional state lines were fit as a usual semirigid rotor (rigid rotor plus distortion effects) and the barrier height was then varied such that the observed splitting between A and E components for each line was obtained. This gave a barrier, $V_3=1572\pm30$ cal/mole.

For a general discussion of the field of internal rotation in molecules the reader is directed to books by Mizushema [5] and Orville-Thomas [6]. Since the initial derivation of the potential barrier of methyl alcohol by Burkhard and Dennison [7], considerable progress has been made in the application of

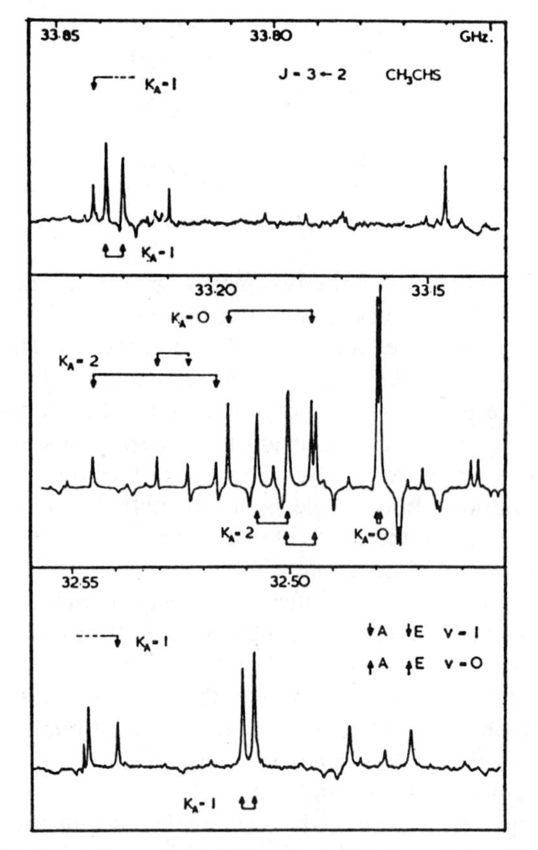

Fig. 12.2 The $J=3\leftarrow 2$ transitions of CH_3CHS under moderate resolution. The $K_A=1E$ torsional satellites are subject to large shifts of the order of 400 MHz and are not shown in the traces. From Kroto and Landsberg [4].

microwave spectroscopy to the elucidation of potential barriers. An excellent review article on the theory and application of microwave spectroscopy to the problem of internal rotation has been given by Lin and Swalen [8], by Dreizler [9], and most recently by Lister et al. [10].

2 POTENTIAL FUNCTION FOR INTERNAL ROTATION

The internal rotation of two parts of a molecule relative to each other can be described by the angle α, called the angle of internal rotation or the torsional angle. For example, in ethane the torsional angle may be taken as the angle between two planes: one defined by the C–C bond and a specifically chosen C–H bond of one CH_3 group, and the other plane defined by the C–C bond and a C–H bond of the other CH_3 group. As the two parts rotate relative to one another, the potential energy will vary as a function of α. In molecules such as CH_3—CH_3 or CH_3—CF_3, there are three equivalent positions into which one methyl group can rotate relative to the other group. The same is true for $CH_3—CHO$, whereas for $CH_3—NO_2$ there are six such equivalent positions, and for $CH_2{=}CH_2$ there are two. The number of equivalent configurations N for a complete internal revolution is obviously dependent on the symmetry of the molecule. It is natural to require that the potential function be a periodic function in α, which repeats itself N times in the interval $\alpha=0$ to $\alpha=2\pi$. Such a periodic function may be represented by a Fourier series expansion. If the potential is taken as an even function of α, a periodic potential function, with period $2\pi/N$, can be expressed in the form:

$$V(\alpha)=a_0+\sum_{k=1}^{\infty} a_k \cos kN\alpha \tag{12.1}$$

where with

$$a_0=-\sum_{k=1}^{\infty} a_k \tag{12.2}$$

$V(\alpha)$ is zero at $\alpha=0$, $\pm 2\pi/N$, $\pm 4\pi/N$, and soon. This potential function for a barrier with N-fold symmetry may be written as:

$$V(\alpha)=\frac{V_N}{2}(1-\cos N\alpha)+\frac{V_{2N}}{2}(1-\cos 2N\alpha)+\cdots \tag{12.3}$$

We will at first be concerned with molecules having a threefold barrier, which have been extensively studied; we shall consider sixfold barriers in Section 8. For a threefold barrier one would have

$$V(\alpha)=\frac{V_3}{2}(1-\cos 3\alpha)+\frac{V_6}{2}(1-\cos 6\alpha)+\cdots \tag{12.4}$$

Usually only the first term of the cosine expansion is retained:

$$V(\alpha)=\frac{V_3}{2}(1-\cos 3\alpha) \tag{12.5}$$

The sinusoidal shape of this periodic hindering potential is depicted in Fig. 12.3.

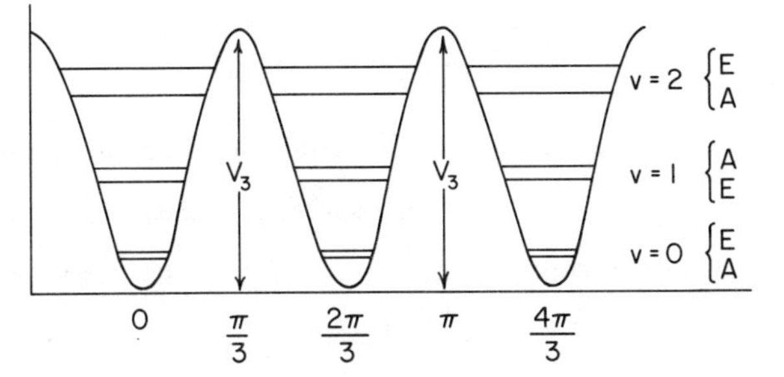

Fig. 12.3 Schematic representation of the potential function and torsional energy levels. A simple cosine potential is depicted with three identical minima and maxima. Each torsional energy level is labeled by the torsional quantum number v. The torsional sublevels are denoted by their symmetry A or E under the C_3 group.

This has a minimum value of zero at $\alpha=0, \pm 2\pi/3, \pm 4\pi/3, \ldots$, and a maximum value of V_3 at $\alpha=\pm\pi/3, \pm\pi, \pm 5\pi/3, \ldots$ This simple potential function has proved to be a rather accurate representation of the actual potential, because in many cases the higher terms in the expansion are very small. Investigation of some threefold-barrier molecules has, in fact, indicated that the V_6 term is much smaller than V_3 [11–14], usually less than 3% of V_3. This is further suggested from results on molecules such as CH_3NO_2 where V_6 is the leading term of the expansion and has been found to be very small (6 cal/mole, see Sections 6 and 8.

3 TORSIONAL ENERGY LEVELS

If the variation of the potential energy with respect to α is taken as given in (12.5), the wave equation for the internal rotation is

$$-F\frac{d^2U(\alpha)}{d\alpha^2}+\left[\frac{V_3}{2}(1-\cos 3\alpha)-E\right]U(\alpha)=0 \tag{12.6}$$

where in this one-dimensional Schrödinger equation, α is the angle of internal rotation, V_3 is the height of the barrier having threefold symmetry which hinders the internal rotation, and $F=\hbar^2/2I_r$ with I_r the reduced moment of inertia for the relative motion of the two groups. Solution of this differential equation yields the torsional eigenvalues E and eigenfunctions $U(\alpha)$.

For the case of two symmetrical coaxial tops, such as CH_3—CH_3 or CH_3—CF_3, the reduced moment for internal rotation is

$$I_r=\frac{I_\alpha I_\beta}{I_\alpha+I_\beta} \tag{12.7}$$

where I_α and I_β are the moments of inertia of the two tops about the axis of internal rotation. The expression for the reduced moment of inertia of more complex cases is given in the next section.

In examination of the torsional energy levels it is helpful to consider first two extremes, V_3 very small and V_3 very large. For a very small barrier such that $V_3\rightarrow 0$ (free rotation), (12.6) reduces to:

$$\frac{d^2U(\alpha)}{d\alpha^2}+\left(\frac{1}{F}\right)EU(\alpha)=0 \tag{12.8}$$

This has the form of a spatial rotator with fixed axis of rotation. The solution is

$$U(\alpha)=Ae^{im\alpha}=A(\cos m\alpha+i\sin m\alpha) \tag{12.9}$$

with A an appropriate normalization factor and with the energy given by

$$E=Fm^2 \tag{12.10}$$

For the function (12.9) to be well behaved, it must satisfy the appropriate boundary condition. Applying the boundary condition

$$U(\alpha)=U(\alpha+2\pi)$$

we see from (12.9) that m is required to take on the values

$$m=0, \pm 1, \pm 2, \pm 3, \ldots$$

The constant A may be evaluated from the normalization condition giving $A=1/2^{1/2}\pi$. The free rotor states, specified by the quantum number m, are doubly degenerate except for the state $m=0$. The functions of (12.9) are also eigenfunctions of the angular momentum operator $p=-i\hbar(\partial/\partial\alpha)$, the eigenvalues of p being $m\hbar$. For the state m the two possible values of p, namely $|m|\hbar$ and $-|m|\hbar$, correspond to the two possible directions of internal rotation.

In the case of a very high barrier, $V_3 \to \infty$, the internal motions will be restricted to small oscillations, torsional oscillations, about the minima of the barrier. The cosine function in (12.6) may be expanded for small values of α as:

$$\cos 3\alpha = 1 - (\tfrac{9}{2})\alpha^2 + (\tfrac{27}{8})\alpha^4 + \cdots \tag{12.11}$$

and thus:

$$V(\alpha) = (\tfrac{9}{4})V_3\alpha^2 - (\tfrac{27}{16})V_3\alpha^4 + \cdots \tag{12.12}$$

For a harmonic approximation we retain only the first term of the expansion. With this potential, the torsional wave equation reduces to:

$$\frac{d^2 U(\alpha)}{d\alpha^2} + \frac{1}{F}\left[E - \tfrac{1}{2}(\tfrac{9}{2}V_3)\alpha^2\right]U(\alpha) = 0 \tag{12.13}$$

which has the same form as the simple harmonic oscillator wave equation. The solutions are well behaved if

$$E = 3(V_3F)^{1/2}(v + \tfrac{1}{2}) \tag{12.14}$$

with

$$v = 0, 1, 2, 3, \ldots$$

The torsional energy levels thus approximate those of a harmonic oscillator. The frequency of torsional oscillation,

$$\nu = \frac{3}{2\pi}\left(\frac{V_3}{2I_r}\right)^{1/2}$$

is related to the barrier, and this relationship provides the basis for determination of the barrier by infrared studies or microwave intensity studies.

In the limit of an infinite barrier each torsional state v is threefold degenerate since the internal motion is torsional oscillation in any one of the three equivalent potential wells. For a finite barrier, the quantum-mechanical tunneling effect leads to a splitting of this threefold degeneracy because the probability of tunneling through the barrier is now finite since V_3 is finite. This is purely a quantum effect since classically if $V_3 > E$ there is no possibility of passage from one potential well to another because the system does not have enough energy to surmount the barrier. Quantum mechanically, however, the molecule may pass from one configuration to another by tunneling through the barrier since

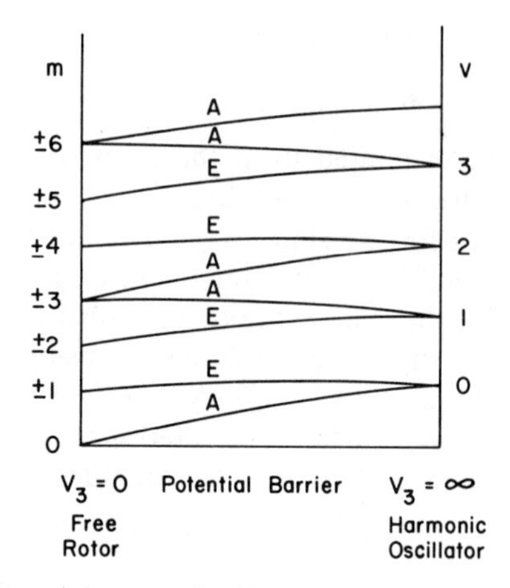

Fig. 12.4 Correlation between the energy levels of free internal rotation specified by the quantum numbers m, and those for harmonic torsional oscillation which are labeled by the quantum numbers v. For barriers between the two extremes some of the degeneracy of the two limiting cases is removed. The degeneracy of the torsional sublevels is indicated by A and E, the E levels being doubly degenerate whereas the A levels are nondegenerate.

the wave functions extend through the classically forbidden regions. The tunneling rate depends on both height and width of the barrier. The effect of tunneling through the potential barrier is to split the triply degenerate torsional level v into two levels, a nondegenerate level designated as an A level and a doubly degenerate level designated as an E level. The connections between the harmonic oscillator limit and the free rotor limit are shown schematically in Fig. 12.4. That each triply degenerate level of the harmonic oscillator limit is split into two levels is apparent. For the free-rotor states, with m a multiple of three, the $\pm m$ degeneracy is removed. For the other values of m this degeneracy remains. The symmetry of the torsional energy levels (the A and E designations) will be considered later.

In Fig. 12.3 the internal rotational energy levels are shown superimposed on the potential curve. Note that the torsional sublevel spacing increases as the torsional energy increases, that is, as v increases. The greater the sublevel splitting, the faster the rate of tunneling. Near the bottom of the well the energy levels approximate those of a harmonic oscillator. When the torsional energy E is greater than the barrier height V_3, the energy levels go over to those of a free rotor.

The qualitative aspects of the torsional energy levels have been presented, but quantitative discussion for intermediate values of the barrier height requires consideration of the solutions of the original torsional wave equation, (12.6). A knowledge of the torsional eigenfunctions and eigenvalues is useful for the evaluation of various perturbation sums that arise in the perturbation treatment of internal rotation, which will be described subsequently.

Because of the choice of the potential function, (12.6) can be transformed to

Mathieu's equation by the substitutions:

$$U(\alpha)=M(x) \qquad 3\alpha+\pi=2x$$
$$E=(\tfrac{9}{4})Fb \qquad V_3=(\tfrac{9}{4})Fs \tag{12.15}$$

giving

$$\frac{d^2M(x)}{dx^2}+(b-s\cos^2 x)M(x)=0 \tag{12.16}$$

where b is an eigenvalue, s is a parameter termed the reduced barrier height, and $M(x)$ a Mathieu function.

The eigenfunctions $U(\alpha)$ of interest must satisfy the boundary condition $U(\alpha)=U(\alpha+2\pi)$ or for the Mathieu functions, $M(x)=M(x+3\pi)$. The Mathieu functions which meet this boundary condition are those with period π in x (period $2\pi/3$ in α) and period 3π in x (period 2π in α). The eigenvalues and eigenfunctions of the torsional wave equation can hence be obtained from existing tables [15–19] on periodic Mathieu functions with the aid of (12.15). The solutions with period π are associated with nondegenerate eigenvalues while those of period 3π are associated with doubly degenerate eigenvalues. For distinction between the eigenvalues ($E_{v\sigma}$ or $b_{v\sigma}$) and the wave functions, the labels v and σ are introduced. The series of degenerate and nondegenerate solutions obtained are ordered in terms of increasing size. The lowest degenerate and nondegenerate eigenvalues are then labeled by $v=0$. The next lowest degenerate and nondegenerate eigenvalues are labeled by the index $v=1$ and so on. For a high barrier, the degenerate and nondegenerate states corresponding to a given v have nearly the same energy, whereas those of different v are more widely separated. The degenerate and nondegenerate energy levels associated with a given v are regarded as the torsional sublevels. As $V_3\rightarrow\infty$, the different energies of the torsional sublevels coalesce, and v becomes the quantum number for the limiting harmonic oscillator state. The usefulness of the label v is thus readily apparent; it is called the principal torsional quantum number. The index $\sigma=0, \pm1$, which gives the symmetry or periodicity of the torsional wave functions, serves to distinguish the torsional sublevels. The $\sigma=0$ levels are the nondegenerate A levels, and the $\sigma=\pm1$ are the degenerate E levels. Appropriate solution of the torsional wave equation, therefore, leads to a set of torsional levels v, each of which is split into two sublevels with one nondegenerate and the other twofold degenerate.

A periodic solution of (12.6) or (12.16) may be represented by a Fourier expansion. A solution of the torsional equation, for example, can be written in the following convenient form

$$U_{v\sigma}(\alpha)=\sum_{k=-\infty}^{\infty} A_k^{(v)}e^{i(3k+\sigma)\alpha} \tag{12.17}$$

where σ is an integer. For the appropriate periodic solutions of a threefold barrier σ takes on the three values -1, 0, and $+1$. As mentioned previously, the index σ gives the symmetry of the torsional wave functions. The torsional

Table 12.1 Character Table for the Group C_3^a

Symmetry Species		E	C_3	C_3^2
A		1	1	1
E	E_1	1	ε	ε^*
	E_2	1	ε^*	ε

[a]Here $\varepsilon = \exp(i2\pi/3)$ where ε^* designates the complex conjugate and $\varepsilon\varepsilon^* = 1$, $(\varepsilon)^2 = \varepsilon^*$, $(\varepsilon^*)^2 = \varepsilon$.

functions may be conveniently classified under the rotational subgroup C_3 of the torsional Hamiltonian. The character table of this group is given in Table 12.1. For classification according to the symmetry species of the group, it is necessary to know how the functions, (12.17), transform under the symmetry operations C_3 and C_3^2. The operations C_3 and C_3^2 on $U_{v\sigma}(\alpha)$ will affect only the angle α, transforming it in the following manner—C_3: $\alpha \to \alpha + 2\pi/3$; C_3^2: $\alpha \to \alpha + 4\pi/3 = \alpha - 2\pi/3$. The identity operation, of course, leaves the function unchanged; the effects of the remaining operations on $U_{v\sigma}(\alpha)$ are

$$C_3\colon\ U_{v\sigma}(\alpha) \to \varepsilon^\sigma U_{v\sigma}(\alpha)$$
$$C_3^2\colon\ U_{v\sigma}(\alpha) \to (\varepsilon^*)^\sigma U_{v\sigma}(\alpha)$$

From the foregoing relations and the character table, it is evident that the $U_{v\sigma}$ with $\sigma = 0$ belong to the nondegenerate species A and give rise to the nondegenerate eigenvalues. The functions with $\sigma = +1$ belong to one part of the doubly degenerate E species, viz., E_1; those with $\sigma = -1$ belong to the other part, E_2, and they give rise to the pairs of degenerate eigenvalues.

Some insight into the nature of the $A(\sigma = 1)$ and $E(\sigma = \pm 1)$ internal torsional states may be obtained from the following considerations. The internal motion for the A states [$U(\alpha)$ periodic in $2\pi/3$] resembles a back-and-fourth oscillation localized in the potential wells. For these states the $\pm m$ degeneracy of the free-rotor states is removed (see Fig. 12.4). The internal motion of the E states [$U(\alpha)$ periodic in 2π] has some of the character of free rotation passing from one potential well to another by tunneling, the $\pm m$ degeneracy of the free rotor which remains for the E states being related to the two possible directions of internal rotation.

In general, for a molecule with N equivalent internal configurations the appropriate periodic solutions [17, 18]

$$U_{v\sigma}(\alpha) = e^{i\sigma\alpha} \sum_{k=-\infty}^{\infty} A_k^{(v)} e^{iNk\alpha} \tag{12.18}$$

are obtained by choice of N integer values of σ, such that $-N/2 < \sigma \leqslant N/2$

with $\sigma \neq -N/2$ for N even. Except for $\sigma=0$ or $\sigma=N/2$ (N even) which give nondegenerate solutions, the remaining values of σ give pairs of degenerate solutions corresponding to $\pm\sigma$. The number of distinct eigenvalues for any v are $(N+2)/2$ for N even and $(N+1)/2$ for N odd. In the limit $V_N \rightarrow \infty$ the sublevels coalesce, leading to an N-fold degeneracy. For finite barriers the N-fold degeneracy is lifted by the effect of tunneling.

Explicit solutions of (12.6) or (12.16) can be obtained by numerical iterative techniques similar to those employed in the determination of the eigenvalues of a rigid rotor (see Appendix B). Substitution of the expansion equation (12.17) into (12.6) and subsequent multiplication by $e^{-i(3k+\sigma)\alpha}$ and integration from 0 to 2π gives

$$A_{k-1}+(\lambda-M_k)A_k+A_{k+1}=0 \tag{12.19}$$

where

$$M_k=\frac{4F}{V_3}(3k+\sigma)^2=\left(\frac{16}{9s}\right)(3k+\sigma)^2$$

$$\lambda=4\left(\frac{E}{V_3}-\frac{1}{2}\right)=\left(\frac{4}{s}\right)\left(b-\frac{s}{2}\right)$$

and

$$k=0, \pm 1, \pm 2, \pm 3, \ldots$$

Here the v labeling of the Fourier expansion coefficients has for the present been omitted, and the substitution of $\cos 3\alpha=(e^{i3\alpha}+e^{-i3\alpha})/2$ and the fact that $\int_0^{2\pi} e^{i(k-k')\alpha}\, d\alpha=0$, unless $k=k'$ with k and integer, have been used for derivation of Eq. (12.19). If this infinite set of linear homogeneous equations is to have a nontrivial solution for the A's, the determinant of the coefficients must vanish. The infinite order determinantal equation has the form:

$$\begin{vmatrix} \ddots & & & \\ M_{k-1}-\lambda & -1 & & \\ -1 & M_k-\lambda & -1 & \\ & -1 & M_{k+1}-\lambda & \\ & & & \ddots \end{vmatrix}=0 \tag{12.20}$$

There will be three separate determinants, one for each value of σ. However, it is apparent that the determinants corresponding to $\pm\sigma$ have the same eigenvalues and thus constitute the degenerate E species solutions with $U_{v,-1}(\alpha)= U_{v1}^*(\alpha)$. The eigenvalues λ, and thus the torsional energies associated with a particular value of σ, may be conveniently obtained by solution of the appropriate determinant by the method of continued fractions [8, 17–20]. Once the eigenvalues are known, the Fourier coefficients of the expansion can be determined by means of the recursion relations between the A's, (12.19). Solving these relations and making use of the normalization requirement, $\int_0^{2\pi} U^*(\alpha)U(\alpha)\,d\alpha=1$, or $2\pi\sum A_k^2=1$, one obtains the torsional eigenfunctions. Both the eigenvalues and the expansion coefficients depend on the reduced barrier height s. Since the values of k are not bounded, a truncated Fourier expansion which leads to a determinant of finite size must be used if the problem is to be mathematically tractable. The number of terms to be retained in the Fourier expansion is dictated by the accuracy desired in the eigenvalues and eigenfunctions.

In the limit of free rotation only one of the Fourier coefficients will be non-vanishing and

$$U_{v\sigma}(\alpha)=\frac{1}{\sqrt{2\pi}}\,e^{i(3k+\sigma)\alpha} \tag{12.21}$$

while the eigenvalues are given by the diagonal elements of (12.20):

$$E_{v\sigma}=F(3k+\sigma)^2 \tag{12.22}$$

Since $(3k+\sigma)$ can have any integral value, a comparison of (12.22) with (12.10) reveals that $m=\sigma$, modulo 3 (see Fig. 12.4). In the high barrier limit the energies are those of a harmonic oscillator and are given by (12.14) as

$$E_v=3(V_3F)^{1/2}(v+\tfrac{1}{2})$$

The energies are independent of σ and hence are threefold degenerate. Each torsional function is a linear combination of three harmonic oscillator functions, each centered about one of the potential minima [8]

$$U_{v0}(\alpha)=\frac{1}{\sqrt{3}}\,[H_v^{(1)}+H_v^{(2)}+H_v^{(3)}] \tag{12.23}$$

$$U_{v1}(\alpha)=\frac{1}{\sqrt{3}}\,[H_v^{(1)}+\varepsilon H_v^{(2)}+\varepsilon^2 H_v^{(3)}] \tag{12.24}$$

$$U_{v,-1}(\alpha)=\frac{1}{\sqrt{3}}\,[H_v^{(1)}+\varepsilon^2 H_v^{(2)}+\varepsilon_v^{(3)}] \tag{12.25}$$

where $\varepsilon=\exp(i2\pi/3)$ and where $H_v^{(1)}$, $H_v^{(2)}$, and $H_v^{(3)}$ are harmonic oscillator functions centered, respectively, about 0, $2\pi/3$, and $4\pi/3$.

Fortunately, the detailed calculations outlined here are usually not necessary because the required perturbation sums, which can be obtained from a knowledge of the torsional eigenfunctions and eigenvalues, have been tabulated for various values of s, and interpolation of these tables is sufficiently accurate for barrier determinations.

Up to this point we have been concerned with the torsional energy levels for a molecule undergoing hindered internal rotation. However, besides these levels, there is a set of energy levels arising from the overall rotation of the entire molecule which will be associated with each of the torsional sublevels. The interaction between overall and internal rotation provides the major mechanism (for asymmetric rotors) by which the splittings of the internal torsional levels are transmitted to the rotational spectrum. Since it turns out that the coupling effect differs for the two torsional sublevels, a given rotational transition associated with the A and E sublevels of a particular torsional state v will appear as a doublet rather than a single line. This doublet separation is a sensitive function of the barrier height, and it is from these splittings that the barrier height may be determined.

The ensuing discussion will be concerned with the derivation of the Hamiltonian that describes a molecule undergoing both internal and overall rotation. From this Hamiltonian, a quantitative discussion of the effects of internal rotation on the rotational spectrum can be made.

4 HAMILTONIAN FOR INTERNAL AND OVERALL ROTATION

The formulation of the internal rotation problem follows the usual procedure which first requires choice of a molecular model from which the classical kinetic and potential energies are developed and subsequently the classical Hamiltonian. The transition to the quantum-mechanical Hamiltonian is then usually accomplished in a straightforward way.

The model chosen for the problem of internal rotation is comprised of a rigid symmetric top (e.g., a CH_3 group) attached to a rigid frame, which may or may not be asymmetric. The two rigid parts rotate about the bond connecting them. The symmetric top (CH_3) is regarded as rotating with respect to the other part which is taken as the framework. The other modes of internal vibration are ignored. This separation of the internal rotation from the remaining $(3N-6-1)$ modes of vibration has proved to be a satisfactory approximation for most of the molecules investigated. The separation is expected to be valid if the torsional frequency is not close to any of the other vibrational frequencies. The entire molecule made up of symmetric top and framework is also rotating and translating in space. The translational motion may be readily separated, and there are then four degrees of freedom—three for overall rotation, which may be described by the three Eulerian angles θ, ϕ, χ of the framework; and one for hindered rotation of the two groups, which is described by the relative angle α between the two groups.

Two methods of approach have been used to handle the problem of internal rotation. These methods differ in choice of coordinate system, and each leads to a somewhat different mathematical formalism. The method introduced by Wilson [21] and Crawford [22] uses the principal axes of the whole molecule as the coordinate system. The axis of the internal rotation coincides with the symmetry axis of the top, but this axis may not be coincident with any one of the principal axes. The terms in the Hamiltonian which describe the interaction between internal and overall rotation are treated by perturbation theory. The boundary condition of invariance under $\alpha \rightarrow \alpha + 2\pi$ applied to the total wave function requires that $U(\alpha) = U(\alpha + 2\pi)$, as applied previously. Physically, this implies that when the original configuration of the system is restored, the wave function describing the system must take on its initial value, that is, be single-valued. This leads to torsional eigenfunctions corresponding to periodic Mathieu functions which have been tabulated. This method is commonly referred to as the principal-axis method (PAM).

An alternate method developed by Nielsen [23] and Dennison et al. [7, 24–26] uses a coordinate system in which the symmetry axis of the top is chosen as one of the coordinate axes. The other two axes are fixed with respect to the framework, and their orientation may be judiciously chosen. One or more coordinate transformations are employed to eliminate or minimize the rotation–torsion coupling; the interactions are then more readily treated by perturbation theory. Some complications, however, arise because the coordinate system chosen is, in general, not a principal axis system and the products of inertia do not vanish. Also the boundary condition of invariance now requires nonperiodic Mathieu functions which have not been tabulated. This method is usually referred to as the internal-axis method (IAM).

In many cases either method is equally applicable, but for slightly asymmetric molecules with light frames, such as methyl alcohol, the internal axis method is to be preferred. The review article by Lin and Swalen [8] discusses these two methods of approach in detail. We shall limit consideration to the principal axis method by which a large class of molecules can be handled. One advantage of the principal axis method is that the necessary perturbation sums have been extensively tabulated; hence analysis of the spectra can be carried out by use of these tables and those developed for rigid rotor molecules. Only the theory for molecules with one internal rotor which is a symmetric top will be treated in detail. Two-top molecules are discussed briefly in Section 10.

For the model of a rigid symmetric top attached to a rigid asymmetric framework (an asymmetric molecule), the principal-axis method uses a coordinate system x, y, z rigidly attached to the framework, the origin of which is located at the center of mass of the molecule and the orientation of which coincides with that of the principal axes of the entire molecule. The model is

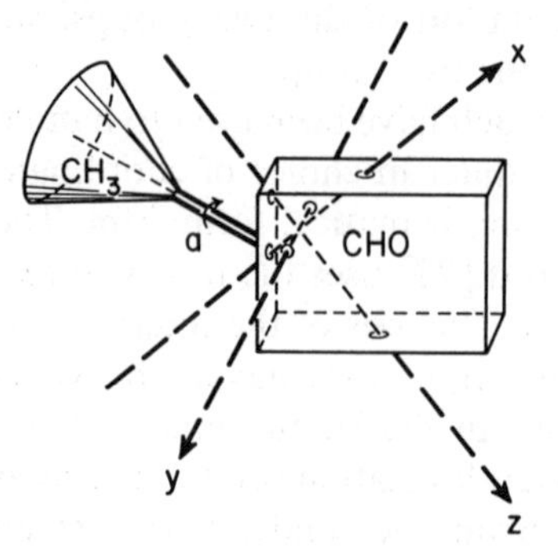

Fig. 12.5 Model used to treat overall and internal rotation.

illustrated in Fig. 12.5. The products of inertia therefore vanish since the coordinate axes are principal axes. Because of the cylindrical symmetry of the top, the principal axes of the molecule are not altered by the rotation of the top relative to the frame, and the moments of inertia of the entire molecule are constants independent of the angle of internal rotation. The moments of inertia, however, would not be independent of α if the internal rotor were not a symmetric top, for example, CH_2DCHO. One of the first molecules whose microwave spectrum was analyzed by means of the PAM is acetaldehyde, studied by Kilb et al. [27].

With the system of coordinates described above, the kinetic energy for a general type of asymmetric molecule (no symmetry apart from that of the CH_3 group) can be written as [22]:

$$T=\frac{1}{2}\sum_g I_g\omega_g^2+\tfrac{1}{2}I_\alpha\dot{\alpha}^2+I_\alpha\dot{\alpha}\sum_g \lambda_g\omega_g \qquad (g=x, y, z) \tag{12.26}$$

where I_g are the principal moments of inertia of the entire molecule, I_α is the moment of inertia of the top about its symmetry axis, ω_g are components along the principal axes of the angular velocity of the framework, $\dot{\alpha}$ is the angular velocity of the top relative to the framework, and λ_g are direction cosines between the symmetry axis of the top and the principal axes of the entire molecule. The first term in the kinetic energy expression represents the energy of overall rotation of the molecule. The second term is the kinetic energy of internal rotation of the top, and the last term represents the coupling between internal and external rotation.

To obtain the Hamiltonian we re-express the kinetic energy as a function of the angular momenta. The angular momenta are defined classically by the relations:

$$P_g=\frac{\partial T}{\partial \omega_g}=I_g\omega_g+\lambda_g I_\alpha\dot{\alpha} \tag{12.27}$$

$$p=\frac{\partial T}{\partial \dot{\alpha}}=I_\alpha\dot{\alpha}+I_\alpha\sum_g \lambda_g\omega_g \tag{12.28}$$

From these relations it is evident that the components P_g of the total angular momentum of the molecule contain a contribution from the internal rotation of the top, whereas p, the total angular momentum of the internal rotor, contains contributions from both internal ($I_\alpha\dot{\alpha}$) and overall rotation. Employing (12.27) and the expression for the kinetic energy, (12.26), we may write

$$2T-\sum\frac{P_g^2}{I_g}=rI_\alpha\dot{\alpha}^2 \tag{12.29}$$

The coefficient rI_α is interpreted as the reduced moment of inertia for internal rotation of the two rigid parts of the molecule with r a reducing factor defined as

$$r=1-\sum_g\frac{\lambda_g^2 I_\alpha}{I_g} \tag{12.30}$$

For the case of two coaxial symmetric tops, $rI_\alpha = I_\alpha(I_z - I_\alpha)/I_z = I_\alpha I_\beta/I_z$, in which I_z is the total moment of inertia of the two tops about the symmetry axis. This is equivalent to the reduced moment expression given previously in (12.7). From (12.27) and (12.28) it is seen that

$$p - \mathscr{P} = rI_\alpha\dot{\alpha} \tag{12.31}$$

with

$$\mathscr{P} = \sum_g \rho_g P_g \qquad \left(\rho_g = \frac{\lambda_g I_\alpha}{I_g}\right) \tag{12.32}$$

The quantity $p - \mathscr{P}$ represents the relative angular momentum of the top and frame. By substituting (12.31) into (12.29), we obtain the kinetic energy in the following convenient form given by Herschbach [28]

$$T = \frac{1}{2}\sum_g \frac{P_g^2}{I_g} + \frac{1}{2}\frac{(p - \mathscr{P})^2}{rI_\alpha} \tag{12.33}$$

If the internal motion is frozen, that is, $\dot{\alpha} = 0$, then the relative angular momentum vanishes, and (12.33) or (12.29) yields the usual expression for the kinetic energy of a rigid rotor in terms of the total angular momentum components. In addition, the angular momenta would no longer include contributions from internal rotation.

Addition of the potential energy to the above kinetic energy expression gives the classical Hamiltonian. The coefficients appearing in (12.33) are constants; therefore the classical kinetic energy becomes a quantum mechanical operator if P_g and p are simply considered as operators. If the angular momentum is measured in units of $\hbar$, the Hamiltonian operator is expressed as

$$\mathscr{H} = \mathscr{H}_r + F(p - \mathscr{P})^2 + V(\alpha) \tag{12.34}$$

in which the potential energy restricting internal rotation is usually assumed to have the form given in (12.5). The $\mathscr{H}_r$ is the usual rigid rotor Hamiltonian and $F(=\frac{1}{2}\hbar^2/rI_\alpha)$ is the reduced rotational constant for internal rotation. The operator $\mathscr{P}$, a reduced angular momentum, is a linear combination of the P_g as defined by (12.32).

The commutation rules for the angular momenta P_g are the standard ones for components of the total angular momentum along the molecule-fixed axis [see (2.58)]. In addition, there is the commutation relation:

$$pP_g - P_g p = 0 \qquad (g = x, y, z) \tag{12.35}$$

This follows from the fact that as a quantum mechanical operator [29]

$$p = -i\left(\frac{\partial}{\partial\alpha}\right)_{\theta,\phi,\chi} \tag{12.36}$$

whereas the corresponding operators for P_g involve the Eulerian angles θ, ϕ, χ but not α. Moreover, it follows from (12.35) that the operators p and $\mathscr{P}$ commute with each other.

5 HIGH-BARRIER APPROXIMATION

The Hamiltonian, (12.34), can be divided into three parts as follows

$$\mathscr{H}=\mathscr{H}_R+\mathscr{H}_T+\mathscr{H}_{TR} \tag{12.37}$$

with

$$\mathscr{H}_R=\mathscr{H}_r+F\mathscr{P}^2 \tag{12.38}$$

$$\mathscr{H}_T=Fp^2+\tfrac{1}{2}V_3(1-\cos 3\alpha) \tag{12.39}$$

$$\mathscr{H}_{TR}=-2F\mathscr{P}p \tag{12.40}$$

where we shall write

$$\mathscr{H}_r=A_xP_x^2+B_yP_y^2+C_zP_z^2 \tag{12.41}$$

in which $A_x=\hbar^2/2I_x$, and so on. The terms $\mathscr{H}_R$ and $\mathscr{H}_T$ are, respectively, the rotational and torsional parts, and $\mathscr{H}_{TR}$ represents the coupling between angular momenta of internal and overall rotation. The internal rotation or torsional Hamiltonian $\mathscr{H}_T$ leads to Mathieu's differential equation and hence to eigenvalues and eigenfunctions discussed in Section 3. The term $\mathscr{H}_R$ is simply a quadratic form in the total angular momentum. The term $F\mathscr{P}^2$ in $\mathscr{H}_R$ introduces terms of the type $P_gP_{g'}$ and P_g^2. The latter terms can be absorbed into the usual rigid rotor expression ($\mathscr{H}_r$) by a modification of the definition of the rigid-rotor rotational constants. As long as the coupling term $\mathscr{H}_{TR}$ is ignored, the Hamiltonian is separable into torsional and rotational parts, and the internal rotation does not directly affect the rotational spectrum. This follows since $\mathscr{H}_T$ is a function only of α and is hence independent of the rotational quantum numbers, whereas $\mathscr{H}_R$ is independent of the internal coordinate α and therefore of the torsional quantum numbers. Inclusion of $-2F\mathscr{P}p$ prevents the Hamiltonian from being separable, since this term depends on both sets of quantum numbers. For molecules with sufficiently large barriers, so that the separation of torsional levels of different v is large compared with the rotational levels, this coupling term can be conveniently treated as a perturbation.

For evaluation of the energy levels, the matrix formulation is employed in which the matrix of $\mathscr{H}$ is constructed in a suitable representation. If the unperturbed Hamiltonian is taken as $\mathscr{H}_R+\mathscr{H}_T$, a convenient basis would be the product functions $\psi_R\psi_T$, with ψ_R a rotational eigenfunction of $\mathscr{H}_R$ with energy E_R and $\psi_T(\equiv U_{v\sigma})$ a torsional eigenfunction of $\mathscr{H}_T$ with energy $E_T(\equiv E_{v\sigma})$. Here the set of rotational quantum numbers is represented by R and the set of torsional quantum numbers by T. In this basis both $\mathscr{H}_R$ and $\mathscr{H}_T$ are diagonal, but $\mathscr{H}_{TR}$ contributes off-diagonal elements. In particular, for the functions $U_{v\sigma}$ the operator p of $\mathscr{H}_{TR}$ is diagonal in σ but has both diagonal and off-diagonal elements in the principal torsional quantum number v [30]. The infinite-order Hamiltonian matrix can be grouped into blocks corresponding to different v. Within each block are the matrix elements of $\mathscr{H}$, all of which are diagonal in v and σ, but not necessarily in R. Associated with each block corresponding to a given v there are different submatrices corresponding to the dif-

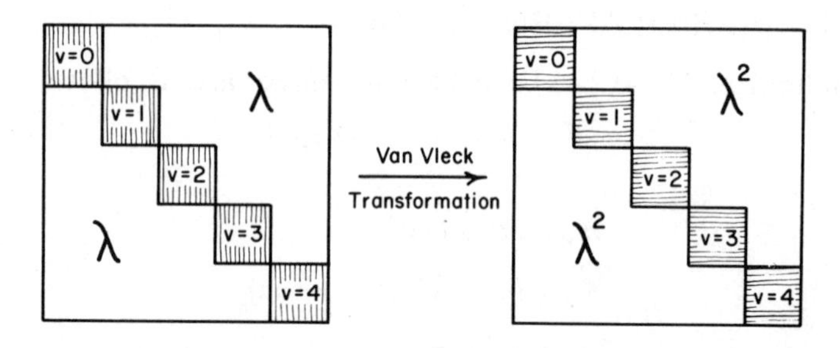

Fig. 12.6 A portion of the energy matrix before and after the Van Vleck transformation. The blocks along the diagonal contain the matrix elements of internal and overall rotation, which are diagonal in the torsional quantum number T, in particular, in v and σ. The unshaded area contains the matrix elements connecting the rotational levels of the various torsional states v. After the transformation these elements are reduced to second order, λ^2, and corrections are introduced within each diagonal block. If the nondiagonal terms in v are now neglected, the transformed diagonal blocks can be diagonalized separately to give the rotational energy levels for each torsional state correct to second order in the separation of rotation and torsion.

ferent torsional sublevels. Blocks of different v are connected by matrix elements of $\mathscr{H}_{TR}$, the nondiagonal elements in v coming from the factor p. A portion of the energy matrix is illustrated in Fig. 12.6. To proceed with the solution further, we must attain some simplification of the matrix. To achieve approximate v diagonalization of this matrix we use a Van Vleck transformation [31, 32]. The effect of this transformation will be to reduce the off-diagonal matrix elements in v from $\mathscr{H}_{TR}$ so that they may be neglected. With neglect of the reduced off-diagonal elements, the transformed matrix may be factored into smaller rotational submatrices, one for each torsional state (see Fig. 12.6). We are then left with the problem of diagonalizing smaller submatrices. The transformation has the effect of modifying the elements diagonal in v by folding the elements off-diagonal in v into the v blocks.

To second order, the matrix elements of $\mathscr{H}$ within a T block, from Appendix C, will be replaced after the Van Vleck transformation by:

$$(R,T|\mathscr{H}|R'T)+\frac{1}{2}\sum_{R'',T''}{}' (R,T|\mathscr{H}'|R'',T'')(R''T''|\mathscr{H}'|R'T)$$
$$\times\left(\frac{1}{E^0_{RT}-E^0_{R''T''}}+\frac{1}{E^0_{R'T}-E^0_{R''T''}}\right) \tag{12.42}$$

where $E^0_{RT}=E_R+E_T$ and $\mathscr{H}'$ is the perturbation giving rise to nondiagonal elements in T. The term $T''=T$ is to be excluded from the sum over R'' and T''. The perturbation term $\mathscr{H}'$ is to be taken as $\mathscr{H}_{TR}(=-2F\mathscr{P}p)$. Since the perturbation is diagonal in σ and hence no degenerate or near-degenerate torsional states are connected, it is a good approximation, for reasonably high barriers, to neglect the energy spacing between rotational levels compared to that between torsional states, i.e., $E^0_{RT}-E^0_{R''T''}\simeq E_T-E_{T''}=E_{v\sigma}-E_{v''\sigma}$. This will allow us to express the energies of a given torsional state in terms of an effective Hamiltonian involving only operators of the total angular momentum.

If the rotational energy differences are neglected, the denominators of (12.42) will no longer involve R, and the summation in R'' and T'' can be separated. Since each matrix element in the numerator can be factored as $(R|\mathscr{P}|R'')\cdot(T|p|T'')$, the R'' summation is simply matrix multiplication giving a factor $(R|\mathscr{P}^2|R')$. Explicitly, we obtain for the matrix elements of a given $v\sigma$ block:

$$(R, v, \sigma|\mathscr{H}_R+\mathscr{H}_T+\mathscr{H}_{TR}|R', v, \sigma)+4F^2(R|\mathscr{P}^2|R')\sum_{v'}{}' \frac{|(v, \sigma|p|v', \sigma)|^2}{\Delta_{vv'}} \tag{12.43}$$

where $\Delta_{vv'}=E_{v\sigma}-E_{v'\sigma}$. This is just the RR'th matrix element of the following effective rotational Hamiltonian operator for the torsional state $v\sigma$:

$$\mathscr{H}_{v\sigma}=\mathscr{H}_r+F[W_{v\sigma}^{(0)}+W_{v\sigma}^{(1)}\mathscr{P}+W_{v\sigma}^{(2)}\mathscr{P}^2] \tag{12.44}$$

where $F=\frac{1}{2}\hbar^2/rI_\alpha$, $\mathscr{P}=\sum \rho_g P_g$ and where

$$W_{v\sigma}^{(0)}=\frac{E_{v\sigma}}{F} \tag{12.45}$$

$$W_{v\sigma}^{(1)}=-2(v, \sigma|p|v, \sigma) \tag{12.46}$$

$$W_{v\sigma}^{(2)}=1+4F\sum_{v'}{}' \frac{|(v, \sigma|p|v', \sigma)|^2}{\Delta_{vv'}} \tag{12.47}$$

with

$$\Delta_{vv'}=E_{v\sigma}-E_{v'\sigma}=\tfrac{9}{4}F(b_{v\sigma}-b_{v'\sigma}) \tag{12.48}$$

and

$$(v, \sigma|p|v', \sigma)=-i\int_0^{2\pi} U_{v\sigma}^*(\alpha)\frac{\partial}{\partial\alpha}\, U_{v'\sigma}(\alpha)\, d\alpha \tag{12.49}$$

This effective rotational Hamiltonian is correct to second order. The W coefficients introduced by Herschbach [28] depend on the barrier, in particular on the ratio V_3/F. The Hamiltonian of (12.44) can also be written in the equivalent form [33, 8]:

$$\begin{aligned}\mathscr{H}_{v\sigma}&=A_{v\sigma}P_x^2+B_{v\sigma}P_y^2+C_{v\sigma}P_z^2+FW_{v\sigma}^{(1)}\sum_g \rho_g P_g\\&\quad+\tfrac{1}{2}FW_{v\sigma}^{(2)}\sum_{\substack{g,g'\\g\neq g'}} \rho_g\rho_{g'}(P_gP_{g'}+P_{g'}P_g)+E_{v\sigma} \qquad (g=x, y, z)\end{aligned} \tag{12.50}$$

where

$$\rho_g=\frac{\lambda_g I_\alpha}{I_g} \tag{12.51}$$

and

$$\begin{aligned}A_{v\sigma}&=A_x+W_{v\sigma}^{(2)}F\rho_x^2\\B_{v\sigma}&=B_y+W_{v\sigma}^{(2)}F\rho_y^2\\C_{v\sigma}&=C_z+W_{v\sigma}^{(2)}F\rho_z^2\end{aligned} \tag{12.52}$$

Table 12.2 Perturbation Coefficients for a Threefold Barrier[a,b]

	Nondegenerate Sublevel (*A symmetry*)			*Degenerate Sublevel* (*E symmetry*)				
s	$n=0$	$n=2$	$n=4$	$n=0$	$n=1$	$n=2$	$n=3$	$n=4$
				Ground Torsional Level ($v=0$)				
8	5.5936	0.280631	−0.0799	5.7972	−0.2617257	−0.137544	0.213762	0.0308
10	6.4056	0.180716	−0.05725	6.5335	−0.1602753	−0.089791	0.124274	0.02676
12	7.1226	0.117586	−0.03948	7.2047	−0.1014166	−0.058664	0.077241	0.019312
16	8.3688	0.052068	−0.01841	8.4047	−0.0437764	−0.026025	0.032531	0.009178
20	9.4499	0.024473	−0.008819	9.4667	−0.0203828	−0.012236	0.015012	0.004408
28	11.3091	0.006261	−0.002281	11.3134	−0.0051860	−0.003130	0.003797	0.001141
36	12.9096	0.001856	−0.000678	12.9108	−0.0015359	−0.000928	0.001123	−0.0003391
40	13.6418	0.001055	−0.000386	13.6425	−0.0008723	−0.0005272	0.0006379	0.0001928
48	15.0029	0.0003643	−0.0001331	15.0031	−0.0003012	−0.0001819	0.0002202	0.0000666
56	16.2539	0.0001354	−0.0000495	16.2540	−0.0001120	−0.00006769	0.0000819	0.00002469
60	16.8457	0.0000845	−0.0000309	16.8457	−0.00006992	0.00004227	0.00005112	0.00001545
68	17.9726	0.0000343	−0.00001252	17.9726	−0.00002835	−0.00001714	0.00002074	0.00000626
72	18.5111	0.00002220	−0.00000812	18.5111	−0.00001838	−0.00001111	0.00001345	0.00000406
80	19.5449	0.00000963	−0.00000352	19.5449	−0.00000796	−0.00000482	0.00000582	0.00000176
88	20.5280	0.00000433	−0.00000158	20.5280	−0.00000358	−0.00000217	0.00000262	0.00000079
92	21.0028	0.00000294	−0.00000107	21.0028	−0.00000243	−0.00000147	0.00000178	0.00000054
100	21.9222	0.00000140	−0.00000051	21.9222	−0.00000115	−0.00000069	0.00000084	0.00000026

First Excited Torsional Level ($v=1$)								
16	24.1805	−1.27334	0.7870	23.4277	0.7977206	0.590647	−0.42680	−0.2421
20	27.2238	−0.654112	0.31265	26.8077	0.4695574	0.319985	−0.294937	−0.13408
28	32.6645	−0.200728	0.07947	32.5300	0.1594403	0.100170	−0.111855	−0.03905
36	37.4427	−0.069411	0.02603	37.3955	0.0566696	0.034699	−0.040897	−0.012993
40	39.6401	−0.042117	0.015628	39.6114	0.0345699	0.021057	−0.0250835	−0.007809
48	43.7320	−0.0162968	0.005989	43.7209	0.0134408	0.0081484	−0.0097997	−0.002995
56	47.4959	−0.0066663	0.002442	47.4914	0.0055072	0.0033332	−0.0040219	−0.0012210
60	49.2765	−0.0043405	0.0015888	49.2735	0.00358727	0.0021702	−0.0026207	−0.0007944
68	52.6667	−0.0018985	0.0006942	52.6654	0.00156961	0.0009493	−0.0011471	−0.0003472
72	54.2864	−0.0012734	0.0004655	54.2855	0.00105298	0.0006367	−0.0007698	−0.0002327
80	57.3951	−0.0005876	0.0002148	57.3947	0.00048597	0.0002938	−0.0003553	−0.0001074
88	60.3509	−0.0002796	0.0001022	60.3507	0.00023119	0.0001398	−0.001690	−0.0000511
92	61.7780	−0.0001948	0.0000712	61.7779	0.00016112	0.0000974	−0.0001178	−0.0000356
100	64,5416	−0.0000964	0.0000352	64.5415	0.00007974	0.0000482	−0.0000583	−0.000017
Second Excited Torsional Level ($v=2$)								
36	58.4614	0.890641	−0.2370	59.1178	−0.8593042	−0.428291	0.721421	0.06245
40	62.3641	0.619322	−0.18380	62.8093	−0.5670974	−0.304883	0.456080	0.07625
48	69.4772	0.291871	−0.09759	69.6812	−0.2522742	−0.145546	0.192538	0.04752
56	75.9067	0.137167	−0.04822	76.0014	−0.1156628	−0.068549	0.086201	0.02400
60	78.9255	0.09449	−0.03365	78.9905	−0.0791721	−0.047238	0.058634	0.016791
68	84.6483	0.045504	−0.016435	84.6795	−0.0378571	−0.022751	0.027842	0.008215
72	87.3744	0.031829	−0.011540	87.3962	−0.02643006	−0.0159144	0.0194013	0.005769
80	92.5971	0.0158271	−0.005763	92.6080	−0.01311438	−0.0079136	0.0096063	0.002882
88	97.5549	0.0080372	−0.002932	97.5604	−0.00665303	−0.0040186	0.0048686	0.001466
92	99.9466	0.0057707	−0.002107	99.9505	−0.00477549	−0.0028885	0.0034936	0.001053
100	104.5751	0.0030175	−0.001102	104.5772	−0.00249627	−0.0015088	0.0018255	0.000552

[a]From Herschbach [28].

[b]The coefficients, $W_{v\sigma}^{(n)}$, are tabulated as functions of the dimensionless parameter $s=\frac{4}{9}(V_3/F)$.

Note that the rotational constants of the rigid molecule $A_x, \ldots,$ are modified by the internal rotation terms arising from $W_{v\sigma}^{(2)} \mathscr{P}^2$.

The perturbation treatment of $\mathscr{H}_{TR}$ can be carried to higher order by application of successive Van Vleck transformations. The effective rotational Hamiltonian for a given torsional state obtained from an nth order perturbation calculation can be expressed simply as [28]

$$\mathscr{H}_{v\sigma} = \mathscr{H}_r + F \sum_n W_{v\sigma}^{(n)} \mathscr{P}^n \tag{12.53}$$

The first term is the rigid asymmetric rotor Hamiltonian, and the last term, which is a power series in $\mathscr{P}$, represents the effect of internal motion on the rotational motion. The $W_{v\sigma}^{(n)}$ are the usual nth order perturbation sums, except for $n=2$ which contains an added contribution arising from the $F\mathscr{P}^2$ term in the original Hamiltonian [see (12.47)].

The dimensionless perturbation coefficients $W_{v\sigma}^{(n)}$ depend on the reduced barrier height s, and these coefficients (except $n=0$) decrease in size with increasing barrier, becoming proportional to the torsional sublevel splittings $(E_{vE} - E_{vA})$ and hence vanishing at the limit of infinite barrier height [28]. At this limit, $\mathscr{H}_{v\sigma}$ reduces to the rigid rotor Hamiltonian plus the energy of a torsional harmonic oscillator, and a usual rigid rotor spectrum would be obtained.

The zeroth order term $W_{v\sigma}^{(0)}$ is just the torsional energy contribution from $\mathscr{H}_T$, and does not affect the rotational spectrum. The first-order term $W_{v\sigma}^{(1)}$ is nonzero only for the degenerate E sublevels. The second-order term $W_{v\sigma}^{(2)}$ differs both in sign and magnitude for the A and E sublevels of a given torsional state v. The following properties of the perturbation coefficients may be listed [28]:

$$W_{v0}^{(n)} = 0 \quad \text{for } n \text{ odd} \tag{12.54}$$

$$W_{v1}^{(n)} = (-1)^n W_{v,-1}^{(n)} \tag{12.55}$$

$$W_{v0}^{(n)} \simeq -2W_{v1}^{(n)} \quad \text{for } n>0 \text{ and even} \tag{12.56}$$

$$\frac{W_{v1}^{(n+2)}}{W_{v1}^{(n)}} \simeq \frac{-(2\pi/3)^2}{[(n+1)(n+2)]} \qquad \text{for } n>0 \tag{12.57}$$

The first two relations are true in general, whereas the latter two relations hold for relatively high barriers. The perturbation coefficients have been tabulated for various values of n and v and for a wide range of the parameter s [18, 19, 28]. The recent tabulation by Hayashi and Pierce [19] lists the perturbation coefficients for $v=0$, 1, 2 and $n=0, 1, \ldots, 6$, d for s values from 4 to 200 with the interval in s being 2 for $s<100$ and 4 for $s>100$. These tables have also been reproduced in the book by Wallrab [121]. A short tabulation of the perturbation coefficients for the $v=0$, 1, and 2 torsional levels are given in Table 12.2. A discussion of the method for evaluating these perturbation coefficients, which avoids the laborious evaluation of the perturbation sums, may be found elsewhere [28, 34]. This is the so-called "bootstrap" procedure which is based on the fact the $W_{v\sigma}^{(n)}$ are not dependent on the molecular asymmetry.

In the derivation of the effective rotational Hamiltonian the rotational energy differences were neglected in comparison with the torsional energy differences. This approximation is adequate for molecules with high barriers but may not be so for those with low and intermediate barriers or small moments of inertia. The error caused by ignoring rotational spacing can be corrected by expansion of the denominators in a Taylor series. This has been discussed by Herschbach [28] and by Stelman [35] and its consideration requires introducing another perturbation coefficient $W_{v\sigma}^{(d)}$ into the effective rotational Hamiltonian. These denominator corrections have no appreciable effect on the determination of the barrier height for high barrier cases [see, e.g., (12.61)] because the contributions to the rotational constants are essentially the same for both torsional sublevels. Nevertheless, they can contribute several megahertz to the rotational constants.

6 HIGH-BARRIER ASYMMETRIC ROTORS

We have seen, by treating the coupling of internal and overall rotation by the degenerate perturbation theory of Van Vleck, that an effective rotational Hamiltonian may be obtained for each torsional level. For a particular torsional state v there will be a set of rotational energy levels associated with each of the two torsional sublevels $\sigma=0$ or A and $\sigma=\pm1$ or $E(\sigma=+1$ or -1 lead to the same energy levels). The two sets of rotational energy levels are obtained by solution of the effective Hamiltonians $\mathcal{H}_{vA}$ and $\mathcal{H}_{vE}$ of (12.53). The number of correction terms to $\mathcal{H}_r$ that must be retained in (12.53) for analysis of the effects of internal rotation varies significantly from molecule to molecule and is dependent on the barrier height and the asymmetry. Also the complexity of the operator $\mathcal{P}$ depends on the molecular geometry. For molecules with a plane of symmetry containing the internal rotor axis, for example, CH_3CHO, some simplification results since the direction cosine λ_y vanishes (y axis perpendicular to the symmetry plane). For two planes of symmetry, for example, CH_3BF_2, both λ_x and λ_y vanish, and $\mathcal{P}$ reduces to $\rho_z P_z$. Even when the molecule has only one plane of symmetry, the internal rotor axis may lie so close to one of the principal axes in the plane that the direction cosine referring to the other principal axis in the plane will be negligibly small and $\mathcal{P}$ will again depend only on the component P_z. It may be mentioned that it is usually convenient to choose the axis of quantization, the z axis, to be the axis making the smallest angle with the top axis, even if this is not the best choice for evaluation of the asymmetric rigid rotor energy levels. Hence in dealing with specific molecules the symbols $A_{v\sigma}$, $B_{v\sigma}$, and $C_{v\sigma}$ [e.g., in (12.50)] may have to be permuted to satisfy the usual convention $A \geqslant B \geqslant C$.

Energy Levels and Spectra of the Pseudorigid Rotor

When barriers are relatively high ($s>30$) [36], terms up to $n=2$ in (12.53) are often sufficient, and the Hamiltonian is then given by (12.50). The cross terms of the form $(P_g P_{g'} + P_{g'} P_g)$ present in this expression can be eliminated as usual by a rotation of the coordinate axes to obtain a new set of principal axes in which these terms vanish. If, for example, only one such cross term were present, a simple 2×2 rotation in the gg' plane may be chosen to remove the term. The orientation of the new coordinate system, referred to as the "effective principal-axis system" will vary with the torsional state since the coefficients of the cross terms contain $W_{v\sigma}^{(2)}$. This transformation alters the rotational constants and the coefficients of the linear terms [8, 27]. Often the modifications produced by such a rotation of coordinate axes is small, and negligible error is introduced by just ignoring the cross terms. The linear terms in P_g, present only for the degenerate E levels since $W_{vA}^{(1)}=0$, can be neglected for a barrier greater than 1000 cal/mole and an asymmetry of $|\kappa|<0.8$ [27]. Therefore, in the cases where the linear terms may be neglected, the effective rotational Hamiltonian may be put into pseudorigid rotor form for both the $A(\sigma=0)$ and $E(\sigma=1)$ sublevels. The Hamiltonians are

$$\mathscr{H}_{vA} = A_{vA} P_x^2 + B_{vA} P_y^2 + C_{vA} P_z^2 \tag{12.58}$$

$$\mathscr{H}_{vE} = A_{vE} P_x^2 + B_{vE} P_y^2 + C_{vE} P_z^2 \tag{12.59}$$

where the effective rotational constants contain contributions from internal rotation, that is,

$$A_{vA} = \frac{\hbar^2}{2I_x} + F\rho_x^2 W_{vA}^{(2)} \tag{12.60}$$

etc. Since the $W_{v\sigma}^{(2)}$ differ for the A and E levels, the effective rotational constants will differ. Thus there is a set of pseudorigid rotor energy levels, $E_{J_\tau v\sigma}$, associated with the A torsional sublevel and with the E sublevel.

The usual selection rules apply for the rotational transitions with the additional restriction that $\Delta\sigma=0$, which results because the dipole moment is independent of the internal rotation angle. Transitions from one internal sublevel to another ($vA \nleftrightarrow vE$) are thus not allowed, and the vA and vE sublevels have separate rigid-rotorlike rotational spectra. Each rotational transition ($J_\tau \rightarrow J'_{\tau'}$) in a torsional state v will hence appear as a doublet rather than as a single line. The A and E lines are interpreted by the Hamiltonians of (12.58) and (12.59), and two separate sets of rotational constants are found to fit the doublets by interpolation of existing tables of rigid rotor energy levels. The analysis is therefore no more involved than that discussed previously in Chapter VII for an ordinary rigid asymmetric rotor except the rigid rotor energy is now a so-called pseudorigid rotor energy computed with the effective rotational constants of (12.60). The barrier may be evaluated from the differences in these pseudorigid rotor rotational constants. For example,

$$\Delta A = A_{vA} - A_{vE} = F\rho_x^2 [W_{vA}^{(2)} - W_{vE}^{(2)}] \tag{12.61}$$

determines the quantity $W_{vA}^{(2)} - W_{vE}^{(2)}$, once the moment of inertia of the methyl group and the direction cosines are established. From this, the barrier height V_3 can be obtained by comparison with the tabulated perturbation coefficients.

From (12.56) the rigid rotor rotational constants which contain the structural information are found to be given by the expressions

$$A_x = \frac{\hbar^2}{2I_x} = \frac{1}{3}(A_{vA} + 2A_{vE}) \tag{12.62}$$

and so on. Furthermore, if (12.56) is a valid approximation, it follows that the shift of the A line from the unperturbed or rigid rotor position is twice that of the E line.

Another procedure is to obtain the barrier height directly from the separation of each doublet [33]. This is particularly useful if the differences in the pseudo-rotational constants are very small. The energy separation between the A and E levels may be expanded as

$$\Delta E = E_A - E_E = \left(\frac{\partial E}{\partial A}\right)\Delta A + \left(\frac{\partial E}{\partial B}\right)\Delta B + \left(\frac{\partial E}{\partial C}\right)\Delta C \tag{12.63}$$

where E_A and E_E are the rigid rotor energy of the levels A and E, respectively, and ΔA, etc., are given by (12.61). Any centrifugal distortion correction is assumed to be the same for both the A and E levels. With the rigid rotor energy expressed by

$$E = \tfrac{1}{2}(A + C)J(J+1) + \tfrac{1}{2}(A - C)E(\kappa) \tag{12.64}$$

we find

$$\frac{\partial E}{\partial A} = \frac{1}{2}\left[J(J+1) + E(\kappa) - (\kappa + 1)\left(\frac{\partial E(\kappa)}{\partial \kappa}\right)\right] \tag{12.65}$$

$$\frac{\partial E}{\partial B} = \frac{\partial E(\kappa)}{\partial \kappa} \tag{12.66}$$

$$\frac{\partial E}{\partial C} = \frac{1}{2}\left[J(J+1) - E(\kappa) + (\kappa - 1)\left(\frac{\partial E(\kappa)}{\partial \kappa}\right)\right] \tag{12.67}$$

The values of $\partial E(\kappa)/\partial\kappa$ may be obtained from tables of $E(\kappa)$. These relations and those of (12.61) will give the energy level differences, ΔE. The difference in the ΔE values for the two levels J_τ, $J'_{\tau'}$ involved in the transition will give the frequency separation of the doublet as a function of the structural parameters, reduced energies, and so on, times the factor $[W_{vA}^{(2)} - W_{vE}^{(2)}]$. Thus, the barrier may be found directly from the observed splittings. If the splittings contain contributions from higher-order effects (see the next section) these contributions would be removed before application of (12.63). It should be noted that in (12.64) the rotational constants have been chosen such that $A \geqslant B \geqslant C$, although this, as indicated previously, is not necessarily the order of $A_{v\sigma}$, $B_{v\sigma}$, and $C_{v\sigma}$; and hence these latter constants must be permuted accordingly.

Higher-order Effects

When the linear terms in P_g are not small compared with the energy spacing between the pseudorigid rotor levels they connect (e.g., the barrier is such that the coefficients, $F\rho_g W_{v\sigma}^{(1)}$, are large or if the asymmetry is small), the energy levels for the E states will not be like that of a rigid rotor. The energy levels may be obtained by diagonalization of the matrix of $\mathscr{H}_{v\sigma}$. With the symmetric rotor functions as basis functions, the Hamiltonian of (12.50) which is correct to second order has the following matrix elements for the torsional state $v\sigma$ [33].

$$(K|\mathscr{H}_{v\sigma}|K) = \tfrac{1}{2}(A_{v\sigma}+B_{v\sigma})J(J+1) + [C_{v\sigma} - \tfrac{1}{2}(A_{v\sigma}+B_{v\sigma})]K^2 + FW_{v\sigma}^{(1)}\rho_z K$$

$$(K|\mathscr{H}_{v\sigma}|K\pm 1) = \tfrac{1}{2}F[W_{v\sigma}^{(2)}\rho_z(\rho_y \pm i\rho_x)(2K\pm 1) + W_{v\sigma}^{(1)}(\rho_y \pm i\rho_x)] \times [J(J+1) - K(K\pm 1)]^{1/2}$$

$$(K|\mathscr{H}_{v\sigma}|K\pm 2) = \tfrac{1}{4}[(B_{v\sigma} - A_{v\sigma}) \pm 2iFW_{v\sigma}^{(2)}\rho_x\rho_y] \times \{[J(J+1) - K(K\pm 1)][J(J+1) - (K\pm 1)(K\pm 2)]\}^{1/2} \quad (12.68)$$

The indices J, M in the matrix elements have been omitted, since $\mathscr{H}_{v\sigma}$ is diagonal in these quantum numbers. Also, the constant term $E_{v\sigma}$ has been dropped because the torsional state does not change in a rotational transition. This matrix differs from that obtained for a rigid rotor in the appearance of the $(K|K\pm 1)$ matrix elements.

For the A levels, $W_{vA}^{(1)}=0$, and the $(K|K\pm 1)$ elements arise only from the quadratic cross terms $(P_g P_{g'} + P_{g'} P_g)$. These terms also contribute to the $(K|K\pm 2)$ elements. They can be neglected or eliminated by choice of a new principal-axis system, as discussed before. As a consequence, the matrix will reduce to that of a rigid rotor with corrected rotational constants, and the energies are obtainable by the usual rigid rotor techniques.

For the E levels the $(K|K\pm 1)$ matrix elements cannot be completely removed even with a rotation of coordinate axes because $W_{vE}^{(1)} \neq 0$, and this complicates the calculation of the $\mathscr{H}_{vE}$ energy levels. The usual factoring obtainable for the rigid rotor matrix is no longer possible, and for a given J one has a $(2J+1)$ by $(2J+1)$ secular determinant to solve. Diagonalization of the secular equation is difficult except for low J. Expressions for some of the low J energy levels are available [8, 12, 37]. Procedures for handling the $n=1$ terms for certain cases have been described [11, 12, 38]. If, for example, the associated direction cosines of the P_x and P_y terms are small enough that these terms may be neglected, then the $(K|K\pm 1)$ matrix elements will arise only from the quadratic cross terms in the angular momentum components, which can be removed by a preliminary transformation, or neglected. The matrix of $\mathscr{H}_{vE}$ with only contributions from the linear term in P_z can now be factored into two submatrices, one for even K and one for odd K, since without the $(K|K\pm 1)$ matrix elements even and odd K are no longer connected. Use of the Wang functions will not, however, lead to additional factoring because of the presence of the linear K term on the diagonal. The eigenvalues of these somewhat smaller secular equations can be more readily determined; for example, see the modified continued fraction procedure of Reference 11. For some of the low J energy levels the matrices are only 2×2 and can be easily solved to give the results of Table 12.3.

Table 12.3 Some Low J Energy Levels[a]

$E_{00}=0$
$E_{10}=A_{v\sigma}+B_{v\sigma}$
$E_{11}=C_{v\sigma}+\frac{1}{2}(A_{v\sigma}+B_{v\sigma})\pm[(F\rho_z W_{v\sigma}^{(1)})^2+\frac{1}{4}(A_{v\sigma}-B_{v\sigma})^2]^{1/2}$
$E_{21}=C_{v\sigma}+\frac{5}{2}(A_{v\sigma}+B_{v\sigma})\pm[(F\rho_z W_{v\sigma}^{(1)})^2+\frac{9}{4}(A_{v\sigma}-B_{v\sigma})^2]^{1/2}$

[a]Energy levels for the Hamiltonian of the E levels in the form $\mathscr{H}_{v\sigma}=A_{v\sigma}P_x^2+B_{v\sigma}P_y^2+C_{v\sigma}P_z^2+FW_{v\sigma}^{(1)}\rho_z P_z$. K, though not a good quantum number, is used to label the energy levels which are designated as E_{JK}.

In many applications, the linear terms can be satisfactorily treated by perturbation theory. In particular, the linear terms are treated as a perturbation of the pseudorigid rotor Hamiltonian ($\mathscr{H}_r$):

$$\mathscr{H}_{vE}=\mathscr{H}_r+\mathscr{H}' \tag{12.69}$$

$$\mathscr{H}_r=A_{vE}P_x^2+B_{vE}P_y^2+C_{vE}P_z^2 \tag{12.70}$$

$$\mathscr{H}'=FW_{vE}^{(1)}\sum_g \rho_g P_g \tag{12.71}$$

where $A_{vE},\ldots,$ are the pseudorigid rotor rotational constants. In the pseudorigid asymmetric rotor representation (basis in which $\mathscr{H}_r$ is diagonal) the P_g have only off-diagonal matrix elements (diagonal in J, off-diagonal in τ) and $\mathscr{H}'$ contributes only in a second- or higher-order perturbation calculation. If $\mathscr{H}'$ is treated with second-order nondegenerate perturbation theory, the rotational energies for the torsional state vE are given by

$$E_{J_\tau}=E_{J_\tau}^0+(FW_{vE}^{(1)})^2\sum_g \rho_g^2 \sum_{\tau'}{}' \frac{|(J,\tau|P_g|J,\tau')|^2}{E_{J_\tau}^0-E_{J_{\tau'}}^0} \tag{12.72}$$

where $E_{J_\tau}^0$ is the pseudorigid rotor energy of the J_τ level. It frequently happens that for a given J near-degeneracies occur between asymmetry doublets J_τ, $J_{\tau'}$ (pairs of adjacent K states); when such levels are coupled by P_g, the nondegenerate perturbation treatment of the linear terms is not applicable. The linear terms can, however, be handled in a manner similar to the way in which the Stark effect for such a situation was treated, that is, by means of a second-order Van Vleck transformation. This leads to a 2×2 secular equation to be solved [39]. The necessary matrix elements and perturbation sums required for solution of the secular equation or application of (12.72) have been tabulated by Dobyns [39] in increments of 0.05 in the asymmetry parameter κ for $J\leqslant 12$.

It has also been pointed out by Rudolph [40] that the required matrix elements of P_g can be expressed in terms of the line strengths, which have been extensively tabulated. In particular, the following relation can be shown to exist [41]

$$\lambda_g(J,\tau;J,\tau')=\frac{2J+1}{J(J+1)}|(J,\tau|P_g|J,\tau')|^2 \tag{12.73}$$

This relation may be used to express (12.72) in terms of the line strengths.

Although it is often sufficient to consider only terms up to $n=2$ in the analysis of internal rotation, higher-order terms in the angular momentum operators become important when the barrier is not sufficiently high. Furthermore, even for a relatively high barrier, higher-order terms may have to be considered. When such terms will be required can be seen by examination of (12.53) in the limit of a symmetric top. In this limiting case $\mathscr{P}^n$ may be expressed as $(\rho K)^n$ where $\rho = I_\alpha/I_z$. Although the $W_{v\sigma}^{(n)}$ terms decrease rapidly as s increases, the convergence of the power series for a given barrier depends on the magnitude of ρK, because for a given barrier the $W_{v\sigma}^{(n)}$ do not converge very rapidly with increasing n. Therefore, the perturbation series converges rapidly only if energy levels have low values of K or if ρ is small. The latter condition is satisfied for molecules with light tops and heavy frames, and hence the PAM method is best suited to this class of molecules. As a rough guide to the number of terms to retain in the Hamiltonian of (12.53), it may be stated that if $\rho K < 0.25$, terms through $n=2$ are sufficient, but if $0.25 < \rho K < 0.55$, terms through $n=4$ should be retained [28].

The matrix elements in a symmetric rotor basis for the Hamiltonian of (12.53) up to $n=4$ are available [28, 42]. The $n=3$ and 4 terms introduce additional matrix elements of the type $(K|K \pm 3)$ and $(K|K \pm 4)$. The solution of such a secular determinant is, however, seldom required since the $n=3$ and 4 terms can usually be treated by perturbation techniques.

For the case where two of the three direction cosines are zero, or negligible, a first-order treatment yields a very simple energy expression. If we choose the asymmetric rigid rotor basis we have

$$E_{v\sigma} = E_r + F \sum_n \rho_z^n W_{v\sigma}^{(n)} \langle P_z^n \rangle \qquad n=2, 4, \ldots \tag{12.74}$$

where all correction terms of (12.53) are treated as first-order perturbations of $\mathscr{H}_r$. To this approximation, all odd n terms vanish. This is particularly convenient when the barrier is high enough that the odd order corrections for the E levels need not be considered. The particularly simple form of the $n=4$ term (and others) in (12.74) arises because $\mathscr{P}$ has been taken to depend significantly only on the P_z component. If this is not the case and all of the $n=4$ terms are required, a term of the type

$$F W_{v\sigma}^{(4)} (\rho_x P_x + \rho_y P_y + \rho_z P_z)^4$$

would appear, one which is comprised of a sum of 81 terms, each a product of four angular momentum components. This term may be written in the equivalent form

$$F W_{v\sigma}^{(4)} \sum \rho_g \rho_{g'} \rho_{g''} \rho_{g'''} P_g P_{g'} P_{g''} P_{g'''}$$

From this form it is immediately evident that these terms have the same form as the centrifugal distortion corrections for an asymmetric rotor (see Chapter VIII). As a consequence, the techniques which have been developed for the handling of centrifugal distortion effects can be employed for the $n=4$ terms with the formal correspondence [28]

$$\frac{\hbar}{4}\tau_{gg'g''g'''}\leftrightarrow F\,W_{v\sigma}^{(4)}\rho_g\rho_{g'}\rho_{g''}\rho_{g'''} \tag{12.75}$$

As before, use of the commutation rules will introduce quadratic terms in the angular momentum components, and hence the rotational constants are modified by terms in $W_{v\sigma}^{(4)}$ in addition to those involving $W_{v\sigma}^{(2)}$ from $\mathscr{P}^2$. For a first-order perturbation treatment with the pseudorigid rotor basis the contribution of the $n=4$ terms to the energy would have the form given in (8.64) for ordinary centrifugal distortion. The distortion coefficients are defined in terms of D_J, and so on, which are now modified as given in Table 12.4. The rotational constants as modified by contributions from both the second- and fourth-order internal rotation terms are also given in Table 12.4. The results of Table 12.4 may also be used to define the reduced pseudocentrifugal distortion constants $\Delta_J,\ldots$. Definitions of the reduced quartic and sextic constants for $\rho_c=0$ may be found elsewhere [43]. Dealing with these pseudocentrifugal distortion terms, $W_{v\sigma}^{(4)}\mathscr{P}^4$, is thus no more complicated than dealing with ordinary centrifugal distortion.

Since the most important contribution usually arises from coupling along one inertial axis, g, the major third-order contribution ($n=3$) will arise from the term P_g^3. Such a term may be included in a perturbation treatment [38, 39]. Fortunately, in many cases the odd n terms, which are more difficult to handle, are not as important as the even n terms. Furthermore, when such terms are important in certain transitions, one can usually find other transitions for the barrier evaluation where the effect of these terms is small.

As mentioned previously for molecules with heavy tops, the internal rotation effects are best treated by the IAM method. A computer program, based on the IAM method, has been described by Woods [44] for treatment of heavy-top molecules such as CF_3CHO. The program is also applicable to molecules with light tops. Extension of the program to molecules with several tops has also been given [45].

Table 12.4 Pseudocentrifugal Distortion Coefficients for Fourth-order Internal Rotation Effects[a]

$$D_J=-(\tfrac{3}{8})(\rho_x^2+\rho_y^2)^2F\,W_{v\sigma}^{(4)}$$
$$D_{JK}=-2D_J-3\rho_z^2(\rho_x^2+\rho_y^2)F\,W_{v\sigma}^{(4)}$$
$$D_K=-D_J-D_{JK}-\rho_z^4F\,W_{v\sigma}^{(4)}$$
$$\delta_J=(\tfrac{1}{4})(\rho_y^4-\rho_x^4)F\,W_{v\sigma}^{(4)}$$
$$R_5=(\tfrac{1}{2})\delta_J-(\tfrac{3}{4})\rho_z^2(\rho_y^2-\rho_x^2)F\,W_{v\sigma}^{(4)}$$
$$R_6=(\tfrac{1}{2})D_J+(\tfrac{1}{4})(\rho_x^4+\rho_y^4)F\,W_{v\sigma}^{(4)}$$
$$A=A_x+F\rho_x^2W_{v\sigma}^{(2)}+F(3\rho_y^2\rho_z^2-2\rho_x^2\rho_y^2-2\rho_x^2\rho_z^2)W_{v\sigma}^{(4)}$$
$$B=B_y+F\rho_y^2W_{v\sigma}^{(2)}+F(3\rho_x^2\rho_z^2-2\rho_x^2\rho_y^2-2\rho_y^2\rho_z^2)W_{v\sigma}^{(4)}$$
$$C=C_z+F\rho_z^2W_{v\sigma}^{(2)}+F(3\rho_x^2\rho_y^2-2\rho_x^2\rho_z^2-2\rho_y^2\rho_z^2)W_{v\sigma}^{(4)}$$

[a] After Herschbach [28].

Continued refinements in the theory of internal rotation of methyl groups have appeared in the literature over the last 10 years, particularly by Dreizler and co-workers and by Bauder and Günthard and co-workers. With the availability of faster and larger capacity computers, powerful direct diagonalization procedures rather than perturbation techniques have been developed [see, e.g., (46–54)].

Rotational Spectra of the Ground Torsional State

From the preceding discussions it is apparent that the appropriate Hamiltonian and procedures for evaluation of the energy levels can range from the relatively simple to the rather complex. In any event, one should remember that the barrier height V_3 is determined by finding the value of $s(=4V_3/9F)$ which in conjunction with the appropriate Hamiltonian satisfactorily accounts for the observed rotational splittings arising from internal rotation. An example of a barrier height calculation is considered later.

Figure 12.7 gives a schematic representation of the energy levels and parallel type ($\Delta K=0$) transitions for different asymmetries and barrier heights. Internal rotation effects in asymmetric rotors lead to doublets consisting of the A and E

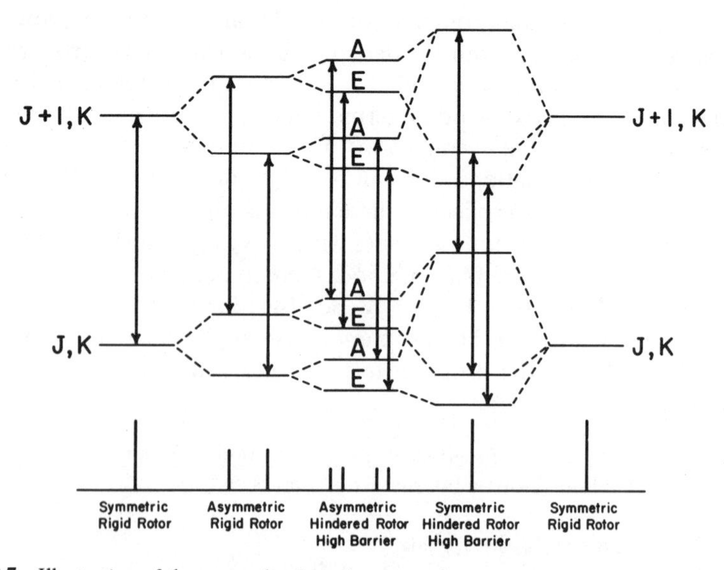

Fig. 12.7 Illustration of the energy levels and rotational transitions ($\Delta K=0$) for a hindered symmetric and asymmetric rotor. For an infinitely high threefold barrier, each J, K level of a symmetric rotor is sixfold degenerate corresponding to the twofold K degeneracy and the threefold torsional level degeneracy. These levels are each split into two asymmetric rotor levels with the introduction of an asymmetry, and two lines are observed. If the barrier is lowered, each asymmetric rotor level is split into two levels, a nondegenerate A level and a degenerate E level, and two doublets are observed. If the asymmetry is removed, the energy levels of a symmetric hindered rotor are obtained. The presence of a high barrier splits each J, K level of a symmetric rotor into three two fold degenerate levels as indicated. The three levels correspond to $\sigma=0, \pm K; \sigma=+1, +K, \sigma=-1, -K; \sigma=+1, -K, \sigma=-1, +K$. However, because of the selection rules only one line is observed.

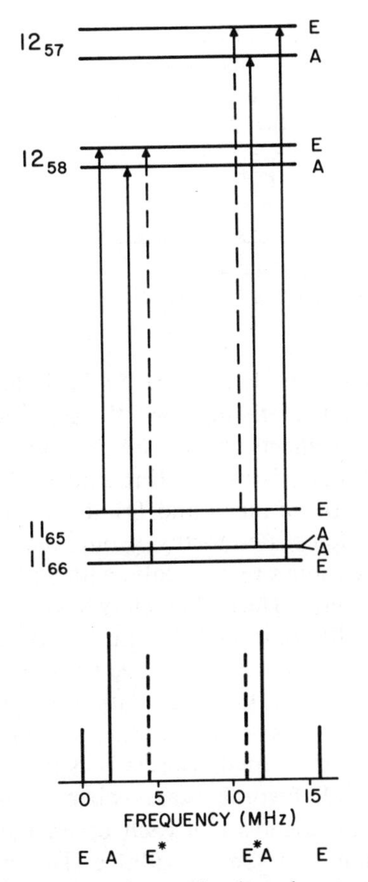

Fig. 12.8 Forbidden transitions in propylene oxide where the asymmetry splitting and internal rotation splittings are comparable. The E^* transitions indicated by dashed lines which are allowed by the general selection rules ($E \leftrightarrow E$) are forbidden for a limiting rigid rotor. From Herschbach and Swalen [11].

lines. These A-E doublets may be closely or widely spaced depending on the barrier height and transitions studied. The A lines will exhibit a pseudorigid rotor spectrum with possible pseudocentrifugal distortion effects if $n=4$ terms are important. This may or may not be true for the E lines, depending on whether the odd n terms are important or not. Furthermore, the odd n terms can result in the observation of additional transitions which are forbidden for a usual rigid rotor as illustrated in Fig. 12.8. General selection rules are summarized in Table 12.5. Derivation of these rules from the symmetry properties of the Hamiltonian are discussed elsewhere [8].

Usually the barrier is determined from the ground torsional state, unless the doublet splittings are unobservable. For the ground torsional state the A line usually, but not always, appears at a higher frequency than the E line. When the barrier is high and the asymmetry large, the Stark effect of a given

Table 12.5 General Selection Rules

Type of Molecule	*Hamiltonian Group*	*Dipole Moment Symmetry*	*Selection Rules*
No symmetry	C_3[a]	A	$A \leftrightarrow A$, $E \leftrightarrow E$
Plane of symmetry	D_3[b]	A_2	$A_1 \leftrightarrow A_2$, $E \leftrightarrow E$

[a]Symmetry species of C_3 are E and A.
[b]Symmetry species of D_3 are E, A_1, and A_2.

rotational transition is essentially the same for both the A and E lines. If, however, the barrier is not sufficiently high or if the asymmetry is small, the E lines can show first-order Stark effects which arise because of the presence of the odd n terms in the Hamiltonian. If the nuclear statistical weights are taken into account, the intensity ratio of the A and E lines is 1:1 for the normal methyl group, whereas for the deuterated methyl group, the ratio is 11:16 with the E line stronger [27]. This applies to molecules with a planar frame and for those with a frame of no symmetry. These characteristics can be useful aids in distinction and assignment of the rotational spectra. It may be pointed out, however, that the barrier has been observed to change upon deuteriation of the methyl group due to vibrational effects (for a tabulation see [43]).

If quadrupole coupling is present in addition to hindered internal rotation, the analysis is further complicated, especially if the contributions are of the same order of magnitude. In favorable cases, where the quadrupole and internal rotation coupling are independent of each other, the internal rotation and quadrupole perturbations are simply additive. This has been found, for example, in the acetyl halides (CH_3COX) [55, 56]. Discussions of magnetic hyperfine interactions [57] in molecules with internal rotation, and quadrupole hyperfine interactions [58] for molecules with a C_s frame and C_{2v} top are available.

Rotational Spectra in Excited Torsional States: V_6 Effects

In addition to the transitions in the ground torsional state ($v=0$), rotational transitions associated with excited torsional states ($v \neq 0$, called satellite lines) are often observable. Because of the interaction of molecular vibrations with the internal rotation and overall rotation, the rotational constants are slightly different in the excited torsional states, and hence the satellite lines are shifted from the ground torsional state line. These satellite lines which accompany each ground state rotational line decrease in intensity because of the Boltzmann factor; however, since the torsional frequencies are low, a number of satellite lines can be observable (see Fig. 12.9). Lines arising from other excited vibrational states are generally much weaker. The Stark pattern for the satellite lines will be essentially the same as that for the corresponding line for the ground torsional state. The splitting of the A and E satellite lines illustrated in Fig. 12.9

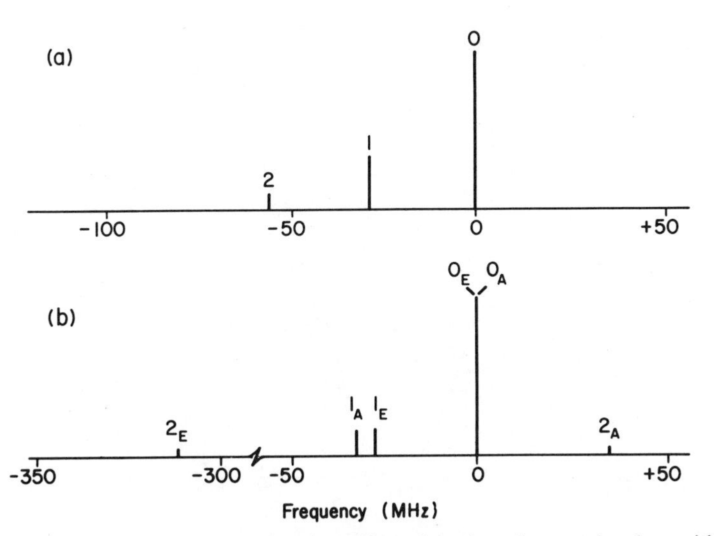

Fig. 12.9 The ground state and torsional satellites of the $1_{01} \rightarrow 2_{12}$ rotational transition of propylene oxide. For this molecule the same barrier is obtained for the first three torsional states. (*a*) The rotational lines of the ground and first two excited torsional states, shown without the internal rotation splitting. (*b*) The same pattern showing the splitting into the v_A and v_E doublets due to the effects of internal rotation. After Herschbach and Swalen [11].

can also be used for determination of the barrier. The analysis, however, can be more difficult because of the contribution of molecular vibrations to the satellite splitting. Fortunately, evidence indicates that the splitting of the A and E lines associated with the torsional state v is, in many cases, due almost entirely to internal rotation, and the vibrational effects may be ignored. In propylene oxide the same barrier is obtained for the first three torsional states [11]. The analysis is then no more difficult than for the ground state, that is, the A and E splittings associated with an excited torsional state v are treated with the working expressions outlined for the ground torsional state. This is particularly advantageous for molecules where the barrier is high enough to make the ground torsional state splittings too small to be observed. Because the perturbation coefficients $W_{v\sigma}^{(n)}$ increase significantly with v, the splittings are magnified in excited torsional states and can become observable, thereby allowing the barrier to be determined [59]. The upper limit for resolvable ground state splitting occurs at about $s=75$ (55 if only parallel transitions are accessible), while for the first and second excited state the upper limit on s is about 130 and 180, respectively [11].

The investigation of excited torsional states is also advantageous for study of the effects of higher terms in the Fourier expansion of the hindering potential. The V_6 constant has been found to be either positive or negative [14]; it modifies the shape of the barrier as demonstrated in Fig. 12.10. If the V_6 term is treated as a perturbation, its contributions change only the values of the perturbation coefficients $W_{v\sigma}^{(n)}$. Perturbation calculations have been carried out to give the

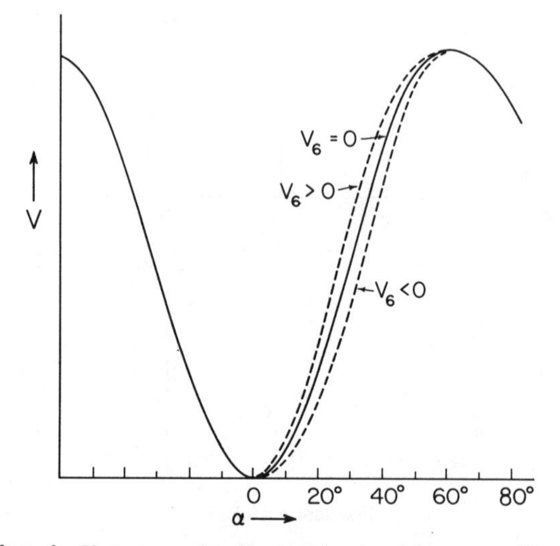

Fig. 12.10 The effect of a V_6 term on the shape of the threefold potential function. The V_6 term does not contribute to either the minimum or maximum of the potential, and the height of the barrier is hence unchanged. A positive V_6 results in a narrower potential well and a broader maximum with the torsional levels becoming more widely separated. A negative V_6 has just the opposite effect. From Fateley and Miller [14].

corrections to $W_{v\sigma}^{(n)}$ caused by a small V_6 term ($V_6/V_3 < 5\%$) [28, 18]. The corrections, however, enter in such a way that V_6 and V_3 cannot be separately determined from data on a single torsional state v. Evaluation of the V_6 term requires data from more than one torsional state. If V_6 is important, it is found that the values of V_3 obtained neglecting V_6 vary systematically with torsional state, and by including the effect of a V_6 term, the variation is effectively removed. Some examples of determinations of both V_3 and V_6 are given in Table 12.6. As apparent, the V_6 constants are small compared to V_3. The variation of V_3 with torsional state, which can be used to determine V_6, may also result from nonrigidity effects, that is, variations in I_α; and if both effects are present, they can be difficult to separate. Physically, as the methyl group rotates, the top is distorted due to a variation in barrier forces closing down slightly [60], for example, at the top of the barrier. Such structural relaxation has been termed torsional flexing [61]. The reduced rotational constant is hence a function of the torsional angle $F(\alpha)$ and may be expressed in terms of a Fourier expansion. The torsional Hamiltonian is thus more complicated with additional kinetic energy terms being introduced. It has been demonstrated by Lees and Baker [62, 63] in the study of CH_3OH that from the spectrum of a single isotopic species it is impossible in first order to separate the effects of V_6 from those of a variation in F because the first-order perturbation corrections due to V_6 and torsional flexing are linear dependent. This nonseparability is reflected in a general way from the fact it is possible to transform [64, 65] the torsional Hamiltonian from a system in which F is a function of torsional angle and V_6

Table 12.6 Examples of V_3 and V_6 Barrier Evaluations

Molecule	V_3(*cal/mole*)	V_6(*cal/mole*)	*Ref.*
CH_3CH_3	2882	20	*a*
CH_3CH_2CN	3226	−172	*b*
CH_3CHO	1145	−31	*c*
CH_3COCN	1198	−13	*d*
$CH_3COCOOH$	972	−10	*e*
(pyridine ring, N) CH_3	258	−12	*f*

[a] Hirota et al. [67].
[b] H. M. Meise, H. Mader, and H. Dreizler, *Z. Naturforsch.*, **31a**, 1228 (1976).
[c] Bauder and Günthard [48].
[d] G. K. Pandey and H. Dreizler, *Z. Naturforsch.*, **33a**, 204 (1978).
[e] R. Meyer and A. Bauder, *J. Mol. Spectrosc.*, **94**, 136 (1982).
[f] H. Dreizler, H. D. Rudolph, and H. Mäder, *Z. Naturforsch.*, **25a**, 25 (1970).

is zero, to one in which F is constant and V_6 is nonzero. A review of this problem has been given by Lees [61], and the importance of considering torsional flexing in the determination of V_6 is pointed out particularly for molecules with light frameworks such as CH_3OH. In this regard, it may be noted that a detailed treatment [66] of a number of isotopic species of CH_3OH, including V_6 and torsional flexing effects, has lead to an essentially zero value for $V_6(=-0.07\pm 0.21$ cm). By measurement of various isotopic forms of ethane [67], and assessing the isotopic effects in the internal-rotation potential constants both V_3 and V_6 have been evaluated (see Table 12.6).

Investigations of the effect on the barrier due to zero-point vibrational averaging and distortion on geometry during internal rotation have been reported [60, 68]. In ethane, calculations indicate that the vibrational averaging correction (374 ± 90 cal/mole), which raises the barrier, is larger and of opposite sign to the distortion correction (-161 ± 20 cal/mole).

Stark Effect

As remarked previously, it is possible for certain E level transitions to exhibit a first-order Stark effect. This is readily demonstrated, as discussed by Lin and Swalen [8]. Consider the Hamiltonian

$$\mathscr{H}=AP_z^2+BP_x^2+CP_y^2+DP_z \tag{12.76}$$

where

$$D=\frac{F W_{vE}^{(1)}\lambda_z I_\alpha}{I_z} \tag{12.77}$$

Because of the presence of the linear P_z term in this Hamiltonian, the Wang functions are no longer the proper zeroth-order wave functions. For example, an eigenfunction for $J=1$ and K odd can be written as

$$\Psi = a\psi(J=1, K=1, M) + b\psi(J=1, K=-1, M) \tag{12.78}$$

where $a \neq b$. The expansion coefficients, which can be evaluated in the usual way, are found to depend on both the asymmetry and the barrier. The average of the Stark operator, $-\mu_z \mathscr{E} \Phi_{Zz}$, may be expressed as follows

$$-\mu_z \mathscr{E} \int \Psi^* \Phi_{Zz} \Psi \, d\tau = \frac{-\mu_z \mathscr{E}(a^2 - b^2)KM}{J(J+1)} \tag{12.79}$$

with $J=1$, $K=1$. Because of the unequal mixing of the K and $-K$ symmetric rotor basis functions the first-order Stark effect term is nonvanishing, and a linear Stark effect can thus result for $K \neq 0$, $M \neq 0$. The first-order Stark effect decreases rapidly as the barrier height increases. This mixing of the rigid rotor wave functions is also responsible for the appearance of forbidden transitions. For similar reasons the spectra of asymmetric rotors with very low barriers and $m \neq 0, \pm 3$ (discussed in Section 8) can show first-order Stark effects.

A general expression for the first-order Stark effect of the $J \rightarrow J+1$, $K_{-1}=1$ transitions has been given by Beaudet [69]. For a given $J \rightarrow J+1$, $K_{-1}=1$ transition of a prolate rotor the frequency splitting between the $+|M|$ and $-|M|$ Stark component is given by

$$\Delta\nu = \frac{8\mu_z \mathscr{E} D|M|}{(B-C)} \left\{ \left[\frac{1}{J^2(J+1)^2} - \frac{1}{J'^2(J'+1)^2} \right] - \left[\frac{1}{J^4(J+1)^4} - \frac{1}{J'^4(J'+1)^4} \right] \frac{8D^2}{(B-C)^2} \right\} + \cdots \tag{12.80}$$

where $J'(=J+1)$ is the higher of the two J levels and D is the coefficient of the linear P_z term defined in (12.77). Since the first-order Stark effect is sensitive to the barrier, via the D constant, it provides a means of determining the barrier. The Stark effect of the $1_{11} \rightarrow 2_{12}$ transition is illustrated in Fig. 12.11 from the work of Beaudet and Wilson [70] on *cis*-fluoropropylene, $CH_3CH{=}CHF$. The A and E transitions were not resolved and four Stark components were found; two linear and two quadratic in the field. One of the second-order Stark components is assigned to the $|M|=1$ component of the A transition, and the two components linear in the electric field are assigned to the $M=+1$ and $M=-1$ components of the E transition. The remaining second-order component is assigned to $M=0$ for both the A and E transitions and is not shown in the figure. From the observed splitting of the $M=+1$ and $M=-1$ components of the E transition the barrier height to internal rotation of the methyl group was determined to be 1037 cal/mole, which may be compared with the value 1057 cal/mole obtained from the line splittings. A general theory of the Stark effect in molecules with one or two symmetric tops based on the concept of isometric groups is described by Bossert et al. [71].

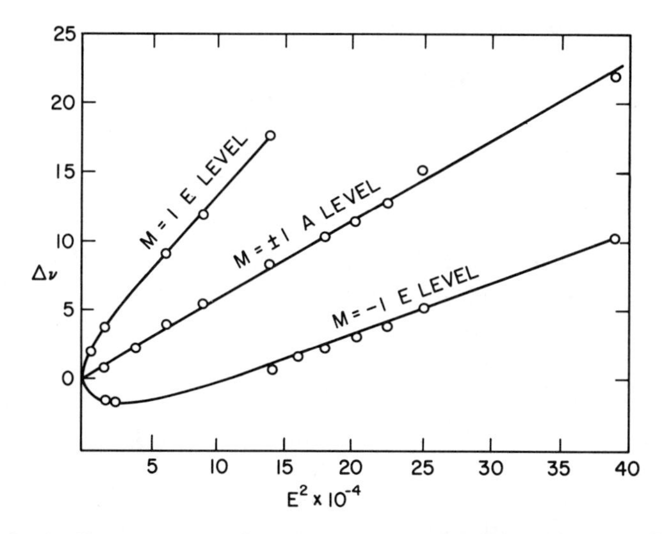

Fig. 12.11 Stark effect of the $1_{11} \rightarrow 2_{12}$ transition of *cis*-$CH_3CH{=}CHF$. From Beaudet and Wilson [70].

Example of a Barrier Calculation

As an illustration of the theory we look at a simple example of a barrier height evaluation using the observed splitting of the $0_{00} \rightarrow 1_{01}$ transition of acetaldehyde, which was found to be $(\nu_A - \nu_E) = 3.0$ MHz. Acetaldehyde is a prolate rotor, and the x, y, z axes are associated with symbols $A_{v\sigma}, \ldots$, as follows: $(x, B_{v\sigma})$, $(y, C_{v\sigma})$, $(z, A_{v\sigma})$ where $A_{v\sigma} > B_{v\sigma} > C_{v\sigma}$ and the xz plane is taken as the plane of symmetry of the molecule. Employing the structural parameters for acetaldehyde (Table 10 of [27]) we find for the moments of inertia (amu Å^2)

$$I_x(I_b) = 49.723 \qquad I_z(I_a) = 8.945$$
$$I_y(I_c) = 55.545 \qquad I_\alpha = 3.123$$

If we take the symmetry axis of the methyl group as coaxial with the carbon–carbon bond the direction cosines are

$$\lambda_z = \cos 23°1' = 0.9204 \qquad \lambda_x = \cos 66°59' = 0.3910$$

with $\lambda_y = 0$, since the $y(c)$ axis is perpendicular to the plane containing the methyl group axis. For the internal rotation parameters we have

$$\rho_x = 0.02456 \qquad \rho_z = 0.32134$$
$$r = 0.6946 \qquad F = 233.0 \text{ GHz}$$

We may apply (12.63)–(12.67) to the splitting of the $0_{00} \rightarrow 1_{01}$ transition, and since $\lambda_y = 0$ we, have $\Delta C = 0$. For the 0_{00} level $E(\kappa) = 0$, $\partial E/\partial A = \partial E/\partial B = 0$, and we have

$$\Delta E(0_{00}) = 0$$

while for the 1_{01} level $E(\kappa)=\kappa-1$, $\partial E/\partial A=0$, $\partial E/\partial B=1$. This gives

$$\Delta E(1_{01})=\Delta B=B_{0A}-B_{0E}$$

The expression for the splitting ($\Delta\nu=\nu_A-\nu_E$) of the transition is therefore

$$\Delta\nu=\Delta B=F\rho_x^2[W_{0A}^{(2)}-W_{0E}^{(2)}]$$
$$=140.54[W_{0A}^{(2)}-W_{0E}^{(2)}]$$

From the observed ground torsional state splitting of 3.0 MHz one obtains the difference in the barrier-dependent perturbation coefficients

$$\Delta W_0^{(2)}=W_{0A}^{(2)}-W_{0E}^{(2)}=0.02135$$

By making use of the following quantities:

s	$W_{0A}^{(2)}$	$W_{0E}^{(2)}$	$\Delta W_0^{(2)}$
22	0.01710	−0.008550	0.025650
23	0.01436	−0.007179	0.021539
24	0.01210	−0.006046	0.018146

a linear interpolation yields $s=23.06$, which gives

$$V_3=\frac{sF}{4.6602}$$
$$=\frac{(23.06)(233.0)}{4.6602}=1153\text{ cal/mole}$$

for the barrier height. The conversion factor 4.6602 gives V_3 in cal/mole when F is expressed in GHz.

7 HIGH-BARRIER SYMMETRIC ROTORS

The method for determining barrier heights described for asymmetric rotors, which utilizes splittings of rotational transitions arising from the direct effects of internal rotation, is not applicable for symmetric rotors. To demonstrate this, let us look at the effective rotational Hamiltonian for a symmetric rotor. For a symmetric top with an internal rotor, illustrated in Fig. 12.12, $B_y=A_x$ and $\mathscr{P}$ reduces to: $\mathscr{P}=\rho P_z$, where $\rho=I_\alpha/I_z$. The Hamiltonian of (12.53) can thus be written as

$$\mathscr{H}_{v\sigma}=A_x(P_x^2+P_y^2)+C_xP_z^2+F\sum_n W_{v\sigma}^{(n)}(\rho P_z)^n \tag{12.81}$$

where $F=\hbar^2 I_z/[2I_\alpha(I_z-I_\alpha)]$. The $W_{v\sigma}^{(n)}$ are the barrier-dependent perturbation coefficients defined previously. The rigid symmetric rotor functions yield a diagonal matrix for this effective rotational Hamiltonian. The rotational energies for the torsional state $v\sigma$ are

$$E_{JKv\sigma}=A_xJ(J+1)+(C_z-A_x)K^2+F\sum_n W_{v\sigma}^{(n)}(\rho K)^n \tag{12.82}$$

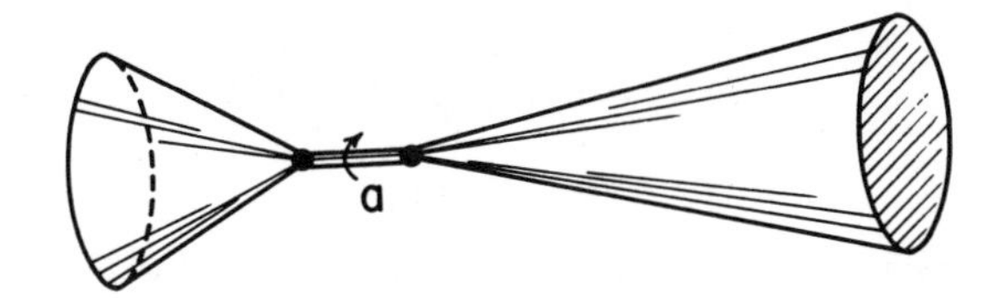

Fig. 12.12 Model for a symmetric rotor with internal rotation.

The first two terms represent the rotational energy of a rigid symmetric rotor, and the last term represents the corrections to the rotational energy due to the effects of internal rotation. The $\sigma=0$ (A-type) levels retain the usual $\pm K$ degeneracy for a symmetric rotor; however, for $\sigma=\pm 1$ (E-type levels the K degeneracy is removed by the coupling of internal and overall rotation. The $\sigma=+1$ or -1 leads to the same energy levels; the rotational levels are doubly degenerate with respect to σ but nondegenerate with respect to K. The three pairs of doubly degenerate rotational levels illustrated in Fig. 12.7 are

$$\sigma=0, \ \pm K$$
$$\sigma=+1, \ +K, \qquad \sigma=-1, \ -K$$
$$\sigma=+1, \ -K, \qquad \sigma=-1, \ +K$$

The selection rules for rotational transitions in the torsional state v are $\Delta J=\pm 1$, $\Delta K=0$ and $\Delta\sigma=0$. The three possible absorption transitions are of the type

$$J, \sigma=0, \ \pm K \rightarrow J+1, \sigma=0, \ \pm K$$
$$J, \sigma=\pm 1, \ \pm K \rightarrow J+1, \sigma=\pm 1, \ \pm K$$
$$J, \sigma=\pm 1, \ \mp K \rightarrow J+1, \sigma=\pm 1, \ \mp K$$

in which the upper or lower signs are taken together. Because of the selection rules it is apparent from (12.82) that the rotational transitions will not be affected by the contributions from internal rotation; hence the three possible transitions are coincident, and no splitting occurs (see Fig. 12.7). Other methods must, therefore, be used for the detection of the effects of internal rotation in the rotational spectra of symmetric top molecules.

One technique, of course, is to introduce an asymmetry by isotopic substitution [37, 72]. Coriolis interaction in a degenerate vibrational mode also allows observation of internal rotation effects [73, 74]. In an excited degenerate vibrational state when both the Coriolis and internal rotation effects are taken into account, the internal rotation effect is found to depend on J and hence to enter into the transition frequency and thus to become directly observable. Other methods which can be utilized are discussed in Section 9.

8 LOW-BARRIER APPROXIMATION

In the previous sections a perturbation treatment of internal rotation was discussed in which the top-frame coupling term was treated as a perturbation.

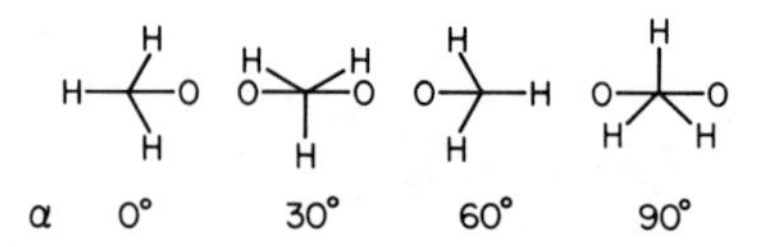

Fig. 12.13 As the methyl group with threefold symmetry is rotated with respect to the frame (NO_2) of twofold symmetry, indistinguishable configurations are obtained at intervals of $2\pi/6$. Hence the barrier has sixfold symmetry.

This method is best suited for molecules with high barriers. As the barrier is lowered or as KI_α/I_z becomes large, the higher-order perturbation terms become more important and have to be included, as discussed previously. If the higher-order corrections for the coupling become excessively large, the internal axis formulation, where the coupling term is initially minimized by a coordinate transformation, is more convenient especially if the asymmetry of the rotor is small. However, for molecules with very low barriers ($s<2$) a different approach is required. For them, the potential barrier is treated as a perturbation. The treatment for a general N-fold barrier has been given by Wilson et al. [29], and the case of a sixfold barrier has been discussed in detail by Lin and Swalen [8].

Both nitromethane, CH_3NO_2, investigated by Tannenbaum et al. [75, 76], and methyl difluoroboron, CH_3BF_2, studied by Naylor and Wilson [77] are example of molecules with very low barriers, the barrier heights being 6.03 and 13.77 cal/mole, respectively. Here the symmetry leads to an internal potential of sixfold symmetry ($N=6$). This is illustrated in Fig. 12.13 for CH_3NO_2. The potential function restricting internal rotation about the symmetry axis of the methyl top has the form

$$V(\alpha)=\frac{V_6}{2}(1-\cos 6\alpha)+\cdots \tag{12.83}$$

The low value found for the barriers to internal rotation in these two molecules is attributable to the higher symmetry [76]. The interaction of one of the N–O bonds in nitromethane with the methyl group gives rise to a barrier to internal rotation which can be written as

$$V_1=\tfrac{1}{2}V_3'(1-\cos 3\alpha)+\tfrac{1}{2}V_6'(1-\cos 6\alpha)+\cdots \tag{12.84}$$

The second N–O bond also, through interaction with the methyl group, contributes a barrier. The angle of torsion for the second N–O bond is $\alpha+\pi$; therefore

$$V_2=\tfrac{1}{2}V_3'[1-\cos 3(\alpha+\pi)]+\tfrac{1}{2}V_6'[1-\cos 6(\alpha+\pi)]+\cdots \tag{12.85}$$

The complete potential barrier due to both N–O bonds in nitromethane is V_1+V_2. Since the contributions are 180° out of phase, a cancellation of terms will result, and the leading term of the potential may be represented by (12.83) with $V_6=2V_6'$. As observed previously, the evidence indicates that the V_3 term of the Fourier series expansion, (12.84), is much larger than succeeding terms.

Therefore, because of the cancellation of this term the barrier can be expected to be much less.

The discussion of the low barrier problem will be confined to molecules such as CH_3BF_2 and CH_3NO_2 with sixfold symmetry. For molecules such as these with the symmetry axis of the methyl group coinciding with one of the principal axes, the Hamiltonian (12.34) takes the following simple form [29]

$$\mathscr{H}=A_xP_x^2+B_yP_y^2+C_zP_z^2+Fp^2-2C_zpP_z+V(\alpha) \tag{12.86}$$

where the potential energy is given in (12.83) and

$$\begin{aligned} A_x&=\frac{\hbar^2}{2I_x}\\ B_y&=\frac{\hbar^2}{2I_y}\\ C_z&=\frac{\hbar^2}{2I_z}+F\left(\frac{I_\alpha}{I_z}\right)^2=\frac{\hbar^2}{2(I_z-I_\alpha)}\\ F&=\frac{\hbar^2}{2rI_\alpha}=\frac{\hbar^2 I_z}{2I_\alpha(I_z-I_\alpha)} \end{aligned} \tag{12.87}$$

Here the z axis is taken as the symmetry axis of the top. The rotational constant C_z is not that of a rigid rotor but is modified by an internal rotation term and represents the rotational constant of the framework part alone rather than of the entire molecule.

In a discussion of the treatment for very low barriers it is convenient to consider first the problem of free rotation, where the methyl group rotates freely, and subsequently to consider the effects of introduction of a small potential barrier.

Energy Levels and Spectra: Free Rotation

For a symmetric rotor where $A_x=B_y$ the Hamiltonian becomes

$$\mathscr{H}=A_x(P_x^2+P_y^2)+C_zP_z^2+Fp^2-2C_zpP_z \tag{12.88}$$

The eigenfunctions of the foregoing Hamiltonian are product functions composed of symmetric rotor wave functions and the torsional functions corresponding to the limiting case of a free internal rotator. They are expressed as

$$\psi_{JKM}(\theta,\phi,\chi)\frac{1}{\sqrt{2\pi}}e^{im\alpha} \tag{12.89}$$

where J, K, M are the usual quantum numbers for symmetric rotors and m is the free internal rotation quantum number which corresponds to the total angular momentum of the top quantized along the axis of internal rotation ($p=m$ in units of $\hbar$). The energies are

$$E_{JKm}=A_xJ(J+1)+(C_z-A_x)K^2+Fm^2-2C_zmK \tag{12.90}$$

with

$$K=0, \pm 1, \pm 2, \ldots, \pm J; \qquad m=0, \pm 1, \pm 2, \ldots, \infty \tag{12.91}$$

All levels are doubly degenerate except for the level with $m=0$, $K=0$.

The selection rules are $\Delta J=\pm 1$, $\Delta K=0$, and $\Delta m=0$. This last restriction follows because the dipole moment is independent of α; therefore, because of the form of the wave function (12.89), the nonvanishing of the dipole matrix element requires the quantum number m to be the same for the two states. Because of these selection rules the rotational transitions are the same as those for a rigid symmetric rotor, and the effect of free internal rotation is not observable. This will not, however, be true for an asymmetric rotor.

For an asymmetric rotor the Hamiltonian (12.86), with the barrier again assumed to be zero, can be divided into two parts as follows:

$$\mathscr{H}=\mathscr{H}^0+\mathscr{H}'$$

$$\mathscr{H}^0=\tfrac{1}{2}(A_x+B_y)(P_x^2+P_y^2)+C_zP_z^2+Fp^2-2C_zpP_z \tag{12.92}$$

$$\mathscr{H}'=\tfrac{1}{2}(A_x-B_y)(P_x^2-P_y^2) \tag{12.93}$$

with the rotational parameters defined as in (12.87). The desired energy levels may be obtained, as usual, by diagonalization of the energy matrix. The appropriate basis for construction of the energy matrix is the $JKMm$ basis. If the asymmetry term $\mathscr{H}'$ were not present, these basis functions, (12.89), yield, as we have seen, a diagonal matrix. The wave functions of $\mathscr{H}^0+\mathscr{H}'$ can be expressed as a linear combination of these basis functions as follows

$$\Psi=\frac{1}{\sqrt{2\pi}}e^{im\alpha}\sum_{K=-J}^{+J}a_K\psi_{JKM} \tag{12.94}$$

since the Hamiltonian is diagonal in the quantum numbers J, M, and m. In this low-barrier representation the nonvanishing matrix elements of the energy matrix are [29, 8]

$$(J, K, m|\mathscr{H}|J, K, m)=\tfrac{1}{2}(A_x+B_y)J(J+1)+[C_z-\tfrac{1}{2}(A_x+B_y)]K^2 + Fm^2-2C_zmK \tag{12.95}$$

$$(J, K, m|\mathscr{H}|J, K\pm 2, m) =\tfrac{1}{4}(B_y-A_x)\{[J(J+1)-K(K\pm 1)][J(J+1)-(K\pm 1)(K\pm 2)]\}^{1/2} \tag{12.96}$$

For each value of J there will be a set of submatrices of order $2J+1$, one submatrix for each value of m. Since the off-diagonal elements do not connect even and odd K values, each submatrix can be further factored into two blocks, one for even K values and one for odd. Because of this factoring, the original wave functions of (12.94) may be divided into two types, one involving a sum over only even K values and the other type a sum over only odd K values.

Except for the presence of the term linear in K on the diagonal, this matrix is similar to that of an ordinary rigid asymmetric rotor. Because of the presence of this term, the Wang transformation does not lead to an additional factoring.

However, for $m=0$ the energy matrix is the same as that for a rigid asymmetric rotor and may be handled in the usual fashion, except that the constant C_z is not the rotational constant of the entire molecule but that of the framework part only. The selection rules for $m=0$ are the same as those for a rigid asymmetric rotor.

The diagonal terms independent of K in (12.95) can be regarded as additive constants for a given J and m. The total energy for the case of free internal rotation can then be written as

$$E=\tfrac{1}{2}(A_x+B_y)J(J+1)+Fm^2+\lambda \tag{12.97}$$

where λ is a root of the secular equation obtained from (12.95) and (12.96) by omission of the diagonal terms independent of K. For small J the secular equations may be readily solved. The secular equation with $J=2$ and even values of $K(K=0, \pm 2)$ has the following form

$$\begin{vmatrix} 4[C_z-\frac{1}{2}(A_x+B_y)]+4mC_z-\lambda & \frac{1}{2}\sqrt{6}(B_y-A_x) & 0 \\ \frac{1}{2}\sqrt{6}(B_y-A_x) & -\lambda & \frac{1}{2}\sqrt{6}(B_y-A_x) \\ 0 & \frac{1}{2}\sqrt{6}(B_y-A_x) & 4[C_z-\frac{1}{2}(A_x+B_y)]-4mC_z-\lambda \end{vmatrix}=0$$

and for odd $K(K=\pm 1)$ it has the form

$$\begin{vmatrix} C_z-\frac{1}{2}(A_x+B_y)+2mC_z-\lambda & \frac{3}{2}(B_y-A_x) \\ \frac{3}{2}(B_y-A_x) & C_z-\frac{1}{2}(A_x+B_y)-2mC_z-\lambda \end{vmatrix}=0$$

It is evident that the secular determinant for $-m$ is identical to that for $+m$, and hence all energy levels for which $m\neq 0$ are doubly degenerate. For small asymmetry and large values of m, the energies may be satisfactorily obtained by second-order perturbation theory, which yields [29]

$$\begin{aligned}\lambda=[C_z-\tfrac{1}{2}(A_x+B_y)]\Bigg(K^2-dmK+&\frac{b^2}{8(2K-dm-2)}\\ &\times[\{J^2-(K-1)^2\}\{(J+1)^2-(K-1)^2\}]\\ &-\frac{b^2}{8(2K-dm+2)}\\ &\times\{[J^2-(K+1)^2][(J+1)^2-(K+1)^2]\}+\cdots\Bigg)\end{aligned} \tag{12.98}$$

in which

$$b=\frac{\frac{1}{2}(A_x-B_y)}{C_z-\frac{1}{2}(A_x+B_y)}$$

and

$$d=\frac{2C_z}{C_z-\frac{1}{2}(A_x+B_y)}$$

For CH_3NO_2 and CH_3BF_2, the dipole moment is along the z axis and the

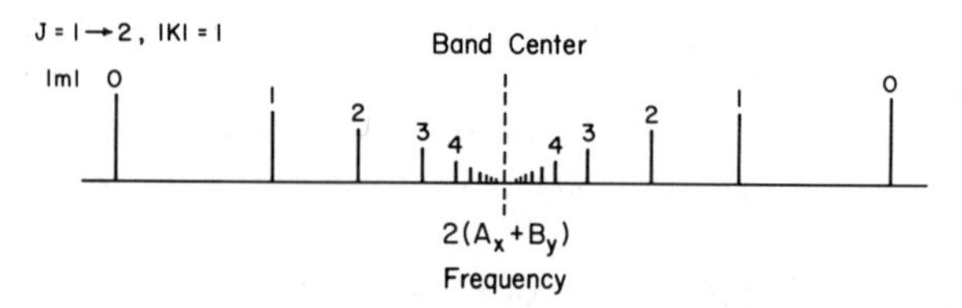

Fig. 12.14 Illustration of the $J=1\rightarrow 2$, $\Delta K=0$, odd K spectrum of an asymmetric rotor with free internal rotation.

parity of K cannot change. The selection rules are [28, 8]

$$\Delta J=0, \pm 1; \qquad \Delta K=0, \pm 2, \pm 4, \ldots ; \qquad \Delta m=0 \tag{12.99}$$

Since these molecules are nearly prolate rotors, K is nearly a good quantum number, and the transitions with $\Delta K=0$ will be stronger than those with larger changes in K.

Transitions of the type $\Delta J=+1$, $\Delta K=0$ and $\Delta m=0$ give rise to a bandlike structure converging to the band head of $(J+1)(A_x+B_y)$ as m increases, the intensities of the lines decreasing with increasing m because of the Boltzmann factor; ultimately the lines of large m will be too weak to observe. This pattern has been observed in the $J=1\rightarrow 2$ transitions of CH_3NO_2 and CH_3BF_2. For the $J=1$, 2 and K odd, the secular equations of order two are easily solvable and give the following expressions for λ

$$\lambda_{J=1}=C_z-\tfrac{1}{2}(A_x+B_y)\pm(4m^2C_z^2+\tfrac{1}{4}(B_y-A_x)^2)^{1/2} \tag{12.100}$$

$$\lambda_{J=2}=C_z-\tfrac{1}{2}(A_x+B_y)\pm(4m^2C_z^2+\tfrac{9}{4}(B_y-A_x)^2)^{1/2} \tag{12.101}$$

The frequencies for the $J=1\rightarrow 2$, $\Delta K=0$ transitions with K odd are

$$\nu_{J=1\rightarrow 2}\,(\text{odd } K)=2(A_x+B_y)\pm\{[4m^2C_z^2+\tfrac{9}{4}(B_y-A_x)^2]^{1/2} - [4m^2C_z^2+\tfrac{1}{4}(B_y-A_x)^2]^{1/2}\} \tag{12.102}$$

Therefore, when K is odd ($K=1$) two lines are expected for each value of m symmetrically spaced about the midfrequency $\nu=2(A_x+B_y)$. For $m=0$ two lines occur as expected for an ordinary asymmetric rotor, that is, the $1_{10}\rightarrow 2_{11}$ and $1_{11}\rightarrow 2_{12}$ transitions.. As m increases, the separation between the pair of lines decreases, converging from both sides to the midfrequency. The spectral pattern is illustrated in Fig. 12.14. For the other possible $J=1\rightarrow 2$, $\Delta K=0$ transition with K even ($K=0$) there will be only one transition for each value of m [77]. This series of transitions will again converge toward $\nu=2(A_x+B_y)$ as m increases.

Energy Levels and Spectra: Hindered Rotation with a Low Sixfold Barrier

The introduction of a small sixfold barrier, which may be treated by perturbation theory, has little effect on the free rotation spectra described previously, except for the $|m|=3$ lines which are split by the barrier. The splittings of these lines are sensitive to the barrier and may hence be used for calculation of the barrier height.

The presence of a sixfold potential barrier in the Hamiltonian introduces off-diagonal matrix elements in m. For the potential given by (12.83) the matrix elements are

$$(J, K, m|V(\alpha)|J, K, m)=\tfrac{1}{2}V_6 \tag{12.103}$$

$$(J, K, m|V(\alpha)|J, K, m\pm 6)=-\tfrac{1}{4}V_6 \tag{12.104}$$

The complete energy matrix is now given by (12.95), (12.96), (12.103), and (12.104). The term $\frac{1}{2}V_6$ which contributes a constant diagonal term may be neglected since it will have no effect on the rotational spectrum. The appearance of off-diagonal elements in m severely complicates the form of the energy matrix. However, approximate diagonalization in m may be obtained by means of a Van Vleck transformation. When the barrier V_6 is small compared with the quantity F, a Van Vleck transformation to second order may be employed [29] to reduce the off-diagonal elements in m to second order; if these elements are neglected, the energy matrix can be factored into blocks corresponding to different values of m. This procedure modifies the diagonal elements by folding in the elements off-diagonal in m. For a given J, the matrix elements of the transformed energy matrix are the same as those given by (12.95) and (12.96) except for an additional diagonal term [29, 8]

$$(J, K, m|\mathscr{H}|J, K, m)=\tfrac{1}{2}(A_x+B_y)J(J+1)+[C_z-\tfrac{1}{2}(A_x+B_y)]K^2+Fm^2 \\ -2C_zmK+\frac{FV_6^2}{32[(C_zK-Fm)^2-9F^2]} \tag{12.105}$$

$$(J, K, m|\mathscr{H}|J, K\pm 2, m) \\ =\tfrac{1}{4}(B_y-A_x)\{[J(J+1)-K(K\pm 1)][J(J+1)-(K\pm 1)(K\pm 2)]\}^{1/2} \tag{12.106}$$

This matrix is similar to that for free rotation. As before, the factoring into even and odd K blocks is obtained, and the $m\neq 0$ levels remain doubly degenerate. The effect of the additional diagonal term is usually such that the rotational spectrum will be essentially the same as for the free internal rotation case [8]. The spectrum is hence virtually insensitive to the low barrier (except for the $|m|=3$ lines).

The above treatment, however, is not applicable to the matrix elements of the barrier connecting the $m=+3$ block with the $m=-3$ block. Both the $m=3$ and $m=-3$ submatrices have to be combined into a single secular equation. The matrix elements of the barrier connecting $m=3$ with $m=9$ and $m=-3$ with $m=-9$ may still be treated by a Van Vleck perturbation technique to reduce these connections. The matrix elements for the $m=\pm 3$ levels then become [29]

$$(J, K, m=\pm 3|\mathscr{H}|J, K, m=\pm 3)=\tfrac{1}{2}(A_x+B_y)J(J+1) \\ +[C_z-\tfrac{1}{2}(A_x+B_y)]K^2+9F\mp 6C_zK\pm\frac{V_6^2}{192(C_zK\mp 6F)} \tag{12.107}$$

$$(J, K, m=3|\mathscr{H}|J, K\pm 2, m=3)=(J, K, m=-3|\mathscr{H}|J, K\pm 2, m=-3)$$
$$=\tfrac{1}{4}(B_y-A_x)\{[J(J+1)-K(K\pm 1)][J(J+1)-(K\pm 1)(K\pm 2)]\}^{1/2} \tag{12.108}$$

$$(J, K, m=3|\mathscr{H}|J, K, m=-3)=-\tfrac{1}{4}V_6 \tag{12.109}$$

For $J=1$ and K odd this gives a submatrix of order four to be solved having the following form [8]:

$$\begin{array}{c} K, m \\ (-1, -3) \\ (1, -3) \\ (-1, 3) \\ (1, 3) \end{array} \begin{array}{c} \begin{array}{cccc} (-1, -3) & (1, -3) & (-1, 3) & (1, 3) \end{array} \\ \begin{bmatrix} a & b & c & 0 \\ b & a' & 0 & c \\ c & 0 & a' & b \\ 0 & c & b & a \end{bmatrix} \end{array} \tag{12.110}$$

where

$$a=\tfrac{1}{2}(A_x+B_y)-5C_z+9F-\frac{V_6^2}{192(6F-C_z)}$$

$$a'=\tfrac{1}{2}(A_x+B_y)+7C_z+9F-\frac{V_6^2}{192(6F+C_z)}$$

$$b=\tfrac{1}{2}(B_y-A_x)$$
$$c=-\tfrac{1}{4}V_6$$

A similar matrix is obtained for $J=2$ and K odd. This matrix is seen to be symmetric under a reflection across both the principal and the secondary diagonal. This symmetry suggests that the $|m|=3$ matrices can be further factored by means of a transformation into two smaller matrices, similar to the way in which the rigid rotor energy matrix was factored by the Wang transformation. For the present case we have

$$\left(\frac{1}{\sqrt{2}}\right)^2 \begin{bmatrix} 1 & 0 & 0 & 1 \\ 0 & 1 & 1 & 0 \\ 0 & -1 & 1 & 0 \\ -1 & 0 & 0 & 1 \end{bmatrix} \begin{bmatrix} a & b & c & 0 \\ b & a' & 0 & c \\ c & 0 & a' & b \\ 0 & c & b & a \end{bmatrix} \begin{bmatrix} 1 & 0 & 0 & -1 \\ 0 & 1 & -1 & 0 \\ 0 & 1 & 1 & 0 \\ 1 & 0 & 0 & 1 \end{bmatrix}$$
$$=\begin{bmatrix} a & b+c & & \\ b+c & a' & \multicolumn{2}{c}{0} \\ & & a' & b-c \\ \multicolumn{2}{c}{0} & b-c & a \end{bmatrix} \tag{12.111}$$

This transformation is equivalent to choosing a new set of basis functions of the type [8]

$$\frac{1}{\sqrt{2}}[\psi(K, m=3)\pm\psi(-K, m=-3)] \tag{12.112}$$

which in the zeroth-order approximation may be expressed as [8]

$$\frac{1}{\sqrt{2}}\left[\psi_{JKM}(\theta, \phi, \chi)\frac{1}{\sqrt{2\pi}}e^{i3\alpha} \pm \psi_{J-KM}(\theta, \phi, \chi)\frac{1}{\sqrt{2\pi}}e^{-i3\alpha}\right] \tag{12.113}$$

These two smaller matrices may be readily solved to give the energy levels. The diagonal elements of the two matrices are connected in one case by the sum of the asymmetry and barrier and in the other by the difference between the asymmetry and barrier. The presence of the low barrier thus splits the double degeneracy when $|m|=3$, and the transitions associated with $|m|=3$ are split by the barrier. For the $J=1\rightarrow 2$ transition with $|m|=3$ and $|K|=1$, there are four lines which appear as two pairs, each symmetrically spaced about the midfrequency as illustrated in Fig. 12.15. The splitting of one of these pairs is approximately 4000 MHz for CH_3BF_2 [77]. For $K=0$, two lines are observed. The splittings of lines such as these are used for calculation of the barrier height.

When a barrier is present, m is no longer a good quantum number. However, if the barrier and the asymmetry are small, m and K are nearly good quantum numbers, and the selection rules are given approximately by

$$\Delta J=0, \pm 1; \qquad \Delta K=0; \qquad \Delta m=0 \tag{12.114}$$

Exact selection rules may be obtained by a group theoretical treatment given elsewhere [29, 8] or by explicit evaluation of the dipole moment matrix elements.

Within the approximation considered here, the splitting of the double degeneracy occurs only for the $|m|=3$ levels. Although the splitting is the largest for $|m|=3$, the degeneracy can be split for those levels where $|m|$ is a multiple of three, if the barrier is sufficiently large (see Fig. 12.4). However, a higher-order perturbation treatment is required to demonstrate the splitting [29].

In summary, the salient features of the low barrier spectrum consists of three sets of lines corresponding to $m=0$, $m\neq 0$, ± 3 and $m=\pm 3$. The transitions associated with $m=0$ will have the pattern of an ordinary asymmetric rotor. These transitions may be conveniently employed to determine the rotational constants. The Stark effect (unless accidental degeneracy occurs) will be a second-order characteristic of nondegenerate levels. For the $m\neq 0$, ± 3 lines, a possible bandlike structure can arise. These lines arising from degenerate rotational levels can exhibit a first-order Stark effect, except for $K=0$, $M=0$. The lines associated with $|m|=3$ levels are widely split by the barrier and allow an accurate barrier determination. They can show rather large second-order Stark effects analogous to the case of nearly degenerate levels of rigid asymmetric rotors.

Several other molecules with low sixfold barriers have been studied. Toluene, *p*-fluorotoluene and 4 methyl pyridine have been found by Rudolph et al. [78], to have barriers of 13.94, 13.82, and 13.51 cal/mole, respectively. For trifluoronitromethane a barrier of 74.4 cal/mole was obtained by Tolles et al. [79]. For CF_3NO_2, with a heavier top (CF_3) and a higher barrier than nitromethane, the

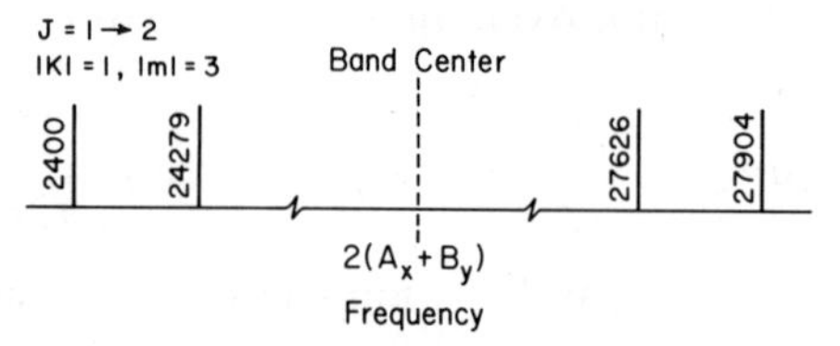

Fig. 12.15 The $J=1 \rightarrow 2$, $|K|=1$, $|m|=3$ lines of CH_3BF_2, which has a low sixfold barrier (frequency in Hz). The splittings of these lines are very sensitive to the barrier height.

Table 12.7 Some Molecules with Sixfold Barriers

Molecule	V_6*(cal/mole)*	*Ref.*
CH_3NO_2	6.03	a
CD_3NO_2	5.19	a
CF_3NO_2	74.4	b
CH_3BF_2	13.77	c
CD_3BF_2	12.2	d
SiF_3BF_2	1.9	e
$C_6H_5CH_3$	13.94	f
$C_6H_5CD_3$	11.8	g
$FC_6H_4CH_3$	13.82	h
$ClC_6H_4CH_3$	13.93	i
$NC_5H_4CH_3$	13.51	j

[a]Tannenbaum et al. [76].
[b]Tolles et al. [79].
[c]Naylor and Wilson [77].
[d]J. Wollrab, E. A. Rinehart, P. B. Rinehart, and P. R. Reed, *J. Chem. Phys.*, **55**, 1998 (1971).
[e]Ogata et al. [81].
[f]Rudolph et al. [78a].
[g]W. A. Kreiner, H. D. Rudolph, and B. T. Tan, *J. Mol. Spectrosc*, **48**, 86 (1973).
[h]Rudolph and Seiler [78b].
[i]G. E. Herberich, *Z. Naturforsch.*, **22a**, 761 (1967).
[j]Rudolph et al. [78c].

spectrum was found to be rather insensitive to the barrier height. In particular, the frequencies of the $|m|=3$ lines were not sensitive to the value of the barrier, as described previously, and hence could not be used for a precise determination of the barrier. However, the Stark effect of one of the transitions studied proved to be sensitive to the barrier height, thereby enabling an accurate barrier determination. Some molecules with sixfold barriers are given in Table 12.7.

The calculation of the energy levels for low-barrier, heavy-top molecules more efficient for higher J levels has been discussed [80] and applied to SiF_3BF_2 [81]. Discussions are available of centrifugal distortion effects [82] in molecules with low barriers such as CH_3NO_2 as well as the rotational Zeeman effect [83, 84].

In the case of low threefold barriers, there are no degenerate levels connected by matrix elements of the barrier, as in the sixfold case for the $m=\pm 3$ levels, and the barrier cannot be as accurately determined. Treatments for the case of low threefold barriers in molecules with one or two internal rotors are available [29, 85–87].

9 OTHER METHODS FOR EVALUATION OF BARRIER HEIGHTS

In addition to the "splitting method," there are three other methods that can be employed for determination of barrier heights. They are: the "torsional satellite-frequency pattern method," the "relative intensity method," and the "avoided crossing method," which will be described briefly. The former methods do not yield as accurate a barrier height as the splitting method, but they are applicable to symmetric rotors as well as asymmetric rotors.

Method Based on Satellite Frequency Pattern

In the foregoing treatment of internal rotation the top and framework were considered to be rigid with only the torsional motion between the top and frame being considered. This approximation predicts that the A-E splittings are zero for symmetric rotors and that all satellite lines of a symmetric rotor would have the same frequency as the ground torsional state line, which is not correct. The lack of rigidity is reflected, for example, in the change of moments of inertia with torsional state, which results in the observation of torsional satellite lines that are shifted from the main line. Since the moments of inertia do not depend directly on the angle of internal rotation, the internal rotation affects the moments of inertia through interactions with the other vibrational modes which, in turn, interact with the overall rotational motion. An analysis of this indirect coupling of internal and overall rotation is of special use for symmetric rotors since it provides a means for determination of the barrier. These higher-order vibration–torsion–rotation interactions have been treated by Kivelson [88] and applied to the $J=0\rightarrow 1$ torsional satellite pattern of the symmetric top CH_3SiH_3 to give the barrier height. The effective rotational constant of a symmetric rotor for the torsional state $v\sigma$ is given by [98, 89]

$$B_{v\sigma}=B+F(Kv\sigma|1-\cos 3\alpha|Kv\sigma)+G(Kv\sigma|p'^2|Kv\sigma)+L(Kv\sigma|p'|Kv\sigma)K \quad (12.115)$$

where $p' = p - (I_\alpha/I_z)P_z$, $|Kv\sigma)$ are the torsional eigenfunctions of the IAM and B, F, G, and L are constants independent of the quantum numbers for overall and internal rotation. The constant B is the usual effective rotational constant for the given vibrational state, G and L may be interpreted as centrifugal distortion constants, and F as a constant arising from the repulsion of the groups as they rotate relative to each other from a position of minimum barrier to maximum barrier. This latter parameter depends on the nature of the forces which give rise to the barrier [90]. The evaluation of the average values of the operators in the above expression has been discussed [88, 28]. These expectation values are dependent on the reduced barrier height. In an analysis the barrier height and the constants B, F, G, and L are adjusted for a fit of the satellite frequencies which are given by

$$\nu = 2B_{v\sigma}(J+1) \tag{12.116}$$

The vibration-torsion interaction also results in a splitting of the satellites into A and E lines, the splitting becoming larger as v increases. This "nonrigidity splitting," however, is usually small for the ground torsional state, amounting to only 0.05 MHz for CH_3SiH_3 [88]. Extension of (12.115) to include higher-order terms is described elsewhere [60] (see also [110]).

Rigorous treatments of vibration–torsion–rotation interactions have been discussed for the case of asymmetric rotors [89, 91–94]. For asymmetric rotors an analysis of nonrigidity effects becomes very difficult because of the complexity of the theoretical treatment and lack of sufficient experimental data. As with symmetric rotors, these nonrigidity effects are responsible for the shift of the satellite lines, and also contribute to the A-E splittings. More tractable treatments [46, 52, 95–97] discuss the interaction between methyl torsion and one low-frequency vibrational mode such as a skeletal torsion or in-plane bending. An alternate approach employing structure relaxation has also been described [48, 51]. Fortunately, the nonrigidity contributions to the A-E splittings associated with the torsional state v are often negligible, especially if low-lying torsional levels are studied. Hence, for asymmetric rotors the A-E splittings are essentially "rigid splittings" arising from only the direct coupling of internal and overall rotation (interaction with other modes of vibration ignored), thereby making possible an accurate determination of the potential barrier by application of the treatment described previously. When internal rotation is not entirely separable from the other vibrations a more detailed analysis is required, or the barrier V_3 should, if possible, be evaluated from ground state torsional splittings.

Relative Intensity Method

Another method, applicable to symmetric and asymmetric rotors, is the intensity method. It is particularly useful for asymmetric rotors when the splittings between the A and E lines are negligibly small. It consists in measurement of the intensities of a rotational transition in the ground state and the excited torsional state. The satellite line is weaker than the main line because of the lower population of the excited torsional state. The intensity ratio at a given temperature is given by the Boltzmann distribution law as

$$\frac{I_v}{I_0}=\left(\frac{g_v}{g_0}\right)e^{-\Delta E/kT} \tag{12.117}$$

where I_0 and I_v are the intensities of a particular rotational transition in the ground and excited state, respectively, with g_0 and g_v the appropriate statistical weights, and $\Delta E(=E_v-E_0)$ is the torsional energy difference between the excited and ground state. From a relative intensity measurement of, for example, the ground and first excited state, ΔE can be evaluated, from which, with the aid of tabulated Mathieu eigenvalues, s may be obtained and thus V_3. For high barriers, where low-lying excited torsional states are studied for which the harmonic oscillator approximation to the energy levels, (12.14), is adequate, an explicit expression for the barrier can be written:

$$V_3=(\tfrac{8}{9})\pi^2\nu^2 I_r \tag{12.118}$$

Here ν is the torsional vibration frequency obtainable from the intensity measurement, and I_r is the reduced moment of inertia.

A number of barriers have been obtained by this method, but there are some problems in its application. For instance, the intensity ratio is not necessarily as simple as indicated in (12.117). Although the major contribution to the intensity ratio comes from the Boltzmann factor, the complete expression involves the ratio of a number of additional quantities, since what is really measured is the ratio of the peak absorption coefficients of the two lines. The intensity ratio thus depends also on the ratio of such things as the line strength, peak absorption frequency, and half-width of the two lines [see (3.27)]. If any of these are significantly different for the two lines associated with the two torsional states, corrections must be made for the differences [98]. Furthermore, even though the A and E splitting may be too small for resolution, it nevertheless may be large enough to affect the measured peak intensity. If so, a correction must be applied to the measured intensity ratio to account for the presence of this unresolved splitting [98]. More important, though, is the problem of measuring intensities of microwave lines with sufficient accuracy. Errors as great as 10 to 20% can appear in the evaluated barrier heights if special care is not taken in the measurement of intensities [98–101].

To effectively eliminate the proportionality factor f in (12.117) depending on the ratio of the half-width, and so on, it is preferable to make intensity measurements over a range of temperatures according to Ruitenberg [102]. Equation 12.117 may be written as

$$\ln\left(\frac{I_v}{I_0}\right)=\ln f-\frac{\Delta E}{kT} \tag{12.119}$$

where f contains the various weight factors. By measurement of the ratio of the peak height for a given rotational transition in the ground and an excited torsional state as a function of temperature, ΔE may be determined from the slope of the line relating the ln of the intensity ratio and the reciprocal temperature. Therefore, a knowledge of f is not required. The main source of error is the fact that the ratio of the effective path length (part of f) of the absorption cell is slightly temperature dependent. This dependence can however be minimized and good results have been obtained with this procedure.

In general, if the motion is approximately harmonic, the solution of (12.6) with a simple N-fold potential, $V(\alpha)=\frac{1}{2}V_N(1\text{-}\cos N\alpha)$, is given by

$$E_v = N\sqrt{FV_N}(v+\tfrac{1}{2}),\ v=0, 1, 2, \ldots \tag{12.120}$$

The N-fold barrier in terms of the torsional frequency is given by

$$V_N = \frac{\nu^2}{N^2 F} \tag{12.121}$$

with F the reduced rotational constant for the internal rotor, and F is in units consistent with ν. Application of this relation for threefold barriers has already been noted. Several molecules with barriers of twofold symmetry have been investigated and the barrier evaluated from the above relation making use of intensity measurements of the torsional satellites to obtain ν. The variation of the inertial defect with torsional state may also be used to evaluate the torsional frequency as described in Chapter XIII, Section 7 and hence V_2. For nitrocyclopropane [103], the torsional frequency is found to be $70 \pm 20\ \text{cm}^{-1}$ and, with $F=1.1\ \text{cm}^{-1}$ for the nitro group, gives $V_2=\nu^2/4F=1100\pm 500\ \text{cm}^{-1}$. Similarly, for nitrosobenzene [104], nitrobenzene [105], benzaldehyde [106], benzoyl fluoride [107], phenyl isocyanate [108], and p-fluoroanisole [109], twofold barriers of $1350\pm 300\ \text{cm}^{-1}$, $1000\pm 500\ \text{cm}^{-1}$, $1713\pm 500\ \text{cm}^{-1}$, $1560\pm 150\ \text{cm}^{-1}$, $1190\pm 350\ \text{cm}^{-1}$, and $1021\ \text{cm}^{-1}$ are found, respectively. It may be noted that if a V_4 term is added to the potential function, as follows from Section 3, the torsional frequency $\nu^2 = F(4V_2+16V_4)$ actually determines the sum rather than V_2.

Avoided-Crossing Method

A new method—the avoided-crossing molecular beam electric resonance technique—for evaluating barrier heights in symmetric tops has been discussed in Chapter VI, Section 4. This method has been extended to observation of "barrier" anticrossings, which breaks the torsional symmetry by Meerts and Ozier [110]. For these anticrossings the values of the crossing field $\mathscr{E}_c$ yield the internal rotation splittings [111] and are independent of $(A-B)$. The torsional level splittings and Stark effect for $|K|=1$, $|M_J|=1$ in methylsilane is shown in Fig. 12.16. The symmetry classification of the torsional levels is according to the irreducible representation Γ of the permutation-inversion group [112, 113, 74] of G_{18}, where $K=0, \sigma=0 \rightarrow A_1$ (J even), A_2 (J odd); $K=0, \sigma=\pm 1 \rightarrow E_4$; $K=\pm 1, \sigma=0 \rightarrow E_1$; $K=\pm 1, \sigma=\pm 1 \rightarrow E_2$; $K=\pm 1, \sigma=\mp 1 \rightarrow E_3$. Anticrossing between E_2 and E_1 was not observed and is believed to be forbidden by symmetry. Several 'barrier" anticrossings have been observed for CH_3SiH_3 in the ground torsional state of the type $(J, K, \sigma, \Gamma)=(J, \pm 1, \mp 1, E_3) \leftrightarrow (J, \mp 1, \mp 1, E_2)$

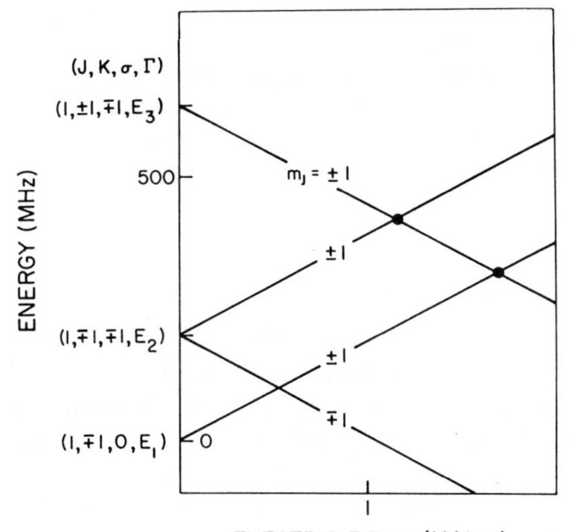

Fig. 12.16 Schematic plot against electric field of the energy levels for the $(K, \sigma)=(\pm 1, \mp 1)$, $(\pm, \pm 1)$, and $(\pm 1, 0)$ states with $J=1$ involved in barrier anticrossings in CH_3SiH_3. The two heavy dots indicate the avoided crossings for which precision measurements of the crossing field have been made. With one exception, upper signs go with upper, and lower signs with lower. If the third intersection were an avoided crossing with the same characteristics as the other two, then the third case would be the exception: upper signs would go with lower and vice versa. However, this third intersection seems to be a true crossing since no anticrossing transitions could be detected. From Meerts and Ozier [110].

and $(J, \pm 1, \mp 1, E_3) \leftrightarrow (J, \mp 1, 0, E_1)$ where the upper (or lower) signs are taken together. In addition, several "rotational" anticrossings have been observed where $\mathscr{E}_c$ does depend on $(A-B)$. These are of the type $(J, K, \sigma)=(1, \pm 1, \pm 1) \leftrightarrow (2, 0, 0)$; $(1, \pm 1, 0) \leftrightarrow (2, 0, 0)$; $(1, \pm 1, \pm 1) \leftrightarrow (2, 0, \pm 1)$; $(1, \pm 1, 0) \leftrightarrow (2, 0, \pm 1)$.

Analysis of all the data with an extended form of (12.116) gave in addition to the rotational constants and several distortion constants, a V_3 barrier of $592.2 \pm 1\ \text{cm}^{-1}$ for $CH_3{}^{28}SiH_3$. This technique can be expected to yield accurate barriers for symmetric-top molecules.

10 MOLECULES HAVING TWO INTERNAL ROTORS WITH C_{3v} SYMMETRY: TWO-TOP MOLECULES

In this section we consider molecules having two separate groups exhibiting restricted internal rotary motions. For simplicity, we include only molecules having threefold symmetric-top internal rotors and shall refer to these as two-top molecules. Those molecules in which the two internally rotating groups are asymmetric rotors are described in Sections 11 and 13.

The theory for two-top molecules has been discussed by various authors [54, 85, 114–120]. Formulation of the theory for them is similar to that for a single top (Sections 4–9). As shown by Swalen and Costain [85] and by Pierce [19, 116], the barrier-dependent perturbation coefficients for a single top are useful for two-top molecules also. Definitions of the appropriate perturbation coefficients are summarized by Wollrab [121]. The two-top problem has been reviewed by Dreizler [46].

As for the single-top molecules, the splitting of the rotational transitions are sensitive functions of the barrier height. For molecules of type $(CH_3)_2X$ the rotational lines are usually split into triplets or quartets.

The potential function for a two-top molecule such as $(CH_3)_2O$ is expressed by

$$V(\alpha_1, \alpha_2)=\frac{V_3}{2}(1-\cos 3\alpha_1)+\frac{V_3}{2}(1-\cos 3\alpha_2)$$
$$+V'_{12} \sin 3\alpha_1 \sin 3\alpha_2+V_{12} \cos 3\alpha_1 \cos 3\alpha_2+\cdots \quad (12.122)$$

Here α_1 and α_2 specify the torsional angles of the two tops. The V_{12} and V'_{12} are the top–top coupling parameters and V_3, the dominate term, is the barrier height. A physical interpretation of the potential coupling constants in terms of two geared methyl groups has been described [122].

An analysis of the rotational spectrum provides values for V_3 and V'_{12} if excited torsional states are studied. However, the evaluation of reliable values for V_{12} has proved to be difficult because of its high correlation with V_3. This may be seen by writing the torsional angle dependence of (12.122) in the harmonic oscillator approximation [123]

$$V(\alpha_1, \alpha_2)\cong\tfrac{9}{4}(V_3-2V_{12})(\alpha_1^2+\alpha_2^2)+9V'_{12}\alpha_1\alpha_2+\cdots \quad (12.123)$$

Hence only an effective barrier $V_3^*=V_3-2V_{12}$ may be obtained. This correlation seems to persist even in higher torsional states where the harmonic approximation should be invalid. Potential constants for some two-top molecules are tabulated in Section 12.

If one of the methyl groups of a molecule with two equivalent tops is deuterated, two nonequivalent tops are obtained, but this usually leads to very small splittings from the deuterated top and the analysis may be often carried out with single top methods. Even in a molecule such as $SiH_3CH_2OCH_3$, the barriers of the methyl and silyl groups may be obtained in the one-top approximation. Extension of the theory to the analysis of the internal rotation structure in the microwave spectrum of molecules with two nonequivalent symmetric tops such as in *N*-methylethyldiamine, $CH_3CH{=}NCH_3$, including effects of top–top coupling has been described [124–126]. In silyl methyl ether, SiH_3OCH_3, top–top coupling effects are readily apparent although the coupling constants could not be extracted [125].

Illustrative values of potential barrier heights for molecules with two equivalent tops and with two nonequivalent tops are tabulated in Section 12.

11 MOLECULES HAVING INTERNAL ASYMMETRIC-TOP ROTORS: TWO-FOLD BARRIERS

The theory and the spectra for molecules with asymmetric-top internal rotors are more complex than those for the three-fold symmetric-top rotors considered in earlier sections of this chapter. Theory for analysis of the microwave rotational spectra for this class of molecules has been developed mostly by Quade and his collaborators [127–131], as well as others [50, 51, 132–134]. Because of their complexity, we make no attempt to describe these developments here. Those seeking to analyze spectra of these types should consult the original works cited.

Three classes of planar molecules with twofold potential barriers have been treated [50, 128]: Class I molecules, for which both the top and frame have C_{2v} symmetry; Class II molecules, for which the top has C_{2v} symmetry, and the frame, C_s symmetry; and Class III molecules, for which the top has C_s symmetry, and the frame, C_{2v} symmetry. The Class II description is chosen for convenience when the C_{2v} part has the smaller inertial constants. The selection rules for these cases have been given [50, 135–137]. Applications of the theoretical methods to phenol, for example, have been reported [135, 138, 139]. The inclusion of the effects of structural relaxation during torsional motion has also been considered in nitrobenzene [140] and nitroethylene [51]. In particular, a quantitative interpretation of the dependence of the rotational constants of nitrobenzene on torsional state has been obtained, indicating that the ONO angle should open up by 1.3° upon rotation of the nitro group from the planar to perpendicular conformation. For nitroethylene an asymmetric nitro group is indicated, with the O_1NO_2 angle opening up by about 3.2° and the N–O_1 and N–O_2 bonds increasing and decreasing by 0.008 Å, respectively, upon rotation from $\theta=0$ to 90°. Microwave studies of H_2O_2 [141] and H_2S_2 [142] have also been made. These molecules have C_2 symmetry with each OH, or each SH, group undergoing rotation relative to the other.

12 SOME OBSERVED PROPERTIES OF MOLECULES THAT UNDERGO INTERNAL ROTATION

Barrier Heights

As illustrations, we give in Table 12.8 a restricted list of barrier heights in molecules having a single internal rotor, a symmetric group having threefold symmetry about the internal axis of rotation. Two-top molecules are illustrated in Tables 12.9 and 12.10. In the earlier edition of this book a rather comprehensive tabulation is given of barriers to internal rotation by threefold symmetric groups as measured with microwave spectroscopy into 1968. Because of the large increase in measured values since that time, a complete updating of this tabulation in the current revision is not feasible. The illustrative values in Tables 12.8–12.10 are chosen from later work. Barrier heights, updated at intervals, may be found in the Landolt-Börnstein tables [143]. In the evaluation of barrier heights the reduced moment of inertia must be known since what is determined is $s(\sim V_3 r I_\alpha)$. The major source of error in the determination of V_3 is not in s, which can be accurately determined, but rather in the uncertainty of I_α. Although r is usually insensitive to small changes in the structure, small

Table 12.8 Some Potential Barriers for Single-top Molecules as Measured with Microwave Spectroscopy

Molecule	V_3*(cal/mole)*	*Ref.*	*Molecule*	V_3*(cal/mole)*	*Ref.*
CH_3CHD_2	2878	*a*	CH_3NH_2	1956	*i*
CH_3CHO	1145	*b*	CH_3NHCl	3710	*j*
CH_3COI	1176	*c*	CH_3PF_2	2300	*k*
$ClCOOCH_3$	1250	*d*	CH_3POF_2	3580	*l*
$CNCOOCH_3$	1172	*e*	CH_3SeH	1001	*m*
$CH_3(I)CH{=}CH_2$	2590	*f*	SiH_3PH_2	1535	*n*
cis-$CH_3CH{=}CHCH{=}CH_2$	741	*g*	$SiH_3CH{=}CH_2$	1488	*o*
trans-$CH_3CH{=}CHC{\equiv}CH$	1903	*h*	$GeH_3CH{=}CH_2$	1238	*p*

[a] E. Hirota, S. Saito, and Y. Endo, *J. Chem. Phys.*, **71**, 1183 (1979).
[b] Bauder and Gunthard [48].
[c] S. Tsuchiya, *J. Mol. Struct.*, **22**, 77 (1974).
[d] D. G. Lister and N. L. Owen, *J. Chem. Soc. Faraday Trans.* 2, **69**, 1036 (1973).
[e] G. Williams, N. L. Owen, and J. Sheridan, *Trans. Faraday Soc.*, **67**, 922 (1971).
[f] P. Groner and A. Bauder, *J. Mol. Spectrosc.*, **74**, 259 (1979).
[g] S. L. Hsu and W. H. Flygare, *J. Chem. Phys.*, **52**, 1053 (1970).
[h] R. G. Ford and L. B. Szalanski, *J. Mol. Spectrosc.*, **42**, 344 (1972).
[i] K. Takagi and T. Kojima, *J. Phys. Soc. Japan*, **30**, 1145 (1971).
[j] A. M. Mirri and W. Caminati, *J. Mol. Spectrosc.*, **47**, 204 (1973).
[k] E. G. Codding, R. A. Creswell, and R. H. Schwendeman, *Inorg. Chem.*, **13**, 856 (1974).
[l] J. R. Durig, K. S. Kalasinsky, and V. F. Kalasinsky, *J. Mol. Struct.*, **34**, 9 (1976).
[m] C. H. Thomas, *J. Chem. Phys.*, **59**, 70 (1973).
[n] R. Varma, K. R. Ramaprasad, and J. F. Nelson, *J. Chem. Phys.*, **63**, 915 (1975).
[o] Y. Shiki, A. Hasegawa, and M. Hayashi, *J. Mol. Struct.*, **78**, 185 (1982).
[p] J. R. Durig, K. L. Kizer, and Y. S. Li, *J. Am. Chem. Soc.*, **96**, 7400 (1974).

changes in the methyl group structure significantly affect the derived V_3 via I_α. Unfortunately, determination of the methyl group structure, as well as the structural parameters of the rest of the molecule, is influenced by zero-point vibration effects (see Chapter XIII). Where the structure of the specific methyl group has not been evaluated, the value assumed for I_α in the calculation of V_3 should be given. A common assumption for I_α is 3.10 amu Å^2. An assessment of the possible errors in the evaluation of barriers is available [144].

Besides the barrier heights and structural parameters, another type of structural information, namely, the equilibrium orientation of the methyl group, has been determined in a number of investigations. When a deuterium atom is substituted for one or two of the hydrogens of a methyl group, the moments of inertia will depend on the angle of internal rotation. The first molecule for which the equilibrium configuration was obtained from its microwave spectra was ethyl chloride. Wagner and Dailey [145] studied monodeuterated ethyl chloride (CH_2DCH_2Cl); since the tunneling rate between potential minima is low compared with the time of a microwave measurement, the spectra of two forms of the molecule were observed, one form in which the deuterium atom is *trans* to the chlorine and one in which the deuterium is in

Table 12.9 Potential Constants for Two-top Molecules

Molecule	V_3(*cal/mole*)	V_{12}(*cal/mole*)	*Ref.*
$(CH_3)_2CH_2$	3294	−158	a
$(CH_3)_2O$	2618	18	b
$(CD_3)_2O$	2572	29	b
$(CH_3)_2S$	2136	34	c
$(CD_3)_2S$	2097	32	c
$(CH_3)_2{}^{80}Se$	1501	—	d
$(CD_3)_2{}^{80}Se$	1493	28	e
$(CH_3)_2PH$	2319	—	f
$(CH_3)_2SiH_2$	1646	−38	a
$(SiH_3)_2S$	523	—	g
$(CH_3)_2CCO$	2065	—	h
$(CH_3)_2CCCH_2$	2025	—	i

[a] A. Trinkaus, H. Dreizler, and H. D. Rudolph, *Z. Naturforsch.*, **28a**, 750 (1973).
[b] Lutz and Dreizler [123].
[c] J. Demaison, B. T. Tan, V. Typke, and H. D. Rudolph, *J. Mol. Spectrosc.*, **86**, 406 (1981).
[d] H. Dreizler, E. Fliege, and G. K. Pandey, *Z. Naturforsch.*, **35a**, 1223 (1980).
[e] G. K. Pandey, H. Lutz and H. Dreizler, *Z. Naturforsch.*, **31a**, 1413 (1976).
[f] J. R. Durig, P. Groner, and Y. S. Li, *J. Chem. Phys.*, **67**, 2216 (1977).
[g] K.-F. Dössel and D. H. Sutter, *Z. Naturforsch.*, **33a**, 500 (1978).
[h] K. P. R. Nair, H. D. Rudolph, and H. Dreizler, *J. Mol. Spectrosc.*, **48**, 571 (1973).
[i] J. Demaison and H. D. Rudolph, *J. Mol. Spectrosc.*, **40**, 445 (1971).

either of two equivalent *gauche* positions (see Fig. 12.17). Rotational constants calculated for an eclipsed configuration could not account for the observed rotational constants. The microwave spectrum thus clearly demonstrates that the equilibrium (or minimum energy) conformation of the methyl group in ethyl chloride is the staggered one. For CH_3CHO, the oxygen is eclipsed with one methyl hydrogen, in CH_3OCHO, the methyl hydrogens stagger the C–O bond, and in $(CH_3)_2O$ each methyl group is staggered with respect to the C–O bond. Additional equilibrium configurations obtained from microwave investigations are tabulated in the earlier edition of this book.

Methyl Group Tilt

An intriguing feature of several molecules which have been studied is that the symmetry axis of the methyl group is not coaxial with the bond axis. This is shown for methyl alcohol in Fig. 12.18. Although this tilt of the axis of rotation has been found to be small, nevertheless the existing evidence indicates that the

Table 12.10 Barriers for Some Molecules with Two Nonequivalent Symmetric Tops

Molecule	V_3*(cal/mole)*	V_3*(cal/mole)*	*Ref.*
CH_3PHCD_3	2390 (CH_3)	2369 (CD_3)	[a]
trans-$CH_3CH{=}NCH_3$	1544 (C—CH_3)	1996 (N—CH_3)	[b]
CH_3OSiH_3	550 (CH_3)	1100 (SiH_3)	[c]
CH_3COOCH_3	285 (C—CH_3)	1215 (O—CH_3)	[d]
CH_3OCD_3	2603 (CH_3)	2565 (CD_3)	[e]
trans-$CH_3OCH_2CH_3$	2702 (O—CH_3)	3300 (C—CH_3)	[f]
trans-$CH_3OCH_2SiH_3$	2520 (CH_3)	2128 (SiH_3)	[g]

[a] J. R. Durig, S. D. Hudson, M. R. Jalilian and Y. S. Li, *J. Chem. Phys.*, **74**, 772 (1981).
[b] Bossert et al. [126].
[c] LeCroix et al. [125].
[d] J. Sheridan, W. Bossert, and A. Bauder, *J. Mol. Spectrosc.*, **80**, 1 (1980).
[e] Durig et al. [53].
[f] M. Hayashi and K. Kiwada, *J. Mol. Struct.*, **28**, 147 (1975).
[g] Y. Shiki, N. Ibushi, M. Oyamada, J. Nakagawa, and M. Hayashi, *J. Mol. Spectrosc.*, **87**, 357 (1981).

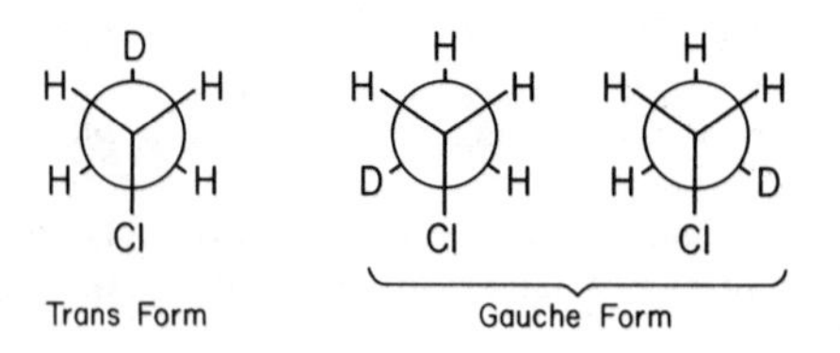

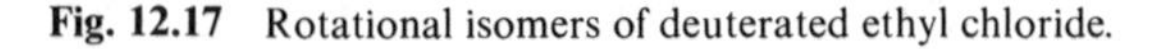

Fig. 12.17 Rotational isomers of deuterated ethyl chloride.

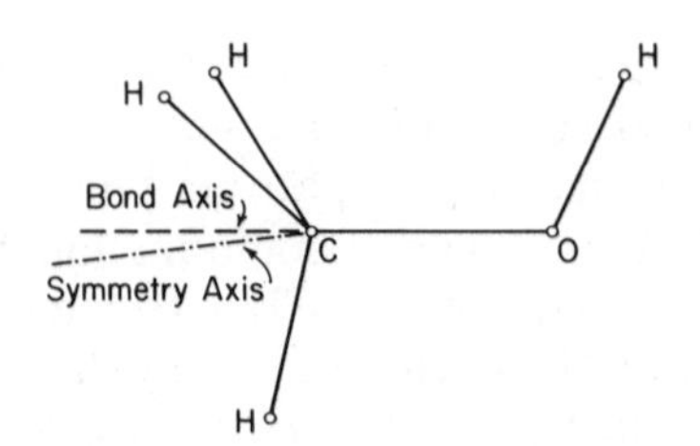

Fig. 12.18 Illustration of the departure of the symmetry axis of the methyl group from the C–O bond axis in methyl alcohol.

effect is real. Evidence for tilted methyl groups has come mostly from an analysis of the internal rotation splittings. Here both the barrier and the direction cosines are fitted to the observed splittings, and the direction cosines so obtained have indicated in several instances that the symmetry axis is tilted with respect to the bond axis. Support for tilts has also come from structural analyses. In the case of dimethyl sulfide [146], for example, the tilt was established from both internal rotation splittings and from structural data determined from isotopic moment-of-inertia information. Both methods are in excellent agreement; the structural calculation indicates that each methyl group is tilted by 2°45′, and the analysis of the internal rotation splitting gives 2°27′. Work on molecules with asymmetrical methyl groups (e.g., CHD_2) has also indicated that the splittings are quite sensitive to the tilting of the methyl group axis [129].

It has been observed by Pierce and Hayashi [146] that tilted methyl groups generally appear when the methyl group is bonded to atoms with unshared electron pair or pairs such as oxygen, sulfur, or nitrogen and, furthermore, that the tilt is towards the unshared pair or pairs. A similar effect has been observed for the methylene group of formaldoxime ($CH_2{=}NOH$) [147] where the methylene group is tilted by about 3° toward the unshared pair on the nitrogen. In the case of nitric acid (HNO_3), a 2° tilt of the NO_2 group away from the hydrogen atom has been reported [148, 149].

For methyl and silyl groups attached to an unsaturated carbon atom, such as in acetone [150], vinyl silane [151], acetyl chloride, and cyanide [150], the tilt is toward the double bond. However, in propylene [152] no such tilt is observed, while for methyl ketene [153] the tilt is away from the double bonds of the ketene group.

In methyl fluorosilane the methyl group tilts toward the two hydrogens in the SiH_2 group by 1°50′, whereas in ethyl chlorosilane and ethyl fluorosilane the ethyl group tilts toward the two hydrogens in the SiH_2 group by 1°21′ and 1°3′, respectively [154]. For ethyl methyl sulfide a tilt of 2°27′ of the ethyl group toward the lone pair electrons on the sulfur atom is found [155].

13 ROTATIONAL ISOMERISM AND RING CONFORMATIONS

For molecules with asymmetric groups connected by a single bond, certain configurations, which differ by a rotation about a single bond, are energetically favored, and two or more rotational isomers of finite lifetime may exist. Each isomer corresponds to a potential energy minima. Since the rotational spectrum is characteristic of the overall geometry of the molecule, geometrically distinct conformers of sufficient abundance give rise to separate distinct spectra, and the various types of information outlined in previous chapters can be obtained for each conformer, including the molecular structure, the methyl group barrier, and the dipole moment. Furthermore, relative intensity measurements of rotational lines associated with the various isomers give information about the energy differences between conformations. Knowledge of the relative stabilities of the isomers and a study of the excited torsional states of the isomers can yield information about the potential that hinders the internal rotation of the asymmetric groups.

Increasing numbers of cyclic structures are being studied with microwave spectroscopy. The ring compounds being investigated are of increasingly large size and complexity. The larger the size of the ring, the more flexible it becomes, and the more probably it has multiple conformations. If some rigidity is introduced into the ring skeleton (e.g., a carbon–carbon double bond), the number of possible conformations is reduced. Microwave spectroscopy can yield information on the stable conformations and the presence or absence of metastable forms; it has been quite successful for elucidation of rotational isomerism and conformations of ring compounds. Ring-puckering potentials, ring conformations, concentrations of rotational isomers, and other information gained from microwave studies of complex molecules may be found in several reviews [156–165].

Identification of the conformers present can be made without a detailed structural study. Often the rotational constants are sufficiently sensitive to different orientations of parts of a molecule that the conformations may be obtained from a study of the common isotopic form of each conformer. Small changes in geometries of different conformations do not prevent the gross conformation from being determined. When the rotational constants are not sufficient, other information obtained from the rotational spectrum can be called upon to aid in confirmation of a particular rotational isomer or ring conformation. The six-membered ring thiane [166] provides a useful example. In Fig. 12.19, the calculated rotational constants from an assumed structure are plotted for thiane over a range of τ values from the chair form (negative values) to the boat form (positive values). For thiane, two possible conformations are clearly possible which satisfy the rotational constant data. One configuration is a chair form with $\tau \cong -60°$, and the other is a boat form with $\tau \cong +15°$. Both these conformations have reasonable interior ring angles and predict "a" and "c"-type transitions, which is consistent with the observed spectrum. Clearly, the rotational constants alone do not allow a unique determination of the conformation of thiane. The dipole moment components μ_a and μ_c, obtained from a study of the Stark effect, will be determined primarily by the relative position of the CSC plane in the principal axis system. This information can be used to confirm the chair form as the correct conformation for thiane. This is demonstrated in Fig. 12.20. From the components $\mu_a = 1.684$ D and $\mu_c = 0.578$ D, the angle between μ and the principal "a" axis can be calculated to be 18.9°. Assuming that the entire dipole moment for the molecule lies in the CSC plane and bisects the CSC angle, one can calculate from reasonable structural parameters the angle between the principal "a" axis and μ for a given conformation. The results of such a calculation are summarized in Fig. 12.20. As can be seen from the figure, the agreement for the chair form, where $\theta = 25.7°$, is much better than the boat form, where $\theta = 37.7°$. From this one can conclude that the conformation for thiane is the chair form.

In certain cases, the intensities of different rotational transitions show the effects of nuclear statistics (Chapter III, Section 4), which in turn reflect the symmetry of the molecule. For example, a study [167] of the relative intensities of certain transitions of dioxene indicate that conformation must have a C_{2v}

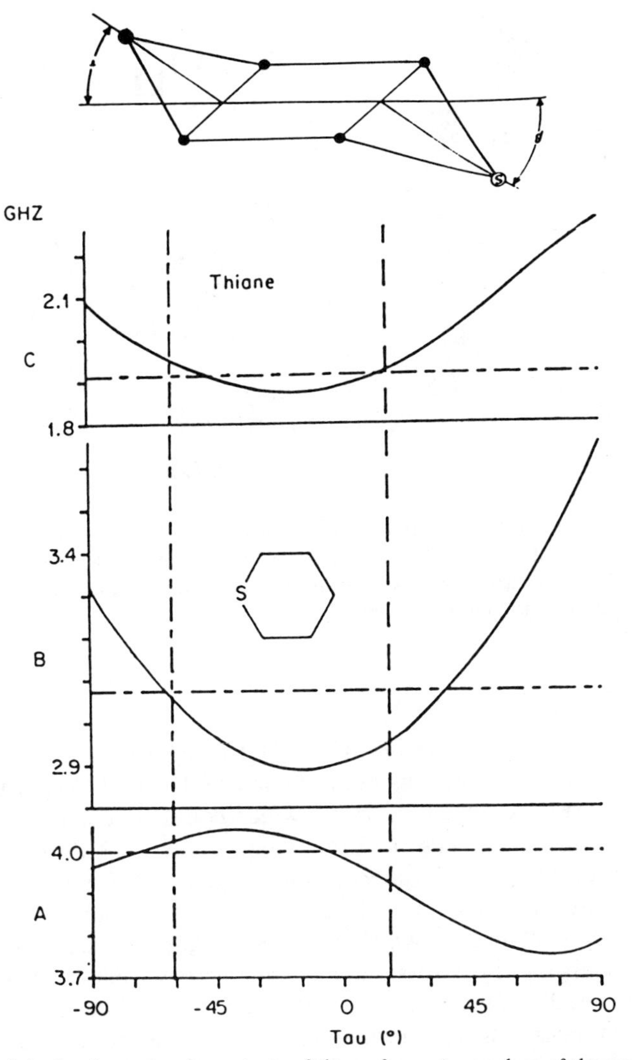

Fig. 12.19 Calculated rotational constants of thiane for various values of the angle $\tau(\beta=50.5°)$. The angles τ and β measure, respectively, the inclination of CCC and CSC planes with respect to the plane of the other four carbon atoms. Positive values of τ correspond to the boat form, negative values to the chair form. Observed rotational constants are indicated by the horizontal broken lines. From Kitchen et al. [166].

or C_2 symmetry consistent with the planar or twisted structure and not with the bent form. The difference in the nuclear quadruopole coupling constants for two conformers can also provide information to aid in establishing the conformation [168]. In 1, 2, 3, 6-tetrahydropyridine, $\overline{CH_2CH_2CH{=}CHCH_2NH}$, quite different coupling constants were obtained for the half-chair axial and half-chair equatorial forms.

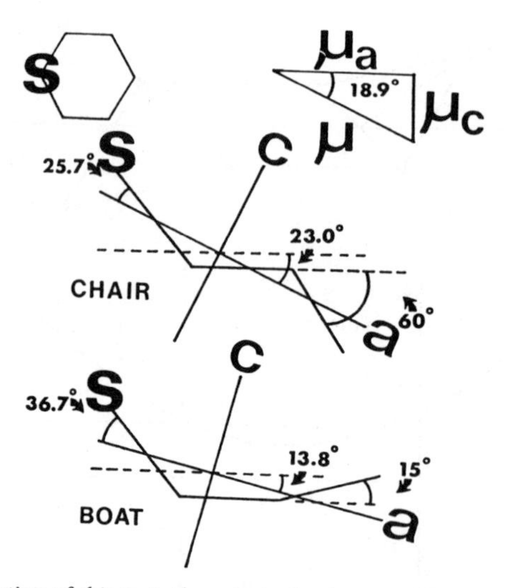

Fig. 12.20 The orientation of thiane in the principal axis system for a chair ($\tau=-60°$) and boat ($\tau=15°$) conformation calculated from an assumed structure. The observed angle between the total dipole moment and the "*a*"-axis is 18.9°, which is in reasonably good agreement with that calculated for the chair conformation. From Kitchin et al. [166].

Examples of Rotational Isomers

The *n*-propyl fluoride molecule ($CH_3CH_2CH_2F$), where internal rotation about two bonds is possible, has been studied by Hirota [169]. Two distinct rotational spectra were observed, one of which could be assigned to the *trans* form and the other to the *gauche* form. These isomers are illustrated in Fig. 12.21. The dihedral angle of the *gauche* form is about 117° from the *trans* ($\theta=0$) position. The ground state lines of both the *trans* and *gauche* isomer were accompanied by satellite lines which could be assigned to excited states of the central C–C(CH_2—CH_2F) torsion and the methyl torsion. Although the ground state lines did not show any splitting, some of the first excited state lines due to methyl torsion were split into doublets; these lines were used for evaluation of the barrier height hindering methyl rotation. The CH_3 barrier was found to be 2.69 kcal/mole for the *trans* isomer and 2.87 kcal/mole for the *gauche* isomer. From measurement of the relative intensities of the excited state lines of the C–C torsion, the energy differences among the first three low-lying levels associated with the *trans* and *gauche* potential wells were found. By a measurement of the temperature dependence of the relative intensity of two nearby transitions, namely the 4_{04}–4_{13} transition of the *trans* isomer and the 7_{16}–7_{25} transition of the *gauche* isomer, the energy difference between the ground states of the two isomers was found to be 0.47 ± 0.31 kcal/mole, with the *gauche* form more stable than the *trans* form.

The potential function for rotation of the CH_2F group may be taken to have the form

$$V=\frac{1}{2}\sum_{n} V_n(1-\cos n\theta) \tag{12.124}$$

with θ the angle of internal rotation about the central C–C bond. By expanding the potential in a Taylor series at each minima and using such information as the energy difference between rotational isomers, the frequency of the C–C torsional mode for the *trans* and the *gauche* isomer, the spacing of the levels associated with the *trans* and *gauche* potential wells, and the variation of the rotational constants with torsional state—the latter two of which give information about the anharmonicity of the wells—and considering only the C–C torsional degree of freedom, one can extract the first seven Fourier coefficients of the foregoing potential function. The first seven potential constants yield the potential curve shown in Fig. 12.22. The potential function is clearly only an approximation to the "real" potential and is subject to errors from the measured

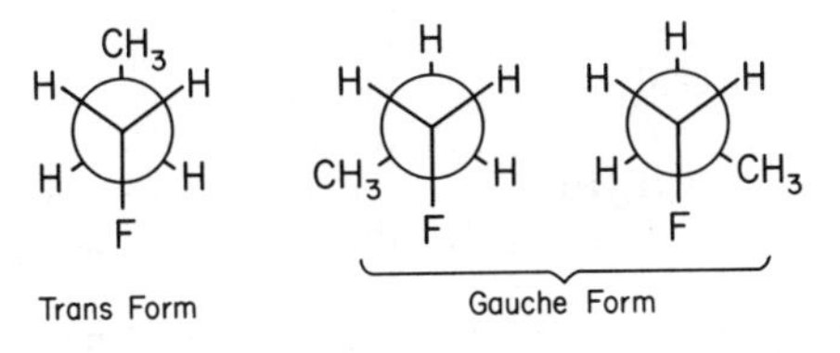

Fig. 12.21 Stable rotational isomers of *n*-propyl fluoride.

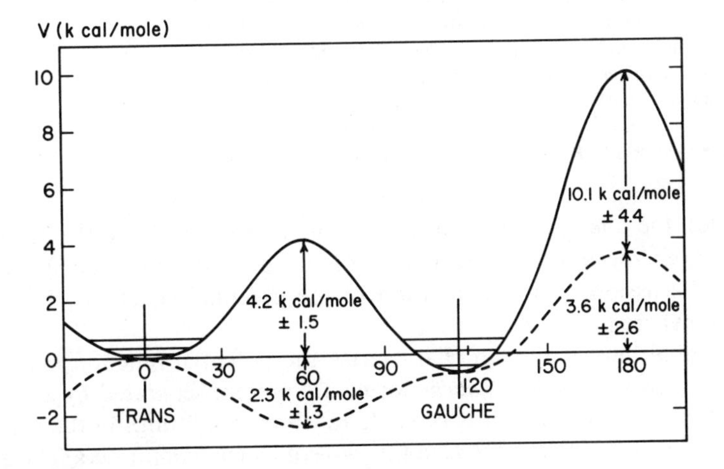

Fig. 12.22 The potential function for rotation of the CH_2F group using the derived potential constants which are (in kcal/mole): $V_1=3.22\pm2.02$, $V_2=-3.05\pm1.72$, $V_3=6.48\pm2.15$, $V_4=-1.25\pm1.17$, $V_5=0.37\pm0.34$, $V_6=-0.76\pm0.54$, $V_7=-0.002\pm0.29$. The V_3 term is the largest; however, the other terms also contribute significantly. The dotted line represents the contribution of all terms except the threefold and sixfold terms. This potential function indicates the barrier at the *cis* ($\theta=180°$) position is higher than the barrier separating the *trans* form from the two equivalent *gauche* forms. From Hirota [169].

Table 12.11 Illustrations of Rotational Isomers Found with Microwave Spectroscopy[a]

Molecule	*Structure*	*Rotational Isomers Observed*	*Energy Difference (E_a-E_b) Between Isomers (cal/mole)*[b]	*Ref.*
Propionyl fluoride	$\underline{CH_3}CH_2C\underline{O}F$	(a) *gauche* (b) *cis*	1290 ± 50	c
Cyclopropane-carbonyl fluoride	CH_2CH_2CHCOF	(a) trans (b) *cis*	572 ± 85	d
n-propyl isocyanide	$\underline{CH_3}CH_2CH_2\underline{NC}$	(a) *trans* (b) *gauche*	283 ± 129	e
Ethyl mercaptan	$\underline{CH_3}CH_2S\underline{H}$	(a) *trans* (b) *gauche*	406 ± 43	f
Monothioformic acid	$\underline{H}COS\underline{H}$	(a) *cis* (b) *trans*	661 ± 17	g
Propyl mercaptan	$\underline{CH_3}CH_2CH_2\underline{SH}$	(a) *trans* (b) *gauche*	389 ± 114	h
Formic acid	$\underline{H}COO\underline{H}$	(a) *cis* (b) *trans*	3903 ± 86	i
Ethyl alcohol	$\underline{CH_3}CH_2O\underline{H}$	(a) *gauche* (b) *trans*	118 ± 14	j

[a]The designation of the isomers observed refers to the relative orientation of the underlined atoms or groups.
[b]Energy difference between ground states.
[c]Stiefvater and Wilson [171].
[d]H. N. Volltrauer and R. H Schwendeman, *J. Chem. Phys.*, **54**, 268 (1971).
[e]M. J. Fuller and E. B. Wilson, *J. Mol. Spectrosc.*, **58**, 414 (1975).
[f]R. E. Schmidt and C. R. Quade, *J. Chem. Phys.*, **62**, 3864 (1975).
[g]B. P. Winnewisser and W. H. Hocking, *J. Phys. Chem.*, **84**, 1771 (1980).
[h]J. Nakagawa and M. Hayashi, *J. Mol. Spectrosc.*, **85**, 327 (1981).
[i]W. H. Hocking, *Z. Naturforsch.*, **31a**, 1113 (1976).
[j]R. K. Kakar and C. R. Quade, *J. Chem. Phys.*, **72**, 4300 (1980).

intensities, the one-vibration approximation, and so on. Nevertheless, knowledge of even the approximate nature of the potential function is valuable. Further discussion on the evaluation of the potential constants, V_n, may be found elsewhere [170–172].

Additional information to characterize the potential function may be obvious when two equivalent *gauche* forms are present separated by a relatively low barrier. In such cases, it is possible to observe splittings in the rotational spectrum due to tunneling between the two equivalent conformers. The general theory to account for these splittings has been given by Meaken et al. [133] and applied to 3-fluoropropane, $CH_2FCH{=}CH_2$. A simple perturbation treatment applicable to such case has also been given [173] and the use of the Stark effect has been discussed [174].

Rotational isomers of some of the molecules for which at least two rotamers have been identified by microwave spectroscopy are listed in Table 12.11. A large number of similar molecules have also been studied with microwave spectroscopy, including many in which only one rotameric form has been observed [156, 159].

Application of Low-resolution Spectroscopy

Under certain conditions, low-resolution microwave spectroscopy (see Chapter VII) can be used to give reliable molecular conformational information quickly. The identity of the conformer can usually be established by transferal of structural data from similar molecules and reasonable structural assumptions in the same way as for a high-resolution study. With large molecules and more than one conformer present, the rotational spectrum can be very complicated yet yield, under low resolution, a tractable spectrum. In such studies, a single rotational constant is obtained for each observable conformer. Since one is usually dealing with large molecules, a large number of structural parameters must be assumed in the calculation of, for example, $(B+C)$ obtained from ${}^{a}R_{01}$ band spectra. Results indicate the agreement between observed and calculated rotational constants is usually 5% or better. In order to more clearly assess the confidence to be attached to the conformer, or dihedral angle, it is useful to obtain an estimate of the error in $(B+C)$ from reasonable estimates of the uncertainties that might be expected in the structural parameters used in the calculation.

When a number of conformers of comparable stability are present, the low-resolution spectra can become very dense, and significant cancellation can occur from overlapping of Stark lobes and absorption bands. Straight chain alkanes, for example, with more than six carbons usually cannot be studied by low-resolution microwave spectroscopy [175]. Trace impurities can also be a more serious problem in such studies. The observation of low-resolution spectra of isotopic species (H and D or ^{35}Cl and ^{37}Cl) can be very useful in the assignment, since isotopic changes in $(B+C)$ can be sensitive to the dihedral angle.

A low-resolution study [176] of crotonyl chloride provided direct evidence of the existence of both the s-*cis* and s-*trans* conformers for this molecule. Both the s-*cis* and s-*trans* forms of crotonyl chloride are slightly asymmetric rotors, and the a-type transitions falling in the R-band region are rather high J transitions. Therefore, both forms, if present, should show bandlike spectra. This is indeed confirmed as illustrated in Fig. 12.23. Note that both the ^{35}Cl and ^{37}Cl species for the *cis* and *trans* forms are observed. The calculated and observed $(B+C)$ and the isotopic shifts in $(B+C)$ are useful in establishing the assignments. The calculated shifts in $(B+C)$ for the s-*trans* and s-*cis* forms are 39.0 and 49.2 MHz, respectively, and the observed shifts are 36.6 ± 2.9 and 48.9 ± 4.6 MHz, respectively. The agreement between the observed and calculated $(B+C)$ is much better for the *trans* form. This perhaps is not surprising since the structural parameters were taken from s-*trans* acryloyl chloride [177]. The much poorer agreement for the *cis* form probably indicates a change in one or more of the molecular parameters in going from the s-*trans* to the s-*cis* form.

A number of conformational studies using this technique have been made by

True and Bohn. Low-resolution studies of the isopropyl compounds [178] of the general form $(CH_3)_2CHOCOX(X=H, F, Cl, CN, \text{or } CF_3)$ show the presence of one rotamer, the syn-*gauche* form. On the other hand, for the similar propargyl derivatives [179] $HC \equiv CCH_2OCOX(X=F, Cl, CN, \text{or } CF_3)$, the $X=CN$ and CF_3 molecules display three band series corresponding to the rotamers syn-anti [$\tau_1(O=COC)=0°$, $\tau_2(COCC)=180°$], syn-*gauche* [$\tau_1(O=COC)=0°$, $\tau_2(COCC) \sim 90°$], and *gauche-gauche* [$\tau_1(O=COC) \sim 60°$, $\tau_2(COCC) \sim 270°$], where τ is a dihedral angle. For $X=F$ and Cl, only the syn-anti and syn-*gauche* rotamers are found. Temperature studies of the band intensities of the different rotamers indicate that the syn-anti and syn-*gauche* forms are of comparable stability while the *gauche-gauche* rotamer is higher in energy for $X=CN, CF_3$. Studies [180] of $CH_3CH_2OCOX(X=Cl, CN, CF_3)$ also show the three rotamers, syn-*anti*, syn-*gauche*, and *gauche-gauche* with the *gauche-gauche* conformer of higher energy.

A series of 1-halogenated alkanes with carbon chains of three to six atoms have been studied by Steinmetz et al. [175]. These molecules have an increasing number of internal rotation centers as the chain lengthens. A number of rotamers are found for each molecule from a study of the low-resolution spectra.

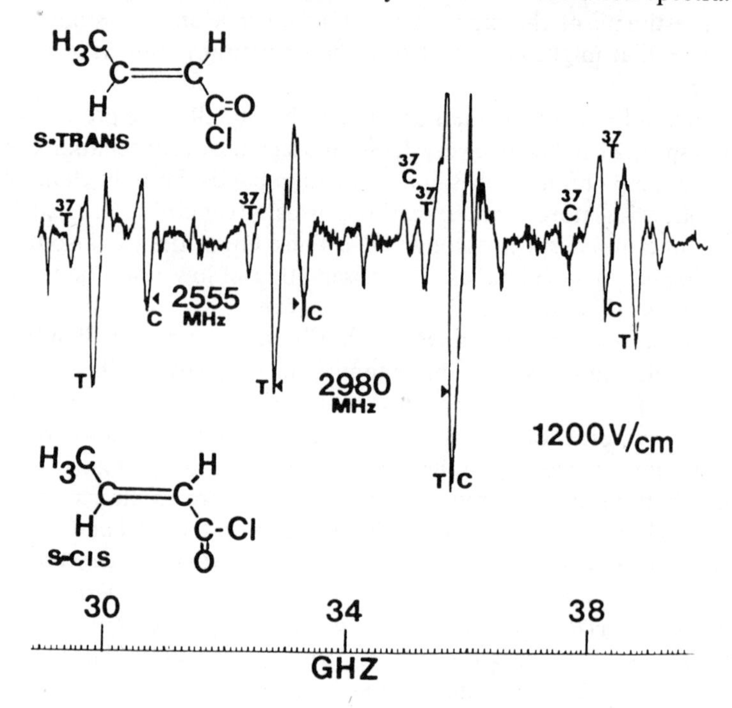

Fig. 12.23 Low-resolution "band" spectra of crotonyl chloride. The stronger bands are the ^{35}Cl peaks. The structural parameters for *s-trans* acryloyl chloride were employed to obtain the calculated values of $(B+C)$. Observed $(B+C)$ values are (in MHz): 2554.5, 2505.6 for *s-cis*-^{35}Cl and ^{37}Cl species; and 2979.9, 2943.3 for *s-trans*-^{35}Cl and ^{37}Cl, while the calculated values are 2480.5, 2431.3 and 2982.0, 2943.0, respectively. From Lum et al. [176].

Effects of Intramolecular Hydrogen Bonding on Conformation

An increasing amount of information is now being obtained by microwave spectroscopy on hydrogen bonds, since the initial work of Costain and Srivastava [181], who studied the bimolecules produced by use of trifluoroacetic acid with formic, acetic, and fluoroacetic acid. Information on intermolecular hydrogen bonds has already been summarized in Chapters V and VI, and other studies may be cited [182–184]. Here we point out the importance of subtle effects of intramolecular hydrogen bonding in determining the conformation of certain molecules where an electronegative atom or group, such as oxygen, nitrogen, a halogen, or electrons of a double, or triple bond or conjugated ring system, is present and properly disposed spatially.

Murty and Curl [185] have studied the microwave spectrum of allyl alcohol, $CH_2{=}CHCH_2OH$. A number of conformations are possible, corresponding to different orientations of the carbinol group $—CH_2OH$ with respect to the ethylenic group $H_2C{=}CH—$ about the C–C bond. For each conformation about the C–C bond there are various possible orientations of the OH group about the C–O bond. The conformer identified for allyl alcohol is such that the O–H hydrogen is brought close to the C=C double bond. Since no significant lengthening of the O–H bond is observed, the interaction is weak, but is nonetheless important in stabilizing the conformation of the most likely dominant form. The H-bonded form is often found to be the more stable one.

In 2-chloroethanol [186], $ClCH_2CH_2OH$, which also has two centers of internal rotation and many possible conformers, the *gauche-gauche* form found (Fig. 12.24) brings the O–H hydrogen close to the chlorine atom, indicating a weak hydrogen bond. The OH···· Cl distance found is 2.609 Å, about 0.4 Å less than the sum of van der Waals radii, and the O–H bond (1.008 ± 0.003 Å) is longer than in a free OH group such as in methanol [187] where $r(\text{O–H}) = 0.96$ Å.

In 2-nitrophenol [188], $OHC_6H_4NO_2$, the most marked structural effect resulting from hydrogen bond stabilization is that the phenolic hydrogen remains in the molecular plane despite a large overlap of van der Waals radii. Similar results are obtained for salicylaldehyde [189].

In a number of molecules with a N–H or O–H bond and a $—C{\equiv}C—$ group, the most stable conformer is one where a maximum hydrogen bonding interaction with the triple bond is possible. This is found in *gauche*-propargyl alcohol, ($HOCH_2C{\equiv}CH$) [190], *gauche*-hydroxy acetonitrile ($HOCH_2C{\equiv}N$) [191], *trans*-amino acetonitrile ($H_2NCH_2C{\equiv}N$) [192], and *trans*-propargyl amine ($H_2NCH_2C{\equiv}CH$) [193, 194].

The rotamer found for methoxyacetic acid [195], CH_2OCH_2COOH, has an intramolecular hydrogen bond formed between the hydroxyl hydrogen and the ether oxygen atom giving a five-membered planar ring configuration. The concentration estimated for this form is between 10 and 30% of the total.

The stable rotamer observed for trifluoroethylamine [196] is the *trans* form where two possible intramolecular hydrogen bonds are possible between the

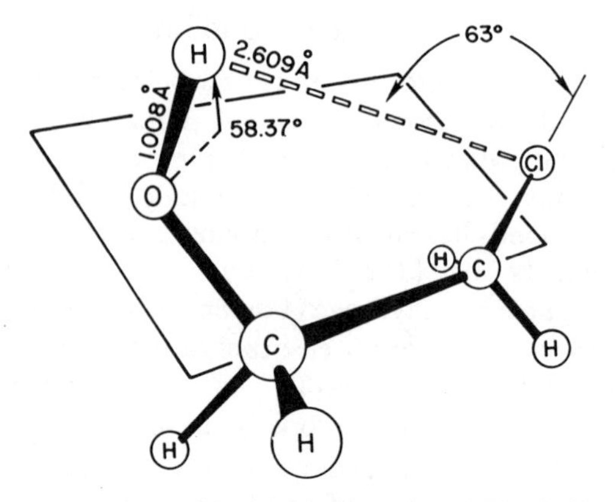

Fig. 12.24 *Gauche-gauche* conformer found for 2-chloroethanol. The O–H and C–Cl are essentially parallel to each other. From Azrak and Wilson [186].

fluorine atoms and the amino hydrogens NH···F, whereas in the *gauche* form only one is possible.

Intramolecular hydrogen bonding has been found involving sulfur. Specifically, OH···S in mercaptoethanol [197] and SH···N in 2-aminoethanethiol [198] where the sulfur atom in the latter molecule serves as a donor, although the H-bond conformer is not the most stable form. Similarly, the H-bond conformer (OH···F) in 3-fluoropropanol [199] is not the most stable form and in 3-chloropropanol [200] no H-bond conformer is found.

14 RING-PUCKERING VIBRATIONS

Since the initial microwave spectral work by Gwinn and his group [201–203] in 1960, important advances have been made in the determination of potential functions for the out-of-plane ring vibrations in four- and five-membered ring molecules. Their initial work on trimethylene oxide confirmed the need for quartic term in the potential function as predicted earlier by Bell [204]. In addition, a quadratic term was also found to be important. In four-membered ring molecules there is one out-of-plane ring vibration, a ring puckering, which is usually the lowest frequency vibration present. This mode can exhibit large amplitude motions. Ring puckering, internal rotation, inversion, and the bending motion of quasilinear molecules are all examples of large amplitude motion. With five-membered ring molecules there are two out-of-plane ring vibrations—a ring puckering and a ring twisting mode. In five-membered rings having a double bond in the ring, the ring is stiffened so that the twisting mode is the higher in frequency. These unsaturated ring compounds then behave like four-membered ring systems. They are called "pseudo-four-membered" ring molecules. Similarly, a six-membered ring molecule with a double bond may be considered a pseudo-five-membered ring molecule. For ring molecules having a double-minimum potential, the planarity is best described by the

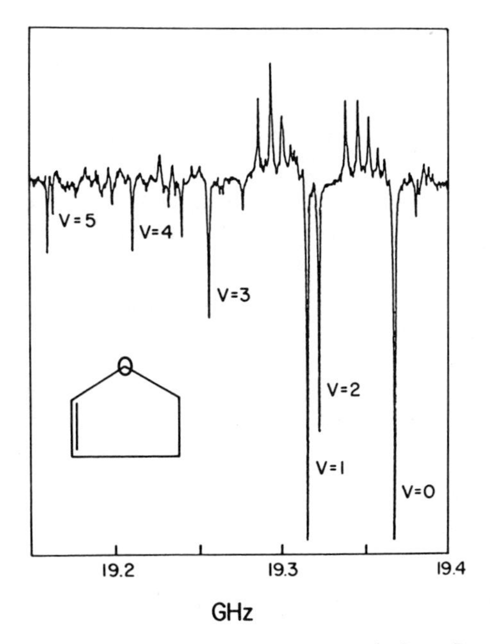

Fig. 12.25 Microwave spectrum of 2,3-dihydrofuran showing the $7_{43} \leftarrow 7_{62}$ transition with $v = 0$–5 excited states of the ring puckering mode. From J. R. Durig, Y. S. Li, and C. K. Tong, *J. Chem. Phys.*, **56**, 5692 (1972).

presence or absence of a barrier at the planar configuration. The size of this barrier provides a measure of the planarity or nonplanarity of the ring. The motion is symmetric about the planar form and governed by a symmetric double minimum potential similar to an inversion (see Fig. 6.6). Both microwave and infrared studies have provided a great deal of information on ring conformations and barriers to planarity which in turn represent a delicate balance between angle strain and internal torsional forces, and hence provide information on these intramolecular forces.

In the microwave spectrum, the presence of a low-frequency puckering vibration is reflected in the appearance of a number of vibrational satellite lines accompanying the ground state transitions. This is illustrated in Fig. 12.25. Relative intensity measurements of these satellites, at a given temperature or at different temperatures provide information on the vibrational energy spacing. The variation of the rotational constants with the puckering vibrational state reflects the large amplitude motion and can provide information on the potential-energy constants. Whereas for most molecules the variation of the rotational constants with vibrational state is usually found to be smooth and nearly linear, the variation for ring molecules with the ring-puckering vibrational quantum number is often quite irregular. To a good approximation, this low-frequency, out-of-plane, mode may be effectively separated from the other higher-frequency modes, and a one-dimensional model may be applied.

Table 12.12 Characteristics of Rotational Spectra for Molecules with Ring Puckering

Potential Function	*Vibrational Energy Separation*	*Rotational Constant Variation with v*	*Spectra*	*Structure*
Single Minimum				
$(a>0, b>0)$	$\Delta E_v > \Delta E_r$	Regular (near linear)	Rigid Rotor	Planar
Double Minimum				
$(a>0, b<0)$ Low barrier ($<100\ \text{cm}^{-1}$)	$\Delta E_v > \Delta E_r$	Zigzagging for low v	Rigid Rotor	Depends on barrier height[a]
Intermediate barrier (100 to 500 cm^{-1})	$\Delta E_v \cong \Delta E_r$	Irregular	Nonrigid Rotor	Nonplanar
High barrier ($>500\ \text{cm}^{-1}$)	$\Delta E_v > \Delta E_r$	Regular	Rigid Rotor	Nonplanar

[a]If the barrier height is small compared to the zero-point energy, the molecule is planar.

The vibrational Hamiltonian may be written as

$$\mathscr{H} = -\frac{\hbar^2}{2\mu}\frac{d^2}{dx^2} + ax^4 + bx^2 \tag{12.124}$$

where x is a one-dimensional large-amplitude ring-puckering coordinate; a, b are potential function constants to be evaluated; and μ is the reduced mass for the ring-puckering vibration and is to a good approximation considered constant independent of x. This potential function takes on different shapes depending on the values of a and b ($a>0, b>0$ single minimum; $a>0, b<0$ double minimum).

With a double minimum potential, the lower vibrational levels can be near degenerate; the usual separation of rotation and vibration is then insufficient, and coupling between the out-of-plane mode and rotation must be introduced. The presence of vibration–rotation coupling which leads to nonrigid rotor-like spectra [205–210] provides information regarding the nature of the ring-puckering vibration. The Stark effect [201, 205, 206, 209, 210] and the observation of transitions between inversion levels [205, 206] can provide additional information. In Table 12.12 we summarize the characteristics of rotational spectra for molecules with ring puckering.

Space does not permit us to describe the theoretical methods for analysis of the effects of ring-puckering vibrations on the spectra. Gwinn and Gaylord [162] and Gwinn and Luntz [211] have discussed the theory and its applications, particularly in microwave spectroscopy. Reviews by Malloy et al. [164] and by Carreira et al. [163] discuss the application of infrared, Raman, and microwave spectroscopy, reviewing the theory and giving specific applications.

15 HINDERED PSEUDOROTATION

In flexible five-membered rings such as tetrahydrofuran (ring structure) there are two low-frequency out-of-plane ring vibrations—a ring-twisting and a ring-puckering mode. When these are close in frequency they may couple, and the vibrational problem must be treated as a two-dimensional problem. When the barrier at the planar configuration is high compared to the barrier between other nonplanar forms, the two vibrations may be effectively separated by transformation to polar coordinates. The two vibrations may be described as a pseudorotation angular mode where in the case of free pseudorotation the phase of the maximum puckering amplitude rotates around the ring, and a radial mode of smaller amplitude which determines the magnitude of the puckering. The angular motion is governed by a Hamiltonian which for a vanishing barrier is that for free internal rotation, hence the term pseudorotation. The simplest molecule for pseudorotation is the nonpolar cyclopentane, which exhibits free pseudorotation. The high density of the pseudorotation energy levels is responsible for the unusually high entropy that is observed [212]. When one of the CH_2 groups in cyclopentane is replaced by, for example, an oxygen atom, the symmetry is reduced. This leads to a barrier to pseudorotation, and the phase of the puckering no longer rotates freely around the ring. The motion is termed hindered pseudorotation. The theory of pseudorotation has been discussed by Harris et al. [213] and the microwave spectrum of tetrahydrofuran [214] and 1, 3-dioxalane [215] have been interpreted in terms of hindered pseudorotation with a low barrier. The microwave spectrum exhibits a rich spectrum due to the high density of the vibrational pseudorotation states. In tetrahydrofuran, nine states were observed below 200 cm^{-1}. Significant rotation–vibration effects can also be observed.

References

1. J. D. Kemp and K. S. Pitzer, *J. Chem. Phys.*, **4**, 749 (1936); *J. Am. Chem. Soc.*, **59**, 276 (1937).
2. L. G. Smith. *J. Chem. Phys.*, **17**, 139 (1949).
3. Thermodynamic value 2875 ± 125 cal/mole, K. S. Pitzer, *Discussions Faraday Soc.*, **10**, 66 (1951); spectroscopic value 3030 ± 300, D. R. Lide, *J. Chem. Phys.*, **29**, 1426 (1958). See also Table 12.6.
4. H. M. Kroto and B. M. Landsburg, *J. Mol. Spectrosc.*, **62**, 346 (1976).
5. S. Mizushima, *Structure of Molecules and Internal Rotation*, Academic, New York, 1954.
6. W. J. Orville-Thomas, *Internal Rotation In Molecules*, Wiley-Interscience, New York, 1974.
7. D. G. Burkhard and D. M. Dennison, *Phys. Rev.*, **94**, 408 (1951).
8. C. C. Lin and J. D. Swalen, *Rev. Mod. Phys.*, **31**, 841 (1959).
9. H. Dreizler, *Fortschr. Chem. Forsch.*, **10**, 59 (1968).
10. D. G. Lister, J. N. Macdonald, and N. L. Owen, *Internal Rotation and Inversion*, Academic, New York, 1978.
11. D. R. Herschbach and J. D. Swalen, *J. Chem. Phys.*, **29**, 761 (1958).
12. L. Pierce, *J. Chem. Phys.*, **29**, 383 (1958).

13. L. Pierce and J. M. O'Reilly, *J. Mol. Spectrosc.*, **3**, 536 (1959).
14. W. G. Fateley and F. A. Miller, *Spectrochim. Acta*, **19**, 611 (1963).
15. *Tables Relating to Mathieu Functions*, Columbia Univ. Press, New York, 1951.
16. G. Blanch and I. Rhodes, *J. Wash. Acad. Sci.*, **45**, 166 (1955).
17. R. W. Kilb, *Tables of Mathieu Eigenvalues and Eigenfunctions for Special Boundary Conditions*, Harvard Univ. Press, Cambridge, Mass., 1956.
18. D. R. Herschbach, "Tables for the Internal Rotation Problem" (Department of Chemistry, Harvard Univ., Cambridge, Mass., 1957); *J. Chem. Phys.*, **27**, 975 (1957).
19. M. Hayashi and L. Pierce, "Tables for the Internal Rotation problem" (Department of Chemistry, Univ. Notre Dame, Notre Dame, Ind., 1961); *J. Chem. Phys.*, **35**, 1148 (1961).
20. The torsional Hamiltonian is invariant under the C_{3v} group ($\alpha \to \alpha + 2\pi/3$, $\alpha \to -\alpha$), and the nondegenerate solutions ($\sigma = 0$) have either even or odd parity under $\alpha \to -\alpha$, i.e.,

$$U(-\alpha) = \pm U(\alpha)$$

It is hence convenient in practice to use trigonometric rather than the exponential expansion (see [19]):

$$U_{v0}(\alpha) = C_0^{(v)}/\sqrt{2} + \sum_{k=1} C_k^{(v)} \cos 3k\alpha, \; v \text{ even}$$

$$U_{v0}(\alpha) = \sum_{k=1} C_k^{(v)} \sin 3k\alpha, \; v \text{ odd}$$

21. E. B. Wilson, Jr., *Chem. Rev.*, **27**, 17 (1940).
22. B. L. Crawford, *J. Chem. Phys.*, **8**, 273 (1940).
23. H. H. Nielsen, *Phys. Rev.*, **40**, 445 (1932).
24. J. S. Kohler and D. M. Dennison, *Phys. Rev.*, **57**, 1006 (1940).
25. E. V. Ivash and D. M. Dennison, *J. Chem. Phys.*, **21**, 1804 (1953).
26. K. T. Hecht and D. M. Dennison, *J. Chem. Phys.*, **26**, 31 (1957).
27. R. W. Kilb, C. C. Lin, and E. B. Wilson, Jr., *J. Chem. Phys.*, **26**, 1695 (1957).
28. D. R. Herschbach, *J. Chem. Phys.*, **31**, 91 (1959).
29. E. B. Wilson, Jr., C. C. Lin, and D. R. Lide, Jr., *J. Chem. Phys.*, **23**, 136 (1955).
30. The matrix elements of the operator p in the basis $U_{v\sigma}(\alpha)$ are: ($\sigma = 0$)

$$U_{v0}(\alpha) = C_0^{(v)}/\sqrt{2} + \sum_{k=1} C_k^{(v)} \cos 3k\alpha, \quad v \text{ even}$$

$$U_{v0}(\alpha) = \sum_{k=1} C_k^{(v)} \sin 3k\alpha, \quad v \text{ odd}$$

$$p_{vv'} = \pm i\pi \sum_{k=1}^{\infty} 3k C_k^{(v)} C_k^{(v')} \begin{cases} + & \text{if } vv' \text{ is odd} \quad \text{even} \\ - & \text{if } vv' \text{ is even} \quad \text{odd} \end{cases}$$

$$p_{vv'} = 0, \quad \text{if } vv' \text{ have the same parity}$$

($\sigma = 1$)

$$U_{v1}(\alpha) = \sum_{-\infty}^{\infty} A_k^{(v)} e^{i(3k+1)\alpha}$$

$$p_{vv'} = 2\pi \sum_{-\infty}^{\infty} (3k+1) A_k^{(v)} A_k^{(v')}$$

The $p_{vv'}$ have been tabulated in [18] and [19] along with matrix elements of other useful operators.

31. J. H. Van Vleck, *Phys. Rev.*, **33**, 467 (1929); O. M. Jordahl, *Phys. Rev.*, **45**, 87 (1934); E. C. Kemble, *Fundamental Principles of Quantum Mechanics*, Dover Publications, New York, 1958, p. 394.

32. For further discussion of the Van Vleck transformation see Appendix C and also E. B. Wilson, Jr., and J. B. Howard, *J. Chem. Phys.*, **4**, 260 (1936). Here the problem of a rotating-vibrating molecule is considered and the Van Vleck transformation is employed to approximately separate rotation and ordinary vibration, thereby obtaining an effective Hamiltonian for a particular vibrational state composed of an effective rigid rotor and distortion Hamiltonian. See also [35].
33. J. D. Swalen and D. R. Herschbach, *J. Chem. Phys.*, **27**, 100 (1957).
34. D. R. Herschbach, *J. Chem. Phys.*, **27**, 1420 (1957).
35. D. Stelman, *J. Chem. Phys.*, **41**, 2111 (1964).
36. Barriers are conveniently classified as high if $s > 30$, intermediate if $s < 15$ and low if $s < 2$. For $(15 < s < 30)$, the odd-order terms in $\mathscr{H}_{vE}$ can cause the E levels to depart from a rigid rotor pattern.
37. R. W. Kilb and L. Pierce, *J. Chem. Phys.*, **27**, 108 (1957).
38. L. Pierce and L. C. Krisher, *J. Chem. Phys.*, **31**, 875 (1959).
39. V. Dobyns, *J. Chem. Phys.*, **43**, 4534 (1965).
40. H. D. Rudolph, *Z. Naturforsch.*, **21a**, 694 (1966).
41. In the symmetric rotor basis the matrix element diagonal in J of the direction cosine Φ_{Zg} is proportional to the matrix element of the angular momentum component P_g (see Table 7.2 and Table 2.1):

$$(J, K, M|\Phi_{Zg}|J, K', M) = \frac{M}{J(J+1)}(J, K|P_g|J, K')$$

The operators are similarly related in the asymmetric rotor basis, that is,

$$(J, \tau, M|\Phi_{Zg}|J, \tau', M) = \frac{M}{J(J+1)}(J, \tau|P_g|J, \tau')$$

By use of this relation and (10.72), (12.73) of the text can be obtained.
42. The second term of the $(K|K)$ matrix element of $\mathscr{P}^4$ given in Appendix B of [28] should be

$$\gamma^2(1/2)(\alpha^2+\beta^2)[(P^2-K^2)(6K^2+1)-4K^2]$$

and the coefficient (1/4) should be replaced by (1/2) in the second term of the $(K|K\pm 1)$ element of $\mathscr{P}^4$ as noted in E. Hirota, *J. Chem. Phys.*, **45**, 1984 (1966).
43. D. Boucher, A. Dubrulle, J. Demaison, and H. Dreizler, *Z. Naturforsch.*, **35** , 1136 (1980).
44. R. C. Woods, *J. Mol. Spectrosc.*, **21**, 4 (1966).
45. R. C. Woods, *J. Mol. Spectrosc.*, **22**, 49 (1967).
46. H. Dreizler, "Rotational Spectra of Molecules With Two Internal Degrees of Freedom," in *Molecular Spectroscopy: Modern Research*, K. N. Rao and C. W. Mathews, Eds., Academic, New York, 1972.
47. H. Dreizler and H. Legell, *Z. Naturforsch.*, **28a**, 1414 (1973).
48. A. Bauder and Hs. H. Günthard, *Chem. Phys.*, **60**, 290 (1976).
49. W. Bossert, A. Bauder, and Hs. H. Günthard, *Chem. Phys.*, **39**, 367 (1979).
50. A. Bauder, E. Mathier, R. Meyer, M. Ribeaud, and Hs. H. Günthard, *Mol. Phys.*, **15**, 597 (1968).
51. P. Nösberger, A. Bauder, and Hs. H. Günthard, *Chem. Phys.*, **4**, 196 (1974).
52. H. Mäder, U. Andersen, and H. Dreizler, *Z. Naturforsch.*, **28a**, 1163 (1973); See erratum F. Scappini, H. Mäder, and H. Dreizler, *Z. Naturforsch.*, **31a**, 1398 (1976).
53. J. R. Durig, Y. S. Li, and P. Groner, *J. Mol. Spectrosc.*, **62**, 159 (1976).

54. P. Groner and J. R. Durig, *J. Chem. Phys.*, **66**, 1856 (1977).
55. L. C. Krisher, *J. Chem. Phys.*, **33**, 1237 (1960); M. J. Moloney and L. C. Krisher, *J. Chem.* Phys., **45**, 3277 (1966).
56. K. M. Sinnott, *J. Chem. Phys.*, **34**, 851 (1961).
57. J. E. M. Heuvel and A. Dymanus, *J. Mol. Spectrosc.*, **47**, 363 (1973).
58. M. Ribeaud, A. Bauder, and Hs. H. Günthard, *J. Mol. Spectrosc.*, **42**, 441 (1972).
59. D. R. Herschbach, *J. Chem. Phys.*, **25**, 358 (1956).
60. C. S. Ewig, W. E. Palke, and B. Kirtman, *J. Chem. Phys.*, **60**, 2749 (1974).
61. R. M. Lees, *J. Chem. Phys.*, **59**, 2690 (1973).
62. R. M. Lees and J. G. Baker, *J. Chem. Phys.*, **48**, 5299 (1968).
63. R. M. Lees, *J. Mol. Spectrosc.*, **33**, 124 (1970).
64. C. S. Ewig and C. O. Harris, *J. Chem. Phys.*, **52**, 6268 (1970).
65. B. P. van Eijck and P. Joosen, *J. Chem. Phys.*, **58**, 5845 (1973).
66. Y. Y. Kwan and C. M. Dennison, *J. Mol. Spectrosc.*, **43**, 291 (1972).
67. E. Hirota, Y. Endo, S. Saito, and J. L. Duncan, *J. Mol. Spectrosc.*, **89**, 285 (1981).
68. B. Kirtman, W. E. Palke, and C. S. Ewig, *J. Chem. Phys.*, **64**, 1833 (1976).
69. R. A. Beaudet, *J. Chem. Phys.*, **40**, 2705 (1964).
70. R. A. Beaudet and E. B. Wilson, Jr., *J. Chem. Phys.*, **37**, 1133 (1962).
71. W. Bossert, J. Ekkers, A. Bauder, and Hs. H. Günthard, *Chem. Phys.*, **27**, 433 (1978).
72. V. W. Laurie, *J. Chem. Phys.*, **30**, 1210 (1959).
73. V. W. Laurie, *J. Mol. Spectrosc.*, **13**, 283 (1964).
74. E. Hirota, *J. Mol. Spectrosc.*, **43**, 36 (1972).
75. E. Tannenbaum, R. D. Johnson, R. J. Myers, and W. D. Gwinn, *J. Chem. Phys.*, **22**, 949 (1954).
76. E. Tannenbaum, R. J. Myers, and W. D. Gwinn, *J. Chem. Phys.*, **25**, **42** (1956).
77. R. E. Naylor, Jr., and E. B. Wilson, Jr., *J. Chem. Phys.*, **26**, 1057 (1957).
78. H. D. Rudolph, H. Dreizler, A. Jaeschke, and P. Wendling, *Z. Naturforsch.*, **22a** 940 (1967); (b) H. D. Rudolph and H. Seiler, *Z. Naturforsch.*, **20a**, 1682 (1965); (c) H. D. Rudolph, A. Dreizler, and H. Seiler, *Z. Naturforsch.*, **22a**, 1738 (1967).
79. W. M. Tolles, E. T. Handelman, and W. D. Gwinn, *J. Chem. Phys.*, **43**, 3019 (1965).
80. T. Ogata, *J. Mol. Spectrosc.*, **54**, 275 (1975).
81. T. Ogata, A. P. Cox, D. L. Smith and P. L. Timms, *Chem. Phys. Lett.*, **26**, 186 (1974).
82. F. Rohart, *J. Mol. Spectrosc.*, **57**, 301 (1975).
83. L. Englebrecht, D. Sutter, and H. Dreizler, *Z. Naturforsch.*, **28a**, 709 (1973).
84. L. Englbrecht and D. Sutter, *Z. Naturforsch.*, **33a**, 1525 (1978).
85. J. D. Swalen and C. C. Costain, *J. Chem. Phys.*, **31**, 1562 (1959).
86. V. W. Laurie and D. R. Lide, Jr., *J. Chem. Phys.*, **31**, 939 (1959).
87. W. H. Kirchhoff and D. R. Lide, Jr., *J. Chem. Phys.*, **43**, 2203 (1965).
88. D. Kivelson, *J. Chem. Phys.*, **22**, 1733 (1954).
89. B. Kirtman, *J. Chem. Phys.*, **37**, 2516 (1962).
90. D. Kivelson, *J. Chem. Phys.*, **23**, 2230 (1955).
91. D. Kivelson, *J. Chem. Phys.*, **23**, 2236 (1955).
92. K. T. Hecht and D. M. Dennison, *J. Chem. Phys.*, **26**, 48 (1957).
93. R. Meyer and Hs. H. Günthard, *J. Chem. Phys.*, **49**, 1510 (1968).
94. C. R. Quade, *J. Chem. Phys.*, **44**, 2512 (1966).
95. H. Dreizler, *Z. Naturforsch.*, **23a**, 1077 (1968).

96. M. Kuhler, L. Charpentier, D. Sutter, and H. Dreizler, *Z. Naturforsch.*, **29a**, 1335 (1974).
97. T. Ikeda, R. F. Curl, Jr., and H. Karlsson, *J. Mol. Spectrosc.*, **53**, 101 (1974).
98. P. H. Verdier and E. B. Wilson, Jr., *J. Chem. Phys.*, **29**, 340 (1958).
99. D. H. Baird and G. R. Bird, *Rev. Sci. Instrum.*, **25**, 319 (1954).
100. A. S. Esbitt and E. B. Wilson, Jr., *Rev. Sci. Instrum.*, **34**, 901, (1963).
101. A. Dymanus, H. A. Dijkerman, and G. R. D. Zijderveld, *J. Chem. Phys.*, **32**, 717 (1960).
102. G. Ruitenberg, *J. Mol. Spectrosc.*, **42**, 161 (1972).
103. A. R. Mochel, C. O. Britt, and J. E. Boggs, *J. Chem. Phys.*, **58**, 3221 (1973).
104. Y. Hanyu, C. O. Britt, and J. E. Boggs, *J. Chem. Phys.*, **45**, 4725 (1973).
105. J. H. Høg, L. Nygaard, and G. O. Sørensen, *J. Mol. Struct.*, **7**, 111 (1970).
106. R. K. Kakar, E. A. Rinehart, C. R. Quade, and T. Kohima, *J. Chem. Phys.*, **52**, 3803 (1970).
107. R. K. Kakar, *J. Chem. Phys.*, **56**, 1189 (1972).
108. A. Bouchy and G. Roussy, *J. Mol. Spectrosc.*, **65**, 395 (1977).
109. D. G. Lister, *J. Mol. Struct.*, **68**, 33 (1980).
110. W. L. Meerts, I. Ozier, *J. Mol. Spectrosc.*, **94**, 38 (1982).
111. I. Ozier and W. L. Meerts, *J. Mol. Spectrosc.*, **93**, 164 (1982).
112. P. R. Bunker, *Mol. Phys.*, **9**, 257 (1965).
113. H. C. Longuet-Higgins, *Mol. Phys.*, **6**, 445 (1963).
114. P. Kasai and R. J. Myers, *J. Chem. Phys.*, **30**, 1096 (1959).
115. R. J. Myers and E. B. Wilson, Jr., *J. Chem. Phys.*, **33**, 186 (1960).
116. L. Pierce, *J. Chem. Phys.*, **34**, 498 (1961).
117. M. L. Sage, *J. Chem. Phys.*, **35**, 142 (1961).
118. H. Drezler, *Z. Naturforsch.*, **20a**, 297 (1965).
119. E. Hirota, C. Matsumura, and Y. Morino, *Bull. Chem. Soc. Japan*, **40**, 1124 (1967).
120. H. Dreizler, *Z. Naturforsch.*, **16a**, 477, 1354 (1961).
121. J. E. Wollrab, *Rotational Spectra and Molecular Structure*, Academic, New York, 1967.
122. H. Dreizler and D. Sutter, *Z. Naturforsch.*, **24a**, 2013 (1969).
123. H. Lutz and H. Dreizler, *Z. Naturforsch.*, **33a**, 1498 (1978).
124. J. Meier, A. Bauder, and Hs. H. Günthard, *J. Chem. Phys.*, **57**, 1219 (1972).
125. C. D. LeCroix, R. F. Curl, P. M. McKinney, and R. J. Myers, *J. Mol. Spectrosc.*, **53**, 250 (1974).
126. W. Bossert, A. Bauder, and Hs. H. Günthard, *Chem. Phys.*, **39**, 367 (1979).
127. J. V. Knopp and C. R. Quade, *J. Chem. Phys.*, **53**, 1 (1970).
128. C. R. Quade, *J. Chem. Phys.*, **47**, 1073 (1967).
129. C. R. Quade and C. C. Lin, *J. Chem. Phys.*, **38**, 540 (1963).
130. J. V. Knopp and C. R. Quade, *J. Chem. Phys.*, **48**, 3317 (1968).
131. C. R. Quade and R. D. Suenram, *J. Chem. Phys.*, **73**, 1127 (1980).
132. D. G. Burkhard, *J. Chem. Phys.*, **21**, 1541 (1953); D. G. Burkhard and J. C. Irwin, *J. Chem. Phys.*, **23**, 1405 (1955).
133. P. Meakin, D. O. Harris, and E. Hirota, *J. Chem. Phys.*, **51**, 3775 (1969).
134. E. Hirota, *J. Mol. Spectrosc.*, **34**, 516 (1970).
135. T. Pedersen, N. W. Larsen, and L. Nygaard, *J. Mol. Struct.*, **4**, 59 (1969).
136. H. Dreizler and K. D. Möller, *Z. Naturforsch.*, **23a**, 1221 (1968).
137. T. Pedersen, N. W. Larsen, and H. Dreizler, *Z. Naturforsch.*, **24a** 649 (1969).
138. C. R. Quade, *J. Chem. Phys.*, **48**, 5490 (1968).
139. E. Mathier, D. Welti, A. Bauder, and Hs. H. Günthard, *J. Mol. Spectrosc.*, **37**, 63 (1971).

140. M. Ribeaud, A. Bauder, and Hs. H. Günthard, *Mol. Phys.*, **23**, 235 (1972).

141. R. H. Hunt, R. A. Leacock, C. W. Peters, and K. T. Heckt, *J. Chem. Phys.*, **42**, 1931 (1965; W. C. Oelfke and W. Gordy, *J. Chem. Phys.*, **51**, 5336 (1969).

142. G. Winnewisser, M. Minnewisser, and W. Gordy, *J. Chem. Phys.*, **49**, 3465 (1968); R. S. Winton and G. Winnewisser, Symposium on Molecular Structure and Spectroscopy, Paper Q3, 1970.

143. Landolt-Börnstein, *Numerical Data and Functional Relationships in Science and Technology*. For a complete reference, see Chapter I, [65].

144. H. Dreizler, "Determination of Barriers to Internal Rotation by Microwave Techniques," in *Critical Evaluation of Chemical and Physical Structural Information*, D. R. Lide and M. A. Paul, Eds., National Academy of Sciences, Washington, D.C., 1974.

145. R. S. Wagner and B. P. Dailey, *J. Chem. Phys.*, **23**, 1355 (1955).

146. L. Pierce and M. Hayashi, *J. Chem. Phys.*, **35**, 479 (1961).

147. I. N. Levine, *J. Mol. Spectrosc.*, **8**, 276 (1962).

148. D. J. Millen and R. J. Morton, *J. Chem. Soc.*, 1960, p. 1523.

149. A. P. Cox and J. M. Riveros, *J. Chem. Phys.*, **42**, 3106 (1965).

150. R. Nelson and L. Pierce, *J. Mol. Spectrosc.*, **18**, 344 (1965).

151. J. M. O'Reilly and L. Pierce, *J. Chem. Phys.*, **34**, 1176 (1961).

152. E. Hirota, *J. Chem. Phys.*, **45**, 1984 (1966).

153. B. Bak, J. J. Christiansen, K. Kunstmann, L. Nygaard, and J. Rastrup-Andersen, *J. Chem. Phys.*, **45**, 883 (1966).

154. M. Hayashi, M. Imachi, and M. Oyamada, *J. Mol. Struct.*, **74**, 97 (1981).

155. M. Hayashi, M. Adachi, and J. Nakagawa, *J. Mol. Spectrosc.*, **86**, 129 (1981).

156. Y. Morino and E. Hirota, *Ann. Rev. Phys. Chem.*, **20**, 139 (1969).

157. H. D. Rudolph, *Ann. Rev. Phys. Chem.*, **21**, 73 (1970).

158. V. W. Laurie, *Acc. Chem. Res.*, **3**, 331 (1970).

159. E. B. Wilson, Jr., *Chem. Soc. Rev.*, **1**, 293 (1972).

160. N. L. Owen, "Studies of Internal Rotation by Microwave Spectroscopy," in *Internal Rotation in Molecules*, W. J. Orville-Thomas, Ed., Wiley-Interscience, New York, 1974.

161. W. J. Lafferty, "Determination of Potential Functions and Barriers to Planarity for the Ring-Puckering Vibrations of Four-Membered Ring Molecules," in *Critical Evaluation of Chemical and Physical Structural Information*, D. R. Lide and M. A. Paul, Eds., National Academy of Sciences, Washington, D.C., 1974.

162. W. D. Gwinn, and A. S. Gaylord, "Spectroscopic Studies of Ring-Puckering Motions," in *MTP International Reviews of Science*, Physical Chemistry Series 2, Vol. 3, Spectroscopy, D. A. Ramsay, Ed., University Park Press, Baltimore, Md., 1976.

163. L. A. Carreira, R. C. Lord, and T. B. Malloy, Jr., "Low-Frequency Vibrations in Small Ring Molecules," in *Topics in Current Chemistry*, F. L. Boschke, Ed., Vol. 82, Springer-Verlag, Berlin, Heidelberg, New York, 1979.

164. T. B. Malloy, Jr., L. E. Bauman, and L. A. Carrier, "Conformational Barriers and Interconversion Pathways in Some Small Ring Molecules," in *Topics in Stereochemistry*, N. L. Allinger and E. L. Eliel, Eds., Vol. 11, Wiley, New York, 1979.

165. J. N. Macdonald and J. Sheridan, "Microwave Spectroscopy," in *Molecular Spectroscopy*, and 6. The Chemical Society, Burlington House, London, 1976, 1977, 1979.

166. R. W. Kitchin, T. B. Molloy, Jr., and R. L. Cook, *J. Mol. Spectrosc.*, **57**, 179 (1975).

167. J. A. Wells and T. B. Malloy, Jr., *J. Chem. Phys.*, **60**, 2132 (1974).

168. S. Chao, T. K. Avirah, R. L. Cook, and T. B. Malloy, Jr., *J. Phys. Chem.*, **80**, 1141 (1976).

169. E. Hirota, *J. Chem. Phys.*, **37**, 283 (1962).

170. J. M. Riveros and E. B. Wilson, Jr., *J. Chem. Phys.*, **46**, 4605 (1967).
171. O. L. Stiefvater and E. B. Wilson, *J. Chem. Phys.*, **50**, 5385 (1969).
172. H. N. Volltrauer and R. H. Schwendeman, *J. Chem. Phys.*, **54**, 260 (1971).
173. B. P. van Eijck, *J. Mol. Spectrosc.*, **82**, 81 (1980).
174. I. Botskor, *J. Mol. Spectrosc.*, **75**, 10 (1979).
175. W. E. Steinmetz, F. Hickernell, I. K. Mun, and L. H. Scharpen, *J. Mol. Spectrosc.*, **68**, 173 (1977).
176. D. K. Lum, L. E. Bauman, T. B. Malloy, Jr., and R. L. Cook, *J. Mol. Struct.*, **50**, 1 (1978).
177. R. Kewley, D. C. Hamphill, and R. F. Curl, Jr., *J. Mol. Spectrosc.*, **44**, 443 (1972).
178. N. S. True and R. K. Bohn, *J. Mol. Struct.*, **36**, 173 (1977).
179. N. S. True and R. K. Bohn, *J. Am. Chem. Soc.*, **99**, 3575 (1977).
180. N. S. True and R. K. Bohn, *J. Am. Chem. Soc.*, **98**, 1188 (1976).
181. C. C. Costain and G. P. Srivastava, *J. Chem. Phys.*, **41**, 1620 (1964).
182. D. J. Millen, *J. Mol. Struct.*, **45**, 1 (1978).
183. E. M. Bellott, Jr. and E. B. Wilson, *Tetrahedron*, **31**, 2896 (1975).
184. A. S. Georgiou, A. C. Legion, and D. J. Millen, *J. Mol. Struct.*, **69**, 69 (1980).
185. A. N. Murty and R. F. Curl, Jr., *J. Chem. Phys.*, **46**, 4176 (1967).
186. R. G. Azrak and E. B. Wilson, *J. Chem. Phys.*, **52**, 5299 (1970).
187. P. Venkateswarlu and W. Gordy, *J. Chem. Phys.*, **23**, 1200 (1955).
188. S. Leavell and R. F. Curl, Jr., *J. Mol. Spectrosc.*, **45**, 428 (1973).
189. H. Jones and R. F. Curl, Jr., *J. Mol. Spectrosc.*, **42**, 65 (1972).
190. E. Hirota, *J. Mol. Spectrosc.*, **26**, 335 (1968).
191. J. K. Tyler and D. G. Lister, *Chem. Commun.*, 1971, p. 1350.
192. H. H. Picket, *J. Mol. Spectrosc.*, **46**, 335 (1973).
193. K. Bolton, N. L. Owen, and J. Sheridan, *Nature*, **217**, 164 (1968).
194. R. Cervellati, W. Caminati, C. D. Esposti, and A. M. Mirri, *J. Mol. Spectrosc.*, **66**, 389 (1977).
195. K.-M. Marstokk and H. Møllendal, *J. Mol. Struct.*, **18**, 247 (1973).
196. I. D. Warren and E. B. Wilson, *J. Chem. Phys.*, **56**, 2137 (1972).
197. E. Sung and M. D. Harmony, *J. Am. Chem. Soc.*, **99**, 5603 (1977).
198. R. N. Nandi, M. R. Boland, and M. D. Harmony, *J. Mol. Spectrosc.*, **92**, **419** (1982).
199. W. Caminati, *J. Mol. Spectrosc.*, **92**, 101 (1982).
200. M. J. Fuller, E. B. Wilson, and W. Caminati, *J. Mol. Spectrosc.*, **96**, 131 (1982).
201. S. I. Chan, J. Zinn, J. Fernandez, and W. D. Gwin, *J. Chem. Phys.*, **33**, 1643 (1960).
202. S. I. Chan, J. Zinn, and W. D. Gwinn, *J. Chem. Phys.*, **34**, 1319 (1961).
203. S. I. Chan, T. R. Borgers, J. W. Russell, H. L. Strauss, and W. D. Gwinn, *J. Chem. Phys.*, **44**, 1103 (1966).
204. R. P. Bell, *Proc. R. Soc. London*, **A183**, 328 (1945).
205. W. C. Pringle, *J. Chem. Phys.*, **54**, 4979 (1971).
206. D. O. Harris, H. W. Harrington, A. C. Luntz, and W. D. Gwinn, *J. Chem. Phys.*, **44**, 3467 (1966).
207. L. H. Scharpen, *J. Chem. Phys.*, **48**, 3552 (1968).
208. A. C. Luntz, *J. Chem. Phys.*, **50**, 1109 (1969).
209. L. H. Scharpen and V. W. Laurie, *J. Chem. Phys.*, **49**, 3041 (1968).
210. S. S. Butcher and C. C. Costain, *J. Mol. Spectrosc.*, **15**, 40 (1965).
211. W. D. Gwinn and A. C. Luntz, *Trans. Am. Crystallogr. Assoc.*, **2**, 90 (1966).

212. J. E. Kilpatrick, K. S. Pitzer, and R. Spitzer, *J. Am. Chem. Soc.*, **69**, 2483 (1947).

213. D. O. Harris, G. G. Engerholm, C. A. Tolman, A. C. Luntz, R. A. Keller, H. Kim, and W. D. Gwinn, *J. Chem. Phys.*, **50**, 2438 (1969).

214. G. G. Engerholm, A. C. Luntz, W. D. Gwinn, and D. O. Harris, *J. Chem. Phys.*, **50**, 2446 (1969).

215. P. A. Baron and D. O. Harris, *J. Mol. Spectrosc.*, **49**, 70 (1974).

Chapter **XIII**

DERIVATION OF MOLECULAR STRUCTURES

1 INTRODUCTION

Microwave spectroscopy provides one of the most widely applicable and accurate methods for evaluation of molecular structures. This structural information is contained in the principal moments of inertia, I_a, I_b, I_c, which are derivable from the spectroscopic constants A, B, C, that is,

$$I_a = \frac{h}{8\pi^2 A} \qquad I_b = \frac{h}{8\pi^2 B} \qquad I_c = \frac{h}{8\pi^2 C}$$

with [1]

$$\frac{h}{8\pi^2}=505{,}376 \text{ amu Å}^2 \text{ MHz}$$

for $I_a, \ldots$, expressed in atomic mass units-angstrom units squared and $A, \ldots$, in megahertz. Of course, before the moments of inertia can be related to the bond lengths and bond angles, the job of assigning the rotational spectrum must be accomplished, that is, correlating the spectroscopic constants and the quantum numbers to the spectral lines. Usually, the spectrum of isotopically substituted molecular species must also be studied to provide sufficient data for the evaluation of the structural parameters. Because, however, of the high sensitivity characteristics of microwave spectroscopy such molecular species can, in many cases, be studied in natural abundance.

Molecular structural parameters, in particular bond lengths, have been the major source of data used to obtain a more detailed understanding of the chemical bond. Explanations of observed trends in structural parameters in terms of such concepts as resonance, hybridization, bond order, conjugation, hyperconjugation, ionic character, and nonbonded interactions, have been vigorously pursued since the introduction of the modern concept of a chemical bond. A study of the factors influencing bond lengths and bond angles requires, of course, an appreciation of the accuracy of the derived molecular parameters and the type of structural parameter obtained. Because of the accuracy of the experimental measurements the moments of inertia may be obtained to six or more significant figures. However, evaluation of structural parameters with an accuracy comparable to that of the moments of inertia has proved to be a difficult problem for spectroscopists. The primary source of the problem is not the uncertainties in Planck's constant or the atomic masses but rather the fact that the dimensions of a molecule are affected by its vibrational energy even in its ground vibrational state. Unless corrections are made for these effects, the meaningfulness of the structural parameters, derived from the observed moments of inertia, is limited to much less than six significant figures. Except for diatomic and some simple polyatomic molecules, the explicit correction of the moments of inertia for vibrational contributions has not been possible because of the difficulties in obtaining the required experimental data. Different procedures have been introduced which correct to various degrees for vibrational effects and which have led to different conceptions of interatomic distance. In particular, five types of bond lengths are defined:

1. r_e, the equilibrium bond length for the hypothetical vibrationless state, evaluated by correction for the effects of vibration including zero-point vibrations, as described in Section 6.

2. r_0, the effective bond length for the ground vibrational state, calculated from the B_0 or I^0 values as described in Section 7.

3. r_s, the substitution bond length, derived from the isotopic substitution method described in Section 8.

4. $\langle r \rangle$ (or r_z), the average bond length associated with the average configuration of the atoms, evaluated by a partial correction for the effects of vibration, as described in Section 9.

5. r_m, the mass dependence bond length, derived from a large number of isotopic species by a first-order treatment of isotopic effects, as described in Section 10.

With the hundreds of molecules that have now been studied by microwave spectroscopy, particularly in the last 10 years, it is no longer possible to provide a reasonably complete compilation of molecular structures in a limited space. In a previous edition of this book [2] a relatively complete compilation to about 1969 was given. Since that time a critical compilation of structures has been given by Harmony et al. [3] covering the period from the late 1940s to halfway through 1977. These authors classify the structural parameters (r_e, r_0, r_s, $\langle r \rangle$) and indicate the range of uncertainty to be associated with the parameters. This is particularly useful as a guide in making realistic evaluations of variations in structural parameters. Another publication [4] lists the best available structural data up to 1974 on more than 1200 molecules.

2 RELATION OF MOLECULAR DIMENSIONS TO MOMENTS OF INERTIA

The components of the moment of inertia tensor (see Chapter II, Section 1) may be arranged into a square matrix

$$\mathbf{I} = \begin{bmatrix} I_{xx} & I_{xy} & I_{xz} \\ I_{xy} & I_{yy} & I_{yz} \\ I_{xz} & I_{yz} & I_{zz} \end{bmatrix} \tag{13.1}$$

which is symmetric, with the diagonal elements, or moments of inertia, defined as

$$\begin{aligned} I_{xx} &= \sum_i m_i(y_i^2 + z_i^2) \\ I_{yy} &= \sum_i m_i(x_i^2 + z_i^2) \\ I_{zz} &= \sum_i m_i(x_i^2 + y_i^2) \end{aligned} \tag{13.2}$$

and the off-diagonal elements, or products of inertia, defined as

$$\begin{aligned} I_{yx} = I_{xy} &= -\sum_i m_i x_i y_i \\ I_{zx} = I_{xz} &= -\sum_i m_i x_i z_i \\ I_{zy} = I_{yz} &= -\sum_i m_i y_i z_i \end{aligned} \tag{13.3}$$

In the foregoing equations for the elements of **I**, m_i is the mass of the ith atom, $x_i y_i z_i$ are the corresponding coordinates measured from the origin of a Cartesian coordinate system fixed in the molecule, and the sum is over all atoms of the molecule. The elements of the inertia matrix depend on both the origin and orientation of the coordinate system. The origin, of course, is to be taken at the center of mass. (This choice allows the translational and rotational motions to be treated separately.) For any choice of origin, it is possible to choose an orientation of the coordinate axis in the molecule such that the inertia matrix is diagonal, that is, the products of inertia vanish. These axes for which the products of inertia vanish are called the principal axes, and the diagonal elements of **I** are known as the principal moments of inertia. Our interest lies in the center-of-mass principal moments of inertia since it is these quantities that are determined from an analysis of the rotational spectrum; we shall designate them as I_x, I_y, I_z.

For a rotation of coordinate axes, the components of a position vector with respect to the "old" and the "new" (rotated) coordinate system are related by a linear transformation. In matrix notation we have simply

$$\mathbf{r} = \mathbf{R}\mathbf{r}' \tag{13.4}$$

or

$$\begin{bmatrix} x \\ y \\ z \end{bmatrix} = \mathbf{R} \begin{bmatrix} x' \\ y' \\ z' \end{bmatrix}$$

where **R** is a square 3×3 orthogonal matrix. The primed symbols refer to the new rotated coordinate system which we may consider here to be the principal axis system of the molecule. Under such an orthogonal transformation of coordinates, the transformed moment of inertia matrix $\mathbf{I}'$ is given by

$$\mathbf{I}' = \tilde{\mathbf{R}}\mathbf{I}\mathbf{R} \tag{13.5}$$

where $\tilde{\mathbf{R}}$ is the transpose of **R**. The new moment of inertia matrix is said to be obtained by a similarity transformation of **I**. If the transformation takes **I** from an arbitrary system to the principal axis system, the resulting matrix, $\mathbf{I}'$, is diagonal. The transformation matrix has a very simple form for the special case for which all but one of the products of inertia vanish, for example, I_{xy}, which is easily obtainable if initially at least one principal axis (here we take this to be the z axis) is known. In this particular case, the transformation matrix corresponds to a simple rotation about the z axis (see Fig. 13.1) that is,

$$\mathbf{R} = \begin{bmatrix} \cos\theta & -\sin\theta & 0 \\ \sin\theta & \cos\theta & 0 \\ 0 & 0 & 1 \end{bmatrix} \tag{13.6}$$

and we may choose the matrix **R** such that it diagonalizes **I** by requiring that

$$\tan 2\theta = \frac{2I_{xy}}{I_{xx} - I_{yy}} \tag{13.7}$$

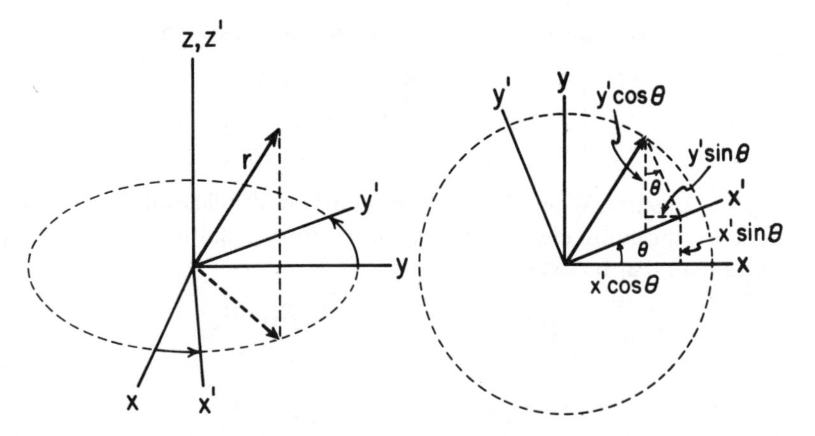

Fig. 13.1 Rotation of the coordinate axes about the z axis by an angle θ takes the unprimed xyz coordinate system into the $x'y'z'$ coordinate system. From the construction, the components of a position vector in the unprimed coordinate system are readily related to those in the primed coordinate system. In particular, the components in the unprimed system are: $(x'\cos\theta - y'\sin\theta, x'\sin\theta + y'\cos\theta, z')$. The transformation matrix R between the primed and unprimed systems is hence given by (13.6) of the text.

The principal moments of inertia may hence be readily calculated by carrying out the matrix multiplication indicated in (13.5) with θ determined from (13.7). The principal moment about the z axis is $I_z = I_{zz}$, and the remaining two principal moments may be expressed by the relation

$$I = \left(\frac{I_{xx} + I_{yy}}{2}\right) \pm \left(\frac{I_{xx} - I_{yy}}{2}\right)(1 + \tan^2 2\theta)^{1/2} \tag{13.8}$$

This procedure is applicable, in many cases, since at least one principal axis may often be located from symmetry considerations.

A principal axis is always associated with an n-fold axis of rotational symmetry, C_n. When $n \geqslant 3$, the molecule is a symmetric top. The methyl chloride molecule, for example, possesses an axis of threefold symmetry, the axis passing through the chlorine and carbon atoms. In addition, for a symmetric top, any axis passing through the center of mass and perpendicular to the symmetry axis will be a principal axis. This is reflected in the fact that two of the principal moments of inertia are equal. Furthermore, if a molecule has a plane of symmetry, then an axis perpendicular to the symmetry plane is a principal axis. The other two principal axes lie in the symmetry plane and might be obtained, as outlined previously, by consideration of a rotation of coordinate axes in the plane of symmetry.

In general, when none of the principal axes is obvious from symmetry, one can choose any convenient reference system x, y, z in the molecule and calculate the elements of the inertia tensor from (13.2) and (13.3); one can then diagonalize the tensor to obtain the principal elements [5]. Let us assume that R_{xa}, R_{ya}, and R_{za} are the cosines of the angles between a principal inertial axis a and the arbitrary axes x, y, z; $R_{xa} = \cos(x, a)$, and so on. Since the off-diagonal elements of the inertia tensor vanish in the principal coordinate system

$$\begin{bmatrix} I_{xx} & I_{xy} & I_{xz} \\ I_{yx} & I_{yy} & I_{yz} \\ I_{zx} & I_{zy} & I_{zz} \end{bmatrix} \begin{bmatrix} R_{xa} \\ R_{ya} \\ R_{za} \end{bmatrix} = I_a \begin{bmatrix} R_{xa} \\ R_{ya} \\ R_{za} \end{bmatrix} \tag{13.9}$$

where I_a is a constant. When the indicated matrix multiplications are performed and the corresponding elements equated, the three resulting equations, upon rearrangement, can be expressed as

$$(I_{xx} - I_a)R_{xa} + I_{xy}R_{ya} + I_{xz}R_{za} = 0 \tag{13.10}$$

$$I_{yx}R_{xa} + (I_{yy} - I_a)R_{ya} + I_{yz}R_{za} = 0 \tag{13.11}$$

$$I_{zx}R_{xa} + I_{zy}R_{ya} + (I_{zz} - I_a)R_{za} = 0 \tag{13.12}$$

These three simultaneous equations in the three unknown cosines R_{xa}, R_{ya}, and R_{za} have nontrivial solutions only if the determinant of the coefficients vanishes. Therefore,

$$\begin{vmatrix} I_{xx} - I_a & I_{xy} & I_{xz} \\ I_{yx} & I_{yy} - I_a & I_{yz} \\ I_{zx} & I_{zy} & I_{zz} - I_a \end{vmatrix} = 0 \tag{13.13}$$

Solution of this cubic equation yields three values of I_a corresponding to the three principal values of the moments of inertia. Each of these values, when substituted for I_a in (13.10)–(13.12), yields a set of three equations which with the auxiliary relation $R_{xa}^2 + R_{ya}^2 + R_{za}^2 = 1$ can be solved for the direction cosines between the x, y, z system and the axis of the principal element which is substituted. The auxiliary relation is required because (13.10)–(13.12) yield only the ratios of the direction cosines.

It will be convenient to have at our disposal expressions for the moments

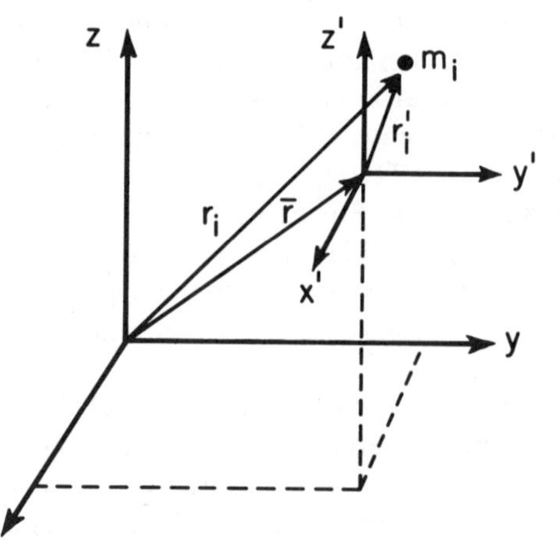

Fig. 13.2 Location of the center of mass and of the ith atom with respect to an arbitrary origin.

and products of inertia with respect to the center of mass even when the coordinates are measured with respect to an arbitrary origin. In Fig. 13.2 the position vector to the ith atom measured from the arbitrary origin is designated by $\mathbf{r}_i$; the vector distance of the center of mass from the arbitrary origin is designated by $\bar{\mathbf{r}}$; and $\mathbf{r}'_i$ represents the distance of the ith atom from the center of mass. The three vectors are related by the equation

$$\mathbf{r}_i = \bar{\mathbf{r}} + \mathbf{r}'_i \tag{13.14}$$

Multiplying by m_i and summing over all atoms, we have for the position vector of the center of mass measured from the arbitrary origin

$$\bar{\mathbf{r}} = \frac{\sum_i m_i \mathbf{r}_i}{\sum_i m_i} \tag{13.15}$$

since by definition of the center of mass (sometimes called the first moment equation)

$$\sum m_i \mathbf{r}'_i = 0 \tag{13.16}$$

Hence, the coordinates of the ith atom measured from the center of mass (x'_i, y'_i, z'_i) in terms of the coordinates measured from the arbitrary origin may be expressed by

$$x'_i = x_i - \frac{\sum_i m_i x_i}{\sum_i m_i} \tag{13.17}$$

and so on. Substitution of the above into the general definition of the moments and products of inertia, (13.2) and (13.3), gives

$$I_{xx}(\text{cm}) = \sum_i m_i (y_i^2 + z_i^2) - \frac{(\sum_i m_i y_i)^2}{\sum_i m_i} - \frac{(\sum_i m_i z_i)^2}{\sum_i m_i} \tag{13.18}$$

and so on, and

$$I_{xy}(\text{cm}) = -\sum_i m_i x_i y_i + \frac{(\sum_i m_i x_i)(\sum_i m_i y_i)}{\sum_i m_i} \tag{13.19}$$

and so on. These expressions give the moments of inertia with respect to the center of mass, with the coordinates measured with respect to an arbitrary origin. If the arbitrary origin is chosen as the center of mass, then only the first terms in I_{xx}, I_{xy}, . . ., are nonvanishing. In any event, use of these equations and subsequent diagonalization of the inertia matrix gives the principal moments of inertia with respect to the center of mass.

When the rotational spectrum of a prospective molecule is being considered, much information about the spectrum to be expected can usually be obtained from moments of inertia evaluated from an assumed structure which is derived from a judicious choice of molecular parameters of similar molecules [6]. When information on similar molecules is not available, a convenient first approximation to the structure may be gained from addition of covalent bond radii given in Appendix F. A comparison of these idealized bond lengths and the actual bond lengths can also give information on the nature of the bonds. The foregoing expressions are particularly useful for calculation of the moments of inertia since, in many cases, it is convenient to choose an origin for the coordinate system other than the center of mass for evaluation of the coordinates of the atoms. For some simple molecules, the moments of inertia may be readily expressed in terms of the bond lengths and bond angles of the molecule. In Table 13.1 explicit expressions for the principal moments of inertia of a few simple molecules are collected. For large molecules, the evaluation of the Cartesian coordinates of all the atoms becomes a rather tedious task. However, Schwendeman [7] has discussed a computer program for calculation of moments of inertia which is particularly convenient, since it uses directly as input data the structural parameters in the form of bond lengths and bond angles rather than Cartesian coordinates.

Table 13.1 Expressions for the Moments of Inertia of Some Simple Molecules[a]

Molecule	*Moments of Inertia*
Diatomic, XY	$I=\left(\dfrac{m_X m_Y}{m_X+m_Y}\right) d_{XY}^2$
Linear, XY_2	$I=2m_Y d_{XY}^2$
Linear, XYZ	$I=\dfrac{1}{M}[m_X m_Y d_{YX}^2+m_Y m_Z d_{YZ}^2+m_X m_Z(d_{YX}+d_{YZ})^2]$
Bent, XY_2[b]	$I_x=2m_Y d_{XY}^2 \sin^2\dfrac{\theta}{2}$
	$I_y=\dfrac{2m_X m_Y}{M} d_{XY}^2 \cos^2\dfrac{\theta}{2}$ $I_z=I_x+I_y$
Bent, XYZ[c]	$I_{xx}=\dfrac{\sin^2(\theta/2)}{M}[m_X(m_Y+m_Z)d_{YX}^2+m_Z(m_X+m_Y)d_{YZ}^2+2m_X m_Z d_{YX} d_{YZ}]$ $I_{yy}=\dfrac{\cos^2(\theta/2)}{M}[m_X(m_Y+m_Z)d_{YX}^2+m_Z(m_X+m_Y)d_{YZ}^2-2m_X m_Z d_{YX} d_{YZ}]$ $I_{xy}=\dfrac{\sin(\theta/2)\cos(\theta/2)}{M}[m_X(m_Y+m_Z)d_{YX}^2-m_Z(m_X+m_Y)d_{YZ}^2]$
Pyramidal, XY_3[d]	$I_x=I_y=2m_Y d_{XY}^2 \sin^2\dfrac{\theta}{2}+\dfrac{m_X m_Y}{M} d_{XY}^2\left(3-4\sin^2\dfrac{\theta}{2}\right)$ $I_Z=4m_Y d_{XY}^2 \sin^2\dfrac{\theta}{2}$

Table 13.1 (Continued)

Molecule	*Moments of Inertia*
Planar, ZXY_2^b	$I_x=2m_Y d_{XY}^2 \sin^2\frac{\theta}{2}$ $I_y=\frac{1}{M}\left[m_Z(2m_Y+m_X)d_{XZ}^2+2m_Y(m_X+m_Z)d_{XY}^2\cos^2\frac{\theta}{2}+4m_Y m_Z d_{XZ} d_{XY}\cos\frac{\theta}{2}\right]$ $I_z=I_x+I_y$
Axial symmetric, $ZXY_3{}^d$	$I_x=I_y=2m_Y d_{XY}^2\sin^2\frac{\theta}{2}+\left(\frac{3m_Y}{M}\right)(m_X+m_Z)d_{XY}^2\left(1-\frac{4}{3}\sin^2\frac{\theta}{2}\right)+\left(\frac{m_Z d_{XZ}}{M}\right)\left[(3m_Y+m_X)d_{XZ}+6m_Y d_{XY}\left(1-\frac{4}{3}\sin^2\frac{\theta}{2}\right)^{1/2}\right]$ $I_z=4m_Y d_{XY}^2\sin^2\frac{\theta}{2}$

[a]The d_{ij} is the bond distance between atoms i and j, M is the total mass of the appropriate molecule.
[b]The x axis corresponds to the C_2 axis, with the z axis perpendicular to the xy plane and with θ as the $Y—X—Y$ bond angle.
[c]The remaining off-diagonal elements of the inertia matrix vanish. Here the principal moments of inertia I_x and I_y must be evaluated from (13.12) of the text. The x axis is parallel to the axis bisecting the $X—Y—Z$ bond angle θ with the z axis perpendicular to the xy plane. Note that $I_z=I_{xx}+I_{yy}=I_x+I_y$.
[d]The z axis is the C_3 symmetry axis and θ is the $Y—X—Y$ bond angle. Note that the acute angle β between the $X—Y$ bond and the symmetry axis is related to the bond angle by $\sin(\theta/2)=(3^{1/2}/2)\sin\beta$.

For the treatment of a general asymmetric rotor it is sometimes convenient to use another symmetric tensor whose elements are defined as [8]

$$P_{xx}=\sum m_i x_i^2 \tag{13.20}$$

and so on, and

$$P_{xy}=\sum m_i x_i y_i \tag{13.21}$$

and so on. The diagonal elements of **P** are called the planar moments of inertia while the off-diagonal elements are equivalent to the products of inertia except for the sign. (The symbol **P** used here should not be confused with angular momentum which has in earlier chapters also been designated by **P**.) The two matrices are related by the matrix equation

$$\mathbf{I}=d\mathbf{E}-\mathbf{P} \tag{13.22}$$

with $d=\sum_i m_i(x_i^2+y_i^2+z_i^2)$ and with $\mathbf{E}$ a 3×3 unit matrix. $\mathbf{I}$ and $\mathbf{P}$ have the same transformation properties; since $d\mathbf{E}$ is invariant with respect to a similarity transformation, that is, a rotation of axes, the roots are related by

$$I_x=d-P_x \tag{13.23}$$

and so on. Furthermore, we may write

$$2d=I_x+I_y+I_z=2(P_x+P_y+P_z) \tag{13.24}$$

so that the following relations between the principal moments of $\mathbf{I}$ and $\mathbf{P}$ may be given as

$$P_x=\tfrac{1}{2}(-I_x+I_y+I_z) \tag{13.25}$$

and so on

$$I_x=P_y+P_z \tag{13.26}$$

and so on. Expressions for P_y and P_z are obtained by cyclic permutation of the subscripts x, y, z. Typical elements of $\mathbf{P}$ computed with respect to the center of mass with an arbitrary origin for the coordinate system have the form

$$P_{xx}(\text{cm})=\sum_i m_i x_i^2-\frac{(\sum_i m_i x_i)^2}{\sum_i m_i} \tag{13.27}$$

and so on, and

$$P_{xy}(\text{cm})=\sum_i m_i x_i y_i-\frac{(\sum_i m_i x_i)(\sum_i m_i y_i)}{\sum_i m_i} \tag{13.28}$$

and so on. When the arbitrary origin of the coordinate system is at the center of mass, the second term of the above expressions vanishes.

3 SINGLE ISOTOPIC SUBSTITUTION: KRAITCHMAN'S EQUATIONS

Kraitchman [8] has given a convenient method for calculation of the position of an atom in a molecule which utilizes the changes in moments of inertia resulting from a single isotopic substitution of the atom. Specifically, the principal axis coordinates of the atom are derived; and once a sufficient number of coordinates have been ascertained, the structure of the molecule may be readily evaluated. In the following discussion, we consider the equations developed by Kraitchman for various types of molecules.

The principal moments of inertia of the parent molecule and of the isotopically substituted molecule will be denoted by I_x, I_y, I_z and I'_x, I'_y, I'_z, respectively. The coordinates x_i, y_i, z_i will be measured from the center-of-mass principal axis system of the parent molecule. Furthermore, the molecule is assumed rigid so that the bond distances and angles are unchanged by isotopic substitution.

For the center-of-mass principal axis system of the parent molecule the elements of **I** are, from (13.18) and (13.19):

$$I_x = I_{xx} = \sum m_i(y_i^2 + z_i^2) \tag{13.29}$$

$$I_y = I_{yy} = \sum m_i(x_i^2 + z_i^2) \tag{13.30}$$

$$I_z = I_{zz} = \sum m_i(x_i^2 + y_i^2) \tag{13.31}$$

$$I_{xy} = -\sum m_i x_i y_i = 0 \tag{13.32}$$

$$I_{xz} = -\sum m_i x_i z_i = 0 \tag{13.33}$$

$$I_{yz} = -\sum m_i y_i z_i = 0 \tag{13.34}$$

If a single isotopic substitution is made for some atom of the molecule, depending on the molecule and the particular atom substituted, both a translation and a rotation of the parent principal axes can result. In many cases at least one of the principal axes of the substituted molecule will remain parallel to a principal axis of the parent molecule. Let the mass of the isotopic atom be denoted by $m + \Delta m$, with m the original mass of the atom. Employing the parent center-of-mass principal axis system with the coordinates of the substituted atom represented by x, y, z, we find from (13.18) and (13.19) that

$$I'_{xx} = I_x + \Delta m(y^2 + z^2) - \frac{(\Delta m y)^2}{M + \Delta m} - \frac{(\Delta m z)^2}{M + \Delta m} = I_x + \mu(y^2 + z^2) \tag{13.35}$$

and similarly that

$$I'_{yy} = I_y + \mu(x^2 + z^2) \tag{13.36}$$

$$I'_{zz} = I_z + \mu(x^2 + y^2) \tag{13.37}$$

$$I'_{xy} = -\mu xy \tag{13.38}$$

$$I'_{xz} = -\mu xz \tag{13.39}$$

$$I'_{yz} = -\mu yz \tag{13.40}$$

where the reduced mass for the isotopic substitution is defined as

$$\mu = \frac{M \Delta m}{M + \Delta m} \tag{13.41}$$

with M the total mass of the parent molecule. These expressions give us the elements of the inertia matrix with respect to the center of mass of the isotopically substituted molecule in terms of the coordinates measured from the center-of-mass principal axis system of the parent molecule. Diagonalization of the inertial matrix yields the principal moments of inertia of the substituted molecule I'_x, I'_y, I'_z which are determined experimentally. These equations are now employed for development of expressions for the coordinates x, y, z of the

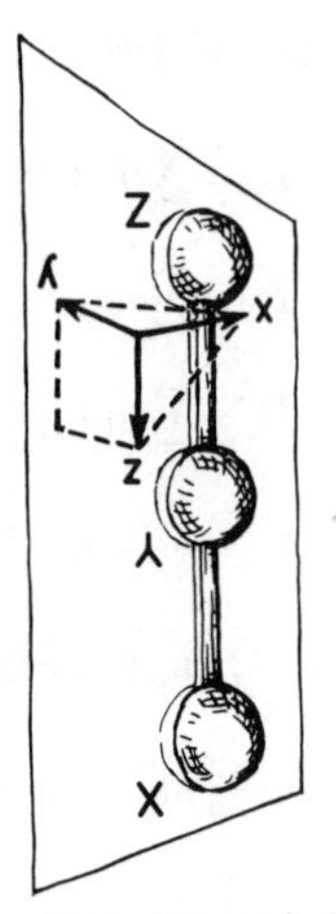

Fig. 13.3 Illustration of the orientation of the coordinate axes for a linear molecule.

isotopically substituted atom in terms of the principal moments I_x, I_y, I_z and I'_x, I'_y, I'_x.

Linear Molecules

Taking the z axis along the molecular axis (see Fig. 13.3) and noting that only the z coordinates of the atoms are nonvanishing, we have $I_z=0$, $I_x=I_y$. The $x'y'z'$ axes may be chosen parallel to the xyz axes; and from (13.35) through (13.40) we have

$$I'_{xy}=I'_{xz}=I'_{yz}=0 \tag{13.42}$$

$$I'_x=I'_{xx}=I_x+\mu z^2 \tag{13.43}$$

$$I'_y=I'_{yy}=I_y+\mu z^2 \tag{13.44}$$

$$I'_z=I_z=0 \tag{13.45}$$

From the equation for I'_x or I'_y, we may obtain the distance of the particular substituted atom from the center of mass of the parent molecule; that is

$$|z|=\left[\frac{1}{\mu}(I'_x-I_x)\right]^{1/2}=\left[\frac{1}{\mu}(I'_y-I_y)\right]^{1/2} \tag{13.46}$$

and $|x|=|y|=0$. The $|z|$ thus obtained from the measured I's represents the distance of the substituted atom from the center of mass of the molecule.

Symmetric-top Molecules: Location of an Atom on the Symmetry Axis

The orientations of the principal axes are not altered by isotopic substitution on the symmetry axis (see Fig. 13.4). If we choose the z axis (and hence z') as the symmetry axis, and taking cognizance of the fact that the x or y coordinate

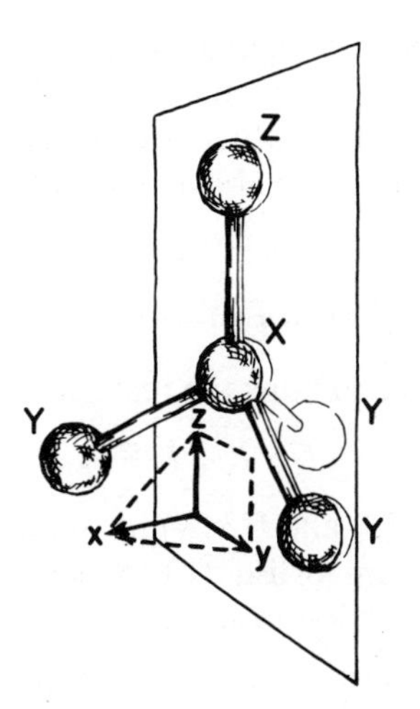

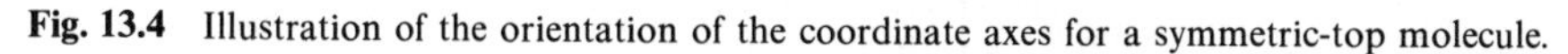
Fig. 13.4 Illustration of the orientation of the coordinate axes for a symmetric-top molecule.

of any atom on the symmetry axis is zero, we obtain for the inertia elements the same expressions as found for a linear molecule, except that $I'_z = I_z \neq 0$. This latter fact, however, has no effect on the determination of the distance $|z|$ of the substituted atom from the center of mass of the parent molecule which is given by (13.46).

Symmetric-top Molecules: Location of an Atom off the Symmetry Axis

Isotopic substitution of an atom which is not on the symmetry axis converts the original symmetric-top to an asymmetric-top molecule. Let the z axis, as before, be along the symmetry axis. In addition, we may choose the orientation of the x and y axes such that the atom to be substituted lies in the yz plane (see Fig. 13.4). With this choise the coordinates of the substituted atom are $(0, y, z)$ and (13.35) through (13.40) give

$$I'_{xy} = I'_{xz} = 0 \tag{13.47}$$

$$I'_{yz} = -\mu yz \tag{13.48}$$

$$I'_x = I'_{xx} = I_x + \mu(y^2 + z^2) \tag{13.49}$$

$$I'_{yy} = I_y + \mu z^2 \tag{13.50}$$

$$I'_{zz} = I_z + \mu y^2 \tag{13.51}$$

Since there are no off-diagonal connections to I'_{xx} we need consider only the submatrix

$$\begin{bmatrix} I_y + \mu z^2 & -\mu yz \\ -\mu yz & I_z + \mu y^2 \end{bmatrix} \tag{13.52}$$

for the substituted molecule. Diagonalization of this matrix yields the two remaining principal moments of inertia, I'_y and I'_z. Since the trace and determinant of a matrix are quantities which are invariant under a similarity transformation, we may derive from the above matrix the equations [9]

$$I_y+\mu z^2+I_z+\mu y^2=I'_y+I'_z \tag{13.53}$$

$$(I_y+\mu z^2)(I_z+\mu y^2)-\mu^2 y^2 z^2=I'_y I'_z \tag{13.54}$$

From (13.49) and (13.53) one obtains the relation

$$I_z=I'_y+I'_z-I'_x \tag{13.55}$$

noting $I_x=I_y$. Solving (13.53) and (13.54) for $|y|$ and $|z|$ and eliminating from these expressions I_z by means of Eq. (13.55), since I_z is usually not obtained from an analysis of a symmetric-top spectrum, one obtains finally

$$|y|=\left[\frac{\Delta I_z}{\mu}\left(1+\frac{\Delta I_y}{I_y-I_z}\right)\right]^{1/2} =\left[\frac{(I'_x-I'_y)(I'_x-I'_z)}{\mu(I'_x-I'_y-I'_z+I_y)}\right]^{1/2} \tag{13.56}$$

$$|z|=\left[\frac{\Delta I_y}{\mu}\left(1+\frac{\Delta I_z}{I_z-I_y}\right)\right]^{1/2} =\left[\frac{(I'_y-I_y)(I'_z-I_y)}{\mu(I'_y+I'_z-I'_x-I_y)}\right]^{1/2} \tag{13.57}$$

and $|x|=0$ for the coordinates of the substituted atom in the center-of-mass principal axis system of the original symmetric top. The determination of an accurate I'_z value for a slightly asymmetric rotor can be difficult. In many cases, I'_z can be evaluated only with a rather large uncertainty and hence, the coordinates derived will be correspondingly limited in accuracy.

To use the first equations given for $|y|$ and $|z|$, ΔI_z may be replaced by $\Delta I_x-\Delta I_y$ [see (13.55) or (13.74) where $\Delta P_x=0$] with an estimate of I_z if the coordinates are not too sensitive to I_z as is often the case. On the other hand, as described in Chapter V, observation of "forbidden" transitions can provide an accurate value of the moment of inertia about the symmetry axis, I_z.

Asymmetric-top Molecules: Planar

If we choose the z axis as perpendicular to the molecular plane (see Fig. 13.5), the z coordinates of all atoms are zero and we have the relation

$$I_x+I_y=I_z \tag{13.58}$$

Substitution of an atom in the plane may result in a rotation of the coordinate axes in the plane; however, the z' axis will remain parallel to the z axis. The coordinates of the substituted atom are taken as $(x, y, 0)$ and (13.35)–(13.40) give

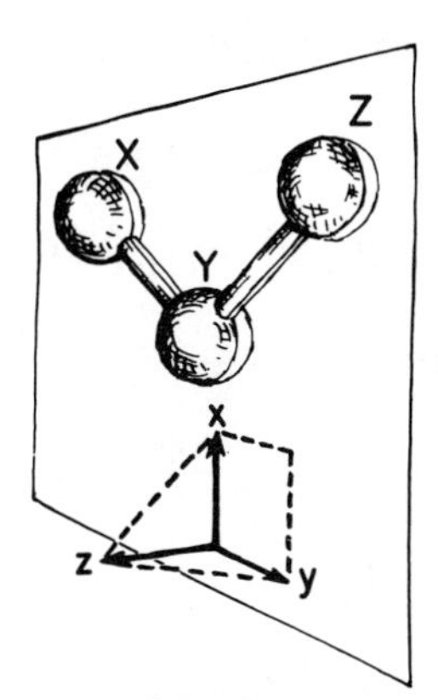

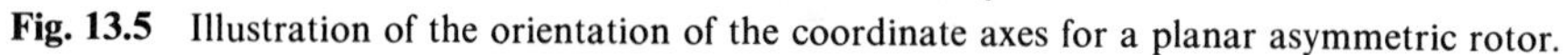

Fig. 13.5 Illustration of the orientation of the coordinate axes for a planar asymmetric rotor.

$$I'_{xz} = I'_{yz} = 0 \tag{13.59}$$

$$I'_{xy} = -\mu xy \tag{13.60}$$

$$I'_{xx} = I_x + \mu y^2 \tag{13.61}$$

$$I'_{yy} = I_y + \mu x^2 \tag{13.62}$$

$$I'_z = I'_{zz} = I_z + \mu(x^2 + y^2) \tag{13.63}$$

These equations are similar to those for a symmetric-top off-axis substitution and may be treated in the same fashion. Expressions like those of (13.53) and (13.54) are readily obtained which can be solved to give the coordinates of the substituted atom in terms of the changes in the moments of inertia. Explicitly

$$\begin{aligned} |x| &= \left[\frac{(I'_y - I_y)(I'_x - I_y)}{\mu(I_x - I_y)}\right]^{1/2} \\ &= \left[\frac{\Delta I_y}{\mu}\left(1 + \frac{\Delta I_x}{I_x - I_y}\right)\right]^{1/2} \end{aligned} \tag{13.64}$$

$$\begin{aligned} |y| &= \left[\frac{(I'_x - I_x)(I'_y - I_x)}{\mu(I_y - I_x)}\right]^{1/2} \\ &= \left[\frac{\Delta I_x}{\mu}\left(1 + \frac{\Delta I_y}{I_y - I_x}\right)\right]^{1/2} \end{aligned} \tag{13.65}$$

and $|z| = 0$ with $\Delta I_x = I'_x - I_x$, and so on.

Asymmetric-top Molecules: Nonplanar

For the general asymmetric rotor, it is convenient to use the tensor **P** rather than **I**. The principal moments of **P** and **I** are related as given in (13.25). For the center-of-mass principal axis system of the parent molecule, the off-diagonal matrix elements of **P** are

$$\begin{aligned} P_{xy} &= \sum m_i x_i y_i = 0 \\ P_{xz} &= \sum m_i x_i z_i = 0 \\ P_{yz} &= \sum m_i y_i z_i = 0 \end{aligned} \tag{13.66}$$

and the diagonal elements are

$$\begin{aligned} P_x &= P_{xx} = \sum m_i x_i^2 \\ P_y &= P_{yy} = \sum m_i y_i^2 \\ P_z &= P_{zz} = \sum m_i z_i^2 \end{aligned} \tag{13.67}$$

Now for the substituted molecule with respect to the same axes, the elements of $\mathbf{P}'$ are, from (13.27) and (13.28),

$$P'_{xx} = P_x + \mu x^2 \tag{13.68}$$

and so on, and

$$P'_{xy} = \mu xy \tag{13.69}$$

and so on. The secular equation for the substituted molecule is hence

$$\begin{vmatrix} P_x + \mu x^2 - P' & \mu xy & \mu xz \\ \mu xy & P_y + \mu y^2 - P' & \mu yz \\ \mu xz & \mu yz & P_z + \mu z^2 - P' \end{vmatrix} = 0 \tag{13.70}$$

which is somewhat simpler than the corresponding equation in terms of $\mathbf{I}$. The roots of this secular determinant are the planar principal moments P'_x, P'_y, P'_z. The coefficients of the polynomial in P' may be related to the roots [10] giving three equations which may be solved to give the coordinates of the substituted atom

$$|x| = \left[\frac{\Delta P_x}{\mu}\left(1 + \frac{\Delta P_y}{I_x - I_y}\right)\left(1 + \frac{\Delta P_z}{I_x - I_z}\right)\right]^{1/2} \tag{13.71}$$

$$|y| = \left[\frac{\Delta P_y}{\mu}\left(1 + \frac{\Delta P_z}{I_y - I_z}\right)\left(1 + \frac{\Delta P_x}{I_y - I_x}\right)\right]^{1/2} \tag{13.72}$$

$$|z| = \left[\frac{\Delta P_z}{\mu}\left(1 + \frac{\Delta P_x}{I_z - I_x}\right)\left(1 + \frac{\Delta P_y}{I_z - I_y}\right)\right]^{1/2} \tag{13.73}$$

where

$$\begin{aligned} \Delta P_x &= (\tfrac{1}{2})(-\Delta I_x + \Delta I_y + \Delta I_z) \\ \Delta P_y &= (\tfrac{1}{2})(-\Delta I_y + \Delta I_z + \Delta I_x) \\ \Delta P_z &= (\tfrac{1}{2})(-\Delta I_z + \Delta I_x + \Delta I_y) \end{aligned} \tag{13.74}$$

and where $\Delta I_x (= I'_x - I_x)$, and so on, are the changes in the principal moments of inertia due to isotopic substitution. The distance of the substituted atom from the center of mass is simply

$$|r| = \left[\left(\frac{1}{2\mu}\right)(\Delta I_x + \Delta I_y + \Delta I_z)\right]^{1/2} \tag{13.75}$$

The expressions (13.71)–(13.73) can also be written in a more symmetrical form by replacing $(I_x - I_y)$ by $(P_y - P_x)$, and so on:

$$|x| = \left[\frac{\Delta P_x}{\mu}\left(1 + \frac{\Delta P_y}{P_y - P_x}\right)\left(1 + \frac{\Delta P_z}{P_z - P_x}\right)\right]^{1/2} \tag{13.76}$$

Expressions for $|y|$ and $|z|$ are obtained by cyclic permutation of P_x, P_y, P_z.

For substitution of an atom in a symmetry plane which is a principal inertial plane (taken as the xy plane) $\Delta P_z = (\Delta I_x + \Delta I_y - \Delta I_z)/2 = 0$, and (13.76) is simplified to

$$|x| = \left[\frac{\Delta P_x}{\mu}\left(1 + \frac{\Delta P_y}{P_y - P_x}\right)\right]^{1/2} \tag{13.77}$$

and likewise

$$|y| = \left[\frac{\Delta P_y}{\mu}\left(1 + \frac{\Delta P_x}{P_x - P_y}\right)\right]^{1/2} \tag{13.78}$$

with $|z| = 0$. These relations can also be derived, starting from (13.70) where elements containing z vanish. These results can be readily shown to be equivalent to (13.64) and (13.65). Other forms are also possible. Since the change in the planar moment perpendicular to the plane (xy) vanishes, the following relations can be written

$$\begin{aligned} \Delta P_x &= \frac{-\Delta I_x + \Delta I_y + \Delta I_z}{2} \\ &= \Delta I_y \\ &= \Delta I_z - \Delta I_x \end{aligned} \tag{13.79}$$

with similar relations for ΔP_y. If equilibrium quantities are used, the relation employed will make no difference. When effective ground state moments of inertia are employed, however, slightly different values for the coordinates will be obtained, and the question of which is the preferred expression arises (see Section 8).

Consider the atom to be located as lying on a symmetry axis, which will be a principal inertial axis, say $z (x = y = 0)$. Hence the changes in the planar moments ΔP_x and ΔP_y vanish. From (13.73), or (13.70) where only the μz^2 term survives, we have

$$|z| = \left(\frac{\Delta P_z}{\mu}\right)^{1/2} \tag{13.80}$$

Alternate forms are readily written since $\Delta I_z=0$ and $\Delta I_x=\Delta I_y$. Thus (13.80) may be reduced to (13.46).

It is possible to write the transformation between the principal axis system of a parent molecule and an isotopic form in terms of the substitution coordinates of an atom as discussed by Rudolph [11]. The transformation is, in general, a rotation plus a translation

$$\mathbf{r}=\mathbf{R}\mathbf{r}'+\mathbf{t} \tag{13.81}$$

where $\mathbf{r}$ represents an arbitrary vector for the parent species and $\mathbf{r}'$ for the isotopic species with

$$\mathbf{R}=-\Delta m\cdot\begin{bmatrix} \dfrac{xx'}{P_x-P'_x} & \dfrac{xy'}{P_x-P'_y} & \dfrac{xz'}{P_x-P'_z} \\ \dfrac{yx'}{P_y-P'_x} & \dfrac{yy'}{P_y-P'_y} & \dfrac{yz'}{P_y-P'_z} \\ \dfrac{zx'}{P_z-P'_x} & \dfrac{zy'}{P_z-P'_y} & \dfrac{zz'}{P_z-P'_z} \end{bmatrix} \tag{13.82}$$

and

$$\mathbf{t}=\frac{\Delta m}{M+\Delta m}\begin{pmatrix} x \\ y \\ z \end{pmatrix} \tag{13.83}$$

For an asymmetric top the six planar moments (parent and single isotopic form) via (13.76) yield the coordinates xyz of the substituted atom relative to the parent. If the roles of the parent and the substituted species are interchanged, the same data provide the coordinates $x'y'z$ of the substituted atom relative to the principal axis frame of the isotopic species. Once the signs of the coordinates have been obtained from other considerations, the signs of the elements in the above transformation are uniquely defined. Then the transformation can be constructed without a knowledge (assumed or otherwise) of the complete molecular structure with only a single isotopic substitution required. The accuracy of the transformation will be dependent on the reliability of the coordinates. For substitution in a principal inertia plane (xy), the off-diagonal terms of R involving z (and z') vanish, and the lower diagonal element is replaced by $-1/\Delta m$. The transformation can have a number of applications [11], such as determination of the orientation of the dipole moment in the molecule from the dipole components (μ_a, μ_b, μ_c) of the parent (μ) and isotopic form (μ'), aid in transformation of the quadrupole coupling tensor to its principal axis system, and aid in the evaluation of small coordinates (See Section 8). In the former application, for example, only the rotational part of (13.82) is required, and the signs of the components of μ and μ' are chosen relative to each other such that $\mu=\mathbf{R}\mu'$ is satisfied.

4 MULTIPLE SUBSTITUTION

Kraitchman's procedure may be extended to multiple isotopic substitution which sometimes proved convenient. Consider, for example, the simultaneous isotopic substitution of the equivalent Y atoms of a bent, triatomic XY_2 molecule. Let the z axis be perpendicular to the plane and the x axis correspond to the twofold symmetry axis (see Fig. 13.5). The orientations of the principal axes are not changed upon double substitution of the Y atoms. The coordinates of the two substituted atoms are $(x, y, 0)$ and $(x, -y, 0)$, and the isotopic mass of each is $m+\Delta m$. Using (13.18) and (13.19) we find

$$I'_{xz}=I'_{xy}=I'_{yz}=0 \tag{13.84}$$

$$I'_x=I'_{xx}=I_x+2\Delta m y^2 \tag{13.85}$$

$$I'_y=I'_{yy}=I_y+\mu_2 x^2 \tag{13.86}$$

$$I'_z=I'_{zz}=I_z+2\Delta m y^2+\mu_2 x^2 \tag{13.87}$$

where

$$\mu_2=\frac{2\Delta m M}{(M+2\Delta m)} \tag{13.88}$$

The coordinates with respect to the parent coordinate system are hence

$$|x|=\left(\frac{\Delta I_y}{\mu_2}\right)^{1/2} \tag{13.89}$$

$$|y|=\left(\frac{\Delta I_x}{2\Delta m}\right)^{1/2} \tag{13.90}$$

and $|z|=0$ with $\Delta I_x=I'_x-I_x$, and so on, and μ_2 given by (13.88). These equations might be applied, for example, to data on SO_2 and $^{18}OS^{18}O$ to give the coordinates of the oxygen atoms.

Another case would be multiple substitution of the equivalent Y atoms in molecules of the class ZXY_3 having C_{3v} symmetry. Let the z axis represent the threefold symmetry axis. If the orientation of the coordinate system is chosen as indicated in Fig. 13.4, one of the Y atoms will be in the yz plane. The coordinates of this atom, with respect to the parent coordinate system, will be $(0, y, z)$. The coordinates of the remaining Y atoms are related by a rotation about the C_3 axis, the z axis. In particular

$$\begin{pmatrix} x_k \\ y_k \\ z_k \end{pmatrix}=\begin{pmatrix} \cos k\omega & -\sin k\omega & 0 \\ \sin k\omega & \cos k\omega & 0 \\ 0 & 0 & 1 \end{pmatrix}\begin{pmatrix} 0 \\ y \\ z \end{pmatrix} \tag{13.91}$$

where $\omega=2\pi/3$ and $k=1, 2$ give the coordinates of the other two Y atoms. Application of the above relation gives for the coordinates $[-(3^{1/2}/2)y, -(1/2)y, z]$ and $[(3^{1/2}/2)y, -(1/2)y, z]$. With these coordinates and the mass of each isotopic atom taken as $m+\Delta m$, (13.18) and (13.19) give

$$I'_{xz}=I'_{xy}=I'_{yz}=0 \tag{13.92}$$

$$\Delta I_x=\Delta I_y=\tfrac{3}{2}\Delta m y^2+\mu_3 z^2 \tag{13.93}$$

$$\Delta I_z=3\Delta m y^2 \tag{13.94}$$

where $\Delta I_x=I'_x-I_x$, and so on, and

$$\mu_3=\frac{3\Delta m M}{M+3\Delta m} \tag{13.95}$$

with M as usual the mass of the parent molecule. Although the foregoing equations may be solved for the $|y|$ and $|z|$ coordinates, the results are of less use, since the moment of inertia about the symmetry axis is usually not determined from the pure rotational spectrum (see, however, Chapter V); but, with the determination of z from the center-of-mass condition employing the z_i coordinates of the on-axis X and Z atoms, the $|y|$ coordinate may be found from (13.93).

Explicit expressions for some other types of molecules, where a set of equivalent nuclei is substituted, have been given by Chutjian [12]. A number of disubstitution cases have been considered by Nygaard [13] and the coordinates are expressed in terms of differences in the planar second moments giving rise to more simplified relations. These useful relations are summarized in Table 13.2. The use of double substitution for location of the coordinates in methyl-like groups such as CCL_3F has been discussed [14]. Evaluation of the hydrogen

Table 13.2 Substitution Coordinates (x, y, z) by Disubstitution for Different Molecular Symmetry[a]

Symmetry	*Isotopic Substitution in the Positions*	*Coordinate Relations* x^2	y^2	z^2	*Example Configuration*
C_{2v}^z	$\pm x,\ 0,\ z$	$\frac{\Delta P_x}{2\Delta m}$	0	$\frac{\Delta P_z}{\mu_2}$	b
C_{2h}^z	$\begin{Bmatrix} x, & y, & 0 \\ -x, & -y, & 0 \end{Bmatrix}$	$\frac{\Delta P_x}{2\Delta m}\left(1+\frac{\Delta P_y}{P_y-P_x}\right)$	$\frac{\Delta P_y}{2\Delta m}\left(1+\frac{\Delta P_x}{P_x-P_y}\right)$	0	c
C_2^z	$\begin{Bmatrix} x, & y, & z \\ -x, & -y, & z \end{Bmatrix}$	$\frac{\Delta P_x}{2\Delta m}\left(1+\frac{\Delta P_y}{P_y-P_x}\right)$	$\frac{\Delta P_y}{2\Delta m}\left(1+\frac{\Delta P_x}{P_x-P_y}\right)$	$\frac{\Delta P_z}{\mu_2}$	d
C_s^{xy}	$x,\ y,\ \pm z$	$\frac{\Delta P_x}{\mu_2}\left(1+\frac{\Delta P_y}{P_y-P_x}\right)$	$\frac{\Delta P_y}{\mu_2}\left(1+\frac{\Delta P_x}{P_x-P_y}\right)$	$\frac{\Delta P_z}{2\Delta m}$	e

[a] $P_x=(-I_x+I_y+I_z)/2$, etc., $\Delta P_x=P'_x-P_x$, etc. (the change in *principal* planar moments by substitution), $P_x-P_y=I_y-I_x$, etc., and $\mu_2=M2\Delta m/(m+2\Delta m)$. From Nygaard [13]. Equivalent atoms indicated.

[b] Example, furan, H_2O.

[c] Example, planar *trans*-1,2-dichloroethylene.

[d] Example, nonplanar H_2O_2 or ethylene ozonide.

[e] Example, acetaldehyde, or aniline.

coordinates of a molecule containing two equivalent methyl groups, for example, $(CH_3)_2-X$, has been described [15]. The coordinates of interest are the symmetric in-plane coordinates $(\pm x_s, 0, z_s)$ and the asymmetric out-of-plane coordinates $(\pm x_a, \pm y_a, z_a)$. The moments of inertia of the normal (6H), hexadeuterated (6D), and trideuterated (3H3D) species are employed.

5 EVALUATION OF STRUCTURES: GENERAL CONSIDERATIONS

When moments of inertia are related to the structural parameters, either of two procedures may be employed. In the first procedure, the molecular structure may be obtained by a best-fitting of calculated and observed moments of inertia of a sufficient number of isotopic species. For example, in the expressions of Table 13.1 which give the moments of inertia in terms of the bond distances and bond angles, the appropriate isotopic masses may be substituted and the molecular parameters found from the solution of the resulting simultaneous equations. Alternately, through the use of the equations considered in Sections 3 through 4 the coordinates of each atom with respect to the principal axes of the parent molecule may be determined, from which the bond distances and bond angles may be easily calculated. One can, of course, choose any one of the molecular species as the "original" or "parent" molecule since within the rigid rotor approximation the structural parameters must be invariant to this choice. For reasons of convenience, however, the normal isotopic species is usually taken as the parent molecule. In application of the equations to a particular molecule the *xyz* axes will be identified in some order with the principal axes *abc*. It should be noted that the distance between any two atoms of a molecule can be evaluated without knowledge of the other distances, provided the moments of inertia are determined for the species with isotopic substitution for both of these atoms as well as for the original unsubstituted species. Furthermore, since the equations for the moments of inertia, the products of inertia, and the first-moment equations provide relations among the coordinates of the parent molecule, it is not necessary to substitute every atom to determine the structure. The minimum amount of data required for complete determination of the structure is the same for both procedures. It may be noted that the number of structural parameters required to characterize a particular molecule is equal to the number of totally symmetric vibrations. Thus by the evaluation of the number of times the totally symmetric species occurs in the Γ_{vib} representation, the number of structural parameters is given [16].

For a linear molecule consisting of N atoms there are N coordinates or $(N-1)$ independent internuclear distances to be evaluated. Analysis of the rotational spectrum of a linear molecule yields one moment of inertia for each isotopic species. The moment-of-inertia equation and the first-moment equation

$$I_x=\sum m_i z_i^2, \qquad \sum m_i z_i=0 \tag{13.96}$$

of the parent molecule provide two equations. Therefore, a minimum of $(N-2)$ isotopic substitutions is required for a complete evaluation of the structure. Furthermore, only one substitution for a given atom yields independent information. Hence at least one substitution for each $(N-2)$ different atoms of the molecule must be made. Knowledge of the coordinates (distance from the center of mass) of two atoms gives the internuclear distance d_{ij}, that is,

$$d_{ij}=|z_i-z_j| \tag{13.97}$$

We must, however, know whether the atoms lie on the same or opposite sides of the center of mass.

For a diatomic molecule knowledge of the moment of inertia I_x is sufficient to determine the distance between the two atoms m_X and m_Y. In particular (see Table 13.1),

$$d=\left[\left(\frac{m_X+m_Y}{m_X m_Y}\right) I_x\right]^{1/2} \tag{13.98}$$

For a linear triatomic molecule XYZ (13.46) and (13.96) allow the calculation of the z_i's and hence of the two bond lengths if a single isotopic substitution on either of its three atoms is made. In particular, if the X atom is isotopically substituted, (13.46) gives the coordinate z_X, and inserting this into (13.96) yields two relations in the two unknowns z_Y and z_Z which may be solved. Equivalently, the moments of inertia for two molecular species gives from Table 13.1 two equations to be solved for the two bond lengths. With more than three atoms, additional isotopic substitutions are required as indicated.

The most common type of symmetric-top molecule, the $Y_pXZ\ldots$ class, has any number p, greater than two, of symmetrically placed off-axis atoms and any number of atoms $X, Z, \ldots$ along the symmetry axis (the z axis). Because of the selection rule only one moment of inertia, for example, I_x, can be found from the spectrum. A single substitution for one of the atoms on the molecular axis allows the distance of that atom from the center of mass to be calculated with (13.46). Substitution of two atoms on the symmetry axis gives the distance between the atoms. Therefore, with N atoms on the symmetry axis N isotopic substitutions in addition to the parent molecule will determine the separations between the atoms. When separations of all the atoms on the symmetry axes are thus obtained, only two unknown parameters remain—the bond angle and the XY bond length. We can obtain these by making a single isotopic substitution on a Y atom and by using (13.56) and (13.57) to calculate the coordinates of the Y atom. Alternately, we may employ the inertial equations. If y_Y represents the perpendicular distance of the Y atoms from the symmetry axis, the z axis, and z_Y the z coordinate distance of the Y_P plane from the center of mass, and z_i the coordinate distance of the ith atom on the symmetry axis from the center of mass, then the moment of inertia can be written as:

$$I_x=pm_Y\left(\frac{y_Y^2}{2}+z_Y^2\right)+\sum_i m_i z_i^2 \tag{13.99}$$

with the first moment equation:

$$pm_Y z_Y + \sum_i m_i z_i = 0 \tag{13.100}$$

in which m_Y represents the mass of Y and m_i the mass of the ith atom located on the symmetry axis. The summation extends over all atoms on the symmetry axis but does not include the atoms off this axis. From the geometry of the molecule we can also write

$$d = [(z_X - z_Y)^2 + y_Y^2]^{1/2} \tag{13.101}$$

and

$$\tan\beta = \frac{y_Y}{|z_X - z_Y|} \tag{13.102}$$

where d is the length of the XY bond and where β is the angle of the bond XY with the symmetry axis. This angle is related to the bond angle YXY, designated θ, by:

$$\sin\tfrac{1}{2}\theta = \sin\frac{\pi}{p}\sin\beta \tag{13.103}$$

With the known z_i values obtained from isotopic substitutions on the symmetry axis, (13.99) and (13.100) can be employed to give y_Y and z_Y and, hence, the bond angle and the XY bond length from (13.101)–(13.103). Conversely, (13.99) through (13.103) are sufficient for calculation of the moment of inertia $I_x(I_b)$ from assumed bond angles and bond lengths. In Table 13.1 explicit expressions for the moments of inertia in terms of the structural parameters have been given for two of the simplest types of symmetric tops. From the foregoing discussion or the expressions of Table 13.1 it is apparent that a minimum of two molecular species must be studied for the XY_3 type and three for the ZXY_3 type. When a simultaneous solution of the inertial equations is to be considered, then it is obviously simpler if symmetric isotopic substitution of the Y atoms is made.

For an asymmetric-top molecule, all three of its principal moments of inertia can be obtained from the pure rotational spectrum. For a planar molecule, however, only two of these are independent because of the relation given in (13.58) which holds for a rigidly planar molecule. For a planar asymmetric rotor, five moment equations of the parent molecule are available:

$$I_x = \sum m_i y_i^2, \qquad I_y = \sum m_i x_i^2, \qquad I_{xy} = -\sum m_i x_i y_i = 0, \qquad \sum m_i x_i = 0,$$
$$\sum m_i y_i = 0 \tag{13.104}$$

where the z axis is taken perpendicular to the plane. If the x and y coordinates of $(N-2)$ atoms are determined by means of (13.64) and (13.65), then the remaining four coordinates of the two unsubstituted atoms may be obtained from the solution of the above relations. Therefore, a minimum of $(N-2)$ isotopic substitutions at different sites is required for the determination of the complete structure of a planar asymmetric rotor having N atoms. For a general, nonplanar, asymmetric-top molecule, all nine moment equations are available, viz., (13.29)–(13.34) and the three center-of-mass equations. Hence $(N-3)$ isotopic substitutions are necessary to fix the complete structure. Expressions for the moments of inertia of some simple asymmetric rotors are given in Table 13.1.

For all types of molecules, if some of the bond lengths or angles are known from symmetry to be equivalent, the minimum number of required substitutions is accordingly reduced. The structures of nonlinear triatomic molecules, XY_2, can be solved completely from the two independent moments of inertia without isotopic substitution. With the molecule in the xy plane and the x axis corresponding to the symmetry axis, we find (see Table 13.1)

$$d=\left[\frac{1}{2m_Y}\left(\frac{M}{m_X}I_y+I_x\right)\right]^{1/2} \tag{13.105}$$

and

$$\tan\tfrac{1}{2}\theta=\left(\frac{m_X I_x}{M I_y}\right)^{1/2} \tag{13.106}$$

where d is the $X-Y$ bond length and θ is the YXY angle and where M $(=m_X+2m_Y)$ is the total mass. For the slightly more complicated case of an XYZ type, such as NOCl, which has three structural parameters, the structure cannot be completely solved from measurements on a single isotopic species. Often some of the dimensions are known with fair precision from electron diffraction, or they can be estimated with confidence from similar molecules. When some molecular parameters are known, fewer substitutions are necessary to complete the structural determination. Substitutions in excess of the minimum number are useful in checking results and simplifying the calculations. Furthermore, additional substitutions are helpful in estimation of the effects of zero-point vibrations when effective moments of inertia are used since the variation of the structural parameters with different choice of data may be observed. Since only the absolute values $|x|$, $|y|$, $|z|$ are obtained, a decision as to the sign of the coordinates must be made from other information. This is really not a serious problem since one usually has a prior knowledge of the arrangement of the atoms in the molecule as well as molecular structure information on the molecule or on similar molecules from other sources, such as infrared or electron diffraction. Also the signs must be such that the center-of-mass conditions $\sum m_i x_i=0$, and so on, and the products of inertia $\sum m_i x_i y_i=0$, and so on, are obeyed (or essentially obeyed for effective rather than equilibrium coordinates). If, for example, the sign of a given coordinate is ambiguous, while others are reasonably obvious, the first-moment equation could be evaluated with both possible signs and the choice made on the basis of how well the relation is satisfied. When the structural calculation is made with both possible signs for an ambiguous coordinate, the derived structure can be used to give a correct sign choice if unreasonable structural parameters are obtained for one

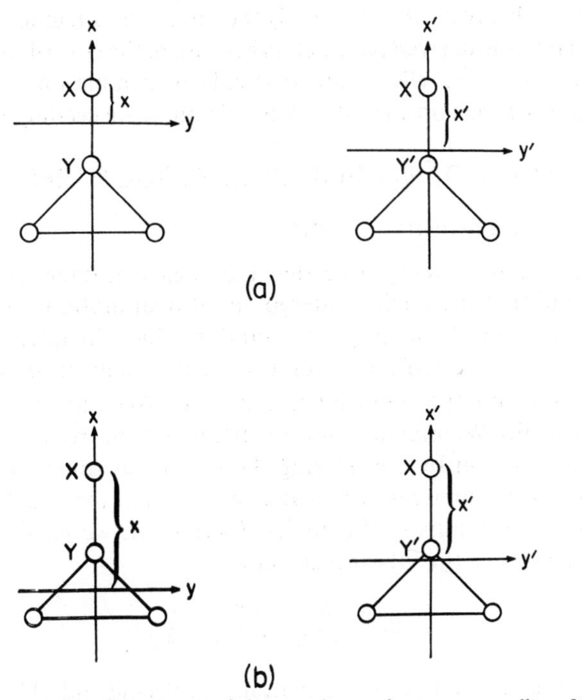

Fig. 13.6 Determining the relative signs of the coordinates by isotopic pulling. Note in (*a*) the coordinate of X increases when referred to molecule $XY' \ldots$ as parent compared to its value relative to $XY \ldots$ as parent. The reverse is true for the center of mass as in (*b*).

sign as compared to another. If sufficient data are available, it is possible to use the changes in the coordinates of an atom with different parent molecules to determine the signs of the coordinates relative to those of another atom. This procedure is referred to as "isotopic pulling" by Pasinski and Beaudet [17], is particularly useful when little is known about the shape of the molecule. Consider the molecule of Fig. 13.6, with $m'_x > m_x$ and $m'_Y > m_Y$. Isotopic $X'Y\cdots$ can be used for evaluation of the x_X coordinate of X in the parent species $XY\cdots$. Likewise, the isotope $X'Y'\cdots$ can be used for evaluation of the coordinate x'_X for the parent species $XY'\cdots$. The center of mass will be shifted toward Y' in the XY' species. Thus, if x_X is positive as shown in Fig. 13.6*a*, we will have $x'_X > x_X$. On the other hand, if X and Y are on the same side of the y axis, as in Fig. 13.6*b*, the $x'_X < x_X$. Therefore, the increase or decrease in the coordinate locates the relative coordinate position of atoms X and Y.

It should be emphasized that here and in previous discussions it has been assumed that the molecule is rigid and therefore that the molecular parameters do not change with isotopic substitution. Hence the equilibrium moments of inertia that correspond to the moments for the vibrationless state where the atoms are at rest should be used in such equations as Kraitchman's equations and the equations of Table 13.1. When effective rather than equilibrium moments of inertia are used, as is usually the case, a certain amount of ambiguity is introduced into the derived parameters because of the contributions from zero-point vibrations. These effects are particularly troublesome for polyatomic molecules and will be considered in more detail in succeeding sections.

6 EVALUATION OF EQUILIBRIUM STRUCTURES

Correction for Vibrational Effects

Previously, we have recognized that the nuclear framework of a molecule is not rigid and that the nuclei undergo vibrational motions about their equilibrium positions simultaneously with rotation. Since the internuclear distance will be changing as the molecule vibrates, the moments of inertia depend in a complicated way on the vibrational state. Consider, for example, a simple diatomic molecule. We may account for the vibration–rotation interaction to a good approximation by considering the rotational constant, characterizing the rather slow rotational motion as an average value, averaged over the rather rapid vibrational motions of the nuclei. Specifically, we have for the effective rotational constant of the vibrational state v:

$$B_v = \frac{h}{8\pi^2 I_b^v} = \frac{h}{8\pi^2 \mu}\left\langle \frac{1}{r^2} \right\rangle \tag{13.107}$$

in which μ is the reduced mass of a diatomic molecule and $\langle 1/r^2 \rangle$ is the appropriate average over the vibrational wave function. Now the instantaneous internuclear distance may be expanded about its equilibrium value in terms of the dimensionless vibrational coordinate:

$$\frac{1}{r^2} = \frac{1}{r_e^2}(1+\xi)^{-2} = \frac{1}{r_e^2}(1-2\xi+3\xi^2-\cdots) \tag{13.108}$$

with $\xi = (r-r_e)/r_e$. The vibrational average of r^{-2} can thus be expressed in terms of the average values of ξ, ξ^2, etc. If the anharmonicity of the potential function is included, that is, the harmonic vibrational wave function is corrected for the presence of a perturbing cubic term ($V' = ha_0 a_1 \xi^3$, $a_0 = \omega_e^2/4B_e$), then the averages are found to be [18]:

$$\langle v|\xi|v\rangle = -a_1\left(\frac{3B_e}{\omega_e}\right)(v+\tfrac{1}{2}) \tag{13.109}$$

$$\langle v|\xi^2|v\rangle = \left(\frac{2B_e}{\omega_e}\right)(v+\tfrac{1}{2}) \tag{13.110}$$

where v is the vibrational quantum number, ω_e is the classical harmonic vibrational frequency, a_1 is the cubic anharmonic constant (dimensionless), and B_e $(=h/8\pi^2\mu r_e^2)$ is the equilibrium rotational constant. Hence taking cognizance of (13.107) through (13.110) we have for the effective rotational constant of a vibrating rotator

$$B_v = B_e - \alpha_e(v+\tfrac{1}{2}) \tag{13.111}$$

correct to terms linear in $(v+\frac{1}{2})$. The coefficient of the $(v+\frac{1}{2})$ dependence is given by the expression

$$\alpha_e = -\left(\frac{6B_e^2}{\omega_e}\right)(1+a_1) \tag{13.112}$$

Equation 13.111 is found to be a good approximation to the effective rotational constant. It is evident that even in the ground vibrational state $(v=0)$, B_0 differs from B_e by $-\alpha_e/2$. Furthermore, even if the vibration were harmonic $(a_1=0)$, a vibrational contribution to the effective rotational constant is realized. Since the anharmonic correction a_1 is usually negative and larger than one, α_e is positive; the effective rotational constant is less than B_e and decreases with an increase in the vibrational state.

For polyatomic molecules, expressions similar to (13.111) hold for the effective rotational constants (see Chapter VII)

$$A_v = A_e - \sum_s \alpha_s^a \left(v_s + \frac{d_s}{2}\right) \tag{13.113}$$

$$B_v = B_e - \sum_s \alpha_s^b \left(v_s + \frac{d_s}{2}\right) \tag{13.114}$$

$$C_v = C_e - \sum_s \alpha_s^c \left(v_s + \frac{d_s}{2}\right) \tag{13.115}$$

Alternately, we may define the effective moments of inertia as

$$I_\alpha^v = I_\alpha^e + \sum_s \left(v_s + \frac{d_s}{2}\right) \varepsilon_s^\alpha \qquad \alpha = a, b, c \tag{13.116}$$

where the rotation–vibration parameter is given by

$$\varepsilon_s^\alpha = \left(\frac{8\pi^2}{h}\right)(I_\alpha^e)^2 \alpha_s^\alpha \tag{13.117}$$

In the foregoing expressions, the sum is over all vibrations of the molecule counting degenerate ones once; $v=(v_1, v_2, v_3, \ldots, v_s, \ldots)$ corresponds to the set of vibrational quantum numbers with v_s the vibrational quantum number for the sth vibration having degeneracy d_s. In asymmetric-top molecules there are no degenerate normal modes of vibration although accidental near-degeneracies are possible. A linear XYZ molecule, for example, has four modes of vibration (see Fig. 5.1), two of which correspond to bending vibrations in two perpendicular planes; these have the same frequency ω_2 and the same α constants and constitute a doubly degenerate vibration $(d_2=2)$. The other two remaining modes are stretching vibrations and are nondegenerate; ω_1, a stretching vibration of the Y–Z bond $(d_1=1)$; ω_3, a stretching vibration of the Y–X bond $(d_3=1)$. The effective rotational constant is thus given explicitly by

$$B_v = B_e - \alpha_1^b(v_1+\tfrac{1}{2}) - \alpha_2^b(v_2+1) - \alpha_3^b(v_3+\tfrac{1}{2}) \tag{13.118}$$

Measurements of B_v for the ground (0, 0, 0) and the first excited vibrational states (1, 0, 0), (0, 1, 0), (0, 0, 1) will allow the evaluation of α_1^b, α_2^b, and α_3^b and hence B_e or I_b^e. Knowledge of the α constants also gives information on the cubic potential constants as is evident from the definition of α_e for a diatomic molecule.

For polyatomic molecules there is an additional Coriolis interaction which contributes to the rotational energy; the quantities α_s^a, etc. depend not only on the harmonic and anharmonic potential constants but also on the Coriolis coupling constants. Because of the additional contributions from Coriolis interactions the effective rotational constants are not simply the average of the inverse of the instantaneous moments. General discussions of the vibrational contributions to the effective moments as well as explicit expressions for the α's of some simple molecules have been given [19] (see also Chapter VII). In addition to the contributions from vibration-rotation, the effective rotational constants contain small contributions from centrifugal distortion (see Chapter VIII) and from electron rotation interactions (see Chapter XI).

Naturally, the preferred measure of the distance between two nuclei is the equilibrium bond distance, which is free from the zero-point vibrational effects, and which has a clear physical meaning, that is, the bond distance corresponding to the minimum of the potential energy curve. Figure 13.7 gives an illustration of the lower portion of the potential energy curve for ^{7}LiF. In principle, the calculation of equilibrium structures is straightforward. Once a sufficient number of vibrational states have been investigated to allow evaluation of the α_i's, the equilibrium rotational constants and hence the equilibrium moments of inertia may be obtained. The equilibrium structure may then be calculated as outlined in the previous section. Since the vibrational effects have been corrected for the errors in r_e, structures are limited only by the experimental uncertainties, uncertainties in fundamental constants such as Planck's constant, and possible uncertainties from electron–rotation effects if these have been ignored.

As we have seen for diatomic and symmetric bent triatomic molecules, only the moments of inertia of a single isotopic species are required for evaluation of the complete structure. For diatomic molecules, then, only one vibrational state need be measured in addition to the ground state if (13.111) holds satisfactorily. If $(v+\frac{1}{2})^2$ dependence is also important, this adds an additional constant γ_e, and measurements on another excited state are necessary. Once B_e is determined, the equilibrium bond distance is given by

$$r_e = \left(\frac{h}{8\pi^2 \mu B_e}\right)^{1/2} = \left(\frac{I_b^e}{\mu}\right)^{1/2} \tag{13.119}$$

where $\mu = m_1 m_2/(m_1 + m_2)$ is the reduced mass [not to be confused with the μ for isotopic substitution defined by (13.41)]. The equilibrium parameters B_e, and r_e of some diatomic molecules are displayed in Tables 4.6 and 4.7.

From measurements of r_0 for two or more isotopes of a diatomic molecule it has been pointed out by Laurie [20] that a good approximation to the equilibrium bond distance may be obtained. From (13.156) and (13.155), neglecting higher-order terms and employing $I_b^e = \mu r_e^2$ and $\omega_e = (1/2\pi)(f/\mu)^{1/2}$ to separ-

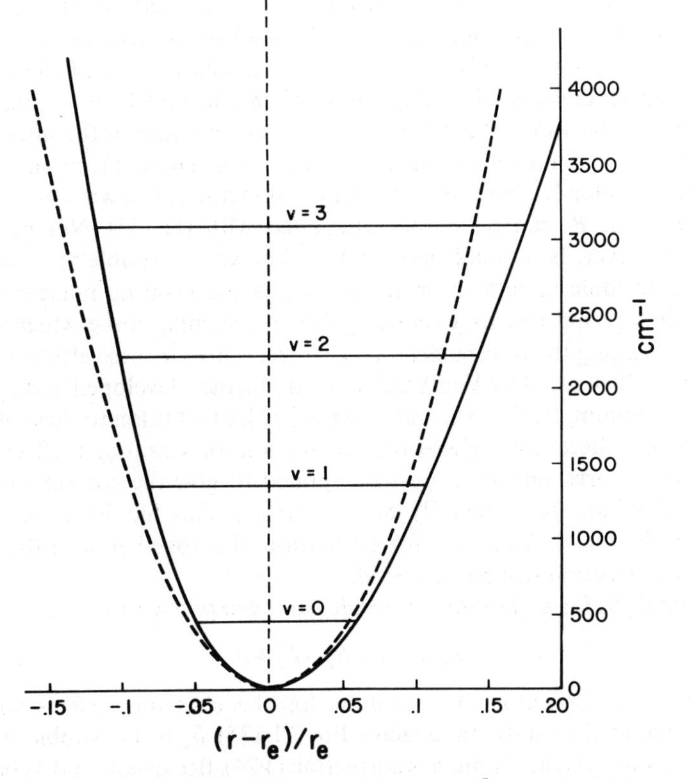

Fig. 13.7 The solid curve represents the potential function of ^{7}LiF evaluated from: $V(\text{cm}^{-1})=(1.5398\times10^5\ \text{cm}^{-1})\times\xi^2(1-2.7006\xi+5.10\xi^2-8.0\xi^3)$. The broken curve represents the harmonic part of the potential function. From E. F. Pearson, Ph.D. dissertation, Duke University, 1968.

ate out the reduced mass dependence, one can obtain

$$r_0=r_e-\left[\frac{3\hbar(1+a_1)}{4r_e f^{1/2}}\right]\mu^{-1/2} \tag{13.120}$$

By plotting r_0 versus $\mu^{-1/2}$ and extrapolating $\mu^{-1/2}$ to zero it is possible to obtain the intercept r_e. From the r_0 values of HF and DF in Table 13.7 one thus obtains $r_e=0.917$ Å, which agrees with the value obtained by direct correction of the rotational constant for the vibrational effects. This procedure, though, has limited applicability since no simple reduced mass function can be defined for a polyatomic molecule. See, however, Sections 9 and 10.

Higher-order Corrections

In addition to the gross vibrational effects just discussed, there are higher-order corrections, described later, that are required for a precise evaluation of B_e and hence of r_e. In the usual derivation of equilibrium structures from observed rotational spectra for molecules in $^1\Sigma$ states, the interaction of rotation with electronic motions is neglected, and the electrons of each bonded atom are assumed to be concentrated at its nucleus. Actually, of course, these assumptions are not strictly true. There are displacements of the electronic charges due to chemical bonding; and, when the molecule is rotating, the electronic angular momentum is not strictly zero because of the interaction of the electronic and molecular rotational motions. These effects are formally treated in Chapter XI, Section 7. When the g factor is known, corrections for these effects on B_e can be simply calculated with (11.135). Not included in (11.135), however, is a small perturbation known as wobble stretching. The nonzero electronic angular momentum causes the rotating nuclear frame to wobble, thus giving rise to a centrifugal-like stretching force which in linear molecules is along the internuclear axis. Corrections for these electronic effects on B_e originally treated by Van Vleck [21] are further developed and applied to CO by Rosenblum, Nethercot, and Townes [22]. In addition to these electronic perturbations, there is a higher-order correction for vibrational effects known as Dunham's correction that must be applied to give the correct value of B_e from the Dunham Y_{01} (generally equated to B_e). This Dunham correction of B_e is given by (4.24). A method for estimation of it together with the wobble-stretching correction will be described.

In general, B_e for a diatomic molecule can be expressed by

$$B_e = Y_{01} + \delta_1 + \delta_2 + \delta_3 \tag{13.121}$$

where the first correction, δ_1, accounts for the electronic effects which are proportional to the rotational g factor Eq. (11.135), δ_2 is the wobble stretching correction, and δ_3 is the Dunham's correction (4.24). Because δ_2 and δ_3 both have $1/\mu^2$ dependence, they can be estimated together from known mass ratios of different isotopic species as follows

$$\frac{B_e}{B'_e} = \frac{B + \delta_2 + \delta_3}{B' + \delta'_2 + \delta'_3} = \frac{\mu'}{\mu} \tag{13.122}$$

where the primed and unprimed quantities represent different isotopic species. With (13.122) and the $(1/\mu^2)$ dependence of δ_2 and δ_3 it can be shown that

$$\delta_2 + \delta_3 = \frac{\varepsilon B'}{(\mu/\mu') - 1} \simeq \varepsilon \left(\frac{BB'}{B' - B} \right) \tag{13.123}$$

in which

$$B = Y_{01} + \delta_1 \tag{13.124}$$

and the isotopic discrepancy is defined by

$$\varepsilon = \frac{B}{B'} - \frac{\mu'}{\mu} \tag{13.125}$$

When δ_3 can be separately evaluated from (4.24), δ_2 can be found from (13.123). When these corrections are important and are not made, then the derived r_e values will not be exactly the same for different isotopic species.

Table 13.3 Rotational Constant Corrections for LiCl, LiF, and LiD (MHz)[a]

Corrections	$^7Li^{35}Cl$	^7LiF	^7LiD
Y_{01}	21,181.004	40,329.808	126,905.36
δ_1	−0.978	−1.410	18.80
δ_3	−0.023	0.081	17.2
$\delta_2+\delta_3$	0.801	1.370	37
B_e	21,180.827	40,329.768	126,961

[a]From Pearson and Gordy [23].

The higher-order corrections described have been applied in studies of CO [22], LiD, LiF, and LiCl [23]. Usually the corrections for the electronic effects are larger than the Dunham correction. The corrections for LiD, LiF, and LiCl are illustrated in Table 13.3. Note that for the light LiD molecule all three terms δ_1, δ_2, and δ_3 are rather large and are essentially equal in magnitude. More detailed discussion of these higher-order effects are available [24, 25].

Equilibrium Structures for Polyatomic Molecules

For a symmetric, bent, triatomic molecule (XY_2) there are three nondegenerate normal modes of vibration: a symmetric stretching ω_1 where the Y atoms move in the same direction along the $X-Y$ bond, an antisymmetric stretching ω_3 where the Y atoms move in opposite directions along the $X-Y$ bond, and a bending vibration ω_2. Therefore, three excited vibrational states must be investigated if the equilibrium constants are to be determined. The equilibrium structure of SO_2 and OF_2 have been determined by Morino et al. [26, 27], from the rotational spectra. The rotational constants for the first excited state of each of the three vibrations ω_1, ω_2, and ω_3 were determined. This information along with the ground state constants gives the constants α_s^a, α_s^b, α_s^c which are collected in Table 13.4. With knowledge of the nine vibration–rotation interaction constants, the equilibrium rotational constants are readily obtained

$$A_e = A_{000} + \tfrac{1}{2}(\alpha_1^a + \alpha_2^a + \alpha_3^a) \tag{13.126}$$

and so on, from which the equilibrium structure can be evaluated by use of (13.105) and (13.106). The equilibrium structures of SO_2 and OF_2 are given in Table 13.5. Within the experimental errors, the relation $I_c^e = I_a^e + I_b^e$ required for a planar molecule is well satisfied. For SO_2, $\Delta^e = I_c^e - I_a^e - I_b^e = -0.0038$ amu Å while for OF_2, $\Delta^e = -0.0035$ amu Å^2. Knowledge of the α constants also allowed the calculation of the cubic force constants [26, 27].

The equilibrium structures for some other symmetric-bent molecules are collected in Table 13.6. That for nitrogen trifluoride (see Table 6.10), trifluoro-

Table 13.4 Vibration–Rotation Constants, Rotational Constants, and Moments of Inertia for SO_2 and OF_2[a]

	Vibration–Rotation Constants (MHz)					
	SO_2			OF_2		
s	α_s^a	α_s^b	α_s^c	α_s^a	α_s^b	α_s^c
1	−31.05	50.14	42.83	38.55	65.77	7.18
2	−1175.90	−2.18	16.00	−699.02	42.36	53.28
3	620.02	34.85	32.88	585.02	69.56	115.25
	Rotational Constants (MHz)					
	A	B	C	A	B	C
Effective	60,778.79	10,318.10	8,799.96	58,782.630	10,896.431	9,167.412
Equilibrium	60,485.32	10,359.51	8,845.82	58,744.90	10,985.28	9,255.27
	Moments of Inertia (amu Å^2)					
	I_a	I_b	I_c	I_a	I_b	I_c
Effective	8.31756	48.9946	57.4470	8.60006	46.3942	55.1444
Equilibrium	8.35515[b]	48.7954[b]	57.1468[b]	8.60553	46.0189	54.6209

[a]Morino et al. [26, 27].
[b]Small corrections from electron-rotation effects have been included for SO_2.

Table 13.5 Equilibrium Structures and Various Ground State Structures of SO_2 and OF_2[a]

	SO_2		OF_2	
Structure	r	θ	r	θ
Equilibrium (r_e)	1.4308	119°19′	1.4053	103°4′
Effective (r_0)	1.4336	119°25′	1.4087	103°19′
Substitution (r_s)	1.4312	119°30′	—	—
Mass dependence (r_m)[b]	1.4307	119°20′	—	—
Average ($\langle r \rangle$)	1.4349	119°21′	1.4124	103°10′

[a]From Morino et al. [26, 27].
[b]From Watson [57].

silane (see Table 6.13) and the methyl halides (see Table 6.12) have been given previously, as well as some equilibrium structures for linear triatomic molecules (Tables 5.9 and 5.10). The equilibrium structure for arsenic trifluoride [28] and phosphorous trifluoride [29] have also been reported, as well as FCN [30] and CH_2F_2 [31]. For NOCl, the equilibrium structure [32] found is N–O = 1.131

Table 13.6 Equilibrium Structures of Some Bent Triatomic Molecules

Molecule	r_e(Å)	θ_e
NF_2[a]	1.3528 ± 0.0001	$103°11' \pm 2'$
O_3[b]	1.2715 ± 0.0002	$117°47' \pm 2$
GeF_2[c]	1.7320 ± 0.0009	97.148°
SiF_2[d]	1.5901 ± 0.0001	$100°46' \pm 1'$
SF_2[e]	1.58745 ± 0.00012	$98.048° \pm 0.013°$
S_eO_2[f]	1.608	113.8°

[a] R. D. Brown, F. R. Burden, P. D. Godfrey, and I. R. Gillard, *J. Mol. Spectrosc.*, **52**, 301 (1974).
[b] J. C. Depannemaecker and J. Bellet, *J. Mol. Spectrosc.*, **66**, 106 (1977).
[c] H. Takeo and R. F. Curl, Jr., *J. Mol. Spectrosc.*, **43**, 21 (1972).
[d] H. Shoji, T. Tanaka, and E. Hirota, *J. Mol. Spectrosc.*, **47**, 268 (1973).
[e] Y. Endo, S. Saito, E. Hirota, and T. Chikaraishi, *J. Mol. Spectrosc.*, **77**, 222 (1979).
[f] E. Hirota and Y. Morino, *J. Mol. Spectrosc.*, **34**, 370 (1970).

± 0.004 Å, N–Cl $= 1.976 \pm 0.004$ Å, and ONCl $= 113°15' \pm 11'$. For more complicated polyatomic molecules the experimental problem of obtaining sufficient data for evaluating equilibrium structures is rather difficult. Not only must the α's for each of the vibrations be obtained, but they must also be measured for a sufficient number of isotopic species to enable a structure evaluation. Furthermore, many of the vibrational frequencies are large, such as those corresponding to bond stretching, and therefore, the populations of these vibrational states are low because of the unfavorable Boltzmann factor. This makes the observation of rotational transitions associated with such states extremely difficult. (Vibrational satellites of low-frequency bending vibrations are, however, usually observable.) As a consequence, it is not usually feasible to acquire enough experimental data to obtain the constants. Theoretical evaluation of the α constants required knowledge of the potential constants, and except for simple molecules little information on the anharmonic constants is available. One is hence forced to use the effective ground state rotational constants to extract structural information.

7 EVALUATION OF EFFECTIVE STRUCTURES

Structures calculated directly from the B_v's observed for a particular vibrational state are called effective structures. Here the structural parameters are chosen to reproduce the effective moments of inertia and the corrections described in Section 6 are not applied. For a diatomic molecule, the effective

Table 13.7 Bond Lengths of some Diatomic Molecules[a]

Molecule	r_e	r_0	$\langle r \rangle$
HF	0.9170	0.9257	0.9326
DF	0.9171	0.9234	0.9284
TF	0.9177	0.9230	0.9272
^{16}OH	0.9707	0.9800	0.9873
^{16}OD	0.9700	0.9772	0.9825
^{12}CH	1.1198	1.1303	1.1388
^{12}CD	1.1188	1.1265	1.1327
H^{35}Cl	1.2745	1.2837	1.2904
H^{37}Cl	1.2746	1.2837	1.2904
D^{35}Cl	1.2744	1.2813	1.2858
D^{37}Cl	1.2744	1.2813	1.2858
T^{35}Cl	1.2746	1.2800	1.2853
T^{37}Cl	1.2746	1.2800	1.2853
^{12}C^{16}O	1.1282	1.1309	1.1323
^{13}C^{16}O	1.1282	1.1308	1.1322
^{12}C^{18}O	1.1282	1.1308	1.1322
^{12}C^{32}S	1.5349	1.5377	1.5392
^{13}C^{32}S	1.5349	1.5376	1.5391
^{12}C^{34}S	1.5349	1.5377	1.5392
^{127}I^{35}Cl	2.3209	2.3236	2.3246
^{127}I^{37}Cl	2.3209	2.3235	2.3245

[a]From Laurie and Herschbach [86]. See [86] for references to the individual molecules.

bond distance r_0 for the ground vibrational state is

$$r_0 = \left(\frac{h}{8\pi^2 \mu B_0} \right)^{1/2} = \left(\frac{I_b^0}{\mu} \right)^{1/2} \tag{13.127}$$

From our previous discussion, it is obvious that any bond distance calculated from B_0 (or I_b^0) will differ from the equilibrium internuclear distance r_e, which must be obtained from B_e (or I_b^e). Furthermore, the distance r_0 does not represent a simple average bond distance, but rather is defined in terms of the reciprocal of the square root of the average inverse square bond distance

$$r_0 = \left\langle \frac{1}{r^2} \right\rangle^{-1/2} \neq r_e \tag{13.128}$$

which is not equal to r_e even when the vibrations are harmonic. Also, since α_e depends on the reduced mass via B_e and ω_e, r_0 can be expected to vary with isotopic substitution. In Table 13.7 the bond lengths of some diatomic molecules are compared. As the effective rotational constant is less than the equilibrium rotational constant, the r_0 distance is larger than r_e. The difference is largest for bonds involving light hydrogen atoms where r_0 is about 0.01 Å larger than r_e. Furthermore, the r_0 values are different for different isotopic species, the variation being particularly large when light atoms such as hydrogen are substituted. Here a big effect on amplitude of vibration results, and the vibrational average over the zero-point vibration is significantly different. For substitution of heavier atoms little effect is observed. On the other hand, the r_e values are intensitive to isotopic substitution. Since the r_e refers to the equilibrium position where the nuclei would be at rest, this type of effect does not enter the calculation of r_e from B_e. Variations, like those for r_0, are also found for the average bond distance, $\langle r \rangle$, which is also given in Table 13.7. This measure of the bond distance will be discussed more fully in Section 9.

For polyatomic molecules r_0 is no longer simply the $\langle r^{-2} \rangle^{-1/2}$ because of Coriolis constributions. Also, isotopic data and the assumption that the effective structural parameters are not affected by mass changes are required for evaluation of the structure. Therefore, when more than a minimum number of isotopic species are available, different r_0 structures can be obtained because the molecular vibrations and hence the effective parameters are modified by isotopic substitution. For the linear molecule HCN, four independent moment of inertia equations may be obtained from the data on $H^{12}C^{14}N$, $H^{13}C^{14}N$, $D^{12}C^{14}N$, and $D^{13}C^{14}N$. These equations may be combined in six different ways to give values of the two bond lengths. The results are given in Table 13.8 along with the r_0 structures of OCS. It is apparent that different r_0 structures are obtained when different isotopic data are used in the calculation. Variations of about 0.01 Å are observed. Obviously the neglect of the zero-point vibrational effects by assumption that r_0 parameters are the same for different isotopic species has seriously limited the accuracy of the derived structural parameters. With more complicated molecules, least-squares fitting techniques are convenient and have been employed. The problem is linearized in the usual way

$$I_i = I_i^0 + \sum_j \left(\frac{\partial I_i}{\partial p_j} \right)_0 \Delta p_j \tag{13.129}$$

where I_i corresponds to the ith experimental moment of inertia, I_i^0 represents the ith moment of inertia calculated from the initial assumed structure, and p_j is the jth structural parameter. The derivatives $\partial I_i / \partial p_j \equiv \Delta I_i / \Delta p_j$ are evaluated by calculating the change in I_i^0 with a small change in p_j while keeping all other parameters constant. The foregoing expression forms the basis of a least-squares analysis in which the corrections Δp_j to the initial structure are obtained. The process is repeated until the corrections are small enough. A discussion of some computational problems is available [33]. Ambiguities caused by zero-point vibrations are, however, still present.

A particular problem arises if the normal least-squares equations are ill-conditioned. One must then resort to obtaining extra data, or must add restrictive conditions, for example, hold a particular bond length constant or evaluate only certain linear combinations of the parameters using techniques for the

Table 13.8 Comparison of r_0 Structures of HCN and OCS

Isotopic Species Used		*Bond Length* (Å) C–H	C–N
	HCN[a]		
$H^{12}C^{14}N$, $H^{13}C^{14}N$		1.0674	1.1557
$H^{12}C^{14}N$, $D^{12}C^{14}N$		1.0623	1.1567
$H^{12}C^{14}N$, $D^{13}C^{14}N$		1.0619	1.1568
$H^{13}C^{14}N$, $D^{12}C^{14}N$		1.0625	1.1566
$H^{13}C^{14}N$, $D^{13}C^{14}N$		1.0624	1.1563
$D^{12}C^{14}N$, $D^{13}C^{14}N$		1.0658	1.1555
	Average	1.0637	1.1563
	Range	0.0055	0.0013
	OCS[b]		
		C–O	C–S
$^{16}O^{12}C^{32}S$, $^{16}O^{12}C^{34}S$		1.1647	1.5576
$^{16}O^{12}C^{32}S$, $^{16}O^{13}C^{32}S$		1.1629	1.5591
$^{16}O^{12}C^{34}S$, $^{16}O^{13}C^{34}S$		1.1625	1.5594
$^{16}O^{12}C^{32}S$, $^{18}O^{12}C^{32}S$		1.1552	1.5653
	Average	1.1613	1.5604
	Range	0.0095	0.0077

[a]From J. W. Simmons, W. E. Anderson, and W. Gordy, *Phys. Rev.*, **77**, 77 (1950); **86**, 1055 (1952).
[b]From C. H. Townes, A. N. Holden and F. R. Merritt, *Phys. Rev.*, **74**, 1113 (1948).

solution of singular systems [34, 35, 36]. When the limits of uncertainties of the structural parameters can be reasonably estimated, the technique of Section 11 can be applied. In some cases, the best that can be done with the available data is to assume some structural parameters from similar molecules and evaluate only certain parameters which ideally do not depend significantly on the assumed parameters (this can be tested). Such structural parameters are useful but less accurate.

The effects of zero-point vibrations are readily manifested when an atom is substituted which lies on a principal axis (e.g., the sulfur atom of SO_2). If the atom lies on, say, the x axis, then in the rigid rotor approximation the corresponding change in the moment of inertia ΔI_x must be zero. In practice, however, small changes are observed because of the effects of vibration. These effects present serious problems to the structural determination of molecules with near-axis atoms or atoms near the center of mass. For such atoms, isotopic

Table 13.9 Comparison of r_0 Structures Obtained from Different Pairs of Moments of Inertia

	SO_2[a]		Cl_2O[b]	
Calculated From	r	θ	r	θ
(I_a^0, I_b^0)	1.4322	119°32′	1.6986	111°6′
(I_a^0, I_a^0)[c]	1.4336	119°36′	1.6994	111°8′
(I_b^0, I_c^0)[d]	1.4351	119°8′	1.7031	110°39′
Range	0.0029	0°28′	0.0045	0°29′

[a]From Morino et al. [27].
[b]From Herberich et al. [61].
[c]Equivalent to I_a^0, $I_b^0+\Delta$.
[d]Equivalent to I_b^0, $I_a^0+\Delta$.

substitution produces only small changes in the moments of inertia, and these changes may be masked by the zero-point vibration effects. In the well known example of N_2O, the observed moment of inertia for $^{15}N^{14}N^{16}O$ is slightly larger than for $^{15}N^{15}N^{16}O$; as a consequence, the application of Kraitchman's equation yields an imaginary coordinate for the central nitrogen atom in $^{15}N^{14}N^{16}O$, an impossible result for a rigid molecule.

Even for a simple, bent, triatomic molecule where two moments of inertia of one molecular species are sufficient to give the complete structure, ambiguities result because of the finite inertial defect Δ. If a molecule is planar in its equilibrium configuration, then

$$I_c^e + I_a^e - I_b^e = 0 \tag{13.130}$$

where c is the axis perpendicular to the plane. However, the effective moments defined by (13.116) have somewhat different vibrational effects associated with the different axes and

$$I_c^v - I_a^v - I_b^v = \Delta \tag{13.131}$$

where Δ is called the inertial defect. Although this quantity is very small compared to the moments of inertia, it nevertheless produces ambiguities in structural calculations. In the case of a bent XY_2 molecule, three different structures are obtainable from the three choices of data available, that is, the three pairs: I_a^0, I_b^0; I_a^0, I_c^0; I_b^0, I_c^0. This is illustrated in Table 13.9 for SO_2 and ClO_2. Such effects are particularly large for the light molecules H_2O, D_2O and T_2O illustrated in Table 13.10. It will be observed that the range in the bond distance decreases as the mass of the vibrating atom increases. On the other hand, it is not clear whether the tendency is for the bond distance to decrease or increase with isotopic substitution. The length, for example, as calculated from I_a^0, I_b^0 increases from H_2O to T_2O, whereas it decreases when calculated from the other two combinations. The trend for the average values also included on the

Table 13.10 Effective Structures of H_2O, D_2O, and T_2O[a]

	Effective Structures[b]					
	H_2O		D_2O		T_2O	
From	r	θ	r	θ	r	θ
(I_a^0, I_b^0)	0.9560	105.1	0.9567	104.9	0.9570	104.9
$(I_a^0, I_c^0)^c$	0.9688	106.3	0.9652	105.7	0.9637	105.5
$(I_b^0, I_c^0)^d$	0.9703	102.9	0.9673	103.3	0.9662	103.5
Range	0.0143	3.4	0.0106	2.4	0.0092	2.0
Average	0.9650 Å	104.8°	0.9631 Å	104.6°	0.9623 Å	104.6°

[a]Data from: (H_2O) F. C. De Lucia, P. Helminger, R. L. Cook, and W. Gordy, *Phys. Rev.*, **A5**, 487 (1972); (D_2O) J. Bellet and G. Steenbeckeliers, *Compt. Rend.*, **271B**, 1208 (1970); (T_2O) F. C. De Lucia, P. Helminger, W. Gordy, H. W. Morgan, and P. A. Staats, *Phys. Rev.*, **A8**, 2785 (1973).

[b]Obtained from different pairs of moments of inertia.

[c]Equivalent to I_a^0, $I_b^0+\Delta$.

[d]Equivalent to I_b^0, $I_a^0+\Delta$.

table are, however, consistent with the lower zero-point vibrational energy of the heavier isotopic species.

Inertial Defect

The problem of calculation of inertial defects has been studied in some detail. The vibration–rotation parameters ε_s^α in (13.116) are separable into harmonic and anharmonic contributions

$$\varepsilon_s^\alpha=\varepsilon_s^\alpha\,(\text{har})+\varepsilon_s^\alpha\,(\text{anhar}) \tag{13.132}$$

(The Coriolis contributions are included in the harmonic part.) Darling and Dennison [37] have shown that the contribution of the anharmonic part of ε_s^α to the inertial defect vanishes exactly for planar molecules and hence Δ is given by

$$\Delta_{\text{vib}}=\sum_s\,(v_s+d_s/2)[\varepsilon_s^c\,(\text{har})-\varepsilon_s^a\,(\text{har})-\varepsilon_s^b\,(\text{har})] \tag{13.133}$$

Therefore, Δ_{vib} correct to terms linear in the vibrational quantum numbers does not depend on the anharmonic potential constants and may be calculated from a knowledge of the harmonic force constants, masses, and geometry of the molecule. Oka and Morino [38] have given a general expression for the inertial defect of a planar molecule. In addition to the contributions from vibrational motion, there are small contributions from centrifugal distortion and electron–rotation interaction, that is

$$\Delta=\Delta_{\text{vib}}+\Delta_{\text{cent}}+\Delta_{\text{elec}} \tag{13.134}$$

where

$$\Delta_{\text{vib}}=\frac{h}{\pi^2 c}\sum_s (v_s+\tfrac{1}{2})\sum_{s'}{}^{'}\frac{\omega_{s'}^2}{\omega_s(\omega_s^2-\omega_{s'}^2)}\left[(\zeta_{ss'}^{(a)})^2+(\zeta_{ss'}^{(b)})^2-(\zeta_{ss'}^{(c)})^2\right]+\frac{h}{\pi^2 c}\sum_t \frac{3}{2\omega_t}\left(v_t+\frac{1}{2}\right) \tag{13.135}$$

$$\Delta_{\text{cent}}=-\hbar^4\tau_{abab}\left(\frac{3}{4}\frac{I_c}{C}+\frac{I_a}{2A}+\frac{I_b}{2B}\right) \tag{13.136}$$

and

$$\Delta_{\text{elec}}=-\frac{m_e}{m_p}\left(I_c g_{cc}-I_a g_{aa}-I_b g_{bb}\right) \tag{13.137}$$

In these equations $\zeta_{ss'}^{(a)}$, and so on, are the Coriolis coupling constants; t runs only over the out-of-plane vibrations; m_e and m_p are the mass of the electron and proton; g_{aa}, and so on, are the rotational magnetic moment g tensor elements. For Δ in amu Å^2, the vibrational frequencies ω_s are expressed in cm^{-1}; the centrifugal distortion constant $\hbar^4\tau_{abab}$ and rotational constants A, B, and C are expressed in MHz; moments of inertia I_a, and so on, in amu Å^2; and $h/\pi^2 c=134.901$. The Coriolis coupling constants are expressed by

$$\boldsymbol{\zeta}^{\alpha}=\mathbf{L}^{-1}\mathbf{C}^{\alpha}(\tilde{\mathbf{L}})^{-1} \tag{13.138}$$

The $\boldsymbol{\zeta}^{\alpha}$ ($\alpha=x$, y, z) matrices are skew-symmetric, that is $\tilde{\boldsymbol{\zeta}}^{\alpha}=-\boldsymbol{\zeta}^{\alpha}$. The $\mathbf{C}^{\alpha}$ matrices depend only on the masses and molecular geometry. Their evaluation has been discussed by Meal and Polo [39]. The matrix $\mathbf{L}$ which connects the normal coordinates and the internal coordinates, $\mathbf{R}=\mathbf{LQ}$, can be determined from a normal coordinate analysis. If the Coriolis coupling constant $\zeta_{ss'}^{(\alpha)}$ is to be nonvanishing, the direct product of the symmetry species of the vibrational coordinates Q_s and $Q_{s'}$ must contain the species of the rotation R_α [40]. For example, a nonlinear XY_2 molecule, which belongs to the point group C_{2v}, has two vibrational modes (ω_1, ω_2) of symmetry A_1 and one (ω_3: antisymmetric stretch) of symmetry B_1. The three rotations R_x, R_y, and R_z under the C_{2v} group belong to the species B_2, B_1, and A_2, respectively. Since we have $A_1\times B_1=B_1$, $A_1\times A_1=A_1$ for the direct products, it follows that the possible nonvanishing Coriolis constants are $\zeta_{13}^{(y)}$, $\zeta_{31}^{(y)}$, $\zeta_{23}^{(y)}$, and $\zeta_{32}^{(y)}$, (y is perpendicular to the molecular plane).

The vibrational contribution to Δ is the most important, although it has been found that for molecules with out-of-plane electrons, for example, the π electrons in the C=O bond of H_2CO, the electronic contribution is not negligible. In the case of ozone the electronic contribution is particularly large, amounting to about 10% of the total inertial defect. The calculation of the electronic contribution to the moments of inertia requires a knowledge of the g tensor of the rotational magnetic moment. Experimental values of the g tensor have been given in Chapter XI. Good agreement between calculated and observed inertial defects has been obtained for a number of simple planar molecules [41], as illustrated in Table 13.11. Note that the inertial defect does not show a large

Table 13.11 Inertial Defect for Some Three- and Four-Atom Molecules in the Ground Vibrational State (amu Å^2)

	Calculated				Observed
Molecule[a]	Δ_{vib}	Δ_{cent}	Δ_{elec}	Δ	Δ
H_2O	0.0460	0.0008	0.0000	0.0467	0.0486
D_2O	0.0627	0.0008	0.0000	0.0635	0.0648
H_2S	0.0631	0.0007	0.0001	0.0639	0.0660
H_2Se	0.0736	0.0008	0^b	0.0744	0.0595
D_2Se	0.1034	0.0008	0^b	0.1042	0.1045
O_3	0.1107	0.0011	−0.0104	0.1014	0.1017
$^{18}OO_2$	0.1148	0.0011	−0.0104	0.1055	0.1046
SO_2	0.1376	0.0004	−0.0037	0.1343	0.1348
$^{34}SO_2$	0.1395	0.0004	−0.0037	0.1362	0.1365
H_2CO	0.0597	0.0016	−0.0052	0.0561	0.0574
HDCO	0.0711	0.0014	−0.0052	0.0673	0.0679
D_2CO	0.0810	0.0014	−0.0052	0.0772	0.0777
F_2CO	0.1554	0.0004	-0.0052^b	0.1506	0.1556
Cl_2CO	0.2628	0.0005	-0.0052^b	0.2581	0.251
$ClNO_2$	0.2154	0.0012	-0.0104^b	0.2062	0.2079
ClF_3	0.202	0^b	0^b	0.202	0.125
BrF_3	0.220	0^b	0^b	0.220	0.260

[a]From Oka and Morino [41].
[b]Assumed.

change with isotopic substitution. This can be useful in distinguishing between an assignment of isotopic species as compared to an excited vibrational state where a larger change can occur.

Both in-plane and out-of-plane vibrations contribute to the inertial defect and Δ_{vib} may be divided into in-plane $\Delta(i)$ and out-of-plane $\Delta(o)$ contributions. Usually the in-plane vibrations make a positive contribution, and the out-of-plane vibrations make a negative contribution to the inertial defect [42]. The in-plane contribution is ordinarily larger, and the total inertial defect is positive, as is observed for the tetratomic molecules in Table 13.11. If one or more of the out-of-plane vibrations is low, much of the in-plane contribution can be canceled with a very small inertial defect resulting, as observed in planar ring molecules. A negative value for the inertial defect is possible if $\Delta(o)$ is greater than $\Delta(i)$, (see later discussion). The values for some planar rings are listed in Table 13.12.

A convenient empirical approximation for calculation of the inertial defect of a planar unsymmetrical $XYWZ$ molecule has been given [41] which requires taking the geometric mean of the inertial defects for the corresponding symmetric molecules:

$$\Delta_{XY} = (\Delta_{XX}\Delta_{YY})^{1/2} \tag{13.139}$$

Table 13.12 Inertial Defects for Some Planar Ring Molecules

Molecule	Δ (*amu* Å^2)	*Molecule*	Δ (*amu* Å^2)
Furan[a]	0.046	Fluorobenzene[f]	0.033
Thiophene[b]	0.065	Chlorobenzene[g]	0.053
Pyrrole[c]	0.076	*m*-Dichlorobenzene[h]	0.098
Pyridine[d]	0.039	Benzonitrile[i]	0.084
2-Chloropyridine[e]	0.055	Phosphabenzene[j]	0.052

[a]B. Bak, D. Christensen, W. V. Dixon, L. Hansen-Nygaard, J. Rastrup-Andersen, and M. Schottländer, *J. Mol. Spectrosc.*, **9**, 124 (1962).
[b]B. Bak, D. Christensen, J. Rastrup-Andersen, and E. Tannenbaum, *J. Chem. Phys.*, **25**, 892 (1956).
[c]B. Bak, D. Christensen, L. Hansen, and J. Rastrup-Andersen, *J. Chem. Phys.*, **24**, 720 (1956).
[d]F. Mata, M. J. Quintana, and G. O. Sørensen, *J. Mol. Struct.*, **42**, 1 (1977).
[e]R. T. Walden and R. L. Cook, *J. Mol. Spectrosc.*, **52**, 244 (1974).
[f]L. Hygaard, I. Bojesin, T. Pedersen, and J. Rastrup-Anderson, *J. Mol. Struc.*, **2**, 209 (1968).
[g]G. Roussy and F. Michel, *J. Mol. Struct.*, **30**, 399 (1976).
[h]M. Onda, O. Ohashi, and I. Yamaguchi, *J. Mol. Struct.*, **31**, 203 (1976).
[i]J. Casado, L. Nygaard, and G. O. Sørensen, *J. Mol. Struct.*, **8**, 211 (1971).
[j]R. L. Kuczkowski and A. J. Ashe, III, *J. Mol. Spectrosc.*, **42**, 457 (1972).

Thus the inertial defect for HFCO is just $(\Delta_{H_2CO} \cdot \Delta_{F_2CO})^{1/2}$. Except for simple molecules the evaluation of the inertial defect, which requires a normal coordinate analysis, is rather tedious. Herschbach and Laurie [42] have considered simple approximations for calculating Δ_{vib} which depend on one or two modes of vibration and give results within 10 to 20% of the experimental values. A particularly simple approximation attributes the major part of the inertial defect to the lowest in-plane vibration,

$$\Delta_{vib} \simeq \frac{4K}{\omega} \tag{13.140}$$

with ω the frequency of the vibration in cm^{-1} and $K = h/8\pi^2 = 16.863$ amu Å^2 cm^{-1}.

Comparison of calculated and observed inertial defects provides the most conclusive evidence for the planarity of molecules. It has become apparent that the observation of a small inertial defect cannot of itself be taken as a decisive test of planarity. For example, in formamide (NH_2CHO) the observation of a small positive inertial defect was originally interpreted as evidence of planarity [42]. However, a subsequent detailed study of various isotopic species by Costain and Dowling [44] revealed rather anomalous behavior in the inertial defects. Specifically, isotopic substitution on the NH_2 group resulted in a decrease in the inertial defect, and for some species there was a negative inertial defect. This behavior suggests that possibly formamide is nonplanar, and all the evidence is, in fact, consistent with the $H_2N—C$ group forming a shallow pyramid. Observed inertial defects also provide information dependent on the harmonic force constants, especially Δ's for excited vibrational states, while calculated inertial defects are useful in the assignment of rotational transitions in vibrationally excited states. Furthermore, evaluation of Δ allows calculation of one of the moments of inertia if the other two are known.

For a nonplanar molecule, the planar moment equation, (13.25), yields

$$2P_\gamma^0 = I_\alpha^v + I_\beta^v - I_\gamma^v = 2P_\gamma^e - \Delta_\gamma \tag{13.141}$$

where, as before, Δ_γ represents the inertial defect often termed a pseudoinertial defect. For a nonplanar molecule, the effective planar moment can be large because of the presence of out-of-plane atoms. In a planar molecule $P_\gamma^e(=\sum m_i\gamma_i^2)$ for the out-of-plane direction vanishes, and (13.131) is obtained. Although the anharmonic contribution to the inertial defect vanishes for a planar molecule, Δ_γ for a nonplanar molecule contains both harmonic and anharmonic vibrational contributions

$$\Delta_\gamma = \Delta_\gamma\,(\text{har}) + \Delta_\gamma\,(\text{anhar}) \tag{13.142}$$

For molecules with a plane of symmetry and with only out-of-plane atoms in symmetrically equivalent pairs, certain generalizations can be made [42]. In this case the vibrations may be classified as symmetric or antisymmetric with respect to the plane of symmetry. The only contribution then to Δ_γ (anhar) arises from the symmetric vibrations of the out-of-plane atoms, whereas for Δ_γ (har) contributions arise from the symmetric and antisymmetric vibrations of the out-of-plane atoms and from antisymmetric vibrations of in-plane atoms. Only the symmetric vibrations of the in-plane atoms do not contribute to Δ_γ (har).

In the structural analysis of nonplanar molecules complications arise because of Δ_γ. These will be considered further in the following sections. Since Δ_c is expected to be small compared to P_c, the quantity $I_c^v - I_a^v - I_b^v$, where c is the out-of-plane direction, will be negative, as follows from (13.141). If, however, the out-of-plane coordinates are very small, i.e., the molecule is only slightly nonplanar, then $I_c^v - I_a^v - I_b^v$ may be positive. For amines and amides that are nonplanar, the quantity $I_c^0 - I_a^0 - I_b^0$ is small and negative, and deuteration of NH_2 usually gives a large decrease in this quantity. On the other hand, for planar amines and amides $I_c^0 - I_a^0 - I_b^0$ is small and positive, and changes on deuteration, though negative, are much smaller [45].

Even for a planar molecule, if a very low-frequency out-of-plane vibration is present, the inertial defect can be larger than usual and negative. The inertial defects for some molecules of this type are listed in Table 13.13, along with the variations of the inertial defect in excited torsional vibrational states. For benzoyl fluoride [46], the variation with the torsional state can be expressed in amu Å² by

$$\Delta = 0.286 - 1.222\,(v_t + \tfrac{1}{2}) \tag{13.143}$$

Table 13.13 Inertial Defects for Some Molecules in the Ground and Excited Torsional States

	Inertial Defect Δ *(amu* Å^2*)*						
Molecule	$v=0$	$v=1$	$v=2$	$v=3$	$v=4$	$\omega_t{}^a$	*Ref.*
C_6H_5NO	-0.146	-0.968	-1.801	-2.589	-3.376	82	b
$C_6H_5NO_2$	-0.4811	-1.8626	-3.1858	-4.4704	—	49	c
C_6H_5CHO	-0.128	-0.966	-3.040	-2.747	—	80	d
C_6H_5COF	-0.325	-1.528	-2.765	-3.963	-5.186	56	e
$FC_6H_4CH{=}CH_2$	-0.775	-2.327	-3.510	-4.606	—	43	f

[a] ω_t in cm^{-1}, $\omega_t = -67.45/\delta\Delta$, where $\delta\Delta$ is taken as the change in the inertial defect between the first excited and ground state.
[b] Hanyu et al. [48].
[c] Høg et al. [49].
[d] Kabar et al. [47].
[e] Kabar [46].
[f] Ralowski et al. [50].

where -1.222 amu Å is the average change of Δ between successive torsional states. This variation will not necessarily be linear unless the vibration is harmonic in nature. This is a good approximation for benzoyl fluoride. Extrapolation of the inertial defect to the vibrationless torsional state ($v_t = -\frac{1}{2}$) gives $+0.286$ amu Å^2 for the residual inertial defect. This value has the correct sign and magnitude for a planar molecule in the ground vibrational state with small amplitude vibrations. The -COF group is thus not bent or twisted out of the plane. If the torsional contribution is subtracted, similar results are obtained for benzaldehyde [47] (0.308 amu Å^2), nitrosobenzene [48] (0.264 amu Å^2), nitrobenzene [49] (0.249 amu Å^2) and *p*-fluorostyrene [50] ($\simeq 0.1$ amu Å^2). Other examples may be cited [51, 52, 53].

As pointed out for nitrosobenzene by Hanyu et al. [48], the variation of the inertial defect can supply information about the torsional vibration. The difference in the inertial defect between two successive singly excited vibrational states (v_s) is, from (13.135),

$$\delta\Delta = \Delta_{v_s+1} - \Delta_{v_s} = \left(\frac{h}{\pi^2 c}\right)\left\{\delta_{st}\frac{3}{2\omega_t} + \sum_{s'}{}' \frac{\omega_{s'}^2}{\omega_s(\omega_s^2 - \omega_{s'}^2)}\left[(\zeta_{ss'}^{(a)})^2 + (\zeta_{ss'}^{(b)})^2 - (\zeta_{ss'}^{(c)})^2\right]\right\} \tag{13.144}$$

where $\delta_{st} = 1$ if s refers to one of the out-of-plane vibrations t, otherwise $\delta_{st} = 0$. Two out-of-plane modes do not interact, $\zeta_{ss'}^{(\alpha)} = 0$, $\alpha = a$, b, and c, while for two interacting in-plane modes, $\zeta_{ss'}^{(a)} = \zeta_{ss'}^{(b)} = 0$, $\zeta_{ss'}^{(c)} \neq 0$, and for an in-plane mode interacting with an out-of-plane mode, $\zeta_{ss'}^{(a)} \neq 0$, $\zeta_{ss'}^{(b)} \neq 0$, $\zeta_{ss'}^{(c)} = 0$ [42]. The torsional mode thus interacts with an in-plane mode, and the $\zeta_{ss'}^{(c)}$ term drops out of (13.144); $\delta\Delta$ between two adjacent excited states of the out-of-plane torsional mode is given by

$$\delta\Delta = \Delta_{v_t+1} - \Delta_{v_t} = \left(\frac{h}{\pi^2 c}\right)\left\{\frac{3}{2\omega_t} + \sum_{s'}{}' \frac{\omega_{s'}^2}{\omega_t(\omega_t^2 - \omega_{s'}^2)}\left[(\zeta_{ts'}^{(a)})^2 + (\zeta_{ts'}^{(b)})^2\right]\right\} \qquad (13.145)$$

If the torsional mode is much lower in frequency than any of the other modes, it couples with $\omega_t^2 < \omega_{s'}^2$, (13.145) reduces to [48]

$$\Delta_{v_t+1} - \Delta_{v_t} = -\left(\frac{h}{2\pi^2 c}\right)\frac{1}{\omega_t} = -\frac{67.45 \text{ amu}\cdot\text{Å}^2 \text{ cm}^{-1}}{\omega_t(\text{cm}^{-1})} \qquad (13.146)$$

since $\sum_{s'} (\zeta_{ts'}^{(a)})^2 = \sum_{s'} (\zeta_{ts'}^{(b)})^2 = 1$ for molecules of C_s symmetry [41]. The difference $\delta\Delta$ is thus predicted to be negative and essentially constant as observed. Equation 13.146 may be applied to find the frequency ω_t. This result is consistent with the general approximate relation [42]

$$\Delta_t \simeq -2\langle Q_t^2\rangle = -\frac{4K(v_t + \frac{1}{2})}{\omega_t} \qquad (13.147)$$

for the contribution of a low-frequency out-of-plane vibration, Q_t, to the inertial defect. The ω_t derived from $\delta\Delta$ are also listed in Table 13.13.

Other qualitative information can be obtained from $\delta\Delta$ [46, 48]. Consider, for example, an interaction between an in-plane mode s and other in-plane modes, then from (13.144)

$$\delta\Delta = \Delta_{v_s+1} - \Delta_{v_s} = -\left(\frac{h}{\pi^2 c}\right)\sum_{s'}{}' \frac{\omega_{s'}^2}{\omega_s(\omega_s^2 - \omega_{s'}^2)}(\zeta_{ss'}^{(c)})^2 \qquad (13.148)$$

where s' runs over the modes that are coupled to s. If the interaction is essentially with a single in-plane vibration higher in frequency, $\omega_s^2 < \omega_{s'}^2$, a positive value of $\delta\Delta$ will be obtained with a minimum value of $(\zeta_{ss'}^{(c)})^2/\omega_s$, $(0 \leqslant |\zeta_{ss'}| \leqslant 1)$.

The planarity of a portion of a molecule is often tested by means of (13.141). This may be illustrated with 1,2-dimethylenecyclobutane [54], studied by Avirah et al. [54] to confirm a planar ring structure with two pairs of out-of-plane methylene hydrogens. The planar moment perpendicular to the plane (ab) is

$$2P_c^0 = 2P_c^e - \Delta_c = 2\sum_i m_i c_i^2 - \Delta_c \qquad (13.149)$$

where the sum is over the masses and perpendicular distances of the out-of-plane atoms from the plane. If the only atoms out-of-plane are the methylene hydrogens, P_c^e will depend only on c_i coordinates of these atoms and in fact on the distance between the H atoms r_{HH}. Thus, except for effects of Δ_c, P_c^0 can be used for direct calculation of the distance between symmetrically equivalent out-of-plane atoms. In the present case, the value of the planar moment of inertia, $P_c = \frac{1}{2}(I_a + I_b - I_c)$, calculated from the ground state rotational constants 3.304 amu Å^2, may be compared to the value of 3.25 amu Å^2 calculated for two out-of-plane methylene hydrogens with C–H = 1.10 Å and ∢HCH = 109°28′, $r_{HH} = 2r_H\sin(\text{HCH}/2)$. A much larger value would be expected if any of the carbon atoms were out of the plane. The variation of the psueudoinertial defect with excited ring-puckering states has also been studied, and the results are as expected for an out-of plane vibration.

Since isotopic substitution of an atom in a symmetry plane will leave P_c^e unchanged, this can be used to test for the presence of such a plane [55]. This is illustrated by cyclopentyl chloride $\overline{CH_2CH_2CH_2CH_2CHCl}$ studied by Loyd et al. [56]. The rotational constants were found to be consistent only with an axial conformation. Whether the ring conformation is twisted or bent can be answered by means of deuterium substitution (α position) data. Consider a molecule with C_s symmetry (no twist). For isotopic substitution in the *ac* plane of symmetry, the change in the planar moment $P_b'-P_b$ for the out-of-plane direction would vanish except for small residual zero-point vibrational effects

$$2\Delta P_b^0 = 2\Delta P_b^e - \Delta\Delta_b = -\Delta\Delta_b \tag{13.150}$$

The observed value of $2\Delta P_b^0$ is -0.026 amu Å^2. Calculation of $2\Delta P_b$ for a twist angle of 10° and 20° gives a value of $+0.059$ and 0.210 amu Å^2, respectively. Thus any substantial twist of the ring leads to positive values of ΔP_b, while a small negative value is observed. The data are hence consistent with an axial-bent structure.

The first-order treatment of isotopic effects to be discussed in Section 10 may be used to define an inertial defect virtually free of vibrational contributions. The substitution inertial defect is defined by

$$\Delta_s = \sum m_i \left(\frac{\partial \Delta_0}{\partial m_i}\right) = \frac{1}{2}\Delta_0 \tag{13.151}$$

where Δ_0 is homogeneous of degree $\frac{1}{2}$ in the masses (see Section 10). The defect Δ_m, defined by [57]

$$\Delta_m = 2\Delta_s - \Delta_0 \tag{13.152}$$

should hence vanish to the first-order approximation made in Section 10. The derivative in (13.151) may be approximated by

$$\frac{\partial \Delta_0}{\partial m_i} = (\Delta_0' - \Delta_0)/\Delta m_i \tag{13.153}$$

where Δ_0 is the inertial defect for the parent and Δ_0' that for isotopic substitution at atom *i*. Substitution of all nonequivalent atoms one at a time allows evaluation of Δ_s via (13.151) and (13.153) and hence Δ_m from (13.152). For $^{32}S^{16}O_2$, $\Delta_0 = 0.1351$, and using data for this parent and $^{16}O^{32}S^{18}O$ and $^{16}O^{34}S^{16}O$, one finds $\Delta_s = 0.0635$ and hence $\Delta_m = -0.0081$ amu Å^2. This is considerably smaller than Δ_0. The defect Δ_m may prove useful as a further test of planarity, though considerable isotopic data are required for its evaluation.

8 EVALUATION OF SUBSTITUTION STRUCTURES

Costain [58] has shown that zero-point vibrational effects tend to cancel in the calculation of interatomic distances with Kraitchman's equations from isotopic substitution, as described in Section 3. Hence, if this method is used for location of each atom in the molecule, more consistent structural parameters can be derived from various combinations of isotopic data and the structure obtained is more nearly equivalent to the equilibrium structure than is the effective ground vibrational structure obtained from simultaneous solution of the moment of inertia equations. This is particularly important for bond lengths involving light atoms [59] such as hydrogen, for which zero-point vibrations are large. Structures derived from the use of effective moments of inertia in Kraitchman's equations are usually designated as substitution structures, and the corresponding interatomic distances are indicated by r_s.

An indication of the relation of the r_s structure to the r_0 and r_e structures can be obtained from the consideration of a diatomic molecule. From (13.116) it follows that

$$I_b^0 = I_b^e + \frac{\varepsilon}{2} \tag{13.154}$$

with

$$\varepsilon = -\left(\frac{6h}{8\pi^2\omega_e}\right)(1+a_1) \tag{13.155}$$

If the binomial expansion is used and the higher order terms are neglected, (13.154) yields

$$r_0 = r_e\left(1+\frac{\varepsilon}{4I_b^e}\right) \tag{13.156}$$

From (13.46), the coordinate of one of the atoms from the center of mass is given by

$$z_s^2 = \frac{1}{\mu}\,\Delta I_b^0 \tag{13.157}$$

where $\Delta I_b^0 = (I_b^0)' - I_b^0$ is the change in the ground state moment of inertia due to isotopic substitution. Therefore, if the contributions from zero-point vibration are included, we have from (13.154)

$$\mu z_s^2 = \Delta I_b^e + \frac{\varepsilon}{2}\left(\frac{\varepsilon'}{\varepsilon} - 1\right) \tag{13.158}$$

It may be easily shown that

$$\frac{\varepsilon'}{\varepsilon} = \left[\frac{(I_b^e)'}{I_b^e}\right]^{1/2} = \left(1+\frac{\Delta I_b^e}{I_b^e}\right)^{1/2} \tag{13.159}$$

Expanding this and inserting it in (13.158) gives approximately

$$\begin{aligned} z_s^2 &= \left(\frac{1}{\mu}\right)\Delta I_b^e\left(1+\frac{\varepsilon}{4I_b^e}\right) \\ &= z_e^2\left(1+\frac{\varepsilon}{4I_b^e}\right) \end{aligned} \tag{13.160}$$

Table **13.14** Evaluation of the Substitution Coordinates for the Atoms of $^{16}O^{12}C^{32}S$

Molecule	*Rotational Parameters,* $B_0(MHz)$	I_a^0 *(amu Å²)*[a]
$^{16}O^{12}C^{32}S$	6081.490	83.12617
$^{18}O^{12}C^{32}S$	5704.83	88.61456
$^{16}O^{13}C^{32}S$	6061.886	83.39500
$^{16}O^{12}C^{34}S$	5932.816	85.20928
	Coordinates (Å)[b]	
O	C	S
−1.68197	−0.52185	1.03835
	$\Sigma m_i z_{si} = 0.03305$ amu Å	

[a]Conversion factor 505,531 amu Å² MHz.
[b]atomic masses employed are: $m(^{12}C)=12.003804$, $m(^{13}C)=13.007473$, $m(^{16}O)=16.000000$, $m(^{18}O)=18.004874$, $m(^{32}S)=31.982236$, $m(^{34}S)=33.97860$.

and finally

$$z_s = z_e\left(1+\frac{\varepsilon}{8I_b^e}\right) \tag{13.161}$$

where the atom is taken to have a positive z coordinate. Determining the remaining coordinate by a second substitution, we find for the bond distance

$$r_s = z_{s_1} - z_{s_2} = \frac{r_e}{2} + \frac{r_e}{2}\left(1+\frac{\varepsilon}{4I_b^e}\right) \tag{13.162}$$

which, from (13.156), yields approximately

$$r_s = \frac{r_e + r_0}{2} \tag{13.163}$$

The r_s distance is between the r_e and r_0 distances and hence closer to the equilibrium value. Experimentally, it is found for polyatomic molecules that the r_s values are usually less than the r_0 values and greater than r_e values with the above relation approximately correct. Exceptions, however, have been observed. In the case of HCN the substitution bond length of C—H has been found to be less than the equilibrium value [60]. Also, r_0 values smaller than r_s values have been obtained; compare the results for the linear molecules OCS and HCN given in Tables 13.8 and 13.15. Similar results ($r_0 < r_s$) have been found in other molecules, for example, Cl_2O [61].

As an illustration of the calculation of substitution structures, we may consider the linear molecule carbon oxysulfide (OCS). If $^{16}O^{12}C^{32}S$ is chosen as the parent molecule, the coordinates of each atom can be determined from data on the isotopic species $^{18}O^{12}C^{32}S$, $^{16}O^{13}C^{32}S$, and $^{16}O^{12}C^{34}S$. The pertinent data are summarized in Table 13.14. With the aid of (13.146) the sulfur coordinate may be evaluated from the increase in I_b^0 resulting from isotopic substitution on the sulfur atom, that is,

$$z_s(S)=\left(\frac{\Delta I_b^0}{\mu}\right)^{1/2}=\left(\frac{2.08311}{1.932064}\right)^{1/2}=1.03835 \text{ Å}$$

The remaining coordinates, calculated in similar fashion, are summarized in Table 13.14. These coordinates give the first substitution structure of OCS listed in Table 13.15. The other r_s structures given in Table 13.15 have been obtained by use of different choices for the parent species along with the appropriate isotopic species required for calculation of the coordinates in each case. It is evident that the r_s parameters are essentially independent of the isotopic species used in the calculation of the structure. Comparison of the r_s structures with the r_0 structures of Table 13.8 shows that the range in the r_0 bond lengths of OCS

Table 13.15 Comparison of r_s Structures of OCS and HCN[a]

Parent Molecule		*Bond Distance* (Å)	
		C—O	C—S
	OCS		
$^{16}O^{12}C^{32}S$		1.16012	1.56020
$^{18}O^{12}C^{32}S$		1.15979	1.56063
$^{16}O^{13}C^{32}S$		1.16017	1.56008
$^{16}O^{12}C^{34}S$		1.16075	1.55963
	Average	1.16021	1.56014
	Range	0.00096	0.00100
	HCN[b]		
		C—H	C—N
$H^{12}C^{14}N$		1.06317	1.15538
$H^{13}C^{14}N$		1.06315	1.15538
$D^{12}C^{14}N$		1.06320	1.15533
$D^{13}C^{14}N$		1.06317	1.15532
	Average	1.06317	1.15535
	Range	0.00005	0.00006

[a]From Costain [58].

[b]The coordinate of the nitrogen atom was determined from the first moment equation, since no ^{15}N isotopic data were available.

and HCN are, respectively, 10 times and a 100 times the range in the r_s parameters. Also, the r_s parameters are less than the average r_0 parameters. Likewise, the r_s bond length for SO_2 (Table 13.5) is less than r_0 and much closer to the equilibrium value. The r_s structures of several other molecules have also been considered by Costain [58] with similar results.

If the substitution coordinates of each atom in the molecule are known, then the substitution moment of inertia I_α^s can be calculated. Since in the substitution procedure there is no longer an attempt to fit the individual moments of inertia, but rather the difference in moments of isotopic species, I_α^s will not agree with I_a^0. In particular, for $^{16}O^{12}C^{32}S$ one finds [58] that

$$I_b^e = 82.882 \text{ amu Å}^2$$
$$I_b^s = 83.008 \text{ amu Å}^2$$
$$I_a^0 = 83.126 \text{ amu Å}^2$$

with the substitution moment almost the average of the equilibrium and effective moments of inertia (See Section 10). For most molecules I_a^s is found to be less than I_a^0.

As a further example we consider the substitution structure calculation of cyclopropenone (C_{2V}) investigated by Benson et al. [62]. The moments of inertia for various isotopic forms of cyclopropenone are listed in Table 13.16. The two ^{13}C species were assigned in their natural abundance $^{13}C_1$ (1%)

Table 13.16 Moments of Inertia for Isotopic Species of Cyclopropenone and Coordinates for the Atoms in Cyclopropenone[a]

	Moments of Inertia[b]				
	Normal Species	$^{13}C_1$	$^{13}C_2$	*Dideuterio*	^{18}O
I_a	15.7704	15.7687	16.1796	20.5950	15.7700
I_b	64.5853	64.6550	65.5537	69.6606	68.8073
I_c	80.4648	80.5350	81.8727	90.3790	84.6881
$I_c - I_a - I_b$	0.1091	0.1113	0.1394	0.1234	0.1108

Center-of-Mass Principal Axis Coordinates

Atom	*a*	*b*	*c*
C_1	0.2660	0	0
C_2	−0.9873	−0.6509	0
C_3	−0.9873	+0.6509	0
O	1.4781	0	0
H_1	−1.6173	−1.5483	0
H_2	−1.6173	+1.5483	0

[a]Moments of inertia in amu Å^2, coordinates in Å. Note $\Sigma m_i a_i = -0.121$ amu Å^2.
[b]Benson et al. [62].

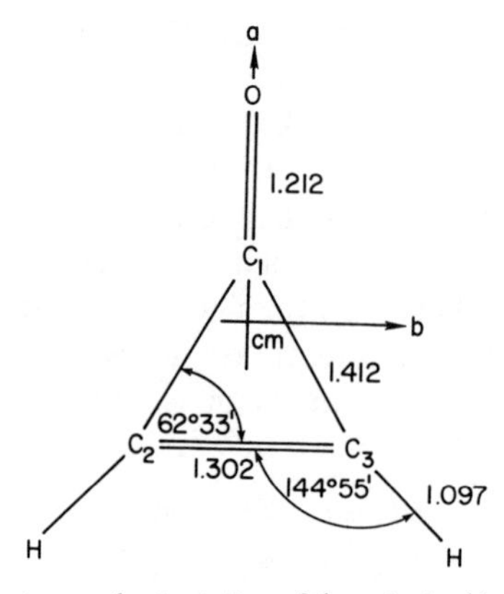

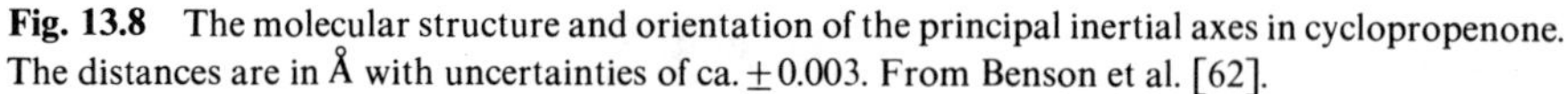

Fig. 13.8 The molecular structure and orientation of the principal inertial axes in cyclopropenone. The distances are in Å with uncertainties of ca. ± 0.003. From Benson et al. [62].

and $^{13}C_2$ (2%), whereas ^{18}O and D_2 species were prepared with essentially 100% purity. The small positive inertial defects also included in the table confirm the planarity of the molecule. All of the out-of-plane c coordinates of the atoms hence vanish. The structure and orientation of the principal inertial axes in cyclopropenone are shown in Fig. 13.8. The coordinates for the carbon and oxygen atoms relative to the common isotopic species may be obtained from Eqs. (13.64) and (13.65)

$$|a| = \left[\frac{\Delta I_b}{\mu}\left(1 + \frac{\Delta I_a}{I_a - I_b}\right)\right]^{1/2} \tag{13.164}$$

$$|b| = \left[\frac{\Delta I_a}{\mu}\left(1 + \frac{\Delta I_b}{I_b - I_a}\right)\right]^{1/2} \tag{13.165}$$

where $\mu = M\Delta m/(M + \Delta m)$ and I_a, I_b are the moments of inertia for the parent molecule of mass M ($=54.01056$) and $\Delta I_a = I_a' - I_a$, $\Delta I_b = I_b' - I_b$ are the differences between the singly isotopically substituted molecule of mass $M + \Delta m$ and the parent. The coordinates of C_2 (or C_3) are obtained by use of the $^{13}C_2$ isotopic species ($\Delta m = 1.00335$, see Appendix E for masses)

$$|a| = \left[\frac{0.9684}{0.98505}\left(1 - \frac{0.4092}{48.8149}\right)\right]^{1/2} = 0.9873$$

$$|b| = \left[\frac{0.4092}{0.98505}\left(1 + \frac{0.9684}{48.8149}\right]^{1/2} = 0.6509$$

The assignment of signs to the coordinates is straightforward in this molecule. The coordinates for C_1 and O atoms are obtained from the $^{13}C_1$ and ^{18}O isotopic species, respectively. Since by symmetry these atoms lie on the symmetry axis, the b coordinate for these atoms vanishes. For a rigid molecule $I_a = I_a'$, but because of zero-point vibrational effects, there are slight differences in the moments of inertia. These vibrational effects would give meaningless b coordinates by straightforward application of (13.164) and (13.165). We therefore set $I_a' = I_a$ to obtain O

$$|a| = \left[\frac{1}{\mu}\,\Delta I_b\right]^{1/2} = \left[\frac{4.2220}{1.93253}\right]^{1/2} = 1.4781$$

$$|b| = 0$$

Likewise for coordinates of C_1

$$|a| = \left[\frac{0.0697}{0.98505}\right]^{1/2} = 0.2660$$

$$|b| = 0$$

It should be pointed out that if the question was whether an atom (e.g., oxygen here) lies on the C_2^a axis of symmetry or not, the observed changes in I with isotopic substitution could be used as evidence. Since the changes must satisfy the relations $\Delta I_a = 0$, $\Delta I_b = \Delta I_c$ in the rigid rotor approximation. The coordinates of the protons (C_2–H_1, C_3–H_2) may be obtained from the dideuterio isotopic species if (13.89) and (13.90) are employed for disubstitution of equivalent atoms in a plane.

$$|a| = \left\{\frac{M + 2\Delta m}{2\Delta m M}\,\Delta I_b\right\}^{1/2} = [0.515396(5.0753)]^{1/2} = 1.6173$$

$$|b| = \left(\frac{\Delta I_a}{2\Delta m}\right)^{1/2} = \left[\frac{4.8246}{2.01255}\right]^{1/2} = 1.5483$$

The final coordinates are included in Table 13.16. Because of zero-point vibrational effects, the coordinates are estimated [62] to be not better than 0.002 Å. The structure derived from the coordinates is shown in Fig. 13.8.

The r_s structures, of course, require that every nonequivalent atom of the molecule be substituted. Often this is not feasible either because of the arduous task involved in the synthesis of the appropriate isotopic species or because of the lack of stable isotopes, such as for the fluorine atom. If all necessary substitutions but one have been made, then the location of the remaining atom might be determined from the observed moments of inertia. Since $I_\alpha^0 \neq I_\alpha^s$, a true substitution coordinate is not obtained. The difference $(I_b^0 - I_b^s)$ must be absorbed by the coordinate of the unsubstituted atom, and the error $\delta z = (I_b^0 - I_b^s)/2mz_s$ is inversely proportional to the mass and coordinate of the atom. The error is greatest if a light atom such as hydrogen is located this way. An alternate procedure is use of the first-moment equations to determine the coordinates, for example,

$$z = -\frac{\sum' m_i z_{si}}{m} \tag{13.166}$$

where z and m are the coordinate and mass of the unsubstituted atom and the sum is over the remaining atoms whose substitution coordinates, z_{si}, are known. The first moment equations are exact for the equilibrium coordinates but hold only approximately for substitution coordinates. This is evident from Table 13.14, where the sum $\sum m_i z_{si}$ for OCS may be compared with the value of zero required for equilibrium coordinates. The error in the coordinate $\delta z = \sum m_i z_{si}/m$, obtained by use of the first-moment equation, is, as before, largest for light atoms, and its use should be restricted to the location of heavy atoms. Light atoms should be located by means of Kraitchman's equations, where [59] $\delta z \sim \delta(\Delta I)/z\Delta m$, and the error is independent of the mass of the substituted atom but increases as z decreases. Since the error, however, is independent of the coordinate value, the first-moment relation is particularly useful for locating the position of atoms close to the center of mass or a principal axis where Kraitchman's equations are unsatisfactory. For example, in the microwave study of benzonitrile by Bak et al. [63], even though a sufficient number of isotopic species were studied (three mono-deuterated species, four mono-^{13}C (ring) species, $C_6H_5{}^{13}CN$, and $C_6H_5C^{15}N$), all the coordinates could not be obtained from Kraitchman's equations. In particular, the equivalent $C(2)$ and $C(6)$ ring carbons (see Fig. 13.9) lie too close to the b axis; their a coordinate was therefore obtained from the first moment equation, $\sum m_i a_i = 0$. In general, when the coordinate of an atom is as small as 0.15 Å. Kraitchman's equations cannot be used for accurate location of the atom [58, 59]. If sufficient data are lacking, the product of inertia relations $\sum m_i x_i y_i = 0$, and so on, can also be called upon for evaluation of coordinates, but the use of moments of inertia directly should be avoided if possible.

It frequently happens that more than one atom is close to a principal axis and the first-moment equations are not sufficient to yield a complete structure. In such cases a double-substitution technique proposed by Pierce [64] can be used. Since second differences of moments of inertia are employed, the vibra-

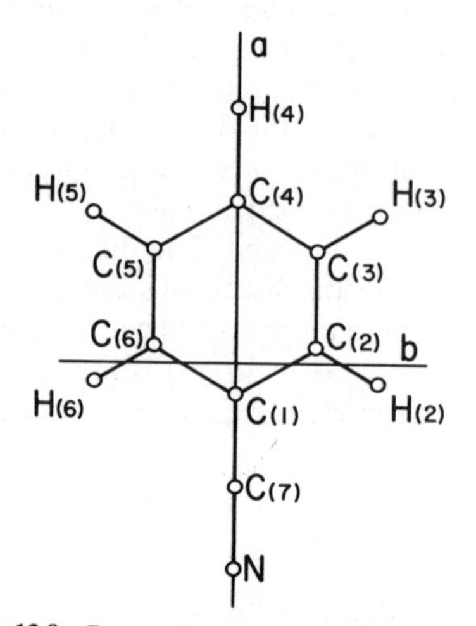

Fig. 13.9 Principal-axis system of benzonitrile.

tional effects tend to be reduced to higher order, and the coordinates calculated in this way are more reliable than those calculated from first differences. As an example of this procedure, consider the location of the central nitrogen atom of nitrous oxide. The method requires four isotopic species, with substitution in two different frameworks, for example, $^{14}N^{14}N^{16}O$, $^{14}N^{15}N^{16}O$ and $^{15}N^{14}N^{16}O$, $^{15}N^{15}N^{16}O$. Let

$$\Delta I_x(\text{NNO}) = I_x(^{14}\text{N}^{15}\text{N}^{16}\text{O}) - I_x(^{14}\text{N}^{14}\text{N}^{16}\text{O}) \tag{13.167}$$

and

$$\Delta I_x(^{15}\text{NNO}) = I_x(^{15}\text{N}^{15}\text{N}^{16}\text{O}) - I_x(^{15}\text{N}^{14}\text{N}^{16}\text{O}) \tag{13.168}$$

From Kraitchman's equation we can write

$$\Delta I_x(\text{NNO}) = \mu z_C^2, \qquad \mu = \frac{M\Delta m}{M+\Delta m} \tag{13.169}$$

where M is the mass of $^{14}N^{14}N^{16}O$, $\Delta m = m(^{15}N) - m(^{14}N)$, and z_C is the central nitrogen coordinate in the principal-axis system of $^{14}N^{14}N^{16}O$. Likewise, we have

$$\Delta I_x(^{15}\text{NNO}) = \mu'(z_C')^2, \qquad \mu' = \frac{M'\Delta m}{M'+\Delta m} \tag{13.170}$$

with M' the mass of $^{15}N^{14}N^{16}O$, Δm as defined previously, and z_C' the coordinate in the $^{15}N^{14}N^{16}O$ framework. When z_C (or z_C') is small, contributions from zero-point vibrations, ignored in the above first-difference equations, may be of comparable magnitude to μz_C^2 or $\mu'(z_C')^2$; hence coordinates calculated from the ΔI^0's can be in serious error. Evidence indicates that if the difference between (13.170) and (13.169) is taken, the vibrational effects are sufficiently reduced to allow calculations of a more reliable coordinate. Taking the difference, we find that

$$\Delta\Delta I_x = \Delta I_x(^{15}\text{NNO}) - \Delta I_x(\text{NNO}) = \mu'(z_C')^2 - \mu z_C^2 \tag{13.171}$$

Now z_C and z_C' are related by the known transformation between the $^{14}N^{14}N^{16}O$ and $^{15}N^{14}N^{16}O$ frameworks. In particular, since in this simple case the orientations of the principal axes are the same, we have from (13.17)

$$z_C' = z_C - \frac{\Delta m z_E}{M+\Delta m} \tag{13.172}$$

where z_E is the coordinate of the end nitrogen in the $^{14}N^{14}N^{16}O$ system. Note that the last term is simply the shift in the center of mass upon substitution of the end nitrogen in $^{14}N^{14}N^{16}O$. Thus the coordinate of the central nitrogen (z_C) in the parent $^{14}N^{14}N^{16}O$ molecule in terms of the second differences of the I_x's is given by the expression

$$\frac{\Delta\Delta I_x}{\mu'} = \left(1 - \frac{\mu}{\mu'}\right) z_C^2 - \left(\frac{2\Delta m z_E}{M+\Delta m}\right) z_C + \left(\frac{\Delta m z_E}{M+\Delta m}\right)^2 \tag{13.173}$$

Table 13.17 Calculation of the Central Nitrogen Coordinate in $^{14}N^{14}N^{16}O$[a]

Method[b]		
I	*II*	*III*
0.0730 Å	0.0725 Å	0.0535 Å
0.0710	0.0725	0.0535

[a]From Pierce [64].
[b]See text for the method of calculation.

Since the end nitrogen atom is far removed from the center of mass, its coordinate z_E may be obtained in usual fashion from first differences in the moments of inertia. In Table 13.17 are given the coordinates for the central nitrogen calculated in three different ways: (I) Use of (13.173) with z_E corresponding to the end nitrogen in one case and the end oxygen in the other; (II) Kraitchman's equation used to determine the oxygen and the end nitrogen coordinate and $\sum m_i z_i = 0$ used to evaluate z_C; (III) Kraitchman's equation in which $^{14}N^{14}N^{16}O$ and $^{14}N^{15}N^{16}O$ species are used. While the first two methods of calculation give coordinates which differ by 0.0005 and 0.0015 Å, Kraitchman's equation gives a coordinate which is in error by 0.02 Å. A more recent application of this double substitution procedure to ClB=S has been reported [65]. A number of isotopic species were studied, and a comparison made of the small boron coordinate as calculated with Methods I through III. As the magnitude of the boron coordinate increases from molecule to molecule, the discrepancy between the methods of calculation decrease as expected.

Equation 13.173 is also applicable to symmetric tops if the atoms substituted lie on the symmetry axis. Applications of this procedure to more complicated cases are discussed elsewhere [66]. In the calculations, very accurate moments of inertia are required, since second differences are employed. The principal drawback to this procedure is the great amount of isotopic data that is necessary, that is, moments of inertia of doubly isotopically substituted molecules.

When sufficient isotopic data are available, more than one isotopic form can be used as parent, and the consistency of the r_s values obtained can serve as an estimate of the uncertainty in these parameters. The use of a different parent can prove useful when a small coordinate of an atom in one parent is significantly large when referred to another isotopic form as parent. An appropriate substitution sequence [11], with the aid of the transformation of (13.81), can also be employed to locate a critically small coordinate where Kraitchman's equations are inapplicable.

When there is more than one poorly determined substitution coordinate and the center-of-mass condition and the product-of-inertia relation are not sufficient, the substitution moments of inertia I_α^s can be employed, as was done by Yamada et al. [67] for HNCS, for which the three substitution b coordinates of the heavy atoms are too small to be reliable. This provides additional relations that the coordinates must satisfy, for example, $I_a^s = \sum m_i[b_s(i)]^2$ for HNCS. It may be shown that I_α^s is physically significant even when the substitution coordinates are very small or imaginary. As discussed in Section 10, to a first-order approximation, the substitution moment of inertia is the average of the equilibrium and effective ground state moment of inertia.

Least-squares adjustment using differences in moments of inertia, ΔI_α^0 with isotopic substitution have been discussed [33, 68, 69] to obtain r_s structures. Less stringent requirements are placed on the type of isotopic data employed. Typke [70] describes a procedure that evaluates directly the Cartesian coordinates rather than the bond lengths and angles. Consider the single substitution of a particular atom in a general asymmetric top. The planar moments of inertia for the parent and isotopic form are related by (13.70) where xyz are the coordinates of the atom relative to the parent inertial system. The essential idea of the procedure is to identify the diagonal elements P_x, P_y, P_z with the experimental values of the parent molecule and the eigenvalues P'_x, P'_y, P'_z of (13.70) with the experimental values for the isotopic molecule. The eigenvalues of (13.70) are hence functions of only the observed planar moments P_x, P_y, P_z and the coordinates xyz of the substituted atom. This provides the basis to obtain the coordinates xyz by a least-squares treatment from

$$P'_x(\text{obs}) = P'_x(\text{calc}) + \left(\frac{\partial P'_x}{\partial x}\right)_0 \Delta x + \left(\frac{\partial P'_x}{\partial y}\right)_0 \Delta y + \left(\frac{\partial P'_x}{\partial z}\right)_0 \Delta z \qquad (13.174)$$

with a similar equation for P'_y and P'_z. From an initial set of values of (x_0, y_0, z_0) for the coordinates of the substituted atom in the parent molecule and the parent planar moments P_x, P_y, P_z, the eigenvalues P'_x, P'_y, P'_z are obtained from (13.70) and employed in the foregoing equations. The coordinates of the unsubstituted atoms do not enter the calculations. The derivatives may be calculated numerically (changes from the initial values of $\Delta x = 0.001$ Å, etc., are usually sufficient). The foregoing equations are then solved for the corrections Δx, Δy, Δz, and the procedure is repeated with the new values $(x_0 + \Delta x)$, and so on. The procedure is iterated until the process converges. The signs of the coordinates are not, as usual, determined but must be fixed from additional considerations. Additional data involving these same coordinates can be readily added to (13.174). One may then proceed to the next atom in the molecule. The center of mass and the inertial products relation can be incorporated into the analysis if desired. Of course, if the atom substituted is on a symmetry plane, the form of (13.70) will be simplified. For any multiple isotopic substitution, the form of (13.70) must be modified appropriately by use of (13.27) and (13.28); the eigenvalues P'_x, and so on, will be functions of the coordinates of all substituted atoms. A simultaneous fit of all the coordinates is obtained with the P''s of an appropriate number of isotopic species. Small coordinates are not accurately determined as in the Kraitchman method and must be fixed to zero or determined from the center-of-mass condition. Such a small coordinate leads to oscillating behavior of the iteration steps without convergence [70]. The

Table 13.18 Substitutional Structures of Water[a]

Species	$(I_a: I_b)$	$(I_a: I_c)$	$(I_b: I_c)$
$H_2{}^{16}O: H_2{}^{18}O: HT^{16}O$	0.9590	0.9607	0.9641
	104.43	106.72	99.46
$H_2{}^{16}O: H_2{}^{18}O: HD^{16}O$	0.9585	0.9613	0.9640
	104.58	106.43	100.74
$D_2{}^{16}O: D_2{}^{18}O: HD^{16}O$	0.9567	0.9622	0.9620
	104.65	104.49	105.30
$D_2{}^{16}O: C_2{}^{18}O: DT^{16}O$	0.9580	0.9604	0.9630
	104.65	105.38	103.00
$H_2{}^{16}O: H_2{}^{18}O: D_2{}^{16}O$	0.9575	0.9618	0.9628
	104.72	105.11	103.90
$H_2{}^{16}O: H_2{}^{18}O: T_2{}^{16}O$	0.9576	0.9612	0.9626
	104.70	105.03	103.93
$D_2{}^{16}O: D_2{}^{18}O: H_2{}^{16}O$	0.9575	0.9618	0.9629
	104.72	105.11	103.89
$D_2{}^{16}O: D_2{}^{18}O: T_2{}^{16}O$	0.9576	0.9606	0.9624
	104.68	104.95	103.95
Average	0.9578 Å	0.9613 Å	0.9630 Å
	104.64°	105.40°	103.02°
Range	0.0023	0.0018	0.0021
	0.29	2.23	5.84

[a]After Helminger et al. [71].

procedure provides equivalent results to those from Kraitchman's equations when the appropriate isotopes are employed.

As with r_0 structures, the presence of a finite inertial defect for planar molecules introduces some uncertainty in the r_s structures. Since differences of moments of inertia are employed in (13.64) and (13.65), the coordinates will be affected by the difference in the inertial defect, $\delta\Delta$, for the isotopic species. However, $\delta\Delta$ is smaller than Δ (see Table 13.11), and the r_s structures obtained from different pairs of ΔI_α^0 tend to be less sensitive to the choice of data than do r_0 structures. This is illustrated in Table 13.18 for water, which has large zero-point vibrational effects. Note that the range in the bond length for different combinations of moments of inertia is smaller than that for the r_0 values of Table 13.10. The bond angles, however, show some rather large inconsistencies when the I_c values are used in the calculations, primarily because of Coriolis contributions to I_c^0 which vanish for I_a^0 and I_b^0. When corrections for these effects are made, much more consistent results are obtained [71]. For structures where double-substitution species were used to locate the hydrogen atom, larger changes in ΔI occur, and the greater consistency in values indicates less severe zero-point vibrational effects. It may be noted that the r_s structural parameters calculated from I_a, I_b are quite close to the equilibrium values.

A quantity of particular importance in the study of internal rotation effects

Table 13.19 Calculation of the H—H Distance[a]

Molecule	r_0(HH)	r_0(DD)$-r_0$(HH)	r_s(HH)	Δ_c
CH_3CHO, CD_3CHO	1.761	0.005	1.776	0.058
CH_3COOH, CD_3COOH	1.733	0.021	1.774	0.146
CH_3COF, CD_3COF	1.724	0.023	1.770	0.160
CH_3COCl, CD_3COCl	1.712	0.028	1.768	0.200
CH_3COCN, CD_3COCN	1.703	0.028	1.766	0.243

[a]From Laurie [20].

is the methyl group structure. The major source of error in the derived barrier heights usually arises from the uncertainty in the methyl group moment of inertia, I_α. Consider a molecule, such as CH_3COX, which possesses a plane of symmetry with a pair of hydrogens as the only out-of-plane atoms. The distance between the H atoms is related to the effective moments from (13.147) as follows

$$I_\alpha^0 + I_b^0 - I_c^0 = m_H r_e^2 - \Delta_c = I_\alpha^e - \Delta_c \tag{13.175}$$

Here r_e is the distance between the two out-of-plane H atoms. If the effects of vibrations (Δ_c) are ignored and the effective distance is defined as

$$m_H r_0^2 = I_a^0 + I_b^0 - I_c^0 \tag{13.176}$$

one finds [20] abnormally large variations in r_0 obtained from the normal and deuterated species. Also r_0 is found to decrease as X becomes larger, as demonstrated in Table 13.19. These variations are due to the neglect of Δ_c. Alternately, the changes in moments of inertia for the normal and deuterated species may be used, that is

$$(m_D - m_H) r_s^2 = \Delta I_a^0 + \Delta I_b^0 - \Delta I_c^0 \tag{13.177}$$

where $\Delta I_a^0 = I_a^0(CD_3) - I_a^0(CH_3)$, and so on. From (13.175) it follows that the vibrational effects are effectively canceled if Δ_c is essentially the same for the CH_3 and CD_3 isotopic species. As is evident from Table 13.19, the r_s values show a much smaller variation of only 0.01 Å as compared with the r_0 values which vary by 0.06 Å. Also, ignoring Δ_c in (13.176) has given an r_0 coordinate that is considerably less than the r_s coordinate where only $\delta\Delta_c$ is ignored. The major contribution to Δ_c arises from low-frequency motions of the heavy in-plane atoms [42, 20], and hence the effect of isotopic substitution of the out-of-plane hydrogen atoms should be small. Values of Δ_c calculated with the assumption that $\Delta_c(CH_3) = \Delta_c(CD_3)$ are also listed in Table 13.19. The Δ_c's are seen to be similar in magnitude to those of planar molecules and to increase as X becomes larger and its associated vibrational frequency decreases. A different treatment of (13.175) is considered in the next section.

At present the r_s structure is the best approximation to the equilibrium structure. Since 1958 a considerable effort has been made to determine structures by the substitution method. Particularly, accurate parameters can be obtained for atoms far removed from the center of mass or a principal axis. Also, accurate partial structures are obtainable when sufficient data are not available for a complete structural determination. Except for unfavorable situations, such as locating near-axis atoms and problems introduced by inertial defects, self-consistency of substitution coordinates to 0.001 Å have been attained, and it may be expected that the r_s bond lengths will be somewhere within 0.002 to 0.005 Å of the equilibrium values. So far, however, it has not been possible to specify with certainty the closeness of r_s structures to r_e structures.

The estimation of the actual reliability of an r_s (or r_0) structure is complicated by the uncertain contribution of the vibrational effects. For physically well-defined r_e and $\langle r \rangle$ structures, the uncertainty in the parameters can be reasonably well assessed in terms of the experimental uncertainties of the quantities that go into the calculation. Because differences in the moments of inertia between the parent and an isotopically substituted molecules are employed in the r_s substitution technique, differences in the zero-point vibrational contributions contribute to the calculations. The vibrational effects cancel to a large degree, but some residual contributions are absorbed by the substitution coordinates. Experimental uncertainties in r_s structures can often be much less important than the residual zero-point vibrational effects. Rudolph [72] has shown that for evaluation of the substitution coordinate of an atom in a principal plane or on a principal axis, the relations (13.77), (13.78), and (13.80) are preferred since they provide coordinates that effectively represent the mean of the values obtained from various other forms of Kraitchmann's equations and take cognizance of the symmetry properties of the substitution.

For ethylene ozonide and furan, nearly symmetric tops (oblate), Kuckowski et al. [73] and Mata et al. [74] have found that the use of Kraitchman's equations for single substitutions can result in amplification of the residual vibrational effects for certain isotopic substitution. If $I_a(I_x) \simeq I_b(I_y)$, certain denominators in (13.71) and (13.72) will be small; if the corresponding numerator is small, the vibrational contributions to the numerator can be significantly amplified in the coordinate calculation. The use of double-substitution species can be useful in minimizing such troublesome vibrational effects [13, 73–76]

A discussion of the estimation of uncertainties, as well as a review of various computational procedures has been given by Schwendeman [33]. Consider the general Kraitchman equation for the a coordinate of the substituted atom relative to the parent

$$a_s = \left\{ \left(\frac{\Delta P_a}{\mu} \right) \left(1 + \frac{\Delta P_b}{P_b - P_a} \right) \left(1 + \frac{\Delta P_c}{P_c - P_a} \right) \right\}^{1/2} \tag{13.178}$$

The uncertainty δa_s and the coordinate a_s are usually dominated by the first factor in the foregoing expression and

$$\delta a_s = \frac{\delta \Delta P_a}{2 \mu a_s} \tag{13.179}$$

Similar relations hold for δb_s and δc_s. The experimental uncertainty can be readily computed from the experimental uncertainty in the planar moments. In the evaluation of rotational constants from the rotational spectrum it is usually good practice to carry out the analysis for the common and the isotopic species in the same way. This will minimize possible experimental errors.

Of additional interest is the estimation of the departure of the coordinates a_s from the equilibrium values a_e. This requires an estimate of the vibrational contribution to $\delta\Delta P_a$. The effective planar moments can be written in terms of the equilibrium planar moments as follows

$$2P_a^e=(I_b^0+I_c^0-I_a^0)+\Delta_a=2P_a^0+\Delta_a \tag{13.180}$$

This gives for the difference

$$\Delta P_a^0=\Delta P_a^e-\frac{\Delta_a'-\Delta_a}{2} \tag{13.181}$$

If the changes in the pseudoinertial defect could be obtained, the equilibrium coordinates could be calculated. The vibrational contribution to the uncertainty is taken as

$$\delta a_s=\pm\left|\frac{\Delta_a'-\Delta_a}{4\mu a_s}\right| \tag{13.182}$$

To obtain Costain's rule for the uncertainties from the equilibrium values one assumes a typical value $\Delta_a'-\Delta_a\simeq 0.006$ amu Å^2 and sets $\mu\simeq 1$, which yields

$$\delta a_s=\pm\frac{0.0015}{|a_s|} \tag{13.183}$$

with a_s in Å. This uncertainty estimate may be increased if the experimental uncertainty in ΔP_a is significant as compared to the estimate of 0.006 amu Å^2. For hydrogen substitution, the uncertainty may be increased [77] by a factor of 2. The estimated uncertainties in the coordinates may be transformed into the uncertainties in the bond distances (or angles). For the bond distance between atoms i and j, the uncertainty estimate is

$$\delta r_{ij}^2=\left(\frac{\partial r_{ij}}{\partial a_i}\right)^2\delta a_i^2+\left(\frac{\partial r_{ij}}{\partial a_j}\right)^2\delta a_j^2+\left(\frac{\partial r_{ij}}{\partial b_i}\right)^2\delta b_i^2+\cdots+\left(\frac{\partial r_{ij}}{\partial c_j}\right)^2\delta c_j^2 \tag{13.184}$$

The derivatives are obtained from the calculated structure. The uncertainty in a bond angle will involve the three coordinates of three atoms $\delta\theta_{ijk}$. For an atom located by means of the first moment or product of inertia relations, these relations can be used to estimate the uncertainty in the derived coordinate δa_i with the uncertainties of the other coordinates δa_j used in the relation. From (13.166), for example,

$$\delta a_i=\sum_{j\neq i}^{N}\frac{m_j\delta a_j}{m_i} \tag{13.185}$$

The vibrational effects tend to give coordinates which are smaller than the equilibrium values and the corresponding uncertainties are not randomly distributed. A procedure for error analysis which takes this into account is discussed by Tobiason and Schwendeman [78]. A method for assessing uncertainties, which takes into account the fact that an average bond distance shrinks when a heavier isotope is substituted for one of the atoms making up the bond, is discussed in the next section.

An empirical correlation between ε and I_0 has been found [79] which may be useful for estimation of the magnitude of the vibrational effects. Caution should be exercised in evaluation of the structure of molecules that undergo vibrations of large amplitude. For HCNO, which can be characterized as a quasilinear molecule (see Chapter V, Section 7), the ground state is consistent with that of a linear molecule. However, the r_s structure for this molecule has a C–H bond length that is short compared to that in other related molecules [80, 81]. All of its molecular vibrations cannot be classified as small amplitude vibrations. In particular, the HCN bending vibration is of sufficiently large amplitude to affect significantly the observed moments of inertia. The short CH bond length in the ground vibrational state may be interpreted as an average projection of the C–H bond length on the CNO axis. The bending potential function has a small potential hump well below the ground vibrational level, at the position corresponding to the linear configuration. Because of this hump the equilibrium structure is a bent one. From its spectrum [82], the quasilinear molecule HNCSe is also found to have an anomalously short N–O bond length of 0.791 Å. For such molecules, corrections to the moments of inertia must be made for the effects of the large amplitude vibrations if the structural parameters are to be meaningful [83, 84]. Similarly, for accurate evaluation of methyl group structures, care must be taken that the torsional contributions to the rotational constants are removed [77].

9 EVALUATION OF AVERAGE STRUCTURES

One disadvantage of the r_s structure is the lack of a well-defined meaning as compared to the equilibrium structure. The possibility of deriving average structures (designated $\langle r \rangle$ or r_z) from the effective moments has been considered by Herschbach and Laurie [85, 86] and by Oka and Morino [87, 88]. Such structures have a clear physical meaning corresponding in particular to the structure for the average molecular configuration for a specific vibrational state. The average structure differs from the equilibrium structure only because of the anharmonicity of the molecular vibrations. The comparison of average parameters for various isotopic species enables the effects of isotopic substitution to be reliably studied. Furthermore, the bond distance r_g (the center of gravity of the probability distribution function) obtained from electron diffraction can be related to $\langle r \rangle$ with only a knowledge of the harmonic force constants [89]. Thus, a reliable comparison between electron diffraction and spectroscopic parameters can be obtained.

The relation between the average distance $\langle r \rangle$ and other interatomic distances such as r_0 and r_e can be illustrated by consideration of a diatomic molecule.

The effective moment of inertia is related to the vibrational average of the reciprocal instantaneous moment ($I=\mu r^2$) as

$$I_b^v=\frac{h/8\pi^2}{B_v}=\langle I^{-1}\rangle^{-1}=\mu\langle r^{-2}\rangle^{-1} \tag{13.186}$$

The effective bond length

$$r_v=\left(\frac{I_b^v}{\mu}\right)^{1/2}=\langle r^{-2}\rangle^{-1/2} \tag{13.187}$$

may hence be written by expansion of r in terms of ξ as

$$r_v=r_e(1+\langle\xi\rangle-\tfrac{3}{2}\langle\xi^2\rangle) \tag{13.188}$$

This is correct to terms linear in the vibrational quantum number (small-oscillations approximation) with the vibrational averages given in (13.109) and (13.110). Similarly, the direct average $\langle r\rangle$ is simply

$$\langle r\rangle=r_e(1+\langle\xi\rangle) \tag{13.189}$$

The deviation of $\langle r\rangle$ from r_e is thus due entirely to the anharmonicity ($\langle\xi\rangle$ vanishes for a harmonic oscillator). In Fig. 13.7 the lower portion of the actual potential function of LiF is compared with the harmonic part of the potential function. The anharmonicity is clearly evident. A comparison of (13.188) and (13.189) gives

$$\langle r\rangle=r_v+\tfrac{3}{2}r_e\langle\xi^2\rangle=r_v+\frac{3B_e}{\omega_e}r_e(v+\tfrac{1}{2}) \tag{13.190}$$

Thus, the average and effective bond distances differ by only a term involving the mean-square harmonic vibrational amplitude, $\langle\xi^2\rangle$. Therefore, the average bond length can be obtained from the effective bond length by application of a correction which depends only on the harmonic part of the potential function. Since the anharmonic term $\langle\xi\rangle$ is usually positive and somewhat larger than the harmonic term $\langle\xi^2\rangle$, we expect

$$\langle r\rangle>r_0>r_e \tag{13.191}$$

The moment of inertia for the average configuration $I^*=\mu\langle r\rangle^2$ is related to the effective moment by the equation

$$I_b^*=I_b^v-\varepsilon_h(v+\tfrac{1}{2}) \tag{13.192}$$

where the correction to I_b^v involves only the harmonic part of ε ($\varepsilon_h=-6h/8\pi^2\omega_e$). By use of (13.190) or the average moment I_b^*, the average bond length may be evaluated. In either case, only a knowledge of the harmonic part of the potential function is required. If the vibration of the molecule is harmonic, then the average moment is equivalent to the equilibrium moment and $\langle r\rangle=r_e$. It should be noted that the moment of inertia for the average configuration is not the same as the average of the instantaneous moment of inertia $\langle I\rangle=\mu\langle r^2\rangle$, since $\langle r^2\rangle\neq\langle r\rangle^2$. The expected order of the various kinds of moments of inertia is

$$\langle I_\alpha\rangle>I_\alpha^*>I_\alpha^v>I_\alpha^e \tag{13.193}$$

A detailed analysis by Herschbach and Laurie [85] has shown that similar considerations also hold for polyatomic molecules. By means of a perturbation treatment to terms linear in $(v+\frac{1}{2})$, it is found that the moments of inertia of the average configuration of the atoms are given by

$$I_\alpha^* = I_\alpha^e + \sum_s \left(v_s + \frac{d_s}{2}\right) \varepsilon_s^\alpha \text{ (anhar)} \tag{13.194}$$

or

$$I_\alpha^* = I_\alpha^v - \sum_s \left(v_s + \frac{d_s}{2}\right) \varepsilon_s^\alpha \text{ (har)} \tag{13.195}$$

where the sum is over all vibrations with degeneracy d_s. As for the diatomic molecule, the moments of inertia for the average configuration can be derived from the effective moments without a knowledge of the anharmonic potential constants; if the vibrations are harmonic, then $I_\alpha^* = I_\alpha^e$, and the average structure will be identical to the equilibrium structure. The anharmonicity of the potential, which is always present to various degrees for real molecules, has the effect of displacing the average configuration of the vibrating nuclei from the equilibrium configuration.

Convenient procedures for evaluation of ε_s^α (har) have been given [85] as well as explicit expressions for some simple molecules [86, 90]. In general [85]

$$d_s \varepsilon_s^{(\alpha)}(\text{har}) = -\left(\frac{2K}{\omega_s}\right)\left[\frac{3}{4}\sum_\gamma \frac{a_s^{\alpha\gamma} a_s^{\alpha\gamma}}{I_\gamma} + \sum_t \zeta_{st}^\alpha \zeta_{st}^\alpha \left(\frac{3\lambda_s + \lambda_t}{\lambda_s - \lambda_t}\right)\right] \tag{13.196}$$

with $K = 16.863$ amu Å^2 cm^{-1}. Other forms are possible since

$$4A_{ss}^{\alpha\beta} = \sum_\gamma \frac{a_s^{\alpha\gamma} a_s^{\beta\gamma}}{I_\gamma} + 4\sum_t \zeta_{st}^\alpha \zeta_{st}^\beta \tag{13.197}$$

where $a_s^{\alpha\beta} = (\partial I_{\alpha\beta}/\partial Q_s)_*$ and $A_{ss}^{\alpha\beta} = (\partial^2 I_{\alpha\beta}/\partial Q_s^2)_*$. In particular, for a linear XY_2 molecule the ground state moment of inertia I_b^* is given by [86]

$$I_b^* = I_b^0 + K\left(\frac{3}{\omega_1} - \frac{1}{\omega_2} - \frac{1}{\omega_3} + \frac{4}{\omega_2 + \omega_3}\right) \tag{13.198}$$

For a linear XYZ molecule [86]

$$I_b^* = I_b^0 - K\left[\frac{1}{\omega_1} + \frac{1}{\omega_2} + \frac{1}{\omega_3} - 4\zeta_{12}^2\left(\frac{1}{\omega_3} + \frac{1}{\omega_1 + \omega_2}\right) - 4\zeta_{23}^2\left(\frac{1}{\omega_1} + \frac{1}{\omega_2 + \omega_3}\right)\right] \tag{13.199}$$

and the Coriolis coupling constants are found from the equations [91]

$$\zeta_{12}^2 + \zeta_{23}^2 = 1 \tag{13.200}$$

$$\lambda_1 \zeta_{23}^2 + \lambda_3 \zeta_{12}^2 = (r_{XY}^2 F_{11} + 2r_{XY} r_{YZ} F_{13} + r_{YZ}^2 F_{33})/I_b \tag{13.201}$$

where F_{11}, F_{33}, F_{13} are the quadratic potential constants for the $X-Y$ bond stretch, the $Y-Z$ bond stretch, and the interaction, respectively.

For a bent XY_2 molecule with b the symmetry axis and with c the out-of-plane axis [86]

$$I_a^* = I_a^0 + 3K\left(\frac{\sin^2\chi}{\omega_1} + \frac{\cos^2\chi}{\omega_2} + \frac{I_a/I_c}{\omega_3}\right) \tag{13.202}$$

$$I_b^* = I_b^0 + 3K\left(\frac{\cos^2\chi}{\omega_1} + \frac{\sin^2\chi}{\omega_2} + \frac{I_b/I_c}{\omega_3}\right) \tag{13.203}$$

$$I_c^* = I_c^0 - K\left[\frac{1}{\omega_1} + \frac{1}{\omega_2} + \frac{1}{\omega_3} - 4\zeta_{13}^2\left(\frac{1}{\omega_2} + \frac{1}{\omega_1+\omega_3}\right) - 4\zeta_{23}^2\left(\frac{1}{\omega_1} + \frac{1}{\omega_2+\omega_3}\right)\right] \tag{13.204}$$

where χ is a parameter determined from ($\zeta_{23} < 0$, $\cos\chi > 0$)

$$\cos^2\chi + 2\zeta_{23}\left(\frac{I_b}{I_c}\right)^{1/2}\cos\chi + \left(\zeta_{23}^2 - \frac{I_a}{I_c}\right) = 0 \tag{13.205}$$

The Coriolis coupling constants which refer to the out-of-plane axis (c axis) are related to the force constants and vibrational frequencies by the relations [39]

$$\zeta_{13}^2 = \frac{\lambda_1 - F_{11}/(G^{-1})_{11}}{\lambda_1 - \lambda_2} \tag{13.206}$$

$$\zeta_{23}^2 = \frac{F_{11}/(G^{-1})_{11} - \lambda_2}{\lambda_1 - \lambda_2} \tag{13.207}$$

where $\lambda_s = 4\pi^2 c^2 \omega_s^2$ with the F and G matrices expressed in terms of the usual symmetry coordinates for a bent XY_2 molecule. Explicit expressions for the average moments of tetrahedral XY_4 molecules and linear X_2Y_2 molecules have also been given in [86]. In the above expressions the harmonic vibrational frequencies ω_s are to be expressed in cm^{-1} with $K = h/8\pi^2 = 16.863$ amu Å^2 cm^{-1} and $\lambda_s = 5.8893 \times 10^{-2}\omega_s^2$. By employment of these expressions the average moments of inertia I_a^* can be evaluated; they in turn can give information on the average positions of the atoms in the molecule.

Laurie and Herschbach [86] have calculated average structures for several simple molecules and pointed out a number of interesting features. In Table 13.7 along with the effective and equilibrium bond lengths the average bond lengths of some diatomic molecules have been given. As expected, $\langle r \rangle > r_0 > r_e$. For the hydrides the $\langle r \rangle$ distance is about 0.02 Å larger than r_e. In the diatomics made up of heavy atoms the difference is smaller. The replacement of H by D, which decreases the amplitude of vibration, causes a shortening in $\langle r \rangle$ of 0.003 to 0.005 Å. Isotopic substitution of the heavier atoms gives a much smaller shortening, of about 0.0001 Å; when the heavy atom is bonded to hydrogen, the shortening is negligibly small since the hydrogen atom does essentially all the vibrating.

The average parameters of some bent XY_2 molecules are illustrated in Table 13.20. It is again observed that replacement of H and D shortens the bond length

Table 13.20 Average Structures of Bent XY_2 Molecules in the Ground Vibrational State

Molecule	$\langle r \rangle$	$\langle \theta \rangle$
H_2O^a	0.9724 Å	104°30′
D_2O^a	0.9687	104°21′
T_2O^a	0.9671	104°16′
H_2S^b	1.3518	92°8′
D_2S^b	1.3474	92°7′
H_2Se^c	1.4754	90°52′
D_2Se^c	1.4711	90°53′
$^{32}SO_2{}^d$	1.4350	119°21′
$^{34}SO_2{}^d$	1.4350	119°21′
$^{35}CLO_2{}^e$	1.4756	117°30′
$^{37}CLO_2{}^e$	1.4755	117°30′
$O_3{}^d$	1.2794	116°44′
$NO_2{}^d$	1.2000	133°48′
$OF_2{}^f$	1.4124	103°10′
$NF_2{}^g$	1.3528	103°11′
$SF_2{}^h$	1.5921	98.20°
$SCl_2{}^i$	2.0153	102.73°

[a]Cook et al. [94].
[b]Cook et al. [93].
[c]Oka and Morino [88].
[d]Laurie and Herschbach [86].
[e]A. H. Clark, *J. Mol. Struct.*, **7**, 485 (1971).
[f]L. Pierce, N. DiCianni, and R. H. Jackson, *J. Chem. Phys.*, **38**, 730 (1963).
[g]R. D. Brown, F. R. Burden, P. D. Godfrey and I. R. Gillard, *J. Mol. Spectrosc.*, **52**, 301 (1974).
[h]W. H. Kirchhoff, D. R. Johnson, and F. X. Powell, *J. Mol. Spectrosc.*, **48**, 157 (1973).
[i]R. W. Davis and M. C. L. Gerry, *J. Mol. Spectrosc.*, **65**, 455 (1977).

by 0.004 Å in H_2O and H_2Se. Replacement of H by T in H_2O gives a shortening of 0.005 Å. Substitution of a heavy atom has a negligible effect. Since for the average structural moments of inertia the inertial defect vanishes [92], different pairs of moments of inertia will give consistent structures, in contrast to the results obtained when effective moments are used. Another interesting isotopic variation found [86] is the decrease in I_b^* (also I_b^0) for SO_2 with replacement of ^{32}S by a heavier isotope. In particular, I_b^* for $^{32}SO_2$, $^{33}SO_2$, and $^{34}SO_2$ is, respectively, 49.0934, 49.0931, and 49.0921 amu Å^2. Here the very small decrease in the average S—O bond causes a decrease in I_b^*. For a rigid molecule like

SO_2 where S is on the b axis, I_b^e would be unaffected. Similar effects have been noted in CO_2 when ^{12}C is replaced by ^{13}C [86].

Average structures in various vibrational excited states have also been evaluated for SO_2 [26] and OF_2 [27]. These are summarized in Table 13.21. It is apparent that the average structures show the characteristics of the excited vibrational states. When the symmetric (v_1) or antisymmetric (v_3) stretching mode is singly excited, the bond lengths increase relative to the ground state, while the bond angles change very little. When the deformation mode (v_2) is excited, the bond angle clearly increases, and there are small changes in the bond lengths. On the other hand, the effective structures of SO_2 do not clearly indicate the variation of the bond angle with vibrational state [26].

It is possible to obtain an estimate of the equilibrium structure from the average structure data. For a diatomic molecule, the average structure is related to the equilibrium structure as follows, with $a_0=\omega_e^2/4B_e$,

$$\langle r\rangle = r_e - \left\{\frac{3a_1}{8}\left(\frac{\hbar}{\pi a_0}\right)^{1/2}\right\}\mu^{-1/2} \tag{13.208}$$

where μ is the reduced mass, and where a_0 (cm^{-1}) and a_1 (dimensionless) are, respectively, the harmonic and cubic potential constants. From the average bond distance and the reduced mass of at least two isotopes, (13.208) can be used to calculate r_e. For an YX_2 molecule, if we assume the Y–X bonds may be treated separately as one-dimensional oscillators, we may apply the above relation. This approximation has been used to derive the equilibrium structure of H_2Se [88] and H_2S [93] and H_2O [94]. The equilibrium structure obtained for H_2O with use of the data in Table 13.20 is $r_e=0.9587$ Å, $\theta_e=103.89°$. In the derivation of θ_e, the same reduced mass dependence has been assumed for

Table 13.21 Average Structures in Excited Vibrational States[a]

Molecule[b]	*State* $v_1v_2v_3$	$\langle r\rangle$	$\delta\langle r\rangle^b$	$\langle\theta\rangle$	$\delta\langle\theta\rangle^b$
SO_2	000	1.4349		119°21′	
	100	1.4388	0.0039	119°24′	3′
	010	1.4346	−0.0003	119°32′	11′
	001	1.4396	0.0047	119°11′	−10′
	020	1.4343	−0.0006	119°44′	23′
OF_2	000	1.4124		103°10′	
	100	1.4183	0.0059	103°2′	−8′
	010	1.4139	0.0015	103°34′	24′
	001	1.4196	0.0072	103°4′	−6′
	020	1.4155	0.0031	103°58′	48′

[a]From Morino et al. [26, 27].
[b]Difference between average parameter in the vibrationally excited state and in the ground state.

$\langle\theta\rangle$ as for $\langle r\rangle$. The extrapolation to $\mu^{-1/2}\rightarrow 0$ is, however, rather long. This result for water may be compared to the more accurate infrared result [95] of r_e= 0.9575 Å and 104.51°.

For molecules like CH_3COX with a single pair of out-of-plane hydrogen atoms, the average out-of-plane distance of the H atoms may be calculated if correction is made for the harmonic part of the inertial defect Δ_y, that is [42]

$$I_a^0 + I_b^0 - I_c^0 = 4m_H\langle y_H\rangle^2 - \Delta_y\,(\text{har}) \tag{13.209}$$

with $c(y)$ the out-of-plane direction. The anharmonic part of the inertial defect simply shifts $\langle y\rangle$ from y_e. If Δ_y (har) is essentially the same for the CH_3 and CD_3 species, then Δy (har) may be evaluated from data on CH_3COX and CD_3COX if the small variation is neglected and if an estimate is made of the difference between $\langle y_H\rangle$ and $\langle y_D\rangle$. Alternately, Δ_y (har) may be estimated from the vibrational frequencies [42]. When corrections are made for Δ_y (har), the $\langle y_H\rangle$ obtained no longer shows the decrease in the methyl group size observed for the effective distance (see Table 13.12). It is found [42] that $\langle y_H\rangle$ is 0.896, 0.891, and 0.896 Å in CH_3CHO, CH_3COF, and CH_3COCl, respectively [obtained by estimation of Δ_y (har) from vibrational data]. These average values of the out-of-plane coordinate compare favorably with that found in methane, namely, 0.896 Å [86].

The consideration of the average structure of a linear XYZ molecule such as HCN reveals a fundamental problem in the calculation of average structures as well as of effective and substitution structures. Here, at least two molecular species are required for determination of the structural parameters; but, as we have seen, the average parameters clearly vary with isotopic substitution. However, the C—H bond distance is not expected to change significantly with a change in mass of carbon, since this will have little effect on the C—H vibration. Also the isotopic variation of C≡N with ^{13}C substitution is expected to be small. Hence, the smallest error is introduced in the structural calculation if either the pair $H^{12}C^{14}N$, $H^{13}C^{14}N$ or $D^{12}C^{14}N$, $D^{13}C^{14}N$ is used. When the structure is evaluated in this way, it is found [86] that $\langle r_{CH}\rangle$ is larger than $\langle r_{CD}\rangle$ by 0.003 Å and that the $\langle r_{CN}\rangle$ is almost the same in HCN and DCN (decrease of 0.004 Å). For H_2CO [90], the derivation of an accurate average structure required an estimate of the variation in the C=O bond with ^{18}O substitution. The difference (0.000123 Å) was estimated by means of an approximate one-dimensional potential function with an assumed anharmonic force constant. More recent calculations on H_2CO have been made by Duncan [96], who used data on seven isotopic species ($H_2^{12}C^{16}O$, $H_2^{13}C^{16}O$, $H_2^{12}C^{18}O$, $D_2{}^{12}C^{16}O$, $D_2{}^{13}C^{16}O$, $D_2{}^{12}C^{18}O$, $HD^{12}C^{16}O$). The data are sufficient for the evaluation of the isotopic effects for all three structural parameters. Corrections were made for the electronic contributions to the moments of inertia. It was assumed that ^{13}C or ^{18}O substitution causes negligible change in the CH(CD) bond lengths or HCH(DCD) bond angles. The average structure and the major isotopic effects are given in Table 13.22. The data for HDCO were not employed in the calculation of the structure; however, the final structure and shrinkage effects were used

Table 13.22 Average Structure of Formaldehyde[a]

r_z(CH)	1.1171 ± 0.0010 Å
r_z(CD)	1.1130 ± 0.0010 Å
r_z(CO)	1.2072 ± 0.0005 Å
α_z(HCH)	$116°14' \pm 6'[\phi_z(\text{HCO})121°53' \pm 3']$
α_z(DCD)	$116°19' \pm 6'[\phi_z(\text{DCO})121°50' \pm 3']$

Isotopic Mass Effects on Structural Parameters

δr_z(CH)	[H–D]	0.0041 ± 0.0005 Å
δr_z(CO)	$[^{12}C-^{13}C]$	0.00003 ± 0.00005 Å
δr_z(CO)	$[^{16}O-^{18}O]$	0.00010 ± 0.00005 Å
δr_z(CO)	$[H_2-D_2]$	0.0005 ± 0.0002 Å [0.00025 Å per D substituion]
$\delta\alpha_z$(HCH)	$[H_2-D_2]$	$-5' \pm 2'$[2.5′ per D substitution]

[a]From Duncan [96].

to calculate the average moments of inertia for this molecule. The agreement with those derived from the harmonic force field is as good as, or better than, ± 0.0006 amu Å^2.

In average structure evaluations, the isotopic changes $\delta\langle r\rangle$ in bond lengths with isotopic substitution are often estimated from the approximate relationships given by Kuchitsu et al. [97, 98]

$$\langle r\rangle = r_e + \frac{3a}{2} u^2 - K \tag{13.210}$$

$$\delta\langle r\rangle = \frac{3a}{2} \delta u^2 - \delta K \tag{13.211}$$

where u^2 is the zero-point, mean-square amplitude of the particular bond in question and K, the corresponding mean-square perpendicular amplitude correction. These quantities may be evaluated from the harmonic force field [99]. Here a is the Morse anharmonicity parameter which may be estimated from similar diatomic molecules [100]. The isotopic effects on bond angles $\delta\langle\theta\rangle$ are small and are usually ignored. Equation 13.210 can be used for estimation of the r_e structure by application of the appropriate correction to the $\langle r\rangle$ structure. These corrections, however, are rather large.

In summary, for even relatively simple molecules, where sufficient harmonic force constants are available for evaluation of I_α^*, if isotopic data must be combined in calculation of the structure, it is necessary to either ignore the isotopic effects or attempt to estimate the isotopic variation in $\langle r\rangle$. This, of course, introduces some uncertainty in the average structural parameters although in favorable cases only small uncertainty. Such isotopic effects also introduce ambiguities in effective and substitution structures, since in the calculation of these structures it is tacitly assumed that the bond lengths are unaffected by isotopic substitution. Because of the isotopic shrinkage of bonds, the substitution method gives coordinates which are too small. (The bond length, however, may be either too short or too long.) In the case of HCN, for example, the "substitution" C—H bond length, which is evaluated by neglect of the shortening caused by deuterium substitution, is found to be some 0.003 Å less than the equilibrium bond length [60]. It has been demonstrated [86] that an isotopic change of 0.0001 Å in a bond length can give an error of 0.002 Å in the substitution coordinate. When an atom is near a principal axis, this effect can be greatly amplified. Unfortunately, the accurate calculation of isotopic changes requires a knowledge of the anharmonic potential constants which are available for only a few polyatomic molecules. The isotopic changes, however, may prove to be characteristic of a bond and hence transferable from simple molecules to more complicated ones. In attempting to estimate the uncertainty in a structure, such as a substitution structure, consideration should be given to the effect of bond shortening and bond angle changes upon isotopic substitution, for example, by consideration of the effect of deuterium substitution where $r(X\mathrm{D})=r(X\mathrm{H})-\delta$ and δ is varied from 0.002 to 0.005 Å. If X is substituted, $\delta=0$. For two heavy atoms, δ is varied between 0.00005 to 0.0001 Å. One procedure [33] would involve calculation of the moments of inertia for all isotopic species with the structure obtained from the experimental moments of inertia. For each isotopic species, the bond distances associated with one or more isotopic substitutions are adjusted by selection of a particular value of δ, and the moments of inertia are recalculated. The difference between the moments of inertia with and without the shrinkage effects is then used for correction of the experimental moments of inertia. Finally, the structure is evaluated from the corrected moments of inertia in the usual way, thus giving an estimate of the uncertainties.

In the absence of equilibrium structures, average structures provide a well-defined structural parameter. Ambiguities introduced by inertial defects of planar molecules are removed, and it is possible to relate $\langle r \rangle$ to the electron diffraction r_g so that a meaningful comparison between the two can be made. This is in contrast to the effective and substitution parameters. On the other hand, the evaluation of average structures requires a knowledge of the harmonic force constants which are usually not available except for relatively simple molecules.

10 EVALUATION OF MASS-DEPENDENCE STRUCTURES

A method of structural calculation has been introduced by Watson [57] which represents an extension of the substitution method but requires a more extensive set of isotopic data. It uses the I^0 values of a number of isotopic species to determine, to first-order, the mass dependence of the vibrational contributions and thus to obtain an estimate of the equilibrium structure. This method is called the "mass-dependence" method and gives the so-called r_m structure.

In the following derivation one exploits the fact that various molecular parameters are homogeneous functions of the masses of degree n, that is,

$$F(\rho m_1, \rho m_2, \ldots \rho m_i, \ldots)=\rho^n F(m_1, m_2, \ldots m_i \ldots) \qquad (13.212)$$

for any value of ρ. Simultaneous isotopic substitution of all the atoms of a molecule $m_i \to \rho m_i$, for example, shows that I_α^e, α_s^α, ε^α, $a_k^{\alpha\beta}$, $\tau_{\alpha\beta\gamma\delta}$, ζ_{kl}^α, and C_{kl}^α are of degree 1, $-\frac{3}{2}$, $\frac{1}{2}$, $\frac{1}{2}$, -2, 0 and -1, respectively [57]. Furthermore, the position of the center of mass and the orientation of the principal inertial axis are unchanged. A more detailed list of molecular parameters and their degree in the masses is given elsewhere [57]. Only in certain cases such as for ozone, $^{16}O_3 \to {}^{18}O_3$, can this isotopic substitution actually be realized, and in these cases the isotopic relations in the foregoing quantities are particularly simple. It is clear that the vibrational contributions to I_α^0 are of a different degree in the masses than I_α^e.

For a linear molecule we may write from (13.116)

$$I_b^0 = I_b^e + \varepsilon \tag{13.213}$$

with

$$\varepsilon = \frac{I_b^e}{2B_e} \sum_s \alpha_s^b d_s \tag{13.214}$$

With isotopic substitution of atom i, we have

$$\Delta I_b^0 = \Delta I_b^e + \Delta\varepsilon \tag{13.215}$$

Introducing the relations

$$[z_e(i)]^2 = \frac{\Delta I_b^e}{\mu_i} \tag{13.216}$$

$$[z_s(i)]^2 = \frac{\Delta I_b^0}{\mu_i} \tag{13.217}$$

with $\mu_i = M\Delta m_i/(M + \Delta m_i)$ and expanding $\Delta\varepsilon$ as a Taylor series in Δm_i gives [57]

$$[z_s(i)]^2 = [z_e(i)]^2 + \left(\frac{\partial \varepsilon}{\partial m_i}\right) + \frac{1}{2M}\left(\frac{\partial^2 (M\varepsilon)}{\partial m_i^2}\right)\Delta m_i + \cdots \tag{13.218}$$

This may be simplified by multiplying by m_i and summing over all atoms i. Thus

$$I_b^s = I_b^e + \sum m_i \left(\frac{\partial \varepsilon}{\partial m_i}\right) + \frac{1}{2M}\sum m_i \left(\frac{\partial^2 (M\varepsilon)}{\partial m_i^2}\right)\Delta m_i + \cdots \tag{13.219}$$

where $I_b^s = \sum m_i [z_s(i)]^2$. The first sum may be evaluated from Euler's theorem on homogeneous functions

$$\sum m_i \left(\frac{\partial \varepsilon}{\partial m_i}\right) = n\varepsilon = \tfrac{1}{2}\varepsilon = \tfrac{1}{2}(I_b^0 - I_b^e) \tag{13.220}$$

where ε is a homogeneous function of degree $n = \frac{1}{2}$. Employing this relation in (13.219) and defining the mass-dependence moment of inertia I_b^m as

$$I_b^m = 2I_b^s - I_b^0 \tag{13.221}$$

gives

$$I_b^m = I_b^e + \frac{1}{M}\sum m_i \left(\frac{\partial^2 (M\varepsilon)}{\partial m_i^2}\right)\Delta m_i + \cdots \tag{13.222}$$

Here Δm_i is the isotopic mass difference for the ith atom employed in deriving the r_s coordinates of the particular parent species, and M (total mass) and m_i refer to the parent species. We see that if linear and higher-order terms in Δm_i are ignored, I_b^m represents an approximation to the equilibrium moment of inertia I_b^e. To the same approximation

$$I_b^s = \frac{I_b^e + I_b^0}{2} \tag{13.223}$$

and the substitution coordinates defined by

$$\begin{aligned}[z_s(i)]^2 &= [z_e(i)]^2 + \frac{\partial \varepsilon}{\partial m_i} \\ &= \frac{\partial I_b^e}{\partial m_i} + \frac{\partial \varepsilon}{\partial m_i} = \frac{\partial I_b^0}{\partial m_i}\end{aligned} \tag{13.224}$$

are well-defined properties of the parent molecule. It is clear, however, that the substitution structure of different parent molecules is mass dependent because $\partial \varepsilon / \partial m_i$ is of degree $-\frac{1}{2}$ in the masses, though the differences are usually small.

To determine an r_m structure one employs various isotopic species to evaluate the substitution coordinate of each atom from (13.217) for a given parent. These coordinates are then employed for evaluation of I^s and hence of I^m from (13.221). Once a sufficient number of I^m for different parent isotopic molecules have been evaluated, the moment-of-inertia equations may be solved for the r_m structure. The validity of this procedure depends on the relation of (13.224) where the vibrational contribution to $[z_s(i)]^2$ is given by $\partial \varepsilon / \partial m_i$. Therefore very small or imaginary coordinates, which can arise when $\partial \varepsilon / \partial m_i$ is large compared to $[z_e(i)]^2$ and negative, should be retained in calculation of I^s. This yields a physically significant substitution moment of inertia regardless of the reliability of the coordinates. Furthermore, first moment equations should not be employed for calculation of z_s coordinates. Ideally, to ensure the correctness of (13.224), the substitution coordinates, (13.218), should be extrapolated to zero values of Δm_i.

For polyatomic molecules it can be shown [57] in a similar manner that first-order estimates of the equilibrium moments of inertia are

$$\left.\begin{aligned} I_\alpha^m &= 2I_\alpha^s - I_\alpha^0 \qquad &(13.225)\\ I_\alpha^s &= \sum_i m_i [\alpha_s(i)]^2 \qquad &(13.226)\end{aligned}\right\} \alpha = a,\ b, \text{ and } c$$

where each nonequivalent atom must be substituted for determination of the I_α^m of a parent molecule. The values of I_α^m for a sufficient number of parent molecules are then solved for the r_m structure. Double-substitution coordinates from substitution of a symmetrically equivalent set of atoms can be employed if care is used to maintain the internal consistency of the method. It has been shown that for bent symmetrical triatomic molecules, (13.225) and (13.226) apply with the coordinates of B located employing the substitution $AB_2 \rightarrow AB_2'$.

They apply for a C_{3v} molecule ABC_3 where ABC'_3 data are employed.

For CO the r_m values are good estimates of r_e, but the spread obtained for the various isotopes is appreciable, though only half the spread of the r_0 values [57]. Using the parent $^{12}C^{16}O$ and the isotopic species ^{13}C and ^{14}C, one finds that the squares of the substitution coordinates are slightly different. Extrapolating them to $\Delta m_c = 0$ yields an improved substitution coordinate in which the higher-order effects in (13.218) are effectively eliminated. The same procedure may be carried out by use of ^{17}O and ^{18}O data for the oxygen substitution coordinate. These data provide an improved I^m value and an r_m bond length that differs from r_e by only 0.00019 Å.

A critical test [101] of the r_m method has been carried out on OCS. It was found that the r_m structure differs significantly from the r_e structure and that the errors are comparable in the r_s and r_m methods. On the other hand, if a corrected set of substitution coordinates is employed for OCS by extrapolation of the coordinates for, say, two different values of Δm_i for each atom i of each parent molecule to $\Delta m_i = 0$, the results are in good agreement with the equilibrium structure. This procedure is clearly of limited applicability and reflects the breakdown of the first-order approximation employed previously. The r_m structure of SO_2, which is in good agreement with the r_e structure, is given in Table 13.5. A detailed study of the r_m structure of OCl_2 and $COCl_2$ has been carried out by Nakata et al. [102, 103]. If averages are taken of the r_m parameters of appropriate isotopic combinations (termed "complimentary sets"), the higher-order effects of (13.222) may be reduced, since these effects will have opposite signs [57, 102]. Parent species which have completely different isotopic make-up so that the signs of Δm_i for any ith atom are opposite, constitute a complimentary set. However, only one meaningful r_m structural parameter can be obtained from the moment of inertia data about a given axis (e.g., two for $COCl_2$) [103].

Unfortunately, the r_m method is of limited applicability. Data of very high precision from a large number of isotopic species are required. The first order approximation $I^m \simeq I^e$ is not sufficient particularly for light atoms such as hydrogen atoms. The method cannot be applied to molecules containing atoms such as fluorine, which cannot be substituted; and even for moderate size molecules, a complete set of reliable r_m parameters cannot be evaluated.

11 DIAGNOSTIC LEAST SQUARES

In calculations of structure, the number of parameters frequently exceeds the number of rotational constants. If one wishes to find a "physically reasonable" set of parameters that reproduce the rotational constants found from the experimental data, it is possible to vary the parameters one at a time until such reasonable parameters are found, but this procedure becomes unrealistic with a large number of parameters. On the other hand, some parameters or some linear combinations of the parameters may not be well determined by the experimental data; yet clearly some linear combinations of the parameters are determined by the data. Curl [35] has, devised a systematic method, referred to as "diagnostic least squares", which is applicable to such situations through use of adjustable parameters weighted by the inverse of their estimated uncertainties. A set of parameters is found that reproduces the experimental data with a minimum distortion in a least-squares sense from the initial set. An outline of this procedure is presented in the following discussion. The technique can also be applied to a combination of data from different sources. An electron diffraction structure, for example, might be adjusted to include the microwave data. In addition to the rotational constants, the dipole moment components can be profitably included when these data are sensitive to the orientation of one part of the molecule relative to another. Another application could involve a set of molecules which constitute a closely related series when one wishes to see if a reasonable set of structural parameters could be obtained that would correlate the microwave data of these molecules. With the assumption that similar structural parameters can be transferred, there may be, in principle, more data than parameters. However, if it is found that the data are not sensitive to all the structural parameters, a diagnostic least-squares analysis can be carried out to obtain a set of structural parameters which adequately fit the spectral data of these molecules. Clearly, the technique can be applied to problems other than calculations of molecular structure.

We consider a set n observables $I_i = I_i(p_j)$, such as moments of inertia, which are considered functions of a set of m parameters p_j. To evaluate the parameters from the observations, the problem is linearized by a Taylor series expansion about the initial set of parameters p_j^0

$$I_i = I_i(p_j^0) + \sum_{j=1}^{m} \left(\frac{\partial I_i}{\partial p_j}\right)_0 \Delta p_j + \cdots, \qquad i = 1, \ldots n \tag{13.227}$$

or

$$\Delta I_i = \sum_{j=1}^{m} \left(\frac{\partial I_i}{\partial p_j}\right)_0 \Delta p_j, \qquad i = 1, \ldots, n \tag{13.228}$$

with $\Delta p_j = p_j - p_j^0$. To proceed where ordinary least squares will not apply, one defines a new set of parameters which have equal uncertainties. Structural parameters are predicted from analogous molecules along with an estimated uncertainty for each parameter. The latter may be done by consideration of the range of values obtained for this parameter in similar molecules. If the range p_j^L to p_j^U is considered a normal distribution with 90% confidence interval, the best estimate of p_j^0 is given by

$$p_j^0 = \frac{p_i^L + p_i^U}{2} \tag{13.229}$$

with estimated standard deviation [35]

$$\sigma_j = \frac{p_j^U - p_j^L}{3.290} \tag{13.230}$$

A new set of weighted parameters is then defined by

$$\Delta P_j = \frac{\Delta p_j}{\sigma_j} \tag{13.231}$$

and (13.228) in terms of these parameters can be written in matrix form as

$$\mathbf{J}\Delta\mathbf{P} = \Delta\mathbf{I} \tag{13.232}$$

where $\Delta\mathbf{P}$ is a column matrix of the m parameter changes ΔP_j, $\Delta\mathbf{I}$ is a column matrix with n elements composed of the difference between the observed and calculated experimental quantities $\Delta I_i = I_i^{obs} - I_i^{cal}$, and $\mathbf{T}$ is the $n \times m$ Jacobian matrix (or design matrix) with elements defined by

$$J_{ij} = \left(\frac{\partial I_i}{\partial P_j}\right)_0 = \left(\frac{\partial I_i}{\partial p_j}\right)_0 \sigma_j \tag{13.233}$$

The derivatives may be evaluated numerically. A small increment is given to the initial value of the jth parameter and the change in the observable I_i is computed to give the derivative. If equivalent parameters are present, the derivatives should reflect this fact. The least-squares normal equations are obtained by multiplication of both sides of (13.232) by $\tilde{\mathbf{J}}$, where the tilde denotes the transpose of the matrix. Thus,

$$\tilde{\mathbf{J}}\mathbf{J}\Delta\mathbf{P} = \tilde{\mathbf{J}}\Delta\mathbf{I} \tag{13.234}$$

where $\tilde{\mathbf{J}}\mathbf{J}$ is a $m \times m$ symmetric matrix (the normal matrix). In ordinary least squares where $\tilde{\mathbf{J}}\mathbf{J}$ is nonsingular, the parameter shifts are given by

$$\Delta\mathbf{P} = (\tilde{\mathbf{J}}\mathbf{J})^{-1}\tilde{\mathbf{J}}\Delta\mathbf{I} \tag{13.235}$$

If the number of adjustable parameters, P_j, exceeds the number of experimentally determined quantities, I_i, or if some linear combinations of the parameters are not well determined by the experimental data, then $\tilde{\mathbf{J}}\mathbf{J}$ may be singular or nearly so. In this case, the matrix inversion leading to (13.235) is not possible. However, some linear combinations of the parameters are determined by the data. The inversion can be effectively obtained by an alternate procedure. The matrix of coefficients of the normal equations may be diagonalized by an orthogonal transformation, $\mathbf{T}$

$$\tilde{\mathbf{T}}\tilde{\mathbf{J}}\mathbf{J}\,\mathbf{T} = \mathbf{\Lambda} \tag{13.236}$$

As a result of the parameter scaling, (13.231), the size of a particular eigenvalue λ_j reflects how well determined is that linear combination of P_j described by the corresponding eigenvalue. Equation 13.234 can then be rewritten as

$$\tilde{\mathbf{T}}\tilde{\mathbf{J}}\mathbf{J}\,\mathbf{T}\tilde{\mathbf{T}}\Delta\mathbf{P} = \tilde{\mathbf{T}}\tilde{\mathbf{J}}\Delta\mathbf{I} \tag{13.237}$$

or

$$\mathbf{\Lambda}\tilde{\mathbf{T}}\Delta\mathbf{P} = \tilde{\mathbf{T}}\tilde{\mathbf{J}}\Delta\mathbf{I} \tag{13.238}$$

and inverting gives

$$\Delta\mathbf{P} = \mathbf{T}\mathbf{\Lambda}^{-1}\tilde{\mathbf{T}}\tilde{\mathbf{J}}\Delta\mathbf{I} \tag{13.239}$$

The shifts in the original parameters are then computed from

$$\Delta p_j = \sigma_j \sum_k T_{jk} \left(\frac{1}{\lambda_k}\right) \sum_l \sum_i T_{lk} J_{il} \Delta I_i \tag{13.240}$$

with $p_j = p_j^0 + \Delta p_j$ and where i runs up to the number of experimentally determined quantities, l runs up to the number of adjustable parameters, k runs over only those values for which $\lambda_k > \sigma^2$, with $\sigma^2 =$ the variance of the experimentally determined quantities.

If the index k runs over all the possible values (i.e., all the $\lambda_k > \sigma^2$), then keeping in mind that one may invert $\tilde{\mathbf{J}}\mathbf{J}$ by diagonalizing, inverting the eigenvalues, and then performing the inverse of the transformation indicated in (13.236), it is seen that (13.240) and (13.235) are equivalent (with $\sigma j = 1$). The difference comes when one or more $\lambda_k < \sigma^2$, for then that eigenvector is not used in computation of the parameter shifts. Once the parameter shifts have been computed, the model calculation is repeated to see if the improved parameters do indeed reproduce the experimental data. This is necessary because a linear least-squares approximation is used and must be tested for convergence.

The parameters obtained will be increasingly unrelated to the data as the eigenvalue decreases. The linear combinations which are better determined by the observations than by the *a priori* estimates will have the corresponding $\lambda_k > \sigma^2$. On the other hand, the *a priori* estimates are expected to be more reliable for $\lambda_k < \sigma^2$. The diagnostic least-squares method thus provides a systematic way of deciding whether the initial assumed parameter is better rather than the least-squares value. If σ is chosen to be too small, the parameters with small dependence on the data are fitted even though they are not determined well by the data. Unreasonable parameter shifts may then occur. Usually the eigenvalues are either much greater or much smaller than σ^2. A problem arises when an eigenvalue approaches σ^2. In such a case, the parameters should be calculated by including and excluding the particular eigenvalue to aid in the decision. The one major factor that must be remembered about this procedure is that the parameters computed are not unique but reproduce the rotational constants within the restraints of the initial estimated parameters and their uncertainties. Confidence intervals for the parameters can also be obtained from the diagnostic least-squares as discussed by Curl [35].

When $\tilde{\mathbf{J}}\mathbf{J}$ is singular, one or more eigenvalues will, in principle, be zero, and only certain linear combinations of the parameters are determined by the data. Because of rounding errors, no eigenvalue will be exactly zero. A parameter scaling technique which allows the eigenvalue size to actually reflect the linear dependence has been developed by Lees [34], and applied in the analysis of the internal rotation parameters V_3 and V_6 in methyl alcohol.

References

1. This conversion factor is based on recent recommended physical constants. In particular, Planck's constant and the conversion factor from amu to grams given in Appendix D has

been employed. In the literature two other values, based on earlier adopted constants, have been extensively used. They are: 505, 480 and 505, 531 amu Å^2 MHz.

2. W. Gordy and R. L. Cook, *Microwave Molecular Spectra*, Wiley-Interscience, New York, 1970, Appendix VIII.
3. M. D. Harmony, V. W. Laurie, R. L. Kuczkouski, R. H. Schwendeman, D. A. Ramsay, F. J. Lovas, W. J. Lafferty, A. G. Maki, *J. Phys. Chem. Ref. Data*, **8**, 619 (1979).
4. J. H. Callomon, E. Hirota, K. Kuchitsu, W. J. Lafferty, A. G. Maki and C. S. Pote, with assistance of L. Buck, and B. Starck, *Structure Data of Free Polyatomic Molecules*, Landolt-Bornstein, Group II: Atomic and Molecular Physics, Vol. 7, Springer-Verlag, Berlin, 1976. See also Chapter I, [65].
5. See also the discussion in Appendix A on the diagonalization of matrices.
6. For a compilation of structural parameters obtained from electron diffraction, microwave spectroscopy, and so on, up to 1955, see L. E. Sutton, D. G. Jenkins, A. D. Mitchell, L. C. Cross, H. J. M. Bowen, J. Donohue, O. Kennard, P. J. Wheatley, and D. H. Whiffen, *Tables of Interatomic Distances and Configuration in Molecules and Ions*, The Chemical Society, London, Special Publication **11**, 1958. Compilations of structural parameters obtained from spectroscopy are given in References 2, 3 and 4.
7. R. H. Schwendeman, *J. Mol. Spectrosc.*, **6**, 301 (1961).
8. J. Kraitchman, *Am. J. Phys.*, **21**, 17 (1953).
9. These relations may also be obtained from the considerations of [10].
10. In the theory of equations it is shown that the coefficients of a polynomial are related to the roots of the polynomial. Specifically, if P'_x, P'_y, P'_z are the roots of a cubic equation in P' then the polynomial equation may be written

$$(P')^3-(P'_x+P'_y+P'_z)(P')^2+(P'_xP'_y+P'_xP'_z+P'_yP'_z)P'-P'_xP'_yP'_z=0$$

Hence by expansion of the secular determinant (13.70) and equation of coefficients of like powers of P' with those in the foregoing equation, the desired relations between the coefficients and roots may be obtained. It will be recognized that two of the relations obtained indicate that the sum of the roots is equal to the trace of the matrix of P' (sum of diagonal elements) and that the product of the roots is equal to the determinant of P'.

11. H. D. Rudolph, *J. Mol. Spectrosc.*, **89**, 430 (1981).
12. A. Chutjian, *J. Mol. Spectrosc.*, **14**, 361 (1964).
13. L. Nygaard, *J. Mol. Spectrosc.*, **62**, 292 (1976).
14. Y. S. Li, K. L. Kizer, and J. R. Durig, *J. Mol. Spectrosc.*, **42**, 430 (1972).
15. G. K. Pandey and H. Dreizler, *Z. Naturforsch.*, **32a**, 482 (1977).
16. J. K. G. Watson, *J. Mol. Spectrosc.*, **41**, 229 (1972).
17. J. P. Pasinski and R. A. Beaudet, *J. Chem. Phys.*, **61**, 683 (1974).
18. See, for example, R. P. Bell and E. A. Guggenheim, *Trans. Faraday Soc.*, **32**, 1013 (1936) and [85].
19. E. B. Wilson and J. B. Howard, *J. Chem. Phys.*, **4**, 260 (1936); B. T. Darling and D. M. Dennison, *Phys. Rev.*, **57**, 128 (1940); H. H. Nielsen, *Phys. Rev.*, **60**, 794 (1941); W. H. Schaffer, *J. Chem. Phys.*, **10**, 1 (1942); A. H. Nielsen, *J. Chem. Phys.*, **11**, 160 (1943); H. H. Nielsen, *Rev. Mod. Phys.*, **23**, 90 (1951); H. H. Nielsen, *Handbuch der Physik*, Vol. 37, Springer Verlag, Berlin, 1959; K. T. Hecht, *J. Mol. Spectrosc.*, **5**, 355 (1960); I. M. Mills, "Vibration–Rotation Structure in Asymmetric and Symmetric-top Molecules," in *Molecular Spectroscopy: Modern Research*, K. N. Rao and C. W. Mathews, Eds. Academic, New York, 1972.
20. V. W. Laurie, *J. Chem. Phys.*, **28**, 704 (1958).
21. J. H. Van Vleck, *J. Chem. Phys.*, **4**, 327 (1936).

22. B. Rosenblum, A. H. Nethercot, and C. H. Townes, *Phys. Rev.*, **109**, 400 (1957).
23. E. F. Pearson and W. Gordy, *Phys. Rev.*, **177**, 52, 59 (1969).
24. J. K. G. Watson, *J. Mol. Spectrosc.*, **45**, 99 (1973).
25. P. R. Bunker, *J. Mol. Spectrosc.*, **42**, 478 (1972).
26. Y. Morino, Y. Kikuchi, S. Saito, and E. Hirota, *J. Mol. Spectrosc.*, **13**, 95, (1964).
27. Y. Morino and S. Saito, *J. Mol. Spectrosc.*, **19**, 435 (1966).
28. J. G. Smith, *Mol. Phys.*, **35**, 461 (1978).
29. Y. Kawashima and A. I. Cox, *J. Mol. Spectrosc.*, **65**, 317 (1977).
30. C. D. Esposti, P. G. Favero, S. Serenellini, and G. Cazzoli, *J. Mol. Struct.*, **82**, 221 (1982).
31. E. Hirota, *J. Mol. Spectrosc.*, **71**, 145 (1978).
32. A. M. Mirri, R. Cervellati, and G. Cazzoli, *J. Mol. Spectrosc.*, **83**, 202 (1980).
33. R. H. Schwendeman, "Structural Parameters from Rotational Spectra," in *Critical Evaluation of Chemical and Physical Structural Information*, D. R. Lide and M. A. Paul, Eds., National Academy of Science, Washington, D.C., 1974.
34. R. M. Lees, *J. Mol. Spectrosc.*, **33**, 124 (1970).
35. R. F. Curl, Jr., *J. Compt. Phys.*, **6**, 367 (1970).
36. G. H. Golub, in *Handbook for Automatic Computation*, Vol II, *Linear Algebra*, J. H. Wilkinson and C. Reinsch, Eds., Springer, Berlin, 1971.
37. B. T. Darling and D. M. Dennison, *Phys. Rev.*, **57**, 128 (1940).
38. T. Oka and Y. Morino, *J. Mol. Spectrosc.*, **6**, 472 (1961).
39. J. H. Meal and S. R. Polo, *J. Chem. Phys.*, **24**, 1119, 1126 (1956).
40. H. A. Jahn, *Phys. Rev.*, **56**, 680 (1939).
41. T. Oka and Y. Morino, *J. Mol. Spectrosc.*, **8**, 9 (1962); **11**, 349 (1963).
42. D. R. Herschbach and V. W. Laurie, *J. Chem. Phys.*, **40**, 3142 (1964).
43. R. J. Kurland and E. B. Wilson, Jr., *J. Chem. Phys.*, **27**, 585 (1957).
44. C. C. Costain and J. M. Dowling, *J. Chem. Phys.*, **32**, 158 (1960).
45. G. B. Little and M. C. L. Gerry. *J. Mol. Spectrosc.*, **71**, 321 (1978).
46. R. K. Kabar, *J. Chem. Phys.*, **56**, 1189 (1972).
47. R. K. Kabar, E. A. Rinehart, C. R. Quade, and T. Kojima, *J. Chem. Phys.*, **52**, 3083 (1970).
48. Y. Hanyu, C. O. Britt, and J. E. Boggs, *J. Chem. Phys.*, **45**, 4725 (1966).
49. J. H. Høg, L. Nygaard, and G. O. Sørensen, *J. Mol. Struct.*, **7**, 111 (1970).
50. W. M. Ralowski, P. J. Mjöberg, and S. O. Ljunggren, *J. Mol. Struct.*, **30**, 1 (1976).
51. R. A. Creswell, *J. Mol. Spectrosc.*, **51**, 111 (1974).
52. A. Bauder, C. Keller, and M. Neuenschwander, *J. Mol. Spectrosc.*, **63**, 281 (1976).
53. M. A. Masson, A. Bouchy, G. Roussy, G. Serratrice, and J. J. Delpuech, *J. Mol. Struct.*, 68, 307 (1980).
54. T. K. Avirah, R. L. Cook, and T. B. Malloy, Jr., *J. Mol. Spectrosc.*, **54**, 231 (1975).
55. B. Bak and S. Skaarup, *J. Mol. Struct.*, **10**, 385 (1971).
56. R. C. Loyd, S. N. Mathur, and M. D. Harmony, *J. Mol. Spectrosc.*, **72**, 359 (1978).
57. J. K. G. Watson, *J. Mol. Spectrosc.*, **48**, 479 (1973).
58. C. C. Costain, *J. Chem. Phys.*, **29**, 864 (1958).
59. If the error in the coordinate arises only from the uncertainty in ΔI, we may write from (13.46) of the text

$$\delta z = [(M+\Delta m)/2zM\Delta m]\delta(\Delta I)$$

Examination of this relation indicates that the error, δz, in the coordinate does not depend on the mass of the atom substituted and that the error decreases with increasing Δm. On the other hand, the closer an atom is to the center of mass (z small), the larger the error in the coordinate.

60. Re-evaluation of the equilibrium structure (see [86]) of HCN has given: r_e(C–H) = 1.0659 Å and r_e(C–N) = 1.1531 Å; the r_s(C–H) value given in Table 13.15 is some 0.003 Å smaller.
61. G. E. Herberich, R. H. Jackson, and D. J. Millen, *J. Chem. Soc.* **(A)**, 1966, p. 336.
62. R. C. Benson, W. H. Flygare, M. Oda, and R. Breslow, *J. Am. Chem. Soc.*, **95**, 2772 1973).
63. B. Bak, D. Christensen, W. B. Dixon, L. Hansen-Nygaard, and J. Rastrup-Anderson, *J. Chem. Phys.*, **37**, 2027 (1962).
64. L. Pierce, *J. Mol. Spectrosc.*, **3**, 575 (1959).
65. C. Kirby and H. W. Kroto, *J. Mol. Spectrosc.*, **83**, 130 (1980).
66. L. C. Krisher and L. Pierce, *J. Chem. Phys.*, **32**, 1619 (1960).
67. K. Yamada, M. Winnewisser, G. Winnewisser, L. B. Szalanski, and M. C. L. Gerry, *J. Mol. Spectrosc.*, **79**, 295 (1980).
68. P. Nösberger, A. Bauder, and Hs. H. Günthard, *Chem. Phys.*, **1**, 418 (1973).
69. C. Hirose, *Bull. Chem. Soc. Japan*, **47**, 976 (1974).
70. V. Typke, *J. Mol. Spectrosc.* **69**, 173 (1978).
71. P. Helminger, F. C. De Lucia, W. Gordy, P. A. Staats, and H. W. Morgan, *Phys. Rev.*, **10A**, 1072 (1974).
72. H. D. Rudolph, *J. Mol. Spectrosc.*, **89**, 460 (1981).
73. R. L. Kuczkowski, C. W. Gillies, and K. L. Gallaher, *J. Mol. Spectrosc.*, **60**, 361 (1976).
74. F. Mata, M. C. Martin, and G. O. Sørensen, *J. Mol. Struct.*, **48**, 157 (1978).
75. R. A. Beaudet and R. L. Poynter, *J. Chem. Phys.*, **53**, 1899 (1970).
76. U. Mazur and R. L. Kuczkowski, *J. Mol. Spectrosc.*, **65**, 84 (1977).
77. B. P. van Eijck, *J. Mol. Spectrosc.*, **91**, 348 (1982).
78. F. L. Tobiason and R. H. Schwendeman, *J. Chem. Phys.*, **40**, 1014 (1964).
79. J. Demaison and L. Nemes, *J. Mol. Struct.*, **55**, 295 (1979).
80. M. Winnewisser and H. K. Bodenseh, *Z. Naturforsch.*, **22a**, 1724 (1967); H. K. Bodenseh and M. Winnewisser, *Z. Naturforsch.*, **24a**, 1973 (1969).
81. M. Winnewisser and B. P. Winnewisser, *J. Mol. Spectrosc.*, **41**, 143 (1972).
82. B. M. Landsberg, *Chem. Phys. Lett.*, **60**, 265 (1979).
83. J. A. Duckett, A. G. Robiette and I. M. Mills, *J. Mol. Spectrosc.*, **62**, 19 (1976).
84. J. A. Duckett, A. G. Robiette, and M. C. L. Gerry, *J. Mol. Spectrosc.*, **90**, 374 (1981).
85. D. R. Herschbach and V. W. Laurie, *J. Chem. Phys.*, **37**, 1668 (1962).
86. V. W. Laurie and D. R. Herschbach, *J. Chem. Phys.*, **37**, 1687 (1962).
87. T. Oka, *J. Phys. Soc. Japan*, **15**, 2274 (1964).
88. T. Oka and Y. Morino, *J. Mol. Spectrosc.*, **8**, 300 (1962).
89. For a discussion of vibrational effects in electron diffraction data, and the relation of r_g to r_z, and so on, see for example: L. S. Bartell, *J. Chem. Phys.*, **23**, 1219 (1955); **38**, 1827 (1963); K. Kuchitsu and S. J. Cyvin, *Molecular Structures and Vibrations*, S. J. Cyvin, Ed., Elsevier, Amsterdam, 1972; V. W. Laurie, "Definitions and General Theory of Interatomic Distances," and K. Hedberg, "Critical Evaluation of Structural Information from Gaseous Electron Diffraction," in *Critical Evaluation of Chemical and Physical Structure Information*, D. R. Lide and M. A. Paul, Eds., National Academy of Science, Washington, D.C., 1974.
90. K. Takagi and T. Oka, *J. Phys. Soc. Japan*, **18**, 1174 (1963).
91. In (13.201) through (13.205) the average configuration moments of inertia should be used although the observed effective moments give sufficient accuracy.
92. For the molecules in Table 13.20 the inertial defect does not quite vanish (see references

quoted therein). The small residual inertial defects are a measure of the errors in the calculations such as those arising from uncertainties in the potential constants.

93. R. L. Cook, F. C. De Lucia and P. Helminger, *J. Mol. Struct.*, **28**, 237 (1975).
94. R. L. Cook, F. C. De Lucia and P. Helminger, *J. Mol. Spectrosc.*, **53**, 62 (1974).
95. W. S. Benedict, N. Gailar, and E. K. Plyler, *J. Chem. Phys.*, **24**, 1139 (1956).
96. J. L. Duncan, *Mol. Phys.*, **28**, 1177 (1974).
97. K. Kuchitsu, *J. Chem. Phys.*, **49**, 4456 (1968).
98. K. Kuchitsu, T. Fukuyama, and Y. Morino, *J. Mol. Struct.*, **1**, 463 (1968); *J. Mol. Struct.*, **4**, 41 (1969).
99. R. Stølevik, H. M. Seip, and J. J. Cyvin, *Chem. Phys. Lett.*, **15**, 263 (1972).
100. K. Kuchitsu and Y. Morino, *Bull. Chem. Soc. Japan*, **38**, 805 (1965).
101. J. G. Smith and J. K. G. Watson, *J. Mol. Spectrosc.*, **69**, 47 (1978).
102. M. Nakata, M. Sugie, H. Takeo, C. Matsumura, T. Fukuyama, and K. Kuchitsu, *J. Mol. Spectrosc.*, **86**, 241 (1981).
103. M. Nakata, T. Fukuyama, K. Kuchitsu, H. Takeo, and C. Matsumura, *J. Mol. Spectrosc.*, **83**, 118 (1980).

Chapter **XIV**

QUADRUPOLE COUPLINGS, DIPOLE MOMENTS, AND THE CHEMICAL BOND

1 INTRODUCTION

In Chapter IX we showed how nuclear quadrupole coupling constants are obtained from the hyperfine structure of molecular rotational spectra. Now we shall relate these constants to the electronic structure of the molecule. The coupling constants obtained from the spectra are of the form

$$\chi_z = eQq_z = eQ\left(\frac{\partial^2 V}{\partial z^2}\right) \tag{14.1}$$

where e is the charge on a proton, Q is the electric quadrupole moment of the nucleus, and $\partial^2 V/\partial z^2$ is the electric field gradient evaluated at the coupling nucleus. Since Q is known for many nuclei (see Appendix E), we shall here consider Q as a known constant and shall concentrate our attention on q, which depends on the electronic structure of the molecule. The reference axis z is fixed in the molecule and is a principal axis of the coupling tensor. In linear or symmetric-top molecules when the coupling nucleus is on the molecular symmetry axis, there is only one observable coupling constant; the reference axis z is the molecular symmetry axis. In asymmetric-top molecules the coupling is often symmetric about a bond to the atom having the coupling nucleus. In the general case, there will be three principal values, χ_x, χ_y, and χ_z, in the coupling tensor of each nucleus; but, as explained in Chapter IX, there are at most only two independent coupling constants. Although the coupling constants are obtained with respect to the principal axes, it is usually convenient for interpretation of the coupling to express the coupling constants with respect to a coordinate system x, y, z, where the z axis is along the bond axis of the coupling atom. This transformation may be accomplished without a knowledge of the off-diagonal elements, for example, χ_{ab}, if the assumption can be made that the x, y, z axes are principal axes of the field gradient tensor (see Chapter IX, Section 4). Since the coupling constants are evaluated in the molecule-fixed reference system, they are independent of rotational state, except for slight centrifugal distortion effects which are generally negligible. Hence coupling constants of molecules measured from microwave rotational spectra are directly comparable to coupling constants measured with pure quadrupole resonance in solids although some differences are expected because of intramolecular interaction in the solids.

The first calculation of a nuclear quadrupole coupling constant in a molecule was that by Nordsieck [1] for HD. Because the bonding in this molecule is essentially through s orbitals, the very small D coupling observed in molecular beam resonance experiments is due to orbital distortion effects. For this reason the theory derived for HD is not applicable to couplings in the more complex molecules usually observed with microwave spectroscopy. Approximate methods for interpretation of nuclear quadrupole coupling in complex molecules in terms of chemical bond properties were developed in the early period of microwave spectroscopy by Townes and Dailey [2–4]. These methods were modified and extended by Gordy [5–7]. Molecular orbital treatment of quadrupole coupling of halogen nuclei in unsaturated organic molecules has been given by Bersohn [8] and by Goldstein [9], who showed that the asymmetry in the coupling is due to π bonding by the halogen. Reviews of quadrupole coupling are available [10–15].

Nuclear quadrupole coupling in many atoms is accurately known from atomic beam resonance experiments. The coupling in others can be calculated with useful accuracy when the quadrupole moment is known. The simplest approach to the interpretation of the couplings in molecules is to relate them to known or calculable atomic couplings. This procedure is most suited to application of the valence bond theories and to LCAO molecular orbital theory which expresses the electronic wave functions of the molecule in linear combinations of the atomic functions. Appropriate methods for interpretation of molecular coupling in terms of known atomic couplings are discussed in the following sections.

Although this chapter is primarily devoted to the interpretation of nuclear quadrupole coupling, we give in Section 14 an elementary interpretation of molecular dipole moments in terms of chemical bond properties. While nuclear quadrupole coupling is sensitive only to the charge distribution near the coupling nucleus, the molecular dipole moment depends strongly on the charge distribution over the entire molecule. Information about the electronic structure of molecules obtained from these two molecular parameters is mostly complementary. In some diatomic molecules for which bond ionic character is derivable from both quadrupole coupling and dipole moments, the results are shown to be consistent.

2 FIELD GRADIENTS OF ELECTRONS IN MOLECULES

An unscreened electronic charge e at distance r from a nucleus will give rise to a potential of e/r at this nucleus. In a coordinate system x, y, z, with origin at the nucleus and with z along the bond axis, $r=(x^2+y^2+z^2)^{1/2}$, the electric field gradient at the nucleus due to this electron is

$$\frac{\partial^2 V}{\partial z^2}=\frac{\partial^2}{\partial^2 z}\left(\frac{e}{r}\right)=e\frac{\partial}{\partial z^2}(x^2+y^2+z^2)^{-1/2}=\frac{e(3\cos^2\theta-1)}{r^3} \tag{14.2}$$

where θ is the angle between r and the reference axis z. To obtain the contribution by an electron of a molecule to the field gradient at the nucleus of one of its atoms, say atom A, it is necessary to average the above quantity over the molecular orbital ϕ_i of the electron. Since ϕ_i is normalized so that $\int \phi_i^*\phi_i\,d\tau=1$, this average may be expressed as

$$q_i^A = e \int \phi_i^* \left(\frac{3 \cos^2 \theta_A - 1}{r_A^3} \right)_i \phi_i \, d\tau \tag{14.3}$$

The total field gradient q^A at the nucleus of atom A due to all the electrons in the molecule is

$$q_{\text{mol}}^A = e \sum_i n_i \int \phi_i^* \left(\frac{3 \cos^2 \theta_A - 1}{r_A^3} \right)_i \phi_i \, d\tau \tag{14.4}$$

where $n_i = 1, 2$, or 0 is the number of electrons in the ith orbital and the summation is taken over all orbitals of the molecule.

Evaluation of the foregoing integral for all the electrons in a typical molecule seems a formidable problem indeed. However, with reasonable approximations and assumptions, the integral can be evaluated for many molecules to useful accuracy.

Let us consider the field gradient at the nucleus of atom A caused by the electrons of a diatomic molecule AB. The molecular orbitals can be expressed as a linear combination of atomic orbitals ψ_i^A and ψ_i^B of atoms A and B.

$$\phi_i = a_i \psi_i^A + b_i \psi_i^B \tag{14.5}$$

Substitution of this function into (14.4) gives the resultant field gradient at nucleus A as

$$q_{\text{mol}}^A = e \sum_i n_i \int (a_i \psi_i^{*A} + b_i \psi_i^{*B}) \left(\frac{3 \cos^2 \theta_A - 1}{r_A^3} \right)_i (a_i \psi_i^A + b_i \psi_i^B) \, d\tau \tag{14.6}$$

where the summation is taken over all electrons of the molecule. Expansion gives

$$\begin{aligned} q_{\text{mol}}^A = {} & e \sum_i n_i a_i^2 \int \psi_i^{*A} \left(\frac{3 \cos \theta_A - 1}{r_A^3} \right)_i \psi_i^A \, d\tau \\ & + 2e \sum_i n_i a_i b_i \int \psi_i^{*A} \left(\frac{3 \cos^2 \theta_A - 1}{r_A^3} \right)_i \psi_i^B \, d\tau \\ & + e \sum_i n_i b_i^2 \psi_i^{*B} \left(\frac{3 \cos^2 \theta_A - 1}{r_A^3} \right)_i \psi_i^B \, d\tau \end{aligned} \tag{14.7}$$

The last term of (14.7) gives the contributions to the field gradient at the nucleus of atom A by all the electrons on atom B. This term is small because of the inverse cube variation of the field gradient with distance and is counterbalanced by an opposite contribution of very nearly the same magnitude from the nuclear charge of B which has been ignored in (14.2). For these reasons the last term, as well as the contribution from the nuclear charge of B, will be neglected. The second term on the right is a contribution from the electronic charge in the overlap region and is zero except for electrons in bonding molecular orbitals and is small as compared with the first term on the right, even for bonding orbitals. For reasons described in Section 7, the cross terms for bonding orbitals in the normalization can be neglected, and the wave functions ϕ_i can be normalized by

$$a_i^2 + b_i^2 = 1 \tag{14.8}$$

This normalisation, in effect, divides overlap charge between the two atomic orbitals. Thus, to a good approximation, the coupling field gradient at nucleus A can be expressed as the summation of the weighted contributions from the electronic charges in the atomic orbitals of atom A. Thus

$$q_{\text{mol}}^A = \sum_i n_i a_i^2 q_i^A \tag{14.9}$$

where

$$q_i^A = e \int \psi_i^{*A} \left(\frac{3 \cos^2 \theta_A - 1}{r_A^3} \right)_i \psi_i^A \, d\tau \tag{14.10}$$

and a_i^2 is a measure of the electronic charge density in the ith atomic orbital ψ_i^A.

Equation 14.9 is easily generalized to polyatomic molecules for which the wave function ϕ_i must be represented by a linear combination of atomic wave functions of N atoms

$$\phi_i = \sum_{j=1}^{N} c_{ji} \psi_{ji} \tag{14.11}$$

where the index j runs over the different atoms. If effects of overlap distortions and of contributions to the field gradient at the kth nucleus by electronic and nuclear charges of other atoms of the molecule are neglected, substitution of this wave function in (14.4) yields the field gradient at the nucleus of atom k

$$q_{\text{mol}}^k = \sum_i \sum_j n_i c_{ki} c_{ji} q_{ki,ji} \tag{14.12}$$

where

$$q_{ki,ji} = e \int \psi_{ki}^* \left(\frac{3 \cos^2 \theta_k - 1}{r_k^3} \right)_i \psi_{ji} \, d\tau \tag{14.13}$$

At most, only overlap terms with atoms bonded to the coupling atom k need be considered, and generally even these can be neglected. Therefore, the field gradient at the kth nucleus in a polyatomic molecule is given quite closely by

$$q_{\text{mol}}^k = \sum_i n_i c_{ki}^2 q_{ki,ki} \tag{14.14}$$

where the c's are normalized by

$$\sum_j c_{ji}^2 = 1 \tag{14.15}$$

and where $q_{ki,ki}$ is the field gradient at the kth nucleus caused by an electron in an atomic orbital ψ_{ki} of the coupling atom k.

It should be realized that the atomic orbitals of the atom bonded in the molecule are not necessarily the same as those of the free atoms. The valence shell atomic orbitals ψ_{ki} might be an admixture of two or more orbitals of the free atom k. Also, the field gradient at the nucleus of atom k might differ slightly from that of the free atom because of changes in nuclear screening. To evaluate the coefficients of c_{ki} from the observed quadrupole coupling or to predict the coupling from the c_{ki} values calculated from molecular orbital theory, one must express the atomic orbitals of the bonded atoms in terms of those of the free atoms. Methods for doing this are described in Sections 3–8.

With ψ_{ki} expressed as a linear combination of hydrogenlike wave functions ψ_{nlm_l}, the integrals involved in (14.14) will vanish unless $l'=l\neq 0$ or $l'=l\pm 2$ and $m_l'=m_l$. Many other terms can be ignored on energetic grounds. Cross terms, for example, involving $s\,(l=0)$ and $d\,(l=2)$ orbitals, although nonvanishing, can be neglected since the integrals are small; also mixing of s and d orbitals is usually small, and hence the integral will be multiplied by a small weighting coefficient. Integrals of the type $n=n'$, $l=l'$ are found to decrease in magnitude with increasing n or l (Section 3). The implications of these and other considerations are summarized later.

The coefficients of c_{ji}^2 of (14.11) represent the weights of the various atomic orbital constituents in the molecular orbital ϕ_i. For polyatomic molecules of high symmetry, such as planar organic ringed groups, molecular orbital methods like those described in texts by Coulson [16] or Pullman and Pullman [17] may be used for calculation of the values of these coefficients which can, in principle, be compared with experimental values obtained from nuclear quadrupole couplings. Unfortunately, such calculations give reliable predictions mainly for π orbital densities on the ring carbons, which have no quadrupole moments. Nitrogen atoms, often bonded in unsaturated ringed systems, have quadrupole moments, but the coupling of ^{14}N is so small as to be difficult to measure accurately in such systems. Quadrupole couplings of halogen atoms attached to planar rings have been treated with molecular orbital theory [8, 18]. Application of molecular orbital theory for prediction of quadrupole coupling for the molecules of low symmetry that are usually observed in microwave spectroscopy is difficult because a high degree of symmetry is required for reduction of unknown parameters of the secular equations. It is evident that quadrupole coupling alone cannot provide very complete information about the molecular orbitals of complex molecules. However, it can give useful information about the bonding to the particular atom for which the quadrupole coupling is measured. Even in the simpler diatomic molecules the unknown parameters are generally more numerous than the measurable coupling constants, and assumptions are necessary in the interpretation of the quadrupole coupling. By correlation of the quadrupole coupling of the same atoms in a number of different molecules, the reasonableness of the approximations and assumptions can be subjected to a consistency test. In this way much has been learned about the nature of chemical bonding.

Simplifying Conditions

Certain reasonable approximations and assumptions can be made that greatly simplify the evaluation of the field gradients of coupling nuclei as expressed by (14.9) or (14.12). (*1*) Electrons in closed subvalence shells can be neglected. Because of spherical symmetry, their resultant contribution is zero except for slight distortion effects such as the Sternheimer corrections [19, 20]. (*2*) Contributions by electronic charge of *s* orbitals of the coupling atom are zero because of spherical symmetry of the *s* orbital. (*3*) Contributions by *d* electrons are negligible in comparison to those of *p* electrons for most molecules. (*4*) Contributions by electronic and nuclear charges of atoms other than those of the coupling nuclei approximately cancel and can generally be neglected. (*5*) The predominant contribution to the field gradient at the nucleus in most molecules is that from *p* electrons of the valence shell of the coupling atom. Other orbitals such as *s* or *d* influence the coupling indirectly by altering the filling of the *p* orbitals via hybridization. (*6*) Contributions by electrons in non-bonding or antibonding electrons in the valence shell of the coupling atoms are the same as those of free atoms except for possible differences in orbital orientation, orbital hybridization, and small effects that are due to change in nuclear screening when there is a formal charge on the coupling atom. (*7*) Contribution by an electron in a bonding molecular orbital i to the field gradient at the kth nucleus is assumed to be that of an electron in the constituent bonding atomic orbital of the coupling atom k times the weighting coefficient c_{ki}^2, which gives the fractional contribution of the atomic orbital to the bonding molecular orbital. (*8*) In calculation of the weighting coefficients c_{ki}^2, either valence bond theory or molecular orbital theory may be applied, with effects of orbital overlap neglected. (*9*) Contributions by electrons in hybridized atomic orbitals are taken to be the sum of the contributions by the fractional constituent orbitals, again with effects of orbital overlap neglected.

With these simplying approximations it is evident that interpretation of the nuclear quadrupole coupling of a bonded atom for which Q and q_{n10} are known (Section 3) depends on the degree and manner in which the p orbitals of the coupling atoms are filled, that is, on the number of unbalanced p electrons in the valence shell (Section 4).

3 FIELD GRADIENTS OF ATOMIC ELECTRONS

The contribution by an electron in an atomic orbital to the field gradient at the atomic nucleus, if effects of electron spin are neglected, can be expressed as

$$q_{nlm_l}=e\int\psi_{nlm_l}^{*}\left(\frac{3\cos^2\theta-1}{r^3}\right)\psi_{nlm_l}\,d\tau \tag{14.16}$$

where e is the electronic charge. The average of r depends only on n and l, and the average of θ only on l and m_l where n, l, and m_l, are, respectively, the principal, the orbital angular momentum, and the magnetic quantum numbers. Therefore

$$q_{nlm_l}=e(l,m_l|3\cos^2\theta-1|l,m_l)\left(n,l\left|\frac{1}{r^3}\right|n,l\right)=-2e\frac{[3m_l^2-l(l+1)]}{(2l-1)(2l+3)}\left\langle\frac{1}{r^3}\right\rangle \tag{14.17}$$

In the usual observations in atomic spectra, l has its maximum projection along the space-fixed reference axis z so that $m_l = l$, and thus the effective value of q is

$$q_{nll} = -\frac{2el}{2l+3}\left\langle\frac{1}{r^3}\right\rangle \tag{14.18}$$

In the bonded atoms where the reference axis is along the σ bond formed by the orbital, m_l equals zero, and the effective value of q is

$$q_{nl0} = \frac{2el}{2l+3}\,\frac{l+1}{2l-1}\left\langle\frac{1}{r^3}\right\rangle \tag{14.19}$$

Comparison of (14.18) and (14.19) shows that

$$q_{nl0} = \frac{l+1}{2l-1}\,q_{nll} \tag{14.20}$$

For p orbitals, $l=1$ and

$$q_{n10} = -2q_{n11} \tag{14.21}$$

For s electrons, l equals zero, and it is evident from (14.18) that the field gradient is zero. The difficulty in evaluation of q comes in obtaining the average $\langle 1/r^3\rangle = (n, l|1/r^3|n, l)$. Because of uncertainty in the wave functions, this average cannot be directly calculated with very useful accuracy except for hydrogen or for highly ionized atoms which have hydrogenlike orbitals. However, other measurable quantities, notably the magnetic hyperfine structure and the doublet fine structure, also depend on $\langle 1/r^3\rangle$ average, and in many complex atoms the desired values of $\langle 1/r^3\rangle$ for p orbitals can be obtained from measured values of these splittings.

Before discussion of the more accurate, semi-empirical methods for obtaining $\langle 1/r^3\rangle$ for p electrons, let us consider the theoretical relation [21, 22] for hydrogenlike orbitals. If all electrons except the one to be considered are removed, the effective nuclear charge number would be Z, the atomic number. When relativistic effects are neglected, we would have

$$\left\langle\frac{1}{r^3}\right\rangle = \frac{Z^3}{a_0^3 n^3 l(l+\frac{1}{2})(l+1)} \tag{14.22}$$

where a_0 is the radius of the first Bohr orbit and n is the total quantum number. Because of the nuclear screening by other electrons in polyelectron alkalilike atoms, the atomic number Z must be replaced by $Z_{\text{eff}} = Z - \sigma$ where σ is a screening constant. An obvious difficulty in applying the relation is in obtaining Z_{eff}. One procedure is to estimate σ from the observed term energies of the atom by use of the relation [21]

$$E_n = -\frac{RZ_{\text{eff}}^2}{n^2} = -\frac{R(Z-\sigma)^2}{n^2} \tag{14.23}$$

in which R is the Rydberg constant. Another procedure is to employ the true Z and alter n in such a way as to get the term energy

$$E_n = -\frac{RZ^2}{n^{*2}} = -\frac{RZ^2}{(n-\mu)^2} \tag{14.24}$$

where μ is the quantum defect or Rydberg correction. A less distorted treatment is to apply corrections to both Z and n, but this requires additional relations for evaluation of the two parameters.

Substitution of (14.22) into (14.18) with Z_{eff} for Z yields

$$q_{nll} = -\frac{4eZ_{\mathrm{eff}}^3}{a_0^3 n^3 (l+1)(2l+1)(2l+3)} \tag{14.25}$$

Without quantitative knowledge of the values of Z_{eff}, some important qualitative deductions can be made from this formula. (*1*) We know that the screening will increase rapidly with increase of n because the higher n orbitals are, on the average, further from the nucleus and hence more screened by the electrons of the inner shells. Therefore, excited states with n values greater than those of the valence shells will have q values considerably less than those of the valence shells not only because of the inverse cube variation with n but because of the large decrease in Z_{eff}. For neutral atoms Z_{eff} approaches unity for n values that are large as compared with the ground state values. (*2*) Evidence indicates that the d electrons are more effectively screened from the nucleus than are the p electrons of the same shell. In addition, the q value for d electrons is reduced from that for p electrons by a factor of $\frac{2}{7}$ by the change in l from 1 to 2. Together, these effects cause the q of d electrons in most atoms to be less by an order of magnitude than that of p electrons of the same shell (same n value). (*3*) Equation 14.25 also shows why a formal charge on the atom increases, and a negative formal charge decreases, the field gradient of the orbitals. Removal of an electron deletes the screening by that electron and thus increases Z_{eff}, whereas addition of an electron correspondingly increases the screening and decreases Z_{eff}. According to Slater's rule [23], an electron in an s or p orbital of the valence shell has a screening constant of 0.35 for other p electrons of this shell.

The most widely applicable method for obtaining $\langle 1/r^3 \rangle$ to useful accuracy for p electrons in complex atoms makes use of measurements of doublet fine structure. The fine structure splitting is a measure of the spin–orbit coupling which, in turn, is a function of $\langle 1/r^3 \rangle$. The well-known formula expressing this relationship [21, 22] is

$$\left\langle \frac{1}{r^3} \right\rangle = \frac{\Delta\nu}{a_0^3 R\alpha^2 (l+\frac{1}{2}) Z_{\mathrm{eff}} H_r} \tag{14.26}$$

in which $\Delta\nu$ is the fine structure doublet frequency in cm^{-1}, a_0 is the radius of the first Bohr orbit, R is the Rydberg constant, α is the fine structure constant, Z_{eff} is the effective atomic number, and H_r is a small relativistic correction. Fine structure doublets of p electrons have been measured for most of the common elements. Obtaining the effective atomic number Z_{eff} is difficult. Barnes and Smith [24] have used the term values to obtain Z_{eff} values for a

Table 14.1 Values of q_{n10} Derived from the Calculations of Barnes and Smith [24]

Atom	*Configuration*	*State*	$\left\langle \left(\frac{a_0}{r}\right)^3 \right\rangle$ [a]	q_{n10} (10^{15} *esu*)[b]
Li	$2p$	^2P	0.0387	−0.100
Be	$2s2p$	^3P	0.173	−0.45
B	$2p$	^2P	0.608	−1.58
C	$2p^2$	^3P	1.23	−3.19
N	$2p^3$		(2.46)	−(6.4)
O	$2p^4$	^3P	4.29	−11.1
F	$2p^5$	^2P	6.55	−17.0
Na	$3p$	^2P	0.243	−0.63
Mg	$3s3p$	^3P	0.77	−2.00
Al	$3p$	^2P	1.28	−3.32
Si	$3p^2$	^3P	2.30	−5.97
P	$3p^3$		(3.46)	−(9.0)
S	$3p^4$	^3P	4.99	−13.0
Cl	$3p^5$	^2P	7.11	−18.6
K	$4p$	^2P	0.434	−1.13
Ca	$4s4p$	^3P	1.12	−2.91
Ga	$4p$	^2P	3.42	−8.88
Ge	$4p^2$	^3P	5.61	−14.6
As	$4p^3$		(7.32)	−(19)
Se	$4p^4$	^3P	9.36	−24.3
Br	$4p^5$	^2P	13.16	−34.2
Rb	$5p$	^2P	0.818	−2.12
Sr	$5s5p$	^3P	1.93	−5.0
In	$5p$	^2P	5.40	−14.0
Sn	$5p^2$	^3P	10.5	−27.3
Sb	$5p^3$		12.2	−30.7
Te	$5p^4$	^3P	14.0	−36.3
I	$5p^5$	^2P	16.8	−43.6
Cs	$6p$	^2P	1.20	−3.12
Ba	$6s6p$	^3P	2.64	−6.85
Tl	$6p$	^2P	9.91	−25.7
Pb	$6p^2$	^3P	13.4	−34.8
Bi	$6p^3$		20.4	−53.0

[a] The values of $\langle (a_0/r)^3 \rangle$ given here are those of Barnes and Smith with the relativistic correction H_r applied.

[b] Coupling per p electron for different nuclear isotopes can be obtained with these q_{n10} values and the Q values of Appendix E. In MHz units, $eQq_{n10}(\text{MHz}) = 72.5\, Q(10^{-24}\ \text{cm}^2)\, q_{n10}(10^{15}\ \text{esu})$. More precise values of eQq_{n10} for some nuclei are given in Table 14.2.

large number of excited and/or ionized states of atoms and by projection of these values have concluded that the relation

$$Z_{\rm eff}=Z-n \tag{14.27}$$

provides a good approximation to $Z_{\rm eff}$. These are similar to the earlier rules [21, 22], $Z_{\rm eff}=Z-2$ for the first period and $Z_{\rm eff}=Z-4$ for higher periods. With values of $\Delta\nu$ obtained from the literature and with $Z_{\rm eff}$ from (14.27), Barnes and Smith calculated values for $\langle 1/r^3\rangle$ for p orbitals of a large number of elements as shown in Table 14.1. From these values the field gradients q_{n10} listed in the last column of Table 14.1 are calculated from the relation

$$q_{n10}=-2.596\left\langle\left(\frac{a_0}{r}\right)^3\right\rangle\times 10^{15}\ \text{esu} \tag{14.28}$$

which is obtained from a combination of (14.26) with (14.19) and substitution of the values for the numerical constants and $l=1$ for p electrons. Values for the p^3 configurations, that is, those for the Group V elements N, P, and so on, which have spherically symmetric charge distributions, are considered most questionable by Barnes and Smith because there is no definitive way to infer the doublet splitting equivalent to that of a single p electron. Consequently, these doubtful values obtained by interpolation are shown in parentheses.

The original tabulation of Barnes and Smith [24] did not include the relativistic correction. They, in effect, tabulated $\langle(a_0/r)^3\rangle H_r$. Although this correction is negligible for first- and second-row elements, it increases with Z and becomes as large as 10% for elements with $Z=60$. Kopfermann [22] provides tabulated numerical values of H_r for all the elements. In Table 14.1 we have used his tabulation to correct the Barnes and Smith values for the relativistic effect. These corrections lower the magnitude of q. The values of q_{n10} listed in Table 14.1 are still not corrected for the Sternheimer polarization [19, 20] which would have the effect of lowering the values slightly more. The accuracy of the field gradients is sufficient, however, for use in most interpretations of molecular quadrupole couplings where other uncertainties are often greater.

The most accurate values of q are those obtained from atomic beam measurements of the combined magnetic and quadrupole coupling constants commonly designated by a and b, respectively, in atoms which have a single unbalanced p electron in the valence shell. Boron in its ground $^2P_{1/2}$ state has the configuration $2s^22p$ with one p electron outside the closed $2s$ subshell. The important halogen atoms have $^2P_{3/2}$ ground states. Each of them lacks the single electron needed to close the p valence shell and hence has a single unbalanced and unpaired electron. The single electron hole in the p valence shell of the halogens gives a field gradient of opposite sign to that of a single p electron outside a closed shell, as in boron. Otherwise, its quadrupole coupling can be treated in a similar manner.

The nuclear magnetic coupling constant of an unpaired non-s electron can be expressed as

$$a = \frac{2\beta\mu_I l(l+1)}{Ij(j+1)} \left\langle \frac{1}{r^3} \right\rangle \tag{14.29}$$

where β is the Bohr magneton, μ_I is the nuclear magnetic moment, I is the nuclear spin, and j is the total electronic angular momentum quantum number. This expression does not include a small relativistic correction calculated by Casimir [25]. Methods for evaluation of the relativistic corrections are described in Kopfermann's monograph [22] on nuclear moments. The nuclear spins and magnetic moments of all abundant isotopes are now known from magnetic resonance (see Appendix E). Hence, when a can be measured for atoms having a single unbalanced electron, the $\langle 1/r^3 \rangle$ value can be evaluated from (14.29). This method gives the most accurate evaluation of $\langle 1/r^3 \rangle$ for the lighter elements. For heavier elements, however, Koster [26] has pointed out that configuration interaction reduces the accuracy of the values. In addition to possible configuration interaction within the valence shell, there is a slight polarization of the inner electron shells by the asymmetric charge distribution of the valence shell predicted by Sternheimer [20]. Some values of field gradients determined from magnetic hyperfine structure are listed in Table 14.2. Some of these values include corrections for configuration interactions and the Sternheimer correction, as indicated.

When the ratio of the magnetic coupling to the quadrupole coupling constant $b = -eQq_{nll}$ can be measured for the same unbalanced electron in accurate atomic beam experiments, precise values of the nuclear quadrupole moment Q can be obtained. The quadrupole coupling in such an atom, in which electron spin is included, is given by

$$b = -eQq_{nll} = -e^2Q \frac{2l}{2l+3} \left\langle \frac{1}{r^3} \right\rangle \tag{14.30}$$

for the state of maximum projection along z ($m_j = j = l + \frac{1}{2}$). Division of (14.30) by (14.29) shows that

$$\frac{b}{a} = -\frac{e^2QIj(j+1)}{\beta\mu_I(l+1)(2l+3)} \tag{14.31}$$

The more accurate values of Q listed in Table 14.2 are those obtained from atomic beam experiments.

For interpretation of molecular quadrupole coupling in terms of chemical bond properties we shall need the quantity eQq_{n10}, which is the nuclear quadrupole coupling of an individual electron in an atomic p orbital of the valence shell n having zero angular momentum, $m=0$, along the reference axis. This corresponds to the orientation of the symmetry axis of the orbital along the reference axis, as occurs for a p orbital employed in formation of a σ bond. The π bonds are formed by p orbitals with $m = \pm 1$ and therefore have their symmetry axes perpendicular to the reference, σ-bond axis.

In atomic experiments the quadrupole coupling constant b measured for a

Table 14.2 Quadrupole Coupling by Atomic p Electrons

Nuclear Isotope	*Nuclear Spin*	Q $(10^{-24}\ cm^2)$	*Electron*	$2(eQq)_{\text{atom}}$ eQq_{n10}(MHz)	*Ref.*
^{10}B	3	+0.0740	2p	−11.83	a
^{11}B	$\frac{3}{2}$	+0.0357	$2p$	−5.39	a
^{14}N	1	+0.1	$2p$	−10	b
^{17}O	$\frac{5}{2}$	−0.026	$2p$	+21	c
^{25}Mg	$\frac{5}{2}$	+0.22	$3p$	−16	d
^{27}Al	$\frac{5}{2}$	+0.155	$3p$	−37.52	e
^{33}S	$\frac{3}{2}$	−0.055	$3p$	+52	c
^{35}Cl	$\frac{3}{2}$	−0.0795	$3p$	+109.74	f
^{37}Cl	$\frac{3}{2}$	−0.0621	$3p$	+86.51	f
^{69}Ga	$\frac{3}{2}$	+0.190	$4p$	−125.0	g
^{71}Ga	$\frac{3}{2}$	+0.120	$4p$	−78.80	g
^{73}Ge	$\frac{9}{2}$	−0.224	$4p$	+224	h
^{75}As	$\frac{3}{2}$	+0.29	$4p$	−400	c
^{79}Br	$\frac{3}{2}$	+0.31	$4p$	−769.76	i
^{81}Br	$\frac{3}{2}$	+0.26	$4p$	−643.03	i
^{115}In	$\frac{9}{2}$	+0.82	$5p$	−899.10	j
^{121}Sb	$\frac{5}{2}$	−0.29	$5p$	+650	c
^{123}Sb	$\frac{7}{2}$	−0.37	$5p$	+830	c
^{127}I	$\frac{5}{2}$	−0.79	$5p$	+292.71	k
^{201}Hg	$\frac{3}{2}$	+0.045	$6p$	−780	l
^{209}Bi	$\frac{9}{2}$	−0.4	$6p$	+1500	m

[a]V. S. Korolkov and A. G. Makhanek, *Opt. i Spektroskopiya*, **12**, 163 (1962); *Opt. Spectrosc. USSR (English Transl.)*, **12**, 87 (1962).
[b]See text, Section 13.
[c]Calculated with Q from Column 3 and q_{n10} from Table 14.1.
[d]A. Lurio, *Phys. Rev.*, **126**, 1768 (1962).
[e]H. Lew, *Phys. Rev.*, **76**, 1086 (1949).
[f]V. Jaccarino and J. G. King, *Phys. Rev.*, **83**, 471 (1951).
[g]G. F. Koster, *Phys. Rev.*, **86**, 148 (1952).
[h]W. J. Childs, L. S. Goodman, and L. J. Kieffer, *Phys. Rev.*, **120**, 2138 (1960).
[i]J. G. King and V. Jaccarino, *Phys. Rev.*, **94**, 1610 (1954).
[j]T. G. Eck and P. Kusch, *Phys. Rev.*, **106**, 958 (1957).
[k]V. Jaccarino, J. G. King, R. A. Satten, and H. H. Stroke, *Phys. Rev.*, **94**, 1798 (1954).
[l]H. G. Dehmelt, H. G. Robinson, and W. Gordy, *Phys. Rev.*, **93**, 480 (1954).
[m]Robinson et al. [58].

single coupling p electron corresponds to the orientation of $m=\pm 1$ and can be expressed as

$$b=-eQq_{n,1,\pm 1}=\tfrac{1}{2}eQq_{n10} \tag{14.32}$$

Therefore

$$eQq_{n10}=2b=2(eQq)_{\text{atom}} \tag{14.33}$$

Table 14.2 gives values of eQq_{n10} thus obtained from the measured atomic constant b. Sometimes eQq_{n10} derived from atomic spectra is indicated as $(eQq)_{\text{at}}$, although this quantity is twice the observable atomic coupling constant b. Q is expressed in units of cm^2, and sometimes q is expressed in units of cm^{-3}. When this is done, the coupling must be represented by e^2Qq since $q(\text{esu}) = eq(\text{cm}^{-3})$. Also, it is apparent that the coupling in frequency units is eQq/h. For convenience, however, h is usually omitted and the coupling constant specified in frequency units simply as eQq.

4 COUPLING BY UNBALANCED *p* ELECTRONS

The field gradient which gives rise to the nuclear quadrupole coupling in most molecules is due primarily to an unequal filling of the p orbitals of the valence shell of the coupling atoms. We shall therefore give a description of the coupling arising from the unbalanced p electronic charge of the valence shell.

The field gradient at the nucleus of an atom caused by a single electron in an atomic p_z orbital is, from (14.16),

$$q_z(p_z) = e\left\langle \frac{1}{r^3} \right\rangle \langle 3\cos^2\theta_z - 1\rangle \tag{14.34}$$

where e is the electronic charge and where the average is taken over the p_z orbital which has its axis of symmetry along z. The field gradient $q_z(p_z)$ of the p_z electron with reference to some other axis g that passes through the nucleus can be obtained by a transformation of the angular-dependent term in (14.34) with the relation

$$\langle 3\cos^2\theta_g - 1\rangle = \tfrac{1}{2}(3\cos^2\alpha - 1)\langle 3\cos^2\theta_z - 1\rangle \tag{14.35}$$

where α is the angle between the symmetry axis of the p_z orbital and the reference axis g. With this relation

$$\begin{aligned} q_g(p_z) &= e\left\langle \frac{1}{r^3} \right\rangle \langle 3\cos^2\theta_g - 1\rangle \\ &= \tfrac{1}{2}e(3\cos^2\alpha - 1)\left\langle \frac{1}{r^3} \right\rangle \langle 3\cos^2\theta_z - 1\rangle \end{aligned} \tag{14.36}$$

Therefore, the contribution to the field gradient in the g direction by an electron in a p_z orbital may be written as

$$q_g(p_z) = \tfrac{1}{2}(3\cos^2\alpha - 1)q_z(p_z) \tag{14.37}$$

When x or y is chosen as the reference axis $\alpha = 90°$, and

$$q_x(p_z) = q_y(p_z) = -\tfrac{1}{2}q_z(p_z) \tag{14.38}$$

Similarly, the gradient of a p_x electron with reference to the y or z axis is

$$q_y(p_x) = q_z(p_x) = -\tfrac{1}{2}q_x(p_x) \tag{14.39}$$

or of a p_y electron

$$q_x(p_y)=q_z(p_y)=-\tfrac{1}{2}q_y(p_y) \tag{14.40}$$

Also,

$$q_x(p_x)=q_y(p_y)=q_z(p_z)=q_{n10} \tag{14.41}$$

where q_{n10} represents the field gradient due to an electron in atomic orbital with $l=1$ (p electron) and $m_l=0$ (angular momentum zero along the reference axis). The subscript n represents the total quantum number of the electron orbital. It is evident that the resultant field gradient q_g due to all the p-shell electrons is

$$q_g=n_x q_g(p_x)+n_y q_g(p_y)+n_z q_g(p_z) \tag{14.42}$$

where n_x, n_y, and n_z represents the number, or fractional number, of electrons in the p_x, p_y, and p_z orbitals respectively. From this relation and from (14.38)–(14.41) it is seen that

$$q_z=-\left(\frac{n_x+n_y}{2}-n_z\right)q_{n10} \tag{14.43}$$

Similarly,

$$q_y=-\left(\frac{n_z+n_x}{2}-n_y\right)q_{n10} \tag{14.44}$$

and

$$q_x=-\left(\frac{n_y+n_z}{2}-n_x\right)q_{n10} \tag{14.45}$$

The quantities within the parentheses represent the unbalanced p electronic charge of the n shell of the coupling atom relative to the x, y, or z axis. It is evident that when the orbitals are symmetrically filled the field gradients are zero with respect to each axis. Thus the completely filled subshells contribute nothing to the coupling, and only the valence shell electrons need be considered.

With (14.43)–(14.45) and with approximations (*1*)–(*9*) from Section 2, we can now express the principal elements in the coupling by

$$\chi_x=eQq_x=-(U_p)_x eQq_{n10} \tag{14.46}$$

$$\chi_y=eQq_y=-(U_p)_y eQq_{n10} \tag{14.47}$$

$$\chi_z=eQq_z=-(U_p)_z eQq_{n10} \tag{14.48}$$

with e the magnitude of the electronic charge and

$$\eta=\frac{\chi_x-\chi_y}{\chi_z}=\frac{3(n_y-n_x)}{n_x+n_y-2n_z} \tag{14.49}$$

where the unbalanced p electrons are

$$(U_p)_x = \frac{n_y + n_z}{2} - n_x \tag{14.50}$$

$$(U_p)_y = \frac{n_z + n_x}{2} - n_y \tag{14.51}$$

$$(U_p)_z = \frac{n_x + n_y}{2} - n_z \tag{14.52}$$

The definition of U_p is such that a positive U_p corresponds to a p-electron deficit and a negative U_p to a p-electron excess along the reference axis. A deficit of one p electron in an otherwise closed p shell with the axis of the electron "hole" oriented along the reference axis would have the value $U_p = 1$; a single p electron outside a closed shell with its symmetry axis oriented along the reference axis would have the value $U_p = -1$. This corresponds to the definition of U_p employed in the earlier books on microwave spectroscopy [10, 11]; it is convenient for analysis of the halide couplings. The quantity $f_g = -(U_p)_g$ has been used [14] to designate an excess p electronic charge along the reference axis $g = x, y, z$.

The quantity q_{n10} used in the discussion of molecules represents the field gradient of an atomic p electron ($l = 1$) of the valence shell n with the symmetry axis of the p orbital oriented along the reference axis ($m = 0$). The quantity eQq_{n10} represents the nuclear quadrupole coupling by such an electron. For many atoms, eQq_{n10} is accurately known from measurements of the atomic hyperfine structure (see Section 3). Measured values are given in Table 14.2. For a single coupling electron, the measured atomic coupling constant is $b = -eQq_{n,1,\pm 1} = \frac{1}{2}eQq_{n10}$.

Field gradients q_{n10} for p electrons can be calculated theoretically by the method described in Section 3; and, when Q is known from experiment (values given in Appendix E), semitheoretical values of eQq_{n10} can be obtained. Usually the coupling constants are given in MHz, Q in cm^2, and q in esu. The following relation is useful in making numerical calculations

$$\chi_g(\text{MHz}) = -72.5(U_p)_g Q(10^{-24}\ \text{cm}^2) q_{n10}(10^{15}\ \text{esu}) \tag{14.53}$$

where $g = x, y, z$ is the reference axis. The quantities U_p and q_{n10} are independent of the isotopic number. When the coupling χ_g is known for one isotopic species, that for other species of the atom can be calculated with the ratios of Q given in Appendix E.

With measured or derived values of eQq_{n10} one can use (14.53) to derive values of U_p from the measured couplings of molecular constants. These can then be compared with values predicted theoretically from various postulated bond structures, as will be explained in later sections.

An orderly relationship between U_p, as measured by $(eQq)_{\text{mol}}/(eQq_{n10})_{\text{atom}}$, and the internuclear distances for the diatomic halides has been found by

Tiemann [27], who has used the relationship to predict the value of $\chi_z = -114$ MHz for the ^{35}Cl coupling in Cl_2.

5 CORRECTIONS FOR CHARGE ON COUPLING ATOM

If the coupling atom has a formal positive charge, the coupling field gradient per p electron will be increased over that of the neutral atom because of the reduced nuclear screening. On the other hand, a negative charge on the coupling atom will decrease the coupling. To a first-order approximation, these effects can be taken into account by multiplication of the q_{n10} for the neutral atom by a correction factor $(1+\varepsilon)$ per unit of positive charge on the coupling atom or by the factor $1/(1+\varepsilon)$ per unit of negative charge. From the difference in q for a number of ionized states of free atoms, Townes and Schawlow [11] have estimated values of ε for a number of different atoms. These values are given in Table 14.3. If there is a fractional positive charge of c^+ electron units on the coupling atom, (14.46)–(14.48) for the coupling are modified in the molecular orbital treatment used here to

$$\chi_g = -(1+c^+\varepsilon)(U_p)_g eQq_{n10} \tag{14.54}$$

If there is a negative charge of c^- electron units, they are modified to

$$\chi_g = -\frac{(U_p)_g eQq_{n10}}{(1+c^-\varepsilon)} \tag{14.55}$$

where $g = x, y, z$. If the coupling atom is bonded by a single bond, as are the halogens, c^+ and c^- will be numerically equivalent to the ionic character of the bond i_c, or to the percentage contribution of the ionic structure in the valence bond resonance method employed by Townes and Dailey [2]. In the valence bond resonance method, the resultant $(U_p)_g$ is the weighted sum of contributions from various resonance structures, that is, $(U_p)_g = \sum_i c_i (U_p^{(i)})_g$ where c_i is the fractional importance of the ith structure and $(U_p^{(i)})_g$ is the corre-

Table 14.3 Screening Constants ε for Correction of p Orbital Couplings for Effects of Charges on the Coupling Atom[a]

Atom	ε	*Atom*	ε	*Atom*	ε	*Atom*	ε
Be	0.90	Mg	0.70	Ca	0.60	Sr	0.60
B	0.50	Al	0.35	Sc	0.30	Ga	0.20
C	0.45	Si	0.30	Ge	0.25		
N	0.30	P	0.20	As	0.15	Sb	0.15
O	0.25	S	0.20	Se	0.20	Te	0.20
F	0.20	Cl	0.15	Br	0.15	I	0.15

[a]Values are from *Microwave Spectroscopy* by C. H. Townes and A. L. Schawlow, Copyright 1955, McGraw Hill, Inc. Used with permission of McGraw Hill Book Company.

sponding number of unbalanced p electrons. The charge correction is applied to the ionic structures by multiplication of the $(U_p^{(i)})_g$ by the factor $1+\varepsilon$ or $(1+\varepsilon)^{-1}$ depending on the sign of the charge on the coupling atom of the ith structure. If the ionic character is such as to produce a closed shell on the coupling atom, as is the case on the halogen in the contributing structures $X(\text{Hal})^-$, the halogen has zero coupling for this structure. Therefore the charge correction has no effect on the overall coupling when it is taken to be the weighted contributions from the various contributing structures in the valence bond resonance treatment. For this reason, the charge correction need not be included in the treatment of the halogen coupling except for ionic structures for which a positive charge is on the coupling atom.

Effects of charges on atoms of the molecule other than the coupling one are usually negligible. For example, an unscreened electronic charge of $\frac{1}{4}e$ on the z axis at a distance of 2.1 Å (the bond length of BrCl) from the coupling nucleus would contribute a field gradient q_z of only 0.026×10^{15} esu. Although such effects of charge are enhanced (on the order of a factor of 10) by polarization of the electron cloud [20, 28] on the coupling atom, the total effect of 0.2×10^{15} esu is still small in comparison with the field gradient of 20×10^{15} esu for a p_z electron in the valence shell of Cl or of a field gradient of 34×10^{15} esu for a similar electron in Br. In the following discussions we shall neglect the effects of charges on neighboring atoms caused by ionic character of covalent bonds.

Because the same charge corrections are applied to all principal elements of the coupling tensor, as indicated by (14.54) and (14.55), these corrections cancel from (14.49) and thus have no effect on the asymmetry parameter $(\chi_{xx}-\chi_{yy})/\chi_{zz}$ of the coupling.

6 HYBRIDIZED ORBITALS

If an electron is in an sp hybridized atomic orbital of a coupling atom A represented by

$$\psi_a = a_s\psi_s + a_p\psi_p \tag{14.56}$$

its field gradient with reference to the z axis will be

$$q_z(a) = e\int \psi_a^*\left(\frac{\partial^2 V}{\partial z^2}\right)\psi_a\,d\tau = a_s^2 q_z(s) + a_p^2 q_z(p) \tag{14.57}$$

However, the average of $(\partial^2 V)/(\partial z^2)$ over the s orbital is zero, $q(s)=0$, because of spherical symmetry. The overlap integral of these orthogonal orbitals is zero, and the normalization is represented by $a_s^2 + a_p^2 = 1$. Therefore (14.57) can be expressed as

$$q_z(a) = a_p^2 q_z(p) = (1-a_s^2)q_z(p) \tag{14.58}$$

If ψ_a is a constituent of a σ bonding orbital ψ_σ, the orbital ψ_σ will be directed along the bond axis chosen as the reference z axis. The p component will then be a p_z orbital, and $q_z(p)=q_z(p_z)=q_{n10}$. If a^2 represents the fractional weight of the component ψ_a in the molecular orbital $\psi_\sigma(=a\psi_a+b\psi_b)$ and if there are two electrons in ψ_σ, the field gradient due to the σ bond will be

$$q_z(\sigma)=2ea^2\int\psi_a^*\left(\frac{\partial^2 V}{\partial z^2}\right)\psi_a\,d\tau=2a^2(1-a_s^2)q_{n10} \tag{14.59}$$

The other hybridized orbital, $\psi'_a=a'_s\psi_s+a'_p\psi_p$, will be normalized if $a_s'^2+a_p'^2=1$. Since the total s character must be unity, that is, $a_s'^2+a_s^2=1$, it follows that $a_p'^2=a_s^2$. This counterhybridized orbital will therefore have a p_z component with weight $a_p'^2=(1-a_p^2)=a_s^2$ which will contribute to q_z. The total number of p_z electrons, n_z, will depend upon the way this counter-hybridized orbital is filled. If it is a nonbonding orbital having the maximum of two electrons, we find that

$$n_z=2a^2(1-a_s^2)+2a_s^2 \tag{14.60}$$

Thus we see that when the counterhybridized orbital has an unshared pair of electrons, as is true for the σ-bonded halogen atoms, the s hybridization increases the p_z electronic charge. If, however, the counterhybridized orbital also forms a bond and has, on the average, less than $2a^2a_s^2$ electrons, n_z would be lowered. The total number of unbalanced p electrons will also depend on the way the p_x and p_y orbitals are filled. If each of them has two unshared electrons, as in the halogens, then we would have

$$(U_p)_z=\frac{n_x+n_y}{2}-n_z=2(1-a^2)(1-a_s^2) \tag{14.61}$$

Since

$$q_z=-(U_p)_z q_{n10}=-2(1-a^2)(1-a_s^2)q_{n10} \tag{14.62}$$

it is evident that under the assumed conditions the magnitude of the coupling is lowered by the factor $(1-a_s^2)$ from that for the nonhybridized p_z bonding orbital. For a pure convalent bond (no ionic character) $a^2=\frac{1}{2}$, and (14.62) becomes

$$q_z=-(1-a_s^2)q_{n10} \tag{14.63}$$

Usually π bonds are formed with pure p orbitals, and hence we shall not consider the hybridization effects of the p_x and p_y orbitals. However, it is evident that if $(n_x+n_y)<4a^2$, the magnitude of q_z would be increased rather than decreased by the above hybridization.

Effects of pd hybridization are calculated in a similar manner. An electron in a pd hybridized orbital

$$\psi_a=a_p\psi_p+a_d\psi_d \tag{14.64}$$

will contribute a field gradient at the nucleus of the atom

$$q_z(a)=a_p^2 q_z(p)+a_d^2 q_z(d) \tag{14.65}$$

where $q(d)$ is the field gradient of a d electron. Although $q(d)$ is not zero, it is an order of magnitude less than $q(p)$ for electrons of the same n shell (see Section 3).

Furthermore, the d contribution is generally expected to be much less than the p contribution, $a_d^2 \ll a_p^2$. Hence, we can neglect the last term of (14.65) and, to a good approximation, can express

$$q_z(a) = a_p^2 q_z(p) = (1 - a_d^2) q_z(p) \tag{14.66}$$

If, as before, the orbital ψ_a is a bonding σ orbital of weight a^2 having a pair of electrons,

$$q_z(\sigma) = 2a^2(1 - a_d^2) q_{n10} \tag{14.67}$$

Although this expression has the same form as (14.59) for sp hybrids, the overall effect of d hybridization on the coupling in common molecules is the opposite to that of s hybridization; that is, whereas s hybridization lowers the magnitude, d hybridization raises it. This difference is caused by the fact that in most common elements forming stable molecules with hybridized p orbitals, the s orbital of the atom has a pair of electrons, but the d orbitals are empty. This is best illustrated by the halogens in which the free atoms have the valence shell configuration ns^2np^5. In the bonded atoms, the counterhybridized sp orbital will have an unshared pair as described above, but the counterhybridized pd orbital would be empty. Thus, if the bonding orbital is a pd hybrid, the total number of the p_z electrons would be

$$n_z = 2a^2(1 - a_d^2) \tag{14.68}$$

When there are unshared pairs in the p_x and p_y orbitals, $n_x = n_y = 2$,

$$(U_p)_z = 2 - 2a^2(1 - a_d^2) \tag{14.69}$$

and

$$q_z = -[2 - 2a^2(1 - a_d^2)] q_{n10} \tag{14.70}$$

which for pure covalent bonds with $a^2 = \frac{1}{2}$ becomes

$$q_z = -(1 + a_d^2) q_{n10} \tag{14.71}$$

Upon comparison with (14.63) it is seen that for the assumed bonding conditions, s and d hybridization are opposed in their effects on the coupling.

If the cross terms between the s and d orbitals are neglected, an spd-hybridized, σ-bonding orbital

$$\psi_a = a_s \psi_s + a_p \psi_p + a_d \psi_d \tag{14.72}$$

will similarly have

$$q_z(\sigma) = 2a^2(1 - a_s^2 - a_d^2) q_{n10} \tag{14.73}$$

One of the other two hybridized orbitals (ψ_a') assumed to be essentially s in character, an sp orbital, will have an amount a_s^2 of p character, while the remaining orbital (ψ_a'') essentially d in character, a pd orbital, will have an amount a_d^2 of p character. These conclusions follow from the assumptions and the normalization conditions. If the counterhybridized s orbital has an unshared pair and the counterhybridized d orbital is empty, as assumed above,

$$n_z = 2a^2(1 - a_s^2 - a_d^2) + 2a_s^2 \tag{14.74}$$

If, as when the p_x and p_z orbitals have unshared pairs,

$$q_z = -[2 - 2a^2(1 - a_s^2 - a_d^2) - 2a_s^2]q_{n10} \tag{14.75}$$

then the coupling would be

$$\chi_z = -2[1 - a^2(1 - a_s^2 - a_d^2) - a_s^2]eQq_{n10} \tag{14.76}$$

For pure covalent bonds with $a^2 = \frac{1}{2}$, this becomes

$$\chi_z = -(1 - a_s^2 + a_d^2)eQq_{n10} \tag{14.77}$$

In these expressions,

a^2 = total weight of the atomic orbital ψ_a in the σ-bonding molecular orbital
a_s^2 = amount of s character of ψ_a
a_d^2 = amount of d character of ψ_a
$1 - a_s^2 - a_d^2$ = amount of p character of ψ_a

It is evident that the hybridization effects on the coupling will depend on the way the orbitals are filled. From these examples, the method of calculation of effects for other electronic configurations should be evident. Methods for evaluation of a^2 are described in Sections 7 and 8. Applications of this theory are given in various sections to follow.

7 PURE COVALENT BONDS—EFFECTS OF ORBITAL OVERLAP

The bonding in the homopolar, diatomic halides consists essentially of a pure covalent, single σ bond with no ionic character to complicate the interpretation. Table 14.4 gives a comparison of the quadrupole couplings of halides Cl_2, Br_2, and I_2 for the free molecules and for condensed molecules in solids. Although the couplings in these molecules cannot be measured with microwave spectroscopy because the molecules are nonpolar, their couplings in the solid state are accurately known from pure quadrupole resonance [29–31] and for the free molecules of Br_2 and I_2 they are accurately known from microwave–optical laser double resonance [32, 33]. The free-molecule coupling in Cl_2 has been reliably inferred [27] from projection of the couplings in Br_2, I_2, and other halides. The solid-state coupling in I_2 is reduced by weak intramolecular bonding [34] revealed by an asymmetry, $\eta = 0.173$, in the coupling of I_2 crystals. The slightly lower coupling for solid Cl_2 and Br_2 as compared with couplings for the free molecules also probably results from weak intramolecular bonding in these solids. For comparison with the molecular couplings, accurate values of the nuclear quadrupole couplings in the free atoms, Cl, Br, and I, are known from

Table 14.4 Observed Nuclear Quadrupole Couplings and Derived Unbalanced p Electrons on the Coupling Atom in Cl_2, Br_2, and I_2

	Observed Coupling, χ_z(MHz)			
Coupling Nucleus	*Free Molecule*	*Ref.*	*Molecular Solid*	*Ref.*
$^{35}Cl_2$	(−114)	a	−108.95	d
$^{79}Br_2$	810.0(5)	b	765	e
$^{127}I_2$	−2452.584(2)	c	−2153	f
	$U_p = -\chi_z(\text{mol})/(eQq_{n10})$			
Cl_2	1.039		0.993	
Br_2	1.052		0.994	
I_2	1.070		(0.931)	

[a]Projected value obtained by Tiemann [27].
[b]Bettin et al. [32].
[c]Yokozeki and Muenter [33].
[d]Livingston [29].
[e]Dehmelt [30].
[f]Dehmelt [31].

atomic beam resonance. They are listed in Table 14.2. It is thus possible to obtain experimental values of the unbalanced p electrons on the bonded halogens in the free molecules as well as in the molecular solids. These experimental values of U_p are given in the lower part of Table 14.4.

It is of interest to compare the values of U_p derived from the observed couplings in Table 14.4 with theoretical U_p values predicted with the assumption of pure p-bonding orbitals and with the neglect of effects of bond overlap distortions on the couplings. From the normalization of (14.8) with $a=b$, it is seen that $a^2=\frac{1}{2}$. Since there are two electrons in the σ bonding orbital, the number of p_z electrons on each coupling atom is $n_z=2a^2=1$. With no hybridization or π bonding, there will be two electrons in each of the nonbonding p_x and p_y orbitals, that is, $n_x=2$ and $n_y=2$. Substitution of these values into (14.52) yields

$$(U_p)_z=\frac{n_x+n_y}{2}-n_z=1 \tag{14.78}$$

for this idealized model of the halogen molecule. The experimental values are

$$(U_p)_z=-\frac{(eQq_z)_{\text{mol}}}{(eQq_{nlm=0})_{\text{atom}}}=-\frac{(q_z)_{\text{mol}}}{(q_{n10})_{\text{atom}}} \tag{14.79}$$

The observed values of U_p in Table 14.4 do not differ grossly from $U_p=1$, predicted with the neglect of the effects of overlap orbitals. However, the very close agreement of the solid state values for Cl_2 and Br_2 with the value predicted for the idealized model is probably fortuitous since the interactions in the solid tend to lower the coupling values. This lowering is most pronounced for I_2 but is evident in the Cl coupling for the substituted methyl halides in solids compared with gases in Fig. 14.3. It is thus reasonable to take the U_p values derived from the coupling of the free molecules as perturbed only by the covalent bonding. These U_p values are seen to be 4 to 7% above the unit value. Since the s hybridization of the bond orbital would reduce rather than increase the coupling in the molecules, it appears that the increase of U_p above unity is due to effects of bond orbital overlap. Although some increase of U_p above unity is expected from orbital overlap distortions, the observed increases are much less than those predicted from normalization of (14.80). As shown below, the simple normalizations of (14.8) or (14.9), usually employed in interpretation of nuclear coupling in molecules, predict values much closer to the observed ones.

Now let us consider possible effects of orbital overlap. Instead of the normalization of (14.8), we employ the normalization of the bonding molecular orbital $\phi = a\psi_a + b\psi_b$:

$$\int \phi^*\phi \, d\tau = a^2 \int \psi_a^*\psi_a \, d\tau + 2ab \int \psi_a^*\psi_b \, d\tau + b^2 \int \psi_b^*\psi_b \, d\tau = 1 \qquad (14.80)$$

With

$$\int \psi_a^*\psi_a \, d\tau = 1 \qquad \int \psi_b^*\psi_b \, d\tau = 1 \qquad \int \psi_a^*\psi_b \, d\tau = S_{ab}$$

this becomes

$$a^2 + 2abS_{ab} + b^2 = 1 \qquad (14.81)$$

For homopolar molecules considered here, $a = b$ and

$$a^2 = \frac{1}{2(1+S_{ab})} \qquad (14.82)$$

For Cl_2, the theoretical value [35] of the overlap integral S_{ab} is 0.34, and with this value (14.82) indicates that $a^2 = 0.37$. Correspondingly, $n_z = 2a^2 = 0.74$, and from (14.78) $(U_p)_z = 2 - 0.74 = 1.26$. Substitution of this value of U_p into (14.79) leads to the predicted $\chi_z(^{35}Cl_2) = -138$ MHz, in poor agreement with the observed value of -109.7 MHz in solids and -114 MHz for free molecules. Inclusion of the overlap term in the normalization requires, however, that we include the cross term, the second term on the right of (14.7) in the evaluation of the field gradient. Inclusion of this term improves the agreement, but only slightly. Precise calculation of this cross term is rather difficult, but its value can be easily approximated by assumption that the overlap charge is concentrated at a distance equal to the covalent radius R from the nucleus. Since there are two electrons in the bonding orbital, the total overlap charge is $2e(2a^2S_{ab})$. From (14.34) with $\theta = 0$ and $3\cos^2\theta - 1 = 2$, the field gradient due to an unscreened charge of this magnitude is

$$q_{ab} \approx \frac{8ea^2S_{ab}}{R^3} \qquad (14.83)$$

With $e=4.80\times10^{-10}$ statcoulombs and $R=0.99\times10^{-8}$ cm, $S_{ab}=0.34$, $a^2=0.37$ for Cl_2, it is seen that $q_{ab}\approx0.5\times10^{15}$ esu. For one unbalanced p electron on Cl, $q_{310}=19.5\times10^{15}$ esu. Thus inclusion of the cross term in (14.7) decreases the magnitude of the coupling by approximately $0.5/19.5=2.5\%$. The predicted value is changed from -138 to -135 MHz, which is still in poor agreement with the observed value. A similar estimate of the cross term for Br_2 with $R=1.14\times10^{-8}$ cm reduces the field gradient by approximately 1%, and the predicted ^{79}Br coupling from 953 to 944 MHz as compared with the observed value of 765 MHz for solids and 810 MHz for free molecules.

Although this estimated correction for the cross term is very approximate, it does indicate that a precise evaluation of this term could not correct for the increase in coupling caused by a lifting of the overlap charge out of the atomic orbital as required by the normalization by (14.81). The relatively good agreement of the observed coupling with that predicted with (14.79) indicates that this normalization gives a more nearly correct representation of the molecular orbital near the nucleus. The implications are that the overlap distortions occur in the outer regions of the orbitals and that the overlap charge signified by the cross term is not produced by the lifting of significant electronic charge density from the area near the nucleus to the outer regions of the orbital, as is implied by the normalization of (14.81). Obviously, such a lifting would be expensive in energy.

The overlap integrals in the pure covalent bonds of these homopolar halides are greater than those of the mixed halides or other polar molecules for which the covalent component of the bonds, and hence the overlap integral S_{ab}, is reduced by significant ionic character. Furthermore, S_{ab} is reduced by a difference in the size of the bonded atoms, which results in less complete overlapping of the bonding orbital of the larger atom. Thus the increase in coupling, only 4 to 7% in these homopolar halides, should be the upper limit of that expected from effects of bond orbital overlap distortions on the halide couplings. For bonds formed between atoms differing appreciably in size, such as HI, or bonds such as AgCl that have significant ionic components, the effects of orbital overlap on the coupling should be negligible.

8 POLAR BONDS—IONIC CHARACTER FROM QUADRUPOLE COUPLING

To explore the effects of ionic character on the coupling, let us consider the halogen coupling in diatomic polar molecules. These provide the simplest cases although possible hybridization effects may cause some uncertainty. If we allow for possible s and d hybridization of the bonding orbital, the coupling is given by (14.76), in which the coefficient a^2 depends on the ionic character of the bond. In a singly bonded diatomic molecule AB the σ-bonding orbital can be represented by the linear combination

$$\psi_\sigma=a\psi_a+b\psi_b \tag{14.84}$$

and the ionic character of the σ bond may be represented by

$$i_c = |a^2 - b^2| \tag{14.85}$$

With normalization of $a^2 + b^2 = 1$ it is evident that

$$a^2 = \frac{1+i_c}{2} \qquad \text{when } a^2 > b^2 \qquad (A \text{ negative}) \tag{14.86}$$

$$a^2 = \frac{1-i_c}{2} \qquad \text{when } a^2 < b^2 \qquad (A \text{ positive}) \tag{14.87}$$

In the first case, $a^2 > b^2$, the negative pole will be on atom A, assumed to be the coupling atom. We consider this case first. Substitution of $a^2 = (1+i_c)/2$ into (14.76) gives the quadrupole coupling

$$\chi_z = -[1 - a_s^2 + a_d^2 - i_c(1 - a_s^2 - a_d^2)]\left(\frac{eQq_{n10}}{1+i_c\varepsilon}\right) \tag{14.88}$$

where a_s^2 is the amount of s character and a_d^2 is the amount of d character of the bonding orbital and where eQq_{n10} is the coupling of an atomic p electron in the orbital ψ_{n10}. If there is s hybridization only,

$$\chi_z = -\frac{(1-i_c)(1-a_s^2)eQq_{n10}}{1+i_c\varepsilon} \qquad \{\text{negative pole on coupling atom}\} \tag{14.89}$$

Here the factor $1/(1+i_c\varepsilon)$ is applied to correct for the change in nuclear screening caused by the negative charge $c^- = i_c$ on the coupling atom as described in Section 5. Approximate values of ε are given in Table 14.3. For Cl, Br, and I, $\varepsilon = 0.15$.

When the positive pole is on the coupling halogen A, the factor $(1+i_c\varepsilon)$ is applied to account for the decreased nuclear screening (Section 5). With this correction, and with $a^2 = (1-i_c)/2$, (14.76) becomes for the positive pole on the coupling atom

$$\chi_z = -[1 - a_s^2 + a_d^2 + i_c(1 - a_s^2 - a_d^2)](1+i_c\varepsilon)eQq_{n10} \tag{14.90}$$

If there is s hybridization only,

$$\chi_z = -(1+i_c)(1-a_s^2)(1+i_c\varepsilon)eQq_{n10} \qquad \{\text{positive pole on coupling atom}\} \tag{14.91}$$

It is evident that there are still more unknown parameters in these equations than there are measurable couplings. This is true even for molecules such as BrCl and ICl in which both atoms have measurable couplings. Since a_s^2 and a_d^2 are not necessarily the same on both atoms, there is a possibility of five unknowns. Assumptions or evaluations of some of the parameters from other experiments are required for solution of these equations.

We make the justifiable assumption that hybridization on a halogen is negligible when the halogen is neutral and when it has a greater electronegativity than has the atom to which it is bonded, that is, when it is negatively charged.

Evidence supporting this assumption for the neutral halogen is provided by the quadrupole coupling for nonpolar Cl_2 and Br_2 as earlier shown and discussed earlier in this section. Furthermore, the couplings in the free molecules, Cl_2, Br_2, and I_2, are slightly greater than those predicted for one unbalanced p electron, not smaller, as would be expected from hybridization. As explained earlier, an increase is expected from orbital overlap. The assumption of negligible effects of hybridization on the negative atom seems required for a reasonable interpretation of the couplings in BrCl and in IBr, for which the ionic characters are small, and for a consistent interpretation of the quadrupole coupling of both halogens in ICl. As explained earlier [7], hybridization on the negative halogen would be quenched by the primary dipole moment, that is, that due to ionic character, whereas hybridization on the positively charged halogen would be induced or supported by the primary moment.

There are inconsistencies in the ionic characters derived from the two halogen couplings in BrCl, IBr, and ICl that cannot be removed by assumption of sp hybridization, but can be removed by inclusion of bond orbital overlap distortions in the calculations, as is explained below. The measured couplings in BrCl are -102.45 MHz for ^{35}Cl and 875.3 MHz for ^{79}Br. With these and with the $(eQq_{n10})_{\rm atom}$ values for ^{35}Cl and ^{79}Br in Table 14.2, one obtains $i_c=0.07$ from substitution of the ^{35}Cl coupling in (14.89), and $i_c=0.12$ from substitution of the ^{79}Br coupling in (14.91) if $a_s^2=0$ in both solutions. Note that the i_c from the Br coupling is twice that from the Cl coupling. This inconsistency is increased by assumption of s hybridization on either atom. There is a comparable inconsistency in the i_c values of IBr predicted in the same way from the I and Br couplings. The covalent component of the bonds in both molecules is large, $\sim 90\%$, with the overlap integrals S_{ab} comparable to those of Cl_2, Br_2, and I_2. Thus to eliminate, or reduce, effects of orbital overlap distortion on the calculations of i_c, it seems reasonable to use, instead of the $(eQq_{n10})_{\rm atom}$, the $(eQq_{n10})_{\rm mol}$ values: 114 MHz for ^{35}Cl, -810.0 MHz for ^{79}Br, and 2452.5 MHz for ^{127}I, as obtained from the couplings listed in Table 14.4 for the free molecules Cl_2, Br_2, and I_2. These values inherently include effects of orbital overlap. With this procedure and with hybridization assumed to be zero on both Cl and Br in BrCl, $i_c=0.09$ is predicted from the ^{35}Cl coupling and $i_c=0.07$ from the ^{79}Br coupling. This inconsistency can now be cleared up by 2% s hybridization on the positive Br with none on the negative Cl. The resulting $i_c=0.09$ is from both coupling atoms. With this procedure the ionic character of IBr predicted from the ^{79}Br coupling is 0.124; from the ^{127}I coupling it is 0.103. A consistent value of $i_c=0.124$ is then obtainable by assumption of 2% s hybridization on the positive I with none on the negative Br. See Section 10.

The values of the ionic character of ICl derived from the observed molecular coupling of Cl and of I are likewise not consistent if the free atomic values of $(eQq_{n10})_{\rm atom}$ are used and the charge correction, $\varepsilon_c=0.15$, is applied with no assumption of hybridization on either element. For example, the $\chi_z(^{35}\text{Cl})=-85.8$ MHz with $(eQq_{n10})_{\rm atom}=109.74$ MHz substituted in (14.89) with $a_s^2=0$ gives $i_c=0.196$, whereas $\chi_z(^{127}\text{I})=-2928$ MHz with $(eQq_{n10})_{\rm atom}=2292.71$ MHz with $a_s^2=0$ substituted in (14.91) gives $i_c=0.234$. Assumption of s hybridization of the bonding orbital on either element widens this inconsistency. If one uses the eQq_{n10} values from the free Cl_2 and I_2 molecules instead of the free atomic values, an assumption of 5.5% hybridization on the I gives the consistent value of $i_c=0.22$. With this ionic character the bond has only 78% covalent character, and one should obtain a more nearly correct eQq_{n10} by interpolating the difference between free-molecule coupling of Cl_2 and of I_2 and the respective atomic eQq_{n10} values in proportion to the covalent ionic character ratios in ICl. This procedure leads to effective values, $eQq_{n10}(^{35}Cl)=113$ MHz and $eQq_{n10}(^{127}I)=2417$ MHz in ICl and yields the consistent value $i_c=0.216$, with only 3.5% s character on the I and none on the Cl. See Section 10.

With no detectable hybridization on the negative halogen in BrCl, ICl, or IBr, it seems improbable that there would be significant hybridization on other halogens at the negative pole of an ionic bond. In Table 14.5 are listed values of i_c that have been calculated from the quadrupole coupling of the negatively charged halogens in the specified diatomic molecules. These values were calculated from

$$i_c=1+\frac{\chi_z}{eQq_{n10}} \tag{14.92}$$

which is obtained from (14.88) if a_s^2 and a_d^2 are set equal to zero and if the screening correction is neglected.

For the free halogen atoms the coupling per unbalanced p electron, $(eQq_{n10})_{\text{atom}}=2b$, is accurately known from atomic beam resonance experiments. See Section 3, Table 14.2. Molecular values of $(eQq_{n10})_{\text{mol}}$, which include effects of orbital overlap, are obtained from the free-molecule couplings in the nonpolar halides, Cl_2, Br_2, and I_2, listed in Table 14.4. These molecular values presumably represent the effective coupling of one unbalanced p_z electron on the covalently bonded halide atom. They make possible a consistent interpretation of the coupling of both halogens of the mixed halides, BrCl, ICl, and IBr, as explained previously. Use of the effective molecular eQq_{n10} would make the predicted i_c values of Table 14.5 slightly smaller, and inclusion of the negative charge correction would make these values slightly greater. Thus the values of i_c in Table 14.5 determined by the simple (14.92) have off-setting effects which tend to increase their accuracy.

For all the halogens eQq_{n10} is accurately known from the atomic coupling constant b measured in atomic beam resonance experiments (Table 14.2) and the molecular couplings χ_z are accurately known from microwave rotational spectra. Thus the error limits of i_c are due primarily to the neglect of orbital overlap and possible hybridization effects. For high values of i_c, 0.75 and above, the predicted values of i_c become relatively insensitive to hybridization and orbital overlap distortions; for moderate ionic character, these effects on the determination of i_c are required to be negligible for a consistent interpretation of the coupling of the two nuclei of BrCl and ICl. Justification for neglect of effects of orbital overlap are provided by the nonpolar molecules Cl_2 and Br_2

Table 14.5 Ionic Character Predicted from Quadrupole Coupling of Negative Halogens on Polar Diatomic Molecules

Molecule AB	*Observed Coupling* $\chi_z(MHz)$[a]	*Predicted Ionic Character* $i_c = 1 + \frac{\chi_z}{eQq_{n10}}$	*Electronegativity Difference* $\lvert x_A - x_B \rvert$
Coupling Atom ^{35}Cl		$eQq_{n10} = 109.74$ MHz	
Cl_2	−108.95[b]	0	0
HCl	−68	0.38	0.85
KCl	0.04	1	2.2
RbCl	0.77	1	2.2
BrCl	−102.4	0.07	0.2
ICl	−85.8	0.22	0.45
AgCl	−37.32	0.66	1.2
AlCl	−8.8	0.92	1.5
InCl	−13.71	0.875	1.5
TlCl	−15.795	0.855	1.5
Coupling Atom ^{79}Br		$eQq_{n10} = -769.76$ MHz	
Br_2	765[b]	0	0
HBr	532	0.31	0.65
LiBr	37.2	0.95	1.85
KBr	10.24	0.99	2.0
AgBr	307	0.60	1.0
GaBr	106	0.86	1.3
InBr	110.6	0.85	1.3
TlBr	126.3	0.84	1.3
Coupling Atom ^{127}I		$eQq_{n10} = 2292.71$ MHz	
HI	−1828.4	0.20	0.4
LiI	−198.15	0.91	1.60
NaI	−259.87	0.89	1.65
KI	86.79	0.96	1.75

[a]Sources for the χ_z values may be found in Landolt-Börnstein, *Numerical Data and Functional Relations in Science and Technology*. For a complete reference, see Chapter 1, [65]. Atomic couplings eQq_{n10} are from Table 14.2; electronegativities, from Appendix G.
[b]For solid-state value, see Table 14.4.

(Section 7) for which the overlap integral is large. For these reasons we believe that the values of ionic character listed in Table 14.5 represent the best available values for these molecules.

We have deliberately selected for consideration here molecules that have low probability for a significant π component in the bonding. Those molecules in which both atoms are halogens are not likely to have significant π bonding because both atoms have closed p shells. Furthermore, large atoms of the higher periods of the atomic table—I, Ga, In, Tl—do not form bonds having significant multiple character because their large covalent radii prevent the required orbital overlap. When the multiple bond character is negligible, as for the molecules listed in Table 14.5, the ionic character i_c of the σ bond orbital represents the total ionic character of the bond. This is not always true. Effects of π character in the bonds are treated in Section 11.

The very small quadrupole couplings observed in essentially ionic molecules of the alkali halides are treated theoretically by Buckingham [36].

9 IONIC CHARACTER—ELECTRONEGATIVITY RELATION

Because we shall need the relationship for interpretation of quadrupole coupling in the more complicated molecules, we digress here to describe the relationship between ionic character of σ bonds and the electronegativity difference of the bonded atoms based on the ionic-character values derived from quadrupole coupling of halogens in simple diatomic molecules (Section 8). The values derived are listed in Table 14.5 along with the corresponding electro-

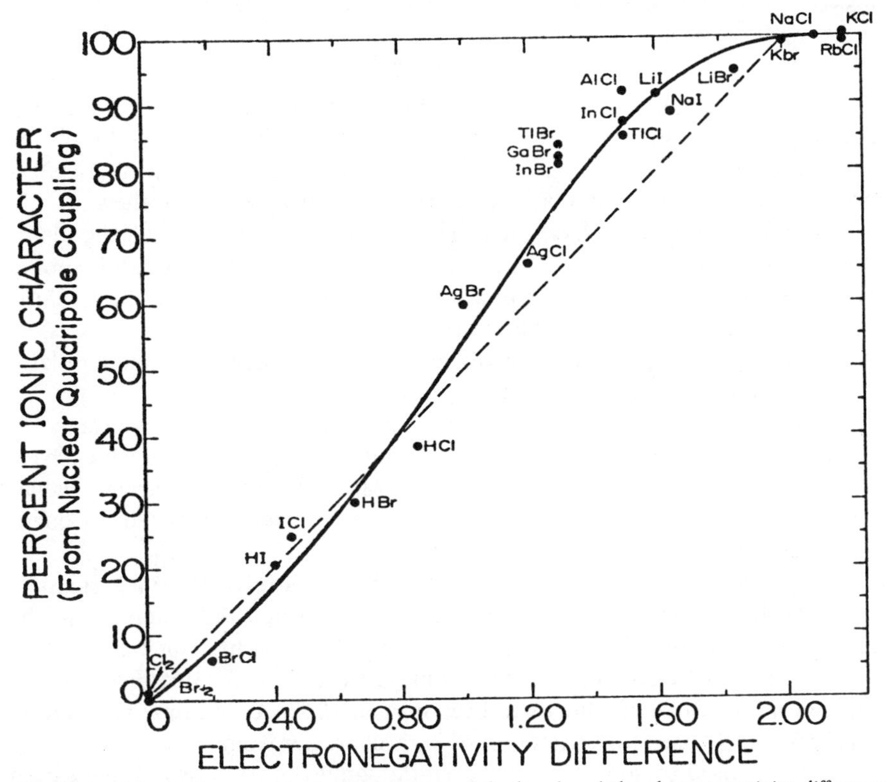

Fig. 14.1 Relation between the ionic character of the bond and the electronegativity difference of the bonded atoms. The ionic character values were derived from nuclear quadrupole coupling of the halogen, as explained in the text.

negativity differences. Figure 14.1 shows a plot of i_c versus $|x_A - x_B|$ with these values. Note that the relationship indicates complete ionicity of the bonds for electronegativity differences of 2 or greater. With similar plots on data earlier available, it was shown [6] that the ionic character-electronegativity relationship can be expressed approximately by the very simple and convenient formula

$$i_c = \frac{|x_A - x_B|}{2} \qquad \text{for } |x_A - x_B| \leqslant 2 \tag{14.93}$$

With complete ionicity, $i_c = 1$ for $|x_A - x_B| > 2$. The dotted line in Fig. 14.1 represents this relationship. For bonds having ionic character of 50% or less or for $|x_A - x_B| \leqslant 1$, it is probably as accurate as the electronegativity values justify. For $|x_A - x_B| > 1$, the quadrupole coupling indicates a somewhat higher ionicity than indicated by (14.93) although this simple relation can still be used for a quick estimate of the approximate ionicity.

The solid curve of Fig. 14.1 represents closely the correlation of the present data. This curve is a plot of the equation

$$i_c = 1.15 \exp[-(\tfrac{1}{2})(2 - \Delta x)^2] - 0.15 \tag{14.94}$$

where

$$\Delta x = |x_A - x_B| \leqslant 2$$

For $\Delta x > 2$, the bonds are completely ionic, $i_c = 1$, and (14.94) does not apply. Although other equations with different constants might be designed to fit the empirical relationship of Fig. 14.1, the relatively simple formula of (14.93) is probably as accurate as can be justified by the electronegativity values.

The relationship of (14.93), or of (14.94), applies only to single bonds. It does not give the total ionic character of a multiple bond, although it should be applicable to a σ component of a multiple bond whenever the effects of the other bond components on the electronegativity difference can be assessed. Applications of the relation to bonds having π components are made in later sections; examples are shown in Table 14.7.

Since a correct evaluation of ionic character is of great importance in chemistry, we shall attempt to clear up some of the difficulties which have caused chemists to be reluctant to accept the rather large ionic-character values predicted from the ionic character-electronegativity relation based on quadrupole couplings.

Many years ago Pauling [37] recognized that there should be a systematic relationship between the ionic character of a bond and the difference in the electronegativity of the bonded atoms. He obtained the first such correlation on the four hydrogen halides with ionic character, estimated from dipole moments, only three of which had been measured. Since that time his relation has been widely used for estimation of ionic character from electronegativities, and it is still used by some workers, even though it is now known to have been based on an incomplete knowledge of dipole moments. Pauling assumed the ionic character to be μ_{obs}/eR, where μ_{obs} is the measured dipole moment, e is the charge on the electron, and R is the interatomic distance. He thus obtained the ionic character values of 0.17 for HCl, 0.114 for HBr, and 0.049 for HI, in contrast to the values given in Table 14.5 of 0.38, 0.31, and 0.20, respectively. It is now known from the work of Mulliken [38] that a large component due to the orbital overlap charge exists in the dipole moments of the hydrogen halides because of the great difference in the size of the bonded atoms. This overlap moment, not taken into account by Pauling, is in the opposite direction to the primary moment $\mu_p = i_c eR$, and for HI is actually larger than the primary moment. Even if the i_c values used by Pauling for HCl, HBr, and HI had been correct ones, these values ranging from 0.049 to 0.17 could not have been safely projected to give values of high ionicity, and the high value $i_c = 0.60$ for HF which he used was not based on a measured dipole moment but on an estimated one. It is to Pauling's credit that he recognized the questionable basis of this relationship and warned that it had only qualitative significance. He justified its presentation on the ground that no other method existed for estimation of even approximate values of ionicity. A reinterpretation of the dipole moments of the hydrogen halides is given in Section 16, where it is shown that the ionic character–electronegativity relation derived from dipole moments is consistent with that obtained above from quadrupole couplings.

10 HYBRIDIZATION FROM QUADRUPOLE COUPLING: POSITIVELY CHARGED HALOGENS AND GROUP III ELEMENTS

Although the nuclear quadrupole evidence does not indicate hybridization on the negative halogen of a polar bond, there is strong quadrupole evidence for *sp* hybridization on the positive element of a polar bond. It is proposed that this difference is due to the fact that hybridization on the positive atom is supported by the primary dipole moment of the polar bond and that hybridization on the negative atom is suppressed by the bond moment. In this section we shall derive from the nuclear quadrupole coupling of the positive atom the amount of such hybridization in a few mixed diatomic halides and in a number of diatomic molecules formed by Group III elements with the halides.

First, we shall consider positively charged halides. In Section 8 it is shown that a consistent interpretation of the two halides in BrCl, IBr, and ICl requires the assumption of small amounts of s character in the bonding orbital of the positive halide, as indicated in Table 14.6. To obtain this consistency, it was necessary also to use for eQq_{n10}, the values obtained from the nonpolar halides, Cl_2, Br_2, and I_2. As explained in Section 8, the molecular eQq_{n10} values must be used to eliminate the effects of bond orbital overlap in these mixed halides that have high covalent character in their σ bonds. We have used the atomic eQq_{n10} values for calculation of the hybridization on the positive atoms of ClF, BrF, and IF which have high ionic components in their bonds. Also, we have used the atomic eQq_{n10} for calculation of the hybridization on other positive atoms in

Table 14.6 Hybridization of Bonding Orbital of Positively Charged Atom of Polar Bonds as Obtained from Nuclear Quadrupole Coupling

Molecule *A – B*	*Hybridized Coupling Atom* *A*	*Coupling* $\chi_z(MHz)$	*Ionic Character of Bond* i_c	*s Character of Bonding Orbital of Atom A*
BrCl	^{79}Br	876.7	0.09[b]	0.02[e]
IBr	^{127}I	−2753.5	0.12[b]	0.02[e]
ICl	^{127}I	−2927.9	0.22[b]	0.05[e]
ClF	^{35}Cl	−145.87	(0.50)[c]	0.18[f]
BrF	^{79}Br	1086.8	(0.58)[c]	0.18[f]
IF	^{127}I	−3438.2	(0.70)[c]	0.20[f]
BF	^{11}B	−4.5	(1.00)[c]	0.28[g]
AlF	^{27}Al	−37.49	(1.00)[c]	0.37[g]
AlCl	^{27}Al	−29.8	0.92[d]	0.27[g]
AlBr	^{27}Al	−27.9	0.90[d]	0.25[g]
AlI	^{27}Al	−25.9	0.86[d]	0.21[g]
GaF	^{69}Ga	−106.5	(1.00)[c]	0.33[g]
GaCl	^{69}Ga	−92.1	0.88[d]	0.27[g]
GaBr	^{69}Ga	−86.5	0.86[d]	0.24[g]
GaI	^{69}Ga	−81.1	0.84[d]	0.21[g]
InF	^{115}In	−723.8	(1.00)[c]	0.33[g]
InCl	^{115}In	−657.5	0.88[d]	0.27[g]
InBr	^{115}In	−633.2	0.85[d]	0.25[g]
InI	^{115}In	−607.1	0.83[d]	0.22[g]

[a]Coupling from Landolt-Börnstein, *Numerical Data and Functional Relations in Science and Technology*. For a complete reference, see Chapter I, [65].
[b]i_c derived from coupling of negative halogen with eQq_{n10} determined from the homopolar molecules Cl_2, Br_2, and I_2, as explained in the text.
[c]i_c determined from electronegativity difference by use of (14.93).
[d]i_c derived from coupling of negative halogen by use of (14.92), with eQq_{n10}(atomic) from Table 14.2.
[e]*s* character a_s^2 derived from (14.91) with coupling in column 3, i_c value in column 4, and eQq_{n10} from the homopolar molecules Cl_2, Br_2, and I_2, as explained in the text.
[f]*s* character derived from (14.91) with coupling of positive halogen in column 3 and $\varepsilon=0.15$ from Table 10.3, eQq_{n10}(atomic) from Table 14.2 and with i_c value from column 4.
[g]*s* character a_s^2 for Group III elements derived from (14.97) with χ_z value in column 3, i_c value in column 4, and eQq_{n10}(atomic) value from Table 14.2 and ε value from Table 14.3.

Table 14.6 for which the i_c is 0.50, or greater. Note that the hybridization is 18 to 20% on the halogens bonded to the highly electronegative *F*.

The Group III elements, B, Al, Ga, and so on, have the valence shell configuration s^2p^1 and normally would be expected to form single bonds with either a pure *p* orbital, an *sp* hybrid, or an *spd* hybridized orbital. We consider the *spd* hybridized bonding orbital, (14.72), of the coupling atom *A*, analyzed in Section 6. For reasons given there, the coupling of the *d*-orbital component is negligible in comparison with that of the *p* component. The significant effect of the d contribution is its reduction of the density of the p orbital component. The p_z density of the bonding orbital is reduced by the hybridization to $a_p^2 = 1 - a_s^2 - a_d^2$. The counterhybridized dp orbital does not contribute to the coupling because it is unpopulated in Group III elements. The counterhybridized sp orbital has a pair of unshared electrons and a p_z density of $a_p^2 = 2(1 - a_s^2)$. The total number of p_z electrons is $n_z = 2a^2(1 - a_s^2 - a_d^2) + 2a_s^2$, as given by (14.74). In the Group III elements the p_x and p_y orbitals are unpopulated, $n_x = 0$, $n_y = 0$. Thus,

$$(U_p)_z = \frac{n_x + n_y}{2} - n_z = -[2a^2(1 - a_s^2 - a_d^2) + 2a_s^2] \tag{14.95}$$

With $a^2 = (1 - i_c)/2$ from (14.87) for the positive atom of an ionic–covalent bond, (14.95) can be expressed as

$$(U_p)_z = -[(1 - i_c)(1 - a_s^2 - a_d^2) + 2a_s^2] \tag{14.96}$$

The coupling of a single-bonded Group III element forming the positive pole of an ionic–covalent bond may then be expressed by

$$\chi_z = [(1 - i_c)(1 - a_s^2 - a_d^2) + 2a_s^2](1 + i_c\varepsilon)eQq_{n10} \tag{14.97}$$

where a_s^2 and a_d^2 signify the s and d character, respectively, of the bonding orbital of the element and $(1 + i_c\varepsilon)$ represents the screening correction for the positive charge.

It is evident from consideration of (14.97) that for the highly ionic bonds generally formed by the Group III elements, the s character of the bonding orbital is likely to have a much greater effect on the coupling than any probable d character. For example, let us consider the coupling χ_z of ^{27}Al in AlBr, which has $i_c = 0.90$. Although this ionic character is high, all the bonds of diatomic molecules of Group III with halogens listed in Table 14.6 have i_c greater than 0.80. With $i_c = 0.90$, (14.97) becomes

$$\chi_z = [0.10 + 1.9a_s^2 - 0.10a_d^2](1 + 0.90\varepsilon)eQq_{n10}$$

If the s and d character were equal in this case, the effects of the s character on the coupling would be 19 times that of the d character. The effect of the s character would be to increase the coupling value, whereas the effect of the equal d character would be to reduce this increase by about 5%. From energy considerations one would expect the d character to be much lower than, rather than equal to, the s character. Consequently, in the calculation of the hybridizations listed in Table 14.6 we have neglected possible effects of d character and have set $a_d^2 = 0$ in (14.97).

There is no reasonable doubt that the Group III elements have significant s hybridization of their valence p orbital when they form single bonds to the more electronegative atoms like the halides. If we set $a_s^2 = 0$ in the calculation of the χ_z coupling of ^{27}Al in AlBr, just described, the magnitude of the calculated $\chi_z(^{27}Al)$ would be only 4.9 MHz as compared with the observed value of 27.9 MHz. Inclusion of any d character would make the calculated value even less than 4.9 MHz. From the electronegativity differences it is reasonably certain that the bonds of GaF and InF are essentially pure ionic bonds. With $i_c = 1.00$ substituted in (14.97), the predicted coupling becomes $\chi_z = 2a_s^2(1+\varepsilon)eQq_{n10}$, which is zero if the s hybridization is zero. An s character of 33%, $a_s^2 = 0.33$, is required to produce the observed couplings in Ga and InF. In effect, this means that the counterhybridized orbital has $\frac{1}{3}p_z$ character. Since this counterhybridized orbital has an unshared pair, the number of p_z-electrons on the Ga or In is $n_z = \frac{2}{3}$ even though the p_z electron density in the bonding orbital is entirely on the F. Since bond orbital hybridization usually occurs because it increases the effectiveness of the orbital overlap and hence increases the strength of the covalent component of the bond, one might ask why there would be such large hybridization in a pure ionic bond. The answer must be that the hybridization is induced entirely by the large primary dipole moment of the ionic bond. See the discussion in Section 16.

Comparison of the last two columns of Table 14.6 shows that the general trend is toward increasing hybridization on the positive atom with increasing ionic character of the bond. The highest amount of s hybridization is for the completely ionic bonds formed by the Group III elements with F. There are, of course, individual differences due to other factors, but the effects of ionic character in inducing hybridization on the positive atom are pronounced. Since there is no orbital overlap, the force for hybridization of the atomic orbitals of a completely ionic bond must be the strong electric field of the primary dipole moment. The primary moment will induce a hybridization moment in such a direction as to lower the overall or resultant moment and hence the potential energy. The negative halogen in these molecules has a closed sp shell, and its polarizability is not great. Significant distortion of these shells would occur through a lifting of p electron charge into d orbitals directed toward the positive pole, but this would lead to significant coupling on the negative halogen, in disagreement with the observed coupling of nearly zero for the halogens. Contrast, for example, the nearly zero coupling of Cl in AlCl or InCl with the very large coupling of both Al and In, almost equivalent to that of one unbalanced p electron. If there were no hybridization on Al or In, removal of the valence p electron would leave in the ionic structure two electrons on the spherically symmetric s shell, which has no quadrupole coupling.

If there were s hybridization on the negative halogen, it would be expected to be in the bonds of low ionic character where the quenching primary moment is small or in bonds such as HI where there is a large orbital overlap moment in the opposite direction (see Section 16). However, the evidence from quadrupole coupling is against such hybridization (see the discussion of BrCl and IBr in Section 8).

11 MEASUREMENT OF π CHARACTER WITH NUCLEAR QUADRUPOLE COUPLING

Nuclear quadrupole coupling provides a useful method for investigation of the π character of chemical bonds. When the coupling is axially symmetric

($\eta=0$), as is always true in linear and symmetric-top molecules, the π character can be detected only through its effects on the magnitude of the coupling. Measurements of the π character in this case requires a knowledge of the coupling to be expected if there were no π character in the bond. When the coupling tensor is made asymmetric by π bonding in one plane (double-bond character), the π component can be measured through the observed asymmetry parameter in the coupling tensor. This method is generally more accurate than that which depends on the magnitude of the coupling.

Symmetrically Bonded Atoms

In linear and symmetric-top molecules there is no observable asymmetry parameter, and one can detect π character only through its effects on the magnitude of the coupling. This method is most applicable to bonds involving a coupling halogen for which the coupling expected for the bond without π character can be estimated from methods derived for single-bonded halogens of diatomic molecules (Sections 7 and 8). We shall illustrate the method for bonds of this type. It is also applicable to symmetrically bonded atoms in asymmetric-top molecules for which the π_x and π_y character is equal and the asymmetry parameter is zero.

In developing the formula for the single-bonded coupling halogen we assumed the p_x and p_y orbitals to have unshared pairs. Thus $n_x=2$ and $n_y=2$. In π bonding either the p_x or the p_y pair, or both, would be shared. There would be, on the average, only one electron in the p orbital forming a complete π bond or $2-\pi_c$ in the halogen orbital which formed a fractional bond of amount π_c. In a symmetrical bond one could not detect the π_x or π_y bonding separately, and hence we designate the total amount of π bonding as π_c. Thus $n_x+n_y=4-\pi_c$. The number of p_z electrons n_z is determined by (14.60) and (14.86). If we assume the coupling to be on a negative halogen atom of a polar bond (the most common case for halogen bonds in organic molecules), the hybridization for the coupling atom will be negligible, and the unbalanced p electrons with reference to the bond axis z will be

$$(U_p)_z=\frac{n_x+n_y}{2}-n_z=2-\frac{\pi_c}{2}-(1+i_\sigma)=\left(1-i_\sigma-\frac{\pi_c}{2}\right) \tag{14.98}$$

With neglect of the small charge correction, the corresponding coupling on a negative halogen will be

$$\chi_z=-\left(1-i_\sigma-\frac{\pi_c}{2}\right)eQq_{n10} \qquad \{\text{negative pole of } \sigma \text{ bond on halogen}\} \tag{14.99}$$

and the π character will be

$$\pi_c=2\left(1-i_\sigma+\frac{\chi_z}{eQq_{n10}}\right) \tag{14.100}$$

in which i_σ is the ionic character of the σ bond only. The total ionic character of the bond including both the σ and π components will be

$$i_c=i_\sigma-\pi_c \tag{14.101}$$

If the resultant dipole moment of the bond is such as to put the positive pole on the coupling atom, we must admit of the possibility of s hybridization. If there is an a_s^2 amount of hybridization on the coupling halogen, (14.91) becomes

$$\chi_z = -\left[(1+i_\sigma)(1-a_s^2) - \frac{\pi_c}{2}\right](1+0.15i_c)eQq_{n10} \qquad \left\{\begin{matrix}\text{positive pole}\\ \text{of } \sigma \text{ bond on}\\ \text{halogen}\end{matrix}\right\} \tag{14.102}$$

where the term $0.15i_c$ is a screening constant for the positive charge on the coupling halogen and where $i_c = i_\sigma + \pi_c$.

In Table 14.7 are given the values for the π character of the C– halogen bond in several linear and symmetric-top molecules as estimated from the quadrupole coupling of the halogen. In all, except ICN and CH_3CCI, (14.100) was used for the calculations, with the ionic character obtained from (14.93). The small possible hybridization in these negative or nearly neutral halogens is neglected. For ICN and CH_3CCI, for which the positive pole of the σ bond is

Table 14.7 Bond Character and Resultant Ionic Character for Various Bonds to Halogens Measured from the Halogen Coupling

Molecule	χ_z(Hal)(MHz) for ^{35}Cl, ^{79}Br, or ^{127}I	*Ionic Character of σ Bond*[a] i_σ	π_c	*Resultant Ionic Character*[a] $i_c = i_\sigma - \pi_c$
CH_3Cl	−74.77	0.35	0	0.35
CH_3Br	577.15	0.25	0	0.25
CH_3I	−1934	0.13	0.04	0.09
SiH_3Cl	−40.0	0.55	0.17	0.38
SiH_3Br	336	0.45	0.22	0.23
SiH_3I	−1240	0.32	0.28	0.04
GeH_3Cl	−46.95	0.57	0	0.57
GeH_3Br	380	0.47	0.06	0.41
ClCN	−83.39	0.10	0.28	+0.18
BrCN	685.6	0	0.22	+0.22
ICN	2418.8	+0.12	0.22	+0.34
HCCCl	−79.67	0.15	0.24	+0.09
HCCBr	646	0.05	0.22	+0.17
CH_3CCCl	−79.6	0.15	0.24	+0.09
CH_3CCBr	647	0.05	0.22	+0.17
CH_3CCI	−2230	+0.07	0.25	+0.32

[a]The positive sign indicates that the positive pole is on the halogen.

on the I, (14.102) is employed with $a_s^2=0$. There is justification for neglect of s character when the positive charge on the halogen is small (see Section 8).

The electronegativity values for the halogen atoms that are needed for calculation of the ionic character from (14.93) are taken from Appendix G, except that group electronegativity values of 2.30 for CH_3, 2.7 for X—C≡C—, and 2.8 for N≡C were used [7].

Among the halogen bonds to the hybride group, only those to SiH_3 show evidence for double-bond character. This bonding can be attributed mainly to contributing structures of the type $H_3Si^-{=}Cl^+$, in which the electron pair on the halogen forms a dative-type π bond with the empty d orbitals of the Si. From bond-length shortening, the Si—Cl and Si—Br bonds are predicted to have about 10% double-bond character, while those for CH_3—Hal and GeH_3—Hal show no bond shortening attributable to π bonding. Such bonds are also possible with Ge and higher-period elements, but π bond character generally decreases with increasing size of the bond atoms because of less effective orbital overlap. Nevertheless, the quadrupole coupling indicates that the π character in the H_3Si—Hal bonds increases slightly from Cl to I. Evidently the decrease in electronegativity from Cl to I is responsible for this effect. When, however, there are other halogens bonded to the central atom, as in $SiCl_4$, $SiBr_4$, and so on, the trend is reversed, that is, the π bonding is greater for the smaller halogen. This reversal is attributable to the difference in effective electronegativity of the group to which the halogen is bonded. For example, the effective electronegativity of the Cl_3Si group is significantly greater than that of I_3Si. The insignificance of π character in the CH_3—halogen bond is due to the fact that no d orbitals are available on the valence shell of C. Slight amounts of π bonding in these molecules might occur, however, through the mechanism of hyperconjugation.

The results in Table 14.7 show, in agreement with consideration of bond lengths [37], that there is a strong tendency for conjugation of a C– halogen bond with an adjacent triple bond. The π bonding of the C halogen can be ascribed to structures of the form

$$N^-{=}C{=}Hal^+ \qquad X{-}C^-{=}C{=}Hal^+$$

which contribute from 22 to 28% to the ground state of the halogen cyanides and acetylenes listed in Table 14.7. Because of the positive charge required on the halogens, structures of this type would be expected to decrease in contributions with increasing electronegativity of the halogen from I to Cl. Other things being equal, however, the π bond character would be expected to decrease with increase in size of the halogen. These two opposing effects are evidently responsible for the nearly equivalent π bond character of the chloride, bromide, and iodide bonds in the cyanides and acetylenes.

The ^{14}N coupling also gives evidence for contributions of structures of the type $N^-{=}C{=}Hal^+$. Because of the small magnitude of the ^{14}N, the prediction of the amount of this contribution from the nitrogen coupling is not as reliable as that from the halogen.

Involvement of *d* Orbitals in Double Bonding

Quadrupole coupling evidence for *d*-orbital involvement in double bonding in SiH_3—Hal and to a lesser extent in GeH_3—Br is shown in Table 14.7. Note that there is no comparable evidence for π bonding in the methyl halides. Since the *d* orbitals of the Group IV elements are normally unoccupied and the *p* shell of a singly bonded halogen is filled, the double bond component is a dative-type bond in which the shared pair is contributed by a *p* orbital of the halogen. The fact that such a double bonded structure as $X_3Si^-{=}Hal^+$ seems to require a positive charge on the very electronegative halogen has caused many to doubt the importance of the structure. Pauling, however, in his well-known treatise on the chemical bond [37] freely postulated such structures to account for bond distances in the silicon halides. The much higher ionic character—of opposite polarity—in the σ component of the bond as revealed by the quadrupole coupling makes this form of double bonding more plausible. For example, (14.93) indicates 55% ionic character for the σ bond in SiH_3Cl, which puts a negative charge of $0.55e$ on the chlorine. The resultant negative charge on the Cl is reduced to $0.38e$ by the 17% π-bonding component. This "π feedback" thus makes the charge on the Cl less negative, but not positive. The rather large positive charge of $0.55e$ in the sp_3 hybrid orbital of the Si significantly reduces the nuclear screening for the *d* orbitals and thus increases their effective electronegativity and tendency for π bonding.

Because of the high electronegativity of the halogens, the group electronegativity of $(Hal)_3Si$— is appreciably greater than that of H_3Si—. The same is true for the Ge groups (see Section 15). However, "π feedback" in the halogen groups puts negative charge back on the central atom and alters its electronegativity. In the spherical halides $Si(Hal)_4$, $Ge(Hal)_4$, and $Sn(Hal)_4$, for which the halogen couplings are known from pure quadrupole resonance, correction for this effect can be taken into account if the number of screening valence electrons is decreased by $c=3(i_\sigma-\pi_c)$ in the calculation of the $(Hal)_3M$— group electronegativity from (14.141). Here $(i_\sigma-i_\pi)$ is the resultant ionic character of each of the halogen bonds. As was shown earlier [7] for the $M(Hal)_4$ molecule, one can then solve the three equations:

$$\chi_z=-\left(1-i_\sigma-\frac{\pi_c}{2}\right)eQq_{n10} \tag{14.103}$$

$$i_\sigma=\frac{x_{Hal}-x_g}{2} \tag{14.104}$$

$$x_g=\frac{0.31[5+3(i_\sigma-\pi_c)]}{r}+0.50 \tag{14.105}$$

for the three unknowns, i_σ, π_c, and x_g, where x_g is the electronegativity of the $(Hal)_3M$— group. Equation 14.105 is described in Section 15. In these equations, χ_z is the halogen coupling, x_{Hal} is the halogen electronegativity (3 for Cl, 2.8 for Br, and 2.55 for I) and r is the covalent radius of M (1.17 for Si, 1.22 for Ge, and 1.40 for Sn). The resulting values of i_σ, π_c, and x_g obtained by this method

Table 14.8 Bond Properties Derived from the Halogen Quadrupole Coupling in Symmetrical $X(\mathrm{Hal})_4$ Molecules in the Solid State

Molecule	χ_z/eQq_{n10}	x_g	i_σ	π_c
$SiCl_4$	0.37	2.07	0.46	0.33
$SiBr_4$	0.46	2.02	0.39	0.30
SiI_4	0.58	1.97	0.29	0.26
$GeCl_4$	0.47	2.07	0.46	0.13
$GeBr_4$	0.54	2.02	0.39	0.13
GeI_4	0.65	1.96	0.30	0.11
$SnCl_4$	0.44	1.94	0.53	0.06
$SnBr_4$	0.50	1.89	0.46	0.08
SnI_4	0.61	1.85	0.35	0.08

are shown in Table 14.8. Although these values are only approximate, they illustrate how quadrupole coupling can be combined with information of other types to provide an understanding of the nature of the bonding in rather complex pentatomic molecules. The π character in these bonds is partly due to hyperconjugation described in Section 12. However, from a consideration of the rather large π_c in SnH_3—Hal, where such hyperconjugation is not significant, it is concluded that the principal mechanisms of the π bonding in the Group IV elements of the third and higher rows involves the d orbitals. The small amount of π character which results from hyperconjugation should not alter significantly the predicted values. Another assumption—that the screening by d electrons is the same as that by s and p electrons—may cause small errors. This assumption would tend to make the predicted x_g values slightly too small.

Figure 14.2 shows a plot of double bond character in the Group IV halides

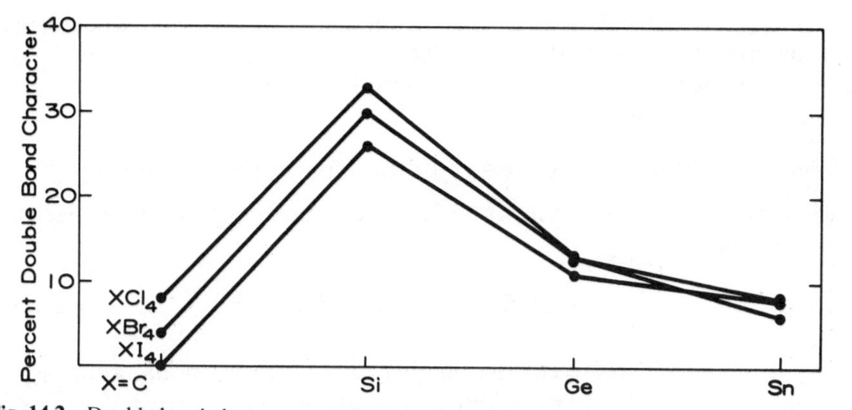

Fig. 14.2 Double bond character in $X(\mathrm{Hal})_4$ molecules as derived from the nuclear quadrupole coupling of the halogen.

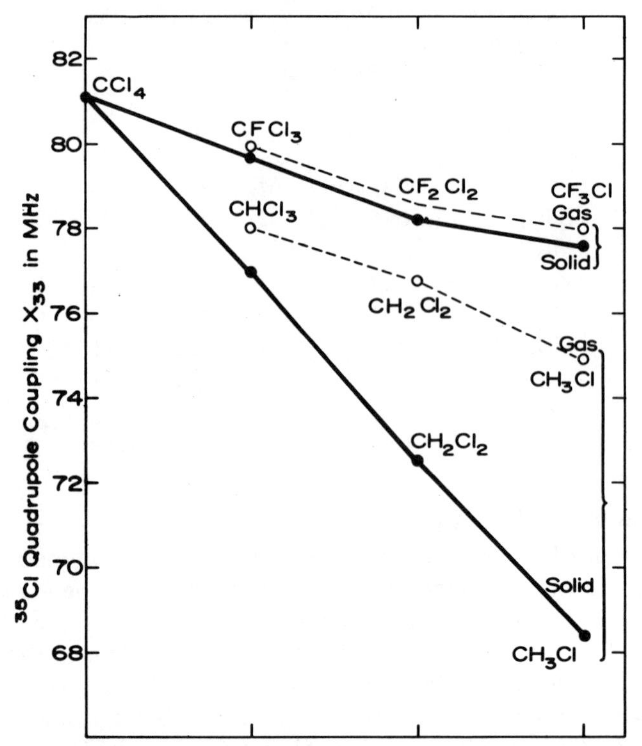

Fig. 14.3 Plots showing effects of chemical constitution and of solid state interactions on Cl quadrupole coupling.

as determined from nuclear quadrupole coupling. The graph demonstrates that the π bonding is most important for Si, which among the Group IV elements is the smallest one having *d* orbitals available in its valence shell. The π bonding in the carbon halides must occur through the mechanism of hyperconjugation; see Section 12.

Although these calculations are made from solid-state quadrupole couplings, it seems probable that for these spherical, nonpolar molecules the solid state values would be very near those for the gaseous state. Evidence for this is given in Fig. 14.3 and Section 12.

π Character from the Asymmetry in the Quadrupole Coupling

The asymmetry parameter in the nuclear quadrupole coupling tensor (Section 4) can be expressed as

$$\eta = \frac{\chi_{xx} - \chi_{yy}}{\chi_{zz}} = \frac{3(n_y - n_x)}{n_x + n_y - 2n_z} \tag{14.106}$$

where n_x, n_y, and n_z represent the number (0 to 2) or fractional number, of electrons in the *p* orbitals with orientations along the respective *x*, *y*, and *z* axes. Since

$$2(U_p)_z = n_x + n_y - 2n_z = -\frac{2\chi_{zz}}{eQq_{n10}} \tag{14.107}$$

Equation 14.106 can be transformed to

$$n_x - n_y = \frac{2}{3}\left(\frac{\chi_{zz}}{eQq_{n10}}\right)\eta \tag{14.108}$$

or to the alternate form

$$n_x - n_y = \frac{2}{3}\left(\frac{\chi_{xx} - \chi_{yy}}{eQq_{n10}}\right) \tag{14.109}$$

With the p_z orbital directed along the reference bond axis, (14.108) provides a means of finding the difference in population of the p_x and p_y orbitals without a knowledge of the σ component—if the atomic coupling per p electron in the bonded atom is known. If the population of one of the orbitals p_x or p_y is known or can safely be assumed, the population of the other can be found from the asymmetry in the coupling.

A typical use of (14.108) is for finding the amount of π bonding in asymmetric-top molecules where the halogen atom is a part of a planar structure over which the π orbital spreads. In such molecules, one p orbital of the halogen normal to the σ bond will be oriented normal to the molecular plane and will participate in the π bonding. The other will be in the plane of the molecule, usually with no orbital overlap for possible bonding. There is no uniformity among microwave spectroscopists in designation of the x and y axes. Some specify the normal to the plane as x and some as y. However, the σ bond axis to the coupling halogen is uniformly designated as the z axis. The halogen p orbital which participates in π bonding, say p_y, shares its two electrons and thus has a reduced n_y value over the other. This reduction in n_y is numerically equivalent to the amount of π_y bond character. Thus we can express (14.108) and (14.109) as

$$\pi_x - \pi_y = \frac{2}{3}\left(\frac{\chi_{zz}}{eQq_{n10}}\right)\eta = \frac{2}{3}\left(\frac{\chi_{xx} - \chi_{yy}}{eQq_{n10}}\right) \tag{14.110}$$

When either π_y or π_x is zero, this equation gives directly the amount of π character of the other orbital. When neither is zero, the asymmetry in the coupling obviously gives the difference in amounts of π_x and π_y character.

Table 14.9 gives the π character of the halogen bonds of a number of asymmetric-top molecules as calculated from the asymmetry in the halogen coupling. In these examples the halogen atom forms a part of a planar group, and we have assumed the π bonding of the halogen orbital in this plane to be zero.

An interesting application of the above theory to nitrogen coupling is provided by $S(CN)_2$, the structure and nuclear coupling of which have been measured with microwave spectroscopy by Pierce et al. [39]. The molecule is planar, with the ^{14}N coupling χ_{zz} of -3.45 MHz along the CN bond. The coupling perpendicular to the molecular plane χ_{yy} is 1.21 MHz, and the coupling in the

Table 14.9 *π Character of Carbon–Halogen Bonds from the Asymmetry in the Nuclear Quadrupole Coupling of the Halogen*

	Coupling Parameter of the Halogen[a]		*π Character of the C–Hal*[b]
Molecule	χ_{zz}	η	π_c
$CH_3CH_2{}^{35}Cl$	−68.8	0.035	0.016
$CH_2{}^{35}Cl_2$	−76.92	0.036	0.017
$CH_2F^{35}Cl$	−70.46	0.103	0.044
$CH_2{=}CH^{35}Cl$	−70.16	0.143	0.061
$CF_2{=}CH^{35}Cl$	−84.3	0.206	0.106
$CH_2C{=}CH^{127}I$	−1877	0.056	0.030
$CH_3C({=}O){-}^{35}Cl$	−59.2	0.271	0.098
$CH_3C({=}O){-}^{79}Br$	464	0.211	0.085
$CH_2{-}CH^{35}Cl$ (ring with CH_2)	−71.40	0.029	0.013
$C_6H_5{}^{79}Br$	567	0.049	0.024

[a]References to source of coupling parameters given in Table 9.3.
[b]Calculated with (14.110).

plane and perpendicular to the CN bond χ_{xx} is 2.27 MHz. Therefore

$$n_x - n_y = \frac{2}{3}\left(\frac{\chi_{xx}-\chi_{yy}}{eQq_{n10}}\right) = \frac{2}{3}\left(\frac{2.27-1.21}{-10}\right) \tag{14.111}$$

or $n_y = n_x + 0.071$. This analysis shows that there is 0.071 e more electronic charge in the p_y atomic orbital (perpendicular to the plane) than there is in the p_x orbital (in the plane). This extra amount can be explained as a 7.1% contribution to the ground state of the molecule from each of the two equivalent structures (II).

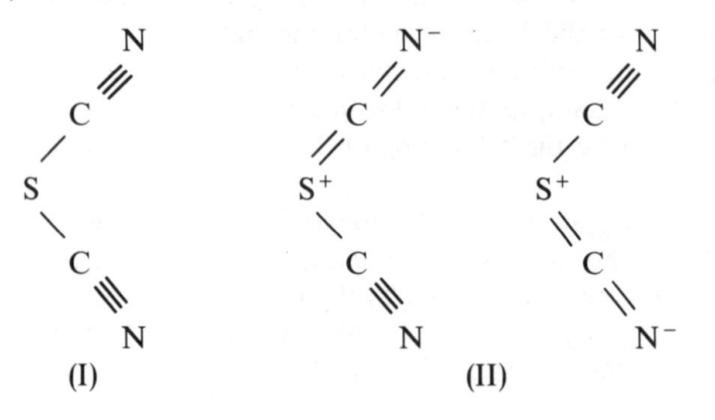

Most of the C– halogen bonds shown in Table 14.9 are adjacent to the C=C double bond, and it is the conjugation with this double bond which is responsible for the large π character of 8 to 10% in the C– halogen bonds. For this configuration the π character is of the order of one-half that for the triple bonds shown in Table 14.7. In C_6H_5Br the conjugation of the C—Br bond with the benzene ring causes 2.4% double bond character in the C—Br bond. It would be interesting to know how this compares with that of C_6H_5Cl, but, so far as we know, the asymmetry in the Cl coupling in this molecule has not been measured.

The π bond character derived from quadrupole couplings gives information about only a part of a molecular orbital which spreads over a planar ring or group involving a number of atoms. Although quadrupole couplings may not provide complete knowledge of such molecular orbitals, they do provide additional observables against which molecular orbital theorists can test their postulates and calculations. The first such use of a quadrupole coupling to test molecular orbital calculations was apparently made by Bersohn [8]. Furthermore, the quadrupole coupling method provides rather definite information about the nature of the specific bond to the coupling atoms, even if knowledge of the entire molecular orbital involved is incomplete.

Much information has been gained about π bonding from pure nuclear quadrupole coupling in solids where coupling by a number of nuclei in the same molecule can be observed. The methods described here are also applicable to interpretation of solid-state coupling, and we shall employ some solid state results in our discussions. However, we make no attempt to give a comprehensive survey of the large amount of data accumulated for solids. Discussions of solids are given by Dehmelt [40], Cohen and Reif [41], Das and Hahn [12], Hooper and Bray [42], and O'Konski [14] among others.

12 HYPERCONJUGATION

Interpretation of the quadrupole coupling of atoms bonded to first-row elements, particularly halogens bonded to carbons, is complicated by hyperconjugation. When other factors influencing the coupling are predictable, quadrupole coupling can be used as a measure of this conjugation. In molecular orbital terminology, the phenomenon, which represents an interaction between the π and σ orbitals, is often designated as configuration interaction. In valence bond resonance theory it can be represented as a conjugation which involves the breaking of a single bond and the consequent formation of a double bond, and vice versa, as illustrated by structures (I) to (III).

$$X{-}M{-}Y \quad \text{(I)} \qquad X^+{=}M \quad Y^- \quad \text{(II)} \qquad X^- \quad M{=}Y^+ \quad \text{(III)}$$

$$^+X{=}^-M{-}Y \quad \text{(IV)} \qquad X{-}M^-{=}Y^+ \quad \text{(V)}$$

The atom X in structure (I) has an unshared pair, which in structure (II) is shared with atom M. To share this pair, atom M employs the orbital used in structure (I) to form the single bond with Y. If Y also has an unshared pair, the reverse conjugation can also occur, as indicated by structure (III). For such conjugation to occur, the orbitals must be sufficiently close and so favorably oriented that the orbital over-lap required for formation of the π-bond component is achieved. If atom M has available an unoccupied orbital of relatively low energy which favorably overlaps those of the unshared pair of X or Y, structures of type (IV) or (V) will also contribute to the ground state of the molecules. When d orbitals of low energy are available, as in Si, structures of type (IV) or (V) apparently are much more important than are the hyperconjugated structures (II) and (III). However, both types of conjugation may occur in such compounds. Hyperconjugation is most important for compounds in which the central atom is carbon and the atoms bonded to it are halogens. Carbon has no empty orbitals in its valence shell to form structures such as (IV) and (V); because of its small size, the orbital overlap required for hyperconjugation can be achieved. The orbital overlap becomes less favorable for hyperconjugation, of course, with the increase in size of the atom bonded to carbon. Hence, hyperconjugation should decrease from the fluorides to the iodides. Because the contributing structures require a positive charge on atom X and a negative charge on atom Y, or vice versa, hyperconjugation also depends sensitively on the relative electronegativity of X and Y.

To account for the abnormally short bond lengths in the carbon halides, Pauling [37] originally postulated hyperconjugated valence bond structures of the type

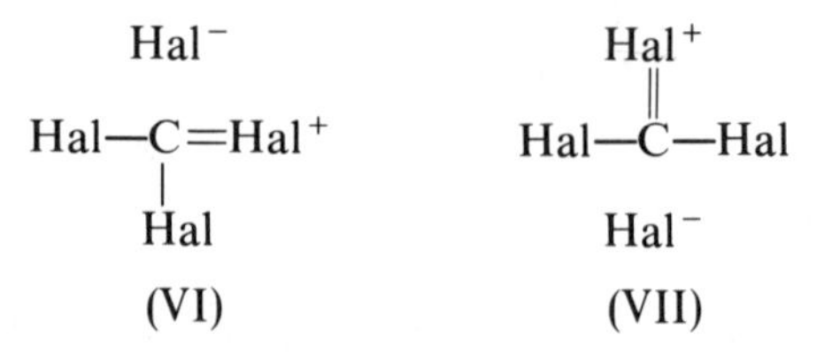

Mulliken et al. [43] later treated hyperconjugation with molecular orbital theory and showed that it occurs to some extent even in pure hydrocarbons.

The molecular orbital theory provides the more satisfactory treatment of conjugated systems where the symmetry and molecular simplicity are such as to make it applicable. However, the conjugated valence bond method is the more practical one for interpretation of quadrupole coupling in the hyperconjugated tetrahedral molecules considered here.

The treatment of quadrupole coupling in hyperconjugated structures is simplest when the atoms X and Y are identical and the contribution from structures (II) and (III) consequently equal. Let us consider the chloromethanes. In CH_3Cl the hyperconjugation is suppressed by the large electronegativity difference between H and Cl, and no significant double bond character is indicated by quadrupole coupling for any of the methyl halides (see Table

14.7). When a hydrogen is replaced by a second chlorine, however, the two equivalent structures (X) and (XI) contribute to the ground state, as do (VIII) and (IX) and other structures (not shown) representing the ionic character and slight hyperconjugation with the CH bonds.

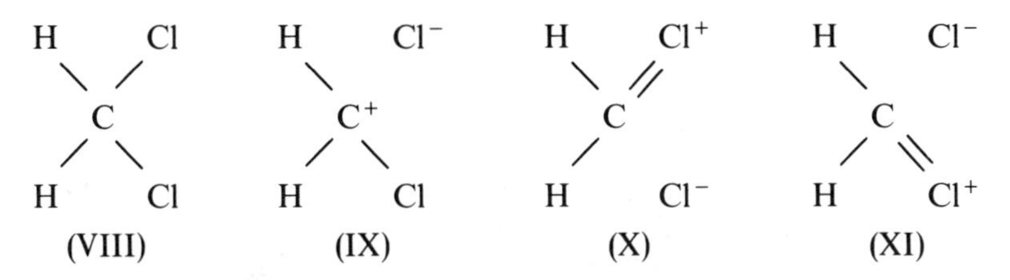

Let us concentrate on the coupling of the upper Cl in the diagram. The coupling of the two chlorines will, of course, be equivalent because of symmetry. The weight of structure (IX) represents the normal ionic character of the σ bond, i_σ. The weight of structure (X) represents the double bond character π_c of the C—Cl bonds. If only structures (VIII), (IX), and (X) contributed, the quadrupole coupling would be given by (14.99) since the hybridization on the Cl is negligible. However, the weight of structure (XI), in which the upper Cl has a closed p shell and hence no coupling, is equal to that of structure (X). Thus the quadrupole coupling given by (14.99) is reduced by the factor $(1-\pi_c)$. The coupling with reference to the C—Cl bond axis is therefore

$$\chi_{zz} = -\left(1 - i_\sigma - \frac{\pi_c}{2}\right)(1-\pi_c)eQq_{n10} \tag{14.112}$$

The ionic character i_σ can be estimated in the normal way, from (14.93), but instead of the usual electronegativity for the carbon we must use the effective electronegativity for the H_2CCl group. It is apparent that the negatively charged Cl in structure (X) or (XI) does not result from the usual ionic character in the Cl bond since there is no countering positive charge on the carbon. It does not depend upon the electronegativity difference between carbon and chlorine.

We have used (14.112) to calculate the π character of the C—Cl bonds for the chloromethane from the χ_{zz} coupling constants, as shown in Table 14.10. To obtain values of i_σ, we used the group electronegativity values for H_2CCl, $HCCl_2$, and CCl_3 given in Section 15 with the normal x value of 3.00 for Cl. Although the group electronegativity values are not precise, these calculations provide a good approximation to the amount of hyperconjugation in these molecules. As expected, they show that the π character in the bonds increases with increasing number of Cl atoms. The coupling for CCl_4 was measured in the solid state, but this should not significantly influence the result.

Note that the value of π_c calculated here for CH_2Cl_2 is greater than that obtained from the asymmetry parameter (Table 14.9). However, both methods agree in the prediction of a small π_c value, from 2 to 3%. The group electronegativity values probably cause the greatest error in the present calculations, for they are difficult to evaluate. The greatest source of error in the method which depends on the asymmetry parameter is the assumption that only one of the

Table 14.10 Double Bond Character π_c of Carbon–Halogen Bonds as Derived from Halogen Quadrupole Coupling (The π_c results from hyperconjugation)

Molecule	χ_{zz}(MHz)	$i_c = \frac{\lvert x_g - x_{\mathrm{Hal}} \rvert}{2}$	π_c of C–Hal
		Gas	
$CH_3{}^{35}Cl$	-74.77[a]	0.35	0
$CH_2{}^{35}Cl_2$	-76.92[a]	0.265	0.03
$CH^{35}Cl_3$	-77.9[a]	0.185	0.09
CH_2FCl	-70.46[a]	0.195 (CCl)	0.044 (CCl) ~0.18 (CF)
		Solid	
$C^{35}Cl_4$	-81.1[b]	0.105	0.08
$C^{79}Br_4$	643[c]	0.115	0.04
$C^{127}I_4$	-2130[d]	0.085	0

[a]Literature source given in Tables 9.2 and 9.3.
[b]Livingston [29].
[c]Calculated from the ^{81}Br resonant frequency observed by A. L. Schawlow, *J. Chem. Phys.*, **22**, 1211 (1954).
[d]H. G. Robinson, H. G. Dehmelt, and W. Gordy, *J. Chem. Phys.*, **22**, 511 (1954).

chlorine orbitals, p_x or p_y, participates in the π bonding. Actually, the asymmetry method gives $\pi_x - \pi_y$, whereas the method based on χ_{zz} gives the total π character, $\pi_x + \pi_y$. Let us assume that both values are correct and choose the x coordinate perpendicular to the Cl—C—Cl plane. Solution of the equations $\pi_x - \pi_y = 0.017$ and $\pi_x + \pi_y = 0.03$ gives $\pi_x = 0.024$ and $\pi_y = 0.006$, a reasonable result.

Equation 14.112 also applies to the bromomethanes and iodomethanes, but coupling values are available only for solid CBr_4 and CI_4. The one value reported in the literature for $HCBr_3$ has been found to be based on an incorrect assignment. With the method described above and with Jaffee's value for the group electronegativity of CBr_3 and of CI_3 (Section 15), we have calculated the π_c values due to hyperconjugation in CBr_4 and CI_4, as shown in Table 14.10. Note that the π_c values decrease from CCl_4 to CI_4, as expected.

Equation 14.112 can also be applied to such mixed halogen compounds as CH_2ClBr. Although an additional π_c parameter is introduced, an additional observable is also introduced when the coupling of both halogens is measurable. So far as we know, coupling for two unlike halogens in such molecules has not yet been measured with microwave spectroscopy.

From consideration of bond lengths [37] and of probable orbital overlap, the π character caused by hyperconjugation is expected to be greater for the

C—F bonds than for those of other halogens. Furthermore, it is evident from comparison of the π_c in the C—Cl bonds of CH_2Cl_2 that fluorine, probably because of its greater electronegativity, induces more ionic character in the C—Cl bonds than does chlorine. Because ^{19}F has no nuclear quadrupole moment, one cannot obtain a direct measurement of the π_c of the C—F bonds. Without taking this hyperconjugation into account, however, one cannot explain the small values observed for other halogens in mixτd fluoromethanes. For example, the group electronegativity of CF_3 (Section 15) is actually greater than that of Cl, and the ionic character of the C—Cl bond in CF_3Cl is such as to make the Cl positive. Therefore, unless there is strong hyperconjugation, one would expect the coupling for ^{35}Cl to be somewhat greater in magnitude than that of one unbalanced p electron, -109 MHz, as observed in Cl_2 (Section 7). Nevertheless, the observed coupling is only -78 MHz, comparable to that in CH_3Cl. The ^{35}Cl coupling is significantly reduced by contributions from the hyperconjugated structures

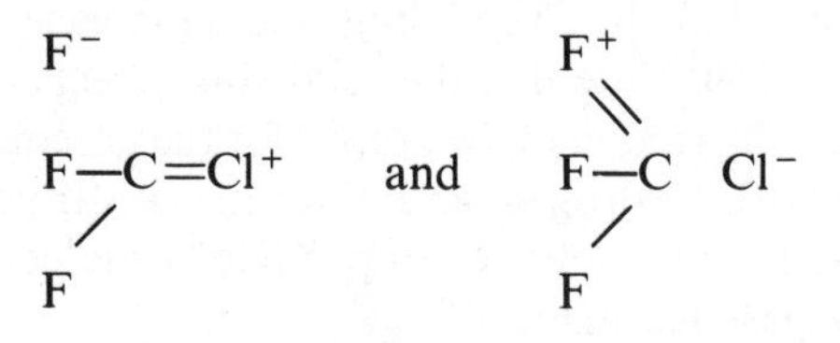

However, it is difficult to estimate the amount of each structure. From the analysis of CH_2FCl given below, it would seem that the influence of the second structure is the greater despite the fact that the positive charge is on the more electronegative F.

An indirect evaluation of the π character in the C—F bond of CH_2FCl can be made from the ^{35}Cl quadrupole coupling. The hyperconjugated structures are principally

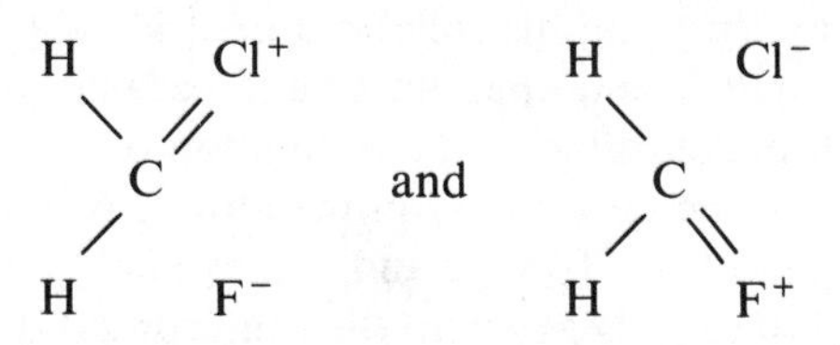

The weight of the first structure which represents the π character in the C—Cl bond is found from the asymmetry parameter to be 0.044. The weight of the second structure is equivalent to the π_c(C—F) and reduces the coupling by the factor, $1-\pi_c$(C—F). The Cl coupling is given by

$$\chi_{zz}=-\left[1-i_\sigma(\text{C—Cl})-\frac{\pi_c(\text{C—Cl})}{2}\right][1-\pi_c(\text{C—F})]eQq_{n10} \quad (14.113)$$

To obtain $i_\sigma(\text{C—Cl})=0.195$ we use (14.93) with x_g equal to 2.61 for CH_2F. Therefore,

$$-70.16=-(1-0.195-0.022)[1-\pi_c(\text{C}-\text{F})](109.46)$$

which yields π_c(C—F)=0.18. This surprisingly large estimate of 18% double bond character in the C—F bond is very approximate, but there can be no doubt that the π character of the C—F bond is significantly greater than that of the C—Cl bond and that the tendency of F to induce hyperconjugation is appreciably greater than that of the other halogens. The greatest uncertainty in this estimate probably arises from the group electronegativity value for CH_2F. If $x_g=2.50$ is used, which we consider the lowest possible value for CH_2F, the value of π_c(C—F)=0.12 is obtained.

Qualitatively, one can see from Fig. 14.3 the effects on the Cl coupling of substitution of increasing numbers of hydrogens and of fluorines in CCl_4. The drop in Cl coupling with hydrogen substitution is expected from the increase in negative charge on the Cl that results from the decreasing electronegativity of the carbon group. Because of additional intermolecular interaction, the decrease is more drastic in the solid state. In contrast, fluorine substitution increases the effective electronegativity of the carbon, and if there were no effects of hyperconjugation, the Cl coupling would increase rather than decrease with fluorine substitution. Note that the solid-state effects are very small in the fluorine substituents. Projection of the graphs for the gaseous and the solid state indicates that for both hydrogen and fluorine substituents the solid-state coupling of the halogen in a spherical-top $X(\mathrm{Hal})_4$ molecule is probably very close to that for the gaseous state.

13 NITROGEN COUPLING

Because of the prevalence of nitrogen among organic chemicals and biochemicals, the ^{14}N quadrupole coupling is potentially a valuable source of information. Much qualitative information has been gained from it, mostly from pure quadrupole resonance of ^{14}N in molecular solids in experiments by O'Konski [14] and by Bray and his collaborators [44–46]. The interpretation of this coupling is made somewhat uncertain because of the complicated bonding on N which makes difficult the estimation of the unbalanced p electrons, also because of some uncertainty in the atomic orbital coupling eQq_{210} for one unbalanced p electron. Townes and Dailey [2, 11] originally estimated this constant to be -10 to -24 MHz. In their interpretation of NF_3, Sheridan and Gordy [47] obtained evidence that the probable magnitude is the lower limit of this range, $eQq_{n10}=-10$ MHz. From NH_3 and $(CH_3)_3N$ couplings, O'Konski [14] calculated the value in the range from -9 to -11 MHz. Kato et al. [48] calculated the value as -11.12 MHz with q_{210} from self-consistent field theory. Lucken [49] obtained the value as -10 MHz. For our analysis in this section we shall employ the commonly used value, $eQq_{210}(^{14}N)=-10$ MHz.

The ^{14}N coupling in free $^{14}N_2$ molecules has now been measured with molecular beam spectroscopy [50] to be $\chi_{zz}(^{14}N)=-2.52(4)$ MHz. With the assumption that the coupling per unbalanced p electron in the molecule is $eQq_{n10}(^{14}N)=-10$ MHz, the σ bonding orbital in the molecule can be predicted to be $(\frac{1}{4})^2 s+(\frac{3}{4})^2 p_z$ and the counterhybridized orbital to be $(\frac{3}{4})^2 s-(\frac{1}{4})^2 p_z$.

Since the bonding is expected to consist of a single σ bond and two π bonds with the counterhybridized orbital having an unshared pair, the number of unbalanced p_z electrons (with this type of bonding) is

$$(U_p)_z = \frac{n_x + n_y}{2} - n_z = \frac{1+1}{2} - \left(\frac{3}{4} + \frac{2}{4}\right) = -\frac{1}{4}$$

and the predicted coupling is

$$\chi_{zz} = -(U_p)_z eQq_{n10} = \tfrac{1}{4} eQq_{n10} = \tfrac{1}{4}(-10) = -2.50 \text{ MHz} \qquad (14.114)$$

in agreement with the observed value of -2.52 MHz.

The bonding of N_2 in the solid state is evidently quite different from that in the free molecules since the solid state couplings are almost twice as great as those in the free molecules. The magnitude of the coupling in solid $^{14}N_2$ has been measured with pure quadrupole resonance by Terman and Scott [51] as 4.650 and as 4.782 MHz in β-quinol clathrate by Meyer and Scott [52]. Generally, one might expect the solid state coupling to be lower rather than appreciably higher than the free molecule coupling. The two values can be brought into approximate agreement by the assumption that in the solids the σ bonding orbital is $(\frac{1}{2})^2 s + (\frac{1}{2})^2 p_z$ and that the unshared pair is in the counterhybridized orbital, $(\frac{1}{2})^2 s - (\frac{1}{2})^2 p_z$. With these assumptions, $(U_p)_z = -\frac{1}{2}$ and the magnitude of the calculated coupling (with eQq_{n10} again assumed to be -10 MHz) is $\chi_{zz} = \frac{1}{2}(-10) = -5$ MHz, in approximate agreement with the higher solid state value of 4.78 MHz. Because intramolecular interactions in the solids may alter the bonding and the coupling in different ways, the predicted orbitals for the free radicals are believed to be the more reliable.

From gaseous microwave spectrocopy the ^{14}N couplings and the molecular structures of the important molecules pyrrole [53] and pyridine [54], as well as N-methyl pyrrole [55] and 4-methyl pyridine [56], have been found. As

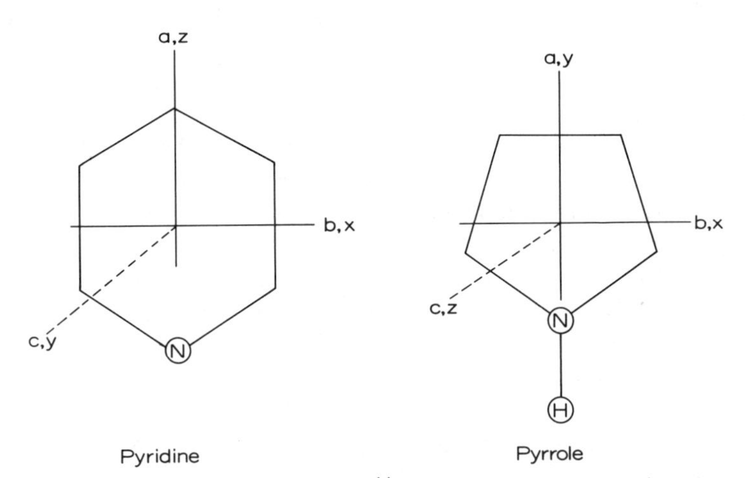

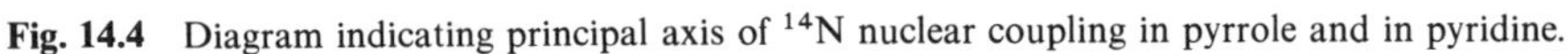
Fig. 14.4 Diagram indicating principal axis of ^{14}N nuclear coupling in pyrrole and in pyridine.

examples, we give interpretation of the coupling in pyrrole and pyridine. Because of their molecular symmetry, the couplings measured represent the principal values of the coupling tensor. The discussion given here and in the following section differs from that of previous sections in that the coupling atom is bonded to more than one atom and the valence shell electrons are referred to axes (x, y, z) that do not necessarily coincide with any bond axis.

The three valence orbitals of N that are in the molecular plane are sp_2 hybrids, and the fourth orbital, perpendicular to this plane, is a pure p orbital. If a_s^2 represents the amount of s character in each of the equivalent σ N—C bonding orbitals, the normalized valence orbitals of the N of pyridine can be expressed as

$$\psi_1=(1-2a_s^2)^{1/2}\psi_s+(2a_s^2)^{1/2}\psi_{p_z}$$

$$\psi_2=a_s\psi_s-\left(\frac{1-2a_s^2}{2}\right)^{1/2}\psi_{p_z}+\left(\frac{1}{2}\right)^{1/2}\psi_{p_x}$$

$$\psi_3=a_s\psi_s-\left(\frac{1-2a_s^2}{2}\right)^{1/2}\psi_{p_z}-\left(\frac{1}{2}\right)^{1/2}\psi_{p_x}$$

$$\psi_4=\psi_{p_y} \tag{14.115}$$

The equivalent orbitals ψ_2 and ψ_3 form the N—C bonds, whereas ψ_4 engages in π bonds and ψ_1 has an unshared pair of electrons. With the measured C—N—C bond angle, the amount of s character a_s^2 can be found from (14.130).

Let us identify the principal coupling axes $z\rightarrow a$, $x\rightarrow b$, $y\rightarrow c$ of pyridine as indicated in Fig. 14.4. The principal coupling elements are $\chi_{zz}=-4.88$ MHz, $\chi_{xx}=1.43$ MHz, and $\chi_{yy}=3.45$ MHz. Because there are two independent coupling parameters, χ_{zz} and η, it is possible to solve the coupling equation for the ionic character, both i_σ and i_π in the σ and π bonds, respectively. From (14.52) and (14.55)

$$\chi_{zz}=\left(n_z-\frac{n_x+n_y}{2}\right)\left(\frac{eQq_{210}}{1+c^-\varepsilon}\right) \tag{14.116}$$

where c^- is the negative charge on the N and where $\varepsilon=0.30$ is the charge-screening correction (Table 14.3). Evidently, $c^-=2i_\sigma+i_\pi$, where i_σ is the ionic character of each σ bond and i_π is that of the π bond. There are two electrons in ψ_1 which from (14.115) has $2a_s^2$ of p_z character. Since the negative pole is on the nitrogen, each of the bonding orbitals ψ_2 and ψ_3 will have $(1+i_\sigma)$ electrons, and the π bonding orbital ψ_4 will have $(1+i_\pi)$ electrons. The observed CNC bond angle of 116°50′ with (14.130) shows that $a_s^2=0.311$. From the foregoing orbitals with these conditions, the p_x, p_y, and p_z orbital population are found to be

$$n_x=1+i_\sigma \qquad n_y=1+i_\pi$$

$$n_z=2(2a_s^2)+2(1+i_\sigma)\left[\frac{1-2a_s^2}{2}\right]=1.622+0.378i_\sigma$$

Substitution of these values into (14.116) with $\chi_{zz}=-4.88$ and $eQq_{210}=-10$, followed by transformation yields

$$i_\sigma+1.557i_\pi=0.323 \tag{14.117}$$

From the asymmetry, (14.109),

$$n_x-n_y=1+i_\sigma-(1+i_\pi)=0.135 \tag{14.118}$$

and hence

$$i_\sigma-i_\pi=0.135 \tag{14.119}$$

Simultaneous solution of (14.117) and (14.119) yields

$$i_\sigma=0.22 \qquad \text{and} \qquad i_\pi=0.08$$

The occupation number of the p_π orbital on the N is therefore 1.08. The negative charge on the nitrogen in units of e is

$$c^-=2i_\sigma+i_\pi=0.52$$

The value of i_σ, with (14.93), indicates an effective electronegativity difference of $2i_\sigma=0.44$ for the σ orbitals of carbon and nitrogen in these compounds.

For treatment of pyrrole we choose the reference z axis perpendicular to the molecular plane and y along the N—H bond. Therefore, $z\rightarrow c$, $y\rightarrow a$, and $x\rightarrow b$. The CNC bond angle is 109°48′, and (14.130) indicates that the s character of the N—C bonding orbitals is 25%. Therefore the valence orbitals of N are

$$\begin{aligned}
\psi_1&=(\tfrac{1}{2})^{1/2}\psi_s+(\tfrac{1}{2})^{1/2}\psi_{p_y}\\
\psi_2&=(\tfrac{1}{4})^{1/2}\psi_2-(\tfrac{1}{4})^{1/2}\psi_{p_y}+(\tfrac{1}{2})^{1/2}\psi_{p_x}\\
\psi_3&=(\tfrac{1}{4})^{1/2}\psi_s-(\tfrac{1}{4})^{1/2}\psi_{p_y}-(\tfrac{1}{2})^{1/2}\psi_{p_x}\\
\psi_4&=\psi_{p_z}
\end{aligned} \tag{14.120}$$

The population of the orbital ψ_1 is $1+i_\sigma(\text{NH})$, that of ψ_2 and of ψ_3 is $1+i_\sigma(\text{NC})$, and that of ψ_4 is $2-\pi_c$. From (14.120) it is evident that $n_x=1+i_\sigma(\text{NC})$, $n_y=1+(\tfrac{1}{2})[i_\sigma(\text{NH})+i_\sigma(\text{NC})]$, and $n_z=2-\pi_c$, where $i_\sigma(\text{NH})$ and $i_\sigma(\text{NC})$ are the ionic character of the N—H and N—C bonds and where π_c is the amount of π bonding by the electrons in the $p_z(\psi_4)$ orbital. The negative charge on the N will be $c^-=2i_\sigma(\text{NC})+i_\sigma(\text{NH})-\pi_c$. With these values and with $\varepsilon=0.30$, the coupling equations (14.116) and (14.109) can be expressed as

$$\frac{\chi_{zz}}{eQq_{210}}=\frac{1-0.75i_\sigma(\text{NC})-0.25i_\sigma(\text{NH})-\pi_c}{1+0.3[2i_\sigma(\text{NC})+i_\sigma(\text{NH})-\pi_c]} \tag{14.121}$$

and

$$\chi_{xx}-\chi_{yy}=(\tfrac{3}{4})[i_\sigma(\text{NC})-i_\sigma(\text{NH})]eQq_{210} \tag{14.122}$$

for which $eQq_{210}=-10$ MHz and the observed couplings are $\chi_{zz}=\chi_{cc}=-2.66\pm0.02$, $\chi_{xx}=\chi_{bb}=1.21\pm0.02$, and $\chi_{yy}=\chi_{aa}=1.45\pm0.02$. There are evidently three unknown quantities, and the two equations cannot be solved without further information. To solve for i_σ(NH) and π_c, we assume that i_σ(NC)=0.25, as predicted by (14.93) with the normal electronegativity difference of 0.5 for C and N. This is a reasonable value of i_σ(NC) since it is somewhat greater than the 0.22 value for the N—C bond of pyridine, where the negative charge on N is greater. With i_σ(NC)=0.25 and with the observed values of the coupling, (14.121) and (14.122) yield the values

$$i_\sigma(\mathrm{NH})=0.22 \qquad \pi_c=0.48$$

The negative charge on the nitrogen is

$$c^- = 2i_\sigma(\mathrm{NC}) + i_\sigma(\mathrm{NH}) - \pi_c = 0.24$$

in electron units, somewhat less than that for pyridine. The group electronegativity of the *N*-pyrrole orbital to which the H is bonded is, from (14.93), $x_g = x_{\mathrm{H}} + 2i_\sigma(\mathrm{NH}) = 2.6$, significantly less than the averaged value of 3 for nitrogen. The p_π orbital occupancy on the nitrogen is 1.52. Although these predictions are, of course, only approximate, the large π bonding of 48% by the "unshared" orbital of the nitrogen is consistent with the observed planar structure of the molecule.

An interpretation of the ^{14}N coupling in $S(CN)_2$ is given in Section 11, and in centrally bonded nitrogen in Section 14.

14 COUPLING BY CENTRALLY BONDED ATOMS

Group V Elements

Let us consider the class of molecule AX_3 in which a Group element—N, As, Sb, or Bi—forms partially ionic σ bonds with three equivalent atoms or groups X. Molecules of this class are NH_3, $(CH_3)_3N$, AsH_3, SbF_3, and so on. The three bond orbitals form the edge of a pyramid with the fourth orbital, which has an unshared electron pair, directed along the axis away from the bonds. If each of the equivalent sp_3 bonding orbitals of atom A has a_s^2 amount of s character, the counterhybridized orbital of the unshared pair will have $3a_s^2$ amount of p character and will be directed along the symmetry axis z. The four normalized orbitals can be expressed as

$$\psi_1 = (1-3a_s^2)^{1/2}\psi_s + (3a_s^2)^{1/2}\psi_{p_z} \tag{14.123}$$

$$\psi_2 = a_s\psi_s - \left(\frac{1-3a_s^2}{3}\right)^{1/2}\psi_{p_z} + \left(\frac{2}{3}\right)^{1/2}\psi_{p_x} \tag{14.124}$$

$$\psi_3 = a_s\psi_s - \left(\frac{1-3a_s^2}{3}\right)^{1/2}\psi_{p_z} - \left(\frac{1}{6}\right)^{1/2}\psi_{p_x} + \left(\frac{1}{2}\right)^{1/2}\psi_{p_y} \tag{14.125}$$

$$\psi_4 = a_s\psi_s - \left(\frac{1-3a_s^2}{3}\right)^{1/2}\psi_{p_z} - \left(\frac{1}{6}\right)^{1/2}\psi_{p_x} - \left(\frac{1}{2}\right)^{1/2}\psi_{p_y} \tag{14.126}$$

where ψ_1 is directed along the symmetry axis and has the unshared pair and where ψ_2, ψ_3, ψ_4 are bonding orbitals having, on the average, $(1+i_\sigma)$ electrons each when the negative charge is on atom A and $(1-i_\sigma)$ when the positive end of the σ bond is on A. The number of p_z electrons will be $2(3a_s^2)$ in the non-bonding orbital. For A at the negative end of the σ bond, the total number of p_z electrons in the three bonding orbitals will be

$$3[(1-3a_s^2)/3](1+i_\sigma)=(1-3a_s^2)(1+i_\sigma)$$

Thus

$$n_z=6a_s^2+(1-3a_s^2)(1+i_\sigma)$$

Since each bonding orbital will have $(1+i_\sigma)$ electrons, it follows from (14.124)–(14.126) that $n_x+n_y=2(1+i_\sigma)$. Therefore

$$(U_p)_z=\frac{n_x+n_z}{2}-n_z=-3a_s^2(1-i_\sigma) \tag{14.127}$$

With the screening correction for the negative charge of $3i_\sigma$ on A, the quadrupole coupling of A can be expressed as

$$\chi_{zz}=\frac{3a_s^2(1-i_\sigma)}{1+3i_\sigma\varepsilon}eQq_{n10} \qquad \{\text{negative pole on coupling atom}\} \tag{14.128}$$

where values of ε are given in Table 14.3.

When the atom or group X has a greater electronegativity than has A (as in NF_3), the positive end of the bonds will be on atom A, and the population of the bonding orbitals will be $1-i_\sigma$. The resulting formula for the coupling will then be

$$\chi_{zz}=3a_s^2(1+i_\sigma)(1+3i_\sigma\varepsilon)eQq_{n10} \qquad \{\text{positive pole on coupling atom}\} \tag{14.129}$$

It is evident that for either polarity of the bonds the coupling vanishes when there is no hybridization of the orbitals. When the polarity of the bonds is such as to put a negative charge on the atom A, its coupling will decrease in magnitude with increase in ionic character and will vanish for completely ionic bonds. When the electronegativity of A is less than that of X (positive polarity on A), the magnitude of the coupling increases with the ionic character of the bonds. The difference in polarity of the bonds is the principal reason that the magnitude of the coupling of ^{14}N in NF_3 is 7.07 MHz, much greater than that in NH_3, 4.10 MHz.

The couplings in Group V hydrides are the simplest to interpret because they are not complicated by π bonding. Since nitrogen has no d orbital, it is possible to get from the bond angle a good estimate of the s hybridization in the bonding orbitals. If θ is the bond angle between two equivalent sp hybrid orbitals, the s character of these orbitals is given by

$$s \text{ character}=a_s^2=\frac{\cos\theta}{\cos\theta-1} \tag{14.130}$$

With $\theta=106°46'$ as the measured bond angle in NH_3, this equation gives the s character as $a_s^2=0.224$. We can now use this value with the observed ^{14}N

coupling of -4.10 MHz to calculate the ionic character of the N—H bonds. With $a_s^2=0.224$, $\varepsilon=0.30$, and $eQq_{210}=-10$ MHz (see Section 13), one obtains from (14.128) $i_\sigma=0.25$ for ammonia. This value of 25% ionic character of the bonds indicates a difference of $2i_\sigma=0.50$ in the electronegativity of H and NH_2. With the value $x_H=2.15$, the group electronegativity of NH_2 is found to be $x_g(NH_2)=2.65$. Since the two hydrogens would be expected to lower the effective electronegativity of N appreciably below its averaged value of 3, the value obtained for NH_2 is a reasonable one.

A treatment similar to that given for NH_3 does not lead to reasonable results for AsH_3 and SbH_3, probably because of *d*-orbital involvement in the hybridization. The *s* character for the bonding orbitals of AsH_3 indicated by the bond angle of 91°50′ is only 3.1%. If this value is used for a_s^2, with the observed coupling $\chi_z(^{75}As)=-160.1$ MHz and $eQq_{310}=-400$ MHz in (14.129), the absurdly large ionic character of 150% is indicated. The group electronegativity of AsH_2 is approximately 2.06 (Table 14.12) and that of H is 2.15. These values with (14.93) indicate that $i_\sigma=0.05$. This value of i_σ with the ^{75}As coupling substituted into (14.129) indicates 12% *s* character in the As bonding orbitals. This amount of *s* character is not in agreement with that, 3.1%, derived from the bond angle. Probably this disagreement arises from *d* contribution to the hybridized orbitals. The nearly orthogonal orbitals can be constructed from *spd* hybrids with appreciable *s* character. However, the introduction of *d* character adds another unknown parameter to make the interpretation uncertain. Nevertheless, we believe that the *s* character of 12% indicated by the quadrupole coupling is near the correct value since the bond angle is more sensitive to small amounts of *d* character than is the quadrupole coupling.

The inconsistency in the *s* character of SbH_3 as indicated by the bond angle and the Sb quadrupole coupling is similar to that in AsH_3. The bond angle of 91°18′ indicates an *s* character of only 2.2%, whereas the ^{121}Sb coupling of 458.7 MHz with $eQq_{n10}=665$ MHz indicates it as 19%. Again, this discrepancy must result from *d* hybridization of the bonding orbitals. A similar discrepancy [57] has been observed between the *s* character in the *s* orbitals of H_2S as derived from the bond angle and from the ^{33}S quadrupole coupling. This discrepancy may likewise arise from *d* character in the hybridized orbitals.

No quadrupole coupling values are available for gaseous bismuth. However, from solid-state pure quadrupole resonance of ^{209}Bi in bismuthtriphenyl [58], the Bi bonding orbital has been estimated as having 8 to 9% *s* character.

Group IV Elements

Because of the prevalence of Group IV elements in both organic and inorganic compounds it is unfortunate that so few of the isotopes of these elements have quadrupole moments. Of the stable isotopes of C, Si, Ge, and Sn, only ^{73}Ge with natural abundance of 7.9% and nuclear spin $I=\frac{9}{2}$ has a quadrupole moment. All other stable isotopes of these elements have spins of zero or $\frac{1}{2}$, and hence $Q=0$. The quadrupole coupling of ^{73}Ge has been measured in only a few molecules.

When a Group IV element has four equivalent tetahedral bonds, as in the spherical-top molecules, the molecular field gradient at the nucleus of the central atom is zero. For this reason there is no observable ^{73}Ge coupling in GeH_4, $GeCl_4$, and so on. When the four bonds to Ge are not equivalent, a field gradient exists at the Ge nucleus which is sufficiently large in many molecules to make the ^{73}Ge coupling measurable. This ^{73}Ge coupling has been measured in GeH_3F $(-92\pm3$ MHz) [59], in GeH_3Cl $(-95\pm3$ MHz) [60], and in GeH_3CH_3 (+3 MHz) [61].

The three equivalent Ge bonding orbitals in GeH_3X can be represented by ψ_2' ψ_3, and ψ_4 of (14.124)–(14.126) where a_s^2 represents the s character of each. The fourth orbital ψ_1, represented by (14.123), bonds to X. Because of the small electronegativity of Ge, 1.8, we assume that the positive pole of each bond is on the Ge. Each of the three equivalent orbitals will have $1-i_\sigma(\text{GeH})$ electrons. Since the p_z character of each of these orbitals is $(1-3a_s^2)/3$, their combined p_z population is $(1-3a_s^2)[1-i_\sigma(\text{GeH})]$. The orbital ψ_1 has a p_z population of $3a_s^2[1-i_\sigma(\text{Ge}X)]$. Also it is evident that $n_x=n_y=1-i_\sigma(\text{GeH})$. Therefore,

$$(U_p)_z=3a_s^2[i_\sigma(\text{Ge}X)-i_\sigma(\text{GeH})] \tag{14.131}$$

This equation shows that regardless of the hybridization the field gradient, and hence the coupling, must vanish when the ionic characters of the bonds are equivalent. The positive pole of each bond is assumed to be on the Ge, and the π character of the bonds is assumed to be negligible. Thus a positive charge of $c^+=3i_\sigma(\text{GeH})+i_\sigma(\text{Ge}X)$ will be on Ge. The coupling equation

$$\chi_z=-(U_p)_z(1+c^+\varepsilon)eQq_{n10} \tag{14.132}$$

with the foregoing values and with $\varepsilon=0.25$ becomes

$$\frac{\chi_z}{eQq_{n10}}=3a_s^2[i_\sigma(\text{GeH})-i_\sigma(\text{Ge}X)]\{1+0.25[3i_\sigma(\text{GeH})+i_\sigma(\text{Ge}X)]\} \tag{14.133}$$

where $eQq_{410}=224$ MHz for ^{73}Ge. For GeH_3Cl the s character a_s^2 is found from the HGeH angle of 111°4′ to be 0.269, and the ionic character of the GeCl bond is found from the ^{35}Cl coupling of -46.95 to be $i_\sigma(\text{GeCl})=0.57$. The ^{73}Ge coupling in this molecule is $\chi_z(\text{Ge})=-95$ MHz. By substitution of these values into (14.133) the value of $i_\sigma(\text{GeH})$ is found to be 0.16. From the relationship between ionic character and electronegativity, (14.93), and with $x_\text{H}=2.15$, the effective electronegativity of these Ge orbitals is indicated to be 1.83. This value seems a bit low, but, of course, the calculations are only approximate. The ionic character of the GeCl bond obtained from the ^{35}Cl coupling in a similar way indicates that $x_g(\text{GeH}_3)=1.86$, in good agreement with the value of 1.87 derived in Section 15.

In GeH_3CH_3 the ^{73}Ge coupling is 3 MHz. The bond angles are nearly tetrahedral, and thus $a_s^2=0.25$. The very small coupling indicates, surprisingly, that the Ge—H and Ge—C bonds have very nearly the same ionic character. To obtain an approximate solution we use the group electronegativity values of 2.30 for CH_3 and 1.87 for GeH_3 (Section 15) and estimate that $i_\sigma(\text{GeC})=0.22$.

This value substituted into (14.133) with $\chi_z=3$ MHz yields i_σ(GeH)=0.23. A calculation of the effective electronegativity of the Ge orbitals bonding the hydrogens results in a value of 1.7. This value is too low, but χ_z is known only approximately, and the hyperconjugated structures are not considered in the treatment above.

The Ge coupling in GeH_3F indicates considerable π bonding in the Ge—F bond through hyperconjugation. Otherwise, its value would be appreciably higher, rather than lower, than that in GeH_3Cl. We shall not attempt a quantitative calculation for this molecule.

Group III Elements: Nature of Boron Bonding

The couplings per p electron, eQq_{n10}, are accurately known for Group III elements from atomic spectroscopy (see Table 14.2). In Section 10 we give an interpretation of the coupling of these elements in diatomic halides where they behave as monovalent elements forming single bonds with hybridized orbitals. Relatively few polatomic molecules involving the Group III elements have been investigated with microwave spectroscopy. Of most interest among them are probably the ones involving boron. As examples we give interpretations of the coupling of ^{11}B in HBS and CH_3BS where the boron forms two σ bonds and one π bond, in BH_3CO where the boron forms four approximately tetrahedral bonds, and in some nonpolar molecules such as $B(CH_3)_3$ where the boron forms three equivalent, planar, sp_2 hybrid bonds.

Since H—B=S is a linear molecule, there is only one observable quadrupole coupling constant for each isotope of boron. These constants have been accurately obtained from measurements of rotational hyperfine structure by Pearson et al. [62]. The value observed for the ^{11}B coupling is $\chi_z(^{11}B)=-3.71(3)$ MHz. This one constant is insufficient for determination of the bond parameters. Nevertheless, with the ionic character–electronegativity relation derived from quadrupole coupling in other molecules, (14.93), and with reasonable assumptions about the bonding based on the known molecular structure, the coupling constant can be accurately predicted. Because of the linear structure it is evident that boron forms σ bonds to the H and S with sp_z hybrids and a localized π bond to the S with a pure p orbital, p_x or p_y. The predicted coupling is not sensitive to the relative amounts of sp hybridization of the two σ bonding orbitals, and for simplicity we assume them to be the equivalent orbitals $(\frac{1}{2})^2(s\pm p_z)$. We then use (14.93) to estimate the ionic character of the bonds from the electronegativity difference of the bonded atoms. It was found, however, that a close fitting of the predicted with the observed coupling could not be obtained when the averaged value of 2.0 from Appendix G was used for the boron electronegativity in the calculation of i_σ. Since it is well known that the effective electronegativity of an s orbital is greater than that of a p orbital of the same valence shell (see Section 15), we assumed that the sp hybrid orbitals have the slightly higher-than-average value of $x_{sp}(B)=2.1$ and that the pure p orbital has the slightly lower-than-average value of $x_p(B)=1.9$ in predicting the ionic character of the bonds with the values $x(H)=2.1$ and $x(S)=2.5$ as given in Table A VII. With these electronegativity values, (14.93) predicts the ionic character values: $i_\sigma(\mathrm{BH})=0$, $i_\sigma(\mathrm{BS})=0.20$, and $i_\pi(\mathrm{BS})=0.30$. These ionic character values allow an accurate prediction of the ^{11}B coupling for the assumed molecular model. For example, with no ionic character there would be, on the average, one electron in the p_x or p_y orbital of the boron and $\frac{1}{2}p_z$ electron in each of the two *sp* hybrids. Since the ionic character is such as to reduce the orbital charge on the B, the *p* electron density in each orbital will be reduced by the amount of ionic character. Thus the estimated unbalanced *p* electron density on the B is

$$(U_p)_z=\frac{n_x+n_y}{2}-n_z=\frac{1-0.30}{2}-[\tfrac{1}{2}+\tfrac{1}{2}(1-0.20)]=-0.55$$

The predicted coupling is

$$\chi_z(^{11}\mathrm{B})=-(U_p)_z eQq_{n10}(1+c^+\varepsilon)$$

where $eQq_{n10}(^{11}\mathrm{B})=-5.39$ MHz from Table 14.2, $\varepsilon=0.50$ from Table 14.3, and $c^+=0.50$ is the sum, $i_\sigma+i_\pi$. Thus the predicted coupling is

$$\chi_z(^{11}\mathrm{B})=0.55(-5.39)(1.25)=-3.71\ \mathrm{MHz}$$

which is in agreement with the observed value. The agreement supports the adjustments made in the electronegativity of boron for the different orbitals. With the averaged value, $x(\mathrm{B})=2.0$, for both pure *p* and *sp* orbitals, the predicted coupling is -3.25 MHz, in only approximate agreement with the observed value.

The ^{11}B coupling in CH_3BS, $-3.71(2)$, measured by Kirby and Kroto [63], is the same as that for HBS; its interpretation is quite similar to that given previously for HBS. The principal difference is in the group electronegativity (see Section 15) of CH_3, $x(CH_3)=2.3$, as compared to that of H. If this group value is used with the same electronegativities as were used previously for B and S, the predicted ^{11}B coupling is -3.61 MHz compared with the observed value, -3.71 MHz. The agreement is as good as can be expected from this type of calculation, and the difference with HBS could arise from the approximate value used for the group electronegativity of CH_3.

The nuclear quadrupole couplings of ^{10}B and ^{11}B in BH_3CO were used for the first measurements [64] of the spins and quadrupole moments of these isotopes. To evaluate the quadrupole moments it was necessary to calculate q_{210} for the atom and to deduce $(U_p)_z$ from the known molecular structure. Now accurate values of the quadrupole moments and of eQq_{210} are available from atomic beam measurements, and these values can be used for improvement of the accuracy of the interpretation. Interestingly, these first interpretations of the quadrupole coupling and structure of BH_3CO showed that the B—C bond is only about half a bond.

The molecular structure of BH_3CO [64] shows that the valence orbitals of boron are sp_3 hybrids. Since the three orbitals forming the B—H bonds are equivalent, they can be represented by orbitals ψ_2, ψ_3, and ψ_4 of (14.124)–

(14.126) with ψ_1 of (14.123) representing the orbital directed along the symmetry z axis which bonds to the C—O. The B—H bond length, 1.194 Å, is that expected from the added covalent radii, 1.20 Å, and hence we assume that it is a normal covalent bond with a small ionic character i_σ and with the positive pole on the less electronegative boron atom. The electrons in ψ_1 will be supplied by the CO and will put negative charge on boron. Let n_1 designate the electron population of ψ_1. The total charge on B can be expressed as $c^- = n_1 - 3i_\sigma$ where i_σ represents the ionic character of the B—H bonds. With a population of $1 - i_\sigma$ in ψ_2, ψ_3, and ψ_4 and of n_1 in ψ_1, the numbers of unbalanced p electrons $(U_p)_z$ can be found in the usual way from the p components of the four orbitals. With the correction for the negative charge screening $c^-\varepsilon$, the coupling equation for B in BH_3CO is found to be

$$\chi_z = \frac{-3a_s^2(1 - i_\sigma - n_1)}{1 + (n_1 - 3i_\sigma)\varepsilon}\, eQq_{210} \tag{14.134}$$

Fortunately i_σ, the ionic character of the B—H bonds, can be found quite closely because of the nearly equal electronegativity of boron (2.00) and of hydrogen (2.15). From (14.93), $i_\sigma = 0.075$. We make no correction in the electronegativity of B because of the small negative charge. The amount of s character $a_s^2 = 0.28$ can be obtained from (14.130) with the known value of 113°52′ for the HBH bond angle. For boron the screening constant ε is 0.50. With these constants, with $\chi_z = 1.55 \pm 0.08$ MHz, and with $eQq_{210} = -5.39$ MHz for ^{11}B, (14.134) can be solved for n_1. The value thus obtained, $n_1 = 0.53$, is equivalent to the bond order of the B—C bond. With the ionic character in the B—H and C—O bonds unspecified, the principal structures contributing to the ground state of this molecule are (I) and (II)

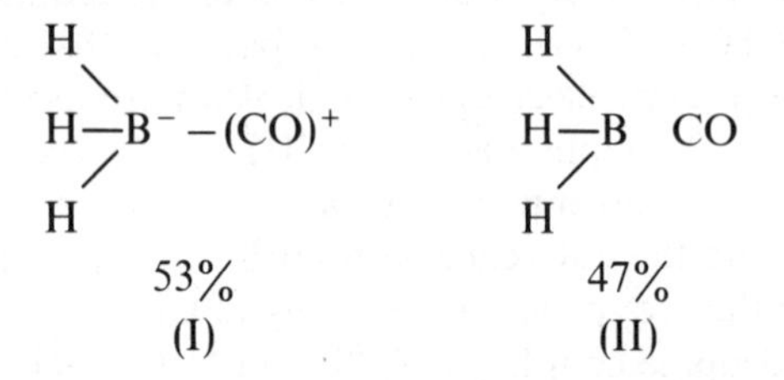

Surprisingly, structure (II), which consists of nonbonded BH_3 and CO molecules, has essentially the same weight as the bonded structure (I). Nevertheless, the high contribution of (II) is in agreement with the C—O bond length in this molecule, 1.131 Å, essentially the same as that in free C—O for which $r_0 =$ 1.1308 Å. It also accounts for the high instability of borine carbonyl which ignites upon exposure to air.

The boron couplings of $B(CH_3)_3$ and $B(C_2H_5)_3$ have been measured in the solid state by Dehmelt [65]. Because the boron and the three carbons to which it bonds are in the same plane for these molecules, we can assume with confidence that the three bonding orbitals of the boron are equivalent sp_2 hybrids. The symmetric z axis is perpendicular to the plane, and the p_z orbital, to a good approximation, is unpopulated, $n_z=0$. Since the carbon groups are more electronegative than boron, the ionic character of the bonds puts a positive charge on the boron. Therefore the population of each bonding orbital of the boron is $1-i_\sigma$, where i_σ is the ionic character of the bond. Since the bond angle is 120°, each bonding orbital has p character of $\frac{2}{3}$, and it is obvious that

$$n_x+n_y=3(\tfrac{2}{3})(1-i_\sigma)=2-2i_\sigma$$

and since $n_z=0$, it is evident that $(U_p)_z=1-i_\sigma$. Consequently,

$$\chi_z=-(1-i_\sigma)(1+3i_\sigma\varepsilon)eQq_{210} \tag{14.135}$$

where $3i_\sigma\varepsilon$ is the screening correction for the positive charge. The measured coupling for $^{11}B(CH_3)_3$ is $\chi_z(^{11}B)=4.87$ MHz. With $\varepsilon=0.50$ and $eQq_{210}(^{11}B)=-5.39$ MHz, (14.135) yields $i_\sigma=9.5\%$. A similar evaluation for $B(C_2H_5)_3$ with the observed $\chi_z(^{11}B)=5.06$ leads to $i_\sigma=7.2\%$ for the ionic character of the bonds in boron triethyl. From the known value of the electronegativity of the methyl group, 2.30, one can estimate the effective electronegativity of B in boron trimethyl with (14.93). It is $x_g=2.30-2i_\sigma=2.11$. The Cl coupling in BCl_3 in the solid state indicates that the B—Cl bonds have appreciable π character [7].

It is of interest that the types of bonds formed by boron differ appreciably in the various combinations in which boron quadrupole couplings have been measured. This diverse bonding is reflected in the boron quadrupole couplings as well as in the observed molecular structures and dipole moments. In BF, HBS, and CH_3BS the ^{11}B coupling is negative, indicating [with the negative eQq_{n10}(atomic)] a negative $(U_p)_z$; in BH_3CO where the boron bonding orbitals are tetrahedral sp_3 hybrids and in molecules such as $B(CH_3)_3$ where the boron bonding orbitals are planar sp_2 hybrids, the coupling is positive, and $(U_p)_z$ is likewise positive. It should be recalled that a negative $(U_p)_z$ corresponds to an excess of p-orbital electron density along the reference z axis and that a positive $(U_p)_z$ corresponds to a relative deficiency of p electron density along this axis.

15 GROUP ELECTRONEGATIVITIES AND QUADRUPOLE COUPLING

As is already evident from previous discussion, electronegativity is a very useful quantity in the interpretation of nuclear quadrupole coupling in molecules. In turn, quadrupole coupling can be used to give information about electronegativity. Electronegativities are also useful for the understanding of molecular dipole moments and for the estimation of interatomic distances. For these reasons we have given a complete table of electronegativity values based on various properties (Appendix G). For multivalent elements, the values in this table represent typical or averaged values of the elements in their most common chemical combinations. For reasons given later, these values must be used with care. When it is available, the effective value of the electronegativity of the element in the specific chemical combination, the "group electronegativity," should be used. A review of group electronegativity has been given by Wells [66].

Electronegativity is not a fixed property of an atom in a molecule, but is one that depends, sometimes appreciably, upon the bonded state. It is a property of a particular orbital employed by the atom, whether *s*, *p*, *d*, or hybridized. For example, the electronegativity of an *sp* hybridized orbital of carbon is greater than that of an sp_3 hybrid. The effective electronegativity of an atom forming bonds to more than one other atom depends upon the electronegativity of these other atoms to which it is bonded. For example, the effective electronegativity of carbon in the group Cl_3C— is appreciably greater than that of carbon in the group H_3C—. The higher electronegativity of the three chlorines is partially reflected in the fourth bond formed by Cl_3C—, and the lower electronegativity of hydrogen is partially reflected in the fourth bond formed by H_3C—. Pauling's original electronegativity scale [37], determined from thermal chemical data, represented essentially an averaged value of carbon in its various bonded states. The Pauling value for carbon, 2.5, is about halfway between that for the H_3C— group (2.30) and that for the Cl_3C— group (2.79).

In the derivation of the relation between electronegativity and ionic character (Section 9), we employed diatomic molecules formed primarily by monovalent atoms. Except when there is a significant positive charge on the halogen, which is rare in organic molecules, the halogens appear to form σ bonds with pure, or nearly pure, *p* orbitals. For these reasons, we can treat the electronegativity of the halogen revealed in its σ bonds as a fixed quantity. When not complicated by significant π bonding, the quadrupole coupling of the halogen can be used for finding the ionic character of the bond formed by the halogen; from the relationship between electronegativity and ionic character (Section 9) the electronegativity of the group relative to that of the halogen can be found. This latter electronegativity is assumed to be a known constant.

Hinze et al. [67] have shown that if the orbital occupancy number n_j is treated not simply as having integral values, 0, 1, or 2, but as a continuously variable quantity within the range $0 \leq n_j \leq 2$, the *j*th orbital electronegativity can be expressed as

$$x_j = \frac{\partial E}{\partial n_j} \tag{14.136}$$

where E is the energy of the charge in the orbital which with n_j continuous must also be considered as a continuous and differentiable function of n_j. They assumed that $E(n_j)$ would be the parabolic function

$$E(n_j) = a + bn_j + cn_j^2 \tag{14.137}$$

and showed that electronegativity values

$$x_j = \frac{\partial E}{\partial n_j} = b + 2cn_j \tag{14.138}$$

calculated with this function give values which are consistent with those of

Table 14.11 Group Electronegatives[a]

Group	x_g	*Group*	x_g	*Group*	x_g
CH_3	2.30	CH_2Cl	2.47	$CHBr_2$	2.49
CH_2F	2.61	$CHCl_2$	2.63	CBr_3	2.57
CHF_2	2.94	CCl_3	2.79	CH_2I	2.38
CF_3	3.29	CH_2Br	2.40	CHI_2	2.44
				CI_3	2.50

[a]From Hinze et al. [67].

Mulliken's definition of electronegativity $X=(I+E)/2$, where I and E are the energies of ionization and electron affinity. Heinze et al. have used (14.138) to calculate the effective electronegativities of atoms in different group-combinations, "group electronegativities." In Table 14.11 we list their effective carbon electronegativities in different groups for singly occupied orbitals. We have employed these values in Sections 11–14. Details of their methods of calculations may be found in the original paper.

A concept developed earlier by one of us [5] can be used with (14.93) for calculation of the effective electronegativity of certain groups. According to this concept, the orbital electronegativity of a bonded atom can be defined as the potential it exerts on an electron in the orbital when the electron is at a distance equal to the covalent radius from the nucleus. Thus

$$x=\frac{(Z_{\text{eff}})e}{r}=\frac{(Z-S)e}{r} \tag{14.139}$$

where r is the covalent radius and S is the screening constant effective at a distance of the covalent radius from the nucleus. At this distance one can assume complete screening by all electrons in the valence subshells and set

$$Z_{\text{eff}}=n-s(n\pm c-1) \tag{14.140}$$

where n is the number of electrons of the valence shell of the neutral atom, c is the magnitude of the charge on the atom (in electron units) with the upper sign taken for a positive charge, and s is the averaged screening constant per valence electron. Screening constants such as those of Slater [23] are not appropriate with this definition unless one uses a value of $\langle 1/r \rangle$ averaged over the atomic orbital instead of the reciprocal covalent radius. In the original work, all electrons of the valence shell were considered to have equally effective screening at a distance of the covalent radius, and the value of s was empirically determined as 0.50 from a fitting of the predicted s values to the Pauling scale. It was found that a shift of origin and a proportionality constant were also necessary for a fitting of the values to the Pauling scale. When a charge c is on the atom, the resulting formula is

$$x = 0.31\,\frac{n+1\pm c}{r} + 0.50 \qquad (14.141)$$

The positive sign is used in (14.141) when the charge is positive, and vice versa. Evidently c depends on the ionic character of the various bonds to the atom, both π and σ bonds, and upon such other effects as π feedback and hyperconjugation discussed in Sections 11 and 12.

The determination of the effective electronegativity of the group is simplest for such hydride groups as CH_3, NH_2, and SH because only ionic character of the σ bonds puts charge on the central atom. If there are three equivalent bonds, as in AB_3 radicals, the charge is $3i_c$, where i_c is the ionic character of each bond. From (14.93) the charge c on A for such groups is

$$c = 3i_c = 3\left(\frac{x_A - x_B}{2}\right)$$

If one assumes that the electronegativity of the fourth orbital of A is the same as that of the three equivalent ones forming the A—B bonds, substitution of this value of c with $x_A = x_g$ into (14.141) and transformation yield

$$x_g = \frac{0.31(n+1) + 0.50 r_A + 0.465 x_B}{r_A + 0.465} \qquad (14.142)$$

where r_A is the covalent radius of A in Angstroms and n is the number of electrons of the valence shell in the neutral atom, that is, $n=4$ for Group IV atoms, and so on. The x_g values for AH_3 radicals in Table 14.12 were calculated with (14.142). The x_g values of the C(Hal)$_3$ group listed in Table 14.12 were also calculated with (14.142); effects of the small amount of π character due to hyperconjugation (Section 12) were neglected. However, for evaluation of the x_g values for the Si, Ge, and Sn halides, listed in Table 14.12, the effects of π character are taken into account as described in Section 11. Calculation of x_g values for PH_2,

Table 14.12 Group Electronegativities Calculated[a] from $(Z_{eff})e/r$

Group	x_g^b	*Group*	x_g^b	*Group*	x_g^b	*Group*	x_g^b
OH	3.24	SH	2.52	SeH	2.33	TeH	2.09
NH_2	2.76 (2.65)[c]	PH_2	2.18	AsH_2	2.06	SbH_2	1.88
CH_3	2.34	SiH_3	1.91	GeH_3	1.87	SnH_3	1.74
CCl_3	2.66[d]	$SiCl_3$	2.07[e]	$GeCl_3$	2.07[e]	$SnCl_3$	1.94[e]
CBr_3	2.59[d]	$SiBr_3$	2.02[e]	$GeBr_3$	2.02[e]	$SnBr_3$	1.89[e]
CI_3	2.50[e]	SiI_3	1.97[e]	GeI_3	1.96[e]	SnI_3	1.85[e]

[a]From (14.141) with c derived from (14.93).
[b]Effective electronegativity of free valence orbital on central atom of the group.
[c]From N quadrupole coupling in NH_3. See Section 14.
[d]Effects of π character neglected.
[e]Correction for π character included. See Section 11.

Table 14.13 Electronegativity of the Methyl Group[a]

Molecule	*From Force Constant*	*From Hal eQq*
CH_3Cl	2.37	2.38
CH_3Br	2.34	2.30
CH_3I	2.28	2.29
	Av. 2.33	2.32

[a]When $x_{Cl}=3.0$, $x_{Br}=2.8$, and $x_I=2.6$. From Gordy [7].

AsH_2, and so on, is like that described for the AB_3 group except that $c=2i_c$. Similarly, for OH, SH, and so on, $c=i_c$.

Although the group electronegativities listed in Table 14.12 represent more closely the effective values of the free valence orbital on the central atom than do the atomic values of Appendix G, these values are only approximately determined. Furthermore, the electronegativity of the group orbital is altered to some extent by the bond it forms. The x_g for the same group is not the same in different molecules. Its range of variation is much less, however, than is that for the same element in groups of different types.

In some molecules, the ionic character of a particular bond can be determined to a good approximation directly from the quadrupole coupling of the bonded atom. In these cases the electronegativity of the particular orbital of the group forming the bond to the coupling atom can be found from the relation between ionic character and electronegativity, (14.93) developed in Section 9.

Table 14.14 Group Electronegativities Derived from Quadrupole Coupling[a]

Molecule	*Group*	x_g
CH_3CH_2Cl	CH_3CH_2—	2.27
SiH_3CH_3Cl	SiH_3CH_2—	2.29
$(CH_3)_3CCl$	$(CH_3)_3C$—	2.24
$CH_3CHClCH_3$	CH_3CHCH_3—	2.28
CH_2—$CHCl$ (ring with CH_2)	CH_2—CH— (ring with CH_2)	2.31
C_6H_5Br	(hexagonal ring)C—	2.30
Pyrrole	(pentagonal ring)N—	2.6

[a]With use of (14.93).

For example, if the small π bond character expected from hyperconjugation in the methyl halides is neglected, an evaluation of the difference in the effective x_g for the CH_3 group in bonding to Cl, Br, and I can be made from the quadrupole coupling of the halogen. Table 14.13 shows the results of such calculations. The values are in good agreement with those obtained from the relation [68] between the force constant and electronegativity, which also provides a method for measurement of group electronegativity. Both types of evaluation indicate a trend toward a decrease in the orbital electronegativity of the group bonded to a halogen from CI to I.

In Table 14.14 are listed group electronegativities of orbitals in specific molecules as estimated from (14.93) with the ionic character determined from quadrupole coupling. The π bonding indicated by the asymmetry parameter in the coupling was taken into account in the evaluation. The values obtained for these diverse hydrocarbon groups are surprisingly close, approximately 2.3 in all of them, even including the benzene radical. Uncertainty in the evaluation is greatest for conjugated groups. The derivation for pyrrole is described in Section 13.

16 INTERPRETATION OF DIPOLE MOMENTS

A straightforward calculation of the electric dipole moment of a molecule implies the averaging of the coordinates of all the charges, both nuclear and electronic, over their respective wave functions, although in practice the nuclear charges can be regarded as fixed-point charges and the inner closed electron shells can be treated as approximately rigid spheres. Since the separated atoms have no dipole moments, it is possible and perhaps simpler to express the molecular dipole moment in terms of the changes in the electron clouds of the

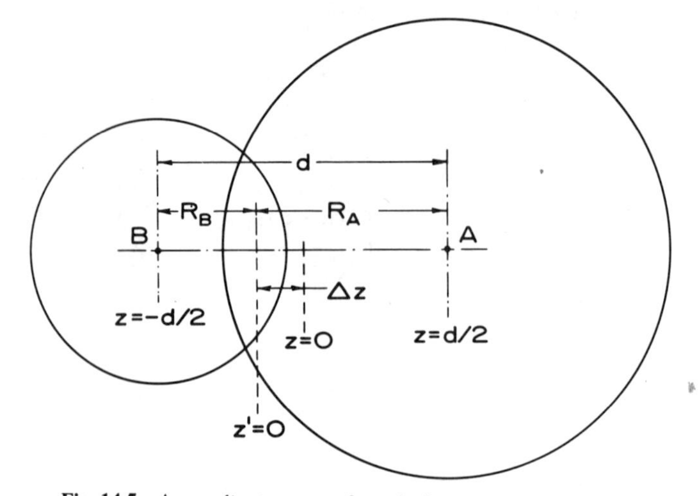

Fig. 14.5 A coordinate system for calculation of bond moments.

atoms upon their becoming bonded. The largest changes generally occur in the orbitals forming the chemical bonds, but the nonbonding electrons are also perturbed by mutual interactions of the bonded atoms.

Only for the very simple molecule LiH was the dipole moment reliably predicted from molecular orbital theory before its measurement. Values of the moment ranging from 5.6 to 6.1 were predicted [66] before the value of 5.888 D was measured [70] with the molecular beam electric resonance method. Afterward the theoretical treatments were refined [71, 72] to give values from 5.8529 to 6.002 D, in good agreement with experiment. Molecular orbital calculations of dipole moments for other first-row diatomic molecules [73, 74] have not proved as successful. Since rigorous molecular orbital methods are of such limited applicability, we shall describe here some approximate concepts for expressing dipole moments in terms of chemical bond properties.

Methods for derivation of dipole moments from microwave spectra are described in Chapter X. Tables of values thus derived for many molecules are given in Chapter X, Section 9.

Bond Moment

Let us consider a diatomic molecule AB bonded by a single covalent σ bond. We choose the coordinate z along the bond axis with the origin at a point halfway between the two nuclei, as indicated in Fig. 14.5. The component of the moment resulting from the two electrons of the bond will be

$$\mu_{\text{bond}} = 2e \int \psi_{\text{mo}} z \psi_{\text{mo}} \, d\tau \tag{14.143}$$

where e is the magnitude of the electronic charge. We express the molecular orbital ψ_{mo} as a linear combination of the atomic orbitals ψ_A and ψ_B of atoms A and B,

$$\psi_{\text{mo}} = a\psi_A + b\psi_B \tag{14.144}$$

and thus

$$\mu_{\text{bond}} = 2e \int (a\psi_A^* + b\psi_B^*) z (a\psi_A + b\psi_B) \, d\tau$$

$$= 2ea^2 \int \psi_A^* z \psi_A \, d\tau + 2eb^2 \int \psi_B^* z \psi_B \, d\tau + 4eab \int \psi_A^* z \psi_B \, d\tau \tag{14.145}$$

To calculate the components due to ionic character and orbital overlap, let us assume that the orbitals ψ_A and ψ_B are unhybridized orbitals. Effects of hybridization are considered later. Since an unhybridized atomic orbital is symmetrical about the nucleus and since A is chosen in the positive z direction, it is evident that

$$\int \psi_A^* z \psi_A \, d\tau = \frac{d}{2} \quad \text{and} \quad \int \psi_B^* z \psi_B \, d\tau = -\frac{d}{2} \tag{14.146}$$

A combination of these expressions with (14.145) yields

$$\mu_{\text{bond}} = ed(a^2 - b^2) + 4eab \int \psi_A^* z \psi_B \, d\tau \tag{14.147}$$

The molecular orbital is normalized by

$$\int (a\psi_A^* + b\psi_B^*)(a\psi_A + b\psi_B) \, d\tau = a^2 + 2abS_{AB} + b^2 = 1 \tag{14.148}$$

where $S_{AB} \equiv \int \psi_A^* \psi_B \, d\tau$ is the overlap integral. The ionic character of the bond is defined by division of the overlap charge equally between the two atoms.

$$i_c = |(a^2 + abS_{AB}) - (b^2 + abS_{AB})| = |a^2 - b^2| \tag{14.149}$$

Hence the first term on the right of (14.147), called the primary moment, is

$$\mu_p = edi_c \tag{14.150}$$

The primary moment results from a transfer of electronic charge between the bonded atoms, putting an excess negative charge on one and leaving an equivalent unscreened nuclear charge on the other. If $a^2 < b^2$, the positive pole of the primary moment will be on atom A.

The last term of (14.147) results from distortions caused by bond-orbital overlap. Its significance was recognized first by Mulliken [75]. This term is difficult to evaluate precisely, but it can be closely approximated in a simple way by a shift in the origin of the coordinate system to the averaged center of the charge in the overlap region. We do this by setting $z = z' + \Delta z$, where z' is measured from the new coordinate origin and where Δz represents the required shift in origin. With the reasonable assumption that the averaged center of the charge in the overlap region is at the point where the two covalent radii meet, as indicated in Fig. 14.5,

$$z = z' + \Delta z = z' - \left(R_A - \frac{d}{2} \right) \tag{14.151}$$

With this transformation, the overlap term becomes

$$\begin{aligned} \mu_s &= 4eab \int \psi_A^* \left[z' - \left(R_A - \frac{d}{2} \right) \right] \psi_B \, d\tau \\ &= 4eab \left[\int \psi_A^* z' \psi_B \, d\tau - \left(R_A - \frac{d}{2} \right) \int \psi_A^* \psi_B \, d\tau \right] \end{aligned} \tag{14.152}$$

Because z' is measured from the center of the overlap region $\int \psi_A^* z' \psi_B \, d\tau = 0$. The overlap charge contribution to the bond moment can be expressed as

$$\mu_s = -4eabS_{AB} \left(R_A - \frac{d}{2} \right) \tag{14.153}$$

where R_A is the covalent radius of the larger atom and S_{AB} is the overlap integral. This term designated by μ_S is often called the bond orbital overlap moment.

The expression can be further reduced by use of (14.148) to eliminate the coefficients a and b. For pure covalent bonds $a=b$ and

$$ab=a^2=\frac{1}{2(1+S_{AB})} \tag{14.154}$$

Equation 14.153 becomes for this case

$$\mu_s=-\frac{2eS_{AB}}{1+S_{AB}}\left(R_A-\frac{d}{2}\right) \quad \{\text{pure covalent bond}\} \tag{14.155}$$

When $a\neq b$ the evaluation is more difficult, but to a good approximation

$$ab\approx\frac{1-i_c}{2(1+S_{AB})} \tag{14.156}$$

and hence

$$\mu_s\approx-\frac{2e(1-i_c)S_{AB}}{1+S_{AB}}\left(R_A-\frac{d}{2}\right) \quad \{\text{ionic–covalent bond}\} \tag{14.157}$$

where i_c is the ionic character of the bond defined by (14.149). Equation 14.157 reduces to that for the pure covalent bond when $i_c=0$. It indicates that μ_s is proportional to the covalent component of the bond. The overlap moment vanishes when $i_c=1$ (complete ionicity) or when the covalent radii of the two bonded atoms are equal, $R_A=d/2$.

The overlap moment is most important for bonds having a large difference in covalent radii, $R_A-d/2$ large, and a small difference in electronegativity, i_c small. Bonds of the small H atom to other elements with comparable electronegativity have μ_s, ~1 D. For example, the S—H bond has [35] $S_{SH}=0.52$, $R_S=1.04$ Å, $d=1.34$ Å, $x_H-x_S=0.35$, and $i_c\approx0.17$. The predicted value of the overlap moment from (14.157) is μ_s(S—H)$=1.0$ D. The negative pole of this moment is toward the H. With the group electronegativity for CH_3 equal to 2.30, the ionic character of the C—H bonds in methane is predicted from (14.93) to be $i_c=0.07$. A similar calculation of the overlap moment for the C—H bond in methane, for which $S_{CH}=0.69$ [35], yields μ_s(C—H)$=0.90$ D. This component is larger in magnitude than the primary moment and of opposite sign, $\mu_p=ied=0.07(4.80)(1.09)=0.36$ D. Hence the resultant negative pole of the bond moment is at the H end of the bond. The hybridization moment considered below does not alter this direction.

Hybridization Moment of Atomic Orbitals

Electrons in a hybridized atomic orbital give rise to a component dipole moment because the averaged center of the hybridized orbital does not coincide with the center of the nuclear charge. Suppose that an electron is in a bond-hybridized sp_z orbital

$$\psi_h=a_s\psi_s+a_p\psi_p \tag{14.158}$$

The reference axis z is along the bond axis, and the origin of the reference system is again at the geometrical center of the molecule, $d/2$ from each nucleus. An electron in the above orbital will have an averaged distance from this origin

$$\langle z\rangle = \int \psi_h^* z \psi_h \, d\tau = a_s^2 \int \psi_s^* z \psi_s \, d\tau + 2a_s a_p \int \psi_s^* z \psi_p \, d\tau + a_p^2 \int \psi_p^* z \psi_p \, d\tau \quad (14.159)$$

and will contribute to the moment the term $e\langle z\rangle$. The components $\int \psi_s^* z \psi_s \, d\tau$ and $\int \psi_p^* z \psi_p \, d\tau$ are symmetrical about the nuclear center and are $d/2$ when z is positive or $-d/2$ when z is negative. However, $\int \psi_s^* z \psi_p \, d\tau$ is not equal to $d/2$ but can be set equal to

$$\int \psi_s^* \left(\frac{d}{2} + z_a\right) \psi_p \, d\tau = \int \psi_s^* z_a \psi_p \, d\tau \quad (14.160)$$

since $\int \psi_s^* \psi_p \, d\tau = 0$. Here z_a is measured from the atomic center. The total contribution to the moment of an electron in this hybridized orbital is the sum of the contributions of the overlapping and nonoverlapping terms

$$e\langle z\rangle = \frac{ed}{2}(a_s^2 + a_p^2) + 2ea_s a_p \int \psi_s^* z_a \psi_p \, d\tau \quad (14.161)$$

Since normalization of the orbital requires that $a_s^2 + a_p^2 = 1$,

$$e\langle z\rangle = \frac{ed}{2} + 2ea_s a_p \int \psi_s^* z_a \psi_p \, d\tau \quad (14.162)$$

The first part, $ed/2$, is simply the contribution to the moment by an electron in an unhybridized orbital with average charge center at the nucleus. This nuclear centered component is included in the primary moment that is due to ionic character. The term $2ea_s a_p \int \psi_s^* z_a \psi_p \, d\tau$ is an increment over and above that caused by an electron with an unhybridized orbital centered at the nucleus.

If n_h represents the number of electrons in the hybridized atomic orbital, the moment will be

$$\mu_h = 2n_h e a_s a_p \int \psi_s^* z_a \psi_p \, d\tau \quad (14.163)$$

If the hybridized orbital is directed toward the molecular center,

$$\langle z_a\rangle = \int \psi_s^* z_a \psi_p \, d\tau \quad (14.164)$$

will be negative when z is positive (atom A in Fig. 14.5). If it is in a counterhybridized sp_z orbital directed away from the molecular center, $\langle z_a\rangle$ will be positive when z is positive. When an s orbital and a p orbital are combined to form two hybrid orbitals, the quantity $a_s a_p \int \psi_s^* z_a \psi_p \, d\tau$ will have the same magnitude, but will be of opposite sign for the two orbitals. If a hybridized sp_z orbital and the counterhybridized sp_z orbital of the same atom have equal electron population, the μ_h of the two orbitals will cancel. The resultant hybridization moment of an atom will depend upon the degree of hybridization of all its orbitals and upon their relative filling. It is evident from the above analysis that the atomic hybridization moment is not included in the primary moment or in the bond overlap moment, although hybridization of the bond orbital causes an increase in the overlap moment through increase of S_{AB}. The atomic hybridization was apparently first recognized by Coulson [16].

That hybridization moments can be very large is demonstrated by Fig. 14.6 which represents a plot given by Coulson of the moment caused by a single electron in an *sp* hybrid orbital of carbon for various degrees of hybridization. For equal mixtures of the *s* and *p* components, the moment is 2.2 D. Hybridization moments for other first-row elements are of comparable magnitude, and those for the second-row elements are even larger. For an sp_3 hybrid orbital of carbon, the moment caused by a single electron is about 0.86 D. Since the C—H bonds of methane have about 7% ionic character, each sp_3 orbital of carbon would have 1.07 electrons and a hybridization moment of 0.92 D with its negative pole directed toward the hydrogen in opposition to the primary moment of 0.36 D, but in the same direction as the overlap moment of 0.90 D calculated previously. Thus the C—H bonds in methane have a resultant moment of about 1.46 D with the negative pole on the hydrogen end of the bond, although the overall moment of the molecule is, of course, zero because of symmetry. In a molecule such as CH_3Cl the approximately tetrahedral orbitals on the carbon are not equally filled, and there is a resultant atomic hybridization moment on the carbon.

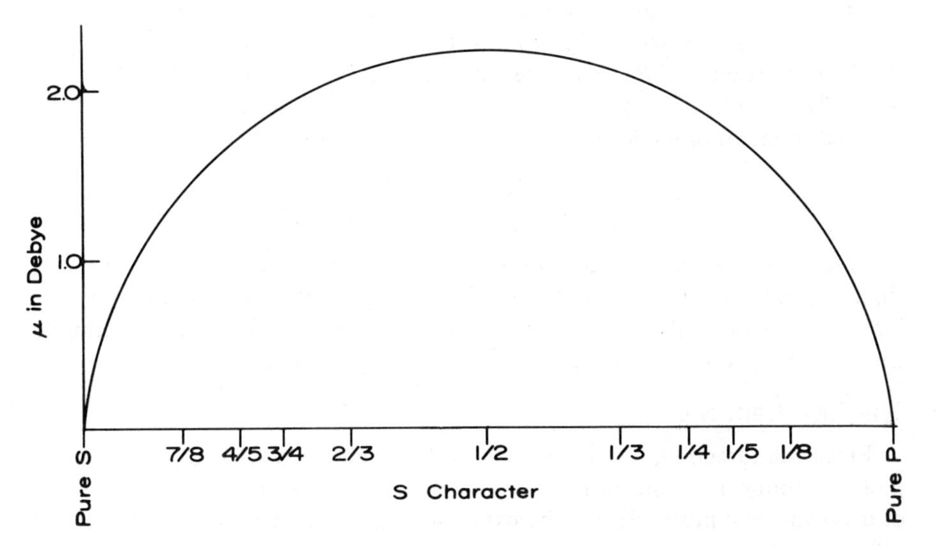

Fig. 14.6 Values of the atomic hybridization moment for a single electron in an *sp* hybridized orbital of carbon for various degrees of hybridization, as calculated by C. A. Coulson, *Trans. Faraday Soc.*, **38**, 433 (1942).

Moment Induced by Internal Fields

The resultant moment arising from ionic character, bond orbital overlap, and so on, produces an electric field that polarizes the electron cloud of the molecule and causes an induced moment that is always of opposite polarity to the resultant or observed moment. A purely theoretical calculation of this induced component is very difficult for it involves slight perturbations of all the orbitals by a nonuniform electric field. However, for ionic or nearly ionic diatomic molecules the induced component can be evaluated semiempirically from measured polarizabilities of the separate ions. For a few semicovalently bonded molecules such as the hydrogen halides, the anisotropic molecular polarizabilities have also been measured. From these constants it is possible to estimate the self-induced moments.

For an ionic molecule AB, the induced moment can be expressed as

$$\mu_{\text{ind}} = \langle E_a \rangle_z \alpha_a + \langle E_c \rangle_z \alpha_b \tag{14.165}$$

where $\langle E_a \rangle_z$ and $\langle E_b \rangle_z$ are the averaged or effective z components of the field acting on ions A and B and where α_a and α_b are their respective polarizabilities. As before, the z coordinate is along the molecular axis. An accurate calculation of the effective field is difficult, but an approximation can be made if it is assumed that the field acting on atom A arises from a charge equivalent to μ_{obs}/d on atom B and that the field acting on atom B arises from an equal charge of opposite sign on atom A. The field at the center of each atom will be in the z direction and will have the value μ_{obs}/d^3, where d is the internuclear distance. However, it seems that a better approximation to the effective polarizing field would be the averaged value of the z component over the cross section of the ion that is perpendicular to the axis and passes through the center of the atom. This average is approximately equal to the z component of the field at a perpendicular distance of $R/2$ from the axis, where R is the ionic radius. This value is $\mu_{\text{obs}}/[d^2+(R/2)^2]^{3/2}$. With this approximation of the averaged field, the induced moment of the ionically bonded diatomic molecule is

$$\mu_{\text{ind}} \approx \mu_{\text{obs}} \left\{ \frac{\alpha_a}{[d^2+(R_a/2)^2]^{3/2}} + \frac{\alpha_b}{[d^2+(R_b/2)^2]^{3/2}} \right\} \tag{14.166}$$

Calculation of the induced moment for covalent bonds is more difficult than that for ionic bonds, but the induced components are generally not very large in such molecules because the inducing moment is less. Later we estimate the induced moments in the hydrogen and alkali halides.

The Resultant Moment

From the foregoing discussion it is evident that the electric dipole moment μ of a diatomic molecule, or in some instances the moment of a particular bond of a polyatomic molecule, can be expressed approximately as the sum of the following components

$$\mu = \mu_p + \mu_s + \mu_h + \mu_{\text{ind}} \tag{14.167}$$

where μ_p is the primary moment resulting from the ionic character of the bond, given by (14.149), μ_s is the bond overlap moment given approximately by (14.157) μ_h is the atomic orbital hybridization moment which for a particular orbital can be calculated from (14.164), and μ_{ind} is the induced moment resulting from the self-polarization of the molecule by the electric fields of the resultant moment. For ionic diatomic molecules, μ_{ind} can be approximated by (14.166). To obtain the total contribution of μ_h to the moment, one must sum the contribution from all hybridized orbitals on both atoms of the bond.

The four components in (14.167) are not entirely independent. For example, the polarization of the molecule by induction must involve hybridization of orbitals although the μ_h component is generally regarded as resulting from the hybridization required to achieve a more effective bond orbital overlap (stronger covalent bond). The hybridization induced by the dipole field likewise reduces the total potential energy, but in a different manner from that of increase in orbital overlap. The two mechanisms are sometimes opposed. For example, in the semicovalent halide bonds X–Hal, an *sp* hybridization of the halide bonding orbital would increase the bond orbital overlap and lower the bond energy. This fact alone might lead one to expect such hybridization; indeed, it has often been postulated. Because the counter hybridized orbital on the opposite side of the halogen from the bond has two electrons whereas the bonding halogen orbital has, on the average, less than two, the resulting hybridization moment would have its positive pole in the direction of the X atom, which in most halide bonds is the direction of the resultant moment. When they are in the same direction, the resultant moment tends to quench the hybridization moment. As has been explained in Section 8, the nuclear quadrupole coupling data indicate little or no such hybridization on the halogen at the negative pole of the moment. When, however, the halogen is at the positive end of the moment, as in Br in FBr, the field of the resultant moment supports the hybridization. Evidence for hybridization on the atom at the positive pole of the bond is cited in Section 10 (see Table 14.6). When the bonds are almost completely ionic, as in the alkali halides, the bond overlap is very small, and μ_s is generally negligible. Also the hybridization moment caused by the covalent bonding is likewise negligible. Although the induced moment in such molecules is large because the primary moment is very large, this induced moment obviously cannot be achieved through hybridization of the *s* and *p* orbitals of a completely filled *sp* shell. Nevertheless, a significant polarization moment could be achieved through a relatively small *pd* admixture.

Ionic Character from Dipole Moments

The relation between ionic character and electronegativity which was derived by Pauling [37] from the dipole moments of the hydrogen halides predicted ionic character values that are considerably lower than those derived later from nuclear quadrupole coupling (see Section 9). The relationship similarly derived with a few additional molecules by Hannay and Smyth [76] also gives similarly low values. This disagreement can be removed [7] by the inclusion of the orbital overlap and polarizability terms in the interpretation of the dipole moments. In deriving his relation Pauling attributed the observed moment entirely to ionic character of the bond. As we shall see below, the considerable bond overlap moment and induced polarization moment of these molecules are in such directions as to reduce the observed moment appreciably below that which would result from bond ionic character alone.

Another development showing that the ionic character values predicted by the Pauling relation are much too low was the very high dipole moments, ~ 10 D, for the alkali halides accurately measured with microwave spectroscopy and with molecular beam resonance. Even if one attributes the observed moments of these molecules entirely to the primary moment, the ionic character values derived are much higher than those predicted from the Pauling or the Hannay-Smyth relation. Nevertheless, many chemists and chemical physicists have been reluctant to accept the high ionic character values indicated by the relationship of (14.93) derived from nuclear quadrupole couplings. Coulson, in his widely used book [16], states, for example, that the assumption of pure p bonding of Cl in HCl cannot give the correct dipole moment without an improbably large ionic contribution. For this reason we shall show here that the relationship between ionic character and electronegativity as derived from dipole moments is the same as that derived from quadrupole couplings when factors such as bond orbital overlap and polarizability are taken into account in the interpretation of dipole moments.

First let us consider the alkali halides listed in Table 14.15, the dipole moments of which have been measured with high accuracy. Their large dipole moments indicate immediately that these molecules are highly ionic. The large polarizabilities of the negative halogen ions, also listed in Table 14.15, indicate the primary moments due to ionic character to be even larger than the observed moments. Because the radius of the halogen ion is larger than that of the alkali ion, any overlap moment would be in the opposite direction to that of the observed moment and would require the primary moment to be still higher. The sp hybridization moment on the halogen cannot be very significant because the sp shell of the halogen is almost completely filled in these molecules. There is little charge to hybridize in the valence shell of the alkali metal. Thus the observed moment must consist essentially of a primary moment and an induced moment of opposite sign. With these considerations, the ionic character of the bond is given by

$$i_c = \frac{\mu_p}{ed} = \frac{\mu_{\text{obs}} - \mu_{\text{ind}}}{ed} = \frac{\mu_{\text{obs}} + |\mu_{\text{ind}}|}{ed} \tag{14.168}$$

From the ionic radii and the polarizabilities listed in Table 14.15 we have estimated μ_{ind} with (14.166) and have then used (14.167) with the observed dipole moments to calculate the ionic character values for the alkali halides that are given in the last column of Table 14.15. With the possible exception of LiI, the high values of i_c show that the omission of components due to covalent bonding is justified. The high i_c values found from (14.168) are in very good

Table 14.15 Ionic Character Derived from Dipole Moments in the Alkali Halides

Molecule	*Polarizability*[a] $\alpha(10^{-24}$ *cm*) Alk$^+$	Hal$^-$	*Observed Moment* μ_{obs}	*Induced Component*[b] μ_{ind}	*Primary Component* $\mu_p=\mu_{obs}-\mu_{ind}$	*Ionic Character* $i_c=\mu_p/ed$
LiF	0.03	1.0	6.328[c]	−1.32	7.65	1.02
LiCl	0.03	3.6	7.1289[d]	−2.39	9.52	0.98
LiBr	0.03	4.7	7.268[e]	−2.56	9.83	0.95
LiI	0.03	7.0	7.4285[f]	−2.84	10.27	0.89
NaF	0.18	1.0	8.1558[g]	−1.17	9.33	1.01
NaCl	0.18	3.6	9.002[h]	−2.07	11.07	0.98
KCl	0.8	3.6	10.2688[d]	−2.06	12.33	0.97
KBr	0.8	4.7	10.6278[i]	−2.23	12.86	0.95

[a]From Landolt-Börnstein, *Numerical Data and Functional Relationships in Science and Technology, Atomic and Molecular Physics, Vol. 1, Atoms and Ions, Part 1, Springer*, Berlin, 1950, p. 401.
[b]Calculated with (14.166) with ionic radii from Pauling [37], p. 350.
[c]L. Wharton, W. Klemperer, L. P. Gold, R. Strauch, J. J. Gallager, and V. E. Derr, *J. Chem. Phys.*, **38**, 1203 (1963).
[d]A. J. Hebert, *Symposium on Molecular Structure and Spectroscopy*, Ohio State Univ., 1966, Paper 0–1.
[e]A. J. Hebert, F. W. Breivogel, and K. Street, *J. Chem. Phys.*, **41**, 2368 (1964).
[f]J. K. Tyler, *J. Mol. Spectrosc.*, **11**, 39 (1963).
[g]C. D. Hallowell, A. J. Hebert, and K. Street, *J. Chem. Phys.*, **41**, 3540 (1964).
[h]L. P. Gold, Thesis, Harvard Univ., 1962.
[i]F. H. de Leeuw, R. van Wachem, and A. Dynamus, *J. Chem. Phys.*, **50**, 1393 (1969).

agreement with those indicated by the very small quadrupole coupling values of the halogens (Table 14.5). There are, of course, no coupling values available for the fluorides.

Table 14.16 lists the ionic character values for the hydrogen halides HCl, HBr, and HI as calculated from the dipole moments observed with the microwave Stark effect for the deuterated species. For reasons given in Sections 7 and 8 we have assumed the bond orbital hybridization on the halogen to be negligible and have used the relation

$$i_c=\frac{\mu_{obs}+|\mu_s|+|\mu_{ind}|}{ed} \tag{14.169}$$

derived from (14.167) for calculation of the ionic character.

For calculation of the μ_s values of Table 14.16 we used (14.157) with the values of S calculated by Mulliken [35] with Slater orbitals. The greatest uncertainty perhaps is in the estimated values of μ_{ind}. Fortunately, anisotropic polarizabilities of the hydrogen halides have been measured. It seems logical to use the bond axis polarizability $\alpha_{\parallel}$ for this calculation, but these values are measured with a uniform electric field, and the inducing field of the dipole

Table 14.16 Ionic Character Derived from Dipole Moments in the Hydrogen Halides

Molecule	S_{AB} [a]	$\alpha_{\parallel}$ [b] (10^{-24} cm^3)	μ_{obs} [c]	μ_s	μ_{ind}	μ_p	$i_c=\mu_p/ed$
HCl	0.49	3.13	1.12	−0.65	−0.83	2.60	0.42
HBr	0.49	4.22	0.83	−0.90	−0.58	2.31	0.34
HI	0.50	6.58	0.445	−1.23	−0.32	2.00	0.26

[a]From $1s$-$np\sigma$ bonds, Mulliken [35].
[b]Parallel component of molecular polarizability, from Landolt-Börnstein, *Numerical Data and Functional Relationships on Science and Technology, Atomic and Molecular Physics, Vol. I, Molecules II*, Springer, Berlin, 1951, p. 511.
[c]For deuterated species, from C. A. Burrus, *J. Chem. Phys.*, **31**, 1270 (1959).

moment is far from uniform. In estimating μ_{ind} we have made the assumption that the measured $\alpha_{\parallel}$ arises from the polarizability of the halogen. This seems to be a good assumption since the polarizability of the semipositive H would be very small. We further assumed that the polarizing field acting on the halogen arises from the pole μ_{obs}/d on the H and that it is approximated by the z component of the field at a distance equal to the covalent radius from the halogen nucleus. The resulting formula is

$$\mu_{ind}=\frac{\mu_{obs}\alpha_{\parallel}}{(d^2+R^2)^{3/2}} \tag{14.170}$$

where d is the internuclear distance and R is the covalent radius of the halogen.

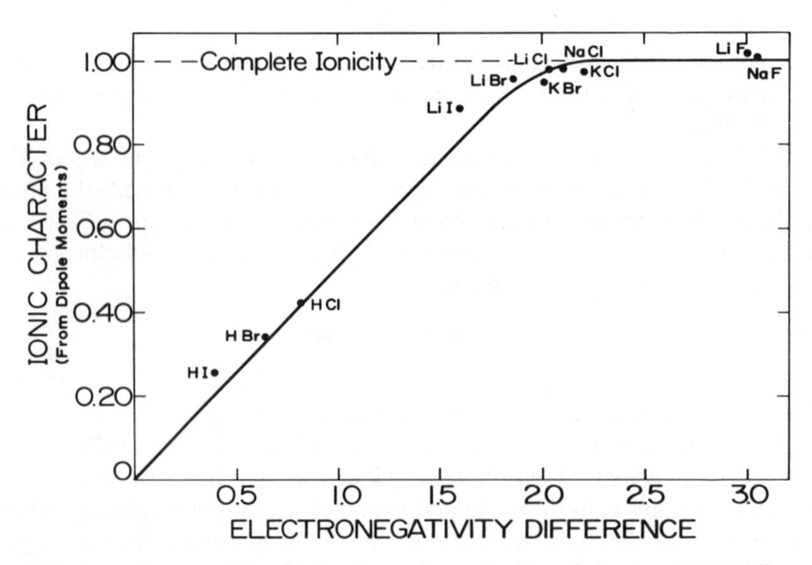

Fig. 14.7 Ionic character (from dipole moments) as a function of electronegativity difference of the bonded atoms.

This formula is similar to that used for estimation of the polarization of the negative ions except that one-half the much larger ionic radius is used instead of the covalent radius.

In Fig. 14.7 the ionic character values obtained from the dipole moments as listed in Tables 14.15 and 14.16 are plotted against the electronegativity difference of the bonded atoms. Within the accuracy of the i_c determinations and electronegativities, the relationship is like that already found from nuclear quadrupole couplings and is described by the formula

$$i_c = \frac{|x_A - x_B|}{2} \quad \text{for } |\Delta x| < 2$$

$$\text{complete ionicity for } |\Delta x| > 2$$

already given by (14.93).

We have not included HF in the plot of Fig. 14.7 because of the uncertainty in the evaluation of the rather large μ_{ind}. The electronegativity difference of $|x_F - x_H| = 1.8$ indicates an almost complete ionicity of the bond, 90%. However, the ionic radius of F^- is 1.33 Å, and yet the measured internuclear distance of HF is only 0.92 Å. This means that if the bond were completely ionic, the proton would actually be buried deep in the negative charge cloud of the F^- ion. If we assume that the proton is at the edge of a semi-ionic radius equal to the observed interatomic distance of 0.92 Å, (14.166) would indicate a polarization moment of $\mu_{\text{ind}} = 1.68$. With this assumption, the ionic character is found from the observed [77] dipole moment 1.8195 D to be $i_c = 0.82$. This value is not unreasonable, but the structure of HF is evidently more complex than this analysis indicates. The value of 1.4 D for its moment predicted from molecular orbital theory [73] is not in good agreement with the accurately observed value.

Polyatomic Molecules

The electric dipole moment of a molecule is determined by the charge distribution over the entire molecule, unlike quadrupole coupling which is determined primarily by the charge distribution in the immediate vicinity of the particular coupling nucleus. For this reason dipole moments of polyatomic molecules are more difficult to interpret quantitatively than are quadrupole couplings. Molecular orbital theory has not yet advanced to the point where molecular dipole moments can be reliably predicted for polyatomic molecules. However, from comparisons of the dipole moments of related polyatomic molecules it has been possible to assign approximate moment values to certain bonds in polyatomic molecules [78–80]. One can then use these empirical bond moments to predict roughly the overall moments of certain molecules. However, bond moments are not vectorially additive because of the changes in bond character from molecule to molecule.

For estimation of the primary bond moments in polyatomic molecules the relationship between ionic character and electronegativity as derived from quadrupole couplings (Fig. 14.1) and confirmed by dipole moments of diatomic molecules (Fig. 14.7) should be of much help when used with the proper group electronegativity values (Section 15). Likewise, the approximation methods described earlier in this section for estimation of other components μ_s, μ_h, and μ_{ind} for diatomic molecules should be useful in the prediction and understanding of the bond moments in polyatomic molecules.

That the hybridization moment can be of much importance in polyatomic molecules is evidenced by the very small moment [81] (0.234 D) observed for NF_3. Because of the large difference in electronegativity of N and F, a large primary N—F bond moment of 3.2 D is expected. The orbital overlap component is negligible because of the nearly equal covalent radii of N and F, but the induced component of each bond ~ 1.0 D is significant because of the large primary moment. There will also be an N hybridization component directed along each bond, but this will be included in the resultant N hybridization moment to be estimated. Exclusive of this hybridization component, the component along each N—F bond will be approximately $3.2-1.0=2.2$ D. Together the three bonds will give rise to an axial component of $\sim 3(2.2) \cos 64^0 \approx 2.9$ D, with the positive pole on the N. The magnitude of the N hybridization moment $\mu_h = \mu_{obs} - 2.9$ is estimated to be ~ 2.7 D or 3.1 D, depending on the direction of the observed moment, which is unknown. This μ_h results from three bonding orbitals having 17% *s* character, and electron populations of 0.5 each, together with a counterhybridized *sp* orbital having 49% *s* character and two unpaired electrons. The overall μ_h is equivalent to that of an 1.5 electron population in the 49% hybridized orbital. The estimated μ_h is reasonable. A similarly hybridized orbital of C with $n_h = 1.5$ would give rise to $\mu_h \approx 3.3$ D (see Fig. 14.6). Although the N orbitals of NH_3 are comparably hybridized, the μ_h of this molecule is appreciably less than that of NF_3 because of the more complete filling of the *sp* shell on the N.

Because of the many uncertainties involved, we shall not attempt further interpretation of dipole moments in polyatomic molecules. In many organic molecules π character in the bonding increases the complexity of the moment. Selected values of moments measured with microwave spectroscopy are given in Tables 10.9–10.11.

References

1. A. Nordsieck, *Phys. Rev.*, **58**, 310 (1940).
2. C. H. Townes and B. P. Dailey, *J. Chem. Phys.*, **17**, 782 (1949).
3. C. H. Townes and B. P. Dailey, *J. Chem. Phys.*, **20**, 35 (1952).
4. C. H. Townes and B. P. Dailey, *J. Chem. Phys.*, **23**, 118 (1955).
5. W. Gordy, *J. Chem. Phys.*, **19**, 792 (1951).
6. W. Gordy, *J. Chem. Phys.*, **22**, 1470 (1954).
7. W. Gordy, *Discussions Faraday Soc.*, **19**, 9 (1955).
8. R. Bersohn, *J. Chem. Phys.*, **22**, 2078 (1954).
9. J. H. Goldstein, *J. Chem. Phys.*, **24**, 507 (1956).

10. W. Gordy, W. V. Smith, and R. F. Trambarulo, *Microwave Spectroscopy*, Wiley, New York, 1954.
11. C. H. Townes and A. L. Schawlow, *Microwave Spectroscopy*, Mc Graw-Hill, New York, 1955.
12. T. P. Das and E. L. Hahn, *Solid State Physics*, Suppl. 1, Academic, New York, 1958.
13. P. Grivet and A. Basssompierre, *Mem. R. Acad. Belg.*, **33**, 217 (1961).
14. C. T. O'Konski, in *Determination of Organic Structures by Physical Methods*, Vol. 2, F. C. Nacnod and W. D. Phillips, Eds., Academic, New York, 1962, pp. 661–726.
15. T. M. Sugden and C. N. Kenney, *Microwave Spectroscopy of Gases*, Van Nostrand, London, 1965.
16. C. A. Coulson, *Valence*, Oxford Univ. Press, Oxford, 1953.
17. B. Pullman and A. Pullman, *Quantum Biochemistry*, Wiley, New York, 1963.
18. J. H. Goldstein, *J. Chem. Phys. Rev.*, **84**, 244 (1951); **86**, 316 (1952); **95**, 736 (1954).
19. R. M. Sternheimer, *Phys. Rev.*, **84**, 244 (1951); **86**, 316 (1952); **95**, 736 (1954).
20. R. M. Sternheimer and H. M. Foley, *Phys. Rev.*, **92**, 1460 (1953); **102**, 731 (1956).
21. H. W. White, *Introduction to Atomic Spectra*, McGraw-Hill, New York, 1934.
22. H. Kopfermann, *Nuclear Moments* (trans. by E. E. Schneider), Academic, New York, 1958.
23. J. C. Slater, *Phys. Rev.*, **36**, 57 (1930).
24. R. G. Barnes and W. V. Smith, *Phys. Rev.*, **93**, 95 (1954).
25. H. B. G. Casmir, *On the Interaction between Atomic Nuclei and Electrons*, Teyler's Tweede Genootschop, E. F. Bohn, Haarlem, 1936.
26. G. Koster, *Phys. Rev.*, **86**, 148 (1952).
27. E. Tiemann, *J. Mol. Struct.*, **97**, 331 (1983).
28. H. M. Foley, R. M. Sternheimer, and D. Tycko, *Phys. Rev.*, **93**, 734 (1954).
29. R. Livingston, *J. Chem. Phys.*, **19**, 1434 (1951).
30. H. G. Dehmelt, *Z. Physik*, **130**, 480 (1951).
31. H. G. Dehmelt, *Naturwiss.*, **37**, 398 (1950).
32. N. Bettin, H. Knöckel, and E. Tiemann, *Chem. Phys. Lett.*, **80**, 386 (1981).
33. A. Yokozeki and J. S. Muenter, *J. Chem. Phys.*, **72**, 3796 (1980).
34. H. G. Robinson, H. G. Dehmelt, and W. Gordy, *J. Chem. Phys.*, **22**, 511 (1954).
35. R. S. Mulliken, *J. Am. Chem. Soc.*, **72**, 4493 (1950).
36. A. D. Buckingham, *Trans. Faraday Soc.*, **58**, 1277 (1962).
37. L. Pauling, *The Nature of the Chemical Bond*, 2nd ed. Cornell Univ. Press, Ithaca, New York, 1948.
38. R. S. Mulliken, *J. Chem. Phys.*, **46**, 497 (1949).
39. L. Pierce, R. Nelson, and C. Thomas, *J. Chem. Phys.*, **43**, 3423 (1956).
40. H. G. Dehmelt, *Discussions Faraday Soc.*, **19**, 263 (1955).
41. M. H. Cohen and F. Reif, in *Solid State Physics*, Vol. 5, F. Seitz and D. Turnball, Eds., Academic, New York, 1957, pp. 321–438.
42. H. O. Hooper and P. J. Bray, *J. Chem. Phys.*, **33**, 334 (1960).
43. R. S. Mulliken, C. A. Reike, and W. G. Brown, *J. Am. Chem. Soc.*, **63**, 41 (1941).
44. P. A. Casabella and P. J. Bray, *J. Chem. Phys.*, **28**, 1182 (1958); **29**, 1105 (1958).
45. E. Schempp and P. J. Bray, *J. Chem. Phys.*, **46**, 1186 (1967).
46. R. A. Marino, L. Guibe, and P. J. Bray, *J. Chem. Phys.*, **49**, 5104 (1968).
47. J. Sheridan and W. Gordy, *Phys. Rev.*, **79**, 513 (1950).
48. Y. Kato, U. Farukane, and H. Takeyama, *Bull. Chem. Soc. Japan*, **32**, 527 (1959).

49. E. A. C. Lucken, *Trans. Faraday Soc.*, **57**, 729 (1961).
50. D. DeSantis, A. Lurio, T. A. Miller, and R. S. Freund, *J. Chem. Phys.*, **58**, 4625 (1973).
51. F. W. Terman and T. A. Scott, *Bull. Am. Phys. Soc.*, **3**, (1958).
52. H. Meyer and T. A. Scott, *Phys. Chem. Solids*, **11**, 215 (1959).
53. L. Nygaard, J. T. Nielsen, J. Kirchheiner, G. Maltesen, J. Rastrup-Andersen, and G. O. Sorensen, private communication.
54. G. A. Sorensen, *J. Mol. Spectrosc.*, **22**, 325 (1967).
55. W. Arnold, H. Dreizler, and H. D. Rudolph, *Z. Naturforsch.*, **23a**, 301 (1968).
56. H. D. Rudolph, H. Dreizler, and H. Seiler, *Z. Naturforsch.*, **22a**, 1738 (1967).
57. C. A. Burrus and W. Gordy, *Phys. Rev.*, **92**, 274 (1953).
58. H. G. Robinson, H. G. Dehmelt, and W. Gordy, *Phys. Rev.*, **89**, 1305 (1953).
59. J. E. Griffths and K. B. McAfee, *Proc. Chem. Soc.*, **1961**, 456.
60. C. H. Townes, J. M. Mays, and B. P. Dailey, *Phys. Rev.*, **76**, 700 (1949).
61. V. W. Laurie, *J. Chem. Phys.*, **30**, 1210 (1959).
62. E. F. Pearson, C. L. Norris, and W. H. Flygare, *J. Chem. Phys.*, **60**, 1761 (1974).
63. C. Kirby and H. W. Kroto, *J. Mol. Spectrosc.*, **83**, 1 (1980).
64. W. Gordy, H. Ring, and A. B. Burg, *Phys. Rev.*, **74**, 1191 (1948); **78**, 512 (1950).
65. H. G. Dehmelt, *Z. Physik*, **134**, 642 (1953).
66. P. R. Wells, in *Progr. Phys. Org. Chem.*, Vol. 6, A. Streitwieser, Jr., and R. W. Taft, Eds., Wiley, New York, 1968, pp. 111–145.
67. J. Hinze, M. A. Whitehead, and H. H. Jaffe, *J. Am. Chem. Soc.*, **85**, 148 (1963).
68. W. Gordy, *J. Chem. Phys.*, **14**, 305 (1946).
69. F. A. Matsen, *J. Chem. Phys.*, **34**, 337 (1961).
70. L. Wharton, L. Gold, and W. Klemperer, *J. Chem. Phys.*, **37**, 2149 (1962).
71. P. Cade and W. Huo, *J. Chem. Phys.*, **45**, 1063 (1966).
72. C. F. Bender and E. R. Davidson, *J. Chem. Phys.*, **49**, 4222 (1968).
73. B. J. Ransil, *Rev. Mod. Phys.*, **32**, 239 (1960).
74. L. C. Allen and A. M. Karo, *Rev. Mod. Phys.*, **32**, 275 (1960).
75. R. S. Mulliken, *J. Chim. Phys.*, **46**, 497 (1949).
76. N. B. Hannay and C. P. Smyth, *J. Am. Chem. Soc.*, **68**, 17 (1946).
77. R. Weiss, *Phys. Rev.*, **131**, 659 (1963).
78. R. J. W. Le Fevre, *Dipole Moments, Their Measurement and Application in Chemistry*, Chemical Publishing Co., New York, 1938.
79. J. W. Smith, *Electric Dipole Moments*, Butterworths, London, 1955.
80. C. J. F. Böttcher, *Theory of Electric Polarisation*, Elesvier, Amsterdam, 1952.
81. S. N. Ghosh, R. Trambarulo, and W. Gordy, *J. Chem. Phys.*, **21**, 303 (1953).

Chapter **XV**

IRREDUCIBLE TENSOR METHODS FOR CALCULATION OF COMPLEX SPECTRA

1 INTRODUCTION

In this section we wish to demonstrate how the irreducible tensor method can be used to extract information on molecular interactions. Direct calculation techniques discussed previously in which wave function expansions are employed to change to a more appropriate basis give rise to particularly complicated expressions. With irreducible tensor methods, such complications are taken care of in the derivation of the relatively simple theorems associated with the theory. Irreducible tensor methods thus provide a highly efficient and powerful way to analyze hyperfine structure arising from plural nuclear coupling, as well as other molecular interactions. Since more applications of these methods are appearing, some familiarity with these techniques is required for an adequate appreciation of the literature on hyperfine interactions, as well as other areas of microwave spectroscopy. The theory of spherical or irreducible tensor operators has been discussed in detail by Rose [1], Edmonds [2], Fano and Racah [3], and Judd [4]. In this section we follow much of the expository discussion given by Cook and De Lucia [5]. Important theories will be simply quoted, and we will concentrate on their application. The reader may consult the previous references for derivations of these theorems.

Discussions and specific applications of irreducible tensor methods to the analysis of complex spectra have been given by Curl and Kinsey [6], Thaddeus et al. [7], and others [8–44], to cite but a few (see also references cited therein). These include discussions applicable to symmetric tops [15, 19, 24, 26, 28, 29, 38, 40, 42] and molecules with unbalanced electronic angular momentum [6, 12, 20, 32, 35, 36, 39]. Techniques for inclusion of the effects of identical particles in a particularly straightforward way have been given by Cederberg [17]. The theory of hyperfine interactions in molecules with internal rotation has been discussed by Heuvel and Dymanus [21], and Ribeaud et al. [43]; that of nuclear hyperfine interactions in spherical tops, by Michelot et al. [34].

By definition, the quantities $T_q^{(k)}(q=-k, -k+1, \ldots, +k)$ constitute an irreducible tensor of rank k (an integer) if they transform like the spherical harmonies $Y_q^{(k)}(\theta, \phi)$ under a rotation. The transformation properties of an irreducible tensor $\mathbf{T}^{(k)}$ under a rotation of the coordinate system is given by [1]

$$T_{q'}^{\prime(k)}=\sum_q T_q^{(k)} D_{qq'}^{(k)}(\alpha\beta\gamma) \tag{15.1}$$

where $T_q^{(k)}$ are the components defined with respect to the old axes, $T_{q'}^{\prime(k)}$ are the components with respect to the new or rotated coordinate system, and $D_{qq'}^{(k)}$ is the $(2k+1)$ dimensional matrix representation of the rotation operator. The $D_{qq'}^{(k)}$ depend on the Euler angles (α, β, γ) that specify the relative orientation of the two frames. By inverting (15.1) we may express $T_q^{(k)}$ in terms of the components, $T_{q'}^{\prime(k)}$. Multiplying by $D_{q',q}^{-1(k)}$ and summing over q' give

$$T_q^{(k)}=\sum_{q'}(-1)^{q-q'}T_{q'}^{\prime(k)}D_{-q,-q'}^{(k)} \tag{15.2}$$

where the unitary character of the $\mathbf{D}^{(k)}$ matrix and the relation

$$D_{q'q}^{-1(k)}=D_{qq'}^{*(k)}=(-1)^{q-q'}D_{-q,-q'}^{(k)} \tag{15.3}$$

have been employed. This relation will be used to express the tensor components $T_q^{(k)}$ in terms of the components $T_{q'}^{\prime(k)}$ with respect to the rotating molecule–fixed principal inertial axis system.

The scalar product of two irreducible tensors $\mathbf{T}^{(k)}$ and $\mathbf{U}^{(k)}$ with components $T_q^{(k)}$ and $U_q^{(k)}$ is defined as

$$\mathbf{T}^{(k)}\cdot\mathbf{U}^{(k)}=\sum_{q=-k}^{k}(-1)^q T_q^{(k)}U_{-q}^{(k)} \tag{15.4}$$

The rule for the product of two irreducible tensor operators is specified by

$$\begin{aligned}T_q^{(l)}&=[\mathbf{T}^{(l_1)}\times\mathbf{T}^{(l_2)}]_q^{(l)}\\&=(2l+1)^{1/2}\sum_{q_1q_2}(-1)^{-l_1+l_2-q}\\&\quad\times\begin{pmatrix}l_1 & l_2 & l\\ q_1 & q_2 & -q\end{pmatrix}T_{q_1}^{(l_1)}T_{q_2}^{(l_2)}\end{aligned} \tag{15.5}$$

which yields irreducible tensors of rank l, where l may be equal to $l_1+l_2\cdots|l_1-l_2|$. Note, in the multiplying of spherical tensors, the ranks are added vectorially not algebraically. The notation $\mathbf{T}^{(l_1)} \times \mathbf{T}^{(l_2)}$ makes clear the origin of the irreducible tensor $T_q^{(l)}$. The symbol above in brackets is a $3j$ symbol and is related to the Clebsch-Gordon coefficients.

The $3j$ symbol has the property that the value is left unchanged for an even permutation of columns and is multiplied by $(-1)^{l_1+l_2+l}$ for an odd permutation of columns or for a sign change of all the elements of the second row. Simple algebraic formulas for some special cases are available [2] as well as tabulated numerical values [45]. They may be readily evaluated from their general definition by means of a computer algorithm. They are nonzero only if the elements of the second row sum to zero, that is, $q_1+q_2-q=0$, and if l_1, l_2, l satisfy the triangular condition, that is, a triangle can be formed with the length of each side specified by a quantum number. A very useful orthogonality relation for the $3j$ symbols is

$$\sum_{m_1 m_2} \begin{pmatrix} j_1 & j_2 & j_3 \\ m_1 & m_2 & m_3 \end{pmatrix} \begin{pmatrix} j_1 & j_2 & j_3' \\ m_1 & m_2 & m_3' \end{pmatrix} = \frac{\delta(j_3, j_3')\delta(m_3, m_3')}{(2j_3+1)} \tag{15.6}$$

To apply the methods of irreducible tensors, the interaction must be written in the form of spherical tensors. In some cases, the interaction can be formulated directly in the form of spherical tensors such as the magnetic and quadrupole interactions [5, 8]. In other cases, it may be convenient to formulate the interaction in Cartesian form. The general hyperfine Hamiltonian in Cartesian tensor form can be expressed as

$$\mathscr{H}_{hfs} = \sum_i^N (\tfrac{1}{6}\mathbf{V}_i : \mathbf{Q}_i + \mathbf{I}_i \cdot \mathbf{C}_i \cdot \mathbf{J}) + \sum_i \sum_{j \neq i} (\mathbf{I}_i \cdot \mathbf{D}_{ij} \cdot \mathbf{I}_j) \tag{15.7}$$

The sum is over all coupling nuclei, and the terms represent, respectively, the electric quadrupole (product of two dyadics), magnetic dipole and spin–spin interactions. This, however, presents no serious limitation to the application of irreducible tensor methods, since the elements of a Cartesian tensor can be cast into forms which transform as spherical tensor operators.

The spherical unit vectors $\mathbf{e}_q^{(1)}$ are defined as [2]

$$\mathbf{e}_0^{(1)} = \mathbf{e}_z, \qquad \mathbf{e}_{\pm 1}^{(1)} = \mp(1/\sqrt{2})(\mathbf{e}_x \pm i\mathbf{e}_y) \tag{15.8}$$

The components of a first-rank irreducible tensor $T_q^{(1)}$ in terms of the Cartesian components (T_x, T_y, T_z) of a vector are given by

$$T_q^{(1)} = \mathbf{e}_q^{(1)} \cdot \mathbf{T} = \mathbf{e}_q^{(1)} \cdot (\mathbf{e}_x T_x + \mathbf{e}_y T_y + \mathbf{e}_z T_z) \tag{15.9}$$

which yields

$$T_0^{(1)} = T_z, \qquad T_{\pm 1}^{(1)} = \mp(1/\sqrt{2})(T_x \pm iT_y) \tag{15.10}$$

These linear combinations transform as a first-rank spherical tensor. Any interaction written in terms of two vectors can hence be readily formulated in spherical tensor notation. The electric dipole interaction with an external field would be written from (15.4) as

$$\mathscr{H}_{\mathscr{E}} = -\boldsymbol{\mu}^{(1)} \cdot \boldsymbol{\mathscr{E}}^{(1)} = \mathscr{E}^{(1)}_{-1}\mu^{(1)}_{1} - \mathscr{E}^{(1)}_{0}\mu^{(1)}_{0} + \mathscr{E}^{(1)}_{1}\mu^{(1)}_{-1}$$
$$\equiv -\mathscr{E}_x\mu_x - \mathscr{E}_y\mu_y - \mathscr{E}_z\mu_z \tag{15.11}$$

with $\mu^{(1)}$ the electrical dipole moment tensor and with the components of the tensors chosen as in (15.10).

A second-rank Cartesian tensor may, in general, be decomposed into three irreducible tensors:

$$\mathbf{T} = \mathbf{T}^{(0)} + \mathbf{T}^{(1)} + \mathbf{T}^{(2)} \tag{15.12}$$

of rank 0, 1, 2. To define the irreducible tensor components in terms of the Cartesian components we first obtain the appropriate unit vectors. The spherical basis vectors may be constructed from (15.5). In particular, the spherical unit vectors obtained by combining $\mathbf{e}^{(1)}_{q_1}$ and $\mathbf{e}^{(1)}_{q_2}$ are given by

$$\mathbf{e}^{(l)}_{q} = (2l+1)^{1/2} \sum_{q_1 q_2} (-1)^q \times \begin{pmatrix} 1 & 1 & l \\ q_1 & q_2 & -q \end{pmatrix} \mathbf{e}^{(1)}_{q_1}\mathbf{e}^{(1)}_{q_2} \qquad (l=0, 1, 2) \tag{15.13}$$

Employing a table of $3j$ symbols, one finds

$$\begin{aligned}
\mathbf{e}^{(0)}_{0} &= (1/\sqrt{3})(\mathbf{e}^{(1)}_{1}\mathbf{e}^{(1)}_{-1} + \mathbf{e}^{(1)}_{-1}\mathbf{e}^{(1)}_{1} - \mathbf{e}^{(1)}_{0}\mathbf{e}^{(1)}_{0}) \\
\mathbf{e}^{(1)}_{0} &= (1/\sqrt{2})(\mathbf{e}^{(1)}_{1}\mathbf{e}^{(1)}_{-1} - \mathbf{e}^{(1)}_{-1}\mathbf{e}^{(1)}_{1}) \\
\mathbf{e}^{(1)}_{\pm 1} &= \pm(1/\sqrt{2})(\mathbf{e}^{(1)}_{\pm 1}\mathbf{e}^{(1)}_{0} - \mathbf{e}^{(1)}_{0}\mathbf{e}^{(1)}_{\pm 1}) \\
\mathbf{e}^{(2)}_{0} &= (\sqrt{2}/\sqrt{3})\mathbf{e}^{(1)}_{0}\mathbf{e}^{(1)}_{0} + 6^{-1/2}(\mathbf{e}^{(1)}_{1}\mathbf{e}^{(1)}_{-1} + \mathbf{e}^{(1)}_{-1}\mathbf{e}^{(1)}_{1}) \\
\mathbf{e}^{(2)}_{\pm 1} &= (1/\sqrt{2})(\mathbf{e}^{(1)}_{0}\mathbf{e}^{(1)}_{\pm 1} + \mathbf{e}^{(1)}_{\pm 1}\mathbf{e}^{(1)}_{0}) \\
\mathbf{e}^{(2)}_{\pm 2} &= \mathbf{e}^{(1)}_{\pm 1}\mathbf{e}^{(1)}_{\pm 1}
\end{aligned} \tag{15.14}$$

To define the irreducible tensor components in terms of the Cartesian components we express $\mathbf{T}$ as a dyadic in terms of its Cartesian components

$$\mathbf{T} = \sum_{g, g'} \mathbf{e}_g \mathbf{e}_{g'} T_{gg'} \qquad (g, g' = x, y, \text{ or } z) \tag{15.15}$$

and with the aid of (15.8) and the relation

$$T^{(l)}_{q} = \mathbf{e}^{(l)}_{q} \cdot \mathbf{T} \tag{15.16}$$

where $l=0, 1, 2$ and $q = -l, -l+1, \ldots, +l$, we find

$$\begin{aligned}
T^{(0)}_{0} &= -(1/\sqrt{3})[T_{xx} + T_{yy} + T_{zz}] & T^{(2)}_{0} &= 6^{-1/2}[3T_{zz} - (T_{xx} + T_{yy} + T_{zz})] \\
T^{(1)}_{0} &= -(i/\sqrt{2})[T_{xy} - T_{yx}] & T^{(2)}_{\pm 1} &= \mp\tfrac{1}{2}[T_{xz} + T_{zx} \pm i(T_{yz} + T_{zy})] \\
T^{(1)}_{\pm 1} &= -\tfrac{1}{2}[T_{zx} - T_{xz} \pm i(T_{zy} - T_{yz})] & T^{(2)}_{\pm 2} &= \tfrac{1}{2}[T_{xx} - T_{yy} \pm i(T_{xy} + T_{yx})]
\end{aligned} \tag{15.17}$$

It is apparent that if the Cartesian tensor is symmetric and traceless, only the second-rank irreducible tensor is nonvanishing. These relations thus allow an interaction involving first- and second-rank Cartesian tensors to be transformed to irreducible tensor form, where irreducible tensor methods may then be employed.

Consider, for example, the case of the spin–rotation interaction. We may write for the interaction in Cartesian coordinates

$$\mathscr{H}_M = \mathbf{I} \cdot \mathbf{C} \cdot \mathbf{J} \tag{15.18}$$

where all elements are referred to space-fixed axes and C is a second-rank Cartesian tensor written as a dyadic. To formulate this interaction so that the irreducible tensor techniques may be employed, we decompose $\mathbf{C}$ as in (15.12) and write the interaction as a sum of scalar products

$$\mathscr{H}_M = \sum_{l=0}^{2} N_l [\mathbf{C}^{(l)} \times \mathbf{J}^{(1)}]^{(1)} \cdot \mathbf{I}^{(1)} \tag{15.19}$$

where the elements of $\mathbf{I}^{(1)}$ and $\mathbf{J}^{(1)}$ are defined by (15.10) and those of $\mathbf{C}^{(l)}$ by (15.17). The irreducible products $[\mathbf{C}^{(0)} \times \mathbf{J}^{(1)}]^{(1)}$, and so on, are obtained from (15.5). The constants N_l are appropriate normalization factors such that when (15.19) is written out explicitly it is equivalent to (15.18). The normalization factors are found to be $N_0 = -1/3^{1/2}$, $N_1 = -1$, and $N_2 = -5^{1/2}/3^{1/2}$. An alternate formulation is also possible in which $\mathbf{I}^{(1)}$ and $\mathbf{J}^{(1)}$ are combined to give irreducible tensors of rank 0, 1, and 2 and the scalar products of these with the corresponding irreducible tensors of $\mathbf{C}$ are formed.

The matrix element equations for irreducible tensor operators, which will be employed in the derivations of the following sections, are briefly summarized below. These theorems are taken from Edmonds [2]. For the basis functions $|\tau jm)$ that are eigenfunctions of $\mathbf{j}^2$ and $\mathbf{j}_z$, the Wigner-Eckart theorem states

$$(\tau' j' m' | T_q^{(k)} | \tau, j, m) = (-1)^{j'-m'} \times \begin{pmatrix} j' & k & j \\ -m' & q & m \end{pmatrix} (\tau' j' \| T^{(k)} \| \tau j) \tag{15.20}$$

where $(\tau' j' \| T^{(k)} \| \tau j)$ is defined as the reduced matrix element, or double-bar matrix element, of $\mathbf{T}^{(k)}$, it is, however, merely a number. The quantum number τ, in general, may be considered a set of quantum numbers required to more clearly specify the state. This factorization theorem shows that the entire projection quantum number dependence of the matrix element is contained in the $3j$ symbol. A knowledge of the matrix element of any one of the components, $T_q^{(k)}$, allows the evaluation of the reduced matrix element.

If the tensor operators $\mathbf{T}^{(k)}$ and $\mathbf{U}^{(k)}$ act, respectively, on subsystems 1 and 2 of a system with a subsystem 1 characterized by the angular momentum $\mathbf{j}_1$ and subsystem 2 by the angular momentum $\mathbf{j}_2$, then the matrix elements of the scalar product of these two tensor operators in the coupled basis $\mathbf{J} = \mathbf{j}_1 + \mathbf{j}_2$ are given by

$$(\tau'_1 j'_1 \tau'_2 j'_2 J' M' | \mathbf{T}^{(k)} \cdot \mathbf{U}^{(k)} | \tau_1 j_1 \tau_2 j_2 J M) = \delta_{JJ'} \delta_{MM'} (-1)^{j_1 + j'_2 + J} \begin{Bmatrix} J & j'_2 & j'_1 \\ k & j_1 & j_2 \end{Bmatrix} \times (\tau'_1 j'_1 \| T^{(k)} \| \tau_1 j_1)(\tau'_2 j'_2 \| U^{(k)} \| \tau_2 j_2) \tag{15.21}$$

where the symbol in braces is the $6j$ symbol, and τ_1 and τ_2 represent additional pertinent quantum numbers characterizing subsystems 1 and 2. Numerical values of the $6j$ symbol have been tabulated [45], and expressions for certain simple cases are available [2]. The value of a $6j$ symbol is unaffected by a permutation of the columns or by an interchange of the upper and lower elements in any two columns. They may be conveniently computed from a general formula due to Racah (see Ref. 2, p. 99). Whereas the $3j$ symbols are involved in the coupling of two angular momenta, the $6j$ symbol is associated with the coupling of three angular momenta. In particular, they are proportional to the expansion coefficients relating the representation of the coupling scheme $\mathbf{j}_2+\mathbf{j}_3=\mathbf{j}'$, $\mathbf{j}'+\mathbf{j}_1=\mathbf{J}$ and the representation of the coupling scheme $\mathbf{j}_1+\mathbf{j}_2=\mathbf{j}'$, $\mathbf{j}'+\mathbf{j}_3=\mathbf{J}$ [see (9.126)]. They are nonzero only if the three numbers of each row and each column satisfy the triangular condition. In applying (15.21) it is important to recognize that $\mathbf{T}^{(k)}$ and $\mathbf{U}^{(k)}$ must be commuting tensor operators.

The relation between the reduced matrix element of a single operator in the coupled and uncoupled representation is given next. For the tensor operator $\mathbf{T}^{(k)}$ acting only on the space of subsystem 1, that is, $\mathbf{T}^{(k)}$ commutes with $\mathbf{j}_2$, the nonvanishing matrix elements are

$$(\tau_1' j_1' \tau_2 j_2 J' \| T^{(k)} \| \tau_1 j_1 \tau_2 j_2 J) = (-1)^{j_1'+j_2+J+k}[(2J+1)(2J'+1)]^{1/2} \times \begin{Bmatrix} j_1' & J' & j_2 \\ J & j_1 & k \end{Bmatrix} (\tau_1' j_1' \| T^{(k)} \| \tau_1 j_1) \tag{15.22}$$

For $\mathbf{U}^{(k)}$ operating only on subsystem 2, that is, $\mathbf{U}^{(k)}$ commutes with $\mathbf{j}_1$, the nonvanishing matrix elements are

$$(\tau_1 j_1 \tau_2' j_2' J' \| U^{(k)} \| \tau_1 j_1 \tau_2 j_2 J) = (-1)^{j_1+j_2+J'+k}[(2J+1)(2J'+1)]^{1/2} \times \begin{Bmatrix} j_2' & J' & j_1 \\ J & j_2 & k \end{Bmatrix} (\tau_2' j_2' \| U^{(k)} \| \tau_2 j_2) \tag{15.23}$$

Let $\mathbf{T}^{(k_1)}$ and $\mathbf{U}^{(k_2)}$ be tensor operators acting on subsystem 1 and 2, respectively. The reduced matrix element of the product $\mathbf{T}^{(k_1)} \times \mathbf{U}^{(k_2)}$ of these two irreducible operators in the coupled representation, in terms of the reduced matrix elements of the individual operators in the uncoupled representation, is given by

$$(\tau'\tau_1' j_1' \tau_2' j_2' J' \| [\mathbf{T}^{(k_1)} \times \mathbf{U}^{(k_2)}]^{(k)} \| \tau\tau_1 j_1 \tau_2 j_2 J) = [(2J+1)(2J'+1)(2k+1)]^{1/2} \times \sum_{\tau''} \begin{Bmatrix} j_1' & j_1 & k_1 \\ j_2' & j_2 & k_2 \\ J' & J & k \end{Bmatrix} (\tau'\tau_1' j_1' \| T^{(k_1)} \| \tau''\tau_1 j_1)(\tau''\tau_2' j_2' \| U^{(k_2)} \| \tau\tau_2 j_2) \tag{15.24}$$

The symbol in braces, known as a $9j$ symbol, can be written in terms of the $3j$ or $6j$ symbols, and can arise when the coupling of four angular momenta are considered. The sum over the additional classifier τ'' is usually not required. On the other hand, for a tensor product which involves operators acting on the same subspace, $\mathbf{jm}$, the reduced matrix elements are related by

$$(\tau' j' \| [\mathbf{T}^{(k_1)} \times \mathbf{U}^{(k_2)}]^{(k)} \| \tau j) = (-1)^{j+j'+k}(2k+1)^{1/2} \times \sum_{\tau'' j''} \begin{Bmatrix} k_1 & k_2 & k \\ j & j' & j'' \end{Bmatrix} (\tau' j' \| T^{(k_1)} \| \tau'' j'')(\tau'' j'' \| U^{(k_2)} \| \tau j) \tag{15.25}$$

Such reduced matrix elements can arise, as, for example, in the application of the Wigner-Eckart theorem or (15.21). The summations in the foregoing equation, where τ'' represents other quantum numbers needed to characterize the state, are usually not required. Derivations of such relationships as given above, applicable to both the rotation group and the point groups, are discussed by Cederberg [17]. The matrix element of a scalar product is a special case of the more general relation of (15.24) with $k_1 = k_2, k = 0$. The $9j$ symbol with two parameters repeated and a single zero element is found to be proportional to a $6j$ symbol. A short list of expressions for some $3j$ and $6j$ symbols which are of particular use here is given in Table 15.1.

An interesting application of the Wigner-Eckart theorem is in the evaluation of the matrix elements of the angular momentum. In particular, we have

$$(J', M'_J | J^{(1)}_q | J, M_J) = (-1)^{J'-M'_J} \begin{pmatrix} J' & 1 & J \\ -M'_J & q & M_J \end{pmatrix} (J' \| J^{(1)} \| J) \tag{15.26}$$

where $J^{(1)}_0 = J_z$; $J^{(1)}_{\pm 1} = \mp 1/2^{1/2}(J_x \pm iJ_y)$ are components referred to the space-fixed coordinate system. In this section the space-fixed coordinates will be designated as (x, y, z), and the body-fixed coordinates will be designated as

Table 15.1 Expressions for Certain $3j$ and $6j$ Symbols

$$\begin{pmatrix} J & J & 0 \\ M & -M & 0 \end{pmatrix} = (-1)^{J-M}(2J+1)^{-1/2}$$

$$\begin{pmatrix} J & J & 1 \\ M & -M-1 & 1 \end{pmatrix} = (-1)^{J-M} \left[\frac{(J-M)(J+M+1) \cdot 2}{(2J+2)(2J+1)(2J)} \right]^{1/2}$$

$$\begin{pmatrix} J & J & 1 \\ M & -M & 0 \end{pmatrix} = (-1)^{J-M} \frac{M}{[(2J+1)(J+1)J]^{1/2}}$$

$$\begin{pmatrix} J & J & 2 \\ M & -M & 0 \end{pmatrix} = (-1)^{J-M} \frac{2[3M^2 - J(J+1)]}{[(2J+3)(2J+2)(2J+1)(2J)(2J-1)]^{1/2}}$$

$$\begin{Bmatrix} a & b & c \\ 0 & c & b \end{Bmatrix} = (-1)^s \left[\frac{1}{(2b+1)(2c+1)} \right]^{1/2}$$

$$\begin{Bmatrix} a & b & c \\ 1 & c & b \end{Bmatrix} = (-1)^{s+1} \frac{2[b(b+1) + c(c+1) - a(a+1)]}{[2b(2b+1)(2b+2)2c(2c+1)(2c+2)]^{1/2}}$$

$$\begin{Bmatrix} a & b & c \\ 2 & c & b \end{Bmatrix} = (-1)^s \frac{2[3X(X-1) - 4b(b+1)c(c+1)]}{[(2b-1)2b(2b+1)(2b+2)(2b+3) \cdot (2c-1)2c(2c+1)(2c+2)(2c+3)]^{1/2}}$$

where $s = a+b+c$, $X = b(b+1) + c(c+1) - a(a+1)$

(x', y', z'). To evaluate the reduced matrix element we choose the simplest case for the LHS of the foregoing equation

$$(J', M'_J|J_0^{(1)}|J, M_J) = (J', M'_J|J_z|J, M_J) = M_J \delta_{JJ'} \delta_{M_J M_{J'}} \tag{15.27}$$

in units of $\hbar$. Evaluation of the $3j$ symbol ($q=0$) from Table 15.1 gives

$$\begin{pmatrix} J & J & 1 \\ M_J & -M_J & 0 \end{pmatrix} = (-1)^{J-M_J} \frac{M_J}{[J(J+1)(2J+1)]^{1/2}}$$

and hence

$$(J\|J^{(1)}\|J) = [J(J+1)(2J+1)]^{1/2} \tag{15.28}$$

Thus, in general

$$(J, M'_J|J_q^{(1)}|J, M_J) = (-1)^{J-M'_J} \begin{pmatrix} J & 1 & J \\ -M'_J & q & M_J \end{pmatrix} [J(J+1)(2J+1)]^{1/2} \tag{15.29}$$

Note $-M'_J + q + M_J = 0$. Replacing the $3j$ symbol by its equivalent yields the well-known matrix elements

$$(J, M_J|J_0^{(1)}|J, M_J) = M_J \tag{15.30}$$

$$(J, M_J \pm 1|J_{\pm 1}^{(1)}|J, M_J) = \mp \frac{1}{\sqrt{2}} [(J \mp M_J)(J \pm M_J + 1)]^{1/2} \tag{15.31}$$

Note in the foregoing expression that the upper (or lower) sign applies throughout. Matrix elements for J_x and J_y may be readily found from these since

$$J_x = \frac{1}{\sqrt{2}} (J_{-1}^{(1)} - J_{+1}^{(1)}) \qquad \text{and} \qquad J_y = \frac{i}{\sqrt{2}} (J_{-1}^{(1)} + J_{+1}^{(1)}) \tag{15.32}$$

This yields (in units of $\hbar$)

$$(J, M+1|J_x|J, M) = \tfrac{1}{2}[(J-M)(J+M+1)]^{1/2} \tag{15.33}$$

$$(J, M-1|J_y|J, M) = \frac{i}{2} [(J+M)(J-M+1)]^{1/2} \tag{15.34}$$

Likewise, the matrix elements of the angular momentum $J_{q'}^{\prime(1)}$ referred to the body-fixed axis in the symmetric rotor basis (JKM_J) may be obtained by use of (15.1)

$$J_{q'}^{\prime(1)} = \sum_q J_q^{(1)} D_{qq'}^{(k)} \tag{15.35}$$

hence,

$$(K'J'M'_J|J_{q'}^{\prime(1)}|KJM_J) = \sum_{qM''_J} (K'J'M'_J|J_q^{(1)}|K'J'M''_J)(K'J'M''_J|D_{qq'}^{(k)}|KJM_J) \tag{15.36}$$

since from (15.29) $J_q^{(1)}$ is diagonal in K and J. The symmetric rotor functions are related to the $\mathbf{D}$ matrix as follows [5]

$$\psi_{JKM_J}(\alpha\beta\gamma)=(-1)^{M_J-K}\,[(2J+1)/8\pi^2]^{1/2}D_{-M_J,-K}{}^{(J)}(\alpha\beta\gamma) \tag{15.37}$$

The matrix elements of $D_{qq'}^{(k)}$ in the symmetric rotor basis may now be obtained from the above identification and are given by [46]

$$(K'J'M'_J|D_{qq'}^{(k)}|KJM_J)=(-1)^{M_J-K}[(2J+1)(2J'+1)]^{1/2} \times\begin{pmatrix} J & k & J' \\ M_J & -q & -M'_J \end{pmatrix}\begin{pmatrix} J & k & J' \\ K & -q' & -K' \end{pmatrix} \tag{15.38}$$

where K is the projection quantum number of the rotational angular momentum $\mathbf{J}$ in the molecule-fixed frame. Therefore,

$$(K'J'M'_J|J_{q'}^{\prime(1)}|KJM_J)=(-1)^t[(2J+1)(2J'+1)^2J'(J'+1)]^{1/2} \times\begin{pmatrix} J & k & J' \\ K & -q' & -K' \end{pmatrix}\sum_{q,M''_J}\begin{pmatrix} J' & 1 & J' \\ M''_J & q & -M'_J \end{pmatrix}\begin{pmatrix} J' & k & J \\ M''_J & q & -M_J \end{pmatrix} \tag{15.39}$$

with $t=J'-M'_J+M_J-K+1$ and $k=1$ and two of the $3j$ symbols have been rearranged. Using the orthogonality property of the $3j$ symbols, (15.6), we obtain

$$(K'JM_J|J_{q'}^{\prime(1)}|KJM_J)=(-1)^{J+1-K}[(2J+1)J(J+1)]^{1/2} \times\begin{pmatrix} J & 1 & J \\ K & -q' & -K' \end{pmatrix} \tag{15.40}$$

Evaluation of this expression gives the matrix elements of the angular moment components in the molecule-fixed axis system for the symmetric rotor basis

$$(KJM_J|J_0^{\prime(1)}|KJM_J)=K \tag{15.41}$$

$$(K\mp 1JM_J|J_{\pm 1}^{\prime(1)}|KJM_J)=\mp\frac{1}{\sqrt{2}}[(J\pm K)(J\mp K+1)]^{1/2} \tag{15.42}$$

Note that the raising and lowering character of the operators is reversed in the two frames, as found in Chapter II.

The components $J_{x'}$ and $J_{y'}$ are defined in terms of $J_{q'}^{\prime(1)}$ as in (15.32), and this yields for the components referred to the molecule-fixed axis system

$$(K-1,\ J,\ M_J|J_{x'}|K,\ J,\ M_J)=\tfrac{1}{2}[(J+K)(J-K+1)]^{1/2} \tag{15.43}$$

$$(K+1,\ J,\ M_J|J_{y'}|K,\ J,\ M_J)=\frac{i}{2}[(J-K)(J+K+1)]^{1/2} \tag{15.44}$$

The phase choice in this section is consistent with both J_x and $J_{x'}$ as real and positive which is consistent with Condon and Shortley [47], but is opposite to the choice used throughout other portions of this text. In application of the techniques discussed in this chapter it is important to use a consistent phase choice throughout.

2 SINGLE COUPLING NUCLEUS: REDUCED MATRIX ELEMENTS

By way of introducing the definitions of the reduced matrix elements for the quadrupole and magnetic interaction, we consider first the case of one coupling nucleus. This case has already been discussed in Chapter IX, Section 4; however, we formulate the problem here in terms of irreducible tensors. The quadrupole and magnetic dipole interactions may be developed in terms of irreducible tensor operators [5, 8]. Since the Cartesian tensor for the quadrupole interaction is symmetric and traceless, only second-rank irreducible tensors are needed to characterize this interaction.

$$\mathscr{H}_{hfs}=\mathbf{V}^{(2)}\cdot\mathbf{Q}^{(2)}+\mathbf{m}^{(1)}\cdot\boldsymbol{\mu}^{(1)} \tag{15.45}$$

where $\mathbf{V}^{(2)}$ is the electric field gradient tensor, $\mathbf{Q}^{(2)}$ the nuclear quadrupole tensor, $\mathbf{m}^{(1)}$ the magnetic field tensor, and $\boldsymbol{\mu}^{(1)}$ the nuclear magnetic dipole tensor. The basis is taken as $|\tau JIFM_F)$ where τ has the usual meaning for an asymmetric rotor and equals K for a symmetric top. From (15.21) for the matrix elements of a scalar product, we find for the quadrupole interaction

$$(\tau' J' IF|\mathbf{V}^{(2)}\cdot\mathbf{Q}^{(2)}|\tau JIF)=(-1)^{J+I+F}\begin{Bmatrix} F & I & J' \\ 2 & J & I \end{Bmatrix}(\tau' J'\|V^{(2)}\|\tau J)(I\|Q\|I) \tag{15.46}$$

All terms off-diagonal in F vanish, and the matrix elements are diagonal and independent of the projection quantum numbers M_F.

In order for the preceding relation to be useful, the reduced matrix element must be evaluated in terms of the usual spectroscopic constants. This is easily accomplished by application of the Wigner-Eckart theorem. The quantum mechanical observable, Q, termed the nuclear quadrupole moment, is defined via the relation

$$(I, M_I=I|Q_0^{(2)}|I, M_I=I)=\tfrac{1}{2}(eQ) \tag{15.47}$$

where $|I, M_I)$ are the nuclear spin basis functions and where $M_I=I$ corresponds to the nuclear state of maximum alignment along z. Application of the Wigner-Eckart theorem, (15.20) to (15.47) gives therefore

$$\tfrac{1}{2}(eQ)=(II|Q_0^{(2)}|II)=\begin{pmatrix} I & 2 & I \\ -I & 0 & I \end{pmatrix}(I\|Q^{(2)}\|I) \tag{15.48}$$

Expressing the $3j$ symbol in terms of its algebraic equivalent from Table 15.1, we find for the reduced matrix element of the nuclear quadrupole tensor

$$(I\|Q^{(2)}\|I)=eQf(I) \tag{15.49}$$

where

$$f(I)=[(2I+1)(2I+2)(2I+3)/8I(2I-1)]^{1/2} \tag{15.50}$$

The reduced matrix element of the field gradient tensor $\mathbf{V}^{(2)}$ is evaluated in a similar manner. The electric field gradient coupling constant $q_{J'J}$ is defined as ($J' \geqslant J$)

$$q_{J'J} = (\tau', J', M'_J = J|(\partial^2 V/\partial z^2)_0|\tau, J, M_J = J) \tag{15.51}$$

and

$$(\tau', J', M'_J = J|V_0^{(2)}|\tau, J, M_J = J) = \tfrac{1}{2} q_{J'J} \tag{15.52}$$

where $|\tau J M_J)$ are the rigid rotor basis functions (e.g., the symmetric rotor or asymmetric rotor functions) with M_J the projection quantum number in the space-fixed frame. It should be noted that in general $q_{J'J}$ depends on the quantum number τ. The Wigner-Eckart theorem yields

$$(\tau' J' J|V_0^{(2)}|\tau J J) = (-1)^{J'-J} \begin{pmatrix} J' & 2 & J \\ -J & 0 & J \end{pmatrix} (\tau' J' \| V^{(2)} \| \tau J) \tag{15.53}$$

Thus

$$(\tau' J' \| V^{(2)} \| \tau J) = q_{J'J} f(J') \tag{15.54}$$

where for $J' \geqslant J$

$$f(J') = \left[2 \begin{pmatrix} J & 2 & J' \\ J & 0 & -J \end{pmatrix} \right]^{-1} \tag{15.55}$$

or explicitly

$$f(J') = \left[\frac{(2J+1)(2J+2)(2J+3)}{8J(2J-1)} \right]^{1/2} \qquad J' = J,$$

$$f(J') = \left[\frac{(2J+2)(2J+3)(2J+4)}{48J} \right]^{1/2} \qquad J' = J+1,$$

$$f(J') = -\left[\frac{2J(2J+1)(2J+2)}{48(J-1)} \right]^{1/2} \qquad J' = J-1,$$

$$f(J') = \left[\frac{(2J+3)(2J+4)(2J+5)}{48} \right]^{1/2} \qquad J' = J+2,$$

$$f(J') = \left[\frac{2J(2J+1)(2J-1)}{48} \right]^{1/2} \qquad J' = J-2 \tag{15.56}$$

where for completeness we have also included the case where $J' < J$ for which $M'_J = M_J = J'$ in the definition of $q_{J'J}$.

In application to specific molecules the quantity $q_{J'J}$, which is defined with respect to axes fixed in space, is conveniently re-expressed in terms of the principal inertial axis system (x', y', z') fixed in the molecule, since as previously pointed out, the field gradients with respect to the molecule-fixed axis system are constants independent of the rotational state. We have from the definition of $q_{J'J}$ and (15.2)

$$\begin{aligned} q_{J'J} &= 2(\tau' J' J | V_0^{(2)} | \tau J J) \\ &= 2 \sum_{q'} (-1)^{-q'} V_{q'}'^{(2)} (\tau' J' J | D_{0,-q'}^{(2)} | \tau J J) \end{aligned} \tag{15.57}$$

In the symmetric-top basis we may write from (15.38)

$$\begin{aligned} q_{J'J} &= (-1)^{J-K-q'} [(2J+1)(2J'+1)]^{1/2} \\ &\quad \times \begin{pmatrix} J & 2 & J' \\ J & 0 & -J \end{pmatrix} \begin{pmatrix} J & 2 & J' \\ K & q' & -K' \end{pmatrix} 2V_{q'}'^{(2)} \\ &= (-1)^{J-K-q'} [(2J+1)(2J'+1)]^{1/2} [2f(J')]^{-1} \\ &\quad \times \begin{pmatrix} J & 2 & J' \\ K & q' & -K' \end{pmatrix} 2V_{q'}'^{(2)} \end{aligned} \tag{15.58}$$

where $q' = K' - K$, $J' \geqslant J$. Note that the latter expression for $q_{J'J}$ also applies to the case $J' < J$ with $f(J')$ as given in (15.56). Usually, the interaction energy is defined as $\mathbf{V}^{(2)} \cdot \mathbf{Q}^{(2)}$ in spherical tensor notation and as $\frac{1}{6}\mathbf{V}:\mathbf{Q}$ in the Cartesian notation. For both expressions to be equivalent, a factor of $6^{-1/2}$ must be included in the definition of the components of $\mathbf{Q}^{(2)}$ and $\mathbf{V}^{(2)}$. The irreducible field gradient tensor components $V_{q'}'^{(2)}$ are thus related to the Cartesian field gradient tensor elements as follows

$$\begin{aligned} 2V_0'^{(2)} &= V_{z'z'} \\ 2V_{\pm 1}'^{(2)} &= \mp(\sqrt{2}/\sqrt{3})(V_{x'z'} \pm iV_{y'z'}) \\ 2V_{\pm 2}'^{(2)} &= 6^{-1/2}(V_{x'x'} - V_{y'y'} \pm 2iV_{x'y'}) \end{aligned} \tag{15.59}$$

In the usual spectroscopic notation, the molecular constants $eQV_{x'x'}$, and so on, are the quadrupole coupling constants denoted by $\chi_{x'x'}$, and so on.

For a symmetric top with the coupling atom on the symmetry axis, only one of the $V_{q'}'^{(2)}$ is nonvanishing, viz., $2V_0'^{(2)} = V_{z'z'} = (\partial^2 V/\partial z'^2) \equiv q$, and the coupling constant for the rotational state J, K is

$$\begin{aligned} q_{JJ} &= (-1)^{J-K}(2J+1) \begin{pmatrix} J & 2 & J \\ K & 0 & -K \end{pmatrix} \begin{pmatrix} J & 2 & J \\ J & 0 & -J \end{pmatrix} q \\ &= \frac{qJ}{(2J+3)} \left[\frac{3K^2}{J(J+1)} - 1 \right] \end{aligned} \tag{15.60}$$

which is the usual definition of q_{JJ} where only diagonal elements in J are considered, see (9.88). The above is also applicable to a linear rotor where $K=0$, and this gives (9.81). For an asymmetric rotor if only diagonal elements in J are to be considered, it is more convenient to return to the definition of q_{JJ} and evaluate it in the asymmetric rotor basis as was done in Chapter IX, Section 4. Otherwise, the symmetric top basis may be conveniently employed, and (15.58) applies. Alternately, (15.57) may be evaluated in the asymmetric rotor basis [see also (9.69)].

Usually the interaction energy is sufficiently small that the rotational quantum numbers may be considered good quantum numbers, and only matrix elements diagonal in J need be considered. This so-called first-order effect is by far the most important although for quadrupole coupling with nuclei of large Q second-order effects can be important, as discussed in Chapter IX, Section 6. In what follows, only the matrix elements diagonal in J will be considered for the magnetic coupling, while for quadrupole coupling general expressions will be given.

For the magnetic spin–rotation interaction we find from (15.21)

$$(\tau JIF|\mathbf{m}^{(1)}\cdot\boldsymbol{\mu}^{(1)}|\tau JIF)=(-1)^{J+I+F}\begin{Bmatrix}F & I & J\\ 1 & J & I\end{Bmatrix}(\tau J\|m^{(1)}\|\tau J)(I\|\mu^{(1)}\|I) \tag{15.61}$$

We now consider the evaluation of the reduced matrix element. Conventionally the observable,

$$(I,\, M_I=I|\mu_z|I,\, M_I=I)$$

is defined as the magnetic moment, $g_I\beta_I I$, and since $\mu_z\equiv\mu_0^{(1)}$ we have

$$(I,\, M_I=I|\mu_0^{(1)}|I,\, M_I=I)=\mu_I=g_I\beta_I I \tag{15.62}$$

where β_I is the nuclear magneton and g_I is the appropriate nuclear g factor. For the molecular magnetic field tensor we may employ the following decomposition theorem for first rank tensors [1]

$$(\tau' J'M'|T_q^{(1)}|\tau JM)=\delta_{J'J}\frac{(\tau' J'M'|J_q^{(1)}(\mathbf{J}^{(1)}\cdot\mathbf{T}^{(1)})|\tau JM)}{J(J+1)} \tag{15.63}$$

which relates the matrix elements of $T_q^{(1)}$ to those of the projection of $\mathbf{T}^{(1)}$ along the angular momentum $\mathbf{J}^{(1)}$. Applying this to $m_0^{(1)}$ gives

$$\begin{aligned}(\tau JJ|m_0^{(1)}|\tau JJ)&=(\tau JJ|J_0^{(1)}(\mathbf{J}^{(1)}\cdot\mathbf{m}^{(1)})|\tau JJ)/J(J+1)\\ &=\frac{(JJ|J_0^{(1)}|JJ)}{[J(J+1)]^{1/2}}\frac{(\tau JJ|\mathbf{J}^{(1)}\cdot\mathbf{m}^{(1)}|\tau JJ)}{[J(J+1)]^{1/2}}\end{aligned} \tag{15.64}$$

The latter factor may be interpreted as the average magnetic field along $\mathbf{J}$

$$\langle H_J\rangle=(\tau JJ|\mathbf{J}^{(1)}\cdot\mathbf{m}^{(1)}|\tau JJ)/[J(J+1)]^{1/2} \tag{15.65}$$

and we have

$$(\tau JM_J=J|m_0^{(1)}|\tau JM_J=J)=\{J/[J(J+1)]^{1/2}\}\langle H_J\rangle \tag{15.66}$$

It is recognized that the left-hand side of this equation is the average field in the z direction, and the above is a relation that is readily derived from the vector model (see Chapter IX, Section 7). The reduced matrix elements can now be found by application of the Wigner-Eckart theorem to (15.62) and (15.66) which yields

$$(I\|\mu^{(1)}\|I)=g_I\beta_I h(I) \tag{15.67}$$

where

$$h(I)=[I(I+1)(2I+1)]^{1/2} \tag{15.68}$$

and similarly,

$$(\tau J\|m^{(1)}\|\tau J)=(C_{J,\tau}/g_I\beta_I)h(J) \tag{15.69}$$

where

$$h(J)=[J(J+1)(2J+1)]^{1/2} \tag{15.70}$$

and the spectroscopic constant is defined by

$$C_{J,\tau}=\frac{g_I\beta_I\langle H_J\rangle}{[J(J+1)]^{1/2}} \tag{15.71}$$

Compare this with (9.137) and (9.142). The $C_{J,\tau}$ depends on the rotational state J, τ and is best expressed in terms of the elements of the magnetic coupling tensor with respect to the principal inertial axis system. The derivation now follows closely that discussed in Chapter IX, Section 7. Referring the elements of $\mathbf{J}^{(1)}$ and $\mathbf{m}^{(1)}$ to the molecule-fixed $x'y'z'$ frame, we can write the scalar product in terms of the Cartesian components of the tensors

$$\begin{aligned}\mathbf{J}^{(1)}\cdot\mathbf{m}^{(1)}&=\sum_{q'}(-1)^{q'}J'^{(1)}_{q'}m'^{(1)}_{-q'}\\&=\sum_{g}J'_g m'_g \qquad (g=x', y', z')\end{aligned} \tag{15.72}$$

For molecules in singlet electronic states the magnetic field arises from the rotation of the molecule and the field components along the molecule-fixed axes may be taken as proportional to the angular velocity, that is,

$$\begin{aligned}m'_g&=(g_I\beta_I)^{-1}\sum_{g'}B_{gg'}\omega'_{g'}\\&=(g_I\beta_I)^{-1}\sum_{g'}C'_{gg'}J'_{g'}\end{aligned} \tag{15.73}$$

since $\omega'_g=J'_g/I_g$. Combining this with (15.72) and (15.65) we have (noting that, in general, only the diagonal elements of $\mathbf{C}$ contribute to the spectrum in first order)

$$C_{J,\tau}=[J(J+1)]^{-1}\sum_{g}C'_{gg}\langle J'^2_g\rangle \tag{15.74}$$

where J'_g are the angular momentum components with respect to the molecule-fixed principal inertial axis system and where the average is taken over the rotational state J, τ. The $C'_{gg'}$ are the coupling constants with respect to the principal axes and are constants independent of the rotational state. This relation agrees with that given previously in (9.143). The quantity $C_{gg}I_g/g_I=B_{gg}$ is invariant under isotopic substitution.

It is of interest to consider the evaluation of the matrix elements of $\mathscr{H}_M$ when defined by (15.18). If for simplicity we consider one coupling nuclei, the matrix elements (diagonal in J) of the terms in (15.19) are from (15.21):

$$\begin{aligned}(\tau'JIF|[\mathbf{C}^{(l)}\times\mathbf{J}^{(1)}]^{(1)}\cdot\mathbf{I}^{(1)}|\tau JIF)&=(-1)^{J+I+F}\begin{Bmatrix}F & I & J\\ 1 & J & I\end{Bmatrix}\\&\quad\times(\tau'J\|[\mathbf{C}^{(l)}\times\mathbf{J}^{(1)}]^{(1)}\|\tau J)(I\|I^{(1)}\|I)\end{aligned} \tag{15.75}$$

The reduced matrix element $(I\|I^{(1)}\|I)=[I(I+l)(2I+l)]^{1/2}=h(I)$ is readily obtained from the Wigner-Eckart theorem. The reduced matrix element of the tensor product which involves operators acting on the same subspace is given by (15.25)

$$(\tau' J\|[\mathbf{C}^{(l)}\times\mathbf{J}^{(1)}]^{(1)}\|\tau J)=-\sqrt{3}\begin{Bmatrix} l & 1 & 1 \\ J & J & J \end{Bmatrix}(\tau' J\|C^{(l)}\|\tau J)(\tau J\|J^{(1)}\|\tau J) \qquad (15.76)$$

As before

$$\begin{aligned}(\tau' J'\|J^{(1)}\|\tau J)&=\delta_{\tau'\tau}\delta_{J',J}[J(J+1)(2J+1)]^{1/2}\\ &=h(J)\end{aligned} \qquad (15.77)$$

Collecting terms we have finally

$$(\tau' JIF|\mathscr{H}_M|\tau JIF)=(-1)^{J+I+F}h(I)h(J)\begin{Bmatrix} F & I & J \\ 1 & J & I \end{Bmatrix} C_J \qquad (15.78)$$

This may be compared to (15.85). Here

$$C_J=-\sqrt{3}\sum_l N_l\begin{Bmatrix} l & 1 & 1 \\ J & J & J \end{Bmatrix}(\tau' J\|C^{(l)}\|\tau J) \qquad (15.79)$$

The $(\tau' J\|C^{(l)}\|\tau J)$ are the spectroscopic constants to be evaluated. For a symmetric rotor basis $(\tau=K)$ we have

$$\begin{aligned}C_J=&-\sqrt{3}(2J+1)\sum_l(-1)^{J+l-K-q'}\\ &\times\begin{Bmatrix} l & 1 & 1 \\ J & J & J \end{Bmatrix}\begin{pmatrix} J & l & J \\ K & q' & -K' \end{pmatrix}(N_l C_{q'}^{\prime(l)})\end{aligned} \qquad (15.80)$$

where $q'=K'-K$ and $C_{q'}^{\prime(l)}$ are the components with respect to the molecule-fixed axes and are defined as in (15.17). These are the molecular constants independent of the rotational state. Equation 15.80 is obtained by use of (15.20) to rewrite the reduced matrix element in terms of an ordinary matrix element, followed by transformation of the $\mathbf{C}^{(l)}$ to principal axes by (15.1), and finally evaluation of the matrix elements in the JKM basis by (15.38). For a symmetric top with coupling atom on the symmetry axis (z') the off-diagonal elements $C_{x'y'}$, and so on, vanish and $C_{x'x'}=C_{y'y'}$. Thus the nonvanishing elements are

$$N_0 C_0^{\prime(0)}=\tfrac{1}{3}(C_{z'z'}+2C_{x'x'}) \qquad (15.81)$$

$$N_2 C_0^{\prime(2)}=-[(10)^{1/2}/3](C_{z'z'}-C_{x'x'}) \qquad (15.82)$$

and hence in (15.80) $q'=0$, $K'=K$. Evaluating (15.80) we find

$$C_J=C_{x'x'}+(C_{z'z'}-C_{x'x'})[K^2/J(J+1)] \qquad (15.83)$$

which is equivalent to (9.158).

If the previous definition of the reduced matrix elements (15.49) and (15.54) are substituted into (15.46), the matrix element for the electric quadrupole interaction is

$$(\tau' J' IF|\mathbf{V}^{(2)}\cdot\mathbf{Q}^{(2)}|\tau JIF)=(-1)^t(eQq_{J'J})f(J')f(I)\begin{Bmatrix} F & I & J' \\ 2 & J & I \end{Bmatrix} \tag{15.84}$$

where $t=J+I+F$, and where $(eQq_{J'J})$ is the quadrupole coupling constant for the rotational state $J\tau$. The quantities $f(J')$ and $f(I)$ are specified by (15.56) and (15.50), respectively. For the magnetic interaction (15.61), (15.67), and (15.69) yield

$$(\tau JIF|\mathbf{m}^{(1)}\cdot\mathbf{\mu}^{(1)}|\tau JIF)=(-1)^r(C_{J,\tau})h(J)h(I)\begin{Bmatrix} F & I & J \\ 1 & J & I \end{Bmatrix} \tag{15.85}$$

where $r=J+I+F$, and $h(J)$ and $h(I)$ are specified by (15.70) and (15.68), respectively. Equations (15.84) and (15.85) yield the usual results for the energy levels $(J=J')$ with replacement of the $6j$ symbol by its algebraic equivalent. See Table 15.1. In general, once a particular value of J, I and F is specified, the $6j$ symbol may be evaluated and hence the corresponding matrix element. This form is particularly convenient because of the ease with which the $6j$ and $3j$ symbols may be evaluated by machine computation.

3 PLURAL, UNEQUAL COUPLING

We now consider explicitly the derivation of the matrix elements for the case of plural nuclear coupling. For N nuclei with spins $I_1, I_2, \ldots I_L, \ldots I_N$ and different coupling energies, it is convenient to employ the coupling scheme

$$\begin{aligned} \mathbf{J}+\mathbf{I}_1&=\mathbf{F}_1 \\ \mathbf{F}_1+\mathbf{I}_2&=\mathbf{F}_2 \\ &\vdots \\ \mathbf{F}_{L-1}+\mathbf{I}_L&=\mathbf{F}_L \\ &\vdots \\ \mathbf{F}_{N-1}+\mathbf{I}_N&=\mathbf{F}_N\equiv\mathbf{F} \end{aligned} \tag{15.86}$$

where $\mathbf{J}$ is the rotational angular momentum and where the nuclear spins are coupled in order of decreasing coupling energy. The basis functions for this coupling scheme are denoted by

$$|\tau JI_1F_1\ldots I_LF_L\ldots I_NF_NM_F)$$

where τ represents any additional quantum numbers, besides the rotational momentum quantum numbers J and M_J, that are necessary to specify the unperturbed rotational state. To be specific, we consider the case of three coupling nuclei. The Hamiltonian is

$$\mathcal{H}_{hfs}=\sum_{i=1}^{3}\mathbf{V}(i)\cdot\mathbf{Q}(i)+\sum_{i=1}^{3}\mathbf{m}(i)\cdot\mathbf{\mu}(i) \tag{15.87}$$

Hereafter the explicit rank designation will be omitted. As indicated by the foregoing equation, the matrix elements of $\mathscr{H}_{hfs}$ will be composed of a sum of contributions from nucleus 1, 2, and 3.

We wish to show first that the electric or magnetic interaction for the Lth nucleus does not depend on the quantum numbers $I_{L+1}F_{L+1}\ldots F_N$ for the coupling scheme of (15.86). For convenience we consider the case of two nuclei. For nucleus 1 we have the following matrix element to evaluate:

$$(\tau' J' I_1 F_1' I_2 F' M_F'|\mathscr{H}(1)|\tau J I_1 F_1 I_2 F M_F)$$

where $\mathscr{H}(1)$ designates, for example, the electric quadrupole or magnetic dipole interaction for nucleus 1. Applying the Wigner-Eckart theorem we have

$$(\tau' J' I_1 F_1' I_2 F' M_F'|\mathscr{H}(1)|\tau J I_1 F_1 I_2 F M_F)=\delta_{M_F M_{F'}}\delta_{FF'}(2F+1)^{-1/2} \times(\tau' J' I_1 F_1' I_2 F'\|\mathscr{H}(1)\|\tau J I_1 F_1 I_2 F) \quad (15.88)$$

where the $3j$ symbol has been evaluated noting that $\mathscr{H}(1)$ is a scalar and hence $k=q=0$. The reduced matrix element may be reduced by use of (15.22) since $\mathscr{H}(1)$ does not operate on I_2:

$$\begin{aligned}&(\tau' J' I_1 F_1' I_2 F\|\mathscr{H}(1)\|\tau J I_1 F_1 I_2 F)\\&=\delta_{F_1F_1'}(2F+1)[(2F+1)(2F_1+1)]^{-1/2}(\tau' J' I_1 F_1'\|\mathscr{H}(1)\|\tau J I_1 F_1)\\&=\delta_{F_1F_1'}\delta_{M_{F_1}M_{F_1'}}(2F+1)^{1/2}(\tau' J' I_1 F_1' M_{F_1'}|\mathscr{H}(1)|J I_1 F_1 M_{F_1})\end{aligned} \quad (15.89)$$

where we again employ the Wigner-Eckart theorem. Therefore we may write

$$(\tau' J' I_1 F_1' I_2 F' M_F'|\mathscr{H}(1)|\tau J I_1 F_1 I_2 F M_F)=\delta_{M_F M_F'}\delta_{FF'}\delta_{F_1F_1'}\delta_{M_{F_1}M_{F_1}'} \times(\tau' J' I_1 F_1' M_{F_1}'|\mathscr{H}(1)|\tau J I_1 F_1 M_{F_1}) \quad (15.90)$$

which shows the matrix elements of $\mathscr{H}(1)$ depend only on the quantum numbers $\tau J I_1 F_1$. In general, the matrix elements of $\mathscr{H}(L)$ are diagonal in all quantum numbers $F_L I_{L+1}\ldots F_N$ and depend only on the quantum numbers $\tau J I_1 F\ldots F_L$. Thus for nucleus 1 we have already discussed the matrix elements; from (15.84) with the appropriate change in notation we have

$$(\tau' J' I_1 F_1|\mathbf{V}(1)\cdot\mathbf{Q}(1)|\tau J I_1 F_1)=(-1)^{t_1}(eQq_{J'J})_1 f(J')f(I_1)\begin{Bmatrix}F_1 & I_1 & J'\\ 2 & J & I_1\end{Bmatrix} \quad (15.91)$$

where $t_1=J+I_1+F_1$, and where $(eQq_{J'J})_1$ refers to nucleus 1. The quantities $f(J')$ and $f(I_1)$ are specified by (15.56) and (15.50), respectively.

For the quadrupole coupling of nucleus 2, (15.21) yields, with $J\equiv F_2, j_2\equiv I_2$, $j_1\equiv F_1$, and $\tau_1\equiv\tau J I_1$,

$$\begin{aligned}&(\tau' J' I_1 F_1' I_2 F_2|\mathbf{V}(2)\cdot\mathbf{Q}(2)|\tau J I_1 F_1 I_2 F_2)\\&=(-1)^{F_1+I_2+F_2}\begin{Bmatrix}F_2 & I_2 & F_1'\\ 2 & F_1 & I_2\end{Bmatrix}(\tau' J' I_1 F_1'\|V(2)\|\tau J I_1 F_1)(I_2\|Q(2)\|I_2)\end{aligned} \quad (15.92)$$

The reduced matrix element of the field gradient of nucleus 2 may be reduced to the desired form by application of (15.22) since $\mathbf{V}(2)$ commutes with $\mathbf{I}_1$. This gives

$$(\tau' J' I_1 F_1' \| V(2) \| \tau J I_1 F_1) = (-1)^{J'+I_1+F_1}[(2F_1+1)(2F_1'+1)]^{1/2} \times \begin{Bmatrix} J' & F_1' & I_1 \\ F_1 & J & 2 \end{Bmatrix} (\tau' J' \| V(2) \| \tau J) \qquad (15.93)$$

The contribution that the rank (k) makes to the phase factor in (15.22) or (15.23) can obviously be omitted when the rank is even. The reduced matrix element on the right-hand side of this equation is given by (15.54). Therefore, taking cognizance of (15.49), (15.54), and (15.93), we may write for (15.92) for nucleus 2

$$(\tau' J' I_1 F_1' I_2 F_2 | \mathbf{V}(2) \cdot \mathbf{Q}(2) | \tau J I_1 F_1 I_2 F_2)$$
$$= (-1)^{t_2} (eQq_{J'J})_2 f(J') f(I_2) [(2F_1+1)(2F_1'+1)]^{1/2} \begin{Bmatrix} J' & F_1' & I_1 \\ F_1 & J & 2 \end{Bmatrix} \begin{Bmatrix} F_2 & I_2 & F_1' \\ 2 & F_1 & I_2 \end{Bmatrix} \qquad (15.94)$$

where $t_2 = J' + I_1 + I_2 + 2F_1 + F_2$.

If there were only two coupling nuclei involved, one would stop here, and F_2 would be replaced by F.

For the quadrupole coupling of nucleus 3, (15.21) yields

$$(\tau' J' I_1 F_1' I_2 F_2' I_3 F_3 | \mathbf{V}(3) \cdot \mathbf{Q}(3) | \tau J I_1 F_1 I_2 F_2 I_3 F_3)$$
$$= (-1)^{F_2+I_3+F_3} \begin{Bmatrix} F_3 & I_3 & F_2' \\ 2 & F_2 & I_3 \end{Bmatrix} (\tau' J' I_1 F_1' I_2 F_2' \| V(3) \| \tau J I_1 F_1 I_2 F_2)(I_3 \| Q(3) \| I_3) \qquad (15.95)$$

Successive application of (15.22) yields for the reduced matrix elements

$$(\tau' J' I_1 F_1' I_2 F_2' \| V(3) \| \tau J I_1 F_1 I_2 F_2) = (-1)^{F_1'+I_2+F_2}$$
$$\times [(2F_2+1)(2F_2'+1)]^{1/2} \begin{Bmatrix} F_1' & F_2' & I_2 \\ F_2 & F_1 & 2 \end{Bmatrix} (\tau' J' I_1 F_1' \| V(3) \| \tau J I_1 F_1) \qquad (15.96)$$

and

$$(\tau' J' I_1 F_1' \| V(3) \| \tau J I_1 F_1) = (-1)^{J'+I_1+F_1}[(2F_1+1)(2F_1'+1)]^{1/2} \times \begin{Bmatrix} J' & F_1' & I_1 \\ F_1 & J & 2 \end{Bmatrix} (\tau' J' \| V(3) \| \tau J) \qquad (15.97)$$

Combination of (15.95)–(15.97) and the definitions of the reduced matrix elements gives for the matrix elements

$$(\tau' J' I_1 F_1' I_2 F_2' I_3 F_3 | \mathbf{V}(3) \cdot \mathbf{Q}(3) | \tau J I_1 F_1 I_2 F_2 I_3 F_3)$$
$$= (-1)^{t_3} (eQq_{J'J})_3 f(J') f(I_3) [(2F_1+1)(2F_1'+1)(2F_2+1)(2F_2'+1)]^{1/2}$$
$$\times \begin{Bmatrix} J' & F_1' & I_1 \\ F_1 & J & 2 \end{Bmatrix} \begin{Bmatrix} F_1' & F_2' & I_2 \\ F_2 & F_1 & 2 \end{Bmatrix} \begin{Bmatrix} F_3 & I_3 & F_2' \\ 2 & F_2 & I_3 \end{Bmatrix} \qquad (15.98)$$

where $t_3 = J' + I_1 + I_2 + I_3 + F_1 + F_1' + 2F_2 + F_3$.

A similar procedure for the magnetic dipole interaction operator yields the following matrix elements for the three coupling nuclei.

$$(\tau J I_1 F_1|\mathbf{m}(1)\cdot \boldsymbol{\mu}(1)|\tau J I_1 F_1)=(-1)^{r_1}(C_{J,\tau})_1 h(J)h(I_1)\begin{Bmatrix} F_1 & I_1 & J \\ 1 & J & I_1 \end{Bmatrix} \tag{15.99}$$

$$(\tau J I_1 F_1' I_2 F_2|\mathbf{m}(2)\cdot \boldsymbol{\mu}(2)|\tau J I_1 F_1 I_2 F_2)=(-1)^{r_2}(C_{J,\tau})_2$$

$$\times h(J)h(I_2)[(2F_1+1)(2F_1'+1)]^{1/2}\begin{Bmatrix} J & F_1' & I_1 \\ F_1 & J & 1 \end{Bmatrix}\begin{Bmatrix} F_2 & I_2 & F_1' \\ 1 & F_1 & I_2 \end{Bmatrix} \tag{15.100}$$

$$(\tau J I_1 F_1' I_2 F_2' I_3 F_3|\mathbf{m}(3)\cdot \boldsymbol{\mu}(3)|\tau J I_1 F_1 I_2 F_2 I_3 F_3)=(-1)^{r_3}(C_{J,\tau})_3$$

$$\times h(J)h(I_3)[(2F_1+1)(2F_1'+1)(2F_2+1)(2F_2'+1)]^{1/2}$$

$$\times\begin{Bmatrix} J & F_1' & I_1 \\ F_1 & J & 1 \end{Bmatrix}\begin{Bmatrix} F_1' & F_2' & I_2 \\ F_2 & F_1 & 1 \end{Bmatrix}\begin{Bmatrix} F_3 & I_3 & F_2' \\ 1 & F_2 & I_3 \end{Bmatrix} \tag{15.101}$$

where $r_1=J+I_1+F_1$, $r_2=J+I_1+I_2+2F_1+F_2+1$, $r_3=J+I_1+I_2+I_3+F_1+F_1'+2F_2+F_3$, and where $h(J)$ and $h(I_i)$ are specified by (15.70) and (15.68), respectively. If two coupling nuclei are present, (15.99) and (15.100) apply, with F_2 replaced by F.

The procedure just outlined may obviously be extended to any number of nuclei. The matrix elements of the electric quadrupole interaction for the Lth nucleus may be written in the form [5]

$$(\tau' F_0' F_1' \ldots F_{L-1}' F_L|\mathbf{V}(L)\cdot \mathbf{Q}(L)|\tau F_0 F_1 \ldots F_{L-1} F_L)$$

$$=(-1)^t(eQq_{J'J})_L f(J')f(I_L)\begin{Bmatrix} F_L & I_L & F_{L-1}' \\ 2 & F_{L-1} & I_L \end{Bmatrix}$$

$$\times\prod_{i=1}^{L-1}[(2F_i+1)(2F_i'+1)]^{1/2}\begin{Bmatrix} F_{i-1}' & F_i' & I_i \\ F_i & F_{i-1} & 2 \end{Bmatrix} \tag{15.102}$$

where the phase factor is determined by

$$t=\sum_{i=1}^{L-1}(F_{i-1}'+I_i+F_i)+(F_{L-1}+I_L+F_L)$$

and $F_0=J$, $F_0'=J'$, and $F_N=F$. As before, $f(J')$ is given by (15.56) where the expression to be used depends on the value of J'. Likewise, $f(I_L)$ is given by (15.50) where the spin of the Lth nucleus is to be substituted.

The matrix elements of the magnetic dipole interaction may be written as [5]

$$(\tau F_0' F_1' \ldots F_{L-1}' F_L|\mathbf{m}(L)\cdot \boldsymbol{\mu}(L)|\tau F_0 F_1 \ldots F_{L-1} F_L)$$

$$=(-1)^r(C_{J,\tau})_L h(J)h(I_L)\begin{Bmatrix} F_L & I_L & F_{L-1}' \\ 1 & F_{L-1} & I_L \end{Bmatrix}$$

$$\times\prod_{i=1}^{L-1}[(2F_i+1)(2F_i'+1)]^{1/2}\begin{Bmatrix} F_{i-1}' & F_i' & I_i \\ F_i & F_{i-1} & 1 \end{Bmatrix} \tag{15.103}$$

where the phase factor is determined by

$$r=(L-1)+\sum_{i=1}^{L+1}(F'_{i-1}+I_i+F_i)+(F_{L-1}+I_L+F_L)$$

and $F_0=F'_0=J$, and $F_N=F$. The factors $h(J)$ and $h(I_L)$ are given, respectively, by (15.70) and (15.68). For both (15.102) and (15.103) in the special case of $L=1$, the product $\prod[\,]^{1/2}\{\,\}$ is set equal to one. The diagonal elements yield the hyperfine energy levels to a first-order approximation. Exact energy levels are obtained by construction of the hyperfine energy matrix, which may be factored into blocks of different F, and subsequent diagonalization. For asymmetric rotors if the symmetric rotor basis is employed, the complete energy matrix of $\mathscr{H}_r+\mathscr{H}_{hfs}$ must be considered. This amounts to inclusion of the matrix elements of the rotational Hamiltonian in the symmetric-top basis along with those given for the hyperfine Hamiltonian in the coupled basis. A consistent phase choice for the matrix elements should be employed.

In this regard, the rigid rotor Hamiltonian can be conveniently written as

$$\mathscr{H}_r=\tfrac{1}{2}(B_{x'}+B_{y'})J(J+1)+[B_{z'}-\tfrac{1}{2}(B_{x'}+B_{y'})]\{J_0'^2-b(J'^2_{+1}+J'^2_{-1})\} \qquad (15.104)$$

with $b=(B_{y'}-B_{x'})/(2B_{z'}-B_{x'}-B_{y'})$.

The study of DC≡CCl by Tack and Kukolich [41] provides a simple example of two nuclei with unequal couplings $\mathbf{J}+\mathbf{I}_{\mathrm{Cl}}=\mathbf{F}_1$, $\mathbf{F}_1+\mathbf{I}_{\mathrm{D}}=\mathbf{F}$. All the hyperfine components of the $J=0\to1$ transition are shown in Fig. 15.1. The

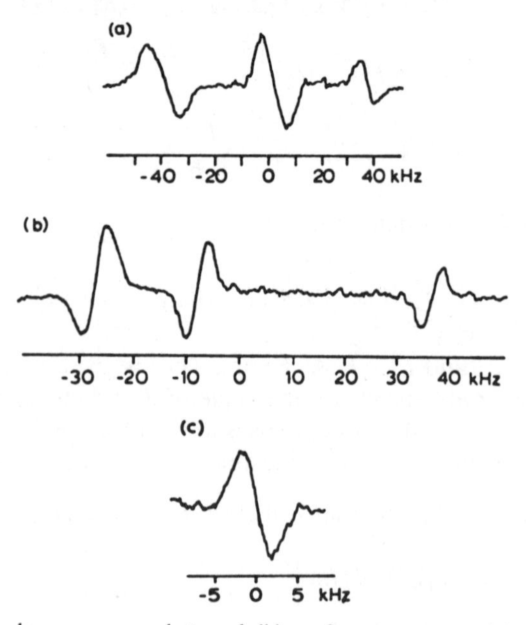

Fig. 15.1 Molecular beam maser resolution of all hyperfine components of the $1\to0$ transitions of ^{35}ClC≡CD. The components are labeled for the $J=1$ level since all hyperfine matrix elements vanish for $J=0$ to first order. (*a*) Recorder tracing of $F_1=\frac{3}{2}$ components. Frequencies in kHz relative to 10,358,017 kHz. (*b*) $F_1=\frac{5}{2}$ components. Frequencies relative to 10,377,921 kHz. (*c*) $F_1=\frac{1}{2}$ components. Frequencies relative to 10,393,892 kHz. From Tack and Kukolich [41].

spectrum was analyzed with a hyperfine Hamiltonian consisting of the deuterium quadrupole interaction and chlorine quadrupole and ^{35}Cl spin–rotation interaction. Second-order quadrupole corrections were calculated for chlorine by perturbation methods (see, e.g., Appendix J) and added to the first-order energies from matrix diagonalization. The coupling constants obtained are: $eQq(^{35}\text{Cl})=79{,}739.5(10)$ kHz, $eQq(\text{D})=202.5(15)$ kHz, and $C_I(^{35}\text{Cl})=1.3(1)$ kHz.

4 PLURAL, EQUAL, OR NEARLY EQUAL, COUPLING

When the coupling energies are the same or nearly the same, the coupling scheme

$$\begin{gathered}\mathbf{I}_1+\mathbf{I}_2=\mathscr{I}_2\\ \mathscr{I}_2+\mathbf{I}_3=\mathscr{I}_3\\ \vdots\\ \mathscr{I}_{L-1}+\mathbf{I}_L=\mathscr{I}_L\\ \vdots\\ \mathscr{I}_{N-1}+\mathbf{I}_N=\mathscr{I}_N\equiv\mathscr{I}\\ \mathbf{J}+\mathscr{I}_N=\mathbf{F}\end{gathered} \tag{15.105}$$

is usually employed. This scheme is appropriate in cases such as NaCl where the nuclei have nearly identical quadrupole coupling [13]. It is also particularly convenient when the coupling nuclei are identical and may be exchanged by symmetry operations since some factoring of the energy matrix is then possible.

The basis functions for the foregoing coupling scheme are denoted by

$$|\tau J I_1 I_2 \mathscr{I}_2 \ldots I_L \mathscr{I}_L \ldots I_N \mathscr{I}_N F M_F)$$

It is convenient for this representation to start with the general N nuclei problem. From (15.21) with $\tau_1\equiv\tau, j_1\equiv J, \tau_2\equiv I_1 I_2 \mathscr{I}_2, j_2\equiv\mathscr{I}_N$ and $J\equiv F$, the matrix element for the electric quadrupole interaction of the Lth nucleus is

$$\begin{aligned}&(\tau' J' I_1 I_2 \mathscr{I}'_2 \ldots I_L \mathscr{I}'_L \ldots I_N \mathscr{I}'_N F|\mathbf{V}(L)\cdot\mathbf{Q}(L)|\tau J I_1 I_2 \mathscr{I}_2 \ldots I_L \mathscr{I}_L \ldots I_N \mathscr{I}_N F)\\ &=(-1)^{J+\mathscr{I}'_N+F}\begin{Bmatrix}F & \mathscr{I}'_N & J'\\ 2 & J & \mathscr{I}_N\end{Bmatrix}(\tau' J'\|V(L)\|\tau J)(I_1 I_2 \mathscr{I}'_2 \ldots \mathscr{I}'_N\|Q(L)\|I_1 I_2 \mathscr{I}_2 \ldots \mathscr{I}_N)\end{aligned} \tag{15.106}$$

The matrix element for the magnetic dipole interaction of the Lth nucleus can also be written as

$$\begin{aligned}&(\tau J I_1 I_2 \mathscr{I}'_2 \ldots I_L \mathscr{I}'_L \ldots I_N \mathscr{I}'_N F|\mathbf{m}(L)\cdot\boldsymbol{\mu}(L)|\tau J I_1 I_2 \mathscr{I}_2 \ldots I_L \mathscr{I}_L \ldots I_N \mathscr{I}_N F)\\ &=(-1)^{J+\mathscr{I}'_N+F}\begin{Bmatrix}F & \mathscr{I}'_N & J\\ 1 & J & \mathscr{I}_N\end{Bmatrix}(\tau J\|m(L)\|\tau J)(I_1 I_2 \mathscr{I}'_2 \ldots \mathscr{I}'_N\|\mu(L)\|I_1 I_2 \mathscr{I}_2 \ldots \mathscr{I}_N)\end{aligned} \tag{15.107}$$

The reduced matrix element of the field tensors which appears in both (15.106) and (15.107) has been given previously, and the problem reduces to the evaluation of the reduced matrix element of the nuclear quadrupole tensor operator and the reduced matrix element of the magnetic dipole tensor operator. Evaluation of these quantities depends on the number of coupling nuclei in the molecule.

We consider explicitly the case of three coupling nuclei. The reduced matrix element of the nuclear quadrupole tensor operator of nucleus 1,

$$(I_1 I_2 \mathscr{I}'_2 I_3 \mathscr{I}'_3 \| Q(1) \| I_1 I_2 \mathscr{I}_2 I_3 \mathscr{I}_3)$$

is brought to the desired form by successive application of (15.22), and the matrix elements are found to be given by

$$\begin{aligned}(\tau' J' I_1 I_2 \mathscr{I}'_2 I_3 \mathscr{I}'_3 F | \mathbf{V}(1) \cdot \mathbf{Q}(1) | \tau J I_1 I_2 \mathscr{I}_2 I_3 \mathscr{I}_3 F)\\ =(-1)^{t_1}(eQq_{J'J})_1 f(J') f(I_1)[(2\mathscr{I}_2+1)(2\mathscr{I}'_2+1)(2\mathscr{I}_3+1)(2\mathscr{I}'_3+1)]^{1/2}\\ \times\begin{Bmatrix} F & \mathscr{I}'_3 & J' \\ 2 & J & \mathscr{I}_3 \end{Bmatrix}\begin{Bmatrix} \mathscr{I}'_2 & \mathscr{I}'_3 & I_3 \\ \mathscr{I}_3 & \mathscr{I}_2 & 2 \end{Bmatrix}\begin{Bmatrix} I_1 & \mathscr{I}'_2 & I_2 \\ \mathscr{I}_2 & I_1 & 2 \end{Bmatrix}\end{aligned} \qquad (15.108)$$

with $t_1 = J + I_1 + I_2 + I_3 + \mathscr{I}_2 + \mathscr{I}'_2 + \mathscr{I}_3 + \mathscr{I}'_3 + F$.

Similarly, for nucleus 2, successive application of (15.22) and (15.23) bring the reduced matrix element of the quadrupole tensor operator in (15.106) to the desired form. This procedure gives as a final result

$$\begin{aligned}(\tau' J' I_1 I_2 \mathscr{I}'_2 I_3 \mathscr{I}'_3 F | \mathbf{V}(2) \cdot \mathbf{Q}(2) | \tau J I_1 I_2 \mathscr{I}_2 I_3 \mathscr{I}_3 F)\\ =(-1)^{t_2}(eQq_{J'J})_2 f(J') f(I_2)[(2\mathscr{I}_2+1)(2\mathscr{I}'_2+1)(2\mathscr{I}_3+1)(2\mathscr{I}'_3+1)]^{1/2}\\ \times\begin{Bmatrix} F & \mathscr{I}'_3 & J' \\ 2 & J & \mathscr{I}_3 \end{Bmatrix}\begin{Bmatrix} \mathscr{I}'_2 & \mathscr{I}'_3 & I_3 \\ \mathscr{I}_3 & \mathscr{I}_2 & 2 \end{Bmatrix}\begin{Bmatrix} I_2 & \mathscr{I}'_2 & I_1 \\ \mathscr{I}_2 & I_2 & 2 \end{Bmatrix}\end{aligned} \qquad (15.109)$$

with $t_2 = J + I_1 + I_2 + I_3 + 2\mathscr{I}'_2 + \mathscr{I}_3 + \mathscr{I}'_3 + F$.

In the case of nucleus 3, (15.23) is used because $\mathscr{I}_2$ commutes with $Q(3)$. This yields specifically

$$\begin{aligned}(I_1 I_2 \mathscr{I}_2 I_3 \mathscr{I}'_3 \| Q(3) \| I_1 I_2 \mathscr{I}_2 I_3 \mathscr{I}_3) = (-1)^{\mathscr{I}_2 + I_3 + \mathscr{I}'_3}\\ \times[(2\mathscr{I}_3+1)(2\mathscr{I}'_3+1)]^{1/2}\begin{Bmatrix} I_3 & \mathscr{I}'_3 & \mathscr{I}_2 \\ \mathscr{I}_3 & I_3 & 2 \end{Bmatrix}(I_3 \| Q(3) \| I_3)\end{aligned} \qquad (15.110)$$

Therefore, from (15.106)

$$\begin{aligned}(\tau' J' I_1 I_2 \mathscr{I}_2 I_3 \mathscr{I}'_3 F | \mathbf{V}(3) \cdot \mathbf{Q}(3) | \tau J I_1 I_2 \mathscr{I}_2 I_3 \mathscr{I}_3 F) = (-1)^{t_3}(eQq_{J'J})_3 f(J') f(I_3)\\ \times[(2\mathscr{I}_3+1)(2\mathscr{I}'_3+1)]^{1/2}\begin{Bmatrix} F & \mathscr{I}'_3 & J' \\ 2 & J & \mathscr{I}_3 \end{Bmatrix}\begin{Bmatrix} I_3 & \mathscr{I}'_3 & \mathscr{I}_2 \\ \mathscr{I}_3 & I_3 & 2 \end{Bmatrix}\end{aligned} \qquad (15.111)$$

with $t_3 = J + I_3 + \mathscr{I}_2 + 2\mathscr{I}'_3 + F$. Note that the matrix elements are diagonal in $\mathscr{I}_2$, as follows from (15.110), in contrast to the results for nucleus 1 and 2.

The matrix elements for the magnetic dipole interactions are summarized below

$$(\tau J I_1 I_2 \mathscr{I}_2' I_3 \mathscr{I}_3' F|\mathbf{m}(1)\cdot \mu(1)|\tau J I_1 I_2 \mathscr{I}_2 I_3 \mathscr{I}_3 F)$$
$$=(-1)^{r_1}(C_{J,\tau})_1 h(J) h(I_1)[(2\mathscr{I}_2+1)(2\mathscr{I}_2'+1)(2\mathscr{I}_3+1)(2\mathscr{I}_3'+1)]^{1/2}$$
$$\times \begin{Bmatrix} F & \mathscr{I}_3' & J \\ 1 & J & \mathscr{I}_3 \end{Bmatrix} \begin{Bmatrix} \mathscr{I}_2' & \mathscr{I}_3' & I_3 \\ \mathscr{I}_3 & \mathscr{I}_2 & 1 \end{Bmatrix} \begin{Bmatrix} I_1 & \mathscr{I}_2' & I_2 \\ \mathscr{I}_2 & I_1 & 1 \end{Bmatrix} \tag{15.112}$$

with $r_1 = J + I_1 + I_2 + I_3 + \mathscr{I}_2 + \mathscr{I}_2' + \mathscr{I}_3 + \mathscr{I}_3' + F$, and

$$(\tau J I_1 I_2 \mathscr{I}_2' I_3 \mathscr{I}_3' F|\mathbf{m}(2)\cdot \mu(2)|\tau J I_1 I_2 \mathscr{I}_2 I_3 \mathscr{I}_3 F)$$
$$=(-1)^{r_2}(C_{J,\tau})_2 h(J) h(I_2)[(2\mathscr{I}_2+1)(2\mathscr{I}_2'+1)(2\mathscr{I}_3+1)(2\mathscr{I}_3'+1)]^{1/2}$$
$$\times \begin{Bmatrix} F & \mathscr{I}_3' & J \\ 1 & J & \mathscr{I}_3 \end{Bmatrix} \begin{Bmatrix} \mathscr{I}_2' & \mathscr{I}_3' & I_3 \\ \mathscr{I}_3 & \mathscr{I}_2 & 1 \end{Bmatrix} \begin{Bmatrix} I_2 & \mathscr{I}_2' & I_1 \\ \mathscr{I}_2 & I_2 & 1 \end{Bmatrix} \tag{15.113}$$

with $r_2 = J + I_1 + I_2 + I_3 + 2\mathscr{I}_2' + \mathscr{I}_3 + \mathscr{I}_3' + F$, and

$$(\tau J I_1 I_2 \mathscr{I}_2 I_3 \mathscr{I}_3' F|\mathbf{m}(3)\cdot \mu(3)|\tau J I_1 I_2 \mathscr{I}_2 I_3 \mathscr{I}_3 F) = (-1)^{r_3}(C_{J,\tau})_3 h(J) h(I_3)$$
$$\times [(2\mathscr{I}_3+1)(2\mathscr{I}_3'+1)]^{1/2} \begin{Bmatrix} F & \mathscr{I}_3' & J \\ 1 & J & \mathscr{I}_3 \end{Bmatrix} \begin{Bmatrix} I_3 & \mathscr{I}_3' & \mathscr{I}_2 \\ \mathscr{I}_3 & I_3 & 1 \end{Bmatrix} \tag{15.114}$$

with $r_3 = J + I_3 + \mathscr{I}_2 + 2\mathscr{I}_3' + F + 1$.

Generalizing the preceding discussion to the case of N coupling nuclei ($N > 1$), we have for the matrix elements of the quadrupole interaction for the Lth coupling nucleus [5]

$$(\tau' J' I_1 I_2 \mathscr{I}_2 \ldots I_L \mathscr{I}_L' \ldots I_N \mathscr{I}_N' F|\mathbf{V}(L)\cdot \mathbf{Q}(L)|\tau J I_1 I_2 \mathscr{I}_2 \ldots I_L \mathscr{I}_L \ldots I_N \mathscr{I}_N F)$$
$$=(-1)^t (eQq_{J'J})_L f(J') f(I_L) \begin{Bmatrix} F & \mathscr{I}_N' & J' \\ 2 & J & \mathscr{I}_N \end{Bmatrix} \Bigg[[(2\mathscr{I}_L+1)(2\mathscr{I}_L'+1)]^{1/2}$$
$$\times \begin{Bmatrix} I_L & \mathscr{I}_L' & \mathscr{I}_{L-1} \\ \mathscr{I}_L & I_L & 2 \end{Bmatrix} \Bigg]$$
$$\times \prod_{i=L}^{N-1} [(2\mathscr{I}_{i+1}+1)(2\mathscr{I}_{i+1}'+1)]^{1/2} \begin{Bmatrix} \mathscr{I}_i' & \mathscr{I}_{i+1}' & I_{i+1} \\ \mathscr{I}_{i+1} & \mathscr{I}_i & 2 \end{Bmatrix} \tag{15.115}$$

where the phase factor is determined by

$$t = J + F + I_N + \mathscr{I}_{L-1} + \prod_{i=L}^{N-1} (I_i + \mathscr{I}_{i+1} + \mathscr{I}_{i+1}') + 2\mathscr{I}_L'$$

and for the magnetic interaction [5]

$$(\tau J I_1 I_2 \mathscr{I}_2 \ldots I_L \mathscr{I}_L' \ldots I_N \mathscr{I}_N' F|\mathbf{m}(L)\cdot \mu(L)|\tau J I_1 I_2 \mathscr{I}_2 \ldots I_L \mathscr{I}_L \ldots I_N \mathscr{I}_N F)$$
$$=(-1)^r (C_{J,\tau})_L h(J) h(I_L) \begin{Bmatrix} F & \mathscr{I}_N' & J \\ 1 & J & \mathscr{I}_N \end{Bmatrix} \Bigg[[(2\mathscr{I}_L+1)(2\mathscr{I}_L'+1)]^{1/2}$$
$$\times \begin{Bmatrix} I_L & \mathscr{I}_L' & \mathscr{I}_{L-1} \\ \mathscr{I}_L & I_L & 1 \end{Bmatrix} \Bigg]$$
$$\times \prod_{i=L}^{N-1} [(2\mathscr{I}_{i+1}+1)(2\mathscr{I}_{i+1}'+1)]^{1/2} \begin{Bmatrix} \mathscr{I}_i' & \mathscr{I}_{i+1}' & I_{i+1} \\ \mathscr{I}_{i+1} & \mathscr{I}_i & 1 \end{Bmatrix} \tag{15.116}$$

where

$$r=J+F+I_N+\mathscr{I}_{L-1}+\sum_{i=L}^{N-1}(1+I_i+\mathscr{I}_{i+1}+\mathscr{I}'_{i+1})+(2\mathscr{I}'_L+1)$$

In the foregoing expressions $\mathscr{I}_0=0$, $\mathscr{I}_1=\mathscr{I}'_1=I_1$, $\mathscr{I}_N=\mathscr{I}$. For the product factor in (15.115) and (15.116), if the upper product limit is less than the initial value of i, this factor is to be set equal to one. Likewise the summation term in the expressions for t and r is to be set equal to zero if the above condition holds. When $L=1$, the factors in large square brackets in (15.115) and (15.116) are to be set equal to one; the last term $(2\mathscr{I}'_L)$ in the expression for t and the last $r(2\mathscr{I}'_L+1)$ in that for r is to be set equal to zero. Note that the matrix elements for a given L (except $L=1, 2$) are diagonal in all $\mathscr{I}$'s below $\mathscr{I}_L$ in the coupling sequence [see (15.105)]. Usually only diagonal elements in J need be considered. Each state may be specified by a given value of J and F. Off-diagonal elements can occur for the intermediate quantum numbers $\mathscr{I}_i$, and these intermediate quantum numbers can be used to further classify the levels.

The complexity introduced with plural nuclear quadrupole coupling may be illustrated by cyanogen azide (N_3CN) with four coupling nuclei, studied by Blackman et al. [22]. The computed and experimental line shapes are shown in Fig. 15.2 for the $1_{11}\rightarrow 2_{12}$ transition. This transition has 864 hyperfine components. It is difficult to unambiguously assign coupling constants to particular nitrogen atoms in the azide groups. Solution of this problem, as well as considerable simplification on the spectrum, can be obtained, however, by study of ^{15}N isotopically substituted species of cyanogen azide.

Some of the simplification possible when there are identical nuclei in the molecule may be illustrated by specialization of the general expression (15.115) to two coupling nuclei. In particular, for the quadrupole interaction we find

$$(\tau'J'I_1I_2\mathscr{I}'F|\mathbf{V}(1)\cdot\mathbf{Q}(1)|\tau JI_1I_2\mathscr{I}F)=(-1)^{t_1}(eQq_{J'J})_1 \times f(J')f(I_1)[(2\mathscr{I}+1)(2\mathscr{I}'+1)]^{1/2}\begin{Bmatrix}F & \mathscr{I}' & J'\\ 2 & J & \mathscr{I}\end{Bmatrix}\begin{Bmatrix}I_1 & \mathscr{I}' & I_2\\ \mathscr{I} & I_1 & 2\end{Bmatrix} \quad (15.117)$$

with $t_1=J+I_1+I_2+\mathscr{I}'+\mathscr{I}+F$, and

$$(\tau'J'I_1I_2\mathscr{I}'F|\mathbf{V}(2)\cdot\mathbf{Q}(2)|\tau JI_1I_2\mathscr{I}F)=(-1)^{t_2}(eQq_{J'J})_2 \times f(J')f(I_2)[(2\mathscr{I}+1)(2\mathscr{I}'+1)]^{1/2}\begin{Bmatrix}F & \mathscr{I}' & J'\\ 2 & J & \mathscr{I}\end{Bmatrix}\begin{Bmatrix}I_2 & \mathscr{I}' & I_1\\ \mathscr{I} & I_2 & 2\end{Bmatrix} \quad (15.118)$$

with $t_2=J+I_1+I_2+2\mathscr{I}'+F$. Here, $\mathbf{I}_1+\mathbf{I}_2=\boldsymbol{\mathscr{I}}$, $\boldsymbol{\mathscr{I}}+\mathbf{J}=\mathbf{F}$.

A study, for example, of (15.117) and (15.118) readily yields an interesting and well-known result when $I_1=I_2$ and $(eQq_{J'J})_1=(eQq_{J'J})_2$. Adding the contributions of nucleus 1 and 2, one finds that the elements of the quadrupole coupling matrix may be written as a factor times $[(-1)^{\mathscr{I}'}+(-1)^{\mathscr{I}}]$. Hence, it follows that matrix elements of the type $(\mathscr{I}\pm 1|\mathscr{I})$ vanish, while those of the type $(\mathscr{I}|\mathscr{I})$ and $(\mathscr{I}\pm 2|\mathscr{I})$ are nonvanishing, and the matrices for a given F thus separate into two submatrices of even and odd $\mathscr{I}$. This is virtually true for

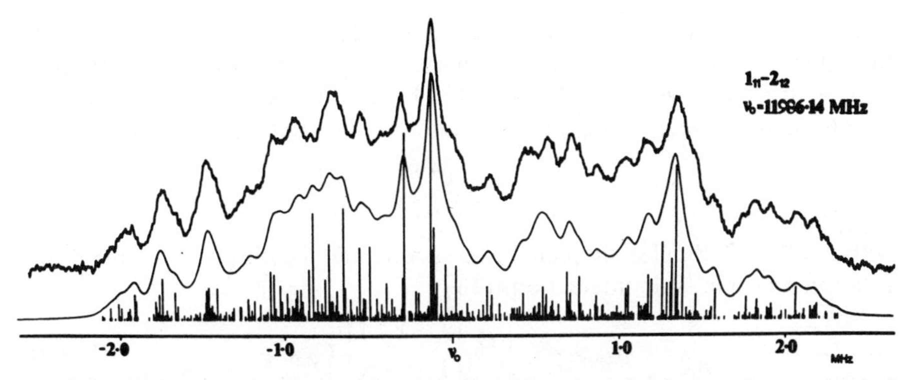

Fig. 15.2 Comparison of experimental and computed line shape for the $1_{11} \rightarrow 2_{12}$ transition of cyanogen azide. The quadrupole coupling constants were obtained by comparing the experimental spectral traces with computer simulated traces. The coupling constants found are: $\chi_{aa}(1) = 4.82$, $\chi_{bb}(1) = -0.70$; $\chi_{aa}(2) = -0.85$, $\chi_{bb}(2) = 0.70$; $\chi_{aa}(3) = -0.75$, $\chi_{bb}(3) = 1.55$; $\chi_{aa}(4) = -2.27$, $\chi_{bb}(4) = 1.15$, where the atom designations indicate the nitrogen position [N(3)-N(2)-N(1)-CN(4)]. From Blackman et al. [22].

NaCl [13] with nonequivalent coupling nuclei and is the case for molecules such as $COCl_2$, D_2S, where two identical quadrupolar nuclei are involved which occupy identical molecular sites. For a given J, the submatrices are conveniently grouped into matrices for a given F with elements within a submatrix labeled by values of $\mathscr{I}$. For the case of $COCl_2(I_1 = I_2 = \frac{3}{2})$ the possible values of the quantum numbers are: $\mathscr{I} = 0$, $F = J$; $\mathscr{I} = 1$, $F = J$, $J \pm 1$; $\mathscr{I} = 2$, $F = J$, $J \pm 1$, $J \pm 2$; $\mathscr{I} = 3$, $F = J$, $J \pm 1$, $J \pm 2$, $J \pm 3$. There are thus seven submatrices (without factoring) corresponding to the various values of F. Symmetry considerations provide further restrictions, and only certain values of $\mathscr{I}$ are allowed for a given rotational state J_{K_{-1}, K_1}. For $COCl_2$, where the rotation C_2^b interchanges identical chlorine nuclei (Fermions), the total wave function must be antisymmetric with respect to this symmetry operation. A total of $(2I_1 + 1)(2I_2 + 1) = 16$ spin functions can be constructed. To divide the possible values of $\mathscr{I}$ into symmetric or antisymmetric spin functions, we note that the lowest $\mathscr{I}$ state is symmetric when I_1 and I_2 are integral or antisymmetric when I_1 and I_2 are half-integral. The remaining symmetries alternate with increasing values of $\mathscr{I}$. Thus, for $I_1 = I_2 = \frac{3}{2}$, the $\mathscr{I} = 0$ (1 spin function) and $\mathscr{I} = 2$ (5 spin functions) give antisymmetric functions, while $\mathscr{I} = 1$ (3 spin functions) and $\mathscr{I} = 3$ (7 spin functions) give symmetric functions. Consider a totally symmetric electronic-vibrational state. For symmetric rotational levels $A(ee)$, $B_b(oo)$, only antisymmetric nuclear states ($\mathscr{I} = 0, 2$) exist, while for the antisymmetric rotational levels $B_a(eo)$, $B_c(oe)$, only symmetric nuclear states ($\mathscr{I} = 1, 3$) exist. Further discussions for the cases $I = \frac{3}{2}$ and $\frac{5}{2}$ are available [48, 49].

As a further specific example we may consider D_2S, where the hyperfine structure of the $1_{10} \rightarrow 1_{01}$ transition has been studied [16]. The symmetry axis is the b axis, and the $1_{10}(B_c)$ and $1_{01}(B_a)$ levels are antisymmetric with respect to exchange of the two identical deuterium nuclei ($I_1 = I_2 = 1$, Bosons). Thus since $\mathscr{I}=0$ (symmetric), 1 (antisymmetric) and 2 (symmetric), $\mathscr{I}$ can take only the value 1 for these rotational levels in order for the total wave function to be symmetric. Specializing (15.117) and (15.118) to the case $J=1$, $I_1=I_2=\mathscr{I}=1$ with $(eQq_J)_1=(eQq_J)_2=eQq_J$, one obtains, for the electric quadrupole energy

$$E_{\text{quad}}=[(-1)^{F+1}]\tfrac{15}{2}\begin{Bmatrix}F & 1 & 1\\ 2 & 1 & 1\end{Bmatrix}eQq_J \tag{15.119}$$

with $F=0$, 1, 2. Similar simplified expressions may be obtained for the spin–rotation and spin–spin interactions [16]

$$E_{\text{mag}}=(-1)^F 6\begin{Bmatrix}F & 1 & 1\\ 1 & 1 & 1\end{Bmatrix}C_{J,\tau} \tag{15.120}$$

$$E_{\text{spin-spin}}=(-1)^F 15\begin{Bmatrix}F & 1 & 1\\ 2 & 1 & 1\end{Bmatrix}(JJ|D_0^{(2)}|JJ) \tag{15.121}$$

It may be noted that the expressions for the electric quadrupole and magnetic dipole energies are equivalent to that for a single spin 1 nucleus except for the sign of the quadrupole energy. The calculated $1_{01}\rightarrow 1_{01}$ transition [28] of D_2O, which is like that for D_2S, is compared with the observed pattern in Fig. 15.3.

The hyperfine interaction by three identical nuclei in symmetric tops is also reduced by the symmetry of the rotational wave function, but its calculation is more involved than that for two nuclei, described previously. The reader is directed to the references cited previously.

The two coupling schemes considered in (15.86) and (15.105) are equivalent in that they lead to the same hyperfine structure, and one coupling scheme or coupling order does not offer any great advantage over the other (excluding symmetry considerations) if the energy matrix is diagonalized. The choice of one coupling scheme over the other rests on the first-order approximation being significantly better in one scheme than in the other so as not to require matrix diagonalization. To this end, other coupling schemes, which are mixtures of the two considered, may prove useful. For example, if two nuclei (spins I_2 and I_3) have comparable couplings which are smaller than that of a third nucleus (spin I_1), the coupling scheme $J+I_1=F_1$, $I_2+I_3=\mathscr{I}$, $\mathscr{I}+F_1=F$ may prove more convenient. This is particularly true if I_2 and I_3 arise from identical particles. Symmetry restrictions will then limit the possible values of $\mathscr{I}$ associated with a given rotational level and these restrictions can more easily be introduced into the matrix elements.

The matrix elements of $\mathscr{H}_{hfs}$ in this representation may be readily evaluated using the procedures already discussed; that for nucleus (1) is, for example, given by (15.84) and (15.85). For nucleus 2 and 3 we would have, for example,

$$\begin{aligned}&(\tau'J'I_1F_1'I_2I_3\mathscr{I}'F|\mathscr{H}_Q(L)|\tau JI_1F_1I_2I_3\mathscr{I}F)=(q_{J'J})_L f(J')\\ &\quad\times(-1)^t[(2F_1+1)(2F_1'+1)]^{1/2}\begin{Bmatrix}F & \mathscr{I}' & F_1'\\ 2 & F_1 & \mathscr{I}\end{Bmatrix}\begin{Bmatrix}J' & F_1' & I_1\\ F_1 & J & 2\end{Bmatrix}\\ &\quad\times(I_2I_3\mathscr{I}'\|Q(L)\|I_2I_3\mathscr{I})\end{aligned} \tag{15.122}$$

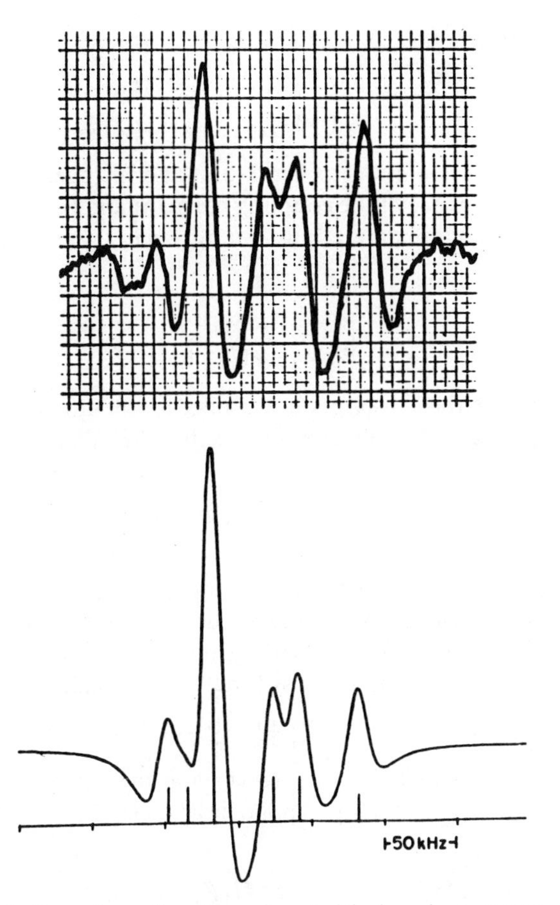

Fig. 15.3 Theoretical hyperfine pattern (lower figure) of the $1_{10} \rightarrow 1_{01}$ transition of D_2O compared with the observed pattern (upper figure). The structure is due to the magnetic and quadrupole coupling by the two identical D nuclei. It was calculated by irreducible tensor methods and observed with a molecular beam maser. The transition falls at the submillimeter wave frequency of 317 GHz. From Garvey and De Lucia [28].

with $t = 2F_1 + \mathcal{I}' + F + J' + I_1$ and $L = 2$ or 3. The reduction of the reduced matrix element follows from (15.22) for $L = 2$ and (15.23) for $L = 3$. Symmetry restrictions for identical particles can be conveniently introduced at this stage in evaluation of the $6j$ symbol arising from this reduction.

5 NUCLEAR SPIN–SPIN INTERACTIONS

We may consider nuclear spin–spin interaction by reference to a specific example, $D^{13}C^{14}N$, studied by Garvey and De Lucia [23]. In addition to the quadrupole and spin–rotation interactions of nitrogen and deuterium, and the spin–rotation interaction of ^{13}C, there is a spin–spin interaction involving the nuclear spins of nitrogen and carbon, and nitrogen and deuterium. The direct spin–spin interaction may be written in Cartesian form as

$$\begin{aligned}\mathscr{H}_{\text{spin-spin}} &= \mathbf{I}_{\mathrm{C}}\cdot\mathbf{D}_{\mathrm{CD}}\cdot\mathbf{I}_{\mathrm{D}}+\mathbf{I}_{\mathrm{C}}\cdot\mathbf{D}_{\mathrm{CN}}\cdot\mathbf{I}_{\mathrm{N}}\\ &= \mathbf{I}_3\cdot\mathbf{D}_{32}\cdot\mathbf{I}_2+\mathbf{I}_3\cdot\mathbf{D}_{31}\cdot\mathbf{I}_1\end{aligned} \tag{15.123}$$

The spin–spin interaction between deuterium and nitrogen is negligibly small due to the $1/r^3$ dependence of the coupling constant and the large D–N separation. Since the largest interaction involves nitrogen, the coupling scheme

$$\begin{aligned}&\mathbf{J}+\mathbf{I}_1=\mathbf{F}_1\\ &\mathbf{F}_1+\mathbf{I}_2=\mathbf{F}_2\\ &\mathbf{F}_2+\mathbf{I}_3=\mathbf{F}_3=\mathbf{F}\end{aligned}$$

is appropriate, where the subscripts 1, 2, 3 refer to the nitrogen, deuterium and carbon atoms, respectively. The elements of the Cartesian coupling tensor are defined by

$$(D_{LK})_{ij}=-\frac{3x_ix_j-\delta_{ij}r^2}{r^5}\frac{\mu_L\mu_K}{I_LI_K} \tag{15.124}$$

where I_L and I_K are the nuclear spins and μ_L and μ_K the corresponding magnetic dipole moments of nuclei L and K. Here $\mathbf{r}$ defines the relative position vector of nucleus K with respect to L. Rewriting the interaction in spherical tensor form, noting that $\mathbf{D}$ is a traceless and symmetric Cartesian tensor, we have

$$\mathscr{H}_{\text{spin-spin}}=-\sqrt{\tfrac{5}{2}}[\mathbf{D}_{32}^{(2)}\times\mathbf{I}_2^{(1)}]^{(1)}\cdot\mathbf{I}_3^{(1)}-\sqrt{\tfrac{5}{2}}[\mathbf{D}_{31}^{(2)}\times\mathbf{I}_1^{(1)}]^{(1)}\cdot\mathbf{I}_3^{(1)} \tag{15.125}$$

where $\mathbf{D}^{(2)}$ is a second-rank spherical tensor and the normalization factor has been chosen such that $D_0^{(2)}=D_{zz}$. Application of (15.21) to the first term of (15.125) gives

$$\begin{aligned}&(JI_1F_1'I_2F_2'I_3F|-\sqrt{\tfrac{5}{2}}[\mathbf{D}_{32}^{(2)}\times\mathbf{I}_2^{(1)}]^{(1)}\cdot\mathbf{I}_3^{(1)}|JI_1F_1I_2F_2I_3F)\\ &=(-1)^{F_2+I_3+F+1}\begin{Bmatrix}F & I_3 & F_2'\\ 1 & F_2 & I_3\end{Bmatrix}(I_3\|I_3^{(1)}\|I_3)\\ &\quad\times\sqrt{\tfrac{5}{2}}(JI_1F_1'I_2F_2'\|[\mathbf{D}_{32}^{(2)}\times\mathbf{I}_2^{(1)}]^{(1)}\|JI_1F_1I_2F_2)\end{aligned} \tag{15.126}$$

The latter tensor product is composed of tensor operators which operate on different subspaces; therefore application of (15.24) yields

$$\begin{aligned}&(JI_1F_1'I_2F_2'\|[\mathbf{D}_{32}^{(2)}\times\mathbf{I}_2^{(1)}]^{(1)}\|JI_1F_1I_2F_2)=[(2F_2+1)(2F_2'+1)3]^{1/2}\\ &\quad\times\begin{Bmatrix}F_1' & F_1 & 2\\ I_2 & I_2 & 1\\ F_2' & F_2 & 1\end{Bmatrix}(JI_1F_1'\|D_{32}^{(2)}\|JI_1F_1)(I_2\|I_2^{(1)}\|I_2)\end{aligned} \tag{15.127}$$

Since $\mathbf{D}^{(2)}$ commutes with $\mathbf{I}_1$ we apply (15.22)

$$\begin{aligned}&(JIF_1'\|D_{32}^{(2)}\|JI_1F_1)=(-1)^{J+I_1+F_1+2}[(2F_1+1)(2F_1'+1)]^{1/2}\\ &\quad\times\begin{Bmatrix}J & F_1' & I_1\\ F_1 & J & 2\end{Bmatrix}(J\|D_{32}^{(2)}\|J)\end{aligned} \tag{15.128}$$

Collection of terms gives

$$(JI_1F_1'I_2F_2'I_3F|-\sqrt{\tfrac{5}{2}}[\mathbf{D}_{32}^{(2)}\times\mathbf{I}_2^{(1)}]^{(1)}\cdot\mathbf{I}_3^{(1)}|JI_1F_1I_2F_2I_3F)$$
$$=(-1)^{F_2+I_3+F+J+I_1+F_1+1}[\tfrac{15}{2}(2F_2+1)(2F_2'+1)(2F_1+1)(2F_1'+1)]^{1/2}$$
$$\times\begin{Bmatrix}F & I_3 & F_2'\\ 1 & F_2 & I_3\end{Bmatrix}\begin{Bmatrix}J & F_1' & I_1\\ F_1 & J & 2\end{Bmatrix}\begin{Bmatrix}F_1' & F_1 & 2\\ I_2 & I_2 & 1\\ F_2' & F_2 & 1\end{Bmatrix}$$
$$\times(J\|D_{32}^{(2)}\|J)(I_3\|I_3^{(1)}\|I_3)(I_2\|I_2^{(1)}\|I_2) \tag{15.129}$$

Similarly, application of (15.21) to the second term of $\mathscr{H}_{\text{spin-spin}}$ and subsequent use of (15.22) and (15.24) gives finally for the matrix elements

$$(JI_1F_1'I_2F_2'I_3F|-\sqrt{\tfrac{5}{2}}[\mathbf{D}_{31}^{(2)}\times\mathbf{I}_1^{(1)}]^{(1)}\cdot\mathbf{I}_3^{(1)}|JI_1F_1I_2F_2I_3F)$$
$$=(-1)^{F_2+I_3+F+F_1'+I_2+F_2}[\tfrac{15}{2}(2F_2+1)(2F_2'+1)(2F_1+1)(2F_1'+1)]^{1/2}$$
$$\times\begin{Bmatrix}F & I_3 & F_2'\\ 1 & F_2 & I_3\end{Bmatrix}\begin{Bmatrix}F_1' & F_2' & I_2\\ F_2 & F_1 & 1\end{Bmatrix}\begin{Bmatrix}J & J & 2\\ I_1 & I_1 & 1\\ F_1' & F_1 & 1\end{Bmatrix}(J\|D_{31}^{(2)}\|J)$$
$$\times(I_1\|I_1^{(1)}\|I_1)(I_3\|I_3^{(1)}\|I_3) \tag{15.130}$$

The reduced matrix element of the rotation-dependent operator is found from the Wigner-Eckart theorem

$$(J\|D^{(2)}\|J)=2f(J)(JJ|D_0^{(2)}|JJ) \tag{15.131}$$

and $(JJ|D_0^{(2)}|JJ)$ can be taken as the spectroscopic constant to be determined. More often the small effects of the spin–spin interaction can be accurately accounted for by calculation of $D_0^{(2)}$ from the molecular geometry. Likewise we have

$$(I_\alpha\|I_\alpha^{(1)}\|I_\alpha)=h(I_\alpha) \tag{15.132}$$

with $h(I_\alpha)$ defined in (15.68).

6 RELATIVE INTENSITIES

In the interpretation of rotational hyperfine spectra the relative intensities of the transitions are required. This may also be readily formulated in terms of irreducible tensor methods. The line strength of the transition $\tau'J'\alpha'F'\to\tau J\alpha F$ is defined by [2]

$$S(\tau'J'\alpha'F'\to\tau J\alpha F)=\sum_{qM_FM_F'}|(\tau'J'\alpha'F'M_F'|\mu_q^{(1)}|\tau J\alpha FM_F)|^2 \tag{15.133}$$

where $\mu_q^{(1)}$ are the components of the irreducible electric dipole moment tensor operator $[\mu_0^{(1)}=\mu_z,\ \mu_{\pm1}^{(1)}=\mp(1/2^{1/2})(\mu_x\pm i\mu_y)]$ and where τJ specifies the pure rotational state. To be able to discuss plural nuclear coupling we let α stand for the set of intermediate momenta in the coupling scheme. From the Wigner-Eckart theorem

$$(\tau' J'\alpha' F' M_F'|\mu_q^{(1)}|\tau J\alpha F M_F)=(-1)^{F'-M_F'}\begin{pmatrix} F' & 1 & F \\ -M_F' & q & M_F \end{pmatrix}(\tau' J'\alpha' F'\|\mu^{(1)}\|\tau J\alpha F) \tag{15.134}$$

Thus

$$S(\tau' J'\alpha' F'\rightarrow\tau J\alpha F)=|(\tau' J'\alpha' F'\|\mu^{(1)}\|\tau J\alpha F)|^2 \sum_{qM_FM_F'}\left|\begin{pmatrix} F' & 1 & F \\ -M_F' & q & M_F \end{pmatrix}\right|^2 \tag{15.135}$$

The last term in this equation can be rewritten as

$$\begin{aligned}\sum_{qM_FM_F'}\left|\begin{pmatrix} F' & 1 & F \\ -M_F' & q & M_F \end{pmatrix}\right|^2 &=\sum_q \sum_{M_FM_F'}\begin{pmatrix} F' & 1 & F \\ -M_F' & q & M_F \end{pmatrix}\begin{pmatrix} F' & 1 & F \\ -M_F' & q & M_F \end{pmatrix}\\ &=\sum_q \sum_{M_FM_F'}\begin{pmatrix} F' & F & 1 \\ -M_F' & M_F & q \end{pmatrix}\begin{pmatrix} F' & F & 1 \\ -M_F' & M_F & q \end{pmatrix}\end{aligned} \tag{15.136}$$

with the aid of the symmetry properties of the $3j$ symbol. The sum over M_F and M_F' is equal to $\frac{1}{3}$. This follows from the orthogonality property of the $3j$ symbols in (15.6). Thus the sum over the three components of q gives unity, and we have finally for (15.133)

$$S(\tau' J'\alpha' F'\rightarrow J\alpha F)=|(\tau' J'\alpha' F'\|\mu^{(1)}\|\tau J\alpha F)|^2 \tag{15.137}$$

Evaluation of the reduced matrix element depends on the coupling scheme.

For a single coupling nucleus α stands for I which does not change during a transition, and (15.22) gives

$$(\tau' J' IF'\|\mu^{(1)}\|\tau JIF)=(-1)^{J'+I+F+1}[(2F+1)(2F'+1)]^{1/2}\begin{Bmatrix} J' & F' & I \\ F & J & 1 \end{Bmatrix} \times(\tau' J'\|\mu^{(1)}\|\tau J) \tag{15.138}$$

The square of the preceding reduced matrix element gives the strength of the hyperfine transition according to (15.137)

$$S(\tau' J' IF'\rightarrow\tau JIF)=(2F+1)(2F'+1)\begin{Bmatrix} J' & F' & I \\ F & J & 1 \end{Bmatrix}^2 |(\tau' J'\|\mu^{(1)}\|\tau J)|^2 \tag{15.139}$$

The reduced matrix element $(\tau' J'\|\mu^{(1)}\|\tau J)$ which represents the intensity of the unsplit rotational transition $\tau' J'\rightarrow\tau J$ may be omitted in evaluation of the relative intensities of the hyperfine transitions. If it is desired, this reduced matrix element may be evaluated by first expressing $\mu_q^{(1)}$ by means of (15.1), in terms of the dipole moment components $\mu_{q'}'^{(1)}$ along the molecule-fixed axes. The matrix elements of $\mu_0^{(1)}$ in the symmetric rotor basis may then be obtained by application of (15.38) and finally the reduced matrix element can be found from the Wigner-Eckart theorem. For asymmetric rotors $|(\tau' J'\|\mu^{(1)}\|\tau J)|^2$ is related to the line strength λ_g introduced in Chapter VII.

We may define a relative intensity of a particular hyperfine line by first summing over F and F' and rearranging the $6j$ symbol

$$\sum_{F,F'} S(\tau' J' I F' \to \tau J I F)$$

$$= \sum_{F'} (2F'+1) \sum_{F} \left[(2F+1) \begin{Bmatrix} 1 & J & J' \\ I & F' & F \end{Bmatrix} \begin{Bmatrix} I & J & F \\ 1 & F' & J' \end{Bmatrix} \right] |(\tau' J' \| \mu^{(1)} \| \tau J)|^2 \quad (15.140)$$

The sum over the term in brackets is $(2J'+1)^{-1}$ by the orthogonality property of the $6j$ symbols. Likewise, the sum over the initial F' states from $|J'-I|$ to $J'+I$ [for $J'>I$ and $J'<I$ number of F' values are $(2I+1)$ and $(2J'+1)$, respectively] is $(2J'+1)(2I+1)$. Hence

$$\sum_{F,F'} S(\tau' J' I F' \to \tau J I F) = (2I+1) |(\tau' J' \| \mu^{(1)} \| \tau J)|^2 \quad (15.141)$$

The relative line strength of a hyperfine line is the ratio of (15.139) to (15.141) [11]

$$s(J' I F' \to J I F) = \frac{S(J' \tau' I F' \to J \tau I F)}{\sum_{FF'} S(J' \tau' I F' \to J \tau I F)}$$

$$= \frac{(2F+1)(2F'+1)}{(2I+1)} \begin{Bmatrix} J' & F' & I \\ F & J & 1 \end{Bmatrix}^2 \quad (15.142)$$

where the notation for s takes into account that it is independent of τ, τ'. Furthermore, it is clear that

$$\sum_{FF'} s(J' I F' \to J I F) = 1 \quad (15.143)$$

As discussed by Rudolph [11], these results can be used to give the unsplit line, corrected for the effects of quadrupole coupling, without actual assignment of the hyperfine structure. The transition frequency of a given hyperfine line $J'\tau' I F' \to J \tau I F$ is (to first-order)

$$\nu = \nu_r + E_Q(J \tau I F) - E_Q(J' \tau' I F') \quad (15.144)$$

where ν_r is the transition frequency $J'\tau' \to J\tau$ if no hyperfine structure were present. If we multiply each hyperfine component by its relative intensity and sum over all components, we have

$$\sum_{FF'} \nu \cdot s = \nu_r \sum_{FF'} s + \sum_{FF'} \{E_Q(J \tau I F) - E_Q(J' \tau' I F')\} s \quad (15.145)$$

By substituting from (15.84) for E_Q, one can show from orthogonality relations of the $6j$ symbol that the sums on the right-hand side vanish identically, yielding [11]

$$\nu_r(J'\tau' \to J\tau) = \sum_{FF'} \nu(J' \tau' I F' \to J \tau I F) s(J' I F' \to J I F) \quad (15.146)$$

Therefore, the unsplit frequency is given to first-order by the intensity-weighted average of the multiplet frequencies. This allows, in principle, the evaluation of ν_r without a detailed analysis of the *hfs* patterns and assignment of individual $F' \to F$ components. The ν_r's can then, of course, be used to give the rotational constants. Partially resolved or coinciding components are properly accounted for by choice of the intensity weighted mean or center-of-gravity of the multiplet. Of course there are problems in application where weak components with large frequency displacements are not observed.

The relative intensities for the two representations discussed previously require evaluation of the reduced matrix element of (15.137). For the case of unequal coupling since $\mathbf{I}_i$ commutes with $\mu^{(1)}$, successive application of (15.22) can be used to bring the reduced matrix element

$$_1 I_2 \ldots I_N F'_N \| \mu^{(1)} \| \tau J I_1 F_1 I_2 \ldots I_N F_N)$$

to the desired form

$$(\tau' J' \| \mu^{(1)} \| \tau J)$$

For N nuclei the relative intensities for unequal coupling are given by [5]

$$S(J'F'_1 \ldots F'_N \to JF_1 \ldots F_N) = |C(J'F'_1 \ldots F'_N \to JF_1 \ldots F_N)|^2 \quad (15.147)$$

where

$$C(J'F'_1 \ldots F_{N'} \to JF_1 \ldots F_N) = (-1)^t \prod_{i=1}^{N} [(2F_i+1)(2F'_i+1)]^{1/2} \begin{Bmatrix} F'_{i-1} & F'_i & I_i \\ F_i & F_{i-1} & 1 \end{Bmatrix} \quad (15.148)$$

and

$$t = \sum_{i=1}^{N} (I_i + F'_{i-1} + F_i) + N \text{ with } F_0 = J,\ F'_0 = J' \text{ and } F_N = F$$

For evaluation of the relative intensities of the hyperfine transitions in the equal or nearly equal coupling scheme, the reduced matrix element

$$(\tau' J' I_1 I_2 \mathscr{I}_2 \ldots I_L \mathscr{I}_L \ldots I_N \mathscr{I}_N F' \| \mu^{(1)} \| \tau J I_1 I_2 \mathscr{I}_2 \ldots I_L \mathscr{I}_L \ldots I_N \mathscr{I}_N F)$$

must first be evaluated. Since the $\mathscr{I}_i$'s commute with $\mu^{(1)}$, (15.22) gives

$$(\tau' J' I_1 F' \| \mu^{(1)} \| \tau J \ldots \mathscr{I}_N F) = (-1)^{J' + \mathscr{I}_N + F + 1} [(2F+1)(2F'+1)]^{1/2} \times \begin{Bmatrix} J' & F' & \mathscr{I}_N \\ F & J & 1 \end{Bmatrix} (\tau' J' \| \mu^{(1)} \| \tau J) \quad (15.149)$$

The resulting relative intensities for equal coupling are [5]

$$S(J' \ldots \mathscr{I}_N F' \to J \ldots \mathscr{I}_N F) = |C(J' \ldots \mathscr{I}_N F' \to J \ldots \mathscr{I}_N F)|^2 \quad (15.150)$$

where

$$C(J' \ldots \mathscr{I}_N F' \to J \ldots \mathscr{I}_N F) = (-1)^t [(2F+1)(2F'+1)]^{1/2} \begin{Bmatrix} J' & F' & \mathscr{I}_N \\ F & J & 1 \end{Bmatrix} \quad (15.151)$$

and $t = J' + \mathscr{I}_N + F + 1$ with $\mathscr{I}_N = \mathscr{I}$. If the intermediate F_i (or $\mathscr{I}_i$) are not good quantum numbers, the first-order intensities given by (15.148) and (15.151) may not be adequate. The matrix of $\mathbf{C}$ must then be transformed by means of a similarity transformation to the basis which diagonalizes the hyperfine Hamiltonian, that is,

$$\mathbf{C}' = \tilde{\mathbf{T}}_i \mathbf{C} \mathbf{T}_f \tag{15.152}$$

where $\mathbf{T}_i$ diagonalizes the energy matrix of the initial state of the transition and $\mathbf{T}_f$ diagonalizes the energy matrix of the final state. The squares of the elements of $\mathbf{C}'$ yield the relative intensities.

7 ELECTRONIC SPIN INTERACTIONS

The oxygen molecule, which has a $^3\Sigma$ ground state, is a classic example of a diatomic molecule having both spin–rotation and electronic spin–spin interactions. Its microwave spectrum, analyzed in some detail in Chapter IV, Section 2, is most simply treated by irreducible tensor methods [50, 51].

Exclusive of centrifugal distortion and nuclear coupling isotopes, the $^3\Sigma$ oxygen Hamiltonian (Chapter IV, Section 2) may be written as

$$\mathscr{H} = B\mathbf{N}^2 + \gamma \mathbf{N}\cdot\mathbf{S} + \tfrac{2}{3}\lambda(3S_{z'}^2 - \mathbf{S}^2) \tag{15.153}$$

where the first term is the rotational part $\mathscr{H}_r$, the second term is the spin–rotation part $\mathscr{H}_{sr}$, and the last term is the spin–spin interaction [50, 51] $\mathscr{H}_{ss}$. Here $S_{z'}$ is the body-fixed component of the electron spin along the internuclear axis, $\mathbf{N}$ is the rotational angular momentum, and $\mathbf{S}$ is the total electron spin operator. Evaluation of the matrix elements of this Hamiltonian allows the energy levels to be evaluated by direct diagonalization of the energy matrix. The appropriate basis functions are $|S, N, J, M_J)$ where the total angular momentum $\mathbf{J} = \mathbf{S} + \mathbf{N}$. The operators $\mathbf{J}^2$, J_z, $\mathbf{N}^2$ and $\mathbf{S}^2$ are diagonal in this basis. It is readily seen that the matrix elements for $\mathscr{H}_r$ and $\mathscr{H}_{sr}$ are given by

$$(S, N, J, M_J|\mathscr{H}_R|S, N, J, M_J) = BN(N+1) \tag{15.154}$$

$$(S, N, J, M_J'|\mathscr{H}_{sr}|S, N, J, M_J) = \tfrac{1}{2}\gamma[J(J+1) - N(N+1) - S(S+1)] \tag{15.155}$$

noting $\mathbf{N}\cdot\mathbf{S} = \tfrac{1}{2}[\mathbf{J}^2 - \mathbf{N}^2 - \mathbf{S}^2]$.

The matrix elements for the spin–spin term may be conveniently evaluated by application of irreducible tensor methods by writing this term in terms of the space-fixed components of $\mathbf{S}$, that is, $\mathscr{H}_{ss} = \tfrac{2}{3}\lambda(\sum \Phi_{iz'}\Phi_{jz'}S_iS_j - \mathbf{S}^2)$, where $\Phi_{iz'}$ is the direction cosine between the body-fixed z' axis and the space-fixed i axis ($i = x, y, z$). In Cartesian tensor form

$$\begin{aligned}\mathscr{H}_{ss} &= \lambda \mathbf{S}\cdot\mathbf{T}\cdot\mathbf{S} \\ &= \lambda \sum_{i,j} S_i S_j T_{ij}\end{aligned} \tag{15.156}$$

where $T_{ij} = \tfrac{2}{3}(3\Phi_{iz'}\Phi_{jz'} - \delta_{ij})$. This may be written in spherical tensor form as

$$\mathscr{H}_{ss} = \sum_{l=0}^{2} N_l[\mathbf{S}^{(1)} \times \mathbf{S}^{(1)}]^{(l)}\cdot\mathbf{C}^{(l)};\ N_l = (-1)^l \tag{15.157}$$

with $\mathbf{S}^{(1)}$ the first rank spherical tensor form of the first rank Cartesian tensor $\mathbf{S}$ defined via (15.10), and $\mathbf{C}^{(l)}$ is the lth rank ($l=0, 1, 2$) spherical tensor form of $\mathbf{T}$ defined via (15.17). In this particular case, $\mathbf{C}^{(l)}=0$ except for $l=2$. Therefore

$$\mathscr{H}_{ss}=\lambda[\mathbf{S}^{(1)}\times\mathbf{S}^{(1)}]^{(2)}\cdot\mathbf{C}^{(2)} \tag{15.158}$$

From (15.21) we have

$$(N', S, J, M_J|\mathscr{H}_{ss}|N, S, J, M_J)=(-1)^{S+N'+J}\begin{Bmatrix} J & N' & S \\ 2 & S & N \end{Bmatrix} \times(S\|[\mathbf{S}^{(1)}\times\mathbf{S}^{(1)}]^{(2)}\|S)(N'\|C^{(2)}\|N) \tag{15.159}$$

Application of (15.25) gives for the reduced matrix element of the electron spin

$$(S\|[\mathbf{S}^{(1)}\times\mathbf{S}^{(1)}]^{(2)}\|S)=\sqrt{5}\begin{Bmatrix} 1 & 1 & 2 \\ S & S & S \end{Bmatrix}[(S\|S^{(1)}\|S)]^2 \tag{15.160}$$

Here $S=1$, and the $6j$ symbol is equal to $\frac{1}{6}$. From the Wigner-Eckart theorem, it follows from the matrix element $(S, M_S=S|S_0^{(1)}|S, M_S=S)=S$ that

$$\begin{aligned}(S\|S^{(1)}\|S)&=[S(S+1)(S+2)]^{1/2}\\ &=\sqrt{6}\end{aligned} \tag{15.161}$$

Similarly, application of the Wigner-Eckart theorem gives

$$(N'\|C^{(2)}\|N)=(-1)^{N'-M_N}(N', M_N|C_0^{(2)}|N, M_N)\begin{pmatrix} N' & 2 & N \\ -M_N & 0 & M_N \end{pmatrix}^{-1} \tag{15.162}$$

where $C_0^{(2)}=(2/6^{1/2})(3\Phi_{zz'}^2-1)$. Evaluation of the direction cosine elements for a linear molecule gives

$$(N'\|C^{(2)}\|N)=-\frac{4}{\sqrt{6}}\left[\frac{N(N+1)(2N+1)}{(2N-1)(2N+3)}\right]^{1/2}\delta_{N',N} +2\left[\frac{(N+1)(N+2)}{(2N+3)}\right]^{1/2}\delta_{N',N+2}+2\left[\frac{N(N-1)}{(2N-1)}\right]^{1/2}\delta_{N',N-2} \tag{15.163}$$

The matrix elements of the Hamiltonian are hence given by the sum of (15.154), (15.155), and (15.159) with cognizance of (15.160), (15.161), and (15.163). The Hamiltonian is diagonal in J, M_J, and S, and since $S=1$ for each value of J, we have $N=J-1$, J and $J+1$ with $J=1, 2, \ldots$. For $J=0$, $N=1$. The energy matrix may hence be factored into a set of 3×3 blocks except for $J=0$ which is 1×1. To take into account distortion effects, the constants B, γ, and λ may be expanded in terms of N. For the diagonal matrix elements, the constants are expanded as in (4.57). For the off-diagonal term $(N'=J+1|N=J-1)$, $\lambda=\lambda_0+\lambda_1(J^2+J+1)$. Elaboration of this theory and calculation of transition strengths are given in the dissertation of Steinbach [51].

The complexity of the spectra of polyatomic molecules with unpaired electron spins is illustrated in Fig. 15.4 for CF_3 which has a doublet electronic ground state with three identical coupling nuclei—the first such molecule analyzed by

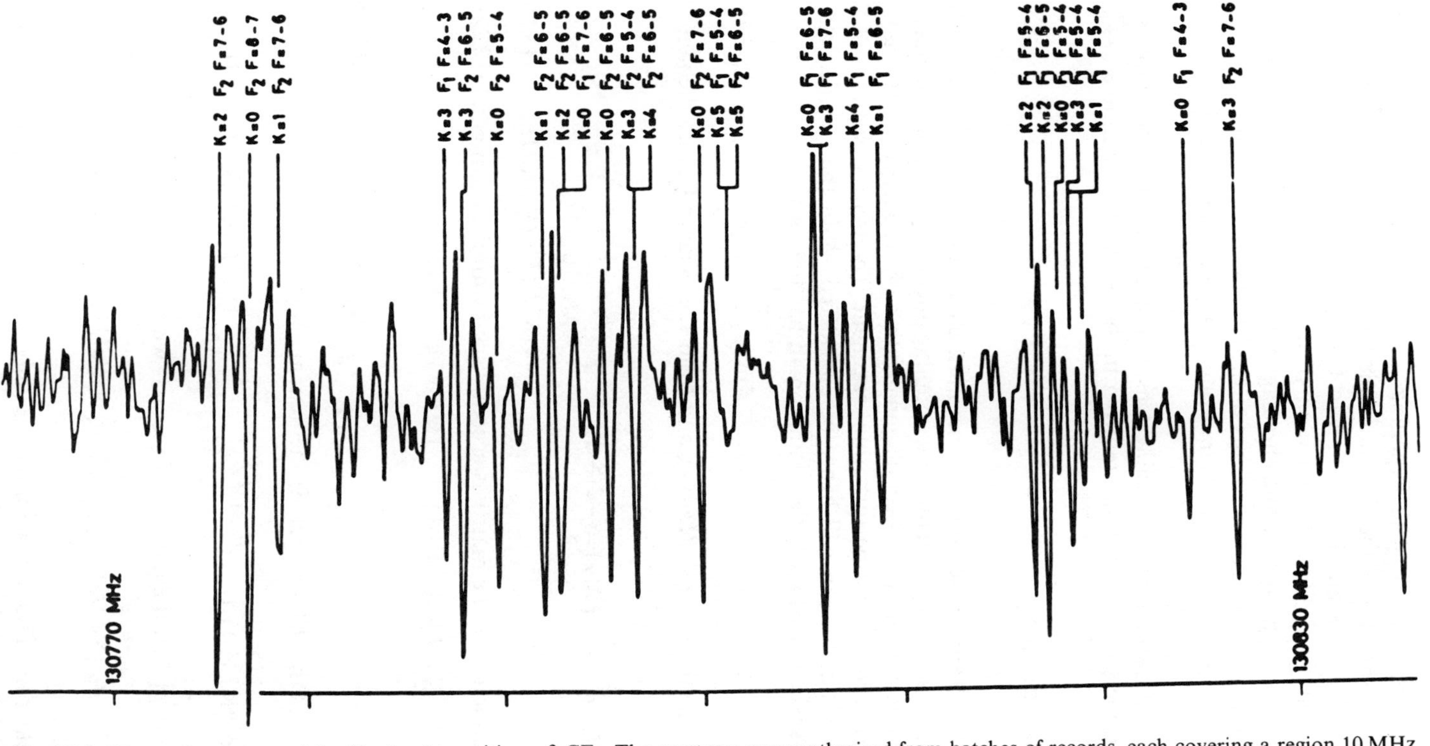

Fig. 15.4 Observed spectrum of the $N=6\leftarrow5$ transition of CF_3. The spectrum was synthesized from batches of records, each covering a region 10 MHz wide and representing the accummulation of 400 scans at a repetition rate of 5 Hz. The spectrum consists of K components split into many lines by the spin–rotation and the hyperfine interactions. From Endo et al. [39].

microwave spectroscopy. The hyperfine Hamiltonian is significantly complicated by the presence of the electronic spin momentum. The coupling scheme employed is $\mathbf{J}=\mathbf{N}+\mathbf{S}$ and $\mathbf{F}=\mathbf{J}+\mathbf{I}_0$ with $\mathbf{I}_0=\mathbf{I}_1+\mathbf{I}_2+\mathbf{I}_3$. A detailed analysis by Endo et al. [39] allowed evaluation of the spin–rotation constants, the Fermi interaction parameter, and the diagonal elements of the dipole–dipole hyperfine interaction tensor. Elements of the latter tensor split the $K=1$ level in first order. A discussion of the Hamiltonian may be found elsewhere [39, 35, 20].

Varied applications of irreducible tensor methods to diatomic molecules with electronic spin momentum are given in the book by Mizushima [44].

8 STARK EFFECT

As a final example we consider the Stark effect for a linear or symmetric top molecule with a single quadrupole coupling nucleus. For a coupling nucleus with large nuclear spin I, a direct diagonalization of the energy matrix may be more appropriate than a perturbation treatment. The matrix elements of the Stark effect Hamiltonian will hence be considered. With the electric field $\mathscr{E}$ along the space-fixed axis z, the Hamiltonian has the form

$$\mathscr{H}_{\mathscr{E}} = -\mathscr{E}\mu_0^{(1)} \tag{15.164}$$

The basis functions are $|KJIFM_F)$, and from the Wigner-Eckart theorem we have

$$(K'J'IF'M_F|\mathscr{H}_{\mathscr{E}}|KJIFM_F) = -\mathscr{E}(-1)^{F'-M_F}\begin{pmatrix} F' & 1 & F \\ -M_F & 0 & M_F \end{pmatrix} \times (K'J'IF'\|\mu_0^{(1)}\|KJIF) \tag{15.165}$$

Further reduction of the matrix is accomplished with the theorem of (15.22)

$$(K'J'IF'\|\mu_0^{(1)}\|KJIF) = (-1)^{J'+I+F+1}[(2F+1)(2F'+1)]^{1/2} \times \begin{Bmatrix} J' & F' & I \\ F & J & 1 \end{Bmatrix} (K'J'\|\mu_0^{(1)}\|JK) \tag{15.166}$$

Application of the Wigner-Eckart theorem again gives

$$(K'J'M_J|\mu_0^{(1)}|KJM_J) = (-1)^{J'-M_J}\begin{pmatrix} J' & 1 & J \\ -M_J & 0 & M_J \end{pmatrix}(K'J'\|\mu_0^{(1)}\|K, J) \tag{15.167}$$

Projection of $\mu_0^{(1)}$ on the molecule-fixed axis with (15.2) yields

$$(K'J'M_J|\mu_0^{(1)}|KJM_J) = \sum_{q'} (-1)^{-q'}\mu_{q'}'^{(1)}(K'J'M_J|D_{0,-q'}^{(1)}|KJM_J) \tag{15.168}$$

For a linear or symmetric-top molecule only, $\mu_0'^{(1)} \equiv \mu$ is nonzero. The matrix element of $\mathbf{D}^{(1)}$ may be evaluated from (15.38); thus $K'=K$ throughout, and

$$(KJ'M_j|\mu_0^{(1)}|KJM_J)=\mu(-1)^{M_J-K}[(2J+1)(2J'+1)]^{1/2}$$
$$\times\begin{pmatrix} J & 1 & J' \\ M_J & 0 & -M_J \end{pmatrix}\begin{pmatrix} J & 1 & J' \\ K & 0 & -K \end{pmatrix} \quad (15.169)$$

Therefore

$$(KJ'M_J|\mu_0^{(1)}|KJM_J)=\mu(-1)^{M_J-K}[(2J+1)(2J'+1)]^{1/2}$$
$$\times\begin{pmatrix} J & 1 & J' \\ K & 0 & -K \end{pmatrix} \quad (15.170)$$

For $J'=J$, $(KJ\|\mu_0^{(1)}\|KJ)=\mu K(2J+1)^{1/2}/[J(J+1)]^{1/2}$, for $J'=J+1$, $(KJ+1\|\mu_0^{(1)}\|KJ)=\mu[(J+1)^2-K^2]^{1/2}/(J+1)^{1/2}$ and for $J'=J-1$, $(KJ-1)\|\mu_0^{(1)}\|KJ)=-\mu(J^2-K^2)^{1/2}/J^{1/2}$. For a linear molecule, $K=0$, which implies that all diagonal elements vanish.

The matrix elements for $\mathscr{H}_{\mathscr{E}}$ are thus given by (15.165) with the auxiliary equation (15.166) and the values of $(J'\|\mu_0^{(1)}\|J)$. To these may be added the matrix elements for the rigid rotor, the magnetic spin-rotation interaction, and the nuclear quadrupole interaction. The zero-field frequencies are found by setting $\mathscr{E}=0$, and the Stark shifts are found by setting $\mathscr{E}$ to the desired field value and diagonalizing the energy matrix for each value of M_F. Since the energy matrix of the total Hamiltonian is no longer diagonal in F with the addition of $\mathscr{H}_{\mathscr{E}}$, a truncated matrix of appropriate dimensions must be chosen.

If there is more than one coupling nucleus, additional quantum numbers will appear in the reduced matrix element of (15.165). Equation (15.22) may be repeatedly used until the reduced matrix element appears in its most reduced form. This will introduce products of $6j$ symbols.

It may be noted that the general form of (15.167) written in terms of $\mu_q^{(1)}(q=0, \pm 1)$ yields the matrix elements of the dipole moment tensor in terms of the constant components along the body-fixed axis with the aid of (15.2). The matrix elements of $\mathbf{D}^{(1)}$ are the matrix elements of the first rank spherical components of the usual direction cosines, $\mathbf{\Phi}$.

References

1. M. E. Rose, *Elementary Theory of Angular Momentum*, Wiley, New York, 1957.
2. A. R. Edmonds, *Angular Momentum in Quantum Mechanics*, Princeton Univ. Press, Princeton, N.J., 1960.
3. U. Fano and G. Racah, *Irreducible Tensorial Sets*, Academic, New York, 1959.
4. B. R. Judd, *Operator Techniques in Atomic Spectroscopy*, McGraw-Hill, New York, 1963.
5. R. L. Cook and F. C. De Lucia, *Am. J. Phys.*, **39**, 1433 (1971).
6. R. F. Curl, Jr., and J. L. Kinsey, *J. Chem. Phys.*, **35**, 17 (1961).
7. P. Thaddeus, L. C. Krishner, and J. H. N. Loubser, *J. Chem. Phys.*, **40**, 257 (1964).
8. N. F. Ramsey, *Molecular Beams*, Clarendon Press, Oxford, England, 1956.
9. W. H. Flygare and W. D. Gwinn, *J. Chem. Phys.*, **36**, 787 (1962).
10. H. P. Benz, A. Bauder, and Hs. H. Günthard, *J. Mol. Spectrosc.*, **21**, 156 (1966).
11. H. D. Rudolph, *Z. Naturforsch.*, **23a**, 540 (1968).
12. A. Carrington, B. J. Howard, D. H. Levy, and J. C. Robertson, *Mol. Phys.*, **15**, 187 (1968).

13. J. W. Cederberg and C. E. Miller, *J. Chem. Phys.*, **50**, 3547 (1969).
14. G. L. Blackman, R. D. Brown, and F. R. Burden, *J. Mol. Spectrosc.*, **36**, 528 (1970).
15. S. G. Kukolich, and S. C. Wofsy, *J. Chem. Phys.*, **52**, 5477 (1970).
16. F. C. De Lucia and J. W. Cederberg, *J. Mol. Spectrosc.*, **40**, 52 (1971).
17. J. W. Cederberg, *Am. J. Phys.*, **40**, 159 (1972).
18. E. Herbst and W. Steinmetz, *J. Chem. Phys.*, **56**, 5342 (1972).
19. J. T. Hougen, *J. Chem. Phys.*, **57**, 4207 (1972).
20. I. C. Bowater, J. M. Brown, and A. Carrington, *Proc. R. Soc. London*, Ser. **A333**, 265 (1973).
21. J. E. M. Heuvel and A. Dymanus, *J. Mol. Spectrosc.*, **47**, 363 (1973).
22. G. L. Blackman, K. Bolton, R. D. Brown, F. R. Burden, and A. Mishra, *J. Mol. Spectrosc.*, **47**, 457 (1973).
23. R. M. Garvey and F. C. De Lucia, *J. Mol. Spectrosc.*, **50**, 38 (1974).
24. R. M. Garvey, F. C. De Lucia, and J. W. Cederberg, *Mol. Phys.*, **31**, 265 (1976).
25. J. T. Murray, W. A. Little, Q. Williams, and T. L. Weatherly, *J. Chem. Phys.*, **65**, 985 (1976).
26. H. K. Bodenseh, W. Hüttner, and P. Nowicki, *Z. Naturforsch.*, **31a**, 1638 (1976).
27. A. Dymanus, "Beam Maser Spectroscopy," in *MTP International Reviews of Science*, *Physical Chemistry Series 2*, Vol. 3 *Spectroscopy*, D. A. Ramsay, Ed., Butterworths and University Park Press, 1976.
28. R. M. Garvey and F. C. De Lucia, *Can. J. Phys.*, **55**, 1115 (1977).
29. S. G. Kukolich, G. Lind, M. Barfield, L. Faehl, and J. L. Marshall, *J. Am. Chem. Soc.*, **100**, 7155 (1978).
30. W. Hüttner, H. K. Bodenseh, P. Nowicki, and K. Morgenstern, *J. Mol. Spectrosc.*, **71**, 246 (1978).
31. K. P. R. Nair, J. Hoeft, and E. Tiemann, *Chem. Phys. Lett.*, **58**, 153 (1978).
32. J. M. Brown and T. J. Sears, *J. Mol. Spectrosc.*, **75**, 111 (1979).
33. K. P. R. Nair, J. Hoeft, and E. Tiemann, *J. Mol. Spectrosc.*, **78**, 506 (1979).
34. F. Michelot, B. Bobin, and J. Moret-Bailey, *J. Mol. Spectrosc.*, **76**, 374 (1979).
35. J. T. Hougen, *J. Mol. Spectrosc.*, **81**, 73 (1980).
36. Y. Endo, S. Saito, and E. Hirota, *J. Chem. Phys.*, **74**, 1568 (1981).
37. H. A. Fry and S. G. Kukolich, *J. Chem. Phys.*, **76**, 4387 (1982).
38. A. M. Murray and S. G. Kukolich, *J. Chem. Phys.*, **77**, 4312 (1982).
39. Y. Endo, C. Yamada, S. Saito, and E. Hirota, *J. Chem. Phys.*, **77**, 3376 (1982).
40. S. G. Kukolich and C. D. Cogley, *J. Chem. Phys.*, **76**, 1685 (1982); **77**, 581 (1982).
41. L. M. Tack and S. G. Kukolich, *J. Mol. Spectrosc.*, **94**, 95 (1982).
42. C. D. Cogley, L. M. Tack, and S. G. Kukolich, *J. Chem. Phys.*, **76**, 5669 (1982).
43. M. Ribeaud, A. Bauder, and Hs. H. Günthard, *J. Mol. Spectrosc.*, **42**, 441 (1972).
44. M. Mizushima, *The Theory of Rotating Diatomic Molecules*, Wiley, New York, 1975.
45. Numerical values have been tabulated by M. Rotenberg, R. Bivens, N. Metropolis, and J. K. Wooten, *The 3j and 6j Symbols* MIT Press, Cambridge, Mass., 1959.
46. Equation 15.38 may be obtained by application of (4.62) of Ref. 1 or (4.6.2) of Ref. 2 and the definition of the symmetric rotor functions.
47. E. U. Condon and G. H. Shortley, *Theory of Atomic Spectra*, Cambridge University Press, Cambridge, England, 1959.
48. G. W. Robinson, and C. D. Cornwell, *J. Chem. Phys.*, **21**, 1436 (1953).
49. J. S. Rigden and L. Nygaard, *J. Chem. Phys.*, **44**, 4603 (1966).
50. W. Steinbach and W. Gordy, *Phys. Rev.*, **A8**, 1753 (1973).
51. W. Steinbach, Ph.D. dissertation, Duke Univ., 1974.

Appendix **A**

NOTES ON MATRIX MECHANICAL METHODS

To facilitate the understanding of certain parts of this book by those not well versed in matrix mechanical methods, we give here a concise description of elements essential to interpretation of microwave spectra.

PROPERTIES OF SPECIAL MATRICES

We start by considering the properties of some special matrices. There are two types of matrices that are of particular importance in quantum mechanics. They are Hermitian matrices and unitary matrices. Before defining these it is useful to define the transpose and Hermitian adjoint matrix.

The transpose of $\mathbf{A}$ denoted by $\tilde{\mathbf{A}}$ is obtained from $\mathbf{A}$ by an interchange of rows and columns thus

$$\tilde{A}_{nm}=A_{mn} \tag{A.1}$$

The transpose of a column matrix is a row matrix.

The Hermitian adjoint $\mathbf{A}^{\dagger}$ of a matrix $\mathbf{A}$ we obtain by transposing $\mathbf{A}$ and taking the complex conjugate of each element, that is,

$$\mathbf{A}^{\dagger}=(\tilde{\mathbf{A}})^{*} \tag{A.2}$$

or in terms of the matrix elements

$$A^{\dagger}_{nm}=(\tilde{A}_{nm})^{*}=A^{*}_{mn} \tag{A.3}$$

where the asterisk signifies complex conjugation.

A matrix $\mathbf{A}$ is said to be Hermitian if it is equal to its Hermitian adjoint, that is,

$$\mathbf{A}=\mathbf{A}^{\dagger} \tag{A.4}$$

or for the elements

$$A_{nm}=A^{*}_{mn} \tag{A.5}$$

The operators of quantum mechanics representing observables are Hermitian operators since such operators yield real eigenvalues and expectation values. The matrix corresponding to a Hermitian operator is likewise a Hermitian matrix. A Hermitian matrix whose elements are real is a symmetric matrix for which

$$\mathbf{A} = \tilde{\mathbf{A}} \tag{A.6}$$

A matrix **A** is unitary if its inverse is equal to its Hermitian adjoint

$$\mathbf{A}^{-1} = \mathbf{A}^{\dagger} \tag{A.7}$$

hence

$$\mathbf{A}\mathbf{A}^{-1} = \mathbf{A}\mathbf{A}^{\dagger} = \mathbf{E} \tag{A.8}$$

where **E** is a unit matrix ($E_{nm} = \delta_{nm}$) also commonly designated **I**. A real unitary matrix

$$\mathbf{A}^{-1} = \tilde{\mathbf{A}} \tag{A.9}$$

is an orthogonal matrix. Transformations between different orthonormal basis systems are accomplished by unitary matrices. Unitary transformations have the property that they preserve the value of a scalar product, which is naturally required since the observable quantities are obtained from scalar products and must hence be independent of the choice of basis system. A unitary transformation is a generalization to complex space of an orthogonal transformation in real space.

MATRIX REPRESENTATION OF WAVE FUNCTIONS AND OPERATORS

In wave mechanics, the dynamical variables or observables are represented by linear operators. Each linear operator, A, has an eigenvalue equation associated with it.

$$A\phi_n = a_n\phi_n \tag{A.10}$$

where ϕ_n is the eigenfunction and a_n the corresponding eigenvalue of the operator A. The most important operator is the energy or Hamiltonian operator $\mathcal{H}$. The solution of its eigenvalue equation, the Schrödinger equation, yields the energy levels (eigenvalues) and energy eigenstates (eigenfunctions) of the quantum system. In matrix mechanics, the quantum mechanical operators are represented by matrices, and the problem of solving the eigenvalue equation (a differential equation) is replaced by the problem of diagonalizing a matrix.

Consider a set of linearly independent functions $\phi_1, \phi_2, \ldots, \phi_n, \ldots$, such as the set of eigenfunctions generated by the solution of an eigenvalue equation. Throughout our discussion we employ a single index to distinguish the various members of the set $\{\phi_n\}$ of basis functions. If more than one quantum number is required in order to specify the basis functions, then the index n must be regarded as standing for the complete set of quantum numbers, viz., $n = i, j, k, \ldots$. We shall assume that the basis functions are orthogonal and normalized, i.e.,

$$(\phi_n|\phi_m) = \int \phi_n^* \phi_m \, d\tau = \delta_{nm} \tag{A.11}$$

where the Kronecker delta is defined by

$$\delta_{nm} = \begin{cases} 0 & \text{if } n \neq m \\ 1 & \text{if } n = m \end{cases} \tag{A.12}$$

Frequently the notation $(\phi_n|\phi_m)$ is further abbreviated to $(n|m)$. The basis functions are said to form a complete set in that any arbitrary wave function may be expanded in terms of them. An arbitrary function ψ can hence be expressed as

$$\psi = \sum_n C_n \phi_n \tag{A.13}$$

where the C's are the expansion coefficients or coordinates of ψ referred to the basis $\{\phi_n\}$ and are defined by

$$C_n = (\phi_n|\psi) \tag{A.14}$$

The function ψ is uniquely specified by enumeration of the coefficients C_n (in general complex). These may be written as a column matrix

$$\begin{bmatrix} C_1 \\ C_2 \\ \vdots \\ C_n \\ \vdots \end{bmatrix} \tag{A.15}$$

In the language of matrix mechanics we refer to this as the matrix representation of ψ referred to the basis $\{\phi_n\}$. The basis system is, of course, arbitrary, and a different basis may be chosen. Then, however, the list of elements in the column matrix will be altered, and a different matrix representation will be obtained.

The scalar product of two functions χ and ψ denoted by $(\chi|\psi)$ is given by the product of a row matrix and a column matrix

$$(\chi|\psi) = \int \sum_n B_n^* \phi_n^* \sum_n C_n \phi_n \, d\tau = \sum_n B_n^* C_n = [B_1^* B_2^* \cdots B_n^* \cdots] \begin{bmatrix} C_1 \\ C_2 \\ \vdots \\ C_n \\ \vdots \end{bmatrix} \tag{A.16}$$

Note that the first factor in the scalar product is the Hermitian adjoint of the column matrix representing χ. If the two functions are orthogonal, the above scalar product will vanish.

The matrix representation of the operator A in the basis $\{\phi_n\}$ is defined by the equation

$$A_{nm} = (\phi_n|A|\phi_m) = \int \phi_n^* A \phi_m \, d\tau \tag{A.17}$$

The array of numbers (called matrix elements) calculated by the above relation may be written as a square matrix **A**, that is,

$$\begin{bmatrix} A_{11} & A_{12} & A_{13} & \cdot & \cdot \\ A_{21} & A_{22} & \cdot & \cdot & \cdot \\ A_{31} & \cdot & \cdot & \cdot & \cdot \\ \cdot & \cdot & \cdot & \cdot & \cdot \end{bmatrix} \tag{A.18}$$

The operator matrix depends on the basis chosen and will have different forms in different basis systems. If the operator represents a measurable quantity, the matrix operator is Hermitian and its elements satisfy (A.5).

The matrix of the sum of two operators $C\ (=A+B)$ is just the sum of the matrices representing the operators A and B in the basis $\{\phi_n\}$. Hence, the elements of **C** are defined by

$$C_{nm}=A_{nm}+B_{nm} \tag{A.19}$$

The matrix of the product operator $C\ (=AB)$ is obtained from the matrix of **A** and **B** by the usual rule for multiplication of two matrices. The matrix elements of the matrix operator **C** are hence given by

$$C_{nm}=\sum_k A_{nk}B_{km} \tag{A.20}$$

Note that all elements are defined with respect to the same basis. Similarly, the matrix for the product of any number of operators is obtained by multiplication of the corresponding matrices. In general, $\mathbf{AB}\neq\mathbf{BA}$ since matrix multiplication is noncommutative. Any matrix **A** does, however, commute with its inverse and with a unit matrix **E**; any two diagonal matrices will also commute. If basis functions are chosen which are simultaneous eigenfunctions of two operators A and B, the corresponding matrices will be diagonal [see (A.33)] and hence commute. This is analogous to the statement that if a function is an eigenfunction of two operators, then the operators commute.

We are now in a position to rewrite any equation of wave mechanics in matrix form. In matrix mechanics, for example, operation on a function ψ with an operator A is represented by the product of a square matrix **A** times the column matrix representing ψ. The result is a new column matrix (or function). With this in mind and taking cognizance of the definition of a scalar product, we may write

$$(\psi|A|\psi)=[C_1^*C_2^*\cdots C_n^*\cdots]\begin{bmatrix} A_{11} & A_{12} & A_{13} & \cdot & \cdot \\ A_{21} & A_{22} & \cdot & \cdot & \cdot \\ A_{31} & \cdot & \cdot & \cdot & \cdot \\ \cdot & \cdot & \cdot & \cdot & \cdot \\ \cdot & \cdot & \cdot & \cdot & \cdot \end{bmatrix}\begin{bmatrix} C_1 \\ C_2 \\ \vdots \\ C_n \\ \vdots \end{bmatrix} \tag{A.21}$$

This gives the average or expectation value of an observable represented by the operator A, for the quantum state ψ. Evaluation of the average value of an observable thus requires a knowledge of the matrix elements of the operator in a given basis and the expansion coefficients of the state ψ in that basis.

CHANGE OF BASIS

Many times it is advantageous to change from one basis to another. Two salient questions then arise: first, how the coordinates of a wave function, that is, its matrix representation, are related in the two basis systems and second, how the matrix of an operator is related in the two basis systems.

Let the connection between the original orthonormal basis functions $\{\phi_n\}$ and the new set of basis functions $\{\phi'_n\}$ be written

$$\phi'_n = \sum_m S_{mn} \phi_m \qquad n = 1, 2, \ldots \tag{A.22}$$

with

$$S_{mn} = (\phi_m | \phi'_n) \tag{A.23}$$

The coefficients S_{mn} define a square matrix $\mathbf{S}$, a transformation matrix, which is a unitary matrix. This insures, for example, that if the original basis functions are orthonormal, then the transformed basis remains orthonormal. If the functions ϕ_n and ϕ'_n are regarded as elements of two column matrices, (A.22) may be written in matrix form as

$$\begin{bmatrix} \phi'_1 \\ \phi'_2 \\ \vdots \\ \phi'_n \\ \vdots \end{bmatrix} = \tilde{\mathbf{S}} \begin{bmatrix} \phi_1 \\ \phi_2 \\ \vdots \\ \phi_n \\ \vdots \end{bmatrix} \tag{A.24}$$

where $\tilde{\mathbf{S}}$ is the transpose matrix of $\mathbf{S}$. The relation between the expansion coefficients of a given wave function ψ in the two different basis systems is found as follows

$$\psi = \sum_n C'_n \phi'_n = \sum_{n,m} S_{mn} C'_n \phi_m = \sum_m C_m \phi_m \tag{A.25}$$

where

$$C_m = \sum_n S_{mn} C'_n \tag{A.26}$$

which gives the old coefficients or coordinates in terms of the new ones. The transformation may be equivalently written as

$$\begin{bmatrix} C_1 \\ C_2 \\ \vdots \\ C_n \\ \vdots \end{bmatrix} = \mathbf{S} \begin{bmatrix} C'_1 \\ C'_2 \\ \vdots \\ C'_n \\ \vdots \end{bmatrix}, \qquad \begin{bmatrix} C'_1 \\ C'_2 \\ \vdots \\ C'_n \\ \vdots \end{bmatrix} = \mathbf{S}^\dagger \begin{bmatrix} C_1 \\ C_2 \\ \vdots \\ C_n \\ \vdots \end{bmatrix} \tag{A.27}$$

Note that if the elements of **S** are real, then both the basis functions and coordinates transform in the same manner.

The basis functions $\{\phi_n\}$ generate a representation of the operator A through the relation

$$A_{nm}=(\phi_n|A|\phi_m) \tag{A.28}$$

while in terms of the basis functions $\{\phi'_n\}$, the operator matrix is defined by

$$A'_{nm}=(\phi'_n|A|\phi'_m) \tag{A.29}$$

The relation between the matrix **A** and **A**′ is found by use of (A.22) in the foregoing equation

$$\begin{aligned} A'_{nm} &= \left(\sum_k S_{kn}\phi_k|A|\sum_l S_{lm}\phi_l\right) \\ &= \sum_{k,l} S^*_{kn}S_{lm}(\phi_k|A|\phi_l) \\ &= \sum_{k,l} (\tilde{S}_{nk})^* A_{kl}S_{lm} \\ &= \sum_{k,l} S^\dagger_{nk}A_{kl}S_{lm} \end{aligned} \tag{A.30}$$

In matrix form

$$\mathbf{A}'=\mathbf{S}^\dagger\mathbf{A}\mathbf{S} \tag{A.31}$$

This gives the matrix representation of the operator referred to the new basis in terms of the matrix representation of the operator in the old basis and of the unitary transformation **S** connecting the two basis systems. The matrix **A** is said to be transformed from the ϕ_n representation to the ϕ'_n representation by a similarity transformation. If the transformation has the property that it converts **A** to diagonal form, then it may be referred to as a principal axis transformation. If **A** is Hermitian, then because of the unitary character of **S**, **A**′ is also Hermitian [1]. Furthermore, the form of an equation is unaltered [2] by a transformation such as **S**, and the eigenvalues of a matrix are invariant to a change of basis [3]. If the transformation matrix is real, as is often the case, then (A.31) reads

$$\mathbf{A}'=\tilde{\mathbf{S}}\mathbf{A}\mathbf{S} \tag{A.32}$$

DIAGONALIZATION OF MATRICES: EIGENVALUES AND EIGENFUNCTIONS

In treating spectra we are especially interested in the eigenfunctions and eigenvalues of Hermitian operators. If the eigenfunctions $\psi_1, \psi_2, \ldots, \psi_n, \ldots$ of a Hermitian operator are known and these functions are used as basis functions for construction of the matrix of this operator, then the matrix will be diagonal. This follows from the orthonormality of the eigenfunctions

$$(\psi_m|A|\psi_n)=a_n(\psi_m|\psi_n)=a_n\delta_{mn} \tag{A.33}$$

The elements a_n of the diagonal matrix are the eigenvalues of the operator A. The diagonal representation of the operator matrix is hence of special interest. Even if the eigenfunctions are initially unknown, it is possible to find a transformation which will diagonalize the operator matrix by means of a similarity transformation. Consider the operator matrix **A** with respect to some orthonormal basis $\phi_1, \phi_2, \ldots, \phi_n, \ldots$. Let the basis functions $\psi_1, \psi_2, \ldots, \psi_n, \ldots$, the eigenfunctions of A, be related to the arbitrary basis functions by

$$\psi_n=\sum_m S_{mn}\phi_m \tag{A.34}$$

The **A** matrix will be transformed from the ϕ representation to the ψ representation by

$$\mathbf{S}^\dagger\mathbf{A}\mathbf{S}$$

and for this basis the transformed matrix must be diagonal with elements $\mathscr{A}_{mn}=a_n\delta_{mn}$ so that

$$\mathbf{S}^\dagger\mathbf{A}\mathbf{S}=\mathscr{A} \tag{A.35}$$

If we premultiply by **S**, since $\mathbf{S}\mathbf{S}^\dagger=\mathbf{S}^\dagger\mathbf{S}=\mathbf{E}$, we have

$$\mathbf{A}\mathbf{S}=\mathbf{S}\mathscr{A} \tag{A.36}$$

and for the matrix elements

$$\sum_m A_{km}S_{mn}=\sum_m S_{km}\mathscr{A}_{mn}=S_{kn}a_n=\sum_m \delta_{km}S_{mn}a_n \tag{A.37}$$

or

$$\sum_m (A_{km}-\delta_{km}a_n)S_{mn}=0 \tag{A.38}$$

Hence, if the transformation diagonalizes the matrix, we are led to a set of simultaneous linear equations defining the S_{mn} in terms of the a_n and the A_{km}. For a given value of n, (A.38) gives a set of linear equations corresponding to the various values of k; these have nontrivial solutions for the $S_{1n}, S_{2n}, \ldots$ if—and only if—the determinant of the matrix of coefficients vanishes

$$|A_{km}-\delta_{km}a_n|=0 \tag{A.39}$$

The roots of this equations (the secular determinant or secular equation) in a_n provide the eigenvalues $a_1, a_2, \ldots$; by substitution of each of the eigenvalues into the set of (A.38) the corresponding **S** matrix may be evaluated. The nth eigenvalue, for example, yields the nth column of **S** that is $S_{1n}, S_{2n}, \ldots$; these are the expansion coefficients of the eigenfunction ψ_n in the basis $\{\phi_n\}$. The columns of $\mathbf{S}$ are also referred to as the eigenvectors of the matrix $\mathbf{A}$. In evaluating the S_{nm} use is made of the normalization condition, that is,

$$\sum_k S_{nk} S^*_{nk} = 1 \tag{A.40}$$

In summary, the diagonalization of the matrix representation of the operator which is tantamount to the solution of the secular equation gives the eigenvalues. The determination of the transformation matrix $\mathbf{S}$ which diagonalizes the operator matrix yields the eigenfunctions. In the text, the secular equation is arrived at from a slightly different point of view, viz., by a rewriting of the eigenvalue equation in matrix form.

Of particular importance is the diagonalization of the Hamiltonian matrix. The diagonalization of $\mathbf{H}$ can be simplified by a knowledge of the operators which commute with the Hamiltonian. If an operator A commutes with $\mathscr{H}$ then

$$[\mathscr{H}, A] = \mathscr{H}A - A\mathscr{H} = 0 \tag{A.41}$$

The matrix representation of this equation based on functions which are eigenfunctions of A, so that $\mathbf{A}$ is diagonal, is

$$\mathscr{H}_{nm}(A_{mm} - A_{nn}) = 0 \tag{A.42}$$

It therefore follows that $\mathscr{H}_{nm}$ vanishes if the states n and m belong to different eigenvalues of $A(A_{nn} \neq A_{mm})$. The $\mathscr{H}$ matrix will hence factor into submatrices with each submatrix involving states with the same eigenvalues for A. In many cases, more then one operator will commute with $\mathscr{H}$. The greatest simplification will result if basis functions are chosen which are simultaneous eigenfunctions of the complete set of commuting operators. The operators which may be considered are not only the commuting observables such as the square of the angular momentum P^2 but also what we might classify as symmetry operators, operators which leave $\mathscr{H}$ invariant and hence commute with $\mathscr{H}$. Consideration therefore of the symmetry of the Hamiltonian also facilitates solution of the diagonalization problem.

References

1. If $\mathbf{A}$ is Hermitian we may write $\mathbf{A}' = \mathbf{S}^\dagger \mathbf{A}^\dagger \mathbf{S}$; taking the Hermitian adjoint we have $(\mathbf{A}')^\dagger = (\mathbf{S}^\dagger \mathbf{A}^\dagger \mathbf{S})^\dagger = \mathbf{S}^\dagger \mathbf{A} \mathbf{S} = \mathbf{A}'$ showing that the Hermiticity is invariant.
2. In the matrix equation $\chi = \mathbf{A}\psi$ if we multiply by $\mathbf{S}^\dagger$ then $\chi' = \mathbf{S}^\dagger \chi = \mathbf{S}^\dagger \mathbf{A}\psi = \mathbf{S}^\dagger \mathbf{A}\mathbf{S}\mathbf{S}^\dagger \psi = \mathbf{A}'\psi'$ showing the transformed equation has the same form.
3. The eigenvalues are solutions of the secular equation $|\mathbf{A} - a\mathbf{E}| = 0$. Since $|\mathbf{AB}| = |\mathbf{A}|\,|\mathbf{B}|$ and $\mathbf{S}^\dagger \mathbf{S} = \mathbf{E}$, we may write

$$|\mathbf{A} - a\mathbf{E}| = |\mathbf{A} - a\mathbf{E}|\,|\mathbf{S}^\dagger \mathbf{S}| = |\mathbf{S}^\dagger|\,|\mathbf{A} - a\mathbf{E}|\,|\mathbf{S}| = |\mathbf{S}^\dagger \mathbf{A}\mathbf{S} - a\mathbf{E}| = |\mathbf{A}' - a\mathbf{E}|$$

Hence the eigenvalues do not depend on the choice of basis.

Appendix **B**

CALCULATION OF THE EIGENVALUES AND EIGENVECTORS OF A HERMITIAN TRIDIAGONAL MATRIX BY THE CONTINUED FRACTION METHOD

When the matrix representation of the Hamiltonian operator has a tridiagonal form

$$\begin{bmatrix} H_{11} & H_{12} & & & & \\ H_{21} & H_{22} & H_{23} & & & \\ & \ddots & \ddots & \ddots & & \\ & & H_{k,k-1} & H_{k,k} & H_{k,k+1} & \\ & & & \ddots & \ddots & \ddots \\ & & & H_{n,n-1} & H_{n,n} & \end{bmatrix} \tag{B.1}$$

the secular equation may be written in continued fraction form. The discussion given by Swalen and Pierce[1] on the continued fraction method will be followed here.

Let $\mathbf{S}$ be a unitary transformation which diagonalizes $\mathbf{H}$, that is,

$$\mathbf{S}^{\dagger}\mathbf{H}\mathbf{S}=\mathbf{\Lambda} \tag{B.2}$$

where $\mathbf{\Lambda}$ is a diagonal matrix. Denoting by $\mathbf{s}$ a column of $\mathbf{S}$, we have for a particular eigenstate the system of homogeneous linear equations

$$(\mathbf{H}-\lambda\mathbf{E})\mathbf{s}=0 \tag{B.3}$$

where λ is an element of $\mathbf{\Lambda}$ and $\mathbf{E}$ is a unit matrix.

To develop the continued fraction about the kth diagonal element of $\mathbf{H}$, let the eigenvector $\mathbf{s}$ be written as

$$\mathbf{s}=N_k\begin{bmatrix}\vdots\\ \sigma_{k-1}\\ 1\\ \sigma_{k+1}\\ \vdots\end{bmatrix} \tag{B.4}$$

The column matrix is an unnormalized eigenvector of $\mathbf{H}$ with the kth element set equal to unity, and N_k is the appropriate normalization factor defined explicitly by the equation

$$|N_k|^{-2}=1+\sum_{\alpha}\left[\sigma_{k+\alpha}^{*}\sigma_{k+\alpha}+\sigma_{k-\alpha}^{*}\sigma_{k-\alpha}\right] \tag{B.5}$$

Here and in following equations when the subscript $k+\alpha$ occurs, the running index α takes on the values $1, 2, \ldots, n-k$; whenever the subscript $k-\alpha$ appears, α takes on the values $1, 2, \ldots, k-1$. The simultaneous equations of relation (B.3) may be written in terms of the elements of the relative eigenvector, the kth equation is

$$H_{k,k-1}\sigma_{k-1}+(H_{kk}-\lambda)+H_{k,k+1}\sigma_{k+1}=0 \tag{B.6}$$

Employing the $(k+\alpha)$th and $(k-\alpha)$th equations we obtain the following recursion relation for the elements of the relative eigenvector

$$\frac{\sigma_{k\pm\alpha}}{\sigma_{k\pm\alpha\mp1}}=\frac{-H_{k\pm\alpha,k\pm\alpha\mp1}}{H_{k\pm\alpha,k\pm\alpha}-\lambda+H_{k\pm\alpha,k\pm\alpha\pm1}(\sigma_{k\pm\alpha\pm1}/\sigma_{k\pm\alpha})} \tag{B.7}$$

Defining the recursion relation

$$R_{k\pm\alpha}=\lambda-H_{k\pm\alpha,k\pm\alpha}-h_{k\pm\alpha\pm1}^{2}/R_{k\pm\alpha\pm1} \tag{B.8}$$

with

$$h_{k\pm\alpha}^{2}=|H_{k\pm\alpha,k\pm\alpha\mp1}|^{2} \tag{B.9}$$

we may write the above recursion relation for the σ's in compact form

$$\frac{\sigma_{k\pm\alpha}}{\sigma_{k\pm\alpha\mp1}}=\frac{H_{k\pm\alpha,k\pm\alpha\mp1}}{R_{k\pm\alpha}} \tag{B.10}$$

The components defined by (B.7) or (B.10) constitute an eigenvector only if (B.6) is satisfied. Note that the limits on α specified above arise because the quantity $R_{k\pm\alpha}$ or $h_{k\pm\alpha}$ is zero when $(k-\alpha)<1$ or $(k+\alpha)>n$.

If the expressions for σ_{k-1} and σ_{k+1} are inserted in the kth simultaneous equation, (B.6), one obtains the continued fraction form of the secular equation

$$\lambda=H_{kk}+\frac{h_{k+1}^{2}}{R_{k+1}}+\frac{h_{k-1}^{2}}{R_{k-1}} \tag{B.11}$$

The above expression represents the kth development (the leading term is H_{kk}) of the continued fraction form of the secular equation. There are n such developments corresponding to the n possible values of k.

To evaluate the eigenvalues of **H**, we may use a first-order iterative procedure. If interest lies in the kth eigenvalue and if the off-diagonal elements are small, then H_{kk} will most nearly approximate the kth eigenvalue. Substitution of this initial approximation to λ on the right-hand side of (B.11) gives an improved approximation to the eigenvalue. Note that the quantities R_{k+1} and R_{k-1} are functions of λ. The improved approximation may now be inserted in (B.11) to give a better approximation, and this process may be continued until the desired accuracy is obtained.

A much more efficient procedure is to use the Newton-Raphson technique, a second-order iterative procedure. Let the function $f_k(\lambda)$ be defined as

$$f_k(\lambda)=\lambda-H_{kk}-\frac{h_{k+1}^2}{R_{k+1}}-\frac{h_{k-1}^2}{R_{k-1}} \tag{B.12}$$

The roots of $f_k(\lambda)$ are the eigenvalues of **H**. There are n functions $f(\lambda)$ which may be defined; the above equation gives the kth development of $f(\lambda)$. If $f_k(\lambda)$ is expanded about the ith approximation $\lambda^{(i)}$ to an eigenvalue, then

$$f_k(\lambda)=f_k(\lambda^{(i)})+f'_k(\lambda^{(i)})(\lambda-\lambda^{(i)}) \tag{B.13}$$

where higher-order terms are neglected. From this, we see that for $f_k(\lambda)$ to be zero the next approximation $\lambda^{(i+1)}$ should be chosen as

$$\lambda^{(i+1)}=\lambda^{(i)}-\frac{f_k(\lambda^{(i)})}{f'_k(\lambda^{(i)})} \tag{B.14}$$

in which

$$\begin{aligned} f'_k(\lambda)&=\frac{df_k(\lambda)}{d\lambda} \\ &=1+\left(\frac{h_{k+1}}{R_{k+1}}\right)^2\left[1+\left(\frac{h_{k+2}}{R_{k+2}}\right)^2(1+\cdots)\right] \\ &\quad+\left(\frac{h_{k-1}}{R_{k-1}}\right)^2\left[1+\left(\frac{h_{k-2}}{R_{k-2}}\right)^2(1+\cdots)\right] \end{aligned} \tag{B.15}$$

As before, with an initial approximation to the eigenvalue, the function $f_k(\lambda)$ and its derivative are evaluated, and (B.14) is employed to give an improved approximation. The iterations may be continued until the change in the eigenvalue is small enough to be ignored. This second-order iterative procedure yields the eigenvalues much more rapidly. Since the sum of the diagonal elements of **H** is equal to the sum of the eigenvalues of **H**, that is,

$$\sum_k H_{kk}=\sum_k \lambda_k \tag{B.16}$$

one of the eigenvalues may be computed without solution of the continued fraction equation.

The derivative $f'_k(\lambda)$ is also related to the normalization constant N_k. By insertion of (B.10) into (B.5) it can be shown that

$$f'_k(\lambda)=1+\sum_\alpha (\sigma^*_{k+\alpha}\sigma_{k+\alpha}+\sigma^*_{k-\alpha}\sigma_{k-\alpha}) \tag{B.17}$$

for any λ. If λ is, in particular, an eigenvalue of **H**, that is, (B.11) is satisfied, then $f'_k(\lambda)$ is related to the normalization constant of the eigenvector

$$|N_k|^2=\frac{1}{f'_k(\lambda)} \tag{B.18}$$

For convergence that is rapid and to the desired root, both the choice of the initial approximation to the eigenvalue and the choice of development are important. For λ' an eigenvalue of **H**, the best development may be defined as the kth if

$$f'_k(\lambda')\leqslant f'_{k\pm\alpha}(\lambda') \qquad \alpha=1, 2, 3, \ldots \tag{B.19}$$

from which it follows that

$$\sigma^*_{k\pm\alpha}\sigma_{k\pm\alpha}\leqslant\sigma^*_k\sigma_k \tag{B.20}$$

for all values of α. Suppose $f(\lambda)$ is developed about some arbitrary position k'. Then, with λ^0 as a good approximation to the true eigenvalue λ', one evaluates $f'_{k'}(\lambda^0)$. If the largest term in $f'_{k'}(\lambda^0)$ is $\sigma^*_{k'+j}\sigma_{k'+j}$, then the best development is the kth, where $k=k'+j$.

Although the discussion has focused on the solution of tridiagonal matrices, it may be applied in other cases since it is possible to transform a general Hermitian matrix into tridiagonal form.[2]

References

1. J. D. Swalen and L. Pierce, *J. Math. Phys.*, **2**, 736 (1961). The reader is also directed to the following additional references on the continued fraction method: (a) G. W. King, R. M. Hainer, and P. C. Cross, *J. Chem. Phys.*, **11**, 27 (1943). (b) M. W. P. Strandberg, *Microwave Spectroscopy*, Wiley, New York, 1954. (c) D. W. Posener, *J. Chem. Phys.*, **24**, 546 (1956). (d) L. Pierce. *J. Math. Phys.*, **2**, 740 (1961).
2. W. Givens, *J. Assoc. Comp. Mach.* **4**, 298 (1957), and Oak Ridge National Laboratory Report ORNL-1574, March 3, 1954.

Appendix **C**

THE VAN VLECK TRANSFORMATION

We wish, in this section, to consider a perturbation technique which has found extensive application. Let the Hamiltonian of interest be divided into two parts

$$\mathscr{H}=\mathscr{H}_0+\lambda\mathscr{H}' \tag{C.1}$$

where $\mathscr{H}'$ is referred to as the perturbing Hamiltonian and $\mathscr{H}_0$ the unperturbed or zero-order Hamiltonian. The parameter λ is introduced for mathematical convenience in keeping track of the various orders of perturbation. It may be considered to vary between 1 and 0 with $\mathscr{H}$ reducing to $\mathscr{H}_0$ when $\lambda=0$; when $\lambda=1$ the full effect of the perturbation is experienced. The eigenfunctions and energy eigenvalues of the unperturbed problem will be denoted by ψ_{tr}^0 and E_{tr}^0, respectively. The E_{tr}^0, as the index r ranges over its values, are a set of near-degenerate or degenerate energy levels while those for the various values of t are assumed to be more widely separated. In general, r and t may each represent a set of quantum numbers. In this basis system, $\{\psi_{tr}^0\}$, $\mathscr{H}_0$ is diagonal

$$(t, r|\mathscr{H}_0|t', r')=\delta_{tt'}\delta_{rr'}E_{tr}^0 \tag{C.2}$$

whereas $\mathscr{H}'$ will, in general, have matrix elements both diagonal and off-diagonal in the two quantum numbers

$$(t, r|\mathscr{H}'|t', r')\neq 0$$

The Hamiltonian matrix is shown schematically in Fig. C.1. To obtain a tractable solution to the energy levels of $\mathscr{H}$ we seek a unitary transformation such that the transformed Hamiltonian matrix, denoted by $\tilde{\mathscr{H}}$, will have no

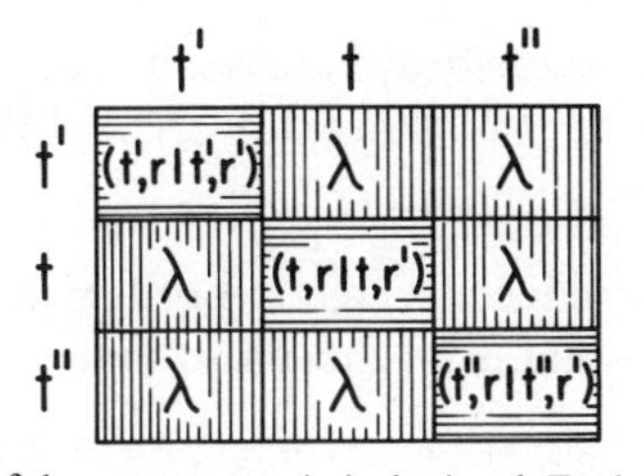

Fig. C.1 A small portion of the energy matrix is depicted. Each diagonal block represents a group of nearly degenerate energy levels. The diagonal blocks are connected by off-diagonal elements arising from the perturbation term in $\mathscr{H}$.

first-order elements connecting states of different t. As a consequence, matrix elements joining different t, that is, connections between the various groups of near-degenerate levels, will be reduced in size and can hence be ignored. This yields a set of smaller submatrices (diagonal in t) to be solved for the energy levels.

The transformed Hamiltonian matrix takes the form

$$\bar{\mathscr{H}} = \mathbf{T}^\dagger \mathscr{H} \mathbf{T} = \bar{\mathscr{H}}_0 + \lambda \bar{\mathscr{H}}_1 + \lambda^2 \bar{\mathscr{H}}_2 + \cdots \tag{C.3}$$

in which the transformed matrix is represented by a power series in the parameter λ, and we restrict ourselves to terms to second-order (λ^2). The transformation matrix may be expressed as

$$\mathbf{T} = e^{i\lambda S} = \mathbf{E} + i\lambda\mathbf{S} - \tfrac{1}{2}\lambda^2\mathbf{S}^2 - \cdots \tag{C.4}$$

where the exponential function is defined by the power series expansion and with $\mathbf{S}$ Hermitian, $\mathbf{T}$ is unitary. We hence have from (C.3)

$$\begin{aligned}
\bar{\mathscr{H}}_0 + \lambda \bar{\mathscr{H}}_1 + \lambda^2 \bar{\mathscr{H}}_2 + \cdots & \\
&= (\mathbf{E} - i\lambda\mathbf{S} - \tfrac{1}{2}\lambda^2\mathbf{S}^2 + \cdots)(\mathscr{H}_0 + \lambda\mathscr{H}')(\mathbf{E} + i\lambda\mathbf{S} - i\lambda^2\mathbf{S}^2 - \cdots) \\
&= \mathscr{H}_0 + \lambda[\mathscr{H}' + i(\mathscr{H}_0\mathbf{S} - \mathbf{S}\mathscr{H}_0)] \\
&\quad + \lambda^2[\mathbf{S}\mathscr{H}_0\mathbf{S} + i(\mathscr{H}'\mathbf{S} - \mathbf{S}\mathscr{H}') - \tfrac{1}{2}(\mathbf{S}^2\mathscr{H}_0 + \mathscr{H}_0\mathbf{S}^2)] + \cdots
\end{aligned} \tag{C.5}$$

For all this to be true for all values of λ, the coefficients of like powers of λ must be equal. Therefore, to second-order

$$\bar{\mathscr{H}}_0 = \mathscr{H}_0 \tag{C.6}$$

$$\bar{\mathscr{H}}_1 = \mathscr{H}' + i(\mathscr{H}_0\mathbf{S} - \mathbf{S}\mathscr{H}_0) \tag{C.7}$$

$$\bar{\mathscr{H}}_2 = \mathbf{S}\mathscr{H}_0\mathbf{S} + i(\mathscr{H}'\mathbf{S} - \mathbf{S}\mathscr{H}') - \tfrac{1}{2}(\mathbf{S}^2\mathscr{H}_0 + \mathscr{H}_0\mathbf{S}^2) \tag{C.8}$$

The first equation tells us that the matrix $\tilde{\mathscr{H}}_0$ is identical to $\mathscr{H}_0$. The remaining relations give the first- and second-order contributions to $\bar{\mathscr{H}}$ in terms of $\mathscr{H}$ and the transformation matrix $\mathbf{S}$.

Writing (C.7) in terms of its matrix elements, we have

$$(t, r|\bar{\mathscr{H}}_1|t', r') = (t, r|\mathscr{H}'|t', r') + i(t, r|\mathscr{H}_0 S - S\mathscr{H}_0|t', r') \tag{C.9}$$

Now we wish to choose the transformation such that no first-order connections exist between the diagonal t block and the rest of the matrix, i.e.,

$$(t, r|\mathscr{H}_1|t', r') = 0 \qquad \text{for } t' \neq t \tag{C.10}$$

Such connections are thereby reduced to second-order; they appear in $\mathscr{H}_2$. This condition from (C.9) requires that $\mathbf{S}$ be chosen so that

$$(t, r|S|t', r')\{(t, r|\mathscr{H}_0|t, r) - (t', r'|\mathscr{H}_0|t', r')\} = i(t, r|\mathscr{H}'|t', r') \tag{C.11}$$

or

$$(t, r|S|t', r') = \frac{i(t, r|\mathscr{H}'|t', r')}{E^0_{tr} - E^0_{t'r'}} \tag{C.12}$$

where we have taken cognizance of the fact that $\mathscr{H}_0$ is diagonal in the basis chosen. Similarly from (C.7) or the Hermitian property of **S** we find

$$(t', r'|S|t, r) = \frac{-i(t', r'|\mathscr{H}'|t, r)}{E^0_{tr} - E^0_{t'r'}} \tag{C.13}$$

The remaining elements of **S** may be conveniently set equal to zero,

$$(t, r|S|t, r') = 0 \tag{C.14}$$

$$(t', r'|S|t'', r'') = 0 \qquad \text{for } t', t'' \neq t \tag{C.15}$$

From this latter choice of the elements of **S** we have

$$(t, r|\bar{\mathscr{H}}_1|t, r') = (t, r|\mathscr{H}'|t, r') \tag{C.16}$$

for the elements of the t-block and

$$(t', r'|\bar{\mathscr{H}}_1|t'', r'') = (t', r'|\mathscr{H}'|t'', r'') \tag{C.17}$$

for $t', t'' \neq t$. The transformation thus affects connections of the type $(t|t')$ only.

With our choice of elements of **S**, the desired effect of reducing elements of the type $(t|t')$ in $\bar{\mathscr{H}}$ to λ^2 has been accomplished. Without the transformation, the $(t|t)$ block could be separated from the rest of the matrix only if terms in $\lambda\mathscr{H}'$ were neglected. See Fig. C.1. In the transformation process, elements of the $(t|t)$ block are modified. We now proceed to evaluate this modification by considering the contribution of $\bar{\mathscr{H}}_2$. The matrix elements diagonal in t are given by

$$\begin{aligned}(t, r)|\bar{\mathscr{H}}_2|t, r') = \sum_{t'', r''} \big[& E^0_{t''r''}(t, r|S|t'', r'')(t'', r''|S|t, r') \\ & + i(t, r|\mathscr{H}'|t'', r'')(t'', r''|S|t, r') - i(t, r|S|t'', r'')(t'', r''|\mathscr{H}'|t, r') \\ & - \tfrac{1}{2}(E^0_{tr'} + E^0_{tr})(t, r|S|t'', r'')(t'', r''|S|t, r')\big]\end{aligned}$$

Employing Eqs. (C.12) and (C.13) and collecting terms gives

$$(t, r)|\bar{\mathscr{H}}_2|t, r') = \frac{1}{2} \sum_{t'', r''}{}' (t, r|\mathscr{H}'|(t'', r'')(t'', r''|\mathscr{H}'|t, r') \left(\frac{1}{E^0_{tr} - E^0_{t''r''}} + \frac{1}{E^0_{tr'} - E^0_{t''r''}} \right) \tag{C.18}$$

The term $t'' = t$ is excluded from the sum. If only one level is present in the t block, viz., $r' = r$, then the above reduces to the second-order energy contribution of ordinary nondegenerate perturbation theory

$$\sum_{t''r''}{}' \frac{(t, r|\mathscr{H}'|t'', r'')(t'', r''|\mathscr{H}'|t, r)}{E^0_{tr} - E^0_{t''r''}} \tag{C.19}$$

where the $E^0_{t''r''}$ are not near E^0_{tr}.

Recapitulating, we can neglect the connections between the $(t|t)$ block and the rest of the matrix, since these have been reduced by the transformation to order λ^2. Diagonalization of this smaller matrix yields the energy levels of the group of near-degenerate or degenerate states, the matrix elements of the $(t|t)$

block being given to second-order by

$$(t, r|\bar{\mathscr{H}}|t, r')=(t, r|\mathscr{H}_0|t, r')+(t, r|\mathscr{H}'|t, r')+\frac{1}{2}\sum_{t'',r''}{}' (t, r|\mathscr{H}'|t'', r'')(t'', r''|\mathscr{H}'|t, r')$$
$$\times\left(\frac{1}{E^0_{tr}-E^0_{t''r''}}+\frac{1}{E^0_{tr'}-E^0_{t''r''}}\right) \qquad (C.20)$$

with $t''\neq t$, $(t, r|\mathscr{H}_0|t, r')=\delta_{rr'}E^0_{tr}$ and $\mathscr{H}'$ the perturbation.

In the text, this perturbation procedure is employed in the discussion of the Stark effect of asymmetric rotors; in the discussion of internal and over-all rotation it is exploited to give approximate diagonalization in the internal rotation quantum number.

Appendix **D**

FUNDAMENTAL CONSTANTS AND CONVERSION FACTORS

Fundamental Physical Constants[a]

Speed of light	c	$(2.997925 \pm 0.000001) \times 10^{10}$ cm/sec
Electronic charge	e	$(4.80298 \pm 0.00007) \times 10^{-10}$ esu
		$(1.60210 \pm 0.00002) \times 10^{-20}$ emu
Avogadro's constant	N	$(6.02252 \pm 0.00009) \times 10^{23}$ molecules/mole
1 amu	$M(^{12}C)/12$	$(1.66043 \pm 0.00002) \times 10^{-24}$ g
Electron rest mass	m_e	$(9.10908 \pm 0.00013) \times 10^{-28}$ g
Proton rest mass	m_p	$(1.67252 \pm 0.00003) \times 10^{-24}$ g
Neutron rest mass	m_n	$(1.67482 \pm 0.00003) \times 10^{-24}$ g
Planck's constant	h	$(6.62559 \pm 0.00016) \times 10^{-27}$ erg sec
Fine structure constant ($e^2/\hbar c$)	α	$(7.29720 \pm 0.00003) \times 10^{-3}$
Bohr magneton ($e\hbar/2m_e c$)	β	$(9.2732 \pm 0.0002) \times 10^{-21}$ erg/gauss
Nuclear magneton ($e\hbar/2m_p c$)	β_I	$(5.05050 \pm 0.00013) \times 10^{-24}$ erg/gauss
Boltzmann's constant	k	$(1.38054 \pm 0.00006) \times 10^{-16}$ erg/deg
Rydberg constant	R_∞	$(1.0973731 \pm 0.0000001) \times 10^{5}$ cm^{-1}
Bohr radius	a_0	$(5.29167 \pm 0.00002) \times 10^{-9}$ cm

Conversion Factors and Equivalents

1 electron-volt/particle	$= 1.60210 \times 10^{-12}$ erg
	$= 8065.73$ cm^{-1}
	$= 23061$ cal/mole
1 calorie (thermochemical)	$= 4.184 \times 10^{7}$ erg
1 erg	$= 1.50930 \times 10^{20}$ Mc
1 statvolt	$= 299.8$ volt
Standard volume, V_0	$= 22413.6$ cm^3/mole
Gas constant, R_0	$= 0.082053$ liter atm/mole deg
	$= 82.055$ cm^3 atm/mole deg
	$= 1.9872$ cal/mole deg
Degree Kelvin (°K)	$T(°K) = T(°C) + 273.15$
Rotational constant	$B(\text{Mc}) = \dfrac{5.05376 \times 10^5}{I_b(\text{amu Å}^2)}$
Stark effect constant	$\mu\mathscr{E} = 0.50344$ (MHz)/(debye volt/cm)
Reduced barrier height	$s = \dfrac{4.6602\ V_3(\text{cal/mole})}{F(\text{GHz})}$

[a] From E. R. Cohen and J. W. M. DuMond, *Rev. Mod. Phys*, **37**, 537 (1965). The unified scale of atomic masses is used throughout ($C^{12} = 12$).

Appendix **E**

ISOTOPIC ABUNDANCES, MASSES, AND MOMENTS

The relative isotopic abundances have been taken from the compilation of G. H. Fuller, *Nuclear Data Tables*, 1959. The masses are after J. H. E. Mattauch, W. Thiele, and A. H. Wapstra, *Nuclear Phys.*, **67**, 1 (1965). The nuclidic masses are for the new unified scale, which is based on the assignment of mass 12 to the isotope ^{12}C. The last place quoted in the masses is uncertain.

The nuclear spins and moments are from a compilation by G. H. Fuller and V. W. Cohen, *Appendix 1 to Nuclear Data Sheets*, May 1965. The zero spins enclosed by parentheses are based on the absence of observable hyperfine structure. A diamagnetic correction has been applied to the nuclear magnetic moments values listed.

Isotopic Abundances, Masses, and Moments

Atomic Number	Element	Symbol	Mass Number	Mass (amu)	Relative Abundance (%)	Spin	Nuclear Magnetic Dipole Moment μ (in units of nuclear magnetons)	Nuclear Electric Quadrupole Moment Q [in units of barns (10^{-24} cm^2)]
1	Hydrogen	H	1	1.00782519	99.9850	$\frac{1}{2}$	2.79278	
			2	2.0141022	0.01492	1	0.85742	0.0028
2	Helium	He	3	3.0160297	1.37×10^{-4}	$\frac{1}{2}$	−2.1276	
			4	4.0026031	99.999863	0		
3	Lithium	Li	6	6.015125	7.42	1	0.82201	−0.0008
			7	7.016004	92.58	$\frac{3}{2}$	3.2564	−0.04
								$\left(\frac{^6\text{Li}}{^7\text{Li}}\right)^a = 0.0176 \pm 0.0010$
4	Beryllium	Be	9	9.012186	100	$\frac{3}{2}$	−1.1776	± 0.03
5	Boron	B	10	10.0129388	19.61	3	1.8007	0.08
			11	11.0093053	80.39	$\frac{3}{2}$	2.6885	0.04
								$\left(\frac{^{10}\text{B}}{^{11}\text{B}}\right)^a = 2.084 \pm 0.002$
6	Carbon	C	12	12.0000000	98.893	0		
			13	13.0033544	1.107	$\frac{1}{2}$	0.7024	
7	Nitrogen	N	14	14.0030744	99.6337	1	0.4036	0.01
			15	15.0001077	0.3663	$\frac{1}{2}$	−0.2831	0.0000845
8	Oxygen	O	16	15.9949150	99.759	0		
			17	16.999133	0.0374	$\frac{5}{2}$	−1.8937	−0.026
			18	17.9991600	0.2039	0		
9	Fluorine	F	19	18.9984046	100	$\frac{1}{2}$	2.6287	

10	Neon	Ne	20	19.9924405	90.92	(0)		
			21	20.993849	0.257	$\frac{3}{2}$	−0.6618	0.09
			22	21.9913847	8.82	(0)		
11	Sodium	Na	23	22.989771	100	$\frac{3}{2}$	2.2176	0.11
12	Magnesium	Mg	24	23.985042	78.70	(0)		
			25	24.985839	10.13	$\frac{5}{2}$	−0.8553	0.22
			26	25.982593	11.17	(0)		
13	Aluminum	Al	27	26.981539	100	$\frac{5}{2}$	3.6414	0.15
14	Silicon	Si	28	27.976929	92.21	(0)		
			29	28.976496	4.70	$\frac{1}{2}$	−0.5553	
			30	29.973763	3.09	(0)		
15	Phosphorus	P	31	30.973765	100	$\frac{1}{2}$	1.1317	
16	Sulfur	S	32	31.9720737	95.0	0		
			33	32.971462	0.760	$\frac{3}{2}$	0.6434	−0.055
			34	33.967865	4.22	(0)		
			36	35.967090	0.0136	(0)		
17	Chlorine	Cl	35	34.968851	75.529	$\frac{3}{2}$	0.82183	−0.079
			37	36.965899	24.471	$\frac{3}{2}$	0.68411	−0.062
								$\left(\frac{^{35}Cl}{^{37}Cl}\right)^a = 1.26878 \pm 0.00015$
18	Argon	A	36	35.967545	0.337	(0)		
			38	37.962728	0.063	(0)		
			40	39.9623842	99.600	(0)		
19	Potassium	K	39	38.963710	93.10	$\frac{3}{2}$	0.3914	0.09
			40	39.964000	0.01181	4	−1.298	−0.07
			41	40.961832	6.88	$\frac{3}{2}$	0.2148	±0.11
								$\left(\frac{^{41}K}{^{39}K}\right)^a = 1.220 \pm 0.002$

Isotopic Abundances, Masses, and Moments

Atomic Number	*Element*	*Symbol*	*Mass Number*	*Mass (amu)*	*Relative Abundance (%)*	*Spin*	*Nuclear Magnetic Dipole Moment μ (in units of nuclear magnetons)*	*Nuclear Electric Quadrupole Moment Q [in units of barns (10^{-24} cm^2)]*
20	Calcium	Ca	40	39.962589	96.97	(0)		
			42	41.958625	0.64			
			43	42.958780	0.145	$\frac{7}{2}$	−1.317	
			44	43.955491	2.06			
			46	45.95369	0.0033			
			48	47.95253	0.185			
21	Scandium	Sc	45	44.955919	100	$\frac{7}{2}$	4.7564	−0.22
22	Titanium	Ti	46	45.952632	7.93			
			47	46.951769	7.28	$\frac{5}{2}$	−0.7884	
			48	47.947950	73.94	0	−1.1040	
			49	48.947870	5.51	$\frac{7}{2}$	−1.1040	
			50	49.944786	5.34			
23	Vanadium	V	50	49.947164	0.24	6	3.3470	
			51	50.943961	99.76	$\frac{7}{2}$	5.148	0.27
24	Chromium	Cr	50	49.946055	4.31			
			52	51.940513	83.76			
			53	52.940653	9.55	$\frac{3}{2}$	−0.4744	−0.03
			54	53.938882	2.38			
25	Manganese	Mn	55	54.938050	100	$\frac{5}{2}$	3.468	0.4
26	Iron	Fe	54	53.939617	5.82			
			56	55.934936	91.66			
			57	56.935398	2.19	$\frac{1}{2}$	0.0905	
			58	57.933282	0.33			

27	Cobalt	Co	59	58.933189	100	$\frac{7}{2}$	4.649	0.4
28	Nickel	Ni	58	57.935342	67.88			
			60	59.930787	26.23			
			61	60.931056	1.19			
			62	61.928342	3.66			
			64	63.927958	1.08			
29	Copper	Cu	63	62.929592	69.09	$\frac{3}{2}$	2.226	−0.18
			65	64.927786	30.91	$\frac{3}{2}$	2.385	−0.17
								$\left(\frac{^{63}Cu}{^{65}Cu}\right)^a = 1.0806 \pm 0.0003$
30	Zinc	Zn	64	63.929145	48.89	(0)		
			66	65.926052	27.81	(0)		
			67	66.92715	4.11	$\frac{5}{2}$	0.8757	0.17
			68	67.924857	18.57	(0)		
			70	69.925334	0.62			
31	Gallium	Ga	69	68.925574	60.4	$\frac{3}{2}$	2.016	0.19
			71	70.924706	39.6	$\frac{3}{2}$	2.562	0.12
								$\left(\frac{^{69}Ga}{^{71}Ga}\right)^a = 1.5867 \pm 0.0004$
32	Germanium	Ge	70	69.924252	20.52	(0)		
			72	71.922082	27.43	(0)		
			73	72.923463	7.76	$\frac{9}{2}$	−0.8792	−0.22
			74	73.921181	36.54	(0)		
			76	75.921405	7.76	(0)		
33	Arsenic	As	75	74.921596	100	$\frac{3}{2}$	1.439	0.29
34	Selenium	Se	74	73.922476	0.87	(0)		
			76	75.919207	9.02	(0)		
			77	76.919911	7.58	$\frac{1}{2}$	0.534	
			78	77.917314	23.52	0		

Isotopic Abundances, Masses, and Moments

Atomic Number	Element	Symbol	Mass Number	Mass (amu)	Relative Abundance (%)	Spin	Nuclear Magnetic Dipole Moment μ (in units of nuclear magnetons)	Nuclear Electric Quadrupole Moment Q [in units of barns (10^{-24} cm^2)]
			80	79.916527	49.82	0		
			82	81.916707	9.19	(0)		
35	Bromine	Br	79	78.918329	50.537	$\frac{3}{2}$	2.106	0.31
			81	80.916292	49.463	$\frac{3}{2}$	2.270	0.26
								$\left(\frac{^{79}\text{Br}}{^{81}\text{Br}}\right)^a = 1.19707 \pm 0.00003$
36	Krypton	Kr	78	77.920403	0.354			
			80	78.916380	2.27			
			82	81.913282	11.56	(0)		
			83	82.914131	11.55	$\frac{9}{2}$	−0.970	0.23
			84	85.911503	56.90	(0)		
			86	85.910616	17.37	(0)		
37	Rubidium	Rb	85	84.911800	72.15	$\frac{5}{2}$	1.3527	0.28
			87	86.909187	27.85	$\frac{3}{2}$	2.7506	0.14
								$\left(\frac{^{85}\text{Rb}}{^{87}\text{Rb}}\right)^a = 2.0669 \pm 0.0005$
38	Strontium	Sr	84	83.913430	0.56			
			86	85.909285	9.86	(0)		
			87	86.908892	7.02	$\frac{9}{2}$	−1.093	0.36
			88	87.905641	82.56	(0)		
39	Yttrium	Y	89	88.905872	100	$\frac{1}{2}$	−0.1373	

40	Zirconium	Zr	90	89.904700	51.46			
			91	90.905642	11.23	$\frac{5}{2}$	−1.303	
			92	91.905031	17.11			
			94	93.906313	17.40			
			95	94.905839	15.72	$\frac{5}{2}$		
			96	95.908286	2.80			
41	Niobium	Nb	93	92.906382	100	$\frac{9}{2}$	6.167	−0.22
42	Molybdenum	Mo	92	91.906810	15.84	(0)		
			94	93.905090	9.04	(0)		
			95	94.905839	15.72	$\frac{5}{2}$	−0.9135	
			96	95.904674	16.53	(0)		
			97	96.906022	9.46	$\frac{5}{2}$	−0.9327	
			98	97.905409	23.78	(0)		
			100	99.907475	9.63	(0)		
								$\left(\frac{^{97}Mo}{^{95}Mo}\right)^{a} > \pm 1$
44	Ruthenium	Ru	96	95.907598	5.51	$\frac{5}{2}$	−0.63	
			98	97.905289	1.87			
			99	98.905936	12.72	$\frac{5}{2}$	−0.63	
			100	99.904218	12.62			
			101	100.905577	17.07	$\frac{5}{2}$	−0.69	
			102	101.904348	31.61			
			104	103.905430	18.58			
45	Rhodium	Rh	103	102.905511	100	$\frac{1}{2}$	−0.0883	
46	Palladium	Pd	102	101.90561	0.96			
			104	103.90401	10.97			
			105	104.90506	22.23	$\frac{5}{2}$	−0.6015	
			106	105.903479	27.33			
			108	107.903891	26.71			
			110	109.90516	11.81			

Isotopic Abundances, Masses, and Moments

Atomic Number	*Element*	*Symbol*	*Mass Number*	*Mass (amu)*	*Relative Abundance (%)*	*Spin*	*Nuclear Magnetic Dipole Moment μ (in units of nuclear magnetons)*	*Nuclear Electric Quadrupole Moment Q [in units of barns* $(10^{-24}\ cm^2)$*]*
47	Silver	Ag	107	106.905094	51.35	$\frac{1}{2}$	−0.1135	
			109	108.904756	48.65	$\frac{1}{2}$	−0.1305	
48	Cadmium	Cd	106	105.906463	1.215			
			108	107.904187	0.875			
			110	109.903012	12.39	(0)		
			111	110.904188	12.75	$\frac{1}{2}$	−0.5950	
			112	111.902763	24.07	(0)		
			113	112.904409	12.26	$\frac{1}{2}$	−0.6224	
			114	113.903360	28.86	(0)		
			116	115.904762	7.58	(0)		
49	Indium	In	113	112.904089	4.28	$\frac{9}{2}$	5.523	0.82
			115	114.903871	95.72	$\frac{9}{2}$	5.534	0.83
50	Tin	Sn	112	111.90484	0.96			
			114	113.902773	0.66			
			115	114.903346	0.35	$\frac{1}{2}$	−0.918	
			116	115.901745	14.30	(0)		
			117	116.902958	7.61	$\frac{1}{2}$	−1.000	
			118	117.901606	24.03	(0)		
			119	118.903313	8.58	$\frac{1}{2}$	−1.046	
			120	119.902198	32.85	(0)		
			122	121.903441	4.72			
			124	123.905272	5.94			

51	Antimony	Sb	121	120.903816	57.25	$\frac{5}{2}$	3.359	−0.29
			123	122.904213	42.75	$\frac{7}{2}$	2.547	−0.37
								$\left(\frac{^{123}Sb}{^{121}Sb}\right)^a = 1.274717 \pm 0.000010$
52	Tellurium	Te	120	119.90402	0.089			
			122	121.903066	2.46			
			123	122.904277	0.87	$\frac{1}{2}$	−.7359	
			124	123.902842	4.61			
			125	124.904418	6.99	$\frac{1}{2}$	−0.8871	
			126	125.903322	18.71	(0)		
			128	127.904476	31.79	(0)		
			130	129.906238	34.48	(0)		
53	Iodine	I	127	126.904470	100	$\frac{5}{2}$	2.808	−0.79
54	Xenon	Xe	124	123.9061	0.096			
			126	125.904288	0.090			
			128	127.903540	1.919			
			129	128.904784	26.44	$\frac{1}{2}$	−0.7768	
			130	129.903509	4.08			
			131	130.905085	21.18	$\frac{3}{2}$	0.6908	−0.12
			132	131.904161	26.89	(0)		
			134	133.905397	10.44	(0)		
			136	135.907221	8.87	(0)		
55	Cesium	Cs	133	132.90536	100	$\frac{7}{2}$	2.579	−0.003
56	Barium	Ba	130	129.90625	0.101			
			132	131.9051	0.097			
			134	133.90461	2.42	(0)		
			135	134.9056	6.59	$\frac{3}{2}$	0.8372	0.18
			136	135. 90430	7.81	(0)		
			137	136.90550	11.32	$\frac{3}{2}$	0.9366	0.28

Atomic Number	*Element*	*Symbol*	*Mass Number*	*Mass (amu)*	*Relative Abundance (%)*	*Spin*	*Nuclear Magnetic Dipole Moment μ (in units of nuclear magnetons)*	*Nuclear Electric Quadrupole Moment Q [in units of barns (10^{-24} cm^2)]*
			138	137.90500	71.66	(0)		
								$\left(\frac{^{137}\mathrm{Ba}}{^{135}\mathrm{Ba}}\right)^a = 1.543 \pm 0.003$
57	Lanthanum	La	138	137.90691	0.089	5	3.707	± 0.8
			139	138.90614	99.911	$\frac{7}{2}$	2.778	0.22
								$\left(\frac{^{138}\mathrm{La}}{^{139}\mathrm{La}}\right)^a \simeq \mp 3.5$
58	Cerium	Ce	136	135.9071	0.193			
			138	137.90583	0.250			
			140	139.90539	88.48			
			142	141.90914	11.07			
59	Praseodymium	Pr	141	140.90760	100	$\frac{5}{2}$	4.5	-0.06
60	Neodymium	Nd	142	141.90766	27.11			
			143	142.90978	12.17	$\frac{7}{2}$	-1.1	-0.6
			144	143.91004	23.85			
			145	144.91254	8.30	$\frac{7}{2}$	-0.71	-0.3
			146	145.91309	17.22			
			148	147.91687	5.73			
			150	149.92092	5.62			
								$\left(\frac{^{143}\mathrm{Nd}}{^{145}\mathrm{Nd}}\right)^a = 1.893 \pm 0.016$

62	Samarium	Sm	144	143.91199	3.09			
			147	146.91487	14.97	$\frac{7}{2}$	-0.90	$< \pm 0.7$
			148	147.91479	11.24			
			149	148.91718	13.83	$\frac{7}{2}$	-0.75	$< \pm 0.7$
			150	149.91728	7.44			
			152	151.91976	26.72			
			154	153.92228	22.71			
63	Europium	Eu	151	150.91984	47.82	$\frac{5}{2}$	3.464	0.95
			153	152.92124	52.18	$\frac{5}{2}$	1.530	2.42
64	Gadolinium	Gd	152	151.91979	0.200			
			154	153.92093	2.15			
			155	154.92266	14.73	$\frac{3}{2}$	-0.27	1.3
			156	155.92218	20.47			
			157	156.92403	15.68	$\frac{3}{2}$	-0.36	1.5
			158	157.92418	24.87			
			160	159.92712	21.90			
65	Terbium	Tb	159	158.92535	100	$\frac{3}{2}$	± 1.7	
66	Dysprosium	Dy	156	155.9239	0.0524			
			158	157.92445	0.0902			
			160	159.92520	2.294			
			161	160.92695	18.88	$\frac{5}{2}$	± 0.42	± 1.1
			162	161.92680	25.53			
			163	162.92876	24.97	$\frac{5}{2}$	± 0.58	± 1.3
			164	163.92920	28.18			
								$\left(\frac{^{163}Dy}{^{161}Dy}\right)^{a} = 1.18 \pm 0.15$
67	Holmium	Ho	165	164.93042	100	$\frac{7}{2}$	4.1	3.0
68	Erbium	Er	162	161.92874	0.136			
			164	163.92929	1.56			
			166	165.93031	33.41			

Isotopic Abundances, Masses, and Moments

Atomic Number	*Element*	*Symbol*	*Mass Number*	*Mass (amu)*	*Relative Abundance (%)*	*Spin*	*Nuclear Magnetic Dipole Moment μ (in units of nuclear magnetons)*	*Nuclear Electric Quadrupole Moment Q [in units of barns (10^{-24} cm^2)]*
			167	166.93206	22.94	$\frac{7}{2}$	−0.56	2.8
			168	167.93238	27.07			
			170	169.93556	14.88			
69	Thulium	Tm	169	168.93425	100	$\frac{1}{2}$	−0.229	
70	Ytterbium	Yb	168	167.9342	0.135			
			170	169.93502	3.03			
			171	170.93643	14.31	$\frac{1}{2}$	0.4930	
			172	171.93636	21.82			
			173	172.93806	16.13	$\frac{5}{2}$	−0.679	3.0
			174	173.93874	31.84			
			176	175.94268	12.73			
71	Lutetium	Lu	175	174.94064	97.41	$\frac{7}{2}$	2.23	5.6
			176	175.94266	2.59	7	3.18	8.0
								$\left(\frac{^{175}Lu}{^{176}Lu}\right)^a = 0.71 \pm 0.01$
72	Hafnium	Hf	174	173.94036	0.18			
			176	175.94157	5.20			
			177	176.94340	18.50	$\frac{7}{2}$	0.61	3
			178	177.94388	27.14	(0)		
			179	178.94603	13.75	$\frac{9}{2}$	−0.47	3
			180	179.9468	35.24	(0)		
								$\left(\frac{^{177}Hf}{^{179}Hf}\right)^a = 0.99 \pm 0.02$

73	Tantalum	Ta	180	179.94754	0.0123			
			181	180.94801	99.9877	$\frac{7}{2}$	2.36	4.2
74	Tungsten	W	180	179.94700	0.135			
			182	181.94830	26.41	(0)		
			183	182.95032	14.40	$\frac{1}{2}$	0.117	
			184	183.95103	30.64	(0)		
			186	185.95444	28.41	(0)		
75	Rhenium	Re	185	184.95306	37.07	$\frac{5}{2}$	3.172	2.6
			187	186.95583	62.93	$\frac{5}{2}$	3.204	2.6
								$\left(\frac{^{185}Re}{^{187}Re}\right)^{a} = 1.056 \pm 0.005$
76	Osmium	Os	184	183.95275	0.018			
			186	185.95387	1.59			
			187	186.95583	1.64	$\frac{1}{2}$	0.067	
			188	187.95608	13.3			
			189	188.95830	16.1	$\frac{3}{2}$	0.6566	0.8
			190	189.95863	26.4			
			192	191.96145	41.0			
77	Iridium	Ir	191	190.96064	37.3	$\frac{3}{2}$	0.18	1.3
			193	192.96301	62.7	$\frac{3}{2}$	0.18	1.2
78	Platinum	Pt	190	189.95995	0.0127			
			192	191.96115	0.78			
			194	193.96273	32.9	(0)		
			195	194.96481	33.8	$\frac{1}{2}$	0.6060	
			196	195.96497	25.3	(0)		
			198	197.96790	7.21			
79	Gold	Au	197	196.96654	100	$\frac{3}{2}$	0.14486	0.58
80	Mercury	Hg	196	195.96582	0.146			
			198	197.966756	10.02	(0)		
			199	198.968279	16.84	$\frac{1}{2}$	0.5027	

Isotopic Abundances, Masses, and Moments

Atomic Number	*Element*	*Symbol*	*Mass Number*	*Mass (amu)*	*Relative Abundance (%)*	*Spin*	*Nuclear Magnetic Dipole Moment μ (in units of nuclear magnetons)*	*Nuclear Electric Quadrupole Moment Q [in units of barns (10^{-24} cm^2)]*
			200	199.968327	23.13	(0)		
			201	200.970308	13.22	$\frac{3}{2}$	−0.5567	0.45
			202	201.970642	29.80	(0)		
			204	203.973495	6.85	(0)		
81	Thallium	Tl	203	202.972353	29.50	$\frac{1}{2}$	1.6115	
			205	204.974442	70.50	$\frac{1}{2}$	1.6274	
82	Lead	ʻLb	204	203.973044	1.48			
			206	205.974468	23.6	(0)		
			207	206.975903	22.6	$\frac{1}{2}$	0.5895	
			208	207.976650	52.3	(0)		
83	Bismuth	Bi	209	208.980394	100	$\frac{9}{2}$	4.080	−0.34
90	Thorium	Th	232	232.03812	100			
92	Uranium	U	234	234.04090	0.0056			
			235	235.04392	0.7205	$\frac{7}{2}$	−0.35	4.1
			238	238.05077	99.2739	0		

[a]Quadrupole moment ratio.

Appendix **F**

BOND RADII

Covalent Radii (Å)[a]

Atom	Single Bond Radius	Double Bond Radius	Triple Bond Radius
H	0.32[b]		
B	0.81	0.71	0.64
C	0.772	0.667	0.603
N	0.74	0.62	0.55
O	0.74	0.62	0.55
F	0.72	0.60	
Si	1.17	1.07	1.00
P	1.10	1.00	0.93
S	1.04	0.94	0.87
Cl	0.99	0.89	
Ge	1.22	1.12	
As	1.21	1.11	
Se	1.17	1.07	
Br	1.14	1.04	
Sn	1.40	1.30	
Sb	1.41	1.31	
Te	1.37	1.27	
I	1.33	1.23	

Van Der Waals Radii (Å)

		H	1.2		
N	1.5	O	1.40	F	1.35
P	1.9	S	1.85	Cl	1.80
As	2.0	Se	2.00	Br	1.95
Sb	2.2	Te	2.20	I	2.15
CH_3	2.0				
C_6H_6	1.70 = half-thickness of benzene molecule				

[a]From L. Pauling, *Nature of the Chemical Bond*, 3rd ed., Cornell Univ. Press, Ithaca, N.Y., 1960. The bond length r_{AB} between

(*continued*)

two atoms A and B is given by

$$r_{AB} = r_A + r_B - \beta |x_A - x_B|$$

where r_A and r_B are the covalent radii of atoms A and B; x_A and x_B are the corresponding electronegativities. A table of electronegativities is given in Appendix G. The empirical constant β is assigned the following values:

$\beta = 0.08$ Å, for bonds involving one (or two) first-row atom(s).

$\beta = 0.06$ Å, for bonds between Si, P, or S and a more electronegative atom not in the first row.

$\beta = 0.04$ Å, for bonds between Ge, As, or Se and a more electronegative atom not in the first row.

$\beta = 0.02$ Å, for bonds between Sn, Sb, or Te and a more electronegative atom not in the first row.

$\beta = 0$, for bonds between C and elements of Groups V, VI, and VII not in the first row.

$\beta = 0.06$ Å, for bonds to hydrogen.[b]

[b]From W. Gordy, W. V. Smith, and R. T. Trambarulo, *Microwave Spectroscopy*, Wiley, New York, 1953.

Appendix **G**

ELECTRONEGATIVITIES OF THE ELEMENTS

			Electronegativities as Derived from		
Atomic Number	*Element*		*Atomic Radius and Nuclear Screening*[a]	*Bond Energies*[b]	*Selected Value*[c]
1	H		2.17	2.1	2.1_5
2	He		—	—	—
3	Li		0.96	1.0	0.95
4	Be		1.38	1.5	1.5
5	B		1.91	2.0	2.0
6	C		2.52	2.5	2.5
7	N		3.01	3.0	3.0
8	O		3.47	3.5	3.5
9	F		3.94	4.0	3.9_5
10	Ne		—	—	—
11	Na		0.90	0.9	0.9
12	Mg		1.16	1.2	1.2
13	Al		1.48	1.5	1.5
14	Si		1.82	1.8	1.8
15	P		2.19	2.1	2.1
16	S		2.58	2.5	2.5
17	Cl		3.00	3.0	3.0
18	A		—	—	—
19	K		0.82	0.8	0.80
20	Ca		1.03	1.0	1.0
21	Sc		1.3	1.3	1.3
22	Ti		1.6	1.6	1.6
		V^{III}	1.4	1.35	1.4
23	V	V^{IV}	1.7	1.65	1.7
		V^{V}	1.9	~1.8	1.9
		Cr^{II}	1.3	1.5	1.4
24	Cr	Cr^{III}	1.5	1.6	1.6
		Cr^{IV}	2.2	~2.1	2.2
		Mn^{II}	1.3	1.4	1.4
25	Mn	Mn^{III}	1.5	~1.5	1.5
		Mn^{VII}	2.6	~2.3	2.5

Atomic Number	Element		Electronegativities as Derived from: Atomic Radius and Nuclear Screening[a]	Bond Energies[b]	Selected Value[c]
26	Fe			1.65	1.7
				1.8	1.8
27	Co			1.7	1.7
28	Ni			1.7	1.8
29	Cu	Cu^{I}		1.8	1.8
		Cu^{II}		2.0	2.0
30	Zn	Zn^{II}		1.5	1.5
31	Ga		1.48	1.6	1.5
32	Ge		1.77	1.7	1.8
33	As		2.04	2.0	2.0
34	Se		2.35	2.3 (2.4)	2.4
35	Br		2.68	2.8	2.8
36	Kr		—	—	—
37	Rb		0.79	0.8	0.8
38	Sr		0.98	1.0	1.0
39	Y		1.21	1.2(1.3)	1.2
40	Zr		1.48	1.4 (1.6)	1.5
41	Nb		1.7	~1.6	1.7
42	Mo	Mo^{IV}	1.6	~1.6	1.6
		Mo^{VI}	2.1	~2.1	2.1
43	Tc	Tc^{V}	1.9		1.9
		Tc^{VII}	2.3		2.3
44	Ru			2.05	2.0
45	Rh			2.1	2.1
46	Pd			2.0	2.0
47	Ag			1.8	1.8
48	Cd			1.5	1.5
49	In	In^{I}	1.36	1.6	1.5
50	Sn	Sn^{II}	1.61	1.65 (1.7)	1.7
		Sn^{IV}		1.8	1.8
51	Sb	Sb^{III}	1.82	1.8 (1.8)	1.8
		Sb^{V}		2.1	
52	Te		2.08	2.1	2.1
53	I		2.36	2.6 (2.5)	2.5_5
54	Xe		—	—	—
55	Cs		0.78	0.7	0.75
56	Ba		0.93	0.85 (0.9)	0.9
57	La		1.2	1.1	1.1
58	Ce	Ce^{III}	1.2	1.05	1.1
59	Pr		1.2	1.1	1.1
60	Nd		1.2		~1.2
61	Il		1.2		~1.2

Atomic Number	Element		*Electronegativities as Derived from* Atomic Radius and Nuclear Screening[a]	Bond Energies[b]	Selected Value[c]
62	Sm		1.2		~1.2
63	Eu		1.1		~1.1
64	Gd		1.2		~1.2
65	Tb		1.2		~1.2
66	Dy		1.2		~1.2
67	Ho		1.2		~1.2
68	Er		1.2		~1.2
69	Tm		1.2		~1.2
70	Yb		1.1		~1.1
71	Lu		1.2		~1.2
72	Hf		1.4	~1.3	1.4
73	Ta	Ta^{III}	1.3	~1.4	1.3
		Ta^{V}	1.7		1.7
74	W	W^{IV}	1.6	~1.6	1.6
		W^{VI}	2.0	~2.1	2.0
75	Re	Re^{V}	1.8		1.8
		Re^{VII}	2.2		2.2
76	Os			2.1	2.0
77	Ir			2.1	2.1
78	Pt			2.1	2.1
79	Au			2.3	2.3
80	Hg	Hg^{I}		1.8	1.8
		Hg^{II}		1.9	
81	Tl	Tl^{I}	1.34	1.5	1.5
		Tl^{III}		1.9	1.9
82	Pb	Pb^{II}	1.56	1.6	1.6
		Pb^{IV}		1.8	1.8
83	Bi		1.8	1.8	1.8
84	Po		2.0	2.0	2.0
85	At		2.2	~2.4	2.2
86	Rn		—	—	—
87	Fa		0.76	~0.7	0.7
88	Ra		0.92	~0.8	0.9
89	Ac		1.1	1.0	1.1
90	Th	Th^{II}	1.0	1.1	1.0
		Th^{IV}	1.4		1.4
91	Pa	Pa^{III}	1.3	~1.4	1.3
		Pa^{V}	1.7		1.7
92	U	U^{IV}	1.5	1.3	1.4
		U^{VI}	1.9		1.9
93	Np		1.1		~1.1
94	Pu		1.3		~1.3
95	Am		1.3		~1.3

Atomic Number	Element	Electronegativities as Derived from: Atomic Radius and Nuclear Screening[a]	Electronegativities as Derived from: Bond Energies[b]	Selected Value[c]
96	Cm	1.3		~1.3
97	Bk	1.3		~1.3
98	Cf	1.3		~1.3

[a]From method derived by W. Gordy, *Phys. Rev.*, **69**, 604 (1946).
[b]Pauling's scale [*J. Am. Chem. Soc.*, **54**, 3570 (1932)] extended by Haissinsky [*J. Phys. Radium*, **7**, 7 (1946)].
[c]Values selected by Gordy and Thomas from those derived by four different methods [*J. Chem. Phys.*, **24**, 439 (1956)].

Appendix **H**

COMPUTATIONAL PROCEDURE FOR THE VIBRATIONAL EIGENVALUE PROBLEM

Relation between internal (S) and normal (Q) coordinates

$$\mathbf{S} = \mathbf{L}\mathbf{Q}$$

Vibrational, kinetic, and potential energy matrices

$$2\mathbf{T} = \tilde{\dot{\mathbf{S}}}\mathbf{G}^{-1}\dot{\mathbf{S}}, \quad 2\mathbf{V} = \tilde{\mathbf{S}}\mathbf{F}\mathbf{S}$$

Vibrational secular determinant (**GF** nonsymmetric)

$$\mathbf{GFL} = \mathbf{L}\boldsymbol{\Lambda} \rightarrow |\mathbf{GF} - \mathbf{E}\lambda| = 0$$

$\boldsymbol{\Lambda}$ diagonal matrix of eigenvalues λ_k, **L** eigenvector matrix, **E** a unit matrix.

* Diagonalize symmetric kinetic energy matrix and normalize orthogonal eigenvector matrix

$$\tilde{\mathbf{L}}_{\mathbf{G}}^{0}\mathbf{G}\mathbf{L}_{\mathbf{G}}^{0} = \boldsymbol{\Lambda}_{\mathbf{G}}, \quad \mathbf{L}_{\mathbf{G}} = \mathbf{L}_{\mathbf{G}}^{0}\boldsymbol{\Lambda}_{\mathbf{G}}^{1/2}[(L_G)_{ij} = (L_G^0)_{ij}(\lambda_G^{1/2})_j]$$

$\boldsymbol{\Lambda}_{\mathbf{G}}$ diagonal matrix of eigenvalues λ_G of **G**. Nonorthogonal transformation $\mathbf{L}_{\mathbf{G}}$ matrix thus defined so

$$\mathbf{L}_{\mathbf{G}}\tilde{\mathbf{L}}_{\mathbf{G}} = \mathbf{G}$$

* Transformation $\mathbf{L}_{\mathbf{G}}$ operates on **F** to give a new secular determinant

$$\mathbf{L}_{\mathbf{G}}\tilde{\mathbf{L}}_{\mathbf{G}}\mathbf{F}\mathbf{L}_{\mathbf{G}}\mathbf{L}_{\mathbf{G}}^{-1}\mathbf{L} = \mathbf{L}\boldsymbol{\Lambda}$$

$$\bar{\mathbf{F}}\mathbf{L}_{\mathbf{F}} = \mathbf{L}_{\mathbf{F}}\boldsymbol{\Lambda} \rightarrow |\bar{\mathbf{F}} - \mathbf{E}\lambda| = 0$$

$$\bar{\mathbf{F}} = \tilde{\mathbf{L}}_{\mathbf{G}}\mathbf{F}\mathbf{L}_{\mathbf{G}} \quad \text{and} \quad \mathbf{L}_{\mathbf{F}} = \mathbf{L}_{\mathbf{G}}^{-1}\mathbf{L}$$

* Diagonalization of the symmetric transformed F-matrix gives the frequencies and the orthogonal eigenvector matrix

$$\tilde{\mathbf{L}}_{\mathbf{F}}\bar{\mathbf{F}}\mathbf{L}_{\mathbf{F}} = \boldsymbol{\Lambda}$$

* The orthogonal eigenvector matrix gives the original nonorthogonal transformation matrix

$$\mathbf{L} = \mathbf{L}_{\mathbf{G}}\mathbf{L}_{\mathbf{F}}$$

* Computational checks

$$\mathbf{L}\tilde{\mathbf{L}} = \mathbf{G}, \ \tilde{\mathbf{L}}\mathbf{F}\mathbf{L} = \mathbf{\Lambda}$$

The eigenvalue problem in the form $\mathbf{FL} = \mathbf{G}^{-1}\mathbf{L}\mathbf{\Lambda}$ or $|\mathbf{F} - \mathbf{G}^{-1}\lambda| = 0$ can be solved by the same sequence of steps. **F** and **G** matrices in terms of symmetry coordinates may, of course, be used.

Appendix I

ENERGIES AND RELATIVE INTENSITIES OF NUCLEAR QUADRUPOLE HYPERFINE STRUCTURE

The function $Y(J, I, F)$ and the relative intensities are tabulated for the most common nuclear spins $I=\frac{1}{2}$, 1, $\frac{3}{2}$, $\frac{5}{2}$, $\frac{7}{2}$ and $\frac{9}{2}$. The $Y(J, I, F)$ is defined as

$$Y(J, I, F)=\frac{\frac{3}{4}C(C+1)-I(I+1)J(J+1)}{2(2J+3)(2J-1)I(2I-1)}$$

where

$$C=F(F+1)-I(I+1)-J(J+1)$$

and F takes on the values

$$J+I, J+I-1, \ldots, |J-I|$$

The quadrupole energies are given by

$$E_Q=\left[\frac{eQq_J(2J+3)}{J}\right]Y(J, I, F)$$

Here the appropriate q_J for a linear, symmetric, or asymmetric rotor must be used (see Chapter IX). The relative intensities [see e.g., (15.142)] have been normalized such that the sum of the intensities of the various hyperfine components of a transition is 100. The intensities for the $J+1\rightarrow J$ transitions are obtained by reversal of the arrows in the entries for $J\rightarrow J+1$. Thus the entries for $F\rightarrow F+1$ correspond to $F+1\rightarrow F$ and those for $F\rightarrow F-1$ to $F-1\rightarrow F$.

			Relative Intensities				
			$J \to J+1$			$J \to J$	
J	F	$Y(JIF)$	$F \to F+1$	$F \to F$	$F \to F-1$	$F \leftrightarrow F+1$	$F \to F$
				$I=\frac{1}{2}$			
0	$\frac{1}{2}$	0	66.667	33.333	0	0.0	0.0
1	$\frac{3}{2}$	0	60.000	6.667	0	0.0	55.556
	$\frac{1}{2}$	0	33.333	0.0	0	11.111	22.222
2	$\frac{5}{2}$	0	57.143	2.857	0	0.0	56.000
	$\frac{3}{2}$	0	40.000	0.0	0	4.000	36.000
3	$\frac{7}{2}$	0	55.556	1.587	0	0.0	55.102
	$\frac{5}{2}$	0	42.857	0.0	0	2.041	40.816
4	$\frac{9}{2}$	0	54.545	1.010	0	0.0	54.321
	$\frac{7}{2}$	0	44.444	0.0	0	1.235	43.210
5	$\frac{11}{2}$	0	53.846	0.699	0	0.0	53.719
	$\frac{9}{2}$	0	45.454	0.0	0	0.826	44.628
6	$\frac{13}{2}$	0	53.333	0.513	0	0.0	53.254
	$\frac{11}{2}$	0	46.154	0.0	0	0.592	45.562
7	$\frac{15}{2}$	0	52.941	0.392	0	0.0	52.889
	$\frac{13}{2}$	0	46.667	0.0	0	0.444	46.222
8	$\frac{17}{2}$	0	52.632	0.310	0	0.0	52.595
	$\frac{15}{2}$	0	47.059	0.0	0	0.346	46.713
9	$\frac{19}{2}$	0	52.381	0.251	0	0.0	52.355
	$\frac{17}{2}$	0	47.368	0.0	0	0.277	47.091
10	$\frac{21}{2}$	0	52.174	0.207	0	0.0	52.154
	$\frac{19}{2}$	0	47.619	0.0	0	0.227	47.392
11	$\frac{23}{2}$	0	52.000	0.174	0	0.0	51.985
	$\frac{21}{2}$	0	47.826	0.0	0	0.189	47.637
12	$\frac{25}{2}$	0	51.852	0.148	0	0.0	51.840
	$\frac{23}{2}$	0	48.000	0.0	0	0.160	47.840
13	$\frac{27}{2}$	0	51.724	0.128	0	0.0	51.715
	$\frac{25}{2}$	0	48.148	0.0	0	0.137	48.011
14	$\frac{29}{2}$	0	51.613	0.111	0	0.0	51.605
	$\frac{27}{7}$	0	48.276	0.0	0	0.119	48.157
15	$\frac{31}{2}$	0	51.515	0.098	0	0.0	51.509
	$\frac{29}{2}$	0	48.387	0.0	0	0.104	48.283

J	F	Y(J, I, F)	Relative Intensities				
			J→J+1			J→J	
			F→F+1	F→F	F→F−1	F↔F+1	F→F
			I=1				
0	1	0.0	55.556	33.333	11.111	0.0	0.0
1	2	0.050000	46.667	8.333	0.556	0.0	41.667
	1	−0.250000	25.000	8.333	0.0	13.889	8.333
	0	0.500000	11.111	0.0	0.0	11.111	0.0
2	3	0.071429	42.857	3.704	0.106	0.0	41.481
	2	−0.250000	29.630	3.704	0.0	5.185	23.148
	1	0.250000	20.000	0.0	0.0	5.000	15.000
3	4	0.038888	40.741	2.083	0.033	0.0	40.179
	3	−0.250000	31.250	2.083	0.0	2.679	28.009
	2	0.200000	23.810	0.0	0.0	2.646	21.164
4	5	0.090909	39.394	1.333	0.013	0.0	39.111
	4	−0.250000	32.000	1.333	0.0	1.630	30.083
	3	0.178571	25.926	0.0	0.0	1.620	24.306
5	6	0.096154	38.462	0.926	0.006	0.0	38.300
	5	−0.250000	32.407	0.926	0.0	1.094	31.148
	4	0.166667	27.273	0.0	0.0	1.091	26.182
6	7	0.100000	37.778	0.680	0.003	0.0	37.677
	6	−0.250000	32.653	0.680	0.0	0.785	31.765
	5	0.159091	28.05	0.0	0.0	0.783	27.422
7	8	0.102941	37.255	0.521	0.002	0.0	37.187
	7	−0.250000	32.813	0.521	0.0	0.590	32.153
	6	0.153846	28.889	0.0	0.0	0.590	28.299
8	9	0.105263	36.842	0.412	0.001	0.0	36.795
	8	−0.250000	32.922	0.412	0.0	0.460	32.414
	7	0.150000	29.412	0.0	0.0	0.460	28.952
9	10	0.107143	36.508	0.333	0.001	0.0	36.474
	9	−0.250000	33.000	0.333	0.0	0.368	32.597
	8	0.147059	29.825	0.0	0.0	0.368	29.456
10	11	0.108696	36.232	0.275	0.001	0.0	36.206
	10	−0.250000	33.058	0.275	0.0	0.302	32.730
	9	0.144737	30.159	0.0	0.0	0.302	29.857
11	12	0.110000	36.000	0.231	0.000	0.0	35.980
	11	−0.250000	33.102	0.231	0.0	0.252	32.830
	10	0.142857	30.435	0.0	0.0	0.252	30.183
12	13	0.111111	35.803	0.197	0.000	0.0	35.787
	12	−0.250000	33.136	0.197	0.0	0.213	32.907
	11	0.141304	30.667	0.0	0.0	0.213	30.454
13	14	0.112069	35.632	0.170	0.000	0.0	35.620
	13	−0.250000	33.163	0.170	0.0	0.183	32.968
	12	0.140000	30.864	0.0	0.0	0.183	30.682

J	F	$Y(J, I, F)$	Relative Intensities				
			$J \to J+1$			$J \to J$	
			$F \to F+1$	$F \to F$	$F \to F-1$	$F \leftrightarrow F+1$	$F \to F$
			$I=1$ (*continued*)				
14	15	0.112903	35.484	0.148	0.000	0.0	35.474
	14	−0.250000	33.185	0.148	0.0	0.158	33.017
	13	0.138889	31.035	0.0	0.0	0.158	30.876
15	16	0.113636	35.354	0.130	0.000	0.0	35.345
	15	−0.250000	33.203	0.130	0.0	0.139	33.056
	14	0.137931	31.183	0.0	0.0	0.139	31.044
			$I=\frac{3}{2}$				
0	$\frac{3}{2}$	0.0	50.000	33.333	16.667	0.0	0.0
1	$\frac{5}{2}$	0.050000	40.000	9.000	1.000	0.0	35.000
	$\frac{3}{2}$	−0.200000	21.000	10.667	1.667	15.000	4.444
	$\frac{1}{2}$	0.250000	8.333	8.333	0.0	13.889	2.778
2	$\frac{7}{2}$	0.071429	35.714	4.082	0.204	0.0	34.286
	$\frac{5}{2}$	−0.178571	24.490	5.224	0.286	5.714	17.286
	$\frac{3}{2}$	0.0	16.000	4.000	0.0	7.000	8.000
	$\frac{1}{2}$	0.250000	10.000	0.0	0.0	5.000	5.000
3	$\frac{9}{2}$	0.083333	33.333	2.315	0.066	0.0	32.738
	$\frac{7}{2}$	−0.166667	25.463	3.023	0.085	2.976	21.769
	$\frac{5}{2}$	−0.050000	19.133	2.296	0.0	3.827	14.745
	$\frac{3}{2}$	0.200000	14.286	0.0	0.0	2.857	11.429
4	$\frac{11}{2}$	0.090909	31.818	1.488	0.028	0.0	31.515
	$\frac{9}{2}$	−0.159091	25.785	1.959	0.034	1.818	23.583
	$\frac{7}{2}$	−0.071429	20.741	1.481	0.0	2.377	18.060
	$\frac{5}{2}$	0.178571	16.667	0.0	0.0	1.786	14.881
5	$\frac{13}{2}$	0.096154	30.769	1.036	0.013	0.0	30.594
	$\frac{11}{2}$	−0.153846	25.888	1.369	0.016	1.224	24.437
	$\frac{9}{2}$	−0.083333	21.694	1.033	0.0	1.612	19.904
	$\frac{7}{2}$	0.166667	18.182	0.0	0.0	1.212	16.970
6	$\frac{15}{2}$	0.100000	30.000	0.762	0.007	0.0	29.890
	$\frac{13}{2}$	−0.150000	25.905	1.010	0.008	0.879	24.882
	$\frac{1}{2}$	−0.090909	22.316	0.761	0.0	1.162	21.041
	$\frac{9}{2}$	0.159091	19.231	0.0	0.0	0.874	18.357
7	$\frac{17}{2}$	0.102941	29.412	0.584	0.004	0.0	29.338
	$\frac{15}{2}$	−0.147059	25.887	0.775	0.005	0.662	25.128
	$\frac{13}{2}$	−0.096154	22.750	0.583	0.0	0.877	21.797
	$\frac{11}{2}$	0.153846	20.000	0.0	0.0	0.659	19.341
8	$\frac{19}{2}$	0.105263	28.947	0.462	0.003	0.0	28.896
	$\frac{17}{2}$	−0.144737	25.854	0.613	0.003	0.516	25.270
	$\frac{15}{2}$	−0.100000	23.068	0.461	0.0	0.685	22.330

J	F	Y(J, I, F)	Relative Intensities				
			J→J+1			J→J	
			F→F+1	F→F	F→F−1	F↔F+1	F→F
			$I=\frac{3}{2}$ (continued)				
	$\frac{13}{2}$	0.150000	20.588	0.0	0.0	0.515	20.074
9	$\frac{21}{2}$	0.107143	28.571	0.374	0.002	0.0	28.534
	$\frac{19}{2}$	−0.142857	25.816	0.497	0.002	0.414	25.353
	$\frac{17}{2}$	−0.102941	23.310	0.374	0.0	0.549	22.722
	$\frac{15}{2}$	0.147059	21.053	0.0	0.0	0.413	20.640
10	$\frac{23}{2}$	−0.108696	28.261	0.309	0.001	0.0	28.233
	$\frac{21}{2}$	−0.141304	25.778	0.412	0.001	0.339	25.401
	$\frac{19}{2}$	−0.105263	23.500	0.309	0.0	0.450	23.021
	$\frac{17}{2}$	0.144737	21.429	0.0	0.0	0.338	21.090
11	$\frac{25}{2}$	0.110000	28.000	0.260	0.001	0.0	27.978
	$\frac{23}{2}$	−0.140000	25.740	0.346	0.001	0.283	25.428
	$\frac{21}{2}$	−0.107143	23.653	0.260	0.0	0.376	23.255
	$\frac{19}{2}$	0.142857	21.739	0.0	0.0	0.282	21.457
12	$\frac{27}{2}$	0.111111	27.778	0.222	0.001	0.0	27.761
	$\frac{25}{2}$	−0.138889	25.704	0.295	0.001	0.239	25.442
	$\frac{23}{2}$	−0.108696	23.778	0.222	0.0	0.318	23.442
	$\frac{21}{2}$	0.141304	22.000	0.0	0.0	0.239	21.761
13	$\frac{29}{2}$	0.112069	27.586	0.191	0.000	0.0	27.573
	$\frac{27}{2}$	−0.137931	25.671	0.254	0.001	0.205	25.447
	$\frac{25}{2}$	−0.110000	23.883	0.191	0.0	0.273	23.596
	$\frac{23}{2}$	0.140000	22.222	0.0	0.0	0.205	22.017
14	$\frac{31}{2}$	0.112903	27.419	0.166	0.000	0.0	27.408
	$\frac{29}{2}$	−0.137097	25.640	0.222	0.000	0.178	25.447
	$\frac{27}{2}$	−0.111111	23.971	0.166	0.0	0.237	23.723
	$\frac{25}{2}$	0.138889	22.414	0.0	0.0	0.178	22.236
15	$\frac{33}{2}$	0.113636	27.273	0.146	0.000	0.0	27.264
	$\frac{31}{2}$	−0.136364	25.611	0.195	0.000	0.156	25.443
	$\frac{29}{2}$	−0.112069	24.047	0.146	0.0	0.207	23.830
	$\frac{27}{2}$	0.137931	22.581	0.0	0.0	0.156	22.425
			$I=\frac{5}{2}$				
0	$\frac{5}{2}$	0.0	44.444	33.333	22.222	0.0	0.0
1	$\frac{7}{2}$	0.050000	33.333	9.524	1.587	0.0	28.571
	$\frac{5}{2}$	−0.160000	17.143	12.190	4.000	15.873	1.905
	$\frac{3}{2}$	0.140000	6.222	9.333	6.667	15.556	6.667
2	$\frac{9}{2}$	0.071429	28.571	4.409	0.353	0.0	27.161
	$\frac{7}{2}$	−0.121429	19.400	6.450	0.816	6.173	11.922
	$\frac{5}{2}$	−0.071429	12.245	6.612	1.143	8.571	3.429
	$\frac{3}{2}$	0.071429	6.857	5.418	1.058	8.000	9.148

J	F	$Y(J, I, F)$	*Relative Intensities*				
			$J \rightarrow J+1$			$J \rightarrow J$	
			$F \rightarrow F+1$	$F \rightarrow F$	$F \rightarrow F-1$	$F \leftrightarrow F+1$	$F \rightarrow F$
			$I=\frac{5}{2}$ *(continued)*				
	$\frac{1}{2}$	0.200000	2.963	3.704	0.0	5.185	1.481
3	$\frac{11}{2}$	0.083333	25.926	2.525	0.120	0.0	25.325
	$\frac{9}{2}$	−0.100000	19.697	3.848	0.265	3.247	15.713
	$\frac{7}{2}$	−0.100000	14.550	4.157	0.340	4.850	9.095
	$\frac{5}{2}$	−0.006667	10.393	3.628	0.265	5.102	4.898
	$\frac{3}{2}$	0.110000	7.143	2.381	0.0	4.286	2.593
	$\frac{1}{2}$	0.200000	4.762	0.0	0.0	2.646	2.116
4	$\frac{13}{2}$	0.090909	24.242	1.632	0.052	0.0	23.932
	$\frac{11}{2}$	−0.086364	19.580	2.532	0.110	1.994	17.164
	$\frac{9}{2}$	−0.107792	15.598	2.785	0.135	3.064	12.121
	$\frac{7}{2}$	−0.037662	12.256	2.463	0.096	3.333	8.571
	$\frac{5}{2}$	0.071429	9.524	1.587	0.0	2.910	6.349
	$\frac{3}{2}$	0.178571	7.407	0.0	0.0	1.852	5.556
5	$\frac{15}{2}$	0.096154	23.077	1.140	0.026	0.0	22.896
	$\frac{13}{2}$	−0.076923	19.373	1.785	0.054	1.347	17.767
	$\frac{11}{2}$	−0.110256	16.138	1.981	0.064	2.098	13.770
	$\frac{9}{2}$	−0.053846	13.350	1.758	0.043	2.314	10.795
	$\frac{7}{2}$	0.050000	10.999	1.122	0.0	2.043	8.780
	$\frac{7}{2}$	0.050000	10.999	1.122	0.0	2.043	8.780
	$\frac{5}{2}$	0.166667	9.091	0.0	0.0	1.299	7.792
6	$\frac{17}{2}$	0.100000	22.222	0.840	0.014	0.0	22.107
	$\frac{15}{2}$	−0.070000	19.160	1.324	0.029	0.970	18.021
	$\frac{13}{2}$	−0.110909	16.440	1.475	0.034	1.522	14.736
	$\frac{11}{2}$	−0.063636	14.051	1.312	0.022	1.691	12.196
	$\frac{9}{2}$	0.036364	11.988	0.833	0.0	1.499	10.372
	$\frac{7}{2}$	0.159091	10.256	0.0	0.0	0.950	9.307
7	$\frac{19}{2}$	0.102941	21.569	0.645	0.009	0.0	21.491
	$\frac{17}{2}$	−0.064706	18.963	1.020	0.017	0.731	18.166
	$\frac{15}{2}$	−0.110860	16.619	1.140	0.020	1.153	15.339
	$\frac{13}{2}$	−0.070136	14.529	1.014	0.013	1.286	13.128
	$\frac{11}{2}$	0.026923	12.692	0.641	0.0	1.142	11.470
	$\frac{9}{2}$	0.152846	11.111	0.0	0.0	0.722	10.390
8	$\frac{21}{2}$	0.105263	21.053	0.511	0.005	0.0	20.998
	$\frac{19}{2}$	−0.060526	18.788	0.809	0.011	0.571	18.134
	$\frac{17}{2}$	−0.110526	16.729	0.906	0.012	0.903	15.735
	$\frac{15}{2}$	−0.074737	14.873	0.806	0.008	1.009	13.780
	$\frac{13}{2}$	0.020000	13.217	0.508	0.0	0.897	12.263
	$\frac{11}{2}$	0.150000	11.765	0.0	0.0	0.566	11.199
9	$\frac{23}{2}$	0.107143	20.635	0.414	0.004	0.0	20.595
	$\frac{21}{2}$	−0.057143	18.634	0.658	0.007	0.458	18.115

J	F	$Y(J, I, F)$	Relative Intensities				
			$J\to J+1$			$J\to J$	
			$F\to F+1$	$F\to F$	$F\to F-1$	$F\leftrightarrow F+1$	$F\to F$
			$I=\frac{5}{2}$ *(continued)*				
	$\frac{19}{2}$	−0.110084	16.799	0.737	0.008	0.726	16.005
	$\frac{17}{2}$	−0.078151	15.129	0.656	0.005	0.813	14.255
	$\frac{15}{2}$	0.014706	13.622	0.413	0.0	0.722	12.858
	$\frac{13}{2}$	0.147059	12.281	0.0	0.0	0.455	11.826
10	$\frac{25}{2}$	0.108696	20.290	0.343	0.003	0.0	20.260
	$\frac{23}{2}$	−0.054348	18.498	0.545	0.005	0.375	18.076
	$\frac{21}{2}$	−0.109611	16.844	0.611	0.005	0.596	16.196
	$\frac{19}{2}$	−0.080778	15.326	0.543	0.003	0.668	14.611
	$\frac{17}{2}$	0.010526	13.944	0.342	0.0	0.594	13.318
	$\frac{15}{2}$	0.144737	12.698	0.0	0.0	0.373	12.325
1	$\frac{27}{2}$	0.110000	20.000	0.288	0.002	0.0	19.977
	$\frac{25}{2}$	−0.052000	18.379	0.459	0.003	0.313	18.029
	$\frac{23}{2}$	−0.109143	16.873	0.514	0.004	0.498	16.335
	$\frac{21}{2}$	−0.082857	15.482	0.458	0.002	0.559	14.887
	$\frac{19}{2}$	0.007143	14.205	0.28-	0.0	0.497	13.684
	$\frac{17}{2}$	0.142857	13.044	0.0	0.0	0.312	12.731
12	$\frac{29}{2}$	0.111111	19.753	0.246	0.001	0.0	19.735
	$\frac{27}{2}$	−0.050000	18.273	0.391	0.003	0.265	17.979
	$\frac{25}{2}$	−0.108696	16.891	0.439	0.003	0.422	16.437
	$\frac{23}{2}$	−0.084541	15.608	0.391	0.002	0.474	15.105
	$\frac{21}{2}$	0.004348	14.421	0.245	0.0	0.421	13.981
	$\frac{19}{2}$	0.141304	13.333	0.0	0.0	0.265	13.069
13	$\frac{31}{2}$	0.112069	19.540	0.212	0.001	0.0	19.526
	$\frac{29}{2}$	−0.048276	18.179	0.338	0.002	0.228	17.928
	$\frac{27}{2}$	−0.108276	16.903	0.379	0.002	0.363	16.514
	$\frac{25}{2}$	−0.085931	15.711	0.337	0.001	0.407	15.280
	$\frac{23}{2}$	0.002000	14.603	0.212	0.0	0.362	14.226
	$\frac{21}{2}$	0.140000	13.580	0.0	0.0	0.227	13.353
14	$\frac{33}{2}$	0.112903	19.355	0.185	0.001	0.0	19.343
	$\frac{31}{2}$	−0.046774	18.095	0.294	0.001	0.197	17.879
	$\frac{29}{2}$	−0.107885	16.909	0.331	0.002	0.315	16.573
	$\frac{27}{2}$	−0.087097	15.797	0.294	0.001	0.353	15.424
	$\frac{25}{2}$	0.0	14.758	0.184	0.0	0.314	14.431
	$\frac{23}{2}$	0.138889	13.793	0.0	0.0	0.197	13.596
15	$\frac{35}{2}$	0.113636	19.192	0.162	0.001	0.0	19.182
	$\frac{33}{2}$	−0.045455	18.019	0.259	0.001	0.173	17.831
	$\frac{31}{2}$	−0.107524	16.912	0.291	0.001	0.276	16.619
	$\frac{29}{2}$	−0.088088	15.870	0.259	0.001	0.310	15.544
	$\frac{27}{2}$	−0.001724	14.892	0.162	0.0	0.275	14.606
	$\frac{25}{2}$	0.137931	13.979	0.0	0.0	0.173	13.806

J	F	$Y(J, I, F)$	Relative Intensities				
			$J \to J+1$			$J \to J$	
			$F \to F+1$	$F \to F$	$F \to F-1$	$F \leftrightarrow F+1$	$F \to F$
			$I = \frac{7}{2}$				
0	$\frac{7}{2}$	0.0	41.667	33.333	25.000	0.0	0.0
1	$\frac{9}{2}$	0.050000	30.000	9.722	1.944	0.0	25.463
	$\frac{7}{2}$	−0.142857	15.278	12.698	5.357	16.204	1.058
	$\frac{5}{2}$	0.107143	5.357	9.643	10.000	16.071	8.929
2	$\frac{11}{2}$	0.071429	25.000	4.545	0.455	0.0	23.636
	$\frac{9}{2}$	−0.096939	16.883	6.926	1.190	6.364	9.470
	$\frac{7}{2}$	−0.081633	10.476	7.483	2.041	9.167	1.905
	$\frac{5}{2}$	0.025510	5.612	6.531	2.857	8.929	0.071
	$\frac{3}{2}$	0.153061	2.143	4.286	3.571	6.000	4.000
3	$\frac{13}{2}$	0.083333	22.222	2.618	0.160	0.0	21.635
	$\frac{11}{2}$	−0.071429	16.827	4.196	0.406	3.365	12.787
	$\frac{9}{2}$	−0.097619	12.311	4.885	0.661	5.276	6.629
	$\frac{7}{2}$	−0.047619	8.598	4.838	0.850	5.952	2.721
	$\frac{5}{2}$	0.035714	5.612	4.209	0.893	5.612	0.638
	$\frac{3}{2}$	0.119048	3.274	3.175	0.694	4.464	0.0
	$\frac{1}{2}$	0.178571	1.488	2.083	0.0	2.679	0.893
4	$\frac{15}{2}$	0.090909	20.455	1.697	0.071	0.0	20.148
	$\frac{13}{2}$	−0.055195	16.485	2.785	0.175	2.074	14.005
	$\frac{11}{2}$	−0.097403	13.054	3.338	0.275	3.365	9.324
	$\frac{9}{2}$	−0.071892	10.124	3.428	0.337	3.977	5.899
	$\frac{7}{2}$	−0.009276	7.660	3.127	0.325	4.012	3.527
	$\frac{5}{2}$	0.065399	5.268	2.494	0.212	3.571	2.012
	$\frac{3}{2}$	0.132653	4.000	1.556	0.0	2.750	1.185
	$\frac{1}{2}$	0.178571	2.778	0.0	0.0	1.620	1.157
5	$\frac{17}{2}$	0.096154	19.231	1.188	0.036	0.0	19.051
	$\frac{15}{2}$	−0.043956	16.120	1.974	0.087	1.404	14.460
	$\frac{13}{2}$	−0.094322	13.374	2.400	0.134	2.318	10.794
	$\frac{11}{2}$	−0.082418	10.974	2.504	0.159	2.797	7.947
	$\frac{9}{2}$	−0.032053	8.900	2.318	0.146	2.893	5.820
	$\frac{7}{2}$	0.036630	7.139	1.865	0.087	2.652	4.329
	$\frac{5}{2}$	0.107143	5.682	1.136	0.0	2.110	3.435
	$\frac{3}{2}$	0.166667	4.545	0.0	0.0	1.273	3.273
6	$\frac{19}{2}$	0.100000	18.333	0.877	0.020	0.0	18.219
	$\frac{17}{2}$	−0.035714	15.789	1.470	0.048	1.012	14.607
	$\frac{15}{2}$	−0.090909	13.510	1.801	0.073	1.689	11.635
	$\frac{13}{2}$	−0.087662	11.484	1.893	0.085	2.060	9.246
	$\frac{11}{2}$	−0.045455	9.700	1.764	0.075	2.156	7.385
	$\frac{9}{2}$	0.018831	8.152	1.421	0.043	1.998	6.015
	$\frac{7}{2}$	0.090909	6.838	0.855	0.0	1.603	5.128
	$\frac{5}{2}$	0.159091	5.769	0.0	0.0	0.962	4.808

J	F	$Y(J, I, F)$	Relative Intensities				
			$J \to J+1$			$J \to J$	
			$F \to F+1$	$F \to F$	$F \to F-1$	$F \leftrightarrow F+1$	$F \to F$
			$I=\frac{7}{2}$ *(continued)*				
7	$\frac{21}{2}$	0.102941	17.647	0.674	0.012	0.0	17.569
	$\frac{19}{2}$	−0.029412	15.502	1.135	0.029	0.764	14.620
	$\frac{17}{2}$	−0.087750	13.558	1.398	0.043	1.283	12.141
	$\frac{15}{2}$	−0.090498	11.808	1.476	0.049	1.576	10.099
	$\frac{13}{2}$	−0.054137	10.245	1.379	0.042	1.659	8.463
	$\frac{11}{2}$	0.006787	8.866	1.111	0.023	1.545	7.213
	$\frac{9}{2}$	0.079670	7.670	0.663	0.0	1.242	6.351
	$\frac{7}{2}$	0.153846	6.667	0.0	0.0	0.741	5.926
8	$\frac{23}{2}$	0.105263	17.105	0.534	0.008	0.0	17.050
	$\frac{21}{2}$	−0.024436	15.256	0.903	0.018	0.597	14.573
	$\frac{19}{2}$	−0.084962	13.563	1.115	0.027	1.007	12.458
	$\frac{17}{2}$	−0.092105	12.024	1.181	0.030	1.242	10.682
	$\frac{15}{2}$	−0.060150	10.634	1.105	0.026	1.312	9.227
	$\frac{13}{2}$	−0.001880	9.391	0.889	0.014	1.225	8.084
	$\frac{11}{2}$	0.071429	8.296	0.528	0.0	0.985	7.254
	$\frac{9}{2}$	0.150000	7.353	0.0	0.0	0.585	6.768
9	$\frac{25}{2}$	0.107143	16.667	0.433	0.005	0.0	16.626
	$\frac{23}{2}$	−0.020408	15.043	0.734	0.012	0.479	14.500
	$\frac{21}{2}$	−0.082533	13.546	0.910	0.018	0.810	12.661
	$\frac{19}{2}$	−0.093037	12.173	0.965	0.020	1.003	11.094
	$\frac{17}{2}$	−0.064526	10.922	0.903	0.017	1.062	9.787
	$\frac{15}{2}$	−0.008403	9.791	0.727	0.009	0.993	8.735
	$\frac{13}{2}$	0.065126	8.781	0.430	0.0	0.798	7.940
	$\frac{11}{2}$	0.147059	7.895	0.0	0.0	0.472	7.422
10	$\frac{27}{2}$	0.108696	16.304	0.359	0.004	0.0	16.274
	$\frac{25}{2}$	−0.017081	14.859	0.609	0.008	0.393	14.417
	$\frac{22}{2}$	−0.080418	13.518	0.756	0.012	0.666	12.794
	$\frac{21}{2}$	−0.093576	12.279	0.802	0.013	0.826	11.393
	$\frac{19}{2}$	−0.067833	11.142	0.752	0.011	0.876	10.208
	$\frac{17}{2}$	−0.013485	10.104	0.604	0.006	0.820	9.235
	$\frac{15}{2}$	0.060150	9.167	0.357	0.0	0.659	8.476
	$\frac{13}{2}$	0.144737	8.333	0.0	0.0	0.389	7.944
11	$\frac{29}{2}$	0.110000	16.000	0.302	0.003	0.0	15.976
	$\frac{27}{2}$	−0.014286	14.698	0.513	0.006	0.382	14.332
	$\frac{25}{2}$	−0.078571	13.484	0.637	0.009	0.557	12.882
	$\frac{23}{2}$	−0.093877	12.357	0.678	0.009	0.692	11.617
	$\frac{21}{2}$	−0.070408	11.314	0.635	0.008	0.735	10.534
	$\frac{19}{2}$	−0.017551	10.356	0.510	0.004	0.688	9.629
	$\frac{17}{2}$	0.056122	9.482	0.300	0.0	0.553	8.905
	$\frac{15}{2}$	0.142857	8.696	0.0	0.0	0.326	8.370

J	F	Y(J, I, F)	Relative Intensities: J→J+1, F→F+1	F→F	F→F−1	J→J: F↔F+1	F→F
			I = 7/2				
12	31/2	0.111111	15.741	0.257	0.002	0.0	15.722
	29/2	−0.011905	14.558	0.438	0.004	0.278	14.249
	27/2	−0.076950	13.449	0.545	0.006	0.473	12.940
	25/2	−0.094030	12.414	0.579	0.007	0.588	11.788
	23/2	−0.072464	11.451	0.543	0.006	0.625	10.790
	21/2	−0.020876	10.561	0.436	0.003	0.585	9.945
	19/2	0.052795	9.744	0.256	0.0	0.470	9.254
	17/2	0.141304	9.000	0.0	0.0	0.276	8.724
13	33/2	0.112069	15.517	0.222	0.001	0.0	15.502
	31/2	−0.009852	14.433	0.378	0.003	0.238	14.170
	29/2	−0.075517	13.413	0.471	0.005	0.406	12.977
	27/2	−0.094089	12.457	0.501	0.005	0.505	11.920
	25/2	−0.074138	11.563	0.470	0.004	0.537	10.996
	23/2	−0.023645	10.732	0.377	0.002	0.504	10.203
	21/2	0.050000	9.964	0.221	0.0	0.404	9.544
	19/2	0.140000	9.259	0.0	0.0	0.237	9.022
14	35/2	0.112903	15.323	0.194	0.001	0.0	15.310
	33/2	−0.008065	14.323	0.330	0.003	0.207	14.096
	31/2	−0.074245	13.379	0.411	0.004	0.353	13.001
	29/2	−0.094086	12.490	0.438	0.004	0.439	12.205
	27/2	−0.075525	11.656	0.410	0.003	0.467	11.164
	25/2	−0.025986	10.876	0.329	0.002	0.438	10.418
	23/2	0.047619	10.152	0.193	0.0	0.351	9.787
	21/2	0.138889	9.483	0.0	0.0	0.206	9.277
15	37/2	0.113636	15.152	0.170	0.001	0.0	15.141
	35/2	−0.006494	14.224	0.290	0.002	0.181	14.026
	33/2	−0.073108	13.345	0.362	0.003	0.309	13.016
	31/2	−0.094044	12.515	0.385	0.003	0.385	12.109
	29/2	−0.076690	11.733	0.361	0.002	0.410	11.303
	27/2	−0.027989	10.999	0.290	0.001	0.384	10.598
	25/2	0.045566	10.314	0.170	0.0	0.308	9.995
	23/2	0.137931	9.677	0.0	0.0	0.181	9.497
			I = 9/2				
0	9/2	0.0	40.000	33.333	26.667	0.0	0.0
1	11/2	0.050000	28.000	9.818	2.182	0.0	23.636
	9/2	−0.1333333	14.182	12.929	6.222	16.364	0.673
	7/2	0.091667	4.889	9.778	12.000	16.296	10.370
2	13/2	0.071429	22.857	4.615	0.527	0.0	21.538

J	F	Y(J, I, F)	Relative Intensities: J→J+1 F→F+1	J→J+1 F→F	J→J+1 F→F−1	J→J F↔F+1	J→J F→F
			$I=\frac{9}{2}$ (continued)				
	$\frac{11}{2}$	−0.083333	15.385	7.161	1.455	6.462	8.084
	$\frac{9}{2}$	−0.083333	9.455	7.879	2.667	9.455	1.212
	$\frac{7}{2}$	0.005952	4.952	6.966	4.082	9.333	0.381
	$\frac{5}{2}$	0.130952	1.796	4.490	5.714	6.286	5.714
3	$\frac{15}{2}$	0.83333	20.000	2.667	0.190	0.0	19.429
	$\frac{13}{2}$	−0.055556	15.111	4.376	0.513	3.429	11.077
	$\frac{11}{2}$	−0.091667	10.989	5.245	0.909	5.495	5.285
	$\frac{9}{2}$	−0.061111	7.576	5.387	1.323	6.364	1.732
	$\frac{7}{2}$	0.005556	4.815	4.913	1.701	6.190	0.136
	$\frac{5}{2}$	0.083333	2.653	3.918	2.000	5.102	0.327
	$\frac{3}{2}$	0.152778	1.048	2.444	2.222	3.143	2.571
4	$\frac{17}{2}$	0.090909	18.182	1.733	0.086	0.0	17.882
	$\frac{15}{2}$	−0.037879	14.631	2.921	0.226	2.118	12.134
	$\frac{13}{2}$	−0.086580	11.539	3.625	0.392	3.526	7.722
	$\frac{11}{2}$	−0.079545	8.876	3.906	0.551	4.308	4.480
	$\frac{9}{2}$	−0.037879	6.612	3.826	0.673	4.545	2.245
	$\frac{7}{2}$	0.020563	4.714	3.448	0.727	4.321	0.854
	$\frac{5}{2}$	0.081169	3.152	2.836	0.679	3.714	0.152
	$\frac{3}{2}$	0.132576	1.891	2.069	0.485	2.800	0.015
	$\frac{1}{2}$	0.166667	0.889	1.333	0.0	1.630	0.593
5	$\frac{19}{2}$	0.096154	16.923	1.215	0.044	0.0	16.746
	$\frac{17}{2}$	−0.025641	14.170	2.078	0.115	1.435	12.490
	$\frac{15}{2}$	−0.080128	11.724	2.626	0.196	2.439	9.052
	$\frac{13}{2}$	−0.085470	9.566	2.892	0.269	3.055	6.344
	$\frac{11}{2}$	−0.057692	7.682	2.910	0.318	3.329	4.275
	$\frac{9}{2}$	−0.010684	6.052	2.712	0.326	3.306	2.755
	$\frac{7}{2}$	0.043803	4.662	2.331	0.280	3.030	1.697
	$\frac{5}{2}$	0.096154	3.497	1.790	0.168	2.545	1.018
	$\frac{3}{2}$	0.138889	2.545	1.091	0.0	1.891	0.655
	$\frac{1}{2}$	0.166667	1.818	0.0	0.0	1.091	0.727
6	$\frac{21}{2}$	0.100000	16.000	0.898	0.025	0.0	15.887
	$\frac{19}{2}$	−0.016667	13.769	1.551	0.065	1.036	12.567
	$\frac{17}{2}$	−0.074242	11.757	1.980	0.109	1.781	9.802
	$\frac{15}{2}$	−0.087121	9.955	2.206	0.147	2.262	7.540
	$\frac{13}{2}$	−0.068182	8.352	2.249	0.169	2.505	5.728
	$\frac{11}{2}$	−0.028788	6.938	2.125	0.168	2.536	4.317
	$\frac{9}{2}$	0.021212	5.706	1.849	0.137	2.378	3.263
	$\frac{7}{2}$	0.073485	4.650	1.429	0.075	2.051	2.533
	$\frac{5}{2}$	0.121212	3.768	0.848	0.0	1.570	2.122
	$\frac{3}{2}$	0.159091	3.077	0.0	0.0	0.923	2.154

J	F	$Y(J, I, F)$	Relative Intensities $J \rightarrow J+1$ $F \rightarrow F+1$	$F \rightarrow F$	$F \rightarrow F-1$	$J \rightarrow J$ $F \leftrightarrow F+1$	$F \rightarrow F$
			$I=\frac{9}{2}$ (*continued*)				
7	$\frac{23}{2}$	0.102941	15.294	0.691	0.015	0.0	15.217
	$\frac{21}{2}$	−0.009804	13.427	1.200	0.039	0.783	12.528
	$\frac{19}{2}$	−0.069193	11.726	1.543	0.065	1.356	10.240
	$\frac{17}{2}$	−0.087104	10.183	1.731	0.087	1.737	8.322
	$\frac{15}{2}$	−0.074284	8.792	1.776	0.098	1.941	6.741
	$\frac{13}{2}$	−0.040347	7.549	1.689	0.095	1.984	5.470
	$\frac{11}{2}$	0.006222	6.448	1.477	0.075	1.879	4.485
	$\frac{9}{2}$	0.058069	5.487	1.141	0.039	1.636	3.771
	$\frac{7}{2}$	0.108974	4.667	0.667	0.0	1.259	3.339
	$\frac{5}{2}$	0.153846	4.000	0.0	0.0	0.735	3.265
8	$\frac{25}{2}$	0.105263	14.737	0.547	0.010	0.0	14.682
	$\frac{23}{2}$	−0.004386	13.137	0.956	0.025	0.612	12.440
	$\frac{21}{2}$	−0.064912	11.666	1.234	0.041	1.066	10.503
	$\frac{19}{2}$	−0.086403	10.320	1.390	0.054	1.373	8.849
	$\frac{17}{2}$	−0.078070	9.095	1.433	0.061	1.543	7.459
	$\frac{15}{2}$	−0.048246	7.986	1.368	0.058	1.586	6.316
	$\frac{13}{2}$	−0.004386	6.993	1.198	0.044	1.510	5.406
	$\frac{11}{2}$	0.046930	6.117	0.924	0.023	1.320	4.723
	$\frac{9}{2}$	0.100000	5.348	0.535	0.0	1.016	4.278
	$\frac{7}{2}$	0.150000	4.706	0.0	0.0	0.588	4.118
9	$\frac{27}{2}$	0.107143	14.286	0.444	0.007	0.0	14.246
	$\frac{25}{2}$	0.0	12.889	0.778	0.017	0.491	12.334
	$\frac{23}{2}$	−0.061275	11.596	1.008	0.027	0.859	10.662
	$\frac{21}{2}$	−0.085434	10.403	1.140	0.036	1.111	9.215
	$\frac{19}{2}$	−0.080532	9.309	1.178	0.040	1.253	7.980
	$\frac{17}{2}$	−0.053922	8.310	1.126	0.037	1.293	6.947
	$\frac{15}{2}$	−0.012255	7.406	0.987	0.028	1.234	6.106
	$\frac{13}{2}$	0.038515	6.595	0.760	0.014	1.081	5.456
	$\frac{11}{2}$	0.093137	5.879	0.437	0.0	0.831	5.006
	$\frac{9}{2}$	0.147059	5.263	0.0	0.0	0.478	4.785
10	$\frac{29}{2}$	0.108696	13.913	0.368	0.005	0.0	13.883
	$\frac{27}{2}$	0.003623	12.676	0.646	0.012	0.403	12.224
	$\frac{25}{2}$	−0.058162	11.523	0.839	0.019	0.707	10.758
	$\frac{23}{2}$	−0.084382	10.454	0.950	0.025	0.916	9.476
	$\frac{21}{2}$	−0.082189	9.466	0.984	0.027	1.037	8.368
	$\frac{19}{2}$	−0.058162	8.557	0.942	0.025	1.072	7.427
	$\frac{17}{2}$	−0.018307	7.727	0.826	0.019	1.025	6.648
	$\frac{15}{2}$	0.031941	6.975	0.635	0.009	0.898	6.030
	$\frac{13}{2}$	0.087719	6.303	0.364	0.0	0.690	5.581
	$\frac{11}{2}$	0.144737	5.714	0.0	0.0	0.396	5.319

J	F	$Y(J, I, F)$	*Relative Intensities*				
			$J \rightarrow J+1$			$J \rightarrow J$	
			$F \rightarrow F+1$	$F \rightarrow F$	$F \rightarrow F-1$	$F \leftrightarrow F+1$	$F \rightarrow F$
			$I=\frac{9}{2}$ *(continued)*				
11	$\frac{31}{2}$	0.110000	13.600	0.310	0.003	0.0	13.576
	$\frac{29}{2}$	0.006667	12.490	0.545	0.008	0.337	12.115
	$\frac{27}{2}$	−0.055476	11.452	0.709	0.014	0.592	10.814
	$\frac{25}{2}$	−0.083333	10.483	0.804	0.017	0.769	9.665
	$\frac{23}{2}$	−0.083333	9.583	0.833	0.019	0.871	8.661
	$\frac{21}{2}$	−0.061429	8.749	0.798	0.017	0.902	7.799
	$\frac{19}{2}$	−0.023095	7.983	0.700	0.013	0.864	7.075
	$\frac{17}{2}$	0.026667	7.282	0.358	0.006	0.757	6.488
	$\frac{15}{2}$	0.083333	6.650	0.307	0.0	0.581	6.043
	$\frac{13}{2}$	0.142857	6.087	0.0	0.0	0.332	5.755
12	$\frac{33}{2}$	0.111111	13.333	0.264	0.002	0.0	13.315
	$\frac{31}{2}$	0.009259	12.328	0.465	0.006	0.285	12.012
	$\frac{29}{2}$	−0.053140	11.384	0.606	0.010	0.502	10.844
	$\frac{27}{2}$	−0.082327	10.499	0.689	0.013	0.654	9.805
	$\frac{25}{2}$	−0.084138	9.672	0.714	0.014	0.742	8.889
	$\frac{23}{2}$	−0.064010	8.903	0.685	0.012	0.769	8.094
	$\frac{21}{2}$	−0.026973	8.190	0.601	0.009	0.737	7.417
	$\frac{19}{2}$	0.022343	7.535	0.461	0.004	0.646	6.858
	$\frac{17}{2}$	0.079710	6.938	0.262	0.0	0.496	6.422
	$\frac{15}{2}$	0.141304	6.400	0.0	0.0	0.282	6.118
13	$\frac{35}{2}$	0.112069	13.103	0.228	0.002	0.0	13.088
	$\frac{33}{2}$	0.011494	12.186	0.402	0.005	0.245	11.916
	$\frac{31}{2}$	−0.011092	11.320	0.524	0.007	0.432	10.858
	$\frac{29}{2}$	−0.081379	10.506	0.596	0.009	0.562	9.910
	$\frac{27}{2}$	−0.084713	9.741	0.619	0.010	0.639	9.068
	$\frac{25}{2}$	−0.066092	9.027	0.594	0.009	0.663	8.330
	$\frac{23}{2}$	−0.030172	8.362	0.521	0.007	0.636	7.696
	$\frac{21}{2}$	0.018736	7.746	0.399	0.003	0.557	7.164
	$\frac{19}{2}$	0.076667	7.181	0.227	0.0	0.427	6.737
	$\frac{17}{2}$	0.140000	6.667	0.0	0.0	0.243	6.424
14	$\frac{37}{2}$	0.112903	12.903	0.199	0.001	0.0	12.891
	$\frac{35}{2}$	0.013441	12.059	0.351	0.004	0.212	11.826
	$\frac{33}{2}$	−0.049283	11.261	0.458	0.006	0.375	10.860
	$\frac{31}{2}$	−0.080496	10.506	0.521	0.007	0.489	9.989
	$\frac{29}{2}$	−0.085125	9.796	0.541	0.008	0.556	9.211
	$\frac{27}{2}$	−0.067802	9.129	0.519	0.007	0.578	8.524
	$\frac{25}{2}$	−0.032855	8.505	0.455	0.005	0.554	7.926
	$\frac{23}{2}$	0.015681	7.925	0.349	0.002	0.486	7.419
	$\frac{21}{2}$	0.074074	7.388	0.198	0.0	0.372	7.003
	$\frac{19}{2}$	0.138889	6.897	0.0	0.0	0.211	6.685

J	F	$Y(J, I, F)$	Relative Intensities				
			$J \rightarrow J+1$			$J \rightarrow J$	
			$F \rightarrow F+1$	$F \rightarrow F$	$F \rightarrow F-1$	$F \leftrightarrow F+1$	$F \rightarrow F$
			$I = \frac{9}{2}$ *(continued)*				
15	$\frac{39}{2}$	0.113636	12.727	0.175	0.001	0.0	12.717
	$\frac{37}{2}$	0.015152	11.946	0.309	0.003	0.186	11.743
	$\frac{35}{2}$	−0.047675	11.205	0.403	0.004	0.329	10.855
	$\frac{33}{2}$	−0.079676	10.503	0.459	0.006	0.429	10.050
	$\frac{31}{2}$	−0.085423	9.839	0.477	0.006	0.488	9.327
	$\frac{29}{2}$	−0.069227	9.214	0.458	0.005	0.507	8.683
	$\frac{27}{2}$	−0.035136	8.627	0.401	0.004	0.487	8.119
	$\frac{25}{2}$	0.013062	8.078	0.307	0.002	0.427	7.634
	$\frac{23}{2}$	0.071839	7.568	0.174	0.0	0.326	7.230
	$\frac{21}{2}$	0.137931	7.097	0.0	0.0	0.185	6.912

Appendix **J**

NUCLEAR QUADRUPOLE SECOND-ORDER CORRECTION ENERGIES FOR LINEAR OR SYMMETRIC-TOP MOLECULES

To obtain $E_Q^{(2)}$, the entries given are to be multiplied by the factor

$$\left(\frac{(eqQ)^2}{B}\right) \times 10^{-3}$$

J	F	$K=0$	$K=1$	$K=2$	$K=3$	$K=4$	$K=5$	$K=6$
					$I=\frac{3}{2}$			
0	$\frac{3}{2}$	−10.4167						
1	$\frac{5}{2}$	−6.0000	−9.4688					
	$\frac{3}{2}$	−2.2500	−10.8750					
	$\frac{1}{2}$	0.0000	−11.7188					
2	$\frac{7}{2}$	−4.0999	−5.6487	−7.2885				
	$\frac{5}{2}$	−2.1866	2.4561	−3.8875				
	$\frac{3}{2}$	10.4167	5.2082	−10.4170				
	$\frac{1}{2}$	0.0000	11.7188	0.0000				
3	$\frac{9}{2}$	−3.0864	−3.8520	−5.3818	−5.3758			
	$\frac{7}{2}$	−1.9290	0.2170	3.4721	1.717119			
	$\frac{5}{2}$	6.0000	3.4466	−2.6042	−7.3244			
	$\frac{3}{2}$	2.2500	5.6668	10.4170	0.0000			
4	$\frac{11}{2}$	−2.4652	−2.8917	−3.9090	−4.7287	−4.0384		
	$\frac{9}{2}$	−1.6904	−0.5944	1.8570	3.1549	−0.8835		
	$\frac{7}{2}$	4.0998	2.8205	−0.4613	−4.0782	−5.2500		
	$\frac{5}{2}$	2.1866	3.5660	6.4915	7.3240	0.0000		
5	$\frac{13}{2}$	−2.0482	−2.3084	−2.9800	−3.7373	−4.0366	−3.1171	
	$\frac{11}{2}$	−1.4935	−0.8679	0.7160	2.3783	2.6536	−0.5096	
	$\frac{9}{2}$	3.0864	2.3715	0.4443	−2.0424	−4.0004	−3.9066	
	$\frac{7}{2}$	1.9290	2.6111	4.2779	5.7901	5.2500	0.0000	
6	$\frac{15}{2}$	−1.7500	−1.9198	−2.3779	−2.9701	−3.4295	−3.4365	−2.4684
	$\frac{13}{2}$	−1.3333	−0.9447	0.0971	1.4220	2.4120	2.2022	−0.3188
	$\frac{11}{2}$	2.4652	2.0288	0.8175	−0.8767	−2.5661	−3.5684	−3.0065
	$\frac{9}{2}$	1.6904	2.0749	3.0805	4.2634	4.8839	3.9066	0.0000

J	F	$K=0$	$K=1$	$K=2$	$K=3$	$K=4$	$K=5$	$K=6$
				$I=\frac{5}{2}$				
0	$\frac{5}{2}$	−5.8333						
1	$\frac{7}{2}$	−2.8930	−6.1475					
	$\frac{5}{2}$	−2.0829	−3.6386					
	$\frac{3}{2}$	−0.5400	−8.2350					
2	$\frac{9}{2}$	−1.8039	−3.1402	−4.8439				
	$\frac{7}{2}$	−1.8039	2.6858	−0.8261				
	$\frac{5}{2}$	4.9478	0.0657	−3.9850				
	$\frac{3}{2}$	−0.1968	5.6991	−5.1118				
	$\frac{1}{2}$	0.0000	−0.8333	−2.0832				
3	$\frac{11}{2}$	−1.2767	−1.9108	−3.2421	−3.5571			
	$\frac{9}{2}$	−1.4773	0.3006	3.1402	−0.4417			
	$\frac{7}{2}$	2.0002	0.7866	−1.6667	−1.7969			
	$\frac{5}{2}$	1.8072	1.7500	0.7640	−3.5926			
	$\frac{3}{2}$	0.5400	1.7341	3.0208	−2.6368			
	$\frac{1}{2}$	0.0000	0.8333	2.0833	0.0000			
4	$\frac{13}{2}$	−0.9753	−1.3193	−2.1601	−2.9227	−2.6497		
	$\frac{11}{2}$	−1.2289	−0.4042	1.4534	2.4952	−0.3603		
	$\frac{9}{2}$	0.9756	0.6127	−0.2797	−1.1141	−0.9101		
	$\frac{7}{2}$	1.5081	1.2013	0.3300	−0.9597	−2.4238		
	$\frac{5}{2}$	0.8856	1.2000	1.7349	1.2654	−2.2500		
	$\frac{3}{2}$	0.1968	0.7918	2.0910	2.6368	0.0000		
5	$\frac{15}{2}$	−0.7835	−0.9894	−1.5290	−2.1682	−2.5167	−2.0281	
	$\frac{13}{2}$	−1.0446	−0.6066	0.5031	1.6706	1.8734	−0.3198	
	$\frac{11}{2}$	0.5225	0.3992	0.0704	−0.3410	−0.6301	−0.5106	
	$\frac{9}{2}$	1.1840	0.9640	0.3562	−0.4822	−1.2888	−1.6970	
	$\frac{7}{2}$	0.8928	0.9656	1.0935	1.0039	0.2435	−1.8228	
	$\frac{5}{2}$	0.2756	0.6230	1.4861	2.3272	2.2500	0.0000	
6	$\frac{17}{2}$	−0.6519	−0.7844	−1.1456	−1.6258	−2.0431	−2.1421	−1.5948
	$\frac{15}{2}$	−0.9052	−0.6482	0.0404	0.9144	1.5623	1.4089	−0.2855
	$\frac{13}{2}$	0.2900	0.2466	0.1258	0.0453	−0.2204	−0.3361	−0.3097
	$\frac{11}{2}$	0.9468	0.8020	0.3961	−0.1836	−0.7938	−1.2316	−1.2380
	$\frac{9}{2}$	0.8283	0.8321	0.8202	0.7229	0.4239	−0.2394	−1.4762
	$\frac{7}{2}$	0.2958	0.5081	1.0693	1.7527	2.1803	1.8228	0.0000

[a]Numerical calculations were made by J. W. Simmons and W. E. Anderson. From W. Gordy, W. V. Smith, and R. F. Trambarulo, *Microwave Spectroscopy*, Wiley, New York, 1953.

AUTHOR INDEX

Numbers in parentheses are reference numbers and indicate that the author's work is referred to although his name is not mentioned in the text. Numbers in *italics* show the pages on which the complete references are listed.

SUBJECT INDEX